Global Positioning System: Theory and Applications
Volume I

Global Positioning System: Theory and Applications
Volume I

Edited by

Bradford W. Parkinson
Stanford University, Stanford, California
James J. Spilker Jr.
Stanford Telecom, Sunnyvale, California

Associated Editors:
Penina Axelrad
University of Colorado, Boulder, Colorado
Per Enge
Stanford University, Stanford, California

Volume 163
PROGRESS IN ASTRONAUTICS AND AERONAUTICS

Paul Zarchan, Editor-in-Chief
Charles Stark Draper Laboratory, Inc.
Cambridge, Massachusetts

Published by the
American Institute of Aeronautics and Astronautics, Inc.
370 L'Enfant Promenade, SW, Washington, DC 20024-2518

Third Printing

Copyright © 1996 by the American Institute of Aeronautics and Astronautics, Inc. Printed in the United States of America. All rights reserved. Reproduction or translation of any part of this work beyond that permitted by Sections 107 and 108 of the U.S. Copyright Law without the permission of the copyright owner is unlawful. The code following this statement indicates the copyright owner's consent that copies of articles in this volume may be made for personal or internal use, on condition that the copier pay the per-copy fee ($2.00) plus the per-page fee ($0.50) through the copyright Clearance Center, Inc., 222 Rosewood Drive, Danvers, Massachusetts 01923. This consent does not extend to other kinds of copying, for which permission requests should be addressed to the publisher. Users should employ the following code when reporting copying from this volume to the Copyright Clearance Center:

1-56347-106-X/96 $2.00 + .50

Data and information appearing in this book are for informational purposes only. AIAA is not responsible for any injury or damage resulting from use or reliance, nor does AIAA warrant that use or reliance will be free from privately owned rights.

ISBN 1-56347-106-X

Progress in Astronautics and Aeronautics

Editor-in-Chief
Paul Zarchan
Charles Stark Draper Laboratory, Inc.

Editorial Board

John J. Bertin
U.S. Air Force Academy

Leroy S. Fletcher
Texas A&M University

Richard G. Bradley
Lockheed Martin Fort Worth Company

Allen E. Fuhs
Carmel, California

William Brandon
MITRE Corporation

Ira D. Jacobsen
Embry-Riddle Aeronautical University

Clarence B. Cohen
Redondo Beach, California

John L. Junkins
Texas A&M University

Luigi De Luca
Politechnico di Milano, Italy

Pradip M. Sagdeo
University of Michigan

Martin Summerfield
Lawrenceville, New Jersey

Dedication

To Anna Marie, Elaine, Virginia, and Tim

Table of Contents

Preface xxxi

Part I. GPS Fundamentals

Chapter 1. Introduction and Heritage of NAVSTAR, the Global Positioning System 3
Bradford W. Parkinson, *Stanford University, Stanford, California*

Background and History 3
   Predecessors 4
   Joint Program Office Formed, 1973 6
Introductory GPS System Description and Technical Design 10
   Principals of System Operation 10
   GPS Ranging Signal 11
   Satellite Orbital Configuration 13
   Satellite Design 14
   Satellite Autonomy: Atomic Clocks 14
   Ionospheric Errors and Corrections 16
   Expected Navigation Performance 16
   High Accuracy/Carrier Tracking 18
History of Satellites 19
   Navigation Technology Satellites 19
   Navigation Development Satellites—Block I 19
   Operational Satellites—Block II and IIA 19
   Replacement Operational Satellites—Block IIR 20
Launches 20
   Launch Vehicles 20
Initial Testing 22
   Test Results 22
   Conclusions 24
Applications 24
   Military 24
   Dual Use: The Civil Problem 24
Pioneers of the GPS 26
   Defense Development, Research, and Engineering—Malcolm Currie and David Packard 26
   Commander of SAMSO, General Ken Schultz 26
   Contractors 26
   Joint Program Office Development Team 27
   Predecessors 27

Future	28
References	28

Chapter 2. Overview of GPS Operation and Design 29
J. J. Spilker Jr., *Stanford Telecom, Sunnyvale, California* and Bradford W. Parkinson, *Stanford University, Stanford, California*

Introduction to GPS	29
Performance Objectives and Quantitative Requirements on the GPS Signal	29
Satellite Navigation Concepts, Position Accuracy, and Requirement Signal Time Estimate Accuracy	31
GPS Space Segment	36
GPS Orbit Configuration and Multiple Access	36
GPS Satellite Payload	38
Augmentation of GPS	40
GPS Control Segment	40
Monitor Stations and Ground Antennas	41
Operational Control Center	42
GPS User Segment	43
GPS User Receiver Architecture	43
Use of GPS	45
GPS Signal Perturbations—Atmospheric/Ionospheric/Tropospheric Multipath Effects	49
Ionospheric Effects	49
Tropospheric Effects	52
Multipath Effects	52
Other Perturbing Effects	54
References	54

Chapter 3. GPS Signal Structure and Theoretical Performance 57
J. J. Spilker Jr., *Stanford Telecom, Sunnyvale, California*

Introduction	57
Summary of Desired GPS Navigation Signal Properties	57
Fundamentals of Spread Spectrum Signaling	59
GPS Signal Structure	67
Multiplexing Two GPS Spread Spectrum Signals on a Single Carrier and Multiple Access of Multiple Satellite Signals	68
GPS Radio Frequency Selection and Signal Characteristics	69
Detailed Signal Structure	73
GPS Radio Frequency Receive GPS Power Levels and Signal-to-Noise Ratios	82
GPS Radio Frequency Signal Levels and Power Spectra	82
Satellite Antenna Pattern	84
Signal Specifications	87
User-Receiver Signal-to-Noise Levels	88
Recommendations for Future Enhancements to the GPS System	93
Detailed Signal Characteristics and Bounds on Pseudorange Tracking Accuracy	94
Cross-Correlation Properties—Worst Case	94
Coarse/Acquisition-Code Properties	97
Bounds on GPS Signal Tracking Performance in Presence of White Thermal Noise	106

Appendix: Fundamental Properties of Maximal Length Shift Registers and Gold Codes 114
References 119

Chapter 4. GPS Navigation Data 121
J. J. Spilker Jr., *Stanford Telecom, Sunnyvale, California*

Introduction 121
   Overall Message Content of the Navigation Data 121
   Navigation Data Subframe, Frame, and Superframe 123
Detailed Description of the Navigation Data Subframe 132
   Subframe 1—GPS Clock Correction and Space Vehicle Accuracy 132
   GPS Ephemeris Parameters—Subframes 2 and 3 136
   Subframes 4 and 5—Almanac, Space Vehicle Health, and Ionosphere Models 139
Time, Satellite Clocks, and Clock Errors 149
   Mean Solar, Universal Mean Sidercal, and GPS Time 149
   Clock Accuracy and Clock Measurement Statistics 152
Satellite Orbit and Position 159
   Coordinate Systems and Classical Orbital Elements 159
   Classical Keplerian Orbits 162
   Perturbation of Satellite Orbit 164
Ionospheric Correction Using Measured Data 169
   Dual-Frequency Ionospheric Correction 169
Appendix 173
References 175

Chapter 5. Satellite Constellation and Geometric Dilution of Precision 177
J. J. Spilker Jr., *Stanford Telecom, Sunnyvale, California*

Introduction 177
   GPS Orbit Configuration, GPS-24 178
   GPS Orbit-Semi-Major Axis 178
   GPS Orbit—Satellite Phasing 180
GPS Satellite Visibility and Doppler Shift 181
   Bound on Level of Coverage for 24 Satellites 182
   GPS Satellite Visibility Angle and Droopler Shift 183
   GPS-24 Satellite Visibility 184
   Augmentation of the GPS-24 Constellation 187
   Constellation of 30 GPS Satellites 187
Coverage Swath for an Equatorial Plane of Satellites 187
   Satellite Ground Traces 189
Geometric Dilution of Precision Performance Bounds and GPS-24 Performance 190
   Bounds on Geometric Dilution of Precision—Two Dimensions 192
   Bounds on Geometric Dilution of Precision—Three Dimensions 197
   Position Dilution of Precision with an Accurate Clock 204
   Position Dilution of Precision for the GPS-24 Constellation 205
References 207
Bibliography 207

Chapter 6. GPS Satellite and Payload 209
M. Aparicio, P. Brodie, L. Doyle, J. Rajan, and P. Torrione, *ITT, Nutley, New Jersey*

Spacecraft and Navigation Payload Heritage 209
  Concept 209
  Relation to Earlier Non-GPS Satellites 209
  Overview of Payload Evolution 209
  On-Orbit Performance History 210
Navigation Payload Requirements 211
  GPS System 211
  GPS Performance 211
  GPS Signal Structure 213
  Payload Requirements 214
Block IIR Space Vehicle Configuration 215
  Navigation Payload Architecture 216
Block IIR Payload Design 216
  Payload Subsystems 216
  Mission Data Unit 223
  L-Band Subsystem 229
Characteristics of the GPS *L*-Band Satellite Antenna 234
  Coverage Area 234
  Antenna Pattern 234
  Antenna Evolution 234
  Crosslinks 236
  Primary and Secondary Functions 237
  Autonomous Navigation 238
Future Performance Improvements 242
  Additional Capabilities 242
References 243

Chapter 7. Fundamentals of Signal Tracking Theory 245
J. J. Spilker Jr., *Stanford Telecom, Sunnyvale, California*

Introduction 245
  GPS User Equipment 245
  GPS User Equipment-System Architecture 246
  Alternate Forms of Generalized Position Estimators 249
  Maximum Likelihood Estimates of Delay and Position 251
  Overall Perspective on GPS Receiver Noise Performance 252
  Interaction of Signal Tracking and Navigation Data Demodulation ... 255
Delay Lock Loop Receivers for GPS Signal Tracking 256
  Coherent Delay Lock Tracking of Bandlimited Pseudonoise Sequences 256
  Noncoherent Delay Lock Loop Tracking of Pseudonoise Signals 272
  Quasicoherent Delay Lock Loop 280
  Coherent Code/Carrier Delay Lock Loop 284
  Carrier-Aided Pseudorange Tracking 287
Vector Delay Lock Loop Processing of GPS Signals 290
  Independent Delay Lock Loop and Kalman Filter 291
  Vector Delay Lock Loop (VDLL) 293

Quasioptimal Noncoherent Vector Delay Lock Loop 298
Channel Capacity and the Vector Delay Lock Loop 305
Appendix A: Maximum Likelihood Estimate of Delay and Position 305
Appendix B: Least-Squares Estimation and Quasioptimal Vector Delay Lock Loops 310
Appendix C: Noncoherent Delay Lock Loop Noise Performance with Arbitrary
  Early-Late Reference Spacing 312
Appendix D: Probability of Losing Lock for the Noncoherent DLL 321
Appendix E: Colored Measurement Noise in the Vector Delay Lock Loop 323
References 325

Chapter 8. GPS Receivers 329
A. J. Van Dierendonck, *AJ Systems, Los Altos, California*

Generic Receiver Description 329
  Generic Receiver System Level Functions 329
  Design Requirements Summary 331
Technology Evolution 335
  Historical Evolution of Design Implementation 335
  Current Day Design Implementation 335
System Design Details 337
  Signal and Noise Representation 338
  Front-End Hardware 340
  Digital Signal Processing 348
Receiver Software Signal Processing 365
  A Signal-Processing Model and Noise Bandwidth Concepts 365
  Signal Acquisition 367
  Automatic Gain Control 368
  Generic Tracking Loops 369
  Delay Lock Loops 372
  Carrier Tracking 378
  Lock Detectors 390
  Bit Synchronization 395
  Delta Demodulation, Frame Synchronization, and Parity Decoding 396
Appendix A: Determination of Signal-to-Noise Density 399
Appendix B: Acquisition Threshold and Performance Determination 402
References 405

Chapter 9. GPS Navigation Algorithms 409
P. Axelrad, *University of Colorado, Boulder, Colorado*, and R. G. Brown,
Iowa State University, Ames, Iowa

Introduction 409
Measurement Models 410
  Pseudorange 410
  Doppler 411
  Accumulated Delta Range 412
  Navigation Delta Inputs 412
Single-Point Solution 412
  Solution Accuracy and Dilution of Precision 413
  Point Solution Example 415

Users Process Models 417
  Clock Model 417
  Stationary User or Vehicle 418
  Low Dynamics 419
  High Dynamics 420
Kalman Filter and Alternatives 420
  Discrete Extended Kalman Filter Formulation 421
  Steady-State Filter Performance 422
  Alternate Forms of the Kalman Filter 423
  Dual-Rate Filter 423
  Correlated Measurement Noise 424
GPS Filtering Examples 424
  Buoy Example 425
  Low Dynamics 427
  Unmodeled Dynamics 427
  Correlated Measurement Errors 430
Summary 430
References 433

Chapter 10. GPS Operational Control Segment 435
Sherman G. Francisco, *IBM Federal Systems Company, Bethesda, Maryland*

Monitor Stations 439
Master Control Station 445
Ground Antenna 447
Navigation Data Processing 449
System State Estimation 457
Navigation Message Generation 463
Time Coordination 464
Navigation Product Validation 465
References 465

Part II. GPS Performance and Error Effects

Chapter 11. GPS Error Analysis 469
Bradford W. Parkinson, *Stanford University, Stanford, California*

Introduction 469
Fundamental Error Equation 469
  Overview of Development 469
  Derivation of the Fundamental Error Equation 470
Geometric Dilution of Precision 474
  Derivation of the Geometric Dilution of Precision Equation 474
  Power of the GDOP Concept 474
  Example Calculations 475
  Impact of Elevation Angle on GDOP 477
Ranging Errors 478
  Six Classes of Errors 478
  Ephemeris Errors 478

Satellite Clock Errors 478
Ionosphere Errors 479
Troposhere Errors 479
Multipath Errors 480
Receiver Errors 480
Standard Error Tables 480
Error Table Without S/A: Normal Operation for C/A Code 481
Error Table with S/A 481
Error Table for Precise Positioning Service (PPS Dual-Frequency P/Y Code) 482
Summary 482
References 483

Chapter 12. Ionospheric Effects on GPS 485
J. A. Klobuchar, *Hanscom Air Force Base, Massachusetts*

Introduction 485
Characteristics of the Ionosphere 485
Refractive index of the ionosphere 488
Major Effects on Global Positioning Systems Caused by the Ionosphere 489
Ionospheric Group Delay—Absolute Range Error 489
Ionospheric Carrier Phase Advance 490
Higher-Order Ionospheric Effects 491
Obtaining Absolute Total Electron Content from Dual-Frequency GPS
  Measurements 493
Ionospheric Doppler Shift/Range-Rate Errors 495
Faraday Rotation 496
Angular Refraction 497
Distortion of Pulse Waveforms 498
Amplitude Scintillation 499
Ionospheric Phase Scintillation Effects 502
Total Electron Content 503
Dependence of Total Electron Content on Solar Flux 504
Ionospheric Models 506
Single-Frequency GPS Ionospheric Corrections 509
Magnetic Storms Effects on Global Positioning Systems 510
Differential GPS Positioning 511
Appendix: Ionospheric Correction Algorithm for the Single-Frequency GPS Users 513
References 514

Chapter 13. Troposheric Effects on GPS 517
J. J. Spilker Jr., *Stanford Telecom, Sunnyvale, California*

Troposheric Effects 517
Introduction 517
Atmospheric Attenuation 520
Rainfall Attenuation 521
Troposheric Scintillation 522
Troposheric Delay 523
Path Length and Delay 524
Troposheric Refraction Versus Pressure and Temperature 528

Empirical Models of the Troposhere 534
    Saastamoinen Total Delay Model 534
    Hopfield Two Quartic Model 534
    Black and Eisner (B&E) Model 536
    Water Vapor Zenith Delay Model—Berman 538
    Davis, Chao, and Marini Mapping Functions 538
    Altshuler and Kalaghan Delay Model 539
    Ray Tracing and Simplified Models 541
    Lanyi Mapping Function and GPS Control Segment Estimate 541
    Model Comparisons 544
Tropospheric Delay Errors and GPS Positioning 544
References 545

Chapter 14. Multipath Effects 547
Michael S. Braasch, *Ohio University, Athens, Ohio*

Introduction 547
Signal and Multipath Error Models 548
    Pseudorandom Noise Modulated Signal Description 549
    Coherent Pseudorandom Noise Receiver 549
    Noncoherent Pseudorandom Noise Receiver 553
    Simulation Results 554
Aggravation and Mitigation 558
    Antenna Considerations 558
    Receiver Design 560
Multipath Data Collection 560
Acknowledgments 566
References 566

Chapter 15. Foliage Attenuation for Land Mobile Users 569
J. J. Spilker Jr., *Stanford Telecom, Sunnyvale, California*

Introduction 569
Attenuation of an Individual Tree or Forest of Trees—Stationary User 571
    Foliage Attenuation—Mobile User 575
    Probability Distribution Models for Foliage Attenuation—Mobile User 576
Measured Models—Satellite Attenuation Data 580
    Measured Fading for Tree-Lined Roads—Mobile Users 581
References 582

Chapter 16. Ephemeris and Clock Navigation Message Accuracy 585
J. F. Zumberge and W. I. Bertiger, *Jet Propulsion Laboratory, California Institute of Technology, Pasadena, California*

Control Segment Generation of Predicted Ephemerides and Clock Corrections 585
Accuracy of the Navigation Message 586
    Global Network GPS Analysis at the Jet Propulsion Laboratory 587
    Accuracy of the Precise Solution 588
    Comparison of Precise Orbits with Broadcast Ephemerides 590

Best Estimate of Trajectory 705
Validation of Truth Trajectory Accuracy 705
Ground Truth 705
Phase I Test (1972–1979) 707
Ground Transmitters 707
Navy Testing for Phase I 710
Tests Between Phase I and Phase II (1979–1982) 711
  Weapons Delivery 711
  Differential Tests 711
Phase II: Full-Scale Engineering Development Tests (1982–1985) 713
Summary 714
Bibliography 715

Chapter 20.   Interference Effects and Mitigation Techniques 717
J. J. Spilker Jr. and F. D. Natali, *Stanford Telecom, Sunnyvale, California*

Introduction 717
  Possible Sources of Interference 719
  Frequency Allocation in Adjacent and Subharmonic Bands 719
Receiver Design for Tolerance to Interference 720
  Receiver Systems 720
  Quantizer Effects in the Presence of Interference 724
Effects of Interference on the GPS C/A Receiver 745
  Effects of the C/A-Code Line Components on Narrow-Band Interference
    Performance 745
  Narrow-Band Interference Effects—Spectra of Correlator Output ... 748
  Interference Effects—Effects on Receiver-Tracking Loops 752
Detection of Interference, Adaptive Delay Lock Loop, Adaptive Frequency Notch
  Filtering, and Adaptive Null Steering Antennas 756
  Adaptation of the Delay Lock Loop and Vector Delay Lock Loop .. 757
  Rejection of Narrow-Band Interference by Adaptive Frequency Nulling Filters .. 757
  Adaptive Antennas for Point Source Interference 759
  Augmentation of the GPS Signals and Constellation 767
Appendix: Mean and Variance of the Correlator Output for an *M*-Bit Quantizer 768
References 771

Author Index 773

Subject Index 775

Comparison of the Precise Clocks with Broadcast Clocks 59
Summary and Discussion 59
Appendix: User Equivalent Range Error 59
References 59

Chapter 17. Selective Availability 60
Frank van Graas and Michael S. Braasch, *Ohio University, Athens, Ohio*

Goals and History ....
Implementation ....
Characterization of Selective Availability ....
    Second-Order Gauss-Markov Model ....
    Autoregressive Model ....
    Analytic Model ....
    Recursive Autoregressive Model (Lattice Filter) ....
    Selective Availability Model Summary ....
References ....

Chapter 18. Introduction to Relativistic Effects on the Global Positioning System ....
N. Ashby, *University of Colorado, Boulder, Colorado*, and
J. J. Spilker Jr., *Stanford Telecom, Sunnyvale, California*

Introduction ....
    Objectives ....
    Statement of the GPS Problem ....
Introduction to the Elementary Principles of Relativity ....
    Euclidean Geometry and Newtonian Physics ....
    Space-Time Coordinates and the Lorentz Transformation ....
    Relativistic Effects of Rotation in the Absence of a Gravitational Field ....
    Principle of Equivalence ....
Relativistic Effects in GPS ....
    Relativistic Effects on Earth-Based Clocks ....
    Relativistics Effects for Users of the GPS ....
    Secondary Relativistic Effects ....
References ....

Chapter 19. Joint Program Office Test Results ....
Leonard Kruczynski, *Ashtech, Sunnyvale, California*

Introduction ....
U.S. Army Yuma Proving Ground (YPG) ....
Reasons for Selection of Yuma Proving Ground ....
Lasers ....
Range Space ....
Joint Program Office Operating Location ....
Satellite Constellation for Test Support ....
Control Segment Responsiveness to Testing Needs ....
Trajectory Determination YPG ....
    Real-Time Estimate ....

Table of Contents for Companion Volume II

Preface xxix

Part III. Differential GPS and Integrity Monitoring

Chapter 1. Differential GPS 3
Bradford W. Parkinson and Per K. Enge, *Stanford University, Stanford, California*

Introduction	3
Standard Positioning Service Users	3
Precise Positioning Service Users	4
Major Categories of Differential GPS	4
Code-Phase Differential GPS	7
User Errors Without Differential GPS	7
Reference Station Calculation of Corrections	10
Application of Reference Correction	11
Analysis of Differential GPS Errors	11
Receiver Noise, Interference, and Multipath Errors for Differential GPS	12
Satellite Clock Errors for Differential GPS	16
Satellite Ephemeris Errors for Differential GPS	17
Ionospheric Errors for Differential GPS	20
Troposhere Errors for Differential GPS	23
Local Area Differential GPS Error Summary	24
Carrier-Phase Differential GPS	27
Attitude Determination	27
Static and Kinematic Survey	28
Near Instantaneous Determination of Integers	30
Radio Technical Commission for Maritime Services Data Format for Differential GPS Data	31
Radio Technical Commission for Maritime Services Message Types 1, 2, and 9	32
Types 18, 19, 20, and 21 Messages	34
Datalinks	34
Groundwave Systems	34
VHF and UHF Networks	36
Mobile Satellite Communications	39
Differential GPS Field Results	41
Short-Range Differential Code-Phase Results	41
Long-Range Differential Code-Phase Results	42
Dynamic Differential Carrier-Phase Results	43
Conclusions	47
Appendix: Differential GPS Ephemeris Correction Errors Caused by Geographic Separation	47
References	49

Chapter 2. Pseudolites 51
Bryant D. Elrod, *Stanford Telecom, Inc., Reston, Virginia* and
 A. J. Van Dierendonck, *AJ Systems, Los Altos, California*

Introduction	51
Pseudolite Signal Design Considerations	52
Previous Pseudolite Designs	52

New Pseudolite Signal Design	53
Integrated Differential GPS/Pseudolite Considerations	57
Pseudolite Siting	57
Pseudolite Time Synchronization	58
User Aircraft Antenna Location	62
Pseudolite Signal Data Message	63
GPS/Pseudolite Navigation Filter Considerations	64
Pseudolite Testing	65
Pseudolite Interference Testing	65
Pseudolite Data Link Testing	67
Navigation Performance Testing	68
Appendix A: Interference Caused by Cross Correlation Between C/A Codes	70
Appendix B: Interference Caused by Pseudolite Signal Level	74
Appendix C: Navigation Filter Modeling with Pseudolite Measurements	76
References	78

Chapter 3. Wide Area Differential GPS 81
Changdon Kee, *Stanford University, Stanford, California*

Introduction	81
Wide Area Differential GPS Architecture and Categories	82
Wide Area Differential GPS Architecture	82
Wide Area Differential GPS Categories	85
User Message Content and Format	87
Error Budget	88
Master Station Error Modeling	88
Ionospheric Time Delay Model for Algorithms A or B	89
Ephemeris and Satellite Clock Errors for Algorithms A, B, or C	92
Simulation of Algorithm B	95
Simulation Modules	95
Ionospheric Error Estimation Results	100
Navigation Performance	101
Summary of Results	104
Test Using Field Data to Evaluate Algorithm C	104
Locations of the Receiver Sites	105
Test Results	105
Latency and Age Concern	111
Conclusion	112
References	114

Chapter 4. Wide Area Augmentation System 117
Per K. Enge, *Stanford University, Stanford, California* and A.J. Van Dierendonck, *AJ Systems, Los Altos, California*

Introduction	117
Signal Design	120
Link Budget and Noninterference with GPS	120
Data Capacity	123
Loop Threshold	124
Ranging Function	124
Nonprecision Approach and Error Estimates	126
Precision Approach and Vector Corrections	128
Vector Corrections	129
Precision Approach Integrity	130
Wide Area Augmentation System Message Format	131
Parity Algorithm	134
Message Type 2 Fast Corrections and User Differential Range Errors	135
Type 25: Long-Term Satellite Error Corrections Message	135
Type 26: Ionospheric Delay Error Corrections Message	136

Type 9: WAAS Satellite Navigation Message ... 137
Applied Range Accuracy Evaluation ... 137
Summary ... 138
Appendix: Geostationary Satellite Ephemeris Estimation and Code-Phase Control ... 139
References ... 142

Chapter 5. Receiver Autonomous Integrity Monitoring ... 143
R. Grover Brown, *Iowa State University, Ames, Iowa*

History, Overview, and Definitions ... 143
Basic Snapshot Receiver Autonomous Integrity Monitoring Schemes and Equivalences ... 145
    Range Comparison Method ... 146
    Least-Squares-Residuals Method ... 147
    Parity Method ... 148
    Maximum Separation of Solutions ... 150
    Constant-Detection-Rate/Variable-Protection-Level Method ... 151
Screening Out Poor Detection Geometries ... 152
Receiver-Autonomous Integrity Monitoring Availability for Airborne Supplemental Navigation ... 155
Introduction to Aided Receiver-Autonomous Integrity Monitoring ... 156
Failure Isolation and the Combined Problem of Failure Detection and Isolation ... 158
    Introductiory Remarks ... 158
    Parity Method and Failure Detection and Isolation ... 158
    Calculation of the P Matrix ... 161
    Failure Detection and Exclusion Algorithm ... 163
References ... 164

Part IV. Integrated Navigation Systems

Chapter 6. Integration of GPS and Loran-C ... 169
Per K. Enge, *Stanford University, Stanford, California* and F. van Graas,
  Ohio University, Athens, Ohio

Introduction ... 169
    Calibration of Loran Propagation Errors by GPS ... 171
    Cross-Chain Synchronization of Loran-C Using GPS ... 171
    Combining Pseudoranges from GPS and Loran-C for Air Navigation ... 171
Loran Overview ... 172
Calibration of Loran Propagation Errors by GPS ... 174
Cross-Rate Synchronization of Loran ... 176
Combining GPS Pseudoranges with Loran Time Differences ... 179
    Navigation Equations ... 179
    Probability of Outage Results ... 182
Summary ... 184
References ... 185

Chapter 7. GPS and Inertial Integration ... 187
R. L. Greenspan, *Charles Stark Draper Laboratories, Cambridge, Massachusetts*

Benefits of GPS/Inertial Integration ... 187
    Operation During Outages ... 189
    Providing All Required Navigation Outputs ... 190
    Reduced Noise in GPS Navigation Solutions ... 190
    Increased Tolerance to Dynamics and Interference ... 191
GPS Integration Architectures and Algorithms ... 191
    Integration Architectures ... 191
    Integration Algorithms ... 194
    Embedded Systems ... 197

Integration Case Studies 199
  GPS/Inertial Navigation Systems Navigation Performance in a Low-Dynamics Aircraft 199
  Using GPS for In-flight Alignment 206
  Integrated Navigation Solutions During a GPS Outage 213
Summary 217
References 218

Chapter 8. Receiver Autonomous Integrity Monitoring Availability for GPS Augmented with Barometric Altimeter Aiding and Clock Coasting 221
Young C. Lee, *MITRE Corporation, McLean, Virginia*

Introduction 221
Methods of Augmentations 222
  Augmented Geometry for Barometric Altimeter Aiding 222
  Barometric Altimeter Aiding with GPS-Calibrates Pressure Altitude Data 223
  Barometric Altimeter Aiding with Local Pressure Input 227
  Augmented Geometry for Clock Coasting 228
  Simultaneous Use of Barometric Altimeter Aiding and Clock 229
Definitions of Function Availability 229
  Navigation Function 229
  Receiver Autonomous Integrity Monitoring Detection Function 229
  Receiver Autonomous Integrity Monitoring Function 230
Results 230
  Parameters of Interest 230
  Discussion of Results 231
Summary and Conclusions 235
Appendix: Statistical Distribution of the Height Gradients 239
References 241

Chapter 9. GPS and Global Navigation Satellite System (GLONASS) 243
Peter Daly, *University of Leeds, Leeds, England, United Kingdom* and Pratap N. Misra, *Massachusetts Institute of Technology, Lexington, Massachusetts*

Introduction to the Global Navigation Satellite System 243
  History of Satellite Navigation Systems 243
  Orbits 244
  History of Launches 247
  Signal Design 248
  Message Content and Format 252
  Satellite Ephemerides 253
  Satellite Almanacs 254
  GPS/GLONASS Onboard Clocks 255
Performance of GLONASS and GPS + GLONASS 258
  Introduction 258
  Requirements of Civil Aviation 259
  Integrated Use of GPS and GLONASS 260
  Performance of GLONASS and GPS and GLONASS 261
Summary 271
Acknowledgments 271
References 271

Part V. GPS Navigation Applications

Chapter 10. Land Vehicle Navigation and Tracking 275
Robert L. French, *R. L. French & Associates, Fort Worth, Texas*

Application Characteristics and Markets 275
  Commercial Vehicle Tracking 275

Automobile Navigation and Route Guidance	277
Intelligent Vehicle Highway Systems	279
Historical Background	281
Early Mechanical Systems	281
Early Electronic Systems	282
Enabling/Supporting Technologies	283
Dead Reckoning	284
Digital Road Maps	286
Map Matching	288
Integration with GPS	291
Mobile Data Communications	292
Examples of Integrated Systems	294
Etax Navigator™/Bosch Travelpilot™	294
Toyota Electro-Multivision	296
TravTek Driver Information System	297
NavTrax™ Fleet Management System	298
References	299

Chapter 11. Marine Applications 303
Jim Sennott and In-Soo Ahn, *Bradley University, Peoria, Illinois* and Dave Pietraszewski, *United States Coast Guard Research and Development Center, Groton, Connecticut*

Marine Navigation Phases and Requirements	303
Marine DGPS Background	304
Global Positioning Systems-Assisted Steering, Risk Assessment, and Hazard Warning System	305
Vessel and Sensor Modeling	308
Vessel Dynamics Model	308
Standardized Sensor Model	310
Combined Ship and Sensor Model	311
Waypoint Steering Functions	312
Filter and Controller Design	313
Sensor/Ship Bandwidth Ratio and Straight-Course Steering	314
Comparative Footprint Channel Clearance Width Distributions	315
Hazard Warning and Risk Assessment Functions	321
Risk Assessment	321
Hazard Warning	323
Summary	323
References	324

Chapter 12. Applications of the GPS to Air Traffic Control 327
Ronald Braff, *MITRE Corporation, McLean, Virginia,* J. David Powell, *Stanford University, Stanford, California* and Joseph Dorfler, *Federal Aviation Administration, Washington, DC*

Introduction	327
Air Traffic Control System	327
General Considerations	329
Operational Requirements	331
Government Activities	333
Air Navigation Applications	334
En Route, Terminal, and Nonprecision Approach Operational Considerations and Augmentations	335
Precision Approach Operational Considerations and Augmentations	344
Other Navigation Operational Considerations	354
Area and Four-Dimensional Navigation	361
Surveillance	362
Current Surveillance Methods	362
Surveillance via GPS	366
Summary of Key Benefits	370
References	370

Chapter 13. GPS Applications in General Aviation 375
Ralph Eschenbach, *Trimble Navigation, Sunnyvale, California*

Market Demographics 375
    Airplanes 375
    Pilots 375
    Airports 377
Existing Navigation and Landing Aids (Non-GPS) 377
    Nondirectional Beacons (NDB) 377
    Very High Frequency Omnidirectional Radio 378
    Distance-Measuring Equipment 379
    Long-Range Radio Navigation 379
    Omega 380
    Approaches 380
Requirements for GPS in General Aviation 384
    Dynamics 385
    Functionality 385
    Accuracy 385
    Availability, Reliability, and Integrity 386
Pilot Interface 387
    Input 387
    Output 388
GPS Hardware and Integration 388
    Installation Considerations 388
    Number of Channels 389
    Cockpit Equipment 389
    Hand-held 389
    Panel Mounts 389
    Dzus Mount 390
Differential GPS 390
    Operational Characteristics 391
    Ground Stations 391
    Airborne Equipment Features 391
Integrated Systems 392
    GPS LORAN 392
    GPS/Omega 392
Future Implementations 393
    Attitude and Heading Reference System 393
    Approach Certification 393
    Collision Avoidance 393
    Autonomous Flight 394
Summary 394
References 395

Chapter 14. Aicraft Automatic Approach and Landing Using GPS 397
Bradford W. Parkinson and Michael L. O'Connor, *Stanford University, Stanford, California*, and Kevin T. Fitzgibbon, *Technical University, São Jose Dos Campos, Brazil*

Introduction 397
    Autolanding Conventionally and with GPS 397
    Simulations Results Presented 398
Landing Approach Procedures 399
    Instrument and Microwave Landing Systems 399
    GPS Approach 401
Aircraft Dynamics and Linear Model 401
    State Vector 401
    Control Vector 402
    Disturbance Vector 402

Measurement Vector	403
Equations of Motion	403
Wind Model	403
Throttle Control Lag	404
Glide-Slope Deviation	404
Autopilot Controller	405
Linear Quadratic Gaussian and Integral Control Law Controllers	405
Regulator Synthesis	405
GPS Measurements	407
Results	409
Cases Simulated	409
Landing with GPS Alone	410
Landing with GPS Plus Altimeter	410
Landing with Differential GPS	411
Landing with Carrier-Phase	411
Linear Quadratic Gaussian vs Integral Control Law	413
Conclusions and Comments	414
Appendix A: Discrete Controllers	415
Appendix B: Discrete Time Optimal Estimator	421
Appendix C: Numerical Values for Continuous System	422
Bibliography	425
References	425

Chapter 15. Precision Landing of Aircraft Using Integrity Beacons 427
Clark E. Cohen, Boris S. Pervan, H. Stewart Cobb, David G. Lawrence, J. David Powell, and Bradford W. Parkinson, *Stanford University, Stanford, California*

Overview of the Integrity Beacon Landing System	427
Centimeter-Level Positioning	428
History of the Integrity Beacon Landing System	429
Doppler Shift and Geometry Change	429
Required Navigation Performance	429
Accuracy	429
Integrity	430
Availability	431
Continuity	431
Integrity Beacon Architecture	432
Droppler Marker	432
Omni Marker	432
Mathematics of Cycle Resolution	434
Observability Analysis	434
Matrix Formulation	435
Experimental Flight Testing	438
Quantification of Centimeter-Level Accuracy	438
Piper Dakota Experimental Flight Trials	439
Federal Aviation Administration Beech King Air Autocoupled Approaches	442
Automatic Landings of a United Boeing 737	444
Flight Test Summary and Observations	447
Operations Using Integrity Beacons	447
Integrity Beacon Landing System Landing Sequence	448
Integrity Beacon Landing System Navigation Integrity	450
Receiver Autonomous Integrity Monitoring	450
System Failure Modes	453
Quantifying Integrity	454
Signal Interference	456
Conclusion	458
References	458

Chapter 16. Spacecraft Attitude Control Using GPS Carrier Phase 461
E. Glenn Lightsey, *NASA Goddard Space Flight Center, Greenbelt, Maryland*

Introduction 461
Design Case Study 462
Sensor Characteristics 464
    Antenna Placement 466
    Sensor Calibration 466
    Multipath 467
    Sensor Accuracy 467
    Dynamic Filtering 468
Vehicle Dynamics 468
    Gravity Gradient Moment 470
    Aerodynamic Moment 471
    System Natural Response 472
Control Design 474
    Control Loop Description 474
    Simulation Results 475
Conclusion 477
Acknowledgments 479
References 479

Part VI. Special Applications

Chapter 17. GPS for Precise Time and Time Interval Measurement 483
William J. Klepczynski, *United States Naval Observatory, Washington, DC*

Introduction 483
Universal Coordinated Time 484
Role of Time in the GPS 485
Translation of GPS Time to Universal Coordinated Time 487
GPS as a Clock in the One-Way Mode 489
Common-View Mode of GPS 490
Melting-Pot Method 498
Problem of Selective Availability 498
Future Developments 499
References 500

Chapter 18. Surveying with the Global Positioning System 501
Clyde Goad, *Ohio State University, Columbus, Ohio*

Measurement Modeling 502
Dilution of Precision 508
Ambiguity Search 509
Use of Pseudoranges and Phase 510
    Review 511
    Three-Measurement Combinations 514
Antispoofing? 516
A Look Ahead 516
References 517

Chapter 19. Attitude Determination 519
Clark E. Cohen, *Stanford University, Stanford, California*

Overview 519
Fundamental Conventions for Attitude Determination 521

Solution Processing	523
Cycle Ambiguity Resolution	524
Baseline Length Constraint	525
Integer Searches	525
Motion-Based Methods	525
Alternative Means for Cycle Ambiguity Resolution	531
Performance	531
Geometrical Dilution of Precision for Attitude	532
Multipath	532
Structural Distortion	533
Troposphere	533
Signal-to-Noise Ratio	534
Receiver-Specific Errors	534
Total Error	534
Applications	535
Aviation	535
Spacecraft	537
Marine	537
References	537

Chapter 20. Geodesy 539
Kristine M. Larson, *University of Colorado, Boulder, Colorado*

Introduction	539
Modeling of Observables	540
Reference Frames	542
Precision and Accuracy	547
Results	550
Crustal Deformation	550
Earth Orientation	553
Conclusions	554
Acknowledgments	554
References	554

Chapter 21. Orbit Determination 559
Thomas P. Yunck, *Jet Propulsion Laboratory, California Institute of Technology, Pasadena, California*

Introduction	559
Principles of Orbit Determination	560
Dynamic Orbit Determination	560
Batch Least Squares Solution	562
Kalman Filter Formulation	563
Dynamic Orbit Error	565
Kalman Filter with Process Noise	565
Orbit Estimation with GPS	567
Carrier-Pseudorange Bias Estimation	567
Kinematic Orbit Determination	569
Reduced Dynamic Orbit Determination	571
Orbit Improvement by Physical Model Adjustment	572
Direct Orbit Determination with GPS	573
Precise Orbit Determination with Global Positioning Systems	574
Global Differential Tracking	574
Fine Points of the Global Solution	576
Precise Orbit Determination	577
Single-Frequency Precise Orbit Determination	580
Extension to Higher Altitude Satellites	584
Highly Elliptical Orbiters	585
Dealing with Selective Availability and Antispoofing	586
Antispoofing	586

Selective Availability	586
Summary	588
Acknowledgments	589
References	589

Chapter 22. Test Range Instrumentation 593
Darwin G. Abby, *Intermetrics, Inc. Holloman Air Force Base, New Mexico*

Background	593
Requirements	595
Test Requirements	595
Training Requirements	596
Range Instrumentation Components	598
Global Positioning Systems Reference Station	598
Data Links	599
Test Vehicle Instrumentation	599
Translator Systems	599
Digital Translators	601
Differential Global Positioning Systems Implementations	604
Existing Systems	604
Department of Defense Systems	604
Commercial Systems	614
Data links	619
Accuracy Performance	619
Position Accuracy	619
Velocity Accuracy	620
Future Developments	621
National Range	621
Kinematic Techniques	622
References	622

Author Index 625

Subject Index 627

Preface

Overview and Purpose of These Volumes

Of all the *military* developments fostered by the recent cold war, the Global Positioning System (GPS) may prove to have the greatest positive impact on everyday life. One can imagine a 21st century world covered by an augmented GPS and laced with mobile digital communications in which aircraft and other vehicles travel through "virtual tunnels," imaginary tracks through space which are continuously optimized for weather, traffic, and other conditions. Robotic vehicles perform all sorts of construction, transportation, mining, and earth moving functions working day and night with no need for rest. Low-cost personal navigators are as commonplace as hand calculators, and every cellular telephone and personnel communicator includes a GPS navigator. These are some of the potential positive impacts of GPS for the future. Our purpose in creating this book is to increase that positive impact. That is, *to accelerate the understanding of the GPS system and encourage new and innovative applications.*

The intended readers and users of the volumes include all those who seek knowledge of GPS techniques, capabilities, and limitations:

- Students attending formal or informal courses
- Practicing GPS engineers
- Applications engineers
- Managers who wish to improve their understanding of the system

Our somewhat immodest hope is that this book will become a standard reference for the understanding of the GPS system.

Each chapter is authored by an individual or group of individuals who are recognized as world-class authorities in their area of GPS. Use of many authors has led to some overlap in the subject matter which we believe is positive. This variety of viewpoints can promote understanding and contributes to our overall purpose. Books written by several authors also must contend with variations in notation. The editors of the volume have developed common notations for the important subjects of GPS theory and analysis, and attempted to extend this, where possible, to other chapters. Where there are minor inconsistencies we ask for your understanding.

Organization of the Volumes

The two volumes are intended to be complementary. Volume I concentrates on fundamentals and Volume II on applications. Volume I is divided into two parts: the first deals with the operation and theory of basic GPS, the second section with GPS performance and errors. In Part I (GPS Fundamentals), a summary of GPS history leads to later chapters which promote an initial understanding of the three GPS segments: User, Satellite, and Control. Even the best of systems has its limitations, and GPS is no exception. Part II, GPS Performance and Error Effects, is introduced with an overview of the errors, followed by chapters devoted to each of the individual error sources.

Volume II concentrates on two aspects: augmentations to GPS and detailed descriptions of applications. It consists of Parts III to VI:

- III. Differential GPS and integrity Monitoring
- IV. Integrated Navigation Systems
- V. GPS Navigation Applications
- VI. Special Applications

Parts III and IV expand on GPS with explanations of supplements and augmentations to the system. The supplements enhance accuracy, availability, or integrity. Of special interest is differential GPS which has proven it can provide sub-meter (even centimeter) level accuracies in a dynamic environment. The last two sections (V and VI) are detailed descriptions of the major applications in current use. In the rapidly expanding world of GPS, new uses are being found all of the time. We sincerely hope that these volumes will accelerate such new discoveries.

Acknowledgments

Obviously this book is a group undertaking with many, many individuals deserving of our sincere thanks. In addition to the individual authors, we would especially like to thank Ms. Lee Gamma, Mr. Sam Pullen, and Ms. Denise Nunes. In addition, we would like to thank Mr. Gaylord Green, Dr. Nick Talbot, Dr. Gary Lennon, Ms. Penny Sorensen, Mr. Konstantin Gromov, Dr. Todd Walter, and Mr. Y. C. Chao.

Special Acknowledgment

We would like to give special acknowledgment to the members of the original GPS Joint Program Office, their supporting contractors and the original set of engineers and scientists at the Aerospace Corporation and at the Naval Research Laboratory. Without their tenacity, energy, and foresight GPS would not be.

B. W. Parkinson
J. J. Spilker Jr.
P. Axelrad
P. Enge

Part I. GPS Fundamentals

Chapter 1

Introduction and Heritage of NAVSTAR, the Global Positioning System

Bradford W. Parkinson*
Stanford University, Stanford, California 94305

I. Background and History

FOR six thousand years, humans have been developing ingenious ways of navigating to remote destination. A fundamental technique developed by both ancient Polynesians and modern navies is the use of *angular measurements* of the natural stars. With the development of radios, another class of navigation aids was born. These included radio beacons, vhf omnidirectional radios (VORs), long-range radio navigation (LORAN), and OMEGA. With yet another technology—artificial satellites—more precise, *line-of-sight* radio navigation signals became possible. This promise was realized in the 1960s, when the U.S. Navy's Navigation Satellite System (known as Transit) opened a new era of navigation technology and capability. However, the best was yet to come.

Over a long Labor Day weekend in 1973, a small group of armed forces officers and civilians, sequestered in the Pentagon, were completing a plan that would truly revolutionize navigation. It was based on radio ranging (eventually with *millimeter precision*) to a constellation of artificial satellites called NAVSTARs. Instead of *angular measurements* to natural stars, greater accuracy was anticipated with *ranging measurements* to the artificial NAVSTARs.

Although it has taken over twenty years to establish that system and to realize its implications fully, it is now apparent that a new *navigation utility* has been created. For under a thousand dollars (price rapidly decreasing), anyone, anywhere in the world, can almost instantaneously determine his or her location in three dimensions to about the width of a street.

This book explains the technology, the performance, and the applications of this new utility—the Global Positioning System (GPS). "*With the quiet revolution of NAVSTAR, it can be seen that these potential uses are limited only by our imaginations.*"[1]

Copyright © 1994 by the author. Published by the American Institute of Aeronautics and Astronautics, Inc., with permission. Released to AIAA to publish in all forms.
*Professor, Department of Aeronautics and Astronautics, and Director, GPS Program.

A. Predecessors

By 1972, the U.S. Air Force and Navy had for several years been competitively studying the possibility of improved navigation from space. These studies became the basis for a new synthesis known as NAVSTAR or the GPS. A brief discussion of the predecessor systems is followed by a description of the Air Force's development program and a summary of the technical design (which is expanded further in succeeding chapters).

1. Applied Physics Laboratory's Transit: Navy Navigation Satellite System

The first operational satellite-based navigation system was called NNSS (Navy Navigation Satellite System), or Transit. Developed by the Johns Hopkins Applied Physics Laboratory (APL) under Dr. Richard Kershner, Transit was based on a user measuring the Doppler shift of a tone broadcast at 400 MHz by polar orbiting satellites at altitudes of about 600 nautical miles (actually, two frequencies were transmitted to correct for ionospheric group delay).

The tone broadcast by Transit was continuous. The maximum rate of change in the Doppler shift of the received signal corresponded to the point of closest approach of the Transit satellite. The difference between "up" Doppler and "down" Doppler can be used to calculate the *range to the satellite at closest approach.* Users with known altitude (e.g. sea level) and the broadcast ephemeris of the satellite could use these Doppler measurements to calculate their positions to a few hundred meters. Of course, corrections had to be made for the user's velocity. Because of this velocity sensitivity and the *two*-dimensional nature, Transit was not very useful for air applications. Another limitation was the intermittent availability of the signals, because mutual interference restricted the number of satellites available worldwide to about five. This limited coverage had unavailability periods of 35 to 100 min.

Originally intended as a system to help U.S. submarines navigate, Transit was soon adopted extensively by commercial marine navigators. Although Transit is still operational, new satellites are no longer being launched, and the Federal Radionavigation Plan has announced the intent to phase it out.

Technology developed for Transit has proved to be extremely useful to GPS. Particularly important were the satellite prediction algorithms developed by the Naval Surface Warfare Center. Transit also proved that space systems could demonstrate excellent reliability. After initial "infant mortality" problems, an improved version exhibited operational lifetimes exceeding its specifications by two or three times. In fact, a number of these valuable spacecraft have lasted more than 15 years.

2. Naval Research Laboratory's Timation Satellites

By 1972, another Navy satellite system was extending the state of the art by orbiting very precise clocks. Known as *Timation,*[2] these satellites were developed under the direction of Roger Easton at the Naval Research Laboratory (NRL). They were used principally to provide very precise time and time transfer between various points on the Earth. In addition, they could provide navigation information. The ranging signals used a technique called *side-tone ranging,* which broadcast a variety of synchronized tones to resolve phase ambiguities.

Initially these spacecraft used very stable *quartz-crystal oscillators;* later models were to orbit the first *atomic frequency standards* (rubidium and cesium). The atomic clocks typically had a frequency stability of several parts in 10^{12} (per day) or better. This frequency stability greatly improves the prediction of satellite orbits (ephemerides) and also extends the time between required control segment updates to the GPS satellites. Timation satellites were flown in inclined orbits: the first two at altitudes of 500 nautical miles and the last in the series at 7500 nautical miles. The third satellite was also used as a technology demonstrator for GPS (see Fig. 1). This pioneering work in space-qualified time standards was an important foundation for GPS.

3. U.S. Air Force Project 621B

The third essential foundation for GPS was a U.S. Air Force program known as 621B. This program was directed by an office in the Advanced Plans group at the Air Force's Space and Missile Organization (SAMSO) in El Segundo, California. By 1972, this program had already demonstrated the operation of a new type of satellite-ranging signal based on pseudorandom noise (PRN). The signal modulation was essentially a repeated digital sequence of fairly random bits—ones or zeros—that possessed certain useful properties. The sequence could be easily generated by using a shift register or, for shorter codes, simply by storing the entire sequence of bits. A navigation user could detect the start

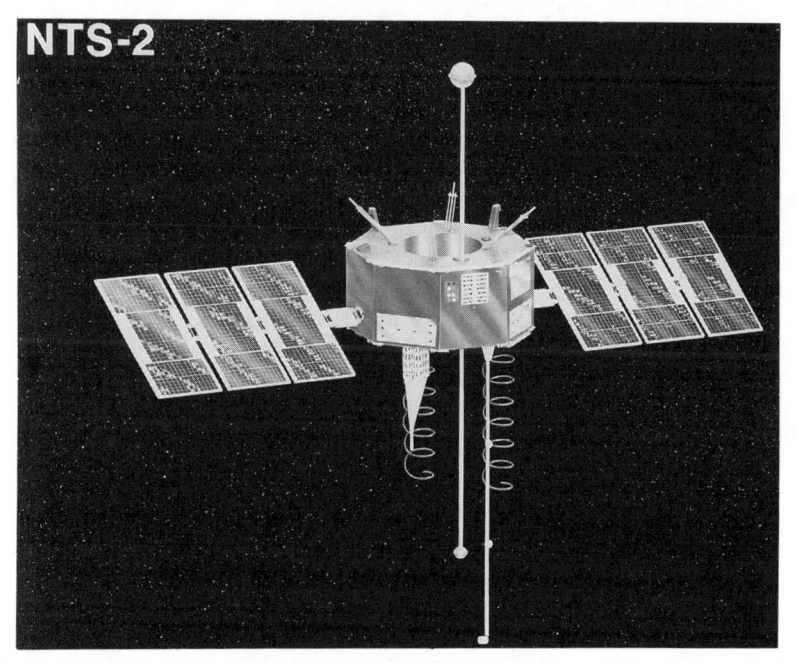

Fig. 1 Navigation technology Satellite II (NTS-II). This satellite was launched as part of the joint program effort to develop reliable spacecraft atomic clocks. NRL called this satellite Timation III (drawing courtesy of NRL).

("phase") of the repeated sequence and use this for determining the range to a satellite. The signals could be detected even when their power density was less than 1/100th that of ambient noise, and *all satellites could broadcast on the same nominal frequency* because properly selected PRN coding sequences were nearly orthogonal. Successful aircraft tests had been run at Holloman AFB to demonstrate the PRN technique. The tests used simulated satellite transmitters located on the floor of the New Mexican desert.

The ability to reject noise also implied a powerful ability to reject most forms of jamming or deliberate interference. In addition, a communication channel could be added by inverting the whole sequence at a slow rate and using these inversions to indicate the ones or zeros of digital data. This slow communication link (50 b/s) allowed the user to receive ephemeris (satellite location) and clock information.

The original Air Force concept visualized several constellations of highly eccentric satellite orbits with 24-h periods. Alternative constellations were nicknamed the *egg-beater, the rotating X and the rotating Y configurations* because of their resulting ground traces. Although these designs allowed the system to be deployed gradually (for example, North and South America first) they had high line-of-sight accelerations. Initially, the concept relied on continuous measurement from the ground to keep the signals time-synchronized. Later, the NRL clock concept was added because the synchronizing link would have been quite vulnerable. The GPS did substitute the Timation clocks later to remove any reliance on continuous ground contact.

B. Joint Program Office Formed, 1973

In the early 1970s, a number of changes in the systems acquisition process had begun to be adopted for the Department of Defense. These changes, recommended by David Packard, were to have a profound effect on NAVSTAR and other major DOD programs. To increase efficiency and reduce interservice bickering, "joint" programs were formed that forced the various services to work together. The GPS was one of the earliest examples. It was decreed to be a Joint Program, with a Joint Program Office (JPO) located at the Air Force's Space and Missile Organization and to have multiservice participation (with the Air Force as the lead service).

The first program director was Dr. (Col.) Bradford W. Parkinson (see Fig. 2), supported by Deputy Program directors—eventually from the Army, Navy, Marine Corps, Defense Mapping Agency, Coast Guard, Air Logistics Command, and NATO. Also continuing their support of 621B was a small cadre of engineers from the Aerospace Corporation under Mr. Walter Melton. Dr. Parkinson was directed to develop the initial concept as a joint development and to gain approval of the Department of Defense to proceed with full-scale demonstration and development.

There have been many speculations on the origin of the names Global Positioning System, and NAVSTAR. The GPS title originated with General Hank Stehling who was the Director of Space for the U.S. Air Force DCS Research and Development (R&D) in the early 1970s. He pointed out to Dr. Parkinson that "navigation" was an inadequate descriptor for the proposed concept. He suggested that "Global

Fig. 2 Joint program included deputy directors from all services. Dr. (Colonel) Parkinson is in discussions with his Navy Deputy, Cdr. Bill Huston of the U.S. Navy. Models of the NTS-II and Phase one GPS satellites are on the table. The civilian is Mr. Frank Butterfield of the Aerospace Corporation (photo courtesy of the U.S. Air Force).

Positioning System" would be a better name. The JPO enjoyed his sponsorship, and this insightful description was immediately adopted.

The title NAVSTAR came into being in a somewhat similar manner. Mr. John Walsh [an Associate Director of Defense Development, Research, and Engineering (DDR&E)] was a key decision maker when it came to the budget for strategic programs in general, including the proposed satellite navigation program. In the contention for funding, his support was not as fervent as the JPO would have liked. During a break in informal discussions between Mr. John Walsh and Col. Brent Brentnall (the program's representative at DOD), Mr. Walsh suggested that NAVSTAR would be a nice sounding name. Colonel Brentnall passed this along as a good idea to Dr. Parkinson, noting that if Mr. Walsh were to name it, he would undoubtedly feel more protective toward it. Dr. Parkinson seized the opportunity, and ever since the program has been known as NAVSTAR the Global Positioning System. Although some have assumed that NAVSTAR was an acronym, in fact, it was simply a nice sounding name* that enjoyed the support of a key DOD decision maker.

---

*We should note that TRW apparently had advocated a navigational system for which NAVSTAR was an acronym (*NAV*igation *S*ystem *T*iming *A*nd *R*anging). This may have been in Mr. Walsh's subliminal memory, but was not part of the process. It was never used as an acronym.

1. Failed Defense System Acquisition and Review Council

Fortunately, the first attempt to gain system approval failed in August 1973. The program that was brought before the Defense System Acquisition and Review Council (DSARC) at that time was not representative of a joint program. Instead it was packaged as the 621B system. Dr. Malcolm Currie,* then head of DDR&E, expressed strong support for the idea of a new satellite-based navigation system, but requested that the concept be broadened to embrace the views and requirements of all services.

2. Synthesis of a New System

With this philosophy, Dr. Parkinson and the Joint Program Office immediately went to work. Over the Labor Day weekend of 1973, he assembled about a dozen members of the JPO on the fifth floor of the Pentagon. He directed the development of a new design that employed the best of all available satellite navigation system concepts and technology. The result was a system proposal that was not exclusively the concept of any prior system but rather was a synthesis of them all. The details of the proposed GPS are outlined below. Its multiservice heritage precluded any factual basis for further bickering, because all contending parties now were part of the conception process. From that point forward, the JPO acted as a multiservice enterprise, with officers from all services attending reviews and meetings that had previously been "Air Force only."

3. Approval to Proceed with GPS

To gain approval for the new concept, Dr. Parkinson began to contact all those with some stake in the decision. After interminable rounds of briefings† on the new approach were given to offices in the Pentagon and to the operating armed forces, a successful DSARC was held on 17 December 1973. Approval to proceed was granted.

The first phase of the program included four satellites (one was the refurbished qualification model), the launch vehicles, three varieties of user equipment, a satellite control facility, and an extensive test program. By June of 1974, the

---

*Dr. Currie was appointed head of DDR&E in early 1973, as part of the incoming administration. He had been living in Los Angeles prior to his appointment and to complete his move he made numerous trips to Los Angeles in the initial months. One legitimate official purpose of these trips was to review programs at SAMSO. After a few trips, he had done all the high-level reviews that were available, so the head of SAMSO, General Schultz, suggested that he receive an indepth review by Dr. (Col.) Parkinson on the space-based navigation concept, then known as 621B. This resulted in a remarkable meeting with the number-three man in all of the U.S. DOD spending about three hours in a small office with a lowly Colonel, talking about engineering, technology, and the wide applications of the proposed system. With his doctorate in Physics, Mal Currie was a keen and quick study. He had a great deal of space experience from his years at Hughes Aircraft. The outcome was that the GPS program enjoyed his steadfast support. Without this key decision maker, the Air Force would have killed the program in favor of additional airplanes. The pivotal (and coincidental) meeting with Dr. Parkinson was destined to be an essential factor in gaining system approval.

†Lt. General Ken Schultz was particularly incensed with the endless presentations that had to be made in the Washington arena. The situation with any bureaucracy is that many can say no, and few (if any) can say yes. To bring the naysayers to neutral, the extended trips from Los Angeles to Washington were necessary for Dr. Parkinson.

satellite contractor, Rockwell International, had been selected, and the program was well underway. The initial types of user equipment included sequential and parallel military receivers, as well as a civil type set for utility use by the military. The development test and evaluation was extensive, with a laser-tracking range set up at the Army's Yuma Proving Ground. An independent evaluation was performed by the Air Force's Test and Evaluation Command.

To maintain the focus of the program the JPO adopted a motto:

> The mission of this Program is to:
> 1. Drop 5 bombs in the same hole, and
> 2. Build a cheap set that navigates (<$10,000),
>   and don't you forget it!

The program developed rapidly; the first operational prototype satellite was launched in February of 1978 (44 months after contract start). By this time, the initial control segment was deployed and working, and five types of user equipment were undergoing preliminary testing at the Yuma Proving Ground. The initial user equipment types had been expanded to include a 5-channel set developed by Texas Instruments and a highly jam-resistant set developed by Rockwell Collins.

4. Needed: A Few More Good Satellites

As stated, only four satellites were initially approved by the DOD, including a refurbished qualification model (see Fig. 3). It became apparent that there was a need for additional satellites, because the minimum number for three-dimensional navigation is four. Any launch or operational failure would have gravely impacted the Phase I demonstration program. Authorization for spare GPS satellites was urgently needed.

The Navy's Transit program inadvertently solved this problem. The chain of events began when Transit requested funds for upgrading certain Transit satellites to a PRN code similar to that used by GPS. The purpose was to provide accurate tracking of the Trident (submarine launched missile) booster during test firings into the broad ocean areas. Dr. Bob Cooper of DDR&E requested a series of reviews to address whether GPS could fulfill this mission.

The GPS solution was to use a signal translator on the Trident missile bus that would relay the GPS modulations to the ground on another frequency. The central issues were whether the ionosphere could be adequately calibrated (because it was a single-frequency system, the ionosphere could not be directly measured), and whether the translated signal could be recorded with sufficient fidelity (it required digitizing at 60 MHz).

During the third and capstone review for Dr. Cooper, Dr. Parkinson (supported by Dr. Jim Spilker and Dr. Jack Klobuchar) was able to present convincing arguments that a GPS solution could solve the Trident problem *provided two additional satellites were authorized.* Dr. Cooper immediately made the decision to use GPS. He directed the transfer of $60M from the Navy to the Air Force, approving two additional satellites, and thereby greatly expanding the Phase one test time as well as significantly reducing the program risk. This little known

Fig. 3 Phase I GPS satellite. It is a three-axis stabilized design with double and triple redundancy where appropriate (drawing courtesy of the U.S. Air Force).

event also eliminated the possibility of an upgraded Transit program competing with the fledgling GPS.

III. Introductory GPS System Description and Technical Design

The operational GPS system of today is virtually identical to the one proposed in 1973. The satellites have expanded their functionality to support additional military capabilities; the orbits are slightly modified, but the equipment designed to work with the original four satellites would still perform that function today. The rest of this volume is devoted to detailed technical descriptions of the system and its applications; the following section provides an overview of the system design.

A. Principles of System Operation

The fundamental navigation technique for GPS is to use *one-way ranging* from the GPS satellites that are also broadcasting their estimated positions. *Ranges are measured to four satellites simultaneously in view by matching (correlating) the incoming signal with a user-generated replica signal and measuring the received phase against the user's (relatively crude) crystal clock.* With four satellites and appropriate geometry, four unknowns can be determined; typically, they are: latitude, longitude, altitude, and a correction to the user's clock. If altitude or time are already known, a lesser number of satellites can be used (see Fig. 4).

Each satellite's future position is estimated from ranging measurements taken at worldwide monitoring stations.* These ranging measurements use the same signals that are employed by a typical user's receiver. Using sophisticated prediction algorithms, the master control station forms estimates of future satellite locations and future satellite clock corrections. For the uploads, which occur daily or (optionally) more frequently, the combined predictions for satellite clock and position have been measured to have an average rms error of 2–3 m. These estimates have demonstrated reasonable errors even after three days (24.3 m of expected ranging error).

B. GPS Ranging Signal

The GPS ranging signal is broadcast at two frequencies: a primary signal at 1575.42 MHz (L_1) and a secondary broadcast at 1227.6 MHz (L_2). These signals are generated synchronously, so that a user who receives both signals can directly calibrate the ionospheric group delay and apply appropriate corrections. However, most civilian users will only use the primary or L_1 frequency.

Potentially, both the signal at the L_1 frequency and the signal at L_2 can *each* have two modulations at the same time (called "phase quadrature"). Current

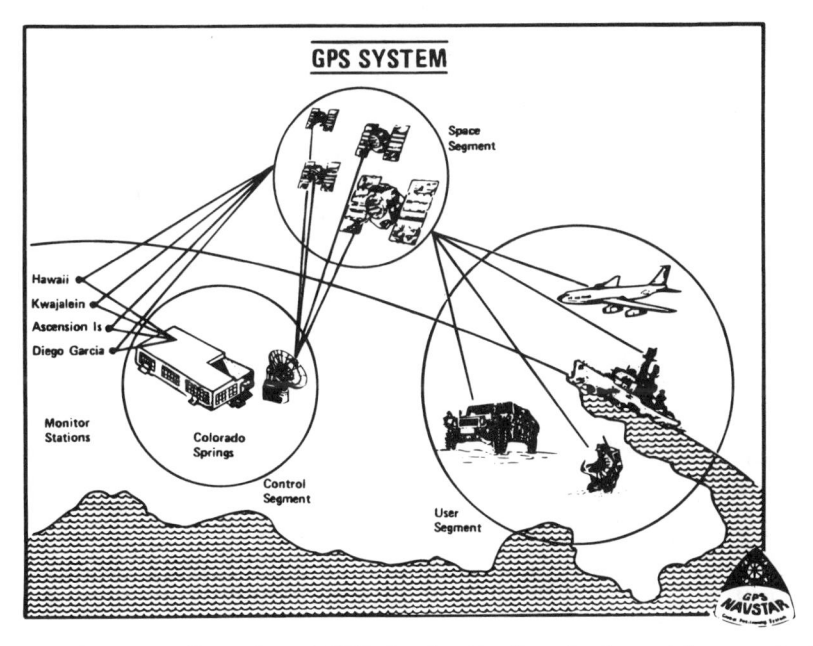

Fig. 4 System configuration of GPS showing the three fundamental segments: 1) user; 2) spacecraft; and 3) ground control (drawing courtesy of the U.S. Air Force).

---

*The Operational Control System (OCS) uses five monitor stations which are located at Colorado Springs, Ascension Island, Diego Garcia, Kwajalein, and Hawaii.

implementation has two modulations on the higher frequency (L_1), but only a single (protected) modulation (see below) on L_2. The two modulations are as follows:

1) C/A or Clear Acquisition Code: This is a short PRN code broadcast at a bit (or chipping) rate of 1.023 MHz. This is the principal civilian ranging signal, and it is always broadcast in the *clear* (unencrypted). It is also used to acquire the much longer P-code. The use of this signal is called the Standard Positioning Service or SPS. It is always available, although it may be somewhat degraded. At this time, and for the projected future, the C/A code is available only on L_1 (some civil users have requested C/A modulation on L_2 to allow ionospheric calibration).

2) P or Precise Code (sometimes called the Protected Code): A very long code (actually segments of a 200-day code) that is broadcast at ten times the rate of C/A, 10.23 MHz. Because of its higher modulation bandwidth, the code ranging signal is somewhat more precise. This reduces the noise in the received signal but will not improve the inaccuracies caused by biases. This signal provides the Precise Positioning Service or PPS. The military has encrypted this signal in such a way that renders it unavailable to the unauthorized user. This ensures that the unpredictable code (to the unauthorized user) cannot be spoofed. This feature is known as antispoof or AS. When encrypted, the P code becomes the Y code. Receivers that can decrypt the Y code are frequently called P/Y code receivers. As a result of the military intent, most civilian users should only rely on the C/A code or SPS.*

1. Selective Availability

In addition, the military operators of the system have the capability to degrade the accuracy of the C/A code intentionally by desynchronizing the satellite clock, or by incorporating small errors in the broadcast ephemeris. This degradation is called *Selective Availability, or S/A*. The magnitude of these *ranging* errors is typically 20 m, and results in rms *horizontal position* errors of about 50 m, one sigma. The official DOD position is that errors will be limited to 100 m, 2 drms, which is about the 97th percentile. A technique known as differential GPS (explained later) can overcome this limitation and potentially provide accuracies sufficient for precision approach of aircraft to landing fields.

2. Data Modulation

One additional feature of the ranging signal is a 50 b/s modulation used as a communications link. Through this link, each satellite transmits its location and the correction necessary to apply to the spaceborne clock.† Also communicated are the health of the satellite, the locations of other satellites, and the necessary information to lock on to the P code after acquiring the C/A code.

---

*There are provisions in the Federal Radionavigation Plan for civilian users with critical national needs to gain access to the P code.

†Although the atomic clocks are extremely stable, they are running in an uncorrected mode. The clock correction is an adjustment that synchronizes all clocks to GPS time.

C. Satellite Orbital Configuration

The orbital configuration approved at DSARC in 1973 was a total of 24 satellites—eight in each of three circular rings with inclinations of 63 deg. The rings were equally spaced around the equator, and the orbital altitudes were 10,980 n.mi. This altitude gave two orbital periods per sidereal day (known as *semisynchronous*) and produced repeating ground traces. The altitude was a compromise among: user visibility, the need to pass over the continental U.S. ground/upload stations periodically, and the cost of the spacecraft launch boosters. Three rings of satellites were initially selected because it would be easier to have orbital spares—only three such spares could easily replace any single failure in the whole constellation. This configuration provided a minimum of six satellites in view at any time, with a maximum of 11. As a result of this redundancy, the system was robust in the sense that it could tolerate occasional satellite outages (see Fig. 5).

Two changes have been made since the original constellation proposal. The inclinations have been reduced to 55 deg, and the number of orbital planes have been increased to six,* with four satellites in each. The number of satellites, including spares, remains 24.

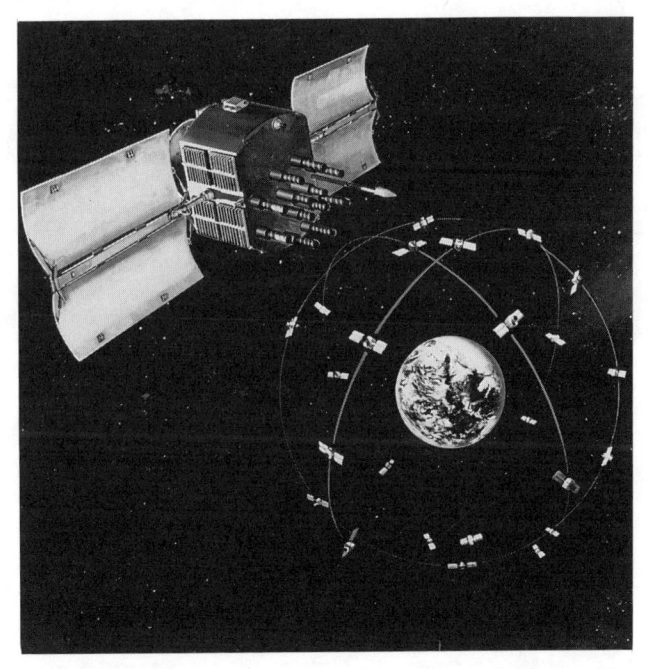

Fig. 5 Original GPS orbital configuration of three rings of eight satellites each. The final operational configuration has the same number of satellites, arranged in six rings of four satellites (drawing courtesy of the U.S. Air Force).

---

*For a number of reasons, the Department of Defense calls the configuration "21 satellites with three orbiting spares." We may find the system eventually having more satellites to increase robustness.

D. Satellite Design

The GPS satellites are attitude stabilized on all three axes and use solar panels for basic power (see Fig. 6). The ranging signal is radiated through a shaped beam antenna—by enhancing the received power at the limbs of the Earth, compensation is made for "space loss." The user, therefore, receives fairly constant power for all local elevation angles.* The satellite design is generally doubly or triply redundant, and the Phase I satellites demonstrated average lifetimes in excess of 5 years (and in some cases over 12).

E. Satellite Autonomy: Atomic Clocks

A key feature of the GPS design is that the satellites need not be continuously monitored and controlled. To achieve this autonomy, the satellites must be predictable in four dimensions: three of *position* and one of *time*. Predictability, in the orbital *position,* is aided because the high-altitude orbits are virtually unaffected by atmospheric drag. Many other factors which affect orbital position must also be considered. For example, variations in geopotential, solar pressure, and outgassing can all have significant effects.

When GPS was conceived, it was recognized that the most difficult technology problem facing the developers was probably the need to fly accurate timing

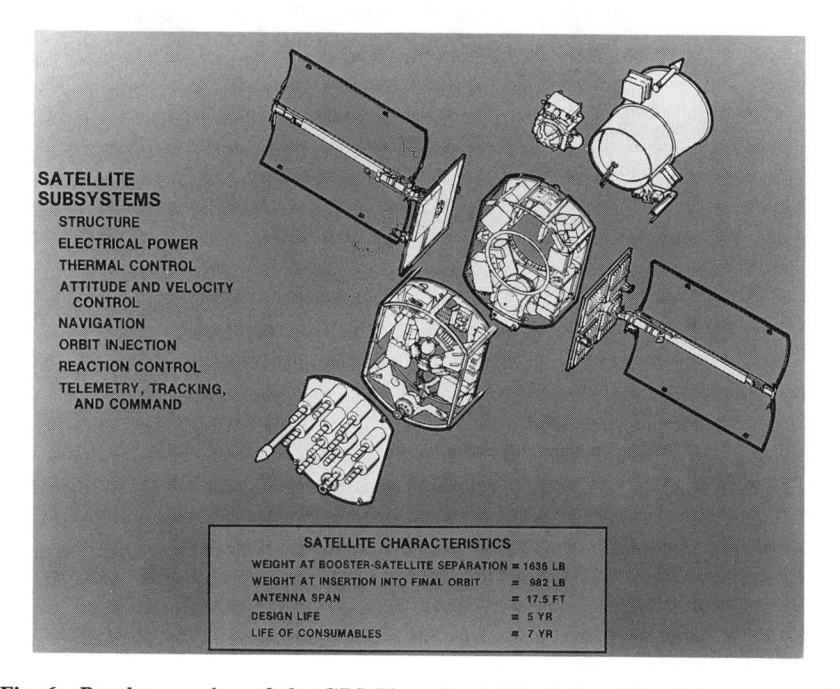

Fig. 6 Breakaway view of the GPS Phase I satellite design (drawing courtesy of the U.S. Air Force).

---

*The requirement for received power on L_1 is -163 dbw into an isotropic, circularly polarized antenna on the primary frequency.

standards, insuring that all satellites' clocks remained synchronized. As mentioned, NRL had been developing frequency standards for space, so this effort was continued and extended.

Payoff of a Good Clock

The basic arithmetic can be understood as follows: A day is about 100,000 s, or 10^5. Light travels about 1 ft per ns (10^{-9} s). If the system can tolerate an error buildup caused by the atomic clock of 5 ft, the stability must be 5 ns per upload (one-half a day). This is about $(5*10^9)/(5*10^4)$ sps, measured over 12 h. Therefore, this requirement is for a clock with about one part in 10^{13} stability,* which can only be met by an atomic standard. Note that there is a roughly constant frequency shift attributable to relativistic effects (both special and general) of about 4.5 parts in 10^{10}, which is compensated by a deliberate offset in the clock frequency.

GPS traditionally has used two types of atomic clocks: rubidium and cesium. Phase one test results for the rubidium cell standard are shown in Fig. 7. A key to outstanding satellite performance has been the stability of the space-qualified atomic clocks, which exceeded the specifications. They have measured stabilities of one part in 10^{13} over periods of 1–10 days.[3]

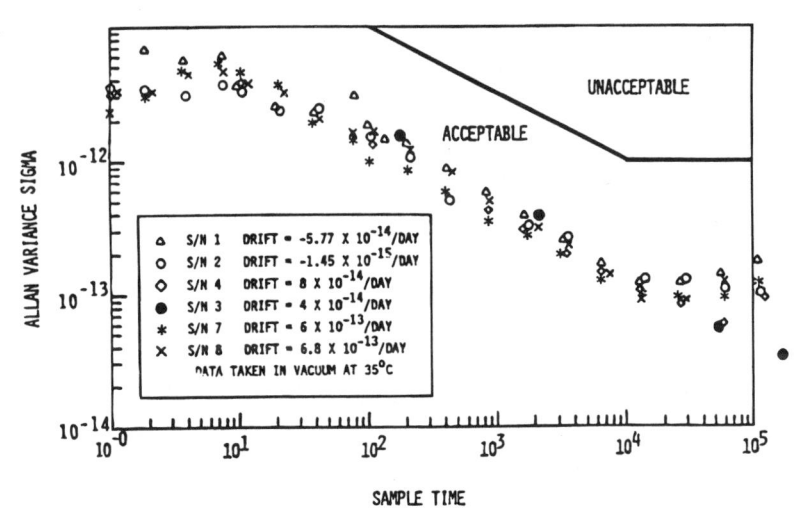

Fig. 7 Space qualified rubidium-cell frequency standard performance. These units were developed by Rockwell as a derivative of a clock designed by Efratom, Inc. (data courtesy of the U.S. Air Force).

---

*Clock stability is traditionally measured with the Allen variance, which shows stability versus averaging time. For short averaging times (1 s) virtually all clocks are dominated by the quartz oscillator, which acts as the short-term flywheel. In Phase one, the clocks were specified at 10^{-12}, measured over 1 day.

F. Ionospheric Errors and Corrections

The free electrons in the ionosphere create a delay in the modulation signal (PRN code). This delay is proportional to the integrated number of free electrons along the transmission path and inversely proportional to the square of the transmission frequency (to first order). The path delay at any elevation angle is often expressed as the product of a zenith delay (elevation equals 90 deg) and an *obliquity* factor that is a function of that elevation angle. This ratio ranges from 1.0 at the zenith (by definition) to about 3.0 at small elevation angles. Typical zenith (or vertical) delays range from a few meters at night to a maximum of ten or twenty meters at about 1400 hours (local solar time). Thus, it is not unusual to find delays of over 30 m at lower elevation angles. Fortunately, these delays are highly correlated between satellites, which helps reduce the calculated horizontal position errors.*

There are two techniques for correcting this error. The first is to use an ionosphere model. The 8-model parameters used to calculate the correction are broadcast as part of the GPS 50 b/s message. This model is typically accurate to a few meters of vertical error.

The second technique uses both broadcast frequencies and the inverse square law behavior to measure the delay directly. By differencing the code measurements on each frequency, the delay on L_1 is approximately 1.546*(difference in delays on L_1 and L_2). This technique is only available to a P/Y-code receiver (because the only L_2 modulation is the P code) or to a codeless (or cross-correlating) receiver.

G. Expected Navigation Performance

The performance capabilities of GPS are primarily affected by two things: 1) the *satellite geometry* (which causes *geometric dilution*); and 2) the *ranging errors*. Under the assumption of uniform, uncorrelated, zero-mean, ranging-error statistics, this can be expressed as follows:

$$\text{RMS position error} = (\text{Geometric dilution})*(\text{rms ranging error})$$

1. Geometric Dilution

The geometric dilution can be calculated for any instantaneous satellite configuration, as seen from a particular user's location. The details of this calculation are explained in Chapters 6 and 11, this volume. For a 21-satellite constellation and a three-dimensional fix, the world median value of the geometric dilution factor (for the nominal constellation) is about 2.7. This quantity is usually called PDOP or position dilution of precision. Typical dilution factors range from 1.5 to 8. The *variations* in this *dilution factor* are typically much greater than the variations in ranging errors.

---

*This correlation of errors due to the ionosphere will mostly show up as a user clock error. While this is not important for many navigation users, it is critical for the precise transfer of time. Such users must employ the dual frequency technique to eliminate this error.

2. Ranging Errors

Ranging errors are generally grouped into the following six major causes:
1) Satellite ephemeris
2) Satellite clock
3) Ionospheric group delay
4) Tropospheric group delay
5) Multipath
6) Receiver measurement errors, including software

Some of these errors tend to be correlated for the same satellite. For example, satellite clock and ephemeris errors tend to be negatively correlated; i.e. they tend to cancel each other somewhat. Other errors tend to be correlated between satellites. For example, the ionospheric and tropospheric group delays always have the same sign, because they are the result of signal paths penetrating the *same blanket of media* with different angles.

With S/A turned off, all errors for *single frequency* SPS are nearly *identical* in magnitude to those for *single-frequency PPS* except for receiver measurement errors (which decrease with increasing bandwidth). *Dual* frequency, which is only available on PPS, can reduce the third error (attributable to the ionosphere) to about 1 m.*

3. Precise Positioning Service (PPS) Accuracy

Ranging errors (including the effects of the satellite clock)[3] for the PPS have been *specified* to be better than 6 m. The product of the average PDOP and the ranging error is the specified three-dimensional accuracy of 16 m spherical error probable (SEP).†

Because each of the five worldwide monitoring stations is continuously measuring the ranging errors to all satellites in view, these measurements are a convenient statistic of the basic, static accuracy of GPS. Table 1 summarizes over 11,000 measurements taken from 15 January to 3 March 1991, during the "Desert Storm" operation of the Gulf War. The S/A feature was not activated during this period. Note that the PPS results presumably are not affected by S/A at all.

During this period, satellite (PRN 9) was ailing but is included in the solution, making the results somewhat worse than would be expected. By dividing the overall SEP by the rms PDOP, an estimate of the effective ranging error can be formed. The average of these results is 2.3 m.‡ This should be compared to the specification of 6 m. Because SEP is smaller than the RMS error, this estimate may be about 15% optimistic.

---

*Multipath errors are generally negligible for path delays that exceed one-and-one-half modulation chips, expressed as a range. Thus, P-code receivers reject reflected signals whose path delay exceeds 150 feet. For the C/A code, the number is 1500 feet, giving a slight advantage to the P code, although it is usually reflections from very close objects that are the main source of difficulty.

†SEP is the radius of the sphere that will contain 50% of the expected errors in three dimensions.

‡This number is probably somewhat better than an average receiver would measure for several reasons. Monitor station receivers are carefully sited to avoid multipath. The receivers are of excellent quality and are not moving. Also, since the monitor station measurements are used to update the ephemeris, there may be some tuning to make the predictions match any peculiarities (e.g. survey errors) at the monitor station locations. Nonetheless, an average ranging error of 2.3 m is an impressive result.

Table 1 PPS measured accuracies; SEP/CEP[a] navigation errors; PPS solutions for the OCS monitor stations during Desert Storm; S/A is off.[4]

Criteria	All	Colorado Springs	Ascension	Hawaii	Diego Garcia	Kwajalein
SEP three-dimensional, m	8.3	7.8	6.8	9.0	9.1	9.0
CEP two-dimensional, m	4.5	4.5	3.8	5.1	4.6	5.0
rms PDOP	3.6	3.9	3.4	3.9	3.4	3.3
Estimated range error, m[b]	2.3	2.0	2.0	2.3	2.7	2.7

[a] CEP is circular error probable, which equals the radius of a circle that would contain 50% of the errors. It is the two-dimensional analog of SEP.
[b] This row is formed by dividing the SEP by the rms PDOP.

Table 2 Expected accuracies for various operating conditions of GPS

	Precise positioning service (PPS)		Standard positioning service (SPS), estimated capability	
	Specification	Measured, static	No S/A	With S/A
Ranging accuracy	6 m	2.3 m	6 m	20 m
CEP (horizontal)	—	4.6 m	12 m	40 m
SEP (three-dimensional)	—	8.3 m[a]	22 m	72 m

[a] The SEP reported for dynamic PPS users was less than 10 m. See Ref. 5.

a. SPS accuracy. Without the degradation of S/A, the SPS would provide solutions with about 50% greater error because of uncompensated ionospheric effects and somewhat greater receiver noise (because of the narrower band C/A code). *It is reasonable to expect that rms horizontal errors for SPS with S/A off would be less than 10 m.*

b. Accuracy summary for code-tracking receivers. Table 2 summarizes the expected accuracies for GPS.

H. High Accuracy/Carrier Tracking

A special feature of GPS, which initially was not generally understood, is the ability to create an extremely precise ranging signal by reproducing and tracking the rf carrier (1575.42 MHz). Because this signal has a wavelength of 19 centimeters (7.5 inches), tracking it to 1/100th of a wavelength provides a precision of about 2 mm. Modern receivers can attain these tracking precisions, but unfortunately, this is not *accuracy*. To provide equivalent accuracy, we must determine which carrier cycle is being tracked (relative to the start of modulation) and compare this with another carrier tracking receiver located at a known position.

INTRODUCTION AND HERITAGE OF NAVSTAR

Surveyors use a technique of double or triple differencing to resolve this *cycle ambiguity*. For dynamic users, the problem is a bit harder. Reflected signals (multipath) and distortions of the ionosphere can be significant errors.

Generally, the carrier-tracking techniques can be used in two ways. For normal use, the carrier tracking can smooth the code tracking and greatly reduce the noise content of the ranging measurement. The other use is in a differential mode for which there are several variations, including surveying, direct measurement of vehicle attitude (with multiple antennas), and various forms of dynamic differential. The GPS control segment uses accumulated delta range (ADR) as the measurement for the monitor stations. This is done with incremental counts of carrier cycles. Later chapters in this volume discuss these techniques in greater detail.

III. History of Satellites

Five groups (or blocks) of satellites have been developed for the GPS program. In chronological order they are: 1) Navigation technology satellites (NTS); 2) navigation development satellites—Block I or NDS; 3) block II satellites; 4) block IIA satellites; and 5) block IIR satellites. In addition, a follow-on group, called block IIF, is being planned.

A. Navigation Technology Satellites

The satellites of the first group were used to explore space technology. They were an extension of the Timation program of the NRL and were known as NTS, or, Navigation Technology Satellites. The first, NTS-1, had been planned as Timation II and was renamed when the JPO was formed. It was launched into a lower orbit than GPS (7500 n.mi.) on the 14th of July 1974. It was the first to fly atomic clocks: two rubidium oscillators were included. The second (and last) of the series, NTS-2, orbited a number of payload components that were identical to the development GPS satellites. This satellite included the first cesium clock in space, the PRN code generator used in the next block, and the first GPS spaceborne computer. The last two items were developed by Rockwell International under the JPO.

B. Navigation Development Satellites—Block I

These pioneering satellites were developed by Rockwell International for the JPO. The initial buy was for four, followed by two additional to support the Trident program. Later, six more were purchased as replacements. Of the 11 satellites that made it into orbit, all achieved initial operational capability. The sole premature failure in these satellites [in all, the satellites launched through the initial Operational capability (IOC)] was caused by the malfunction of a refurbished Atlas-F booster at Vandenberg. The first launch was on the 22nd of February 1978, 44 months after contract award. Designed for 3-year lifetimes, several operated for over 10 years.

C. Operational Satellites—Block II and IIA

In all, 29 satellites have been produced and are on orbit or ready to launch. The first satellite in this series was declared operational on the 10th of August

1989. These satellites were initially launched at a rate of about six per year. Initial operational capability was declared at the end of 1993, with full operational capability (FOC) attained by the end of 1994.

D. Replacement Operational Satellites—Block IIR

These are enhanced performance GPS satellites being developed by Martin Marietta (after buying out the division from GE, which bought it from RCA). The contract was awarded on 21 June 1989 for 20 satellites, with options for 6 more. The first delivery is planned for 1995. These satellites have enhanced autonomy, including the capability to meet a degraded range error specification of up to 180 days since the last ground control segment upload. They also have increased hardening against natural and man-made radiation.

IV. Launches

A. Launch Vehicles

The 12 original (Phase I) satellites were to be launched on refurbished Atlas-F ICBMs (see Fig. 8). The initial plan was to use the McDonnell-Douglas Delta for the next series of launches (Phase II). About 1979, this was changed, and the shuttle was decreed to be the booster of choice for Air Force missions. The

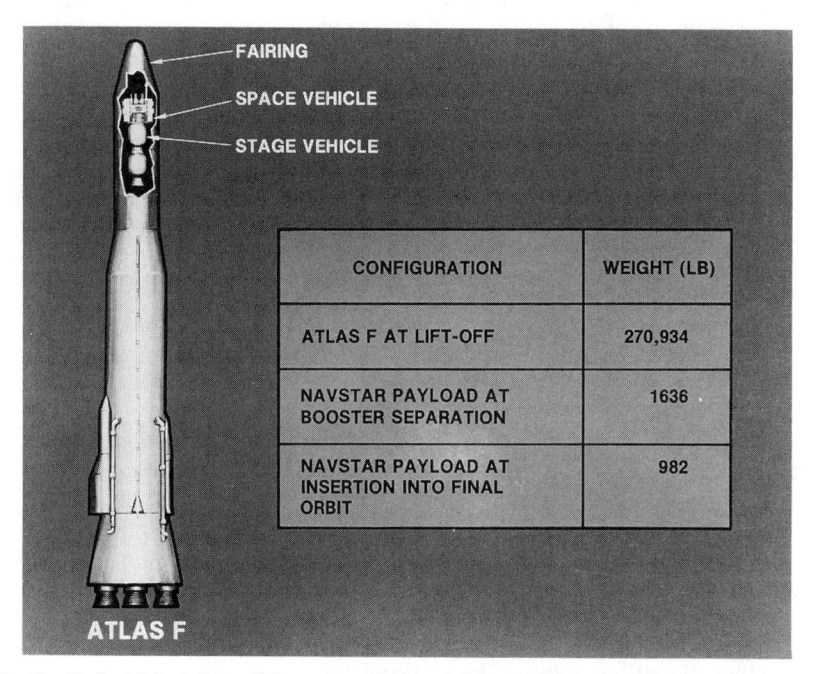

CONFIGURATION	WEIGHT (LB)
ATLAS F AT LIFT-OFF	270,934
NAVSTAR PAYLOAD AT BOOSTER SEPARATION	1636
NAVSTAR PAYLOAD AT INSERTION INTO FINAL ORBIT	982

Fig. 8 Refurbished Atlas F launch vehicle used for the first 12 launches. The stage vehicle was a tandem (stacked) solid rocket configuration (drawing courtesy of the U.S. Air Force).

INTRODUCTION AND HERITAGE OF NAVSTAR

Table 3 History of initial GPS launches through IOC

Block II seq.	SVN	PRN, code	Internat. ID	NASA catalog number[a]	Orbit plane pos'n.[b]	Launch date, UTC	Clock[c]	Available	Decommissioned
Block I									
—	01	04	1978-020A	10684	—	78-02-22	—	78-03-29	85-07-17
—	02	07	1978-047A	10893	—	78-05-13	—	78-07-14	81-07-16
—	03	06	1978-093A	11054	—	78-10-06	—	78-11-13	92-05-18
—	04	08	1978-112A	11141	—	78-12-10	—	79-01-08	89-10-14
—	05	05	1980-011A	11690	—	80-02-09	—	80-02-27	83-11-28
—	06	09	1980-032A	11783	—	80-04-26	—	80-05-16	91-03-06
—	07				—	81-12-18	Launch failure		—
—	08	11	1983-013A	14189	—	83-07-14	—	83-08-10	93-05-04
—	09	13	1984-059A	15039	C-1	84-06-13	Cs	84-07-19[d]	—
—	10	12	1984-097A	15271	A-1	84-09-08	Rb	84-10-03	—
—	11	03	1985-093A	16129	C-4	85-10-09	Rb[e]	85-10-30	—
Block II									
II-1	14	14	1989-013A	19802	E-1	89-02-14	Cs	89-04-15	—
II-2	13	02	1989-044A	20061	B-3	89-06-10	Cs	89-08-10	—
II-3	16	16	1989-064A	20185	E-3	89-08-18	Cs	89-10-14	—
II-4	19	19	1989-085A	20302	A-4	89-10-21	Cs	89-11-23	—
II-5	17	17	1989-097A	20361	D-3	89-12-11	Cs	90-01-06	—
II-6	18	18	1990-008A	20452	F-3	90-01-24	Cs	90-02-14	—
II-7	20	20	1990-025A	20533	B-2	90-03-26	Cs	90-04-18	—
II-8	21	21	1990-068A	20724	E-2	90-08-02	Cs	90-08-22	—
II-9	15	15	1990-088A	20830	D-2	90-10-01	Cs	90-10-15	—
Block IIA									
II-10	23	23	1990-103A	20959	E-4	90-11-26	Cs	90-12-10	—
II-11	24	24	1991-047A	21552	D-1	91-07-04	Cs	91-08-30	—
II-12	25	25	1992-009A	21890	A-2	92-02-23	Rb	92-03-24	—
II-13	28	28	1992-019A	21930	C-2	92-04-10	Cs	92-04-25	—
II-14	26	26	1992-039A	22014	F-2	92-07-07	Cs	92-07-23	—
II-15	27	27	1992-058A	22108	A-3	92-09-09	Cs	92-09-30	—
II-16	32	01[f]	1992-079A	22231	F-1	92-11-22	Cs	92-12-11	—
II-17	29	29	1992-089A	22275	F-4	92-12-18	Cs	93-01-05	—
II-18	22	22	1993-007A	22446	B-1	93-02-03	Cs	93-04-04	—
II-19	31	31	1993-017A	22581	C-3	93-03-30	Cs	93-04-13[g]	—
II-20	37	07	1993-032A	22657	C-4	93-05-13	Cs	93-06-12	—
II-21	39	09	1993-042A	22700	A-1	93-06-26	Cs	93-07-20	—
II-22	35	05	1993-054A	22779	B-4	93-08-30	Cs	93-09-28	—
II-23	34	04	1993-068A	22877	D-4	93-10-26	Cs	93-11-22	—
II-24	—	—	—	—	—	—	—	—	—

[a] NASA Catalog Number is also know as NORAD or U.S. Space Command object number.
[b] No orbital plane position listed = satellite no longer operational.
[c] Clock: Rb = rubidium; Cs = cesium
[d] The power supply of PRN 13 has insufficient capacity to maintain L_1/L_2 transmissions during eclipse season. During this period, the L_1/L_2 transmissions of PRN 13 may be turned off for up to 12 h a day.
[e] PRN 03 is operating on Rb clock without temperature control.
[f] The PRN number of SVN 32 was changed from 32 to 01 on 93-01-28.
[g] Corrective maintenance performed on PRN 31 on 93-06-16 seems to have fixed the L_2 intermittent-lock problem.

Block II satellites were designed to that interface. After the Challenger accident, this decision was reconsidered, and the Delta II has since been used as the GPS launch vehicle. The history and recent status of launches is shown Table 3.

V. Initial Testing

The objective of the Phase I approval of GPS was to validate the total system concept. A major stumbling block in obtaining Phase I approval was a classic bureaucratic "catch 22." The issues raised were the following: 1) How could user equipment development be approved when it wasn't clear they would work with the satellites? but ... 2) How could the satellites be launched without ensuring they would work with the user equipment? Pursued to a superficial conclusion, nothing could be done at all. The solution was adapted from the 621B program. A system of solar-powered GPS transmitters was deployed on the desert floor at the Yuma test ground. These transmitters all radiated one of the unique orthogonal GPS codes (at the approved frequencies), which were synchronized to each other and to the satellites as they were launched. These transmitters were called pseudolites (from pseudosatellites). They provided a geometry that approximated that of the satellites, although the signals were coming from negative elevation angles. The user equipment could be verified to work with satellite transmitters prior to launch.* As satellites were launched, psuedolites could be dropped from the test system; when four satellites were on orbit, the equipment was completely debugged and able to verify the claims that had been made at DSARC. This approach solved the logical impasse. The pseudolite concept was later expanded as a technique to improve accuracy and integrity for civil landing of aircraft.

This approach satisfied the doubters, and in fact significantly strengthened the program. By 1978, when the first NDS satellite was launched, the main varieties of user equipment had been validated quantitatively and qualitatively.

A. Test Results

Initial testing of user equipment included seven different types that were integrated into 11 types of land, sea, and air vehicles plus manpack testing. Literally hundreds of tests were run. Two results are presented here. Figure 9 shows the summary of integrated tests. Values of SEP† range from 6 to 16 m. The later testing in the A-6 is probably more representative of total system performance.

The blind bombing results, Fig. 10, reflect a substantial improvement because of the GPS. Radar bombing is the usual alternative in poor visibility. The results are particularly significant because the probability of a hit is usually inversely

---

*In fact a further cross check had been conceived and implemented by (then Major) Gaylord Green who initially ran the satellite development for the JPO. The satellite transmitter was activated during ground testing and shown to allow lock up by the phase one user equipment. Col. Green later returned to the JPO and completed a distinguished career as the Director of the GPS program.

†Spherical error probable is the radius of the sphere that contains 50% of the measured errors. The A-6, B-52, and F-16 are aircraft. The SSN is a submarine, the CV is an Aircraft Carrier. The UH-60 is a helicopter, and MV is the manpack/vehicular set test. Data for these two sets of results come from the JPO publication YEE-82-009B of September 1986, titled "User's Overview" (unclassified).

SPHERICAL ERROR PROBABLE (METERS)

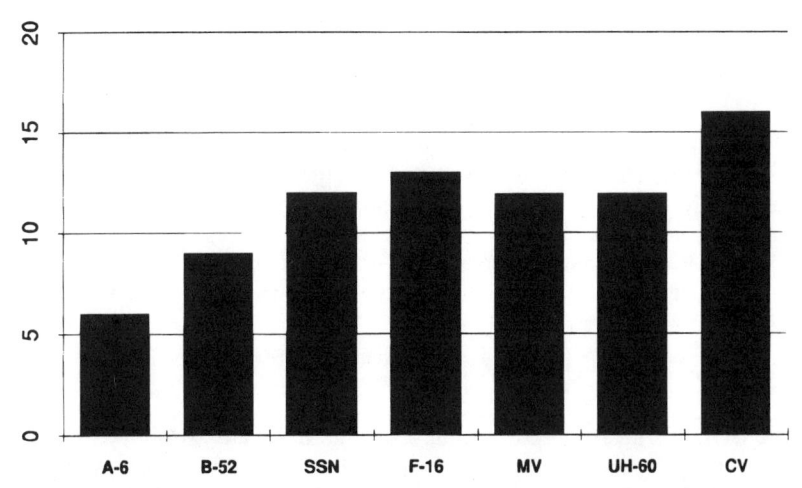

Fig. 9 Test results for seven vehicles using integrated GPS. The earliest tests are on the right and the later on the left. The improved accuracy, in part, reflects system maturity (data courtesy of the U.S. Air Force).

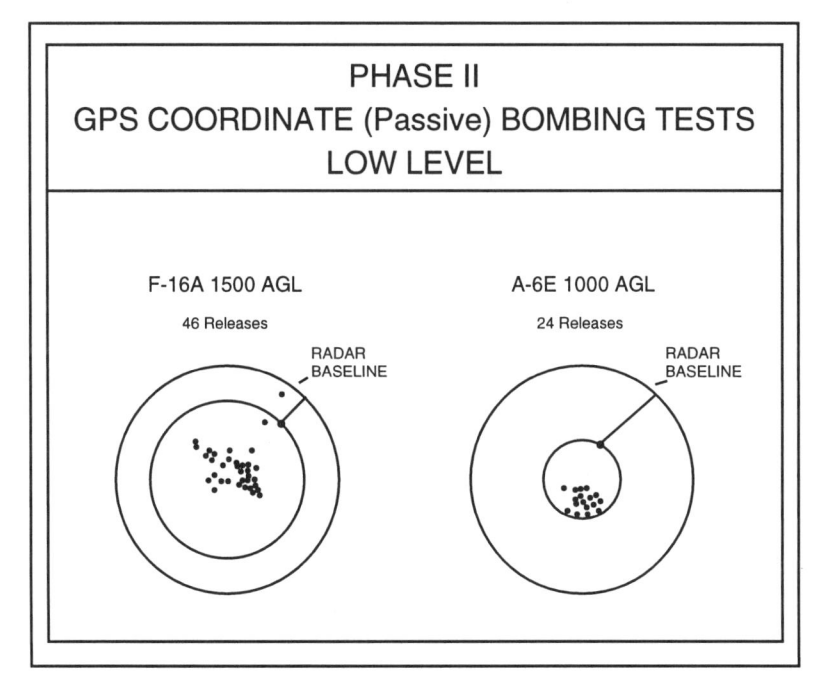

PHASE II
GPS COORDINATE (Passive) BOMBING TESTS
LOW LEVEL

F-16A 1500 AGL — 46 Releases

A-6E 1000 AGL — 24 Releases

RADAR BASELINE

Fig. 10 GPS blind bombing results compared to radar bombing baselines (data courtesy of the U.S. Air Force).

proportional to the *square of the miss radius*. Indeed these results more than satisfied the original part of the motto that called for "five bombs in the same hole."

B. Conclusions

During its initial tests, the Global Positioning System more than met the original design objectives. The impact was not fully appreciated by the operational forces of the military until the Desert Storm battles showed the value of GPS as a force multiplier. The second half of the motto "build a cheap set ($10,000 1973 dollars) that navigates has been overtaken by the civil rush to build integrated chip sets that have driven the costs of GPS down to less than $300 in 1973 dollars. The decision to use a digitally formatted signal has also been vindicated.

VI. Applications

A. Military

The DOD's primary purposes in developing GPS were as follows: 1) its use in precision weapon delivery; and 2) providing a capability that would help reverse the proliferation of navigation systems in the military. Military applications include mine sweeping, aircraft landing, and infantry operations (to name just a few). The Desert Storm campaign was almost a boutique war to demonstrate the effectiveness of GPS. The tactical commanders were finally able to experience the power that comes from precise knowledge of position in a common coordinate frame. It was ironic that the majority of receivers being used were developed by civil companies, with no help from military sponsorship.

B. Dual Use: The Civil Problem

From the beginning of the GPS, it was recognized that the proposed GPS system would provide utility for many more users than the U.S. military. The code structure was arranged to have a precise, protected modulation (the P code), which could be encrypted, and a clear acquisition modulation (the C/A code), which could be exploited by civil users. The earliest presentations always included descriptions of the usefulness to the worldwide civil community. The applications of the GPS for the civilian community are extensive. Initially the GPS was used for accurate time transfer and survey, because these applications could accept the limited initial coverage. Later uses span marine, air, land, and even space. Civil sets currently outnumber military by more than ten to one. That ratio will probably increase as civil set cost decreases. Other chapters discuss in detail many of these applications for these civil receivers.

Table 4 summarizes some of these major civil applications.

This proliferation has led to legitimate fears that the GPS system would be used against its builders, the U.S. Military. Initially it was felt that the P code would demonstrate accuracies about seven times better than C/A.* Therefore, civil receivers would be inherently less accurate. Technology invalidated that assumption. By using carrier aiding, the noise in the C/A receivers could be

---

*The P code has a chip rate ten times higher, but the C/A code has approximately twice as much power.

Table 4 Some civil applications of GPS

Air navigation	Nonprecision approach and landing
	Domestic en route
	Oceanic en route
	Terminal
	Remote areas
	Helicopter operations
	Aircraft attitude
	Collision avoidance
	Air traffic control
Land navigation	Vehicle monitoring
	Schedule improvement
	Minimal routing
	Law enforcement
Marine navigation	Oceanic
	Coastal
	Harbor/approach
	Inland waterways
Static positioning and timing	Offshore resource exploration
	Hydrographic surveying
	Aids to navigation
	Time transfer
	Land surveying
	Geographical information systems
Space	Launch
	In-flight/orbit
	Reentry/landing
	Attitude measurement
Search and Rescue	Position reporting and monitoring
	Rendezvous
	Coordinated search
	Collision avoidance

smoothed to the point that receiver measurement was an insignificant error source. Anticipating the need to withhold full accuracy from an enemy, the system design had included the ability to degrade the accuracy of the satellite clock or the broadcast satellite location. This so-called selective availability is an important feature for the potential protection of the free world. However, Dr. Parkinson had argued that this capability should not be used all the time, because it could be defeated by various forms of differential (locally corrected) GPS. Degrading the signal continuously would lead to rapid introduction of improved differential techniques. It was felt to be better if S/A were only used when an urgent need was determined.

Alleviating fears of enemy use, any civil differential technique could also be countered in time of hostility by using local area jamming of the more susceptible C/A code. The military would continue to rely on the more jam resistant P code for combat operations.

This civil problem is only partially resolved. The international and civil communities have been pushing for less restrictive civil use. The U.S. government has now agreed to provide the signal worldwide, for the foreseeable future, at the 50 m (one sigma) error level. In addition, they have agreed to give 10 years notice, should they not be able to continue to meet the commitment. Moreover, there is now a joint *military/civil* task force overseeing the operation of GPS with representatives from the departments of defense and transportation.

VII. Pioneers of the GPS

The GPS owes its existence to many foresighted and self-sacrificing people. The following list is not complete; it is hoped that those not mentioned will not feel offended. As is often the case, the engineers who took a concept and made it a reality tend to be forgotten. The writer would like to personally thank the outstanding, and dedicated men and women of the initial JPO who truly made GPS possible with their heroic efforts.

A. Defense Development, Research, and Engineering—Malcolm Currie and David Packard

A staunch and essential supporter from early 1973 was Dr. Malcolm Currie, then Deputy Secretary of Defense for Research and Engineering. In the early years, the GPS suffered because it did not have a single operational Armed Forces command that felt space-based navigation was an operational necessity. Most of the affected commands felt it was desirable, but hoped it would be sponsored (and funded) by someone else. Dr. Currie could visualize the value and threw his support into the bureaucratic fray. Without his intercession, the GPS would have been canceled before the first satellites flew.

Another pioneer—although he may not have been aware of his impact on the GPS program—was David Packard (previously Deputy Secretary of Defense for Research and Engineering). Mr. Packard had brought significant reforms to the DOD decision-making process. This streamlining included brief (7-page) decision coordinating papers (DCP) and crisp decisions after a meeting of the DSARC. Without these reforms, gaining program approval would have been a much longer and more arduous task.

B. Commander of SAMSO, General Ken Schultz

As Program Director, Dr. Parkinson was extremely fortunate to work for a general who also had been a Program Director. Lt. General Kenneth Schultz was tough and fair and knew how difficult it was to run a large program. He laid down the objectives but did not attempt to control the process totally. Along the way, he taught the JPO many essential things about keeping a program on track, from procurement to personnel.

C. Contractors

The principal Phase I hardware contractors are listed in Table 5. Literally hundreds of hardworking, capable engineers and managers produced the Phase I success.

Table 5 Principal contractors

Contractor	Development effort	Comments
Rockwell International	Development satellites	Initially three plus a flying Qual model; two more added later
General Dynamics	Control segment and direction of Magnavox	Also developed the inverted range for testing
Magnavox	User equipment	Included monitor receivers for the control segment
Texas Instruments	User equipment	An alternate, competitive receiver source
Collins Radio of Rockwell International	Jam resistant user equipment	Actually under contract to the U.S. Air Force Flight Dynamics Lab
Stanford Telecommunications	Signal structure	Also instrumental in obtaining extra satellites

D. Joint Program Office Development Team

It was members of this team who synthesized the design of GPS in 1973, prior to signing contracts with any of the support contractors. From the beginning, a conscious decision was made *not* to use an integrating contractor. The overall integration was to be handled by the Joint Program Office in cooperation with various contractors. For this to work, the JPO had to manage the technical tradeoffs and all major systems interfaces.

Fortunately, Dr. Parkinson had been given strong support in directing the effort. This included careful selection of the Air Force officers who had to make the management and technical approach succeed. Of the approximately 35 military officers involved, 6 held Ph.D.s in engineering, and virtually all the others held Master's degrees. Many had experience in running large programs, and some (who had been at the Air Force's Central Inertial Guidance Test Facility) were extremely skilled in devising and running test programs for navigation systems. With the extensive travel demands, it was essential that Dr. Parkinson had a strong and effective deputy to smooth administration of the complex program. That role was filled effectively by then Lt. Col. Steve Gilbert during the initial phase of development and later by Lt. Col. Don Henderson. In addition, the JPO was supported by a small, but effective, cadre of engineers from the Aerospace Corp., which was initially headed by Walt Melton, who had done much of the pioneering work on 621B.

E. Predecessors

The Air Force 621B program had been developed in the Plans Directorate of the Air Force's Space and Missile Systems Organization with strong support from the Aerospace Corp. Two individuals who pioneered the most essential pieces of other GPS technology were Roger Easton of the NRL, who developed the initial clock technology, and Dr. Richard Kirschner of the ARL, who had developed Transit.

VIII. Future

The quiet revolution of NAVSTAR GPS has just begun. Given that the number of active satellites in the constellation now has reached 24, the use of GPS surely will expand. As that expansion continues, the demand will be for greater integrity, which will lead to a modest increase in the number of satellites. The major issues awaiting resolution are the following:

1) Sufficiency of the number of satellites
2) Expansion of the backup control segment capability
3) Resolution of the international request for some civilian control of the system

An expanded dialogue between the military operators and civilian users has begun. The next great wave of progress will be *differential* GPS systems, which squeeze the expected dynamic errors down to less than 1 m. All users will benefit as this new "navigation utility" comes into full operation and usefulness over the next 20 years.

References

[1] Parkinson, B. W., "Overview," *Global Positioning System,* Vol. I, Institute of Navigation, Washington, DC, 1980, p. 1.

[2] Easton, R. L., "The Navigation Technology Program," *Global Positioning System,* Vol. I, Institute of Navigation, Washington, DC, 1980, pp. 15–20.

[3] Bowen, R. et al., "GPS Control System Accuracies," *Global Positioning System,* Vol. III, Institute of Navigation, Washington, DC, 1986, p. 250.

[4] Sharrett, Wysocki, Freeland, Brown, and Netherland, "GPS Performance: An Initial Assessment," *Proceedings of ION GPS-91,* Institute of Navigation, Washington, DC, 1991.

[5] Anon., *Proceedings of ION-89,* Institute of Navigation, Washington, DC, 1989, p. 19.

Chapter 2

Overview of GPS Operation and Design

J. J. Spilker Jr.*
Stanford Telecom, Sunnyvale, California 94089
and
Bradford W. Parkinson†
Stanford University, Stanford, California 94305

I. Introduction to GPS

THE Global Positioning System (GPS) consists of three segments: the space segment, the control segment, and the user segment, as shown in Fig. 1. The *control segment* tracks each satellite and periodically uploads to the satellite its prediction of future satellite positions and satellite clock time corrections. These predictions are then continuously transmitted by the satellite to the user as a part of the navigation message. The *space segment* consists of 24 satellites, each of which continuously transmits a ranging signal that includes the navigation message stating current position and time correction. The *user receiver* tracks the ranging signals of selected satellites and calculates three-dimensional position and local time.

This chapter is designed to provide a summary discussion of the GPS. Later chapters develop the details. All segments of the system, along with a detailed discussion of the signal and the multiple applications of the GPS, are covered in separate chapters.

II. Performance Objectives and Quantitative Requirements on the GPS Signal

The key performance objectives of the GPS system can be summarized as follows:

1) High-accuracy, real-time position, velocity, and time for *military users* on a variety of platforms, some of which have high dynamics; e.g., a high-performance

---

Copyright © 1994 by the authors. Published by the American Institute of Aeronautics and Astronautics, Inc., with permission. Released to AIAA to publish in all forms.
*Ph.D., Chairman of the Board.
†Professor, Department of Aeronautics and Astronautics, and Director, GPS Program.

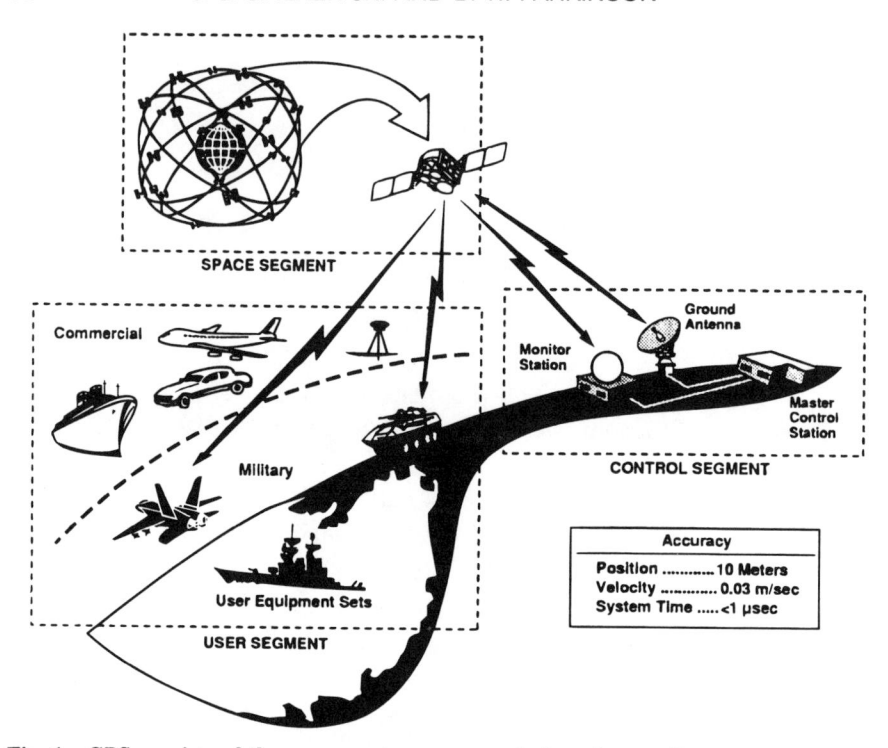

Fig. 1 GPS consists of three segments: space, control, and user; the user segment contains both civil and military users.

aircraft—high accuracy translates into 10-m three-dimensional rms position accuracy or better; velocity accuracy < 0.1 m/s.

2) Good accuracy for *civil users*—the real-time civil user accuracy objective is considered to be 100 m (at about the 95th percentile) or better in three dimensions. In the future, this accuracy may be improved by reducing or eliminating the deliberate degradation of the ranging signal.

3) Worldwide, all weather operation, 24 h a day.

4) Resistance to intentional (jamming) or unintentional interference for all users—enhanced resistance to jamming for military users.

5) Capability for highly accurate geodetic survey to centimeter levels using radio frequency carrier measurements—capability for high-accuracy time transfer to 100 ns or better.

6) Affordable, reliable user equipment—users cannot be required to carry high-accuracy clocks; e.g., atomic frequency standards, or sophisticated arrays of directional antennas that must be pointed at the satellites.

In addition to these performance requirements for the user, the GPS must also employ a cost-efficient space segment, must live within constraints of available bandwidth and frequency allocations, and have a control segment capable of measuring the satellite orbits, clocks, and uploading data to the spacecraft for retransmission to the users.

GPS OPERATION AND DESIGN

This chapter provides a summary of the GPS and describes how these accuracy and other performance requirements translate into requirements on the GPS signal. For example, user position accuracy translates into accuracy requirements for the time measurement performed by the GPS user receiver. These requirements also affect the radio frequency frequency selection. Because the satellites must be limited in transmit power, and there are many perturbing physical phenomena and geometrical considerations as discussed later, all of these selections have been made with care.

A. Satellite Navigation Concepts, Position Accuracy, and Requirement Signal Time Estimate Accuracy

As an elementary step in discussing the use of satellites for real-time navigation, consider the single idealized navigation satellite and a single user, as shown in Fig. 2. Assume that the user is fixed in inertial space on a nonrotating Earth. Assume also that the satellite has information as to its precise position vs. time and contains a perfect clock. Imagine that both position and time are displayed in lights on the side of the spacecraft so that they are observable to the user who has a telescope and camera. For this example of the principal of satellite ranging, the satellite and user coordinates are both expressed in an Earth–centered-inertial (ECI) nonrotating coordinate system with the origin at the Earth's center (see Fig. 3) and we neglect atmospheric and relativistic effects.

In this example, the user camera periodically photographs the satellite clock and position indicator and compares the satellite clock reading t_s with a simultaneous

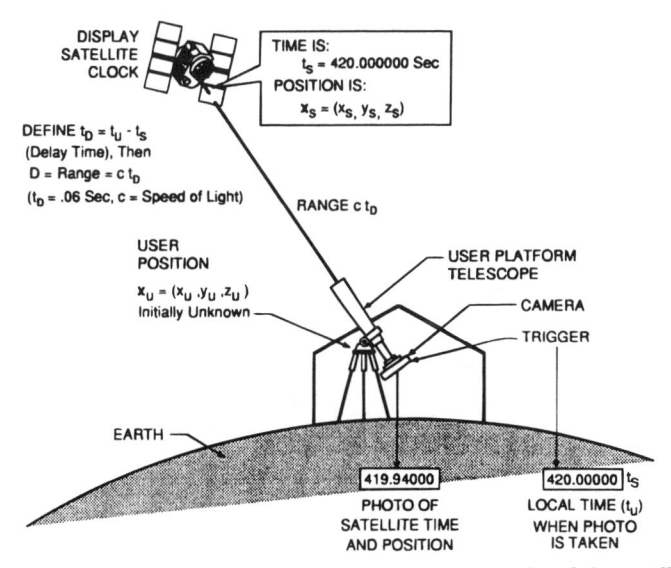

Fig. 2 Satellite and user clock timing concepts—photographs of the satellite clock are taken by the user. Coordinates are expressed in an Earth–centered-inertial (ECI) coordinate system. The true system time is t_u at the time of reception. A perfect user clock is assumed. The satellite position can be alternately denoted by x_s in Cartesian coordinates or r_s in radius vector coordinates.

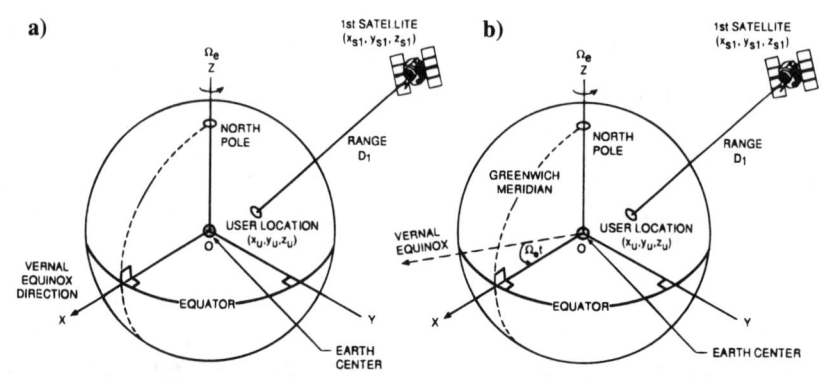

Fig. 3 Earth–centered-inertial and Earth-centered, Earth-fixed (ECEF) coordinates: a) the ECI coordinates are nonrotating with the *x*-axis aligned with a vector from the Sun to the Earth position at the vernal equinox (the first day of autumn); b) ECEF coordinates rotate with the Earth with the *x*-axis on the Greenwich meridian.

reading of the local user clock t_u. Also for this example, both clocks are assumed to be exactly synchronized to system time, and the user is assumed to be stationary. The photograph reveals that, at the receive time instant when the user clock shows $t_u = 420.00000$ s, the image received at the camera from the satellite showed the satellite position at $x_s(t_s)$ [or $r_s(t_s)$ in radius vector coordinates] and $t_s = 419.94000$; i.e., it shows the satellite clock and position 0.060 s earlier. Thus, there is a measured delay caused by the finite speed of light c of $t_D = t_u - t_s = 0.06$ s and a range to the satellite $D = c\, t_D$ measured at time $t = t_u$. Therefore, the receiver's location at time t_u is somewhere on a sphere of radius D centered at $x_s(t_s)$. This simple example does not address the impact of the satellite's velocity nor the possibility of a dynamic user. Note also that in order to convert this result to meaningful user coordinates; e.g., Earth-centered, Earth-fixed (ECEF) coordinates, we must account for the fact that the Earth is rotating through 360 deg in inertial space per sidereal day. (A sidereal day is approximately 23 h 56 min 4 s of mean solar time.)

Clearly, if we were able to perform the same type of measurement with three satellites simultaneously, we could locate the user position in three dimensions at the intersection of three spheres and perform the desired real-time navigation. However, doing so requires accurate, synchronized time at the user terminal. Assume now that the user clock has an unknown bias error b_u and thus the user clock reads $t'_u = t_u + b_u$, where t_u is the "true" system time at the user's time of reception. By adding a ranging measurement to a fourth satellite, the solution can be found for both $x(t_u)$ and true user clock time t_u (or user clock error Δt_u), at the time at which the measurement is taken (see Fig. 4).

The difference between satellite clock time and user clock time when the user clock is not precise is termed *"pseudorange."* For the ith satellite, this range difference is denoted as ρ_{iT} where subscript T denotes true pseudorange. The *true pseudorange* ρ_{iT} to satellite i, in the idealized error-free condition, is the true range plus the user clock bias correction b_u expressed in seconds and is expressed as $\rho_{iT} = c(t_u - t_{si}) + c\, b_u$. However, in the real measurement, there are random

GPS OPERATION AND DESIGN 33

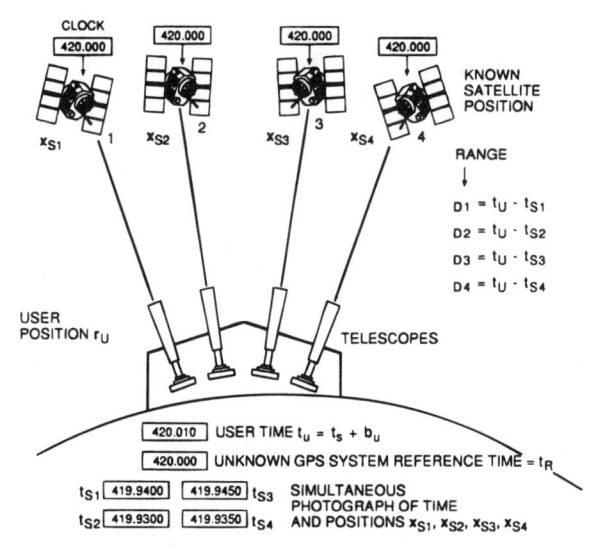

Fig. 4 Three-dimensional user position and clock bias measurement for a user with an accurate user clock. Four pseudorange equations are needed to solve for the four unknowns, user x, y, and z coordinates and user clock bias b_u. Note that the satellite positions are all observed by the user at slightly different times t_{si}.

noise effects, various other bias errors, propagation errors (and relativistic effects) so that the *measured pseudorange* ρ_i is $\rho_i = \rho_{iT} + \Delta D_i - c\,\Delta b_i + c\,(\Delta T_i + \Delta I_i + v_i + \Delta v_i)$ where Δb_i is the satellite bias clock error (s); ΔD_i is the satellite position error effect on range; v_i is the receiver measurement noise error for satellite i (s); ΔI_i is the ionospheric delay error (s); ΔT_i is the tropospheric delay error (s); and Δv_i is the relativistic time correction (s). Thus, the actual user clock reads $t'_u = t_u + b_u$, and the actual clock of satellite i reads $t'_{si} = t_{si} + \Delta b_i$, where t_u and t_{si} are the true system times at the user at the time of reception and at satellite i at the time of its signal transmission, respectively.

We define *true pseudorange* $\rho_{iT} = |x_{si} - x_u| + cb_u = |r_{si} - r_u| + cb_u = D_i + cb_u$. Note that pseudorange defined in this manner is not directly an observable, and each of these quantities may vary with time. Note also that the techniques discussed here permit real-time satellite navigation and position measurement. This technique is distinct from systems that must observe the changes in satellite Doppler shift over some period of time as the satellite passes overhead to determine position. Techniques of this latter type clearly cannot autonomously determine the position of a rapidly moving vehicle.

Although the preceding examples of Figs. 2 and 4 used optical measurements, exactly the same process can be performed using microwave signaling with coded signals. Examination of the fine detail of the received signal code provides exactly the same information as a photograph of a clock, because the structure of the code can be interpreted as time counts of a clock. One simple radio frequency analogy to the optical clock display is a signal that is modulated by a 10-stage binary counter that counts the number of precisely timed 1 μs clock cycles up to $2^{10} = 1024$ and then repeats. Examination of the last 10 bits of the binary waveform then gives coarse time analogous to hours and minutes, and examination of the exact timing of the clock transition gives the finer resolution of time. As compared to a system that simply monitors periodic sequences of identical 1 μs pulses, the counter, in effect, reduces the level of ambiguity in the measurement from 1 to 1024 μs, etc.

The preceding simplified description of the GPS positioning calculations briefly introduces a number of significant effects, which are treated in detail in later chapters of this volume. These include the following:

1) *User motion*—The user is generally moving in inertial space, and we must account for the user motion between the time the signal is transmitted from the satellite and the time when the signal is received.

2) *Atmospheric effects*—Excess delay caused by the wave traveling through the atmosphere (troposphere and ionosphere) must be estimated or measured.

3) *Relativistic effects*—There are a number of effects caused by satellite and user motion, Earth rotation, and the Earth's gravitational field, all of which can be significant.

1. User Receiver Measurements—Pseudorange and Carrier Phase

Next we examine how a receiver makes the measurements on the signal waveform or the radio frequency carrier to form an accurate estimate of the pseudorange ρ_i. Figure 5 illustrates some of the range measurements that can be

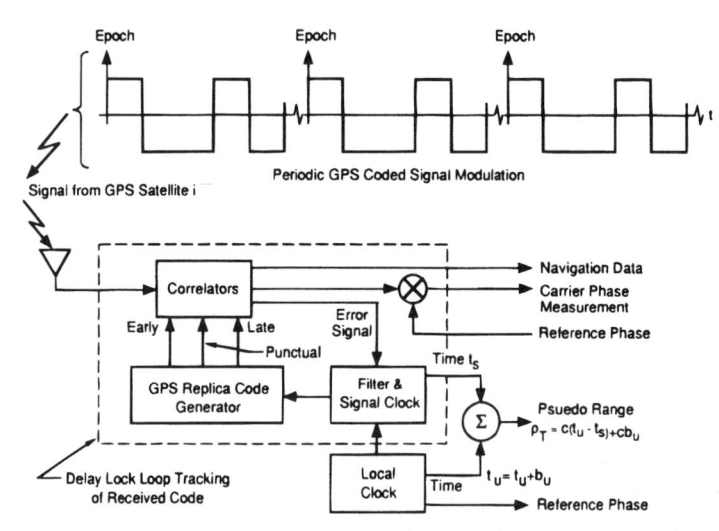

Fig. 5 Pseudorange measurement using a delay–lock-loop (DLL). The GPS replica code generates early and late reference signals that are both fed to a correlation device that produces an estimate of whether the early or late signal provides the best match. The resulting error signal is then used to control the signal clock in a tracking mode. The differences between the two clocks $t'_u - t_s$ is then multiplied by c to form pseudorange $c(t'_u - t_s)$. Similar measurements are made on the received carrier phase. The user receiver has a clock bias offset b_u in seconds. (Satellite clock error is ignored here.)

made on a signal waveform. Pulses of radio frequency energy can be modulated with a special pulse code sequence that has a distinct beginning (or epoch). The GPS civilian [coarse acquisition (C/A)] signal repeats such a sequence every millisecond. The C/A code is a random-like or pseudonoise (PN) binary sequence of 1023 chips. The pseudorange $t_u - t_s$ can then be recovered in a special delay lock loop receiver designed to track and to detect the code from that satellite, as shown in Fig. 5. The measurement relies on the unique code properties of each satellite signal. These properties enable the receiver to measure pseudorange to each satellite separately. The delay lock technique creates an internal replica of the known modulation sequence and adjusts the internal epoch until it exactly matches the received signal in delay. This matching is performed by cross-correlating the received and internal signals and finding the start time that maximizes the correlator output. The satellite clock time at the time of transmission is then subtracted from the user clock time to recover the measured pseudorange.

If the user receiver clock t_U and the satellite clock are both synchronized to GPS reference time,* then range $= D = c(t_u - t_s)$. For purposes of this discussion, atmospheric and other propagation path delay perturbations are neglected. Pseudorange is the same measurement when the user receiver clock t_U has an unknown and possibly time varying clock bias b_u expressed in seconds. The pseudorange is then $\rho = D + cb_u$, as shown in Fig. 5. This technique is known as *code ranging* to distinguish it from carrier phase measurements.

Most GPS receivers can also reconstruct the GPS radio frequency carrier at 1575.42 MHz and use this sinusoid as a ranging signal. This measurement is very precise (typically subcentimeter or a fraction of the 19-cm wavelength), but its *accuracy* is limited by the difficulty of resolving which cycle is being received (called the cycle ambiguity or $n\lambda$ problem). Carrier phase accuracy corresponds to the equivalent carrier phase noise expressed in distance which would be on the order of 2 mm if the carrier phase could be measured to 1% of the wavelength. However, the initial value of the carrier phase is completely ambiguous, and we must resort to the use of various phase differencing techniques. Nonetheless, it is possible to measure changes in phase both very precisely and without significant cycle slipping over many seconds. This carrier phase measurement can then allow us to make very precise position measurements. The use of carrier phase measurements for various surveying and aircraft landing applications is discussed in later chapters in this volume. Table 1 gives a rough estimate of the measurement accuracy, bias, and precision for GPS carrier and code measurements.

a. Relating pseudorange accuracy and positioning accuracy—dilution of precision (DOP). Positioning accuracy reflects the final capability of most GPS receivers. Although it is related to *ranging* accuracy, they are not the same. The relationship between them is a function of the geometry of the selected satellites; that is, the directions of arrival of the satellite signals. To achieve a positioning accuracy requirement of 10 m, the ranging accuracy and geometry must both combine to acceptable values. For example, if each individual pseudorange measurement has a statistically independent error of zero mean with the same rms

---

*All time intervals can be expressed in equivalent distance in meters by multiplying by the speed of light c.

Table 1 Rough order of magnitude estimates of GPS
code and carrier phase measurements in meters[a]

Measurement	Rough measurement accuracy	
	Bias	Precision
Code	≈ 5 m	≈ 2 m
Carrier phase	$n\lambda$	≈ 0.002 m

[a]There is assumed to be no selective availability (SA) degradation. The value of n in the carrier phase ambiguity must be determined.

value of σ (caused by all effects), then the rms position errors are given by the following:

Position error = DOP $*$ σ, where DOP is a multiplier determined by the geometry and is typically between 1 and 100.

The quantity DOP is calculated from the unit vectors to each of the satellites, as shown in Chapters 5 and 11, this volume. Generally, if the DOP rises above six, the satellite geometry is not very good. There are several measures of positioning accuracy. For the current nominal constellation, the worldwide median position dilution of precision (PDOP) (50th percentile) is approximately 2.5. As another example, the horizontal error (in the x and y coordinates) is given by rms error horizontal = HDOP (horizontal dilution of precision) $*$ σ, where rms denotes the root mean square error. The vertical error (in the z coordinate) is found from rms error vertical = VDOP (vertical dilution of precision) $*$ σ.

The speed of light is approximately $c \cong 3$ m/ns and $1/c \cong 3.3$ ns/m. If the satellite geometry produces an HDOP of 3 (and a horizontal error less than 10 m is required), then HDOP $*$ σ = $3\sigma \leq 10$ m $*$ 3.3 ns/m. Thus, the required ranging accuracy is then $\sigma \leq 11$ ns. In a similar manner, the rms position error (in three dimensions x, y, and z) is estimated by the relationship rms position error = PDOP $*$ σ.

Table 2 summarizes some of the key definitions of parameters in the GPS user position calculations and error sources.

III. GPS Space Segment

A. GPS Orbit Configuration and Multiple Access

The discussions of the basic concepts for GPS and analysis of the geometric dilution of precision have shown that the user must make measurements on four or more satellites simultaneously to provide real-time three-dimensional navigation. Thus, the satellite orbital constellation must provide a user anywhere in the world simultaneous access to measure pseudorange to four or more satellites at any time, 24 h a day. Furthermore, as shown later in the DOP analysis, the satellites should be widely spaced in angle. Measurements on each of the four or more satellites must be made simultaneously or nearly simultaneously without mutual interference if we are to solve for position. This capability is termed

Table 2 Summary of notation for GPS position and pseudorange parameters

Parameter	Description
x_{si} or r_{si}	Position of satellite i in either x, y, z or radius vector coordinates, respectively. The position of the satellites at the time of transmission is $x_{si}(t_{si})$ or $r_{si}(t_{si})$.
x_u or r_u	User position in either x, y, z or radius vector coordinates. User position may also vary with time. User position at time of reception is $x_{ui}(t_{Ri})$ or $r_{ui}(t_{Ri})$.
t_{si}	True time of transmission from satellite i. This parameter is the true time of transmission and may not be exactly the same as indicated by the satellite.
t'_{si}	Actual satellite clock reading $t'_{si} = t_{si} + \Delta b_i$
Δb_i	Satellite bias clock error (expressed in s)
t	GPS system time
t'_u	Actual user clock time at time of reception of signal $t'_u = t_u + b_u$
t_u	True user time at time of reception
b_u	User clock bias—can vary with time
ρ_{iT}	True value of pseudorange $\rho_{iT} = |x_{si} - x_u| + b_u = |r_{si} - r_u| + b_u$
ρ_i	Measured pseudorange with various error contributors
$n(t)$	Receiver thermal noise
Δb_i	Satellite bias clock error (expressed in s)
ΔD_i	Satellite position bias error effect on range
v_i	Receiver pseudorange measurement noise error for satellite i, s
ΔI_i	Ionospheric delay (expressed in s)
ΔT_i	Tropospheric delay (expressed in s)
Δv_i	Relativistic time correction (expressed in s)
c	Velocity of light
DOP	Dilution of precession
VDOP, HDOP, PDOP	Vertical, horizontal, and position geometric dilutions of precession
GDOP	Geometric dilution of precession includes both position and time error effects
ϕ_i	Measured carrier phase offset as received from satellite i
ϕ_{iT}	True received carrier phase offset $\phi_{iT} = \omega_o(t_{si} - t_u)$ at frequency ω_o
λ	Wavelength of GPS carriers can have value λ_{L_1} or λ_{L_2} for GPS L_1 or L_2 frequencies.
n	Carrier phase cycle count ambiguity
D_i	True range from satellite i to user $|x_{si} - x_u| = |r_{si} - r_u| = c(t_u - t_{si})$

multiple access. Multiple access signaling permits measurements to be made on a signal from one satellite without signals from other satellites interfering with that measurement.

From a user performance standpoint, satellite orbit altitude selection has several effects:

1) The higher the orbit altitude, the greater the fraction of the Earth visible by a single satellite

2) Within limits, power flux density on the Earth is nearly independent of orbit altitude because the satellite antenna beamwidth can be selected (widened or narrowed) to provide full Earth coverage.

3) A low-orbit altitude with its corresponding short visibility time leads to a larger number of signal acquisitions and satellite–satellite handovers by the user receiver, and larger Doppler shift must be tolerated in the receiver.

The selected satellite orbital constellation contains 24 satellites, the GPS-24, in six orbit planes. There are four satellites in each of the six planes, as shown in Fig. 6. The satellites have a period of 12 hours sidereal time* and a semimajor axis of 26,561.75 km. A satellite with an orbit period of 12-h sidereal time produces a ground track (projection on the Earth's surface) which repeats over and over. For GPS the longitude crossing at the equator is kept fixed to within ± 2° by the GPS Control Segment. This orbit has 63% of the radius of a geostationary satellite orbit with a 24-h period. The GPS semimajor axis is the orbit radius of the circular GPS orbits, and thus the GPS satellites have an altitude of 20,162.61 km above the Earth's equatorial radius[1] of 6378.137 km. The altitude of the GPS orbit obviously is well above the atmosphere and not subject to atmospheric drag. Other perturbations such as solar pressure and lunar and solar gravitational orbit perturbations can be significant. The satellites are inclined with respect to the equator by 55 deg (the initial GPS satellites had a 63-deg inclination). Table 3 summarizes the approximate parameters of the GPS orbit. The satellite orbital constellation is described in detail in Chapter 5, this volume.

B. GPS Satellite Payload

The key role of the satellites is to transmit precisely timed GPS signals at two *L*-band frequencies† 1.57542 GHz and 1.2276 GHz. These signals must have embedded in them, in the form of navigation data, both the precise satellite clock time as well as satellite position so that a user receiver can determine both satellite time t_{si} and satellite position r_{si} at the time of transmission. These navigation data are uploaded from the GPS control segment (CS) to each satellite and then stored in memory in the satellite for readout in the satellite navigation data stream. Figure 7 shows a simplified view of the GPS satellite payload. The GPS upload station sends the satellite the ephemeris information regarding the satellite orbit

---

*A sidereal day is defined as the time for the Earth to complete one revolution on its axis in ECI space and consists of 24 mean sidereal hours where 1 mean sidereal day is slightly shorter than a mean solar day. One sidereal day is 23 h, 56 min, 4.009054 s or 86,164.09054 s of mean solar time. One mean sidereal day is equal to 0.997269566 mean solar day or one mean solar day is equal to 1.002737909 mean sidereal day.

†These signals, as well as the L_3 signal at 1381.05 MHz, are all selected and filtered so as to minimize interference with the radio astronomy bands (see Chapter 3, this volume).

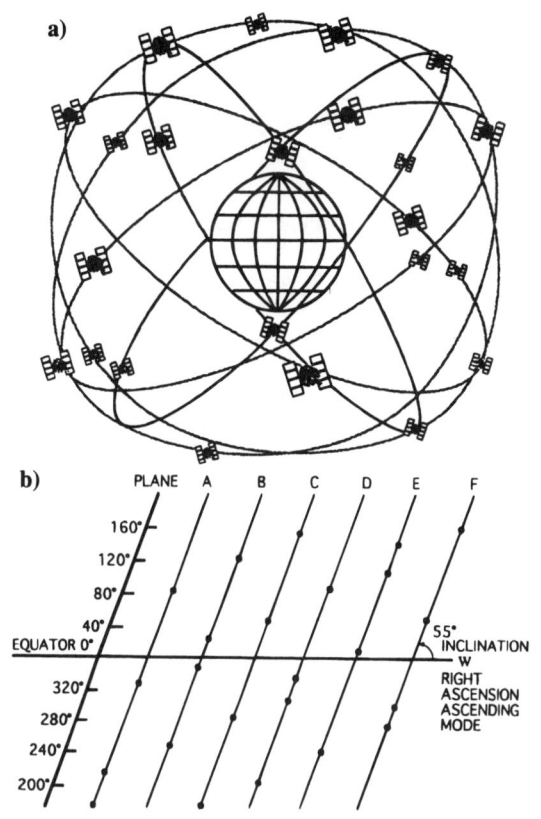

Fig. 6 GPS satellite constellation: a) the six orbit planes shown in ECI coordinates; and b) satellite positions on each of the six orbit planes. The GPS constellation has satellites in six equally spaced orbit planes. The present GPS-24 satellites constellation shown in b) contains 24 satellites. The uneven satellite phasing in each plane is designed to minimize the effect of satellite outage.

and the exact position in that orbit vs. time. Included is a satellite clock correction that calibrates the offset of the satellite clock relative to GPS system time. These data are uploaded to the satellite through an S-band telemetry and command system.

One of the keys to GPS satellite performance is the stability of the GPS satellite clocks. Each satellite carries redundant atomic oscillators of high stability. These atomic clocks are stabilized using either rubidium or cesium atoms in gaseous form.

The atomic clocks along with appropriate frequency synthesizers then synchronize the GPS signal generators and also control the radio frequency center frequencies of the two L-band frequencies. The signals are then amplified and filtered to remove signal power outside the allocated frequency bandwidth of 20 MHz for each of the two L-band signals. The signals also are modulated by the navigation data that carry the satellite position and time information to the user.

Table 3 Approximate GPS satellite parameters

Orbit plane	Six equally spaced ascending nodes at 120 deg
Orbit radius r_{cs}	26,561.75 km semimajor axis
Orbit velocity (circular) (ECI)	$= \sqrt{\dfrac{\mu}{r_{cs}}} = 3.8704$ km/s
Eccentricity	Nominally zero, but generally less than $e = 0.02$
ω_s angular velocity	1.454×10^{-4} rad/s
Period[a]	12 h mean sidereal time
Inclination	$i = 55$ deg nominal

[a]The period of an orbit in seconds of mean solar time is $T_p = (2\pi/\sqrt{\mu})a^{3/2}$ where a is the semimajor axis in meters and μ is the Earth's gravitational parameter $\mu = 3.986005 \times 10^{14}$ m^3/s^2. For 12-h mean sidereal time period, $a = 26561.75$ km including a minor correction for a nonspherical Earth. A new model, Joint Gravity Model #2, is being proposed by NASA and the University of Texas, which gives $\mu = 3.986004415 \times 10^{14}$ m^3/s^2, and $R_e = 6378.1363$ km for the Earth's mean equatorial radius.

(The reader is referred to Chapter 6, this volume on the GPS satellite payloads for a more thorough description.) One of the limitations on GPS clock accuracy is selective availability. Selective availability is a clock dither that can be imposed on the GPS signals to restrict unauthorized (nonmilitary) access to the full accuracy of the system.* Chapter 1 of the companion volume describes differential GPS that can be used to improve civil accuracy.

C. Augmentation of GPS

As discussed in later chapters in this volume and the companion volume, we can augment the GPS satellite system with other ranging signal sources:

1) Ground transmitters or pseudolites that transmit GPS signals and other information to support GPS—a special form is the integrity beacon used for aircraft landing

2) Additional satellites can either carry transponders that can relay GPS-type signals from synchronized ground transmitter uplinks or have navigation payloads similar to the basic GPS satellites. The Federal Aviation Administraton's Wide Area Augmentation System (WAAS) is an example of an augmentation of GPS. This system adds geostationary relay satellites.

3) Differential and wide area differential ground stations. These ground stations transmit correction information to appropriately equipped GPS receivers to improve the accuracy of the receivers.

IV. GPS Control Segment

The GPS CS has several objectives:
1) Maintain each of the satellites in its proper orbit through infrequent small commanded maneuvers.

---

*It is expected that selective availability effects will eventually be eliminated. A recent report of the National Research Council[5] has recommended that selective availability be turned to zero.

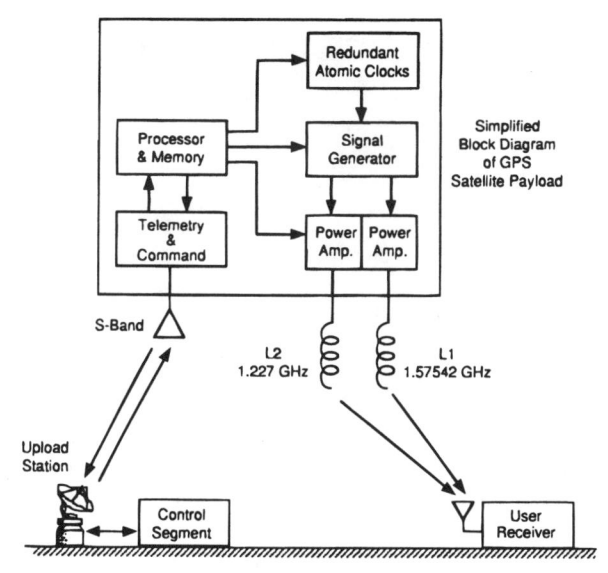

Fig. 7 Simplified GPS satellite payload functional diagram.

2) Make corrections and adjustments to the satellite clocks and payload as needed.
3) Track the GPS satellites and generate and upload the navigation data to each of the GPS satellites.
4) Command major relocations in the event of satellite failure to minimize the impact.

Although each of these objectives is important, this discussion concentrates on the third objective.

The Operational Control Segment began operation in 1985 and consists of five monitor stations, four ground antenna upload stations, and the Operational Control Center. Each of these facilities is shown in Fig. 8. The sites have been selected to provide a significant separation in longitude between each of the monitor stations. Each of these sites, except the site in Hawaii, also contains a ground antenna upload station.

A. Monitor Stations and Ground Antennas

Each of the five monitor stations contains multiple GPS tracking receivers designed to track both the L_1 and L_2 codes and carriers for each of the satellites in view. The monitor stations also contain redundant cesium standard clocks for the GPS receivers to use as a reference oscillator and also to time tag each of the measurements. The measurements of code clock delay and carrier phase for each satellite in view are then sampled, time tagged, and multiplexed in a datastream to send back to the operational control center. Each of the four ground antenna (GA) upload stations has the capability of uploading navigation data to the satellites on an S-band T T&C link. As discussed earlier, the visibility region around each GA extends approximately ±72 deg in Earth angle about the GA.

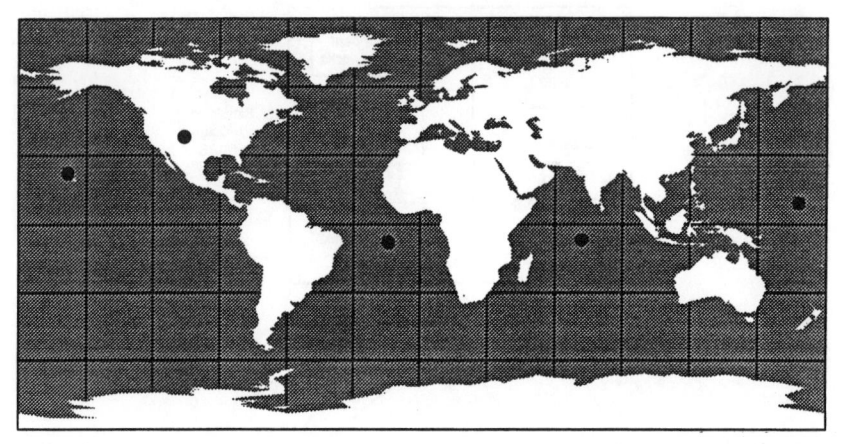

Fig. 8 GPS control segment. There are monitor stations at Hawaii, Colorado Springs, Ascension Island, Diego Garcia in the Indian Ocean, and at Kwajalein Island in the West Pacific.

B. Operational Control Center

The operational control center receives the multiplexed pseudorange measurements and carrier measurements from each satellite in the L_1 and L_2 carriers. The Kalman filter processor in the OCS then estimates the ephemerides, clock error, and other navigation data parameters; e.g., satellite health, for each satellite. The objective of the OCS is to format navigation data for a minimum of 14 days of updates. Navigation data are then transmitted to the upload ground antennas. Each satellite can be given a fresh upload three times a day, approximately eight hours apart. However, normally there is only one upload per day. (See Chapter 16, this volume.) Each upload contains many pages of navigation data that are then fed to the GPS spacecraft processor. The GPS satellite processor then reads out the appropriate set of navigation data for the specific time period appropriate to the time of transmission.

Satellite clock errors in the navigation data are the dominant source of user range error (URE) when the time since the last upload reaches several hours.* Even with three uploads per day, computer simulations run in 1985 show that satellite clock predictability based on clock specification values can limit the GPS user range accuracy to 3m (1σ) 10 h after the satellite ephemeris and clock prediction upload using the specified Allan variance for the satellite clock (2×10^{-13} at $\tau \geq 61,200$ s and 7×10^{-12} at $\tau \leq 50$ s).[2] However, actual satellite clocks perform better than the specification. The specified URE is computed as the following rms sum: URE = [(radial perturbation)^2 + 0.0192 (in-track perturbation)^2 + 0.0192 (crosstalk perturbation)^2]$^{1/2}$, which is a representative error projection to the user.

The simulations showed the total URE had an rms value for 10-h updates of 4.2m (1σ), which was dominated by the clock error component for the specified

---

*This statement assumes that both L_1 and L_2 are available. If not, ionospheric modeling errors are often the largest.

clock and is well within the specified 6m maximum. By way of comparison, the simulated URE for 0 h, 3 h, 6 h, and 24 h predictions were 1.4m, 2.4m, 3.2m, and 8.4m (1σ), respectively. Most (95%) of the URE at 10 h is a result of the accumulated clock noise. When the simulated clock stability was reduced to the typical observed value (rather than specified levels) of 1×10^{-13}, the URE at 10 h decreased to 2.3m (1σ). For further information on expected errors see Chapters 11 and 16, this volume. A discussion of the GPS control segment is found in Chapter 10, this volume.

V. GPS User Segment

There are a great many applications for the GPS system. New applications seem bounded only by the imagination. This section lists some of the more common modes of operation. All of these are discussed in more detail elsewhere in this volume. Although the primary purpose of GPS was a military application, civil users are already more prevalent. The section begins however by describing the fundamental user system architecture.

A. GPS User Receiver Architecture

A generalized view of a typical GPS user system is depicted in Fig. 9. This section discusses the basic configuration of each of these elements. Although most receivers employ only a single GPS antenna, the generalized GPS receiver begins with one or more antenna/low noise amplifiers. More than one antenna/amplifier may be employed in order to achieve the following:

1) Accommodate maneuvering of the user platform; e.g., an aircraft banking

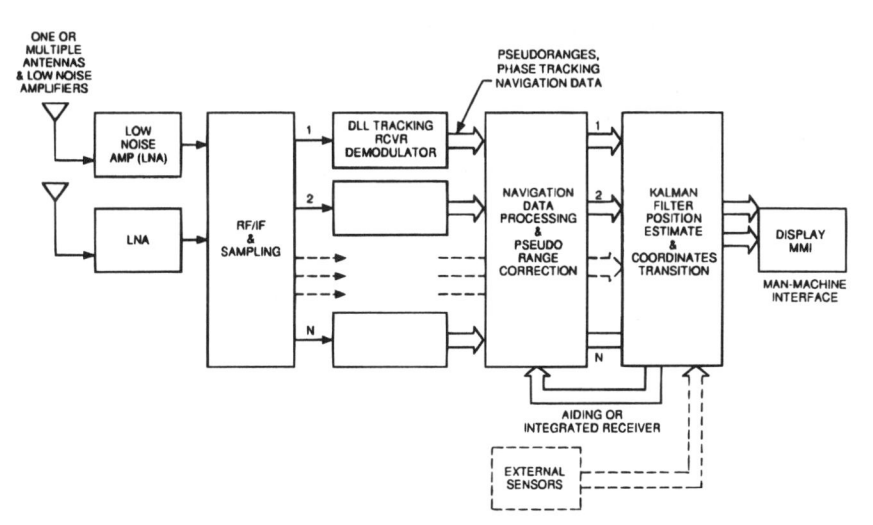

Fig. 9 Generalized GPS user system configuration with separate receiver and position estimating functions.

and thereby avoid blocking some of the satellites with a wing
2) Increase the antenna gain
3) Discriminate against interfering jammers through the use of multiple narrow beam antennas or adaptive antennas
4) Measure attitude

The antenna beams can be steered electronically or mechanically, if necessary. The most common GPS receivers employ only a single omnidirectional (really hemispherical antenna). The output of the antenna is then fed to a radio frequency filter/low-noise amplifier combination in order to amplify the signal and to filter out potential high-level interfering signals in adjacent frequency bands that might either saturate the amplifier or drive it into a nonlinear region of operation. The filters must be selected with low loss and sufficient bandwidth and phase linearity to minimize the distortion of the desired C/A- or P(Y)-code (precise nonstandard code) signals. The signal then passes through serial stages of radio frequency amplification, downconversion, and intermediate frequency (IF) amplification and sampling/quantizing. The sampling and quantizing of the signal can either be performed at IF or at baseband. In either case, in-phase and quadrature (I, Q) samples are taken of the received signals plus noise. At the present state-of-the-art, the functions of radio frequency amplification, downconversion, IF amplification, and A/D sampling can be implemented with a single MMIC (monolithic microwave integrated circuit) chip.

The I, Q samples are then fed to a parallel set of DLLs each of which tracks a different satellite signal and recovers the carrier, which is bi-phase modulated with both the GPS codes and the GPS navigation data. The DLL[2-4] and associated demodulators provide estimates of the pseudorange, carrier phase, and navigation data for each satellite. Typically, the number of parallel tracking DLL varies from 2 to 16 and can possibly track all of the satellites in view at both L_1 and L_2 frequencies simultaneously (see Chapters 7 and 8, this volume). Generally, at least five satellites are tracked as a minimum, either in parallel or in time sequence. At the present state of the art, a 10-channel receiver with 10 parallel DLLs can be implemented on one CMOS chip.

The parallel measurements of pseudoranges and carrier phase along with the navigation data for each satellite are then sent to the navigation data processor where the position of each satellite is calculated from the navigation data in subframes 2,3 at the time of each pseudorange measurement (see Chapters 3 and 4, this volume, for detailed discussion of the GPS signal and navigation data). The pseudorange and phase data are then corrected for the various perturbations, including satellite clock errors, Earth rotation, ionosphere delay, troposphere delay, relativistic effects, and equipment delays. The corrected pseudorange data, phase or accumulated phase [accumulated delta range ADR] measurements along with other sensor data are then processed by the Kalman filter, which estimates user position and velocity state vector. As discussed in Chapter 9, this volume, it is also possible to integrate the Kalman filter with the DLL instead of performing these operations independently. The output of the Kalman filter estimator provides position, velocity, and time estimates relative to the user antenna phase center. These coordinates are usually computed in ECEF coordinates and are then transferred by appropriate geodetic transformation to a local coordinate set convenient to the user.

The Kalman filter may also receive inputs from various other sensors; e.g., barometric altimeters, dead-reckoning estimates of attitude, heading, speed, inertial navigation ring laser gyros, or other navigation aids. This Kalman filter estimate of user position can also be used in a differential mode or kinematic survey mode with other GPS units where at least one unit is at a known reference point, in order to provide precision geodetic survey, more accurate airborne or shipborne navigation, or a common view mode, precision differential time transfer (see Chapter 9, this volume and Chapter 1 of the companion volume). The position, velocity, and time information can then be used with other user-provided information to provide tracks of user positions vs. time, display position on a map, to show way-points to a desired destination, or to satisfy a wide variety of other applications.

B. Uses of GPS

A partial listing of the uses of GPS includes the following:
1) Aircraft navigation—GPS and differential GPS, commercial and general aviation aircraft
2) Land mobile navigation—automobiles, trucks, and buses
3) Marine vessel navigation—GPS and differential GPS
4) Time transfer between clocks
5) Spacecraft orbit determination
6) Attitude determination using multiple antennas
7) Kinematic survey
8) Ionospheric measurement

See the companion volume for in-depth discussions of many of these applications.

1. Various Applications of GPS Receivers

a. Airborne GPS receivers. Airborne GPS receivers provide three-dimensional real-time navigation. These receivers must receive GPS signals from a minimum of four satellites to solve for four unknowns (x,y,z,T) because the airborne receiver clock generally is imprecise. A fifth satellite is needed for satellite handover because periodically new satellites are coming into view while another satellite is going out of view. In addition, the receivers also generally provide three-dimensional velocity estimates and clock drift estimates by making Doppler measurements on the carrier. The GPS receiver may process these satellite signals either in parallel or in time sequence. Aircraft in banked turns may suffer blockage of one or more satellites because of wing or other obstructions. Generally, receivers operate by tracking many more than the minimum four or five satellites in parallel; e.g., 8 to 12, and the clock stability may be sufficient to flywheel through momentary satellite blockage. The GPS receiver may be augmented by inertial navigation systems and/or other navigation aids to provide a hybrid GPS/inertial solution. For higher accuracy, the GPS receiver may operate in a differential mode wherein a GPS receiver at an appropriate known site; e.g., an airport, transmits differential corrections for GPS errors. In the Wide Area

Augmentation System (WAAS), corrections are transmitted to users in the form of satellite clock and position corrections, and ionospheric delay estimates which are valid over a wider geographic region. Further applications are covered in Chapters 12–15 of the companion volume.

   b. Land mobile navigation. Land vehicles may require either two- or three-dimensional positioning. Generally, the altitude is varying, but typically, its variation is at a much slower relative rate compared to its horizontal motion in contrast to an aircraft. In principle, land vehicles can operate at least for a time with only two or three satellites because of the slowly varying vertical component. A fourth satellite could provide less frequent periodic updates of the altitude. However, land vehicles also are subject to blockage or shadowing of one or more satellites by trees, hills, or man-made obstructions. Another augmentation is the use of a magnetic or gyroscopic heading indicator and a wheel counter (see Chapter 10 of the companion volume).

   c. Marine navigation. Navigation on the ocean or large bodies of water is usually at an altitude that varies only with the tides and any roll, pitch, and yaw-induced motion of the GPS antenna aboard the ship. For most purposes, only two dimensions are unknown. Thus, three satellites are adequate to solve for position and two satellites can suffice if a third is employed for periodic updates of the ship receiver clock. Again, however, typical receivers may operate on all satellites in view, and differential navigation can be employed for greater accuracy in harbors. This application is discussed in Chapter 11 of the companion volume.

   d. Time transfer using GPS. Time can be transferred from a reference station to a clock of known location with a single satellite. Four or more satellites can be used initially if the exact location of the remote clock is unknown initially. Greater accuracy can be obtained with the civil signal using "common view" time transfer wherein two clocks at different locations are both within line of sight of the same GPS satellite. In this instance, receivers at each site are tracking the same GPS satellite simultaneously. Satellite clock errors along with any clock dither caused by SA cancel in this mode. Some fraction of the satellite position and ionospheric errors also cancel, depending on the relative separation between the two locations. This cancellation has a residual that depends (approximately) on the size of the angle between the two locations (see Chapter 17 of the companion volume).

   e. Spacecraft GPS receivers. The position/orbit of a near-Earth satellite can be determined by placing a GPS receiver onboard. If the user satellite is below the altitude of GPS, then the satellite can receive the GPS signals from satellites in view above and to the sides where not shadowed by the Earth. If the user satellite is above the altitude of GPS, for example, at geostationary orbit, the user satellite can still receive the GPS signal as it passes on either side of the Earth's shadow. The GPS satellite signals are transmitted to Earth by an antenna pattern slightly greater than the Earth angle, and therefore they extend beyond the Earth's limb. The range to the GPS satellite is, of course, approximately equal to the sum of the GPS altitude and the synchronous satellite altitude, and the GPS signal is accordingly weaker. Furthermore, the GDOPs usually will not be as good as for an Earth-based user. The weaker signal can be compensated by the gain of a directional antenna (at some cost). This approach is discussed in Chapter 21 of the companion volume.

f. Differential GPS (DGPS). If two GPS receivers operate in relatively close proximity, many of the errors inherent in two GPS position solutions are common to both solutions. For example, satellite clock time errors and a significant fraction of the satellite ephemeris, ionospheric, and other errors cancel when we seek a differential or relative position solution. Thus, if one receiver is at a known, fixed position it can transmit pseudorange correction information to other receivers in the area so they can achieve higher relative position accuracy. In the future, it is expected that many GPS receivers will operate in the differential mode as differential corrections become more available. A simplified version of DGPS operation is shown in Fig. 10. The DGPS reference station transmits pseudorange correction information ($\approx$250 bps) for each satellite in view on a separate radio frequency carrier. Because there may be a number of DGPS stations in a network, the data would typically include an almanac giving the location of other DGPS reference stations so the user can use the closest station. Differential GPS is discussed in Chapter 1 of the companion volume.

Differential GPS normally is limited to separations between users and reference stations to approximately 100 km. To carry out a similar differential GPS operation over a wider region a concept known as wide area differential GPS (WADGPS) or wide area augmentation system (WAAS) has evolved. Wide area differential GPS employs a set of monitor stations spread out geographically and a central control or monitor station in somewhat the same mold as the GPS control segment, but simpler. The WADGPS upload station then would relay GPS satellite position and clock and atmosphere corrections via separate geostationary relay satellites.

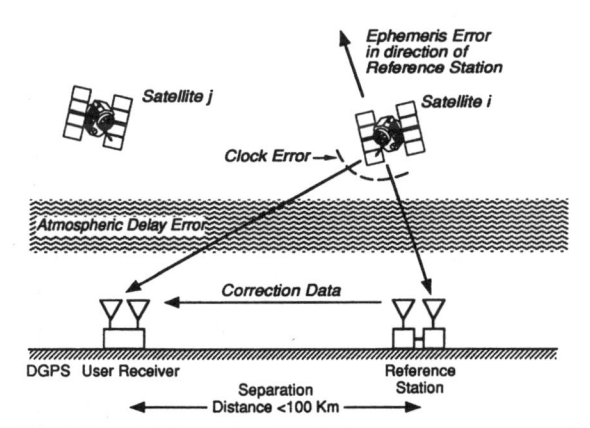

Fig. 10 Simplified view of differential GPS. This correction can completely eliminate satellite clock error offsets but ephemeris and atmospheric corrections differ for the user from the reference station by an amount that depends on the separation distance.

The corrections would be in real time with delays of less than 30 s. Wide area DGPS is discussed in Chapters 3 and 4 of the companion volume.

    g. GPS survey. Global positioning system survey operates with double differencing operations similar to those described above, but carrier measurements are employed to get resolutions on the order of a fraction of the carrier wavelength. For example, the GPS carrier wavelength at 1.57542 GHz is approximately 19 cm. If a carrier phase cycle can be measured to 1%, the differential range error is only 2 mm. Certainly there are ambiguities in the carrier phase measurements, and these must be resolved with various double-differencing techniques.

    The basic concept of double differencing is illustrated in simplified form is Fig. 11 where carrier phase measurements are differenced for a single satellite and the double differenced for two (or more) satellites. Obviously, this difference can be carried out at multiple time intervals to resolve the ambiguity. Survey is discussed in detail in Chapters 18 and 20 in the companion volume.

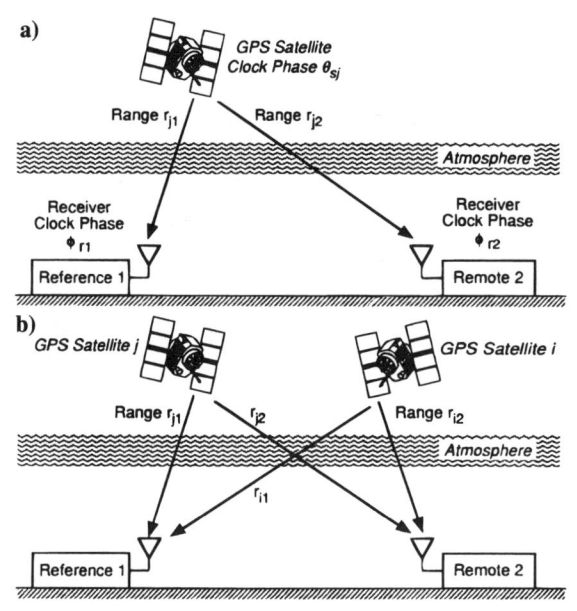

Fig. 11 Single and double carrier phase differencing for GPS and GPS-kinematic survey: a) *Single-difference receiver.* The phase received in receiver 1 is the satellite phase θ_{sj} minus the range delay effect $\omega_o r_{j1}/c$ minus the receiver reference clock phase ϕ_{r1}; namely, $\phi_{1j}(t) = \theta_{sj} - \omega_o r_{j1}/c - \phi_{r1}$. The first difference is then $\phi_{2j}(t) - \phi_{1j} = \Delta_j(t) = \omega_o(r_{j1} - r_{j2})/c + \phi_{r1} - \phi_{r2}$ where the first difference cancels out the satellite clock phase. Also, most of the atmospheric effects cancel if the separation distance is sufficiently small; b) *double-difference receiver.* The second difference is the difference between two first differences for satellite i and j; namely, $\delta_{ij}(t) = \Delta_j(t) - \Delta_i(t) = \omega_o[(r_{j1} - r_{j2}) - (r_{i1} - r_{i2})]/c$ where both the receiver carrier phase offsets cancel.

h. Attitude Determination. In addition to position measurement, GPS can also be employed in an interferometric mode with multiple user antennas to determine vehicle attitude orientation. The GPS codes can isolate each satellite signal, which can then be employed in a differential phase measurement mode with two antennas to measure angular offset for the axis of the rotation of each antenna pair and the direction of the satellite signal. Usually this technique employs carrier phase measurements. Because the baselines are short, simplified techniques to resolve the $n\lambda$ uncertainty can be used. Attitude determination is further explored in Chapter 19 of the companion volume.

i. Hybrid GPS receivers. In many applications, it makes sense to integrate GPS receiver measurements with inertial measurement systems or other navigation aids. The two sets of measurements can provide better capability than either could alone. For example, temporary shadowing or other interruption of the GPS satellite signals by a momentary obstruction or interference can be accommodated by an inertial system that can allow the navigation systems to continue to operate without interruption. (See Chapters 6–9 of the companion volume.)

VI. GPS Signal Perturbations—Atmospheric/Ionospheric/Tropospheric Multipath Effects

The GPS signal frequencies L_1 and L_2 at 1.57542 GHz and 1.2276 GHz are sufficiently high to keep the ionospheric delay effects relatively small, yet not so high as to cause too large a path loss with the use of small omnidirectional antennas (which do not require pointing). In addition, the signal frequency is not so high as to cause any significant path loss attributable to rainfall attenuation. Nonetheless, the atmosphere does cause small but nonnegligible effects. As the GPS signal passes through the atmosphere from the satellite to the user, the signal encounters a number of propagation effects, the magnitude of which depends on the elevation angle of the signal path and the atmospheric environment where the user is located. These effects include the following:

1) Ionospheric group delay and scintillation
2) Group delay caused by wet and dry atmosphere—the troposphere and stratosphere
3) Atmospheric attenuation in the troposphere and stratosphere

There are also effects caused by multipath signals from reflective surfaces and scattering. These effects are discussed fully in Chapters 12, 13, and 14, this volume; the discussion below simply introduces the key principles.

A. Ionospheric Effects

The ionosphere is a region of ionized gases that varies widely from day to day and with solar conditions and also has a large diurnal fluctuation. The presence of the ionosphere changes the velocity of propagation v according to the refractive index $n = c/v$. The cumulative effect also depends on the angle of penetration through the ionosphere as shown in Fig. 12. The refractive index $n(r)$, in turn, varies along the propagation path r. The lower extent of the ionosphere is above 75–100 km, and the ionosphere peak electron content is somewhere in

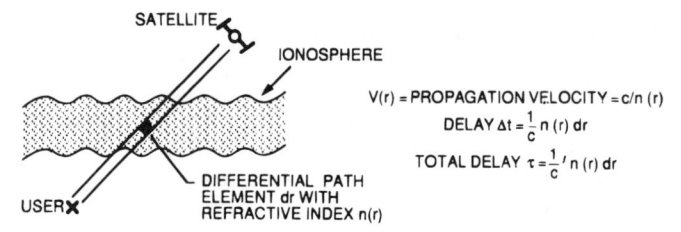

Fig. 12 Ionospheric delay along path through the medium.

the vicinity of 200–400 km. The peak ionospheric electron content can vary by as much as two orders of magnitude between day and night. A fundamental difference between the refractive index for the ionosphere and that for the troposphere is that the refractive index for the ionosphere varies with frequency because of the ionized gases.

The ionosphere can cause two primary effects on the GPS signal. The first is a combination of *group delay* and *carrier phase advance,* which varies with the exact paths and the electron density through which the satellite to user signal traverses the ionosphere. The second effect is ionospheric *scintillation,* which can at some latitudes cause the received signal amplitude and phase to fluctuate rapidly with time. Both effects depend the on the radio frequency and influence the GPS signal design. There are other effects, Faraday rotation and ray bending changing the angle of arrival, but these effects are not significant for purposes here.

At GPS frequencies in the 1.6 GHz frequency region, the ionospheric zenith path delay tends to vary with time in a diurnal fashion, as shown in Fig. 13 and might vary from 2 to 50 ns. As can be seen, however, the diurnal variation fluctuates quite markedly from day to day in a manner that seems difficult to predict. To first order, the ionospheric delay $\Delta \tau$ varies inversely with frequency squared $\Delta \tau \cong A/f^2$. As shown later, this relationship permits us to make dual frequency measurements at L_1 and L_2 to estimate the ionospheric delay. Note that because the lower extent of the ionosphere is typically well above the Earth's surface (see Fig. 14), the angles ϕ of entrance and exit of a satellite observed at 0 deg elevation angle by a user on the ground are well above 0 deg.

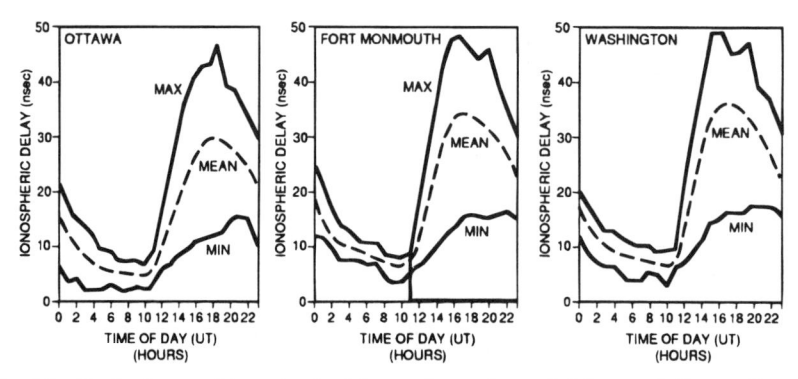

Fig. 13 Typical mean ionospheric delay and envelope of delay variation vs time of day during March 1958. Satellite at zenith $f = 1.6$ GHz.[5]

GPS OPERATION AND DESIGN 51

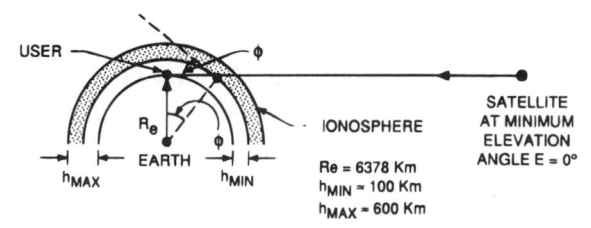

Fig. 14 Angle of incidence and exit of the ionosphere φ.

Thus, at lower elevation angles, the signal path transits through a larger extent of the ionosphere as shown in Fig 14. The delay at any elevation angle can be described as the ratio of actual delay to the vertical delay at $E = 90$ deg. This ratio is defined as the obliquity factor Q as shown in Fig. 15.

Note that the obliquity factor can be as high as three for low elevation angles. Because the ionosphere extends over moderate altitudes, 0.1–0.3 Earth radii, the satellite user signal path does not penetrate the ionosphere at very low elevation angles. This ionospheric effect is unlike the troposphere effects where the troposphere extends down to the Earth's surface. For an upper limit of the ionosphere of 600 km, the minimum angle of entrance to the ionosphere from the satellite is 24 deg. If the lower limit of the ionosphere is at $h = 100$ km, then the angle of exit is 10 deg. Height differences of 160–220 NM (160 NM $\cong$ 296 km) do not cause the obliquity factor to vary greatly at low elevation angles, as shown in Fig. 15. If the zenith ionospheric delay is 50 ns, then for an obliquity factor of three, the ionospheric delay at the lower elevation angles can be as much as 150 ns or approximately 45 m at 1.6 GHz. Clearly, this amount of unknown excess propagation delay is not consistent with a 15-m position accuracy objective for GPS and must be compensated for in some manner, either by modeling, measurement, or operation in a differential mode. Intuitively common ranging errors tend to affect the clock much more than horizontal position.

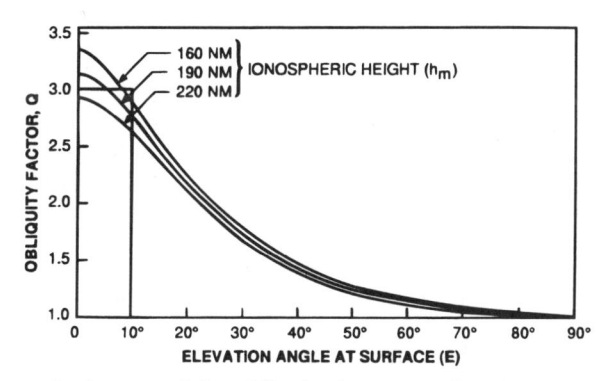

Fig. 15 Ionospheric group delay obliquity factor as a function of the relationship surface elevation angle.[5] The obliquity factor Q is the ratio of the ionospheric path delay for a satellite at elevation angle E to the delay function satellite at zenith (note 1 NM = 1.852 km).

B. Tropospheric Effects

The troposphere/stratosphere can produce a variety of propagation effects on radio waves from the satellite including the following:
1) Atmospheric attenuation
2) Tropospheric scintillation
3) Tropospheric refraction caused by the wet and dry atmosphere that produce excess delay in the signal

1. Tropospheric Group Delay

The troposphere is a region of dry gases and water vapor that extends up to approximately 50 km. This region has an index of refraction, $n(h) = 1 + N(h) \times 10^{-6}$, that varies with altitude. The index of refraction is slightly greater than unity, and hence, causes an excess group delay in the signal waveform beyond that of free space. This region is not ionized and is not frequency dispersive because the excess group delay $\Delta\tau$ is constant with frequency for frequencies below 15 GHz and is approximately equal to the following:

$$\Delta\tau = \int_{\text{path}} N(h)dh \times 10^{-6}$$

The excess group delay is normally on the order of 2.6 m for a satellite at zenith, and it can exceed 20 m at elevation angles below 10 deg. Thus, it must be modeled and removed if high accuracy positioning and time transfer are to be achieved. Dry atmospheric effects that are relatively easily modeled account for approximately 90% of the tropospheric excess delay. The wet atmosphere, although only about 10% of the total, is highly variable and very difficult to model. Detailed discussions of the ionospheric and tropospheric effects on GPS are given in Chapters 12 and 13, this volume.

C. Multipath Effects

1. Multipath

Some of the most difficult navigation problems are for aircraft. Aircraft navigation and three-dimensional navigation in general are also prime motivations for the GPS system. It is important to examine potential multipath effects that can be present in aircraft navigation. An aircraft flying at altitude h has a multipath environment with ground or sea surface reflections, as shown in Fig. 16. If the satellite elevation angle is E, the reflected ray is delayed with respect to the direct ray by $\Delta R = c\Delta\tau = 2h \sin E$. If, as an example, $h = 1$ km and $E = 10$ deg, then the delay difference in the reflected signal is $\Delta\tau = 1.16$ μs. The sea surface reflection cannot be avoided easily by antenna design if we must operate with satellites at low or moderate elevation angles. (The antenna pattern must allow for aircraft banking.) Furthermore, the reflected signal amplitude from the sea surface can at times be nearly as large or sometimes even larger than the direct ray. As shown later in Chapters 3 and 7, the GPS receiver can effectively reject most of the multipath signal if the differential delay $\Delta\tau > 1.5$ μs for the C/A code and 0.15 μs for the P(Y)-code. Note the region of potential

GPS OPERATION AND DESIGN

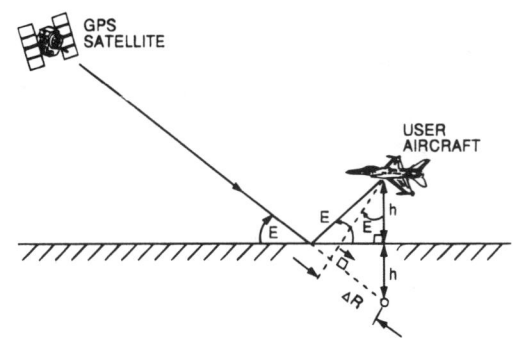

Fig. 16 Multipath effects. Multipath delay varies with elevation angle E and user altitude h. Delay = $\Delta R = 2h \sin E$.

mutipath delay problems for the C/A code is then

$$1.5 \; \mu s > \frac{2b}{c} \sin E = \Delta \tau$$

or

$$h \sin E < (1.5 \; \mu s)c = 448.5 \text{ m}.$$

Because the satellite is moving, the multipath will, in general, be time varying. A reflected multipath signal has the following Doppler shift

$$\Delta f = \Delta \dot{\phi} = \frac{2\dot{h} \sin E + 2h \dot{E} \cos E}{\lambda}$$

for the example shown in Fig. 16. More generally, the multipath may consist of a whole array of scatterers/reflectors, as shown in the impulse response of Fig. 17b. If the number of reflectors is sufficiently large and they are modeled as

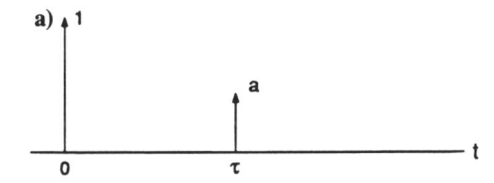

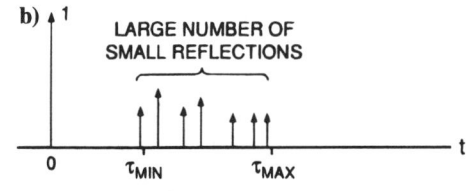

Fig. 17 Multipath channel impulse response with a) a single specular reflection; and b) a large number of smaller reflections.

independent random reflections, then the summed multipath reflection is approximately Gaussian and has a Rayleigh distribution in amplitude.

Within the constraints of available bandwidth (≈ 2 MHz per channel for the usual civilian receivers) and limitations on receiver complexity, the GPS signal is designed to resist the interference from multipath signals for delay differences that exceed 1 μs. Of special importance are multipath signals with a delay difference corresponding to the aircraft sea surface reflections described above. The capability for multipath discrimination of the selected GPS signal is examined in Chapter 14, this volume. We should recognize however, that there can also be reflections from nearby metallic or conducting surfaces (e.g., aircraft wings or stabilizers) that cannot be discriminated against by choice of signal within the bandwidth constraint because the delay difference is too small.

D. Other Perturbing Effects

In addition to the satellite clock and ephemeris errors of the GPS control segment and the atmospheric effects of the ionosphere, troposphere, and multipath, there are several other effects that are important for at least some users. Each of these is briefly discussed below.

1. Relativistic Effects

The Global Positioning System is perhaps the first widely used system where relativistic effects are not negligible. For example, referring back to Figs. 2 and 4, the position and time were all discussed in gravity-free inertial space with stationary users and a nonrotating Earth. In fact, there are several relativistic effects that are nonnegligible. These effects include the following: 1) gravitational field of the Earth and Earth rotation; and 2) velocities of satellite and user. The major effects cause an average increase in the satellite clock frequency as observed by a user on Earth. These effects are partially accommodated by purposely setting the satellite clock slightly low in frequency prior to launch. This topic is discussed in Chapter 18, this volume.

2. Foliage Attenuation

One of the major classes of users for GPS are ground mobile users. If these users are traveling along a road or highway, there is the possibility of obstructions or tree foliage attenuations of the GPS segments. This topic is discussed in Chapter 15, this volume.

3. Selective Availability

To reduce the potential for GPS to be used in hostilities toward the United States, the accuracy of the GPS signal for civil users can be purposely reduced by a capability called selective availability, which is discussed in Chapter 17, this volume.

References

[1]Kaplan, G. H., "The IAU Resolutions on Astronomical Constants, Time Scales, and the Fundamental Reference Frame," U.S. Naval Observatory Circular 163, Washington, DC, Dec. 1981.

[2]Spilker, J. J., Jr., and Magill, D. T., "The Delay Lock Loop—An Optimum Tracking Device," *Proceedings of the IRE,* Vol. 49, Sept. 1961.

[3]Spilker, J. J., Jr., *Digital Communications by Satellite,* Prentice-Hall, Englewood Cliffs, NJ, 1977, 1995.

[4]Spilker, J. J., Jr., "GPS Signal Structure and Performance Characteristics," *Navigation,* Summer 1978, pp. 121–146; also published in *Global Positioning System Papers,* Institute of Navigation, Washington, DC.

[5]National Research Council, *The Global Positioning System—A Shared National Asset,* National Academy Press, Washington, DC, 1995.

Chapter 3

GPS Signal Structure and Theoretical Performance

J. J. Spilker Jr.*
Stanford Telecom, Sunnyvale, California 94089

I. Introduction

THIS chapter discusses the details of the Global Positioning System (GPS) signal structure, its specifications, and general properties. The chapter begins with a review of the general performance objectives and quantitative requirements of the signal. Because the GPS signal falls into a broad category of signals known as spread spectrum signaling, the fundamentals of spread spectrum signaling are introduced. The chapter continues with a detailed description of the GPS signal structure for both the precision (P code) and civil coarse/acquisition (C/A code) signals. The various minor distortions and imperfections permitted by the GPS satellite–user interface specification are also discussed. Although the general format of the navigation data is summarized, details of the navigation message are given in the next chapter. The radio frequency signal levels and relevant signal-to-noise ratios are discussed next. The chapter concludes with a discussion of the signal performance characteristics including: 1) C/A- and P-code cross-correlation properties for multiple access, and 2) performance bounds on the C/A- and P-code pseudorange tracking accuracy for the optimal delay–lock-loop tracking receivers.

A brief summary of Galois fields that are the mathematical basis for maximal length and Gold sequences is given in the Appendix.

A. Summary of Desired GPS Navigation Signal Properties

Based on the navigation accuracy and system requirements and the relevant physics/communication theory discussed in the previous chapter, the system level accuracy requirements can now be stated and translated into signal measurement accuracy requirements. User position and velocity accuracy objectives translate into requirements on pseudorange and other GPS signal measurements and infor-

---

Copyright © 1994 by the author. Published by the American Institute of Aeronautics and Astronautics, Inc., with permission. Released to AIAA to publish in all forms.
*Ph.D., Chairman of the Board.

mation on satellite position and clock time at the time of transmission that must be available to the user receiver. This required information is summarized in Fig. 1.

The accuracy of the pseudorange measurements can be related to the desired accuracy of position by the various **dilutions of precision** (DOP); e.g., position dilution of precision (PDOP), and horizontal dilution of precision (HDOP). If it is assumed that PDOP $\approx$ 3 then a 10-m rms user position error translates into a pseudorange accuracy* of $\leq$ 10 m/3 = 3.33 m or 11 ns. Likewise, a civil user needing 100-m real-time accuracy translates to a pseudorange accuracy of roughly 110 ns. In addition, the GPS signal should also possess the following properties:

1) Tolerance to signals from other GPS satellites sharing the same frequency band; i.e., multiple access capability

2) Tolerance to some level of multipath interference—there are many potential sources of multipath reflection; e.g., man-made or natural objects or the sea surface for an aircraft flying over water

3) Tolerance to reasonable levels of unintentional or intentional interference, jamming, or spoofing by a signal designed to mimic a GPS signal

4) Ability to provide ionosphere delay measurements—dual frequency measurements made at L_1, L_2 frequencies must permit accurate estimation of the slowly changing ionosphere

1. Flux Density Constraints

In addition to the requirements stated above for the GPS signal, there are requirements that the GPS signal received on the Earth be sufficiently low in power spectral density so as to avoid interference with terrestrial microwave line-of-sight communication. For example, a line-of-sight microwave terminal carrying a large number of 4-kHz voice channels potentially can receive interference from a GPS satellite signal that might be observed within the antenna pattern of the receive microwave antenna. Thus, the power flux density of the GPS

GPS INFORMATION SOUGHT

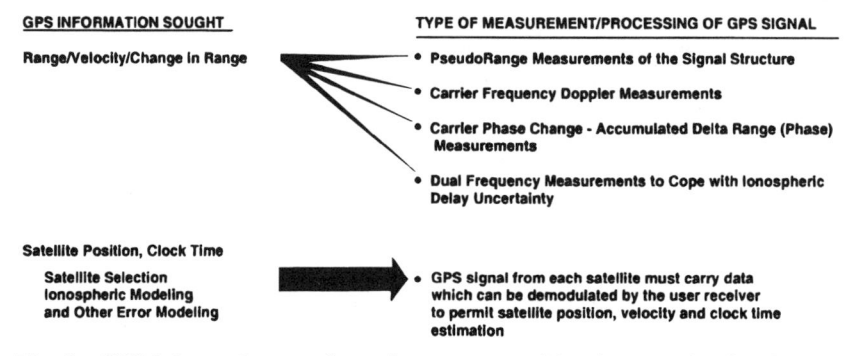

Range/Velocity/Change in Range

TYPE OF MEASUREMENT/PROCESSING OF GPS SIGNAL
- PseudoRange Measurements of the Signal Structure
- Carrier Frequency Doppler Measurements
- Carrier Phase Change - Accumulated Delta Range (Phase) Measurements
- Dual Frequency Measurements to Cope with Ionospheric Delay Uncertainty

Satellite Position, Clock Time
Satellite Selection
Ionospheric Modeling
and Other Error Modeling

- GPS signal from each satellite must carry data which can be demodulated by the user receiver to permit satellite position, velocity and clock time estimation

Fig. 1 GPS information sought and measurement/signal processing by the user receiver.

---

*This statement assumes that there are no other error effects besides pseudorange. There are other errors; e.g., satellite position and clock time errors, but at the moment only the GPS pseudorange error effects are considered.

SIGNAL STRUCTURE AND THEORETICAL PERFORMANCE

satellite signal in a 4-kHz band is constrained so as to remain below a certain level, thus eliminating the possibility of interference on one or more voice channels. Because the 24 GPS satellites orbit the Earth in 12-h orbits, many microwave radio locations will observe GPS satellites at low-elevation angles at one time or another. For this reason, the International Telecommunication Union (ITU) has set flux density regulations on the power that can be generated by a satellite-to-Earth link. In the 1.525–2.500 GHz band, the flux density limit for low-elevation angles is -154 dBW/m^2 for any 4/kHz frequency band[1]. Because the constraint is on power flux spectral density rather than total radiated power, a satellite can radiate more total power and stay within the flux density limit if the signal energy is spread out fairly uniformly over a wider spectral band. The Global Positioning System uses spread spectrum signals to achieve this goal wherein the signal spectra are spread out over a much wider bandwidth than their information content in order to permit use of higher power levels and, of course, to achieve sufficiently precise ranging accuracy. For a unity (0 dB) gain antenna, the aperture area is $A = \lambda^2/4\pi$, and for GPS L_1 at 1.57542 GHz, where the wavelength is $\lambda = 0.1904$ m and $A = 2.886 \times 10^{-3}$ m^2 or -25.4 dB relative to 1 m^2. Thus, this flux density limit transblates to a power level to a unit gain antenna at 1.54542 GHz of $-154 - 25.4 = -179.4$ dBW in any 4-kHz frequency band.

In addition to constraints for line-of-sight microwave radio, there are also constraints to protect radio astronomy. Radio astronomy makes use of the 1420.4 MHz spectral line of neutral atomic hydrogen (the 1400–1427 MHz band is assigned for radio astronomy) and the OH radical molecule with lines at 1612.232, 1665.402, 1667.359, 1720.530 MHz.[2] Thus, the GPS satellite signal is specially filtered to avoid interference with these bands.

B. Fundamentals of Spread Spectrum Signaling

Spread spectrum signaling in its most fundamental form is a method of taking a data signal $D(t)$ of bandwidth B_d that is modulated on a sinusoidal carrier to form $d(t)$, and then spreading its bandwidth to a much larger value B_s where $B_s >> B_d$. The bandwidth spreading can be accomplished by multiplying the data-modulated carrier by a wide bandwidth-spreading waveform $s(t)$. A simplified spread spectrum system is shown in Fig. 2. The figure shows a conventional biphase modulated transmitter* on the far left, followed by a spectrum-spreading operation, an additive noise and interference transmission channel, and the spread spectrum receiver processor. A binary data bit stream $D(t)$ with values $D = \pm 1$ and clock rate f_d is first modulated on a carrier of power P_d to form the narrow bandwidth signal:

$$d(t) = D(t)\sqrt{2P_d} \cos \omega_o t \qquad (1)$$

This narrow bandwidth signal of bandwidth B_d is then spread in bandwidth by a binary pseudorandom signal $s(t)$ where $s(t) = \pm 1$ and has a clock rate f_c that greatly exceeds the data bit rate; i.e., $f_c >> f_d$. For random data and spreading

---

*In general the data signal $D(t)$ can be multilevel and complex. However, this discussion is restricted to binary real $D(t) = \pm 1$.

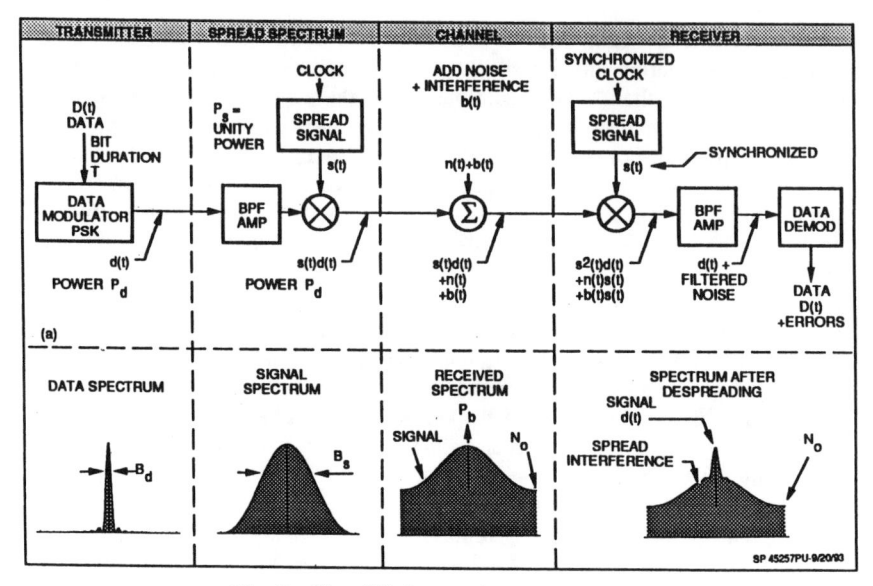

Fig. 2 Simplified spread spectrum link.

codes, the data $D(t)$ and spreading waveforms $s(t)$ have the following power spectral densities, respectively, as shown in Fig. 2.

$$G_d(f) = \frac{1}{f_d}[(\sin \pi f/f_d)/(\pi f/f_d)]^2 \quad \text{and} \quad G_s(f) = \frac{1}{f_c}[(\sin \pi f/f_c)/(\pi f/f_c)]^2$$

Because the timing of the data and clock transitions are synchronous, the spread spectrum product $D(t) s(t)$ has exactly the same spectrum as that of $s(t)$ alone. The spread spectrum signal then has the following form:

$$s_o(t) = s(t)d(t) = s(t)D(t)\sqrt{2P_d} \cos \omega_o t \tag{2}$$

where the spreading signal is as follows:

$$s(t) = \sum_{n=-\infty}^{\infty} S_n p(t - nT_c) \tag{3}$$

and $p(t)$ is a rectangular unit pulse over the interval $\{0, T_c = 1/f_c\}$ and S_n is a random or pseudorandom sequence $S_n = \pm 1$. In general, $p(t)$ can represent a filtered pulse, and different spreading waveforms $s_i(t)$ with coefficients S_{in} in Eq. (3) can separately modulate in-phase and quadrature carrier components. For this example, we restrict the discussion to rectangular pulses and biphase modulation.

This form of spread spectrum is termed **direct sequence-spread spectrum** (DS-SS), and it is one of many different forms of spread spectrum. Other forms* include spreading by pseudorandom frequency hopping, termed frequency hop-

---

*There are other means for bandwidth spreading, such as low rate error correction coding, that do not employ an independent spreading waveform.[3]

spread spectrum (FH-SS), and various hybrid forms of DS-SS and FH-SS. We restrict our discussion here to DS-SS because it provides a means to recover precise timing, and at the same time, it permits recovery of the pure rf carrier. The ability to recover pure carrier is key to precision differential delay and Doppler measurements that provide accuracies on the order of 1% of a carrier wavelength.

The DS-SS signal in Fig. 2 next passes through a channel (with zero channel delay for simplicity and without loss of generality) with additive white noise $n(t)$ of power spectral density N_0 and interference $b(t)$ to form the received signal $r(t) = s(t) d(t) + n(t) + b(t)$ where $b(t)$ is a pure tone interference of power P_b. In the receiver, an identical and precisely time-synchronized replica spreading signal $s(t)$ is generated and correlated (multiply and filter) with the received noisy signal. The fact that the received replica must be accurately time synchronized is shown later to be the exact property that enables the system to extract accurate timing and ranging information. That is, the signal has a narrow autocorrelation envelope of width inversely proportional to the clock rate f_c. The receiver multiplier converts the desired signal $d(t) s(t)$ to $d(t) s^2(t) = d(t)$ because $s^2(t) = 1$; that is, it compresses the spread spectrum signal to its original narrow bandwidth with only the data modulation remaining. The noise spectral density is still N_0 because convolving white thermal noise with a continuous constant envelope spread spectrum signal is still white Gaussian noise. The narrowband interference $b(t)$ has now been spread to look like $s(t)$ and has bandwidth B_s, similar to the manner in which the narrowband signal $d(t)$ was spread in the transmitter. Filtering this multiplier output through a bandpass filter passes the narrowband signal $d(t)$ relatively undistorted, however only a fraction of the noise and interference power passes through the bandpass filter with power $N_0 B_d$ and $P_b (B_d/B_s)$, respectively.

Demodulation of this filtered output then produces a bit error rate that is determined by this noise and interference level. Note that if there is only thermal noise and no other interference is present, then the receiver output is exactly the same in terms of signal power and noise density as if there had been no spectral spreading at all. That is, the effects of the spreading and despreading by the binary pseudorandom code $s(t)$ in the transmitter and receiver cancel. Thus, the use of properly synchronized spread spectrum signaling neither improves nor degrades the signal performance against a thermal noise background.

However, the performance against a tone interference of fixed power is greatly improved because the interference power level is reduced by the ratio of the clock rates f_c/f_d. The ratio, f_c/f_d, of the PN chip rate f_c to the data bit rate f_d is termed the *processing gain of* the spread spectrum system and is a key parameter because it determines what fraction (f_d/f_c) of the interference power passes through to the output. Whereas the thermal noise power increases in direct proportion to radio frequency bandwidth, the interference power is fixed and independent of bandwidth. In fact, spread spectrum signaling is effective against a much broader class of interference than simple tone interference.

1. Direct Sequence-Spread Spectrum Signals—Autocorrelation Function and Spectrum

A noise-like pseudorandom binary spreading sequence $s(t)$ bears a close resemblance to a random sequence. A purely random binary sequence is generated by a coin-flipping operation where the outcome is equally probable ±1. This sequence at clock rate f_c can generate the waveform $s(t)$ of (3), which is shown in Fig. 3a, and it has a triangular autocorrelation function and $(\sin \pi \tau f / \pi \tau f)^2$ shaped power spectral density as shown in Fig. 3b,c. As is shown in a later paragraph, a close approximation (pseudorandom) to a random sequence can be generated by using suitable feedback shift registers. Thus, we can generate a replica waveform at the receiver and suitably time synchronize this replica to the received signal.

2. Multiple Access Performance of Spread Spectrum Signaling

It is often desired to provide a method by which multiple signals can simultaneously access exactly the same frequency channel with minimal interference between them. Spread spectrum signaling has the capability to provide a form of multiple access signaling called code division multiple access (CDMA) wherein multiple signals can be transmitted in exactly the same frequency channel with limited interference between users, if the total number of user signals M is not too large. This multiple access capability is important for GPS because a user receiver may receive simultaneously 10 GPS signals from 10 different satellites, wherein all signals occupy the same frequency channel and are continuous (i.e., not time gated). For example, assume that there are M signals, all of exactly the same power P_s received at a receiver antenna. If all M signals are received with exactly the same code clock delay, it is possible to select a certain number of signals that are completely orthogonal, and thus cause no multiple access interference provided $M \leq f_c T_d$ where $f_d = 1/T_d$ is the data bit rate. However, in many communication/ranging tasks orthogonal signaling is not possible because the signal sources—the GPS satellites in our example—cannot possibly all be equally distant from each user. Good multiple access performance is still possible, however, by selecting the different GPS spread spectrum pseudorandom codes to be nearly uncorrelated for all possible time offsets.

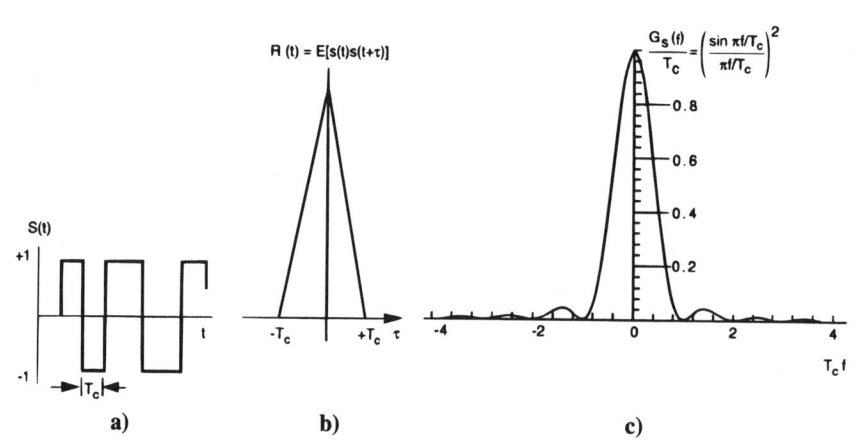

Fig. 3 Random binary sequence, autocorrelation function, and power spectral density—a) waveform, b) autocorrelation function, and c) power spectral density. The clock rate is $f_c = 1/T_c$.

SIGNAL STRUCTURE AND THEORETICAL PERFORMANCE

Examine two multiple access signals $s_i(t)$ and $s_j(t)$, which are both uncorrelated pseudorandom codes with identical spectra $G_s(f)$ and are both transmitted on the same frequency channel and received with independent random timing. The receiver of Fig. 2 for the first signal; e.g., $s_i(t)$, cross-correlates the received additive signals with the desired reference signal code $s_i(t)$. Ignoring the data modulation, carriers, and noise for the moment, the correlator output is then $s_i(t - \tau_1)[s_i(t - \tau_1) + s_j(t - \tau_2)] = 1 + s_i(t - \tau_1) s_j(t - \tau_2)$. The unity term is the desired component, and the spread spectrum multiple access interference term is $s_i(t - \tau_1) s_j(t - \tau_2)$. For random time offsets between the two signals and power level P_s, the multiple access interference spectrum is defined as $G_{ma}(f)$ and is obtained by convolving the individual spectra (see Fig. 4):

$$G_{ma}(f) = P_s \int G_s(v)G_s(v - f)dv \quad \text{and} \quad G_{ma}(0) = P_s \int G_s^2(v)dv$$

Assume that the processing gain is large; i.e., $f_c/f_d \gg 1$. Then only the multiple access interference spectrum near $f = 0$ is significant because the correlation filters can have a bandwidth on the order of f_d. The convolved spectrum at $f = 0$ can be computed to be:

$$G_{ma}(0) = P_s \int_0^\infty \left(\frac{\sin \pi f/f_c}{\pi f/f_c}\right)^4 df$$

$$G_{ma}(0) = \left(\frac{2}{3}\right)(P_s/f_c)$$

Note that if the multiple access signal has transitions coincident with the reference signal; that is, a nonrandom time offset, the multiple access interference

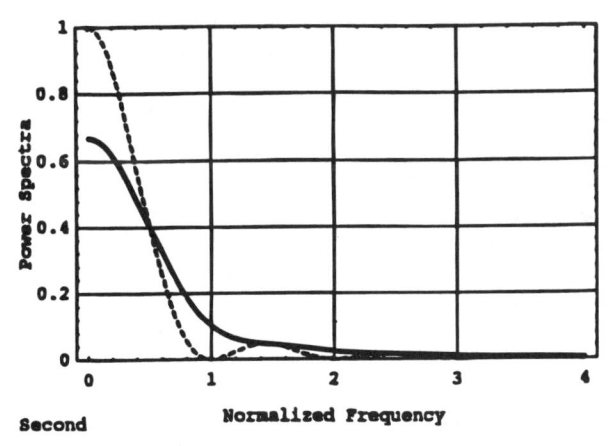

Fig. 4 Original spectrum $(\sin \pi/\tau f)^2$ (dashed line) and the convolved spectrum $G_{ma}(f)$ (solid line) at the correlator output, for a normalized clock rate $f_c = 1$ for an unfiltered pseudorandom signal. Note that the multiple access power spectral density at $f = 0$ decreases to 2/3.

is not spread, and the factor of 2/3 does not appear. Note also, that this result (Fig. 4) assumes that the $[(\sin \pi f/f_c)/\pi f/f_c]^2$ spectrum includes all of its sidelobes and is not filtered. If the signal is filtered to include only the main lobe the factor of 2/3 increases to approximately 0.815. If the signal spectrum is rectangular the factor is unity. As shown later, the transmitted GPS C/A code is nearly unfiltered and contains sidelobes out to the 10th. The GPS P code has the same bandwidth but 10 times the clock rate, and thus only contains the mainlobe.

Because there are $M-1$ interfering multiple access signals, and there is only one desired signal for each tracking receiver, the net effect of the $M-1$ multiple access signals is to increase the effective noise spectrum in the vicinity of the desired data modulation from a value N_o with no multiple access interferences to an equivalent noise density:

$$N_{oeq} = N_o + \frac{2}{3}(M-1)P_s/f_c = N_o[1 + \frac{2}{3}(M-1)P_s/f_cN_o] \qquad (4)$$

Table 1 summarizes the equivalent noise density relationships for spread spectrum multiple access signals for the complete $(\sin x/x)^2$ spectrum, $(\sin x/x)^2$ mainlobe only, and rectangular spectra.

Thus, if all sidelobes are included, the effective noise density is increased by the factor $1 + (2/3)(M-1)P_s/f_c$, and the effective energy per bit E_b to equivalent noise density ratio decreases to the following:

$$\frac{E_b}{N_{oeq}} = \frac{P_s T_d}{N_o\left[1 + \frac{2}{3}(M-1)P_s/f_cN_o\right]} = \frac{P_s}{N_of_d}\frac{1}{\left[1 + \frac{2}{3}(M-1)P_s/f_cN_o\right]} \qquad (5)$$

The quantity E_b is the energy per bit $E_b = P_s/f_d$. The quantity E_b/N_{oeq} determines the output error rate. For biphase modulated (antipodal) signaling the E_b/N_{oeq} needs to be on the order of 10 if no error correction coding is employed.* If the performance of the system is not to be degraded by more than 3 dB relative to thermal noise performance, then from (5) the number of equal power multiple access signals is limited by the following:

$$M < \frac{3}{2}\left(\frac{N_of_c}{P_s}\right) + 1 = \frac{3}{2}\left(\frac{N_of_c}{E_sf_d}\right) + 1 = \frac{3}{20}\left(\frac{f_c}{f_d}\right) + 1 \quad \text{for} \quad \frac{E_b}{N_{oeq}} = 10 \qquad (6)$$

the limit on M increases as the spread spectrum clock rate of f_c increases.† Again, it is pointed out that the factor of 3/2 applies only for the unfiltered $(\sin x/x)^2$ spectrum.[4] Note that for GPS there are often $M = 9$, 10 GPS signals in view. See the later chapter on the GPS constellation. Furthermore, note that if the

---

*Although GPS does not employ error correction coding, it should be pointed out that the use of spread spectrum signaling generally does permit the use of low rate error correction codes that can allow operation at low values of E_b/N_o without suffering from bandwidth expansion because the spread spectrum signaling is already broadening the spectrum by itself.

†For example, with $f_c = 10^6$ and $f_d = 50$, this result, (6), becomes:

$$M < (3/20)(10^6/50) + 1$$
$$M < 3 \times 10^3 + 1$$

SIGNAL STRUCTURE AND THEORETICAL PERFORMANCE

Table 1 Equivalent noise density for M equal power spread spectrum signals of different power spectral densities[a]

Spread spectrum signal spectra	Multiple access equivalent noise density at $F = 0$
$G_s(f)$, $-\infty < f < \infty$	$N_o \left[1 + \dfrac{(M-1)P_s}{N_o} \right] \displaystyle\int_{-\infty}^{\infty} G_s^2(f)\, df$
$\dfrac{1}{f_c}\left[\dfrac{\sin \pi f/f_c}{\pi f/f_c}\right]^2$, $-\infty < f < \infty$ All sidelobes	$N_o\left[1 + \left(\dfrac{2}{3}\right)\dfrac{(M-1)P_s}{f_c N_o}\right]$
$\dfrac{1}{f_c}\left[\dfrac{\sin \pi f/f_c}{\pi f/f_c}\right]^2 \left(\dfrac{\pi}{2 \sin \text{integral}(2\pi)}\right)$, $-f_c < f < f_c$ Mainlobe only	$N_o\left[1 + \dfrac{\pi(M-1)P_s}{3 f_c N_o}\right.$ $\left.\left(\dfrac{2 \sin \text{integral}(4\pi) - \sin \text{integral}(2\pi)}{(2 \sin \text{integral}(2\pi))^2}\right)\right]$ $= N_o\left[1 + \dfrac{(M-1)P_s}{f_c N_o}(0.815497)\right]$
Rectangular spectrum $\dfrac{P_s}{2 f_o}$, $-f_o < f < f_o$	$N_o\left[1 + \dfrac{(M-1)P_s}{f_o N_o}\right]$

[a] The reference signal spectra $G_s(f)$ are all normalized to unity signal power. Each of M received signals has power P_s.

desired signal is 1/10th of the power of the other signals, the value of M permitted decreases approximately by a factor of 10. As shown later, in some instances the DS-SS code has a relatively short period (e.g., the GPS C/A code), the spectrum of the spreading code has line components spaced at the epoch rate, and the performance is not quite as good as indicated above.

3. Generation of the Spreading Signal Using Linear Feedback Shift Registers

Figure 5 shows a simple four-stage linear feedback shift register with taps after stages 1 and 4, which are modulo-2 added to form a short period maximal length pseudorandom or pseudonoise PN sequence. The sequence of shift register state vectors is shown in Fig. 5b, where the initial state vector is as follows:

a) n = 4 FOUR STAGE FEEDBACK SHIFT REGISTER

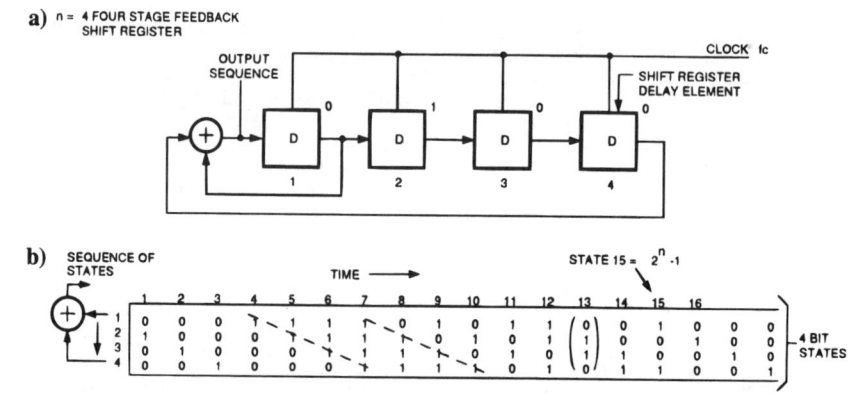

b) SEQUENCE OF STATES

Fig. 5 Generation of a PN sequence using a maximal length feedback shift register and the $2^4 - 1 = 15$ sequence of code generator states. The state vectors each have four binary components, as shown in b.

$$s_1 = \begin{pmatrix} 0 \\ 1 \\ 1 \\ 1 \end{pmatrix}$$

The state vector components are defined by the state of each of the four binary shift register delay elements. As long as this shift register is not set to the "all zero" state, it will cycle through all $2^4 - 1 = 15$ state vectors in a periodic manner. Only specific tap combinations produce a sequence of length $2^4 - 1 = 15$. In general, an n stage linear feedback shift register (LFSR) with proper taps generates all 2^n of the n-bit states except the all-zero state, thus it cycles through each of the possible state vectors. The use of the word "linear" refers to the restriction on the logic to modulo-2 adders. Thus, there are $2^n - 1$ states in the period of the sequence, and the sequence has a period $2^n - 1$.

Figure 6 shows the PN sequence at the output of a selected stage. The autocorrelation of the PN sequence where $s(t) = \pm 1$ is the following:

$$R(i) = (1/15) \int_0^{15} s(t)s(t + i)dt$$

and for rectangular pulses has the form shown in Fig. 6c when expressed as a function of continuous time.

The maximal length sequence can be designed to have an arbitrarily long period $2^n - 1$ by increasing the number of stages n. As n increases, the maximal length sequence becomes more random in appearance or pseudorandom, and its spectrum approximates the $(\sin x/x)^2$ spectrum of Fig. 3. Because of the sharp (narrow in time) peak of the autocorrelation function for a high clock rate signal, the waveform also can be used for very accurate measurements of time and range or pseudorange. Obviously, this characteristic is key for GPS. An introduction

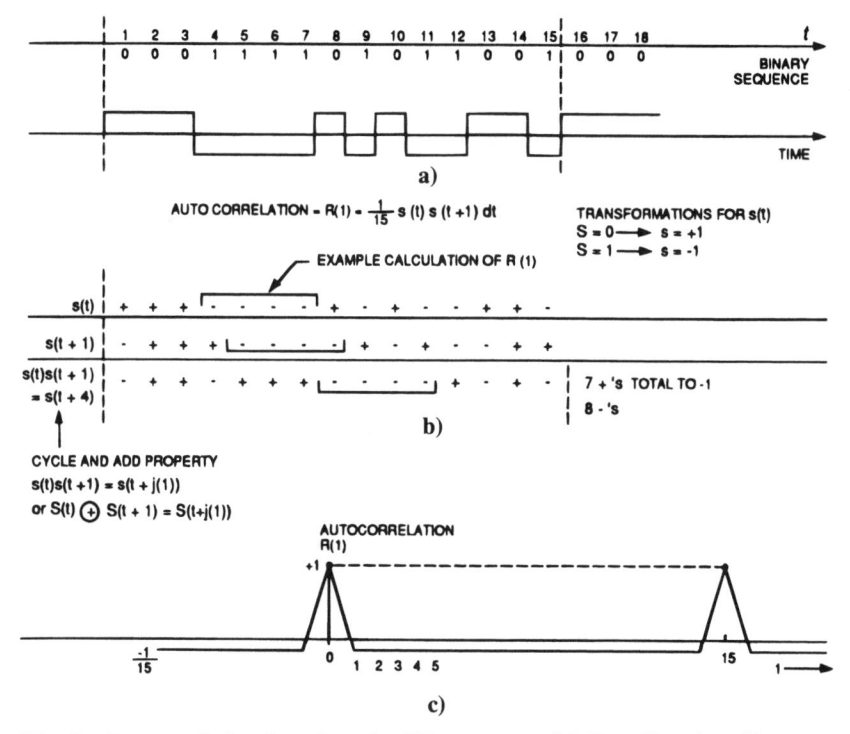

Fig. 6 Autocorrelation function of a PN sequence with $P = 2^n - 1 = 15$ states.

to the more detailed analysis of PN codes and Galois field algebra for PN codes is given later in the chapter as an appendix.

II. GPS Signal Structure

In this section, the structure of the GPS signal is described in detail. The general properties of the civil and precision direct sequence spread spectrum signals and codes are discussed and related to many of the system requirements. Detailed analyses of the performance of the signal are given in the next section.

As stated earlier, the GPS system must provide authorized users with a 10-m or less rms position error, which for a PDOP $\approx$ 3 translates to a required accuracy of pseudorange measurement on the order of 11 ns. The Global Positioning System chose to accomplish this required accuracy with a 10.23-Mcps precision P code. Two other GPS objectives, rapid acquisition of the P-code and providing a lesser but still revolutionary three-dimensional accuracy for the civil user are achieved by the use of the civil coarse/acquisition (C/A)-code, which has a 1.023-Mcps chip rate and a code period of 1023 chips. Civil users do not have access to the P(Y) code when the P code is in the antispoof (AS) Y-code mode. The somewhat unusual code rates of 1.023 Mcps and 10.23 Mcps are selected so that the period of the C/A code corresponds exactly to 1 ms for time-keeping purposes.

A. Multiplexing Two GPS Spread Spectrum Signals on a Single Carrier and Multiple Access of Multiple Satellite Signals

Two important questions to deal with are how to multiplex the two codes, C/A and P, on a single carrier and how to provide the multiple access of the various GPS signals that are to be received from the different satellites within the available frequency band.

The GPS L_1 signal has two spread spectrum signals, civil, C/A, and precision, P, multiplexed onto a single radio frequency carrier. In addition, the signals from multiple satellites must share the same frequency channel. The Global Positioning System multiplexes the civil and precision code on a single carrier in phase quadrature and then employs CDMA so that the different satellite signals can share the identical frequency band. Each satellite P signal occupies the entire available bandwidth to maximize timing accuracy. Table 2 shows the multiplexing and multiple access alternatives considered during the original design of the GPS signal.

Time multiplexing of the two signals, civil (coarse) and precision on one carrier; i.e., transmitting a portion or all of the period of the civil signal followed by a portion of the long military signal, was a possible choice for GPS. However, a time multiplexed signal would not have permitted continuous phase measurement of the carrier because the civil user does not have access to the military precision signal. The ability to perform precision carrier phase measurements was always considered to be of key value to the GPS system. Time gating of the shorter C/A civil signal would also change its autocorrelation characteristics and results in a less desirable cross-correlation performance.

The alternative selected for GPS is to modulate the civil C/A signal on the in-phase component of the L_1 carrier and modulate the precision P signal on a quadrature phase (90 deg rotated), thus providing a constant envelope modulated carrier even if the two signals have different power levels. The GPS signal then has the form (neglecting data modulation) $XP_i(t) \cos \omega_o t + XG_i(t) \sin \omega_o t$ where XP_i represents the P-code and XG_i represents the C/A code. Data biphase modulates both inphase quadrature components identically.

The selected multiple access technique for GPS is CDMA wherein the signals are separated through the use of codes with good cross-correlation properties. Code division multiple access in some systems has a so-called "near–far" problem when substantial differences exist in the received signal levels from different

Table 2 Alternative multiplexing and multiple access techniques considered for the global positioning system

Methods of multiplexing onto a single carrier	Time multiplex civil and military codes		In phase and quadrature multiplex
Multiple access methods	Frequency division multiple access	Time division multiple access	Code division multiple access

SIGNAL STRUCTURE AND THEORETICAL PERFORMANCE 69

transmitters. However, with GPS, the satellites are all at roughly the same range, and the received signal levels normally do not vary greatly. (Exceptions occur when the signal from a given satellite is blocked momentarily by an aircraft wing tip, or, if on the ground, by tree foliage.) The choice of a specific family of codes for GPS that provide the desired code division multiple access performance is discussed in detail in later paragraphs.

An alternative multiple access technique considered was frequency division multiple access (FDMA). Frequency division multiple access of the satellite signals, which was subsequently selected for the GLONASS navigation satellites, has the advantage that the civil signals can be truly uncorrelated by offsetting the carriers in frequency by the bandwidth of the civil PN code. However, this approach occupies a larger bandwidth for a given code bandwidth, a disadvantage that the GLONASS developers diminished by operating the civil signal at roughly half the clock rate of the GPS signal. The GLONASS civil signal operates with a single 511-bit length PN code at 511 kbps, and spaces the carriers in frequency by 562.5 kHz.*[5] GLONASS is discussed more extensively in a later chapter. Decreasing the C/A-code clock rate for the same power flux density on the ground would provide a somewhat lesser accuracy. The other aspect of the frequency division approach felt to be a disadvantage was that the user receiver would have to operate with several frequency offsets if many satellites were to be tracked simultaneously. It was believed that the frequency division multiple access operation had a cost implementation disadvantage for the state of the art at that time (1973–1974).

B. GPS Radio Frequency Selection and Signal Characteristics

During the design phase of the GPS system, various frequency bands were considered for the GPS signal. Although a strict optimization of the frequency is not meaningful because only certain frequency bands could be made available, several considerations were important in selecting the GPS frequency band. Some of these are noted in Table 3.

The use of *L*-band gives acceptable received signal power with reasonable satellite transmit power levels and Earth coverage satellite antenna patterns, whereas the *C*-band path loss is roughly 10 dB higher because the path loss is proportional to f^2 for an omnidirectional receive antenna and fixed transmit antenna beamwidth and range. The large ionospheric delay and fluctuation in delay weighs against uhf as does the difficulty in obtaining two large ($\approx$20 MHz) bandwidth frequency assignments in the uhf band (two frequency bands are necessary for ionospheric correction). Thus, *L*-band was selected, and dual frequencies permit ionospheric group delay measurements. The signal bandwidths at both center frequencies are 20 MHz.

1. Global Positioning System Signal Characteristics

Thus, the GPS signal consists of two components, Link 1 or L_1, at a center frequency of 1575.42 MHz and Link 2, L_2, at a center frequency of 1227.6 MHz.

---

*The center frequencies of the channels are $1602 + 0.5625\,n$ MHz, where $n = 0, 1, 2, \ldots, 24$. See Chapter 9 in the companion volume.

Table 3 Global positioning system transmission frequency band selection considerations

Performance parameter	uhf ≈400 MHz	L-band (1–2 GHz)	C-band (4–6 GHz)
Path loss for omnidirectional receive antenna-loss $\sim f^2$	Path loss lowest of the three	Acceptable	Path loss ≈ 10 dB larger than at L-band
Ionospheric group delay, $\Delta R \sim 1/f^2$	Large group delay, 20–1500 ns	Group delay 2–150 ns at 1.5 GHz	Group delay ≈ 0–15 ns
Other considerations	Galactic noise ≈ 150°K at 400 MHz	—	Rainfall/atmospheric attenuation can be significant in 4–6 GHz band 0.1 to 1 dB/km at 100 mm of rain/hour

Each of the center frequencies is a coherently selected multiple of a 10.23 MHz master clock. In particular the link frequencies are the following:

$$L_1 = 1575.42 \text{ MHz} = 154 \times 10.23 \text{ MHz}$$
$$L_2 = 1227.6 \text{ MHz} = 120 \times 10.23 \text{ MHz} \tag{7}$$

Similarly, all of the signal clock rates for the codes, radio frequency carriers, and a 50 bps navigation data stream are coherently related.

The frequency separation between L_1 and L_2 is 347.82 MHz or 28.3%, and it is sufficient to permit accurate dual-frequency estimation of the ionospheric group delay. (The ratio of $L_1/L_2 = 77/60 = 1.2833$.) The ionospheric group delay varies approximately as the inverse square of frequency, and thus measurement at two frequencies permits calculation of the ionospheric delay. The ionospheric group delay correction is obtained by subtracting the total L_1 group delay τ_{GDL_1} from the total L_2 group delay τ_{GDL_2} in order to cancel the true pseudorange delay. This difference $\Delta\tau$ is then (neglecting random noise for the moment) the following:

$$\Delta\tau = \tau_{GDL_2} - \tau_{GDL_1} = \frac{A}{f_{L_2}^2} - \frac{A}{f_{L_1}^2} = \frac{A}{f_{L_1}^2}\frac{1}{1.54573} = \frac{\tau_{iono}}{1.54573} \tag{8}$$

or

$$\tau_{iono} = A/f_{L_1}^2 = 1.54573 \, \Delta\tau$$

where τ_{iono} is the ionospheric group delay at L_1, and $\Delta\tau$ is the measurable difference between total propagation delays at L_1 and L_2. Thus, the frequencies L_1 and L_2 are separated far enough in frequency so that the ratio is only a factor 1.54573. Ionospheric effects and models are discussed both in Chapter 4 and in considerable detail in Chapter 12, this volume.

SIGNAL STRUCTURE AND THEORETICAL PERFORMANCE

As discussed in Chapter 18, the relativistic effects are not negligible in GPS but are partially compensated for in the satellite by offsetting the 10.23 MHz master clock rate to a slightly lower number before launch. As the signal approaches the Earth from the satellite, the frequency increases slightly because of relativity by approximately the same factor as the offset, and for a stationary user on the Earth's surface, the GPS signal clock appears to have a frequency very close to the desired 10.23 MHz. Henceforth, when reference is made to the desired 10.23 MHz, the frequency will always be this slightly offset frequency as far as the satellite clocks are concerned when observed prior to launch. The actual frequency of the satellite clocks before launch is 10,229,999.995453 MHz or an offset of $\Delta f = 4.57$ mHz below 10.23 MHz. The fractional frequency offset is -4.46×10^{-10} (see Ref. 6).

The L_1 signal is modulated by both a 10.23 MHz clock rate precision P signal and by a 1.023 MHz civil C/A signal to be used by the civil user. The transmitted signal spectra for both L_1 and L_2 are shown in Fig. 7. The binary modulating signals are formed by a P code or a C/A code that is modulo-2 added to the 50 bps binary data D, to form* P$\oplus$D and C/A $\oplus$ D, respectively. The P code also can be converted to a secure antispoof Y code at the same clock rate, and is labeled the P(Y) code. The L_1 signal has an in-phase component of its carrier that is modulated by the P signal, P$\oplus$D, and a quadrature (within ± 100 m rad) carrier component that is modulated by C/A$\oplus$D. The peak power spectral density of the C/A signal exceeds that of the P code at L_1 by approximately 13 dB because it is nominally 3 dB stronger and has 1/10 the chip rate and bandwidth. The in-phase and quadrature waveforms and phasor diagram of the L_1 signal are shown in Fig. 8.

The L_2 signal is biphase-modulated by either the P code or the C/A code. Normal operation would provide P- or Y-code [labeled P(Y)] modulation on the L_2 signal. There may or may not be data modulation on L_2 dependent on ground command.

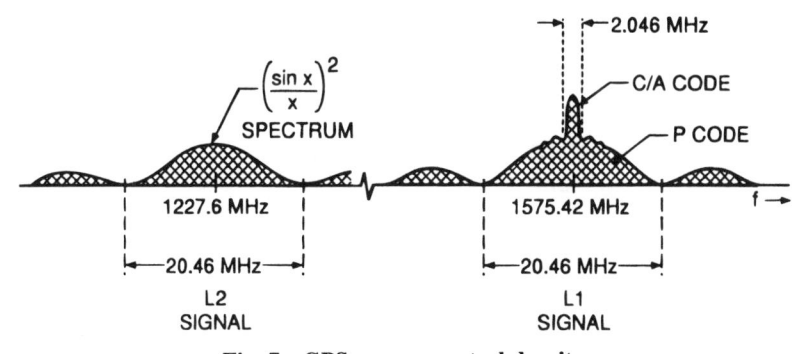

Fig. 7   GPS power spectral density.

---

*The symbol $\oplus$ stands for modulo-2 addition of 0, 1 numbers, which is equivalent to multiplication of $+1$, -1 numbers, respectively.

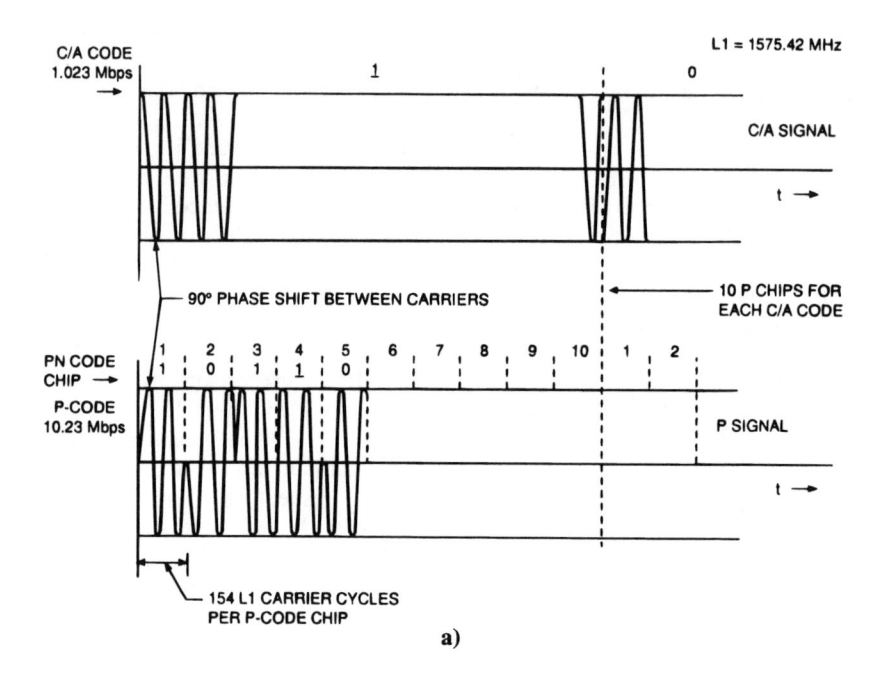

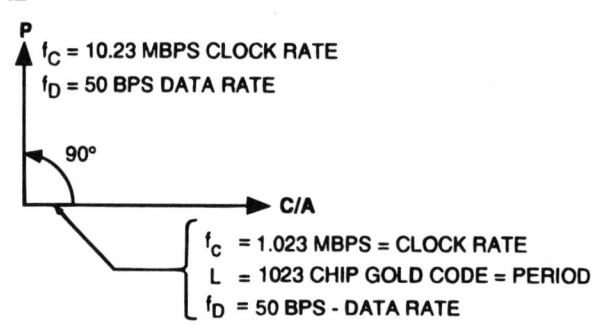

Fig. 8 GPS L_1 signal waveform and phasor diagram. The P code for satellite i is labeled XP_i, and the C/A code for satellite i is labeled XG_i. a) Radio frequency waveforms for the P signal and C/A signal (carrier not to scale). b) Phasor diagram.

The GPS satellite can also transmit a third L-band carrier modulated by the C/A code at an L_3 frequency of 1381.05 MHz = 135 × 10.23 MHz. This signal is utilized only in a time-gated mode for a Nudet (Nuclear Detonation) Detection System (NDS) and is not used in the GPS navigation receiver.

C. Detailed Signal Structure

The L_1 signal contains both in-phase and quadrature signals. The signal transmitted (see Fig. 8) by the satellite i is then as follows:

$$S_{L_{1i}}(t) = \sqrt{2P_c}XG_i(t)D_i(t)\cos(\omega_1 t + \phi) + \sqrt{2P_p}XP_i(t)D_i(t)\sin(\omega_1 t + \phi) \tag{9}$$

where ω_1 is the L_1 frequency as defined above, ϕ represents a small phase noise and oscillator drift component and P_c and P_p are the C/A and P signal powers, respectively. Oscillator frequency stability is obtained using redundant cesium and rubidium atomic frequency standards. (The first satellite in the GPS series NTS-2, had a long-term clock stability better than 2×10^{-13} and later satellites have improved stability $\approx 3 \times 10^{-14}$). The P code $XP_i(t)$, is a ± 1 pseudorandom sequence with a clock rate of 10.23 Mbps, and a period of exactly 1 week. Each satellite i transmits unique C/A and P codes. The binary data $D_i(t)$, also has amplitude ± 1 at 50 bps and has a 6 s subframe and a 30 s frame period. The C/A code XG_i is a unique Gold code of period 1023 bits and has a clock rate of 1.023 Mbps. Thus, the C/A code has a period of 1 ms.

In GPS, the C/A-code strength is nominally 3 dB stronger than the P code on L_1. As already mentioned above, the code clocks and transmitted radio frequencies are all coherently derived from the same on-board satellite frequency standard. The rms clock transition time difference between the C/A and P code clocks is required to be less than 5 ns. Both C/A and P codes are of a class called product codes; i.e., each is the product of two different code generators clocked at the same rate where the delay between the two code generators defines the satellite code i (see Fig. 9). The specific component codes forming the product code for P and C/A are quite different, but the principle is similar. The clock interval for the C/A code $T_{cc} = 10T_c$ where T_c is the P-code clock interval in the figure.

1. P Code—Precision Code

The P code for satellite i is the product of 2 PN codes, $X1(t)$ and $X2(t + n_iT)$, where $X1$ has a period of 1.5 s or 15,345,000 chips, and $X2$ has a period of 15,345,037 or 37 chips longer. Both sequences are reset to begin the week at the same epoch time. Both $X1$ and $X2$ are clocked in phase at a chip rate $f_c = 1/T_c = 10.23$ MHz. Thus, the P-code is a product code of the following form:

$$XP_i(t) = X1(t)X2(t + n_iT), \quad 0 \leq n_i \leq 36 \tag{10}$$

where $X1(t)$ and $X2(t)$ are binary codes of value ± 1 and $XP_i(t)$ is reset at the beginning of the week. The delay between $X1(t)$ and $X2(t)$ is n_i code clock intervals of T_c s each (see Fig. 10). The $X1$ and $X2$ codes are each generated as the products of two different pairs of 12-stage linear feedback shift registers)

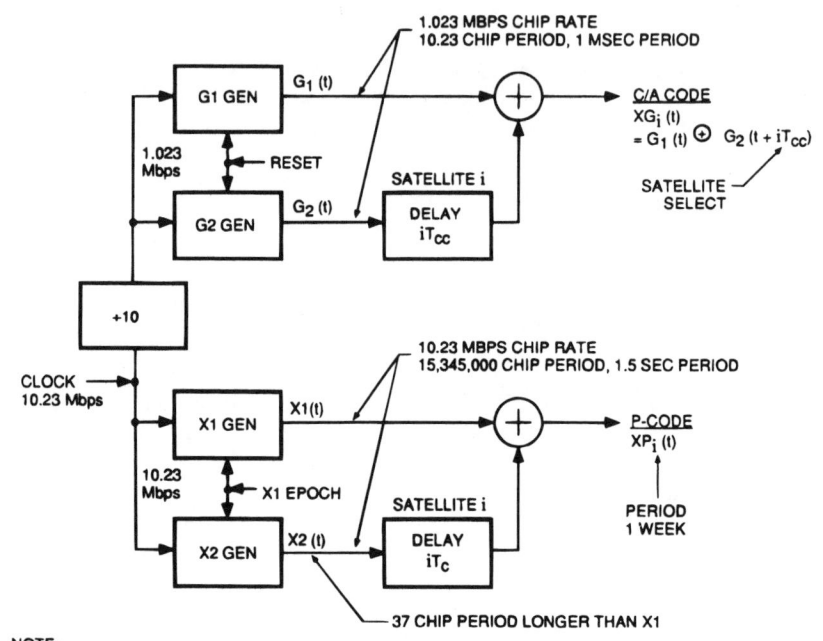

Fig. 9 GPS code generators for satellite *i*. Both the C/A- and P-codes are product codes.

X1A and X1B and X2A and X2B with polynomials specified in the GPS-ICD-200[6] as follows:

$$X1A: 1 + X^6 + X^8 + X^{11} + X^{12}$$
$$X1B: 1 + X + X^2 + X^5 + X^8 + X^9 + X^{10} + X^{11} + X^{12} \quad (11)$$
$$X2A: 1 + X + X^3 + X^4 + X^5 + X^7 + X^8 + X^9 + X^{10} + X^{11} + X^{12}$$
$$X2B: 1 + X + X^3 + X^4 + X^8 + X^9 + X^{12}$$

See the Appendix for a discussion of code polynomials, Galois fields, and shift registers.

For now, suffice it to say that these polynomials give the feedback tap positions of the 12-stage shift registers, X1A, X2A, X1B, X2B, of Fig. 10. Recall the introductory discussion of Fig. 5. The X1A and X1B codes have different relatively prime periods as do the X2A and X2B codes. A 12-stage maximal length shift register produces a code period of $2^{12} - 1 = 4095$. If two code generators are short cycled to give relatively prime periods less than or equal to 4095, then the product code can have a period in the vicinity of 1.6×10^7, the product of the two periods. For GPS, the two product codes have been short cycled to relatively prime periods of 15,345,000 and 15,345,037 for the X1 and X2 respectively. Likewise, the product of X1 and X2 codes generates a new code that has a period that is the product of the periods, unless it is short cycled.

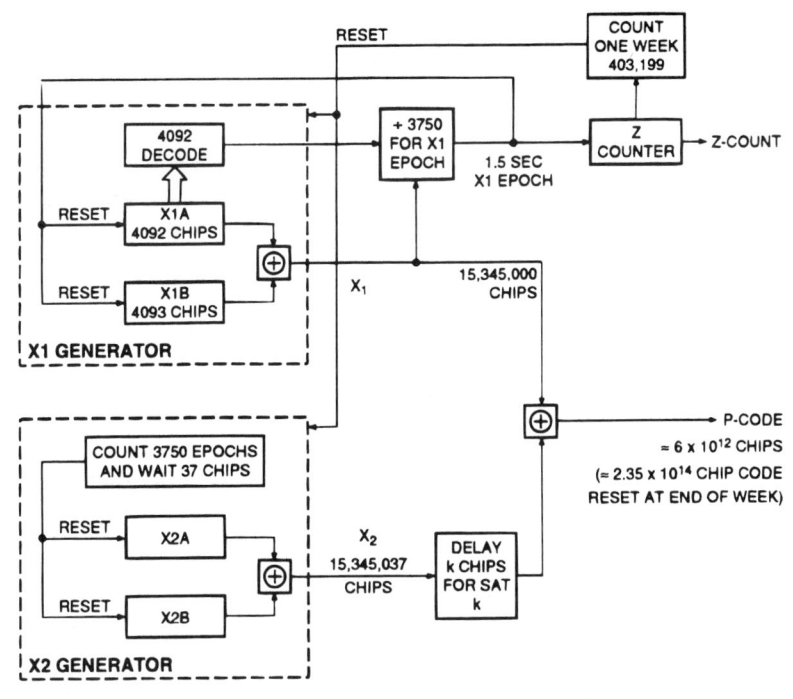

SIGNAL STRUCTURE AND THEORETICAL PERFORMANCE 75

Fig. 10 Simplified P-code generator block diagram.

The product of $X1$ and $X2$ codes clocked together act somewhat like two gears with a number of teeth on each gear corresponding to the periods of $X1$ and $X2$ as shown in Fig. 11. If we imagine that the teeth of both gears are coded in black and white according to the respective binary chips in the PN sequence, then for relatively prime code lengths for $X1$ and $X2$ the period of this product code is equal to the product of the two individual codes periods (the number of teeth on each gear wheel). Each satellite has a unique code offset $n_i T$ between code $X1$ and code $X2$, which makes the P-code unique as well. The increase in code period for $X2$ by 37 relative to $X1$ allows the values of n_i to range over 0–36 without having any significant segment of a P code of one satellite match that of another. Thus, we have 37 different pseudorandom P codes.

For a different view of the P code, note that the period of a product of $X1$ and $X2$ codes, each of which has a relatively prime period, is the product of the periods; i.e. $(15,345,000) \times (15,345,037) = 2.35469592765 \times 10^{14}$. Thus, if the P-code were allowed to continue without being reset, each P code would continue without repetition for slightly more than 38 weeks. In effect, this overall period has been subdivided so that each of 37 possible GPS satellites or ground transmitters (pseudolites) gets a 1-week period code, which is nonoverlapping with that of any other satellite.

A long period code such as the P code is difficult to acquire without acquisition aids. For example, a receiver correlator must be timed to within roughly one P-code chip or roughly 0.1 μs and clocked in synchronism in order to correlate at

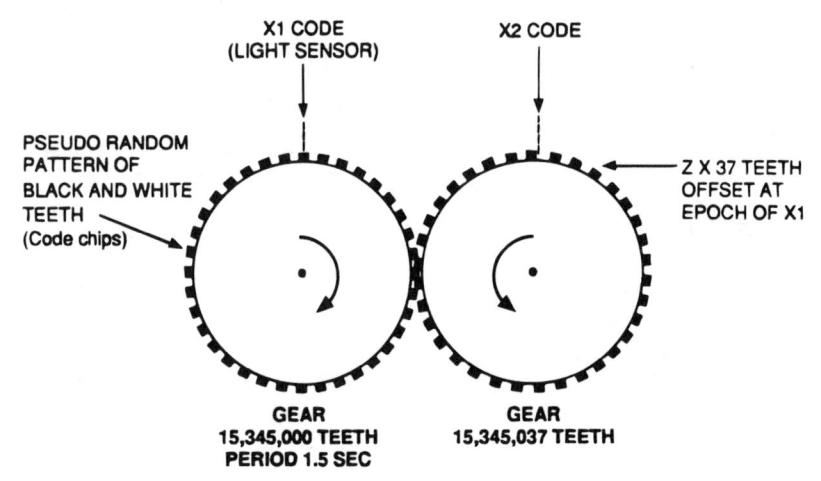

Fig. 11 P-code–subcode characteristics as represented pictorially by two gear wheels with light sensors for both gear wheels.

Labels in figure: X1 CODE (LIGHT SENSOR); X2 CODE; PSEUDO RANDOM PATTERN OF BLACK AND WHITE TEETH (Code chips); Z × 37 TEETH OFFSET AT EPOCH OF X1; GEAR 15,345,000 TEETH PERIOD 1.5 SEC; GEAR 15,345,037 TEETH.

all. Note that the period of the $X1$ code is exactly 1.5 s, i.e., $1.5 \times 10.23 \times 10^6 = 15{,}345{,}000$, and there are this same number of code chip time bins to search.

A timing Z-count is defined in Fig. 12 as the number of 1.5 s $X1$ epochs since the beginning of the week. There are four $X1$ epochs per data subframe of 6 s. To help the receiver to acquire the long period P code, the 50 bps datastream contains an updated handover-word (HOW) for each 6-s subframe. The HOW when multiplied by 4, equals the Z-count at the beginning of the next 6-s subframe. Thus, if we have acquired timing from the relatively short C/A code and know the subframe epoch times and the HOW words, we can instantly acquire the P-code at the next subframe epoch. Figure 13 summarizes the timing relationships between $X1$, $X2$ epochs, and the Z-count and HOW words.

2. Antispoof P(Y) Code, Nonstandard Codes, and Selective Availability

The P code is a long, 1-week period code; however, it is published in GPS-ICD[6] and is available to potential spoofers or jammers. (A spoofer generates a signal that mimics the GPS signal and attempts to cause the receiver to track the wrong signal.) For this reason, the GPS system has the option to replace the P code with a secure Y code available only to authorized U.S. Government users. The Y code is employed when the "antispoof" or AS mode of operation is activated. The Y code is a secure version of the published P code that operates at the same clock rate as the P code, but has a code available only to authorized users. The main purpose of the Y code is to assure that an opponent cannot spoof the Y-code signal by generating a Y-code replica.

Nonstandard C/A and Y codes (NSC and NSY codes) are used in place of the C/A and P(Y) codes to protect the user from a malfunction in the spacecraft. They are only used for a malfunctioning satellite. These codes are not used in navigation receivers.

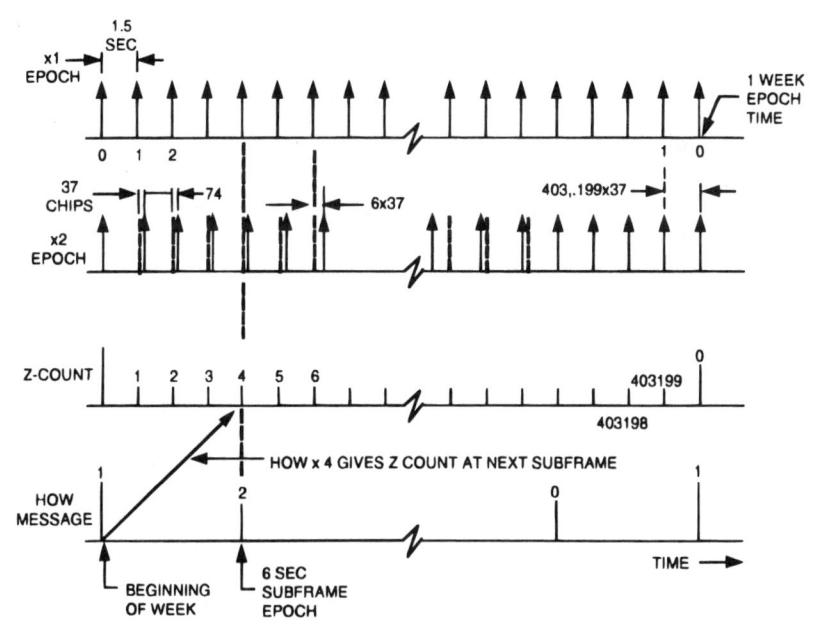

SIGNAL STRUCTURE AND THEORETICAL PERFORMANCE   77

Fig. 12 Timing diagram for the P-code components X_1, X_2, and the Z-count and HOW message relationship (not to scale). The HOW message is carried in the 50-bps datastream.

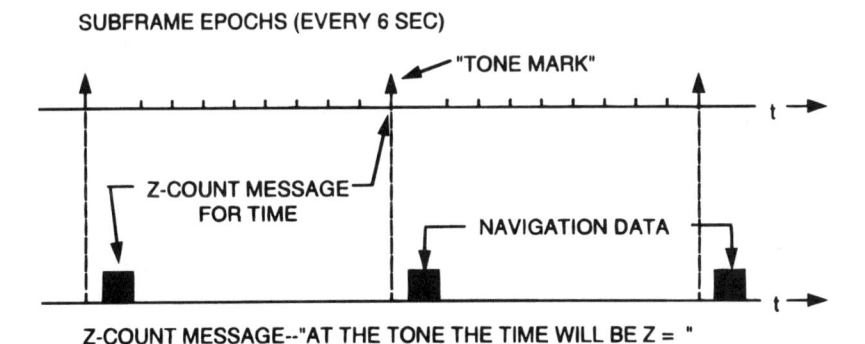

Fig. 13 GPS received signal time and the Z-count navigation data that are used to help acquire the P-code once the C/A code is acquired. The Z-count also aids in time ambiguity resolution for the C/A code.

Selective availability is a purposeful degradation of the GPS signal by the U.S. Government that can be imposed to restrict the full accuracy of the GPS system to authorized military users. Selective Availability (SA) is discussed in a later chapter in this volume. RMS position accuracy with SA is less than or equal to 100 m. SA purposely dithers the GPS clock in a pseudorandom manner. The clock dither has been reported by Allan and Dewey[7] to have a decorrelation time of 300–400 s. For observation times shorter than 300 s the clock dither can be modeled as a random walk phase modulation; for longer observation times it can be modeled as white noise phase modulation. As discussed in Chapter 2, this volume, recommendations have been made to turn selective availability to zero.

3. Coarse/Acquisition (C/A) Code—Civil GPS Signal

The C/A code for the civil user is a relatively short code with a period $2^{10} - 1 = 1023$ bits or 1-ms duration at a 1.023 Mbps bit rate. The code period is purposely selected to be relatively short so as to permit rapid acquisition. That is, there are only 1023 code chip time bins to search. The C/A codes are selected to provide good multiple access properties for their period. The C/A codes for the various satellites are taken from a family of codes known as Gold codes that are formed by the product of two equal period 1023 bit PN codes $G1(t)$ and $G2(t)$ (see Appendix and Ref. 8). Thus, this product code is also of 1023 bit period and is represented as follows:

$$XG(t) = G1(t)G2[t + N_i(10T_c)] \tag{12}$$

where N_i determines the phase offset in chips between $G1$ and $G2$. Note that C/A-code chip has duration $10T_c$ s where T_c is the P-code chip interval. There are 1023 different offsets N_i, and hence 1023 different codes of this form.* Each code $G1$, $G2$ is generated by a maximal-length linear shift register of 10 stages. The $G1$ and $G2$ shift registers, are set to the "all ones" state in synchronism with the $X1$ epoch. The tap positions are specified by the generator polynomial for the two codes:

$$G1: G_1(X) = 1 + X^3 + X^{10}$$
$$G2: G_2(X) = 1 + X^2 + X^3 + X^6 + X^8 + X^9 + X^{10} \tag{13}$$

Because each Gold code has a 1-ms period, there are 20 C/A-code epochs for every databit. The 50-bps data clock is synchronous with both the C/A epochs and the 1.5 s $X1$ epochs of the P code. Figure 14 shows a simplified block diagram of the C/A-code generator. The unit is comprised of two 10-stage feedback shift registers clocked at 1.023 Mbps having feedback taps at stages 3 and 10 for $G1$ and at 2, 3, 6, 8, 9, 10 for $G2$, as indicated by the polynomials of (13). The various delay offsets are generated by tapping off at appropriate points on the $G2$ register and modulo-2 adding the two sequences together to get the desired delayed version of the $G2$ sequence using the so-called "cycle-and-add" property of the linear maximal length shift register (LMLSR). Maximal length shift register codes have the property that the addition of two time offset ("cycled") versions of the same code gives a shifted

---

*There actually are 1025 different Gold codes of this period and family. The codes, $G1(t)$ and $G2(t)$, by themselves, are the other two codes.

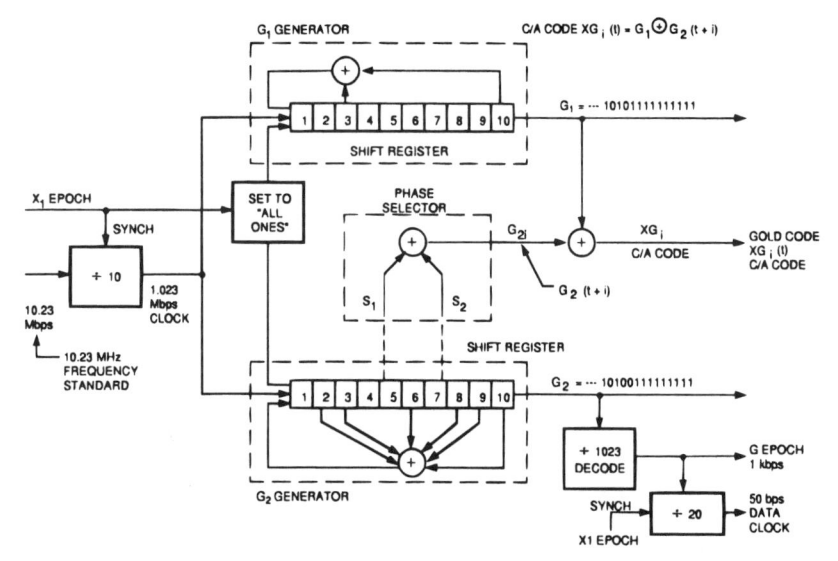

SIGNAL STRUCTURE AND THEORETICAL PERFORMANCE 79

Fig. 14 C/A-code generator block diagram showing the 1-ms G epoch and data clock generators that are all in synchronism.

version of the same code; hence the name "cycle and add" property. This property is discussed later in the chapter. The code tap positions for various codes are given in Table 4. Note that there are 45 possible tap positions in Fig. 14, but as shown in Table 4, only 37 codes are defined in GPS-ICD-200.[6] The general relationship between code taps and code phase is analyzed in the Appendix. Epochs of G code occur at 1 Kbps and are divided down by 20 to get the 50 bps data clock. All clocks are in phase synchronism with the $X1$ clock, as shown in Fig. 14.

The recursive equations for the $G1$ and $G2$ sequences that correspond to the $G1$, $G2$ polynomials of (2–8) are as follows:

$$G1(i) = G1(i - 10) \oplus G1(i - 3)$$
$$G2(i) = G2(i - 10) \oplus G2(i - 9) \oplus G2(i - 8) \quad (14)$$
$$\oplus\ G2(i - 6) \oplus G2(i - 3) \oplus G2(i - 2)$$

Table 4 gives the first bits of each of the 37 C/A codes in octal form. For reference, the first 31 bits of the first Gold code for SV#1 are {1, 1, 0, 0, 1, 0, 0, 0, 0, 1, 1, 1, 0, 0, 1, 0, 1, 0, 0, 1, 0, 0, 1, 1, 1, 1, 0, 0, 1, 0, 1}.

Notice that the first 10 bits of this code match exactly those shown in Table 4.

It is also possible to generate each Gold code with a single 20-stage shift register (not maximal length) by simply using a code generator that corresponds to a polynomial that is the product of the $G1(x)$ and $G2(x)$ polynomials (modulo-2). This code generator still produces codes of length 1023. Different codes are formed by starting the shift register in the correct state. However, that form does not allow one generator to generate all codes easily. It is also possible simply to delay one code generator, the $G2$ generator, relative to $G1$ simply by changing the starting state of $G2$.

Table 4 GPS code phase assignments for various spacecraft ID numbers
(taken from GPS-ICD-200)

SV ID No.	GPS PRN signal No.	Code phase selection, C/A, (G2₊)	X2₊	Code delay chips C/A	P	First 10 chips octal[a] C/A	First twelve chips octal P
1	1	2⊕6	1	5	1	1440	4444
2	2	3⊕7	2	6	2	1620	4000
3	3	4⊕8	3	7	3	1710	4222
4	4	5⊕9	4	8	4	1744	4333
5	5	1⊕9	5	17	5	1133	4377
6	6	2⊕10	6	18	6	1455	4355
7	7	1⊕8	7	139	7	1131	4344
8	8	2⊕9	8	140	8	1454	4340
9	9	3⊕10	9	141	9	1626	4342
10	10	2⊕3	10	251	10	1504	4343
11	11	3⊕4	11	252	11	1642	4343
12	12	5⊕6	12	254	12	1750	4343
13	13	6⊕7	13	255	13	1764	4343
14	14	7⊕8	14	256	14	1772	4343
15	15	8⊕9	15	257	15	1775	4343
16	16	9⊕10	16	258	16	1776	4343
17	17	1⊕4	17	469	17	1156	4343
18	18	2⊕5	18	470	18	1467	4343
19	19	3⊕6	19	471	19	1633	4343
20	20	4⊕7	20	472	20	1715	4343
21	21	5⊕8	21	473	21	1746	4343
22	22	6⊕9	22	474	22	1763	4343
23	23	1⊕3	23	509	23	1063	4343
24	24	4⊕6	24	512	24	1706	4343
25	25	5⊕7	25	513	25	1743	4343
26	26	6⊕8	26	514	26	1761	4343
27	27	7⊕9	27	515	27	1770	4343
28	28	8⊕10	28	516	28	1774	4343
29	29	1⊕6	29	859	29	1127	4343
30	30	2⊕7	30	860	30	1453	4343
31	31	3⊕8	31	861	31	1625	4343
32	32	4⊕9	32	862	32	1712	4343
——[c]	33	5⊕10	33	863	33	1745	4343
——[c]	34[b]	4⊕10	34	950	34	1713	4343
——[c]	35	1⊕7	35	947	35	1134	4343
——[c]	36	2⊕8	36	948	36	1456	4343
——[c]	37[b]	4⊕10	37	950	37	1713	4343

[a] In the octal notation for the first 10 chips of the C/A code as shown in this column the first digit (1) represents a "1" for the first chip and the last three digits are the conventional octal representation of the remaining 9 chips. (For example, the first 10 chips of the C/A code for PRN Signal Assembly No. 1 are: 1100100000).
[b] C/A codes 34 and 37 are common
[c] PRN sequences 33 through 37 are reserved for other uses (e.g. ground transmitters).
⊕ = "exclusive or"

We might ask why not simply take 37 different maximal length shift register codes and use them in place of the Gold codes? After all, 2 of the Gold codes of this family of length 1023 are the maximal length codes themselves. The answer is that the other maximal length codes of length 1023 would not guarantee uniformly low cross-correlation sidelobes for all other needed satellite codes and all possible delay offsets.

4. L_2 Signal

The L_2 signal is biphase-modulated by either the P code or the C/A code, as selected by ground command. The same 50 bps datastream modulates the L_2 carrier as transmitted on L_1. Thus, the L_2 signal is represented in the normal P format as follows:

$$S_{L_{2_i}}(t) = \sqrt{2P_2} X P_i(t) D_i(t) \cos(\omega_2 t + \phi_2) \tag{15}$$

where $\sqrt{2P_2}$ represents the L_2 signal amplitude at the satellite, $XP_i(t)$ is the P code for the ith satellite, which is clocked in synchronism with the L_1 codes. As with the L_1 signal, both L_2 carrier and code are synchronous with one another. The L_2 signal can also be modulated with the P-code without the data. This feature permits the precision receiver tracking loops to be reduced further in IF bandwidth, and thereby can improve the noise/interference performance.

Because the L_2 signal is biphase-modulated, it is possible to recover the L_2 carrier without knowledge of the P code by simply squaring the signal or cross-correlating L_1 with L_2 with a delay offset that matches the L_1 to L_2 ionospheric delay difference. These types of codeless recovery of the L_2 carrier can then be used in estimating the ionosphere delay (see Chapter 4, this volume). There is added noise degradation in this codeless carrier recovery because of the nonlinear squaring operation or noisy cross-correlation operation. However, the information bandwidth of the ionosphere is sufficiently small that if we have already tracked the L_1 code, the noise bandwidth required to track the L_2–L_1 ionospheric difference is also very small, and noise effects can be kept small by narrow bandwidth filtering.

5. GPS Data Format

Table 5 summarizes the signal and data characteristics just discussed. There are five subframes of 6 s each for a total frame period of 30 s. One of the key points to be made in the signal structure discussion is that acquisition by a receiver of the relatively short period C/A code plus the recovery of a single full subframe of data permits us to acquire the P code with minimal or zero search. Knowledge of the C/A epoch plus the data subframe epoch and the HOW word gives the exact phasing of the P(Y) code. Navigation solutions require, as a minimum, reception of data subframes 1, 2, 3 containing clock-correction and ephemeris data and, in general, require reception of a full 30- s frame of data. The navigation data are discussed in detail in the next chapter.

6. Codes for GPS Augmentation

The GPS satellites are augmented at times by pseudolites or ground transmitters that may transmit frequency offset or a low-duty factor pulsed GPS-type signal

Table 5 Summary of GPS signal parameters and data formats

Parameter	C/A Signal	P Signal
Code clock (chip) rate	1.023 Mbps	10.23 Mbps
Code period	1023	$= 6 \times 10^{12}$; 1 week
Data rate	50 bps	50 bps
Transmission frequency	L_1	L_1, L_2

Data format—frame and subframe structure

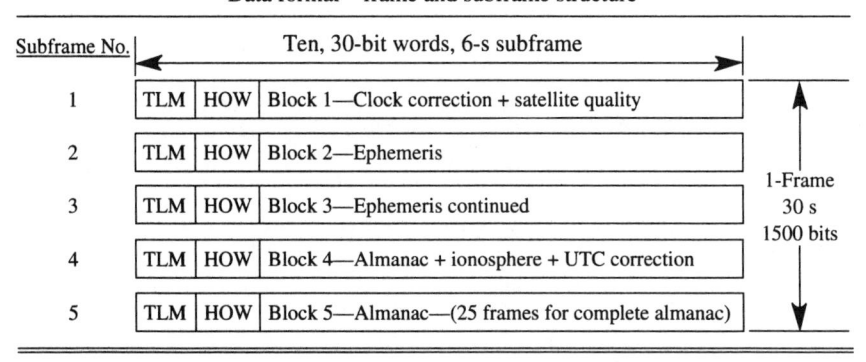

Subframe No.			Ten, 30-bit words, 6-s subframe	
1	TLM	HOW	Block 1—Clock correction + satellite quality	
2	TLM	HOW	Block 2—Ephemeris	1-Frame
3	TLM	HOW	Block 3—Ephemeris continued	30 s
4	TLM	HOW	Block 4—Almanac + ionosphere + UTC correction	1500 bits
5	TLM	HOW	Block 5—Almanac—(25 frames for complete almanac)	

Each Telemetry (TLM) word contains an 8-bit Barker word for synchronization. The Handover Word (HOW) contains a 17-bit Z-count for handover from the C/A code to the P code.

and use codes different from those employed in the GPS satellites to avoid confusion. GPS pseudolites are discussed in detail in a later chapter.

Also planned is the augmentation of the GPS satellites with geostationary overlay satellites using different Gold codes. These signals might be generated either by the satellite or generated and synchronized on the ground and broadcast to the GPS users via geostationary satellite transponders. Table 6 lists the Gold codes selected by INMARSAT for future transmission by the INMARSAT satellites.[9] Table 6a lists the codes for the Wide Area Augmentation System (WAAS) of the Federal Aviation Administration. GPS augmentation is also discussed in later chapters on the GPS wide area differential GPS (WADGPS) and Wide Area Augmentation System in Volume II.

III. GPS Radio Frequency Receive GPS Power Levels and Signal-to-Noise Ratios

A. GPS Radio Frequency Signal Levels and Power Spectra

The minimum specified received signal strength for a user receiver employing a 0 dBIC antenna is given below in Table 7 for a satellite at elevation angles above 5 deg. As shown in the next subsection, the actual minimum varies with elevation angle to the satellite because of the shaped satellite antenna pattern. The signal power spectral densities for the P and C/A signal components are shown in Fig. 15. Figure 15 also shows the measured radiofrequency power spectral density of the L_1 signal. Note the narrowband high-power density C/A signal in the center of the signal spectrum. Note that these spectra are

SIGNAL STRUCTURE AND THEORETICAL PERFORMANCE 83

Table 6 Final INMARSAT C/A-code selection

Order	PRN	Delay	Initial $G2$ state (octal)
1	201	145	1106
2	205	235	1617
3	208	657	717
4	206	235	1076
5	202	175	1241
6	207	886	1764
7	209	634	1532
8	211	355	341

Table 6a Wide Area Augmentation System (WAAS) PRN ranging C/A codes (Note that these codes include the INMARSAT codes of Table 6)

PRN Code #	Display (Chips)	First 10 WAAS Chips (Octal)[a]
115	145	0671
116	175	0536
117	52	1510
118	21	1545
119	237	0160
120	235	0701
121	886	0013
122	657	1060
123	634	0245
124	762	0527
125	355	1436
126	1012	1226
127	176	1257
128	603	0046
129	130	1071
130	359	0561
131	595	1037
132	68	0770
133	386	1327

[a] The first digit represents a 0 or 1 in the first chip. The next three digits are the octal representation of the remaining nine chips.

Table 7 GPS minimum received signal power levels at output of a 0 dBIC antenna with right-hand circular polarization[a]

Link	GPS signal component (minimum strength) specified		Expected maximum does not exceed this level with 0.6 dB atmospheric loss	
	P	C/A	P	C/A
L_1	−163 dBW	−160 dBW	−155 dBW	−153 dBW
L_2	−166 dBW	−166 dBW	−158 dBW	−158 dBW

[a] The satellite is assumed to be at an elevation angle ≥ 5 deg.

the transmitted spectra. In normal operation, the thermal white noise of the receiver significantly exceeds the signal spectral density, and the signal is not visible using a spectrum analyzer. Recall that the maximum power spectral density for a pseudonoise signal with a continuous $[(\sin \pi f/f_c)/\pi f/f_c)]^2$ shaped spectrum is P_s/f_c. Thus, if $P_s = -160$ dBW, the maximum power density is −160 dBW − 60.1 dB = −220.1 dBW/Hz.

It is useful to compare this number with the recommended power flux density limit of the CCIR[1] cited earlier in this chapter in Sec. I.A. If an effective antenna aperture area* of $A = \lambda^2/4\pi = 2.8856 \times 10^{-3} \mathrm{m}^2$ or −25.4 dBm² at 1.57542 GHz is assumed for a unity gain antenna, then the power flux density per Hz is $P_s/f_s A = -194.7$ dBW/Hz-m². The total flux density in a 4-kHz band is then $4 \times 10^3 P_s/f_s A = -158.7$ dBW/m², which is within the level recommended by the CCIR[1] of −154 dBW/m².

B. Satellite Antenna Pattern

The radio frequency received signal levels are transmitted from the satellite by shaped pattern antennas (see Fig. 16) to compensate partially for the increased path loss to the user at low-elevation angles. The GPS Block II satellite antenna is an array of helices on the face of the satellite. The edge of Earth is approximately 13.87 deg off the satellite antenna boresight, i.e., the Earth subtends an angle of approximately 27.74 deg from the GPS satellite altitude. The satellite antenna pattern extends somewhat beyond the edge of the Earth, as shown in Fig. 16. Thus, even a GPS receiver in another satellite can receive signals from GPS satellites to perform satellite positioning, provided that it is not blocked by the Earth's shadow and is not too far off the main lobe of the satellite antenna pattern (see Fig. 17).

The transmitted signal from the satellite is right-hand circularly polarized with an ellipticity (offset from perfectly circular) no worse than 1.2 dB for L_1 and 3.2 dB for L_2 within an angle of ±14.3 deg from boresight.† Because the user antenna can be at various orientations with respect to the satellites, the satellite received power is specified under the following conditions; a) the signal is measured at the output of a 3 dBIC (isotropic) linearly polarized receiving antenna; b) the space

---

*The relationship between antenna gain G and effective aperture area A for an ideal lossless antenna is $G = 4\pi A/\lambda^2$.[10]

†NAVSTAR GPS Space Segment/Navigation User Segment Interface Control Document, ICD-GPS-200, IRN-200B-005, Dec. 16, 1991.

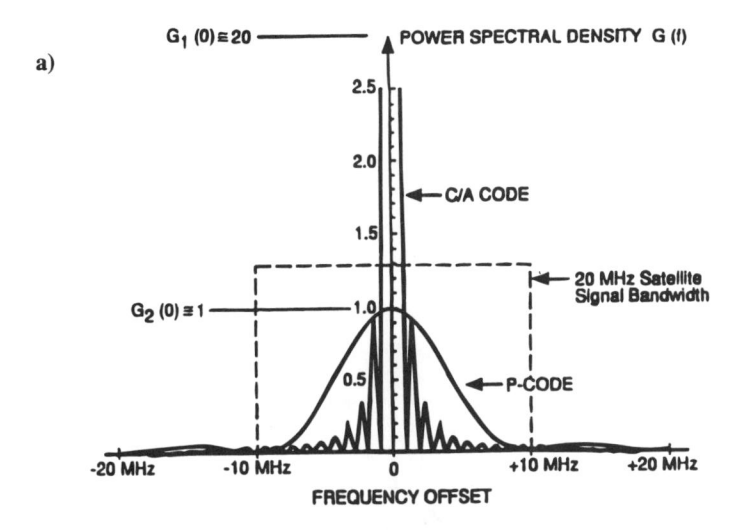

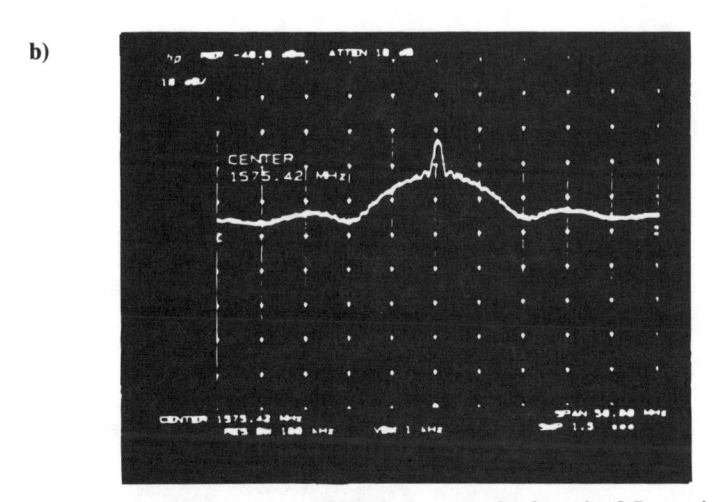

Fig. 15 Radio frequency spectrum plot and photograph of received L_1 carrier with C/A and P QPSK modulation; a) Power spectra of carriers with bit rates of 1.023 Mb/s and 10.23 mb/s. The ratio of C/A power to P-code signal power is 3 dB in this figure, b) Photograph of signal generated by Stanford Telecom GPS signal simulator 7200. Spectrum scales: horizontal, 5 MHz/division; vertical, 10 dB/division (courtesy Stanford Telecom).

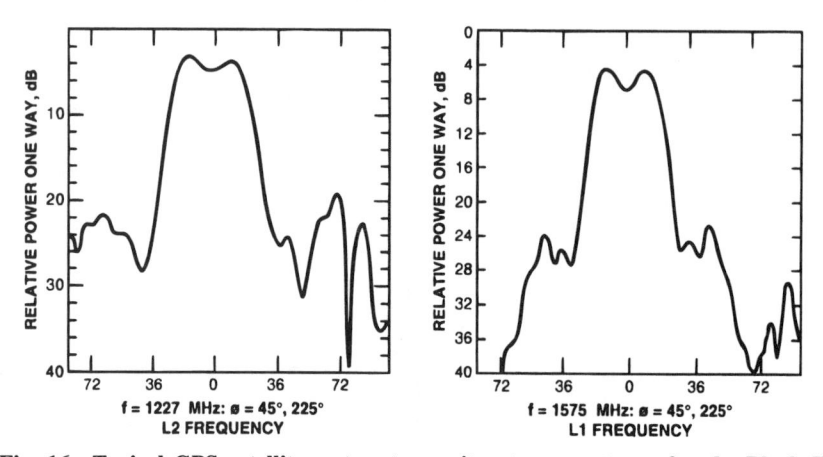

Fig. 16 Typical GPS satellite system transmit antenna patterns for the Block II satellites.

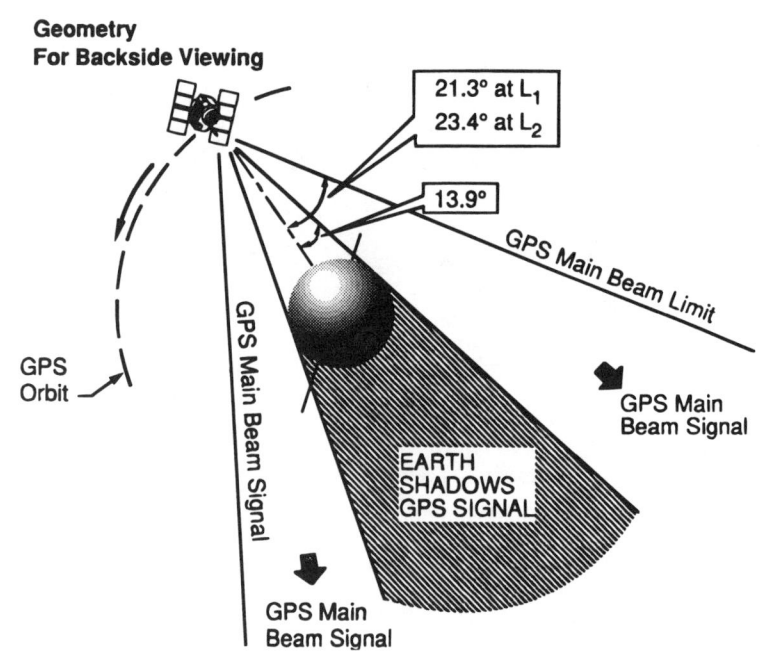

Fig. 17 GPS satellite main beam relative to Earth (not to scale). User satellites can navigate using GPS provided they are in the main beam of the GPS antenna but outside the Earth's shadow.

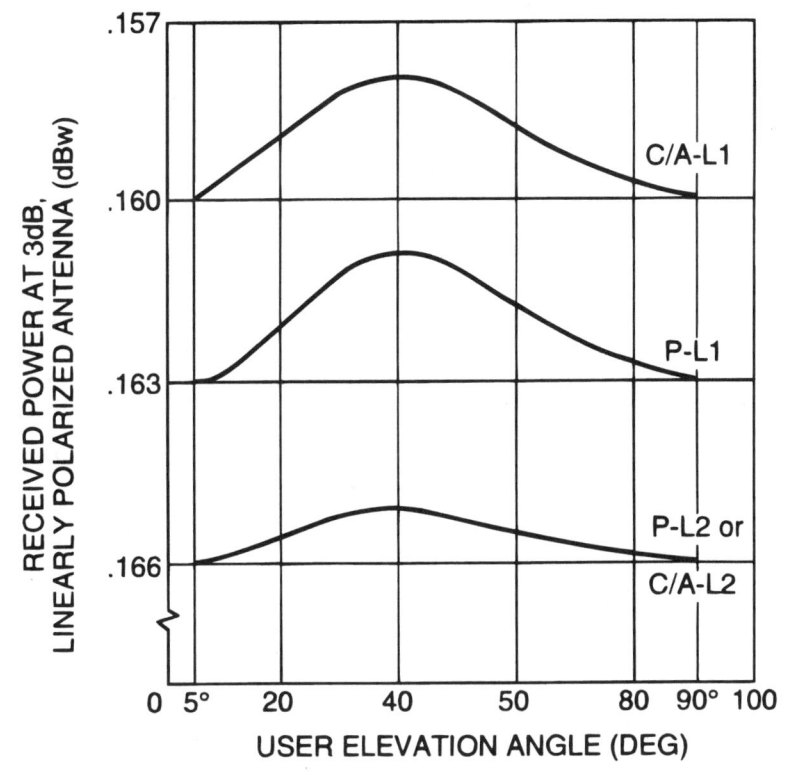

SIGNAL STRUCTURE AND THEORETICAL PERFORMANCE

Fig. 18 GPS user received minimum signal levels vs elevation angle as stated in the GPS document GPS-ICD-200.[6]

vehicle is above a 5 deg elevation angle; c) the received signal levels are observed within the 20 MHz frequency allocation; d) the atmospheric path loss is 2.0 dB; and e) the space vehicle's attitude error is 0.5 deg (toward reducing signal level). The specified minimum received signal power vs. elevation angle for these conditions is shown in Fig. 18. Note that the specified received signal level peaks at 40 deg elevation angle at a level that is approximately 2 dB above the nominal −160 dBW level for the C/A code on L_1. As shown in Chapter 13 this volume, atmospheric path loss is generally less than 2 dB, except at low-elevation angles. In addition, the satellites are designed so that these numbers are met at the end of the satellite's life. Beginning-of-life power levels are generally higher. Thus, these numbers are somewhat conservative.

C. Signal Specifications

1. Signal Correlation Loss

The GPS C/A and P(Y) signals are filtered to a bandwidth of 20.46 MHz. There are two such 20.46 MHz frequency bands centered at L_1 and L_2. The GPS space segment GPS-ICD-200[6] defines a maximum correlation loss of 1.0 dB from that of an ideal signal and ideal receiver due to satellite signal generation and filtering imperfections and waveform distortion. The loss is apportioned as follows: 1) space vehicle modulation imperfection 0.6 dB; and 2) user equipment receiver waveform distortion 0.4 dB caused by the 20.46 MHz filter.

Note that because the C/A code is only 1.023 Mbs, there is no significant filtering of the C/A signal out to the 10th sideband. However, only the mainlobe of the P(Y) code is transmitted.

2. Other GPS Signal Specifications

1) Carrier phase noise. The phase noise spectral density of the satellite carrier (unmodulated) is defined to be sufficiently small so that a receiver phase–lock-loop of 10-Hz one-sided closed-loop noise bandwidth can track this carrier (with its phase noise) with an accuracy of 0.1 rad rms (excluding thermal noise effects).

2) L_1 phase quadrature accuracy and signal coherence. The L_1 C/A and P(Y) carrier components are modulated in phase and quadrature with a phase offset of less than ± 100 mrad from 90 deg. On the L_1 channel, the code chip transitions between the two modulating signals [P(Y) or C/A] will have an average time difference less than 10 ns.

3) Group delay uncertainty. The effective uncertainty of the group delay in the satellite transmission will not exceed 3.0 ns (two sigma).

4) Spurious transmissions. In-band spurious transmissions from the satellite are to be at least 40 dB below the L_1, L_2 carriers.

5) Signal polarization. The GPS signal is right-hand circularly polarized with an ellipticity for L_1 no worse than 1.2 dB and 3.2 dB for L_2 within an angular range of ±14.3 deg from boresight. (Perfectly circular polarization has an ellipticity of 0 dB.)

D. User-Received Signal-to-Noise Levels

It is useful to estimate the received signal-to-noise density levels, and then to translate these ratios into expected carrier phase jitter and code delay jitter. As a first step, the noise density levels must be computed in terms of the receiver noise figure.

1. Receiver Noise Power Density

The receiver antenna output is fed to a transmission line and bandpass filter and then to the low-noise amplifier, as shown in Fig. 19. Because of the potential for line losses, the low-noise amplifier is generally kept in close physical proximity to the antenna. The bandpass filter must similarly have low loss but provide adequate filter selectivity to attenuate adjacent channel interference. The effective noise temperature and one-sided noise spectral density are related to the receive signal power for a given satellite L_1 or L_2 signal, as shown in Fig. 19. The received noise density (one-sided) is $N_0 = kT_{eq}$ where the equivalent noise temperature is defined as follows:

$$T_{eq} = T_A/L + (L - 1)T_0/L + T_R \; °K \qquad (16)$$

SIGNAL STRUCTURE AND THEORETICAL PERFORMANCE

```
        ▽ P_c, T_A   ┌─────────┐            LNA
        ─────────────│XMISSION │─ C, T_R, N_0 ─▷─
                     │LINE &   │
                     │BANDPASS │
                     │FILTER   │
                     └─────────┘
```

$C = P_c/L$

$N_0 = kT_{eq}$

$T_{eq} = \dfrac{T_A}{L} + \dfrac{(L-1)}{L} T_0 + T_R \,°K$

LOSS = L

P_c = RECEIVED SIGNAL POWER AT ANTENNA OUTPUT

C = RECEIVED SIGNAL POWER AT PREAMPLIFIER INPUT

N_0 = RECEIVED NOISE DENSITY AT PREAMPLIFIER INPUT

L = POWER TRANSMISSION LOSS BETWEEN THE ANTENNA AND THE PREAMPLIFIER $L > 1.0$

T_A = ANTENNA NOISE TEMPERATURE

T_0 = AMBIENT TEMPERATURE OF THE TRANSMISSION LINE

T_R = LOW NOISE AMPLIFIER RECEIVER NOISE TEMPERATURE = $T_0 (F-1)$

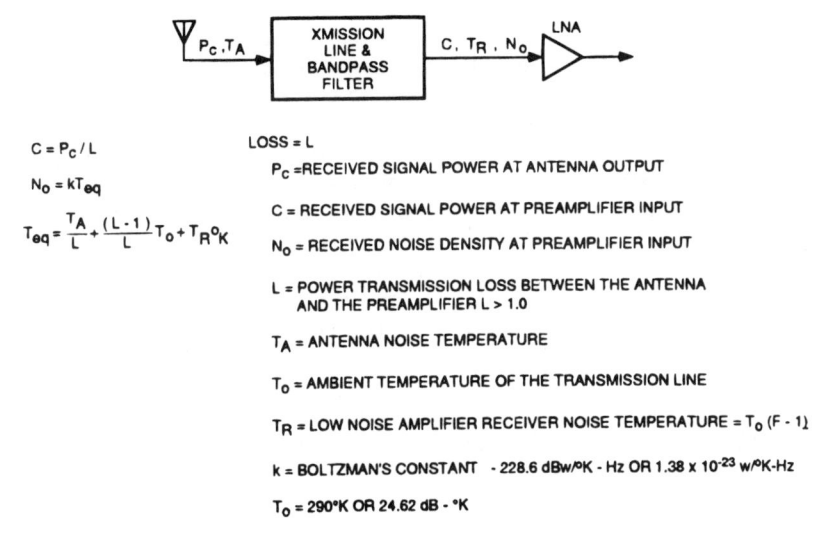

k = BOLTZMAN'S CONSTANT -228.6 dBW/°K-Hz OR 1.38×10^{-23} w/°K-Hz

$T_0 = 290°K$ OR 24.62 dB - °K

Fig. 19 Low noise amplifier configuration and received C/N_0 computation. Typical noise figure might be in the range 1.0 dB to 1.5 dB. If $F = 1.0334$ (or 1.25 dB), then $F - 1 = 0.334$ and $(F - 1) T_0 = 97°$ K for $T_0 = 290°$ K.

where the receiver noise temperature is $T_R = T_0 (F - 1)$, F is the receiver noise figure (not in dB), T_0 is the ambient temperature, and T_A is the noise temperature of the antenna (depends on sky noise and antenna sidelobes and backlobes that are looking at the warm Earth). All temperatures are in °K. The value of Boltzman's constant is $k = -228.6$ dBW/°K-Hz. If, for example, the LNA receiver noise figure is 1.0 dB then $F = 1.259$, $F - 1 = 0.259$, and $T_R = (F - 1) 290$ °K $= 75.1$ °K. For a typical example of small line/filter loss, $L = 1.1$, and $T_A = 130$ °K, the equivalent noise temperature and noise density for this example are as follows: $T_{eq} = 130$ °K/1.1 $+ (0.1/1.1) 290$ °K $+ 75.1$ °K $= 219.6$ or 23.42 dB °K; $N_0 = kT_{eq} = -205.2$ dBW/Hz. Note that the peak signal power spectral density is P_s/f_c. Note also that T_{eq} is heavily influenced by T_A for low-noise figure receivers. The C/A signal power density for the specified $P_s = -160$ dBW and $f_c = 1.023 \times 10^6$ Hz is -220.1 dBW/Hz, which compares to the noise density in the example of -205.2 dBW/Hz. Thus, the C/A signal spectrum, even at its spectral peak, is 14.9 dB below the noise power spectral density. The P-signal power spectral density is roughly 13 dB below this number or 27.9 dB below the thermal noise density. Thus, a single GPS signal normally is not visible to a spectrum analyzer.

2. Received Carrier-to-Noise Density Ratio—C/N_0

The C/N_0 at a receiver is one of the key parameters that determines receiver performance. At the receiver, the minimum expected carrier power at L_1 for the C/A code is $P_c = -160$ dBW. As shown in Fig. 19, the effective carrier power C is this level minus the effect of line losses in the transmission line if the antenna has a 0 dB isotropic gain 0 dBIC. Thus, $C = P_c/L$ where L is the loss.

In addition to the line loss, there may also be loss caused by foliage attenuation (atmospheric path losses at low-elevation angles have already been included in the 2-dB specification value), or antenna gain positive or negative with respect to the 0 dBIC used in the specification.

If the received signal power for the L_1 C/A signal is $C = -160$ dBW and $N_0 = -205.2$ dBW-Hz as in the example above, then the carrier power-to-noise density ratio is $C/N_0 = 45.2$ dB-Hz. At a higher elevation angle with a signal level of $C = -158$ dBW, the received signal-to-noise density increases to $C/N_0 = 47.2$ dB-Hz. In practice, as previously mentioned, the satellite power for the Block II satellites often exceeds the specified level because the satellite power is expected to degrade with time, and newer satellites are designed to have a higher output power than their end-of-life specification, perhaps by as much as 6 dB. In addition, the downlink power assumes a 2.0-dB atmospheric path loss,* which in fact may be less than 0.3 dB. Thus, the actual C/N_0 can perhaps be as much as 7.3 dB higher, which would give a $C/N_0 = 51.5$ dB-Hz for the C/A signal.

3. Simplified Estimate of Delay Measurement Noise on Code and Carrier

It is of interest to get a rough idea of the accuracy of the range estimate for both code and carrier measurements, although these accuracies are computed more precisely later. Assume for the moment that both the spreading code and the data modulation have been stripped off the carrier, and one is left with a pure carrier signal plus bandlimited white noise:

$$r(t) = \sqrt{2P_s} \sin(\omega_0 t + \phi) + N_s(t)\sin(\omega_0 t + \hat{\phi}) + N_c(t)\cos(\omega_0 t + \hat{\phi}) \quad (17)$$

If we multiply this received waveform with a near quadrature coherent reference with phase $\hat{\phi}$; namely, $2\cos(\omega_0 t + \hat{\phi})$, the phase detector output at baseband (low-pass filtered) is then as follows:

$$r(t)2\cos(\omega_0 t + \hat{\phi}) \cong \sqrt{2P_s}\sin(\phi - \hat{\phi}) + N_c(t) \cong \sqrt{2P_s}(\phi - \hat{\phi}) + N_c(t) \quad (18)$$

for $|\phi - \hat{\phi}| \ll 1$. The noise power spectral density (one-sided) of $N_c(t)$ is $2N_0$. If the phase error $\phi - \hat{\phi}$ is filtered to a bandwidth B_L, then the output phase differrence plus phase noise can be written in normalized form $\phi - \hat{\phi} + \phi_n$ where the phase noise ϕ_n has rms value $\sigma_\phi^2 = (N_0 B_L/P_s) = 1/(\text{SNR}_L)$. Convert this phase error to an equivalent rms time error:

$$\sigma_T = \frac{1}{2\pi f_o} \frac{1}{\sqrt{\text{SNR}_L}} \quad (19)$$

As described above, the GPS carrier phase noise is specified to be sufficiently low so that it can be tracked with a 10-Hz noise bandwidth. If a carrier-tracking loop of bandwidth $B_L = 10$ Hz is employed and the C/N_0 is 45.2 dB, then the signal-to-noise ratio SNR_L in the loop, all else being linear is $\text{SNR}_L = C/N_0 B_L = 45.2 - 10 = 35.2$ dB because $\sigma_\phi = 2\pi f_o \sigma_T$ at frequency f_0. Using

---

*ICD-GPS-200, Sheet 42a of 115, Dec. 2, 1991.

Eq. (19), we can show that this SNR corresponds to an rms phase noise σ_ϕ corresponding to the following:

$$\sigma_\phi^2 = (\sigma_T 2\pi f_o)^2 \cong \frac{1}{\text{SNR}_L} \quad \text{or} \quad c\sigma_T = \frac{\lambda}{2\pi} \frac{1}{\sqrt{\text{SNR}_L}} \qquad (20)$$

for sufficiently high $\text{SNR}_L \gg 1$, where $c\sigma_T$ is the rms differential distance error (see Fig. 20). If $\text{SNR}_L = 35.2$ dB or 3311 and the wavelength $\lambda = 19.05$ cm for L_1, then the output phase error equivalent in distance is $c\sigma_T \cong 0.5$ mm.

Thus, the carrier-tracking phase noise measurement error caused by thermal noise is only 0.5 mm. Other errors caused by carrier phase noise and dynamic tracking errors would likely increase this number to several mm, but the carrier phase measurement error is still very small. However, as with any phase measurement, there is an ambiguity in the absolute number of phase cycles. Thus, only differential phase measurements are possible. Nonetheless, carrier phase measurements are extremely valuable. Carrier phase can be tracked over a considerable time, and if no phase cycles are slipped, the carrier phase delay can be

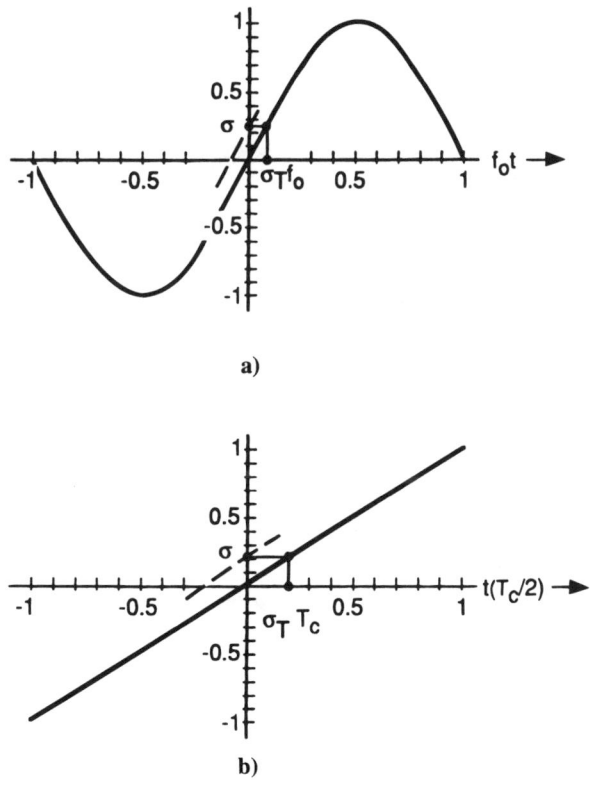

Fig. 20 Simplified view of delay measurement noise on a) carrier, b) code where $f_0 \approx 1.51542$ GH$_z$. The wavelength at L_1 is $\lambda = c/f_0 \approx 19.05$ cm and $T_c = 1/f_c = 1.023$ MH$_z$.

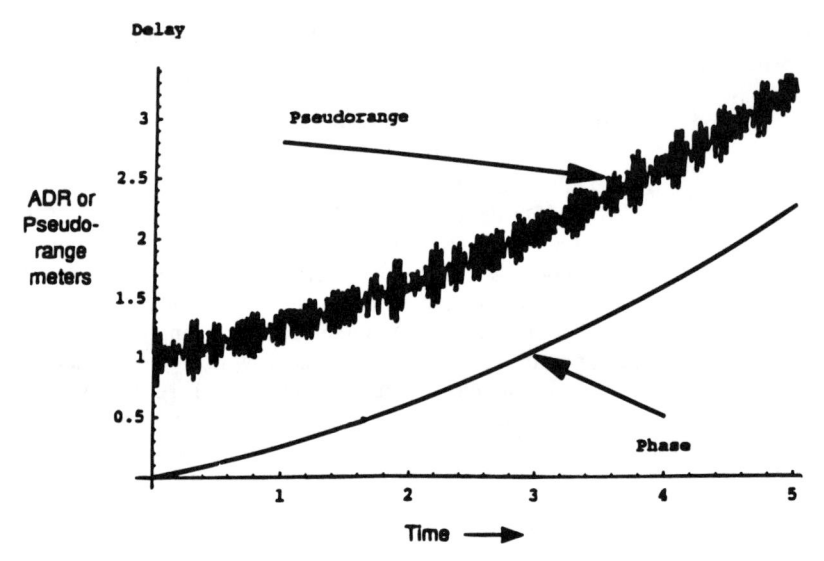

Fig. 21 Example of rms noise on carrier phase and code measurements, both converted to meters (not to scale).

tracked for several hours, i.e., during the time of satellite visibility. Carrier phase measurements made over time and converted to distance are termed *Accumulated Delta Range* (ADR). An example plot of the carrier phase measurement vs. time is shown as the ADR or phase curve in Fig. 21. Note that the carrier phase only has meaning in a differential sense. Initial carrier phase measurements have an ambiguity in the number of phase cycles and initial phase and thus are set arbitrarily at zero. Both the code and carrier phase delay measurement curves have exactly the same shape, except for noise and a bias offset and ionospheric delay changes.

4. Code-Tracking Measurement Noise

Measurements on the code would follow a similar approach as that for the carrier, except that the slope of the code tracking is on the order of T_c where $f_c = 1/T_c = 1.023$ MHz for the C/A code as opposed to λ for the carrier. Absolute rather than differential delay measurements of code delay are made here, however, and in any event, the code delay must be estimated before carrier phase can be extracted. If, as a simple example, the received code-tracking signal is coherently converted to baseband and multiplied by a delayed replica $s(t + T_c/2 + \hat{\tau})$ we have the following:

$$\sqrt{2P_s}s(t + \tau)s(t + T_c/2 + \hat{\tau}) + N_{c1}(t) \cong \sqrt{2P_s}\left(\frac{\tau - \hat{\tau}}{T_c}\right) + N_{c1}(t)$$

$$+ \text{ constant for } |\tau - \hat{\tau}| \ll T_c \quad (21)$$

where we have approximated $s(t)$ by its Taylor's series expansion for small

SIGNAL STRUCTURE AND THEORETICAL PERFORMANCE

$|\tau - \hat{L}|$. A crude estimate of the rms delay error is then $\sigma_\tau = T_c 1/\sqrt{\text{SNR}_L}$, where SNR_l, is the signal-to-noise ratio in the tracking loop filter of bandwidth B_L, namely $\text{SNR}_L = P_S/N_0 B_L$. If the code-tracking loop bandwidth is 3 Hz, all remains linear, and $C/N_0 = 45.2$ dB-Hz, then the SNR in the code tracking loop is $\text{SNR}_L = 45.2 - 4.8 = 40.4$ dB.

For this simple nonoptimum code-tracking measurement, the rms tracking error is as follows:

$$\sigma_\tau \equiv T_c \frac{1}{\sqrt{\text{SNR}_L}} = 9.32 \text{ ns} \quad \text{or} \quad c\sigma_\tau \equiv 2.8 \text{ m} \tag{22}$$

for $\text{SNR}_L \gg 1$ and a value of $T_c = 0.9775 \mu s$ for the 1.023 Mbs C/A code. Figure 21 shows the approximate relative rms errors of the code and carrier measurements. Code-tracking errors are larger than carrier-tracking errors roughly by the ratio of carrier frequency to code clock frequency. These estimates of delay measurement noise are only intended to be crude estimates to give a rough idea of the true accuracy. As shown in the next section and later chapters, the rms code-tracking error can be reduced significantly below this number by using various types of delay-lock-loops for tracking the code.

5. Multiple Access Noise

It was shown earlier in Eq. (5) that the multiple access effect of M equal power spread spectrum signals degrades the equivalent noise density by a factor of $1 + (2/3)(M - 1)P_S/f_C N_0$. For the example, just given where $C/N_0 = P_S/N_0 = 45.2$ dB, $f_c = 1.023$ MHz, if we set $M - 1 = 10$, that is $M = 11$ satellites are in view, then the degradation in equivalent noise power spectral density is $1 + (2/3)(10)(3.3 \times 10^4)/1.023 \times 10^6 = 1.216$ or only 0.8 dB degradation relative to the thermal noise density. The P-code multiple access, of course, would have an $f_c = 10.23$ MHz, and the multiple access interference degradation in noise density is less than 0.1 dB. In reality, the C/A code spectrum is not continuous because of its periodiocly, and therefore this analysis gives only a rough approximation. The more accurate result is one topic in the next section.

It should be pointed out that the factor of 2/3 in the equation above appears because the code delay offsets are assumed to be random, and all of the code spectral sidelobes are included, as is approximately true for the C/A code. For the P code with only the first sidelobe included, the factor of 2/3 becomes 0.815, as discussed earlier, because only the main lobe of the P-code signal is used. (Note also that if all of the codes momentarily had their clock transition aligned perfectly, the worst case, the factor of 2/3 becomes 1.)

E. Recommendations for Future Enhancements to the GPS Signal

For future enhancement of the military and civil GPS system performance, the author makes the following recommendations:

1) Increase the user-received GPS L_1, C/A and P(Y) signal levels in the GPS specification by approximately 4 dB to be consistent with the actual power levels in the Block 2 satellites. Add a second civil signal, L_5, near L_2.

2) Add a C/A(YC) signal to the L_2 channel in phase quadrature to the P(Y) signal. The Block 2 satellites can either transmit P(Y) or C/A but not both. This L_2 C/A(YC) signal could be transmitted simultaneously with the P(Y) code. This 1.023-MHz clock rate C/A(YC) code can be transmitted either as pure C/A code or, for use by authorized users only, as an antispoof YC code. There are several potential advantages in this signal:
- It provides an antispoof acquisition aid for military users on the L_2 channel with the 1.023-MHz YC code. For example, a programmable matched filter could be used by an authorized user to acquire the YC code very quickly.
- Civilian codeless GPS receivers could use the L_2 YC signal for much improved ionospheric delay correction.

3) Increase the power levels of this L_2 signal with both the C/A(YC) and P(Y) signals to the same power levels of the L_1 channel as described above. This change would permit authorized users to operate primarily on the L_2 channel if they so desired, and they would only need the L_1 channel for ionospheric correction.

IV. Detailed Signal Characteristics and Bounds on Pseudorange Tracking Accuracy

The previous paragraph defined the structure of the GPS signal. In this section, detailed spectra and the multiple access characteristics of the C/A and P codes are analyzed, and the reasons for their selection are discussed. In particular, the "cross-correlation" properties between signals from different satellites are examined for the P and C/A signals both with and without Doppler offset. Second, the bounds on performance for optimal tracking of the C/A and P codes are analyzed and discussed. A brief discussion of the finite field algebra as it pertains to PN sequences and Gold codes is given as an Appendix to this chapter.

A. Cross-Correlation Properties—Worst Case

The key multiple access performance parameter of the GPS signals is the generalized cross-correlation performance. Any GPS receiver must perform a correlation operation if it is to extract the signal timing and recover the data for the desired satellite. Figure 22 shows the typical received signal and correlation receiver for an example where two satellites are in view; an arbitrary satellite h, the desired signal, and a second satellite j, which in this example represents an interfering satellite. If there are M satellites in view, then there are $M-1$ interfering satellites. Of course, we must realize that a complete GPS receiver contains a bank of perhaps 10 parallel receiver correlators. For one of these parallel correlators, the roles are reversed, and satellite j is the desired signal, and satellite h is the interference.

The interfering satellite signal in Fig. 22 is time offset by $k_j T_c$ and Doppler offset by frequency f_d. Clearly, if there is a Doppler offset, this value of time offset is changing with time, and the value $k_j T_c$ only represents the delay at some snapshot in time. As discussed previously, an integer number of T_c s offset is the worst case. In general, the offset will be noninteger and random, and the interference in the average diminishes by a factor of 2/3 for the C/A code. With the P

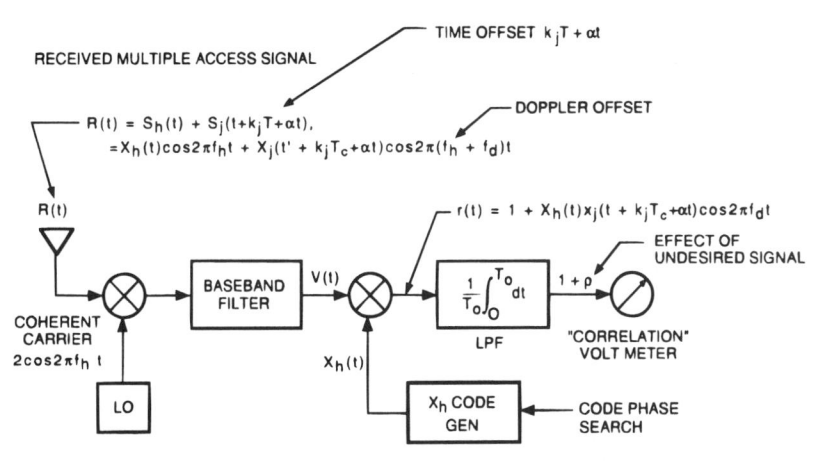

SIGNAL STRUCTURE AND THEORETICAL PERFORMANCE

Fig. 22 Multiple access interference in P-code user receiver. The received signal consists of the desired signal plus a time offset, Doppler-shifted multiple access signal from another satellite.

code, which is filtered to the first null, the factor of 2/3 becomes 0.815. For our purposes, we assume the worst case phasing, and the factor of 2/3 becomes unity.

The two received signals are assumed to be modulated by binary codes $X_h(t)$ and $X_j(t)$, respectively. These signals can represent either the P code, the C/A code, or an arbitrary signal. For the moment, the data modulation is ignored, and the signals from different satellites are assumed to be of equal strength. Noise effects are additive and can be considered separately. The block diagram of Fig. 22 shows a coherent correlation operation, where the coherent carrier is multiplied with the received signal, and the resultant baseband output is multiplied by a phase-synchronized replica of the desired code $X_h(t)$. The output of the multiplier is then integrated for some time T_m to produce the "correlation" output. If T_m is equal to or a multiple of the period of the waveforms or approaches infinity, the output will be the true correlation, otherwise it will be a partial correlation function.[11] For the GPS C/A code, the databit interval corresponds to 20 C/A-code periods, so this effect is small. With the P code, the code period is much longer than a databit period, but the clock rate is 10 times higher, and partial correlation effects are still small.

The output of the normalized "correlation" meter would be exactly unity if there is no cross-correlation ρ between X_h and X_j codes. However, in general, there will be some finite cross-correlation either positive or negative, and $1 + \rho \neq 1$. This nonzero cross-correlation can cause interference in the receiver-tracking operation or possibly cause false lock in a receiver code search and acquisition operation if $|\rho|$ is sufficiently large; e.g., $|\rho| > 0.3$. The effect can be made more severe by the user receiver antenna pattern, which might have more gain in the direction of the interfering satellite, or if there is less space loss for that satellite. For example, if the interfering satellite is at the zenith and in the direction of maximum antenna gain while the desired signal is at a 5-deg elevation angle, the difference in received signal levels can favor the interfering signal by more than 6 dB. Finally, of course, if there are M satellites in view,

96   J. J. SPILKER JR.

M-1 of them represent interfering signals. If the interference were severe and some signals markedly stronger than others, it is possible to acquire the strongest signals first and cancel out its effect on the others by subtraction. Fortunately, with GPS, this complication is not necessary.

1. P Code

Figure 23 shows the amplitude spectra and signal-to-interference ratio (multiple access gain) computed for the P codes where we have assumed the following:

1) The desired and multiple access interference signals are received at equal power and the same Doppler offset.
2) Both are received with 50 bs data.
3) The two signals are clocked in synchronism.

The output spectrum $G_w(f)$ for the multiplier output waveform $W(t)$ then takes the form shown in Fig. 22b.

Because the P(Y) codes, in effect, are random (the period of the P code is 1 week) and are also modulated by random data, the cross-correlation spectrum is continuous with no discrete line components. The desired component is $D_h(t)$,

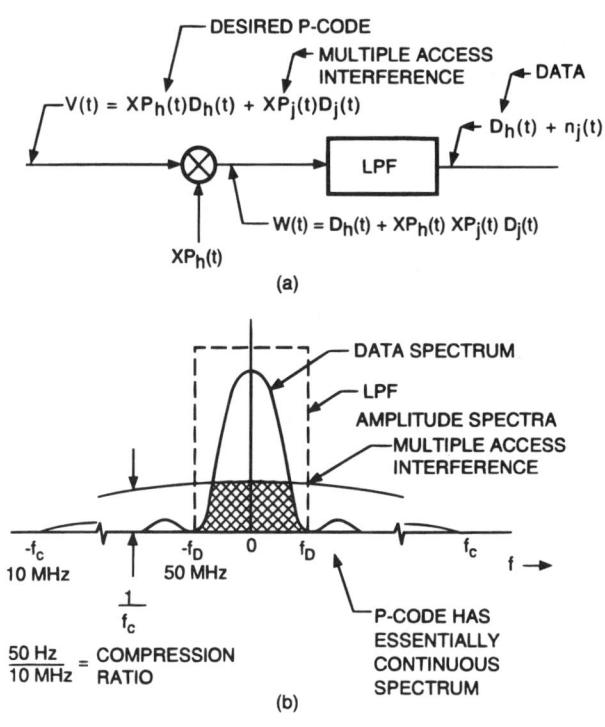

Fig. 23 Multiple access gain for a long period P-code signal. There is no thermal noise in this example, but the delay offset is an integer number of clock cycles. In GPS, $f_D = 50$ Hz, and $f_c = 10.23$ MHz for the P code. The box labeled LPF is a low-pass filter.

SIGNAL STRUCTURE AND THEORETICAL PERFORMANCE

which here is assumed to be a random binary datastream. The interfering multiple access component has the spectrum of the product of the two different synchronous PN codes, which is for all purposes of interest here, a pseudorandom bit stream with a continuous spectrum. Thus, both desired and undesired signals have (sin $x/x)^2$ spectra but vastly different bandwidths; the desired component, the 50 bps data, has a bandwidth to the null of $f_d = 50$ Hz, and the interfering component has a worst case bandwidth to the null of $f_c = 10.23$ MHz. As shown earlier, the degradation in E_b/N_0 caused by this worst case multiple access interference for $M - 1 = 10$ signals is $1 + 10 P_s/N_0 f_c$. If $P_s = -163$ dBW and $N_0 = -205.2$ dB-Hz and $f_c = 10.23$ MHz, then the worst case degradation for 10 interfering signals is a factor $1 + 10 P_s N_0/f_c = 1 + 0.016 = 1.016$, which represents a negligible degradation of 0.07 dB.

B. Coarse/Acquisition-Code Properties

The multiple access properties of the 1023-chip period Gold codes are substantially different from the 1-week period P-code signal in two respects:

1) The C/A code is periodic with a 1-ms period, and thus has line components spaced by 1 kHz rather than the continuous spectrum of the P code. The C/A-code spectrum consists of discrete line components not uniform in height. Likewise, C/A-code cross-correlation sidelobes are periodic, are not of equal height, and are much larger than those of the P code. The data modulation randomizes the C/A spectrum but only with the 50-Hz databit stream compared to the 1-kHz line component spacing.

2) The cross-correlation property is dependent in a significant way on both Doppler offset, as well as code offset. The P code, on the other hand, has multiple access properties essentially independent of Doppler and time offset.

It is also important to point out that the multiple satellite signals are generally received with different value of delay and Doppler frequency offset. For example, a satellite at zenith is at the point of closest approach 20,183 km and has no Doppler; whereas a satellite on the horizon has maximum range 25,783 km (not quite as large as the orbit radius), thus there is a potential range difference 5600.9 km or a 18.68-ms delay difference because the speed of light $c = 2.99792 \times 10^5$ km/s. Furthermore, the satellite on the horizon has a radial velocity that is maximum in magnitude and either positive or negative. Both of these effects (delay and Doppler offsets) make the multiple access problem more difficult (see Fig. 24a). Figure 24b shows the contours of constant Doppler shift for fixed users at various locations on the rotating Earth for a GPS satellite temporarily at 0,0 latitude, longitude. The maximum Doppler shift is approximately ±6 kHz. The contours are shaped both by satellite orbital motion and the rotation of the Earth. Thus, the various GPS signals must be processed with multiple access interference with differing delay and Doppler frequency offsets and not simply a fixed delay offset.

1. Coarse/Acquisition Signal Spectrum

The amplitude spectral density of a maximal length PN code is shown in Fig. 25. If the PN-code period is 1023 bits and the clock rate is 1.023 Mbs, then the PN code amplitude spectrum is a set of line components with a uniform (sin kf/kf)

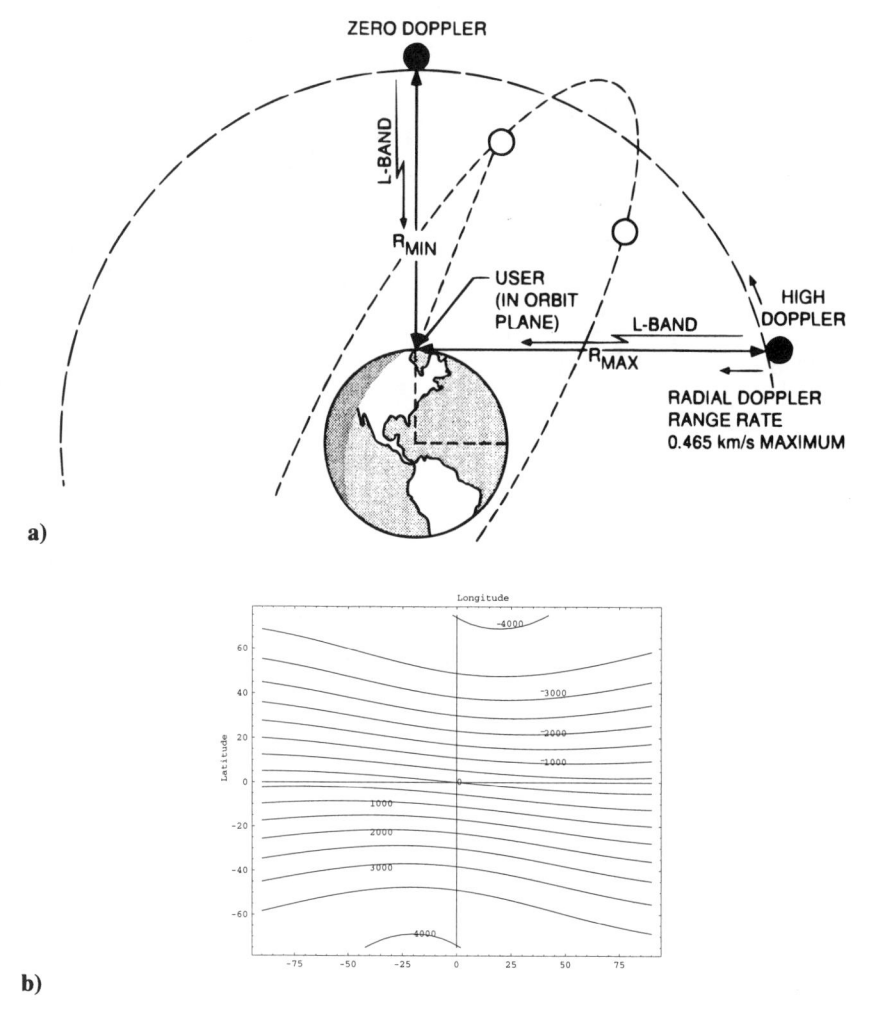

Fig. 24 GPS range and Doppler for a user; a) range variation ≤ 5600 km, maximum range variation occurs if user is in orbit plane, Doppler variation within ± 0.465 km/s (1 nm = 1.852 km). For acceptable multiple access performance, signals must tolerate both delay and Doppler offsets; b) contours of constant Doppler shift for satellite at 0° latitude, 0° longitude vs user position. The maximum Doppler shift is approximately ±4.02 kHz. The largest negative Doppler shift for a fixed user is in the upper right-hand at 20° offset in longitude and at the highest latitude 75° in latitude. The contours are shaped by the rotation of the Earth. The satellite has a 55-deg inclination angle.

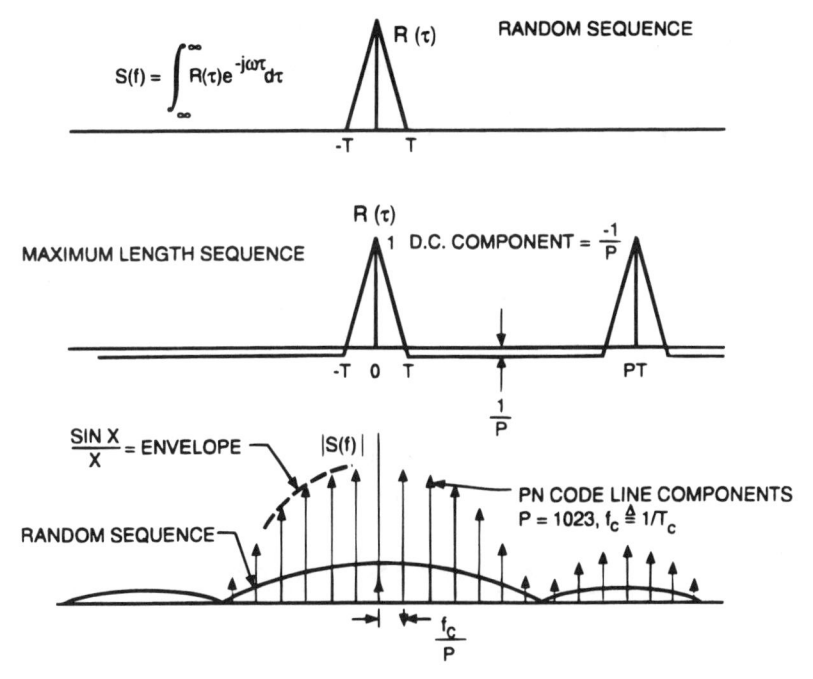

SIGNAL STRUCTURE AND THEORETICAL PERFORMANCE 99

Fig. 25 Autocorrelation function and amplitude spectrum of short PN code (magnitude of the Fourier transform).

variation in amplitude level. The line components are, of course, separated in frequency by the inverse code period rate $f_c/P = 1$ kHz apart where $f_c = 1.023 \times 10^6$ is the code clock rate, and $P = 1023$ is the period.

Gold codes of the same period are composed of a similar set of line components at exactly the same frequencies. However, in this instance, the line components do not follow the same smooth $(\sin x/x)^2$ envelope, although the frequency spacing is the same. Figure 26 shows an example spectrum for a Gold code period $P = 1023$. Note that instead of a line component power 30 dB down from the total signal power P_s, as would be true if we had 1000 line components of equal power, the line component power level can vary significantly above and below this level, depending on the code delay difference and Doppler offset. The spectral contents for the long maximal length PN sequence in a 1.023-kHz band is in fact 30 dB down from the total power near the center of the spectrum.

Figure 27 shows the exact power levels for each of the line components for the GPS C/A code for SV#1. Note that most all of these line components are below -25 dB below the total signal power, but one has an amplitude of approximately -23 dB. Note also that because this C/A code is a balanced code with only one more 1 than 0, the line at $f = 0$ is down 60.2 dB and is essentially zero. A balanced Gold code of length $2^n - 1$ has 2^{n-1} ones. The number of balanced Gold codes is $2^{n-1} + 1 = 512 + 1 = 513$ for $n = 10$.

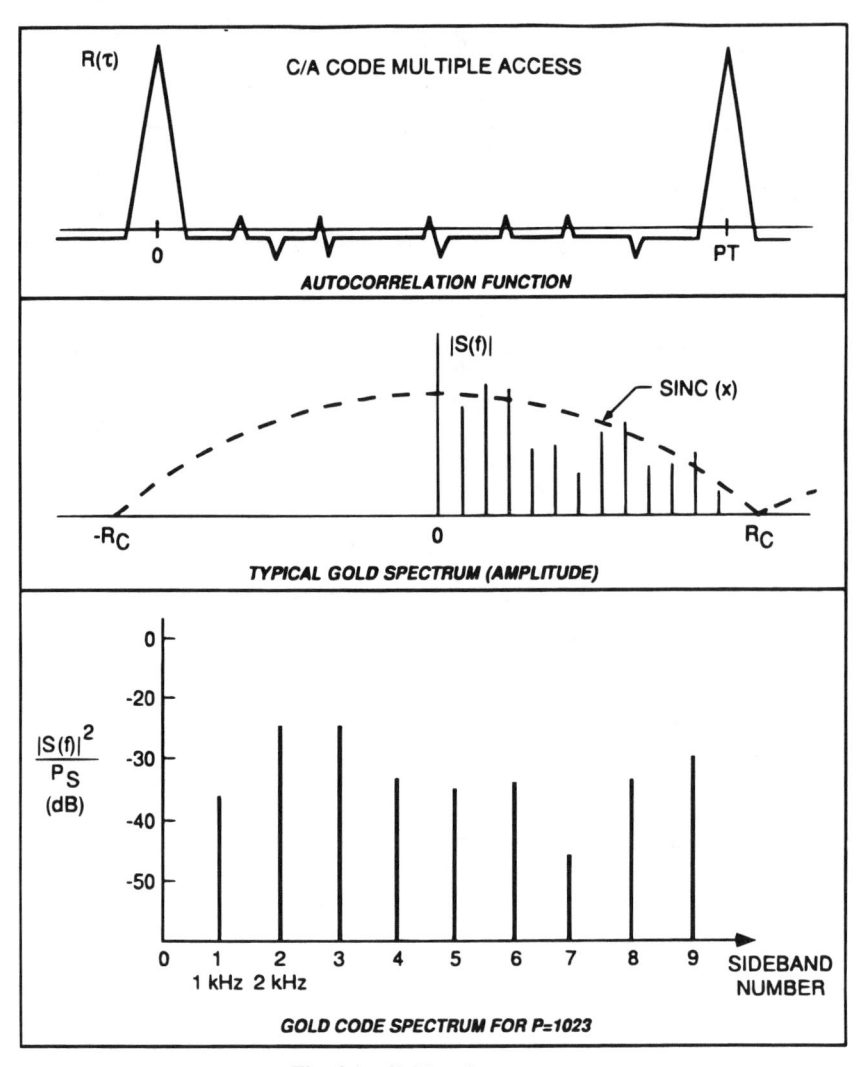

Fig. 26 Gold code spectra.

2. *Gold Code Cross-Correlation Properties*

The Gold codes selected for the C/A signal[8] are a family of codes formed as the product (or modulo-2 sum) of two different *properly paired* maximal length linear feedback shift registers (LFSR), both of the same period $P \triangleq 2^n - 1$ (see the Appendix for the analysis). Table 8 illustrates some of the cross-correlation properties of a Gold code. The two PN sequences $x_i(t)$, $x_j(t)$ are specially selected from the set of LFSR sequences having the same period, as discussed later. As discussed in the Appendix, the cycle and add (delay and multiply) property of maximal length PN codes shows that the product of a code and itself offset in

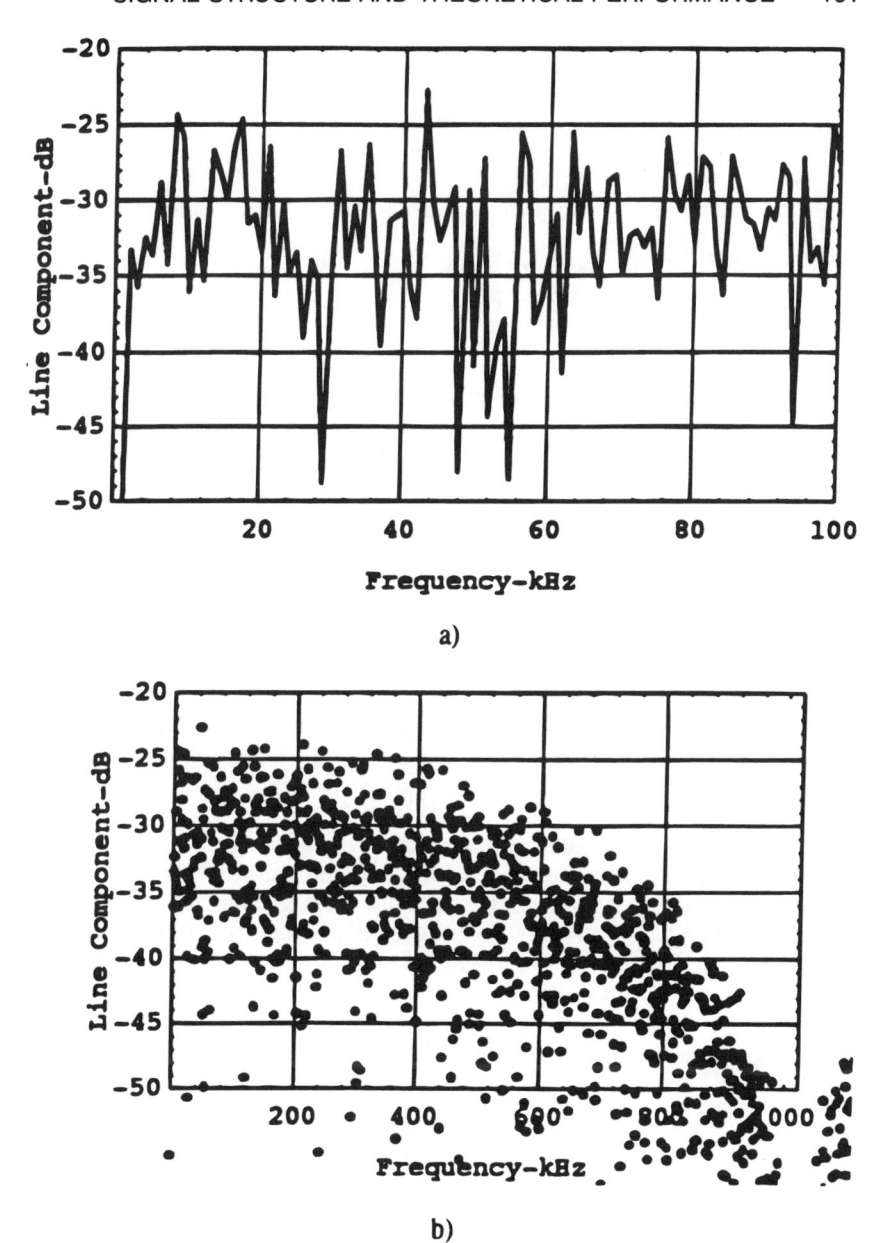

Fig. 27 Line component power relative to the total signal power in dB for the C/A code for satellite SV #1. The line component values have been connected by straight lines in (a); a) the first 100 line components. Each component is spaced at 1-KHz intervals; b) complete set of line components out to the first null in the C/A spectrum. Note that instead of a relatively smooth envelope of line components for a maximal length code, the Gold codes have different spectra, and most have uneven components, some above and some below −30 dB.

Table 8 Gold code cross-correlation properties for no Doppler offset

Product codes: Product of 2 maximal length PN codes $x_i(t)$ and $x_j(t)$, of same period
$G_k(t) = x_i(t)x_j(t + k)$ where x_i have period P
P different values of k. Hence P different codes plus the codes x_i, x_j by themselves provide a total of $2^n + 1$ codes
Family of codes generated for different values of k with low cross-correlation between any pair for all delay offsets
Cross correlation for zero delay offset (using the cycle and add property of PN sequences)

$$\overline{G_k(t)G_m(t)} = \overline{x_i(t)x_j(t + k)x_j(t + m)} = \overline{x_j(t + k)x_j(t + m)}$$
$$= \overline{x_j(1 + n)} = -1/P$$

where the time average is simply $\overline{x_i(t)} = (1/P) \sum_{m=1}^{P} x_i(t + m)$

For other values of code shift $G_k(t)G_1(t + n)$, the cross-correlation is bounded by the time average

$$\overline{G_k(t)G_m(t + q)} = \overline{x_i(t)x_j(t + k)x_i(t + q)x_j(t + q + m)}$$
$$= \overline{x_i(t + r)x_j(t + r + s)} = \overline{G_s(t + r)} = \overline{G_s(t)}$$

time yields the same code with yet a different offset, namely $x_i(t) x_i(t + k) = x_i(t + j)$. Using Table 8, it is easily seen that the cross-correlation between any two different Gold codes of the same family $G_k(t)$ and $G_e(t)$ with no time offset is simply $- 1/P$; i.e., the same as for the PN-code autocorrelation. More generally, however, the cross-correlation between two codes gives the following:

$$\overline{G_k(t)G_e(t + n)} = \overline{G_s(t + r)} = \overline{G_s(t)} \tag{23}$$

where $\overline{G_s(t)}$ is simply the time average of another code in the same family. For balanced Gold codes, the number of ones and zeros differs only by one and $\overline{G_s(t)} = -1/P$. However, even though the Gold codes used in GPS are balanced, the cross-correlation operation yields another Gold code that will not always be balanced. Table 9 summarizes the quantitative cross-correlation with zero Doppler offset for Gold codes of length $P = 2^n - 1$.

Table 9 Cross-correlation properties of Gold codes for n odd and n even (no Doppler offset)

Code period	n = number of shift register stages	Normalized cross-correlation level	Probability of level
$P = 2^n - 1$		$-[2(n + 1/2) + 1]/P$	0.25
	n—odd	$-1/P$	0.50
		$[2(n + 2/2) - 1]/P$	0.25
$P = 2^n - 1$	n—even	$-[2(n + 2/2) + 1]/P$	0.125
	and	$-1/P$	0.75
	$n \neq 4i$	$[2(2 + n/2) - 1]/P$	0.125

SIGNAL STRUCTURE AND THEORETICAL PERFORMANCE

Table 10 summarizes the some of the properties of three types of sequences; linear maximal length shift register sequences, nonlinear maximal length shift registers (contains one more state in the period because the code includes the "all zero" state), and the Gold codes. Note that there are $2^n + 1$ Gold codes in a family of codes of period $2^n - 1$. There are all shift offsets k allowed as indicated in Table 10 plus the two PN components by themselves $x_i(t)$, $x_j(t)$. Thus, the advantage of the Gold codes is not simply a low cross-correlation between all members of the family but that there are a large number of codes all of similarly good properties. Linear maximal length sequences do not in general possess this uniformly low cross-correlation property. The two PN codes used to form the Gold code are, of course, an exception; however, these are only two such codes, not a large family.[11,12]

The Euler's ϕ function of Table 10 can be defined as follows. If the integer n is expressed as

$$n = \prod_i p_i^{a_i}$$

where the p_i are primes, then[13]

$$\phi(n) = \prod_i p_i^{a_i-1}(p_i - 1)$$

and we can define the recursion relationships $\phi(pm) = (p - 1)\phi(m)$, $\phi(p^k m) = p^{k-1}\phi(pm)$, Euler's $\phi(n)$ function is equal to the number of positive integers less than n and relatively prime to n.

Assume next that the received Gold code has been modulated on a radio frequency carrier and is received with a fixed Doppler shift relative to the local oscillator in the receiver. Thus, the received signal has the form

$$G_i(t+\tau)\cos(\omega_o t + \omega_d t + \phi) \tag{24}$$

where delay $\tau \triangleq \tau_0 + \alpha t$ and $\omega_d \triangleq \omega_0 \alpha$ represents the signal with a constant

Table 10   Generalized properties—maximal length codes and gold codes[a]

	Linear shift Registers	Nonlinear shift registers	Gold codes, 2 registers of length n
Period of n-stages	$P = 2^n - 1$	2^n	$2^n - 1$
Cycle and add property	Yes	No	Yes
Autocorrelation function	Two-level	Cannot be two-level	4-level
Number N of codes of period	$N = \phi(2^n - 1)/n$ $\leq (2^n - 2)/n$ For $2^n - 1 = 1023$,	$\approx 2^{2^n}/2^n$	$2^n + 1$
2^n-1 or 2^n	$N = 60$		1025

[a] $\phi(m)$ is the number of integers relatively prime to m, $\phi(m)$ is the Eulers function.

rate of change of delay or a fixed Doppler shift. Cross-correlation by in-phase and quadrature correlators matched to a code $G_j(t)$ with no Doppler shift yields the two products:

In-phase $C_I(t) = G_i(t + \tau)\cos(\omega_0 t + \omega_d t + \theta)[2G_j(t)\cos\omega_0 t]$
$= G_i(t + \tau) G_j(t)\cos(\omega_0 t + \theta)$
$= G_m(t + k + \tau)\cos(\omega_d t + \theta)$,

and

Quadrature $C_Q(t) = G_i(t + \tau)\cos(\omega_0 t + \omega_d t + \theta)[2G_j(t)\sin \omega_0 t]$
$= G_i(t + \tau) G_j(t)\cos(\omega_0 t + \theta)$
$= G_m(t + k + \tau)\sin(\omega_d t + \theta)$

where the double frequency terms have been eliminated, and we have used the fact that the product of two Gold codes is another Gold code with a different delay. In complex notation, the time average over a period of $G_m(t)$ of this product then becomes the following:

$$\overline{G_i(t)} = \overline{G_m(t + k + \tau)e^{j(\omega_d t + \theta)}} \tag{25}$$

Thus, the cross-correlation output simply offsets the line components of the Gold code by the Doppler offset.

Represent the periodic Gold code by its Fourier series:

$$G_m(t) = \sum a_n \cos(2\pi nt/T) + b_n \sin(2\pi nt/T)$$
$$= \sum c_n \cos(2\pi nt/T + \phi_n) = Re \sum c_n e^{j(2\pi nt/T + \phi_n)} \tag{26}$$

where for unit power in the code $\overline{G_m^2(t)} = 1 = (1/2) \sum c_n^2$. From Eqs. (25) and (26) the correlator product output for the in-phase and quadrature components can only have a nonzero time average if $2\pi n/T = \omega_d$; i.e., the Doppler offset matches a specific line component, say the qth line component, in this instance q kHz. Thus, the time average of the in-phase and quadrature output components are then (in complex notation) the following:

$$\overline{C(t)} = \overline{c_q \cos(2\pi qt/T + \phi_n)\cos(\omega_d t + \theta)} + \overline{jc_q \sin(2\pi qt/T + \phi_n)\cos(\omega_d t + \theta)}$$
$$= (1/2) c_q[\cos(\phi_n - \theta) - j \sin(\phi_n - \theta)]$$

Or in complex notation the time average of the correlator output is as follows:

$$\overline{C(t)} = (1/2) c_q e^{j(\theta_n - \theta)}$$

and the squared envelope ($I^2 + Q^2$) has an average power of $c_q^2/4$ where q represents an integer number of kHz of doppler shift. The power in the qth line component of the code itself is $c_q^2/2$. Thus, the cross-correlation with Doppler shift picks out that line component in the appropriate Gold code corresponding to the Doppler and passes it directly to the output (after a factor of 2 reduction in power). The output of the correlator for the desired signal is as follows:

$$\overline{C(t)} = \overline{G_i^2(t)}\cos \theta + \overline{G_i^2(t)}\sin \theta$$
$$= \cos \theta + j \sin \theta = e^{j\theta}$$

which has unity average power in the square-law envelope correlator output.

SIGNAL STRUCTURE AND THEORETICAL PERFORMANCE   105

Thus, if the Doppler offset is an integral multiple of the line component spacing, the correlator output for the interfering signal corresponds to the amplitude of the corresponding code line component. Recall that even with a stationary user, the Doppler offset between satellites can vary by as much as approximately ±6 kHz. Thus, we might encounter any of the first 12 line components.

Table 11 summarizes the cross-correlation results for both zero Doppler and the worst possible Doppler for three different Gold code periods. Note that the zero Doppler cross-correlation changes by 6 dB at every 2-stage increase in the number of shift registers. Thus, the zero Doppler cross-correlation decreases by 6 dB by going from period $P = 511$ to $P = 2047$ codes, whereas an increase from 511 to 1023 causes no improvement if there is no Doppler shift.

It is easily seen, however, that the zero Doppler condition is not of greatest importance. When worst case Doppler shifts are considered, the peak cross-correlation changes by 3 dB with each doubling in code period. The 1023 chip period was chosen for GPS as a compromise between low cross-correlation and receiver acquisition time. If the code period doubles, the code acquisition time doubles also. Note one additional point. If there is a Doppler shift between two signals, then the delay difference is by definition changing between codes. For example, if there is a 1-kHz Doppler shift at L_1, then the C/A-code clock rate differs by 1 kHz/1540 = (1/1.54) Hz = 0.649 Hz. Thus, the two codes will shift in relative delay by one C/A-code chip every 1.54 s. Thus, the worst case code sidelobes with Doppler shift are only temporary in nature. However, they persist for times that are long compared to a data bit at 50 bps.

Figure 28 shows the cumulative probability of various cross-correlation interference levels for the GPS C/A code for various Doppler shifts from $f_d = 0$ to ± 5 kHz. It should be pointed out that the worst case cross-correlation sidelobes for Doppler shift of Table 11 occur with only a low probability as is apparent from the variation in line component amplitudes shown earlier in Fig. 27.

Note that the 4-kHz Doppler gives the worst cross-correlation sidelobe over this range; however, the other Doppler shifts give similar results. These cumulative averages are formed by averaging results for all 1023 of the Gold codes of period 1023 in the GPS family. All possible code time offsets are considered for each Doppler offset for all possible pairs of codes in this family. It should also be pointed out that with multiple interfering GPS signals, there is an averaging effect; not all of the signals will produce a worst case Doppler delay offset at the same time. In fact, as mentioned above the worst case cross-correlation with Doppler is rather rare, and some signals will produce a lower than average component, and the results tend to average to the random sequence result of

Table 11 Peak cross-correlation sidelobes for Gold codes

Parameter	Code period		
	511	1023	2047
Peak cross-correlation (any Doppler shift)	−18.6 dB	−21.6 dB	−24.6 dB
Peak cross-correlation (zero Doppler)	−23.8 dB	−23.8 dB	−29.8 dB
Probability of worst case or near worst case cross-correlation (zero Doppler)	0.5	0.25	0.5

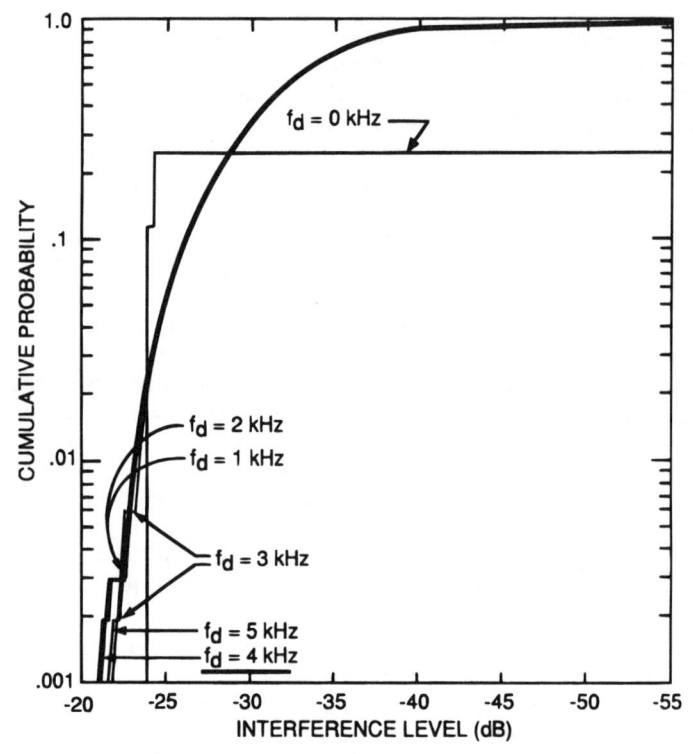

Fig. 28 Cumulative probability of interference level for 1023-bit Gold code at 1.023 Mb/s (Courtesy H. Chen, Stanford Telecom).

approximately −30 dB. Likewise, they will not have an integer number of chips offset delay, and the spectrum is widened, as was shown earlier with the 2/3 factor in the multiple access noise calculation shown in Table 1.

C. Bounds on GPS Signal Tracking Performance in the Presence of White Thermal Noise

The objective of this section is to determine the performance bounds on the minimum mean square pseudorange tracking error in the presence of white Gaussian receiver noise. Recall that the GPS satellite signals are filtered to a 20.46-MHz radio frequency bandwidth. Thus, many sidelobes of the C/A code are included. Note that the C/A signal of power P_s has a power spectral density (one-sided) of the following form:

$$G_s(f) = \left[\frac{\sin \pi(f - f_o/f_c)}{\pi(f - f_o)/f_c}\right]^2 P_s/2f_c, f > 0 \qquad (27)$$

where f_o is the L_1 frequency, $f_c = 1.023$ MHz, and P_s is the power. A plot of

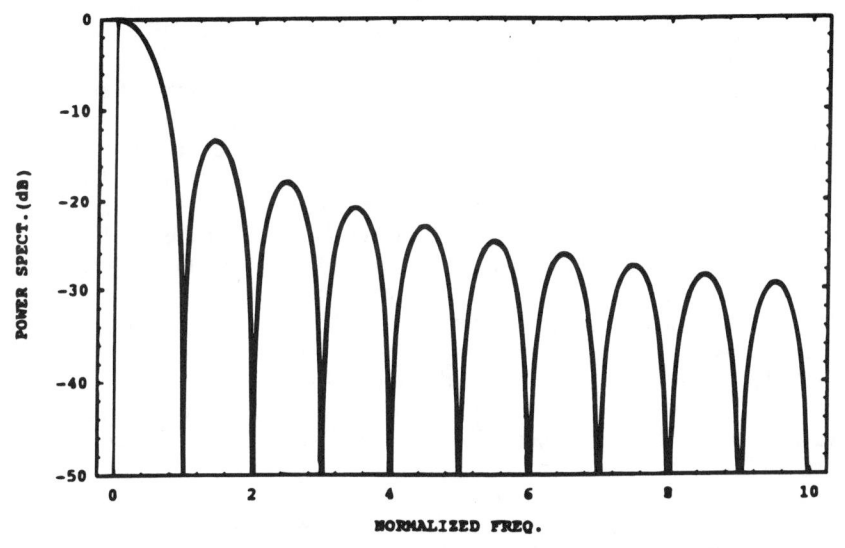

SIGNAL STRUCTURE AND THEORETICAL PERFORMANCE 107

Fig. 29 Plot of the one-sided C/A-code power spectral density in dB out to the tenth spectral sidelobe corresponding to a 20.46 MHz radio frequency bandwidth. Frequency is normalized to the C/A-code clock rate $f_c = 1.023$ MHz.

$G_s(f)$ is given in Fig. 29 out to the tenth spectral sidelobe corresponding to the 20.46-MHz bandwidth. This bandwidth is just sufficient to pass the main lobe of the P code. As is shown below, the optimal estimator for delay for the 20.46-MHz C/A signal in the presence of white thermal noise utilizes the full 20.46-MHz bandwidth. In this subsection comparisons are made between the performance of the optimal estimators for the following: 1) C/A-code signal using the full 20.46-MHz bandwidth; 2) C/A-code signal using only the 2.046 MHz bandwidth of the C/A-code main lobe; and 3) P code using the 20.46 MHz bandwidth.

1. Near-optimal Delay–Lock-Loop Estimator and Performance Bounds

The near-optimal delay–lock-loop estimator of pseudorange τ is discussed in detail in a later chapter on signal-tracking theory and is shown in simplified form in Fig. 30. The detailed discussion of the delay–lock-loop and its operation is

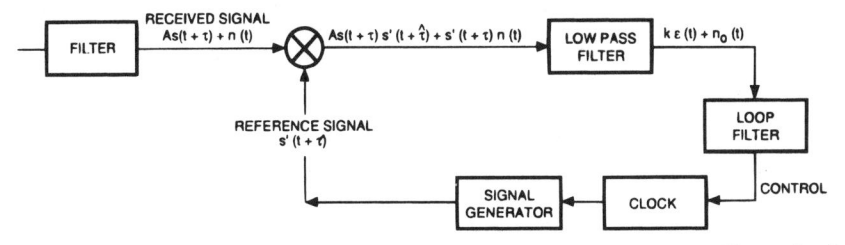

Fig. 30 Simplified block diagram of the near-optimal delay–lock-loop. The optimal reference signal is the differentiated signal $s'(t)$. The delay error is $\epsilon \triangleq \tau - \hat{\tau}$.

left for later chapters. For the moment, simply assume that this delay–lock-loop provides the near-optimal estimate of delay given that the delay error $\epsilon = \tau - \hat{\tau}$ is sufficiently small compared to the width of the linear region of the differentiated autocorrelation function $R'(\epsilon)$ of the signal. The quantity $R'(\epsilon)$ is the differentiated autocorrelation function, and is assumed to exist everywhere. For simplicity, the signal $s(t)$ represents the signal from a single satellite. Vector delay–lock-loop estimators which process many signals in parallel are discussed in Chapter 7.

Notice that the optimum delay estimator cross-correlates the received base band signal not with the signal waveform $s(t)$ but with the differentiated reference waveform $s'(t)$, which is assumed to be finite everywhere.[11,14] Note that for the simple example where the signal is $\cos \omega_o t$ the differentiated reference is $-\omega_o \sin \omega_o t$ and the, delay–lock-loop reduces to the simple phase–lock-loop. As examples using the full 20.46 MHz of the signal available for both C/A and P signals, Fig. 31 shows the spectra for the reference waveforms $s'(t)$ for the optimal C/A-code and P-code delay–lock-loops. Note that the reference spectra are of the form $(\sin \pi f/f_o)^2$ rather than $[(\sin \pi f/f_o)/(\pi f/f_o)]^2$, the spectrum of the received signal.

The output of the multiplier of Figure 31 is processed by some minimal amount of low-pass filtering just sufficient to reduce the self-noise to a relatively small level. Self-noise is the broadband noise formed in the process of multiplying the received signal $s(t + \tau)$ with the reference $s'(t + \hat{\tau})$.[10] For a pure sinusoidal carrier, the product $\sin \omega_o(\tau + \hat{\tau}) \cos \omega_o(\tau - \hat{\tau})$ yields $R'(\tau - \hat{\tau}) = \sin \omega_o (\tau - \hat{\tau})$ (ignoring terms at $2\omega_o$), and there is no self-noise. However, with PN-modulated carriers there is a pseudorandom term in addition to $R'(\tau - \hat{\tau})$. In a sufficiently small region of delay error $\epsilon = (\tau - \hat{\tau})$, the signal $s(t + \tau)$ can be represented by a Taylor's series expansion about $s(t + \hat{\tau})$ and the low-pass output of the multiplier is as follows:

$$As(t - \tau)s'(t + \hat{\tau}) + s'(t + \tau)n(t) \cong A\{s(t + \hat{\tau})s'(t + \hat{\tau}) + \epsilon[s'(t + \hat{\tau})]^2\} + n_1(t)$$

$$\cong A\epsilon[s'(t + \hat{\tau})]^2 + n_1(t) \qquad (28)$$

where the received signal power is $A^2 = P_s$, and $n(t)$ and $n_1(t)$ are both white Gaussian noise of power spectral density N_o. The power in $s'(t)$; i.e., $E[s'(t)]^2 = P_{s'}$ is normalized to unity.

More generally, the output of the low-pass filter can be approximated by the following:

$$AR'(\epsilon) + (R''(0))^{1/2}n_0(t) + \text{small self-noise term} \qquad (29)$$

where the first term represents its expected value of the product $s(\tau - \tau)s'(t - \hat{\tau})$, and we use the following relationships: $E[s(t)s(t + \epsilon)] = R(\epsilon)$ and $E[s(t)s'(t + \epsilon)] = R'(\epsilon) = dR(\epsilon)/d\epsilon$; i.e., R and R' are the autocorrelation and differentiated autocorrelation functions, respectively. We assume that $R'(\epsilon)$ and $R''(\epsilon)$ exist everywhere; i.e. the signals are of finite bandwidth. Because $s'(t)$ is independent of $n(t)$, and $n(t)$ is white Gaussian noise, $n_o(t)$ is also white Gaussian noise, although it is nonstationary and is amplitude-modulated by $s'(t)$. The slope of $R'(\epsilon)$ at $\epsilon = 0$ is $R_s''(0)$ and corresponds to the slope in the earlier equation from the Taylor's series expansion. The power in $s'(t)$ is $P_{s'} = -R_{s'}(0)$.

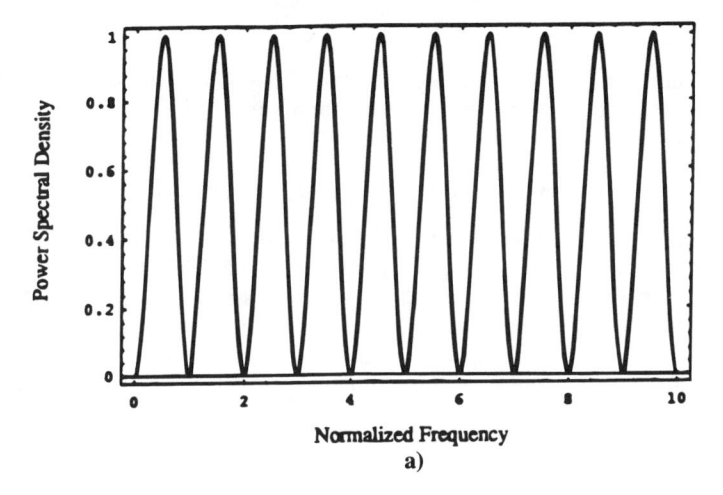

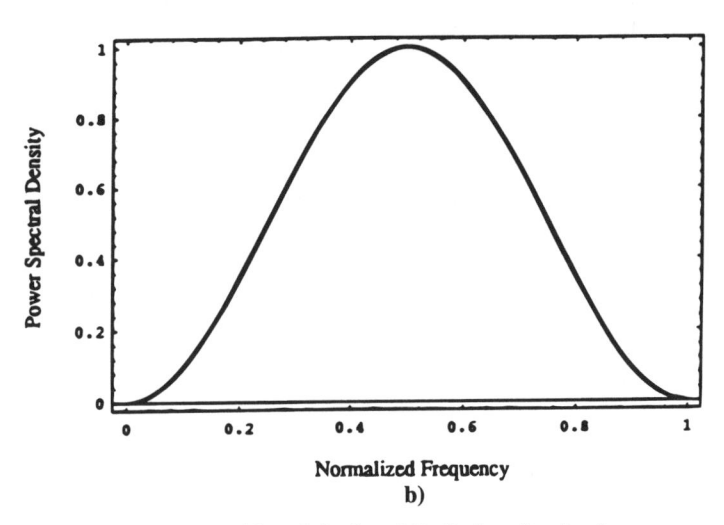

SIGNAL STRUCTURE AND THEORETICAL PERFORMANCE 109

Fig. 31 Power spectra densities of the band-limited optimal reference waveforms $s'(t)$ for the C/A code (a) and the P code (b). The frequencies are both normalized to the respective clock rates; the maximum frequency in both examples is 20.46 MHz.

2. *Estimate of Delay Error*

In the finite width quasilinear region of the differentiated autocorrelation function, the first term in the output can be approximated by the slope at $\epsilon = 0$ as follows:

$$\sqrt{P_s}R_s''(0)\{\epsilon(t) + [-R_s''(0)]^{1/2}n_o(t)/P_s^{1/2}R_s''(0)\} \tag{30}$$

where $n_o(t)$ is then also white noise of power spectral density N_o and $R_s''(0)$ is the slope of the delay–lock-loop discriminator curve. The $\epsilon(t)$ term represents the useful signal for the delay estimate. If it is assumed that the output is to be filtered by a filter of effective closed-loop bandwidth B, and the signal-to-noise ratio is sufficiently high to maintain quasi-linear operation, then the effective delay error variance is approximated by the following:

$$\sigma_\epsilon^2 \cong [-R_s''(0)]N_oB/[R_s''(0)]^2P_s = \{(P_s/N_oB)[-R_s''(0)]\}^{-1}$$
$$\cong 1/\{(\text{SNR})[-R_s''(0)]\} \tag{31}$$

for $\text{SNR} > (\text{SNR})_{Threshold}$. The exact variance for this nonlinear problem must be found as the solution to the Fokker-Planck stochastic differential equation. Thus, the approximate rms delay error $\sigma_\epsilon \cong 1/\{\text{SNR}[-R_s''(0)]\}^{1/2}$ where P_s/N_oB is a normalized signal-to-noise ratio $= \text{SNR}$. For fixed input signal power and noise density, the error variance performance is inversely proportional to the second derivatives of the signal autocorrelation function.

Recall that the autocorrelation function and the signal power spectral density $G_s(f)$ are related by the Fourier transform and that similar relationships can be written for the derivatives of $R_s(\tau)$, as follows:

$$R_s(\tau) = \int_{-\infty}^{\infty} G_s(f)e^{j\omega\tau}\,df, \quad R_s(0) = \int_{-\infty}^{\infty} G_s(f)\,df = P_s = 1$$

$$R_s'(\tau) = \int_{-\infty}^{\infty} j\omega G_s(f)e^{j\omega\tau}\,df, \quad R_s'(0) = \int_{-\infty}^{\infty} j\omega G_s(f)\,df \tag{32}$$

$$R_s''(\tau) = \int_{-\infty}^{\infty} (-\omega^2)G_s(f)e^{j\omega\tau}\,df,$$

$$R_s''(0) = -\int_{-\infty}^{\infty} \omega^2 G_s(f)\,df = -2\int_0^{\infty} \omega^2 G_s(f)\,df$$

The quantity

$$\int_{-\infty}^{\infty} \omega^2 G_s(f)\,df = -R_s''(0)$$

is also known as the square of the Gabor bandwidth $\overline{\Delta\omega^2}$ Ref. 15. In a related problem for radar pulse time estimation, Helstrom[16] has shown that the optimum estimate of arrival time for a received signal in the presence of white noise is related to the Gabor bandwidth $\overline{\Delta\omega^2}$ by $\sigma_\tau^2 = 1/(d^2\,\overline{\Delta\omega^2})$ where $d^2 \triangleq E/N_o$ is the received signal pulse energy-to-noise density ratio. Thus, the time delay estimate

SIGNAL STRUCTURE AND THEORETICAL PERFORMANCE

for the optimum delay–lock-loop gives the analogous result because $\overline{\Delta\omega^2} = -R''_s(0)$, namely

$$\sigma_\epsilon^2 = \frac{1}{\text{SNR}\overline{\Delta\omega^2}} \tag{33}$$

for SNR sufficiently high. The SNR must be sufficiently high in order for the quasi-linear assumption to remain valid. Note that the average slope $-R''(0)$ diminishes rapidly with increasing noise variance. Thus, these estimates of rms delay error are lower bounds as consistent with the Cramer-Rao bounds.

3. Cramer-Rao Bound

The above estimate of delay error variance relates closely to the Cramer-Rao lower bound on the variance of any unbiased estimator. Define a set of N scalar measurements at time i as follows: $x(i) = s(i, \tau) + n(i)$ where $i = 0, 1, \ldots, N - 1$, where $n(i)$ are white Gaussian noise terms of zero mean and covariance matrix $E[n^T n] = \sigma^2 I$ and τ is the parameter to be estimated. The Cramer-Rao lower bound on the variance in the estimate of τ is then as follows:[17]

$$\sigma_{\hat\tau}^2 \geq 1/I(\tau) \tag{34}$$

where $I(\tau)$ is the Fisher information (a matrix if τ is a vector parameter). Define the conditional probability of x given τ as $p(x/\tau)$, which for this example is Gaussian. The Fisher information is then as follows:

$$I(\tau) = -E[\partial^2[\ln(p(x|\tau))]/\partial \tau^2] = E[(\partial[\ln(p(x|\tau))]/\partial \tau)^2] \tag{35}$$

When $p(x/\tau)$ is viewed as a function of τ for an observed x, it is termed the likelihood function. The Fisher information is in effect a measure of the sharpness of the peak of the likelihood function. For this Gaussian noise example the likelihood function is as follows:

$$p(x|\tau) = \left(\frac{1}{\sigma\sqrt{2\pi}}\right)^N \exp\left[-\frac{1}{2\sigma^2}\sum_{i=0}^{N-1}[x(i) - s(i, \tau)]^2\right]$$

The Fisher information is as follows:

$$I(\tau) = -\frac{1}{\sigma^2}\sum_{i=0}^{N-1}\left[\frac{\partial s(i|\tau)}{\partial \tau}\right]^2 \tag{36}$$

More generally, if τ is a vector parameter with components τ_i, the Cramer-Rao bound states the following:

$$\sigma_{\hat\tau_i}^2 \geq [I^{-1}(\tau)]_{ii} \tag{37}$$

where

$$[I(\tau)]_{ij} = -E\left[\frac{\partial^2[\ln(p(x|\tau))]}{\partial \tau_i \partial \tau_j}\right]$$

If $s(t, \tau)$ is a band-limited signal with a bandwidth W, and the signal samples are taken every $\Delta = 1/2W$ s, and τ is the delay that $s(i, \tau) = s(i\Delta - \tau)$, then the Cramer-Rao bound becomes the following:

$$\sigma_\tau^2 \geq \frac{\sigma^2}{\sum_{i=0}^{N-1} \left|\frac{\partial s}{\partial \tau}\right|_{t=i\Delta}^2} = \frac{1}{\frac{E_s}{N_0/2} \overline{\Delta\omega^2}} \tag{38}$$

where the signal energy over the time interval $(0, N\Delta)$ is as follows:

$$E_s = \int_0^{N\Delta} s^2(t) \, dt$$

and $\overline{\Delta\omega^2}$ is the mean square bandwidth

$$\overline{\Delta\omega^2} = (2\pi)^2 \frac{\int f^2 |S(f)|^2 \, dt}{\int |S(f)|^2 \, dt}$$

where $S(f)$ is the Fourier transform of $s(t)$. Thus, the estimate of delay error variance of Eq. 33 is the lower bound and is approximated only at sufficiently high signal-to-noise ratio. Note further that the Cramer-Rao bound must be used with caution because an unbiased estimator that achieves this bound is not always realizable (see Refs. 17 and 18 and the chapter in this volume on Signal Tracking Theory.

4. Delay Estimate RMS Error for Optimal Delay–Lock-Loops

To compare the rms errors for the three GPS tracking problems of interest, C/A code with 2.046-MHz bandwidth, 20.46-MHz bandwidth, and P code with 20.46 MHz, the value of the Gabor bandwidth, or $R_s''(0)$, must be computed for each example. Unity signal power is assumed in all three examples. Note that we approximate the C/A code with a continuous spectrum. Note also that the optimal reference signal $s'(t)$ for this ideal bandlimited example is a series of $(\sin kt)/kt =$ shaped impulses of pseudorandom sign. The results are as follows:

Coarse/Acquisition-Code 2.046-MHz Bandwidth

$$\overline{\Delta\omega^2} = \frac{2}{f_c} \int_0^{f_c} \left(\frac{\sin \pi f/f_c}{\pi f/f_c}\right)^2 \omega^2 df = \frac{2}{f_c} (2)^2 f_c^2 \int_0^{f_c} (\sin \pi f/f_c)^2 df \frac{\pi}{f_c} \left(\frac{f_c}{\pi}\right) \tag{39}$$

$$\overline{\Delta\omega^2} = \frac{2}{f_c} \frac{2^2 f_c^3}{\pi} \int_0^\pi \sin^2 x \, dx = \frac{2}{f_c} \frac{2^2 f_c^3}{\pi} \left(\frac{\pi}{2}\right) = 4 f_c^2$$

where $x \triangleq \pi f/f_c$.

Coarse/Acquisition-Code 20.46-MHz Bandwidth (Unity Power)

$$\overline{\Delta\omega^2} = \frac{2}{f_c} \int_0^{10 f_c} \left(\frac{\sin \pi f/f_c}{\pi f/f_c}\right)^2 \omega^2 \, df = \frac{2}{f_c} \left(\frac{4 f_c^3}{\pi}\right) \int_0^{10\pi} \sin^2 x \, dx = 40 f_c^2 \tag{40}$$

SIGNAL STRUCTURE AND THEORETICAL PERFORMANCE 113

Table 12 Comparison of various delay estimation rms delay errors for C/A-code signal 3 dB stronger than the P-code

Signal Type	Root-mean-square delay error, σ_ϵ	Relative root-mean-square delay error, $\sigma_\epsilon/\sigma_{\epsilon o}$
C/A code—2.046-MHz BW	$\dfrac{1}{\text{SNR}^{1/2}}\sqrt{\dfrac{1}{4f_c^2}} = \dfrac{1}{\text{SNR}^{1/2}}\dfrac{1}{2f_c}$	$\sqrt{1.107}\,\dfrac{10}{\sqrt{2}} = 7.440$
C/A code—20.46-MHz BW	$\dfrac{1}{(\text{SNR})^{1/2}}\sqrt{\dfrac{1}{40f_c^2}} = \dfrac{1}{\text{SNR}^{1/2}}\dfrac{1}{2f_c\sqrt{10}}$	$\sqrt{1.107}\,\sqrt{5} = 2.352$
P code—20.46-MHz BW	$\dfrac{1}{\text{SNR}^{1/2}}\sqrt{\dfrac{1}{400f_c^2(1.107)}} = \dfrac{1}{\text{SNR}^{1/2}}\dfrac{(1/1.052)}{20f_c}$	1

The reference rms error $\sigma_{\epsilon o}$ is that for the P-code optimal receiver.

Precision-Code 20.46-MHz Bandwidth (Unity Power)

$$\overline{\Delta\omega^2} = \frac{2}{f_c}\left(\frac{1+\Delta}{10}\right)\int_0^{10f_c}\left(\frac{\sin \pi f/10f_c}{\pi f/10f_c}\right)^2 \omega^2 df$$

$$= \frac{4\cdot 2}{f_c}\left(\frac{10f_c^2}{10}\right)(1+\Delta)\int_0^{10f_c}(\sin \pi f/10f_c)^2 df\,\frac{\pi}{10f_c}\left(\frac{10f_c}{\pi}\right)$$

$$\overline{\Delta\omega^2} = \frac{2(1+\Delta)}{f_c}\frac{2^2}{10}\frac{(10f_c)^3}{\pi}\int_0^\pi \sin^2 x\,dx$$

$$= \frac{2(1+\Delta)}{f_c}\frac{2^2}{10}\frac{(10f_c)^3}{\pi}\frac{\pi}{2} = 400 f_c^2(1+\Delta) \quad (41)$$

The quantity $1+\Delta = 1.107$ is needed to normalize the P code to unity power because the spectral sidelobes have been removed prior to transmission at unity power. Table 12 compares the three types of receivers for C/A signals which are twice the power of that for the P-code signal. Notice that the optimum P-code estimator is a factor of $7.07 \times 1.05 = 7.44$, better than the best C/A-code estimator when the C/A-code receiver is constrained to operate with a 2.046-MHz bandwidth. However, the P-code receiver is only a factor of 2.352 better than the optimal C/A-code delay-lock estimator, which utilizes the full 20.46 MHz available.

Thus because of the improved processing of the optimal C/A code, its performance bound, including the effect of the higher power and hence higher SNR in the C/A signal, is not far below that of the P code receiver. It should be pointed out, however, that the P-code receiver has definite performance advantages relative to nonwhite noise interference and security in the P(Y) code.

These results, give the bounds on the best delay estimate possible for a given tracking loop bandwidth B. As discussed in later chapters, the bandwidth B required is related to the dynamics of the delay variation with time. We have also assumed that each satellite signal is processed independently, one at time;

i.e. in a scalar delay–lock-loop. A later chapter discusses the vector delay–lock-loop, and it is shown that we can improve on this performance substantially by processing these satellite signals in parallel in vector fashion, and effectively increase the SNR in these equations by a vector-combining operation.

Appendix: Fundamental Properties of Maximal Length Shift Registers and Gold Codes

The GPS C/A codes are Gold codes that are products (modulo-2 sums) of maximal length feedback shift register codes, and it is appropriate to discuss some of the fundamental properties of these codes. This appendix introduces the elements of finite field algebra as applicable to maximal length codes, illustrates the cycle and add property used in the C/A-code generator, and discusses the relationship between the maximal length code pairs used to generate the Gold codes.[7]

A. Galois Fields, Primitive Polynomials, and Vector Representation of Maximal Length Sequences

The state vectors of the linear feedback shift registers used in GPS are elements of a finite field termed the Galois field (GF). A finite field is a set of elements including 0 and 1 that can be added, subtracted, multiplied, or divided (except by 0) to give a unique result in the field.[12] All finite fields are commutative. Although Galois fields in general have p^n elements where p is prime and n is an integer, this discussion is restricted to the binary p. Thus, there are $p^n = 2^n$ elements in the field. For the GPS C/A code, $n = 10$. This section gives only a simplified discussion of the algebra sufficient to address the specific GPS objectives. For more detailed analysis, the reader is referred to Refs. 13 and 19–21.

Define the elements of the GF(2^n) field as $0, \alpha^0, \alpha^1, \alpha^2, \ldots, 2^{2n-2}$. Each of the α^i can also be represented as the state vector of an n-stage binary feedback shift register. For the C/A code a 10-element state vector where element $\alpha^0 = 1$ is also equal to the element $\alpha^{2^n-1} = \alpha^{1023} = 1$. Likewise $\alpha^{2^n} = \alpha$, etc.; namely, the elements are all expressed as powers of α and because $\alpha^{2n-1} = 1$, α is termed a primitive element of GF(2^n). For example, if $2^n - 1 = 15$, as shown in Table A.1, $\alpha^0 = 1 = \alpha^{15}, \alpha = \alpha^{16}$, etc.

As shown below, the properties of maximal length sequences are based on the theory of finite field algebra and irreducible primitive polynomials. Define a polynomial $f(x)$ to be irreducible if $f(x)$ cannot be expressed as the product of two polynomials over the field; that is, it has no divisors except for multiples of itself and scalars. If α is defined as an element in the field GF(2^n) and α is a root of the polynomial $f(x)$; i.e., $f(\alpha) = 0$, then α^2, α^{2^i} are also roots of the irreducible polynomial $f(x)$ because $f(\alpha^2) = [f(\alpha)]^2 = 0$. An irreducible polynomial contains an odd number of nonzero terms. Furthermore, if the irreducible polynomial has a primitive element α (where $\alpha^{2^n-1} = 1$) that is a root, then the polynomial is termed a primitive polynomial and corresponds to the polynomial for a maximal length feedback shift register. For example, the GPS polynomials

$f(x) = 1 + x^3 + x^{10}$ and $f(x) = 1 + x^2 + x^3 + x^6 + x^8 + x^9 + x^{10}$ are primitive polynomials. There are actually 60 distinct primitive polynomials of degree 10, and each corresponds to a separate maximal length sequence.

Consider an example finite field composed of 16 binary 4-ruples, as shown in Table A.1. Defining a finite field addition table does not require the definition of the polynomial, but the generation of a multiplication table does. This finite field GF (2^4) of Table A.1 has its multiplication defined by the minimal polynomial $m_\alpha(x) = 1 + x^3 + x^4$. The vector representation corresponds to the coefficients of the polynomial representation. Note also that if we examine the first bit in the vector, the sequence of this first bit (excluding the all zero state) gives a maximal length sequence. The same can be said for the sequence of each second bit in the vector, etc. This GF (2^n) finite field is defined by a minimal polynomial $m_\alpha(x) = 1 + x^3 + x^4$, which has the element α as one of its roots: i.e., $m_\alpha(\alpha) = 0$. The minimal polynomial of an element α in GF (2^n) is the monic polynomial of smallest degree with coefficients in GF (2); namely, degree $= 1$, that has α as a root. Addition and multiplication of polynomials is performed modulo-$[m_\alpha(x), 2]$. For the GF (2^n) to correspond to a maximal length sequence of length $2^n - 1$, the polynomial must be irreducible and primitive (has a primitive element of α where all nonzero elements of GF (2^n) are of the form α^i). If α^i represents the present state of a cyclic code at time i (e.g., the vector representation of Table A.1), then the next state is $\alpha(\alpha^i) = \alpha^{i+1}$.

Figure A.1 shows a specific implementation of a 4-stage feedback shift register. As shown below, this 4-stage feedback shift register generates the same sequence as that in Table A.1 (except for the all zero state). The difference equation for

Table A.1 Various representations of a finite field of elements α^i in GF (2^4) with polynomial $m_\alpha(x) = 1 + x^3 + x^4$

Element	Polynomial representation	Vector representation
0	0	(0000)
$\alpha^{15} = 1 = \alpha^0$	$0 + 0 + 0 + 1$	0001
$\alpha^{16} = \alpha$	$ + + \alpha + $	0010
α^2	$ + \alpha^2 + + $	0100
α^3	$\alpha^3 + + + $	1000
α^4	$\alpha^3 + + + 1$	1001
α^5	$\alpha^3 + + \alpha + 1$	1011
α^6	$\alpha^3 + \alpha^2 + \alpha + 1$	1111
α^7	$ + \alpha^2 + \alpha + 1$	0111
α^8	$\alpha^3 + \alpha^2 + \alpha + $	1110
α^9	$ + \alpha^2 + + 1$	0101
α^{10}	$\alpha^3 + + \alpha + $	1010
α^{11}	$\alpha^3 + \alpha^2 + + 1$	1101
α^{12}	$ + + \alpha + 1$	0011
α^{13}	$ + \alpha^2 + \alpha + $	0110
α^{14}	$\alpha^3 + \alpha^2 + + $	1100

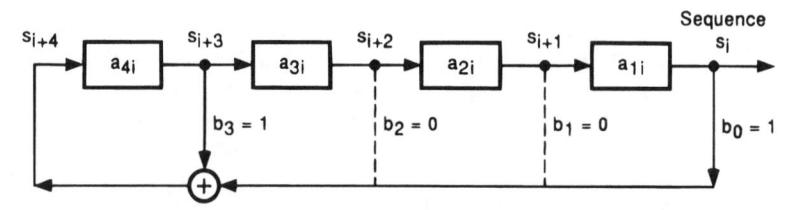

Fig. A.1 Block diagram of a 4-stage linear feedback shift register with a primitive polynomial $f(x) = 1 + x^3 + x^4$, which generates the same sequence as the sequence GF(2^4), of Table A.1. The shift register generates a sequence of period $2^n - 1 = 2^4 - 1 = 15$. (See also Fig. 5, this chapter.)

The sequence of Fig. A.1 at iteration or discrete time i is as follows:

$$s_{i+4} = s_i + s_{i+3} = \sum_{i=0}^{n-1} b_0 = s_{i+j} \tag{A.1}$$

where s_1 is the binary output sequence as shown, and its companion matrix or the state vectors, a_i is as follows:

$$a_{i+1} = \begin{bmatrix} a_{4_{i+1}} \\ a_{3_{i+1}} \\ a_{2_{i+1}} \\ a_{1_{i+1}} \end{bmatrix} = \begin{bmatrix} b_3 & b_2 & b_1 & b_0 \\ 1 & 0 & 0 & 0 \\ 0 & 1 & 0 & 0 \\ 0 & 0 & 1 & 0 \end{bmatrix} \begin{bmatrix} a_{4i} \\ a_{3i} \\ a_{2i} \\ a_{1i} \end{bmatrix} \overset{\Delta}{=} Ta_i$$

$$= \begin{bmatrix} 1 & 0 & 0 & 1 \\ 1 & 0 & 0 & 0 \\ 0 & 1 & 0 & 0 \\ 0 & 0 & 1 & 0 \end{bmatrix} a_i \tag{A.2}$$

The transition or companion matrix T can easily be seen to correspond to the state of the shift register of Fig. A.1.

It is easy to show that the PN sequence generated by the shift register of Fig. A.1 generates the same PN sequence as that of the sequence of binary 4-ruples taking, for example, the far left binary symbol from each vector representation binary 4-ruple) of α^1 in Table A.1]. These binary sequences are the same (aside room a possible delay offset), although the vector sequences may not be the same.

B. Shift and Add Property and GPS Code Selection

It is also easy to show the modulo 2 summation of two time-offset versions of the same maximal length sequence generates another offset version of the same sequence. This property is known alternatively as "Cycle and Add," "Shift and Add," or "Shift and Multiply." For example, taking an unshifted (no delay) version of the sequence α^i and adding to it another version of the sequence α^{i+1} we obtain a time-shifted version of the same sequence: $(1 + \alpha^k)\alpha^i = \alpha^k + \alpha^{i+k} = \alpha^j \mod[m_\alpha(x), 2]$ because $1 + \alpha^k$ is an element in GF(2^n), and the product of two elements in GF(2^n) yields another element in GF(2^n) for primitive polynomials.

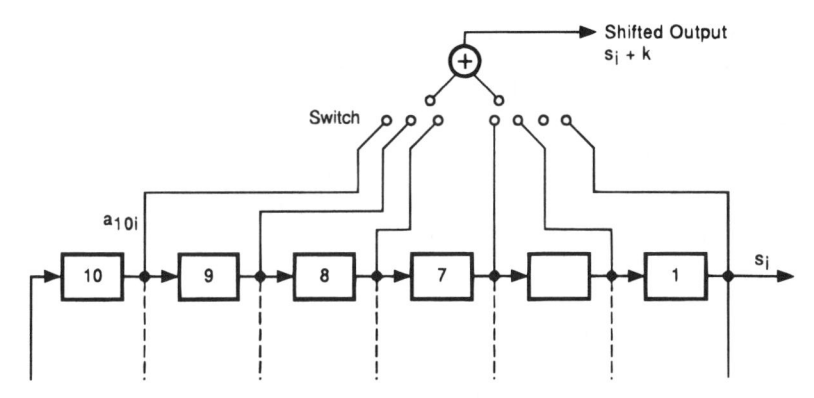

Fig. A.2 Generation of a time-shifted version of the global positioning systems PN sequence s_i using the shift and add spoperty of the PN sequence.

The GPS C/A codes are generated from two different maximal length shift registers each of 10 stages and each with a primitive polynomial. They each generate different PN sequences of length $2^n - 1 = 2^{10} - 1$. One of these employs a shift and add operation to delay one sequence with respect to the other, as shown earlier in Fig. 14. The switch shown has $n(n-1)/2 = 10(9)/2 = 45$ possible tap combinations (sampling without replacement) see Fig. A.2. The resulting delay offset for each of the tap combinations can easily be determined by computing the polynomials for each of the $2^{10} - 1 = 1023$ elements and identifying the $(\alpha^i + \alpha^1)\alpha^n = \alpha^j$ offset using algebra $\text{Mod}[m_\alpha(x), 2]$ to generate a table similar to that shown in Table A.1. Table 4 shown earlier gives the code phase selection laps j, k, for the section GPS Gold codes using this "Shift and Add" property. Clearly these 45 tap combinations only generate 45 of the 1025 possible Gold codes in this family. Although the satellite PN generator is fixed, it is easy to modify PN generators in the user receivers by adding modulo-2 adders, selecting a different initial state, or offsetting the delay by other means so that a larger number or all of the 1025 Gold codes can be generated with the same two shift registers.

C. Sequence Decimation

If α is a root of the primitive minimal polynomial $m_\alpha(x)$, then, as already stated, α^2 and α^{2^i} $\text{Mod}[m_\alpha(x)]$ are also. The α^{2^i} are conjugates (analogous to $+\sqrt{-1}$ and $\sqrt{--1}$ being complex conjugates, they are both roots of the same minimal polynomial $x^3 + 1 = 0$.[13]) of α. Thus, a decimated sequence α^{2^i} and α^{4^i}, etc., are also elements of $GF(2^n)$ and thus all produce exactly the same sequence, although perhaps time offset. Thus, decimation by 2: i.e., taking every other bit in a maximal length PN sequence or every 4th bit or very 2^{2^i} th bit gives exactly the same PN sequence except for a possible time offset. Taking other decimations; e.g., $\alpha^{3 \cdot 2^i}$ (cyclotomic cosets), on the other hand, gives other sequences (some of which are also maximal length), and these decimations correspond to other minimal polynomials: i.e., $m_\beta(x) = m_{\alpha^r}(x)$, where $\beta = \alpha^r$. For example, for the GF(2^4) polynomial, $m_\alpha = 1 + x + x^4$ the cyclotomic cosets and their minimal polynomials and powers of α are as shown in Table A.2.

Only cyclotomic cosets C_1 and C_7 correspond to primitive polynomials. Notice that from one primitive polynomial it is trivial to find one other. That is, the element α^{-1} corresponds to another primitive polynomial $m_{\alpha_{-1}}(x)$ simply because the order of its coefficients are reversed, as is the case for the C_1 and C_7 minimal polynomials.

D. Gold Codes

To generate the Gold codes, two PN sequences of length $2^n - 1$ are selected, which have a special relationship to one another. If one sequence generator has a polynomial $m_\alpha(x)$ with linear span n, the other should have a minimal polynomial $m_{\alpha^r}(x)$ where the exponent, $r = 2^{(n+2)/2} + 1$, for n even, and $r = 2^{(n+2)/2} + 1$, for n odd. Thus, for the PN sequence of GPS with polynomial $m_\alpha(x) = 1 + x^3 + x^{10}$, the question is what is the matching PN code minimal polynomial in order to form a Gold code? As a simple example, if the $m_\alpha(x) = 1 + x + x^4$, then the value of $r = 7$, and we need to find the minimal polynomial corresponding to $\beta \triangleq \alpha^7$. The elements in this new field $\text{Mod}[m_\alpha(x), 2]$ are then as follows:

$$\beta^0 = 1, \qquad \beta^3 = \alpha^{21} = \alpha^2 + \alpha^3,$$
$$\beta^1 = \alpha^7 = 1 + \alpha + \alpha^3, \qquad \beta^4 = \alpha^{28} = 1 + \alpha^2 + \alpha^3$$
$$\beta^2 = \alpha^4 = 1 + \alpha^3,$$

By definition, β is a root of the unknown minimal polynomial we are seeking. This polynomial can be expressed as $1 + b_1\beta + b_2\beta^2 + b_3\beta^3 + \beta^4 = 0$ and solved for the b_i to obtain $m_\beta(x) = 1 + x^3 + x^4$. This was the polynomial used in the earlier example of Table A.1. The same procedure results in the transformation for the GPS codes for $\beta = \alpha^{65}$ to find the Gold code polynomial pair for GPS. Select $m_\alpha(x) = 1 + x^3 + x^{10}$ as one polynomial, and the matching polynomial is easily computed to be $m_\beta(x) = 1 + x^2 + x^3 + x^6 + x^8 + x^9 + x^{10}$. This pair of polynomials, as shown earlier, is the pair used for the GPS Gold codes.

Table A.2 Power α^i and corresponding minimal polynomials for GF(2^4)

Cyclotomic coset	Powers of α	Minimal polynomial
C_1	(1, 2, 4, 8)	$1 + x + x^4$
C_7	(7, 14, 13, 11) = (7, 14, 28, 56 mod 15)	$1 + x^3 + x^4$
C_3	(3, 6, 12, 9)	$1 + x + x^2 + x^3 + x^4$
C_5	(5, 10)	$1 + x + x^2$
C_0	(0)	$1 + x$

References

[1]Anon., "Protection of Terrestrial Line of Sight Radio-Relay System Against Interference from the Broadcasting Satellite Service (Sound) in the Band 1427-1530 MHz," International Telecommunications Union, ITU, CCIR, International Radio Consultative Committee, CCIR-Rept. 941.

[2]Ponsonby, J. E. B., "Spectrum Management and the Impact of the GLONASS and GPS Satellite Systems on Radio Astronomy," *Journal of Royal Institute of Navigation*, Vol. 44, No. 3, 1992.

[3]Enge, P., and Sarwate, "SSMA Performance of Orthogonal Codes; Linear Receivers," *IEEE Transactions on Communication*, Dec. 1987.

[4]Viterbi, A. J., "On Dispelling the 'Factor of 3' Myth in Direct Sequence CDMA," private communication, 1993.

[5]Ivanov, N., Salichev, V., "Status, Plans, and Policy of the GLONASS for Civil Application," *Proceedings of ION GPS-91*, Institute of Navigation, Institute of Navigation, Washington, DC, Sept. 1991.

[6]Anon. *Interface Control Document GPS-ICD-200, with IRN-200B-PR001*, Department of the Air Force, July 1, 1992.

[7]Allan, D. W., and Dewey, W., "Time Domain Spectra of SA," Institute of Navigation, GPS-93, Sept. 1993.

[8]Gold, R., "Optimal Binary Sequences for Spread Spectrum Multiplexing," *IEEE Transactions on Information Theory*, Oct. 1967, pp. 619–621.

[9]Nagle, J. R., Van Dierendonck, A. J., and Hua, Q. D., "INMARSAT-3 Navigation Signal C/A Code Selection and Interference Analysis," *Navigation*, Vol. 39, Winter 1992–93.

[10]Kraus, J. D., *Antennas*, McGraw-Hill, New York, 1988.

[11]Spilker, J. J., Jr., *Digital Communications by Satellite*, Prentice Hall, Englewood Cliffs, NJ, 1977 and 1995, Chapter 17, "Satellite Timing Concepts," and Chapter 18, "Delay Lock Tracking of Pseudonoise Signals."

[12]Sarwate, and Pursley, M., "Cross-Correlation Properties of Pseudorandom and Related Sequences," *Proceedings of the IEEE*, New York, Institute of Electrical and Electronic Engineers, May 1980.

[13]Berlekamp, E. R., *Algebraic Coding Theory*, Aegean Park Press, Laguna Hills, CA, 1984.

[14]Spilker, J. J., Jr., and Magill, D. T., "The Delay Lock Discriminator—An Optimum Tracking Device," *Proceedings of the IRE*, Vol. 49, Sept. 1961, pp. 1404–1416.

[15]Gabor, 1953.

[16]Helstrom, C. W., "Statistical Theory of Signal Detection," Macmillan, New York, 1960, p. 214.

[17]Kay, S. M., "Statistical Signal Processing-Estimation Theory," Prentice Hall, Englewood Cliffs, NJ, 1993.

[18]White, S. C., and Beaulieu, N. C., "On the Application of the Cramer-Rao and Detection Theory Bounds to Mean Square Symbol Timing Recovery," *IEEE Transactions on Communication*, Oct. 1992.

[19]Golomb, S., *Shift Register Sequences*, Aegean Park Press, Laguna Hills, CA, 1982.

[20]Pless, V., *Introduction to the Theory of Error Correcting Codes*, Wiley, New York, 1989.

[21]Simon, M. K., et al., *Spread Spectrum Communications*, Vol. 1, Computer Science Press, Rockville, MD, 1985.

Chapter 4

GPS Navigation Data

J. J. Spilker Jr.*
Stanford Telecom, Sunnyvale, California 94089

I. Introduction

GPS signal carries with it data from the satellite that the user receiver needs to solve for position, velocity, and time. This chapter describes these GPS navigation data in some detail and gives the background analysis useful in understanding their functions. The data formats correspond to the ICD-GPS-200.[1] The first section gives a complete overview of the entire navigation data format and overall frame structure. The second section describes in some detail the format and algorithms of ICD-GPS-200 for each of the subframes. The final sections present some of the physical and mathematical bases for the algorithms of Sec. II. Detailed discussion of the effects of relativity, the ionosphere, and troposphere is reserved for later chapters. A table of physical constants used in this chapter and elsewhere is given in the Appendix.

A. Overall Message Content of the Navigation Data

The 50 bits/s datastream provides the user with several key sets of data required to obtain a satisfactory navigation, geodetic survey, or time transfer solution. These navigation data are uploaded to each satellite by the GPS Control Segment (CS) for later broadcast to the user. Uploads occur once per day, or more often, if needed to keep the user range error (URE) within specification. The navigation data provide the information shown in Table 1. The form of the information is also described in the table.

1. Perturbing Factors in the Navigation Measurements

Various perturbations affect the relationship between measurements made on the received signal and the true range to the satellite from the user. Some of these are discussed in the preceding chapter. The remaining effects are discussed here and in the next chapter. Let us begin by defining (refer to Chapter 2, this volume) the "true" geometric pseudorange $\rho_{iT}(t)$, which assumes perfect

---

Copyright © 1994 by the author. Published by the American Institute of Aeronautics and Astronautics, Inc., with permission. Released to AIAA to publish in all forms.
*Ph.D., Chairman of the Board.

Table 1 Requirements and characteristics of GPS navigation data

Requirement	Information provided by GPS navigation data
Precise satellite position at time of transmission	Satellite ephemeris using a modified Kepler model (sinusoidal perturbations) in an Earth–centered-inertial (ECI) coordinate frame with transformation to Earth-centered, Earth-fixed (ECEF) coordinates
Precise satellite time at time of transmission	Satellite clock error models and relativistic correction
P(Y) code[a] acquisition from C/A code[b]	A handover word (HOW) is transmitted that keeps track of the number of P(Y) code 1.5-s ($X1$ subsequence) periods thus far in the week. These data can aid in P(Y) code acquisition.
Select the best set of satellites for lowest appropriate GDOP[c] within elevation angle constraints (requires approximate knowledge of satellite position)	Moderate accuracy almanac that gives approximate position, time, and satellite health for the entire GPS constellation
Time transfer information	GPS time to universal coordinated time (UTC) time conversion data
Ionospheric corrections for single-frequency users	Approximate model of ionosphere vs time and user location
Quality of satellite signals/data	User range accuracy (URA)—a URA index N is transmitted that gives a quantized measure of space vehicle accuracy available to the civil (unauthorized) user

[a]P(Y) code, precise, antispoof code.
[b]C/A code, coarse/acquisition code.
[c]GDOP, Geometric Dilution of Precision.

knowledge of satellite clock time and position, assumes further the absence of any atmospheric propagation delay or relativistic effects, and uses a nonrotating ECI coordinate system. The "true" geometric pseudorange $\rho_{iT}(t)$ at time t for satellite i is then defined as follows:

$$\rho_{iT}(t) \stackrel{\Delta}{=} c(t_u - t_{si}) + cb_u = |\mathbf{x}_{si} - \mathbf{x}_u| + cb_u = |\mathbf{r}_{si} - \mathbf{r}_u| + cb_u = D_i + cb_u \tag{1}$$

where c is the speed of light, $\mathbf{x}_{si}(t - D_i(t)/c)$ is the true satellite position at the true time of transmission, $t - D_i(t)/c$, in ECI coordinates $\mathbf{x}_u(t)$ is the unknown user position also in ECI coordinates where t is the time of reception, and $t_u(t)/c$ is the user clock reading at the time of reception. Pseudorange is expressed in meters. User position is expressed either in Cartesian coordinates as $\mathbf{x}_u$ or in spherical coordinates as $\mathbf{r}_u$. The received satellite signal indicates the satellite clock time at the time of transmission* $t_{si}(t - D_i(t)/c)$, and $b_u(t)$ is the user clock time offset $b_u(t) = [t_u(t) - t_{GPS}(t)]$ expressed in seconds. The geometric transit time of the signal is $D_i(t)$ expressed in meters. The coordinates can be rotated to ECEF coordinates, but the rotation of the Earth during the transit time $D_i(t)$ must also be taken into account.

The true pseudorange $\rho_{iT}(t)$ is not directly an observable, but instead must be observed with various perturbations. The measured pseudorange $\rho_i(t)$ is equal to the true pseudorange plus various perturbing factors, as shown below [refer again to Chapter 2, Eq. (2)].

$$\rho_i = \rho_{iT} + \Delta D_i - c\Delta b_i + c(\Delta T_i + \Delta I_i + v_i + \Delta v_i) \tag{1}$$

where

Δb_i = satellite bias clock error, s
ΔD_i = satellite position bias error effect on range, m
v_i = receiver measurement noise error for satellite i, s
ΔI_i = ionospheric excess delay, s
ΔT_i = tropospheric excess delay, s
Δv_i = relativistic time correction, s

To be precise one also must account for small second-order effects caused by motion of the satellite during the time interval caused by ionospheric and tropospheric excess delay when computing the satellite position at time of transmission.

Each of the perturbations in the pseudorange measurement equation above must be either estimated, measured, or computed using the navigation datastream or other information. Satellite position information itself, of course, must be estimated from the navigation datastream. Table 2 identifies which type of data is involved for each perturbing parameter. Figure 1 (modified from ICD-GPS-200)[1] illustrates in specific terms how each of the corrections is applied to the pseudorange measurements. Each of these parameters is discussed either in this chapter or in other chapters of this book. Note that the tropospheric corrections are not discussed in ICD-GPS-200.[1] However, tropospheric correction models are discussed in detail in Chapter 13, this volume.

B. Navigation Data Subframe, Frame, and Superframe

The navigation data are transmitted in a 50 bits/s stream that is modulo-2 added to the C/A and P(Y) codes on the L_1 frequency and may or may not be carried on the L_2 P(Y) code, depending on the satellite mode. The data bit stream is synchronous with the 1-kHz C/A code epochs. The databits are formatted into 30-bit words, and the words are grouped into subframes of 10 words that are 300 bits in length and 6 s in duration. Frames (or pages) consist of 5 subframes of 1500 bits and 30 s in duration, and a superframe consists of 25 frames and has a duration 12.5 min. The general frame and subframe format is shown in Fig. 2. Much of the data repeat every frame and some data; e.g. the 8-bit preamble repeat every subframe. Periodically, the navigation dataframes are updated. New navigation data set (4-h curve fit for normal operations) cutovers occur only on

---

*The time indicated is the "proper" time indicated by the satellite clock at the time of transmission.

Table 2 Relationship between measured pseudorange and other parameters

	Parameters	Source of information for parameter
True pseudorange	$\rho_{iT} = |x_{si} - x_u| + cb_u$	Estimated by the GPS receiver
Pseudorange measurement	$\rho_i(t)$	Measured by GPS navigation receiver
Satellite position for satellite i	$x_{si}(t)$	Satellite position, calculated from the navigation data
User position	$x_u(t)$	Unknown user position, to be estimated
Ionospheric excess path delay	$\Delta I_i(t)$	Measured using dual-frequency measurements, or for single-frequency user, modeled using navigation data
Tropospheric excess path delay	$\Delta T_i(t)$	Estimated using approximate equations of various levels of accuracy, some of which rely on pressure and humidity measurements.
Satellite clock error	$\Delta b_i(t)$	Computed using navigation data. There can remain a residual error caused by selective availability for unauthorized users or users without differential GPS connection
User clock bias	b_u	Unknown user clock bias, to be estimated. Once estimated, it may vary only slowly depending on quality of user clock
Satellite position error bias error effect on range	$\Delta D_i(t)$	Unknown bias error cause by errors in GPS control segment estimate

one hour boundaries except for the first data set after a new upload, which may be cut in at any time during the hour. Block II satellite data sets for subframes 1, 2, and 3 are transmitted for periods of two hours before update. Block I satellite subframes 1, 2, and 3 are transmitted for periods of one hour before update.

Words 1 and 2 of each subframe are used for synchronization (preamble), handover word, and C/A code time ambiguity removal. The remaining words, 3–10, of subframe 1 provide clock correction information for the space vehicle clock and space vehicle health and user range accuracy (URA) measures. Subframes 2 and 3 contain ephemeris data that allow estimation of the transmitting satellite's position. Subframes 1, 2, and 3 have the same format from frame to frame, however, subframe 4 and 5 have 25 pages or different sets of data and contain the almanac that gives the approximate satellite ephemeris, clock correction, and space vehicle status for all of the satellites. The almanac data permit the user to select the best set of satellites or simply to determine which satellites are in view. Subframe 4 also contains ionospheric modeling and UTC-GPS clock correction information. A detailed view of the frame structure is shown in Fig. 3. Note that each 30-bit word includes six parity check bits that permit the user receiver to check for errors in the received datastream. A detailed discussion of each of these subframes in Fig. 3 can be found in Sec. II of this chapter, and their mathematical bases are given in the later sections.

GPS NAVIGATION DATA

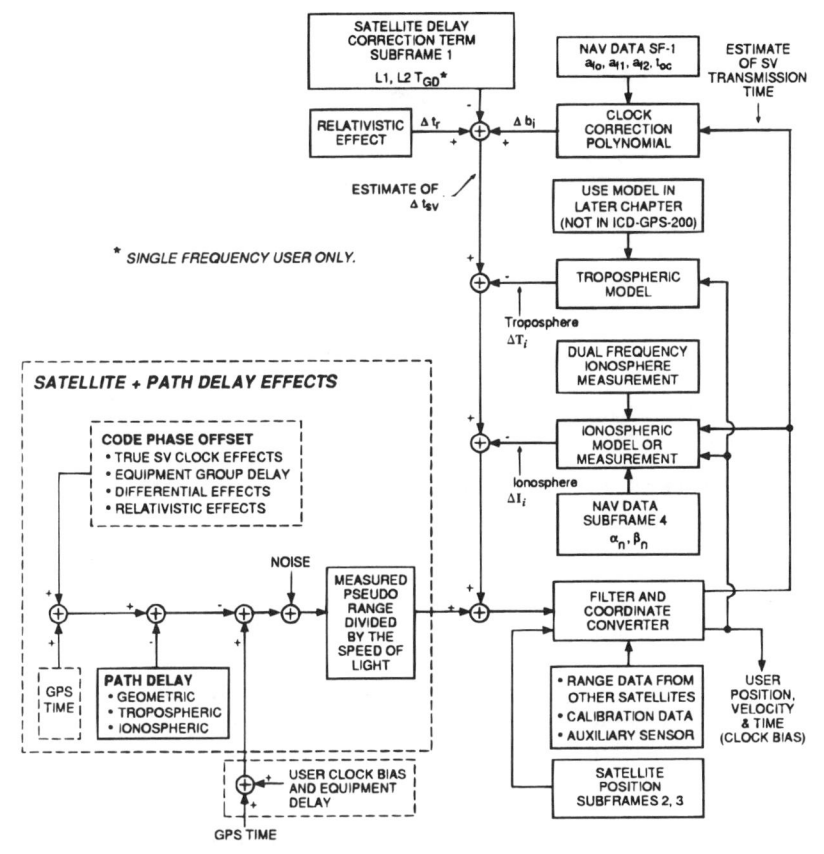

Fig. 1 Navigation data and correction parameters for pseudorange estimate from the pseudorange measurement. This diagram is a modified version of that contained in ICD-GPS-200.[1] The notation *SF* represents the navigation data subframe.

The subframes, frames, and 25-frame superframes are all synchronous with the 1.5-s, X1 epochs of the P code. Recall that the full P code has a 1-week period. The superframe also begins at the beginning of each week. Subframes begin at the beginning of the week and are numbered consecutively from the beginning of the week to aid in C/A to P(Y) code acquisition/handover. The timing relationships, shown in Fig. 4 and Fig. 5, illustrate the relationships between X1 epochs, time of week (TOW), handover word, Z-count, data bits, frames, and the chip durations.

The timing starts at the beginning of each week, defined as midnight Saturday night–Sunday morning in GPS time, which is referenced to UTC time kept by the U.S. Naval Observatory. GPS zero time point is defined as midnight on the night of January 5, 1980/morning of January 6, 1980. Note that the Z-count rolls over 1024 weeks later every 19+ years. UTC time is nominally referenced to time at the Greenwich meridian. GPS time differs from UTC in that GPS time does not exhibit the leap second that is sometimes inserted in UTC. GPS time,

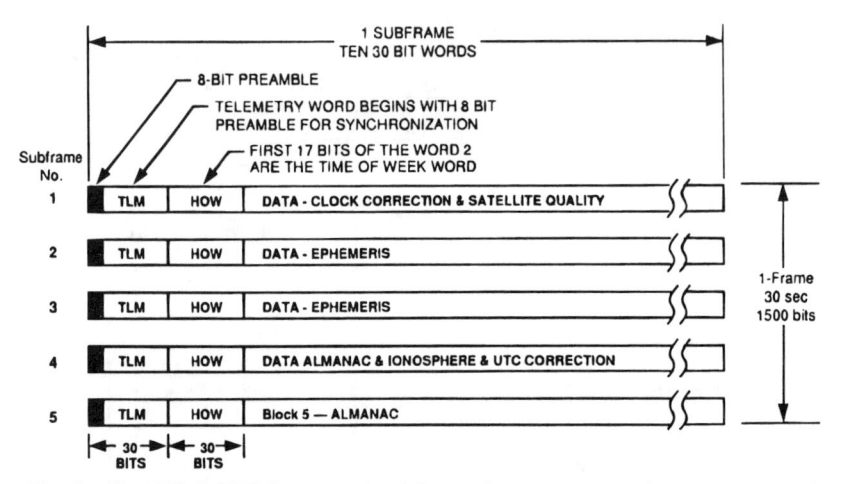

Fig. 2 Simplified GPS frame and subframe format. A superframe consists of 25 frames of 30 s each or 750 s or 12.5 min, and provides a complete almanac.

however, is kept to within 1 μs of UTC (modulo-1 s) by the GPS Control Segment. As years pass, the GPS time will differ from UTC by an integer number of seconds.

The number of $X1$ epochs (1.5 s each) since the GPS zero time point modulo-1024 weeks is a 29-bit number called the Z-count. The 19 least significant bits of the Z-count are referred to as the *time of week* (TOW) count, which is defined as the number of $X1$ epochs (1.5 s each) since the transition from the previous week. A truncated version, the 17 most significant bits of the TOW word is defined as the handover word, which ranges from 0 to 100,799, corresponds to the number of 6-s subframes since the beginning of the week, and is contained in the L-band downlink datastream. The HOW removes any timing ambiguity caused by the 1-ms C/A-code period and aids in C/A- to P-code handover/acquisition. The ten most significant digits of the 29-bit Z-count represent the number of weeks since the GPS zero time point.

As shown in Fig. 6, each 10-word subframe begins with a telemetry (TLM) word, which in turn begins with an 8-bit preamble, as shown for synchronization (a modified Barker sequence). Other parts of the TLM contain data needed by authorized users, as defined in GPS-ICD-205 and/or GPS-ICD-207.

1. Subframe Synchronization

The 8-bit modified Barker word at the beginning of the TLM word (or the 8-bit modified Barker word plus the two zeros in positions 29 and 30 of the HOW word) provides a synchronization pattern for subframe synchronization by the GPS receiver. However, random data patterns elsewhere in the frame can provide an identical pattern. Furthermore, there is a ± 1 sign ambiguity in demodulating any biphase modulated data signal. Thus, the 8-bit or the 8-bit plus 2-bit pattern still has a probability of a "false alarm" or random bit pattern match with the sync pattern of 2×2^{-8} or 2×2^{-10} for 1/128 or 1/512, respectively. Thus,

GPS NAVIGATION DATA

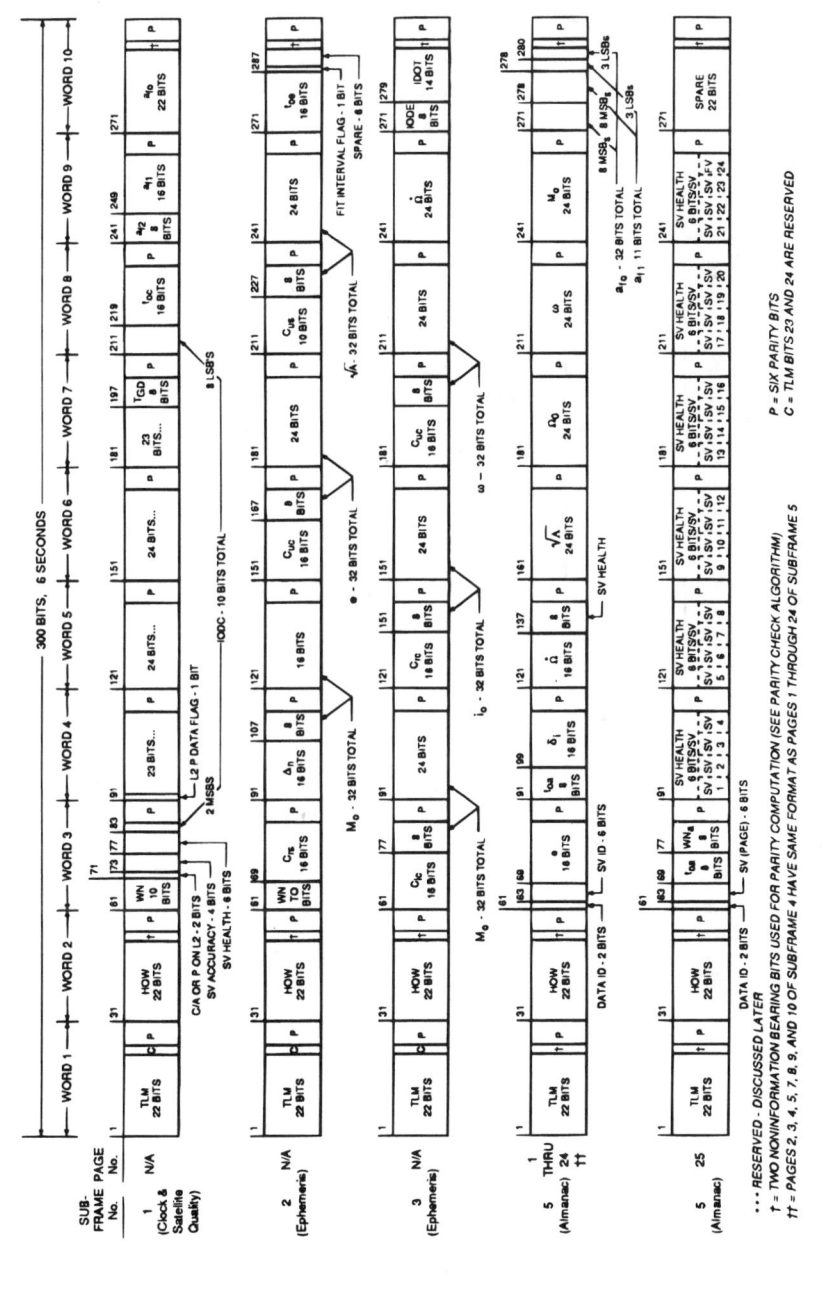

Figure 3 GPS data format from ICD-GPS-200.[1] Bit 1 is transmitted first. Within each word the most significant bits are transmitted first.

Figure 3 (continued) GPS data format from ICD-GPS-200.[1] Bit 1 is transmitted first. Within each word the most significant bits are transmitted first.

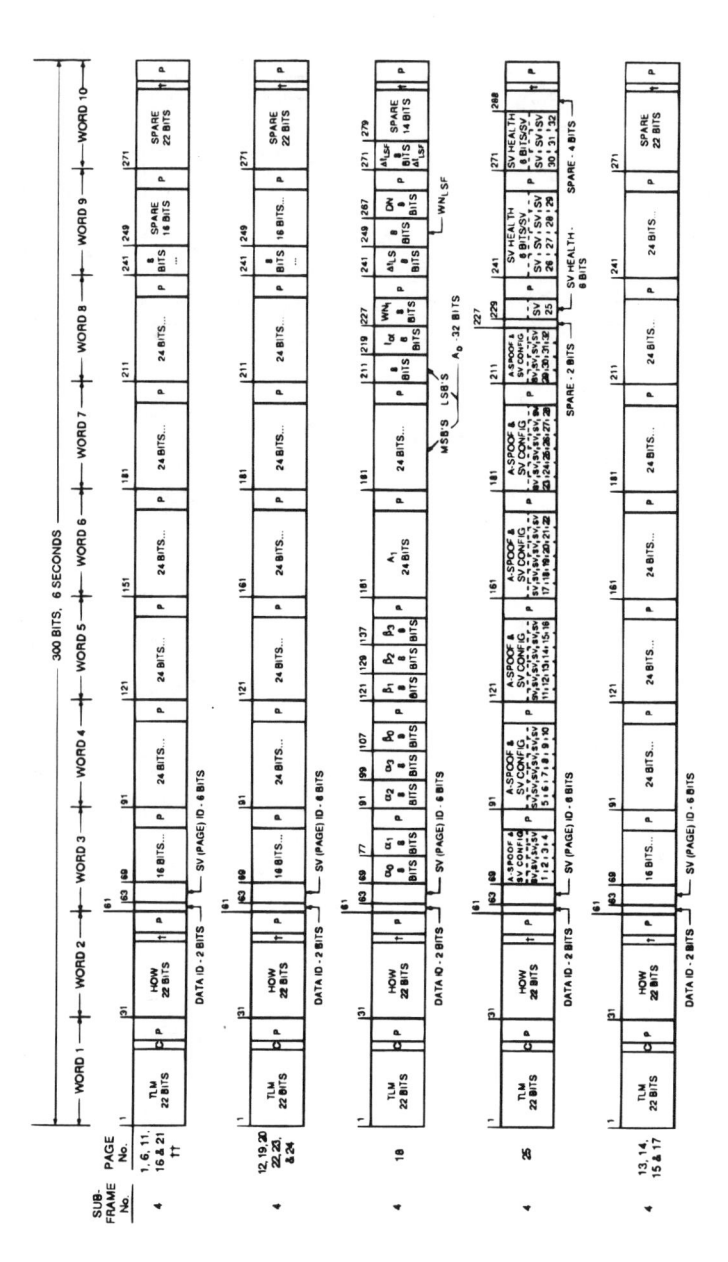

GPS NAVIGATION DATA

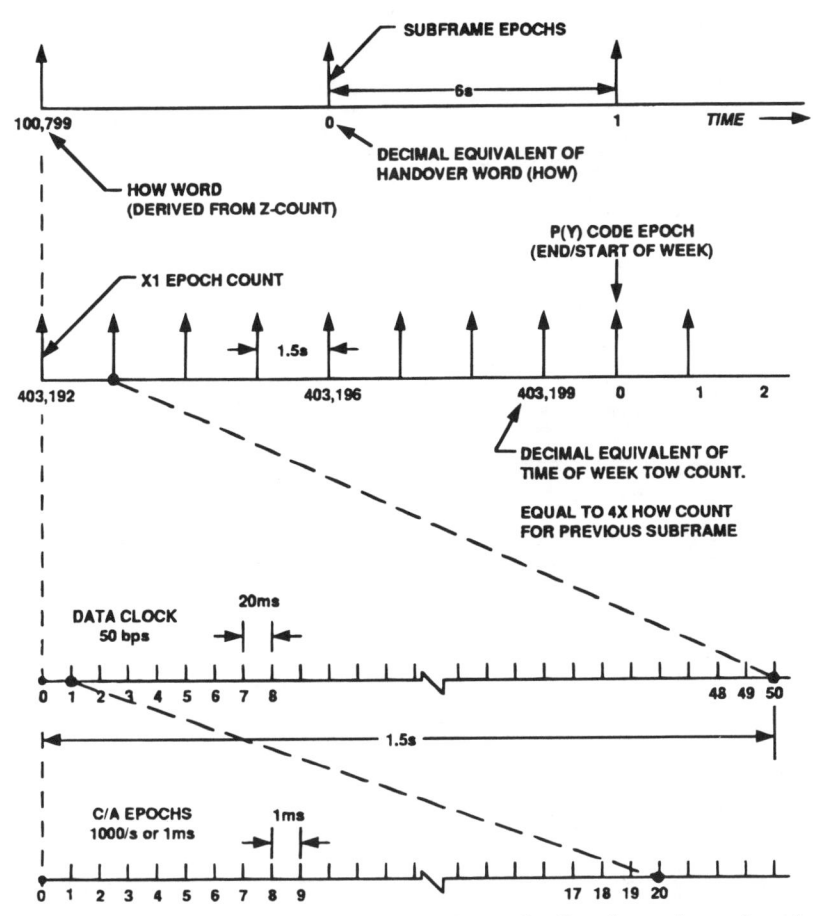

Fig. 4 Timing relationships between C/A-code epochs, P-code epochs, and navigation data.

the modified Barker word by itself yields too high a false alarm rate to be acceptable. However, we can also check the 17-bit truncated TOW message at the beginning of the HOW word to see that it increments by one and only one from subframe to subframe as a means for confirming the subframe synchronization.

2. Parity Check Algorithm

Although the navigation data are normally received at a relatively high signal-to-noise ratio and correspondingly low bit error rates $P_b < 10^{-5}$, it is important to have a parity check algorithm to reject words with any errors in them. Each 30-bit word plus the last two bits of the previous word is encoded into an extended Hamming (32,26) block code of $n = 32$ symbols and $k = 26$ "information" bits, where only 24 of the bits are true information bits d_i, $i \leq 24$.

If the parity transmission bits D_i (computed from the equation below and the H matrix of Fig. 7 or Table 3) do not match the six received parity bits D_i for

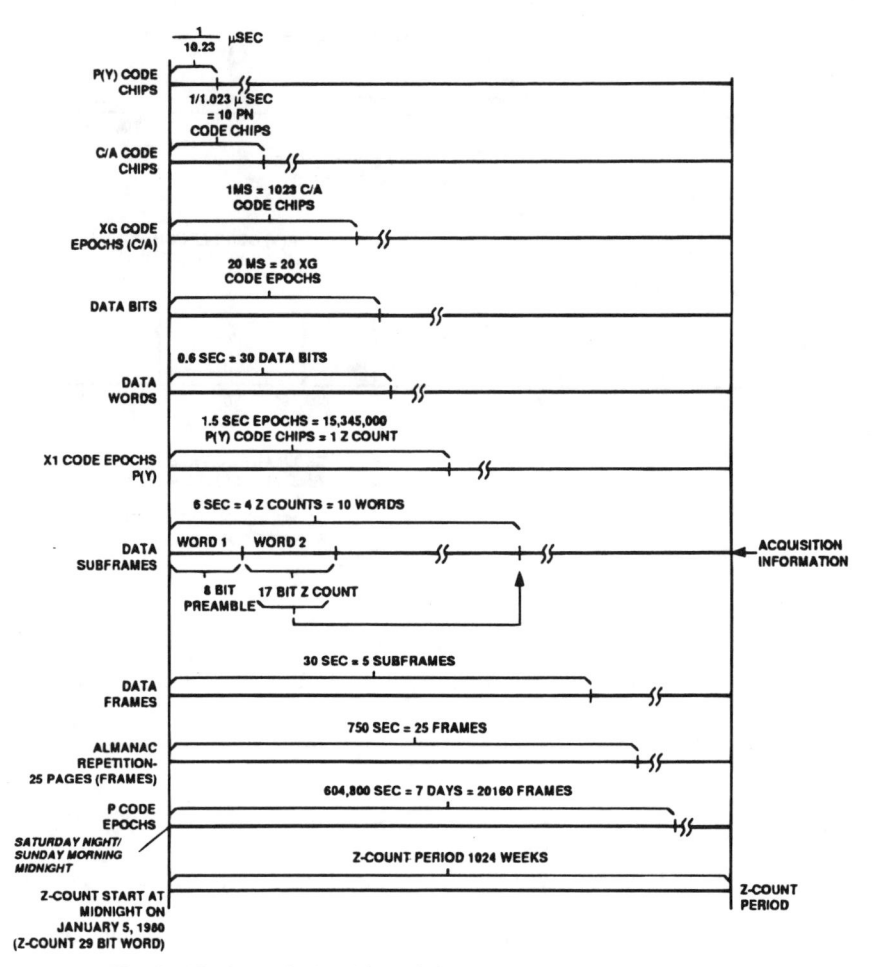

Fig. 5 Timing relationships of C/A codes, and data to Z-count.

D_{25}, D_{26}, D_{27}, D_{28}, D_{29}, D_{30}, then the information bits d_i for $i \leq 24$ are rejected. The parity check equation for the six parity check bits is $\boldsymbol{p} = \boldsymbol{Hd}$ where $\boldsymbol{H}$ is the (24 × 6) matrix of Fig. 7 and $D_i = d_i + D_{30}^*$ for $i \leq 24$, the received databit vector is $\boldsymbol{d} = (D_{29}^*, D_{30}^*, d_1, d_2, \ldots, d_{24})$, and $\boldsymbol{p}$ represents the vector $(D_{25}, D_{26}, D_{27}, D_{28}, D_{29}, D_{30})$.

In making this calculation, the $d_i = D_i \oplus D_{30}^*$, $i \leq 24$ are computed first, then the parity check bits D_i, for $25 \leq i \leq 30$ are computed and checked. In Fig. 7, note that each row in the parity matrix is a cyclic shift of the previous row, except for the last row. The last row is a check of all the previous rows. Note that the sum of all of the row vectors in the matrix is the all "1" row vector.

The Hamming code of length n with parameter m is of the form $(2^m - 1, 2^m - m - 1, 3) = (n, k, 3)$ is of minimum weight 3 for all m. For $m = 5$, the code is (31, 26, 3). The number of information bits in the code word is k. The

GPS NAVIGATION DATA

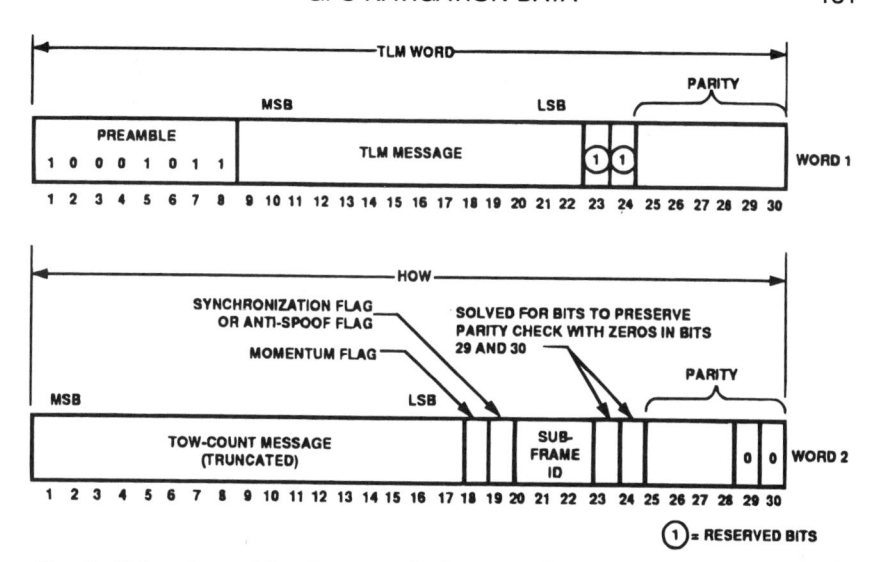

Fig. 6 Telemetry and handover words formats referenced from ICD-GPS-200.[1]

normal Hamming code (31,26) is a perfect code, which means that $2^{n-k} = 1 + n$, that is, $2^{31-26} = 2^5 = 1 + 31$ and has distance 3 for double error detection capability. It takes a minimum of three errors to cause an undetected error.

For GPS the distance 3 Hamming code is converted to a minimum distance 4 code by appending an added parity bit (the 32nd bit) that checks all of the other symbol bits including the other parity bits. In effect, this change adds the last row to the parity check matrix shown in Fig. 7. The code is then shortened to 30 bits by deleting two of the databits.

The GPS parity check code is an extended Hamming code (32,26) and has distance 4. Therefore, it takes certain patterns of four errors to cause an undetectable error.[2-4] If the error probability is p, then the probability of an undetectable error is approximately equal to $p_u \cong 1085\,p^4 - 29295\,p^5 + 403403\,p^6 + 0\,(p^7)$, and it is negligibly small for moderately low error probabilities $p < 10^{-3}$.

$$H = \begin{pmatrix} 1 & 1 & 1 & 0 & 1 & 0 & 0 & 0 & 1 & 1 & 1 & 1 & 1 & 0 & 0 & 1 & 1 & 0 & 1 & 0 & 0 & 1 & 0 \\ 0 & 1 & 1 & 1 & 0 & 1 & 1 & 0 & 0 & 0 & 1 & 1 & 1 & 1 & 1 & 0 & 0 & 1 & 1 & 0 & 1 & 0 & 0 & 1 \\ 1 & 0 & 1 & 1 & 1 & 0 & 1 & 1 & 0 & 0 & 0 & 1 & 1 & 1 & 1 & 1 & 0 & 0 & 1 & 1 & 0 & 1 & 0 & 0 \\ 0 & 1 & 0 & 1 & 1 & 1 & 0 & 1 & 1 & 0 & 0 & 0 & 1 & 1 & 1 & 1 & 1 & 0 & 0 & 1 & 1 & 0 & 1 & 0 \\ 1 & 0 & 1 & 0 & 1 & 1 & 1 & 0 & 1 & 1 & 0 & 0 & 0 & 1 & 1 & 1 & 1 & 1 & 0 & 0 & 1 & 1 & 0 & 1 \\ 0 & 0 & 1 & 0 & 1 & 1 & 0 & 1 & 1 & 1 & 1 & 0 & 1 & 0 & 1 & 0 & 0 & 0 & 1 & 0 & 0 & 1 & 1 & 1 \end{pmatrix}$$

Fig. 7 Parity matrix H for the extended Hamming (32,26) code where $d_1, d_2, \ldots d_{24}$ are the source databits and $D_1, D_2, \ldots D_{30}$ are the bits transmitted by the global positioning system satellite. The notation $D_{29}*, D_{30}*$ represents the last 2 bits transmitted in the previous 30-bit word. Note that each row in H is simply a cyclic shift of the previous row except for the row D_{30}.

Table 3 Parity encoding equations for each 30-bit word (from GPS-ICD-200)[1]

$D_1 = d_1 \oplus D_{30}*$
$D_2 = d_2 \oplus D_{30}*$
$D_3 = d_3 \oplus D_{30}*$
$\cdot \quad \cdot$
$\cdot \quad \cdot$
$\cdot \quad \cdot$
$D_{24} = d_{24} \oplus D_{30}*$
$D_{25} = D_{29}* \oplus d_1 \oplus d_2 \oplus d_3 \oplus d_5 \oplus d_6 \oplus d_{10} \oplus d_{11} \oplus d_{12} \oplus d_{13} \oplus d_{14} \oplus d_{17} \oplus d_{18} \oplus d_{20} \oplus d_{23}$
$D_{26} = D_{30}* \oplus d_2 \oplus d_3 \oplus d_4 \oplus d_6 \oplus d_7 \oplus d_{11} \oplus d_{12} \oplus d_{13} \oplus d_{14} \oplus d_{15} \oplus d_{18} \oplus d_{19} \oplus d_{21} \oplus d_{24}$
$D_{27} = D_{29}* \oplus d_1 \oplus d_3 \oplus d_4 \oplus d_5 \oplus d_7 \oplus d_8 \oplus d_{12} \oplus d_{13} \oplus d_{14} \oplus d_{15} \oplus d_{16} \oplus d_{19} \oplus d_{20} \oplus d_{22}$
$D_{28} = D_{30}* \oplus d_2 \oplus d_4 \oplus d_5 \oplus d_6 \oplus d_8 \oplus d_9 \oplus d_{13} \oplus d_{14} \oplus d_{15} \oplus d_{16} \oplus d_{17} \oplus d_{20} \oplus d_{21} \oplus d_{23}$
$D_{29} = D_{30}* \oplus d_1 \oplus d_3 \oplus d_5 \oplus d_6 \oplus d_7 \oplus d_9 \oplus d_{10} \oplus d_{14} \oplus d_{15} \oplus d_{16} \oplus d_{17} \oplus d_{18} \oplus d_{21} \oplus d_{22} \oplus d_{24}$
$D_{30} = D_{29}* \oplus d_3 \oplus d_5 \oplus d_6 \oplus d_8 \oplus d_9 \oplus d_{10} \oplus d_{11} \oplus d_{13} \oplus d_{15} \oplus d_{19} \oplus d_{22} \oplus d_{23} \oplus d_{24}$
where $d_1, d_2, \ldots d_{24}$, are the source data bits;
the symbol (*) is used to identify the last 2 bits of the previous word of the subframe;
$D_{25}, \ldots D_{30}$ are the computed parity bits;
$D_1, D_2, D_3, \ldots D_{29}, D_{30}$, are the bits transmitted by the space vehicle (SV);
and $\oplus$ is the modulo-2 or "exclusive-or" operation.

II. Detailed Description of the Navigation Data Subframes

The previous section defined the overall format of the 50-bs navigation data. This section describes in detail each of the elements of the navigation data for each of the five subframes. The next section provides some of the analytical background for the satellite clock and ephemeris calculations.

A. Subframe 1—GPS Clock Correction and Space Vehicle Accuracy Measure

Subframe 1 contains the data to be used in the algorithms described below to provide the space vehicle clock correction. It also contains data to give an estimate of the effect of space vehicle accuracy on user range accuracy (URA).

1. GPS Clock Correction Data Formats—Subframe 1

The user receiver needs to correct the GPS satellite clock errors. The user receiver must have an accurate representation of GPS system time t at the time of transmission for the GPS signal it now is receiving from satellite i. The satellite clock correction Δt_{sv} is obtained using coefficients broadcast from the satellite after being uploaded by the GPS control segment. The control segment actually uploads several different sets of coefficients to the satellite, of which each set is valid over a given time period. The data sets are then transmitted in the downlink datastream to the users in the appropriate time intervals. Subframe 1, words 8, 9, 10, shown previously in Fig. 3, contain the data needed by the users to perform corrections of the space vehicle clock. These corrections represent a second-order polynomial in time. Specifically, bits 9–24 of word 8, bits 1–24 of word

9, and bits 1–22 of word 10 provide four clock correction parameters, t_{oc}, a_{f2}, a_{f1}, a_{f0}, which are described in the following paragraphs.

The GPS time t (the space vehicle SV clock time) needed to solve for user position is $t = t_{SV} - \Delta t_{SV}$, where t_{SV} is the SV pseudorandom noise (PRN) code phase time at the time of transmission and is easily determined by the GPS receiver. The satellite clock correction term is approximated by a polynomial $\Delta t_{SV} = a_{f0} + a_{f1}(t - t_{oc}) + a_{f2}(t - t_{oc})^2 + \Delta t_R$, where a_{f0}, a_{f1}, and a_{f2} are the polynomial correction coefficients corresponding to phase error, frequency error, and rate of change of frequency error; the relativistic correction is Δt_R; and t_{oc} is a reference time (in s) for clock correction. Table 4 describes the parameters in number of bits, scale factors, and units.

The relativistic correction must be computed by the user. A first-order effect described in the GPS ICD[1] gives the relativistic correction for an Earth-centered, Earth-fixed (ECEF) observer and a GPS satellite of orbit eccentricity e. This relativistic correction varies as the sine of the satellite eccentric anomaly E_k as follows:

$$\Delta t_R = F e \sqrt{A} \sin E_k = 2 \mathbf{R} \cdot \mathbf{V}/c^2$$

where

$F = -2\sqrt{\mu}/c^2 = -4.442807633 \times 10^{-10}$ $s/m^{1/2}$
$\mu = 3.986005 \times 10^{14}$ m^3/s^2 value of Earth universal gravitational parameter
$c = 2.99792458 \times 10^8$ m/s
$\mathbf{R}$ = instantaneous position vector of the space vehicle
$\mathbf{V}$ = instantaneous velocity vector of the space vehicle
e = space vehicle orbit eccentricity

Table 4 Subframe 1 parameters for clock correction and other data

Parameter	No. of bits	Scale factor, LSB[a]	Effective range[b]	Units
Code on L_2	2	1	—	N/A
Week no.	10	1	—	Week
L_2 P data flag	1	1	—	Discretes
SV accuracy	4	—	—	(see text)
SV health	6	1	—	N/A
T_{GD}	8[c]	2^{-31}	—	s
Issue of data clock (IODC)	10	—	—	(see text)
t_{OC}	16	2^4	604,784	s
a_{f2}	8[c]	2^{-55}	—	s/s^2
a_{f1}	16[c]	2^{-43}	—	s/s
a_{f0}	22[c]	2^{-31}	—	s

[a]Least significant bits.
[b]Unless otherwise indicated in this column, effective range is the maximum range attainable with indicated bit allocation and scale factor.
[c]Parameters so indicated shall be two's complement, with the sign bit (+ or −) occupying the most significant bits (MSB).

E_k = eccentric anomaly of the satellite orbit
A = semimajor axis of the satellite orbit

R and V are expressed in the same inertial coordinate system. Chapter 18, this volume, discusses the relativistic effects and the derivation of this equation in more detail.

In addition, as discussed in Chapter 18, the other relativistic effects are as follows:

1) Increase in the received clock frequency by a fixed user on the surface of the Earth's geoid by a fraction $\Delta f/f = +4.46 \times 10^{-10}$. This effect is compensated by purposely setting the 10.23 MHz satelite clock low by $\Delta f = +4.56 \times 10^{-3}$ Hz. Thus, the satellite clock is set to 10,229,999.99543 Hz before launch to ensure that the received GPS signals arrive at the Earth geoid at the correct frequencies.

2) We may also have to account for any significant velocity of the user relative to the Earth or displacement in altitude (gravitational potential) from the surface of the geoid. For example, if the user is at an altitude above the geoid, the fractional increase in the received satellite frequency is not as large. Some of these effects simply may be accounted for by a modification in the user clock bias offset and may not significantly affect user position estimates, because they are approximately the same for all satellites.

We must also account for the rotation of the Earth during the time of transit of the GPS signals from satellite to user. It has already been pointed out that the satellite position has been computed at the time of transmission, whereas the user receiver is computing position at a slightly later time. The Earth has rotated during this transit time, and this rotation must be taken into account. Note that these times of transit for different satellites are not all identical. This effect is a simple effect of the finite velocity of light.

2. $L_1 - L_2$ Correction—Single-Frequency Users

The $L_1 - L_2$ delay correction term is calculated by the GPS Control Segment (CS) to account for the group delay difference in the space vehicle transmission between L_1 and L_2 signals based on measurements made on the SV prior to launch in the factory test. The GPS CS uses a two-frequency ionospheric correction and estimates the a_{fo} satellite clock correction term based on the dual-frequency measurement. Thus, the user employing both L_1 and L_2 in the ionospheric correction need make no further correction. However, the user who employs only L_1 must modify the space vehicle clock correction by $(\Delta t_{sv})_{L_1} = \Delta t_{sv} - T_{GD}$, where T_{GD} is provided in the subframe 1 data by bits 17–24 of word 7 (see Table 4). For the user who employs only L_2 the space vehicle clock correction is $(\Delta t_{sv})_{L_2} = \Delta t_{sv} - \Gamma T_{GD}$ where $\Gamma = (f_{L_1}/f_{L_2})^2 = (1575.42/1227.6)^2 = (77/60)^2$.

The value of the correction term T_{GD} is not equal to the SV group delay differential but rather $T_{GD} = (t_{L_1} - t_{L_2})/(1 - \Gamma)$, where $t_{L_1} - t_{L_2}$ is the SV differential group delay for the satellite. Thus, the value of T_{GD} is not equal to the mean SV group delay differential but to that delay multiplied by $1/(1 - \Gamma)$.

3. Subframe 1—Space Vehicle Accuracy—User Range Accuracy Index

Bits 13 through 16 of word 3 subframe 1 give the user range accuracy (URA) index, of the SV for the user who does not have access to the full accuracy of

Table 5 Table of user range accuracy index N, vs the user range accuracy interval in meters

URA index	URA, m
0	0.0–2.4
1	2.4–3.4
2	3.4–4.85
3	4.85–6.85
4	6.85–9.65
5	9.65–13.65
6	13.65–24.0
7	24.0–48.0
8	48.0–96.0
9	96.0–192.0
10	192.0–384.0
11	384.0–768.0
12	768.0–1536.0
13	1536.0–3072.0
14	3072.0–6144.0
15	>6144.0[a]

[a](No accuracy prediction is available. Unauthorized users are advised to use the SV at their own risk.)

Table 6 Ephemeris data definitions[1]

Symbol	Definition
M_0	Mean anomaly at reference time
Δ_n	Mean motion difference from computed value
e	Eccentricity of the orbit
$(A)^{1/2}$	Square root of the semimajor axis
$(OMEGA)_0$	Longitude of ascending node of orbit plane at reference time
I_0	Inclination angle at reference time
ω	Argument of perigee
OMEGADOT	Rate of right ascension
IDOT	Rate of inclination angle
C_{uc}	Amplitude of the cosine harmonic correction term to the argument of latitude
C_{us}	Amplitude of the sine harmonic correction term to the argument of latitude
C_{rc}	Amplitude of the cosine harmonic correction term to the orbit radius
C_{rs}	Amplitude of the sine harmonic correction term to the orbit radius
C_{ic}	Amplitude of the cosine harmonic correction term to the angle of inclination
C_{is}	Amplitude of the sine harmonic correction term to the angle of inclination
T_{oe}	Reference time for ephemeris
IODE	Issue of data (ephemeris)

GPS. (These nonmilitary users are termed the unauthorized users in ICD-GPS-200.[1]) The URA itself (as opposed to the index) is given in meters. The URA index N is an integer in the range of 0–15 and has the relationship to the URA of the SV shown in Table 5.

4. Issue of Data-Clock

The issue of data clock (IODC) indicates the issue number of the data set for clock correction, which provides a means for detecting any change in the clock correction parameters. This information is carried in bits 23 and 24 of word 3 MSB and bits 1–8 of word 8 in subframe 1.

B. GPS Ephemeris Parameters—Subframes 2 and 3

The purely elliptical Kepler orbit is precise only for a simple two-body problem where the mutual gravitational attraction between the two bodies is the only force involved. In the actual GPS satellite orbit, there are many perturbations to the ideal orbit, including nonspherical Earth gravitational harmonics; lunar, solar gravitational attraction; and solar flux. Thus, the GPS orbit is modeled as a modified elliptical orbit with correction terms to account for these perturbations: 1) sin, cos perturbations to the a) argument of latitude, b) orbit radius, and c) angle of inclination; and 2) rate of change of a) right ascension, and b) inclination angle.

Furthermore, the parameters for this model are changed periodically to give a best fit to the actual satellite orbit. In normal operations, the fit interval is 4 hours. Subframes 2 and 3 provide 375 bits of information for the modified Keplerian model. Table 6 shows ephemeris model parameters including the sinusoidal perturbations to the orbit radius, the angle of inclination and argument of latitude; the rate of change of inclination angle, angular rate of change of the right ascension; and the basic Keplerian parameters. The scale factors for these parameters are given in Table 7.

1. Calculation of Satellite Position

By demodulating and extracting the navigation data in subframes 2 and 3, the user can calculate the satellite position vs. time. The issue of data-ephemeris IODE is a number provided in both subframes 2 and 3 for purposes of comparison and for comparison with the 8 LSB of the IODC term in subframe 1. It should be pointed out that the two IODE numbers in subframes 2 and 3 must match and should also correspond to the IODC for the clock in subframe 1; otherwise, a data set cutover has occurred, and the user must collect new data.

2. Curve Fit Intervals for the Ephemeris Data

Bit 17 in word 10 of subframe 2 is a fit interval flag that indicates whether the GPS CS used a least squares fit over a 4-h period or a longer 6-h period; a "0" bit is transmitted for fit periods greater than 4 h. For data sets with a 4-h fit interval (transmitted during the first approximately 1-day period after upload), the curve fit procedures provide a URE contribution for the predicted SV ephemeris of

GPS NAVIGATION DATA

Table 7 Ephemeris parameters[1]

Parameter	Number of bits[a]	Scale factor, LSB	Effective range[b]	Units
IODE	8	—	—	(See text)
C_{rs}	16[c]	2^{-5}	—	m
Δn	16[c]	2^{-43}	—	Semicircles
M_0	32[c]	2^{-31}	—	Semicircles
C_{uc}	16[c]	2^{-29}	—	rad
e	32	2^{-33}	0.03	Dimensionless
C_{us}	16[c]	2^{-29}	—	rad
$(A)^{1/2}$	32	2^{-19}	—	$m^{1/2}$
t_{oe}	16	2^{4}	604,784	s
C_{ic}	16[c]	2^{-29}	—	rad
$(OMEGA)_0$	32[c]	2^{-31}	—	Semicircles
C_{is}	16[c]	2^{-29}	—	rad
i_0	32[c]	2^{-31}	—	Semicircles
C_{rc}	16[c]	2^{-5}	—	m
ω	32[c]	2^{-31}	—	Semicircles
OMEGADOT	24[c]	2^{-43}	—	Semicircles/s
IDOT	14[c]	2^{-43}	—	Semicircles/s

[a]See Fig. 3 for complete bit allocation in subframe.
[b]Unless otherwise indicated in this column, effective range is the maximum range attainable with indicated bit allocation and scale factor.
[c]Parameters so indicated shall be two's complement, with the sign bit (+ or −) occupying the MSB.

less than 0.35 m, one sigma. These URE component values apply when the data set is transmitted, as well as for a period of 3 h thereafter. The longer, less accurate, 6-h fit interval normally is not used. It is employed if the upload does not occur daily, and the same uploaded data set must apply for a 2nd day through the 14th day after upload. For data sets with a 6-h fit interval, the curve fit provides a URE of less than 1.5 m, one sigma. These URE values apply during transmission and for 2 h thereafter.

The equations in Table 8 give the space vehicle antenna phase center position in WGS-84[5] Earth-centered, Earth-fixed reference frame (including correction for the Earth's rotation with the x'_k to x_k matrix transformation). The ECEF coordinate system is defined as follows:

Origin = Earth center of mass (Geometric center of the WGS-84[5] ellipsoid)
z axis = Parallel to the direction of the Conventional International Origin (CIO) for polar motion as defined by the Bureau International de l'Heure (BIH) on the basis of the latitudes adopted for the BIH stations (Rotation axis of the WGS-84[5] ellipsoid)
x axis = Intersection of the WGS-84[5] reference meridian plane and the plane of the mean astronomic equator, the reference meridian being parallel to the zero meridian defined by the Bureau International de l'Heure on the basis of the longitudes adopted for the BIH stations
y axis = Completes a right-handed, Earth-centered, Earth-fixed orthogonal coordinate system measured in the plane of the mean astronomic equator 90° east of the *x* axis (*x, y* axis of the WGS-84[5] ellipsoid)

Table 8 Elements of ephemeris model equations[1]

$\mu = 3.986005 \times 10^{14}$ m³/s²	WGS-84[5] value of the Earth's universal gravitational parameter
$\dot{\Omega}_e = 7.2921151467 \times 10^{-5}$ rad/s	WGS-84[5] value of the Earth's rotation rate
$A = (\sqrt{A})^2$	Semimajor axis
$n_0 = \sqrt{\mu/A^3}$	Computed mean motion–rad/s
$t_k = t - t_{oe}$ [a]	Time from ephemeris reference epoch
$n = n_0 + \Delta n$	Corrected mean motion
$M_k = M_o + n t_k$	Mean anomaly
$\pi = 3.1415926535898$	GPS standard value for π
$M_k = E_k - e \sin E_k$	Kepler's equation for the eccentric anomaly E_k (may be solved by iteration), rad
$v_k = \tan^{-1}\left\{\dfrac{\sin v_k}{\cos v_k}\right\} = \tan^{-1}\left\{\dfrac{\sqrt{1-e^2}\sin E_k/(1-e\cos E_k)}{(\cos E_k - e)/(1 - e\cos E_k)}\right\}$	True anomaly v_k as a function of the eccentric anomaly
$E_k = \cos^{-1}\left\{\dfrac{e + \cos v_k}{1 + e\cos v_k}\right\}$	Eccentric anomaly
$\Phi_k = v_k + \omega$	Argument of latitude
$\delta u_k = C_{us} \sin 2\Phi_k + C_{uc} \cos 2\Phi_k$	Argument of latitude correction ⎫ Second
$\delta r_k = C_{rs} \sin 2\Phi_k + C_{rc} \cos 2\Phi_k$	Radius correction ⎬ harmonic
$\delta i_k = C_{is} \sin 2\Phi_k + C_{ic} \cos 2\Phi_k$	Inclination correction ⎭ perturbations
$u_k = \Phi_k + \delta u_k$	Corrected argument of latitude
$r_k = A(1 - e\cos E_k) + \delta r_k$	Corrected radius
$i_k = i_0 + \delta i_k + (\text{IDOT}) t_k$	Corrected inclination
$x_k' = r_k \cos u_k$ $y_k' = r_k \sin u_k$	Satellite position in orbital plane
$\Omega_k = \Omega_o + (\dot{\Omega} - \dot{\Omega}_e) t_k - \dot{\Omega}_e t_{oe}$	Corrected longitude of ascending node
$x_k = x_k' \cos \Omega_k - y_k' \cos i_k \sin \Omega_k$ $y_k = y_k' \sin \Omega_k + y_k' \cos i_k \cos \Omega_k$ $z_k = y_k' \sin i_k$	Satellite position in Earth-centered, Earth-fixed coordinates

[a] t is GPS system time at time of transmission; i.e., GPS time corrected for transit time (range/speed of light). Furthermore, t_k shall be the actual total time difference between the time t and the epoch time t_{oe} and must account for beginning or end of week crossovers. That is, if t_k is greater than 302,400 s, subtract 604,800 s from t_k. If t_k is less than 302,400 s, add 604,800 s to t_k.

From Table 8 note that it is the mean anomaly $M_k = M_0 + n t_k$ that varies linearly with the time interval t_k. However, the solution of the satellite position requires knowledge of the eccentric anomaly E_k, which does not vary linearly with time unless the eccentricity $e = 0$.

The eccentric anomaly E_k must be solved for by iterative calculations not given in Table 8. However, some of these techniques are briefly discussed in Sec. III of this chapter. Note again, that the model of Table 8 is not merely a simple elliptical orbit. Second harmonic (sinusoidal) corrections are made for the argument of latitude, radius, and inclination of the Kepler model for satellite position. These corrections are then introduced to provide the corrected position for the satellite in the orbital plane, the corrected radius, argument of latitude, and inclination. Finally, the x, y, z coordinates for the satellite position are transferred to Earth-centered, Earth-fixed coordinates to be used in the final computations of the user position.

3. Geometric Range

The ICD-GPS-200[1] also states "The user must also account for the effects due to Earth rotation during the time of signal propagation so as to evaluate the path delay in an inertially stable system." This effect is discussed later in Chapter 18 on relativistic effects.

C. Subframes 4 and 5—Almanac, Space Vehicle Health, and Ionosphere Models

1. Almanac Data

Almanac data are used for satellite selection purposes and as aids to acquisition; the almanac can also be used to give approximate Doppler and delay information. Almanac data are used by P(Y) code users in order to perform direct P(Y) code acquisition (if they choose not to acquire the C/A signal first and then handover to the P(Y) code). The almanac data provides a truncated, reduced precision set of the ephemeris parameters described earlier in Table 6. Almanac data provide approximate ephemeris information for up to 32 satellites along with the associated health data for each satellite. Almanacs are provided only for valid satellites or perhaps for a satellite that is about to become active. Where there is no satellite data to fill an almanac data slot, dummy alternating "0s" and "1s" are transmitted to aid in synchronization.

Subframes 4 and 5 each carry 25 pages of information, one new page per frame repetition. We term a 25-frame segment a superframe. Thus, a GPS receiver must demodulate 25 frames over a period of 25×30 s or 12.5 min in order to receive all 25 pages of the subframe 4 and 5 almanac data. Of particular interest are the pages shown in Table 9.

Table 9 Key elements of pages in subframes 4 and 5

Pages	Subframe 4	Pages	Subframe 5
2,3,4,5,7,8,9,10	Almanac data for SV 25 through 32	1–24	Almanac data for SV 1–24
18	Ionosphere and UTC data	25	SV health for SV 1–24, almanac reference time and reference week number
25	Anti-spoof flag SV configuration for 32 SV, SV health for SV 25–32	—	—
Other pages	Reserved + special messages, spares	—	—

Table 10a Almanac parameters and accuracy

Age of data time since transmission	Almanac accuracy
1 day	900 m
1 week	1200 m
2 weeks	3600 m

The almanac data are much less accurate than the detailed ephemeris data of subframes 2 and 3. However, the almanac data are valid for longer periods of time and do not require frequent updates. Approximate one sigma almanac accuracy varies as a function of the time since the time of transmission approximately as shown in Table 10a.

The almanac parameters and their scale factors are shown in Table 10b. The algorithm for these parameters is the same ephemeris algorithm as discussed for subframes 2 and 3. Where the almanac does not include a parameter; e.g., sinusoidal corrections, these parameters are set to zero. For the inclination angle, a nominal value of 0.30 semicircles is implicit, and only a parameter δ_i, the correction to the inclination, is transmitted.

In addition, the almanac provides truncated clock correction a_{f0}, a_{f1} parameters for the algorithm discussed in subframe 1. The Almanac time correction provides time to within 2 μs of GPS time using the first order polynomial $t = t_{SV} - \Delta t_{SV}$ where t is GPS time, t_{SV} is the space vehicle clock time (PRN code phase at message transmission time) and $\Delta t_{SV} = a_{f0} + a_{f1} t_k$, where t_k is the time from epoch.

The almanac data occupy almost all bits of words 3–10 of each page of subframe 5 (pages 1–24) and subframe 4 (pages 2–5 and 7–10). The exceptions are the first 8 bits of word 3 (data ID and SVID), bits 17–24 of word 5 (SV health), and the 50 bits of parity. The "0" SVID, binary all zeroes, is used to

Table 10b Almanac parameters

Parameter	Number of bits	Scale factor LSB	Effective range	Units
ϵ	16	2^{-21}		dimensionless
t_{oa}	8	2^{-12}	602,112	s
δ_1	16[a]	2^{-19}	——	semicircles
OMEGADOT	16[a]	2^{-38}	——	semicircles/s
$(A)^{1/2}$	24	2^{-11}	——	$m^{1/2}$
(OMEGA)0	24[a]	2^{-11}	——	semicircles
ω	24[a]	2^{-23}	——	semicircles
M_0	24[a]	2^{-23}	——	semicircles
a_{f0}	11[a]	2^{-20}	——	s
a_{f1}	11[a]	2^{-38}	——	s/s

[a]Parameters so indicated shall be two's complement, with the sign bit (+ or −) occupying the MSB.

identify a dummy satellite. When not all satellite slots are needed for different satellites, the same satellite almanac data may be repeated in more than one page. Space vehicle ID is given by bits 3–8 of word 3.

3. Space Vehicle Health

Subframes 4 and 5 also contain two types of space vehicle health data: 1) each of the 32 pages that contain satellite clock/ephemeris data also provide an 8-bit SV health status for that particular satellite; and 2) the 25th page of subframe 4 and 5 jointly contain a satellite summary consisting of 6-bit health status words for up to 32 space vehicles.

The first three most significant bits of the 8-bit health words give the health of the navigation data for that space vehicle; e.g., an indication that the Z-count in the HOW word is good or bad. The five LSB of the 8-bit words and the 6-bit words give the health of the space vehicle signal components, as described in Table 11.

4. Translation of GPS Time to UTC Time

GPS time is based on atomic standard time, and the time broadcast from the satellite is continuous (modulo-1 week) without the leap seconds of UTC, because the introduction of leap seconds would throw the P-code receivers out of lock at the time when they are introduced. Nonetheless, GPS time is maintained by the GPS CS to be within 1 μs of UTC (USNO) time (modulo-1 s) and provides correction parameters in the GPS navigation message. Thus, the GPS provides an important time transfer function. The UTC–GPS translation parameters are shown in Table 12.

The correction parameters to convert GPS time broadcast by the satellite to UTC are contained in the 24 MSB of words 6–9 plus the 8 MSB of word 10 in page 18 of subframe 4. The bit length scale factors are shown in Table 12.

The information contains the parameters required to relate GPS time to UTC and provides notice to the user of any delta time in the recent past or the near future due to leap seconds Δt_{LS} and the week number WN_{LSF} at which that leap second becomes effective. The above relationships apply for the vast majority of the time. However, when the user is operating at a time near the time for a leap second change, special adjustments are required.

The algorithm defining the relationship between GPS time and UTC using the navigation data in subframe 4 is as follows:[1]

$t_{UTC} = (t_E - \Delta t_{UTC})$ [modulo-86,400 s]

where t_{UTC} is in ss and

$\Delta t_{UTC} = \Delta t_{LS} + A_0 + A_1(t_E - t_{ot} + 604{,}800 (WN - WN_t))$, s

t_E = GPS time as estimated by the user on the basis of correcting t_{SV} for factors given in the Subframe 1 clock correction discussion as well as for ionospheric and SA (dither) effects

Δt_{LS} = delta time due to leap seconds

A_0 and A_1 = constant and first-order terms of polynomial

t_{ot} = reference time for UTC data (see Table 13)

Table 11   Codes for health of space vehicle signal components

MSB				LSB	Definition
0	0	0	0	0	All signals OK
0	0	0	0	1	All signals weak[a]
0	0	0	1	0	All signals dead
0	0	0	1	1	All signals have no data modulation
0	0	1	0	0	L_1 P signal weak
0	0	1	0	1	L_1 P signal dead
0	0	1	1	0	L_1 P signal has no data modulation
0	0	1	1	1	L_2 P signal weak
0	1	0	0	0	L_2 P signal dead
0	1	0	0	1	L_2 P signal has no data modulation
0	1	0	1	0	L_1 C signal weak
0	1	0	1	1	L_1 C signal dead
0	1	1	0	0	L_1 C signal has no data modulation
0	1	1	0	1	L_2 C signal weak
0	1	1	1	0	L_2 C signal dead
0	1	1	1	1	L_2 C signal has no data modulation
1	0	0	0	0	L_1 and L_2 P signal weak
1	0	0	0	1	L_1 and L_2 P signal dead
1	0	0	1	0	L_1 and L_2 P signal has no data modulation
1	0	0	1	1	L_1 and L_2 C signal weak
1	0	1	0	0	L_1 and L_2 C signal dead
1	0	1	0	1	L_1 and L_2 C signal has no data modulation
1	0	1	1	0	L_1 signal weak[a]
1	0	1	1	1	L_1 signal dead
1	1	0	0	0	L_1 signal has no data modulation
1	1	0	0	1	L_2 signal weak[a]
1	1	0	1	0	L_2 signal dead
1	1	0	1	1	L_2 signal has no data modulation
1	1	1	0	0	SV *Is* temporarily out (do not use this SV during current pass[a])
1	1	1	0	1	SV *Will Be* temporarily out (use with caution[a])
1	1	1	1	0	Spare
1	1	1	1	1	More than one combination would be required to describe anomalies (except those marked by[a])

[a]Three to six-dB below specified power level due to reduced power output, excess phase noise, SV attitude, etc.

WN = current week number (derived from subframe 1)
WN_t = UTC reference week number

Note that the number of seconds in a day is 86,400.

The estimated GPS time is in seconds relative to end/start of week. The reference time for UTC data t_{ot} is referenced to the start of that week whose number is given in word 8 of page 18 in subframe 4, which represents the 8 LSB of the week number. The user must account for the truncated nature of the week number (see ICD-GPS-200[1]).

When the effectivity time of the leap second event is in the past relative to the user's current time, the relationship presented above is valid except that Δt_{LSF}

GPS NAVIGATION DATA   143

Table 12 GPS-UTC clock correction parameters from subframe 4, page 18[a]

Parameter	Number of bits	Scale factor LSB	Effective range	Units
A_0	32[b]	2^{-30}	—	s
A_1	24[b]	2^{-50}	—	s/s
Δt_{LS}	8[b]	1	—	s
t_{ot}	8	2^{12}	602,112	s
WN_t	8	1	—	weeks
WN_{LSF}	8	1	—	weeks
DN	8[c]	1	7	days
Δt_{LSF}	8[b]	1	—	s

[a]The notations DN and WN stand for day number and week number, respectively.
[b]Parameters so indicated shall be two's complement, with the sign bit (+ or −) occupying the MSB.
[c]Right justified.

Table 13 Reference times for block II satellite vehicles to be used for various clock, ephemeris, almanac, and UTC correction polynomials.[a]

		Hours after first valid transmission time			
Fit interval, h	Transmission interval, h	t_{oc} clock	t_{oe} ephemeris	t_{oa} almanac	t_{ot} UTC
4	2	2	2	—	—
6	4	3	3	—	—
8	6	4	4	—	—
14	12	7	7	—	—
26	24	13	13	—	—
50	48	25	25	—	—
74	72	37	37	—	—
98	96	49	49	—	—
122	120	61	61	—	—
146	144	73	73	—	—
144 (6 days)	144	—	—	84	84
144 (6 days)	4080	—	—	84	84

[a]This table describes the nominal selection that is expressed modulo-604,800 s in the navigation message.

is substituted for Δt_{LS}. The exception to the above algorithm occurs whenever the user's current time falls within the timespan of $DN + 3/4$ to $DN + 5/4$ where DN is the day number. In this time interval, proper accommodation of the leap second event with a possible week number transition is provided by the following expression for UTC: $t_{UTC} = W[\text{modulo-}(86400 + \Delta t_{LSF} - \Delta t_{LS})]$, s, where $W = (t_E - \Delta t_{UTC} - 43200) [\text{modulo-}86400] + 43200$, s; and the definition of Δt_{UTC} (as given in the paragraph above) applies throughout the transition period. Note that when a leap second is added, unconventional time values of the form 23:59:60.xxx are encountered. Some user equipment may be designed to approximate UTC by decrementing the running count of time within several seconds after the event, thereby promptly returning to a proper time indication. Whenever a leap second event is encountered, the user equipment must consistently implement carries or borrows into any year/week/day counts.

5. Subframe 4 Ionospheric Delay Corrections

Ionospheric group delay can cause a significant error in the measured pseudorange group delay by perhaps as much as 300 ns during the daytime at low elevation angles and a lesser error, perhaps 5–15 ns, delay at nighttime and varies in roughly a diurnal pattern. Ionospheric delay effects are discussed in detail in Chapter 12, this volume. The two-frequency user with access to the P(Y) code can correct for most of this delay error by measurement.

The single-frequency user has several alternatives:
1) Ignore the ionosphere and accept the ionospheric error,
2) Use a model of the ionosphere,
3) Use a single-frequency carrier/code differential ionospheric measurement scheme,
4) Use a dual-frequency codeless technique either by cross-correlating L_1 and L_2 channels or by squaring the L_2 P(Y) code to recover the pure carrier as shown at the end of this chapter. (Carrier frequency estimates from L_1 and the almanac can be used to reject the multiple access interference caused by other carriers in the estimation process.)

The single-frequency user can get an approximate correction by using a model of the ionosphere with model parameters transmitted in the downlink datastream. However, the user should be advised that the ionosphere varies in a manner that is difficult to predict, hence the model provides only an approximate correction (for perhaps 70% of the ionosphere delay). In addition, the single-frequency user must correct for the L_1–L_2 delay differential T_{GD} in the satellite that is not needed by the two-frequency user.

The ionospheric group delay model developed for GPS by Klobuchar (see Chapter 12, this volume) essentially employs a half cosine approximation, as shown in Fig. 8. The ionospheric group delay is modeled essentially as follows:

$$I(t) = F(E)\left[5 \times 10^{-9} + \text{AMP(L)} \cos[2\pi(t - 50{,}400)/\text{PER(L)}] \text{ for day}\right]$$

$$= F(E)(5 \times 10^{-9}) \text{ for night} \qquad (3)$$

where $F(E)$ is the obliquity factor that gives a larger I for lower elevation angles E. The parameter AMP(L) is the amplitude of the half cosine for daytime, which is a function of the geomagnetic latitude L of the Earth projection of the ionosphere contact point and PER(L) is the period of the half cosine.

The small constant level of delay equal to 7 ns in this example of Fig. 8 is meant to represent the delay at night. As the sun rises and sets, the ionospheric model gives rise to the cosine-shaped pulse for daytime.

The half-cosine model in the actual GPS ionospheric group delay model is represented by the first three terms in the series expansion $\cos x \cong 1 - (x^2/2) + (x^4/24)$, $|x| < 1.57 \approx \pi/2$, as shown in the algorithm of Table 14.

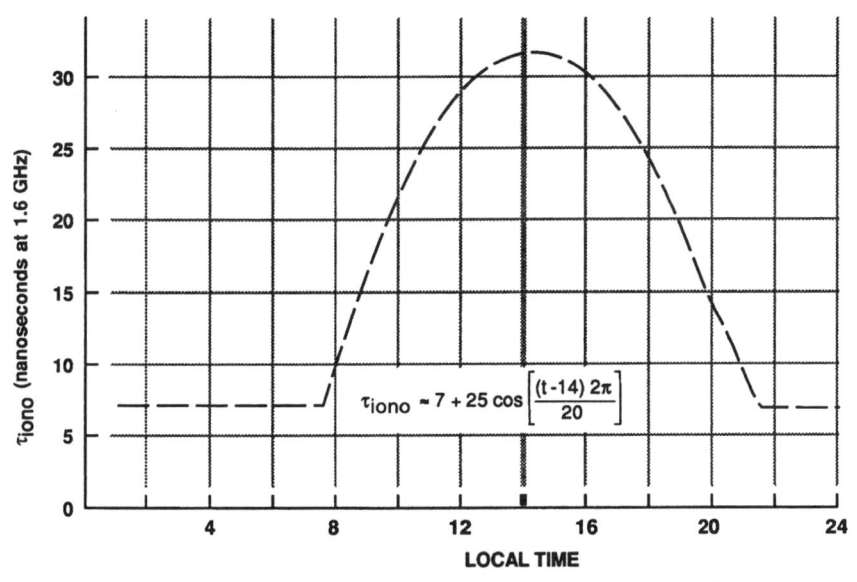

Fig. 8 Example ionospheric group delay cosine fit model.

$$\tau_{iono} \approx 7 + 25 \cos\left[\frac{(t-14)\,2\pi}{20}\right]$$

The specific ionospheric parameters for GPS are given by subframe 4, page 18 for the single-frequency L_1 (or L_2 user) for use in the algorithm given below. These data occupy bits 9–24 of word three plus the 24 MSB of words four and five. The scale factors are shown in Table 15.

Notice that with this definition of the obliquity factor F at elevation angle $E = 0$-deg gives $F = 3.382032$. Thus, the obliquity factor is slightly greater than three at low elevation angles, and $F = 1.0004$ for $E = 0.50$ semicircles (90 deg).

The GPS ionospheric obliquity factor from the model is plotted in Fig. 9. The obliquity factor of 3.382 at 0-deg elevation angle does not become as great at low elevation angles as the tropospheric obliquity factor, because the ionosphere occupies an altitude range of approximately 50–500 km, and as a consequence of this altitude, even a ray path at 0 deg elevation angle on Earth enters the ionosphere at a steeper elevation angle than 0 deg, whereas the troposphere has maximum effect right at the Earth's surface (see Chapter 13, this volume).

The approximate behavior of the variation of ionospheric delay vs. elevation angle can be estimated by referring to Fig. 10, which shows a uniform spherical shell of ionosphere extending from one altitude h_{min} to an upper altitude h_{max}. The upper and lower limits shown are only examples, the ionosphere is in reality not uniform, and in any event, the upper and lower extent of any model would vary with time. In contrast to the troposphere, the ionosphere does not extend to the Earth's surface. Thus, even if the user elevation angle at the Earth's surface is $E = 0$ deg, the angle of incidence ϕ to the ionosphere for this simple model is greater than zero and for this example: $\phi = \cos^{-1}\{[1/(1+\delta)]\cos E\} = 13.455$ deg for $E = 0$ deg, where $\delta = (h_{min}/R)$, and for this example $h_{min} = 180$ km and $\delta = 0.02822$. The length L_I through the ionosphere for this simple model is easily determined to be as follows:

Table 14 Ionospheric model

The ionospheric correction model is given by the following:

$$I = T_{iono} = \begin{cases} F^* [(5.0 * 10^{-9}) + (AMP)(1 - (x^2/2) + (x^4/4))], & |x| < 1.57 \text{ day} \\ F^* (5.0 * 10^{-9}), & |x| \geq 1.57 \text{ night} \end{cases}$$

where $I = T_{iono}$ is referred to the L_1 frequency; if the user is operating on the L_2 frequency, the correction term must be multiplied by $\gamma = 1.646944444$. (T_{iono} is the notation of ICD-GPS-200.)[1]

$$AMP = \begin{cases} \sum_{n=0}^{3} \alpha_n \phi_m^n, & AMP \geq 0 \\ \text{if } AMP < 0, \text{ set } AMP = 0 \end{cases} \text{(s)} \quad \text{Amplitude}$$

$$x = \frac{2\pi (t - 50400)}{PER}, \text{ (rad)}$$

$$PER = \begin{cases} \sum_{n=0}^{3} \beta_n \phi_m^n, & PER \geq 72000 \\ \text{if } \quad PER < 72{,}000, \text{ set } PER = 72{,}000 \end{cases} \text{(s)} \quad \text{Period}$$

$F = 1.0 + 16.0 [0.53 - E]^3$, the obliquity factor
α_n and β_n are the satellite transmitted data words with $n = 0, 1, 2,$ and 3.

Other equations that must be solved are as follows:

$\phi_m = \phi_i + 0.064 \cos(\lambda_i - 1.617)$ (semicircles),

$\lambda_i = \lambda_u + \dfrac{\Psi \sin A}{\cos \phi_i}$ (semicircles),

$$\phi_i = \begin{cases} \phi_u + \Psi \cos A \text{ (semicircles)}, & |\phi_i| \leq 0.416 \\ \text{if } \phi_i > 0.416, \text{ then } \phi_i = +0.416 \\ \text{if } \phi_i < -0.416, \text{ then } \phi_i = -0.416 \end{cases} \text{(semicircles)},$$

$\Psi = \dfrac{0.0137}{E + 0.11} - 0.022$ (semicircles),

$t = 4.32 * 10^4 \lambda_i + $ GPS time (s), $t = $ local solar time (s)

where $0 \leq t < 86{,}400$, therefore: if $t \geq 86{,}400$ seconds, subtract 86,400 seconds;
 if $t < 0$ seconds, add 86,400 seconds.

The terms used in computation of ionospheric delay are as follows:

Satellite Transmitted Terms

α_n = the coefficients of a cubic equation representing the amplitude of the vertical delay, 4 coefficients—8 bits each

β_n = the coefficients of a cubic equation representing the period of the model, four coefficients—8 bits each

(Table 14 continues on next page.)

Table 14 Ionospheric model (continued)

Receiver Generated Terms

- E = elevation angle between the user and satellite, semicircles
- A = azimuth angle between the user and satellite, measured clockwise positive from the true north, semicircles
- ϕ_u = user geodetic latitude, semicircles, WGS-84[5]
- λ_u = user geodetic longitude, semicircles, WGS-84[5]

GPS time receiver computed system time

Computed Terms

- x = phase, rad
- F = obliquity factor, dimensionless
- t = local time, s
- ϕ_m = geomagnetic latitude of the Earth projection of the ionospheric intersection point, mean ionospheric height assumed 350 km, semicircles
- λ_i = geodetic longitude of the Earth projection of the ionospheric intersection point, semicircles
- ϕ_i = geodetic latitude of the Earth projection of the ionospheric intersection point, semicircles
- Ψ = Earth's central angle between user position and Earth projection of ionospheric intersection point, semicircles

Table 15 Ionospheric parameters from subframe 4, page 18

Parameter	Number of bits	Scale factor, LSB	Effective range[a]	Units
α_0	8[b]	2^{-30}	——	s
α_1	8[b]	2^{-27}	——	s per semicircle
α_2	8[b]	2^{-24}	——	s per semicircle^2
α_3	8[b]	2^{-24}	——	s per semicircle^3
β_0	8[b]	2^{11}	——	s
β_1	8[b]	2^{14}	——	s per semicircle
β_2	8[b]	2^{16}	——	s per semicircle^2
β_3	8[b]	2^{16}	——	s per semicircle^3

[a]Effective range is the maximum range indicated by the bit allocation and scale factor.
[b]Parameters so indicated shall be two's complement, with the sign bit (+ or −) occupying the MSB.

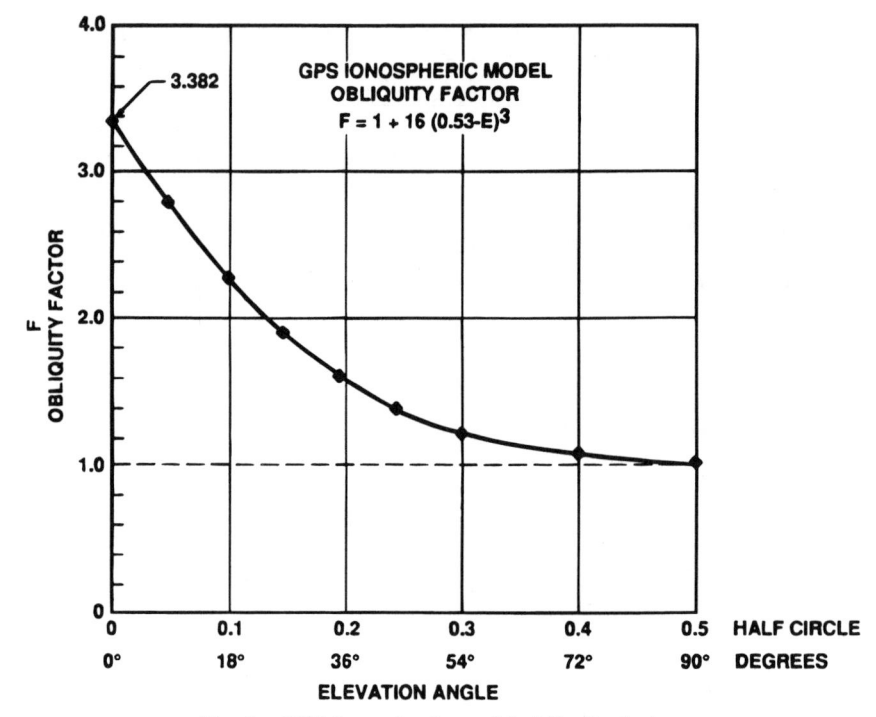

Fig. 9 GPS ionospheric model obliquity factor.

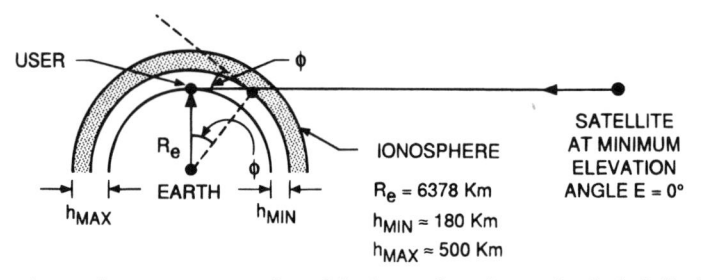

Fig. 10 Approximate representation of the ionosphere by a spherical shell of width $h = h_{max} - h_{min}$. The angle of incidence of the ionosphere d relative to the elevation angle E is given by $\phi = \cos^{-1}[(\cos E)/(1 + \delta)]$ where $\delta = h_{min}/R_e$.

$$L_l = \frac{2h(1 + \Delta/2)}{\sin\phi + \sqrt{\sin^2\phi + 2\Delta + \Delta^2}}$$

where $\Delta = (h_{max} - h_{min})/R_e = 0.05017$. For a uniform density ionosphere, the ratio of L for some arbitrary elevation angle E to that at zenith $E = 90$ deg corresponds to the obliquity factor. For the simplified model of Fig. 10, the obliquity factor for $E = 0$ deg is the ratio of the two values of L_l which gives the following: $[L_l(E = 0 \text{ deg})/L_l(E = 90 \text{ deg})] = 2.0502/0.62899 = 3.228$, which is fairly close to the value for the GPS model of Fig. 9 of 3.382.

III. Time, Satellite Clocks, and Clock Errors

Precision satellite clocks and time measurements are the keys to the accuracy of GPS. In this section, different concepts of time are examined, clock stability is discussed in statistical terms, and clock stability as it relates to GPS navigation is summarized.

Time is an abstract concept that man has created in order to order events subjectively at a given location. "Time is defined so that motion looks simple."[6] For scientific purposes, this means that we need a uniform time scale, "two time intervals are equal if it can be shown that equal processes took place during these two time intervals."[7] The following paragraphs describe several different time scales relevant to GPS.

A. Mean Solar, Universal Mean Sidereal, and GPS Time

Early forms of time keeping employed time as indicated by the sundial or the apparent solar time. Ptolemy (ca. 100–178 A.D.) noted the irregularity of the solar day and defined mean solar time by assuming a mean movement of the Sun relative to an observer on Earth. In this way, a clock and the mean solar time were in approximate agreement. The difference between apparent solar time and mean solar time varies with the seasons and location and is approximated by the "Equation of Time," which in centuries past, was sometimes printed on sundials. Because the Earth is tilted on its axis by 23.45 deg with respect to the ecliptic (plane of rotation of the Earth about the Sun), the apparent rotation rate of the Sun about the Earth is not constant throughout the year. This time offset, the "Equation of Time," can vary by roughly ±16 min, and the two times are approximately equal four times per year (see, for example, Ref. 8). Furthermore, the Earth's orbit around the Sun is not perfectly circular (see Fig. 11). The slight ellipticity of the Earth's orbit $\epsilon = 0.0167$[9] also causes a variation of the apparent solar day by approximately 4.7 s.

1. Universal Time

We can make the appropriate correction for these effects to obtain mean solar time. If this correction is made at the Greenwich meridian in England, we have UT0 the first universal time scale. More precisely, UT0 is based on a mathematical expression for the right ascension of the fictitious mean Sun, and the clock time of transit of any celestial object with known position by an observatory yields

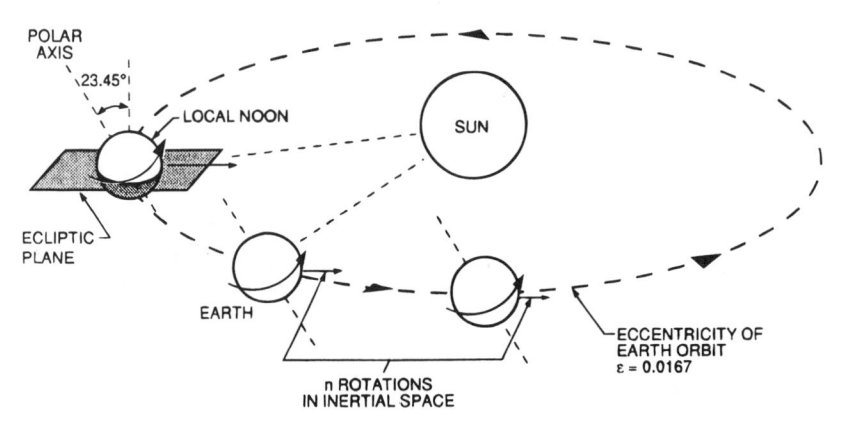

Fig. 11 Rotation of the Earth about the Sun. The Earth must rotate slightly more than 360 deg on its axis to reach solar noon than it does to rotate once on its axis in inertial space. The helio-centric-eliptic coordinate system has its origin at the center of the Sun. Thus, a sidereal day is slightly shorter (by approximately 4 min) than a solar day. The plane of the Earth's orbit around the Sun is the ecliptic plane. (Mean distance to the Sun = 1.495×10^8 km. Orbit speed $\cong$ 29.79 km/s.)

UT0 after correction for longitude, aberration, parallax, nutation, and precession.[7] However, UT0 is not strictly uniform, as discussed in the following paragraphs.

A sidereal day is defined as the time that it takes the earth to rotate once on its axis relative to inertial space; i.e. relative to the stars, whereas a solar day is the time that the earth takes to rotate so that the sun crosses the local meridian, as shown in Fig. 11. Since the earth revolves once about the sun in a solar year (365.25 mean solar days), there is a difference between these two time scales by approximately one extra rotation of the Earth per year. The mean solar day is approximately 4 min longer than a mean sidereal day (24 h = 86,400 mean solar s). One day of mean solar time consists of 1.002737093 days of mean sidereal time. (Note that $1 \times 1/365.25 = 1.00273785$.) A mean sidereal day consists of 24 sidereal hours or 23 h, 56 min, 4.0954 s or 0.9972695664 days or 86164.09054 s of mean solar time. Thus, a sidereal day is shorter than a solar day by approximately 4 min.

In addition to the relatively simple Earth axis tilt and Earth orbit eccentricity effects, there are also effects of Earth precession of its axis caused by the Earth's oblateness, effects from the Sun–Earth gravitation, lunar gravitation effects that cause a nodding or nutation effect of approximately 9 s of arc, and several polar wander effects. These all affect solar time relative to a precise clock time and are summarized in Fig. 12 (Ref. 9). In addition, there seem to be small, shorter-term periodic effects ($\approx$0.3 cycle/year) and a general downward trend in the Earth's spin rate. All of these effects can have some relevance to time and position measurement.[10]

Universal time represents a family of time scales based on the Earth's rotation on its axis, and UT0 was discussed previously. The coordinates of an observatory used to generate UT0 are subject to small changes caused by slow movements of the Earth rotation axis polar variation. Universal Time UT1 is a true navigator's time scale, which has been corrected for polar variation. UT1 is not uniform

NODDING OR NUTATION
PERIOD 18.6 YR
WITH 9 SECONDS OF
ARC MOTION

PRECESSION
PERIOD
26,000 yr

POLAR AXIS
OF ROTATION

PLANE OF
EARTH'S ORBIT
(ECLIPTIC PLANE)

VERNAL
EQUINOX MOVES
WESTWARD

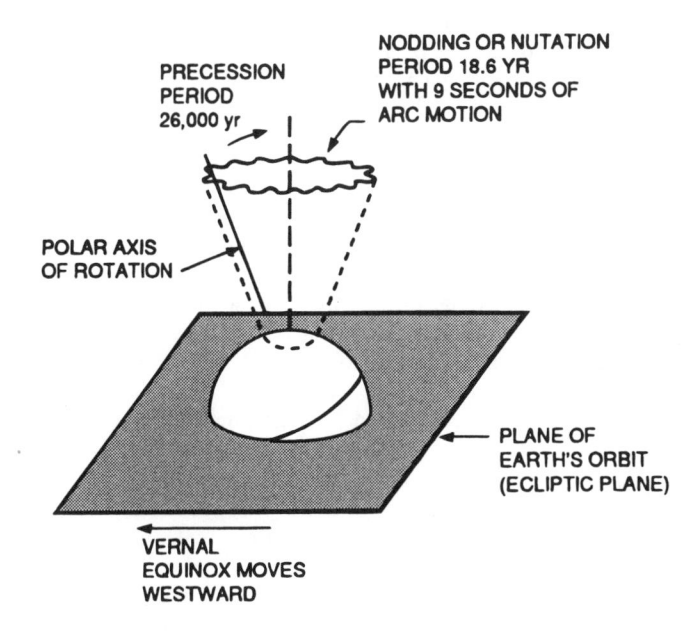

POLAR VARIATIONS (PV) IN ADDITION TO PRECESSION & NUTATION

- 430 DAY PERIOD MEANDERING MOTION-CHANDLER WOBBLE
  ≈ 0.1 ARC SECOND TOGETHER ≈ ± 9 METERS

- 12 MONTH SEASONAL MOTION ≈ 0.1 ARC SECOND

- SECULAR DRIFT 0.002-0.003 ARC SECOND/YEAR

SEASONAL VARIATIONS IN ROTATION PERIOD (SV)

CHANGE IN LENGTH OF DAY ≈ 1 PART IN 10^8 OR 10^{-3} SEC
PLUS A GENERAL TREND OF DECELERATION IN SPIN RATE
OBSERVED IN THE LAST 100 YEARS.

Fig. 12 Precession, nutation, and polar motion of the Earth rotation axis and change in rotation rate. The precession of the equinox is 1.39697128 deg/century. The obliquity of the ecliptic and Epoch 2000 is 23.4392911 deg.

because of small changes caused by both regular seasonal variation and irregular and unpredictable changes in the Earth's rotation period. A further smoothing is performed to remove the seasonal variation to form UT2, which still has small unpredictable (10^{-8}) variation and long-term drift (10^{-10}/year) effects. Both UT1 and UT2 are stable to within approximately 3 ms in a day.

2. *Ephemeris Time*

Ephemeris time is based on the Earth's orbital motion about the Sun, whereas solar time is based on the rotation rate of the Earth on its axis. Astronomers make use of ephemeris time, which is defined on the basis of the Tropical Year, that is the time for the Sun to return to its starting point (the vernal equinox) with respect to the distant stars. One second of ephemeris time is defined as 1/31556925.97474 of the Tropical Year 1900. The Tropical Year is 365.24220 mean solar days.

3. Universal Coordinated Time

With the availability of high-precision atomic and superconductive clocks (superconducting resonators), atomic time scales are now in existence with time stabilities better than one part in 10^{13} from year to year. The accuracy of time measurement now exceeds that of any other physical measurement. UTC is an atomic clock time scale coordinated by the Bureau International de l'Heure in Paris. The GPS system is referenced to UTC (USNO), the U.S. time standard kept by the U.S. Naval Observatory. The UTC (USNO) and UTC(OP) Paris Observatory are regularly coordinated by common view GPS time transfer.[10] UTC differs from a pure atomic clock time in that it occasionally introduces leap seconds. These leap seconds are introduced to keep this atomic time scale in approximate step with the Earth's rotation. The leap second adjustment can cause a particular minute to have 59 or 61 s instead of 60. UTC is, by international agreement, kept to within 0.9 s of the navigator's time scale UT1. Leap seconds are usually added or deleted on June 30 or December 31. Leap seconds have been implemented since 1972.

4. GPS Time

The GPS control segment contains a set of high-accuracy cesium beam atomic clocks, both at the central control station site and at the various remote monitor station sites. The composite set of clocks constitutes a highly accurate GPS system time to which the satellite clocks are compared and corrected by upload commands to the satellite. The GPS satellites themselves carry redundant cesium and rubidium atomic clocks. Thus, the GPS system time is an atomic clock time similar to but not the same as UTC time. One marked difference is that GPS time does not introduce any leap seconds. To do so would throw the GPS P(Y)-code receivers using the system out of lock. Thus, introducing leap seconds is out of the question for GPS, and the introduction of UTC leap seconds causes the GPS time and UTC time to differ by a known integer number of cumulative leap seconds. Other than the leap second effect however, the GPS control segment attempts to keep the GPS time to within 1 µs of UTC (USNO) time (modulo-1s).

B. Clock Accuracy and Clock Measurement Statistics

One of the keys to obtaining high-accuracy position and time keeping with GPS is the use of redundant atomic clocks on the GPS satellites themselves and in the GPS control segment. These clocks provide the means for estimating both the satellite orbits as well as the correction parameters for the satellite clocks. Both of these quantities are estimated on the ground by the control segment, transmitted to the satellites, and then the appropriate datablocks containing ephemeris model and clock correction terms are retransmitted to the users. One objective of the GPS CS is to attempt to estimate both the clock error and to be able to perform piecewise extrapolation of the clock error in the future. Before the details of the clock corrections are described, it is important to discuss clock accuracy and the statistical measures of clock performance. These statistical measures also apply to the user receiver clock.

A clock generator produces a waveform that in general can be represented as follows:

$$A \sin\left[(\omega_0 t + \phi_0) + \left(\phi_e + \omega_e t + 2\pi d \frac{t^2}{2}\right) + x(t)\right] \quad (4)$$

where we neglect as unimportant any minor amplitude modulation. The quantity $(\omega_0 t + \phi_0)$ represents the desired clock phase, the next term, $(\phi_e + \omega_e t + 2\pi d\, t^2/2)$, represents the causal phase error effects where ϕ_e is a fixed phase error offset, ω_e is a frequency error that may be a function of the clock environment; e.g., temperature, magnetic field, acceleration, and pressure. The term, $(2\pi d\, t^2/2)$, represents a frequency drift or aging effect, and the final term $x(t)$ represents a random phase error. In analyzing these effects, we separate the causal errors ($\phi_e + \omega_e t + 2\pi d\, t^2/2$) from the random phase error $x(t)$. Some typical environmental (causal) effects are listed in Table 16 for various types of high-performance atomic clocks.[7,12-14] By contrast, the aging rate of a crystal oscillator might be on the order of 10^{-10}/day or larger, much higher than that of atomic-based clocks.

1. Allan Variance of Clock Frequency Noise

For the remainder of this section, the discussion on oscillator phase noise refers only to the random component of the phase term $x(t)$ and its derivative $y(t)$, which is assumed to exist and is represented by $y(t) = dx(t)/dt$.

The frequency noise $y(t)$ assumed to be a stationary random process with a power spectral density (one-sided) defined as $G_y(f)$, where $G_y(f)$ is nonzero only for a finite frequency range $f_{min} < f < f_{max}$, and $G_y(f) = 0$ for $f \geq f_{max}$ or $f < f_{min}$ where $f_{min} = 1/T_0$, and T_0 is the observation interval. All real electronic

Table 16 Environmental and systematic drift (aging) effect in example clocks[a]

	Atomic hydrogen maser	Cesium atomic beam controlled oscillator	Rubidium gas cell controlled oscillator
Temperature drift	$<1 \times 10^{-13}$/°C	$\approx 2 \times 10^{-13}$/°C	$\approx 10^{-12}$/°C
Systematic drifts	None detectable $<10^{-12}$/yr	None detectable $<3 \times 10^{-12}$/yr	Aging $\approx 1 \times 10^{-11}$/month
Magnetic field effect on frequency	1,420,405,751.768 Hz + 2750 H_c^2 Hz	9,192,631,770 Hz + 427.18 H_c^2 Hz	Higher magnetic sensitivity

[a]The magnetic field is H_C in oersteds.

processes have some practical upper-frequency limit f_{max}. It is further assumed that the total power is finite in the following frequency noise spectrum:

$$\int_0^\infty G_y(f) \, df = S_y^2$$

The frequency noise power spectral density is one important measure of frequency stability. Another measure, as shown next, is expressed in the time domain.

Define an incremental average frequency $\bar{y}(t_k + \tau)$ as the average value of frequency $y(t)$ over the time interval $(t_k, t_k + \tau)$ as follows:

$$\bar{y}(t_k + \tau) \stackrel{\Delta}{=} \frac{1}{\tau} \int_{t_k}^{t_k+\tau} y(t) \, dt = \frac{x(t_k + \tau) - x(t_k)}{\tau} = \frac{x(t_{k+1}) - x(t_k)}{\tau} \quad (5)$$

where $t_{k+1} \stackrel{\Delta}{=} t_k + \tau$. This term $\bar{y}_k$ represents the first difference in phase x_k, or the average slope of $x(t)$ over this same time interval. Define a rate of change in incremental average frequency $\Delta(t)$ as follows: $\Delta(t) \equiv d\bar{y}(t)/dt$. Further define an average rate of change in incremental average frequency $\bar{\Delta}(t_k)$, where $\bar{\Delta}(t_k)$ is also the second difference in phase $x(t)$:

$$\bar{\Delta}(t_k + \tau) = \frac{1}{\tau} \int_{t_k}^{t_k+\tau} \Delta(t) \, dt = \bar{y}(t_k + \tau) - \bar{y}(t_k)$$

$$= \frac{x(t_{k+2}) - x(t_{k+1}) - [x(t_{k+1}) - x(t_k)]}{\tau^2}$$

$$= \frac{x(t_{k+2}) - 2x(t_{k+1}) - x(t_k)}{\tau^2} \quad (6)$$

Thus $\bar{\Delta}(t)$ is the average value of the slope of $\bar{y}$, for the same interval. An estimate of the clock phase error at time t based on data at time $t - \tau$ can be approximated by the Taylor's series $x(t) \cong x(t - \tau) + y(t - \tau)\tau + \frac{1}{2} \dot{y}(t - \tau)\tau^2 + ,\ldots$, for small τ for finite sufficiently small τ, y, $\dot{y}$, etc. By using the definition given above in Eq. (5) for incremental average frequency for time increments of τ, the expression for $x(t)$ can be written in terms of the definition, $\bar{y}(t) \stackrel{\Delta}{=} [x(t) - x(t - \tau)]/\tau$, as follows:

$$x(t) = x(t - \tau) + \bar{y}(t)\tau \quad (7)$$

(Note the similarity in form only to the first two terms of Taylor's series.) If each $\bar{y}$ increment is a random white Gaussian process with fixed mean μ and variance $\sigma_{\bar{y}}^2$, then the phase $x(t)$ is a random walk with uncorrelated segments and the expected value $E[x(t)] = t\mu$ and $\sigma_x^2 = t\sigma_{\bar{y}}^2$; that is, the phase $x(t)$ is a nonstationary process with a variance that grows linearly with time. Recall from elementary statistics[15-17] that the unbiased estimate of the variance of a random variable z_k from N samples is as follows:

$$M_2 = \frac{1}{N-1} \sum_{n=1}^{N} \left[z_n - \frac{1}{N} \sum_{k=1}^{N} z_k \right]^2 \quad (8)$$

This estimate converges with large N provided that $M_4 = E(Z_k - \bar{Z}_k)^4$ exists.

With this background, the sample variance of the incremental frequency $\bar{y}$ is defined by analogy as follows:

$$\langle \sigma_y^2(N, \tau) \rangle = \left\langle \frac{1}{N-1} \sum_{n=1}^{N} \left[\bar{y}_n - \frac{1}{N} \sum_{k=1}^{N} \bar{y}_k \right]^2 \right\rangle \quad (9)$$

where $\langle \rangle$ represents the time average. For $N = 2$ in Eq. (9), this sample variance is called the Allan variance, which is defined as follows:

$$\sigma_y^2(\tau) \triangleq \langle \sigma_y^2(2, \tau) \rangle = \left\langle \frac{(\bar{y}_{k+1} - \bar{y}_k)^2}{2} \right\rangle \quad (10)$$

Replace the time average by a summation over time to get an unbiased estimate of σ_y^2, which converges for large m. We can make m successive measurements where the beginning of each τ s averaging interval is separated by T s where $T \geq \tau$. If we set $T = \tau$ so that each τ s averaging interval begins at the end of its predecessor then the following results:

$$\sigma_y^2(\tau) = \left\langle \frac{(\bar{y}_{k+1} - \bar{y}_k)^2}{2} \right\rangle = \frac{1}{2} \langle [\bar{\Delta}(t_k)]^2 \rangle$$

$$\cong \frac{1}{m} \sum_{k=1}^{m} \frac{(\bar{y}_{k+1} - \bar{y}_k)^2}{2}, \quad \text{for} \quad m \to \infty \quad (11)$$

The Allan variance[13] can also be written using Eq. (6) in terms of the phase variable $x(t)$:

$$\sigma_y^2(\tau) = \frac{1}{2} \frac{\langle (x_{k+2} - 2x_{k+1} + x_k)^2 \rangle}{\tau^2} \quad (12)$$

$$\cong \frac{1}{2m} \sum_{k=1}^{m} \left[\frac{x_{k+2} - 2x_{k+1} + x_k}{\tau^2} \right]^2 \quad \text{for} \quad m \to \infty$$

where the time average has been approximated by averaging over m nonoverlapping time increments. This second difference is depicted in Fig. 13. Clearly, if the phase variation is precisely linear with time; i.e., constant frequency, the Allan Variance is zero. If τ is small, then the Allan variance measures short-term stability of the oscillator. If τ is large; e.g., 24 h, then Allan variance gives the long-term stability.

There is a relationship between the power spectral density of the frequency variable $y(t)$ and the Allan variance. Clock noise power spectral density can be quite generally represented by the following[12]:

$$G_y(f) = h_{-2}f^{-2} + h_{-1}f^{-1} + h_0 + h_1 f + h_2 f^2, \text{ for frequencies } 0 < f < f_h \quad (13)$$

where the $1/f^2$ term represents random walk in frequency y, the $1/f$ term represents the flicker noise in frequency y, the f^0 term represents white noise in y (or random walk in phase x), the f term is flicker noise in phase x, and the f^2 term is white noise in the phase x.

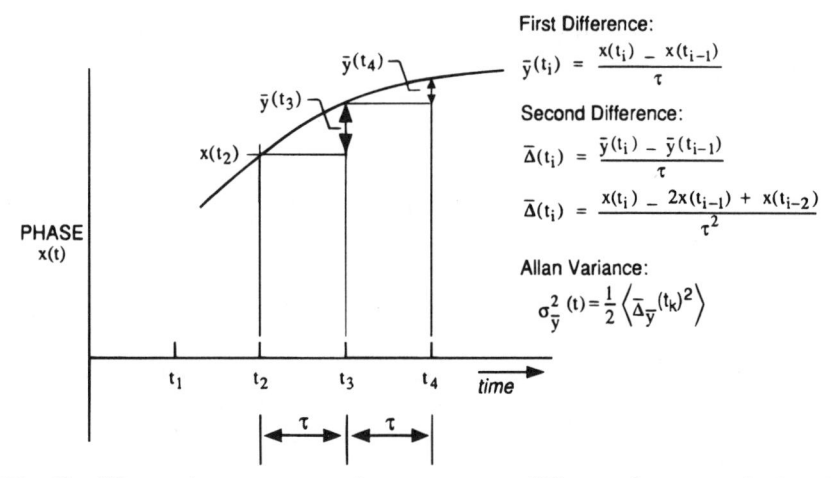

Fig. 13 Allan variance measures the mean square difference between x (t_{i-1}) and a straight line fit between $x(t_{i-2})$ and $x(t_i)$. The second difference of the random phase function $x(t)$ is $\bar{\Delta}(t_i)$, and the Allan variance is $\sigma_{\bar{y}}^2(\tau)$.

The transformation between the clock noise spectral components and the Allan variance for $N = 2$ expressed vs sample interval time τ is as follows[12]:

$$\sigma_{\bar{y}}^2(\tau) = 2\int_0^\infty G_y(f)\left[\frac{\sin \pi f\tau}{\pi f\tau}\right]^2 \left\{1 - \frac{\sin^2 2\pi f\tau}{4\sin^2 \pi f\tau}\right\} df = 2\int_0^\infty G_y(f)\left[\frac{\sin^2 \pi f\tau}{\pi f\tau}\right]^2 df \quad (14)$$

For $N = 2$ and $G_y(f) = h_\alpha f^\alpha$ this equation reduces to the following[13]:

$$\sigma_{\bar{y}}^2(\tau) = \frac{2h_\alpha}{(\pi\tau)^{\alpha+1}} \int_0^\infty u^{\alpha-2} \sin^4 u \, du \quad (14a)$$

Although the transformation from $G_y(f)$ to $\sigma_{\bar{y}}^2(\tau)$ is unique, the reverse transformation is not unique.

Using Eq. (14), the Allan variance for these frequency noise spectral density terms provides a time domain representation vs τ as follows:

$$\sigma_{\bar{y}}^2(\tau) = h_{-2}\frac{(2\pi)^2}{6}|\tau| + h_{-1} 2 \ln 2$$
$$+ \frac{h_0}{2}\frac{1}{|\tau|} + h_1 \frac{1}{(2\pi\tau)^2}\{3[2 + \ln(2\pi f_h|\tau|)] - \ln 2\}$$
$$+ h_2 \frac{3f_h}{(2\pi\tau)^2} \quad (15)$$

Thus, if the frequency noise spectral density is $G_y(t) \sim f^\alpha$, then the Allan variance terms are $\sigma_{\bar{y}}^2 \sim |\tau|^\mu$ where α and μ are related, as in Table 17. A measure of time stability measure is the corresponding variance in phase x and is defined as follows:

GPS NAVIGATION DATA

Table 17 Relationship between frequency noise spectral density and Allan variance variation with time interval t

Type of noise	Frequency noise spectral density, $G_s(f) \sim f^\alpha; f < f_h$ α	Allen variance, frequency, $\sigma_y^2(\tau) \sim |\tau|^\mu$ μ	Phase variance, time variance, $\sigma_x^2(\tau) \sim |\tau|^\nu$ ν
Random walk FM	−2	1	3
Flicker FM	−1	0	2
White FM	0	−1	1
Flicker phase	+1	a	a
White phase	+2	−2	0

[a] The flicker phase noise term contains a $(1/\tau^2) \ln 2\pi f_h |\tau|$ term for Allan variance and a $\ln 2\pi f_h |\tau|$ term for the phase variance.

$$\sigma_x^2(\tau) = \tau^2 \sigma_y^2 = h_{-2} \frac{(2\pi)^2}{6} |\tau|^3 + h_{-1} 2\tau^2 \ln 2$$

$$+ \frac{h_0}{2} |\tau| + \frac{h_1}{(2\pi)^2} \{3[2 + \ln(2\pi f_h |\tau|)] - \ln 2\} + h_2 \frac{3 f_h}{(2\pi)^2} \quad (16)$$

Figure 14 plots the Allan variance $\sigma_y^2(\tau)$ vs τ for some of the individual components of clock noise for arbitrary component levels h_i using Eq. (15).

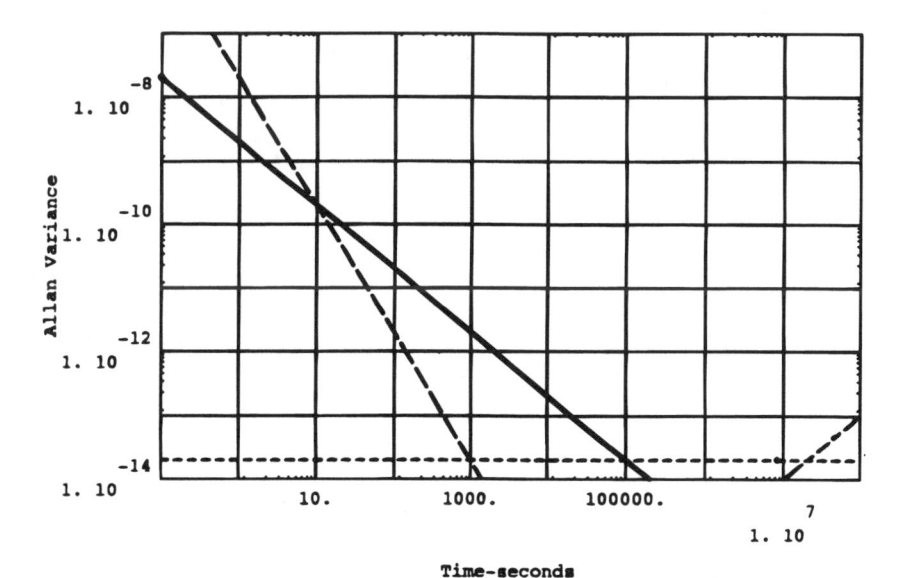

Fig. 14 Plot of examples of some of the components of the Allan variance: 1) white phase noise (large dashes); 2) white frequency noise (solid curve); 3) flicker frequency noise (small dashes); and 4) random walk frequency noise (alternate dashes). Refer to the value of μ in Table 17.

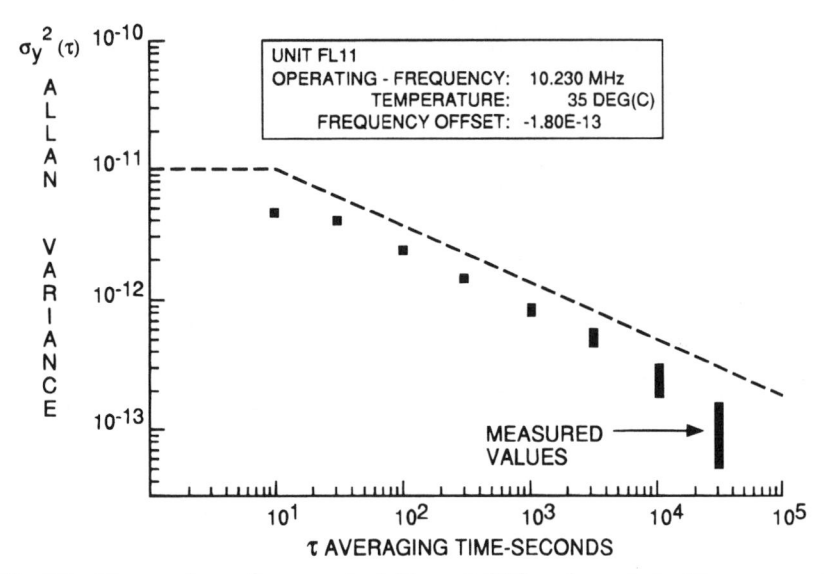

Fig. 15 Allan variance for an actual Phase I GPS cesium clock. The ranges of measured points are indicated by the bars. The dashed curve is the specification curve. (Courtesy Frequency and Time System.)

Figure 15 shows an example of the Allan variance plotted vs τ for one of the actual Phase I GPS satellite cesium frequency clocks taken before launch. Note that the Allan variance of the clock decreases to as low as 10^{-13} for time intervals τ on the order of 1 day.

The long-term stability of cesium clocks has been measured for both Block I cesium and the higher stability Block IIA satellites at the U.S. Naval Observatory.[18] Stability estimates have been made using both the broadcast ephemerides and the postprocessed precise ephemerides from the Defense Mapping Agency.[5] The results are shown in Table 18. The only significant difference between the broadcast and postprocessed ephemerides results is for the Block IIA GPS#23, which showed a somewhat better stability at 1 day with the postprocessed ephemerides and indicates some radial error contribution at 1 day in the broadcast ephemerides. In both clock measurements, the frequency stability was still trending downward at a 10-day sample time interval. Both satellite clocks demonstrated

Table 18 Frequency stability (Allan variance) estimates of GPS NAVSTAR on-orbit cesium standards for 1-day and 10-day frequency stabilities

Ephemerides sample time	Block I—GPS NAVSTAR #9		Block IIA—GPS NAVSTAR #23	
	Broadcast ephemerides	Post-processed precise	Broadcast ephemerides	Post-processed precise
1 day	1.7×10^{-13}	1.8×10^{-13}	1.2×10^{-13}	0.84×10^{-13}
10 day	5.4×10^{-14}	6.2×10^{-14}	2.9×10^{-14}	3.4×10^{-14}

GPS NAVIGATION DATA

a better 1-day stability than the specified value of 2×10^{-13}. The accuracy of the clocks and the broadcast ephemerides is discussed more fully in a later chapter.

IV. Satellite Orbit and Position

User position calculations require precise knowledge of the satellite position at the time of transmission. As already described, this information is carried in the downlink datastream in subframes 2 and 3 as ephemeris parameters from which satellite position can be computed. The ephemeris parameters describe the orbit during a given curve fit time interval. The orbit parameters correspond to a modified Keplerian orbit. As a preliminary to these discussions, it is appropriate to review the classical orbital parameters and the Kepler model. The effects of various perturbing factors on the orbit are then discussed along with computational methods for solving the nonlinear equations.

A. Coordinate Systems and Classical Orbital Elements

The satellite position is solved using Earth–Centered-Inertial (ECI) coordinates and then must be translated to the user's coordinate system, which is Earth-centered, Earth-fixed (ECEF). Figure 16a shows the nonrotating ECI coordinate system, which is fixed with respect to the stars. GPS satellite orbits are determined in ECI coordinates and then converted to ECEF. Figure 16b shows the rotating ECEF coordinates that ultimately must be used to solve for user position on the real Earth.

1. Inclined Orbits

The classical orbital elements of an ideal elliptical orbit are shown in Fig. 17. The satellite orbit is inclined with respect to the equatorial plane by angle i. The GPS Block II satellite orbit planes are inclined at 55 deg. The older Block I

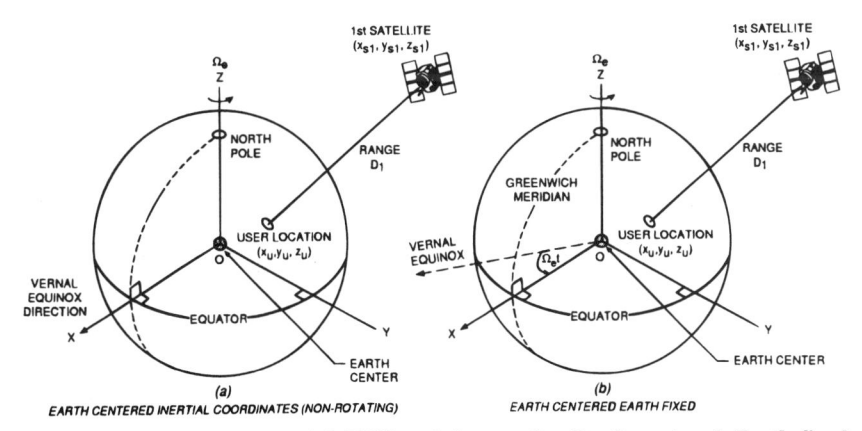

Fig. 16 Earth-centered inertial (ECI) and the rotating Earth-centered, Earth-fixed coordinate systems.

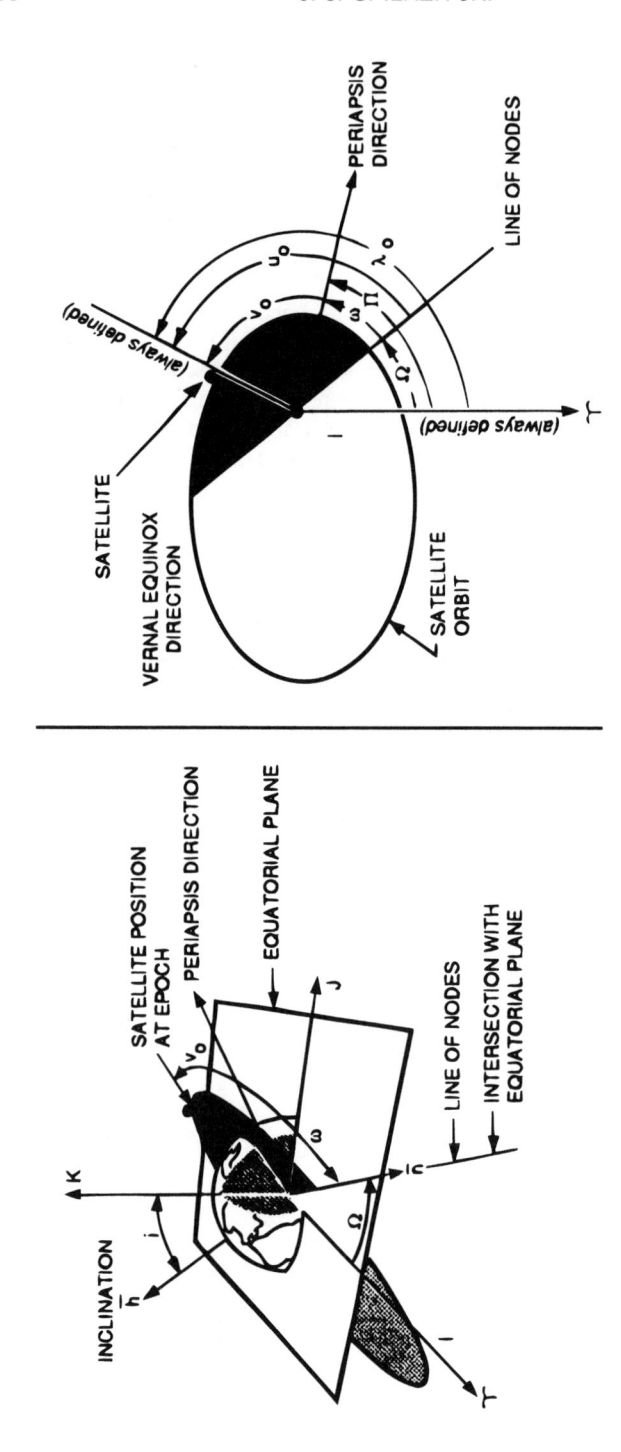

Fig. 17 Classical orbital elements for a satellite in an inclined elliptical orbit. The figure on the right is drawn in the plane of the orbit. The portion of the ellipse above the equatorial plane is shown in black in both figs. The satellite phase at time t_0 in the orbit referenced to the periapsis direction is given by v_0 the true anomaly.

satellites are inclined at 63 deg. In the figure, the satellite orbital plane is shown as it intersects the equatorial plane; the intersection line is the line of nodes. The GPS constellation consists of six orbit planes where the ascending nodes, as expressed in astronomical coordinates, are equally spaced by 360 deg/6 = 60 deg. Clearly, each satellite plane has a line of ascending node at the same angle where another plane has its line of descending nodes because they are 180 deg apart in astronomical ECI coordinates.

Table 19 describes the classical orbital elements for the inclined elliptical orbit of Fig. 17. The parameters might be summarized as follows:

1) The size and shape of an orbit are defined by the semimajor axis a and the eccentricity e.

2) The position of the satellite on the elliptical orbit is described by the true anomaly v. Relationships are given later between the true anomaly v and the eccentric and mean anomalies.

3) The orientation of the elliptical orbit with respect to inertial space is given by the following parameters: a) the angle to the line of ascending nodes Ω, b) the orbit inclination i, and c) the argument of the perigee (periapsis) ω.

With this background, the next objective is to determine the equations of motion of the satellite in the orbit—Kepler's equations.

Table 19 Classical orbital elements

Symbol	Orbital element	Definition
a	Semi-major axis	A constant defining the size of the conic orbit
e	Eccentricity	A constant defining the shape of the conic orbit
i	Inclination	The angle between the K unit vector and the angular momentum vector H
Ω	Longitude of the ascending node	The angle in the equatorial plane, between the I unit vector and the point where the satellite crosses through the equatorial plane in a northerly direction (ascending node) measured counterclockwise when viewed from the north side of the equatorial plane
ω	Argument of periapsis	The angle in the plan of the satellite's orbit, between then ascending node and the periapsis point, measured in the direction of the satellite's motion
T	Time of periapsis passage	The time when the satellite was at periapsis

B. Classical Keplerian Orbits

Figure 18 illustrates the position of the satellite in the elliptical orbit. Kepler's laws for the ideal two-body problem restated in terms of satellites for purposes of this discussion state that 1) the satellite moves in an elliptical orbit with Earth at the focus as shown, 2) the line joining the satellite with the Earth's center sweeps out equal areas A in equal times, 3) the period T_0 of the orbit given by Table 20 summarizes some of the key relationships for Keplerian orbits (Refs. 9, 19): $T_0 = (2\pi/\sqrt{\mu}) a^{3/2}$. Note that in the discussion of the ephemeris parameters of subframes 2 and 3 in an earlier section, the eccentric anomaly symbol was E_k. The symbol E is also used elsewhere and in the ionosphere discussion of subframe 4 and 5 for the elevation angle to the satellite. Nonetheless, E is commonly used for the eccentric anomaly, and the reader is advised to use caution.

In classical Keplerian analysis, the mean anomaly M is defined as a variable that varies linearly with time $M = n(t - T)$ where n is the mean motion $n = \sqrt{\mu/a^3}$, and a is the orbit semimajor axis. The position of the satellite on the elliptical orbit (see Fig. 18), however, must be obtained using either the eccentric anomaly E or the true anomaly (or polar angle) v where Kepler's equation is as follows: $M = n(t - T) = E - e \sin E$.

A plot of M vs E for small eccentricity e would show a straight line ($M = E$) plus a small sinusoidal ripple of amplitude ($e \sin E$). The true anomaly v is given by $v = \cos^{-1}[(e - \cos E)/(e \cos E - 1)]$, where v is in the same quadrant as E. For a perfectly circular orbit ($e = 0$) the satellite velocity is

$$v_c = \sqrt{\mu/r_c} = 631.348\sqrt{r_c} \text{ km/s} \qquad (17)$$

where v_c is the satellite velocity in circular orbit and r_c is the circular orbit radius in km.

A great deal of mathematics has been directed at solutions to Kepler's equation $E = M + e \sin E$. A series representation of the eccentric anomaly can be viewed

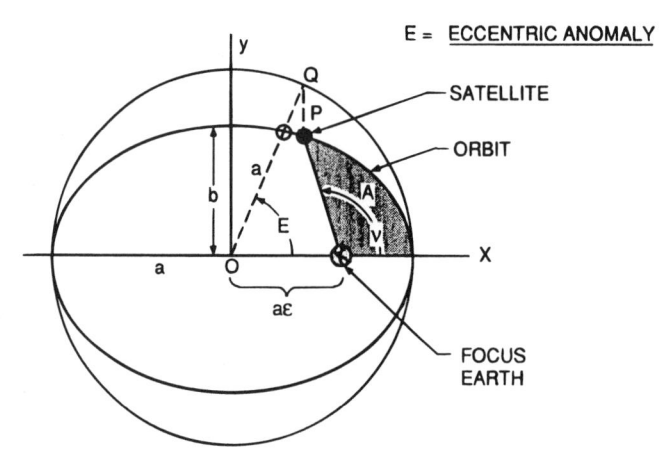

Fig. 18 Time-of-flight and eccentric anomaly E for an elliptic orbit, as viewed in the orbit plane

GPS NAVIGATION DATA

Table 20 Key relationships for Keplerian orbits

Kepler's equation $M \triangleq E - e \sin$ $E = n(t - T_o), 0 \le t \le T$ E = eccentric anomaly M = mean anomaly	Series representation of E $$E(t) = M(t) + 2\sum_{j=1}^{\infty} \frac{1}{(j+1)} J_j(j\epsilon)\sin[jM(t)]$$
$n \triangleq \sqrt{\mu/a^3}$ = mean motion T_o = orbit period = $2\pi\sqrt{a^3/\mu}$ μ = Earth's gravitational parameter e = eccentricity of the orbit v = true anomaly t = time	$\sin v(t) = \dfrac{\sqrt{1 - e^2} \sin E_k(t)}{1 - e \cos E_k(t)}$ $\cos v(t) = \dfrac{\cos E_k(t) - e}{1 - e \cos E_k(t)}$
Specific mechanical energy = $(v^2/2) - (\mu/r) = -\mu/2a$ = sum of kinetic and potential energies Radius from focus to satellite $r = a(1 - e^2)/(1 + e \cos v) = a(1 - e \cos E)$ Equation of an ellipse $(x^2/a^2) + (y^2/b^2) = 1$ For circular orbits $\epsilon = 0$, orbit velocity = $v_c = \sqrt{\mu/a}$	

as either a Fourier series or a Bessel function series with appropriate coefficients:

$$E = M + 2\sum_{j=1}^{\infty} \frac{2}{j} J_j(je) \sin(jM) \qquad (18)$$

Bessel, in fact, invented the Bessel function at least in part to solve Kepler's equation with a series solution. This expression, Eq. (18), can be obtained by defining $X(M) \triangleq E - M$ and expanding $E - M = e \sin E$ in a Fourier series[21]:

$$X(M) = E(M) - M = e \sin[X(M) + M] = \sum_{j=1}^{\infty} b_j \sin(jM)$$

where

$$b_j = \frac{2}{\pi} \int_0^{\pi} X(M) \sin(jM) dM$$

The Bessel function $J_j(z)$ integral or Neuman expansion relationships can be used in this expression[22]:

$$J_n(z) = \frac{1}{\pi} \int_0^{\pi} \cos(n\theta - z \sin \theta) \, d\theta$$

and

$$\cos(z \sin \theta) = J_0(z) + 2J_2(z)\cos 2\theta + 2J_4(z)\cos 4\theta + \cdots$$
$$\sin(z \sin \theta) = 2J_1(z)\sin \theta + 2J_3(z)\sin 3\theta + 2J_5(z)\sin 5\theta + \cdots$$

1. Numerical Methods of Solving Kepler's Equation

There is an enormous body of literature on numerical methods of solving Kepler's equation. Space permits mention of only two. The first is a method of successive substitution by Ref. 23. Using the relationship $E - M = X(M) = \epsilon \sin(X + M)$, an iterative expression for $X(M)$ can be written as follows: $X_{n+1}(M) = e \sin(M + X_n(M)), n = 0, 1, 2 \ldots$.

For $X_o = 0$ this series converges absolutely and uniformly with a residual $|X_{n+1}(M) - X_n(M)| \leq e^{n+1} |\sin(M)|$. We can also use a version of Newton's method of iteration, by defining the iterative value of M, namely $M_n \triangleq E_n - e \sin E$; e.g., as follows:

$$E_{n+1} = E_n - \frac{M_n - M}{1 - e \cos E_n} = E_n - \frac{E_n - e \sin E_n - M}{1 - e \cos E_n}$$

for $n = 1, 2, \ldots, p$

As n increases and $M_n \to M$, the iterative steps decrease in size toward zero. These iterations can be made to converge more rapidly by using an initial value,[10] which is a linear interpolation of the maximum and minimum values of E; that is, $E = M, M + e$ for $E = \{0, \pi\}$. Thus, the initial value of E_n used in the iteration is the average of the following: $E = M + e \sin M$ and $E = M + e \sin(M + e)$.

A linear interpolation of these two can give a better initial setting for E_o as the following:

$$E_o = M + \frac{e \sin M}{1 - \sin(M + e) + \sin M}$$

For a more complete discussion of methods of solution of Kepler's equation the reader is referred to Refs. 19–21 and Ref. 24.

C. Perturbation of Satellite Orbit

A satellite under the influence of a truly spherical Earth with no external effects at all would follow precisely* the Kepler orbit. However, the Earth is not truly a sphere,† and there are other effects caused by the moon and sun.[25] The oblateness of the Earth is the dominant perturbing effect, and higher-order spherical harmonics also contribute to the Earth's gravitational potential. The shape of the Earth in its

---

*There are relativistic effects on the satellite orbit, but these are negligible for our purposes here.

†There are two models of the Earth's actual physical shape, an equipotential ellipsoide of revolution, which is used here, and the equipotential surface of the Earth, the geoid. The geoid is defined in WGS-84 as "that particular equipotential surface of the Earth that coincides with mean sea level over the oceans and extends hypothetically beneath all land surfaces. In a mathematical sense the geoid is also defined as so many meters (+ N) or below (− N) the ellipsoid.

simple model is a nearly spherical ellipsoid of radius $R_e = 6378.137$ km at the equator and a flattening of 1/298.257,[26] as shown in Fig. 19.

On the surface of the ellipsoid, the x and z coordinates vary with geodetic latitude* ϕ as follows:

$$x = \frac{R_e \cos \phi}{\sqrt{1 - \epsilon^2 \sin^2 \phi}}, \quad z = \frac{R_e(1 - \epsilon^2)\sin \phi}{\sqrt{1 - \epsilon^2 \sin^2 \phi}} \quad (19)$$

where ϵ is the Earth's eccentricity $\epsilon = 0.0818192$. An observer at height h above the ellipsoid has coordinates x and z, as shown in Fig. 19:

$$x = \left(\frac{R_e}{\sqrt{1 - \epsilon^2 \sin^2 \phi}} + h\right)\cos \phi \quad \text{and} \quad z = \left(\frac{R_e(1 - \epsilon^2)}{\sqrt{1 - \epsilon^2 \sin^2 \phi}} + h\right)\sin \phi \quad (20)$$

The general representation of the Earth's gravitational potential is given in Table 21 and is referenced to the gravitational spherical harmonics of Fig. 20. The Earth has zonal, tesseral, and sectoral spherical harmonics. However, for purposes of this discussion only the J_2 term is used because the higher harmonics are on the order

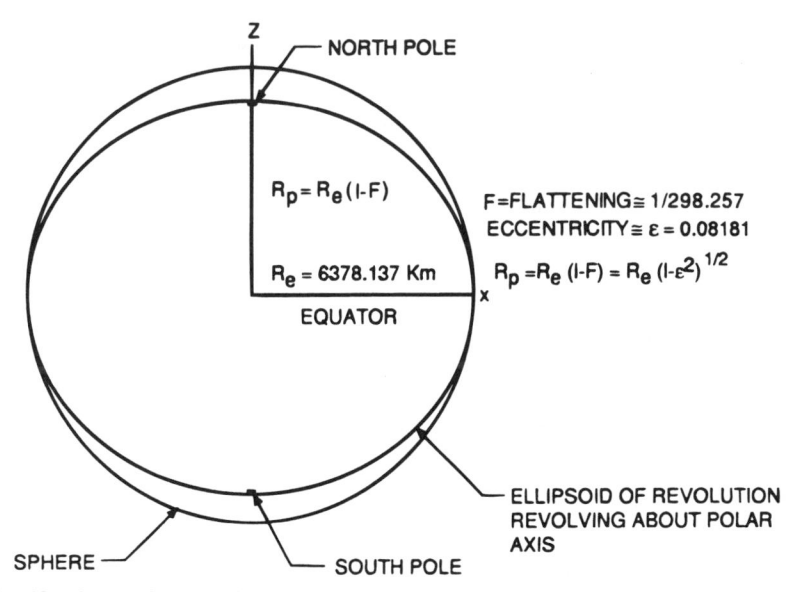

Fig. 19 Approximate ellipsoid of revolution model of Earth's oblateness—ellipticity exaggerated for effect. The ellipticity of the equator itself is very small $\cong (a_{max} - a_{min})/a_{min} = 1.6 \times 10^{-5}$ where a is the radius.

---

*Geocentric latitude is defined as the angle between the equatorial plane and the line drawn from the center of the Earth to a point on the ellipsoid.

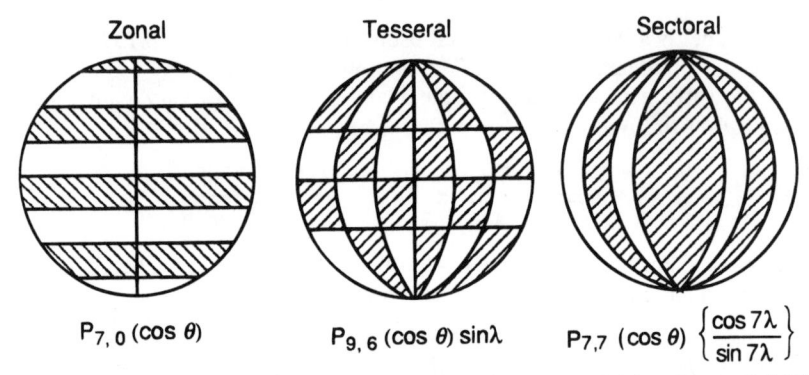

Fig. 20 Examples of spherical harmonics of the Earth's gravitational potential. The shaded and unshaded regions represent positive and negative relative potential (disturbing function).[27] The functions p_{nm} (cos ϕ) are the associated Legendre polynomials. The even numbered zonal harmonics are symmetric about the equatorial plane; e.g., $J_2 = J_{20}$, $J_4 = J_{40}$, and the odd zonal harmonics are antisymmetric; $J_3 = J_{30}$, etc.

Labels under spheres: Zonal: $P_{7,0}(\cos\theta)$; Tesseral: $P_{9,6}(\cos\theta)\sin\lambda$; Sectoral: $P_{7,7}(\cos\theta)\begin{Bmatrix}\cos 7\lambda \\ \sin 7\lambda\end{Bmatrix}$

Table 21 Generalized gravitational potential function Φ of the Earth.[5, 20, 28] Notice that harmonics higher than J_2 are less than 1/400 of the J_2 value.[a]

$$\Phi(r,\phi,\lambda) = \frac{\mu}{r}\left[1 - \sum_{n=2}^{\infty} J_n\left(\frac{R_e}{r}\right)^n P_n(\sin\phi) + \sum_{n=2}^{\infty}\sum_{m=1}^{\infty} J_{nm}\left(\frac{R_e}{r}\right)^n P_{nm}(\sin\phi)\cos[m(\lambda - \lambda_{nm})]\right]$$

where
- r = geocentric radius to satellite
- ϕ = geocentric latitude
- λ = geographic longitude
- μ = gravitational potential of the Earth
- R_e = equatorial radius of the Earth
- J_n = J_{no} = harmonic coefficients
- J_{nm} = tesseral harmonic coefficients, $m \neq 0$
- λ_{nm} = equilibrium longitude for J_{nm}

The zonal harmonics are (5th and higher order are on the order of 10^{-7} or less):
- J_2 = 1082.6300×10^{-6}
- J_3 = $-2.5321531 \times 10^{-6}$
- J_4 = $-1.6109876 \times 10^{-6}$

The higher order cosine and sine tesseral harmonics are on the order of 10^{-6} or less.

[a]The actual WGS-84 relationship is

$$\Phi = \frac{\mu}{r}\left[1 + \sum_n \sum_m \left(\frac{R_e}{r}\right)^n \bar{P}_{nm}(\sin\phi)(C_{nm}\cos m\lambda + S_{nm}\sin m\lambda)\right] \text{ where } C_{20} = -J_2\sqrt{5}.$$

of 1/400 of the J_2 term or less. The approximate potential function* representation is[22]

$$\Phi(r, \phi, \lambda) = \frac{\mu}{r}\left[1 - \frac{J_2}{2r^2}(1 - 3\sin^2\phi)\right]$$

where J_2 is a coefficient determined by experimental observations and ϕ is the geographic latitude. The quantity J_2 is the Earth quadrapole moment and $(1 - 3\sin^2\phi)$ is the Legendre polynomial $P_2(\sin\phi)$.

In a simplified view, the oblateness of the Earth acts as if there were a ring of mass around the equator, which causes a torque on the orbiting satellite and causes it to precess (see Fig. 21).

The precession rate of the angle of the ascending node and the argument of the perigee is as shown[26,29] in Table 22. Note the appearance of the term $4 - 5\sin^2 i$ in the equation for the argument of the perigee in Table 22. The zeros of this equation at $i = 63.4349488$ and 116.565051 deg are inclination angles (sometimes called the critical inclination) for which first-order perturbation theory predicts the argument of perigee will neither advance nor retard.

In addition to this secular precession of the orbit there are also short-term oscillating and long-term oscillating effects on the orbital angle, as shown pictorially in Fig. 22. The principal short period effects are caused by J_2, nonzonal, and lunar/solar effects. Table 23 lists some relative magnitudes of various disturbing forces on the orbit perturbation.

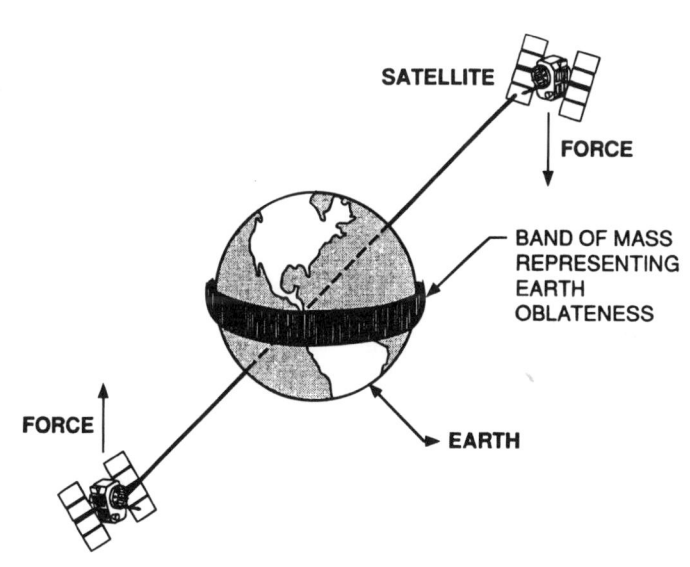

Fig. 21 Representation of the Earth oblateness by a ring of mass around the Earth. The gravitational force of this ring causes a precession of the satellite orbit plane.

---

*Note that the specific potential energy at a point is $V(r,\phi,\lambda) = -\Phi(r,\phi,\lambda)$.

Table 22 Secular precession rate of the angle of the ascending node Ω and the argument of the perigee, in general, and for a global positioning system of semimajor axis 26,600 Km and zero eccentricity, inclination $i = 60$ deg[a]

Secular precession angle rate	Lunar, solar and J_2 effect equations	Precession rate, deg/s		
		J_2 Effect	Moon	Sun
$\dot{\Omega}$	$= -1.5nJ_2(R_{e/a})^2 (\cos i)/(1 - \epsilon^2)^{-2}$ $\cong -2.06474 \times 10^{14} a^{-7/2} (\cos i)/(1 - \epsilon^2)^{-2}$	$-.033$	$-.00085$	-0.0038
	Sun: $-.00154 (\cos i)/n$ Moon: $-.00338 (\cos i)/n$			
$\dot{\omega}$	$= 0.75nJ_2(R_{e/a})^2 (4 - 5 \sin^2 i)(1 - e^2)^{-2}$ $= 1.03237 \times 10^{14} a^{-7/2} (4 - 5 \sin^2 i)(1 - e^2)^{-2}$	0.008	0.00021	0.00010
	Sun: $0.00077 \dfrac{(4 - 5 \sin^2 i)}{n}$ Moon: $0.00169 \dfrac{(4 - 5 \sin^2 i)}{n}$			

[a]These results for the sun and moon effects are approximations in that they neglect the changing orientation between the orbital plane and the Moon's orbital plane and the ecliptic plane.[21,30]

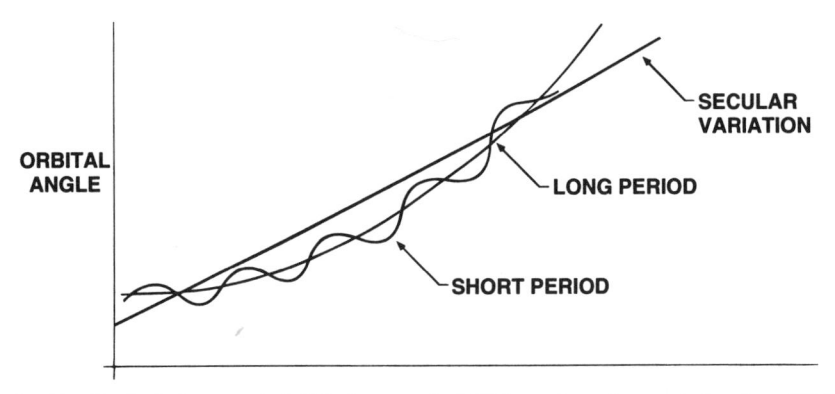

Fig. 22 Typical short period (SP), long period (LP), and secular variation in orbital parameters; e.g., ascending node angle Ω. In general, the secular variations trend linearly with time, short-term variations have a period less than or equal to the orbit period, and longer period fluctuations have a period longer than the orbit period.[21,30]

Table 23 Table of disturbing forces on the satellite orbit and the approximate magnitude of the effect (A is surface area, M cgs is satellite mass cgs)[24]

Disturbing force	Factor	Long period/secular relative magnitude
J_2	10^{-3}	1
J_n $(n > 2)$	10^{-6}	10^{-3}
Nonzonal	10^{-5} J_{22}, 10^{-6} others	10^{-4}
Lunar/solar	$3.1/n^2 \sim 10^{-5}$ synchronous 10^{-7} low altitude	10^{-4}
Solar radiation	$4.6 \times 10^{-8}(a/R_e)^2$ $(A/M) \sim 10^{-8}$	10^{-2} mean anomaly 10^{-5} other elements

For the GPS orbits, these perturbing forces on the satellite have appropriate values, as shown in Table 24. These are dominated by the second zonal harmonic (oblateness of the Earth), and lunar and solar gravity.

Table 24 Summary of the approximate perturbing forces for GPS space vehicles

Source	Maximum perturbing acceleration, m/s^2	Maximum excursion growth in one hour, m	Dominant perturbation period, h
Earth-mass attraction	5.65×10^{-1}	—	12
Second zonal harmonic	5.3×10^{-5}	300	Secular + 6
Lunar gravity	5.5×10^{-6}	40	Secular + 12
Solar gravity	3×10^{-6}	20	Secular + 12
Fourth zonal harmonic	10^{-7}	0.6	3
Solar radiation pressure	10^{-7}	0.6	Secular + 3
Gravity anomalies	10^{-8}	0.06	Various
All other forces	10^{-8}	0.06	Various

V. Ionospheric Correction Using Measured Data

The use of an ionospheric model employing parameters from subframe 4 has already been discussed. The reader is also referred to Chapter 12, this volume, where the ionospheric physics and effects on GPS are discussed in some detail. However, the user may desire accuracy greater than that obtainable using the model. This section describes the fundamentals of standard dual-frequency correction, codeless dual-frequency corrections, and a single-frequency correction method that makes use of the fact that carrier phase delay and code group delay vary in an equal and opposite manner with changes in the ionosphere.

A. Dual-Frequency Ionospheric Correction

Greatest accuracy in ionospheric correction is obtained by the user who can operate on both L_1 and L_2 signals. The dual-frequency user measures pseudoranges

pseudoranges R_{L_1} and R_{L_2} at frequencies $L_1 = f_1$, $L_2 = f_2$. The ionospheric group delay group caused by the ionosphere is approximately inversely related to frequency squared as described in the previous chapter.

$$R_{L_1} \cong R + A/f_1^2, \qquad R_{L_2} \cong R + A/f_2^2$$

where R is the path delay without the ionospheric delay error, and A is dependent on the integrated electron content of the ionosphere—an unknown. The corrected pseudorange is then obtained by using a weighted difference of the two pseudoranges as follows:

$$R = R_{L_1} - (R_{L_2} - R_{L_1})/(\Gamma - 1) = R_{L_1} - \frac{\delta R}{(\Gamma - 1)} = R_{L_1} - 1.545728(\delta R)$$

where $\delta R \triangleq R_{L_2} - R_{L_1}$. It is important to point out that R_{L_1} and R_{L_2} have independent additive noise errors in their measurements; i.e., the measurements of R_{L_1} and R_{L_2} are as follows:

$$\hat{R}_{L_1} = R + A/f_1^2 + n_1(t), \qquad \hat{R}_{L_2} = R + A/f_2^2 + n_2(t)$$

where n_1 and n_2 have rms values σ_{n1}, σ_{n2} When $\hat{R}_{L_1}$ and $\hat{R}_{L_2}$ are subtracted, the weighted difference term $\delta \hat{R}$, that is, the delay correction term is as follows:

$$\delta \hat{R}/(\Gamma - 1) = 1.545728 \, [A((1/f_2^2) - (1/f_1^2)) + n_2(t) - n_1(t)]$$
$$= \delta \hat{R}/(\Gamma - 1) = 1.545728[A((1/f_2^2) - (1/f_1^2)) + n_2(t) - n_1(t)]$$

has a variance $\sigma^2 = 2.389 \, (\sigma_{n2}^2 + \sigma_{n1}^2)$.

This noise error is substantially larger (by more than 3.8 dB because $\sigma_{n2}^2 > \sigma_{n1}^2$ than the noise error on the L_1 pseudorange by itself. Fortunately, however, the ionosphere drift varies only very slowly with time (except for the small scintillation effects), and the noise can be heavily filtered to produce a very accurate estimate of the slowly varying ionospheric delay.

1. Codeless Dual-Frequency Ionospheric Corrections

Civil users of GPS may not have access to the P(Y) code on the L_2 channel. In this situation, the user can still make use of dual-frequency ionospheric measurements with codeless techniques that operate by:

- squaring the L_2 channel to recover the L_2 carrier, or
- cross-correlating the L_1 and L_2 channels in a delay lock fashion with a noisy reference.

Figure 23 shows these two techniques in simplified block diagram forms. In Fig. 23a, the receiver channel employs a squaring operation. The received L_2 signal generally contains a summation of GPS P(Y)-code signals of the following form:

$$\sum_i s_i(t + \tau_i) \sin(\omega_0 t + \phi_i)$$

where $\phi_i(t) = \omega_0 t(t + \tau) + \phi_{Ii}$ and ω_0 is the L_2 carrier frequency. The doppler term $\omega_0 \dot{\tau}_i(t)$ varies in a different manner for each satellite, and ϕ_{Ii} is the ionosphere

GPS NAVIGATION DATA

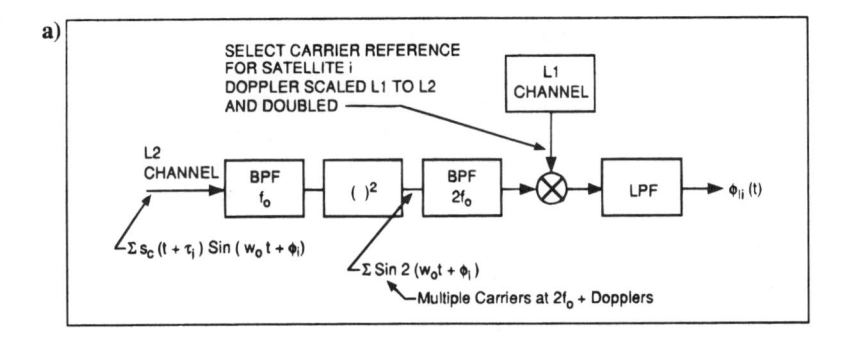

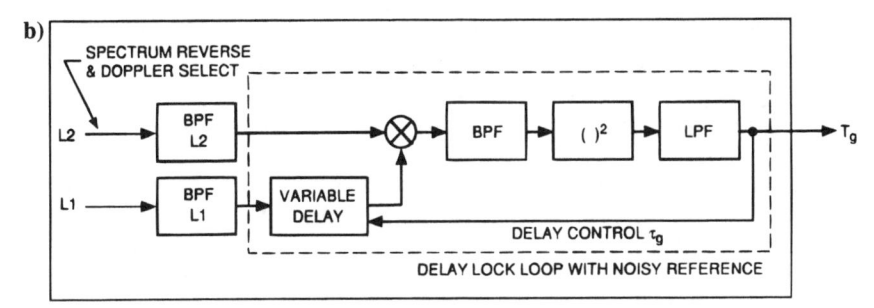

Fig. 23 Simplified block diagrams of codeless and partially codeless L_2 ionospheric correction techniques; a) squaring loop, and b) $L_1 - L_2$ cross-correlation loop.

in phase shift. Squaring the summation of these signals removes the spreading code since $s_i^2 = 1$ and results in the sum of pure carriers plus noise (including self-noise which is produced from the products $s_i(t)s_j(t)$ of the uncorrelated P(Y) codes from different satellites and is negligible compared to thermal noise effects). Each of the satellite signals has a different Doppler profile that can be used to select the desired signal and can be determined precisely from the L_1 C/A carrier. The individual carriers can then be tracked by making use of the recovered carriers from the L_1 channel. The only difference between the L_1 Doppler from the L_2 doppler, aside from the L_1 to L_2 frequency ratio, is the ionospheric phase shift, which is recovered as shown. Note that the L_2 Doppler in Fig. 23a is doubled in the squaring operation.

In Fig. 23b, the L_1 channel is cross-correlated with the L_2 channel after a variable delay is introduced in a delay–lock-loop tracking system[31] to recover the weighted ionospheric group delay τ_{iono}. (Note that not all satellites have the same obliquity factor and hence not all the same ionospheric delay.) The individual satellite signal ionospheric delay can be tracked by reversing the spectrum of L_2 so that the Doppler offsets add rather than subtract and then selecting the desired satellite signal through the use of the known Doppler for the desired signal recovered in the L_1 C/A channel.

2. Single-Frequency Ionospheric Correction and the Ionospheric Model

The single-frequency user must either ignore the ionosphere and accept the ionospheric error, use a model of the ionosphere, or use a single-frequency carrier/code differential ionospheric measurement scheme. The basis for the single-frequency correction is that the code and carrier are phase shifted by the ionosphere by an equal and opposite amount. Thus, if we measure the code delay change and the carrier delay change, the difference between these two quantities is twice the ionospheric group delay change because the conventional Doppler shift cancels. Similarly, if we simply add the two delays, the ionospheric effect cancels, as shown in the next paragraph.

3. Ionosphere Group and Phase Delay—Example with Two Sinusoids

The ionospheric group delay varies inversely with frequency squared, as shown in Fig. 24; namely, $\tau(\omega) = k/\omega^2$ where $k = (2\pi)^2 A$. Because group delay $\tau(\omega) \triangleq d\phi(\omega)/d\omega$, where $\phi(\omega)$ is the phase delay, the corresponding phase delay varies inversely with frequency as $\phi(\omega) = -k/\omega$.

As a simple example, apply a signal $s(t)$, which is the sum of two pure sinusoids centered at frequency ω_0 to this channel: $s(t) = \cos(\omega_0 + \Delta)t + \cos(\omega_0 - \Delta)t = (2 \cos \Delta t) \cos \omega_0 t = B(t)\cos \omega_0 t$ where it will be assumed that $\Delta \ll \omega_0$. Because each of the two carriers is phase shifted by $\phi(\omega)$, the output $r(t)$ of the channel is easily shown to be

$$r(t) = 2 \cos \Delta[t - (k/\omega_0^2)]\cos\{\omega_0[t + (k/\omega_0^2)]\}$$

$$= B[t - (k/\omega_0^2)]\cos \omega_0 [t + (k/\omega_0^2)].$$

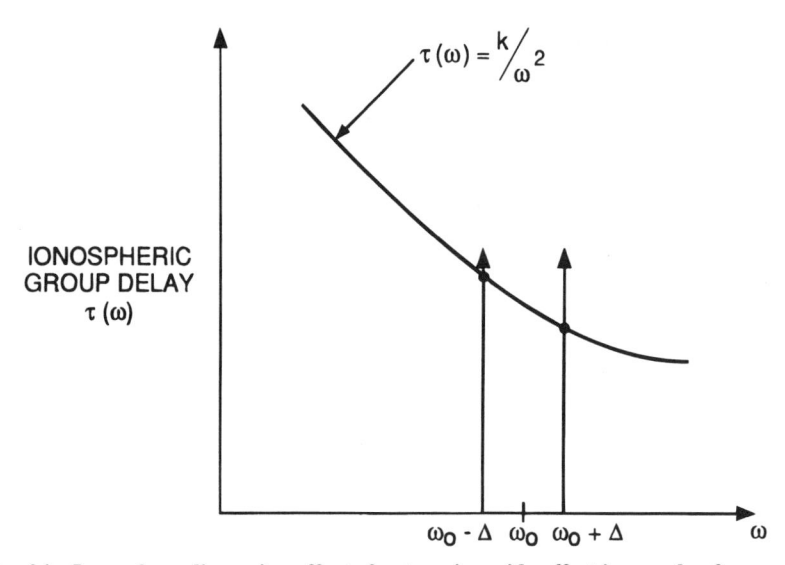

Fig. 24 Ionosphere dispersion effects for two sinusoids offset in angular frequency by 2Δ.

GPS NAVIGATION DATA

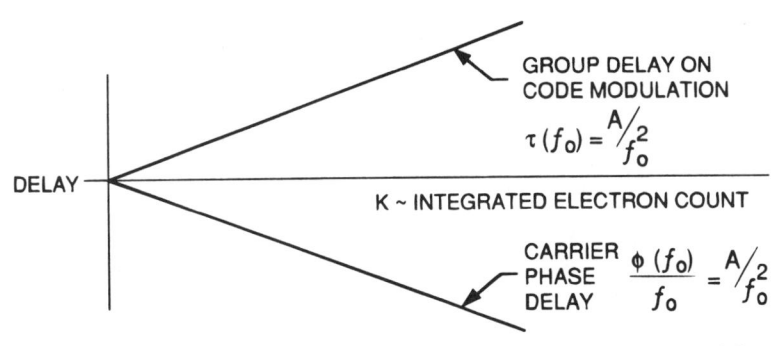

Fig. 25 Impact of ionosphere on pseudorange and phase delay. The group delay and phase delay are equal and opposite for a given ionosphere electron count parameter A.

Note that the envelope term $B(t)$ is delayed by the group delay while the carrier phase is advanced. Thus, the group delay effect on the envelope and phase delay effect on the carrier are exactly equal and opposite, as shown in Fig. 25. The same effect would be observed if $B(t)$ were some more complex waveform; e.g., a coded GPS signal. The code would be delayed while the carrier phase advances (see also Refs. 32 and 33).

This equal and opposite phase and group delay must be carefully taken into account because the ionospheric delay changes with time in any receiver design. For example, carrier tracking of the received signal can be used to aid in tracking the signal envelope modulation. However, changes in the ionosphere causes these two measurements to shift in opposite directions. On the other hand, the equal and opposite relationship can, in some situations, be employed to cancel the effects of changes in the ionosphere. For example, we can use a single-frequency receiver capable of simultaneously tracking both envelope modulation (group delay) and carrier phase (phase delay). The average of these two measurements, properly scaled, is immune to changes in the ionosphere. Similarly, if we measure both carrier and code delay and subtract them, the difference is twice the ionospheric delay. Note, however, that because the carrier phase has an initial ambiguity, this approach only applies to changes in the ionospheric delay and not to its absolute value.

Appendix

Table A1 Physical constants[a]

Symbol	Constant	Source
$a = R_e$	Semimajor axis of Earth's reference ellipsoid = 6378137.0 m	WGS-84[5]
c	Velocity of light 2.99792458×10^8 m/s	ICD-GPS-200[1]
C/A-code	Coarse/acquisition-code @ $f_c = 1.023$ Mb/s	ICD-GPS-200[1]
e	Eccentricity of the Earth's orbit = 0.017	
ϵ	Earth ellipsoid eccentricity $\epsilon = 0.0818191908426$	WGS-84[5]
F	Earth flattening $1/298.257$, $F = 1 - (1 - \epsilon_e^2)^{1/2} = 0.00335281066474$; $1/F = 298.257223563$	IAU (1976)

(Table A1 continues on next page)

Table A1 Physical constants (continued)[a]

Symbol	Constant	Source
G	Obliquity of the ecliptic, at standard epoch 2000	
J_2	Earth quadripole moment or dynamical form-factor for Earth = 0.00108263	IAU (1981)[34]
k	Boltzman's constant 1.380658×10^{-23} J/K	
L_1	L_1 GPS frequency 1575.42 MHz	ICD-GPS-200[1]
L_2	L_2 GPS frequency 1227.6 MHz	ICD-GPS-200[1]
μ	3.986005×10^{14} m^3/s^2 value of Earth's gravitational constant. This value of μ includes the Earth's atmosphere. If the Earth's atmosphere is excluded the WGS-84 value of μ is $\mu' = 3.9860015 \times 10^{14}$ m^3/s^2	WGS-84[5]
M_d	Mean molar mass of dry air 28.9644 kg/kmol	
M_e	Mass of the Earth 5.9733328 WGS-84, includes Earth's atmosphere $\times 10^{24}$ kg	
M_m	Mass of the moon 7.35×10^{22} kg	
M_s	Mass of the Sun 1.9891×10^{30} kg	
M_w	Molar mass of water = 18.0152 kg/kmol	
p	General precession in longitude, per Julian century, at standard epoch 2000	
P-Code	Precision-code @ f_c = 10.23 Mb/s, also P(Y)-code	ICD-GPS-200[1]
π	GPS value of π = 3.1415926535898	ICD-GPS-200[1]
1 Atmosphere	1013.25 mbars	
1 kJ	10^3 kg m^2/s^2 = 10^{10} erg	
R	Universal gas constant 8.31434 kJ-/kmole · K	
R_e	Equatorial radius for Earth (IAG 1975); [IUGG value 1980] = 6378137 m	WGS-84[5]
R_{me}	Moon–Earth distance 3.84×10^8 m	
R_{se}	Sun–Earth distance 1.50×10^{11} m	
0 deg C	273.16 K	
$\dot{\Omega}_e$	Earth's rotation rate $7.2921151467 \times 10^{-5}$ rad/s, WGS-84[5] value of the Earth's rotation rate; rotation rate of Earth vector viewed in ECI coordinate expressed in solar time s WGS-84 also gives the angular velocity of the Earth in a precessing reference frame as $\dot{\Omega}_e^* = 7.291158553 \times 10^{-5}$ rad/s (sidereal period of rotation = 86,164.1004 s)[b]	SCD-200 WGS-84 value[5]
$2\pi/\dot{\Omega}_e$	1 sidereal day = 86,164.0989038 s	SCD-200 WGS-84 value[5]
	1 solar day = 86,400 s	
	1 nautical mile = 1852 m (exact)	
	1 international foot = 0.3048 m (exact)	

[a] A more recent gravity model than WGS-84 has been developed jointly by NASA and the University of Texas. This gravity model, Joint Gravity Model #2 or JGM-2, is used in Chapter 18, this volume. This value of μ for this model is $\mu = 3.986004415 \times 10^{14}$ m^3/s^2.
[b] WGS-84 also gives the angular velocity of the Earth in a processing reference frame as

$$\dot{\Omega}_e^* = 7.291158553 \times 10^{-5} \text{ rad/s}$$

References

[1] Anon. "GPS Interface Control Document," ICD-GPS-200, IRN-200B-PR-OOJ, Rev. B-PR, U.S. Air Force, July 1, 1992.

[2] Hamming, R. W., *Coding and Information Theory*, Prentice-Hall, Englewood Cliffs, NJ, 1986.

[3] Clark, G. C., and Cain, S. B., *Error Correction Coding for Digital Communications*, Plenum, New York, 1981.

[4] Poli, A., and Hugnet, L., *Error Correcting Codes—Theory and Applications*, Prentice-Hall, Hempstead, England, 1992.

[5] Anon., "Department of Defense World Geodetic System, 1984 (WGS-84), Its Definition and Relationships with Local Geodetic Systems," DMA-TR-8350.2, Defence Mapping Agency (DMA), Sept. 1987.

[6] Misner, C. W., Thorne, K. S., and Wheeler, J. A., *Gravitation*, W. H. Freeman, San Francisco, CA, 1973.

[7] Winkler, G. M. R., "Timekeeping and its Applications," *Advances in Electronics and Electron Physics*, Vol. 44, 1972, Academic Press, New York.

[8] D. Howse, *Greenwich Time and the Discovery of the Longitude*, Oxford University Press, Oxford, 1980.

[9] Bate, R. R., Mueller, D. D., and White, J. E., *Fundamentals of Astrodynamics*, Dover, New York, 1971.

[10] Smith, D. G., *Cambridge Encyclopedia of Earth Sciences*, Crown Publishers, Cambridge Univ. Press, New York, 1981.

[11] Allan, D. W., Granveaud, M., Klepczynski, W. J., and Lewandowski, W., "GPS Time Transfer with Implementation of Selective Availability," *Proceedings of the Precision Time and Time Interval* (PTTI), 1990.

[12] Cutler, L. S., and Searle, C. L., "Some Aspects of the Theory and Measurements of Frequency Fluctuations in Frequency Standards," *Proceedings of the IEEE*, Feb. 1966.

[13] Spilker, J. J., Jr., *Digital Communications by Satellite*, Prentice Hall, Englewood Cliffs, 1977, 1995.

[14] Barnes, J. A., et al., "Characterization of Frequency Stability," National Bureau of Standards, TN 394, U.S. Dept. of Commerce, Washington, DC, 1970.

[15] H. Cramer, *Mathematical Methods of Statistics*, Princeton Univ. Press, Princeton, NJ, 1946.

[16] Orfanidis, S. J., *Optimal Signal Processing*, McGraw Hill, New York, 1988.

[17] E. Kreyszig, *Advanced Engineering Mathematics*, Wiley, New York, 1993.

[18] McCaskill, T. B., et al., "Effect of Broadcast and Precise Ephemerides on Estimates of the Frequency Stability of GPS Navstar Clocks," *International Technical Meeting*, Institute of Navigation, Washington, DC, Sept. 1993.

[19] Battin, R. H., *An Introduction to the Mathematics and Methods of Astrodynamics*, AIAA Education Series, AIAA, Washington, DC, 1987.

[20] Chobotov, V. A., (ed.), *Orbital Mechanics*, AIAA, Washington, DC, 1991.

[21] Escobar, P. R., *Methods of Orbit Determination*, Krieger, Malabar, FL, 1976.

[22] Whitaker, E. F., and Watson, G. N., *A Course of Modern Analysis*, Cambridge Univ. Press, 1958.

[23] Kurth, R., *An Introduction to the Mechanics of the Solar System*, Pergamon, New York, 1959.

[24] Taff, L. G., *Celestial Mechanics*, Wiley, New York, 1985.

[25] Kaula, W. K., "Theory of Satellite Geodesy," Blaisdell, 1966.

[26] Larson, W. J., and Wertz, J. R., (eds.), *Space Mission Analysis and Design,* 2nd ed., Kluwer Academic, Dordrecht, The Netherlands, 1992.

[27] Burkard, R. K., "Geodesy for the Layman," U.S. Dept. of Commerce, AD-670156, Feb. 1968.

[28] Leick, A., *GPS Satellite Surveying,* Wiley, New York, 1990.

[29] Goldstein, H., *Classical Mechanics,* Addison-Wesley, Reading, MA, 1980.

[30] Boyden, D. G., "Space Mission Geometry," *Space Mission Design and Analysis,* edited by J. R. Wertz and W. J. Larson, Kluwer, Dordrecht, The Netherlands, 1992.

[31] Spilker, J. J., Jr., and Magill, D. T., "The Delay Lock Discriminator—An Optimum Tracking Device," *Proceedings of the IRE,* Sept. 1961, pp. 1403–1416.

[32] Muhleman, D. O., and Johnston, I. D., "Radio Propagation in the Solar Gravitational Field," *Physical Review Letters,* Aug. 1966.

[33] MacDoran, P. F., "DRVID Charged Particle Measurements with a Binary-Coded Sequential Acquisition Ranging System," JPL Space Programs Summary, March 1970.

[34] Kaplan, G. H., "The IAU Resolutions on Astronomical Constants, "Time Scales and the Fundamental Reference Frame," U.S. Naval Observatory Circular 163, 1981.

Chapter 5

Satellite Constellation and Geometric Dilution of Precision

J. J. Spilker Jr.*
Stanford Telecom, Sunnyvale, California 94089

I. Introduction

THE GPS satellite constellation is selected to satisfy many different conditions to provide worldwide three-dimensional navigation. The satellite constellation has been placed in an inclined orbit at 55 deg (formerly 63-deg inclination for earlier Block I satellites) to provide full Earth coverage, whereas a purely equatorial orbit cannot provide coverage above approximately 72° latitude at the GPS orbit altitude. The requirement for a minimum of four simultaneous pseudorange measurements with good geometry directly leads to a constellation of many satellites. A nongeosynchronous orbit was selected to permit the use of carrier phase/Doppler measurement profiles in addition to pseudorange measurements of code phase/delay. This objective was one of the reasons for the selection of the 12 sidereal hour orbit.

This chapter discusses the operational satellite constellation of 24 satellites termed the GPS-24 (closely related to the previous Primary 21) constellation. The specific characteristics of this constellation are discussed in terms of the statistics of the numbers of satellites visible at different user latitudes and various elevation angles, the satellite ground tracks, and the signal Doppler shifts. The potential for the future augmentation of this GPS constellation by adding satellites in either the same planes or as geostationary satellites is also discussed.

The chapter then continues with an analysis of the *geometric dilution of precision* (GDOP) and the related DOPs: PDOP, HDOP, VDOP, TDOP for position, horizontal, vertical, and time dilutions of precision. The bounds on the minimum value of these DOPs are analyzed for various numbers of satellites for both two- and three-dimensional problems with various constraints on elevation angle and user clock stability. Further discussion and development of the DOP concept is found in Chapter 11, this volume. The chapter concludes by describing the GDOP for the GPS-24 satellite constellation with a single satellite outage.

Copyright © 1994 by J. J. Spilker Jr. Published by the American Institute of Aeronautics and Astronautics, Inc., with permission. Released to AIAA to publish in all forms.
*Ph.D., Chairman of the Board.

II. GPS Orbit Configuration, GPS-24

As mentioned earlier, the operational constellation is defined as GPS-24. The selected satellite orbital constellation contains 24 satellites in six orbit planes. There are four satellites in each of the six planes, which are shown in Fig. 1. The satellites have a period of 12 h *sidereal time** and a semimajor axis of 26,561.75 km.† A sidereal day is defined as the time for the Earth to complete one revolution on its axis in Earth–Centered-Inertial (ECI) space and consists of 24 sidereal hours where 1 sidereal day is slightly shorter than a mean solar day (see Chapter 4, this volume). One sidereal day is 23 h, 56 min, 4.009054 s or 86,164.09054 s of mean solar time. One mean sidereal day is equal to 0.997269566 mean solar day. The satellites are inclined with respect to the equatorial plane by 55 deg. Table 1 summarizes the nominal parameters of the GPS orbit.

A. GPS Orbit-Semi-Major Axis

The nominal period T_{po} of an orbit around the Earth is as follows:

$$T_{po} = \frac{2\pi}{\sqrt{\mu}} a_o^{3/2} \quad \text{and} \quad a_o = (T_{po}^2 \mu/(2\pi)^2)^{1/3} \tag{1}$$

where a_o is the semimajor axis in m, and μ and π are defined for GPS using

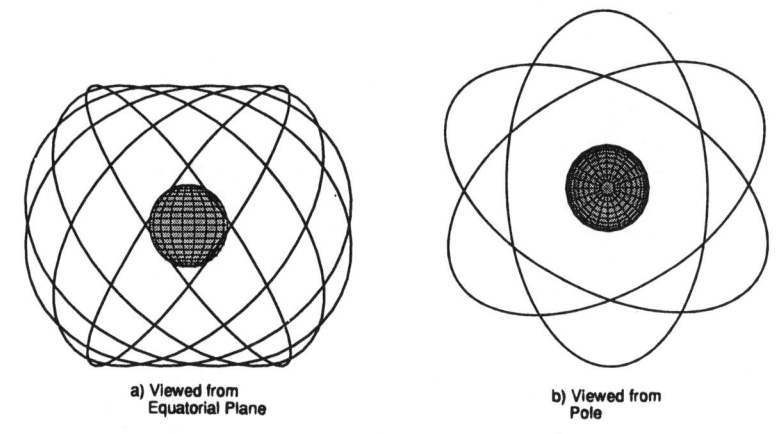

a) Viewed from Equatorial Plane

b) Viewed from Pole

Fig. 1 GPS-24 satellite constellation: a) the six orbit planes inclined at 55 deg shown in Earth–centered-inertial (ECI) coordinates viewed from the equatorial plane. The GPS constellation has four satellites in six equally spaced orbit planes. Note that the symmetrical satellite orbit planes are superimposed in part b. The present GPS-24 satellite constellation contains four nonequally spaced satellites in six orbit planes. The satellite phasing is designed to minimize the effect of satellite outage. The satellite phasing is shown in Fig. 2.

---

*A satellite with a 12-h orbit in sidereal time rotates once in inertial space in 12 h of sidereal time. The semimajor axis for a 12-h sidereal orbit is 26,561.75 km and has an altitude above the Earth equatorial radius of 6378.137 km of 20,183.6 km.

†Strictly speaking, the GPS satellites have a 12-h orbit (sidereal time) wherein the ground tracks repeat to within ± 2° in longitude. If the ground track drifts off from its nominal value by more than 2°, the GPS control segment makes a minor orbit correction.

Table 1 Approximate GPS satellite parameters and physical constants

Parameter	Value	
Orbit plane spacing	6 equally spaced ascending nodes at 120 deg	
Orbit radius r_{cs}	26,561.75 km semimajor axis	
Orbit velocity (circular) (ECI)	$\cong \sqrt{\mu/r_{cs}} = 3.8704$ km/s	
Eccentricity	Nominally zero but generally less than $e = 0.02$	
ω_s angular velocity	$2 \times 7.29211 \times 10^{-5}$ rad/s	
Period	12-h mean sidereal time	
Inclination	$i = 55$ deg nominal	
Velocity of light c	2.99792458×10^8 m/s	GPS-ICD-200[a]
Earth's gravitational parameter μ	3.986005×10^{14} m^3/s^2	GPS-ICD-200 WGS-84 value[b]
Pi, π	3.141592653898	GPS ICD-200
Earth's rotation rate $\dot{\Omega}_e$ in inertial space	$7.2921151467 \times 10^{-5}$ rad/s	ICD-200 WGS-84 value
One sidereal day $2\pi/\dot{\Omega}_e$	86164.0989038 s	WGS-84 (calculated from $\dot{\Omega}_e$)
One solar day	86400 s	
J_2 harmonic	1.08268×10^{-3}	

[a] GPS-ICD 200, GPS Interface Control Document.
[b] WGS-84, 1984 World Geodetic System.

WGS-84 and GPS-ICD-200 values as shown in Table 1. The values of a_o and T_{po} neglect the effect of Earth oblateness. If the period of the orbit is set as 1/2 of a sidereal day, (12 h sidereal time), then the nominal value of the semimajor axis a_o is 26,561.765 km. However, because of Earth oblateness, the orbit period must be corrected slightly.*

The orbit mean motion (angular) of the satellite with the J_2 correction for Earth oblateness is as follows[1]:

$$\bar{n} = \sqrt{\frac{\mu}{a^3}} \left[1 + \frac{3}{2} \frac{J_2 R_e^2}{p^2} \left(1 - \frac{3}{2} \sin^2 i\right)(1 - e^2)^{1/2} \right] \quad (2)$$

where a is the corrected semimajor axis at epoch; e = eccentricity; $p = a(1 - e^2)$; R_e is the Earth equatorial radius; i is the orbit inclination; and $J_2 \cong 1082.68 \times 10^{-6}$. For $e \cong 0$, this equation reduces to the following:

$$\bar{n} \cong \sqrt{\frac{\mu}{a^3}} \left[1 + \frac{3}{2} \frac{J_2 R_e^2}{a^2} \left(1 - \frac{3}{2} \sin^2 i\right) \right] \} \quad \text{for} \quad e \cong 0$$

$$\cong \sqrt{\frac{\mu}{a^3}} (1 + \Delta) \quad (2a)$$

where $\Delta = (3/2) J_2 [R_e/a]^2 [1 - (3/2) \sin^2 i]$ for $e \cong 0$.

---
*There is also a correction for solar pressure, which is dependent on the spacecraft size and shape, that is neglected here.

The period of the orbit corresponds to $\bar{n}T_p = 2\pi$, and then is given by the following:

$$\bar{n}T_p = 2\pi, \quad \text{or} \quad T_p = 2\pi/\bar{n} \cong 2\pi\sqrt{\left(\frac{a^3}{\mu}\right)/[1 + \Delta]} \tag{3}$$

for $T_p = 12$ h sidereal time, and $\Delta \ll 1$. Solve Eq. (3) for the semimajor axis a to obtain the following:

$$a \cong \left[\mu\left(\frac{T_p}{2\pi}\right)^2 (1 + \Delta)^2\right]^{1/3} \tag{4}$$

noting that Δ is inversely proportional to a^2.

For the GPS orbit with an orbit inclination of 55 deg and zero eccentricity, $\Delta \cong -6.098 \times 10^{-7}$. Thus, the second-order effect on Δ by the variation of a can be neglected, and the corrected value of the semimajor axis is as follows:

$$a \cong \left[\frac{(1 + \Delta)^2}{(2\pi)^2} \mu T_p^2\right]^{1/3} = \left[\frac{\mu T_p^2}{(2\pi)^2}\right]^{1/3} (1 + \Delta)^{2/3}$$

$$\cong \left[\frac{\mu T_p^2}{(2\pi)^2}\right]^{1/3} \left(1 + \frac{2}{3}\Delta\right) \quad \text{for} \quad \Delta \ll 1 \tag{5}$$

For GPS, the J_2 correction is relatively small and only corresponds to a 0.0108 = km reduction in the semimajor axis.*

Thus, the corrected value of a is 26561.754 km.

B. GPS Orbit–Satellite Phasing

Table 2 lists the orbital parameters for the GPS-24 satellite constellation. The satellites all have a design eccentricity of $e = 0$ and an inclination $i = 55$ deg. The angle Ω is the right ascension of the ascending node measured in inertial coordinates from the vernal equinox. The *lan* is the longitude of the ascending node in ECEF coordinates at the epoch time 7,1,0,0,0 of year 1993. The quantity M is the mean anomaly, which is the satellite phase for a circular orbit. If the 24 satellites were equally spaced in each orbit plane, there would be a 360 deg/ 4 = 90 deg separation in M.

In Table 2, note that the longitude of the ascending node *lan* is expressed in Earth-centered, Earth-fixed (ECEF) rotating coordinates. The right ascension of ascending node Ω for each of the six planes *A, B, C, D, E,* and *F* as expressed in ECI coordinates is the same for all four satellites in each individual plane and are offset by 60 deg from one another. All of the phase angles and longitudes of the ascending node *lans* (expressed in ECEF coordinates) are different. Thus, there are 24 different ground tracks. Units are in km and deg. The satellite phases in each of the planes are shown in Fig. 2. Notice that in inertial space, the satellite planes are equally spaced in longitude relative to the vernal equinox, but the

---

*Note that at the earlier Block I satellite inclination of 63 deg, the semimajor axis reduction is 29.29 times as large.

Table 2 GPS-24 satellite constellation

	ID	a	i	Ω	M	ΔM	lan
1	A3	26561.75	55.0	272.847	11.676	103.55	179.63
2	A4	26561.75	55.0	272.847	41.806	31.13	14.69
3	A2	26561.75	55.0	272.847	161.786	119.98	74.68
4	A1	26561.75	55.0	272.847	268.126	106.34	127.85
5	B1	26561.75	55.0	332.847	80.956	130.98	94.27
6	B2	26561.75	55.0	332.847	173.336	92.38	140.46
7	B4	26561.75	55.0	332.847	204.376	31.04	155.98
8	B3	26561.75	55.0	332.847	309.976	105.6	28.78
9	C1	26561.75	55.0	32.847	111.876	100.08	169.73
10	C4	26561.75	55.0	32.847	241.556	129.68	54.57
11	C3	26561.75	55.0	32.847	339.666	98.11	103.62
12	C2	26561.75	55.0	32.847	11.796	32.13	119.69
13	D1	26561.75	55.0	92.847	135.226	100.07	61.40
14	D4	26561.75	55.0	92.847	167.356	32.13	77.47
15	D2	26561.75	55.0	92.847	265.446	98.09	126.51
16	D3	26561.75	55.0	92.847	35.156	129.71	11.37
17	E1	26561.75	55.0	152.847	197.046	130.98	152.31
18	E2	26561.75	55.0	152.847	302.596	105.55	25.09
19	E4	26561.75	55.0	152.847	333.686	31.09	40.63
20	E3	26561.75	55.0	152.847	66.066	92.38	86.82
21	F1	26561.75	55.0	212.847	238.886	103.54	53.23
22	F2	26561.75	55.0	212.847	345.226	106.34	106.40
23	F3	26561.75	55.0	212.847	105.206	119.98	166.39
24	F4	26561.75	55.0	212.847	135.346	30.00	1.46

Units are in km for a, and degrees elsewhere, identification ID = Plane/slot; a = Semimajor axis, size in km; i = Inclination, deg; Ω = Right ascension of ascending node, deg; M = Mean anomaly, deg; lan = Longitude of ascending node, deg; ΔM = Phase difference in mean anomaly to adjacent satellite in the same plane.

satellites themselves in each plane are not equally spaced. In fact, two of the satellites in each plane are spaced by between 30.0 and 32.1 deg. If the remaining two satellites were equally spaced in the remaining 330 deg, that would put the spacing of the other satellites at approximately 330 deg/3 = 110 deg. The actual separation varies from 92.38 to 130.98 deg. The spacing has been optimized to minimize the effects of a single satellite failure on system degradation. As discussed later in the chapter, even with a single satellite failure, the PDOP does not exceed six for more than 25 min/day for a user constraint on satellite elevation angles of 5 deg or more.

II. GPS Satellite Visibility and Doppler Shift

Some important characteristics of the GPS satellite constellation as they relate to user receiver navigation performance are the number of satellites in view and the range of Doppler shifts. It is critical that at least four satellites be in view, but it is highly desirable that five or more be in view at all times. When one

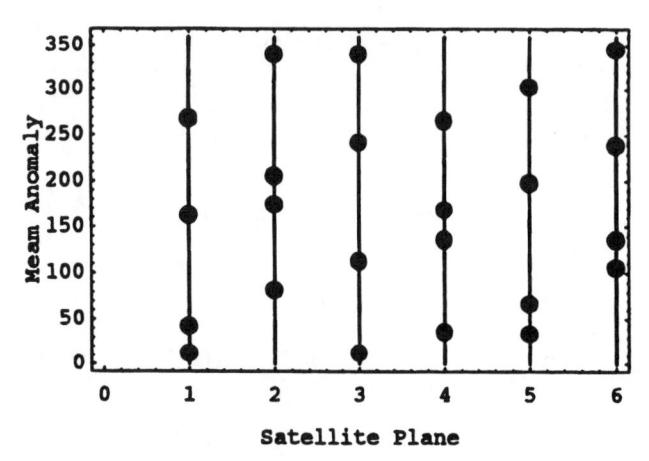

Fig. 2 GPS-24 satellite mean anomaly (satellite phase) in each of the six planes at the epoch 7,1,0,0,0 in 1993 (ECI space). Time is in universal clock time (UTC).

satellite is going out of view, the user receiver must begin to transition to another satellite as its replacement. Furthermore, four satellites by themselves may not provide a sufficiently low GDOP because of poor geometry at certain times.

A. Bound on Level of Coverage for 24 Satellites

A general class of circular orbit satellite constellations with equally spaced satellites and orbit planes has been defined by Walker.[2] In this family of constellations, there are T total satellites in P uniformly spaced planes of circular satellites, each plane at inclination angle i with respect to the equatorial plane. There are T/P uniformly spaced satellites in each plane. The relative phasing between satellites in adjacent planes is given by F, which is in units of 360 deg/T; i.e., if a satellite in one plane is just crossing through the equatorial plane in the northerly direction, the adjacent plane satellite is offset by an angle (360 deg/T)F below the equatorial plane. Thus, the constellation can be described by the notation ($T/P/F$), and the inclination angle by i. For example, if there are $T = 18$ satellites, and the relative phasing between satellites is $F = 2$, then the relative phasing between satellites in adjacent planes as they pass through the equator is (360 deg/T)F = (360/18)2 = 40 deg.

The degree of worldwide multiple satellite coverage can be determined by computing the maximum Earth angle separation between satellites for a cone that contains N satellites. That is, in any conical section of the sphere of satellites there must be N satellites in view for a conical angular separation of 2β deg where β deg is determined by the minimum allowed user elevation angle E (see Fig. 3). This N-satellite visibility must be maintained at all user coordinates and at all times.

Walker[2] has shown that continuous worldwide coverage with at least six satellites in view everywhere is possible with 24 satellites in six planes using a 24/6/1 constellation at an inclination angle of 57 deg for users with a minimum elevation angle of 7 deg. The maximum Earth angle separation between satellites of $\beta = 69.9$ deg for this constellation corresponds to $E \approx 7$ deg minimum elevation

SATELLITE CONSTELLATION AND GEOMETRIC DILUTION 183

```
EARTH RADIUS ................... ≈ 6378.1 km
SATELLITE ALTITUDE .......... 20,184 km
SATELLITE ORBIT RADIUS... 26,561.7 km
```

ELEVATION ANGLE E°

CONE OF VISIBILITY

h = 20184 km

6378 km
EARTH RADIUS

β = 71.2
FOR ELEVATION
ANGLE E = 5°

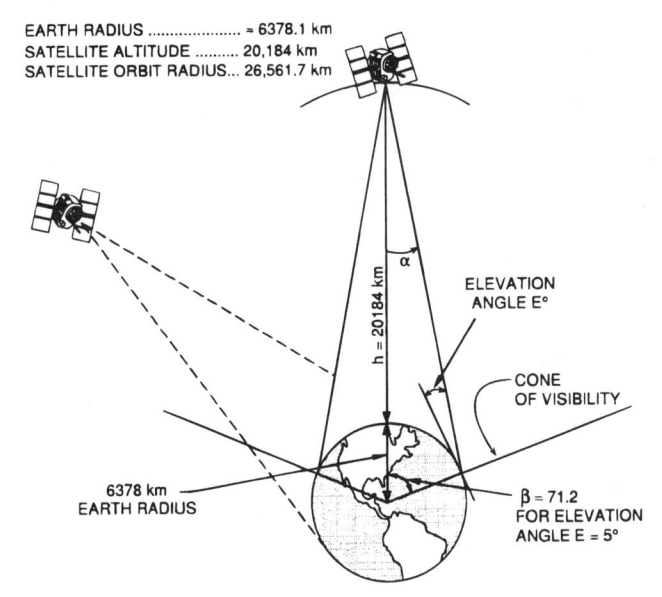

Fig. 3 Ground coverage by GPS satellites. The user is restricted to receiving satellites above a minimum elevation angle E.

angle (see Fig. 4a). Sevenfold coverage is achievable with a (24/8/4) constellation with a smaller minimum elevation angle of 3 deg is too low for normal operation. However, this constellation's seven-satellite minimum coverage is obtained with the maximum Earth angle separation between satellites of $\beta = 75.8$ deg and the corresponding minimum elevation angle of 3 deg which is too low for normal operation. It should also be noted that the maximum satellite visibility does not necessarily correspond to the smallest GDOP nor to the best performance.

The selected GPS-24 satellite constellation is shown later to give fivefold visibility. Although it does not have as good a full constellation satellite visibility as the (24/6/1) constellation, the GPS-24 satellite constellation has instead been selected on the basis of best coverage if a single satellite becomes inoperative.

B. GPS Satellite Visibility Angle and Doppler Shift

The half angle subtended by the users as viewed from the satellite is termed α as shown earlier in Fig. 3. Each GPS satellite broadcasts to the Earth with an antenna coverage pattern that somewhat exceeds the angle $\alpha = 13.87$ deg subtended by the Earth, as shown in Fig. 3. That is, the satellite antenna pattern extends beyond the edge of the Earth, in order to provide coverage to user satellites not shadowed by the Earth. Each satellite covers users within $\pm \beta$ deg of the subsatellite point, where β is determined by the minimum elevation angle. The subsatellite point can be defined by: A line drawn between the Earth's center, and the satellite intersects the Earth's surface at the subsatellite point. (For a spherical Earth, the satellite is at the zenith of a user at the subsatellite point.) Thus, the fraction of the Earth covered

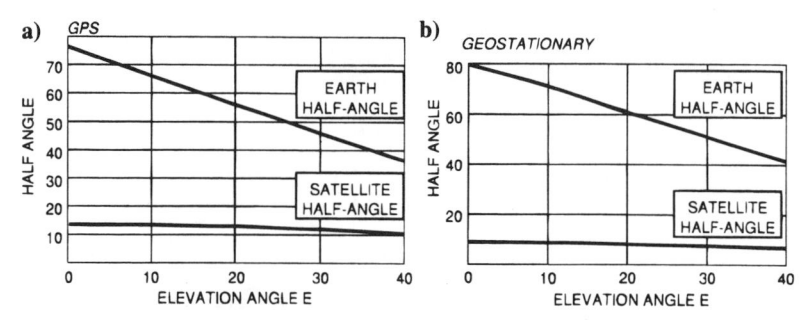

Fig. 4 Half-angles subtended from Earth center β and from the satellite α as a function of the elevation angle E. The half-angles are given both for the GPS 12-h orbit (a), and for the 24-h geostationary orbit (b).

depends on the constraint on the minimum elevation angle from the user to the satellite. Within this elevation angle constraint E, the value of α is as follows:

$$\alpha = \sin^{-1}\left[\left(\frac{R_e}{R_e + h_s}\right)\cos E\right], \quad \begin{cases} \alpha = 13.87 \text{ for } E = 0 \text{ deg} \\ \alpha = 13.82 \text{ for } E = 5 \text{ deg} \end{cases} \quad (6)$$

where R_e is the Earth radius; h_s is the satellite altitude; $h_s + R_e = a$; and a is the orbit radius or semimajor axis because we have assumed a circular orbit and spherical Earth.

The Earth half-angle β = 90 deg − α − E = 71.2 deg for E = 5 deg, and β is termed the visibility half-angle. Figure 4a shows the visibility half-angles β as a function of elevation angle E for GPS altitude orbits. For completeness, the visibility half-angles are shown in Fig. 4b for geostationary 24-h orbits, because there is a consideration of augmenting the GPS orbit satellites with one or more geostationary orbit satellites. The fraction of the spherical shell of satellites visible to the user is $(1/2)(1 - \cos \beta)$. On the average, if all 24 satellites were always equally spaced on a sphere (which they cannot be), we would expect to see a fraction of the satellites corresponding to the fraction of the sphere subtended by the β deg cone; namely, $24(1/2)(1 - \cos \beta) = 8.1$ satellites for E = 5 deg of the total 24 satellites.

An example of the actual GPS satellite visibility region for a user at a fixed point on Earth is shown* on a Mercator projection map in Fig. 5. The visibility region is defined by the subsatellite points above the dashed line on the figure.

C. GPS-24 Satellite Visibility

A set of satellite visibility statistics for the operational GPS-24 satellite constellation is shown in Figs. 6 and 7 for various user latitudes. The results have been

---

*The parametric equation for the locus of latitude and longitude at the edge of satellite visibility (subsatellite point) for a user at latitude φ and longitude θ is as follows:

latitude = $\sin^{-1}[\sin (71 \text{ deg}) \sin b \cos \phi + \cos (71 \text{ deg}) \sin \phi]$;
longitude = $\tan^{-1}[\sin (71 \text{ deg}) \cos \phi/(\cos (71 \text{ deg}) \cos \phi - \sin (71 \text{ deg}) \sin b \sin \phi)] + \theta$,

where b is the parametric angle $b = (-\pi, \pi)$ and 71 deg is the assumed coverage half-angle from the Earth's center.

SATELLITE CONSTELLATION AND GEOMETRIC DILUTION 185

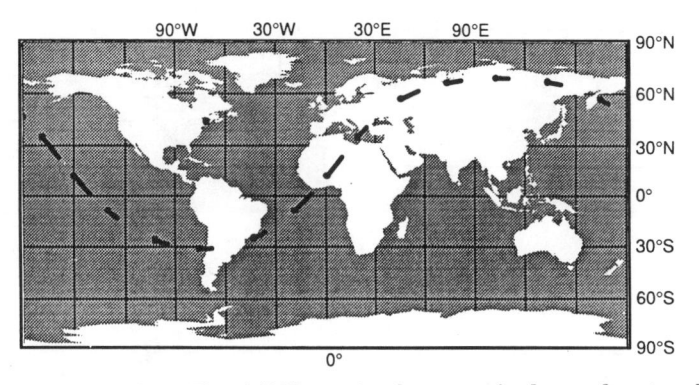

Fig. 5 Example GPS satellite visibility region for a user in the northeastern United States at latitude 40.1°N, longitude 74.5°W (see the dot) with an Earth visibility half-angle of 71 deg. Satellites with a subsatellite point above the dashed curve are in view of the user.

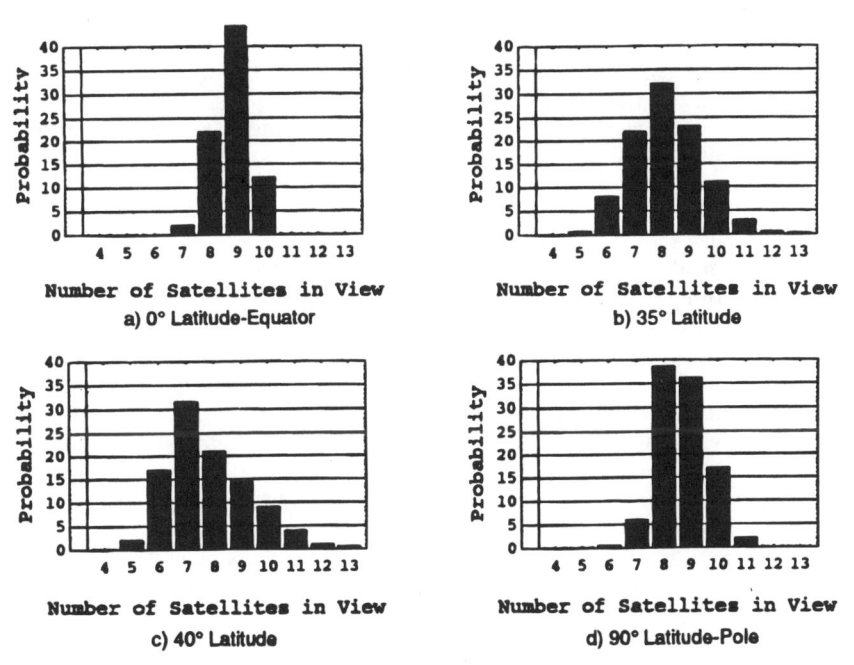

a) 0° Latitude-Equator

b) 35° Latitude

c) 40° Latitude

d) 90° Latitude-Pole

Fig. 6 GPS satellite visibility for the GPS-24 satellite constellation for a 5-deg elevation mask angle. The bar charts correspond to user latitudes as indicated. Note that there are always at least five satellites in view and at least seven more than 80% of the time.

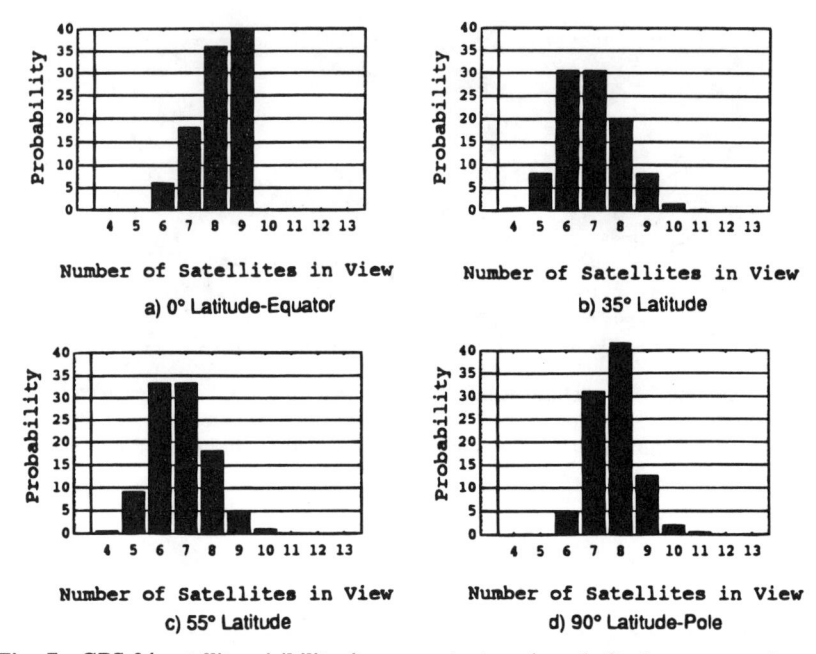

Fig. 7 GPS-24 satellite visibility in percent at various latitudes, averaged over longitude and a 24-h time span. The 35° user latitude corresponds to the approximate worst latitude where momentarily there are only four satellites in view (approximately 0.4% of the time) at this high elevation angle.

averaged over all user position longitudes at that latitude. Figure 6 shows the satellite visibility for a 5-deg elevation mask angle. The majority of the time there are at least seven satellites in view. Compare this result with the previously presented result for equally spaced satellites on a sphere where the average number of satellites on a sphere is 8.1. There are a minimum of five satellites visible for this 5-deg elevation mask angle. Note that a moderately high altitude aircraft can view satellites down to 0-deg elevation angle, and hence have even better satellite visibility. Both sets of results are based on 100 time samples over 24 h and 16 longitude samples at each latitude. It should also be pointed out that a uniform spacing of GPS satellites in an orbit such as the (24/6/1) constellation would lead to better visibility statistics, but greater sensitivity to a satellite outage.

In Fig. 7, a 10-deg elevation mask angle has been used, and satellite visibility has again been averaged over 24 h in time and a set of longitudes at 11.25-deg increments. Note that at both very high and very low latitudes, a minimum of six satellites are always in view and as many as 9–11. However, for a 10-deg elevation mask angle in the vicinity of 35 to 55° latitude, only four satellites are in view a small fraction, <0.5%, of the time. Clearly whenever only four satellites are in view, a failure of one of these satellites would cause an outage. Furthermore, even if four satellites are in view, the satellite/user geometry might correspond to a high GDOP. Thus, although these outages are not very likely and only occur at certain short intervals of time, methods of augmentation of the GPS-24 constellation are of interest.

A word of caution is in order for users at high latitudes. Note that the maximum satellite elevation angle for a user at the pole is

$$E = \tan^{-1}(a \cos (55 \text{ deg})/a \sin (55 \text{ deg}) - R_e) = 44.7 \text{ deg}$$

because the satellite inclination angle is 55 deg. Thus, there are no satellites anywhere at or near the zenith, and the GDOP is degraded, although satellite visibility is good. If the minimum elevation angle is 5 deg, then the usable range of satellite elevation angles is only 39.7 deg.

D. Augmentation of the GPS-24 Constellation

A number of methods can be used to increase the minimum number of satellites in view by adding satellites to the constellation:

1) Add another satellite to each plane to create a (30/6/X) constellation with five satellites in each plane instead of four.

2) Add a ring of GPS satellites at the same altitude but in the equatorial plane. Because the GPS visibility outage regions appear most likely at the midlatitude range 30–50°, a ring of three evenly spaced GPS satellites can add another satellite in view for users anywhere in the midlatitudes. However, sparing is made more difficult by adding yet another plane of satellites.

3) Add several geostationary satellites. These satellites also cover the midlatitude region, and because they are of higher altitude than the GPS satellites, fewer are required for the same degree of coverage. Second, we may be able to put GPS-like payloads on host geostationary communications or weather satellites.

4) Another augmentation being considered is the use of satellites with inclined planes on lower altitude 6-h orbits.

E. Constellation of 30 GPS Satellites

As an example of 30-satellite visibility, the coverage statistics for the (30/6/1) constellation are computed (not necessarily the optimum 30-satellite constellation) and shown in Fig. 8, which illustrates the visibility statistics for a 5-deg elevation mask angle. *Note that here there are a minimum of eight satellites in view at all tested user latitudes.* (55° latitude also showed a minimum of eight satellites in view.) Thus, there seem to be several real advantages in a 30-satellite constellation; namely, visibility is markedly improved; single-satellite outages are easily tolerated; and as is shown later in the chapter on signal-tracking theory, the added power of the other satellites in view can improve receiver performance.

III. Coverage Swath for an Equatorial Plane of Satellites

One of the methods for augmenting GPS is to add an equatorial plane of satellites at either GPS altitude of geostationary altitude. A plane of n equally spaced satellites with separation angle $2\phi = 360/n$ between subsatellite points generates a swath of continuous coverage, as shown in Fig. 9. If the coverage region half-angle for a single satellite for the desired minimum elevation angle

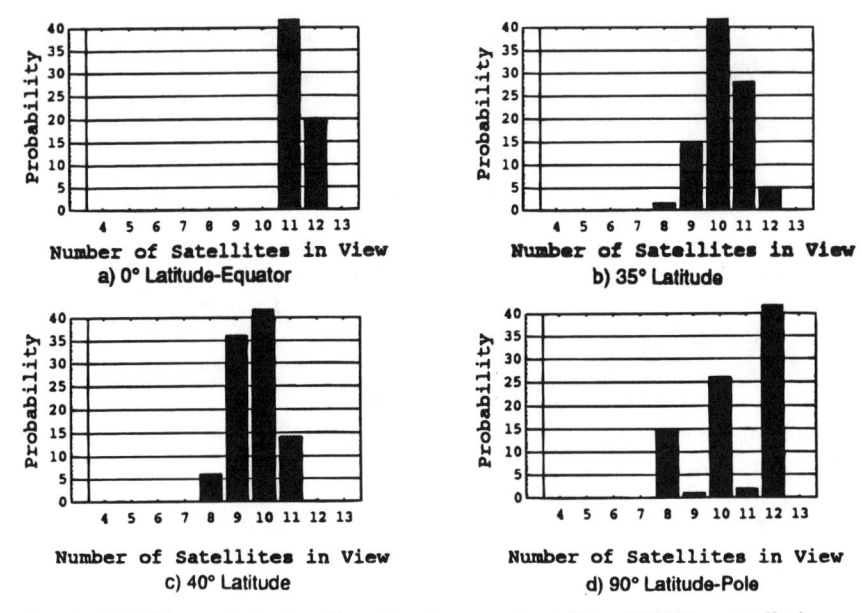

Fig. 8 Visibility statistics for 30 uniformly spaced satellites (30/6/1) constellation vs latitude for a 5-deg elevation mask angle. The user latitudes are as indicated.

E is β as measured from Earth's center, then the coverage swath angle λ in latitude is given by $\lambda = \cos^{-1}[\cos \beta / \cos \phi]$ for satellites in the equatorial plane. If the satellites are at GPS altitude and $\beta = 71$ deg for $E = 5$ deg, then the coverage swath latitudes are as shown in Table 3. Table 3 also gives the "swath for a geosynchronous orbit" at radius 42,162 km.

Thus, it is clear that 3, 4, or 5 satellites in the equatorial plane at GPS altitude produce a continuous coverage swaths of ± 49.3, ± 62.6, ± 66.3 deg, respectively, in latitude about the equator. The operational GPS satellite constellation already has a high degree of redundancy in satellite visibility at the higher latitudes

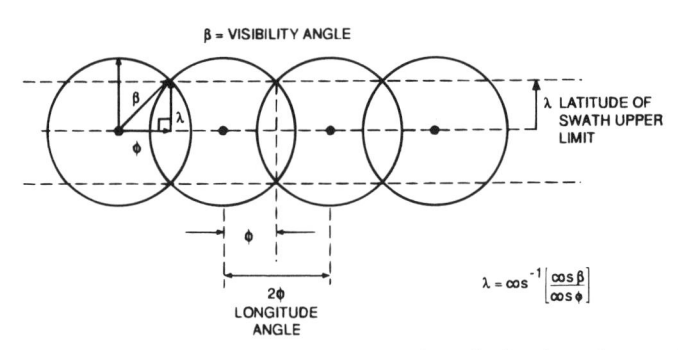

Fig. 9 Visibility coverage swath for n satellites in a single plane (equatorial) on a spherical Earth.

Table 3 Coverage swath width latitude λ for $n = 3,4,5$ satellites in an equatorial plane at GPS altitude and geosynchronous altitude (separation $2\phi = 360°/n$)

Number of satellites n		GPS		Geosynchronous	
	Separation 2ϕ	Coverage β	Swath λ	Coverage β	Swath λ
3	120 deg	71 deg	49.4 deg	77 deg	63.3 deg
4	90 deg	71 deg	62.6 deg	77 deg	71.5 deg
5	72 deg	71 deg	66.3 deg	77 deg	73.9 deg
6	60 deg	71 deg	67.9 deg	77 deg	74.9 deg

because the GPS orbits are inclined at 55 deg (see Fig. 6). Momentary periods of low satellite visibility that might occur in the event of a catastrophic satellite failure tend to occur at the midlatitudes, as shown earlier. Thus, it is clear that an equatorial ring of satellites at either GPS or geostationary altitudes with GPS-like payloads could provide an extra degree of redundancy for the GPS system. More specifically, the GPS-24 constellation has momentary periods of only four-satellite visibility for user elevation angle constraints at 10-deg elevation angle and user latitudes in the 30–60-deg region. These regions could be easily covered by three equatorial satellites at either geostationary or GPS altitudes to provide a minimum of five satellite visibility at 10-deg elevation angle or six-satellite visibility at 5-deg elevation angle. Even better would be an additional six satellites in either the (30/6/1) constellation at 55-deg inclination or a set of six equatorial satellites.

A. Satellite Ground Traces

From a user's standpoint, it is sometimes useful to show the ground trace that each of these satellites makes as its subsatellite point moves with time. The ground trace is the line generated on the Earth's surface by the line joining the satellite and the Earth's center as both the satellite moves in its orbit and the Earth rotates. Because the satellites have precisely a 12-h (sidereal time) orbit, each satellite traces out exactly the same track on the Earth's surface each sidereal day.* The Earth, of course, rotates once in inertial space each sidereal day underneath the satellite orbit. Thus the satellite produces a ground track (the locus of points directly below the satellite on the surface of the Earth), which exactly repeats every day. If we stand on a ground track of a satellite, we see the same satellite appear overhead at the zenith every day. In fact, a user at any fixed point sees exactly the same pattern of satellites every day. However, because the user's clock time is mean solar time rather than sidereal time of the satellite period, the user sees this satellite pattern appear approximately four minutes earlier each day (235.9 s).

Define a reference GPS satellite in a circular orbit at inclination i which crosses the equator at $t = 0$ at longitude θ_j. If another satellite of interest is in the same

---

*As pointed out earlier, the GPS satellite ground tracks are held to within ±2° in longitude by the GPS Control Segment at the time of the writing.

plane but offset in phase (the mean anomaly) by ϕ_i, then it can be shown that this satellite ground track has latitude that varies with time as follows:

$$\text{Lat} = \sin^{-1}[\sin i \sin(\omega t + \phi_i)] \tag{7a}$$

where the period of the orbit $T = 2\pi/\omega$.

The longitude likewise varies with time as follows:

$$\text{Long} = \phi_j - \Omega_e t + \tan^{-1}[\tan(\omega t + \phi_i)\cos I] \tag{7b}$$

where Ω_e is the Earth rotation rate.*

As a simple example, if there were four equally spaced satellites, and one of these crossed the equator at 0° longitude, the ground traces would appear as shown in Fig. 10. Notice that each satellite ground track repeats and makes two complete cycles as it moves from -180 to $+180°$ in longitude, as we would expect from the 12-h period. It can be shown that a satellite in another plane, but offset in phase, can have exactly the same ground track. For example, if one satellite crosses the equator in a positive direction at $t = 0$, another satellite crossing the equator in a negative direction at the same time but 180° offset in longitude would have exactly the same ground track. In general, two satellites have the same ground track if $\Delta\phi = 2\Delta\theta(\text{mod } 2\pi)$. That is, the offset in satellite phase is equal to twice the offset in longitude of their planes modulo 2π. Thus, it is possible for satellites in different orbit planes in inertial space to have the same ground track if the phasing is selected accordingly.

Clearly, with the GPS-24 constellation the satellites do not all have the same ground tracks because the longitude of the ascending nodes (measured in the rotating Earth coordinates) are different (mod π or 180°). Thus, there are 24 separate ground tracks, one for each satellite. However, each of their ground traces have exactly the same shape as those shown here except for the position of the longitude of the ascending node.

IV. Geometric Dilution of Precision Performance Bounds and GPS-24 Performance

The previous section discussed satellite orbits and visibility, and previous and later chapters discuss the pseudorange measurement accuracy. However, it is really user position that is of greatest interest, and as described earlier, the general relationship between the errors in the pseudorange measurements by the user to the user position accuracy is described by the GDOP. The GDOP generally assumes that the measured pseudorange errors are independent with zero mean, and all *measurement errors* have the same rms value σ. (GDOP is also discussed in Chapter 11, this volume, wherein a different approach is taken but leads, of course, to the same result.) The following are the GDOP parameters:

$$\text{GDOP} = \frac{1}{\sigma}\sqrt{\sigma_x^2 + \sigma_y^2 + \sigma_z^2 + \sigma_b^2} \text{ for three dimensions plus time}$$

$$= \text{Geometric Dilution of Precision}$$

---

*For computational purposes, it is sometimes useful to approximate the $\tan^{-1}(\ )$ relationship for $I = 55$ deg of Eq. (6b) by $\text{Long}(t,\phi,\theta) \cong 2\pi t/12 + \phi - 0.271 \sin 2(2\pi t/12 + \phi) + 0.0367 \sin 4(2\pi t/12 + \phi) - 0.00663 \sin 6(2\pi t/12 + \phi) + 0.00135 \sin 8(2\pi t/12 + \phi) - 2\pi t/2y + \theta)$.

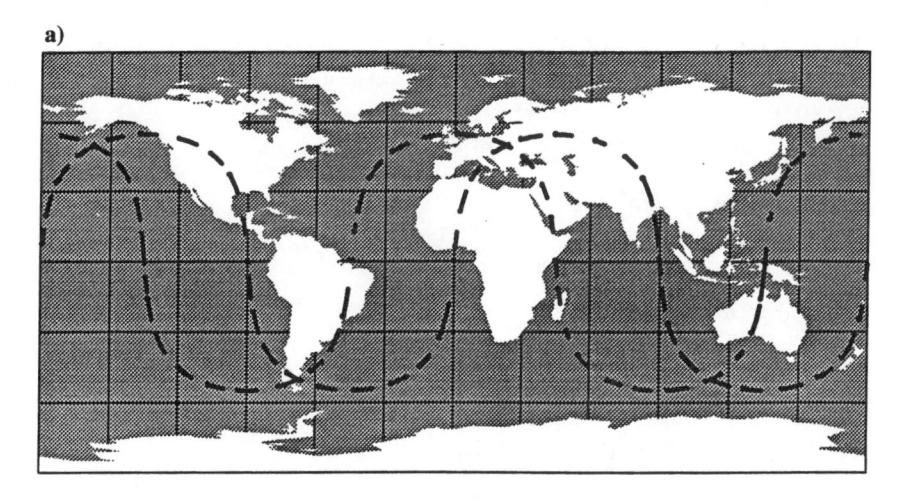

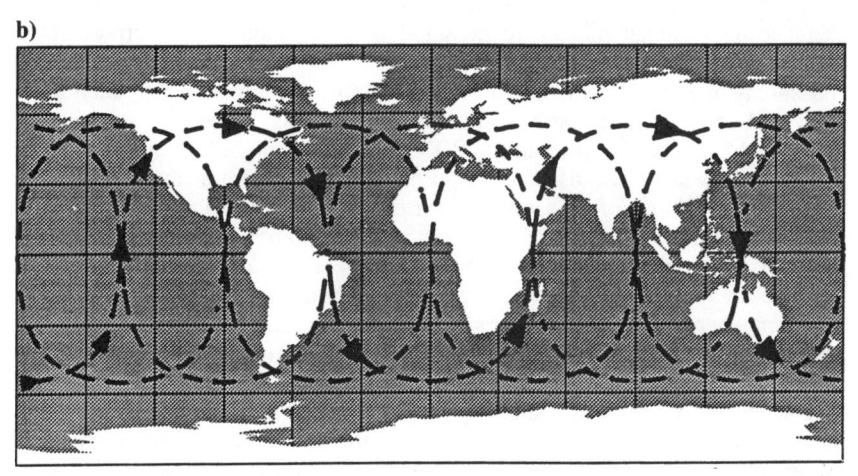

Fig. 10 Example of ground traces for four equally spaced satellites at 55-deg inclination in a single orbit plane at inclination 55 deg where one of the satellites crosses the equator at longitude 0°. Each satellite travels from west to east along its ground track: a) ground traces for two satellites, one of which crosses the equator at 0° longitude (longitude of ascending node is 0°), and the second is in the same orbit plane but spaced by 90 deg; b) ground traces for all four equally spaced satellites. Each satellite follows the ground trace moving from west to east. Note that the GPS satellites are not equally spaced in the GPS-24 constellation.

$$\text{PDOP} = \frac{1}{\sigma}\sqrt{\sigma_x^2 + \sigma_y^2 + \sigma_z^2}$$

= Position Dilution of Precision

$$\text{HDOP} = \frac{1}{\sigma}\sqrt{\sigma_x^2 + \sigma_y^2} \qquad (8)$$

= Horizontal Dilution of Precision

$$\text{VDOP} = \sigma_z/\sigma$$

= Vertical Dilution of Precision

$$\text{TDOP} = \sigma_b/\sigma$$

= Time Dilution of Precision

where σ_x, σ_y, σ_z are the rms *errors in the estimated user position* coordinates x, y, z, and the user clock bias error estimate has an rms value σ_b, expressed in distance units. All parameters are measured in units of meters. The purpose of this section is to define the bounds of GPS position estimates for a given pseudorange measurement error. Parameters to be considered include number of satellites in view and constraints on the minimum elevation angle.

It should also be pointed out that PDOP constraints do not always tell the full story if used alone because the rms pseudorange errors are not generally the same for all satellites; for example, some measurement errors vary as a function of elevation angle. Nonetheless, the various DOPs provide a useful measure of performance that is dependent on the relative geometry of the satellites relative to the user. As an illustrative example, GDOP bounds are also computed for a situation where measurement errors contain both an independent and a correlated component that increases with decreasing elevation angle. This correlated component mimics to some extent the effects of the atmosphere.

The section concludes by showing maps of GDOP performance with a single satellite outage. These GDOP outage maps are shown as potential momentary outage areas when PDOP exceeds some numerical constraint. The GPS-24 satellite constellation with a single satellite outage is the reference constellation.

A. Bounds on Geometric Dilution of Precision—Two Dimensions

Although most users seek three-dimensional position, users at sea level operate essentially on a two-dimensional surface. The two-dimensional example also provides a simple introduction to the GDOP concept and calculations. Furthermore, some of the three-dimensional DOPs are shown to be related to the two-dimensional DOPs.

In addition to developing the equations for the various DOPs, we also seek to find the optimum satellite geometry and the bounds on GDOP. The best position solution for a given measurement error variance is the one that minimizes GDOP. GDOP, in turn, is minimized for an assumed user location by optimizing the geometry of the satellites. Figure 11 shows the geometry for a two-dimensional (x, y) position solution for a user with unknown clock bias b. Thus, there are

SATELLITE CONSTELLATION AND GEOMETRIC DILUTION   193

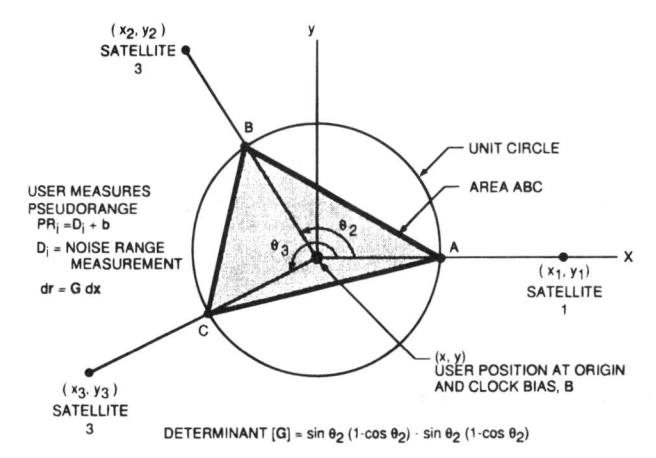

Fig. 11 Two-dimensional satellite geometry and GDOP for three "satellites." With no loss in generality, we can set $\theta_1 = 0$. The optimum geometry is for equally spaced satellites in two dimensions. For three satellites, this relationship yields satellites with direction vectors at the tips of an equilateral triangle.

three unknowns: x, y, and b. Pseudoranges are measured for three satellites, which are then used to solve for the three unknowns. With no loss in generality, the x-coordinate can be aligned in azimuth with one of the satellites. Because the user does not have absolute time, only the pseudorange $PR_i = D_i + cb_u$ can be measured where D_i is the actual range from the satellite to the user. For simplicity in notation, we replace cb_u with b. The measured pseudorange is denoted as $\rho_i = PR_i + n_i$. The three satellites have angles $\theta_1 = 0$ deg, and θ_2, and θ_3 are measured from the x axis. A differential change in user position dx, dy, db results in a differential change in each pseudorange $d\rho_i = \dfrac{\partial \rho_i}{\partial x} dx + \dfrac{\partial \rho_i}{\partial y} dy + \dfrac{\partial \rho_i}{\partial b} db$.
Apply this differential change to all three pseudoranges and add a measurement noise dn_i to give the following matrix equation:

$$d\boldsymbol{\rho} = \begin{pmatrix} d\rho_1 \\ d\rho_2 \\ d\rho_3 \end{pmatrix} = \begin{bmatrix} 1 & 0 & 1 \\ \cos\theta_2 & \sin\theta_2 & 1 \\ \cos\theta_3 & \sin\theta_3 & 1 \end{bmatrix} \begin{pmatrix} dx \\ dy \\ db \end{pmatrix} + \begin{pmatrix} dn_1 \\ dn_2 \\ dn_3 \end{pmatrix} = \boldsymbol{G} d\boldsymbol{x} + d\boldsymbol{n} \quad (9)$$

and form an estimate $d\hat{\boldsymbol{x}} = \boldsymbol{G}^{-1} d\boldsymbol{\rho}$ where $\boldsymbol{G}$ is the direction cosine matrix,* $d\boldsymbol{\rho}$ is the differential measured pseudorange vector, and $d\boldsymbol{x}$ is the differential user position vector. To solve for GDOP and HDOP, the values of σ_x^2, σ_y^2, σ_b^2 must be found. We can invert the Eq. to obtain the following (assume $\boldsymbol{G}$ to be nonsingular):

---

*Both $\boldsymbol{H}$ and $\boldsymbol{G}$ are used in the literature to represent the direction cosine or geometry matrix. Here we use $\boldsymbol{G}$.

$$d\mathbf{x} = \begin{pmatrix} dx \\ dy \\ db \end{pmatrix} = \frac{1}{|\mathbf{G}|} \begin{bmatrix} g'_{11} & g'_{12} & g'_{13} \\ g'_{21} & g'_{22} & g'_{23} \\ g'_{31} & g'_{32} & g'_{33} \end{bmatrix} \begin{pmatrix} dPR_1 \\ dPR_2 \\ dPR_3 \end{pmatrix} = \mathbf{G}^{-1}(d\boldsymbol{\rho} - d\mathbf{n}), \quad (10)$$

and form an estimate $d\hat{\mathbf{x}} = \mathbf{G}^{-1} d\boldsymbol{\rho}$ where the g'_{ij} are the terms in the adjoint matrix and $d\mathbf{n}$ is the measurement noise (or error). For small errors, the mean square error in position x, y, b is as follows:

$$\sigma_x^2 + \sigma_y^2 + \sigma_b^2 = \text{tr}[\text{cov } \mathbf{x}] \quad (11)$$

where $E[dx\, dx^T] = \text{cov}(\mathbf{x}) = E[\mathbf{G}^{-1}\, d\mathbf{n}(\mathbf{G}^{-1}\, d\mathbf{n})^T] = \mathbf{G}^{-1}\, \text{cov}(\mathbf{n})\, \mathbf{G}^{-T}$ and $tr[\ ]$ is the trace of $[\ ]$.

If all of the measurement noises dn_i are independent, zero mean, and of equal variance σ^2 then the following obtains:

$$\text{cov}(\mathbf{n}) = \sigma^2 \mathbf{I}, \quad \text{and} \quad \text{cov}(\mathbf{x}) = (\mathbf{G}^T \cdot \mathbf{G})^{-1} \sigma^2 \quad (12)$$

where $\mathbf{1}$ is the identity matrix. Thus, GDOP can be expressed as follows:

$$\text{GDOP} = \frac{1}{\sigma} \sqrt{\sigma_x^2 + \sigma_y^2 + \sigma_b^2} = \sqrt{\text{tr}(\mathbf{G}^T\mathbf{G})^{-1}} \quad (13)$$

Notice that $tr[(\mathbf{G}^T\mathbf{G})^{-1}] = \left[\dfrac{1}{|\mathbf{G}|^2}\right] \sum_{i,j} (g'_{ij})^2$ where $g'_{ij}/|\mathbf{G}|$ are the elements of $\mathbf{G}^{-1}$ if $\mathbf{G}^{-1}$ exists.

In general, it can be shown that the GDOP is approximately minimized by maximizing the denominator of Eq. (10), the determinant of $\mathbf{G}$, namely $|\mathbf{G}|$, because the numerator varies for these GDOP problems less strongly with geometry than does the $|\mathbf{G}|$. It can be shown that the determinant of $|\mathbf{G}|$ is directly proportional to the area of the triangle ABC circumscribed by the unit circle. This situation is illustrated in Fig. 11. It is clear that the area ABC is maximized by equally spacing the satellites to form an equilateral triangle; i.e., $\theta_2 = 120$ deg, $\theta_3 = 240$ deg. In fact, this equal spacing does minimize the GDOP and does not just maximize the determinant. For this equally spaced condition, the following obtains:

$$[\mathbf{G}^T\mathbf{G}]^{-1} = \begin{bmatrix} 2/3 & 0 & 0 \\ 0 & 2/3 & 0 \\ 0 & 0 & 1/3 \end{bmatrix} \quad (14)$$

Thus, GDOP $= \sqrt{5/3} = 1.291$; HDOP $= \sqrt{4/3} = 1.1547$; and TDOP $= \sqrt{1/3} = 0.5773$ for three satellites optimally positioned in two dimensions.

1. Least-Squares Estimators: The Gauss-Markov Theorem

Consider the more general problem of estimating a vector $d\mathbf{x}$ of k dimensions given an observation vector $d\boldsymbol{\rho}$ of m dimensions where $k \leq m$, and where the measurements $d\boldsymbol{\rho}$ have observation noise $d\mathbf{n}$:

$$d\boldsymbol{\rho} = \mathbf{G}\, d\mathbf{x} + d\mathbf{n} \quad (15)$$

where $\mathbf{G}$ is a known $m \times k$ matrix. Obviously, when $m > k$, the solution is overdetermined. The observation noise dn is zero mean and has the following covariance matrix:

$$E[dn dn^T] = R, \quad \text{and} \quad E[dn] = 0$$

The Gauss-Markov theorem states that the minimum variance, unbiased estimate of dx, is given by (Ref. 3):

$$d\hat{x} = (G^T R^{-1} G)^{-1} G^T R^{-1} d\rho \tag{16}$$

with the assumption that $(G^T R^{-1} G)$ is nonsingular. For uncorrelated measurement noise where $R = \sigma^2 I$ this estimate reduces to the following:

$$d\hat{x} = (G^T G)^{-1} G^T d\rho \tag{17}$$

with the assumption that $(G^T G)$ is nonsingular. If G is nonsingular, the n the following obtains:

$$(G^T G)^{-1} G^T = G^{-1}, \quad \text{and} \quad d\hat{x} = G^{-1} d\rho \tag{17a}$$

These relationships can also be written in the form of a generalized inverse. A widely used notation for one of several generalized inverses, the so-called (1) inverse is B^-. This generalized inverse is defined[4] by its behavior;

$$B B^- B = B \tag{18}$$

Thus, we can write both Eqs. (16) and (17) as follows:

$$d\hat{x} = G^- d\rho \tag{19}$$

where $G^- = (G^T R^{-1} G)^{-1} G^T R^{-1}$ for Eq. (16), and $G^- = (G^T G)^{-1} G^T$ for Eq. (17) are the generalized inverse expressions. Both G^- expressions satisfy Eq. (18).*
For the special case where G is a square, nonsingular matrix and $R = \sigma^2 I$, the generalized inverse reduces to $G^- = G^{-1}$. Note also that $(G^T G)^{-1}$ is symmetric.

The estimate of Eq. (19) is unbiased; i.e., the expected value of the error is zero, and the error covariance matrix is as follows:

$$\Sigma = E[(d\hat{x} - dx)(d\hat{x} - dx)^T] = (G^T R^{-1} G)^{-1} \tag{20}$$

For $R = \sigma^2 I$ where I is the identity matrix, this expression reduces to the following:

$$\Sigma = \sigma^2 (G^T G)^{-1} \tag{20a}$$

where $(G^T G)^{-1}$ is symmetric, as is $(G^T R^{-1} G)^{-1}$ because R is symmetric.

The DOP terms are then the appropriate sums of terms on the diagonals of the Σ matrix:

$$\text{GDOP} = \frac{1}{\sigma^2} \text{tr}(\Sigma) = \text{tr}[(G^T G)^{-1}] \tag{21}$$

2. Minimum Two-Dimensional GDOP for More Than Three Satellites

In general, of course, we may use more than three satellite measurements to solve for two-dimensional position. If m satellite measurements are employed,

---

*Thus relationship is easily shown by substituting G^- for B^- in Eq. (18) and using the usual matrix relationships $(AB)^{-1} = B^{-1} A^{-1}$, $(AB)^T = B^T A^T$, and $A^T = A$ if A is symmetric.

the measurement vector ρ then has m dimensions and $d\rho = G\ dx + dn$. where G is an $m \times 3$ matrix for a two-dimensional space x, y, plus b (or the G is a $m \times 4$ matrix generalized to the three-dimensional x, y, z space $+ b$). The minimum variance unbiased estimate of dx for zero mean noise dn with covariance matrix R is found by applying the Gauss-Markov theorem of Eq. (16) namely, $d\hat{x} = G^-\ d\rho \triangleq (G^T R^{-1} G)^{-1} G^T R^{-1}\ d\rho$.

For five satellites operating in two dimensions on the x, y plane, the question is: What selection of the five satellites minimizes GDOP? The answer is shown in Fig. 12 where contours of constant GDOP are plotted, and the minimum is found at $\theta_1 = 72$ deg, $\theta_5 = -72$ deg, $\theta_2 = 144$ deg, $\theta_4 = -144$ deg, $\theta_0 = 0$ deg. Again, the x axis has been aligned with satellite S_1, and satellites S_2 and S_5 and satellites S_3 and S_4 have been assumed to be symmetrical with respect to the x-axis. Note that the satellite direction vectors form a regular pentagon, which again maximizes the area formed by the satellite unit vectors circumscribed on the unit circle. It can be shown that regular polygons maximize the area inside the unit circle and simultaneously minimize GDOP, not only for three satellites but for larger numbers of satellites as well.

The covariance matrix for the error in the estimate of dx is as follows:

$$\Sigma = [G^T G]^{-1} = \begin{bmatrix} 2/5 & 0 & 0 \\ 0 & 2/5 & 0 \\ 0 & 0 & 1/5 \end{bmatrix} \quad (22)$$

The minimum values of two-dimensioned GDOP measures for five satellites are as follows:

$$\text{GDOP} = \sqrt{1} = 1$$
$$\text{PDOP} = \text{HDOP} = \sqrt{0.8} = 0.8944 \quad (23)$$
$$\text{TDOP} = \sqrt{0.2} = 0.447$$

Note that the GDOP for five satellites has been reduced with respect to the three

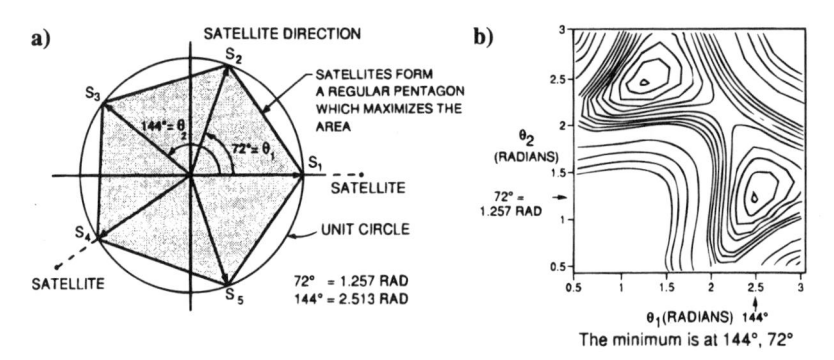

Fig. 12 a) Optimum two-dimensional satellite selection for five satellites satellite geometry: and b) contours of constant GDOP vs satellite position (angles). Minimum GDOP = 1.0, contours near minimum are spaced by 0.01, the minimum GDOP is at 144 deg, 72 deg.

SATELLITE CONSTELLATION AND GEOMETRIC DILUTION 197

satellites GDOP by a divisor $\sqrt{N/3} = \sqrt{5/3}$, the square root of the ratio of the number of satellites, as we might expect.

B. Bounds on Geometric Dilution of Precision—Three Dimensions

The more general problem of interest is the solution to the user coordinates for the three-dimensions position problem, x, y, z plus b user clock bias. The matrix Eq. (16) and (20) applied above to the two-dimensional problem apply directly, however. The satellite geometry for the three-dimensional problem with four satellites is shown in Fig. 13. One satellite is positioned on the z axis and another in the x, z plane intersection with no loss in generality (the z plane is not necessarily at the zenith for this initial general problem). The differential equation for this situation is as follows:

$$d\boldsymbol{\rho} = \begin{pmatrix} d\rho_1 \\ d\rho_2 \\ d\rho_3 \\ d\rho_4 \end{pmatrix} = \begin{bmatrix} g_{11} & 0 & g_{13} & 1 \\ g_{21} & g_{22} & g_{23} & 1 \\ g_{31} & g_{32} & g_{33} & 1 \\ 0 & 0 & 1 & 1 \end{bmatrix} \begin{bmatrix} dx \\ dy \\ dz \\ db \end{bmatrix} + \begin{pmatrix} dn_1 \\ dn_2 \\ dn_3 \\ dn_4 \end{pmatrix} = \boldsymbol{G}\, d\boldsymbol{x} + d\boldsymbol{n} \quad (24)$$

Again, it can be shown that there is the approximate inverse relationship of GDOP to determinant $|G|$, and minimizing the determinant also minimizes the position error variance.

In Fig. 13, the unit vectors to each of the satellites describe a tetrahedron (four faces) with nodes $ABCD$. It is readily shown that the determinant $|G|$ is directly proportional to the volume of this tetrahedron; namely, $|G| = 6V$, where V is the

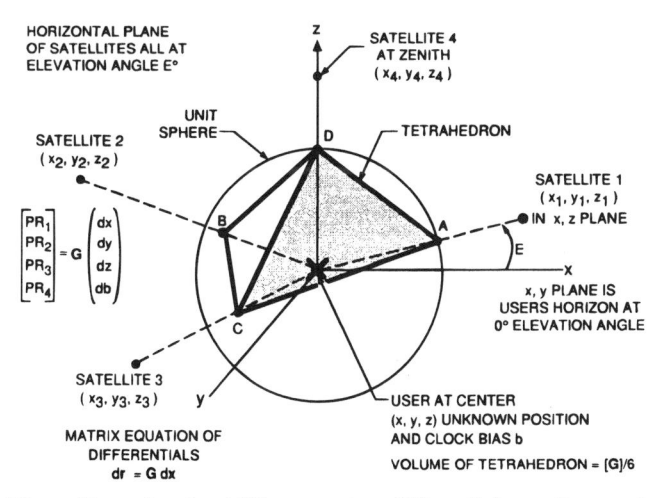

Fig. 13 Three-dimensional satellite geometry, differential equation, and volume of tetrahedron. The unit vectors A, B, C, D are all on the surface of the unit sphere pointing toward satellites 1, 2, 3, 4, respectively.

tetrahedron volume.* As we maximize the volume, the GDOP decreases also almost everywhere.[6] Thus, the satellites should be selected to maximize the volume of the tetrahedron, and that selection, in fact, does minimize GDOP. The volume is maximized by placing one satellite at the zenith of the user and the other satellites equally spaced in a plane perpendicular to the user link to the satellite at zenith. Thus, the base of the tetrahedron is an equilateral triangle. If there are no constraints on elevation angle, the tetrahedron volume is maximized if satellites 1, 2, 3 are all equally spaced at 120 deg and placed at $E = -19.47$ deg for $E = \sin^{-1}(-1/3)$; i.e., 19.47 deg below the horizon to provide a regular tetrahedron with four equal faces. Of course, that positioning is impractical unless the user is in a satellite above the Earth permitting this large a negative elevation angle.

If satellites 1, 2, 3 are equally spaced by 120 deg in the azimuth plane and at an elevation angle E and, the other satellite is at zenith, then the differential equation is as follows:

$$d\boldsymbol{\rho} = \begin{pmatrix} d\rho_1 \\ d\rho_2 \\ d\rho_3 \\ d\rho_4 \end{pmatrix} = \begin{bmatrix} \cos E & 0 & \sin E & 1 \\ -\frac{1}{2}\cos E & \sqrt{\frac{3}{4}}\cos E & \sin E & 1 \\ -\frac{1}{2}\cos E & -\sqrt{\frac{3}{4}}\cos E & 1 & 1 \\ 0 & 0 & \sin E & 1 \end{bmatrix} \begin{pmatrix} dx \\ dy \\ dz \\ db \end{pmatrix} + d\boldsymbol{n} \quad (25)$$

The equation for the error covariance matrix for an assumption of uncorrelated, equal variance measurement errors; i.e., $\mathbf{R} = \sigma^2 \mathbf{I}$ is the following symmetric matrix:

$$\boldsymbol{\Sigma} = \begin{bmatrix} \sigma_x^2 & 0 & 0 & 0 \\ 0 & \sigma_y^2 & 0 & 0 \\ 0 & 0 & \sigma_z^2 & \Sigma_{zb} \\ 0 & 0 & \Sigma_{zb} & \sigma_b^2 \end{bmatrix} \quad (26)$$

where

$\sigma_x^2/\sigma^2 = \sigma_y^2/\sigma^2 = (2/3)\sec^2 E$
$\sigma_z^2/\sigma^2 = (4/3)/[\cos(E/2) - \sin(E/2)]^4$
$\sigma_b^2/\sigma^2 = (5 - 3\cos 2E)/6[\cos(E/2) - \sin(E/2)]^4$ and
$\Sigma_{zb}/\sigma^2 = -(1 + 3\sin E)/3 [\cos(E/2) - \sin(E/2)]^4$

The bounds on GDOP, PDOP, HDOP, VDOP, TDOP derived from Eq. (27) are shown in Table 4 along with their values at $E = 0$ deg.

---

*The volume of a parallel-piped with three edge vectors is equal to the triple scalar product $(\boldsymbol{abc}) = \boldsymbol{a} \cdot (\boldsymbol{b} \times \boldsymbol{c})$, and the volume of the tetrahedron is 1/6 the volume of the parallel-piped.[5]

Table 4 Values of the various three-dimensional DOPs for four satellites in the optimum constellation vs minimum elevation angles and the value at elevation angle of $E = 0$ deg

DOP	DOP value at $E = 0$ deg	DOP expressions vs elevation angle E
GDOP	$\sqrt{3}$	$[(4/3) \sec^2 E + [(13/6) - (1/2)\cos 2E]/[\cos(E/2) - \sin(E/2)]^4]^{1/2}$
PDOP	$2\sqrt{2/3}$	$[(4/3) \sec^2 E + (4/3)/[\cos(E/2) - \sin(E/2)]^4]^{1/2}$
HDOP	$2/\sqrt{3}$	$(2/\sqrt{3}) \sec E$
VDOP	$2/\sqrt{3}$	$(2/\sqrt{3})/[\cos(E/2) - \sin(E/2)]^2$
TDOP	$1/\sqrt{3}$	$(1/\sqrt{6})(5 - 3\cos 2E)^{1/2}/[\cos(E/2) - \sin(E/2)]^2$

These results are plotted in Fig. 14 as a function of the elevation angle E. Note that these GDOP results are for the optimum satellite positions within that elevation angle constraint. In general, of course, we do not have one satellite at zenith and three others equally spaced in azimuth and at the same minimum elevation angle. Nevertheless, it is interesting to note that the GDOP, PDOP, etc. all decrease uniformly as we decrease E from 90 deg toward $E = 0$ deg (the horizon). In fact, HDOP for three satellites on the horizon is the same as the optimum two-dimensional HDOP = $\sqrt{4/3}$ = 1.154. Note also that $E = 0$ deg corresponds to the optimum HDOP for any elevation angle, whereas the GDOP and PDOP continue to decrease with decreasing negative E as the tetrahedron

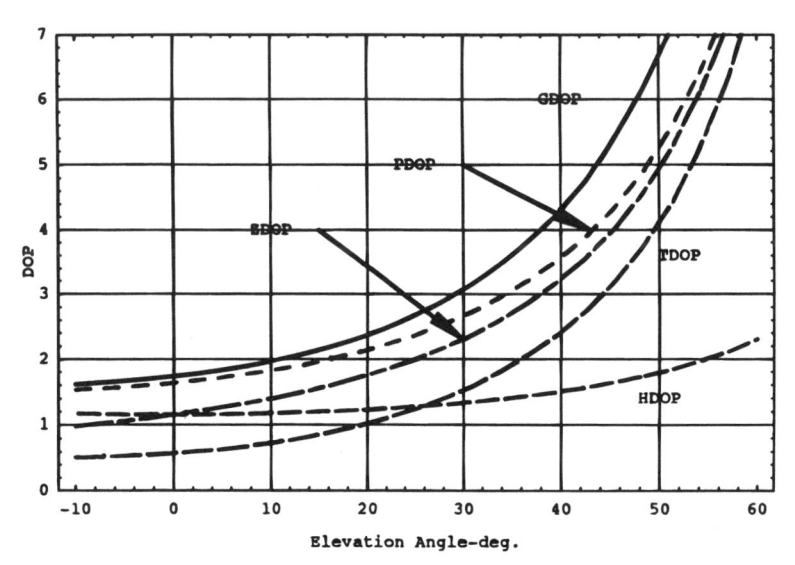

Fig. 14 Bounds on the minimum value of GDOP for three dimensions x, y, z, plus user clock bias b as a function of elevation angle mask for four satellites, three at elevation angle E in deg and a fourth at zenith.

volume increases until $E = -19.47$ deg where they are both at the minimum. The minimum GDOP at this negative elevation angle is $\sqrt{5/2} = 1.5811$. Note that a high altitude aircraft or a low orbiting satellite can operate with negative elevation angles.

In general, we find that for satellite constellations with 23–24 satellites, the GDOP numbers vary with time and are more typically in the range of 2–3 for four-satellite solutions. Thus, for a 15-m position error, the pseudorange accuracy must be less than or equal to 5 m, a value on which we originally based our requirements for the signal waveform.

1. Effects at Low Elevation Angles—Independent Plus Correlated Measurement Errors

For a user near or on the Earth's surface, operation at low elevation angles can lead to increased random errors caused by multipath and ionospheric/tropospheric delay effects on the pseudorange measurements. These effects are magnified at low elevation angle by various obliquity factors* and lead to degraded performance when the elevation angle decreases below 5 deg. On the other hand, the atmospheric measurement errors have a component highly correlated for satellites at the same elevation angle.

As an illustration of this type of effect assume that pseudorange measurements are made to three equally spaced satellites at elevation angle E, plus one satellite at the zenith. Assume as a simplified example that the measurement noise for satellite i is of the following form:

$$n_i = n_{ri} + n_a f(E) \qquad (27)$$

where the n_{ri} are independent noise terms of variance $7\sigma^2/8$, and n_a has variance $\sigma^2/8$, and for this simplified model, the obliquity factor is approximated as follows:

$$f(E) = 1.1/(\sin E + 0.1), \qquad E > 0 \text{ deg} \qquad (28)$$

and $f(90 \text{ deg}) = 1$, $f(0 \text{ deg}) = 11$. The two noise terms n_{ri} and n_a are assumed to be independent of one another. The total measurement error variance is then equal to σ^2 at $E = 90$ deg. The n_{ri} term can be likened to a constant variance random noise term. The n_a terms can correspond to an atmospheric error contribution of value $n_a f(E)$, which, although random, has a variance that depends on elevation angle. Both noise terms are assumed to have zero mean. (In effect the mean value has been subtracted in the receiver.)

The resulting measurement error covariance matrix then varies with elevation angle, and for this simplified example, it has the following values:

---

*The obliquity factors account for the increased effective path lengths through the ionosphere and troposphere as elevation angle decreases. See the later chapters on the ionospheric and tropospheric effects in this volume.

$$R(E) = \sigma^2 \begin{bmatrix} 1 & 0.125 & 0.125 & 0.125 \\ 0.125 & 1 & 0.125 & 0.125 \\ 0.125 & 0.125 & 1 & 0.125 \\ 0.125 & 0.125 & 0.125 & 1 \end{bmatrix} \quad \text{at} \quad E = 90 \text{ deg}$$

$$= \sigma^2 \begin{bmatrix} 2.89 & 2.20 & 2.20 & 0.50 \\ 2.20 & 2.89 & 2.20 & 0.50 \\ 2.20 & 2.20 & 2.89 & 0.50 \\ 0.50 & 0.50 & 0.50 & 1 \end{bmatrix} \quad \text{at} \quad E = 60 \text{ deg} \quad (29)$$

Note that it has been assumed that the error term $n_a f(E)$ that varies with elevation angle is random and correlated for all satellites. The output error covariance in the position estimate is as follows:

$$\Sigma = (G^T R^{-1} G)^{-1} \quad (30)$$

and then leads to an effective DOP that can be normalized to σ^2, the total measurement variance at $E = 90$ deg for this example. The GDOP, VDOP, TDOP, and HDOP are shown in Fig 15. Observe that for this illustrative example, the GDOP does not drop below 2.6 and begins to increase significantly for elevation angles below 6 deg. Observe also that because of the correlation between the

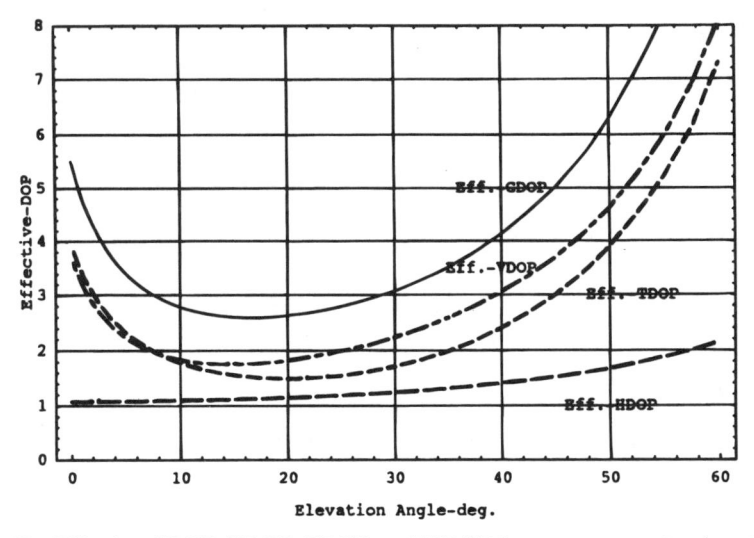

Fig. 15 Effective GDOP, VDOP, TDOP, and HDOP for measurement noise with a correlated component that increases VDOP in variance with decreasing elevation angle in accord with the simplified model of Eq. (27).

measurement error at low elevation angles, there is no impact on the HDOP. The major effect is on GDOP, VDOP, PDOP, and TDOP.

Although this illustrative example is only a crude approximation to the actual effects at low elevation angle, it does indicate some of the typical impacts of low elevation angle operation. Figure 15 should be compared to Fig. 14 where no elevation angle dependent noise is present.

2. Three-Dimensional GDOP—More than Four Satellites in View

The more general GDOP problem is: What are the bounds on GDOP vs elevation mask angle for the overdetermined situation where more than four satellites are in view and being measured? Because there are often 6–8 satellites in view, sometimes 8–11, as shown in the previous section, and many modern GPS receivers are capable of continuously tracking 6–12 satellites, this question is of more than academic interest. The solution to the problem follows exactly the reasoning in the overdetermined two-dimensional problem discussed earlier. Measurement noise errors are assumed to be independent and of equal variance.

Consider the problem where six satellites are in view, and six pseudoranges are being measured. Constrain the minimum elevation angle to a value E. Initially, let us assume that the minimum elevation angle $E = 0$ deg. Let 6-m satellites be equally spaced on the horizon, and the other m are equally spaced at some other elevation angle E_z.* Computer analyses and optimization give the results in Table 5. As is shown, the optimum GDOP occurs with two satellites at zenith, however the GDOP changes little if the high elevation angle satellites are reduced to 60 deg if there are three high elevation angle satellites.

Note that the optimum constellation gives a GDOP $= \sqrt{2}$ for six satellites vs a GDOP $= \sqrt{3}$ for four satellites for $E = 0$ deg. Thus, the minimum GDOP decreased directly as the inverse square root of the number of satellites, just as it did in the two-dimensional example. The optimum constellation forms a four-sided pyramid, as shown in Fig. 16a. Note that with six satellites, one-third at are zenith, and two-thirds on the horizon in proportion to the number of coordinates x,

Table 5 GDOP results for six satellites in different configurations

Three satellites in the horizon and three equally spaced at elevation angle E_z	GDOP $= 1.48$ at $E_z = 90$ deg, (for $E_z > 60$ deg GDOP ≤ 1.6)
Four satellites equally spaced on the horizon and two equally spaced at elevation angle (*Optimum configuration, four on horizon, two at zenith*)	GDOP $= \sqrt{2} = 1.41$ at $E_z = 90$ deg, (for $E_z \geq 70$ deg GDOP $\leq \sqrt{2.1} = 1.45$)
Five satellites on the horizon and one at the zenith.	GDOP $= 1.45$ at $E_z = 90$ deg

*Note that the maximum elevation angle for a user at the pole is $E = \tan^{-1}[(R_s \sin(55 \deg) - R_e)/R_s \cos(55 \deg)) = 45.27$ deg for $R_s = 26{,}561$ km. Thus, for polar users, there is a constraint on the maximum elevation angle as well.

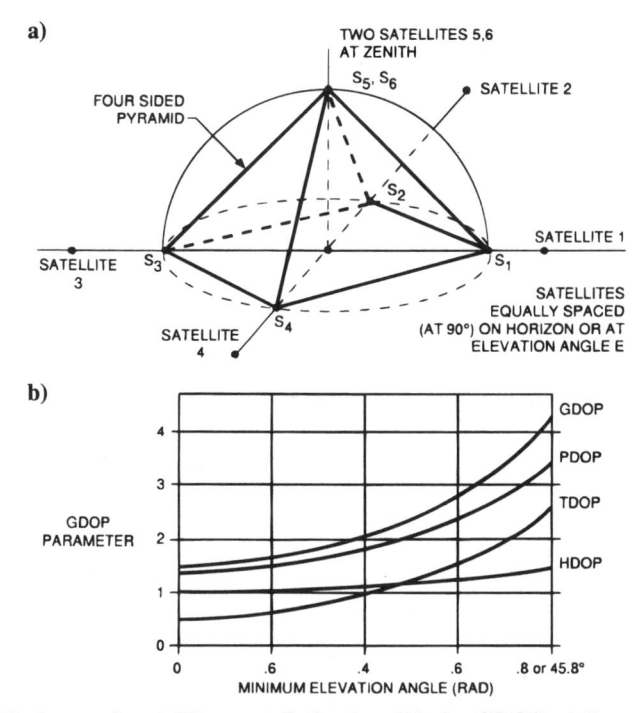

Fig. 16 Optimum six-satellite constellation to minimize GDOP: a) the satellites are in the directions of a four-sided pyramid circumscribed in a hemisphere; b) optimum GDOP for six satellites with an elevation angle mask angle E constraint. The minimum GDOP $= \sqrt{2} = 1.41$ for six satellites for an elevation angle mask of 0 deg.

y, and z in the respective planes. The bounds on GDOP, PDOP, HDOP, TDOP vs. elevation angle mask angle are shown in Fig. 16b.

Clearly, we can expand the inverse square root relationship further by simply taking one of the satellite constellations known to be optimum for N satellites and simply putting two satellites in the same position where one existed previously to obtain a $2N$ constellation. For example, take the optimum four-satellite constellation and put two satellites where only one existed previously for an eight-in-view scenario. Each satellite in effect has double the power or, equivalently, the noise σ has decreased by $\sqrt{2}$. Thus, the GDOP has decreased by $\sqrt{2}$ also.

Finally, it should be pointed out again that the real satellite pseudorange measurements contain some correlated error components; for example, atmospheric effects that increase with decreasing elevation angle, as discussed in the previous subsection. There are then at least the three following effects for users on or near the Earth's surface, as pointed out earlier in Fig. 15:

1) Some of the errors increase with decreasing elevation angle especially for E below 10 deg. Both tropospheric and ionospheric errors fall in this category. Multipath effects can also be important.

2) Because of the correlation in some of the error terms, the noise variances in the estimate do not necessarily average as well with increasing numbers of satellites, especially if several satellites are closely spaced in angle.

3) Correlation of the measurement error terms with satellites at the same elevation angle cancels some of the effective HDOP error.

C. Position Dilution of Precision with an Accurate Clock

In some situations, the user clock is quite stable in frequency. Once the clock bias is estimated in the receiver, the user clock estimate is often very accurate for the short term and initially has only a small variance σ_B^2. This variance grows slowly with time. For this situation, the user position can be computed using a three-dimensional model where the various pseudorange measurement noise terms now have both an independent thermal noise component and a clock bias component common to all pseudorange measurements. (For another approach, see Ref. 8.)

If the estimator operates with a pseudorange measurement noise that has the following covariance matrix:

$$E[d\mathbf{n} \quad d\mathbf{n}^T] = \sigma^2 \mathbf{I}, \quad \text{and} \quad d\mathbf{\rho} = \mathbf{G}d\mathbf{x} + d\mathbf{n} \tag{31}$$

then the estimate for differential change in user position is as follows[7]:

$$d\hat{\mathbf{x}} = [\mathbf{G}^T\mathbf{G}]^{-1}\mathbf{G}^T d\mathbf{\rho} \tag{32}$$

If the user clock bias is known to within some rms value σ_b, then we can simply subtract the bias b from the pseudorange measurement $\mathbf{\rho}$. Then we can write the user position error covariance as follows:

$$d\mathbf{\rho} = \mathbf{G}d\mathbf{x} + \mathbf{I}\,db + d\mathbf{n}, \tag{33}$$

and

$$[d\mathbf{x} \quad d\hat{\mathbf{x}}^T] = \sigma_n^2[\mathbf{G}^T\mathbf{G}]^{-1} + \sigma_b^2[\mathbf{G}^{-1} \quad \mathbf{I}\mathbf{I}^T \quad \mathbf{G}^{-T}] \tag{34}$$

where $\mathbf{I}$ is the unit vector, σ_n^2 is the noise variance, and σ_b^2 is the residual clock bias variance. The PDOP is defined by the following:

$$\sigma_n \text{PDOP} = \sigma_n \operatorname{tr}[\mathbf{G}^T\mathbf{G}]^{-1} + \sigma_b \operatorname{tr}[\mathbf{G}^{-1} \quad \mathbf{I}\mathbf{I} \quad \mathbf{G}^{-T}] \tag{35}$$

This expression can be rewritten as follows:

$$\text{PDOP} = \left[\text{PDOP}_n + \frac{\sigma_b}{\sigma_n}\text{PDOP}_b\right] \tag{36}$$

As a simple example, assume that there are three satellites, one at zenith, one in the x direction, and one in the y direction with an elevation angle constraint E. Assume also that the receiver has accurately estimated the ionosphere and troposphere and has corrected for selective availability. The PDOP_n and PDOP_b

for noise and clock bias then vary as shown in Fig. 17. In the optimum satellite configuration for $E = 0$, the optimum PDOP = $\sqrt{3}\,[1 + \sigma_b/\sigma_n]$, and for a good user clock and a limited time since the last clock update; i.e., a small ratio σ_b/σ_n, the PDOP = $\sqrt{3}$. Typically, we expect that immediately after an accurate four-satellite user position solution, the value of σ_b/σ_n is small ($<<1$). As time progresses, σ_b/σ_n grows slowly in a manner dependent on the quality of the user clock as, discussed in Chapter 4, this volume.

D. Position Dilution of Precision for the GPS-24 Constellation

Position dilution of precision outage maps and statistics have been computed for the GPS-24 satellite constellation. With all 24 satellites in operation, there are always at least 5 satellites in view, as shown previously for an elevation angle of 5 deg, and there are no PDOP outages for PDOP ≤ 6 as the requirement.

If a single GPS satellite fails, however, there can be momentary PDOP outages as shown, in Fig. 18 for a constraint PDOP ≤ 6. The outage regions are momentary, typically on the order of 5 min, but with a maximum of 24 min/day. Notice that the main outage regions of interest are those between 30 and 50° latitude, and there is another region between 55 and 75° latitude. These are regions within the coverage area of a geostationary satellite and also could be covered by a GPS altitude satellite in the equatorial plane.

Results are also shown in Fig. 19 for an average single satellite outage for a constraint PDOP < 10 and the GPS-24 constellation (P. Massatt, private communication. Courtesy Aerospace Corp. and the U.S. Air Force GPS Joint Program Office, July 11, 1994). These results indicate that there are relatively small regions and short time intervals of PDOP > 10, except for some polar regions. The worst case outages are less than 15 min per day, and generally much less. As discussed earlier, even these momentary outages can be eliminated by additional GPS satellites or by augmenting GPS with geostationary satellites.

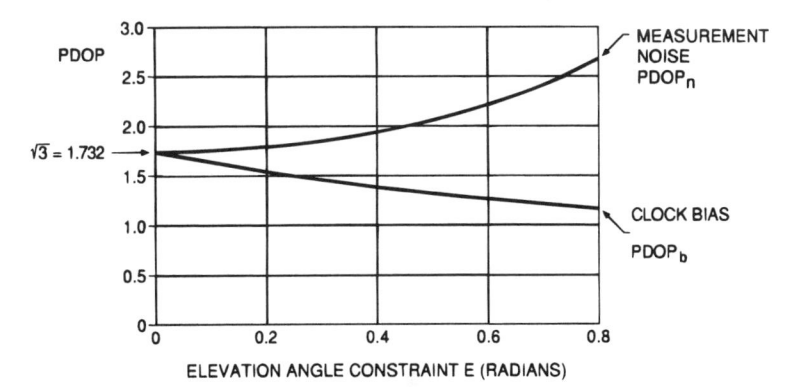

Fig. 17 PDOP components for three satellites with an accurate user clock of clock bias σ_b with measurement noise σ_n and elevation angle constraint E.

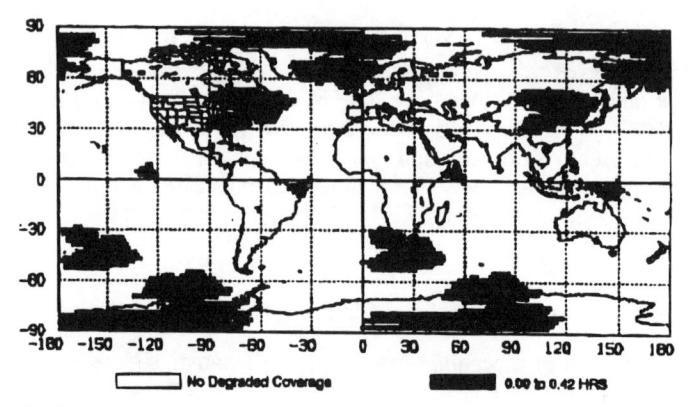

Fig. 18 Regions with short term degraded coverage with 23 operational satellites of the GPS-24 constellation—PDOP ≤6 constraint. An average satellite outage is assumed (courtesy U.S. Air Force). (Average case of 24 satellite constellation degraded coverage with 1 SV non operational at PDOP ≦ 6. Degraded coverage region is region that violates this constraint. Time interval of degraded coverage is less than 25 min/day.)

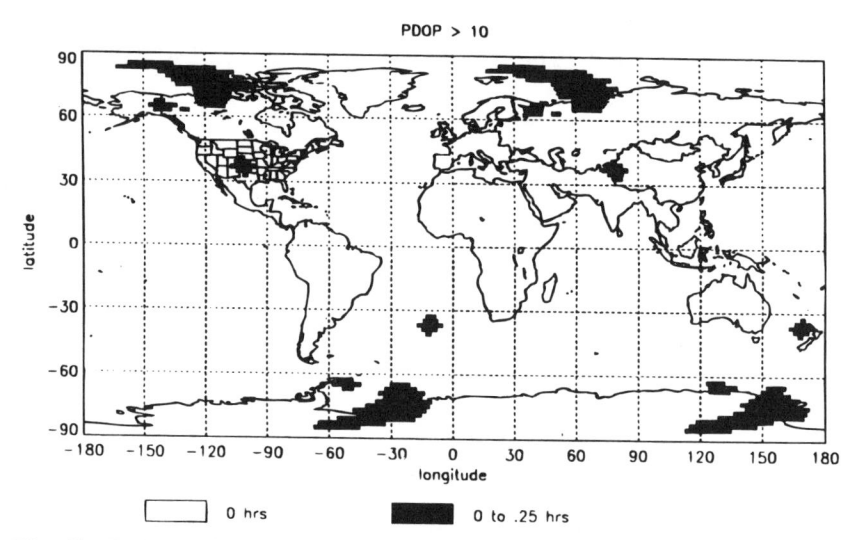

Fig. 19 Regions of short term degraded coverage for a single (average) satellite outage in the GPS-24 constellation (a failure of the $F3$ satellite) for PDOP > 10 considered an outage. (P. Massatt, personal communication. Courtesy Aerospace Corp and U.S. Air Force Joint Program Officer, July 11, 1994).

References

[1] Chobotov, V. A. (ed.), *Orbital Mechanics,* AIAA Education Series, AIAA, Washington, DC, 1991.

[2] Walker, J. G., "Circular Orbit Patterns Providing Continuous Whole Earth Coverage," Royal Aircraft Establishment, TR 77044, 1977.

[3] Stark, H., and Woods, J. W., *Probability, Random Processes, and Estimation Theory for Engineers,* Prentice-Hall, Englewood Cliffs, NJ, 1986.

[4] Barnett, S., *Matrices—Methods and Applications,* Oxford University Press, Oxford, 1990.

[5] Kreyszig, E., *Advanced Engineering Mathematics,* John Wiley, New York, 1993.

[6] Kihara, M., and Okada, T., "A Satellite Selection Method and Accuracy for the Global Positioning System," *Navigation,* 1984.

[7] Sturza, M. A., "GPS Navigation Using Three Satellites and a Precise Clock," *Navigation,* 1983, pp. 146–156.

[8] Copps, E. M., "An Aspect of the Role of Clocks in a GPS Receiver," *Navigation,* Fall 1984.

[9] Massatt, P., Private Communication, Aerospace Corp., July 1994.

Bibliography

Bate, R. R., Mueller, D. D., and White, J. E., *Fundamentals of Astrodynamics,* Dover Publications, New York, 1971.

Bowen, R., Swanson, P. L., Winn, F. B., Rhodus, N. W., and Feess, W. A., "Global Positioning System–Operatioal Control Systems Accuracies," Institute of Navigation, National Technical Meeting, 1985.

Escobal, P. R., *Methods of Orbit Determination,* Krieger Publishing Co., Malabar, FL, 1976.

Green, G., Massatt, P. D., and Rhodus, N. W., "The GPS 21 Primary Satellite Constellation," Institute of Navigation, Satellite Division, International Technical Meeting, 1988.

Lacarell, G., et al., "Flowers and Satellites," *Royal Institute of Navigation,* Vol. 44, No. 1, pp. 122–126.

Lee, N. S. "Accuracy Limitations in Hyperbolic Multilateration Systems," *IEEE Trans. AES,* Jan. 1975.

Massatt, P., and Rudnick, K., "Geometrical Formulas for the Dilution of Precision Calculations," *Navigation,* Winter 1990–1991, pp. 3179–391.

Millikan, R. J., and Zoller, C. J., "Principle of Operation of Navstar and System Characteristics," *Navigation,* 1980.

Philips, A., "Geometrical Determination of PDOP," *Navigation,* Winter, 1984–1985.

Phlong, W. S., Elrod, B. D., "Availability Characteristics of GPS and Augented Alternatives," Institute of Navigation, 1993 National Technical Meeting, Jan. 1993.

Rhodus, W., Private Communication, August 1992.

Rider, L., "Optimum Redundant Earth Coverage Using Polar Orbits," Aerospace Corp., Report, TOR-0066, March 1970.

Schuchman, L., Elrod, B. D., Van Dierendonck, A. J., "Applicability of an Augmented GPS for Navigation in the National Airspace System," Proceedings of the IEEE, Nov. 1989.

Walker, J. G., "Continuous Whole-Earth Coverage by Circular-Orbit Satellite Pattern," Royal Aircraft Establishment, *Journal British Interplanetary Society,* Technical Report 70211, Nov. 1970.

Walker, J. G., "Satellite Constellation," *Journal British Interplanetary Society,* Vol. 37, 1984.

Chapter 6

GPS Satellite and Payload

M. Aparicio, P. Brodie, L. Doyle, J. Rajan, and P. Torrione*
ITT, Nutley, New Jersey 07110

I. Spacecraft and Navigation Payload Heritage

A. Concept

THE launch of Sputnik I by the USSR in October 1957 was the beginning of the age of navigational satellites. That event was the culmination of centuries of navigation based on the known position of natural heavenly bodies and ever-improving clock accuracy. Observations of the signals transmitted by Sputnik I established the idea that accurate timing signals coming from artificial satellites of known position could aid navigators substantially. In effect, the artificial satellite replaced the functions of the sun and star tables with the known ephemeris of the satellite. In addition, the satellite's accurate timing signals replaced the ship's chronometer. The satellite-based system could provide continuous, worldwide coverage with few satellites; it could be an all-weather system, and it could provide extraordinary position location accuracy.

B. Relation to Earlier Non-GPS Satellites

The challenge of Sputnik I led to the U.S. satellite system Transit I. Begun in December 1958, its goal was continuous, worldwide, all-weather coverage. Transit was placed in operation January 1964. During the late 1960s, the growing need for accurate navigation among the U.S. strategic and tactical forces, combined with the rapid reduction in the cost of computers and processors (user equipment) established the need for, and potential feasibility of, a highly accurate tri-service navigational satellite system. The Air Force and Navy began independent programs called, respectively, 621B and TIMATION. NAVSTAR/GPS is the program that combined features of 621B and TIMATION. The Air Force's signal structure and frequencies were used, while the Navy's orbital configuration was chosen.

C. Overview of Payload Evolution

Two prime contractors under the management of the U.S. Air Force have taken the responsibility of taking the GPS satellite from concept into practice. Rockwell

---

Copyright © 1995 by the authors. Published by the American Institute of Aeronautics and Astronautics, Inc., with permission. Released to AIAA to publish in all forms.
*Aerospace/Communications Division.

International was the developer of the initial or Block I spacecraft and the Production Block II and IIA spacecraft, and Martin Marietta is the current developer of the GPS Block IIR (replenishment) spacecraft. ITT Corporation developed the critical navigational payload elements on all these satellites. The pseudorandom noise signal assembly (PRNSA), which constitutes the navigation payload on the original eight GPS development spacecraft consists of the following components: baseband processor; L_1/L_2 synthesizer; L_1 modulator; L_2 modulator; L_1 high-power amplifier; L_2 high-power amplifier; and diplexer. It also included delivery of GPS peculiar support equipment (GPSE) for ground station and prelaunch testing. Four more PRNSAs were subsequently added for a total of 12 GPS Block I deliveries.

The development Block I program resulted in 10 GPS spacecraft that were successfully launched from 1978 through 1985. One Block I spacecraft was destroyed because of launch failure. The 12th Block I was a qualification unit that was not flown. The Block II production spacecraft launches began in 1989. The Block II changes to the basic navigation service of the Block I spacecraft consisted of a gradually degrading navigation service for a period of 14 days, if the Control Segment (GPS CS) became inoperable. Block IIA spacecraft added an autonomous momentum management capability that functioned for a period of 180 days without ground contact.

The navigation message uploads to the Block II/IIA spacecraft are performed on a daily basis by the GPS CS. Block IIR satellites incorporate a ranging capability in the satellite crosslink. This ranging capability, combined with an onboard Kalman filter, gives the Block IIR autonomous navigation Autonov capability. Using the crosslink range measurements, the Block IIR spacecraft estimates the error in the Kepler orbital parameters. This enables the Block IIR satellites to support full navigation accuracy of 16-m spherical error probable (SEP), without CS contact for periods of up to 180 days. The comparable error of a Block IIA at the end of 180 days is of the order of kilometers.

D. On-Orbit Performance History

Figure 1 summarizes the historical development of the GPS space segment. Block IIR spacecraft are presently in development and will replace the Block I, II, and IIA satellites as they are declared nonfunctional. There are two basic ways in which satellites become nonfunctional. The first is unplanned and involves an on-orbit failure causing loss of the spacecraft, "caused by," "and" failure of two or more redundant components. The second is depletion of such life-limiting items as thruster fuel and degradaton of the solar arrays or batteries.

The GPS achieved a major milestone in December 1993 when it established initial operational capability or IOC. At IOC, the Air Force Space Command achieved operation of a full constellation of 24 GPS satellites. Shortly thereafter in February 1994, the Federal Aviation Agency (FAA) declared GPS operational for aviation use.

Figure 2 summarizes the launch and on-orbit performance history of the Block's I, II, and IIA space vehicles (SV). The last Block IIA SV is expected to be launched in the first quarter of 1996. Block IIR satellites will begin launches in

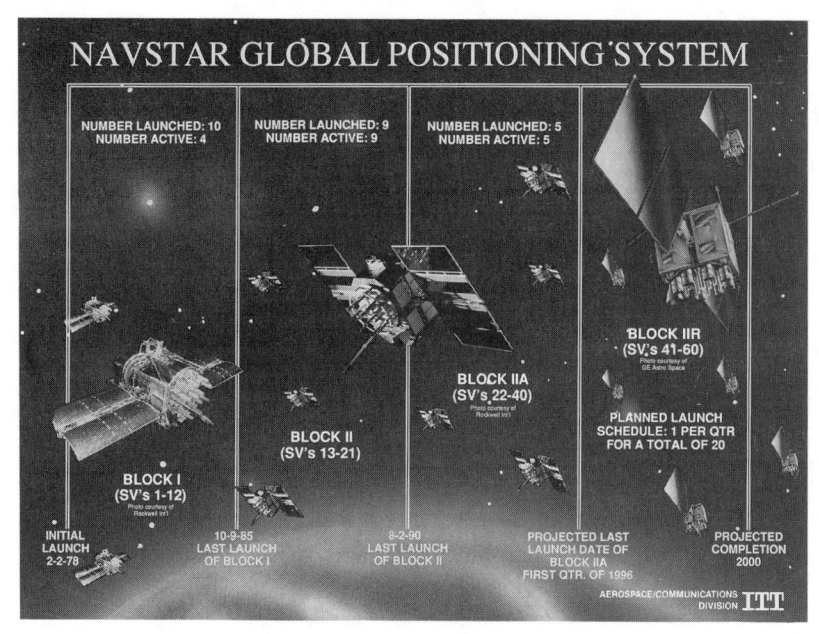

Fig. 1 Historical development of the space segment.

1996. The Air Force plans to begin the Block IIF (Follow-on) program in 1995 with expected launches beginning in 2001.

II. Navigation Payload Requirements

A. GPS System

GPS is a highly accurate, passive, all-weather, 24-h, worldwide, common-grid navigation system. The navigation payload must supply a continuous, precision, high-integrity signal to support this requirement.

The GPS has also enabled specialized users to enhance system accuracy, availability, and integrity with augmentation and sensor integration. The augmentations include differential stations providing differential corrections to GPS measurements; the GPS integrity channel (GIC) providing faster integrity information via geostationary satellites and pseudolites, enabling faster cycle ambiguity resolution and precise localized navigation. The sensor integrations include GPS integrations with inertial navigation systems (INS) and low-cost multisensors to provide accurate navigation comparable to INS at much lower cost. These combined systems provide superior short-term accuracy of GPS and control the error growth during GPS signal outage (caused by masking and aircraft maneuver) by utilizing the long-term stability of an INS.

B. GPS Performance

The GPS is designed to provide a precision positioning service with a 16-m SEP. This performance is achievable by a dual-frequency precision code user

SV/PRN No.	GPS Block	Launch Date	On-Orbit Available Date	Nav Lost Date	Deactivation Date	Total Months Available	Mission Life Rqmt (Months)	Exceeded Life by (Months)	Failure Mechanism
1/4	I	22-Feb-78	29-Mar-78	25-Jan-80	17-Jul-85	22	48	-26	Clock
2/7	I	13-May-78	14-Jul-78	30-Jul-80	16-Jul-81	24	48	-24	Clock
3/6	I	06-Oct-78	09-Nov-78	19-Apr-92	25-Apr-92	161	48	113	Clock
4/8	I	11-Dec-78	08-Jan-79	27-Oct-86	14-Oct-89	93	48	45	Clock
5/5	I	09-Feb-80	27-Feb-80	28-Nov-83	28-Nov-83	45	48	-3	Wheel
6/9	I	26-Apr-80	16-May-80	10-Dec-90	06-Mar-91	127	48	79	Wheel
7/	I	18-Dec-81 - Satellite destroyed during launch (booster failure)							
8/11	I	14-Jul-83	10-Aug-83	04-May-93	04-May-93	117	48	69	EPS Degr.
9/13	I	13-Jun-84	19-Jul-84	28-Feb-94	28-Feb-94	115	48	67	Clock
10/12	I	08-Sep-84	03-Oct-84			119	48	71	
11/3	I	09-Oct-85	30-Oct-85	27-Feb-94	27-Feb-94	100	48	52	Not ID'd
12/	II	Qual Unit - will not be launched							
13/2	II	10-Jun-89	11-Jul-89			62	60	2	
14/14	II	14-Feb-89	15-Apr-89			65	60	5	
15/15	II	01-Oct-90	15-Oct-90			47	60		
16/16	II	18-Aug-89	14-Oct-89			59	60		
17/17	II	11-Dec-89	06-Jan-90			56	60		
18/18	II	24-Jan-90	16-Feb-90			55	60		
19/19	II	21-Oct-89	26-Nov-89			58	60		
20/20	II	24-Mar-90	18-Apr-90			53	60		
21/21	II	02-Aug-90	22-Aug-90			49	60		
22/22	IIA	03-Feb-93	30-Mar-93			18	60		
23/23	IIA	26-Nov-90	10-Dec-90			45	60		
24/24	IIA	04-Jul-91	30-Aug-91			37	60		
25/25	IIA	23-Feb-92	24-Mar-92			30	60		
26/26	IIA	07-Jul-92	23-Jul-92			26	60		
27/27	IIA	09-Sep-92	30-Sep-92			24	60		
28/28	IIA	10-Apr-92	25-Apr-92			29	60		
29/29	IIA	18-Dec-92	05-Jan-93			20	60		
30	IIA	FY 95					60		
31/31	IIA	04-Apr-93	13-Apr-93			17	60		
32/1	IIA	22-Nov-92	11-Dec-92			21	60		
33	IIA	FY 95					60		
34/4	IIA	26-Oct-93	29-Nov-93			10	60		
35/5	IIA	30-Aug-93	28-Sep-93			12	60		
36/6	IIA	10-Mar-94	28-Mar-94			6	60		
37/7	IIA	13-May-93	12-Jun-93			15	60		
38	IIA	FY 96					60		
39/9	IIA	26-Jun-93	20-Jul-93			14	60		
40	IIA	FY 97					60		

Data is current as of 06-Sep-94

Total On-Orbit Availability (to Date): 1,751 months / 145.9 years
On-Orbit Mission Life above Program Rqmts (to Date): 37.5 years

Fig. 2 The GPS satellite on-orbit performance history.

under dynamic environment. The current policy is to encrypt the precision code so that it is available only to authorized users. The user ranging error (URE) is derived from the Block II requirement that the system meet a navigational accuracy of 16-m SEP. The derived user ranging error (URE) is 6.6 m (one sigma). The URE budget is shown in Table 1. The velocity accuracy achievable by the system in a dynamic environment is 0.1 m/s. The time transfer accuracy that can be achieved by the system is at least 100 ns. The system also provides a standard positioning service (SPS) of 100-m 2d rms for unauthorized users. The C/A code by itself can provide an accuracy of about 25-m SEP. However, the accuracy is degraded to a 100-m level by man-made degradation of the navigation signal and parameters.

The total root-sum-square errors (rss) allowed for the space segment is approximately 3.5 m (one sigma). An rss is taken, because these errors are uncorrelated

GPS SATELLITE AND PAYLOAD

Table 1 System error budget (Block II)

Error sources	User range error (1σ), m
Space segment	
Clock and navigation subsystem stability	3.0
L-Band phase uncertainty	1.5
Predictability of SV parameters	1.0
Other	0.5
Control segment	
Ephemeris prediction and model implementation	4.2
Other	0.9
User segment	
Ionospheric delay compensation	2.3
Tropospheric delay compensation	2.0
Receiver noise and resolution	1.5
Multipath	1.2
Other	0.5
Total (root-sum-square) rss URE	6.6

or statistically independent. Achieved performance for the satellite is 2.2 m (one-sigma) using the rubidium atomic frequency standard (AFS), and 2.9 m (one-sigma) for the cesium AFS. Both performance measures are taken 24 h after the last update from the ground. For normal operations, the ground updates the satellite every 24 h.

C. GPS Signal Structure

The GPS satellites provide precise ranging signals at two frequencies, L_1 and L_2. The satellite transmits precise ranging information using precision (P) code and transmits coarse range using the coarse/acquisition (C/A) code. The C/A code is a Gold code of register size 10, which has a sequence length of 1023. The clock rate of the C/A code is 1.023 MHz, and the code period is 1 ms. The P code is clocked at 10.23 MHz. Each satellite uses a different member of the C/A Gold code family. The P code is over 37 weeks long but is short-cycled on a weekly basis. Different satellites use a different 1-week segment of the P code. The short length of the C/A code allows user equipment that has a low-cost clock with time uncertainty of the order of seconds to search the entire code phase of the C/A code quickly and acquire and track the C/A signal. Tracking the C/A code enables the receiver to demodulate the navigation data. The navigation message is a 50 b/s datastream arranged in 25 pages, each page containing 5 subframes, with each subframe containing 10 words of 30 bits each. The navigation datawords are encoded with (32, 24) Hamming parity providing single error correction and double error detection capabilities. The navigation data on the C/A code has, in addition to the ephemeris and satellite clock correction and double error detection information, the hand-over information that enables the receiver to acquire and track the P code. Specialized receivers with direct time transfer capability can directly search, acquire, and track the P code.

Dual-frequency transmission of ranging signals by the GPS satellite enables user equipment with dual-frequency capability to measure ranges at the L_1 and

L_2 frequencies. These measurements allow the user equipment to accurately compensate for the propagation delay through the ionosphere.

The signal levels guaranteed to the users near the surface of Earth are shown in Table 6-2 for various signal components. The C/A-code power level is 3 dB higher than that of P code to enable fast initial acquisition of the C/A code. The L_2 power is considerable lower, because it typically is not used for initial acquisition. The receivers in general search, acquire, and track the L_1 C/A signal. Based on range measurements made while tracking L_1 C/A, a limited and somewhat slower search is made on the L_2 signal to enable L_2 acquisition and subsequent tracking.

D. Payload Requirements

The link budget covers the three segments of the system and satisfies the power density levels guaranteed to the users. The obvious trade-off in the link budget is the antenna gain vs payload transmit power. This trade-off is made by evaluating the cost, size, power, and weight.

Key requirements are derived from the receive bands of the receivers onboard the satellites and from such specialized frequency bands as a radio astronomy band. An additional requirement minimizes the phase noise of the space-borne L-band transmitters.

Other key performance parameters include: group delay variation on the L-band transmitter chain, which has an impact on the payload URE; the uncertainty in the differential group delay between L_1 and L_2, which has an impact on the accuracy to which the ionospheric corrections can be made; and gain flatness, which influences the symmetry of the code autocorrelation function at the receiver.

As in most complex systems, high reliability is an important factor in the GPS payload design. High reliability provides an assurance that the system is providing an accurate signal to meet the user's navigational needs.

The design of the spacecraft incorporates subassembly and component redundancy; environmental controls; the use of flight-proven, high-reliability piece parts (e.g., transistors, integrated circuits, relays); and proven manufacturing and test procedures and practices. Particular design and test emphasis is placed on ensuring that a failure within a component will neither degrade the performance of the components within the spacecraft nor propagate throughout the system.

Single point failures have been absolutely minimized. Detailed reliability analyses have been performed on all spacecraft elements to determine the effectiveness of design trade-offs. Failure modes, effects, and criticality analyses (FMECA)

Table 2 L-band rf power

	L_1		L_2
	I	Q	I
	C/A	P(Y)	P(Y)/C/A
Output power at the antenna input, dbw	+14.3	+11.3	+8.1
Power near the Earth's surface	−160.0	−163.0	−166.0

have been performed on all elements of the SV. The analyses are performed concurrently with the design efforts, thus producing designs that reflect the analyses conclusions and recommendations. The GPS spacecraft is required to meet a 7.5-year life and is designed for a 10-year life.

III. Block IIR Space Vehicle Configuration

As noted earlier, the Block IIR spacecraft will replace the Block II and IIA spacecraft as they become nonoperational. The contract to develop the Block IIR system was awarded to Martin Marietta (then General Electric Astro Space) and to ITT Aerospace/Communications Division (then ITT Defense Communications Division) in 1987. The team consists of more than 20 subcontractors nationwide. The launch of the first Block IIR spacecraft is planned for the last quarter of 1996. An exploded view of the Block IIR SV is shown in Fig. 3.

The GPS navigation signal is generated and transmitted by the total navigation payload (TNP). The AFS is the heart of the TNP. It provides the precision timing needed to achieve GPS accuracy. The TNP includes two rubidium atomic frequency standards and a cesium atomic frequency standard.

The L-band system consists of the three transmitter chains for three radio frequencies denoted as L_1, L_2, and L_3. The L_1 and L_2 frequencies are used for the navigation mission of the GPS, and the L_3 frequency is used by the nuclear detonation detection system (NDS or Nudet), also located onboard the GPS SV.

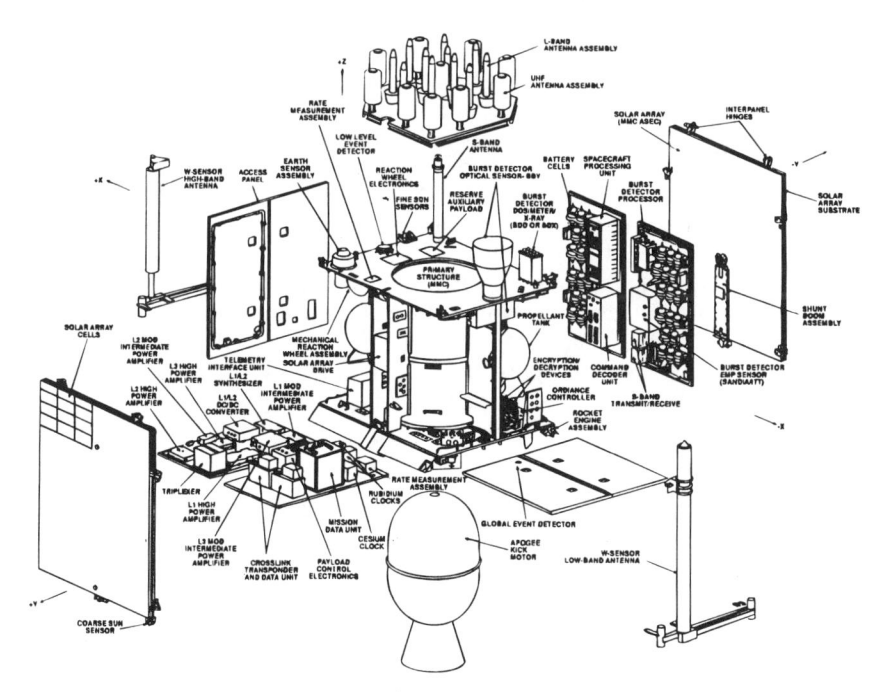

Fig. 3 The GPS Block IIR space vehicle.

Within the TNP, are two major functional capabilities that provide the Block IIR with a giant leap in mission capabilities as compared to Block II/IIA. These are referred to as the time-keeping system (TKS) and the autonomous navigation (AutoNav) capability.

The subsystems of TNP, TKS, and the AutoNav capabilities are explored in detail in the following sections.

The central body of the Block IIR SV is a cube of approximately 6 ft on each side. The span of the solar panel is about 30 ft. The lift-off weight of the spacecraft is 4480 lb and the on-orbit weight is about 2370 lb. The Block IIR spacecraft will be launched by the Delta launch vehicle. The spacecraft has 16 thrusters and a standard set of bus equipment such as telemetry, tracking, and command (TT&C), payload control electronics (PCE), a spacecraft processing unit (SPU), and an attitude reference system. The Block IIR spacecraft are designed to meet 10-year mission life and 4-year storage life. They are designed to be ready to launch with 60-days notice. The SV is designed to operate autonomously for at least 180 days without ground intervention; whereas, the Block I and II satellites' accuracy degrades without ground contact.

A. Navigation Payload Architecture

The TNP components are shown mounted on the two payload panels of the spacecraft shown on the lower left of Figure 6-3. This assembly consists of all the components needed to generate the signals that provide the navigational capability of the GPS. The antennas that radiate the signals to Earth are shown on the top face in the figure. This face is always accurately pointed toward the Earth to ensure uniform and stable illumination of the Earth by the navigational signals. Placing the TNP on two adjacent and stand-alone panels allows easy integration and testing of the spacecraft.

The mission data unit (MDU), atomic frequency assembly, and L-band subsystem produce the navigational signals. The crosslink transponder and data unit (CTDU) provides direct satellite-to-satellite communication and ranging. This allows the satellites to operate autonomously (i.e., without ground control segment time and ephemeris updates) at full accuracy for at least 180 days.

The MDU is the brains of the TNP. It integrates all mission functions, such as ephemeris calculations, encryption, NDS data, pseudorandom code generation, and autoNav, as well as monitoring the health of specific TNP components. The MDU software consists of approximately 25,000 lines of code written in Ada running on a MIL-STD 1750A radiation-hard processor at 16 MHz. A block diagram of the TNP is shown in Fig. 4.

IV. Block IIR Payload Design

A. Payload Subsystems

The L-band subsystem also includes bi- and quadriphase shift keyed modulators, (BPSK and QPSK) which place the MDU information with the pseudorandom C/A and P codes (1.023 Mbs and 10.23 Mbs, respectively) on the three L-band carriers. These three carriers are at 1227.60 MHz (L_2), 1381.05 MHz (L_3), and 1575.42 MHz (L_1). The three carriers are amplified by bipolar transistor

GPS SATELLITE AND PAYLOAD

MISSION DATA UNIT

- 1750A Central Processor
- ADA HOL Used Throughout
- Clock Frequency Synthesis from Multiple Standards
- Integral Baseband Processor
- Full Message Encoding and Message Processing
- Real Time Kalman Filter Navigation and Clock State Estimation

CROSSLINK TRANSPONDER DATA LINK

- RF Receive Transmit of Digital Data
- Precision Inter Satellite Ranging
- Frequency Hopped TDMA
- Full Frame Modulation and Mode Control

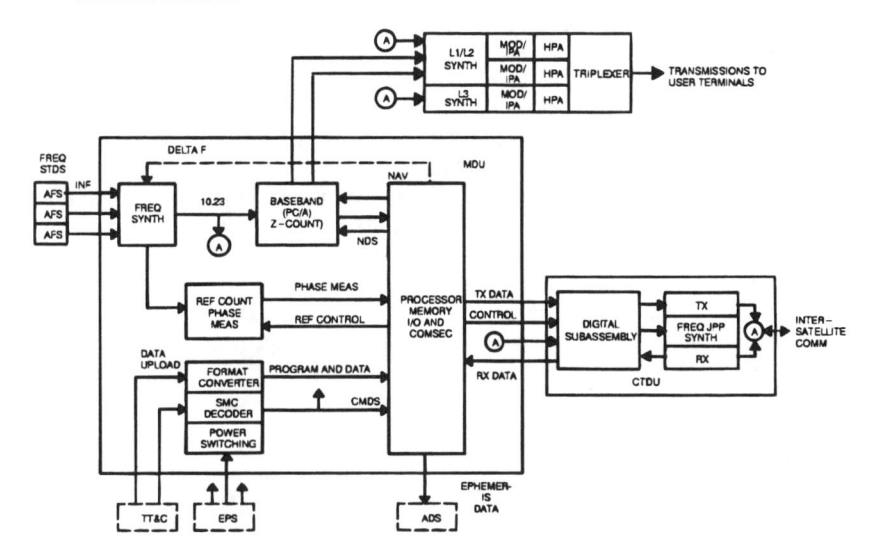

TIME STANDARD ASSEMBLY

- Multiple Atomic Frequency Standards for Reliability
- Accommodates Various Clock Types (Cs, Rb)
- RAD–Hard Upset Proof Design
- Synthesized High Stability GPS Timing Signals
- Automated Integrity Monitoring

L-BAND SUBSYSTEM

- 25–50 Watt Transmitter
- Bandwidth 20 MHz
- Radiation Hardened
- L_1: 1 or 10 MChip/s Quadraphase
  L_2, L_3: 1 or 10 MChip/s Biphase
- Space Proven Design Operational on Block I and Block II

Fig. 4 Complete system from atomic reference clocks through transmitted navigation messages to users.

amplifiers using microwave integrated circuits technology and sent to the triplexer, which combines them and delivers them to the spacecraft antennas for radiating to users on the Earth. The L_3 signal is filtered through an L_3 astronomy filter.

A principal element of the spacecraft is the atomic frequency standard or atomic clocks. The key requirement is to maintain an accuracy of 6 ns with respect to GPS time. To do this, cesium and rubidium atomic frequency standards are integrated into the TKS of the navigation payload. To ensure operation if one of these critical components fails, two rubidium standards and one cesium standard provide redundancy for the time standard assembly (TSA). The Block

IIR TNP unit is shown in Fig. 5 undergoing prequalification tests at the ITT A/CD facility in Clifton, New Jersey.

The following sections discuss the key design issues associated with each element. Table 3 summarizes the functions of each payload element. Each of these is described in detail in the following sections.

1. Atomic Frequency Standards

a. Introduction. Placing a very stable time reference in a position where maximum user access can be achieved is the basis for modern satellite navigation. At the speed of propagation of electromagnetic signals in the atmosphere, 1 ns of phase uncertainty of the signal, as measured at the user, is roughly equivalent to 1 ft or 1/3 m in position uncertainty. The only instruments that can maintain the phase uncertainty within the required limits for GPS; i.e., 9 ns, one sigma for 24, are atomic frequency standards. This phase stability is equivalent to an uncertainty of 1 mm out of the distance from the Sun to the Earth.

Atomic frequency standards* are amazingly accurate and stable devices. The history of frequency standards and clocks spans the spectrum from sand and water clocks to the modern chronometer and present day AFS. The fundamental problem of clock use has remained the same: "How do we measure the time

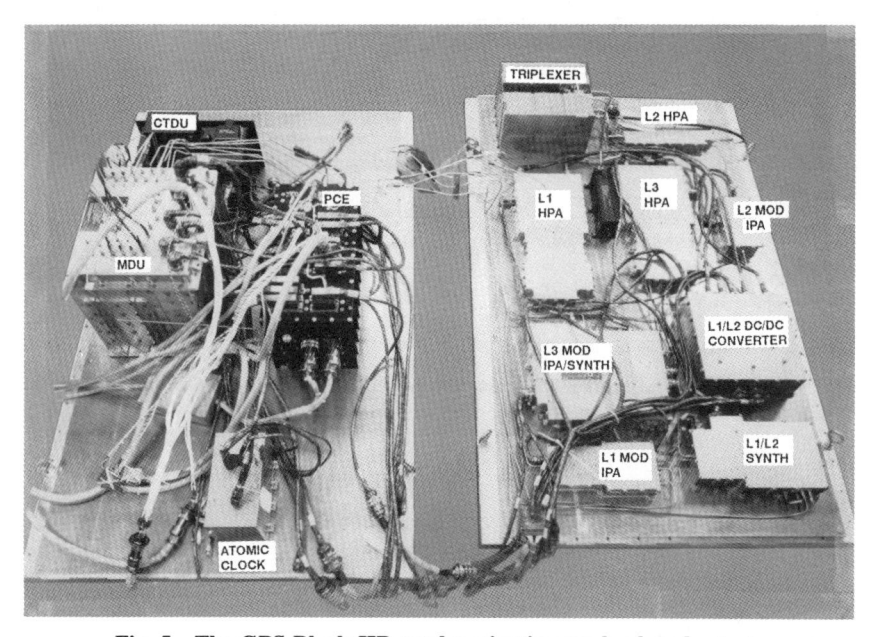

Fig. 5 The GPS Block IIR total navigation payload under test.

---

*The primary difference between clocks and frequency standards is that clocks keep a record of how many repetitions of a periodic phenomenon have occurred since an event; whereas, frequency standards provide only the periodic phenomenon to the external world.

Table 3 GPS Block IIR payload element functionality

Element	Function
Atomic frequency standards	Provide accurate, stable and reliable timing for all GPS signals. (This is the most important function, because accurate navigation depends primarily upon accurate time.)
Onboard processing	Generate navigational messages, perform ephemeris calculations, and data encryption, generate P and C/A code, monitor health of payload, provide clock error corrections
Software	Implement onboard processing functions. Reprogrammable from ground stations
L-band system	Generate and modulate the L_1, L_2, and L_3 signals and combine them for transmission to Earth
Crosslinks	Provide satellite-to-satellite communications and ranging
Auto navigation	Provide accurate autonomous operation without regular communications from Earth

interval between two events?" The answer has also remained the same; "Count periods of a stable phenomenon." The technological history of clocks and frequency standards is a search for more accurate and stable phenomena and implementation of the measuring device, ranging from the rising and setting of the sun, through mechanical pendulums, to transitions of electrons between energy bands.

The principle of operation of atomic frequency standards[12,14,15] in its most fundamental form is: "The coupling of the output frequency to a periodic natural phenomenon whose period (time to complete a cycle) is essentially invariant." The natural phenomenon used is the change in energy of the outermost electron of an atom of a given element or compound. When this change in energy occurs, the atoms release or absorb the energy at a precisely determined frequency, hence the term atomic frequency standard.

There are different types of AFS, classified by the element or compounds they use for the electron energy transition and the way they couple this information to external devices. Among the most successful AFS types are: hydrogen masers; ammonia frequency standard; cesium frequency standard; rubidium frequency standard and maser; beryllium frequency standard; and mercury electromagnetic ion trap frequency standard.

All AFS types have three common functions: 1) preparation of the outermost electron population into a known state; 2) injection of an electromagnetic signal that causes the energy transition, (higher precision of the frequency of the injected signal causes more atoms to transition); and 3) interrogation and sorting of the resulting energy state of the resulting outermost electron atomic or molecular population. The last step generates an error signal that is used to tune a voltage control oscillator (VCXO) to that natural frequency. The VCXO provides the clock signal with the correct output frequency. Specific examples of this process are the rubidium and cesium AFS. These two types of AFS are of significant importance and are the only types in use in the GPS space segment. The diagrams shown (Fig. 6 and 7) are of the resonator sections of the cesium and rubidium AFS. They are those elements of the AFS that perform the three operations described above.

b. Operation of the cesium AFS. For the GPS cesium AFS, cesium atoms are emitted from a heated cesium reservoir (Fig. 6). Then the three operations are performed. The first is the preparation or sorting of the energy states of the electron in the atomic or molecular population of the cesium atoms. This is accomplished using a magnetic field tuned to the natural frequency of the magnetic dipole of the atoms that contain the electrons in the ground state. This field is produced by a magnet known as the "A" magnet that deflects atoms into the cavity only if they contain electrons in the ground state. This operation creates a relatively pure population of atoms that have their outermost electrons in the ground state.

The second operation is the electromagnetic stimulation of the outermost electron of the cesium atoms. The electrons are shifted from the ground energy state to the next energy state or hyperfine state. This stimulation takes place, if and only if, the electromagnetic field is oscillating extremely close to the specific frequency. That frequency is 9, 192, 631, 770 Hz. If the electromagnetic stimulus is not very close to the oscillating frequency, few transitions to the hyperfine energy state take place, and most electrons remain in the ground energy state. If the electromagnetic stimulus is at the right frequency, many electrons will make the hyperfine transition. This transition changes the properties of the atoms that contain the electrons. In particular, this transition changes the magnetic dipole of the host atoms and allows sorting of the atoms.

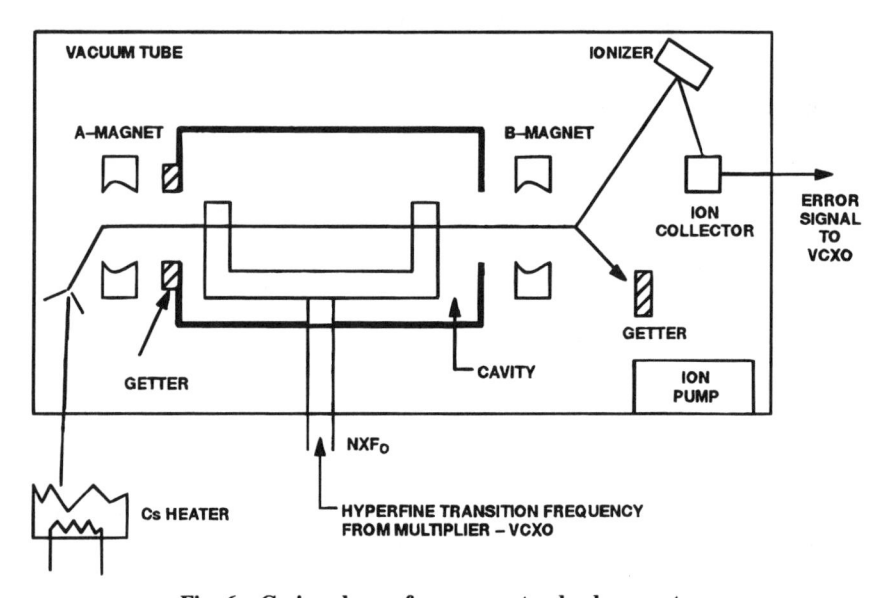

Fig. 6 Cesium beam frequency standard resonator.

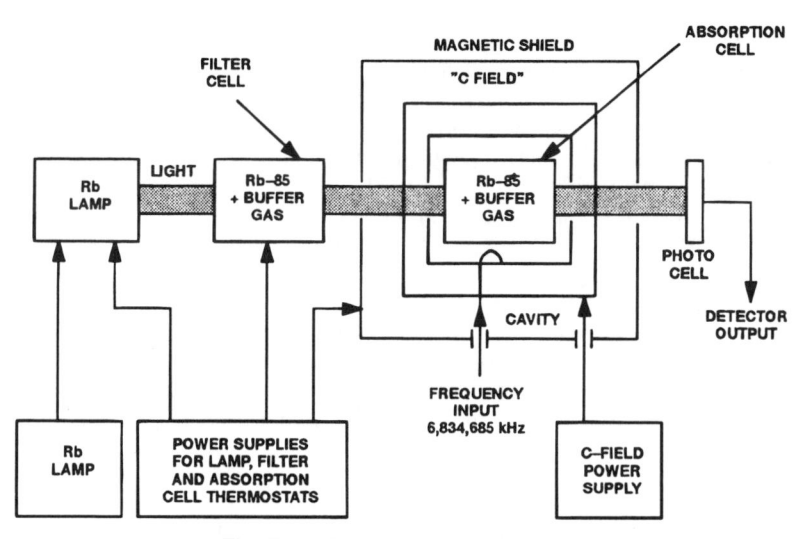

Fig. 7 Rubidium atomic resonator.

The third operation, which is the interrogation and sorting of the atomic population, is performed with the magnet placed at the other end of the cavity. This magnet, the "B" magnet, is tuned so that it deflects atoms that contain electrons in the hyperfine energy state. If the atoms coming out of the cavity contain electrons in the hyperfine energy state, those are then deflected. This means that the electromagnetic stimulus is precisely on frequency if it is successfully causing many energy transitions. The deflected atoms are sent in a path that causes them to strike an ionizer rod. The ionizer rod very efficiently converts the neutral atoms into ions. The ions continue their trajectory into an electron multiplier that turns the ion flow into a current flow proportional to the number of atoms that have been deflected. The current is an indication of the frequency of the electromagnetic stimulus that was injected in the cavity. From the electron multiplier detector output, an error signal is derived, which corrects the VCXO in a control loop. The VCXO oscillation frequency is the actual clock output frequency of the device.

c. Operation of the rubidium AFS. For the rubidium AFS, three similar operations are performed (see Fig. 7). Initially, electrons are in the ground energy state. There will always be a fraction of the total population of atoms that naturally contain outermost electrons in the ground state. In a working standard, this is normally 0.1% of the total population.

The outermost electrons are then excited to the hyperfine energy state. This excitation is carried out by injecting an electromagnetic signal into the cavity that contains the rubidium vapor (the so-called absorption cell). If this stimulus is oscillating at precisely the right frequency 6, 834, 682, 608 Hz (for a zero magnetic field), many electrons in the ground energy state in the atomic population change into the hyperfine energy state. This transition in energy state changes the optical absorption properties of the host atoms.

The third operation is interrogation of the atomic population for energy changes. This optical interrogation process begins at the lamp. The lamp is a discharge device filled with the rubidium isotope 87 and excited by an rf source at about 100 MHz. The lamp emits a spectrum of electromagnetic frequencies, the majority in the visible range. The electromagnetic emissions are passed through a filter cell filled with the rubidium isotope 85. The filter cell stops all but two visible frequencies from reaching the absorption cell. Atoms containing the outermost electrons that have made the hyperfine energy state transition from the ground state will absorb the light. The absorption of the light by the host atoms creates a minimum current output at the optical detector located at the other end of the absorption cell. The minimum occurs only if the electrons in the population of host atoms make the hyperfine energy transition. The optical detector or photodetector generates a current proportional to the amount of light that reaches it. The current is used to shift the VCXO oscillation frequency to the desired clock output frequency.

d. Performance. The most important requirements for operating an AFS in a space-borne platform for a navigational application are: phase stability; reliability; low power; low weight; low volume; and high tolerance to the space environment. To meet these requirements, special space-qualified units are used in the GPS satellites. The most difficult requirements to meet are phase stability and reliability. Although the AFS units used by the National Bureau of Standards exhibit excellent performance, they are too large and heavy for spacecraft applicaton. The reliability of AFS has traditionally been low because of the nature of their high sensitivity to the environment and their complexity. Balancing these unique requirements and constraints to produce units accurate and reliable enough for space poses unique technological challenges.

Phase stability is measured in terms of the time-averaged integral of the fractional frequency stability ($\sigma_y(\tau)$). The fractional frequency stability is measured by the Allan variance $\sigma^2_y(\tau)$, which is the square of the Allan deviation $\sigma_y(\tau)$. The Allan variance is defined by the following equation:

$$\sigma_y^2(\tau) = \frac{1}{2N-1} \sum_{i=0}^{N-1} (Y_{i+1} - Y_i)^2$$

where Y_i = fractional frequency at time interval i (actual frequency/nominal desired frequency, and N = number of fractional frequency samples.

The Allan deviation is a measure or statistical estimate of the noise contribution to the frequency instability of frequency standards. The noise processes that have experimentally been found to have significant effects in precision frequency standards are: white phase modulation noise (PM); white frequency modulation noise; flicker phase modulation noise; flicker frequency modulation noise; and random walk of phase noise.

Table 4 is a comparison of some of the critical parameters for the GPS Blocks I, II, IIA, and IIR.[15] As mentioned previously, reliability, power, weight, volume, and tolerance to the space environment are the critical parameters for AFS in the GPS context.

GPS SATELLITE AND PAYLOAD

Table 4 GPS Blocks I, II, IIA, and IIR atomic frequency standards specifications

AFS type	Reliability requirement	Size, in.	Weight, lb	Power consumption, W
Rb GPS Block I	0.763 for 5.5 years	L = 5.00 W = 6.00 H = 7.50	13	24.75
Cs GPS Block I[a]	0.663 for 5.5 years	L = 5.30 W = 15.10 H = 7.80	28	22.00
Rb GPS Block II	0.763 for 5.5 years	L = 5.00 W = 6.00 H = 7.50	13	24.75
Cs GPS Block II	0.663 for 5.5 years	L = 5.30 W = 15.10 H = 7.80	28	22.00
Rb GPS Block IIA	0.763 for 5.5 years	L = 5.00 W = 6.00 H = 7.50	13	24.75
CS GPS Block IIA	0.663 for 5.5 years	L = 5.30 W = 15.10 H = 7.80	28	22.00
CS GPS Block IIA second source	0.750 for 5.5 years	L = 5.30 W = 15.10 H = 7.80	28	22.00
Rb GPS Block IIR	0.763 for 7.5 years	L = 8.50 W = 5.60 H = 6.20	14	15
Cs GPS Block IIR	0.775 for 7.5 years	L = 16.50 W = 5.50 H = 5.25	22	26

[a]Only the last four satellites of Block I carry cesium atomic frequency standards.

Table 4 shows that the greatest improvements from Block I, II, and IIA to IIR are the reliability, weight, and power consumption of the rubidium standards. Of these parameters, the most relevant change is the power consumption from 24.75 W to 15 W. Predicted reliability has been increased significantly for both standards. The reliability parameter is of extreme importance in the space mission environment, where the life of components and subsystems limit the mission's life.

B. Mission Data Unit

1. Onboard Processing

Onboard processing is performed in the MDU. The MDU and the frequency synthesizer unit (FSU) are housed together in one physical package. The MDU provides storage of navigation data as uploaded by the CS via the TT&C subsystem of the SV. The MDU combines these data with internally generated ranging codes and sends the resulting navigation message to the L-band system (LBS) for transmission to the ground. The MDU also is capable of altering these navigation signals when necessary to deny full GPS navigation accuracy to unauthorized users [i.e., it provides selective availability (SA)]. The FSU plays a role in both the generation of the navigation signals and selective availability, and it is controlled by the MDU processor. The MDU has the capability of autonomously operating for 180 days without receiving navigation data updates from the CS. When the MDU operates in this mode, it computes the SV ephemeris and clock correction data by processing rf ranging performed between spacecraft. The MDU then updates the contents of the navigation message sent to the users.

In addition to the function of providing navigation data to GPS users, the MDU receives data from the NDS. These data are encoded by the MDU and transferred to the ground with navigation data via the LBS. The data are also transmitted via the uhf crosslink to other in-view satellites for retransmission to the ground via the L-band. The MDU/FSU operates in the space radiation environment and can operate through or recover from specified transient nuclear events without a permanent impact on navigation functional accuracy or the NDS data-processing function.

The following is a summary of MDU/FSU functions:

1) provide storage of messages uploaded by the CS;
2) process, format, and generate navigation data;
3) provide precise timing for other payload components;
4) generate pseudorandom noise (PRN) codes for navigation;
5) provide SA; i.e., alter the navigation downlink when necessary to deny full GPS navigation accuracy to unauthorized users;
6) perform antispoof (AS); i.e., alter the navigation downlink on CS command to allow the authorized user full GPS accuracy through hostile environments;
7) operate through certain specific transient nuclear events without a permanent impact on the navigation functional accuracy;
8) recover from nuclear radiation transient including logic upset without assistance;
9) autonomously operate for 180 days without update from the CS; i.e., modify navigation data on a periodic basis by processing ranging data from other in-view satellites and exchange navigation data with other SVs and enable autonomous determination of ephemeris and clock corrections;
10) operate for 14 days, with at least the same navigation accuracy as the Block II vehicles are required to meet, without updates from the CS or other satellites;
11) encode NDS data received from the burst detection processor (BDP).
12) insert current SA data and SV ephemeris data into the NDS datamessage for transmission to other GPS satellites and the nuclear detonation user segment (NDUS).
13) update the SA data element (rapid turn-on) via the CTDU upon ground command.
14) perform a graceful turn-on and turn-off function to a known acceptable condition.
15) provide telemetry, diagnostics, and self-check capabilities.

2. Software

The computer program in the Block IIR navigation payload is large and complicated, as compared to those used in other spacecraft. This program is referred to as the mission processor (MP) software. The requirement to function autonomously for 180 days dictates that the spacecraft will have to perform many functions currently performed by the CS. In addition to ephemeris and clock parameters estimation, this includes integrity monitoring, curve fitting of the navigation parameters, user range accuracy (URA) estimation, formatting the navigation message, selective availability, universal coordinated time (UTC) steering, and unassisted recovery from upsets.

Another factor contributing to the complexity of the software is the variety of hardware and software interfaces that the processor must deal with concurrently. These interfaces are shown in Fig. 8. Abbreviations for the interfaces that connect directly to the MP are defined in Table 5. Each of these corresponds to one or more I/O ports.

The interfaces connect the MP to the other subsystems shown in Fig. 8, including the SPU; reserve auxiliary payload (RAP); TT&C; CTDU; BDP; FSU; LBS; AFS; hop sequence generator (HSG); COMSEC; watchdog monitor; and error detection and correction (EADC).

The MP is programmed entirely in Ada. This affected the overall approach to developing the software. The Ada tasking model is used to accommodate the wide variety of processing deadlines. Ada portability and modularity are exploited to allow testing of the code in a variety of test environments.

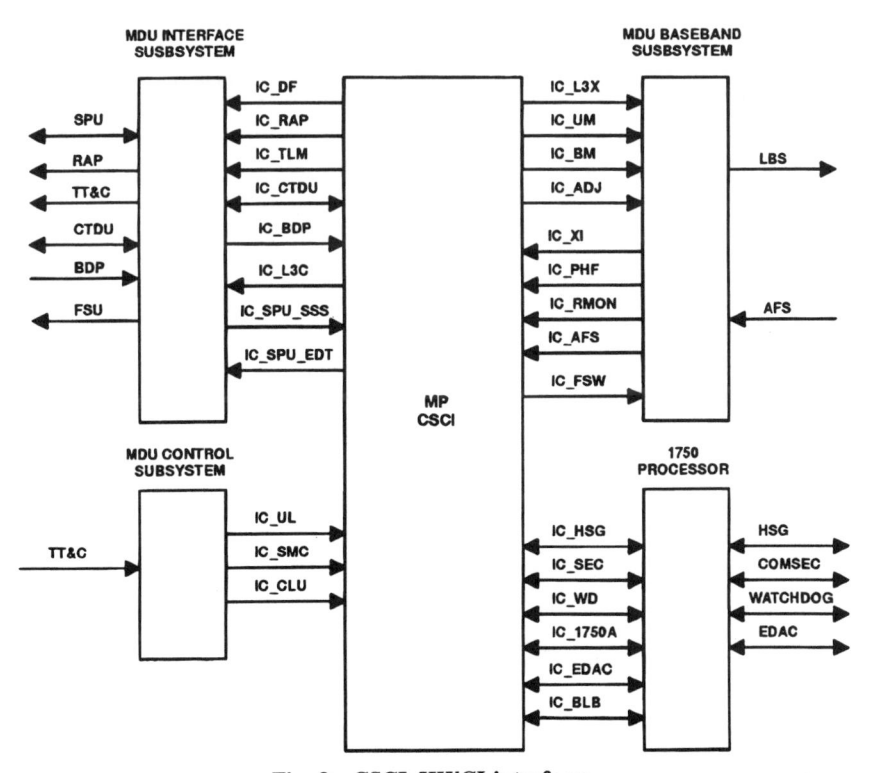

Fig. 8 CSCI–HWCI interfaces.

Table 5 Mission processor CSCI external Interfaces

PUID	Name
IC-UL	Upload data interface
IC-SMC	Serial magnitude command interface
IC-CLU	Clear upload interface
IC-TLM	Telemetry interface
IC-CTDU	CTDU interface
IC-BDP	BDP data interface
IC-L3X	L3 transmit data interface
IC-L3C	L3 on/off command interface
IC-BM	Baseband modulation interface
IC-UM	User message interface
IC-ADJ	Adjustment interface
IC-XI	X1/Z-count interface
IC-PHF	Phase feedback interface
IC-RMON	Reference monitor interface
IC-DF	Delta F command interface
IC-AFS	Atomic frequency standards interface
IC-FSW	AFS switch interface
IC-BLB	Blind bus interface
IC-WD	Watchdog monitor interface
IC-HSG	Hop sequence generator interface
IC-SEC	Comsec interface
IC-1750A	MIL-STD-1750A processor interface
IC-SPU-SSS	Spacecraft subsystem status interface
IC-SPU-EDT	Ephemeris data/time interface
IC-RAP	RAP interface
IC-EDAC	Error detector and correction interface
IC-LLED	Low-level event detector interface

3. In-Space Reprogrammability

The operational flight code can be completely reprogrammed from the ground. Upon cold start, the processor executes a program that is resident in PROM (programmable read-only memory). This program then uplinks the operational program over the S-band datalink. The PROM program has sufficient diagnostic capability to verify proper operation of the processor, memory, and data interfaces needed to upload and execute the flight program.

In addition to the capability to upload the entire program, partial uploads are possible. Certain parts of the program have been segmented so that they can be modified without uploading the entire program.

4. Software Partitioning Facilitates Test

An important feature of the onboard software is its ability to test certain components in accelerated time. This feature is most significant for AutoNav. A special test bed, the autonomous navigation emulator (ANE), has been developed by ITT to support this testing both before and after launch.

In addition to testing of AutoNav on simulators and several hardware configurations is planned prior to launch, extensive on-orbit certification testing has been planned before AutoNav becomes operational. The SV and the ANE have been designed to interact to support this testing. The Block IIR SV can operate in a mode similar to that of the Block I and II vehicles while this testing is performed in the background. Thus, AutoNav certification testing can be performed while the vehicle is fully operational in the GPS constellation.

The ANE relies extensively on the modularity and code portability of the onboard AutoNav code. Because the same source code is used in both the simulator and the flight processor simulation, fidelity of the onboard algorithm is assured.

5. Rate Monotonic Scheduling

Block IIR flight software makes full use of pre-emptive scheduling which is provided by the Ada tasking model. Many real-time systems, such as the Block II onboard computer program, use what is commonly called a cyclical executive. Although the design of a cyclical executive is simple, it imposes design constraints that can make the design of the application code very complicated. This is particularly true if the processor has many different asynchronous responses with a wide range of response times. In these case, pre-emptive scheduling can both significantly improve the effective processor throughput and simplify the application code design.

In rate monotonic scheduling, the faster a task must respond to synchronization events, the higher the priority. Certain design constraints are also necessary to ensure that a high-priority task never waits for a low-priority task to arrive at a synchronization point. This design and analysis technique provides a rigorous method of guaranteeing response times in a pre-emptive scheduling environment.

6. Time-Keeping System

The TNP TKS is an integrated collection of assemblies and components that distribute precise clock and frequency to the TNP and the NDS.[13] It contains an AFS, a FSU, and system and reference clock generators. Its operation is controlled by the MP. The MP is a MIL-STD-1750A component that is a part of the MDU.

Figure 9 is a system block diagram of the TKS. The switch matrices provide the means for using redundant VCXOs and AFSs. They select one VCXO and one AFS to be on-line. The AFS input switch matrix permits a back-up AFS to be powered on for warm-up purposes without affecting normal operation. The back-up AFS can be switched on-line by commands from the MP to replace the original on-line AFS. The VCXO switch matrix permits VCXO redundancy management by the CS; only one VCXO can be powered up at a given time.

The 10.23 MHz (nominal) VCXO is used to generate the system clock frequency of 10.23 MHz; its frequency is controlled by a delta-F command generated by the MP. This frequency is then used to generate a system clock, which is a 1.5-s time interval called an "epoch." The epoch is the base unit of time utilized

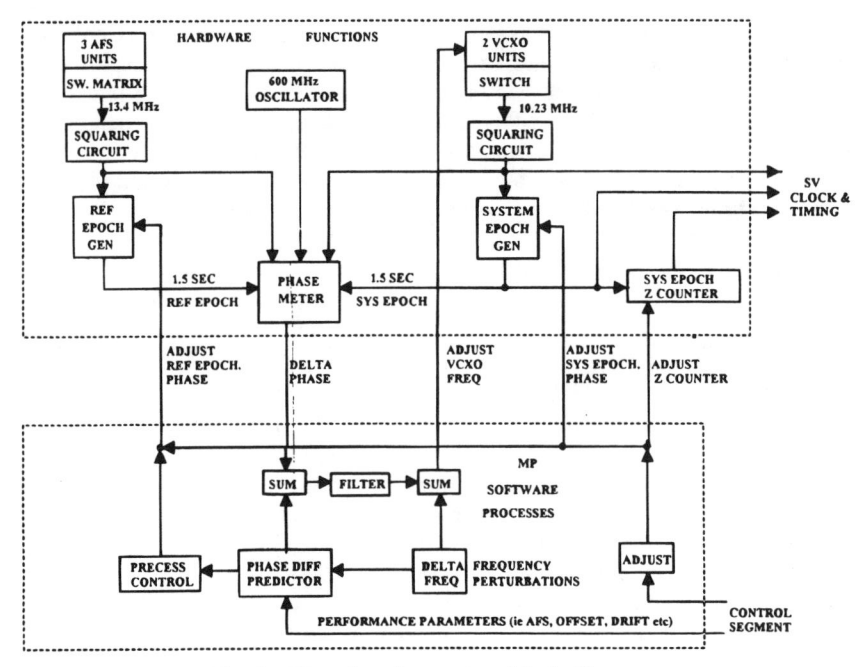

Fig. 9 Time-keeping system block diagram.

by the TNP and transmitted to the users. During initialization, the system epoch is synchronized to GPS time by adjusting TKS calibration and by placing clock corrections in the navigation message sent to the user. The 10.23 MHz output frequency is deliberately perturbed by algorithmic changes applied to the delta-F command by the MP. These perturbations restrict use of the full capabilities of the TNP to authorized users who can correct for these perturbations.

The 13.4-MHz (nominal) AFS is used to drive a reference clock generator that also generates a 1.5-s epoch (known as the reference epoch) as the base unit of time. Because the system epoch is driven by the 10.23 MHz VCXO, it is asynchronous to the reference epoch. The MP removes this asynchronism by adjusting the reference generator and by keeping track of the asynchronism.

Short-term stability of the system clock relative to GPS time is maintained by the stability of the VCXO. Long-term synchronism is maintained by coupling the system clock to the more stable reference clock. The phase of the system clock is compared to the phase of the reference clock by a phase meter. Phase differences cause the MP to adjust the delta-F command until the phase difference between the two epochs is brought to its target value. The target phase difference is initially about 100 ms and is maintained in software to account for precession, drift, and planned perturbation. Discrete commands from the CS are used to apply power and to reconfigure redundant TKS components.

The MP contains the algorithms and logic necessary for proper operation of the TKS. It controls initialization, operational adjustments, SA, EDAC, and

GPS SATELLITE AND PAYLOAD

interfaces to the control segment. Interfaces between TKS hardware and the MP are via parallel and serial registers.

C. L-Band Subsystem

1. Requirements and Description

The payload of the GPS Block I and II navigation satellite is the PRNSA.[1,2] This digital and rf equipment continuously transmits coded navigational data to all users. The rf signals are modulated with two codes: one for precision measurements (P code), and one for carrier acquisition, and less precise navigational measurements (C/A code). The total satellite system comprises components that receive updated positioning information from ground stations, store and forward this data as coded information, modulate the data on the generated carrier frequencies, and amplify the signals to levels required for ground reception. A block diagram of the GPS Block IIR payload is shown in Fig. 10.

The baseband processor is the component that processes the telemetry and ephemeris data for modulation on the rf carrier. Periodically, the processor's navigation message is updated as it passes over the ground control station.

The processor outputs a 50 b/s datastream to the baseband where it is modulo-2 added to two generated PRN codes. The P code is at a 10.23 MBs rate, and the C/A code is at a 1.023 MBs rate. The P-code signal can be encrypted to provide SA and prevent unauthorized use of this precision signal. The code clocks are in synchronism with the PRNSA 10.23 MHz frequency standard. The

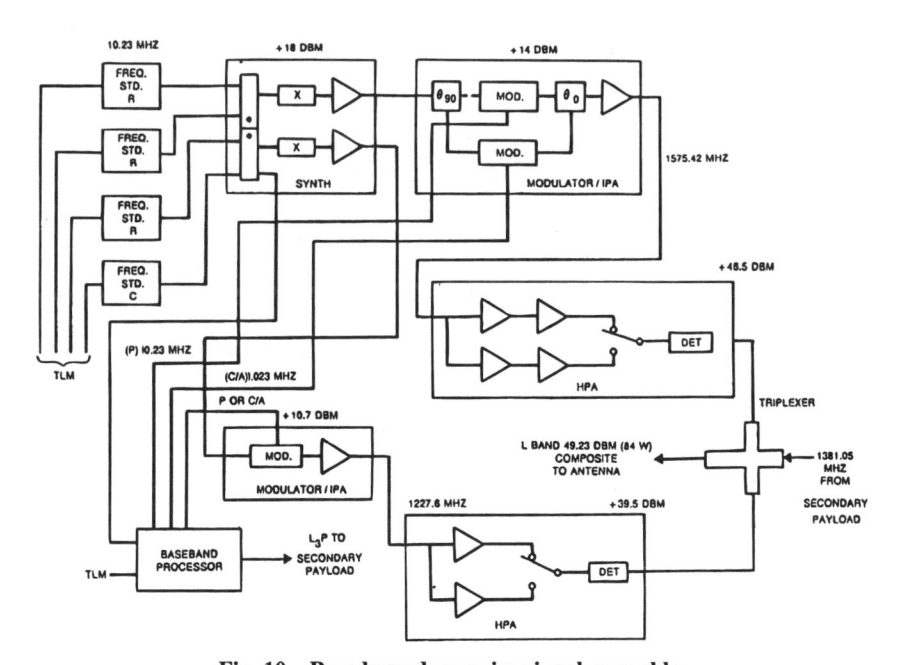

Fig. 10 Pseudorandom noise signal assembly.

baseband processor provides a P-code output and a C/A-code output to the L_1 modulator; and a single output, either C/A or P code, to the L_2 modulator.

The coded outputs are phase-modulated on to an L_1 carrier of 1575.42 MHz and an L_2 carrier of 1227.6 MHz. Data are sent over both rf to permit high-accuracy users to make ionospheric corrections. The two carrier frequencies are generated by the synthesizer component of the PRNSA. The 10.23-MHz frequency standard input is multiplied to the final frequency using step recovery diodes. The L_1 frequency multipliers are X14 and X11, followed by a three-stage rf amplifier producing the L_1 carrier (i.e., 10.23 MHz × 14 × 11 = 1575.42 MHz). The L_2 frequency multipliers are X12 and X10 followed by a 2-stage rf amplifier producing the L_2 carrier (i.e., 10.23 MHz × 12 × 10 = 1227.6 MHz).

The L_1 modulator quadraphase modulates the L_1 rf carrier from the synthesizer with C/A and P codes from the baseband unit. The L_1 carrier input is split into two paths. One carrier is delayed by 90 deg and biphase modulated with the C/A code in a balanced mixer. The second carrier is biphase modulated with P code in a balanced mixer. The two carriers are then recombined, resulting in a quadraphase modulated carrier.

The resultant modulated L_1 carrier is amplified to a nominal level of 50 W in a high-gain solid-state bipolar power amplifier (L_1 HPA). The output amplifier stage consists of four transistors in parallel. The four transistors are mutually balanced for phase and amplitude balance. They are then combined at the amplifier output. The L_2 modulator is of similar design except that it biphase modulates the L_2 carrier with either C/A or P code selectable by ground command.

The L_2 HPA is of similar design to the L_1 HPA and develops 10 W of rf power in a single transistor output stage.

The basic design of all the rf transistor stages in the modulators and high-power amplifiers employs microwave integrated circuit (MIC) technology with alumina substrates on aluminum carriers.

The outputs of the L_1 and L_2 rf amplifiers are combined in a diplexer (a triplexer in later designs when the L_3 NDS was installed) designed to add the two rf signals in a low-loss junction and provide the spectrum shaping for transmission.

The diplexer delivers the combined rf (L_1 and L_2) power to the antenna system, which consists of a phased array assembly of helical antennas producing a shaped beam for total Earth coverage.

Figure 11 shows the GPS Block I SV under test at Rockwell International and clearly shows the 12 helical antennas operating in an array on the spacecraft near side. The arced solar panels are shown in their open position on both sides of the space craft.

The PRNSA payload for the spacecraft is shown in Fig. 12. All components except the passive diplexer are at least doubly redundant. The baseband processor has triple redundancy.

The L-band (the frequency band between 1 and 2 GHz) subsystem for GPS IIR consists of 10 components that make up the three transmitters at L_1 (1575.42 MHz), L_2 (1227.60 MHz), and L_3 (1381.05 MHz) frequencies. The L-band components are: (1) L_1/L_2 frequency synthesizer; (2) L_1 modulator/intermediate-power amplifier (MOD/IPA); (3) L_2 MOD/IPA; (4) L_1 HPA; (5) L_2 HPA; (6) L_1/L_2 dc/dc converter; (7) L_3 frequency synthesizer/MOD/IPA/converter; (8) L_3 HPA; (9) L_3 astronomy band-reject filter; and (10) Triplexer.

Fig. 11 The GPS Block I satellite undergoing tests at Rockwell International.

A block diagram of the L-band subsystem is shown in Fig. 13. Each transmitter chain is configured as the following major functions: (1) an L-band synthesizer locked to a 10.23 MHz AFS reference; (2) a modulator (QPSK or BPSK) and IPA; and (3) a high-power amplifier that generates the required transmit power.

The triplexer filters each modulated carrier and combines the three into a single output. It has extremely stable performance over temperature changes to minimize group delay variations. Variations in group delay would degrade URE.

Each active function is redundant with cross strapping between functions to provide very high reliability (see Fig. 13). The LBS hardware has been designed to provide high performance, minimal changes in performance caused by aging and radiation, rapid recovery from nuclear upset, and high producability. All this hardware is radiation-hardened by design and shielding techniques and provides continuous operation through severe nuclear environments.

2. Synthesizers

Frequency synthesizers can be designed in a wide variety of ways including phase–lock-loops, multiplier chains, and numerically controlled oscillators. The choice depends upon the required spurious level, phase noise, power consumption, size, and weight.

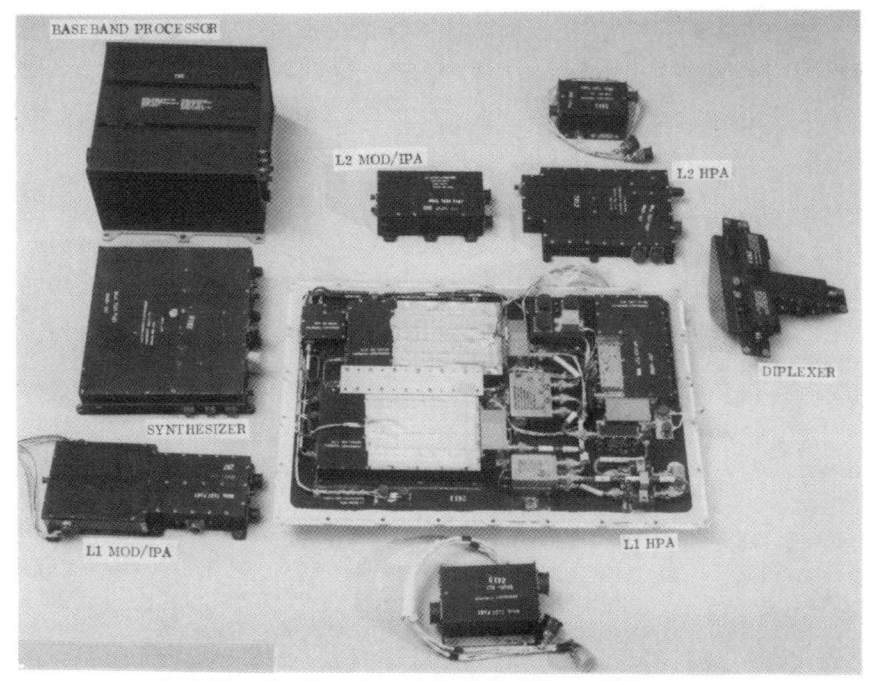

Fig. 12 Pseudorandom noise subassembly.

3. Modulators/Intermediate-Power Amplifiers

Modulators that inject the modulation at baseband and the rf carrier at final frequency are used on the GPS. The designs directly modulate the L-band carrier and include an output amplifier that provides highly stable output power. These designs have the following features:

1) Final frequency (L-band) BPSK (L_2 and L_3) or QPSK (L_1) modulators that provide very accurate phase states to maintain the URE of GPS.

2) each stage of the MOD/IPA rf amplifier includes active biasing to maintain the required performance over environment, radiation, voltage, and aging.

4. High-Power Amplifiers

High-power amplifiers provide the final rf amplification for transmission to Earth. High efficiency is critical, because spacecraft and power are limited resources. In the GPS IIR system, the high-power amplifiers provide the final transmit power prior to output filtering. Highlights of these HPA design features are:

1) high-efficiency bipolar transistors are used because of their efficient high-voltage, low-current operation;

2) hybrid combined output stages to obtain the final transmit power while maintaining transistor junction temperatures below 125°C for high reliability. High-power amplifiers' output powers are 50 W (L_1), 10 W (L_2), and 20 W (L_3).

A photograph of the LBS engineering development model is shown in Fig. 14.

GPS SATELLITE AND PAYLOAD 233

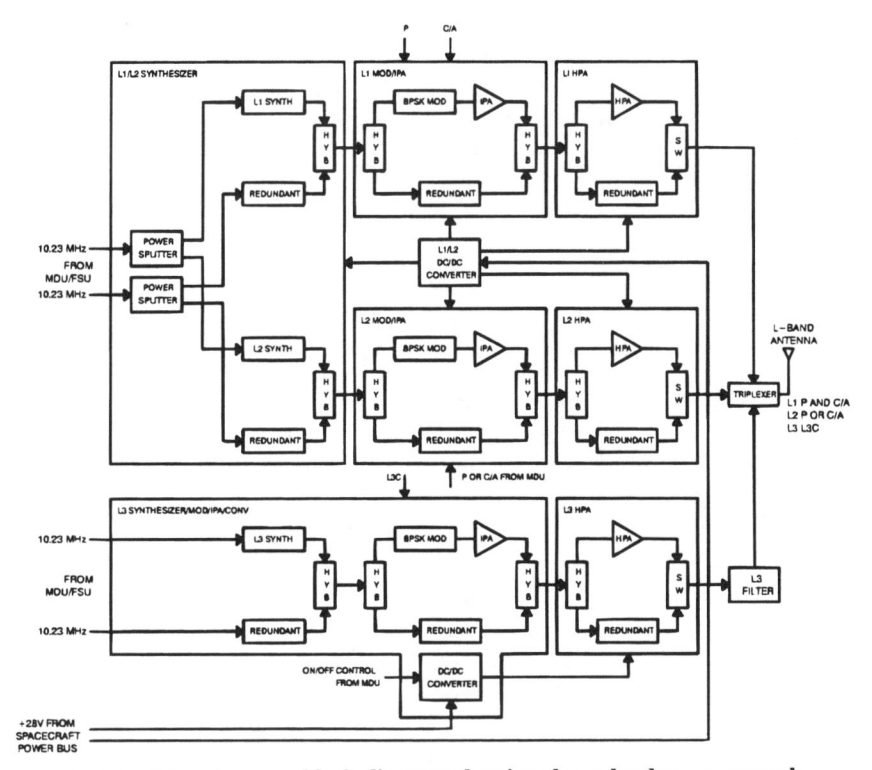

Fig. 13 *L*-band system block diagram showing the redundancy approach.

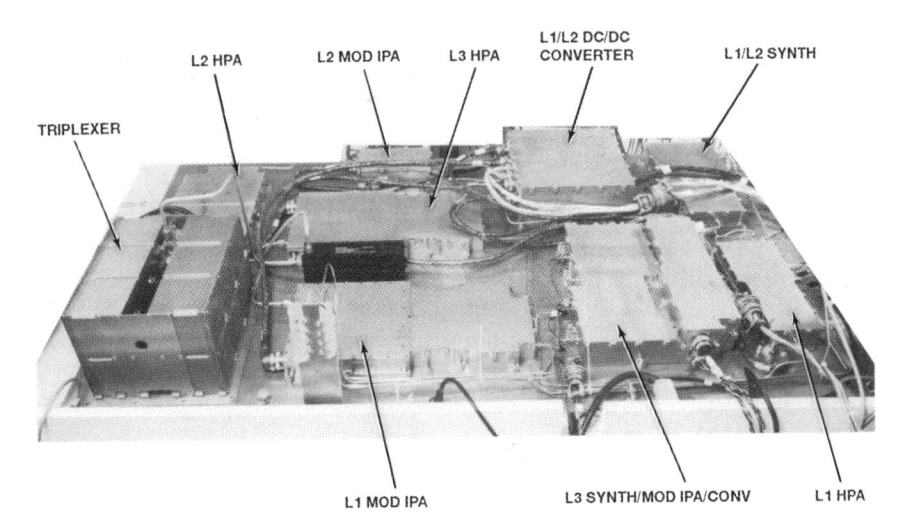

Fig. 14 *L*-band system engineering development model.

V. Characteristics of the GPS *L*-Band Satellite Antenna

The *L*-band antenna onboard the GPS satellite is designed to radiate the composite *L*-band signals to the users on and near Earth. It provides a nearly constant signal level to the user receivers over the whole Earth coverage with circular polarization at L_1, L_2, and L_3 frequencies.

A. Coverage Area

This *L*-band antenna is a broadband, fixed-beam antenna providing whole Earth coverage at L_1, L_2, and L_3 frequencies. The GPS SV is a three-axis stabilized satellite, with an attitude control system that keeps the *L*-band antenna pointed toward the center of the Earth. From the GPS satellite altitude, the view angle from edge-to-edge of Earth is about 27.7 deg. The total pointing error of the satellite is specified to be less that ± 0 0.15 deg, with 99% probability. Thus, a fixed-beam antenna with adequate gain over 28 deg can be used in the GPS satellite.

B. Antenna Pattern

The goal of the antenna is to illuminate the Earth's surface in view of the satellite with an almost uniform signal strength. The path loss of the signal is a function of the distance from the antenna phase center to the surface of the Earth. The path loss is minimum when the satellite is directly overhead (satellite is at 90 deg elevation), and is maximum at the edge of Earth coverage (satellite at the horizon). The difference in path length between these two extremes is about 500 km. The difference in path loss caused by this variation in path length is about 2.1 dB. The ideal antenna pattern required to illuminate the surface of the Earth uniformly is shown in Fig. 15. It is symmetric about the axis from the satellite to the center of the Earth. Beyond the 28-deg view angle, the ideal antenna will radiate near-zero rf energy. A practical antenna with a smooth antenna pattern is designed so that the variations in signal strength over the 28-deg view angle is minimized, and the total radiated energy over the 28-deg angle is maximized.

C. Antenna Evolution

The initial concept and design of the antenna array was performed by Rockwell International, Space Systems Division, for Block I, II, and IIA satellites.[16] The Block I antenna has a circular peak antenna gain (in the plane perpendicular to the central axis), with a dip in antenna gain at the antenna boresight. The peak is located approximately 10 deg from the boresight. This antenna is a phased array design. It is comprised of 12 helical elements, arranged in two concentric circles on the Earth-facing satellite panel. The footprint is shown in Fig. 16. The inner circle is composed of four equally spaced elements, and the outer circle contains eight elements, also equally spaced. Each element is a monofilar, axial mode helix. The helix element design provides a wide bandwidth and circular polarization with minimal element-to-element interaction. The relative radii of the inner ring and outer ring control the angular location of the near-circular

GPS SATELLITE AND PAYLOAD

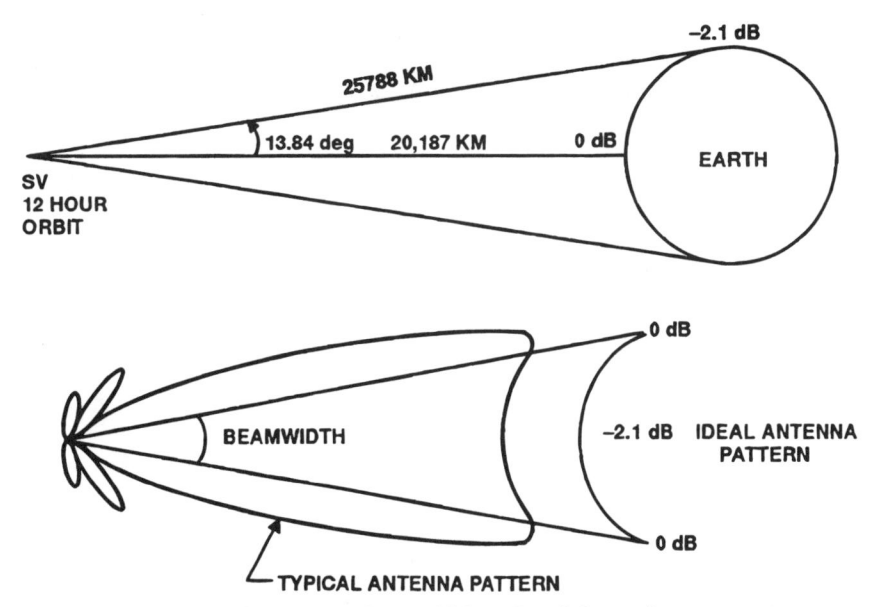

Fig. 15 Antenna beamwidth and path loss values.

antenna peak. The elements in the inner ring are fed in-phase with 90% of total power, and the outer ring elements are fed 180 deg out of phase with 10% of total power. The antenna feed network is a passive device that provides the proper phase and amplitude distribution to the elements. The resultant pattern produces a beamwitdth of approximately 28 deg, with a dimple in gain at the boresight, as shown in Fig. 17. This design gives the users the maximum signal when the satellite is approximately at an elevation of 40 deg. The dimpled shape is achieved by 180-deg relative phasing between the inner and outer elements. The depth of the dimple with respect to the peak is controlled by the ratio of power between the inner and outer ring elements. Because there are only four elements in the inner ring, the antenna pattern is not perfectly symmetrical in the plane perpendicular to the boresight. The GPS antenna array generates a right-hand circularly polarized (RHCP) signal over the frequency range of 1200–1600 MHz. The GPS antenna is an entirely passive design that exhibits high reliability.

Keeping the signal illumination uniform on the ground reduces the total radiated power required at the spacecraft. The system advantages of this design concept are that it reduces the size, weight, and power consumption of the spacecraft navigation payload. This also reduces the amount of dc power (and hence, number of solar cells) required on the satellite, which simplifies satellite complexity and cost.

In Block II/IIA, the cylindrical ground plane for each element was changed to a conical design. This design provides significant reduction in side and backlobe radiation. This reduction, in addition to providing better antenna efficiency, is less susceptible to degradation from adjacent and forward bulkhead obstructions.

Several changes were made in the Block IIR design. The ratio of inner and outer radii and the rf power feed ratio were changed. In Block IIR, the 180-deg

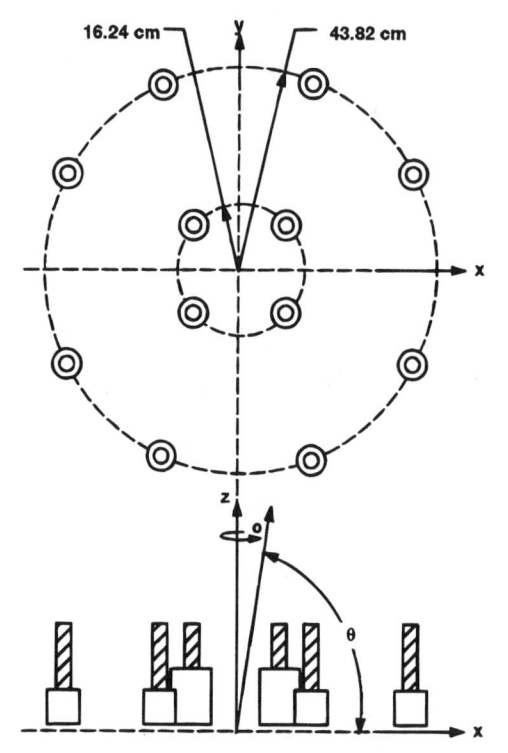

Fig. 16 GPS block I *L*-band antenna.[16]

relative phase between inner and outer elements is achieved with a 90-deg electrical phase shift from the stripline beam-forming network and a 90-deg mechanical rotation of the outer elements relative to the inner elements. This new Block IIR design was developed by Martin Marietta, Astro Space Division. A view of the Block IIR satellite highlighting the antenna form is shown in Fig. 18. The antenna design parameters are tabulated in Table 6, for Block I, II/IIA, and IIR satellites.

D. Crosslinks

Each GPS Block IIR satellite has a crosslink transponder data unit. Simply stated, this CTDU has both data and ranging modes for direct GPS satellite-to-satellite communications.

The purpose of the CTDU is twofold. The first is to supply a precise ranging signal for AutoNav. and to exchange the AutoNav state vector among Block IIR spacecraft. The second is a data exchange peculiar to the NDS system. AutoNav supplies spacecraft-computed navigation parameters to users. The AutoNav state vector consists of both Keplerian orbit parameters and clock states. To the user, AutoNav is a transparent feature, because the L_1 and L_2 signals remain unchanged from those of Block I and Block II. AutoNav supplies the Block IIR spacecraft

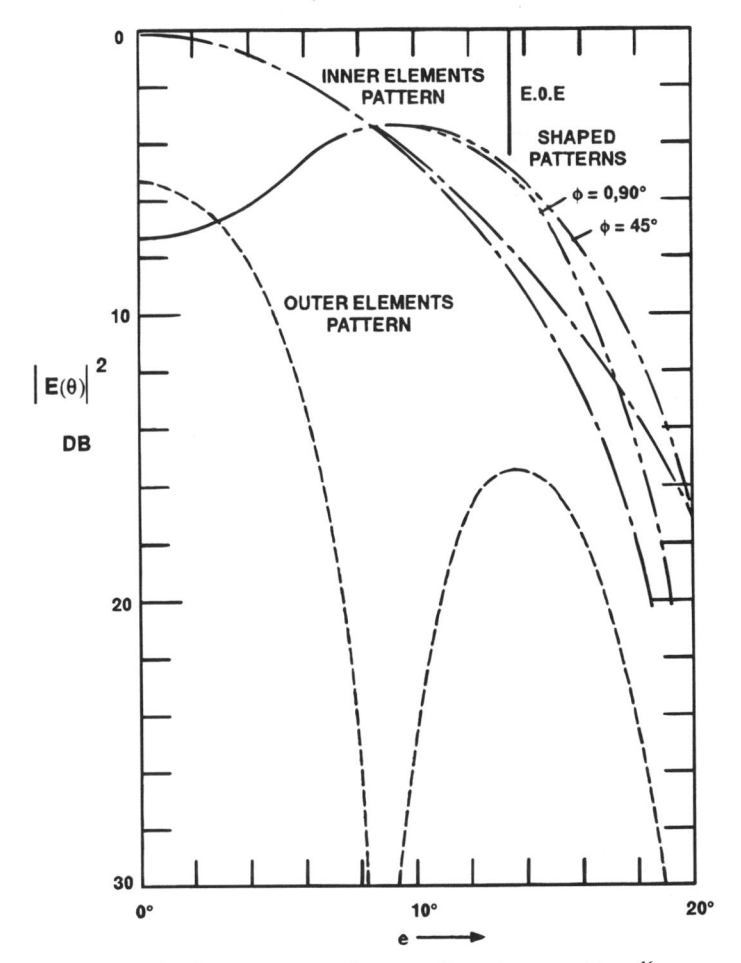

Fig. 17 Inner, outer, and composite antenna pattern.[16]

the ability to operate autonomously with respect to the CS for as long as 6 months while maintaining a URE of less than 6 m.

Two CTDUs are provided in each Block IIR SV. Each CTDU has an integral power converter and exhibits modular or slice construction, as shown in Fig. 19. Switching between CTDUs can be initiated by the GPS CS after determination of a failure or an operation out of specification.

The CTDU is a time division multiple access (TDMA) frequency-hopped, spread-spectrum communication system in the UHF band. It incorporates a 5-MChip/s pseudorandom code and generates a 108-W rf output power.

E. Primary and Secondary Functions

Primary functions of the CTDU are as follows: transfer data between satellites on a frequency-hopped carrier; perform one-way ranging between satellites on

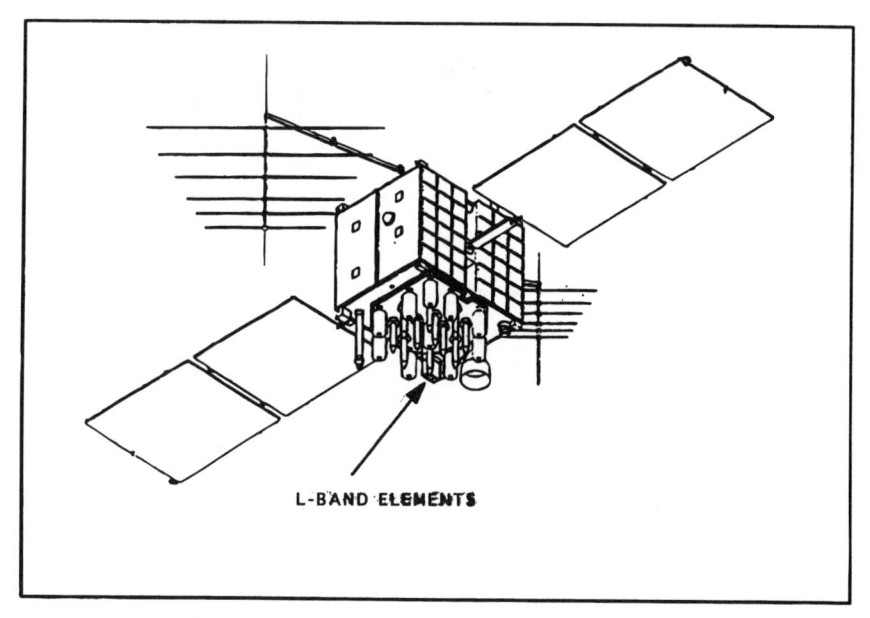

Fig. 18 *L*-band transmitting array for the GPS spacecraft.

two carrier frequencies; and measure internal (CTDU) delay. The secondary functions are the following: acquire and discriminate cross-link transmission; recover symbol, hop, and frame timing; and generate acquisition preamble in the transmit mode.

From an architectural point of view, the CTDU is divided into several distinct functions, as shown in Fig. 20. The rf converter takes the received frequency-hopped signal, and downconverts it to if with the aid of the digital frequency synthesizer. A dedicated digital signal processor digitally converts the if signal and performs detection, tracking, and demodulation of the received signal. The GPS timing signals and references are supplied through an interface to the MDU.

F. Autonomous Navigation

The current Block I and Block II GPS satellites broadcast a navigation message that is uploaded daily by the CS. The clock and ephemeris parameters in the message are predictions based on the control segment's current estimate. In addition to this mode of operation, the Block IIR SV will have the onboard capability to estimate the ephemeris and clock autonomously and generate the navigation message. This capability is referred to as autonomous navigation or AutoNav,[3-9] and its development was motivated by four considerations.

1. Survivability

The CS is considered to be the most vulnerable link in the system. By providing the capability to meet accuracy requirements for 180 days without the CS, this

GPS SATELLITE AND PAYLOAD

Table 6 Summary of GPS *L*-band antenna designs

Elements	Block I 12[a]	Block II/IIA 12	Block IIR 12
Arrangement (geometry)	Two concentric circles, 4 el[b] inner, 8 el outer	Two concentric circles, four el inner, eight el outer	Two concentric circles, four el inner, eight el outer
Inner diameter	15.24 cm	15.24 cm	18.03 cm
Outer diameter	43.82 cm	48,82 cm	47.5 cm
Helix radius	3.56 cm Uniform	3.56 cm tappered over last two turns	3.4 cm tappered over last two turns
Helix length	5.1.18 cm	62.10 cm	52.6 cm
Helix ground shield	Cylindrical cup	conical cup	conical cup
Power inner ring	90%	90%	90.5%
Power inner ring	10%	10%	9.5%
Phase inner ring	0 deg	0 deg	0 deg
Power outer ring	180 deg	180 deg	180 deg
Polarization	RHCP[c]	RHCP	RHCP
Bandwidth	1200—1600 MHz	1200–1600 MHz	1200–1600 MHz

[a]Number of elements.
[b]el = elements.
[c]Right-hand circular polarization.

vulnerability is mitigated. The solution also allows for the permanent loss of any monitor stations.

2. Reduced Upload Requirements

With AutoNav, the CS uploads fewer data to the vehicle. This relieves scheduling requirements on the ground stations.

3. Integrity

Crosslink ranging provides an independent reference against which the vehicle can compare its ephemeris and clock.

4. Accuracy

Although this was not a stated requirement, the prospect of improved accuracy played a part in the decision to implement AutoNav. Accuracy improves, because AutoNav can update the ephemeris and clock predictions as often as four times an hour. In the current system, these are updated every day.

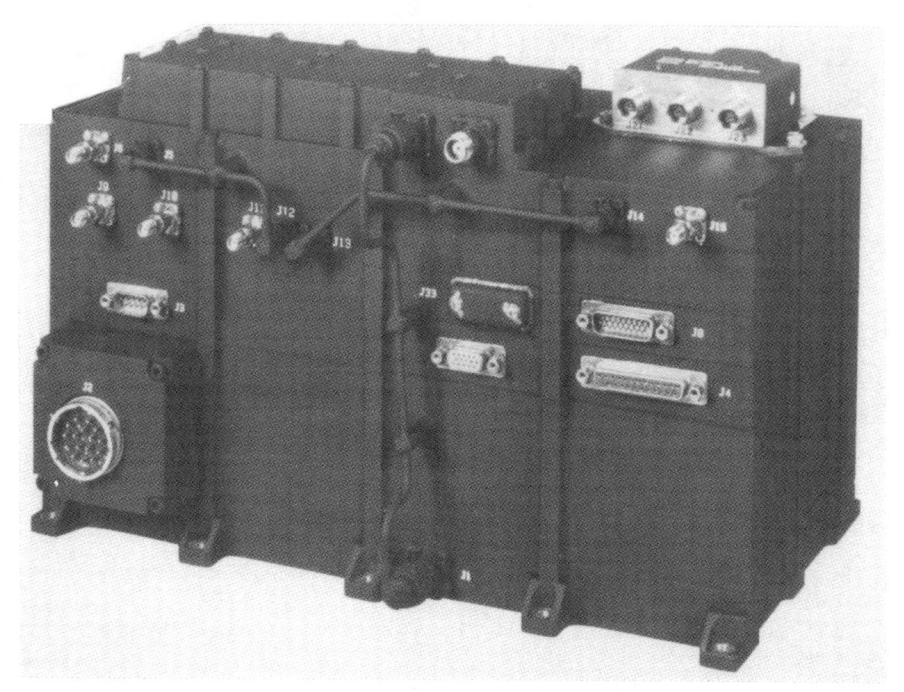

Fig. 19 Crosslink transponder data unit developed for ITT by Motorola.

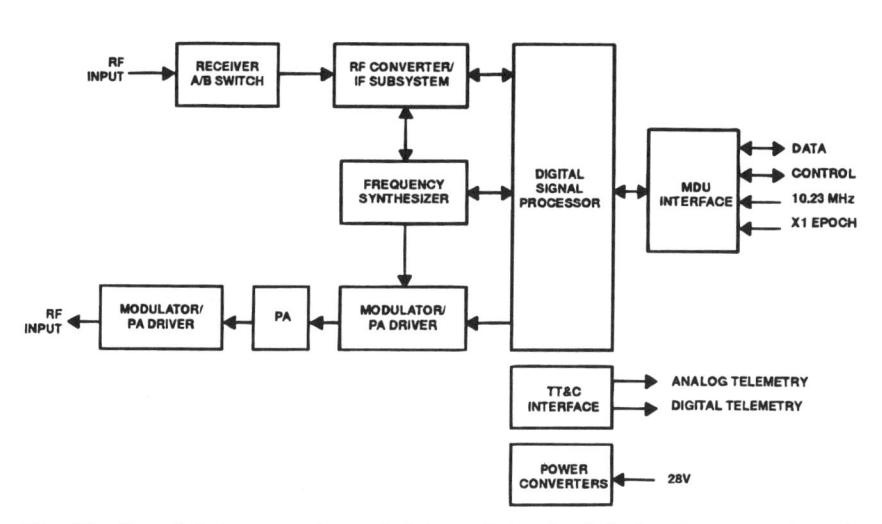

Fig. 20 Crosslink transponder and data unit top level (redundancy not shown) functional description.

AutoNav estimates the ephemeris and clock from intersatellite pseudorange measurements performed by a uhf crosslink. In addition to using the UHF crosslink for ranging, AutoNav utilizes the crosslink to exchange messages between the SVs. In this way, the SVs make both clock offset and distance measurements between each pair. Each SV then uses these measurements to update onboard Kalman filters for its own clock and ephemeris.

AutoNav uses a uhf crosslink that is backward compatible with the Block II crosslink used for relaying NDS communications. The crosslink performs the same NDS function on the Block IIR satellite. Because of this, Autonav is accomplished with minimal additional onboard equipment. However, the need to be backward compatible with Block II put significant constraints on the design. AutoNav had to use the inherited link protocol without interfering with Block II vehicles, which will be on orbit at the same time. In this protocol, each satellite is assigned a 1.5-s time slot within a 36-s TDMA frame. On each frame, the satellite transmits on its assigned slot and listens on the other 23 slots. AutoNav is performed in a cycle of several TDMA frames.

A block diagram of AutoNav is shown in Fig. 21. The CTDU performs the crosslink communications and ranging. Data from the CTDU are sent to a processor that performs the other functions shown. The results of the processing are formatted into the navigation message, which is part of the L-band signal transmitted to the users.

To begin each cycle, all the satellites broadcast a ranging signal in the same TDMA frame. Two frequencies are used so that delay caused by the plasmasphere can be corrected. On subsequent frames, they broadcast their pseudorange measurements so that 2-way ranging is available. In addition, such other parameters

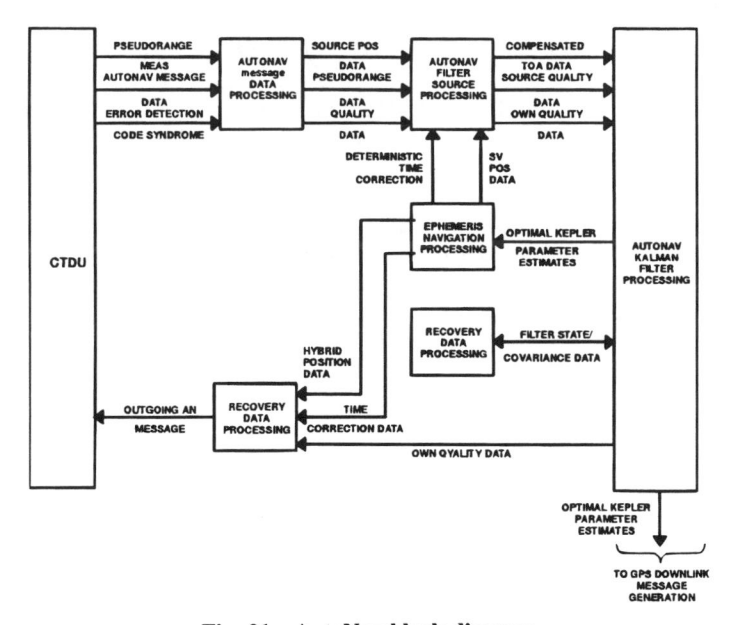

Fig. 21 AutoNav block diagram.

as the ephemeris are broadcast in an outgoing message. These derived measurements are absorbed by separate clock and ephemeris Kalman filters.

Depending up on the geometry, AutoNav will meet system accuracy and integrity requirements when between four and seven Block IIR vehicles are on orbit. Because the Block IIRs will be launched to replace old or failed Block II SVs, the precise geometry cannot be predicted. Prior to this, the first few Block IIR satellites will be operated in the Block II-like mode, while AutoNav is running in the background. In this mode of operation, AutoNav can be tested on-orbit, and accuracy can be verified. When a sufficient number of Block IIRs are on orbit, the satellites will be commanded to broadcast the AutoNav-generated navigation messages on the L-band to the users. In this hybrid mode, the GPS time between the Block II and the Block IIR AutoNav ensemble must be synchronized by the CS.

AutoNav has been designed to be totally transparent to the user. The user receiver will not distinguish between AutoNav and non-AutoNav satellites. However, authorized users will notice improved accuracy.

With AutoNav, the spacecraft will perform many of the functions currently performed by the CS. In addition to ephemeris and clock estimation, these includes integrity monitoring, curve fitting the navigation message, URA estimation, formatting the navigation message and UTC steering.

VI. Future Performance Improvements

As is by now apparent, the key element of the GPS is accurate time, and the keepers of GPS's accurate time are the FS. Improvements in the 3-m (one-sigma) URE allotted to the atomic clock and navigation subsystem (see Table 1) to 0.5 m seems feasible. When coupled with improved ephemeris estimation and the minimization of any clock/ephemeris correlation, the URE can be reduced from 6.6 m to 4 m or better. ITT is working actively with its atomic clocks suppliers, EG&G for the rubidium clocks and Kernco for the cesium clocks, to achieve these goals by applying advanced and innovative concepts to the problem.

In addition, investigations funded by the Naval Research Laboratory and the Air Force are always looking for new and innovative methods to improve all aspects of payload performance including reliability, size, weight, cost, and efficiency. Technologies such as gallium arsenide rf transistors, novel rf filtering techniques, multichip module packaging techniques, advanced processor and software development tools and methods, and advanced system engineering tools, are continuously being evaluated and developed.

A. Additional Capabilities

1. Integrity

When so many users depend upon GPS for critical functions, the integrity of the signal from a GPS satellite must be beyond question. If the signal is not acceptable, that fact must be broadcast widely and quickly. There are many independent methods of evaluating and locally communicating the existence of a nonacceptable signal. However, a worldwide system using GPS assets seems feasible. Earth-based accuracy measurement, uplinked to GPS satellites and relayed among all GPS satellites via an improved crosslink would distribute integrity information among all satellites. A multiplexed integrity channel incorporated in L_2 or L_1 would distribute the integrity information to all GPS receivers. This multiplexed channel would be backward compatible to existing receivers, so they would continue to operate normally. However, to take advantage of this new integrity channel, modifications to existing receivers would be required.

2. Other Platforms

GPS-like signals need not emanate from GPS satellites only. Secondary payloads incorporating the TNP could be flown on any satellite. ITT has recently completed a study for INMARSAT on the feasibility of incorporating navigation signals on satellites that are geostationary and satellites in 6-hour orbits.

The distribution of GPS signals from other space-based platforms will improve the availability of the system for all users, because at any one time many more signals will be available to a user. Other platforms include ground-based navigational transmitters. Placed near airports, these GPS "pseudolites" could provide increased accuracy for aircraft approaches.

The entire GPS community—industry, academia, the Air Force, the FAA, the International Civil Aeronautics Organization, and the user community—are continuously working to improve the GPS payload, satisfy all its customers and to improve the system's capabilities.

References

[1] Feit, L., and Domanico, P., "Navigation Signal Random Noise Generator Design and Performance Verification," ITT, EASCON Proceedings, IEEE 81CH1724-4, 1981.

[2] Domanico, P., "A Technical Note on GPS NAVSTAR Pseudo Random Noise Signal Assembly (PRNSA)," ITT, presented at The National Telesystems Conf., IEEE 83CH1975-2, 1983.

[3] Siegel, B., and Ananda, M., "The Next Generation Global Positioning System Block III Space Vehicle Concept," AAS/AIAA Astrodynamics Specialist Conf., August 1985.

[4] Ananda, M., et al., "Autonomous Navigation of the Global Positioning System Satellites," AIAA Guidance and Control Conf., August 1984.

[5] Codik, A., "Autonomous Navigation of GPS Satellites: A Challenge for the Future," Institute of Navigation, January 1985.

[6] Menn, M., "Autonomous Navigation for GPS via Crosslink Ranging," IBM Corp., IEEE CH2365, May 1986.

[7] Ananda, M. P., Bernstein, H., Cunningham, K. E., Feess, W. A., and Stroud, E. G., "Global Position System (GPS) Autonomous Navigation," IEEE Position Location and Navigation Symposium, Las Vegas, NV, Nov. 8, 1990.

[8] Bernstein, H., Bowen, A. F., and Gartside, J. N., "GPS User Position Accuracy with Block IIR Autonomous Navigation (AUTONAV)," Aerospace Corp., Institute of Navigation Meeting GPS-93, Sept. 1993.

[9] Brower, Chen, Doyle, Klein, Sloander, and Wiederholt, "GPS Block IIR Autonomous Navigation Emulator," Proceedings of the Institute of Navigation, Sept. 1991.

[10] Leinbaugh, D. W., "Guaranteed Response Times in a Hard-Real-Time Environment," *Transactions on Software Engineering*, Vol. SE-6, No. 1, Jan. 1980.

[11]Sha and Goodenough, "Real-Time Scheduling Theory and Ada," *Computer*, April 1990.

[12]Rawicz, H. C., Epstein, M. A., and Rajan, J. A., "Timekeeping System for GPS Block IIR," ITT Corp., 24th Annual PTTI Meeting, Dec. 1992.

[13]Baker, A., "GPS Block IIR Time Standard Assembly (TSA) Architecture," ITT, 22nd Annual PTTI Conf., Dec. 1990.

[14]Vanies, J., and Audoin, C., *The Quantum Physics of Atomic Frequency Standards*, Adam Hilger, Philadelphia, 1988.

[15]Van Mille, M. J., "Cesium and Rubidium Frequency Standard Status and Performance on The GPS Program," PTTI Conf., 1993.

[16]Czopek, F., and Shollenberger, Lt. S., "Description and Performance of the GPS Block I and II L-Band Antenna and Link Budget," 6th International Technical Meeting, The Institute of Navigation, Sept. 1993.

Chapter 7

Fundamentals of Signal Tracking Theory

J. J. Spilker Jr.*
Stanford Telecom, Sunnyvale, California 94089

I. Introduction

A. GPS User Equipment

THE GPS basic concepts, signal structure, navigation data, satellite constellation, satellite payload, and geometric dilution of precision (GDOP) concepts have all been introduced in previous chapters. This chapter and the next two discuss the user equipment. These chapters provide the foundation for much of the material in the companion volume on GPS applications.

Figure 1 illustrates the basic configuration of a GPS user equipment in its typical form. Generally, the user equipment performs two functions: the first, track the received signals, usually with some form of delay lock loop (DLL) so as to measure the pseudorange and usually the pseudorange-rate or accumulated delta range (ADR), a carrier measurement, as well. This chapter discusses the fundamentals of signal tracking theory as applied to the GPS signals. The next chapter, Chapter 8, discusses means for implementation of the GPS receiver functions. The following chapter, Chapter 9, discusses the navigation algorithm used to convert these receiver measurements into the desired output; namely, user position, velocity, and user clock bias error. This present chapter on signal tracking theory also describes a technique for combining the two functions, receiver tracking and navigation algorithm into one combined integrated tracking system.

In this chapter, the received signals are assumed to be received with stationary additive white Gaussian noise (AWGN) that is representative of the thermal noise at the receiver frontend. For most of this chapter, the received radio frequency signal-to-noise ratio is small, typical of most GPS receiver applications. Thus, the effects of self-noise or multiple access noise caused by the other GPS signals is generally small and is usually neglected. However, in the last section on the vector delay lock loop (VDLL), it is shown that the quasioptimal detector that tracks all signals is designed to remove much of this multiple access noise.

Copyright © 1994 by the author. Published by the American Institute of Aeronautics and Astronautics, Inc., with permission. Released to AIAA to publish in all forms.
*Ph.D., Chairman of the Board.

```
                    GPS
                 Satellites
                    \|/
                     V
                     |
                     v
            ┌─────────────┐     ┌─────────────┐
            │     GPS     │====>│  Navigation │====>
            │   Receiver  │     │  Algorithms │<---
            └─────────────┘     └─────────────┘
                     ^
                     |
              ┌─────────────┐
              │    User     │
              │    Clock    │
              └─────────────┘
```
— Pseudo-range, range rate estimates

Estimates of user position, velocity, clock bias

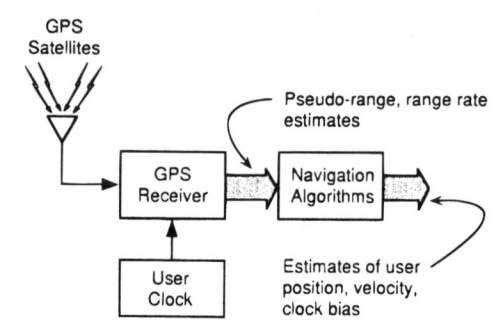

Fig. 1 Simplified configuration of the GPS user equipment.

B. GPS User Equipment-System Architecture

A generalized view of a basic GPS user equipment system is depicted in Fig. 2. The satellite signals are received by one or more antennas/low-noise amplifiers.* The output of the antenna is fed to a radio frequency bandpass filter/low-noise amplifier combination in order to amplify the signal and to filter out potential high-level interfering signals in adjacent frequency bands. Otherwise, these potential interfering signals might either saturate the amplifier or drive it into a nonlinear region of operation. The radio frequency filters must be selected with low loss in order to maintain a low noise temperature and also must have sufficient bandwidth and phase linearity to minimize the distortion of the desired C/A- or P(Y)-code signals. The signal then passes through serial stages of radio frequency amplification, downconversion, IF amplification and filtering, and sampling/quantizing. The sampling and quantizing of the signal can be performed either at intermediate frequency (IF) or at baseband. In either approach, in-phase and quadrature (I, Q) samples are taken of the received signals plus noise. At the present state of the art, we can implement the functions of radio frequency amplification, downconversion, IF amplification, and A/D sampling with a single monolithic microwave integrated circuit chip (MMIC). Filtering and reference frequency generation may require additional circuitry.

The I, Q samples are then fed to a parallel set of DLLs each of which tracks a different satellite, measures pseudorange, and recovers the carrier which is bi-phase modulated with the GPS navigation data.† The DLL 1–3 and associated

---

*Although most receivers employ only one antenna, some use more than one antenna/amplifier in order to:
- Accommodate maneuvering of the user platform; e.g., an aircraft banking and thereby avoid blocking some of the satellites with a wing. For example, one antenna can be at the top of the aircraft and others can be at the sides.
- Provide increased antenna gain. Each higher gain antenna can be pointed at single or clustered groups of satellites. The antenna beams can be steered electronically or mechanically, if necessary.
- Discriminate against interfering signals or multipath. Multiple narrow beam antennas or adaptive antennas can be employed. Some antennas use special ground planes to reduce multipath. Null steering antennas can be employed.

†An alternative time sequencing approach can be employed in which a DLL is sequenced over several satellites, dwelling on each satellite for a short period of time. There is, however, some performance degradation with this sequencing approach.

FUNDAMENTALS OF SIGNAL TRACKING THEORY 247

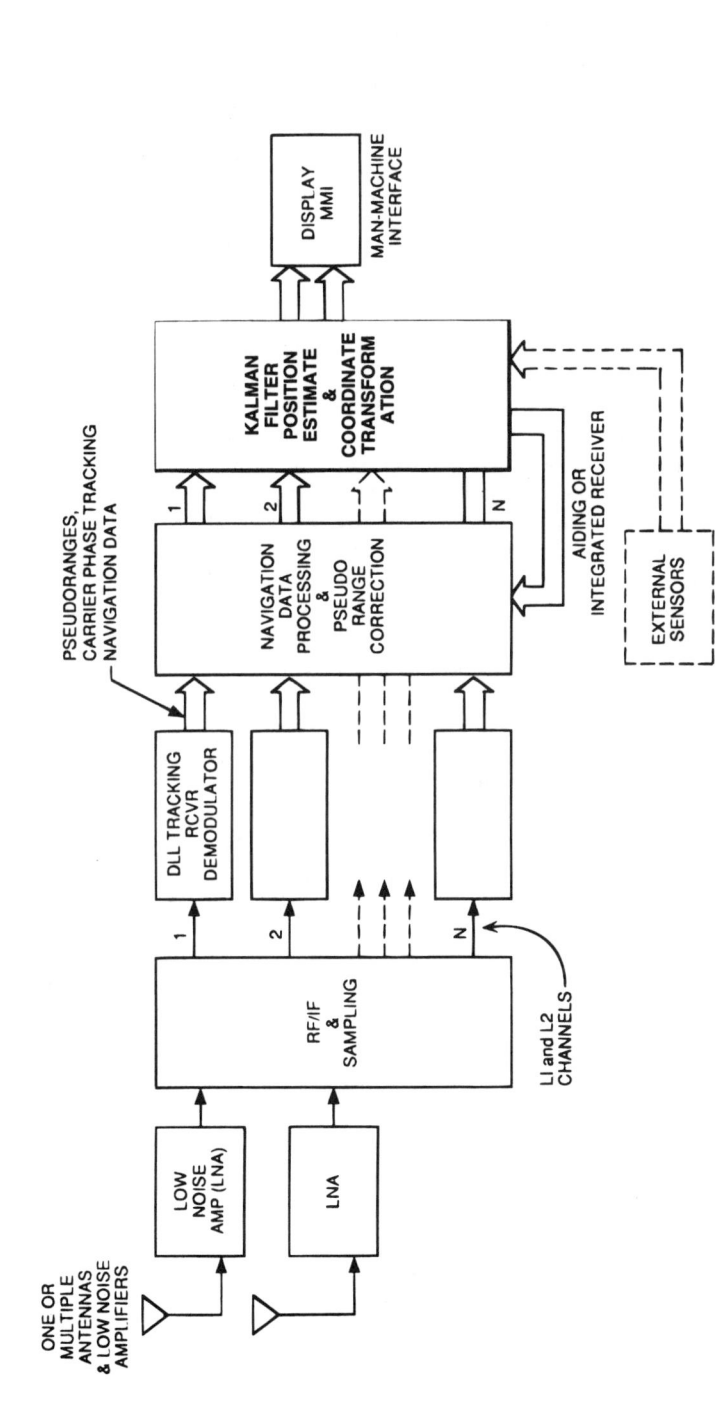

Fig. 2 Generalized GPS user system configuration with separate delay estimating receiver and position-estimating functions. There are N satellite signals tracked by the parallel DLL ($2N$ signals if both L_1 and L_2 channels are used).

demodulators provide estimates of the pseudorange, carrier phase, and navigation data for each satellite and are usually implemented using digital processing. As discussed in Chapter 5, this volume, on the GPS satellite constellation, there may be as many as ten satellites in view at one time. Typically the number of parallel tracking DLL varies from 2 to 16, and it is possible to track all of the satellites in view at both L_1 and L_2 frequencies. Generally, at least five satellites are tracked as a minimum, either in parallel or in time sequence. At the present state of the art, an L_1 10-channel receiver with 10 parallel DLLs can be implemented in digital form on one CMOS chip. Semiconductor technology is increasing significantly every 18 months, and the processing power available for GPS receivers is expanding in the same manner.

The receiver system with its parallel signal DLL and computer processors must carry out the operations of satellite selection, signal search, tracking, and data demodulation, as shown in Table 1.

This chapter concentrates on the signal-tracking task. The other tasks are described in detail in Chapter 8, this volume. The parallel pseudoranges, navigation data, and, in the more sophisticated receivers, the carrier phase measurements are fed to the navigation data processor where the position of each satellite is calculated, and pseudorange and clock corrections are made. As a first step in this operation, the pseudorange and carrier phase measurements are corrected for the various perturbations, including satellite clock errors, Earth rotation effects, ionosphere delay, troposphere delay, relativistic effects, and equipment delays.

The corrected pseudorange, phase, or accumulated phase (ADR) measurements along with other sensor data are then fed to the extended Kalman filter (EKF) or similar filter. The output of the EKF estimator provides position, velocity, and time estimates relative to the user antenna phase center (see Chapter 9, this volume.) There may or may not be information from additional sensors; e.g., altimeters, inertial measurement units (IMUs), or dead-reckoning instruments. If so, these measurements are also fed to the EKF. Data from carrier tracking and some of these sensors can also be used to aid the DLL tracking operation itself. As discussed in the companion volume, this Kalman filter estimate of user position can also be used in a differential mode with other GPS receivers where at least one receiver is at a known reference point for geodetic sensing, more accurate airborne or shipborne navigation, or in common-view mode for precision differential time transfer. The user position is usually computed in Earth-centered, Earth-fixed (ECEF) coordinates and are then transferred by appropriate geodetic

Table 1 Simplified sequence of operations in a GPS receiver system

1) Select the satellites to be tracked among those in view. Approximate satellite position can be determined using the Almanac, and the selection criteria can be based on GDOP.

2) Search and acquire each of the GPS satellite signals selected.

3) Recover navigation data for each satellite.

4) Track the satellites under whatever conditions of user dynamics are present and measure pseudorange and range-rate and/or ADR.

transformation to a desired local coordinate set or map display convenient to the user. As pointed out in a previous paragraph and discussed in more detail later, it is also possible to integrate the EKF with the DLL instead of performing these operations independently. The author has termed the integrated system the vector delay lock loop (VDLL) because it processes the signals in parallel as a vector operation. The VDLL can be interpreted as a further extension of the EKF.

C. Alternate Forms of Generalized Position Estimators

Figure 3 shows the received GPS signals from N satellites, each with delay $\tau_i(x)$ and signal amplitude a_i. The user position is x. The C/A or P received signal is expressed as follows;

$$r(t, x) = \sum_{i=1}^{N} a_i d_i(t) s_i[t - \tau_i(x)] \cos[\omega_o(t - \tau_i) + \phi_i] + n(t) \quad (1)$$

where $s_i(t)$ represents the radio frequency signal* transmitted by satellite i with amplitude a_i, delay τ_i, and phase ϕ_i, and d_i is the binary data modulation. Both the delay and phase vary with time in accord with user dynamics relative to the satellite i. The signal s_i can represent either the C/A or P codes at either L_1 or L_2 frequencies†. The noise $n(t)$ is assumed to be stationary AWGN.‡ The delay

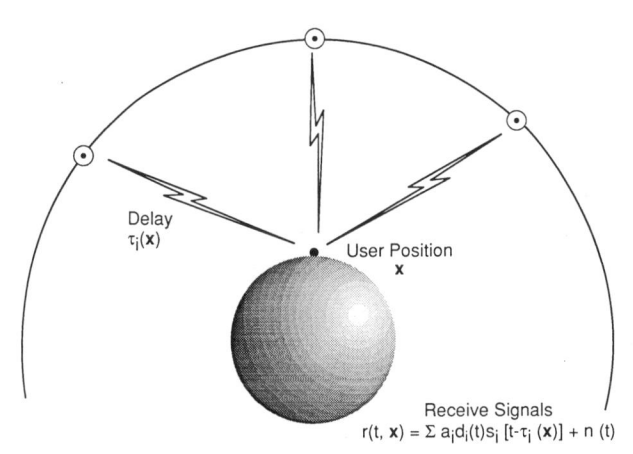

Fig. 3 Simplified representation of received GPS signals for a user at position x and path delays $\tau_i(x)$. The signals are received in the presence of white Gaussian noise $n(t)$.

---

*In reality, of course, the signal has both a C/A code and a P code in-phase quadrature. Both of these pseudonoise (PN) signals can be used simultaneously in delay estimation by a simple generalization of these results.

†More precisely, of course both C/A and P codes should be shown in phase quadrature, and both signals can be tracked in one receiver.

‡Although $n(t)$ is generally white Gaussian thermal noise, there clearly can be signal level fluctuations caused by multipath, attenuation caused by blockage from physical obstruction, and satellite-user motion/geometry. The delay $\tau_i(x)$ is also perturbed by atmospheric delay effects, selective availability, and multipath, which are separate colored noise effects discussed later.

τ_i and phase ϕ_i are functions of the user position vector x. In general, the position vector $x(t)$ is a function of time and is assumed to have some limited set of dynamics governed by a process model.

There are at least two general forms of position estimators that can be configured for the GPS system and many types within each of those forms. The most general estimator, shown in Fig. 4a, shows a single estimator that produces in one step an estimate of the user position vector x. The second more restricted form of Fig. 4b first processes the received signal so as to estimate each of the satellite pseudoranges τ_i (and range-rate, etc.) and then generates an estimate of position $\hat{x}$ based on the $\hat{\tau}_i$. Each of the estimates of τ_i are performed completely independently in this second form.

Assume for the moment that the user position is an unknown constant vector and that the received signal samples at discrete times t_k are $r(t_k,x)$ and are represented over the time interval $t_1, t_2, \ldots t_k$ as the following received vector:

$$r(x) = [r(t_1, x), r(t_2, x), \ldots, r(t_K, x)] \qquad (2)$$

Then the estimate in Fig. one 4a can be represented as follows:

$$\hat{x}_a = F_a[r(x)] \qquad (3)$$

The estimate of Fig. 4b is the following more restricted type of estimate;

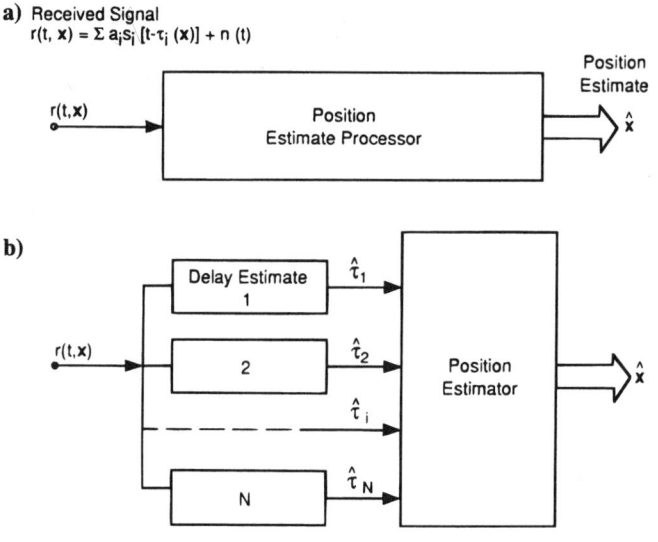

$$\hat{x}_b = F_b[r(x)] = F_{bx}\{\hat{\tau}[r(x)]\} \qquad (4)$$

Fig. 4 Two forms of generalized GPS position estimate processors. The first form of processor a) estimates the position directly without an independent intermediate delay estimate. This estimator may also produce an estimate of delay, because with GPS a delay estimate is needed for recovery of the navigation data. The second form b) first estimates the delay using independent parallel estimators for each τ_i and then estimates position as a separate process.

where each of these τ_i delay (pseudorange) estimates in (b) are made independently. That is, no use is made of the fact that the delays τ_i may be correlated by the geometry of the transmission paths. Most present GPS receivers fall into this latter more restricted class.

An obvious potential disadvantage of the two-step approach of Fig. 4b can be illustrated by a hypothetical example where there are a large number of equal power satellite signals, say $N = 100$, and only one coordinate, a scalar x, to be estimated. Further assume that the satellite path delays are simple offsets of one another, namely, $\tau_i(x) = x + C_i$, where the C_i are known delay offsets. This signal model obviously represents an extreme example of an overdetermined estimation problem. Clearly, an estimate of x, based on the totality of 100 equal power received signals can, in general, produce a better estimate by making use of the linear relationship of the 100 different τ_i and the full power of the 100 signals than would an estimate based on processing each individual signal independently, and then combining the 100 independent estimates of delay. For example, a receiver could offset each of the signals by the known delays C_i add the signals coherently,* and estimate the delay using the composite signal that now has a signal-to-noise ratio 100 times as large. We can envision a situation where the signal-to-noise ratio of each individual signal is too small to process independently (i.e., below threshold); whereas, the delay in the composite signal is easily measured.

Section II of this chapter discusses receivers of the form of Fig. 4b wherein each delay estimate is made independently. Quasioptimum forms of the receivers (scalar delay lock loops) are discussed and related to more conventional delay lock loop tracking systems. The chapter concludes (Sec. III) with a discussion of quasioptimum forms of receivers of the more general form of Fig. 4a, which the author has defined as the vector delay lock loop (VDLL).[4]

D. Maximum Likelihood Estimates of Delay and Position

As shown in Appendices A and B, the DLL is a quasioptimal iterative form of two different statistically optimum delay estimators, the maximum likelihood estimator and the least mean square error estimator. In this subsection, we review a class of estimates termed the maximum likelihood estimate.[5-10] The parameter to be estimated can represent either delay or user position, is assumed to be constant over some time interval of interest, and slowly varies from interval to interval. When cast in an iterative approximation form, these estimators take the configuration of the DLL or the VDLL.

Assume that we make independent observations of a set of scalar random variables r_i, each of which depends on an unknown vector parameter x that we want to estimate. Define r as a vector representing a set of K random variables r_i, where the vector has a probability density $p(r)$. The conditional probability density of r is conditioned on a certain value for the unknown vector parameter x, and is defined as $p(r|x)$. The conditional density, $p(r|x)$, is termed the likelihood function when the variable x is assumed to be the unknown random variable (rather than the opposite). In particular, suppose that we make K independent

---
*The noise also adds for each signal, but not coherently.

observations of the random variables r_i at discrete times t_k to form an observation vector $\mathbf{r} = (r_1, r_2, \ldots r_K)$, where the observations include additive Gaussian noise. The unknown x is assumed to be constant over this K sample interval. We define the maximum likelihood estimate as that estimate $\hat{x}$, which maximizes $p(\mathbf{r}|x)$; namely,* the following:

$$p(\mathbf{r}|\hat{\mathbf{x}}) \geq p(\mathbf{r}|\mathbf{x}_o) \qquad (5)$$

where x_o is any other estimate of x.

Appendix A derives the maximum likelihood estimates for (1) delay $\tau(t_k)$; (2) a scalar position variable $x(t_k)$ where delay is a nonlinear function of x; and (3) a vector position variable $x(t_k)$, where all signals are received in the presence of white Gaussian noise.

Figure 5 shows the block diagrams of the estimators (delay lock loops) that provide iterative closed-loop approximations to maximum likelihood for the two forms of estimators of Fig. 4 with time compared to the noise effects. Thus, the effective bandwidth of the closed-loop tracking filters can be small because it need only track the dynamics of user motion rather than the wide bandwidth of the signal itself. The estimator of Fig. 5a is configured as a closed loop operation in a form similar to what has been called the delay lock loop in earlier papers by the author and has the same form as the delay estimator of Fig. 4b.† The estimator of Fig. 5c is in a form that corresponds to Fig. 4a and is termed a vector delay lock loop.

The reader is referred to Appendix A for details. Suffice it to say here that each of these forms of delay lock loops serves to track the variable to be estimated by generating a correction term that is directly proportional to the error in the estimate. Furthermore, each tracking loop begins by forming the product of the received signal with a differentiated version of the signal component; namely, $s_i'(t)$.

A third closely related estimate is the least mean square error estimate for the AWGN channel. As shown in Appendix B, the least-mean-square estimate of a parameter for signals received is produced by a receiver that can also be configured in an iterative closed-loop form very similar to that for the maximum likelihood estimate when certain linearizing assumptions are made in both estimators. Thus, both estimators configured in iterative form are versions of delay lock loops.

E. Overall Perspective on GPS Receiver Noise Performance

Although this chapter discusses only the fundamentals of GPS signal tracking, it is important to place the noise performance that is discussed here in perspective with the noise effects on the other operations that must be performed in the GPS

---

*Another type of optimal estimate is the maximum a posteriori estimator (MAP),[11] which maximizes $p(x|r)$; i.e., the most probable value of x given the observation vector r. The MAP can be written $p(x|r) = \dfrac{p(r|x)\,p(x)}{\int p(r|x)\,p(x)\,dx}$. Thus, use of the MAP estimate requires knowledge of the a priori probability density $p(x)$ as well as $p(r|x)$. However, in many problems, the a priori probability $p(x)$ is assumed to be unknown.

†For a pure sinusoidal signal, Viterbi[12] showed that the MAP estimator for a Gaussian phase variable can be configured in the form of a phase locked loop. For a pure sinusoidal signal, the delay lock loop reduces to a phase locked loop. See also Ref. 10.

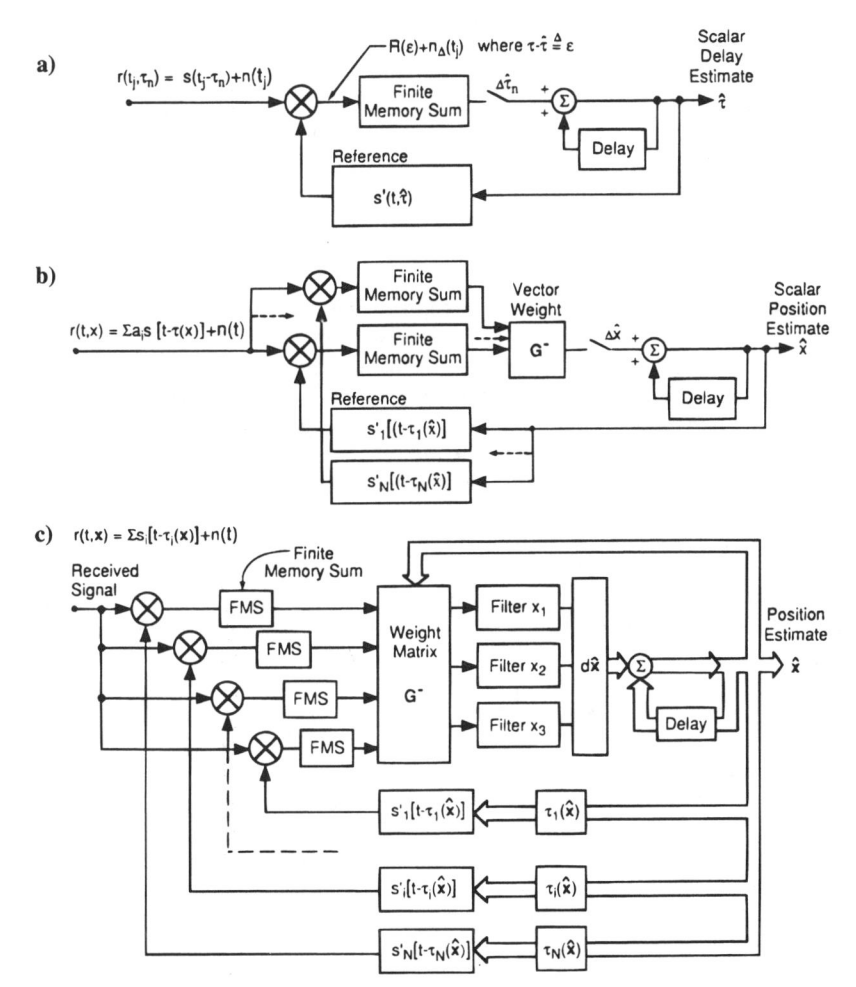

Fig. 5 These various forms of the delay-lock tracking loop provide estimates are iterative closed-loop approximations to the maximum likelihood estimates of τ or x provided that the changes in x occur in steps every K samples and are sufficiently small. Also assume that the loop begins with an initial estimate $\hat{\tau}$ or $\hat{x}$, which is close to the true value of τ or x. Three different generalizations of the delay lock estimator are shown for various levels of complexity of the estimation task. In part a, only a single received signal is present, and the parameter to be estimated is a scalar τ. In part b, multiple signals are received, and the quantity estimated is a scalar position variable x. In part c, multiple signals are received, and the parameter to be estimated is a position vector $x = (x_1, x_2, x_3, \ldots x_m)$. The G^- matrix is the generalized inverse that gives the minimum mean square error estimate of dx given $d\tau$. Different loop filters process each of the position variable components.

receiver; namely, GPS signal search and acquisition, and navigation data demodulation.

It is also important to illustrate some contrasts in the multiple access performance of GPS used for signal tracking with the multiple access performance of code division multiple access (CDMA) used for communications. The primary differences are twofold. First, the GPS signals are transmitted with only moderate power and are generally received with low gain, nearly omnidirectional antennas. Consequently the received radio frequency signal-to-noise ratios that result are low. Thus multiple access and self-noise are generally (although not always) of secondary importance to thermal noise effects. Second, with GPS, the primary objective is to estimate the user position and velocity, rather than transmit and receive a multiplicity of communications signals. Thus, as the number of satellites increases, the information sought does not increase significantly.* Contrast that objective with that of a CDMA communications network wherein each user signal carries with it digitized information at a substantial datarate, the transmission of which is the key objective of the communications system. In a CDMA system, maximum channel capacity is achieved when the signal-to-thermal noise ratio is large, thus multiple access noise effects dominate and limit the number of signals that can be transmitted (because each signal carries additional, rather than redundant, information).

Search and acquisition of the GPS signal is discussed in the next chapter. However, since this chapter on signal tracking assumes that the initial value of delay error has been reduced to a small value, it is well to spend a moment to make plausible that that assumption is realistic. The required C/N_o must be reasonable.

GPS signal search and acquisition is often accomplished by sequencing a reference C/A code over each of the 1023 chips of the C/A code in fractional chip (often in 1/2 chip) steps, in each of several Doppler frequency offset increments. This time-frequency search is completed when the signal component is detected by a noncoherent square law detector. The noncoherent detector operates using a relatively narrow IF bandwidth W, and acquisition occurs when the detector produces an output that exceeds some threshold level. The threshold level is set, in turn, to produce some acceptable level of false alarms, perhaps on the order of 5 in 10^3. If threshold is exceeded several times; e.g., three times out of five, "lock" is declared. If each time-frequency cell is examined for T_r s; i.e., the search rate is $S_r = 1/T_r$. For a false alarm probability $P_{FA} = 5 \cdot 10^{-3}$, and a probability of detection $P_d = 0.9$, the required ratio of carrier power $C = P_s$ to noise density, N_o (one-sided) is approximately $C/N_o S_r \cong 22$ or 13.4 dB if the IF signal-to-noise ratio is unity or more.[3,13] Thus if† $C/N_o = 33$ dB-Hz, and the IF bandwidth is 2000 Hz, then the maximum search rate is $S_r < 90$ or 45 chips/s if 1/2 chip steps are used. If the maximum residual Doppler phase oscillator drift is ± 1 kHz, then only one IF channel of 2 kHz must be searched.

---

*There is, of course, more navigation data to be recovered but these data are at a very low datarate of 50 bps.

†Typical GPS C/N_o are larger; $C/N_o > 40$ dB-Hz. The rule of thumb $C/N_o S_R = 22$ for signal search can also be written $E_R/N_o = 22$ where $E_R = C/S_R$ is the energy per search time interval $1/S_R$. Note the similarity in form with the bit error rate constraint $E_b/N_o = 10$ where E_b is the energy per bit.

FUNDAMENTALS OF SIGNAL TRACKING THEORY   255

Thus search and acquisition do not pose any severe constraints on C/N_o for these search rates. Search and acquisition are discussed more fully in the next chapter.

The navigation data demodulation places a similar requirement on C/N_o. Coherent demodulation of the binary phase-shift keyed (BPSK) waveform produces a theoretical bit error probability $P_E = \text{erfc}\sqrt{2E_b/N_o} = \text{erfc}\sqrt{2C/N_o f_d}$ where E_b is the energy per bit $= P_s T_d = CT_d$, where $T_d = 1/f_d$, and f_d is the 50 bps datarate. If $C/N_o f_d = 10$, then the output error probability is approximately 10^{-5}, which is quite satisfactory for good receiver performance. This ratio corresponds to a $C/N_o = 500$ or 27 dB, which is less than the 33 dB-Hz just cited in the search and acquisition example.

Thus, we can proceed with the discussion of tracking knowing that the C/N_o requirement for acquisition in a reasonable time is consistant with available GPS signal levels (see Chapter 3, this volume), and data can be demodulated at sufficiently low error rate.

F. Interaction of Signal Tracking and Navigation Data Demodulation

As described above, the complete GPS receiver has two closely related and interacting tasks; the receiver must track the delay of each of the received signals and must also coherently (or differentially coherently) detect the 50 bps navigation databit stream, as shown in Fig. 6. Each of these tasks can support the other. For example, the extraction of the navigation data relies on a reasonably accurate delay estimate of each signal so that a reference waveform properly aligned in time; i.e., punctual, can be used to remove the GPS spread spectrum PN code and recover the BPSK signal. The BPSK signal itself can be coherently demodulated by recovering an estimate carrier phase $\phi_i(t)$ and then extracting the navigation data $d_i(t + \tau)$. The recovered carrier phase and navigation data estimates can, in turn, be used as a coherent reference to eliminate the carrier and data in the

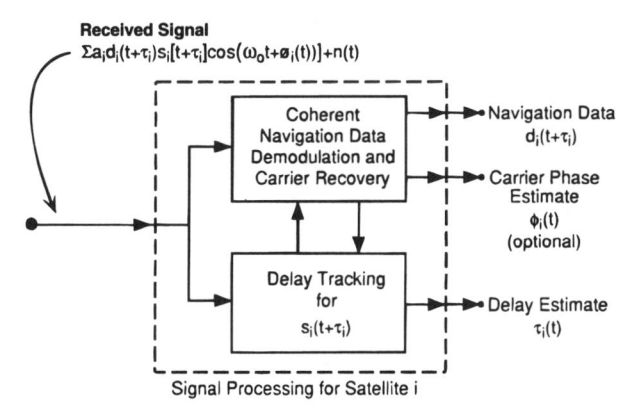

Fig. 6 Two separable, but interacting, signal processing tasks in the GPS receiver, navigation data demodulation, and signal delay tracking. The outputs of each of these tasks can aid the operation of the other.

received signal to permit coherent tracking. Alternatively, this carrier frequency estimate can be used as an aid in removing Doppler shift from the received signal prior to the delay tracking operation, or as an aid in code tracking.

II. Delay Lock Loop Receivers for GPS Signal Tracking

This section describes several alternate methods for tracking the GPS signal. Both code tracking and, to a lesser extent, carrier tracking are described. The signal is generally assumed to be received in the presence of stationary additive white Gaussian noise (AWGN). In this chapter, the signals $s_i(t)$ are always assumed to be bandlimited finite rise time signals with a continuous autocorrelation function $R(\epsilon)$ and continuous, finite, differentiated autocorrelation function $R'(\epsilon) \triangleq \partial R(\epsilon)/\partial \epsilon$. The discussion begins with the quasioptimal delay lock loop that employs a differentiated signal as the reference.[1,3] This delay lock loop is then related to the early-late gate delay lock loop.[2,3,14] It is shown that for a signal represented by a PN trapezoidal pulse sequence with finite risetime, the two forms are identical. Although some of the discussion of tracking systems is expressed in terms of continuous time systems, the reader should be aware that implementations are generally performed in discrete time digital operations. Thus, read $s(t_k)$ for $s(t)$ where the samples are taken at discrete times t_k with independent samples taken at the Nyquist rate.

A. Coherent Delay Lock Tracking of Bandlimited Pseudonoise Sequences

Consider a coherent tracking receiver as shown in Fig. 7 where for the moment we focus on a single received PN signal of power $P_s = A^2/2$ plus bandpass AWGN of spectral density N_o (one-sided).* The coherent receiver downconverts the signal and removes the data modulation using information and control signals from a yet to be described coherent demodulation channel (see Sec. I.F and II.C). The output of the coherent downconverter is then passed through a low-pass

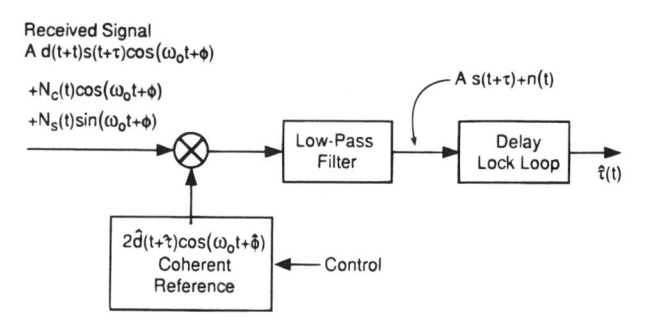

Fig. 7   General form of the coherent delay lock loop receiver system.

---

*Throughout much of this section, the functions are often expressed in continuous time t. The reader should be aware that these results are easily translated to complex discrete time functions at sample time t_k sampled at the Nyquist rate.

filter that maintains the signal spectrum but removes the components at frequency $2f_o$. The output of the coherent downconverter, assuming sufficiently accurate estimates of carrier phase ϕ and data d, is then a pure PN signal plus white Gaussian noise $n(t)$. It should be pointed out that coherent tracking loops of this form have definite practical limitations in that a carrier tracking cycle slip can cause the code tracking loop to lose lock. Nonetheless, the coherent tracking loop serves a useful purpose as a bound on performance and as an introduction to the other forms of the DLL.

The output of the multiplier low-pass filter combination in Fig. 7 for perfect data demodulation and carrier phase recover, namely, $\hat{d}(t) = d(t)$ and $\phi = \hat{\phi}$, is then $A s(t + \tau) + N_c(t)$. The radio frequency signal power is $P_s = (A^2/2)$ for constant envelope signals and somewhat less than that for finite rise time trapezoidal waveforms. The ratio of the signal power in the multiplier output $A^2 = 2P_s$ to the noise density* $2N_o$ of the $n(t) = N_c(t)$ term is then $A^2/2N_o = P_s/N_o$, the same as at the radio frequency input. The baseband pseudonoise PN signal is represented by the following:

$$s(t) = \sum_i P_i p(t - iT)$$

where P_i is a random or pseudorandom binary sequence of numbers ± 1, and $p(t)$ is a bandlimited finite rise time pulse waveform of approximate pulse width T. The PN clock rate is $f_c = 1/T$.

The original paper on the "Delay-Lock Discriminator"[1] pointed out that the optimal tracking system for tracking the delay of a signal $s(t + \tau)$ can be approximated by a tracking loop that first multiplies the received signal plus noise by the differentiated signal, as shown in Fig. 8. The true delay τ and

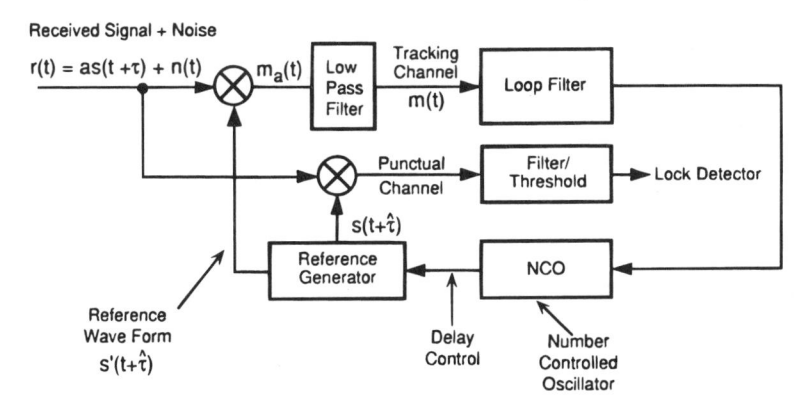

Fig. 8 Delay lock tracking of signal $s(t + \tau)$ where τ varies with time. The punctual channel reference $s(t + \hat{\tau})$ is used to detect that the receiver is locked-on. The bandwidth of the closed-loop tracking must match the dynamics of the delay variation.

*The power of the radio frequency finite bandwidth AWGN, $N_c(t) \cos\omega_o(t) + N_s(t) \cos\omega_o(t)$ with center frequency ω_o is $P_n = (1/2)E[N_c^2(t) + N_s^2(t)] = E[N_c^2(t)]$. If the one-sided spectral density of the radio frequency noise is N_o, then the one-sided spectral density of $N_c(t)$ is $2N_o$, noting that $N_c(t)$ has half the bandwidth of the radio frequency noise.

the delay estimate $\hat{\tau}$ are assumed to be sufficiently close that first-order linear approximations are accurate; i.e., the delay error $\epsilon \Delta \tau - \hat{\tau}$ is sufficiently small relative to the width of the autocorrelation function. The received signal, $s(t + \tau)$, can then be expanded in a Taylor series referenced to the signal with a delay estimate $\hat{\tau}$ as follows:

$$s(t + \tau) \cong s(t + \hat{\tau}) + \epsilon s'(t + \hat{\tau}) + \frac{\epsilon^2}{2} s''(t + \hat{\tau}) +, \ldots, \text{ for small } \epsilon \quad (6)$$

where the partial derivative of the signal waveform with respect to time is $s' \Delta (\partial/\partial t) s$, etc.

The output of the upper multiplier, the tracking channel, in Fig. 8 is $m_a(t)$, which can be expanded using Eq. (6) as follows:

$$m_a(t) = s'(t + \hat{\tau})[As(t + \tau) + n(t)] \quad (7a)$$

$$m_a(t) \cong As(t + \hat{\tau})s'(t + \hat{\tau}) + A\epsilon(t)s'(t + \hat{\tau})^2$$
$$+ (A\epsilon^2(t)/2)s''(t + \hat{\tau})s(t + \hat{\tau}) +, \ldots, + n(t)s'(t + \hat{\tau}) \quad (7b)$$

As before, $n(t)$ is assumed to be stationary AWGN is independent of $s(t)$. The signal is also assumed to be stationary. Define $R(\epsilon) = E[s(t + \epsilon)s(t)]$ as the autocorrelation function of the signal $s(t)$. Note that the expected value of the multiplier output is as follows:

$$E[m_a(t)] = E(s(t + \tau)s'(t + \hat{\tau})) = E[s(t + \hat{\tau} + \epsilon)s'(t + \hat{\tau})] = R'(\epsilon) = D(\epsilon) \quad (8)$$

where $R'(\epsilon) = D(\epsilon)$ is the differentiated autocorrelation function where $R'(\epsilon) = (\partial/\partial\epsilon)R(\epsilon)$, and $D(\epsilon)$ is termed the delay lock loop discriminator characteristic. The slope of the discriminator characteristic at $\epsilon = 0$ is $D'(0)$ and defines the loop gain of the DLL. Define $P_d = E[s'(t)]^2 = -R_s''(0) = -D'(0)$ as the power of the differented signal,* and assume that the clock-rate f_c of the signal $s(t)$ is large compared to the closed-loop noise bandwidth† B_n of the linearized equivalent tracking loop, which is defined later. We can represent the product $s(t)s'(t + \epsilon)$ by its expected value plus a "self-noise" term‡; namely,

$$s(t + \tau) s'(t + \hat{\tau}) = D(\epsilon) + [s(t + \tau) s'(t + \hat{\tau}) - D(\epsilon)] = D(\epsilon) + n_{sn}(t)$$

where $n_{sn}(t)$ is a wide bandwidth self-noise term

$$n_{sn}(t) \Delta [s'(t + \hat{\tau})s(t + \tau) - D(\epsilon)],$$

---

*The slope of the discriminator characteristic is $D'(0) = R_s''(0)$ and is negative. Note that the dimension of $s'(t)$ is s^{-1} and the dimension of $(s')^2$, and hence, P_d is s^{-2}.

†If the closed-loop transfer function is $H(j\omega)$ then the closed-loop noise bandwidth is

$$B_n = \frac{1}{2\pi} \int_0^\infty |H(j\omega)|^2 \, d\omega / |H(0)|^2$$

‡Self-noise is defined as the difference between the expected value of the product $s(t + \tau) s'(t + \hat{\tau})$ and its actual value. The self-noise is a broadband noise-like waveform.[3,4]

most of which is removed by the low-pass and tracking loop filters, and thus, can often be neglected.

The output of the upper multiplier $m_a(t)$ in Fig. 8 is next filtered with a low-pass filter. This filter has bandwidth that is small compared to f_c but large compared to the dynamics of the delay. Thus, it performs a limited amount of averaging. Hence, it is useful to represent the low-pass filtered multiplier output $m(t)$ of Eq. (7) as follows:

$$m(t) = AD(\epsilon) + [n(t)s'(t + \hat{\tau}) + An_{sn}(t)]|_{\text{low pass}} = AD(\epsilon) + n_s(t)$$

$$m(t) \cong -A\epsilon(t)P_d + [n(t)s'(t + \hat{\tau}) + An_{sn}(t)]$$

$$+ (\epsilon^2, \epsilon^3 \text{ terms})|_{\text{low pass}}$$

$$\cong -A\epsilon(t)P_d + n_s(t) \qquad \text{for small } \epsilon \text{ and } B_n << B \qquad (9)$$

where $D(\epsilon) \approx D'(0)\epsilon = -P_d\epsilon$ for small ϵ, self-noise effects are neglected, and it is assumed that the ϵ^2 and higher-order ϵ terms are negligible for small ϵ and $n_s(t) = n(t)s'(t + \hat{\tau})|_{\text{low pass}}$. Thus, the last form of Eq. (9) represents the first two terms in the Taylors' series of Eqs. (6) and (7b). The first component of $m(t)$ in Eq. (9) represents the delay lock loop discriminator characteristic* $D(\epsilon)$.

Thus, for small ϵ, the output of the low-pass filter of Fig. 8 produces a correction term directly proportioned to delay error ϵ, and thereby, enables the loop to track the delay $\tau(t)$. To begin the tracking operation we must initialize (lock up) the DLL with a moderately accurate initial estimate of τ obtained by some search procedure (see Chapter 8, this volume). Note also that for a pure sine wave signal $s(t) = \sin \omega t$, the differentiated signal $s'(t) = \omega \cos \omega t$, and the delay lock loop for this special case simplifies to a conventional phase lock loop. In this case, the filtered product $s(t + \tau)s'(t + \hat{\tau}) = \omega \sin \omega\epsilon \cong \omega^2\epsilon$ for $\omega\epsilon << 1$ (neglecting the 2ω terms), and the discriminator characteristic is $D(\epsilon) = \omega \sin (\omega\epsilon)$. Digital phase lock loops are employed in the carrier-tracking operation.

1. Trapezoidal Pseudonoise Waveform

It is useful to approximate the finite rise time PN waveform by a symmetrical trapezoidal waveform† $s(t)$ of zero mean (equiprobable zero and ones) and rise time δT, as shown in Fig. 9. This finite rise time PN waveform is a more realistic representation of the GPS signal than the ideal rectangular shape. Thus, the DLL reference waveform $s'(t)$ is a ternary waveform that is a sequence of narrow pseudorandom pulses with pulse width equal to the rise time δT. It can be seen that the synchronized product $s(t)s'(t)$ with zero delay offset has zero long-term average ($D(0) = 0$), as we would expect. For a finite rise time pseudorandom

---

*Note that with this definition of $D(\epsilon)$, the slope at $\epsilon = 0$ is negative, and we must put a sign reversal in the loop in order to track the delay variable.

†To be consistent with the assumption of a bandlimited waveform with a continuous derivative, assume that the corners of the trapezoid are slightly rounded.

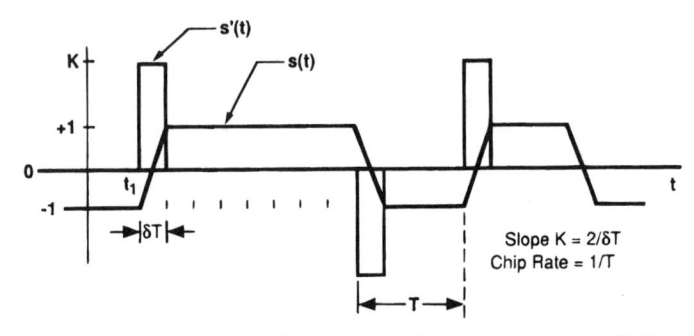

Fig. 9 Plot of PN sequence with a trapezoidal waveform and finite rise time $\delta T < T/2$. The differentiated waveform $s'(t)$ is also shown. The peak value of $s'(t)$ is $K = 2/\delta T$. Where there is no transition $s'(t) = 0$. Because there are transitions half the time, the duty factor of $s'(t)$ is $\delta T/2T$.

sequence of chip duration T, the multiplier output integrated over one chip interval T with zero delay error also is zero; namely

$$\int_{NT}^{NT+T} s(t)s'(t)dt = 0$$

Note that for the trapezoidal waveform, the only portion of the waveform useful for time delay measurement is during the transient rise time δT of the waveform, which occurs at the beginning of the chips 50% of the time. This characteristic is in marked contrast to a communication channel where essentially all of the waveform is useful. Thus, the noise is time-gated to a duty factor of $\delta T/2T$ without significant loss of tracking information.

Thus, the multiplier output of Eq. (9) can be approximated as follows:

$$m(t) \cong AD[\epsilon(t)] + n(t)s'(t)|_{\text{low pass}}$$
$$\cong -AP_d\epsilon(t) + n(t)s'(t)|_{\text{low pass}} \quad \text{for small } \epsilon \quad (10)$$

where the multiplier output noise is time-gated by $s'(t)$, which has a duty factor* of $\delta T/2T$. The discriminator function for this trapezoidal PN sequence is quasilinear $D(\epsilon) \cong -[2\epsilon/\delta T - (\epsilon/\delta T)^2]$ for $|\epsilon| < \delta T/2$. Figure 10 shows a plot of example trapezoidal pulse waveforms, their autocorrelation functions $R(\epsilon)$, and differentiated autocorrelation functions $R'(\epsilon) = D(\epsilon)$ for unit amplitude triangular and trapezoidal pulse waveforms. The power in the differentiated signal is as follows:

$$-D'(0) = P_d = E[s'(t)^2] = K^2 \frac{\delta T}{2T} = \left(\frac{2}{\delta T}\right)^2 \frac{\delta T}{2T} = \frac{2}{T(\delta T)} \quad (11)$$

where $D'(0) = -P_d$, and $K = 2/\delta T$ is the slope of $s(t)$ at a transition, and it is assumed that the PN signal has transitions one-half the time. The rise time δT is inversely related to the one-sided 3-dB bandwidth of the baseband wave-

---

*The duty factor of each pulse is $\delta T/T$, but its pulses occur only in half of the chip intervals.

FUNDAMENTALS OF SIGNAL TRACKING THEORY

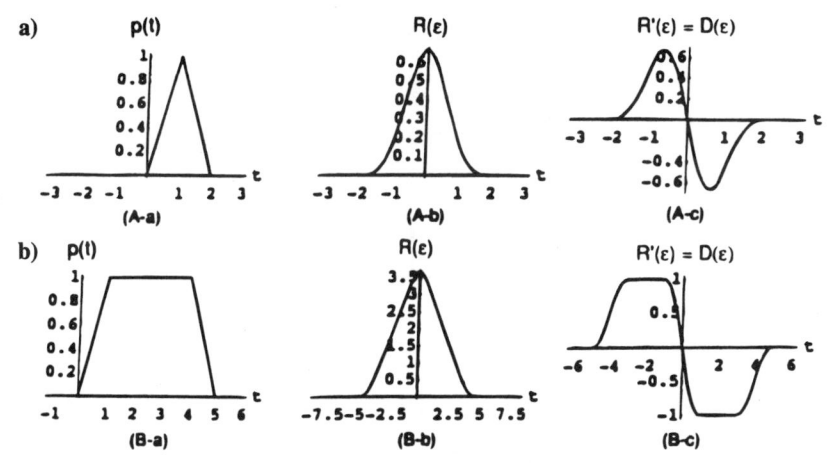

Fig. 10 Triangular- and trapezoidal-shaped pulses with the same normalized $\delta T = 1$, their autocorrelation functions and the quasioptimal delay lock discriminator functions $D(\epsilon) = R'(\epsilon)$. In part a, the PN sequence chip rate is normalized to 1.0, and the rise time of the PN pulse is also 1.0 leading to a triangular waveform shown in A-a with $\delta T/T = 1.0$. The autocorrelation function $R(\epsilon)$ is shown in A-b, and the differentiated waveform or quasioptimal delay lock loop discriminator characteristic $R'(\epsilon)$ is shown in A-c. Figure 10b shows the trapezoidal waveform B-a, autocorrelation function B-b, and discriminator characteristic $R'(\epsilon)$) in B-c, for a PN pulse where the rise time $\delta T = 1.0$ is 25% of the pulse width $T = 4.0$; i.e., $\delta T/T = 0.25$.

shaping filter B by the approximation* $\delta T \cong 0.44/B$. If, for example, the bandwidth $B = 1/T$, then $\delta T = 0.44T$. If, on the other hand, the bandwidth is considerably larger and $B = 10/T$, as is approximately the situation for the transmitted C/A code on the GPS L_1 channel, then $\delta T = 0.044T$. Thus, the pulse width of the quasioptimum reference waveform is very narrow for short rise-time pulses. Even for the P code, the rise time is significantly less than T.

2. Delay Lock Loop Discriminator Curve

Assume again that the signal waveform is the trapezoidal shaped PN sequence. The reference waveform is then a PN sequence of finite width pair of rectangular pulses shown in Fig. 11a where $T = 4$, $\delta T = 1$. The coherent delay lock loop with a differentiated reference has a discriminator curve $D(\epsilon) = R'(\epsilon)$, as shown in Fig. 11b (and also Fig. 10B-c).

Thus, for small values of $\delta T/T$, this delay lock loop operates approximately as a piecewise-linear "bang-bang" servo system, and the discriminator curve

---

*The trapezoidal waveform can result by filtering a rectangular shaped PN sequence with a finite memory integrator filter with impulse response $h(t) = 1/\delta T$ for $0 \leq t \leq \delta T$, and $h(t) = 0$ otherwise. This filter has a frequency transfer function $|H(j\omega)| = \sin(\pi f \delta T)/\pi f \delta T$. This signal spectrum has a first zero in the frequency response at $f = 1/\delta T$. At frequency $f = 0.44/\delta T$, the spectral response is down 2.97 dB. Thus, we define the bandwidth measure $B = 0.44/\delta T$. Clearly, the actual relationship between δT and bandwidth of the waveshaping filter is dependent on the exact signal waveform and the definitions of bandwidth and rise time.

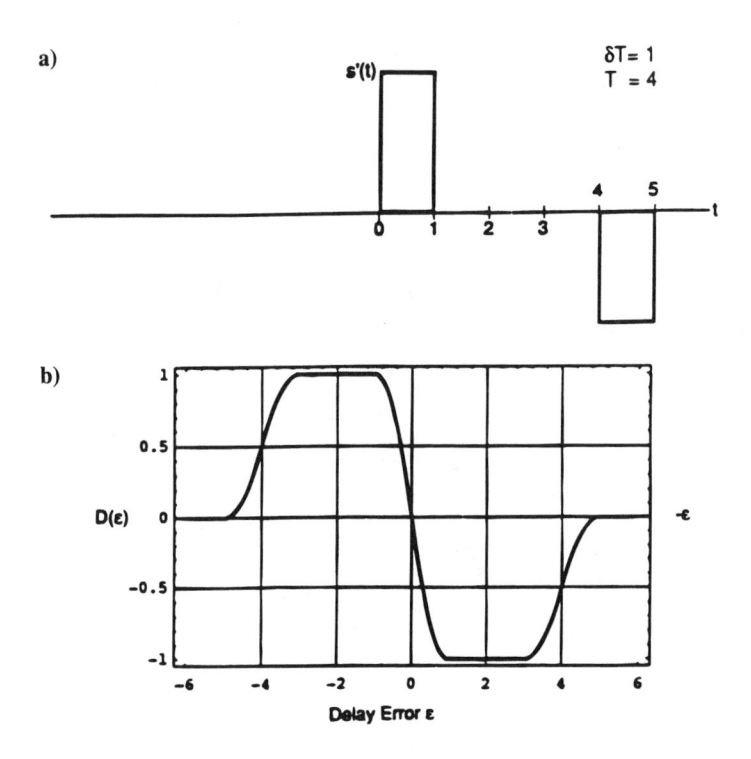

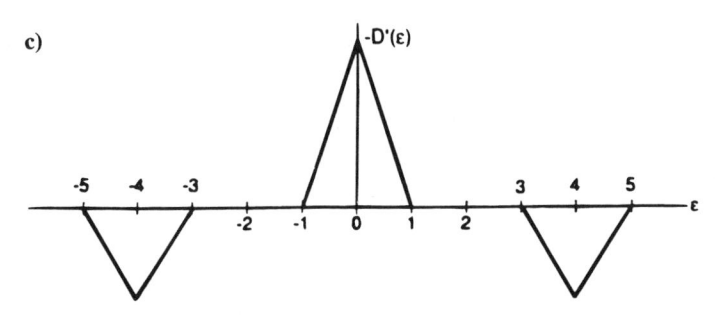

Fig. 11 Coherent delay lock loop discriminator curve $D(\epsilon)$ for a finite rise-time PN signal with a trapezoidal waveform for the optimal delay lock loop. The slope of $D(\epsilon)$," namely, $D'(\epsilon)$, is also shown. The differentiated reference s' is shown in part a. The discriminator curve $D(\epsilon)$ is shown in part b, and the derivative of the discriminator curve (loop gain) $D'(\epsilon)$ is shown in part c. The rise time $\delta T = 1$ and the pulse period $T = 4$ in this example. Thus, the normalized rise time is $\delta T/T = 0.25$. The slope $-D'(0) = 2/(T\delta T)$.

approximates a square wave "doublet." The slope $D'(\epsilon)$ of the discriminator function is as follows:

$$D'(\epsilon) = \frac{\partial}{\partial \epsilon} D(\epsilon) = -\frac{\partial}{\partial \epsilon} \int s(t+\epsilon)s'(t)dt = -\int s'(t+\epsilon)s'(t)dt \quad (12)$$

and is zero for $|\epsilon/T| > 1 + \delta T/T$. If $s'(t)$ is a quasirectangular-shaped "doublet" of Fig. 11a then the slope of $D(\epsilon)$ is a triangular-shaped "triplet," as shown in Fig. 11c.

3. Coherent Early-Late Gate Delay Lock Loops

The earliest published delay lock loop specifically for binary PN signals used two reference signals, an early reference and a late reference signal, each binary PN signal with nearly rectangular shape[2] (see Fig. 12a).* Often the difference between the early signal and the late signal is set equal to a chip width T s. The low-pass filters in the two legs of Fig. 12a simply remove broadband noise and self-noise, as discussed in the previous section. Define the (nearly) zero rise-time binary PN reference waveform as $s_o(t)$. Clearly, the difference between the early and late binary reference signals shown in Fig. 13 is identical to the ternary signal at the bottom of Fig. 13 if the waveforms $s_o(t)$ are rectangular with zero rise-time.

The low-pass filtered output $m(t)$ for the early-late version of the DLL can be written from Fig. 12 as follows:

$$m(t) = [As(t + \hat{\tau}) + n(t)][s_o(t + \hat{\tau} - \Delta/2) - s_o(t + \hat{\tau} + \Delta/2)]|_{\text{low pass}}$$

$$\cong A[R_s(\epsilon - \Delta/2) - R_s(\epsilon + \Delta/2)] + n(t)s_\Delta(t + \hat{\tau})|_{\text{low pass}} = AD_\Delta(\epsilon)$$

$$+ n(t)s_\Delta(t + \hat{\tau})|_{\text{low pass}}$$

$$\cong AD'_\Delta(0)\epsilon + n(t)s_\Delta(t + \hat{\tau})|_{\text{low pass}} \quad (13)$$

where $R_s(\epsilon) = E[s(t + \epsilon)(s_o(t))]$ is the crosscorrelation between the reference $s_o(t)$ and the finite rise-time received waveform $s(t)$, $s_\Delta(t) = s_o(t + \Delta/2) - s_o(t - \Delta/2)$ and the DLL discriminator function is $D_\Delta(\epsilon) = R_s(\epsilon + \Delta/2) - R_s(\epsilon - \Delta/2)$. Note that this early-late DLL multiplier output, Eq. (13), is in exactly the same form as that for the quasioptimal DLL and is identical if $s_\Delta(t) = s'(t)$.

Thus, for the trapezoidal PN waveform, the $s'(t)$ ternary pulse sequence of Fig. 11 is exactly equivalent to the properly scaled difference between two time displaced zero rise-time PN sequences, as shown in Fig. 13. Thus, the differentiated delay lock loop of Fig. 12b is also equivalent to an early-late gate delay lock loop with delay offsets of $\pm\delta T/2$, as shown in Fig. 12a. In this chapter, the difference in delay between the early and late reference signals is defined as Δ.

For a signal with bandwidth $B = 10/T$ then if rise time is related to bandwidth by $\delta T = 0.44/B$, the quasioptimal delay differences $\Delta = \delta T = 0.044T$. This offset is significantly smaller than the commonly used delay difference $\Delta = T$.

---

*The original delay lock loop paper in 1961 defined the differentiated reference that gives an early-late spacing equal to the rise-time for a trapezoidal waveform. However, at that time, the state of the art made a longer delay spacing more practical.

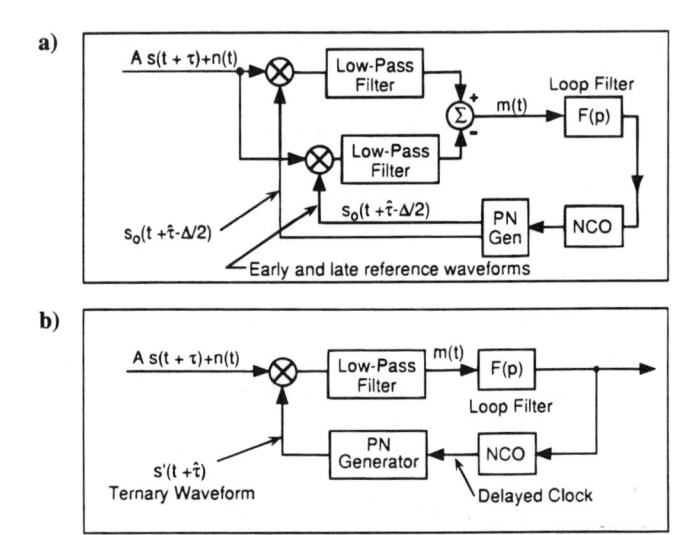

Fig. 12 The early-late delay lock loop with a delay difference between early and late reference signals of δT (part a). For a trapezoidal PN waveform, this loop is mathematically equivalent to the loop shown in part b, the optimal delay lock loop. The early-late delay difference Δ is set at $\Delta = \delta T$, where δT is the pulse rise-time. The NCO is a number-controlled oscillator used to control the clock phase of the PN generator. Part b shows the optimal delay lock loop for the trapezoidal or ternary waveforms of Fig. 9 utilizes a differentiated reference that is a ternary waveform.

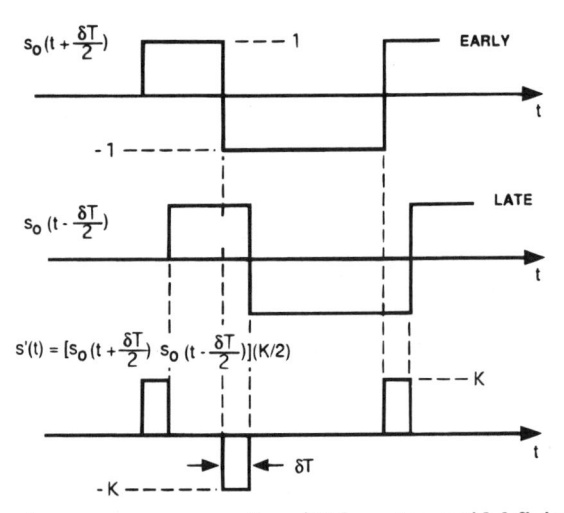

Fig. 13 The pulse sequence representing $s'(t)$ for a trapezoidal finite rise-time δT waveform of Fig. 11 is equal to the difference between two zero rise-time reference PN sequences; namely, $s'(t) = [s_o(t + \Delta/2) - s_o(t - \Delta/2)](K/2)$ with only a scale factor $K/2$ difference if the delay difference is $\Delta = \delta T$.

The smaller delay offset gives a higher accuracy but a slightly smaller threshold acquisition range and a substantially smaller quasilinear region. As discussed later in the subsection on transient performance of the DLL, there are certain disadvantages if the early-late spacing Δ becomes too small.* Note, however, that if the rise-time of the PN signal δT is equal to the chip interval; i.e., $\delta T = T$, then the signal becomes the triangular wave of Fig. 10, and the quasioptimal early-late gate delay lock loop delay offset is the commonly used plus or minus half-chip setting.[2] Thus, the commonly used delay lock loop for an early-late spacing of $T/2$ is optimum only for the triangular wave PN signal.†

If the received signal has an ideal rectangular pulse shape, the discriminator characteristic $D(\epsilon)$ for the early-late reference signal with early–late offsets of $\pm T/2$; i.e., $\Delta = T$ is piecewise linear, as shown in Fig. 14, and has a normalized one-sided width $(\epsilon/T) = 1.5$, and thus, gives a nearly 50% wider acquisition range than the DLL with narrow correlator spacing for $\delta T/T \ll 1$. However, as shown later, the noise performance of the DLL that uses an approximation to the differentiated reference, a PN sequence of narrow pulses or equivalently a narrow early-late spacing close to the rise-time of δT, gives better noise performance than that for the early-late gate DLL with the wide early-late spacing, $\Delta = T$.

It is informative to compare the noise performance of two different delay lock loop discriminator characteristics that can be used for the same received trapezoidal pulse waveform of Fig. 9. Consider a pulse of width $T = 4$ for this waveform with a rise-time $\delta T = 1$ for a ratio $\delta T/T = 0.25$. The two different reference waveforms are shown in Fig. 15. Figure 16 shows the discriminator characteristic for the quasioptimal reference waveform $s'(t)$ as the solid curve where its discriminator characteristic $D(\epsilon) = R'(\epsilon)$. The dashed curve is the discriminator characteristic for an early-late gate delay lock loop with an early-late separation of $\Delta = T$. Note that both discriminator curves have the same slope at $\epsilon = 0$. However, the effective noise spectral density is four times as large for the loop with the $\Delta = T$ separation because the effective noise is time gated by the duty factor. Thus, the noise performance of the quasioptimal loop is superior for this trapezoidal signal for small delay error.

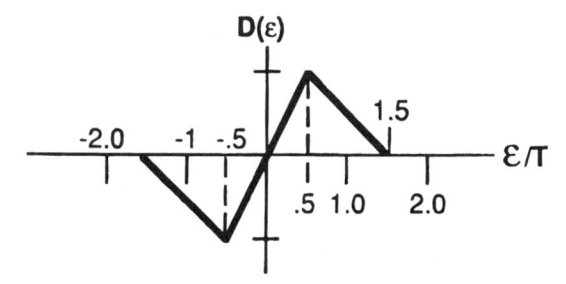

Fig. 14 Delay lock loop discriminator curves for offset reference signals $s(t \pm T/2)$; i.e., $\Delta = 1$, and an ideal rectangular received PN signal.

---

*During acquisition, we may have to increase the loop gain to improve the pull-in performance.
†Note that there are more recent papers that discuss the use of narrower time separations between the early and late gate signals. See also the next chapter.[15,16]

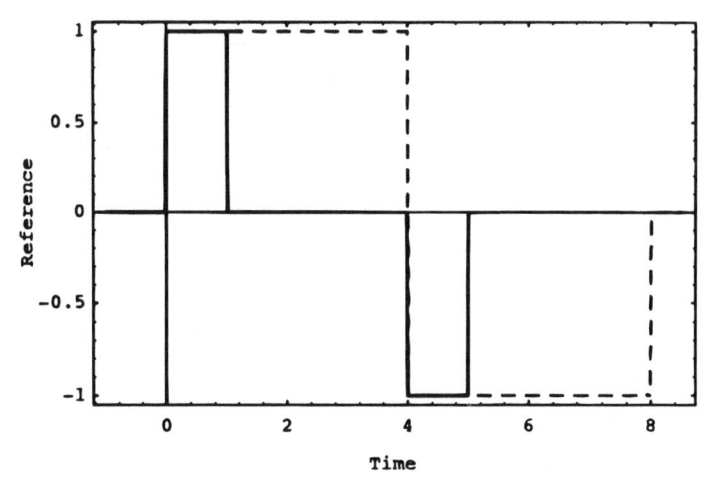

Fig. 15 Quasioptimal reference waveform $s'(t)$ for the trapezoidal shaped PN pulse waveform, $\Delta = \delta T = T/4$ (solid curve), and a reference with an early-late separation of T. The received Trapezoidal pulse waveform has a pulse duration of $T = 4$ and a $\delta T = 1$ rise-time. The dashed curve shows the commonly used early-late reference waveform with an early-late separation of $\Delta = T$ s.

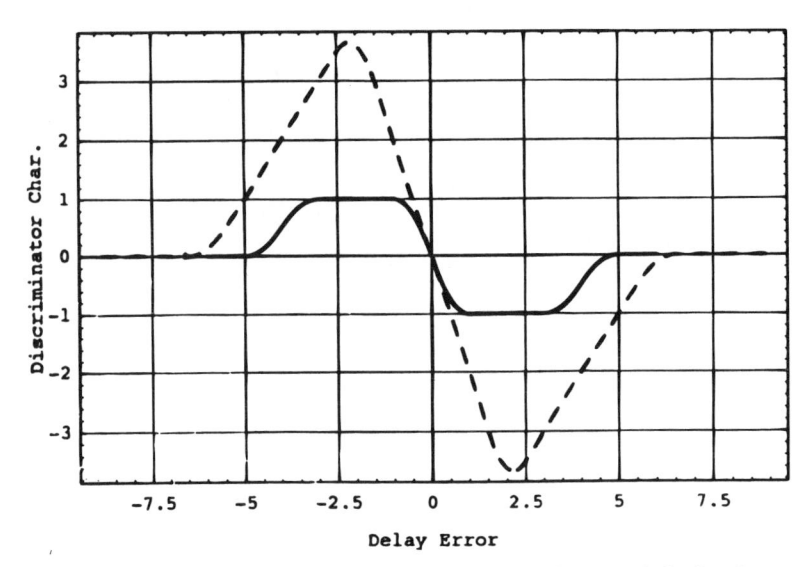

Fig. 16 Quasioptimal and early-late gate discriminator characteristics for the trapezoidal waveform of Fig. 10b with $T = 4$ and $\delta T = 1$ for a rise-time ratio $\delta T/T = 0.25$. The solid curve, quasioptimal, is obtained using the differentiated reference $s'(t)$. The dashed curve is obtained using an early-late gate rectangular reference with an early-late separation of T. Note that the slope of both discriminator characteristics at the origin is approximately equal to $D(0) \cong -2$.

On the other hand, the discriminator characteristic for the quasioptimal receiver, if multiplied by a gain factor of 2 to equate noise effects has double the loop gain but still does not have as large a peak nor quite as wide a tracking region. The discriminator characteristic drops to zero at a delay error of $\epsilon = 1.25T$ for the quasioptimal detector; whereas, the early-late gate detector discriminator with $\Delta = T$ drops to zero at a delay error $\epsilon = 1.5T$. Note, also, that the quasioptimal delay lock loop has a constant correction (as in a "bang-bang servo,")[17] for delay errors larger than $2\delta T$ in magnitude but smaller than T; whereas, the $\Delta = T$ early-late delay lock loop has a wider quasilinear range. Thus, although quasioptimal delay lock loop, indeed, has superior noise performance, the early-late gate delay lock loop with a T s separation is expected to have better initial acquisition performance. Thus, we can envision a DLL where the early-late separation is set at T during initial acquisition and decreases during normal tracking to $\Delta = \delta T$ or some other value $\Delta < T$ (see Ref. 16).

Because the linear region of this quasioptimal delay lock loop can be quite narrow if δT is much less than T, appropriate account must be taken of the saturation effects in designing the tracking loop bandwidth. Notice that if $\epsilon > \Delta$, the correction or steering voltage of the discriminator function does not go to zero; however, the effective loop gain diminishes either with increased delay estimate noise variance or loop-tracking transient error. Thus, the loop-tracking gain and noise bandwidth may have to be increased somewhat to take this saturation effect into account since the expected value of the gain diminishes with increasing delay error variance. For example, the loop gain may have to be increased during acquisition and then reduced to its optimal tracking values.

4. Linearized Equivalent Circuit of the Delay Lock Loop

The expression for the multiplier output (low-pass filtered) $m(t)$ from Eq. (10) can be written as follows:

$$m(t) \cong +A\epsilon(t)D'(0) + n(t)s'(\tau + \hat{\tau})|_{\text{low pass}} = -AD'(0)(\tau + \hat{\tau}) + n_s(t)$$

$$\cong -AP_d(\tau - \hat{\tau}) + n_s(t) \qquad (14)$$

In this latter form, the output $m(t)$ of the DLL can be generated by the subtractor in the linearized closed loop of Fig. 17 where the equivalent input to the linearized equivalent is now approximately equal to $AP_d\tau(t) + n_s(t)$.

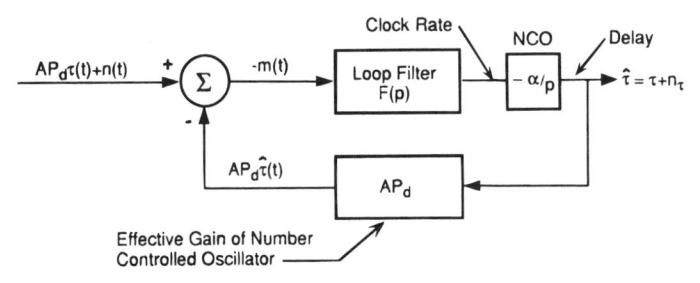

Fig. 17 Equivalent circuit of a coherent base band delay lock loop for a trapezoidal-shaped received signal operating in the quasilinear range $\epsilon < \delta T$.

Thus, in the "linear" region of the discriminator characteristic, the differentiated reference delay lock loop has an equivalent circuit, shown in Fig. 17. The $\Delta = T$ early-late DLL has a similar linearized equivalent with a slightly different loop gain as does the noncoherent DLL of the next subsection. The NCO has as its input a frequency control word and produces as its output clock phase or delay $\hat{\tau}(t)$. Thus, the NCO acts as an integrator with transfer function α/p after the loop filter. Operator notation is employed here where $p = d/dt$.

If the loop filter is $F(p)$, and the NCO (integration) has a gain $-\alpha/p$, then the closed-loop operation of Fig. 12b can be represented by the following differential equation:

$$\hat{\tau}(t) = \left(\frac{\alpha}{p}\right) F(p)[-m(t)] = \left(\frac{\alpha}{p}\right) F(p)[-aD'(0)[\tau(t) - \hat{\tau}(t)] + n_s(t)] \quad (15a)$$

Solve for $\hat{\tau}(t)$ to obtain the following:

$$\hat{\tau}(t) = \left[\frac{-AD'(0)\frac{\alpha}{p}F(p)}{1 - AD'(0)\frac{\alpha}{p}F(p)}\right]\left[\tau(t) + \frac{n_s(t)}{-AD'(0)}\right] = H(p)\left[\tau(t) + \frac{n_s(t)}{-AD'(0)}\right] \quad (15b)$$

where $H(p)$ represents the closed-loop response and has noise bandwidth B_n, and $D'(0) = -P_d$. This expression for the estimate can be separated into transient delay errors $\tau_T(t)$ and noise-induced delay errors $\tau_n(t)$; i.e., $\hat{\tau}(t) = \tau(t) + \tau_T(t) + \tau_n(t)$. The transient delay error can be written as $\tau(t) + \tau_T(t) = H(p)\tau(t)$, and the noise induced delay error is as follows:

$$\tau_n(t) = H(p)\left[\frac{n_s(t)}{-aD'(0)}\right] \quad (16a)$$

The closed-loop response for this linearized model of the DLL for a noise-free input is then given by the following differential equation:

$$\hat{\tau}(t) = H(p)\tau(t) = \left[\frac{AP_d F(p)(\alpha/p)}{1 + AP_d F(p)(\alpha/p)}\right]\tau(t) \quad (16b)$$

where $H(p)$ represents the closed-loop response. The closed-loop filter response must be adequate to track the dynamics of the delay variable τ. For the early-late DLL with a delay separation Δ, the same expression applies if $D'(0)$ is replaced by $D'_\Delta(0)$ and $n_s(t) = [s_o(t + \Delta/2) - s_o(t - \Delta/2)]n(t)|_{\text{low pass}}$.

5. Delay Lock Loop Transient Tracking Performance

The closed-loop bandwidth of the linearized DLL must be designed to be wide enough to tolerate the dynamics of the delay $\tau(t)$ of the received signal with a small transient error but not so wide as to degrade noise performance. For GPS, these dynamics depend both on the satellite motion and especially on the user receiver dynamics. Consider a DLL operating in its quasilinear mode of operation with a code-tracking loop filter of the following form:

$$F(p) = \left[\frac{(1 + \sqrt{2}\, p/p_o)}{(1 + g_o p/p_o)}\right] \tag{17}$$

where we have selected the loop parameters for critical damping of a second-order closed-loop response function $H(p)$. By inserting Eq. (17) into Eq. (16b) and assuming a large g_o, the closed-loop transfer function can be expressed in the following form:

$$\frac{\hat{\tau}(t)}{\tau(t)} = H(p) = \frac{AP_d F(p)(\alpha/p)}{1 + AP_d F(p)(\alpha/p)} = \frac{\alpha A P_d (1 + \sqrt{2}\, p/p_o)}{\frac{g_o p^2}{p_o} + \alpha A P_d (1 + \sqrt{2}\, p/p_o)}$$

$$= \frac{1 + \sqrt{2}\, p/p_o}{\frac{g_o p_o}{\alpha A P_d}\left(\frac{p}{p_o}\right)^2 + 1 + \sqrt{2}(p/p_o)} = \frac{1 + \sqrt{2}\, p/p_o}{1 + \sqrt{2}\, p/p_o + (p/p_o)^2} \tag{18}$$

where the gain is set to give $g_o p_o/\alpha A P_d = 1$. In general, the dc loop gain depends on the correlator spacing, signal level, and loop amplifier gain. The equivalent closed-loop noise bandwidth B_n for this filter is $B_n = 1.06\, p_o$. That is, if the input noise density is N_o (one-sided) and the loop has unity gain at $f = 0$, then the output noise power is $N_o B_n$ for white noise. Thus, the output noise increases in direct proportion to B_n. The value of p_o should be small for best noise performance but not so narrow as to degrade tracking performance.

For a dynamic user platform, the closed-loop bandwidth must be sufficiently wide to track the transient variation of delay vs time which is caused by user-satellite range dynamics without creating large transient errors. The steady-state transient delay error ϵ_{ss} in response to a step of acceleration $\ddot{\tau}_s$ for this loop is as follows:*

$$\frac{\epsilon_{ss}}{T} = \frac{1.12}{(2B_n)^2}\frac{\ddot{\tau}_s}{T} \tag{19}$$

As an example, if the delay acceleration is $A_g g$ where the acceleration $g = 32.1578$ ft/s$^2 = 9.8017$ m/s^2 (note that 1 ft $= .3048$ m), $c = 2.99792458$ m/s, and 1 g corresponds to $\ddot{\tau}_s = 3.2695 \cdot 10^{-8}$ s/s^2. If the PN chip rate is $f_c = 10^6$, $T = 10^{-6}$s, then if we set $\epsilon_{ss}/T = 0.01$ then the following results:

$$\frac{\epsilon_{ss}}{T} = \frac{1.12}{(2B_n)^2}\frac{\ddot{\tau}_s}{T} = \frac{1.12 A}{4 B_n^2}\left[\frac{3.2693 \cdot 10^{-8}}{10^{-6}}\right] = \frac{9.154 \cdot 10^{-3}\, A}{B_n^2} \tag{20a}$$

or

$$B_n = 0.9568\sqrt{AT/\epsilon_{ss}} = 0.9568\sqrt{A}\ \text{Hz for}\ \epsilon_{ss}T = .01 \tag{20b}$$

If the normalized acceleration is $A_g = 5$, then the required noise bandwidth is

---

*There is no steady-state transient error for a second-order loop with constant velocity input.

$B_n = 2.139$ Hz. Figure 18 shows plots of the tracking error vs. loop noise bandwidth for acceleration steps of $0.61\ g = 6\text{m/s}^2$ and $1.22\ g = 12\ \text{m/s}^2$ for a PN chip rate of 10^6 Hz.

Note that in order for these transient error relationships to hold, the loop must be in the quasilinear region of the discriminator curve. For an early-late spacing of Δ, then, the steady-state transient error must be less than Δ. Because $B_n \sim \sqrt{T/\epsilon_{ss}}$, as ϵ_{ss} decreases, the value of B_n increases. Thus, if we set a narrow correlator spacing $\Delta \ll T$, the value of ϵ_{ss} may have to decrease, and hence, increase the noise bandwidth. There are two constraints on ϵ_{ss} that set a minimum to the value of B_n. First ϵ_{ss} must be small enough in an absolute sense to give the desired accuracy as given by Eq. (2). Second, ϵ_{ss} must be small enough relative to Δ to give quasilinear operation on which Eq. (19) is based. For example, if we set $\epsilon_{ss}/\Delta \leq 0.2$, then from Eq. (20a) we obtain the following:

$$B_n \geq 0.204\sqrt{A_g T/\Delta} \tag{20c}$$

For very small Δ/T, the minimum value of B_n can be constrained by the value of delay spacing Δ. Thus, the performance advantage of the smaller value of Δ may not be quite as great because the value of B_n may have to increase somewhat.

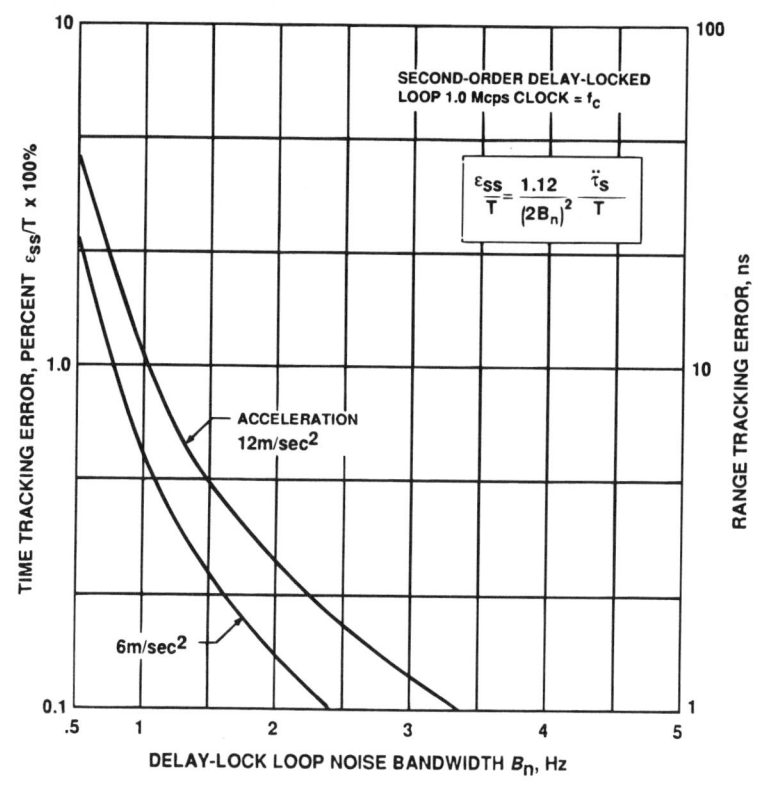

Fig. 18 Normalized steady-state tracking error of a DLL caused by acceleration steps of $c\ddot{\tau} = 6\text{m/s}^2, 12\text{m/s}^2$ vs closed-loop noise bandwidth B_n.

6. Mean-Square Noise Error

From Eq. (16a), the noise induced delay error is as follows:

$$\tau_n(t) = H(p)\left[\frac{n_s(t)}{-AD'(0)}\right],$$

where $n_s(t) = n(t)s'(t + \hat{\tau})|_{\text{low-pass}}$. If the noise spectral density of $n_s(t)$ is N_{so} (one-sided), then the output delay noise variance is as follows:

$$\sigma_\epsilon^2 = N_{so}B_n/A^2[D'(0)]^2 \qquad (21)$$

If σ_n is the rms value of $H(p) n_s(t)$, then the rms delay error is $\sigma_\epsilon = \sigma_n/aD'(0)$.

The noise $n(t)$ has noise density (one-sided) of $2N_o$. Recall that for a trapezoidal received waveform of rise-time δT, that $s'(t)$ is a sequence of pulses of positive and negative amplitude $\pm K = \pm 2/\delta T$ that have an average duty factor of $\delta T/2T$. Thus, the noise density of $n_s(t)$ (after low-pass filtering to smooth the transients) has a value at low frequencies of

$$N_{so} = 2N_o(2/\delta T)^2 \, (\delta T/2T) = 2N_o(2/T\delta T)$$

for the DLL with a differentiated signal as the reference. The corresponding value of the gain is $-D'(0) = 2/(T\delta T)$. Thus, the mean-square delay error from Eq. (21) is then as follows:

$$\sigma_\epsilon^2 = \frac{N_{so}B_n}{A^2[D'(0)]^2} = \frac{2N_o\left(\frac{2}{T\delta T}\right)B_n}{A^2\left[\frac{2}{T\delta T}\right]^2} = \frac{2N_oB_n}{A^2}\frac{T\delta T}{2}$$

and

$$\left(\frac{\sigma_\epsilon}{T}\right)^2 = \frac{2N_oB_n}{A^2}\left(\frac{\Delta T}{2T}\right) \qquad (22a)$$

For a trapezoidal pulse waveform with a small rise-time $\delta T \ll T$, the received signal power at radio frequency is $P_s \cong A^2/2$, and thus the following results:

$$\left(\frac{\sigma_\epsilon}{T}\right)^2 \cong \frac{N_oB_n}{2P_s}\left(\frac{\delta T}{T}\right) \qquad \text{for } \sigma_\epsilon/T \ll 1 \qquad (22b)$$

which decreases as the rise-time diminishes, because the noise is time-gated by the ratio $\delta T/2T$.

If, on the other hand, the received signal has essentially a zero risetime so that the received signal waveform is $s_o(t + \tau)$ and the early-late reference separation is Δ, then the multiplier output from Eq. (13) is
$m(t) = AD'_\Delta(0)\epsilon + n(t)s_\Delta(t + \hat{\tau})|_{\text{low pass}} = AD'_\Delta(0)\epsilon + n_s(t)$.

For this example, $s(t) = s_o(t)$, the zero rise-time signal. Thus $D'_\Delta(0) = -2/T$, and $s_\Delta(t)$ takes on the values ± 2 and has a duty factor $\Delta/2T$. The signal component is then $-2A\epsilon$, and the noise density of $n_s(t)$ is $2N_o(2^2)\Delta/2T = 4N_o\Delta/T$. The output delay noise variance is as follows:

$$\sigma_\epsilon^2 = \frac{4N_oB_n}{A^2[D'(0)]^2}\frac{\Delta}{T} = \frac{4N_oB_nT^2}{2P_s(4)}\left(\frac{\Delta}{T}\right) = \frac{N_oB_nT^2}{2P_s}\left(\frac{\Delta}{T}\right),$$
(23a)

$$\text{and } \left(\frac{\sigma_\epsilon}{T}\right)^2 = \frac{N_oB_n}{2P_s}\frac{\Delta}{TA} \quad \text{for } \sigma_\epsilon/T \ll 1$$

and for the commonly used value $\Delta = T$ the normalized mean square delay error is as follows*:

$$(\sigma_\epsilon/T)^2 = N_oB_n/2P_s \tag{23b}$$

Two comments should be made. For a finite rise-time, trapezoidal signal, noise performance, (σ_ϵ/T), improves as Δ decreases as long as $\Delta \geq \delta T$, the rise-time. The optimum occurs at $\Delta = \delta/T$. The performance actually degrades if Δ is made less than the rise-time. Second, the value of Δ should be significantly larger than the transient error ϵ_{ss} of Eq. (22). This constraint, in turn, can cause the minimum required value of B_n to increase as Δ decreases, because $B_n \sim 1/\sqrt{\epsilon_{ss}}$ and from Eq. (20c), $B_n \geq 0.303\sqrt{A_gT/\Delta}$ can be one of the constraints on B_n.

B. Noncoherent Delay Lock Loop Tracking of Pseudonoise Signals

The previous discussion assumed a coherent downconversion of the radio frequency signal to baseband and removal of the data demodulation prior to the DLL tracking. As mentioned in Sec. I.A, coherent tracking has limitations for GPS in that accurate code tracking is necessary for the carrier-tracking operation to begin. Furthermore, cycle slips in the carrier-tracking operation can cause loss of code lock. Thus, there is considerable interaction between carrier and code tracking, and coherent DLL tracking is somewhat fragile and not often used for GPS. In this section we assume that the signal enters a noncoherent DLL at IF with a residual Doppler offset, data modulation, and that the carrier phase is unknown. In this instance, we revert to early-late correlation rather than using a single ternary reference, because we must achieve sufficient correlation to produce a high signal component in each of the noncoherent detectors in the IF. An extension of the coherent early-late DLL (Fig. 19a) replaces the baseband correlators with bandpass correlators as shown in Fig. 19b.[18] The early-late delay spacing is set at Δ. Although a correlator spacing of $\Delta = T$ is often employed, smaller spacings of $\Delta = \delta T$ give better noise performance for a trapezoidal waveform with rise-time δT, as discussed in the previous section on the coherent DDL.† Here there is an even greater advantage than in the coherent DLL, because a smaller delay spacing Δ yields a higher signal-to-noise ratio in each of the IF channels, and hence, there is a lesser small signal suppression effect caused by the square law detector nonlinearity when the IF signal-to-noise ratio diminishes below unity.

---

*Note that even the early-late DLL with a $\Delta = T$ delay separation gates off the noise half the time; i.e., when there is no transition.
†Under the assumption of small delay error.

FUNDAMENTALS OF SIGNAL TRACKING THEORY 273

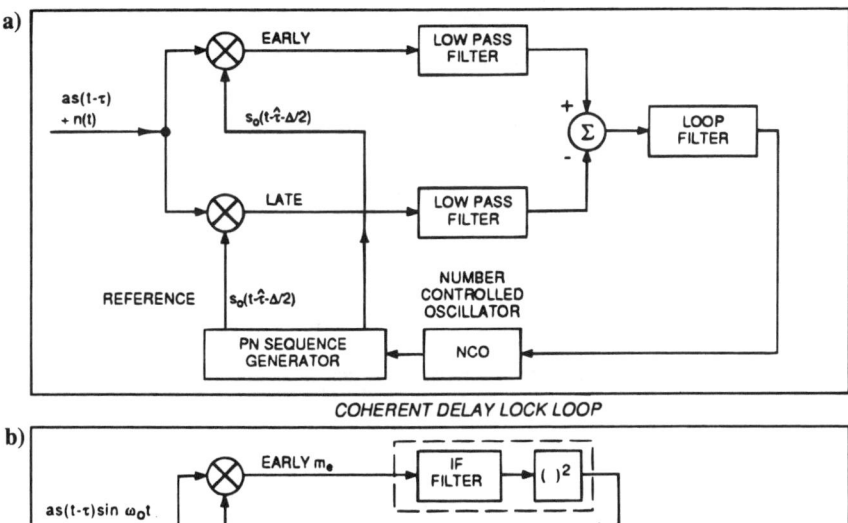

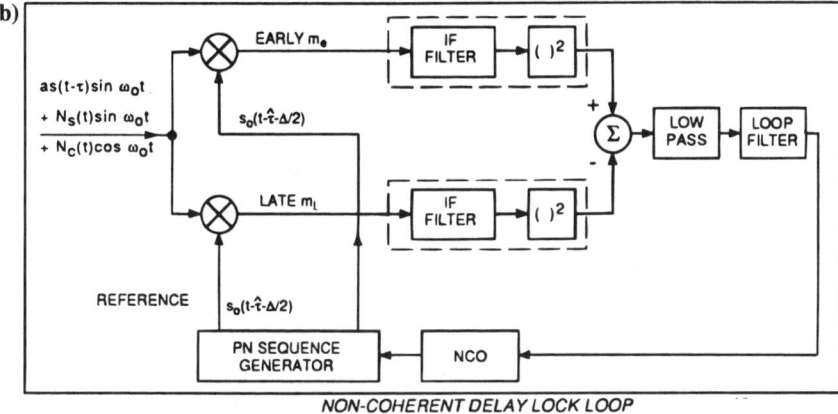

Fig. 19 Baseband and noncoherent versions of the early-late DLL. The baseband version is shown in part a. The noncoherent form in part b operates on a modulated carrier and simply replaces each low pass filter of part a with an IF filter followed by a square-law envelope detector.

1. Performance of Second-Order Noncoherent Delay Lock Loop

Figure 20 is a simplified block diagram of the noncoherent DLL and the related punctual channel that can be used for data demodulation and code lock detection. The signal is assumed to have both data and carrier modulation; i.e., the signal is assumed to be $Ad(t)s(t)\cos(\omega_o t + \phi)$. For GPS, the IF bandpass filters in each leg of the early and late IF channels (or their baseband I/Q digitized sample equivalents) must have sufficient bandwidth to accommodate residual frequency offset of the modulated signal caused by residual satellite/user Doppler shifts (uncorrected GPS satellite Doppler is typically $\approx \pm 6$ kHz) plus residual user clock drift, bias frequency offsets, and the data modulation (BPSK signal modulated by the 50 bps navigation data).

Residual Doppler shift is generally the dominant requirement on the IF bandwidth, because the GPS data modulation is only 50 bps. Much of the residual

274 J. J. SPILKER JR.

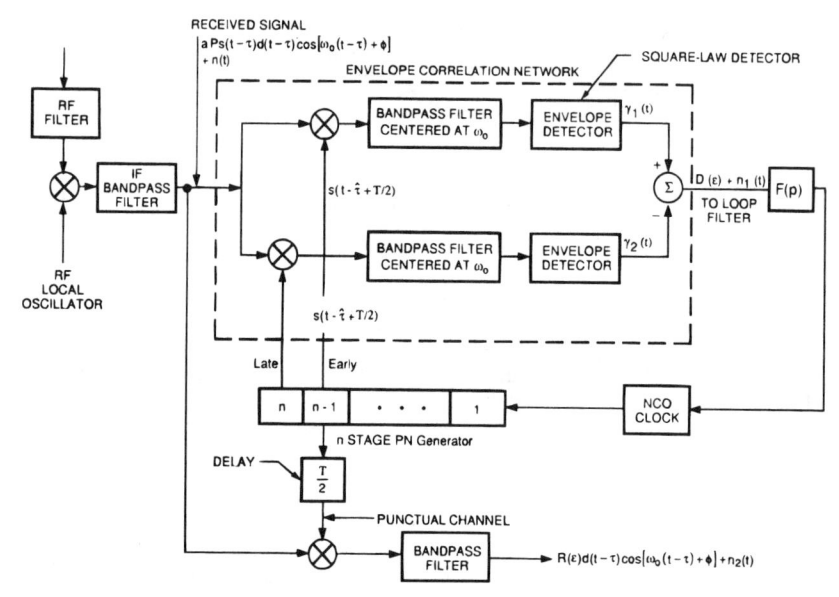

Fig. 20 Noncoherent DLL using envelope correlation for a data-modulated PN signal on an radio frequency carrier (Ref. 18, pages 416–426). The delay error is $\epsilon \triangleq \tau - \hat{\tau}$. The local oscillator performs downconversion and a limited amount of Doppler frequency removal.

IF frequency offsets can be removed by some form of automatic frequency control (AFC) signal processing. Doppler estimation also can be performed as part of the acquisition and tracking process. Both of these operations can permit reduction in the IF bandwidth.*

Assume as a first example that the PN signal is rectangular with essentially zero rise-time. The discriminator characteristic for this noncoherent DLL is then as shown in Fig. 21 for values of $\Delta/T = 0.125, 1.0$, and 2.0. In this instance $D(\epsilon) = R^2(\epsilon + \Delta/2) - R^2(\epsilon - \Delta/2)$ where the squared autocorrelation function terms appear because of the square-law detector, and Δ is the early-late separation. Note that if the delay difference is increased beyond $\Delta/T = 1$ the loop gain at $\epsilon = 0$ degrades. As can be seen, the loop gain for $\Delta/T = 0.25$ has increased, but the peak amplitude of $D(\epsilon)$ has decreased to approximately 0.4375 at $\epsilon/T = \Delta/2T = 0.125$, and thereby limits its pull-in range.

The slope of the discriminator characteristic increases as Δ decreases, and as shown later for finite rise-time trapezoidal signals the optimum noise performance occurs at $\Delta = \delta T$ for quasilinear operation. There is one other advantage of a smaller early-late delay spacing $\Delta < T$. At zero delay error $\epsilon = 0$, the signal

---

*For example, the initial IF bandwidths can be set sufficiently wide to accommodate the full Doppler shift uncertainty; e.g., the bandwidth might be set in the 5–10 kHz range. After search, acquisition, and AFC tracking, the IF bandwidth can be adaptively reduced to a significantly smaller value. (Recall also from Chapter 4, this volume, that the P code on the L_2 channel may have no navigation data modulation.)

FUNDAMENTALS OF SIGNAL TRACKING THEORY 275

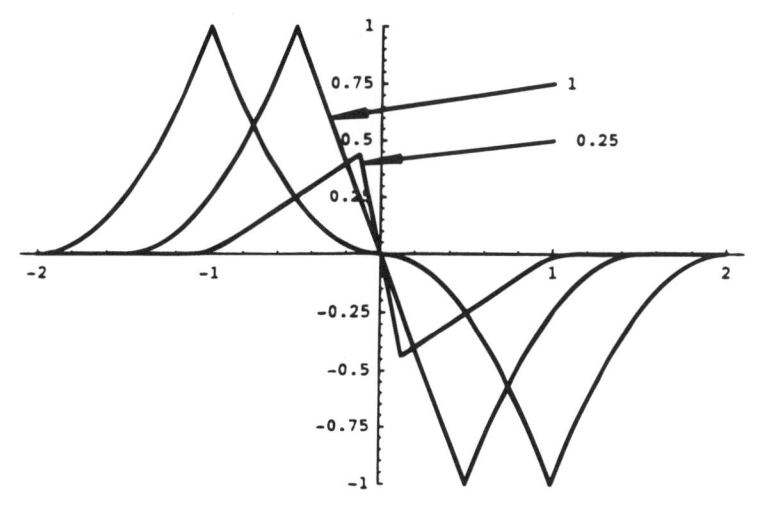

Fig. 21 Discriminator characteristic of a noncoherent DLL with an envelope correlator (square-law detector) for normalized code separations Δ/T of 2, 1, 0.25, plotted as ϵ/T. A zero rise-time received signal is assumed. The spacing $\Delta/T = 2$ gives zero slope at $\epsilon = 0$ and thus gives undesirable performance.

power in each of the bandpass filter outputs is proportional to $A^2 R^2(\Delta/2)$ and as Δ decreases, the signal component power increases. For example, if $\Delta = T$, the squared autocorrelation $R^2(T/2) = 1/4$; whereas, for $\Delta = T/4$, it increases to $R^2(\Delta/8) = 0.766$, a 5.4 dB improvement.

2. Noise Performance of the Noncoherent Delay Lock Loop

The noise performance of the noncoherent DLL with an early-late spacing of $\Delta = T$, and a nearly rectangular input waveform is calculated in Appendix C. For an input-received carrier-to-noise ratio* C/N_o, the rms random noise-induced delay error σ_ϵ produced by the noncoherent DLL of Fig. 20 with a delay separation of $\Delta = T$, a closed-loop noise bandwidth B_n and IF bandwidth ω is given by the following:

$$\frac{\sigma_\epsilon}{T} = \left[\frac{B_n}{2C/N_o}\left(1 + \frac{2W}{C/N_o}\right)\right]^{1/2} = \left[\frac{1}{2(\text{SNR})}\left(1 + \frac{2}{(\text{SNR})_{\text{IF}}}\right)\right]^{1/2} \quad (24)$$

where the tracking bandwidth signal-to-noise ratio (SNR) = $C/N_o B_n$, and the IF $(\text{SNR})_{\text{IF}} = C/N_o W$. The second term in Eq. (16), the $1/(\text{SNR})_{\text{IF}}$ term, represents square-law small signal suppression effect of the noise and becomes important if $(\text{SNR})_{\text{IF}} \leq 1$. If the SNR_{IF} in the IF is > 5, then the square-law effect, noise · noise term, is small, and the rms random delay error is approximately $\sigma_{\epsilon/T} \cong [B_n/(2C/N_o)]^{1/2}$. In this situation, the degradation of the noise performance caused by the square-law nonlinearity is small, and the performance is the same as that for the coherent DLL.

---

*The carrier power C is synonymous to signal power $P_s = A^2/2$.

Figure 22 shows the variation in σ_ϵ/T vs C/N_o for various values of IF bandwidth W. For the DLL discriminator function to remain in the quasilinear mode, the total noise plus maximum transient delay error ϵ_{max} must be small compared to T; e.g., $2\sigma_\epsilon + \epsilon_{max} < 0.5T$. That is, if the total rms value of transient plus thermal noise delay error reaches the region of $3\Delta/2$, the limit of the DLL discriminator curve, the loop has a moderate probability of losing its "locked-on" state, and threshold occurs. Threshold effects are discussed more precisely in a later section.

3. Noise Performance for the Noncoherent Delay Lock Loop vs Early-Late Spacing Δ

The previous paragraph showed the performance for an early-late spacing of $\Delta = T$, where T is the chip duration and the reference is a zero rise-time waveform. It is easy to show that for a waveform with a rise-time $\delta T < T$, noise performance can be improved by using a smaller early-late spacing. Appendix C computes the delay noise variance σ_ϵ^2 of the output of the noncoherent DLL as a function of the early-late spacing Δ for trapezoidal wave shape PN sequences with rise time δT.

Appendix C shows that the output of the noncoherent DLL has a noise variance, σ_n^2, which is closely related to the delay error variance σ_ϵ^2 as described below. The variance σ_n^2 contains both signal · noise and noise · noise terms because of the square-law nonlinearity.

$$\sigma_n^2 = 2A^2 R_s^2(\Delta/2) N_o B_n(\Delta/2T) + 2N_o^2 W B_n(\Delta/2T)(1 - \Delta/2T) \qquad (25)$$

where the $(\Delta/2T)$ and $(1 - \Delta/2T)$ terms appear because of the effective time-gating of the noise by the reference waveform. The second term in Eq. (25) is the noise · noise term and is significant if the IF signal-to-noise ratio (which is approximately $C/N_o W$), falls to unity or below.

The key performance measure is the mean-square delay error σ_ϵ^2, where σ_ϵ^2 is related to σ_n^2 in Eq. (25) by the following:

$$\left(\frac{\sigma_\epsilon}{T}\right)^2 = \frac{\sigma_n^2}{\left[\frac{C}{2} D_s'(0) T\right]^2} \qquad (26a)$$

The discriminator slope at $\epsilon = 0$; namely, $D_s'(0)$, is plotted in Fig. 23 vs the early-late spacing Δ. The rise-time of the signal used in Fig. 23 assumes that $\delta T/T = 0.25$. As shown, the value of $D_s'(0)$ attains its maximum magnitude (negative) at an early-late spacing of $\Delta = \delta T$, the spacing equal to the rise-time. If $\Delta = \delta T$ the optimal setting for noise performance, then the slope of the discriminator function is

$$D_s'(0) = \frac{-4}{T}\left(1 - \frac{\delta T}{2T}\right)$$

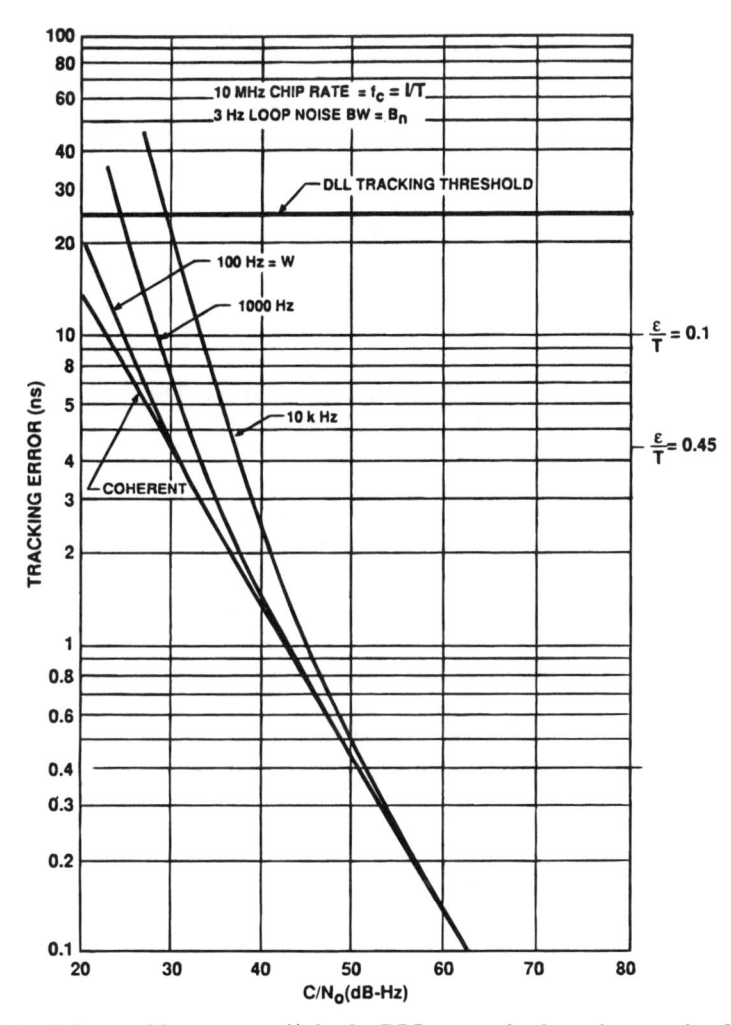

Fig. 22 Noise-tracking error σ_ϵ/Δ in the DLL vs received carrier-to-noise density for a noise bandwidth of $B_n = 3$ Hz and various values of $B_{IF} = W$. Note the small signal suppression effects caused by IF nonlinearities (the square-law detection in the IF) if the SNR in the IF goes to unity or below. For a 1 kHz IF bandwidth, the $(SNR)_{IF}$ is unity at a $C/N_o = 30$ dB.

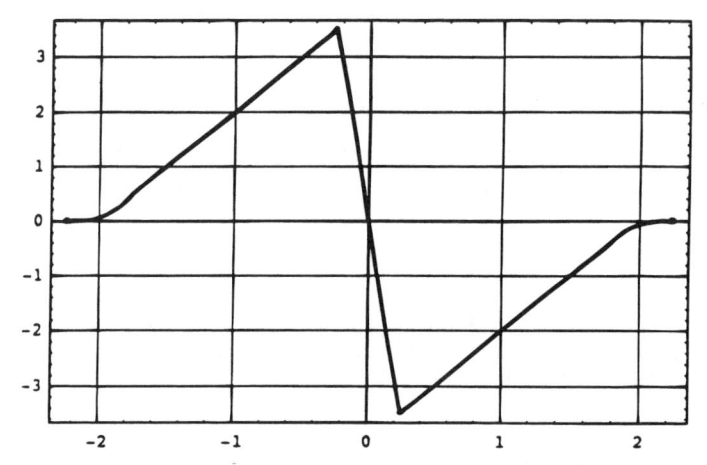

Fig. 23 Slope of the noncoherent DLL discriminator curve $D_s'(0)$ plotted vs normalized early-late delay spacing Δ/T for a trapezoidal PN waveform with normalized rise-time $\delta T/T = 0.25$. Note that the slope decreases to zero at $\Delta = 0$. The slope has its maximum amplitude at $\Delta = \delta T$.

The normalized delay error variance is then as follows;

$$\left(\frac{\sigma_\epsilon}{T}\right)^2 \cong \frac{N_o B_n}{2C}\left(\frac{\delta T}{T}\right) \quad \text{for large IF signal-to-noise ratio} \quad (26b)$$

That is, the noise variance is reduced relative to that for the $\Delta = T$ early-late separation by the duty factor $\delta T/2T$ just as shown for the coherent DLL in Eq. (25b). For an arbitrary value of early-late delay spacing and high IF signal-to-noise ratio, the normalized delay error variance is as follows;

$$\frac{1}{N_o B_n}\left(\frac{\sigma_\epsilon}{T}\right)^2 \cong \frac{2A^2 R_s^2(\Delta/2)(\Delta/2T)}{\left[\frac{A^2}{4}D_s'(0)T\right]^2} = \frac{2A^2 R_s^2(\Delta/2)(\Delta/2T)}{\left[\frac{A^2}{4}4R_s(\Delta/2)R_s'(\Delta/2)\right]^2}$$

$$\cong \frac{2(\Delta/2T)}{A^2[R_s'(\Delta/2)]^2 T^2} = \frac{2}{A^2 T^2(2T}\frac{\Delta}{[R_s'(\Delta/2)]^2} \quad (26c)$$

Thus, the normalized noise-induced delay error variance is proportional to $\Delta/[R_s'(\Delta/2)]^2$ for high IF signal-to-noise ratio. Figure 24 shows the variation of $\Delta/[R_s'(\Delta/2)]^2$ with Δ for $\delta T/T = 0.25$. As shown, this quantity is minimized at a value of $\Delta = \delta T$. Clearly, too small a value of Δ leads to a large noise error and a larger delay spacing; e.g., $\Delta = T$ increases the normalized delay error variance by a factor of approximately 4 for a 2-to-1 increase in σ_ϵ.

The value of σ_ϵ/T at the optimum setting is approximated by the following:

$$\left(\frac{\sigma_\epsilon}{T}\right)^2 \frac{1}{N_o B_n} = \frac{(\Delta/2T)}{A^2[R_s'(\Delta/2)]^2 T^2} \quad (26d)$$

For conditions of very short rise-time $\delta T \ll T$, we must also consider the effect

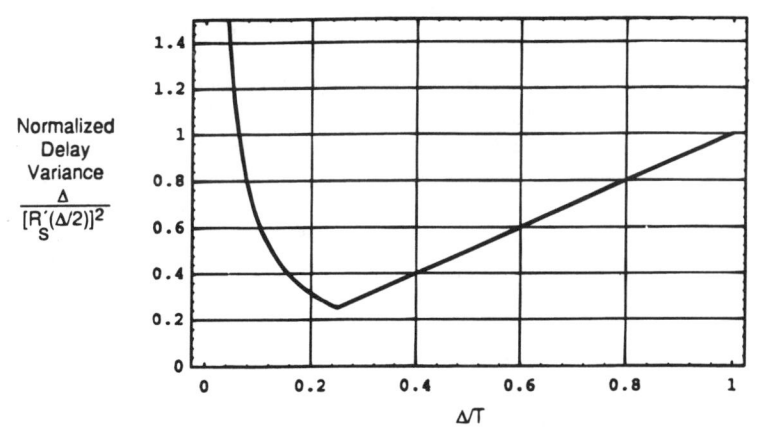

Fig. 24 Normalized variation in delay error noise variance vs normalized early-late delay spacing Δ/T for a normalized rise-time $\delta T/T = 0.25$. Note that the minimum delay error variance occurs at $\Delta/T = \delta T/T = 0.25$.

of very small Δ on the transient performance of the DLL. As previously pointed out, the required loop bandwidth is inversely related to the square root of steady-state transient error caused by a step of acceleration; i.e., $B_n \sim 1/\sqrt{\epsilon_{ss}}$. Thus, if the transient error must be reduced to a small fraction of Δ for quasilinear operation of the DLL, then one of the constraints on the noise bandwidth must increase with decreasing Δ, and the noise-induced tracking error may increase also.

4. Quasioptimal Noncoherent Delay Lock Loop

Further improvement can be obtained by using as the reference signal a punctual signal plus or minus the differentiated signal; namely,

$$s(t + \hat{\tau}) \pm s'(t + \hat{\tau}) \pm s'(t + \hat{\tau}).$$

The two correlator outputs then have expected values

$$E\{s(t + \tau)[s(t + \hat{\tau}) + s'(t + \hat{\tau})]\} = R(\epsilon) + R'(\epsilon),$$

and

$$E\{s(t + \tau)[s(t + \hat{\tau}) + s'(t + \hat{\tau})]\} = R(\epsilon) - R'(\epsilon).$$

Squaring these two terms and taking the difference yields

$$[R(\epsilon) + R'(\epsilon)]^2 - [R(\epsilon) + R'(\epsilon)]^2 = 4R(\epsilon)R'(\epsilon) = D(\epsilon).$$

The advantage is that the first component, the punctual signal, is matched to the received signal at the estimated value of delay. This is the same DLL discriminator curve we find later in the quasicoherent DLL in Sec. II.C.

5. Probability of Losing Lock for the Noncoherent Delay Lock Loop

A more precise estimate of the threshold effect caused by thermal noise can be computed by analyzing the stochastic differential equation of a first-order noncoherent DLL.[14,19,20] A brief summary of the derivation is given in Appendix D.

Figure 25 shows the normalized mean time to lose lock $\bar{T}$ in the DLL (for a noncoherent second-order DLL) vs rms delay jitter.[19] An early-late spacing of $\Delta = T$ has been assumed for the DLL. Radio frequency filtering of bandwidth $2/T$ has been assumed that slightly rounds the discriminator characteristic. Note that the time $\bar{T}$ is shown on a logarithmic scale. As can be seen from the graph, the mean time to lose lock decreases greatly as $\sigma_\epsilon/T > 0.25$ (where $\log_{10} 2B_n\bar{T} \cong 3$). This result is consistent with the previous estimate of threshold where we approximated $2\sigma_\epsilon = T/2$.

C. Quasicoherent Delay Lock Loop

The noncoherent DLL of the previous section has at least two disadvantages.

1) There is a noise-squared component that degrades performance if the IF signal-to-noise ratio is below unity: the so-called squaring loss.

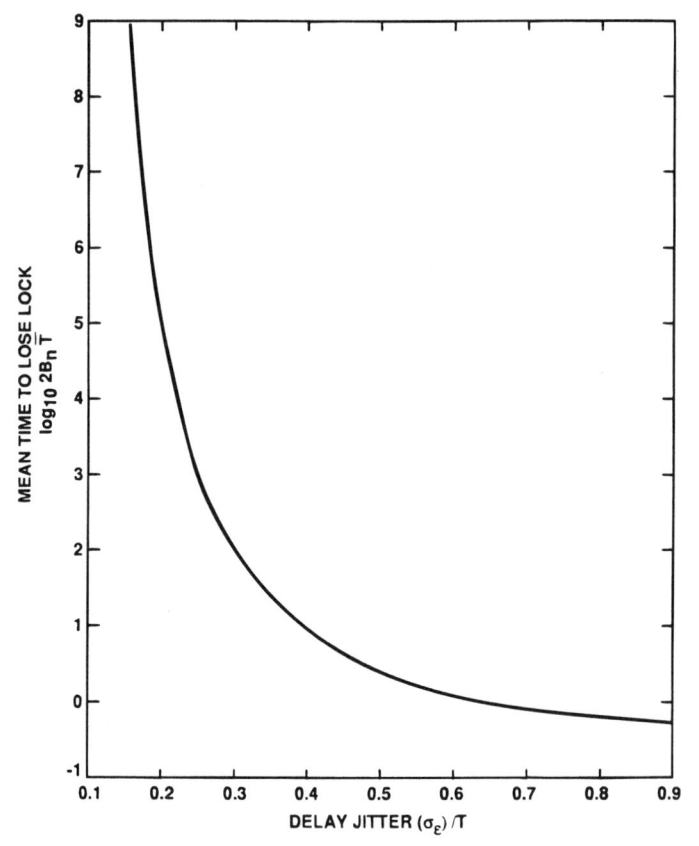

Fig. 25 Normalized mean slip time $\bar{T}$ vs rms delay jitter σ_ϵ/T for a second-order noncoherent DLL (plot of computed points from Ref. 20).

FUNDAMENTALS OF SIGNAL TRACKING THEORY

2) The early and late channel outputs of the IF filter/square-law device are not at a cross-correlation peak. Thus, the useful signal power component is not as high as it could be, although that effect is minimized if the delay spacing is decreased significantly below $\Delta = T$.

The quasicoherent delay lock loop[35] (QCDLL) of Fig. 26 improves noise performance and eliminates the second of these disadvantages. Furthermore, it does so with no more complexity than in the noncoherent DLL, eliminates possible problems with gain offsets in the early-late channels, and improves the noise performance. This QCDLL* has two correlators. The punctual upper channel correlator is intended to provide an estimate of the binary data waveform (without making hard decisions) and to remove residual Doppler shift. The punctual channel also provides the needed data demodulation and lock detector functions for the receiver. The reference for the upper correlator is the punctual signal $s(t + \hat{\tau})$ where $\hat{\tau}$ is the best estimate of the actual delay. The bandpass filter has a bandwidth somewhat greater than the data signal to allow the data to pass relatively undistorted and tolerate residual Doppler shift. It has a fixed delay T_o. The tracking channel multiplier output likewise is filtered to an identical bandwidth and delay so that the data-modulated tracking error term $m(t)$ can pass through relatively undistorted.

The product of the two waveforms after the bandpass filters, then, effectively removes the data modulation (because $d^2(t) = 1$) and residual Doppler. Thus, only the tracking error remains, plus the additive noise. Note also that the noise terms in the two channels are essentially uncorrelated, because they are generated by different reference functions s and s' and, as shown later, do not produce the square-law noise effects.† The product of n_0 and n_2, however, does have a noise · noise power effect. The upper channel output can also be applied to a bandpass filter (low-pass filter), square-law device and threshold detector to give an indication of the "locked on" condition. "Locked on" means that the delay error is sufficiently small and remains small; i.e., $\epsilon(t) < T$. The bandpass filter in the

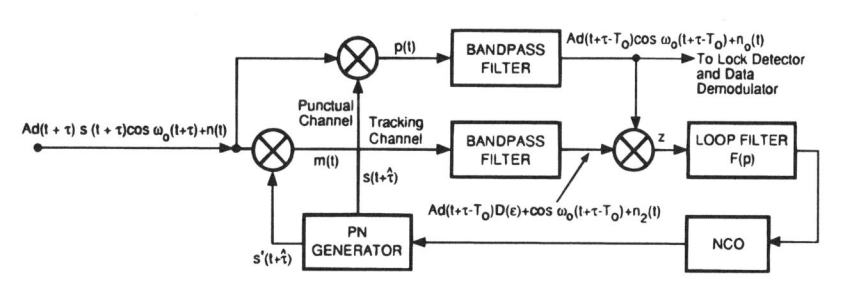

Fig. 26 Carrier- and data-modulated signal and a quasicoherent DLL.

---

*A variation of the QCDLL was shown in Ref. 15. It is termed the modified code tracking loop (MCTL) and employs an early-late reference signal with a delay spacing of one chip.

†The autocorrelation function of squared Gaussian noise $n^2(t)$ has autocorrelation function $R_n^2(\tau)$, which is $R_n^2(\tau) = R_n^2(0) + 2 R_n^2(\tau)$, where $R_n(\tau)$ is the autocorrelation function of the noise input to the square-law device. Note the factor of 2 in front of the second term. On the other hand, the product of two independent Gaussian noise terms $n_0(t)$ and $n_2(t)$, has an autocorrelation function $R_{n0n2}(\tau) = R_{n0}(\tau) R_{n2}(\tau)$.

punctual channel can be a bandpass finite memory integrator of memory T_o, where T_o is the period of one databit, and $T_o \gg T$. The output of the multiplier in the punctual channel is as follows:

$$\begin{aligned} p(t) &= [Ad(t + \tau)s(t + \tau)\cos(\omega_o t + \tau) + n(t)]s(t + \hat{\tau}) \\ &= Ad(t + \tau)s(t + \tau)\cos(\omega_o t + \tau)s[(t + \hat{\tau}) + n_o(t)] \\ &= Ad(t + \tau)R(\epsilon)\cos(\omega_o t + \tau) + [n_o(t) + Ad(t + \tau)n_{sn}(t)\cos(\omega_o t + \tau)] \end{aligned} \quad (27)$$

where the $R(\epsilon)$ is the autocorrelation function of the PN signal $s(t)$. Again, we have approximated the output by its expected value plus a self-noise term $n_{sn}(t)$. The filtered output of the punctual channel is approximated by
$p(t) \cong Ad(t + \tau - T_o)R(\epsilon)\cos(\omega_o t + \tau + \omega_o T_o) + n_o(t)|_{\text{bandpass}}$. The bandpass filter removes much of the thermal noise and self-noise

$$n_{sn}(t) = [s(t + \tau)s(t + \hat{\tau}) - R(\epsilon)],$$

and thus, the filter output can be represented by

$$R(\epsilon) d(t - \tau - T_o) + n_p(t).$$

The output of the first multiplier in the lower tracking channel of Fig. 26 is represented by the following:

$$m(t) = [Ad(t + \tau)s(t + \tau)\cos(\omega_o t + \tau) + n(t)]s'(t + \hat{\tau}) \quad (28)$$

Delay is introduced in the lower channel to match the delay and cancel the phase shift of the upper channel filter. Both of these operations, as well as the correlators, can be implemented digitally, and the two bandpass filters (implemented in in-phase/quadrature form) made identical. The filtered output of the lower channel is approximately $d(t + \tau - T_o)R'(\epsilon) + n_s(t)$. The output $z(t) + n_m(t)$ of the second multiplier has effectively removed the data, but it has added some additional noise caused by additive noise in the punctual channel not removed by the data filter.

Note that the closed-loop DLL filter response has also been affected by the delay T_o, and this effect must be included in calculating the closed-loop transfer function $H(p)$ or its discrete time equivalent. The closed-loop bandwidth must be small compared to the inverse of the delay to prevent instability. The output of the second multiplier in the lower channel is then $z(t - T_o)$ where the following occurs:

$$\begin{aligned} z(t - T_o) &\cong [Ad(t + \tau - T_o)R'(\epsilon) + n_s(t)][Ad(t + \tau - T_o)R(\epsilon) + n_p(t)] \\ &\cong A^2 R(\epsilon)R'(\epsilon) + n_p(t)Ad(t + \tau - T_o)R'(\epsilon) \\ &\quad + n_s(t)Ad(t + \tau - T_o)R(\epsilon) + n_s(t)n_p(t) \\ &\cong A^2 R(\epsilon)R'(\epsilon) + n_m(t) \\ &\cong A^2 R''(0)\epsilon(t) + n_m(t) \qquad \text{for } \epsilon \ll T \end{aligned} \quad (29)$$

where $d^2(t) = 1$, because the data $d(t) = \pm 1$, and we assume $R(\epsilon) \cong 1$ for $\epsilon \ll T$. The equivalent noise is

$$n_m(t) = Ad(t + \tau - T_o)[R'(\epsilon)n_p(t) + R(\epsilon)n_s(t)] + n_s(t)n_p(t),$$

and $n_s(t)$ is filtered time-gated thermal noise because of the multiplication by $s'(t)$. This QCDLL is a generalization of the Costas loop where for a pure sine wave signal $\cos\omega_o t$, the differentiated reference is $s'(t) = -\omega_o \sin\omega_o t$. Both of these loops operate much as the coherent DLLs at high signal-to-noise ratios but degrade because of the noise · noise term as the signal-to-noise ratio in the bandpass filter declines toward unity. The primary advantage of the QCDLL of Fig. 26 over the noncoherent loop is that the punctual channel generally has a higher crosscorrelation than either the early or late channels. This advantage yields a relatively smaller noise · noise effect and generally better performance. Furthermore, the hardware for the punctual channel is already required for data demodulation, so there is no hardware penalty in implementing the QCDLL. Furthermore, it also has improved multipath performance.

1. Quasicoherent Delay Lock Loop Discriminator Characteristic

As an example, assume that the received signal is a finite rise-time trapezoidal PN signal with a rise-time δT, as shown earlier in Fig. 9. The output of the multiplier from Eq. (29) is $z(t - T_o) = A^2 R(\epsilon)R'(\epsilon) + n_m(t)$. For simplicity in implementation, assume that the reference signal is $s'(t + \hat{\tau})(1/\delta T)$ so that the peak amplitude is $s'(t + \hat{\tau}) = \pm 1$ and is zero otherwise. The output is then $2z/\delta T$, and the effective multiplier output for a normalized signal level $A = 1$ is then $z(t - T_o)/\delta T = R(\epsilon)R'(\epsilon)/\delta T + n_m(t)/\delta T = -D(\epsilon) + n_m/\delta T$, where $D(\epsilon)$ is the discriminator characteristic.

Figure 27 shows a plot of $D(\epsilon)$ for several values of δT, namely, $\delta T/T = 0.125, 0.25, 0.5, 1.0$. Note that the noise power output for this system decreases in proportion to the duty factor $\delta T/2T$ of the reference waveform $s'(t)/\delta T$. The slope

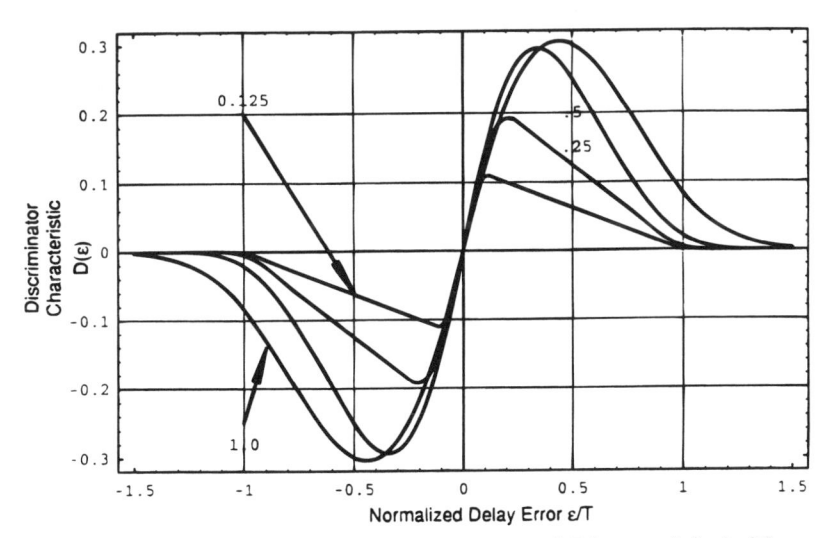

Fig. 27 Discriminator characteristics $D(\epsilon) = -R(\epsilon)R'(\epsilon)/\delta T$ (reversed sign) of the near optimum QCDLL for various values of PN pulse rise-time $\delta T/T$ of **0.125, 0.25, 0.5, 1.0**. The tracking channel reference is a pulse "doublet" of unit amplitude and width δT. Note that all slopes at the origin are equal.

of the discriminator characteristic $D'(0)$ at $t = 0$, on the other hand, remains constant. Thus, the output signal-to-noise improves directly in proportion to $2T/\delta T$.

Note that the peak amplitude of the discriminator decreases as the value of δT decreases. Thus, during the initial acquisition and pull-in operation, we can increase the loop gain, an action that increases the maximum correction voltage that drives the NCO. Thus, during acquisition, the loop gain can be increased, and as the delay estimate error ϵ decreases, the loop gain can be decreased in order to adjust the closed-loop bandwidth to its optimum value.

It should also be pointed out that the QCDLL generalizes to track simultaneously both in-phase and quadrature codes of quadrature-modulated PN signals such as GPS, where the in-phase and quadrature carrier components are modulated by separate PN codes, C/A and P.

D. Coherent Code/Carrier Delay Lock Loop

In GPS, the PN signal is not only data modulated and modulated on a carrier, but the carrier frequency is coherent with the code clock rate and delayed by the pseudorange in the same manner. The quasioptimal DLL then must track a combination of both the PN code phase and the higher-frequency carrier phase. The received signal component in this example is $S(t + \tau) = s[t + \tau(t)] \cos \omega_o[(t + \tau(t) + \phi]$. Both the PN code phase and the carrier phase are coherently related; i.e., $s(t)$ has a PN chip rate $f_c = 1/T$, and the carrier frequency $f_o = \omega_o/2\pi = M(1/T)$ is an integral multiple of the code chip rate f_c. For GPS, $M = 154$ for the L_1 carrier relative to 10.23 MHz P code and $M = 120$ for the L_2 carrier. This relationship between code and carrier delay is perturbed slightly by changes in the ionospheric delay, and we must also take this effect into account. Excess code delay and carrier phase delay are equal and opposite effects of the ionosphere (See Chapters 4 and 12, this volume). For the moment, however, this effect is ignored.

The optimal reference waveform for this signal is the differentiated signal $S'(t)$ where the following result applies:

$$S'(t + \tau) \triangleq \frac{\partial S}{\partial \tau} = s'(t + \tau)\cos \omega_o(t + \tau) - \omega_o s(t + \tau)\sin \omega_o(t + \tau) \quad (30)$$

Thus, the differentiated reference S' contains two terms and requires two separate cross-correlation operations, which are then weighted and summed together in accord with expression (30).

The product of the received signal component S with the reference waveform times 2 yields the following baseband output:

$$S(t + \tau)2S'(t + \hat{\tau}) = s(t + \tau)s'(t + \hat{\tau})\cos(\omega_o\epsilon) - \omega_o s(t + \tau)s(t + \hat{\tau})\sin \omega_o\epsilon \quad (31)$$

where the $2\omega_o$ terms have been removed by low-pass filtering. The expected value of this product waveform for fixed ϵ is as followed:

$$D(\epsilon) = R'(\epsilon)\cos \omega_o\epsilon - \omega_o R(\epsilon)\sin \omega_o\epsilon \quad (32)$$

where $R(\epsilon) = E[s(t)s(t + \epsilon)]$ and $R(0) = 1$.

FUNDAMENTALS OF SIGNAL TRACKING THEORY

Because the value of ω_o is larger than $R'(\epsilon)$ for any bandlimited radio frequency signal the discriminator characteristic has a shape similar to that shown in Fig. 28. For GPS, however, where $M = 154$ for the L_1 P-code, there would be many more sinusoidal cycles within the envelope than that shown in Fig. 28. Use a Taylor's series expansion of $s(t)$, and assume that ϵ is sufficiently small. The expected value of the product then yields the following:

$$\begin{aligned} D(\epsilon) &= E[S(t + \tau)2S'(t + \hat{\tau})] \\ &\cong \epsilon R''(0) + \omega_o^2 \epsilon \\ &= (R''(0) + \omega_o^2)\epsilon \quad \text{for small} \quad \epsilon \end{aligned} \quad (33a)$$

and for a trapezoidal PN waveform with rise-time δT, the following obtains:

$$D(\epsilon) \cong \epsilon\left[\left(\frac{2}{T\delta T}\right) + \omega_o^2\right] \quad \text{for small} \quad \epsilon \quad (33b)$$

where $R''(0) = -P_d$, and P_d is the power in the differentiated code signal $s'(t)$,

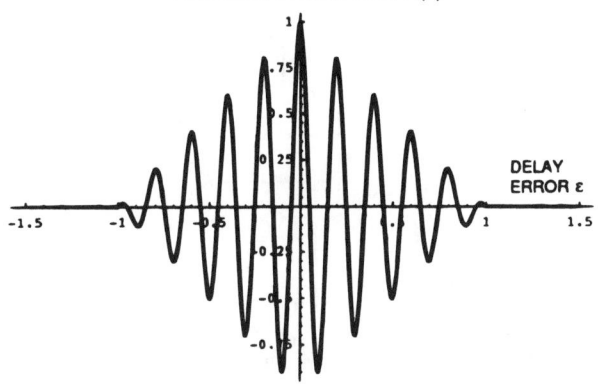

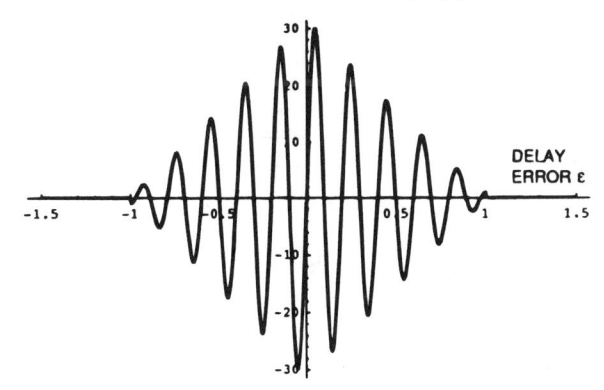

Fig. 28 Autocorrelation function $R(\epsilon)$ and DLL discriminator characteristic $D(\epsilon)$ for a signal $s(t)\cos\omega_o t$. Because $\omega_o \gg -R''(0)$ in this example

$$D(\epsilon) \cong -\omega_o R(\epsilon) \sin \omega_o \epsilon.$$

and the power of $s(t)$ is $R(0) = 1$. Recall that for a unit amplitude trapezoidal signal with rise-time δT and chip interval T, the power in the differentiated signal is $P_d = 2/(T\delta T)$. Because the carrier frequency is generally high compared to the slope of the PN signal; i.e., $\omega_o^2 \gg P_d = 2/(T\delta T)$, the ω_o^2 term dominates.

The quasioptimal tracking loop for this signal is as shown in Fig. 29. The quasioptimal DLL can be viewed as providing an optimally weighted summation of the code DLL and the carrier loop (phase lock loop) correlators. The code loop is shown in a dashed box in Fig. 29. The tracking and control loop weights the output of the two loop filters shown. Because the carrier is tracked, the code loop operates with a coherent carrier reference. In principle, except for ionospheric delay changes, there need be only one loop filter operating on the weighted signal discriminator curve once the system has achieved lock.* Because the code clock and carrier are related by a factor M, a single NCO can control both. A phase stepper can be used to step the carrier in integer phase cycles to attempt to track code phase at the optimal (peak) value of the autocorrelation $R(\epsilon)$. For this loop to operate correctly, the carrier phase error must be small compared to $\pi/2$ radians.

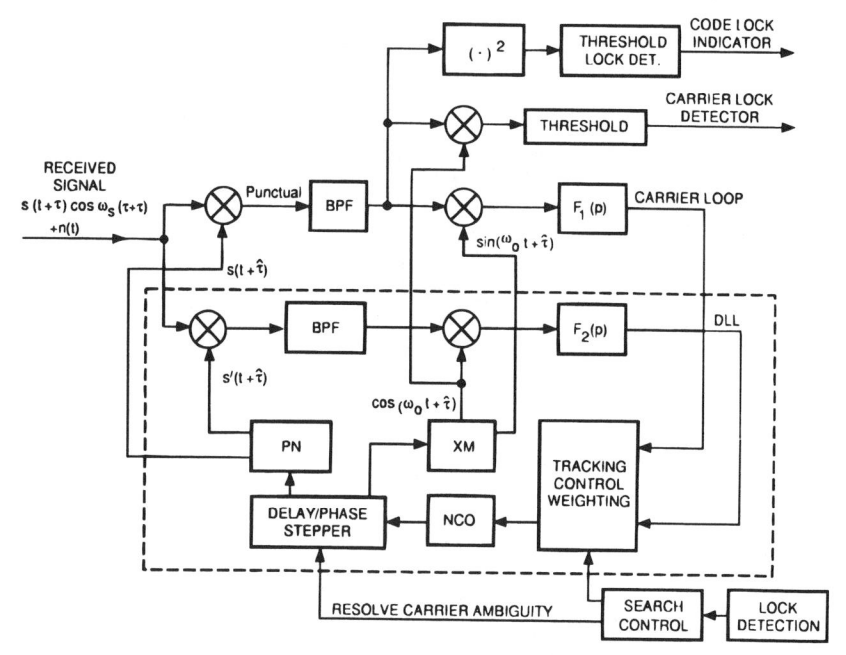

Fig. 29 Quasioptimal DLL for a PN signal modulation on a carrier $s(t) \cos \omega_o t$ where the PN code clock and carrier frequency are coherently related.

---

*As time passes, changes in the ionosphere can cause the carrier and code loops to diverge, and the code loop must control the tracking, or code correlation can be lost. Thus, we cannot simply track the carrier and expect to remain at the minimum of the code discriminator characteristic over a long time period because the ionosphere delay changes with time. However, this delay varies slowly with time (see Chapter 12, this volume), and carrier tracking still is very useful.

Clearly, there are many ambiguities in such a loop if $\omega_o/2\pi \gg 1/T$, as is true for GPS, and the autocorrelation function and differentiated autocorrelation function generally have many more ambiguities than shown in Fig. 28. The number of ambiguities approximates $M/2$, and for GPS, complete ambiguity resolution is often not possible. In general, there would be many carrier cycles at spacing $1/f_o$ within the autocorrelation function envelope (which has width $\approx 2T$). Thus, the DLL can lock on at any of the ambiguous negative-going (or positive-going, depending on the sign of the loop gain) zero crossings, of which there can be several depending on the ratio of $\omega_o/2\pi$ to $1/T$. Thus, carrier ambiguity resolution, as discussed above, is important if the full accuracy is to be obtained. Otherwise, the carrier tracking is simply considered as aiding the code-tracking loop and provides range rate and ADR measurements. However, it is possible to achieve accuracies on the order of 1% or better for the code, with or without carrier aiding. Higher accuracies can be achievable using the delay separation $\Delta = \delta T$, as discussed earlier, for the quasioptimal DLL. As mentioned earlier, even the P code has a rise-time δT, which is significantly less than the chip interval T. Thus, full ambiguity resolution may be possible under some circumstances for the P code, if the full signal bandwidth is utilized and due account is taken of the ionospheric effects.

Initial search and acquisition of the code $s(t)$ can occur by sequencing the code states until a noncoherent detection operation finds the approximately correct code delay. Carrier lock can then be found by sweeping the carrier phase. Carrier tracking can commence independent of carrier ambiguity resolution. Fine tuning of the code to resolve carrier ambiguity can then occur by incrementally slipping or adding cycles of the carrier phase, creating much finer jumps in the code phase so as to maximize the filtered output of a coherent punctual channel or to minimize the error in the tracking channel. The code tracking can be accomplished in the code loop by digitally stepping the carrier phase in incremental 360 deg phase steps to minimize the code tracking error. Again, we must use caution to avoid carrier cycle slips, a serious limitation of this technique.

E. Carrier-Aided Pseudorange Tracking

The carrier-tracking operation with phase-locked loops can also be employed as an aid to code tracking. The following measured phase shift

$$\phi(t) = \omega_o[\tau(t) - \tau_I(t)] + \phi_o + \phi_n(t) \tag{34}$$

has an unknown initial phase ϕ_o, phase noise $\phi_n(t)$; $\tau(t)$ is the range delay; and τ_I is the ionospheric delay. The measured pseudorange group delay is as follows:

$$\rho(t) = \tau(t) + \tau_I(t) + n_\epsilon(t) \tag{35}$$

where $n_\epsilon(t)$ is the measured pseudorange tracking noise where the τ_I effects are equal and opposite on carrier and code delay. Figure 30 shows simulated scaled phase delay $\phi(t)/\omega_o$ and pseudorange $\rho(t)$. As shown and already discussed for the optimal DLL, the noise in the phase measurement of delay is generally much smaller than that of the pseudorange by the ratio $[2/(T\delta T)]/\omega_o^2 = 2/[\omega_o^2 T\delta T]$. However, the phase variable has an obvious uncertainty in its initial value (unless the ambiguity can be removed), the initial value has been removed by simple

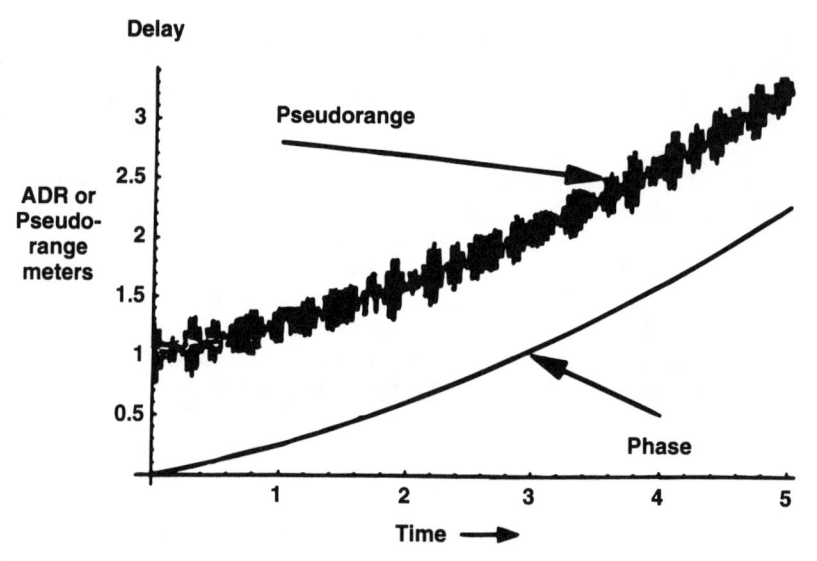

Fig. 30 Example of rms noise on carrier phase and code where both converted to meters (not to scale). Note that the carrier phase only has meaning in a differential sense and is usually termed accumulated delta range or ADR. Initial carrier phase measurements have an ambiguity in the number of phase cycles and initial phase, and here the initial phase is set to zero. Both code and carrier measurements have exactly the same shape, except for differences in the noise, a bias offset, and ionospheric delay charger.

differencing. With the exception of the bias offset and the ionospheric delay changes with time, the pseudorange delay and phase delay curves have exactly the same shape.

Thus, we can subtract the scaled phase difference variable $[\phi(t) - \phi(0)]/\omega_o$ from the code group delay $\rho(t)$ to yield a delay difference $\Delta(t)$. This difference between code pseudorange and normalized carrier phase delay from Eqs. (34) and (35) is as follows:

$$\Delta(t) = \rho(t) - \left[\frac{\phi(t) - \phi(0)}{\omega_0}\right] = 2\tau_I(t) + n_\epsilon(t) - \left(\frac{\phi_n(t) + \phi_0}{\omega_0}\right)$$

$$+ [\tau(0) - \tau_I(0)] + \left(\frac{\phi_n(0) + \phi_0}{\omega_0}\right)$$

$$\Delta(t) = \tau(0) + n_\epsilon(t) - \left(\frac{\phi_n(t) - \phi_n(0)}{\omega_0}\right) + 2\tau_I(t) - \tau_I(0) \qquad (36a)$$

If there is no change in the ionosphere delay; i.e., $\tau_I(t) = \tau_I(0)$, and the phase noise is negligible, then this delay difference is approximated by the following:

$$\Delta(t) \cong \tau(0) + \tau_I(0) + n_\epsilon(t) \qquad (36b)$$

FUNDAMENTALS OF SIGNAL TRACKING THEORY

The difference $\Delta(t)$ can be averaged over time to estimate $\tau(0) + \tau_I(0)$ provided $\tau_I(t)$ does not change significantly. If we average $\Delta(t)$ over a relatively long period, a good estimate of the initial value of $\tau(0) + \tau_I(0)$ can be obtained. The only limit is the potential for cycle slips in the phase measurement and ionospheric delay changes. The major effects of the ionosphere are somewhat diurnal, and thus, change very slowly (see Chapters 4 and 12, this volume), and hence, permit rather long averaging times. Some or all of these ionospheric effects can be compensated or measured by alternate means. Thus, the initial value of the delay $\tau(0)$ can be estimated accurately.

1. Simultaneous Measurement of L_1 and L_2 Carriers

Simultaneous reception of the L_1 and L_2 carriers can be used to aid in resolving carrier ambiguities as well as in the measurement of the ionosphere. This approach to ambiguity resolution is similar to that employed in a sidetone ranging signal. In a sidetone ranging system, we transmit a parallel set of tones, as follows:

$$\sum \sin[(\omega_0 + \Delta\omega_i)t + \phi_0] \tag{37}$$

where the offsets $\Delta\omega_i$ are used to resolve the carrier phase ambiguities in the largest $\Delta\omega_i$ frequency.[3] Likewise, a small enough carrier frequency offset can be used to aid in the resolution of the ambiguity of the carrier phase $\omega_0 t$ itself. For example, the L_1-L_2 frequency separation is

$$\Delta f = 1575.42 \text{ MHz} - 1227.6 \text{ MHz} = 347.82 \text{ MHz},$$

which is 22.08% of the L_1 carrier frequency. Thus, there is a ratio of approximately 4.53 between L_1 and Δf. The wavelengths of the various components in GPS are shown in Table 2. In addition, we should note that if the P-code rise-time

$$\delta T = 0.2T, \text{ then } c\delta T = 586.52 \text{ cm}.$$

The ratio of the L_1-L_2 frequency difference, 347.82 MHz, to 10.23 MHz is exactly 34 to 1. If we can resolve the P-code group delay to better then 1/34 or 3% of a chip width, then we can resolve the ambiguities in the phase offset between L_1 and L_2. (We must, of course, properly account for the offset between L_1 and L_2 caused by the ionosphere in the process.) This "wide-laning" technique is discussed in Chapter 18 in the companion volume.

Table 2 Wavelength of various components in the GPS signal

Signal	Wavelength
L_1 carrier wavelength	$c/L_1 = 19.04$ cm
$\Delta f = L_1-L_2$ wavelength	$c/(L_1 - L_2) = 86.25$ cm
P-code wavelength	$c/10.23$ MHz $= 2932.6$ cm
C/A-code wavelength	$c/1.023$ MHz $= 29326$ cm

III. Vector Delay Lock Loop Processing of GPS Signals

A conventional GPS receiver consists of several parallel scalar DLLs, each of which independently estimates the individual pseudoranges.* The parallel set of measured pseudoranges (plus Doppler or ADR measurements on the carrier) are then fed to a Kalman filter estimator, as shown in Fig. 31. The reader is referred to Chapter 9, this volume and to Brown and Huang[21] for a more detailed discussion of extended Kalman filters for GPS position/time estimation.

However, as shown in this configuration, each DLL effectively is producing an independent estimate for each of the N pseudoranges for each of the N satellites. As discussed earlier in Sec. I, not all of their measurements are truly independent, although they are treated as such by the DLL, and if there are more than four measurements being made and four or fewer unknowns, the system is overdetermined. Furthermore, the geometry of the satellite-user paths generally prevents the measurements from being truly independent.

Some reasons why the system may be overdetermined and pseudoranges correlated are shown in Table 3. Earlier in Sec. I (and in Appendix A) it was shown that the maximum likelihood estimator of position does not estimate the individual signal delays independently and then solve for position, but rather performs both functions in a single vector-processing loop, the vector DLL.

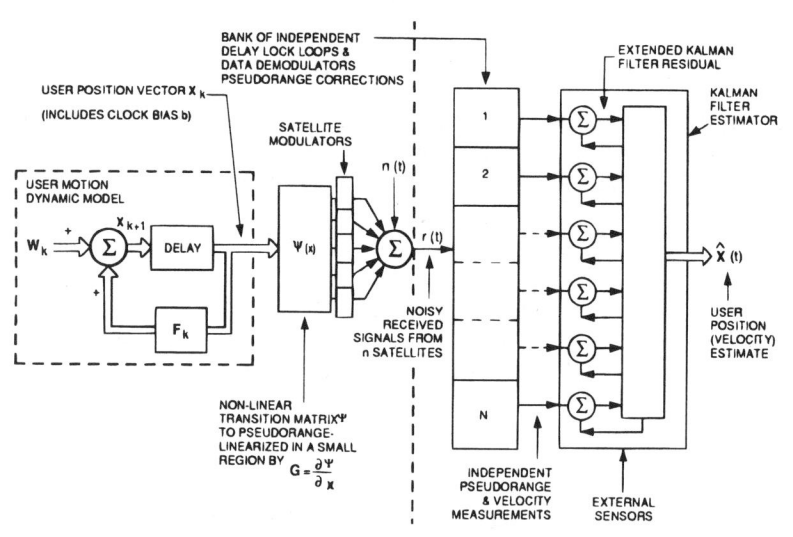

Fig. 31 Conventional receiver configuration. Separated parallel, independent pseudorange measurements are followed by a user position estimator. The DLLs each act independently of the others in estimating pseudorange in this configuration.

---

*The only major exception is when both L_1 and L_2 are estimated for the same satellite. For this problem, often a single L_2 channel is sequenced over several satellites with aiding from the L_1 channel, because the L_1 and L_2 channels for a given satellite differ only by a slowly varying ionospheric delay.

Table 3 Possible causes of redundancy in pseudorange measurements

1) There may be many, $N > 4$, pseudoranges and carrier phases measured (perhaps eight satellite measurements) in parallel; whereas, there are only four unknown user position coordinates (x,y,z,b). Thus, in principle, the total signal power received from all satellites may be adequate to perform the position estimate although no one single satellite may not have enough power to produce an accurate pseudorange estimate. With the use of VLSI digital DLL, as many as $N = 8$–12 GPS satellites can be tracked continuously using a single chip.

2) The user clock bias b may be relatively stable and of much lower dynamics than the x,y,z user coordinates. Thus, there may be a relatively low information rate in the clock phase noise/drift $(b,\dot{b})$ and very little signal power needed to track b accurately.

3) For some users; e.g., on a ship or vehicle on relatively flat land, the vertical coordinate remains approximately constant at sea level or is changing slowly with time.

A. Independent Delay Lock Loops and Kalman Filter

As a preamble to the discussion on the vector delay-lock loop (VDLL) and in order to define the variables, it is useful to review briefly the operation of a conventional system that first estimates the pseudoranges independently and then estimates the position vector. In Fig. 31, the user state vector (position/velocity, clock bias) is defined by the process equation at discrete time t_k, as follows

$$x_{k+1} = F_k x_k + w_k \tag{38}$$

where F_k represents the dynamics of the user platform and clock, and w_k is the Gaussian random driving variable. Consider as an example a user position that varies in a random walk manner (white noise velocity) rather than with abrupt changes in velocity. We can model this position process as a four-dimensional (4-D) user state vector of the following form*:

$$x_k^T = [x_k, y_k, z_k, b_k] \tag{39}$$

where x, y are horizontal positions, z is the altitude, and b is the user clock bias.

The forcing variable w_k samples are independent Gaussian samples with zero mean and covariance $E(w_k^T w_j) = Q_k \delta_{jk}$. The measured (indirectly) observable, the pseudorange vector of N satellite measurements, is expressed by the following

$$\rho(x_k) = \Psi_k(x_k) + n_k \tag{40}$$

where the noise samples are independent Gaussian samples of zero mean and covariance $E[n_k^T n_j] = R_k \delta_{jk}$. (The constraint of independent noise samples (white noise) is removed later.)

Define the estimate $\hat{x}_{k|k-1}$ as the estimate of x_k given the previously observed measurements of $\rho(x_k)$ at times $t_0, t_1, \ldots, t_{k-1}$. Linearize Eq. (40) by expanding $\Psi(x_k)$ in a Taylor's series about the estimated value; namely,

---

*A 4-vector is assumed here for simplicity; more common would be an 8-vector with position, velocity, clock error, and clock drift rate for a user position with white noise acceleration. The reader is referred to Chapter 9, this volume for greater depth.

$$\Psi(x_k) = \Psi(x_{k/k-1}) + G_k \Delta x_k \tag{41}$$

where the Jacobian matrix G_k is

$$G_k = \left[\frac{\partial \Psi(x_k)}{\partial x_k}\right], \text{ and } \Delta x_k = x_k - \hat{x}_{k/k-1}$$

(Note that many texts use H for this matrix; here we use G_k as the "geometric" matrix, in keeping with many GPS references.) Then, by substituting Eq. (41) into Eq. (40), the linearized pseudorange vector can be written as follows:

$$\rho(x_k) = \Psi_k(\hat{x}_{k/k-1}) + G_k \Delta x_k + n_k \tag{42}$$

where it is assumed that $\Psi(x)$ is sufficiently smooth and that each of the satellite signals are separately observable with independent noise. Obviously, this assumption is only an approximation, because the signals are all observed in a multiple access environment, and there is some interference (multiple access noise) between different PN signals. However, they are largely separable because the PN codes have low cross correlation. Each of these signals are then fed to a set of separate parallel DLL receiver systems each operating independently of the others. The linearized Eq. (42) can be used to generate the extended Kalman filter (EKF) estimate of x_k as depicted in Fig. 32. Note that $G_k = G_k(x_k)$ is not constant with time if the geometry changes.

The EKF estimates are then as follows:

$$x_{k/k} = \hat{x}_{k/k-1} + K_k[\rho(x_k) - \Psi_k(\hat{x}_{k/k-1})]$$

and

$$\hat{x}_{k+1/k} = F_k \hat{x}_{k/k} \tag{43a}$$

where K_k is the Kalman gain matrix. These expressions can also be written as follows:[10,11]

$$\hat{x}_{k+1/k} = F_k \hat{x}_{k/k-1} + F_k K_k[\rho(x_k) - \Psi_k(\hat{x}(_{k/k-1}))] \tag{43b}$$

Define $\hat{\rho}_{k/k-1} \triangleq \Psi_k(\hat{x}_{k/k-1})$ where the Kalman gain matrix is as follows:

$$K_k = P_{k/k-1} G_k^T [R_k + G_k P_{k/k-1} G_k^T]^{-1} \tag{44}$$

and must be continually updated with time. The estimate error $\epsilon_k - \hat{x}_{k/k-1}$ has a covariance matrix $E(\epsilon_k^T \epsilon_k) = P_{k/k-1}$ estimated as follows:

$$\hat{P}_{k/k-1} = F_{k-1} \hat{P}_{k-1/k-1} F_{k-1}^T + Q_{k-1} \tag{45}$$

Both the Kalman gain and the estimation error covariance must be updated on line, because the basic Eq. (41) is nonlinear, and the geometric function matrix (the Jacobian matrix) changes with time as the user/satellite geometry changes with time.* For GPS, many of these changes are rather slow, however.

By the way of reference to Chapter 5, this volume, it should be pointed out that the Kalman filter for estimating a constant Δx_o with fixed measurement

---

*For a detailed discussion of Kalman, Schmidt-Kalman, and adaptive Magill–Kalman filters see Chapter 9, this volume, and Refs. 6, 10, 11, 21, 23–25.

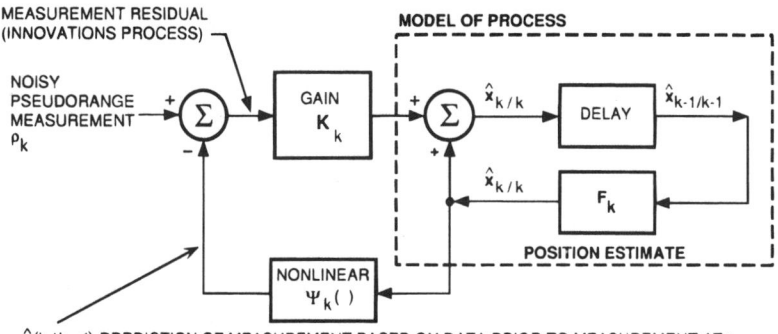

Fig. 32 Extended Kalman filter estimate of user position vector x_k based on pseudorange measurements ρ_k. The pseudorange estimate is $\hat{\rho}_{k/k-1} = \Psi_k(\hat{x}_{k/k-1})$, and it is assumed that $\rho - \hat{\rho}$ is sufficiently small.

statistics (rather than dynamic position vector) is identical to the least-squares solution, based on the same observation vector $\Delta\rho_o$, namely,*

$$\Delta\hat{x}_o = (G_o^T R_o^{-1} G_o)^{-1} G_o^T R_o^{-1} \Delta\rho_o = G^- \Delta\rho_o \tag{46}$$

where G_o is the constant linearized geometric matrix about a small region of uncertainty of x and G^- is a generalized inverse. Note that G_o is only approximated as a constant matrix over a short period of time, because the satellites are moving.

B. Vector Delay Lock Loop (VDLL)

The VDLL is a further generalization of the EKF and is a quasioptimal extension of the EKF estimation process. The VDLL closes the loop all the way back to the signal correlators instead of having two separate sets of shorter loops (delay lock loops and the EKF loop)† (see Refs. 22 and 26.) The VDLL is based on the fact that the pseudorange measurement Eq. (40) $\rho(x_k) = \Psi_k(x_k) + n_k$ is not directly an observable. Instead, the received signal is a single scalar observable, and each satellite-received signal component $s_i[t - \tau_i(x(t))]$ has two layered nonlinearities; namely, $s_i[\tau_i]$ and $\tau_i(x)$ and not just a single nonlinearity. The scalar observable signal at baseband is as follows:

$$r_k = \sum_{i=1}^{N} a_i s_i[k - \tau_i(x_k)] + n_k \tag{47}$$

where the noise samples are independent and Gaussian with zero mean. The objective is to perform a quasioptimal modified extended Kalman estimate of a vector user position x_k from this scalar observable.

To derive the VDLL configuration, we write the Taylor's series expansion for the signal from satellite i as follows:

---

*Note that the measurement vector can be easily generalized to include multiple samples over time as well as multiple satellites by increasing the dimension of ρ_o.

†A related integrated system is described in Refs. 22 and 26. See also Ref. 27.

$$s_i[t - \tau_i(x(t))] = s_i(t - \tau_i(\hat{x})) + s_i'(t - \tau_i(\hat{x}))g_i^T(t)(x - \hat{x}) +, \ldots,$$

where

$$g_i^T \triangleq \left(\frac{\partial \tau_i}{\partial x}, \frac{\partial \tau_i}{\partial y}, \frac{\partial \tau_i}{\partial z}, \frac{\partial \tau_i}{\partial B}\right), \frac{\partial s_i}{\partial x} = s_i'[t - \tau_i(x)]\frac{\partial \tau_i}{\partial x}, \; s_i'(t - \tau_i) = \frac{\partial s_i}{\partial \tau_i}(t - \tau_i) \quad (48)$$

The vector $g_i(t)$ has components that vary slowly with time as the satellite-user geometry changes, as discussed earlier. Note the similarity in this expression to the Taylor's series expansion in Sec. II, except that this expansion is about x rather than τ.

Form the product of the received signal $r(t)$ with each of N differentiated signal reference waveforms $s_i'\ [t - \tau_j(\hat{x})]$, (or equivalent early-late waveforms) to obtain the observation vector

$$v(t) = \begin{bmatrix} r(t)s_1'(t - \hat{\tau}_1) \\ r(t)s_2'(t - \hat{\tau}_2) \\ \cdots \\ \cdots \\ r(t)s_N'(t - \hat{\tau}_N) \end{bmatrix} = \begin{bmatrix} a_1 D(\Delta\tau_1) \\ a_2 D(\Delta\tau_2) \\ \cdots \\ \cdots \\ a_N D(\Delta\tau_N) \end{bmatrix} + \begin{bmatrix} n_1(t) \\ n_2(t) \\ \cdots \\ \cdots \\ n_N(t) \end{bmatrix} \cong AP_d \begin{bmatrix} \Delta\tau_1 \\ \Delta\tau_2 \\ \cdots \\ \cdots \\ \Delta\tau_N \end{bmatrix} + n \quad (49)$$

for small $\Delta\tau_i$ where $\Delta\tau_i = \tau_i - \hat{\tau}_i$; $\Delta\tau_i = \Delta\tau_i(x)$ and $D(\Delta\tau_i)$ is the delay lock discriminator characteristic, $D(\Delta\tau_i) = E[s(t)s'(t + \Delta\tau_i)] = R_s'(\Delta\tau_i) \cong R_s''(0)\Delta\tau$ for $\Delta\tau << T$, where $R_s''(0) = D'(0) = -P_d$, and P_d is the power in the differentiated signal. Notice that in Eq. (49), the noise terms n_{ik} are now different from one another because of the multiplication by $s_i'(t_k - \tau_i)$. The signal waveforms $r(t_k)$, $s(t_k)$ are sampled at a very high rate $f_s > 2B$, which for GPS, represents a sampling rate of either several MHz (C/A) or several 10s of MHz (P(Y)). These sample rates are generally much higher than one need sample the user position–velocity state vector. Thus, as shown earlier in Fig. 32, the output of the multipliers sampled at rate f_s is first filtered; e.g., by an integrate and dump filter operating at a much lower (decimated) output sample rate f_o. This output sample rate f_o is perhaps a factor of more than 1000 or 10,000 lower than the IF sample rate. This filtered output then gives outputs $D(\tau_{ik} - \hat{\tau}_{ik}) + n_{ik}$ where $D(\epsilon)$ is the DLL discriminator characteristic, and $\tau_{ik} - \hat{\tau}_{ik}$ is the delay error for satellite i at the sample times t_k. Thus t_k. Thus, although the correlators operate at sample rate f_s consistant with the bandwidth of the PN signal, the rest of the VDLL operates at a much lower sample rate.

Use the Jacobian geometric matrix $(\partial \tau_i/\partial x_j) = G$ to obtain an approximation for Eq. (49). The following linearized matrix equation* is produced (for the four-dimensional user state vector):

---

*Note that there are higher-order filters based on additional terms in the Taylors' series expansion. Examples of these higher-order filters include iterated EKF and Gaussian second-order filters.[28] Higher-order terms can be important for applications such as pseudolites where the range to the transmitter is relatively small, and there is considerable curvature in $\Psi(x)$.

FUNDAMENTALS OF SIGNAL TRACKING THEORY

$$v(t_k) = \begin{bmatrix} r(t_k)s_1'(t_k - \hat{\tau}_1) \\ r(t_k)s_2'(t_k - \hat{\tau}_2) \\ \vdots \\ r(t_k)s_N'(t_k - \hat{\tau}_N) \end{bmatrix} \cong A_k P_d \begin{bmatrix} g_1(t_k) \\ g_2(t_k) \\ \vdots \\ g_N(t_k) \end{bmatrix} \epsilon + n + , \ldots,$$

$$\cong A_k P_d \begin{bmatrix} \frac{\partial \tau_1}{\partial x} & \frac{\partial \tau_1}{\partial y} & \frac{\partial \tau_1}{\partial z} & \frac{\partial \tau_1}{\partial b} \\ \frac{\partial \tau_2}{\partial x} & \frac{\partial \tau_2}{\partial y} & \frac{\partial \tau_2}{\partial z} & \frac{\partial \tau_2}{\partial b} \\ \vdots & \vdots & \vdots & \vdots \\ \frac{\partial \tau_N}{\partial x} & \frac{\partial \tau_N}{\partial y} & \frac{\partial \tau_N}{\partial z} & \frac{\partial \tau_N}{\partial b} \end{bmatrix} \begin{bmatrix} x - \hat{x} \\ y - \hat{y} \\ z - \hat{z} \\ b - \hat{b} \end{bmatrix} + n$$

$$\cong P_d A_k G(t_k) \epsilon_k + n_k \tag{50}$$

where ϵ is the user position vector error $\epsilon_k = x_k - \hat{x}_{k/k-1}$ in vector notation. We have also assumed that the amplitude vector A can vary with time as A_k. As before, G_k has some dependence on x_k.

Write Eq. (50) in discrete time notation as follows;

$$v_k = \begin{bmatrix} r(t_k)s_1'(t_k + \hat{\tau}_1) \\ r(t_k)s_2'(t_k + \tau_2) \\ \ldots \\ r(t_k)s_N'(t_k + \tau_N) \end{bmatrix} = P_d A_k G_k [x_k - \hat{x}_{k/k-1}] + n_k \tag{51}$$

where the noise samples are independent in time, and the noise has a covariance matrix $E[n_k^T n_j] = R_k \delta_{jk}$.

The amplitude vector A_k is a known diagonal matrix with elements $[a_i]$ representing the relative signal strength of each of the received signals.* Although the signal levels for each satellite also vary with x, it is assumed here that A, and hence the a_i terms, have small dependence on x and time.

Equation (51) is now in the form where we can easily generate the standard Kalman estimator by assuming the following process and measurement equation:

$$x_k = F_k x_{k-1} + w_k \tag{52}$$

$$v_k = P_d A_k G_k [x_k - \hat{x}_{k/k-1}] + n_k \tag{53}$$
$$= H_k x_k + n_k - H_k \hat{x}_{k/k-1}$$

where we define $P_d A_k G_k = H_k$. The Kalman estimator equation can then be written as follows:
Predictor

---

*The matrix A can also be estimated as part of the VDLL.

$$\hat{x}_{k+1/k} = F_k \hat{x}_{k/k}$$

Correction update

$$\hat{x}_{k/k} = \hat{x}_{k/k-1} + K_k[v_k - H_k \hat{x}_{k/k-1}] \tag{54}$$

where K_k is the following standard Kalman gain equation;

$$K_k = P_{k/k-1} + H_k^T[R_k + H_k P_{k/k-1} H_k^T]^{-1} \tag{55}$$

and P and R are the error and noise covariance matrices, respectively. The VDLL of Figs. 33 and 34 implements this quasioptimal estimate. The NCOs act as the integrators included in the design.

There are several potential advantages of this VDLL receiver if the number of satellites exceeds the number of dimensions to be estimated; i.e. the system is overdetermined, as shown in Table 4. In principle, the use of signals from $N > 4$ satellites in parallel may provide enough *total* signal power to track successfully and to obtain accurate position estimates using the VDLL receiver under the same conditions where the signal strength from each individual satellite is

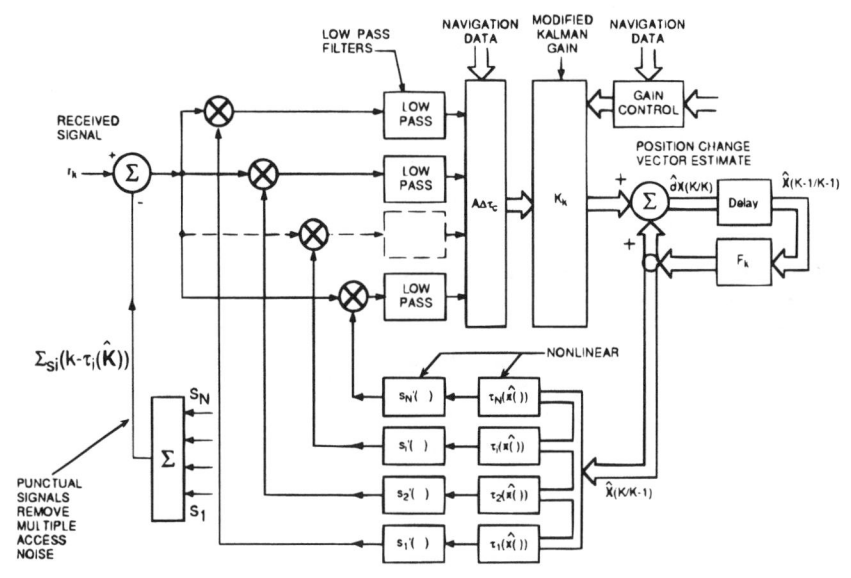

Fig. 33 Vector delay lock loop for a baseband PN signal. This DLL produces a quasioptimal position vector estimate from a scalar input containing multiple signals. The summation term subtracted from r_k at the initial part of the processing removes self-noise effects that are assumed to be small. The low-pass filters that follow the multipliers are designed to remove self-noise and multiple access noise effects and reduce the processing bandwidths in later portions of the VDLL without being so narrow in bandwidth as to affect the dynamics of tracking the user motion. The block that follows the low-pass filters labeled $\Delta\tau_c$ corrects the pseudorange-type information using satellite position, clock error, selective availability (if available), ionospheric corrections, and converts these measurements into a vector pseudorange format.

Table 4 Potential advantages of the VDLL in improving noise performance

1) Noise is reduced in all of the tracking channels making them less likely to enter the nonlinear region and fall below threshold. The effective bandwidths of the position-tracking loop filters are now governed by the dynamics of receiver motion and any correlation between τ_i and τ_j is now utilized. This noise reduction operation can be improved further if there are both a large number of satellites tracked, and the user clock bias b has relatively low dynamics (corresponds to an even higher redundancy in the number of satellite signals received).[a]

2) The VDLL can operate with momentary blockage of one or more satellites. The signal level from one or more of the satellites may be momentarily blocked by an obstruction; e.g., the wing tip of an aircraft, a freeway overpass, a building, or a grove of trees along a highway. Thus, the a_i for that satellite decreases, or equivalently, the relative noise level on that measurement increases. If sufficient satellites remain in view or the blockage is short enough, the remaining satellites may be able to give a sufficiently good position estimate that there is never any loss of lock in any of the delay lock loops. (Loss of lock is a condition defined as one where the position error ϵ in the estimate is large enough so that there is no useful correction signal generated.)

For example, satellites may be sequentially blocked for a short period of time throwing the corresponding independent DLL out of lock, while the VDLL can remain in lock throughout the momentary outage.

3) The VDLL filtering can be better optimized than the conventional sequential DLL-Kalman filter combination which is constrained to be the product of two stable closed-loop tracking filters in series. Thus, we would expect to see improved performance in the processor performance for highly dynamic user motion.

[a] It can be argued that as long as the tracking filters and estimators operate as linear filters, it makes no difference how they are configured, and the serial combination of the independent DLL followed by the EKF is still optimum. However, that statement is not valid if there is any nonlinearity, and the independent DLL moves into the nonlinear region or if there are realizability constraints on the closed-loop filters.

so low (or the interference levels are so high) that none of the individual scalar DLL can remain in lock when operating independently. Thus, we can visualize VDLL systems that can operate successfully, when the conventional independent parallel DLL approach fails completely! In Chapter 5, this volume, the effective GDOP was shown to decrease as the inverse square root of the number of satellites, as might be expected. Thus, if the number of satellites increases by a factor of 2, with proper geometry, the signal power received from each satellite could decrease by a factor of 2, and we could achieve the same position accuracy. Thus, the addition of more satellites can, indeed, allow a user with a degraded antenna/receiver (or a higher level of interference) to operate satisfactorily when the same antenna/receiver system utilizing separate, independent DLL for each satellite may not be able to track even a single satellite.

As shown in Figs. 33 and 34, the VDLL requires that computations be made at a relatively high speed; i.e., the computation delay must be small compared to the inverse closed-loop noise bandwidth $1/B_n$ in order to keep the loop stable when the effective closed-loop processing bandwidth is sufficiently wide to track the user position dynamics. High-speed, low-cost computer chips make this task feasible now, whereas in the past, it would have been too complex.

There are variations in this vector delay lock receiver depending on 1) the dimensionality of the user state vector; e.g., 4–8 dimensions; and 2) the dimensionality of the measurements; e.g., pseudoranges or pseudoranges plus Doppler/rate of change measurements on the carrier and other sensor inputs. Just as discussed in the previous section on the scalar delay lock loop, coherent, noncoherent, and quasicoherent forms of the VDLL can be configured.

1. Time Variation in Kalman Gain and G Matrix for the VDLL

In general, both the Kalman gain and the G matrix vary with time. The G matrix, of course, varies as the satellite position changes, and the direction vectors to the user position change. Likewise, the extended Kalman filter gain matrix changes with satellite position and signal strength. The usual considerations for avoiding Kalman filter divergence are also important for this quasioptimal vector delay lock estimator. Thus, we would expect to consider use of standard techniques for iterating the Kalman filter gain and position estimation and simplifying the computations, such as shown in Table 5.

2. Acquisition for the Vector Delay Lock Loop

Although the acquisition process is left for discussion in the next chapter, that discussion primarily applies to the scalar DLL. If we attempted to acquire the N GPS signals in parallel with a VDLL receiver, it would be necessary to search the user state vector $(x,y,z,b,\dot{x},\dot{y},\dot{z},\dot{b})$ in increments small enough to accommodate the DLL discriminator characteristic width, and the IF bandwidth of any noncoherent detectors. Because these parameters must be searched in at least a 4-D position space (and possibly a 4-D position velocity space) in parallel over the entire uncertainty space, direct search and acquisition, except with very small initial uncertainty ranges, may often be impractical within an acceptable acquisition time. Thus, we might expect that the search and acquisition process should independently search for lock with each of the DLL operating in parallel until at least four satellites are acquired and their pseudoranges measured. At that point, we can solve the simultaneous equations for user position by a surface searching technique. Once the position is estimated with sufficient accuracy, the position estimating Kalman or other similar filter can be put in operation, and the closed-loop VDLL operation can begin without searching the remaining satellites, if the other errors are small. Thus, the initial acquisition process is very similar to that used for the scalar DLL/Kalman estimator approach.

C. Quasioptimal Noncoherent Vector Delay Lock Loop

The coherent and noncoherent DLL of Sec. II have their counterparts in the quasioptimal VDLL and the noncoherent VDLL. Figure 35 illustrates one possible configuration of the noncoherent VDLL. Each of N separate noncoherent correlators has early-late delay separation matched to the signal rise-time δT as, discussed

FUNDAMENTALS OF SIGNAL TRACKING THEORY 299

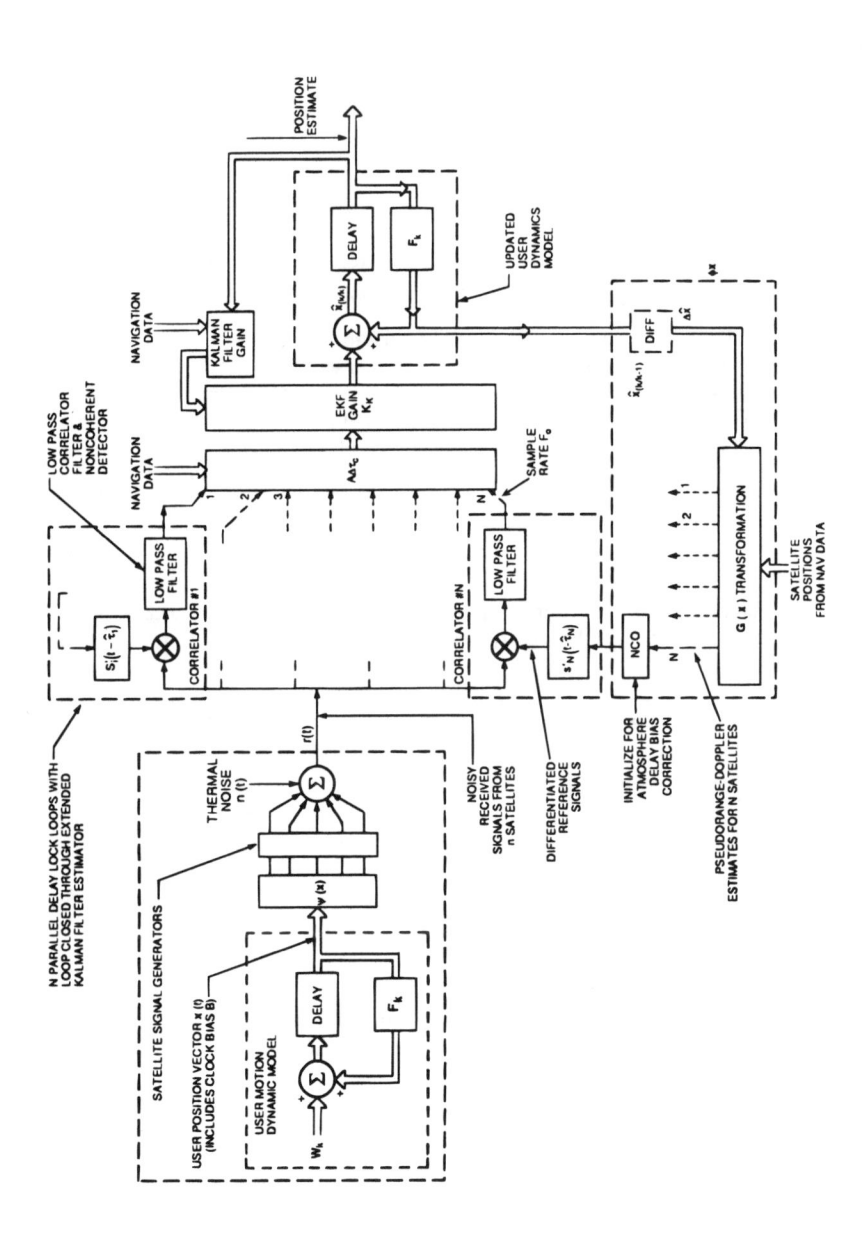

Fig. 34 Quasioptimal vector delay lock/position estimator. The reference signal here shown as a differentiated signal can also be implemented as an early-late discriminator in coherent and noncoherent forms and in a quasicoherent form. The more complete quasioptimal estimators continues a subtractor that precedes.

Table 5 Standard techniques for simplifying the Kalman computations

1) Covariance matrix factorization techniques (in order to reduce order of magnitude of computations); e.g., Cholesky, lower/upper LU, upper diagonal (UD) triangular; i.e., factorization into a Toeplitz matrix.

2) Various ϵ techniques (due to Schmidt) to avoid unrealistically small values for the error covariance and, hence, Kalman filter gain. Under these small gain conditions the filter stops paying much attention to new data.[28]

3) Schmidt–Kalman filter. This variation of the Kalman Filter is a method for dealing with colored noise and biases by partitioning the state vector to reduce computation load.[21] We can include additional components (e.g., bias errors) to the user state vector but estimate only the needed components.

4) Adaptive Kalman filter gain utilizing tests on the innovations sequence. An innovations process can be defined as a sequence of random variables $z_k - E[z_k|z_{k-1}, z_{k-2}, \ldots, z_0]$, which represents the new information contained in the new sample. The residual differences between the observations/measurement and the predictions of the observation/measurement is the innovation sequence. The Kalman filter with the optimal gain should whiten the innovations sequence, and this test can be used as a criterion for optimality of the filter gain K.

---

in Sec. II. The closed-loop filter operates in a quasisteady-state mode,* as did the filters of Sec. II, except that the G^- matrix changes with geometry and signal strength. However, rather than closing each loop independently, the residual error terms $\epsilon_i(t)$ are all weighed by the optimal generalized inverse matrix operation $G^-(t) = [G^T R^{-1} G]^{-1} G^T R^{-1}$ where G is the slowly varying direction cosine matrix dependent on the relative position of the user and the satellites, and R is the measurement error covariance matrix that varies as the signal strength of each of the satellite signals.

The loop filters $F_i(p)$ and integrators are selected either from the process model as a steady-state closed-loop version of the Wiener filter, or by reference to the maximum steps of acceleration anticipated, as discussed earlier in Sec. II of this chapter. The output position vector estimate $x_T = (x,y,z,B)$ is then transformed back by G to the pseudoranges for each of the satellites by the linear direction cosine matrix $G(t)$. Both G and G^-, of course, vary slowly with time and must be periodically updated. The updates must be done sufficiently often and smoothly so that the vector loop is not thrown out of lock in the process.

1. One-Dimensional Example of Vector Delay Lock Loop

In the simplest example of this VDLL, consider a one-dimensional unknown position x. The VDLL receives signals from N satellites positioned along the x-axis, and thus, is highly overdetermined for $N \gg 1$. This form of vector delay lock receiver then has a configuration reminiscent of diversity combining in a

---

*In many examples, the Kalman filter converges rather rapidly to a steady-state operation except for the time variation in the G matrix and signal strength.

FUNDAMENTALS OF SIGNAL TRACKING THEORY

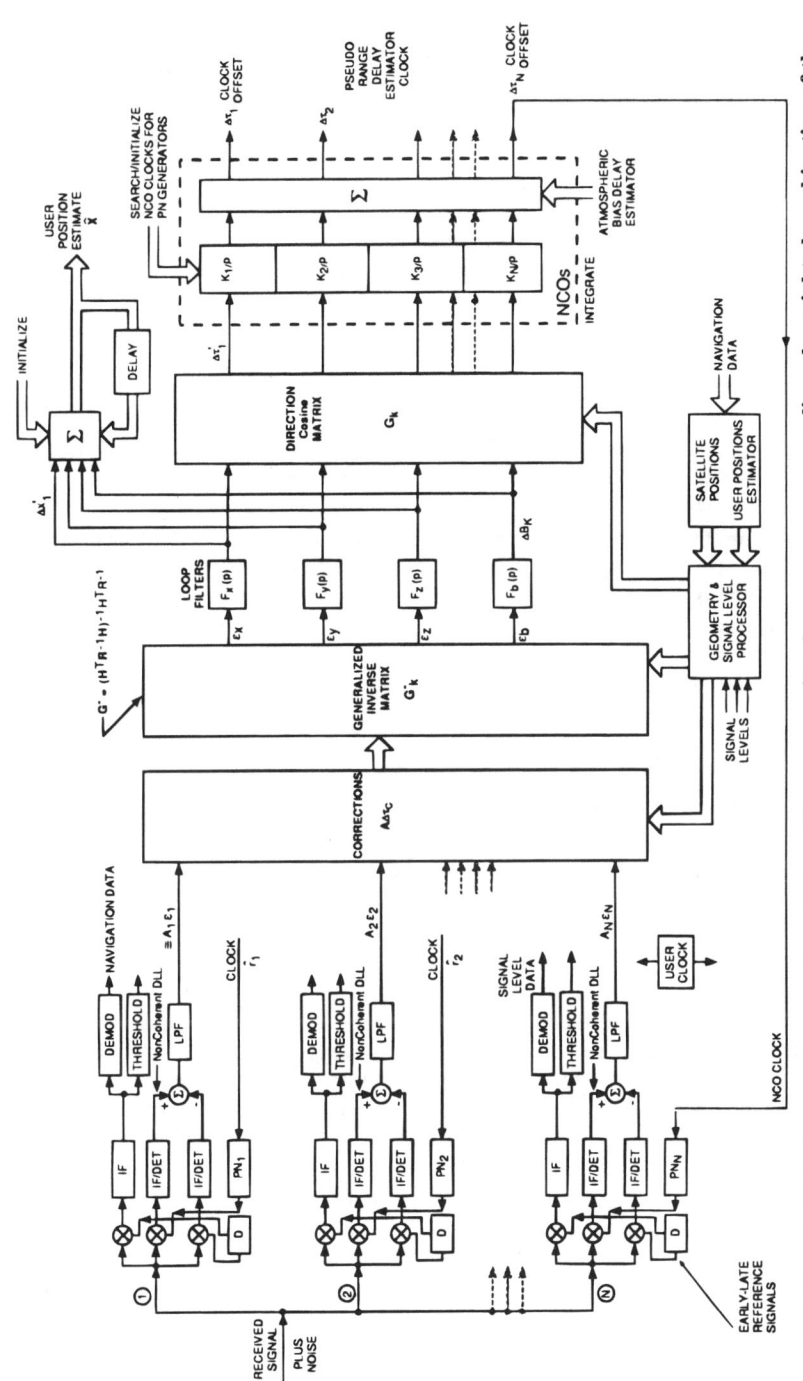

Fig. 35 Quasioptimal VDLL in a noncoherent form. The error residuals ϵ_x, ϵ_y, ϵ_z, ϵ_B are a linearly weighted combination of the individual correlator outputs.

receiver for a fading channel.[29] Strictly speaking, the C/N_o does not increase linearly with the number of satellites in view because the additional satellites also add to the amount of self-noise, and hence, to the equivalent noise density. The equivalent noise density for random PN streams increases as

$$N_{oeff} = N_o + \Sigma P_i/f_c \tag{56}$$

where P_i is the received signal power for the ith satellite, and f_c is the PN clock rate. Thus, the ratio of the total carrier power C_{Total} to equivalent noise density N_{oeff} at the center frequency is approximately (see Chapter 3, this volume):

$$\frac{C_{total}}{N_{oeff}} = \frac{\Sigma P_i}{N_o + \Sigma P_j/f_c} = \frac{1}{(N_o/C_{total}) + 1/f_c} \tag{57}$$

However, in most instances with GPS, the multiple access noise of the $\Sigma P_i/f_c$ terms are not dominant because the received C/N_o levels are generally relatively small.

For the previous example, it is clear that the noise bandwidth of the x component tracking loop need be no wider than that for a single independent DLL; however, the effective signal power is tripled if three satellites are in view along the x-axis, as compared to one. Thus, the threshold signal-to-noise ratio for a single satellite can drop by a factor of N, if all N signals are of equal received power. Of equal importance, one or more of the N satellite signals can be severely attenuated, and the VDLL can, under the right conditions, still track accurately.

2. Minimum Mean Square Error Estimate for Two Dimensions

Consider that the task of estimating a position vector $x^T = (x,y,b)$ with a fixed but unknown position and a slowly varying clock bias b. As an example, consider a VDLL operating in two dimensions x,y of user position uncertainty and a slowly varying clock bias uncertainty b; that is, the user z coordinate is fixed and known. If there are six satellites total, as shown in Fig. 36 (two at the $\pm$ x-axis; two at the $\pm$ y-axis; and two at the $+$ z-axis), then the geometric matrix G, the Jacobian matrix, is shown in the figure.

Figure 37 shows the implementation of a VDLL with steady-state rather than Kalman filtering, where the three loop filters $F_x(p)$, $F_y(p)$, $F_b(p)$ are fixed except for a time-varying gain, which is dependent on received signal level and geometry. The correlators in Fig. 37 can be either an early–late noncoherent correlator or a coherent correlator (multiplier followed by a finite memory integrator). As discussed in Sec. III, the outputs of each correlator for a differentiated reference are $a_i D(\epsilon_i) + n_i$ for each of the six signal channels where $\epsilon_i = \tau_i - \hat{\tau}_i = \psi(x) - (\hat{x})$.

For small error in $\Delta x = x - \hat{x}$ we can approximate

$$\Psi(x + \Delta x) = \Psi(x) + G\Delta x$$

and

$$\tau - \hat{\tau} = \Psi(x) - \Psi(\hat{x}) \cong G\Delta x \tag{58}$$

FUNDAMENTALS OF SIGNAL TRACKING THEORY

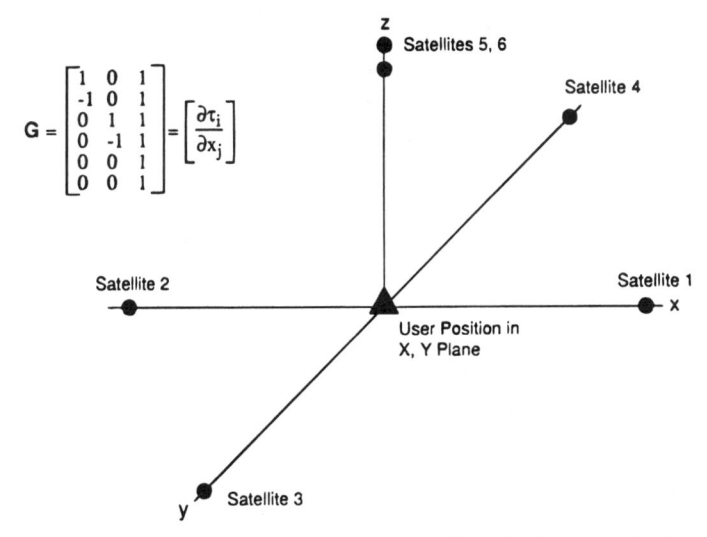

$$G = \begin{bmatrix} 1 & 0 & 1 \\ -1 & 0 & 1 \\ 0 & 1 & 1 \\ 0 & -1 & 1 \\ 0 & 0 & 1 \\ 0 & 0 & 1 \end{bmatrix} = \left[\frac{\partial \tau_i}{\partial x_j}\right]$$

Fig. 36 Example satellite configuration, six satellites. It is assumed the user has only two coordinates of position uncertainty plus a slowly varying clock bias.

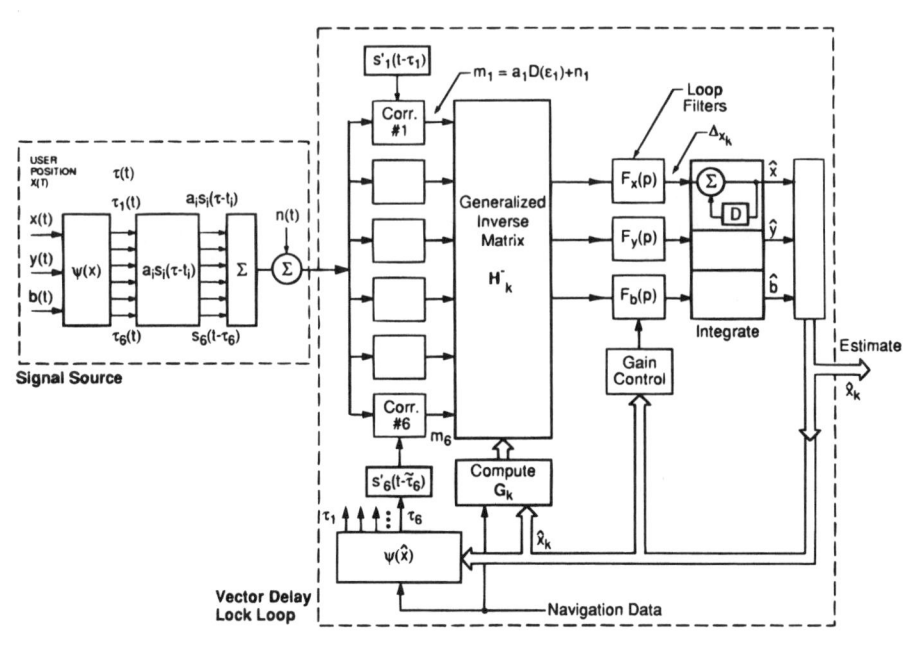

Fig. 37 Example VDLL with steady-state rather than Kalman-type filtering for two-dimensional position plus clock bias estimate. Some of the filters vary with user position estimate and $\Psi(\hat{x})$ can be approximated by an iterative expression using G and $d(\hat{x})$. Early-late noncoherent correlators can be used. Pseudorange corrections are added prior to the matrix H^-.

Thus, each channel of the correlator output is approximated by

$$m_i = a_i D(\epsilon_i) + n_i$$
$$= a_i D'(0)\epsilon_i + n_i$$
$$= a_i D'(0)[\Psi_i(x) - \Psi_i(\hat{x})] + n_i$$
$$= a_i D'(0) G_i \Delta x + n_i$$

or in vector form the set of the six correlator outputs are as follows:

$$\mathbf{m}_i = AD'(0)G\Delta x + n = H\Delta x + n \tag{59}$$

Define $H^- = [H^T R H]^{-1} H^T R^{-1} = [1/D'(0)] [(AG)^T R A G]^{-1} (AG)^T R^{-1}$. Then the estimate of Δx at time t_k is as follows:

$$\Delta \hat{x}_k = H_k m_k \tag{60}$$

as shown in Fig. 37. Note that G varies with user position and time.

For a unit matrix measurement noise covariance $R = 1$ and equal signal strength $A = 1$, the generalized inverse linear weight matrix is then as follows:

$$H^- = G^- = \begin{bmatrix} 0.5 & -0.5 & 0 & 0 & 0 & 0 \\ 0 & 0 & 0.5 & -0.5 & 0 & 0 \\ 1/6 & 1/6 & 1/6 & 1/6 & 1/6 & 1/6 \end{bmatrix} = (G^T G)^{-1} G^T$$

Note the similarity in this process to predetection diversity combining.[29] The covariance matrix at the output of the G^- estimator is as follows

$$E[\boldsymbol{\epsilon}^T \boldsymbol{\epsilon}] = [G^T G]^{-1} = \begin{bmatrix} 1/2 & 0 & 0 \\ 0 & 1/2 & 0 \\ 0 & 0 & 1/6 \end{bmatrix} \tag{61}$$

Assume a steady-state VDLL of the form shown in Fig. 35 with three loop filters x, y, b. The closed-loop transient performance of this tracking loop is identical in tracking the x-coordinate to that described previously for the conventional DLL. If a critically damped second-order loop is employed, the closed-loop transfer function is as follows:

$$\hat{x}(t)/x(t) = H(p)$$
$$= KAF(p)/(p/p_o + KAF(p)) \tag{62}$$
$$= (1 + \sqrt{2}\, p/p_o)/[1 + \sqrt{2}\, p/p_o + (p/p_o)^2]$$

The noise bandwidth is $B_n = 1.06\, p_o$. Because of redundancy, the effective noise in the VDLL for this example is, however, reduced by a factor of 2. That is, the effective noise term $n(t)/\sqrt{2}$ has only half as much power as the noise at the x,y input channel and 1/6 as much for the b channel [see Eq. (61)]. The effective signal-to-noise ratio is improved by a factor of 2, and the threshold performance improves accordingly, as we might expect from the 2-to-1 redundancy.

The VDLL concept can be specialized to the form of a vector phase–lock-loop for carrier tracking. In this example, the parallel delay lock correlators are replaced by parallel phase lock correlators. For carrier tracking, there might be cycle ambiguities, but vector processing can still be employed.

3. Colored Noise

In many tracking applications, the effective noise in the delay measurement is not white but colored because of such effects as atmospheric noise, selective availability, and ephemeris errors. Appendix E describes the modifications to the VDLL for a colored noise environment.

D. Channel Capacity and the Vector Delay Lock Loop

It is the total power received from all GPS satellites relative to a fixed noise density that determines the channel capacity C. However, the only information sought for GPS is the user position with a desired level of accuracy and position update rate, *not* the individual pseudoranges. Because the user position vector typically is of limited dynamics, the information rate required is quite small.

The channel capacity C is defined as the maximum rate at which a user can transmit information through a channel and recover that information at the receive end with a vanishingly small probability or error. As the number of signals increases, the total received signal power increases. That is, the total multiple access channel capacity C also increases with the number of signals N, as follows[30]:

$$C = B \log_2\left(1 + \sum_{i=1}^{N} P_i/N_o B\right)$$

where each received signal has power P_i, bandwidth B, and there is a single noise term of one-sided noise density N_o. Then as N increases for equal power signals, the channel capacity also increases approximately in proportion to $\log_2 N$ for $P_i/N_o B \gg 1$.

On the other hand, the total information rate sought from the channel; e.g., the x,y,z coordinates of the user position to a desired accuracy level is a constant independent of N. Assume independent parallel Gaussian information sources with zero mean, variance σ_i^2, and squared error distortion metric

$$D_i = E[(x_i - \hat{x}_i)^2].$$

For example, the desired information rate (rate distortion function) $0 \leq D_i \leq \sigma_i^2$ for three coordinates x,y,z is as follows:

$$R(D) = W \sum_{i=1}^{3} \log_2(\sigma_i^2/D_i)$$

where $2W$ is the rate at which measurements are sought (sample rate $2W$) and σ_i^2/D_i is the measure of desired accuracy. Thus, it is clear that with N satellite signals with good geometry, it should be possible to operate with signal power/satellite that can decrease as N increases, or alternatively, to tolerate more noise link interference as long as $R(D) < C$ by a sufficient margin. The VDLL is a technique to make use of this capacity increase.

Appendix A:
Maximum Likelihood Estimate of Delay and Position

Assume that the observed received scalar signal samples, $r(t_j) = s(t_j,x) + n(t_j)$, taken at time t_j are independent Gaussian random variables with mean $s(t_j,x)$.

The noise samples $n(t_j)$ are assumed to be independent stationary, Gaussian random variables with zero mean, variance σ^2, and are independent of the signal $s(t_j,x)$. The signal samples $s(t_j,x)$ are some known nonlinear function of an unknown parameter x, which we want to estimate. Assume also that $s(t,x)$ represents a constant power, bandlimited signal and that the signal autocorrelation function $R_s(\epsilon)$ and all of its derivatives exist and are continuous and invariant with time. Consider first that $x = \tau$ and is a scalar.

Assume that there are K independent samples, that the unknown is the scalar delay τ and that τ remains constant over this time interval. The joint probability of the K observed samples conditioned on the unknown value of τ is as follows*:

$$p(r|\tau) = \left(\frac{1}{\sigma\sqrt{2\pi}}\right)^K \exp-\left\{\frac{1}{2\sigma^2} \sum_j [r(t_j) - s(t_j, \tau)]^2\right\} \quad (A1)$$

The *maximum likelihood estimate* of τ is that value of τ that maximizes $p(r|\tau)$ for a given observed r. A necessary condition for the existence of the maximum likelihood estimate Eq. (A1) is that the derivative of $p(r|x)$ with respect to τ evaluated at $\tau = \hat{\tau}$ must be zero. This condition also corresponds to the condition that the derivative of the log of the probability; i.e., the *log likelihood* must also be zero; namely,

$$\left.\frac{\partial p(r|\tau)}{\partial \tau}\right|_{\tau=\hat{\tau}} = \left.\frac{\partial \ln p(r|\tau)}{\partial \tau}\right|_{\tau=\hat{\tau}} = 0 = \frac{\partial}{\partial \tau} \sum_j [r(t_j) - s(t_j, \tau)]^2|_{\tau=\hat{\tau}} \quad (A2)$$

where we have omitted the constant multiplier in Eq. (A2), and $\hat{\tau}$ is the maximum likelihood estimate.

Assume that the noise-free received signal energy

$$E_s = \sum_j [s(t_j, \tau)]^2$$

is independent† of τ. With this assumption the constraint (A2) can be written as follows:

$$\frac{\partial p(r|x)}{\partial x} = 0 = \sum_j r(t_j) \frac{\partial}{\partial \tau} s(t_j, \tau)|_{\tau=\hat{\tau}} = \sum_j r(t_j) s'(t_j, \tau)|_{\tau=\hat{\tau}} \quad (A3)$$

where $s'(t_j,\tau) \triangleq (\partial/\partial \tau) s(t_j,\tau)$ is the differentiated signal.

Thus, the maximum likelihood estimate $\hat{\tau}$ satisfies the relationship that the summation of the product of the received signal samples $r(t_j)$ with the differentiated reference signal samples $s'(t_j,\tau)$ must be equal to zero. Of course, if there are multiple solutions of this equation, we must select the value corresponding to the maximum value of $p(r|\tau)$.

Assume, now, that we have found the initial value of the maximum likelihood estimate $\hat{\tau}$ for the first time interval of K samples and that $\hat{\tau} \cong \tau$ for that interval. Also assume that the parameter τ changes slowly in small increments from τ_0 to

---

*We can also represent the random noise process over this finite interval as a Karhunen–Loeve series. However, we take the discrete time representation[12] and assume that K is large.

†This assumption is equivalent to an absence of self-noise. If that condition is not valid, the self-noise is removed by subtraction by a punctual signal.

τ_1, etc. from one time interval of K samples to the next. Specifically, assume that the value of τ during the first K time samples t_1 to t_K is τ_0 and that $\hat{\tau}$ has converged to τ_0 for that interval. During the next K samples, τ takes on a slightly different value, τ_1, where $\tau_1 - \tau_0 = \epsilon$ is a small change in a sense, yet to be defined. Notice that the expected value of the product $r(t,x + \epsilon)s'(t,x)$ in Eq. (A3), is equal to the following:

$$E[r(t, \hat{\tau} + \epsilon)s'(t, \hat{\tau})] = E[s(t, \hat{\tau} + \epsilon)s'(t, \hat{\tau})]$$

$$= R_s'(\epsilon) \cong R_s''(0)\epsilon \quad \text{for small } \epsilon \quad (A4)$$

The sum in Eq. (A3), properly scaled, approximates Eq. (A4) for sufficiently large K.

Define the differentiated autocorrection functions of the signal as follows:

$$R_s'(\epsilon) = \frac{\partial R_s(\epsilon)}{\partial \epsilon}, \quad \text{and} \quad R_s''(\epsilon) = \frac{\partial^2 R_s(\epsilon)}{\partial \epsilon^2}$$

where $R_s(\epsilon)$ is the autocorrelation function of $s(t)$. The condition of small ϵ assures that ϵ is within the quasilinear region of the differentiated autocorrelation function $R_s'(\epsilon)$. The quantity $R_s'(\epsilon)$ has also been termed the DLL discriminator function. The expected value of the product in Eq. (A3) gives a correction term proportional to the parameter change ϵ. Thus, for large K, once the parameter τ has been estimated sufficiently accurately to arrive at an estimate $\hat{\tau}$ close to the true value, the estimator can be configured in the form of a closed-loop tracking function. Then small incremental changes in the parameter τ can be tracked to provide a piecewise continuous approximation to the maximum likelihood estimate as shown in Fig. 5a in the text.

Therefore, if τ_1 increases (positively) slightly with respect to τ_0, ϵ momentarily increases by an identical or proportional amount. Hence, the average of the product in Eq. (A3) over the K sample increases by an amount proportional to $R_s'(\epsilon) \cong R_s''(0) \epsilon = R_s''(0) [\tau_1 - \tau_0]$. If the multiplied output $s'(\tau + \hat{\tau}) r(t + \tau)$ is followed by some type of low-pass averaging filter, the filter output (closed-loop) serves to increase the value of the estimate, and thus, decrease the tracking error. Thus, the loop tracks the variation in delay τ and serves to minimize Eq. (A3), the constraint equation for maximum likelihood estimate.

More specifically, assume that the increments

$$j = nK + i$$

where

$$i = \{0, K - 1\}$$

i.e., there are many intervals n of length K samples each. Assume that at the end of interval n_o, the estimate of τ_{n_o} exactly matches the true value of τ_{n_o}; namely, $\tau_{n_o} = \hat{\tau}_{n_o}$. The output of the summation where the new value of τ is τ_{n_o+1} is as follows:

$$\Sigma_{n_o+1} = \sum_{j=1}^{K} r(t_j)s'[t_j + \hat{\tau}_{n_o}] = \sum_{j=1}^{K} [s[t_j + \hat{\tau}_{n_o+1}] + n(t_j)]s'[t_j + \hat{\tau}_{n_o}]$$

$$= \sum_{j=1}^{K} s[t_j + \tau_{n_o+1}]s'[t_j + \hat{\tau}_{n_o}] + n_{\Delta n_o+1}$$

$$= \sum_{j=1}^{K} s[t_j + \Delta\tau_{n_o+1} + \hat{\tau}_{n_o}] s'[t_j + \hat{\tau}_{n_o}] + n_{\Delta n_o+1}$$

$$= R_s'(\Delta\tau_{n_o+1}) + n_{sn_o+1} + n_{\Delta n_o+1}$$

$$= R_s''(0)\Delta\tau_{n_o+1} + n_{\tau_{n_o+1}} \qquad (A5)$$

where $n_{s_{n_o+1}}$ is a small self-noise term that decreases to zero as K becomes large, and $n_{\Delta n}$ is a noise term after summation of independent random noise samples for the n interval, and $\Delta_{\tau n}$ is the change in τ.

The equation for Σ can be expressed in the form of a closed-loop DLL block diagram, as shown in Fig. A1 and in Fig. 5a. Note that this block diagram only approximates the maximum likelihood estimate, because for the true maximum likelihood estimate, $\Sigma_{(n_o+1)} = 0$. However, for sufficiently large values of K and slow variations of τ, which lead to correspondingly small values of $\Delta\tau$, the DLL is a good approximation to the maximum likelihood estimate.*

Estimate of Scalar Position from a Set of Measurements

As a further generalization, assume that a summation of N equal power signals are received as in GPS, each with a delay that is a function of a scalar position variable x. The received signal is then of the following form:

$$r(t, x) = \sum_{i=1}^{N} s_i[t - \tau_i(x)] + n(t) \qquad (A6)$$

where each of the signals $s_i(t)$ are uncorrelated, and the τ_i are different delay functions $\tau_i(x)$. The received noise $n(t)$ is Gaussian and has zero mean and variance σ^2. The conditional probability density for the signal at K independent sample times t_j is then as follows:

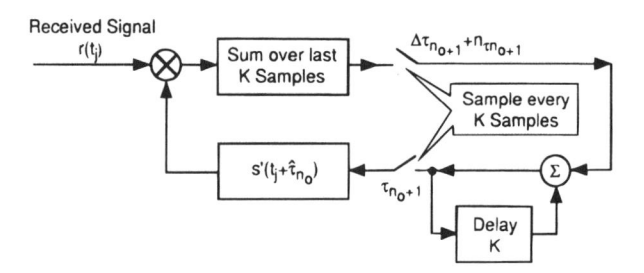

Fig. A1 Delay lock loop block diagram representing Eq. (A5), which approximates an iterative version of the maximum likelihood estimate.

---

*The reader is also referred to Ref. 12, who analyzed the maximum a posterior estimators which for the example of a phase modulated sinusoid take the form of a phase locked loop; i.e., a special case of the DLL. This tracking loop also bears a similarity to the linear sequential least-squares error estimator discussed in Ref. 10 (see Appendix B).

FUNDAMENTALS OF SIGNAL TRACKING THEORY

$$p(r|x) = \left(\frac{1}{\sigma\sqrt{2\pi}}\right)^K \exp\left(-\frac{1}{2\sigma^2}\sum_{j=1}^{K}\left\{r(t_j) - \sum_i s_i[t_j - \tau_i(x)]\right\}^2\right) \quad (A7)$$

Again, the summation of the received signal samples is the energy

$$E_s = \sum [s_i(t - \tau_i)]^2$$

and is assumed to be independent* of x; that is there is assumed to be negligible self-noise. A necessary condition for the maximum likelihood estimate is the constraint equation which is as follows:

$$\frac{\partial p(r|x)}{\partial x} = 0 = \frac{\partial}{\partial x}\sum_j\left\{r(t_j) - \sum_i s_i[t_j - \tau_i(x)]\right\}^2\bigg|_{x=\hat{x}}$$

$$= 0 = \sum_j r(t_j) \sum_i s_i'[t_j - \tau_i(x)]\tau_i'(x)|_{x=\hat{x}}$$

$$= 0 = \sum_{i=1}^{N}\frac{\partial \tau_i(x)}{\partial x}\left\{\sum_{j=1}^{K} r(t_j)s'[t_j - \tau_i(x)]\right\}\bigg|_{x=\hat{x}} = \sum_{i=1}^{N}\frac{\partial \tau_i}{\partial x}\Sigma_i(x)|_{x=\hat{x}}$$

where

$$\Sigma_i(x) \triangleq \sum_{j=1}^{K} r(t_j)s'[t_j - \tau_i(x)], \quad \text{and} \quad s'(t, x) = \frac{\partial}{\partial \tau} s(t, \tau),$$

$$\tau'(x) = \frac{d}{dx}\tau(x), \quad \text{and} \quad \frac{\partial s}{\partial x} = \frac{\partial s}{\partial \tau}\frac{d\tau}{dx} \quad (A8)$$

In this estimator, the reference is the sum of the differentiated signals, each with a separate scale factor $\partial \tau_i/\partial x$ related to the geometry. Thus, the optimum estimator seeks to adjust $\hat{x}$ so as to set the sum in Eq. (A11) to zero and weights the satellite measurements with the slope $\partial \tau_i/\partial x$. Assume that the derivatives $\partial \tau_i(x)/\partial x$ are nonzero.† This equation has a number of solutions, however the maximum likelihood estimate‡ occurs when each $\Sigma_i(x) = 0$.

It should be pointed out that if the term $[\Sigma s_i(t + \tau_i)]^2$ is not a constant independent of τ_i, then the constraint equations become the following:

$$\{r(t_j) - \sum s_i[t_j - \tau_i(\hat{x})]\}s_i'(t_j - \tau_i(\hat{x}))\frac{\partial \tau_i}{\partial x_k} = 0 \quad (A9)$$

and the $r(t_j) - \Sigma s_i[t_j - \tau_i(\hat{x})]$ term serves to remove self-noise effects by subtraction of the estimate of the sum of the signal terms.

---

*Generally, the assumption that E_s is independent of x requires zero crosscorrelation between signals from other satellites and is only an approximation (see Refs. 9 and 20 and Chapter 3 of this volume).

†If any component is zero, that satellite signal carries no useful information and can be eliminated from the calculation.

‡Clearly, if the satellites are positioned so that half of the $\partial \tau_i/\partial x$ are $+C$, and the other half are at $-C$, then Eq. (A11) could be satisfied if all $\Sigma_i \cong$ nonzero constant, which is not a maximum likelihood condition (see Fig. 5b).

Estimate of Vector Position from a Set of Measurements

As a further generalization, assume that there are several parameters x_k to be estimated; i.e., the unknown position is a vector variable. A simple extension of Eq. (A.8) then yields the following:

$$\frac{\partial p(r|x)}{\partial x_k} = 0 = \sum_j r(t_j) \sum_i s_i'[t_j - \tau_i(\hat{x})] \frac{\partial \tau_i}{\partial x_k}\bigg|_{x=\hat{x}}, \quad k = 1, 2, 3, 4$$

$$= 0 = \sum_{i=1}^N \frac{\partial \tau_i}{\partial x_k} \left[\sum_{j=1}^N r(t_j) s_i'(t_j - \tau_i(\mathbf{x})) \right]_{x=\hat{x}} = \sum_{i=1}^N \frac{\partial \tau_i}{\partial x_k} [\Sigma_i(\mathbf{x})]_{x=\hat{x}}$$

where

$$\Sigma_i(\mathbf{x}) = \sum_{j=1}^N r(t_j) s_i'[t_j - \tau_i(\mathbf{x})]_{x=\hat{x}} \tag{A10}$$

and $x = (x_1, x_2, x_3, x_4)$ is the unknown vector. As before, the estimate can be generated iteratively to generate an approximate minimum for each of the signal component summations Σ_i over the K sample intervals. The estimator for the position vector $\hat{x}$ is then shown in Fig. 5c. In this example for N parallel received signals, the received signal is separately multiplied by N differentiated reference signals. We still must obtain an estimate of dx given dτ, however, because it is x that we want to estimate. The outputs of the multipliers are then weighted and combined by a weighting matrix which is the generalized inverse G^- in order to generate the estimates of dx_1, dx_2, dx_3, etc. The matrix G^- gives the minimum mean-square error estimate of dx given dτ.[10] The optimum matrix G^- is related to the geometric matrix G and the covariance matrix, R, of the noise in the estimate of dτ_i by the following:

$$G^- = [G^T R G]^{-1} G^T R^{-1} \tag{A11}$$

The output is then filtered by separate filters for each of the dx_1, dx_2, dx_3, dx_4 components. The weight matrix G^- is dependent on x and can vary slowly with time as the geometry changes. More precise models that account for the dynamics of the position vector process are discussed in Sec. III devoted to the VDLL that uses Kalman-type filtering.

Appendix B:
Least-Squares Estimation and Quasioptimal Vector Delay Lock Loops

For a position estimation problem with stationary Gaussian noise, constant power signals, and a position variable represented by a stationary Gaussian process, we can generate a least-square estimator that is slightly different configuration of the VDLL.

Consider a vector differential pseudorange observation $\Delta\rho$:

$$\Delta\rho = G\Delta x + n \tag{B1}$$

where the vector set of observation residual $\Delta\rho = \rho - \hat{\rho}$ of dimension N are related by a linear matrix G to a constant unknown vector $\Delta x = x - \hat{x}$ of

FUNDAMENTALS OF SIGNAL TRACKING THEORY

dimension k, $N \geq k$, and there is additive noise $\boldsymbol{n}$ in each observation. These observations represent the bank of correlator outputs as before. Assume that the noise has zero mean $E(\boldsymbol{n}) = 0$ and that it has covariance matrix $E[\boldsymbol{nn}^T] = \sigma^2 \boldsymbol{I} = \boldsymbol{R}$. It is well known that the best estimate in the minimum least-squares sense; i.e., minimizing $E[(\Delta \boldsymbol{x} - \Delta \hat{\boldsymbol{x}})(\Delta \boldsymbol{x} - \Delta \hat{\boldsymbol{x}})^T]$ is to set the estimate $\Delta \hat{\boldsymbol{x}}$ to be as follows:

$$\Delta \hat{\boldsymbol{x}} = (\boldsymbol{G}^T \boldsymbol{R}^{-1} \boldsymbol{G})^{-1} \boldsymbol{G}^T \boldsymbol{R}^{-1} \Delta \boldsymbol{\rho} = \boldsymbol{G}^- \Delta \boldsymbol{\rho} \tag{B2}$$

where it is assumed that the matrix $(\boldsymbol{G}^T \boldsymbol{R}^{-1} \boldsymbol{G})^{-1}$ exists. The mean-squared error in this estimate is $E[(\boldsymbol{x} - \hat{\boldsymbol{x}})(\boldsymbol{x} - \hat{\boldsymbol{x}})^T] = [\boldsymbol{G}^T \boldsymbol{R}^{-1} \boldsymbol{G}]^{-1}$, and the generalized inverse is $\boldsymbol{G}^-$.

Note that expression (B2) implies all of the $\boldsymbol{\rho}$ coordinates measured from different satellites are of equal strength and accuracy. That assumption is removed shortly. This estimate, Eq. (B2), if written in iterative form can be approximated by the feedback loop of Fig. B1 and the estimator gives a measure of each coordinate error produced by the loop $\Delta \hat{x}_i \Delta x_i - \hat{x}_i$. The $\Delta \hat{x}_i$ components are then individually filtered. The user motion in each coordinate and the noise* are assumed to be independent of one another. Differential corrections are made to $\hat{\boldsymbol{\rho}}$ so as to minimize each of the $\Delta \hat{x}_i$ in this closed-loop tracker. The loop filters $F_{xi}(p)$ are each designed to track the dynamics of motion in that particular coordinate, much as discussed in Sec. II of this chapter. The closed-loop response can be made to approximate the realizable Wiener filter for the process.

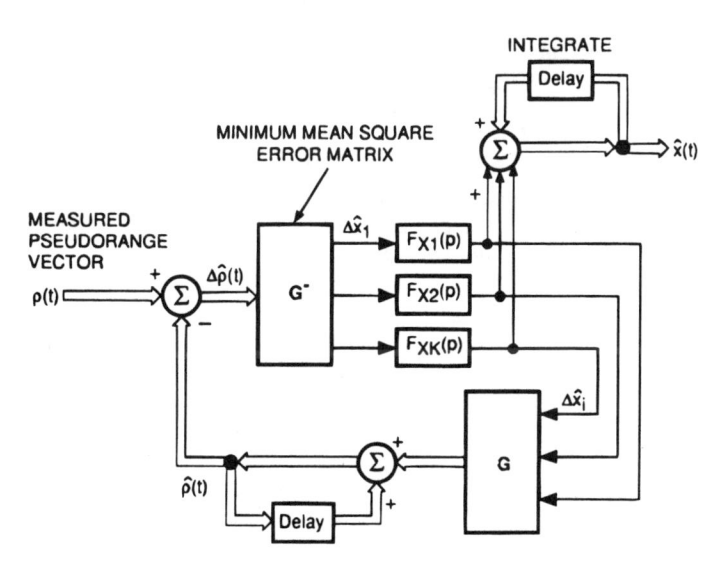

Fig. B1 Simplified form of minimum mean square error position vector estimator configured in iterative form as part of a VDLL. The loop operates so as to make adjustments to ρ that minimize each of the $\Delta \hat{x}_i$. The dynamics of each of the loop filters are designed to be consistent with the dynamics of user motion in that particular coordinate.

---

*In general, the noise may not always be completely independent from coordinate to coordinate.

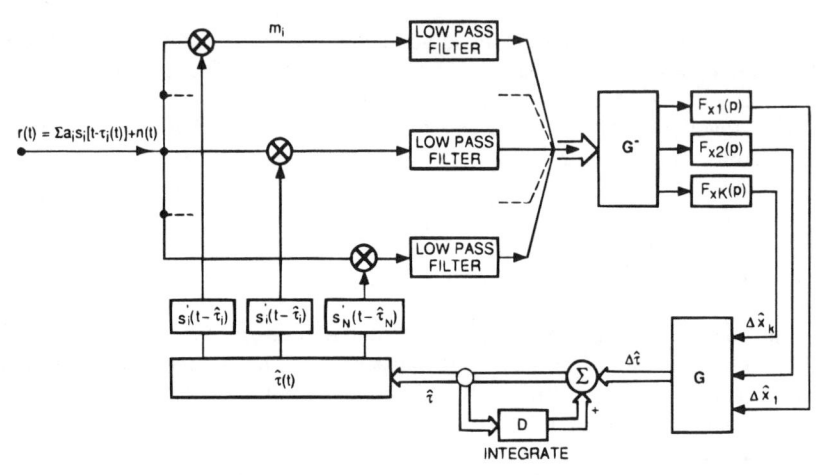

Fig. B2 Vector delay lock loop for baseband PN signals.

Figure B2 shows the next extension of this tracking operation where the input is a baseband received signal:

$$r(t) = \sum_{i=1}^{N} a_i s_i[t - \tau_i(t)] + n(t) \tag{B4}$$

where the a_i represent the signal strengths of the various components. After multiplication by the differentiated reference $s_i'\,[t - \hat{\tau}_i(t)]$ and integration, the output can be approximated for small delay error by the following:

$$\Delta m = A\Delta\tau + n = A[\tau(x) - \hat{\tau}(\hat{x})] + n(t) \cong AG\Delta x + n(t) \tag{B5}$$

where A is a diagonal matrix of signal amplitudes a_i, and $\Delta\tau(x)^T = [\Delta\tau_1(x); \Delta\tau_2(x), \ldots, \Delta\tau_k(x)]$, etc. Because the measurements here are weighted by the signal strength, we use the symbol Δm instead of $\Delta\rho$. The noise components are all assumed to be independent with zero mean and $E[n^T n] = R$ is the noise covariance matrix. The optimal least-squares estimator for Δx for this received signal is then the same as before, except that G is replaced by AG; namely, $\Delta\hat{x} = G^- \Delta m$, where the matrix $G^- = \left((AG)^T R^{-1} AG\right)^{-1} (AG)^T R^{-1}$ is again assumed to exist.

Appendix C:
Noncoherent Delay Lock Loop Noise Performance with Arbitrary Early–Late Reference Spacing

The noncoherent DLL discriminator characteristic is discussed in Sec. II.B. The objective of this Appendix is to analyze the noise performance of a noncoherent DLL with arbitrary early-late delay separation of Δ, where $0 < \Delta < T$. The received signal waveform is assumed to be a trapezoidal-shaped PN sequence with a rise-time δT. The waveform is received with carrier and data modulation as well as stationary additive white Gaussian noise in the IF frequency band with one-sided noise density N_o. The received waveform is as follows:

$$Ad(t + \epsilon)s(t + \epsilon)\cos(\omega_o t + \phi) + N_c(t)\cos(\omega_o t + \phi)$$
$$+ N_s(t)\sin(\omega_o t + \phi) \quad \text{(C1)}$$

where for small δT, the signal power* is $P_s \cong A^2/2$, and the noise spectral density of $N_c(t) \cos(\omega_o t + \phi)$ and $N_s(t) \sin(\omega_o t + \phi)$ are each $N_o/2$. The two noise components $N_c(t)$ and $N_s(t)$ are independent white Gaussian noise terms. Thus, the noise density of $N_c(t)$ and $N_s(t)$ are each N_o. The IF output of the early and late multipliers of Fig. 19b with an early–late spacing of Δ is as follows:

$$m_e(t) = s_o(t + \Delta/2)[Ads(t + \epsilon)\cos(\omega_o t + \phi) N_c \cos(\omega_o t + \phi)$$
$$+ N_s \sin(\omega_o t + \phi)]$$
$$m_L(t) = s_o(t - \Delta/2)[Ads(t + \epsilon)\cos(\omega_o t + \phi) + N_c \cos(\omega_o t + \phi)$$
$$+ N_s \sin(\omega_o t + \phi r)] \quad \text{(C2)}$$

It is useful to represent the early and late reference waveforms in terms of common and difference waveforms:

$$s_o(t + \Delta/2) = s_c(t) + s_d(t)$$
$$s_o(t - \Delta/2) = s_c(t) - s_d(t) \quad \text{(C3a)}$$

where $s_c(t)$ is composed of those portions of the waveform that are common to both early and late waveforms, and $s_d(t)$ is the portion that has opposite sign in each (see Fig. C1). The values of s_o are ± 1, and s_c, s_d are ternary with values $\pm 1, 0$.

The common waveform is nonzero only where the two waveforms $s_o(t + \Delta/2)$ and $s_o(t - \Delta/2)$ have the same sign. The product of the two waveforms is zero; i.e., $s_d(t)$ is nonzero only where $s_c(t)$ is zero, and vice versa. The difference waveform is nonzero only in intervals where there is a transition in the bit pattern.

$$s_c(t) = \frac{1}{2}[x_o(t + \Delta/2) + s_o(t - \Delta/2)]$$
$$s_d(t) = \frac{1}{2}[s_o(t + \Delta/2) - s_o(t - \Delta/2)] \quad \text{(C3b)}$$

and because s_c, s_d are ternary and defined as above, $s_c(t)$ is nonzero only at those instants of time when $s_d(t) = 0$, and the product $s_c(t) s_d(t) = 0$ at all t. The duty factor of $s_d(t)$ is $\Delta/2T$, because $s_d(t)$ is nonzero only for those intervals where there is a transition in $s_o(t)$; i.e., half the time for equally likely ± 1 in $s_o(t)$.

We express the PN waveform as follows:

$$s(t) = \Sigma P_i p(t - iT) \quad \text{(C4)}$$

where P_i is a pseudorandom binary pattern ± 1 and $p(t - iT)$ is a trapezoidal pulse with rise-time δT and duration of approximately T. Likewise, define $s_o(t)$ as follows:

---

*For the trapezoidal signal pulse shape with equally probable $s_o \pm 1$. The actual power is $P_s = (1 - 1/3 \, \delta T/T) \, a^2/2$.

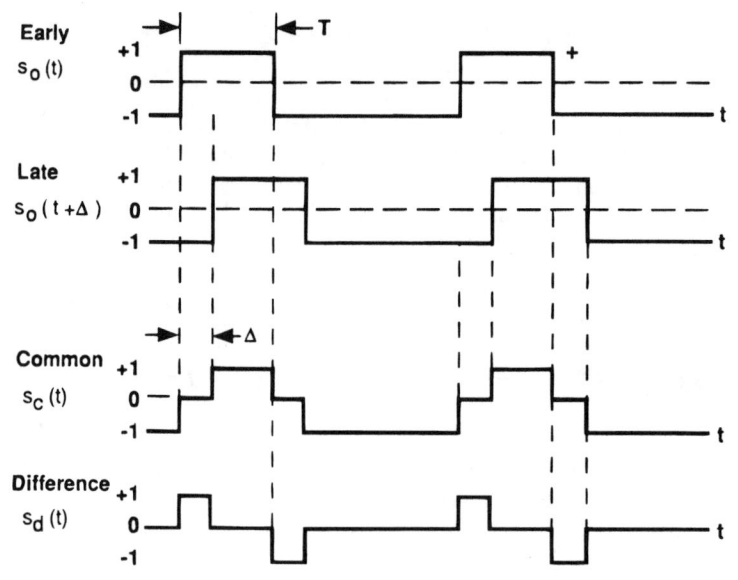

Fig. C1 Decomposition of early and late reference signals into common s_c and difference s_d ternary reference signals. Note that s_d is nonzero only when s_c is zero and $s_d s_c = 0$.

$$s_o(t) = \sum P_i p_o(t - iT) \qquad (C4)$$

where $p_o(t)$ is a zero rise-time pulse of duration T. Both $p(t)$ and $p_o(t)$ are symmetrical about $T = 0$. The multiplier outputs can then be expressed as follows:

$$m_e(t) = [s_c(t) + s_d(t)][Ads(t + \epsilon)\cos(\omega_o t + \phi) + N_c \cos(\omega_o t + \phi)$$
$$+ N_s \sin(\omega_o t + \phi)]$$

$$m_L(t) = [s_c(t) - s_d(t)][Ads(t + \epsilon)\cos(\omega_o t + \phi) + N_c \cos(\omega_o t + \phi)$$
$$+ N_s \sin(\omega t + \phi)] \qquad (C5)$$

Note that the product noise terms $s_d N_c$, and $s_c N_c$ are independent, because s_c and s_d sample the noise at different nonoverlapping time instants; the same is true for $s_c N_s$, and $s_d N_s$.

The outputs of the multipliers are then filtered by IF bandpass filters of bandwidth W, which is much smaller than the bandwidth of the PN sequence and give signal components that are close to their expected value. The signal is assumed to be centered in the bandwidth of the IF filter. The filtered IF outputs M_e and M_L are, thus, as follows:

$$M_e(t) \cong Ad[R_c(\epsilon) + D_d(\epsilon)]\cos(\omega_o t + \phi) + N_{cc} \cos(\omega_o t + \phi)$$
$$+ N_{cd} \cos(\omega_o t + \phi) + N_{sc} \sin(\omega_o t + \phi) + N_{sd} \sin(\omega_o t + \phi)$$

$$M_L(t) \cong Ad[R_c(\epsilon) - D_d(\epsilon)]\cos(\omega_o t + \phi) + N_{cc} \cos(\omega_o t + \phi)$$

FUNDAMENTALS OF SIGNAL TRACKING THEORY

$$- N_{cd}\cos(\omega_o t + \phi) + N_{sc}\sin(\omega_o t + \phi) - N_{sd}\sin(\omega_o t + \phi)$$

where the expected values are defined

$$E[s_c(t)s(t+\epsilon)] = R_c(\epsilon)$$

$$E[s_d(t)s(t+\epsilon)] = D_d(\epsilon) \quad \text{(C6)}$$

which are the autocorrelation and discriminator terms. The term N_{cc} represents the filtered version of the product $s_c(t)N_c(t)$, etc. Because the noise product $s_d(t)N_c$ and $s_d(t)N_s$ are both time-gated with duty factor $\Delta/2T$, the noise terms N_{sd}, N_{cd} both have noise density $N_o\Delta/2T$ and bandwidth W, the IF bandwidth. Likewise, the noise density of the N_{sc}, N_{cc} terms are $N_o[1 - \Delta/2T]$ and also has bandwidth W. However, the filtered terms N_{cc}, etc. are no longer time-gated.

The filtered multiplier outputs of Eq. (C6) can be written in the following form:

$$M_e = B + C$$

$$M_L = B - C \quad \text{(C7)}$$

From Eq. (C6) the B and C terms are as follows:

$$B = AdR_c(\epsilon)\cos(\omega_o t + \phi) + N_{cc}\cos(\omega_o t + \phi) + N_{sc}\sin(\omega_o t + \phi)$$

$$C = AdD_d(\epsilon)\cos(\omega_o t + \phi) + N_{sd}\cos(\omega_o t + \phi) + N_{sd}\sin(\omega_o t + \phi) \quad \text{(C8)}$$

The difference in the squared outputs in the envelope detectors of the early and late channels can be written as follows:

$$M_e^2 - M_L^2 = (B + C)^2 - (B - C)^2 = 4BC \quad \text{(C9)}$$

Thus, the B^2 and C^2 terms cancel. The low-pass filter at the output of $M_e^2 - M_L^2$ operation eliminates the $\cos(\omega_o t + \phi)\sin(\omega_o t + \phi)$ terms and the $2\omega_o t$ terms and results in the following:

$$\frac{1}{2}[M_e^2 - M_L^2|_{\text{low pass}}] = 2BC|_{\text{low pass}}$$

$$= A^2 R_c(\epsilon)D_d(\epsilon) + AR_c(\epsilon)N_{cd} + AD_d(\epsilon)N_{cc}$$

$$+ (N_{cc}N_{cd} + N_{sc}N_{sd}) \quad \text{(C10)}$$

The spectral components in this output are shown in Fig. C2. The output

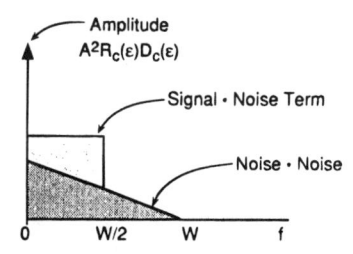

Fig. C2 Spectral components in the output of the square-law detector after differencing. The bandwidth W is the IF bandwidth.

signal-to-noise is then followed by a second low-pass filter that corresponds to the linearized closed-loop DLL noise bandwidth B_n, which set by the loop filter $F(p)$.

The signal component can be written as

$$\frac{1}{2}(M_e^2 - M_L^2)_{\text{low pass}} = A^2 R_c(\epsilon) D_d(\epsilon) \tag{C11}$$

The terms $R_c(\epsilon)$ and $D_d(\epsilon)$ can be expressed in terms of the cross-correlation function of the rectangular signal $s_o(t)$ with the actual signal $s(t)$ in the following form:

$$R_c(\epsilon) = E\left[s_c(t)s(t+\epsilon)\right] = \frac{1}{2}E\left\{\left[s_o\left(t+\frac{\Delta}{2}\right) + s_o\left(t-\frac{\Delta}{2}\right)\right]s(t+\epsilon)\right\}$$

$$= \frac{1}{2}\left[R_s\left(\epsilon + \frac{\Delta}{2}\right) + R_s\left(\epsilon - \frac{\Delta}{2}\right)\right] \tag{C12a}$$

where $R_s(\epsilon) \triangleq E[s_o(t)s(t+\epsilon)]$ and $R_o(\epsilon) \triangleq E[s_o(t)s_o(t+\epsilon)]$. Likewise, the term $D_d(t)$ can be expressed as follows:

$$D_d(t) = E[s_d(t)s(t+\epsilon)B_s = \frac{1}{2}E\left\{\left[s_o\left(t+\frac{\Delta}{2}\right) - s_o\left(t-\frac{\Delta}{2}\right)\right]s(t+\epsilon)\right\}$$

$$= \frac{1}{2}\left[R_s\left(\epsilon - \frac{\Delta}{2}\right) - R_s\left(\epsilon - \frac{\Delta}{2}\right)\right] \tag{C12b}$$

and, thus, multiplying Eq. (C12a) and (C12b) gives the following:

$$R_c(\epsilon)D_d(\epsilon) = \frac{1}{4}\left[R_s^2\left(\epsilon + \frac{\Delta}{2}\right) - R_s^2\left(\epsilon - \frac{\Delta}{2}\right)\right] = \frac{1}{4}D_s(\epsilon)$$

where $D_s(\epsilon)$ is defined as the noncoherent DLL discriminator characteristic; namely, $D_s(\epsilon) \triangleq \left[R_s^2\left(\epsilon + \frac{\Delta}{2}\right) - R_s^2\left(\epsilon + \frac{\Delta}{2}\right)\right]$. Thus, the signal component amplitude is as follows:

$$\frac{1}{2}(M_e^2 - M_L^2)_{\substack{\text{low-pass} \\ \text{signal}}} = A^2 D_s(\epsilon)/4 \cong P_s D_s(\epsilon)/2$$

$$\cong A^2/4\, D_s'(o)\epsilon \quad \text{for} \quad \epsilon \le \Delta \tag{C13}$$

where $dD_s(\epsilon)/d\epsilon|_{\epsilon=0} = D_s'(o)$, and the signal power at the input for this trapezoidal

FUNDAMENTALS OF SIGNAL TRACKING THEORY

waveform is equal to $P_s = A^2/2(1 - 1/3(\delta T/T)) \cong A^2/2$ for small δT. The signal and reference waveform origins are defined so that $R_c(\epsilon)$ is symmetrical about $\epsilon = 0$, with a maximum at $\epsilon = 0$ for this trapezoidal waveform.

The signal times noise term for $\epsilon = 0$ can then be written from Eq. (C10), (C12) as follows:

$$n_{sn} = \frac{1}{2}\left[M_e^2 - M_L^2\right]_{\substack{\text{low-pass}\\ \text{noise} \cdot \text{signal}}}$$

$$= \frac{A}{2}\left[R_s\left(\epsilon + \frac{\Delta}{2}\right) + R_s\left(\epsilon - \frac{\Delta}{2}\right)\right]N_{cd}(t)$$

$$+ \frac{A}{2}\left[R_s\left(\epsilon + \frac{\Delta}{2}\right) - R_s\left(\epsilon - \frac{\Delta}{2}\right)\right]N_{cc}(t) \quad \text{(C14)}$$

$$= AR_c\left(\frac{\Delta}{2}\right)N_{cd}(t) \quad \text{for} \quad \epsilon = 0$$

because the second term cancels at $\epsilon = 0$ and $R_s(\Delta/2) = R_s(-\Delta/2)$.

The power spectral density of the signal times noise term $n_{sn}(t)$ at baseband is then $N_{osn} = 2A^2R_s^2(\Delta/2)N_o(\Delta/2T)$ for $\epsilon = 0$, where the factor of 2 appears because of the spectral foldover of the IF spectrum when converted to baseband. The signal times noise power in bandwidth B_n is then as follows:

$$P_{sn} = 2A^2R_s^2(\Delta/2)N_o(\Delta/2T)B_n \quad \text{for} \quad \epsilon = 0 \quad \text{(C15)}$$

The noise · noise in the low-pass difference is the following:

$$\frac{1}{2}[M_e^2 - M_L^2]_{\substack{\text{low-pass}\\ \text{noise} \cdot \text{noise}}} = N_{cc}N_{cd} + N_{sc}N_{sd}$$

The noise power spectral density at the origin ($t = 0$) for the noise times noise term $N_{cc}N_{cd} + N_{sc}N_{sd}$

$$N_{nn} = 2N_o^2(\Delta/2T)[1 - \Delta/2]W \quad \text{(C16)}$$

and the output noise · noise power is as follows:

$$P_{nn} = 2N_o^2(\Delta/2T)[1 - \Delta/2T]WB_n \quad \text{(C17)}$$

Then the total output noise power is the sum of Eqs. (C15) and (C17):

$$P_n = 2A^2R_s^2(\Delta/2)N_o(\Delta/2T)B_n + 2N_o^2(\Delta/2T)[1 - \Delta/2T]WB_n \quad \text{(C18)}$$

If the received signal is rectangular with $\delta T = 0$, then $s(t) = s_o(t)$, and $R_s(\epsilon) = R_o(\epsilon)$, and

$$P_n = 2A^2[1 - \Delta/2T]^2(\Delta/2T)N_oB_n + 2(N_oW)(1 - \Delta/2T)(\Delta/2T)N_oB_n \quad \text{(C19a)}$$

because $R_o(\epsilon) = 1 - |\epsilon/T|$ for $|\epsilon| < T$. Because $A^2/2 = P_s$, the signal power, and

N_oW is the IF noise power P_{nl}, the total noise power in the output for $\delta T = 0$ is as follows:

$$P_n = 4P_s(\Delta/2T)(1 - \Delta/2T)(N_oB_n)[(1 - \Delta/2T) + (P_{nl}/2P_s)] \quad \text{(C19b)}$$

Note that the noise power decreases in direct proportion to the time-gating duty factor $\Delta/2T$, as discussed earlier. However, performance improvement in output delay noise power does not diminish indefinitely with Δ for finite rise-time signals. As pointed out earlier, the optimum result comes at $\Delta = \delta T$, the rise-time of the waveform.

To translate this noise power of Eqs. (C18) and (C19) to rms delay error, we must divide the noise power by the squared slope in the effective discriminator characteristic; namely, $\left[\dfrac{d}{d\epsilon}[A^2D_s(\epsilon)/2]\right]^2$ from Eq. (C13). For small error, the slope is $D_s'(0)$. Thus, the mean-square delay noise normalized to T^2 is given by the following:

$$(\sigma_\epsilon/T)^2 = \dfrac{2P_s(\Delta/2T)(1 - \Delta/2T)N_oB_n\left[(1 - \Delta/2T) + \dfrac{P_{nl}}{P_s}\right]}{\left[\dfrac{P_s^2}{4}[D_s'(0)]^2T^2\right]}$$

$$= \dfrac{2(\Delta/2T)(1 - \Delta/2T)N_oB_n\left[(1 - \Delta/2T) + \dfrac{P_{nl}}{P_s}\right]}{[P_sD_s(0)^2T^2/4]} \quad \text{(C20)}$$

For the example where $\Delta = T$, $D_s'(0) = 1/(.5T) = 2/T$. This expression reduces to the following:

$$(\sigma_\epsilon/T)^2 = \dfrac{2\left(\dfrac{1}{2}\right)\left(1 - \dfrac{1}{2}\right)N_oB_n\left[(1 - 1/2) + \dfrac{P_{nl}}{P_s}\right]}{P_s(4/T^2)(T^2/4)} = \left(\dfrac{N_oB_n}{2P_s}\right)\left[1 + 2\dfrac{P_{nl}}{P_s}\right] \quad \text{(C21)}$$

where P_s/P_{nl}, is the IF signal-to-noise ratio.

Discriminator Characteristic for the Noncoherent DLL vs Early-Late Separation

As shown in Eq. (C20), the noncoherent DLL discriminator characteristic and the slope of that discriminator curve at $\epsilon = 0$ are key parameters of the DLL. These parameters are calculated vs early-late delay spacing Δ for both received signals with both zero rise-time pulse and trapezoidal PN pulses with a rise-time of δT s.

Zero Rise-Time Received Signal

The autocorrelation function and differentiated autocorrelation function for a PN waveform $s_o(t)$ with chip duration T s and zero rise-time are shown in Fig. C3.

FUNDAMENTALS OF SIGNAL TRACKING THEORY

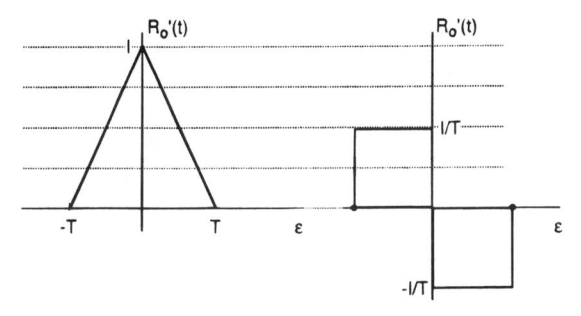

Fig. C3 Autocorrelation function and differentiated autocorrelation function for a zero rise-time PN waveform of $s_o(t)$ chip duration T.

The discriminator characteristic for a noncoherent DLL for a zero rise-time received signal $s_o(t + \epsilon)$ and reference waveforms $s_o(t + \Delta/2)$ and $s_o(t - \Delta/2)$, is as follows:

$$D_o(\epsilon) = R_o^2(\epsilon + \Delta/2) - R_o^2(\epsilon - \Delta/2) \quad (C22)$$

where Δ is the early-late separation. The slope of this discriminator characteristic is then as follows:

$$D_o'(\epsilon) = \frac{d}{d\epsilon}[D_o(\epsilon)] = 2[R_o(\epsilon + \Delta/2)R_o'(\epsilon + \Delta/2) - R_o(\epsilon - \Delta/2)R_o'(\epsilon - \Delta/2)] \quad (C23)$$

and at $\epsilon = 0$

$$D_o'(0) = 4R_o(\Delta/2)R_o'(\Delta/2)$$

$$= 4[1 - \Delta/2T][-1/T] = \frac{4}{T}\left[1 - \frac{\Delta}{2T}\right] \quad \text{for} \quad 0 < \Delta < T \quad (C24)$$

Trapezoidal Waveform Received Signal with Rise-Time δT

If the received signal is a finite rise-time trapezoidal waveform with rise-time δT, shown in Fig. C4, and the reference waveforms are early and late versions

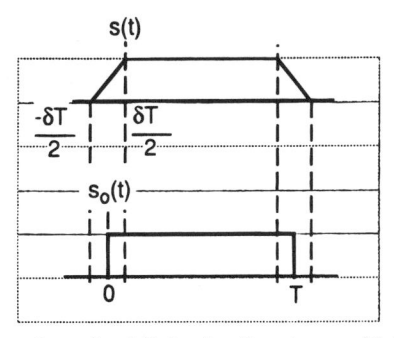

Fig. C4 Reference s_o and received finte rise-time trapezoidal signals $s(t)$. The rise-time is δT.

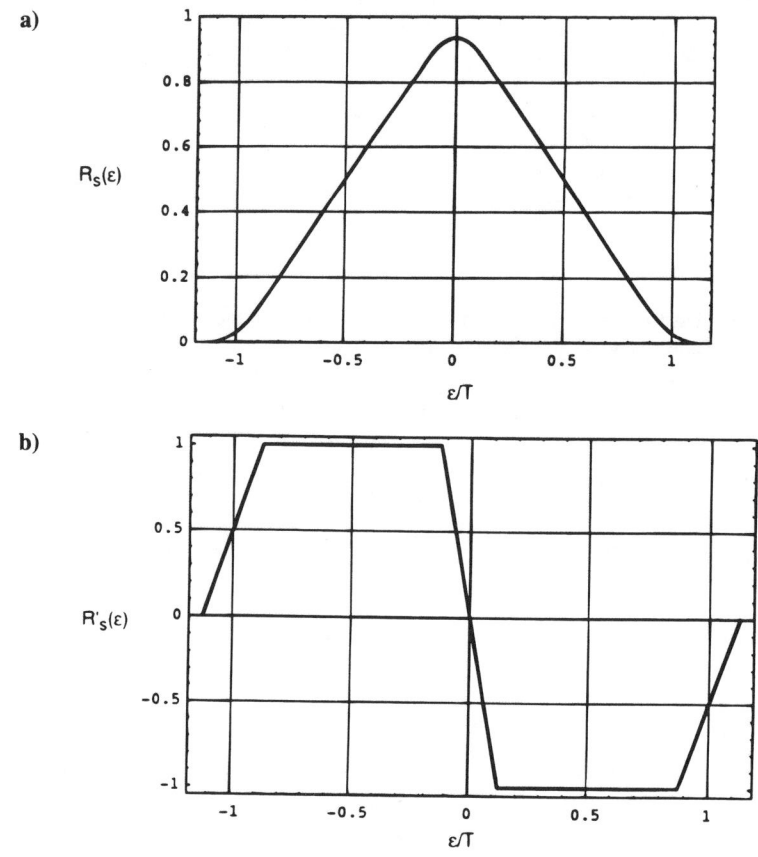

Fig. C5 Cross-correlation and differentiated correlation functions for a trapezoidal waveform: a) cross-correlation function $R_s(\epsilon)$ between the trapezoidal signal $s(t)$ with normalized rise-time $\delta T/T = 0.25$ and $s_o(t)$ with zero rise-time; b) differentiated correlation function $R'_s(\epsilon) = (d/d\epsilon) R_s(\epsilon)$ for the trapezoidal waveform with $\delta T/T = 0.25$.

of $s_o(t)$, then the crosscorrelation $R_s(\epsilon)$ between $s(t + \epsilon)$ and $s_o(t)$ is shown in Fig. C5 for $T = 1.0$ and $\delta T/T = 0.25$ where $R_s(\epsilon)$ is defined as follows:

$$R_s(\epsilon) = E[s_o(t)s(t + \epsilon)] \tag{C25}$$

For the trapezoidal-shaped waveform with rise-time δT, the expression for $R_s(t)$ is as follows:

$$R_s(\epsilon) = 1 - \left(\frac{\delta T}{4T} - \frac{\epsilon^2}{T\delta T}\right) \quad \text{for} \quad |\epsilon| < \delta T/2;$$

$$= 1 - \frac{|\epsilon|}{T}, \quad \delta T/2 < |\epsilon| < T = \delta T/2; \tag{C26}$$

$$\left[1 - \left(\frac{T - \delta T}{2}\right)\right]\left[\left(\frac{\epsilon - T - \delta T}{2}\right)^2 (\delta T)^2\right], \quad \frac{T - \delta T}{2} < \epsilon < \frac{T + \delta T}{2}$$

and the differentiated correlation waveform

$$R'_s(\epsilon) = \frac{\partial}{\partial \epsilon} R_s(t) = \frac{-2\epsilon}{T\delta T} \quad \text{for } |\epsilon| < \delta T/2$$

$$= -1/T \quad \delta T/2 < \epsilon < T - \delta T/2$$

$$= \left[1 - \left(\frac{T - \delta T}{2}\right)\right] 2\left(\frac{\epsilon - T - \delta T}{2}\right)/(\delta T)^2, \quad \frac{T - \delta T}{2} < \epsilon < \frac{T + \delta T}{2} \quad (C27)$$

as shown in Fig. C5b.

Thus, the discriminator curve in $D_s(\epsilon)$ for the finite rise-time waveform is related to the cross-correlation function $R_s(\epsilon)$ by the following:

$$D_s(\epsilon) = R_s^2(\epsilon + \Delta/2) - R_s^2(\epsilon - \Delta/2) \quad (C28)$$

The slope of the discriminator characteristic at $\epsilon = 0$ is then as follows:

$$D'_s(\epsilon) = 4R_s(\Delta/2)R'_s(\Delta/2) \quad (C29)$$

The slope of the discriminator characteristic $D_s'(0)$ is then equal to the following:

$$D'_s(0) = -4R_s(\Delta/2)R'_s(\Delta/2)$$

$$= 4\left(1 - \frac{\delta T}{4T} - \frac{(\Delta/2)^2}{T\delta T}\right)\left(\frac{-2(\Delta/2)}{T\delta T}\right) \quad \text{for } |\Delta| < \delta T$$

$$= 4(1 - |\Delta|/2T)(-1/T) \quad \delta T < \Delta < T - \delta T/2$$

$$= -4(1 - |\Delta|/2T)(1/T) \quad -(t - \delta T/2) < \Delta < -\delta T/2 \quad (C30)$$

At $\Delta = \delta T$, we have the following:

$$D'_s(0) = -4/T(1 - \delta T/2T) \quad (C31)$$

This slope is as plotted in normalized form in Fig. 23. As can be seen, the peak slope, the optimum, occurs at an early–late separation Δ equal to the signal rise-time (i.e., $\Delta = \delta T$).

Appendix D:
Probability of Losing Lock for the Noncoherent DLL

Assume an input signal of the form $\sqrt{2P_s}\, d(t)s(t)\cos(\omega_0 t + \phi) + n(t)$, where d and s are binary ± 1, P_s is the average signal power, and the noise is white Gaussian noise. Then, the stochastic differential equation of the noncoherent DLL with the early–late delay difference of $\Delta = T$ is

$$\hat{\tau}(t) = 2KP_s \frac{F(p)}{p}\left[D(\epsilon) + \frac{n_s(t)}{2P}\right] \quad (D1)$$

where $n_s(t) = N_1(t) + N_2(t)$ represents the noise output of the square-law detector $N_1(t)$ representing the $n(t) \cdot$ reference term, and $N_2(t)$ represents the $n(t) \cdot n(t)$ term. The $D(\epsilon)$ and $\epsilon = \tau - \hat{\tau}$ are the normalized (to unity gain at $\epsilon = 0$) DLL discriminator characteristic, and normalized tracking error for zero rise-time signals, respectively, and K is the NCO gain.

Define $g(\epsilon, t)$ as the probability density function for normalized delay error at time t. Define $\bar{T}$ as the mean time for the DLL to lose lock, which in this example, occurs at $|\epsilon| = (3/2)$, because the discriminator curve becomes zero at $|\epsilon| \geq (3/2)$. Set the initial condition as zero delay error at $t = 0$. Thus, the conditions on $q(\epsilon, t)$ are $q(t, 0) = \delta(\epsilon)$; $q(\tau, t) = 0$ for $|\epsilon| \geq 3/2$; and, $q(\tau, t) = q(-\epsilon, t)$, where ϵ is normalized to T. As time increases in the limit, $q(\epsilon, \infty) = 0$.

The probability density $q(\epsilon,t)$ for the delay error ϵ varies with time t and satisfies the Fokker Planck equation.[3,14,20,31]

$$\frac{\partial q(\epsilon, t)}{\partial t} = \frac{-\partial}{\partial \epsilon}[A_1(\epsilon)q(\epsilon, t)] + \frac{1}{2}\frac{\partial^2}{\partial t^2}[A_2(t)q(\epsilon, t)] \quad (D2)$$

where

$$A_1(\epsilon) = \lim_{\Delta t \to 0}\left[\frac{E(\Delta\epsilon | \epsilon)}{\Delta t}\right] = -2\alpha P_s K D_n(\epsilon),$$

$$A_2(\epsilon) = \lim_{\Delta t \to 0}\left[\frac{E[(\Delta\epsilon)^2 | \epsilon]}{\Delta t}\right] = \frac{N_o K^2}{2}$$

where $\Delta\epsilon$ is the incremental delay error in time Δt, and $D_n(t)$ is the normalized discriminator characteristic

$$D_n(t) = D(\epsilon)/\left|\frac{dD}{d\epsilon}\right|_{\epsilon=0}$$

The probability that the delay error ϵ has *not* yet reached the threshold limit which for this DLL is $\epsilon = \pm (3/2)$, at the time t is as follows:

$$\phi(t) = \int_{-3/2}^{-3/2} q(\epsilon, t)d\epsilon \leq 1 \quad (D3)$$

Likewise, define the probability that the threshold condition is reached at time t as $\psi(t) = 1 - \phi(t)$ where $\psi(0) = 0$, and $\psi(\infty) = 1$.

Consequently, the mean time to lose lock; i.e., the first passage time to reach the threshold value of ϵ is as follows:

$$\bar{T} = \int_0^\infty t\frac{\partial \psi}{\partial t}dt = \int_0^\infty -t\frac{\partial \phi}{\partial t}dt = \int_0^\infty \phi(t)dt$$

$$= \int_0^\infty \int_{-3/2}^{3/2} q(\epsilon, t)d\epsilon\, dt \quad (D4)$$

where we have integrated by parts and used the fact that $\phi(t)$ must decrease faster than $1/t$ to obtain this result. Integrate the Fokker–Planck equation $(0,\infty)$

to obtain the following:

$$-\delta(\epsilon) = 2\alpha P K \frac{d}{d\epsilon}[D_n(\epsilon)Q(\epsilon)] + \frac{N_o K^2}{2}\frac{d^2}{d\epsilon^2}[Q(\epsilon)]$$

where

$$Q(\epsilon) \stackrel{\Delta}{=} \int_0^\infty q(\epsilon, t)dt$$

and $Q(-3/2) = Q(3/2) = 0$. The solution to this equation; i.e., Green's function for this problem, can be inserted in Eq. (24) to obtain the expected time to lose lock. This result can be expressed in terms of the closed-loop noise bandwidth B_n, the normalized tracking error variance of the loop $\sigma\epsilon^2$, and the IF bandwidth W, and can be shown to be the following[20]:

$$2\bar{T}B_n = \frac{1}{4\sigma_\epsilon^2}\int_0^{3/2}\int_\epsilon^{3/2} e^{\left[\frac{-g(\epsilon)}{\sigma_\epsilon^2}\right]} e^{\left[\frac{g(\epsilon')}{\sigma_\epsilon^2}\right]} d\epsilon\, d\epsilon'$$

where

$$\sigma_\epsilon^2 = \frac{B_n N_o}{2P_s}\left[1 + \frac{2N_o W}{P_s}\right] \tag{D5}$$

where P_s is the signal power, and $g(\epsilon)$ is related to the noise-free DLL discriminator curve $D(\epsilon)$ by

$$g(\epsilon) = \frac{1}{2}\int_0^\epsilon D(\epsilon')d\epsilon' \tag{D6}$$

Note that $g(\epsilon)$ is an even function of ϵ. (Note that Ref. 20 uses G for g.) The normalized expected time to lose lock is given in the text in Fig. 25.

Appendix E:
Colored Measurement Noise in the Vector Delay Lock Loop

In Sec. III, the effective noise samples have been assumed to be independent from time instant to time instant. Although the thermal noise samples at the receiver input are generally independent, there are perturbations to the delay τ_i that are not white, and have an impact on the Kalman gain calculations. More precisely, the received signal is of the following form:

$$\sum a_i s_i(t - \tau_i(\mathbf{x}) - \Delta\tau_i) + n(t) \tag{E1}$$

where $\Delta\tau_i$ represents colored noise errors caused by atmospheric delay fluctuations,* selective availability (SA) or satellite ephemeris errors,† and $n(t)$ is white thermal noise. Atmospheric delay error varies slowly with time, and definitely is not white noise. As before, we represent the position process equation as $\mathbf{x}_{k+1} = F_k \mathbf{x}_k + \mathbf{w}_k$.

---

*See Chapters 12 and 13, this volume, on Ionospheric Effects and Tropospheric Effects, respectively.
†See Ref. 32.

After the signal is correlated with the differentiated delayed reference in a given channel, the output for satellite signal j is as follows:

$$s_j(t - \tau_j(x) - \Delta\tau_j)s_j'(t - \tau_j(\hat{x})) + n_{rj}(t) \tag{E2}$$

where the random noise component $n_{rj}(t)$ is also white. We neglect the self-noise or lump it in $n_{rj}(t)$. For $\tau_j(x) \approx \tau_j(\hat{x})$, we have the approximation for the multiplier output with its colored noise:

$$m_j(t) \cong a_{jk}R_s''(0)[\epsilon_j + \Delta\tau_j] + n_{rj}(t)$$
$$= a_{jk}R_s''(0)\epsilon_j(t) + n_{rj}(t) + \Delta\tau_j R_s''(0) \tag{E3}$$

Represent $\Delta\tau_j R_s''(0)$ as a Markov process noise $n_{\Delta jk}$. At discrete sample times t_k, the multiplier output is then as follows:

$$m_{jk} = a_j R_s''(0)\epsilon_{jk} + [n_{rjk} + n_{\Delta jk}] \tag{E4}$$

and in vector form as follows:

$$\mathbf{m}_k = A_k R_s''(0)\boldsymbol{\epsilon}_k + \mathbf{n}_{rk} + \mathbf{n}_{\Delta k} = A_k R_s''(0)\boldsymbol{\epsilon}_k + \mathbf{n}_k \tag{E5}$$

where $\mathbf{n}_{rk}$ represents independent noise samples and $\mathbf{n}_{\Delta k}$ is colored measurement noise, and the sum of the two random noise perturbations is represented by $\mathbf{n}_k$, which is a Markov process;

$$\mathbf{n}_k = C_{k-1}\mathbf{n}_{k-1} + \boldsymbol{\eta}_{k-1} \tag{E6}$$

where $\boldsymbol{\eta}_k$ is white Gaussian noise. Finally, assume that estimate $\hat{x}$ is sufficiently close to x that the linear geometry matrix G_k can be employed. Thus, the observation is expressed by the following:

$$\mathbf{m}_k = A_k R_s''(0) G_k \Delta\mathbf{x}_k + \mathbf{n}_k$$
$$= H_k \Delta\mathbf{x}_k + \mathbf{n}_k \tag{E7}$$

where $\mathbf{n}_k$ is a Gauss–Markov measurement noise which is *not* white and $\mathbf{n}_k$ has zero mean and a covariance matrix R_k at time t_k.

Let us change variable by offsetting $\mathbf{m}_k$ to define a new quantity $\overline{\mathbf{m}}_k$ so that the measurement noise in $\overline{\mathbf{m}}_k$ is white in order to use the standard extended Kalman filter theory (see Refs. 11, 23, 33, 34). Define the new variable and use Eq. (E8) to obtain the following:

$$\overline{\mathbf{m}}_{k+1} \triangleq \mathbf{m}_{k+1} - C_k \mathbf{m}_k = H_{k+1}\Delta\mathbf{x}_{k+1} + \mathbf{n}_{k+1} - C_k[H_k \Delta\mathbf{x}_k + \mathbf{n}_k] \tag{E8}$$

If we write $\Delta\mathbf{x}_{k+1} = F_k \Delta\mathbf{x}_k + G_k \mathbf{w}_k$ and use Eq. (E.5), then this expression becomes the following:

$$\overline{\mathbf{m}}_{k+1} = H_{k+1}(F_k\Delta\mathbf{x}_k + G_k\mathbf{w}_k) + (C_k\mathbf{n}_k + \boldsymbol{\eta}_k) - [H_k\Delta\mathbf{x}_k + \mathbf{n}_k]$$
$$= (H_{k+1}F_k - B_kH_k)\Delta\mathbf{x}_k + \boldsymbol{\eta}_k + H_{k+1}G_k\mathbf{w}_k = \overline{H}_{k+1}\Delta\mathbf{x}_k + \boldsymbol{\eta}_k \overline{G}_k\mathbf{w}_k \tag{E9}$$

where $\overline{H}_k = (H_{k+1}F_k - C_kH_k)$ and $\overline{G}_k = H_{k+1}G_k$ and set $\overline{\mathbf{m}}_o = \mathbf{m}_o = H_o\delta\mathbf{x}_o + \mathbf{n}_o$.

Note that the additive noise term in the expression for $\overline{\mathbf{m}}_{k+1}$ is $\boldsymbol{\eta}_k + \overline{G}_k\mathbf{w}_k$ and is now white for this definition of $\overline{\mathbf{m}}_{k+1}$. Note, however, that the measurement

FUNDAMENTALS OF SIGNAL TRACKING THEORY

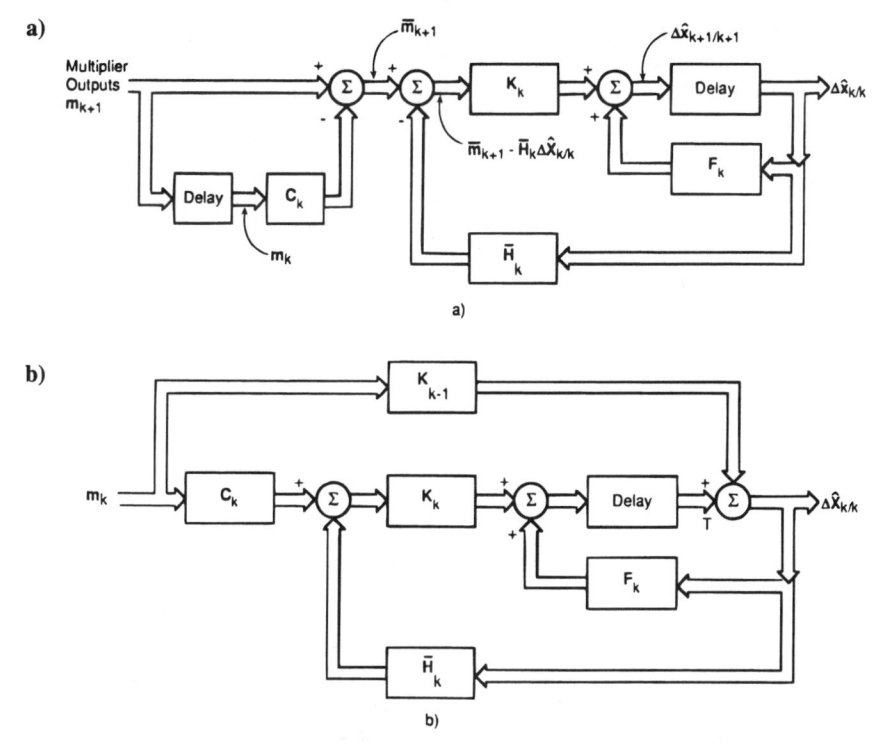

Fig. E1 Two configurations of the block diagram of the modified extended Kalman filter in the VDLL optimized for Gauss–Markov measurement (colored) noise in the multiplier outputs. Colored noise can appear because the atmospheric delay error and selective availability or bias-type errors in the delay variable are not independent from sample to sample. The modified multiplier output term $\overline{m}_k$ has white measurement noise.

noise for $\overline{m}_{k+1}$ now is no longer independent of the process noise w_k. However, the dimensionality of $\overline{m}_k$ is exactly the same as for m_k. Thus, the standard Kalman filter algorithm can be developed for this process in estimating x_k given the samples 0, 1, 2, ... k, namely $\hat{x}_{k/k}$ if the proper definitions are employed. The resulting VDLL filters are then shown in Fig. E1. The estimate is $\Delta\hat{x}_{k+1/k+1} = F_k\hat{x}_{k/k} + K_k(\overline{m}_{k+1} - \overline{H}_k\Delta\hat{x}_{k/k})$.

References

[1] Spilker, J. J., Jr., and Magill, D. T., "The Delay Lock Discriminator—An Optimum Tracking Device," *Proceedings of the IRE,* 1961, pp. 1403–1416.

[2] Spilker, J. J., Jr., "Delay Lock Tracking of Binary Signals," IRE Trans. SET, 1963, pp. 18.

[3] Spilker, J. J., Jr., *Digital Communications by Satellite,* Prentice Hall, Englewood Cliffs NJ, 1977.

[4] Spilker, J. J., Jr., "Vector Delay Lock Loop Tracking—Position Estimation," Paper presented at the IEEE Communication Theory Workshop, FL, April, 1993. See also U.S.

Patent 5,398,034 March, 1995, Vector Delay Lock Loop Processing of Radiolocation Transmitter Signals.

[5]Cramer, *Mathematical Methods of Statistics*, Princeton Univ. Press, Princeton, NJ, 1946.

[6]Stark, H., and Woods, J. W., "Probability and Estimation Theory for Engineers," Prentice-Hall, Engelwood Cliffs, NJ, 1986.

[7]Anderson, T. W., *An Introduction to Multivariate Statistical Analysis*, Wiley, New York, 1958.

[8]Feller, W., "Probability Theory and its Applications," Wiley, New York, 1950.

[9]Stiffler, J. J., "Theory of Synchronous Communications," Prentice Hall, Englewood Cliffs, NJ, 1971.

[10]Kay, S. M., "Statistical Signal Processing—Estimation Theory," Prentice Hall, Engelwood Cliffs, NJ, 1993.

[11]Anderson, B. D. O., and Moore, J. B., *Optimal Filtering*, Prentice Hall, Englewood Cliffs, NJ, 1979.

[12]Viterbi, A. J., *Principles of Coherent Communications*, McGraw-Hill, New York, 1966.

[13]Helstrom, C. W., *Statistical Theory of Signal Detection*, Pergamon Press, New York, 1960.

[14]Meyer, H., and Ascheid, G., *Synchronization in Digital Communications*, Vol. 1, *Phase, Frequency Locked Loops and Amplitude Control*, Wiley, New York, 1990.

[15]Simon, M. K., Omura, J. K., Scholtz, R. A., and Levitt, B. K., *Spread Spectrum Communications*, Vols. 1, 2, 3, Computer Science Press, Rockwell, MD, 1985.

[16]Van Dierendonck, A. J., Fenton, P., and Ford, T., "Theory and Performance of Narrow Correlator Spacing in a GPS Receiver," *Navigation*, 1992.

[17]Bellman, R., *Adaptive Control Processes—A Guided Tour*, Princeton University Press, Princeton, NJ, 1961.

[18]Gill, W. J., "A Comparison of Delay Lock Loop Implementations," *IEEE Transactions on Aerospace and Electronic Systems*, 1966.

[19]Holmes, J. K., "Delay Lock Loop Mean Time to Lose Lock," *IEEE Transactions on Communications*, 1978, pp. 1549–1556.

[20]Holmes, J. K., *Coherent Spread Spectrum Systems*, Krieger, Melbourne, FL, 1990.

[21]Brown, R. G., and Hwang, P. Y. C., *Introduction to Random Signals and Applied Kalman Filtering*, John Wiley, New York, 1992.

[22]Copps, E. M., et al., "Optimal Processing of GPS Sensors," *Navigation*, 1980, pp. 171–182.

[23]Gelb, A. (ed.), *Applied Optimal Estimation*, MIT Press, Cambridge, MA, 1974.

[24]Magill, D. T., "Noise Theory of Tracking Correlators," *IEEE International Conference on Communication*, 1965.

[25]Sorenson, H. W., *Kalman Filtering: Theory and Applications*, IREEE Press, New York, 1985.

[26]Sennott, J. W., and Senffner, D., "The Use of Satellite Geometry for Prevention of Cycle Slips in a GPS Processor," *Navigation*, 1992.

[27]Iltis, R. A., Mailaender, L., "An Adaptive Multiuser Detection with Joint Amplitude and Delay Estimation," *IEEE Transactions on Communication*, 1994.

[28]Jazwinski, A. H., *Stochastic Processes and Filtering Theory*, Academic Press, New York, 1970.

[29]Wozencraft, J. N., and Jacobs, I. M., *Principles of Communications Engineering*, Wiley, New York, 1967.

[30]Cover, T. M., and Thomas, J. A., *Elements of Information Theory,* Wiley, New York, 1991.

[31]Doob, J. L., *Stochastic Processes,* Wiley, New York, 1953.

[32]Chou, H., "An Anti-SA Filter for Nondifferential GPS Users," *Proceedings of ION GPS-90,* (Colorado Springs, CO), Institute of Navigation, Washington, DC, Sept. 19–21.

[33]Sage, A. P., and Melsa, J. L., *Estimation Theory with Applications to Communications and Control,* McGraw-Hill, New York, 1971.

[34]Bryson, A. E., Jr., and Henrikson, L. J., "Estimation Using Sampled-Data Containing Sequentially Correlated Noise," *Journal of Spacecraft and Rockets,* 1968.

[35]Spilker, J. J., Jr, U.S. Patent No. 5,477,195, "Near Optimal Quasi-Coherent Delay Lock Loop (QC DLL) for Tracking Direct Sequence Signals and CDMA," SN 08/353, 208, Dec. 19, 1995.

Chapter 8

GPS Receivers

A. J. Van Dierendonck
AJ Systems, Los Altos, CA 94024

I. Generic Receiver Description
A. Generic Receiver System Level Functions

A SYSTEM level functional block diagram of a generic GPS receiver is shown in Fig. 1. The generic receiver consists of the following functions: 1) antenna; 2) preamplifier; 3) reference oscillator; 4) frequency synthesizer; 5) downconverter; 6) an intermediate frequency (IF) section; 7) signal processing; and; 8) applications processing.

In a general sense, not all GPS receivers perform navigation processing. Many perform time transfer or differential surveying, or simply collect measurement data. Thus, the last function is more appropriately called applications processing, thus covering a broad set of applications.

The antenna may consist of one or more elements and associated control electronics, and may be passive or active, depending upon its performance requirements. Its function is to receive the GPS satellite signals while rejecting multipath and, if so designed, interference signals. The preamplifier generally consists of burnout protection, filtering, and a low-noise amplifier (LNA). Its primary function is to set the receiver's noise figure and to reject out-of-band interference.

The reference oscillator provides the time and frequency reference for the receiver. Because GPS receiver measurements are based on the time-of-arrival of pseudorandom noise (PRN) code phase and received carrier phase and frequency information, the reference oscillator is a key function of the receiver. The reference oscillator output is used in the frequency synthesizer, from which it derives local oscillators (LOs) and clocks used by the receiver. One or more of these LOs are used by the downconverter to convert the radio frequency (rf) inputs to intermediate frequencies (IFs) that are easier to process in the IF section of the receiver.

The purpose of the IF section is to provide further filtering of out-of-band noise and interference and to increase the amplitude of the signal-plus-noise to a workable signal-processing level. The IF section may also contain automatic

---

Copyright © 1995 by the author. Published by the American Institute of Aeronautics and Astronautics, Inc., with permission. Released to AIAA to publish in all forms.

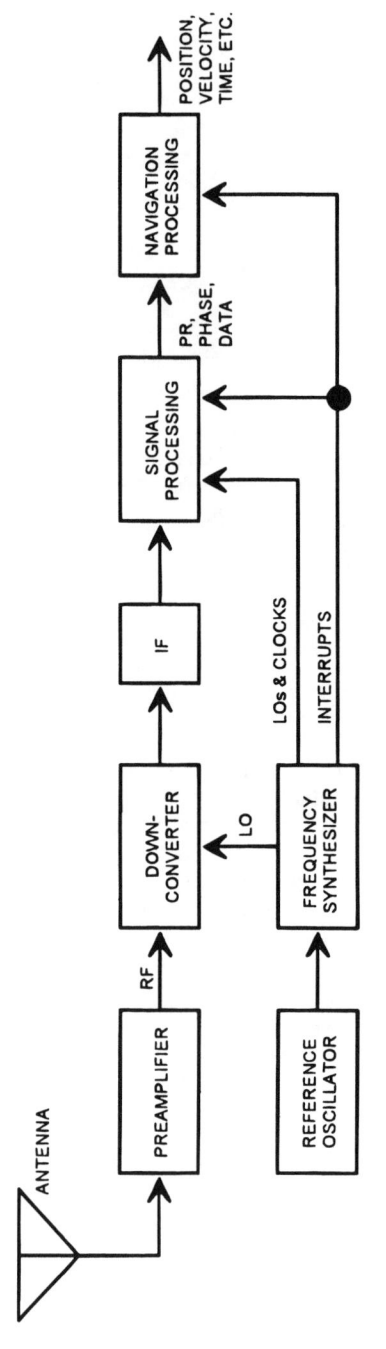

Fig. 1 Generic GPS receiver functional block diagram.

gain control (AGC) circuits to control that workable level, to provide adequate dynamic range, and to suppress pulse-type interference.

The signal-processing function of the receiver is the core of a GPS receiver, performing the following functions:

1) splitting the signal-plus-noise into multiple signal-processing channels for signal-processing of multiple satellites simultaneously;

2) generating the reference PRN codes of the signals;

3) acquiring the satellite signals;

4) tracking the code and the carrier of the satellite signals;

5) demodulating the system data from the satellite signals;

6) extracting code phase (pseudorange) measurements from the PRN code of the satellite signals;

7) extracting carrier frequency (pseudorange rate) and carrier-phase (delta pseudorange) measurements from the carrier of the satellite signals;

8) extracting signal-to-noise ratio (SNR) information from the satellite signals; and

9) estimating a relationship to GPS system time.

The outputs of the signal-processing function are pseudoranges, pseudorange rates and/or delta pseudoranges, signal-to-noise ratios, local receiver time tags, and GPS system data for each of the GPS satellites being tracked, all of which are used by the applications-processing function. Most of this chapter is devoted to the details of the signal-processing function.

The applications-processing function controls the signal-processing function and uses its outputs to satisfy application requirements. These requirements vary with application. Although GPS is primarily a satellite navigation system, the applications of a GPS receiver are diverse. Some of the other applications with significantly differing processing requirements are as follows:

1) time and frequency transfer;

2) static and kinematic surveying;

3) ionospheric total electron content (TEC) and amplitude and phase scintillation monitoring;

4) differential GPS (DGPS) reference station receivers; and

5) GPS satellite signal integrity monitoring.

The common link between these diverse applications is that they all use the same signal-processing measurements in one form or another. However, because of bandwidth and accuracy requirements imposed by these various applications, the requirements on signal-processing function also differ. In general, GPS receivers do not meet the signal-processing and applications-processing requirements for all the applications. Special processing is required for some of the applications.

B. Design Requirements Summary

In addition to the variation in signal- and applications-processing requirements with application, the receiver front-end functions will also have varying design requirements with application. In the following, these variations are noted by function, with details provided later in the chapter.

1. Antenna

The parameters that dictate the antenna requirements are as follows: gain vs azimuth and elevation, multipath rejection, interference rejection, phase stability and repeatability, profile, size, and environmental conditions. The gain requirements are a function of satellite visibility requirements and are closely related to multipath rejection, and somewhat related to interference rejection. The goal is to have near uniform gain toward all satellites above a specified elevation angle, but, at the same time, reject multipath signals and interference typically present at low elevation angles. These are usually conflicting requirements. Some multipath rejection can also be achieved by reducing the left-hand, circularly polarized (LHCP) gain of the antenna without reducing the right-hand, circularly polarized (RHCP) gain. This is because the satellite signals are RHCP signals; whereas, reflected multipath signals usually tend to be either linearly polarized (LP) or even LHCP, depending upon the dielectric constant of the reflecting surface.

Interference rejection can also be achieved using a phased-array antenna, where the relative phase received from each antenna is controlled to "null" out the interference in the combined reception. This type of antenna is called a controlled-reception pattern antenna (CRPA), which is usually used only for military applications.

Phase stability and repeatability are important in differential surveying applications when differential carrier-phase accuracy is important. In this case, orientation of the antenna is important, taking advantage of phase repeatability.

Antenna profile is important in dynamic applications, such as for aircraft and missiles. Normally, requiring a low profile for those applications must be traded off against a desired gain pattern or other desired parameter.

Environmental conditions dictate the type of material used for the antenna and whether or not a radome is required. Some materials change their dielectric properties as a function of temperature.

2. Preamplifier

The preamplifier generally consists of burnout protection, filtering, and an LNA. The parameters that dictate the preamplifier requirements are as follows: the unwanted rf environment as received through the antenna, losses that precede and follow the preamplifier, and desired system noise figure (or noise temperature) as derived from overall receiver performance requirements. The gain of the preamplifier is not a system-level requirement, per se, but a derived requirement that satisfies that system level requirement.

The unwanted rf environment as received through the antenna affects the preamplifier in two ways. Either it could cause damage to the preamplifier electronics, or cause saturation of the preamplifier and circuitry that follows. Of course, except for damage prevention, we can do nothing to suppress the rf environment, as passed by the antenna, at frequencies that are in the bandwidth of the desired GPS signal. (Actually, there are techniques such as adaptive equalizers, usually applied at IF. These techniques are not discussed in this chapter.) That environment is considered to be either jamming or unintentional interference. However, suppression of the rf environment out of the desired GPS signal band can be accomplished by filtering, either before, during, and/or after amplification. When it is accomplished, it is based upon a trade-off between system noise figure requirements and filter insertion loss and bandwidth efficiency. Suppression of in-band and out-of-band damaging interference is usually accomplished with diodes that provide a ground path for strong signals. In the case of lightning protection, more complex lightning arrestors are sometimes used.

The system noise figure is set using an LNA that provides enough gain to cause any losses inserted after the LNA to have a negligible effect. Losses inserted prior to the LNA add directly to the system noise figure and are not affected by the LNA.

3. Reference Oscillator

The requirements on reference oscillators for GPS receivers have changed considerably over the years, mainly because of their expense. A high-quality oscillator can be the most significant cost item of a modern receiver. Thus, there have been compromises made on oscillator performance. Also, the oscillator's performance is not as critical in the modern multichannel receivers, especially in most commercial applications. However, there are some commercial and military applications where reference oscillator performance is critical. Typical requirements applied to reference oscillators are as follows:

1. *Size*—Stable oven-controlled crystal oscillators (OCXOs) and rubidium oscillators can be relatively large. Temperature-compensated crystal oscillators (TCXOs) are relatively small. Larger oscillators have more temperature inertia.

2. *Power*—Oven-controlled crystal and rubidium oscillators consume significant power.

3. *Short-term stability* caused by temperature, power supply, and natural characteristics. Short-term stability affects the ability to estimate and predict time and frequency in the receiver.

4. *Long-term stability* caused by natural characteristics, including crystal aging

5. *Sensitivity to acceleration*—g force and vibration sensitivity. Vibration causes phase noise, and dynamic g forces affect the ability to estimate time and frequency in the receiver.

6. *Phase noise*—high-frequency stability. Phase noise degrades the signal-processing performance of the receiver.

4. Frequency Synthesizer

Mostly, the requirements placed on the frequency synthesizer are derived requirements and the receiver designer's choice. Its design is based on the designer's *frequency plan,* which defines the receiver's IF frequencies, sampling clocks, signal processing clocks, etc. The frequency plan requires careful analysis to ensure adequate rejection of mixer harmonics, LO feed-through, unwanted sidebands and images. A key design parameter for the synthesizer is the minimization of phase noise generated in the synthesizer. Phase noise generated at the reference oscillator frequency is multiplied by the ratio of the rf frequency to its frequency through the synthesis process. Thus, the design of the synthesizer is critical to the performance of the GPS receiver.

The frequency synthesizer may also be required to generate local clocks for signal processing and interrupts for applications processing. This requirement might be assigned to signal processing. These local clocks comprise the receiver's time base.

5. Downconverter

The downconverter mixes LOs generated by the frequency synthesizer with the amplified rf input to IF frequencies, and, if so designed, IF frequencies to lower IF frequencies. This process implements the frequency plan, which, again, is the receiver designer's choice. The outputs of the mixers include both the lower and upper sideband of the mixing process, either of which can be used as an IF frequency. The unwanted sideband, LO feed-through, and harmonics are rejected by filtering at the IF. The unwanted image is filtered at rf before the mixing process. Because all of these processes are a function of the frequency plan, the requirements placed upon the downconverter are also derived requirements.

6. Intermediate Frequency Section

The requirements on the IF section are as follows:
1) Final rejection of out-of-band interference, unwanted sidebands, LO feed-through, and harmonics. The bandwidth of this rejection is a trade-off against correlation loss caused by filtering. In addition, the rejection of wide-band noise is required to minimize aliasing in a sampling receiver.
2) Increase the amplitude of the signal-plus-noise to workable levels for signal processing and control that amplitude as required for signal processing (AGC).
3) Suppress pulse-type interference.
4) Depending upon design, convert the IF signal to a baseband signal composed of in-phase (I) and quadraphase (Q) signals.

7. Signal Processing

As previously stated, signal processing may include the generation of the local clocks and interrupts. However, its prime requirement is to provide the GPS measurements and system data from selected satellites required to perform the navigation or other applications function. How this requirement is met constitutes the signal–processing-derived requirements and is based upon the receiver designer's choice. The functions of the signal processing are listed in the beginning of this section.

8. Applications Processing

The requirements for applications processing are to control the signal processing to provide the necessary measurements and system data and to use those measurements and system data to perform one or more of a variety of GPS applications. The processing requirements for these applications are covered elsewhere in this volume.

II. Technology Evolution
A. Historical Evolution of Design Implementation

In this section, the historical evolution of GPS receiver design implementation is described, dating back to the mid-1970s, the start of GPS Phase I concept validation. This does not include the development effort during the 621B program, predecessor to the GPS.

The first seven GPS receivers were mostly developed concurrently. These were the Phase I sets—the X Set, the Y Set, the Z Set, the Manpack developed by Magnavox, the High Dynamic User Equipment (HDUE) developed by Texas Instruments, the Advanced Development Model (ADM) developed by Collins Radio, and a satellite-monitoring receiver developed by Stanford Telecommunications, Inc. Three of these receivers, the Z Set, Manpack, and the satellite-monitoring receiver, had analog baseband signal processing and used processors only for the applications-processing function. The other three used microprocessors of one form or another to perform some of the signal processing. Needless to say, these receivers were all quite large, except for the medium sized Z Set and the Manpack. However, these two receivers had only one channel with one correlator, using this channel in a sequencing mode.

The rf sections of these receivers used discrete components, such as transistors, including large cavity filters. Circuit isolation was difficult. Consequently, there were a number of IF stages (three or four or more) to distribute the gain over a number of frequencies. This made the frequency synthesizers very complicated. Furthermore, the high-frequency LOs were generated using cavity multipliers.

The lower-frequency analog signal-processing sections were made up of operational amplifiers and other discrete components. The digital portions, such as the code generators and clocks, consisted of medium-scale integrated (MSI) circuits, at best. CMOS circuitry was in its infancy at that time, so higher-power consumption circuitry was used. The X Set used a bit-slice process controller for its baseband signal processing, which was state of the art at that time. Computers ranged from Hewlett Packard minicomputers to DEC large-scale integrated (LSI) computers. A structured version of Fortran was used by Magnavox as the applications-processing language; whereas, assembler language was used for the baseband signal processing.

The evolution into the Phase II GPS receivers was not that dramatic, although analog baseband signal processing was completely replaced with microprocessors and some MSI gave way to LSI. The evolution into the Phase III GPS receivers included some digital gate arrays and more powerful microprocessors. However, multiple IF stages and complicated frequency plans were still common. Signal processing was still not accomplished digitally, and the receivers were generally still quite large and expensive. Some of these receivers are still being produced today (the RCVR-3A airborne set and the RCVR-3S shipboard set). The operational control segment (OCS) monitor stations still use large hybrid analog/digital baseband receivers. However, these receivers are no longer in production.

B. Current Day Design Implementation

Except for the RCVR-3A and the RCVR-3S, all GPS receivers in production today are *probably* all true digital signal-processing receivers. I use the word *probably* because there are so many different GPS receivers in existence today, that one could never know. However, the consensus is that, in order to produce inexpensive receivers, they have to perform true digital signal processing. One reason for this is because almost all new GPS receivers have at least 4 full tracking channels; whereas, some have up to 24 to 36 channels for dual-frequency processing. Even the new military receivers, which usually lag in technology, have up to six tracking channels. The evolution of CMOS very-large scale integration (VLSI) technology has caused an explosion in the capabilities of new GPS receivers. The two technologies were made for each other. The processing speed of the new CMOS matches the signal-processing requirements for GPS receivers. Five or more channels on a chip are the norm in new GPS receivers.

The front-end electronics of GPS receivers have also experienced a dramatic evolution with the introduction of monolithic microwave integrated circuits (MMIC), stripline filter techniques, chip capacitors and resistors, high-speed digital integrated circuits, surface acoustic wave (SAW) filters, and surface mount printed circuit boards (PCBs).

1. Radio Frequency Electronics

The use of gallium arsenide (GaAs) MMIC for GPS receivers has been attempted and used on occasion, but with limited success. Most modern receivers use silicon bipolar technology because of efficiency and cost. Although the Defense Advanced Research Projects Agency (DARPA) sponsored Rockwell Collins on a program for the development of gallium arsenide (GaAs) MMIC for GPS receivers, Rockwell abandoned its use in favor of silicon bipolar in their miniature airborne GPS receiver (MAGR) design for those very reasons.[11] The GPS frequencies are not high enough to warrant the use of highly integrated GaAs circuitry. Most rf designers agree that the use of GaAs is usually limited to a front-end FET, and little more than a marketing "buzz word." Any advantage gained in using GaAs with respect to noise figure and power dissipation is lost because of circuit-matching problems at the GPS frequencies.

To an extent, the same is true for large-scale MMIC, even if silicon bipolar is used. However, in this case, the use of large-scale MMIC is a trade-off between size and circuit efficiency. Large-scale MMIC is certainly smaller, but may not be as efficient as smaller-scale MMIC in terms of power dissipation, cost, and out-of-band rejection. That is because the large-scale MMIC is designed for a wide range of applications, not just for a GPS receiver. Furthermore, in the rf and IF sections of a receiver, filters must be injected between stages of amplification and mixing for out-of-band rejection and gain stability. Efficient filtering cannot be accomplished in the MMIC chips.

Times are changing, however. GEC Plessey has introduced a silicon bipolar MMIC chip, the GP1010, that was developed specifically for GPS receivers.[2] Its input is at rf at the GPS L_1 frequency, although already amplified and filtered via a preamplifier. Intermediate frequency filtering is not included on the chip. Its output is a stream of 1.5-bit samples for input to digital signal processing. The chip also contains the frequency synthesizer and AGC, but not the loop filters.

2. Frequency Synthesizer Electronics

The use of cavity multipliers has been replaced with phase–lock-loops (PLLs) using voltage-controlled oscillators (VCOs) and high-speed digital divider circuits (prescalers). Chips are commercially available that contain programmable prescalers. However, because of the clocking speeds required for the first LO provided by these frequency synthesizers, sometimes the frequency of the VCO and clocking speed of the prescaler are half what is required, and a frequency doubler or a second-harmonic mixer is used to achieve the L-band LO. However, these components are still small and quite simple. In the end, we have small, efficient frequency synthesizer electronics that are significantly smaller and lower power than technology used 5–10 years ago. As described above, the GEC Plessey GP1010 chip contains the VCO and prescalers, but it is not programmable.[2]

3. Down Conversion and Intermediate Frequency Electronics

The conversion from rf to IF has generally become a silicon bipolar MMIC implementation, where the MMIC includes mixing and a stage or two of amplification. The VCO for the LO may also be included. Once at IF, surface acoustic wave (SAW) filters provide a small, efficient means of final filtering to minimize unwanted out-of-band signals and noise.

4. Reference Oscillators

Unfortunately, the development of good, stable reference oscillators has not kept up with the pace of the development of the other sections of a GPS receiver. There have been improvements, but, depending upon the ultimate stability required, the oscillator can be the most expensive and largest component in the receiver. For example, for very good stability, not including that achieved with relatively large and expensive atomic oscillators, a crystal oscillator must be ovenized to minimize frequency excursions that are the main cause of frequency instability. Unfortunately, to have temperature inertia, mass is required. Thus, although the sizes of these oscillators have been reduced over the years, they are still relatively large. In fact, these smaller oscillators are not usually as stable as the older, larger, oscillators.

The advancement of digital signal processing and multiple receiver channels has allowed the use of TCXOs in most commercial applications. This has prompted some improvement in TCXO performance over what had been the norm in the past. However, that improvement has not been dramatic, although the size and cost of the TCXOs have been reduced significantly.

III. System Design Details

In this section, the block diagram of Fig. 1 is expanded, and design details are presented, starting with the hardware and following with the software part of the digital signal processing. Only the implementation of modern digital receivers is presented along with some trade-offs between different modern implementations. Specifically, two receiver designs that are familiar to the author are used as examples—the Rockwell Collins miniature airborne GPS receiver

(MAGR), an example of a military receiver, and the NovAtel GPSCard™, an example of a commercial receiver. A functional overview of the MAGR is shown in Fig. 2.[1] Details of these functions may vary from receiver to receiver, but they exist in one form or another in all modern receivers. Although the MAGR is a dual-frequency receiver that receives both L_1 and L_2 signals and both the C/A- and P-codes, processing that is common for those signals is presented only once.

A. Signal and Noise Representation

Before describing the remainder of the receiver operations, it is appropriate to provide a representation of the received signal and noise in both the time domain and the frequency domain. The signal is represented as follows:

$$s(t) = AC(t)D(t)\cos[(\omega_0 + \Delta\omega)t + \phi_0] \tag{1}$$

where A = signal amplitude; $C(t)$ = PRN code modulation (± 1); $D(t)$ = 50 bps data modulation (± 1); $\omega_0 = 2\pi f_0$ = carrier frequency (L_1 or L_2); $\Delta\omega = 2\pi\Delta f$ = frequency offset (Doppler, etc.); and ϕ_0 = nominal (but ambiguous) carrier phase.

For the purpose of the processing described herein, it is necessary to represent only one component of the L_1 signal, because one of the two components (in-phase or quadrature) does not correlate with the receiver channel's reference code in channel tracking the desired code. L_2 processing is identical to that of the L_1 signal except that L_2 acquisition and tracking are usually aided with information obtained by tracking the more powerful L_1 signal.

In the frequency domain, the spectral density of the signal is the spectral density of the PRN code centered at $\pm(\omega + \Delta\omega)$. At baseband, this spectral density is as follows:

$$S_s(\omega) = \frac{A^2 T_c}{2} \frac{\sin^2(\omega T_c/2)}{(\omega T_c 2)^2} \tag{2}$$

where T_c is the PRN code chip width, or the inverse of the PRN code chipping rate. Although the C/A-code spectrum is a line spectrum, this representation suffices for most of the processing described herein. Exceptions are noted as they arise.

Signal power in a $2B$ Hz two-sided bandwidth is given by the following:

$$P_s = \frac{1}{2\pi} \int_{-2\pi B}^{2\pi B} S_s(\omega) d\omega \tag{3}$$

If $B = \infty$, $P_s = A^2/2$. If $B = 1/T_c$, $P_s = 0.9 A^2/2$, resulting in a 0.45 dB signal loss.

Ambient noise is represented in the frequency domain as white noise with a constant spectral density $N_0/2$. If passed through a unity-gain bandpass filter with a two-sided noise bandwidth $2B$ Hz centered at frequency ω_0, the spectral density appears as illustrated in Fig. 3. The resulting noise power in the two-sided bandwidth is as follows:

$$P_n = 2N_0 B \tag{4}$$

and a signal-to-noise ratio in a $2B$ two-sided noise bandwidth of

GPS RECEIVERS

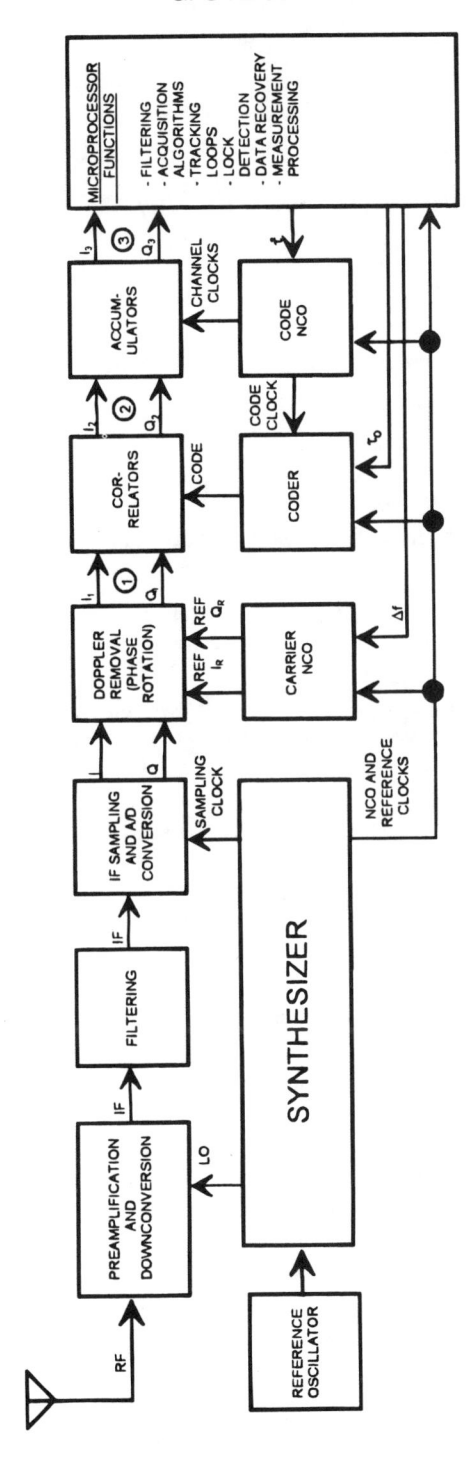

Fig. 2 Functional block diagram of the MAGR.

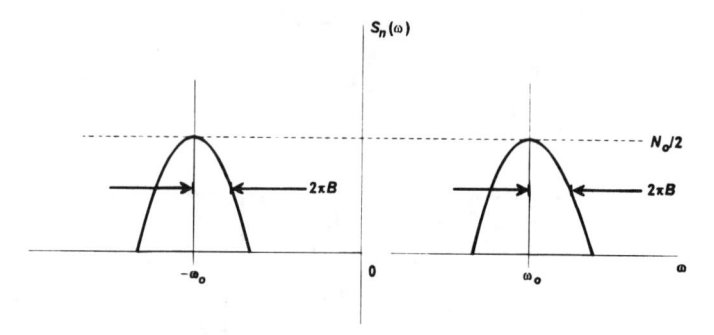

Fig. 3 Spectral representation of ambient noise.

$$\frac{P_s}{P_n} = \frac{A^2}{4N_0 B} \tag{5}$$

The output of the bandpass filter has two equivalent time domain representations[3]:

$$n(t) = x(t)\cos\omega_0 t - y(t)\sin\omega_0 t \tag{6}$$

$$n(t) = r(t)\cos[\omega_0 t + \varphi(t)] \tag{7}$$

where $x(t)$ and $y(t)$, the in-phase and quadraphase components, respectively, are bandlimited Gaussian processes with the properties defined as follows:

$$E[x(t)] = E[y(t)] = 0$$
$$E[x^2(t)] = E[y^2(t)] = E[n^2(t)] = P_n \tag{8}$$
$$E[x(t)y(t)] = 0$$

Furthermore, $r^2(t) = x^2(t) + y^2(t)$ has a chi-squared (or Rayleigh) distribution with two DOF, and

$$\varphi(t) = \tan^{-1}\left[\frac{x(t)}{y(t)}\right] \tag{9}$$

is uniformly distributed between 0 and 2π radians.

B. Front-End Hardware

The hardware is comprised of the front-end electronics and part of the digital signal processing. The other part of the digital signal processing is implemented in software.

1. Antenna

The following requirements were placed upon the Fixed Reception Pattern Antenna (FRPA3) for Rockwell's military receivers[4]:

1) *rf:* "The GPS antenna shall accept the GPS navigation signals at both the L_1 and L_2 frequencies and output them to the GPS antenna electronics."[4] This implies either a wide-band antenna that accepts both frequencies and all frequencies between them, or implies two antennas packaged in a single unit. Both types of antennae are available, because some are naturally narrow band while others are wide band.

2) *Antenna gain:* "The GPS antenna shall provide a minimum gain of -2.5 dBic to a RHCP signal over a 160° solid angle cone of coverage (above 10° elevation angle) for signals in both the L1 and L2 bandwidths. Gain shall be measured at the L1 and L2 carrier frequencies at the prevailing ambient temperature using a standard gain horn for comparison. The combined effects of environmental temperature range and bandwidth ... shall not cause the gain to be less than -3.3 dBic. These gain requirements apply when using a test ground plane."[4] (This ground plane matches the housing of the antenna electronics for shipboard installation.)

3) *Other antenna specifications:* Other FRPA3 specifications are as follows:

a) *Voltage standing wave ratio (VSWR):* $\leq 2:1$ (referenced to 50 ohms over L_1 and L_2 bands)

b) *dc impedance @ signal interface:* 0 ohms

c) *Connector:* Single TNC female receptacle

d) *Size:* Less than 6 in. in diameter and 1.75 in. in height, including connector

The dc impedance of 0 ohms provides lightning protection. More than one vendor supplies the antenna and not necessarily through Rockwell Collins. Typical gain patterns from one vendor (Sensor Systems, Inc. S67-1575-14) are shown in Fig. 4. This model has two narrowband antennas packaged in a single unit. The Sensor Systems antenna has a diameter of 3.5 in. and a height of 0.565 in.[5]

The following technical specifications are given for a Sensor Systems' commercial aeronautical antenna (S67-1575-16), which was designed to meet ARINC 743A characteristics[5,6]:

1) *rf:* 1575 ± 2 Mhz with a VSWR $\leq$ 1.5:1 or 1575 ± 10 MHz with a VSWR $\leq 2:1$.

2) *antenna gain:* The gain pattern is specified as follows:

> -1 dBic to 75 deg from vertical

> -2.5 dBic to 80 deg from vertical

> -4.5 dBic to 85 deg from vertical

> -7.5 dBic at 90 deg from vertical

The RHCP gain pattern is shown in Fig. 5.

2. Receiver Front End

The receiver front end consists of filtering and limiting, an LNA, a frequency synthesizer, downconversion, and conversion to baseband. Initial filtering, limiting, and LNA can be housed with the antenna to comprise an integrated antenna electronics. This is an optional configuration of the MAGR and is usually the case in commercial receivers. In the case of the MAGR, the following requirements are imposed:[7]

1. *Preselector filtering:* The preselector filtering is required to reject out-of-band interference and to limit the noise bandwidth of the antenna electronics.

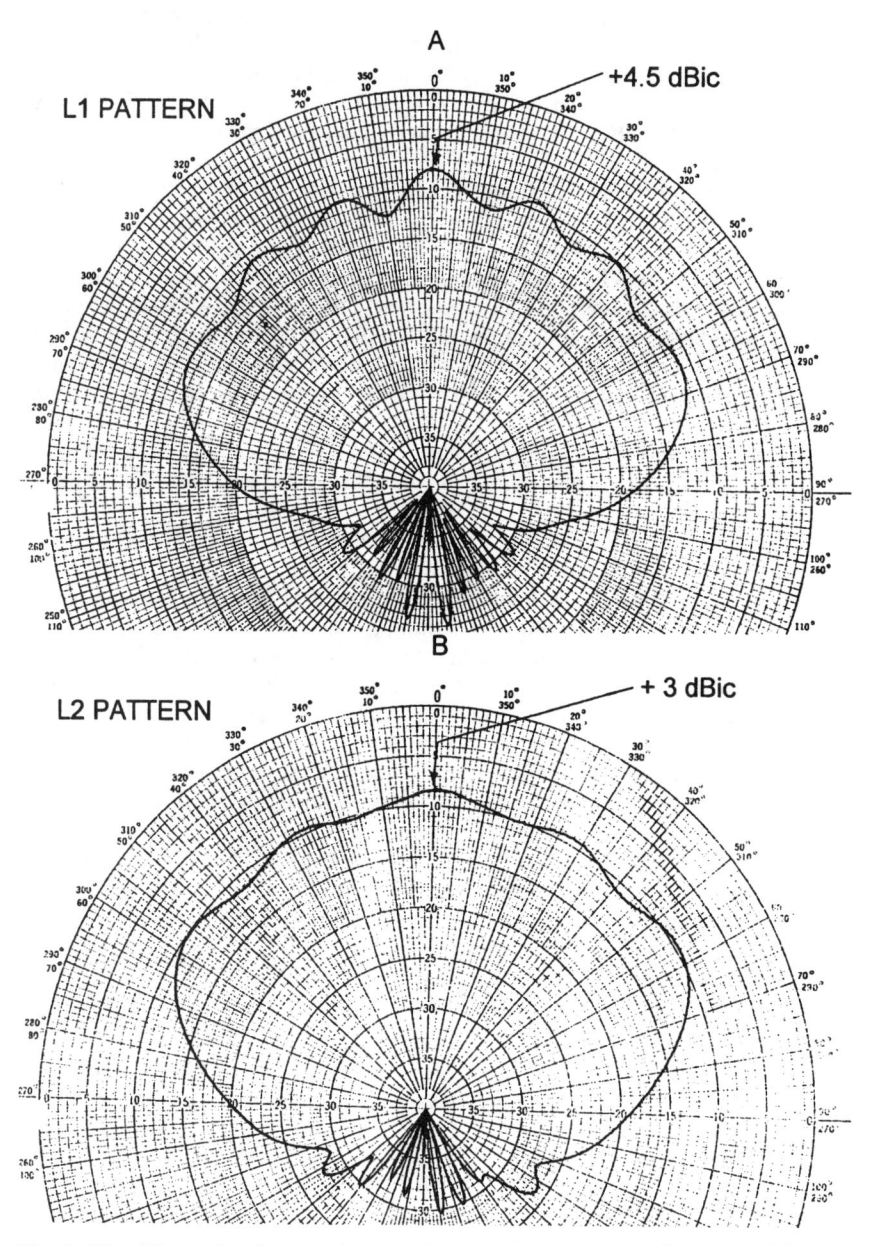

Fig. 4 Fixed Reception Pattern Antenna 3 antenna gain patterns (courtesy of Sensor Systems, Inc.).

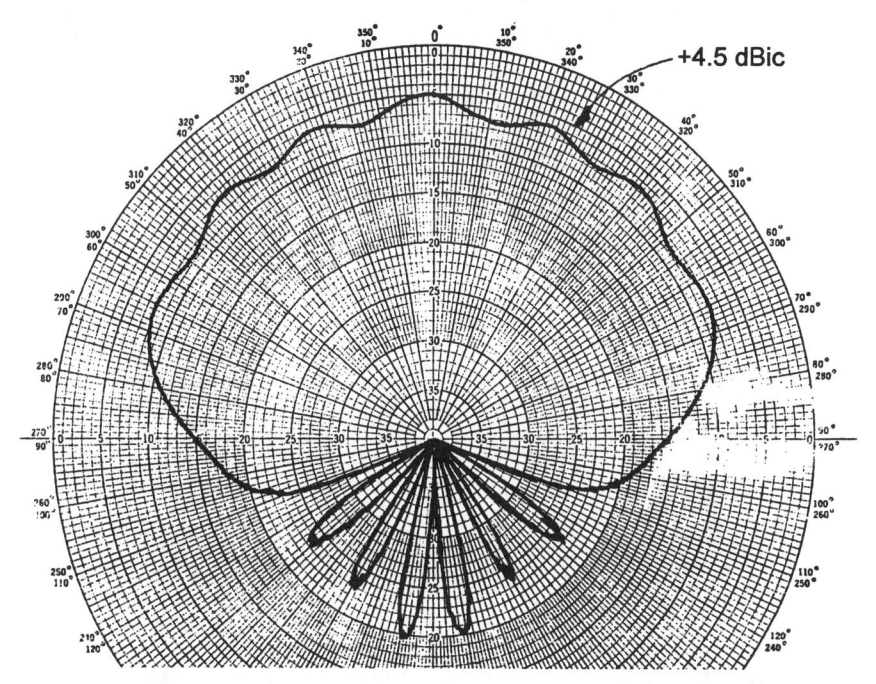

Fig. 5 Commercial L_1 Antenna Gain Patterns (courtesy of Sensor Systems, Inc.).

The filtering shall be dual-band centered at L_1 and L_2, with a noise bandwidth of 80 MHz in each band. The insertion loss shall be sufficiently low to meet overall gain and noise figure requirements.

2. *Burnout protection:* The antenna electronics shall not incur damage or performance degradation after being subject to a peak signal power density of 69 kW/m^2 for not more than 10 μs, or a continuous signal power density of 348 W/m^2 in either band of frequencies.

3) *Gain and noise figure:* The antenna electronics, including interconnecting cabling, preselector, and protection circuitry, shall have a minimum overall gain of 23 dB, with a noise figure of 4 dB at the input to the MAGR receiver.[8] The maximum overall gain shall be 33 dB.

The Sensor Systems' commercial ARINC 743A antenna described is also available with an internal preamplifier with 26 ±3 dB gain with an internal interference rejection filter and an LNA (S67-1575-52). A single TNC connector carries both the L_1 signal and dc power to the LNA (+4 to +24 VDC at 25 mA, maximum). Out-of-band rejection is 35 dB at 1625 MHz.

3. Noise Figure Computations

Figure 6 provides a model of a receiver front end for the purposes of computing the receiver and system noise figure and noise temperature. In general, the system noise figure (in dB) is related to system noise temperature (in Kelvin) as follows[9]:

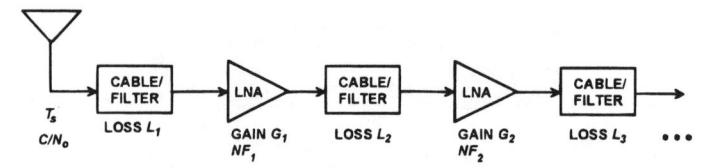

```
         ┌────────┐       ┌────────┐       ┌────────┐
  Y      │ CABLE/ │       │ CABLE/ │       │ CABLE/ │
──┬──────┤ FILTER ├─▷─────┤ FILTER ├─▷─────┤ FILTER ├──→
 Tₛ      │ LOSS L₁│  LNA  │ LOSS L₂│  LNA  │ LOSS L₃│  • • •
 C/N₀    └────────┘ GAIN G₁└────────┘GAIN G₂└────────┘
                    NF₁              NF₂
```

Fig. 6 Noise figure computation model.

$$\mathrm{NF} = 10 \log_{10}\left(1 + \frac{T_{\mathrm{sys}}}{T_0}\right) \qquad (10)$$

where $T_0 = 290$ K $= 24.6$ dB$-$K.

The corresponding noise density, in W/Hz, is $N_0 = K_B T_{\mathrm{sys}}$; where $K_B = -228.6$ dBW/K$-$Hz $= 1.380 \times 10^{-23}$W/K$-$Hz, is the Boltzmann constant.[9] Based upon the model given in Fig. 6, the system noise temperature is computed as follows:[10,11]

$$\begin{aligned}
T_{\mathrm{sys}} &= T_s + T_R \\
&= T_s + T_0[L_1 - 1 + L_1[\mathrm{NF}_1 - 1 + G_1^{-1}[L_2 - 1 \\
&\quad + L_2[\mathrm{NF}_2 - 1 + G_2^{-1}[L_3 - 1 + \cdots]]]]]
\end{aligned} \qquad (11)$$

where T_s is the source (antenna) temperature; T_R is the receiver noise temperature; and the L_i, NF_i, and G_i are loss, amplifier noise figure, and amplifier gain of each stage i, respectively, all given in ratio. This formula is known as the *Friis Formula*. Note that the loss and noise figure of the first stage affects the system noise figure directly; whereas, any losses after the first amplification are reduced proportional to that gain. Thus, the first stage is said to *set* the system noise figure or noise temperature.

Note, also, that the source temperature adds directly to the system noise temperature. Normally, this is the antenna sky temperature, which is relatively low with respect to the ambient noise temperature T_0. One exception is when the receiver is connected directly to a GPS signal generator or simulator, in which the source temperature is the ambient noise temperature (290 K), which is a worst-case situation. Normally, adjustments to the signal power must be made to compensate for this. Note that in this case, if the first stage gain is high enough, the noise density is simply as follows:

$$\begin{aligned}
N_0 &= K_B T_0 \mathrm{NF}_1 \\
&= -228.6 + 24.6 + \mathrm{NF}_1(\mathrm{dB}) \\
&= -204 \text{ dBW/Hz} + \mathrm{NF}_1(\mathrm{dB})
\end{aligned} \qquad (12)$$

Sometimes this equation is erroneously used for "real-world" computations, providing pessimistic analysis results. A source temperature of 75–100 K is typical, depending upon the antenna pattern and the amount of ground temperature observed.[9,12]

4. Synthesizers and Frequency Plans

Figure 7 presents the block diagram of two of the MAGR's custom silicon bipolar chips that make up its synthesizer.[1] One chip (*L*-band chip) also includes

GPS RECEIVERS   345

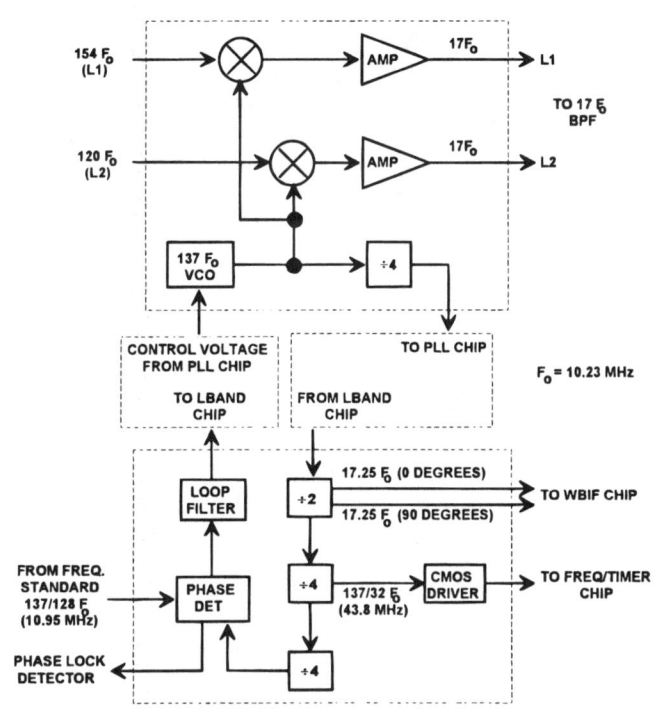

Fig. 7 Miniature airborne GPS receiver L-band/phase-lock-loop chips and frequency plan.

the downconverter function of the receiver. This synthesizer design reflects the concept used in modern GPS receivers, where a nonstandard reference oscillator frequency is used. Older receivers used such frequencies as 5 or 10 MHz, or 5.115 or 10.23 MHz, which also became standards because of GPS. Note that the MAGR uses 10.949296875 MHz as its reference.

The MAGR synthesizer generates a common local oscillator at $137F_0$ (1401.51 MHz) for both the L_1 and L_2 frequencies by phase-locking the LO voltage controlled oscillator, divided by 128, to the 10.95 MHz reference ($F_0 = 10.23$ MHz). It also generates common in-phase and quadraphase LOs for conversion of the IF frequencies to baseband at 17.25 F_0 (176.4675 MHz) and a CMOS clock at 43.7971875 MHz, which is used for clocking the digital signal-processing circuitry. Note that the baseband LOs do not match the IF frequecies at 17 F_0. This leaves a residual frequency offset at baseband, which is part of the frequency planning scheme. The overall effects of this residual offset is treated as Doppler. These effects are described later. However, along with the nonstandard reference frequency, allowing for such an offset also simplifies the frequency plan and the synthesizer design.

The GPSCard™ uses the same concept for its frequency plan and synthesizer as illustrated in Fig. 8. Its reference frequency is 20.473 MHz, which is also its digital signal-processing clock. The GPSCard™ utilizes a commercial synthesizer

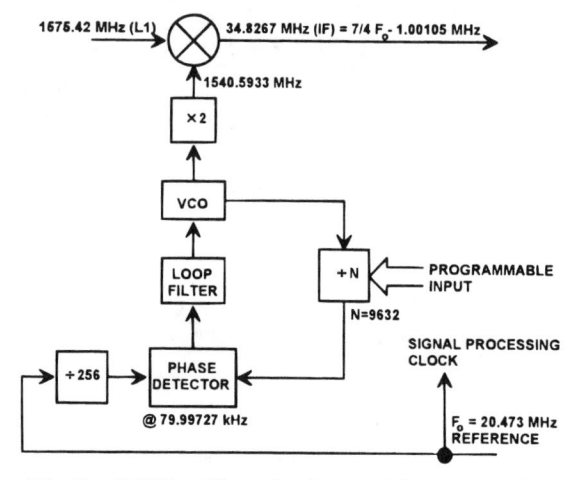

Fig. 8 GPSCard™ synthesizer and frequency plan.

chip with a programmable divider (prescaler) rather than custom chips. The key difference between this frequency plan and that of the MAGR is that the GPSCard™ uses IF sampling for its conversion to baseband, which is described later. Note that there is also a residual frequency offset.

5. Mixing Operations and Intermediate Frequency Filtering

Downconversion from rf to IF, and, in the case of the MAGR, conversion from IF to baseband, are accomplished by mixing the incoming signal and noise (rf or IF) with an LO. This process is illustrated in Fig. 9.

If the local oscillator is represented as $LO_1(t) = 2\cos\omega_1 t$ with power of two units, then the output of the mixer is this LO multiplied by the sum of Eq. (1) and Eq. (7), or $s_{IF}(t) + n_{IF}(t) = 2[s(t) + n(t)]\cos\omega_1 t$ + harmonics + LO feedthrough + image noise. Ignoring for the moment the harmonics, LO feedthrough, and image noise and using the product of cosines, for the signal we have $s_{IF}(t) = AC(t)D(t)\{\cos[(\omega_0 + \omega_1 + \Delta\omega)t + \phi_0] + \cos[(\omega_0 - \omega_1 + \Delta\omega)t + \phi_0]\}$ consisting of upper and lower sideband components, each with a power of

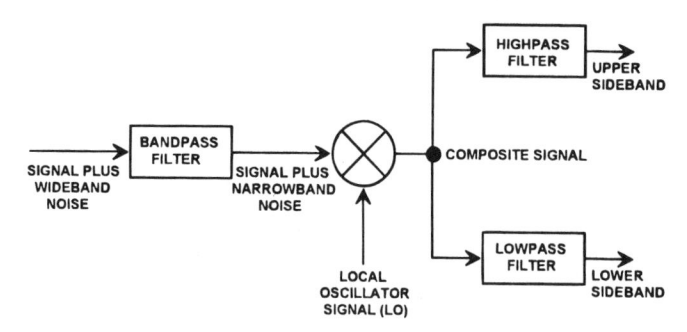

Fig. 9 Mixing operations.

$A^2/2$. Only the lower sideband is wanted. Therefore, the upper sideband is eliminated via a low-pass filter, resulting in the IF frequency $\omega_{IF} = \omega_0 - \omega_1$. Similarly, for the noise, in terms of Eq. 7, $n_{IF}(t) = r(t) \cos[\omega_{IF}t + \varphi(t)]$.

Harmonics and LO feedthrough are removed via a well-designed frequency plan and the use of bandpass filters at IF. Harmonics are generated by the mixer, because it is a nonlinear device. LO feedthrough (at ω_1) is the LO leaking through the mixer. In addition, these filters also provide the final rejection of out-of-band interference and image noise, primarily because it is easier to obtain narrow-band filtering at the lower IF frequencies.

In-band image noise is either at the frequency $\omega_0 - 2\omega_1$ (upper sideband component at the lower sideband), or at the frequency $\omega_0 - 2\omega_{IF}$ (mixes to the negative of the lower sideband). Both would mix to the IF frequency and, thus, would not be filtered at IF. It is necessary to filter noise at these frequencies prior to mixing at rf so that they do not exist. In order to avoid the use of narrowband filters at rf for this purpose, the frequency plan should be designed to prevent image noise close to the IF frequency.

6. Conversion to Baseband

Conversion to baseband is the process of converting the IF signal to that of in-phase and quadraphase components of the signal envelope, but still modulated with residual Doppler. However, as pointed out in the discussions of synthesizers and frequency plans, in most modern receivers, an intentional residual frequency offset may still exist. There are two methods for achieving this conversion—by analog mixing or by a technique known as IF (or pass-band) sampling. Because the latter (used in the GPSCard™) is a sampling process, its description is delayed to the next section.

The MAGR uses the former method, and does so in its wide-band IF chip (silicon bipolar), which is shown in Fig. 10.[1] Also shown are the AGC and the analog-to-digital (A/D) converters. This conversion to baseband is realized by mixing the IF signal with two LOs, one of which is shifted 90° in phase with respect to the other (in quadrature). The low-pass filters reject the upper sidebands. The in-phase and quadraphase combined unity power LOs are, respectively,

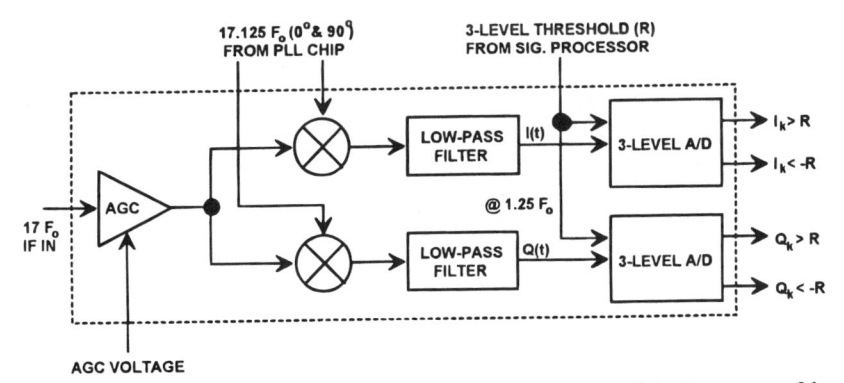

Fig. 10 Miniature airborne GPS receiver wideband intermediate frequency chip.

as follows:

$$LO_{2I}(t) = \sqrt{2}\cos\omega_2 t \qquad (13)$$

$$LO_{2Q}(t) = \sqrt{2}\cos\left(\omega_2 t + \frac{\pi}{2}\right) = -\sqrt{2}\sin\omega_2 t \qquad (14)$$

The resulting analog in-phase and quadraphase baseband signal components are then as follows:

$$I_s(t) = \frac{A}{\sqrt{2}} C(t)D(t)\cos(\Delta\omega_B t + \phi_0) \qquad (15)$$

$$Q_s(t) = \frac{A}{\sqrt{2}} C(t)D(t)\sin(\Delta\omega_B t + \phi_0) \qquad (16)$$

where the residual frequency offset is

$$\Delta\omega_B = \omega_{IF} - \omega_2 + \Delta\omega \qquad (17)$$

The relationship of these I and Q levels at this point and the signal power, related to that defined in Eq. (3), is $P_s = E[I_s^2(t) + Q_s^2(t)] = A^2/2$.

Under the assumption that this residual frequency offset is quite small with respect to the bandwidth of the low-pass filters, and that their single-sided bandwidths are essentially B Hz, the baseband noise components are simply $I_n(t) = x(t)/\sqrt{2}$ and $Q_n(t) = y(t)/\sqrt{2}$, and the noise power, in terms of Eqs. (4) and (8), is $P_n = E[I_n^2(t) + Q_n^2(t)] = 2N_o B$. The MAGR AGC shown in Fig. 10 does not operate on signal power, because the signal is still below the noise level at this point in the receiver. It is a very wideband AGC whose time constant is such that it suppresses pulse interference (time constant < 1 μs).

C. Digital Signal Processing

Digital signal processing consists of precorrelation sampling, Doppler removal, PRN coders, correlators, number-controlled oscillators (NCOs), postcorrelation filtering, and various receiver clocks.

1. Precorrelation Sampling

As is evident for the MAGR in Fig. 10, modern GPS receivers all become digital prior to correlation and Doppler removal. However, this is where the commonality ends. They differ in sample rates, sample quantization, and, as discussed previously, some receivers convert to baseband as part of the sampling process, known as IF sampling.

Intermediate frequency sampling is illustrated in Fig. 11. The concept is to sample the IF signal at a rate at which the I and Q samples are obtained directly. Suppose the sample rate is as follows:

$$SR = \frac{4f_{IF}}{N} \qquad (18)$$

where f_{IF} is the IF frequency being sampled, and N is an odd number. Then, the samples would be taken at

GPS RECEIVERS

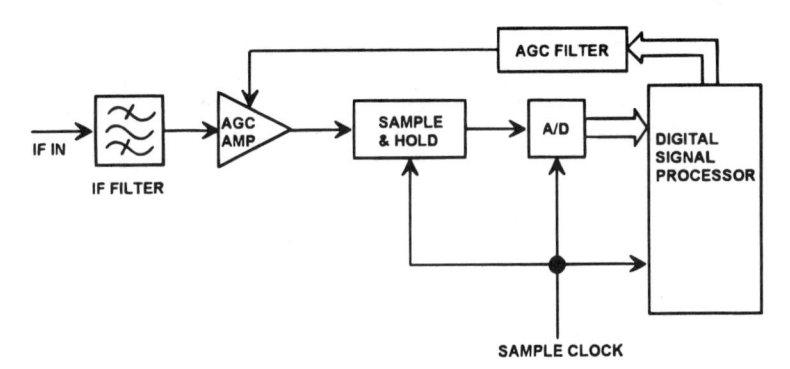

Fig. 11 Intermediate frequency sampling process.

$$t_k = \frac{kN}{4f_{IF}} \text{ sec}; \quad k = 0, 1, \ldots \tag{19}$$

Sampling the IF signal at these times yields the following:

$$s_k = s_{IF}(t_k) = AC(t_k)D(t_k)\cos\left[2\pi(f_{IF} + \Delta f)\frac{kN}{4f_{IF}} + \phi_0\right]$$

$$= AC_k D_k \cos\left[\frac{\pi kN}{2}\left(1 + \frac{\Delta f}{f_{IF}}\right) + \phi_0\right]$$

$$= AC_k D_k \cos\left[\frac{\pi kN}{2} + \phi_k\right] \tag{20}$$

where Δf is an intentional frequency offset plus that attributable to Doppler, C_k and D_k are the code and data at time t_k, and

$$\phi_k = \phi_0 + \frac{\pi kN\Delta f}{2f_{IF}} = \phi_0 + \Delta\omega t_k = \phi_0 + \Delta\phi_k \tag{21}$$

is the baseband phase of the sample attributable to the nominal phase and frequency offset at time t_k. If the sample rate is offset from that of Eq. (18), $\Delta\omega$ in Eq. (21) simply becomes $\Delta\omega_B$, analogous to Eq. (17). If we ignore the Δf of Eq. (20) for the moment, note that the IF signal is sampled at exactly successive 90 deg phases, producing the following sequence of samples:

$$\sqrt{2}[I_{sk}, Q_{sk}, -I_{sk}, -Q_{sk}, I_{sk}, Q_{sk}, -I_{sk}, -Q_{sk}, \ldots] \tag{22}$$

or

$$\sqrt{2}[I_{sk}, -Q_{sk}, -I_{sk}, Q_{sk}, I_{sk}, -Q_{sk}, -I_{sk}, Q_{sk}, \ldots] \tag{23}$$

depending upon the value of N. The Δf results in a time-varying $\Delta\phi_k = \Delta\omega t_k$, causing a phase rotation of the samples that is removed after the sampling process.

This IF sampling process is sometimes called *pseudo sampling*, because the I and Q samples do not occur at the same time. For large frequency offsets with respect to the sampling rate, but still within the Nyquist rate, this will induce an additional phase shift in the Q samples causing aliasing to a negative frequency offset. However, this is not a problem for this GPS receiver application for reasonable Δf values.

The digital signal-processor can simply invert the sign on half the samples and sort them into in-phase and quadraphase samples. This sign inversion is actually an advantage in that, if there are any dc biases present in the sampling process, they will eventually cancel in subsequent signal processing. Also, because the I and Q samples are generated in the same circuitry, there are no gain and phase imbalances between them, except as noted above. This can occur in the analog baseband conversion process. Also, only one A/D converter is required, although it must sample at twice the rate.

There are disadvantages, however, other than the double sample rate. First, the aperture time of the sampling process must be small with respect to the period of the IF frequency. That is why a sample-and-hold circuit is shown in Fig. 11. If the A/D is flash, and the IF frequency is low enough, this circuit is not required. For IF frequencies as high as they are in the MAGR, certainly a silicon bipolar flash A/D would be required for this process. How quick the sample must be is debatable, other than it must be fast with respect to the IF frequency. $\text{Sin}(x)/x$, where x is proportional to the product of the frequency and the aperture time, attenuation occurs if the aperture time is too long, but then this attenuation also occurs on the noise, which is also at the IF frequency. Thus, there would be no loss in signal-to-noise ratio to a point.

The second disadvantage is minor. That is, as described above, the sample rate must be high enough so that there is no significant delay between the I and Q samples. This delay should be small with respect to a pseudonoise (PN) chip so that most of the time the I and Q samples do not straddle chip transitions. Thus, there is some loss associated with sampling right at twice the Nyquist frequency (four times the code chipping rate), but it is minimal. It has an effect similar to filter phase distortion.

This delay between the I and Q also has an effect on interference that may be present in the IF bandwidth. For example, consider an interference signal that is offset in frequency from the center of the IF band by a large amount, but is still in band. The sampling process also generates I and Q of the interference. The delay between these I and Q can be significant if the frequency offset is large, but, for normal receiver processing, this is acceptable. The effect is that some of the interference energy is folded over to the other side of center. Energy cannot be created, so this is of no consequence. Both sides of the center frequency are spread by the code correlation process.

As previously stated, the GPSCard™ uses IF sampling. Note from Fig. 8 that its IF frequency and sample frequency are such that N is 7 with a frequency offset of -1.00105 MHz. There are also at least 10 individual I and Q samples per chip, so the loss caused by the second disadvantage is negligible. Furthermore, the sample rate is not an integer multiple of the chipping rate, so that the I and Q sample times will never stay synchronous with the chip transitions. Because the IF frequency is relative low with a period of approximately 30 ns, the CMOS A/D aperture time poses no problem. Five levels of the A/D are used for a 2.5-bit quantization.

The MAGR performs baseband sampling, as shown in Fig. 10 using two 1.5-bit quantization (three levels: −L, 0, and +L) A/D converters. The result is the same as IF sampling, where

$$I_{sk} = \frac{A}{\sqrt{2}} C_k D_k \cos \phi_k \qquad (24)$$

$$Q_{sk} = \frac{A}{\sqrt{2}} C_k D_k \sin \phi_k \qquad (25)$$

with the exception that the Q sample is one sample later in the case of IF sampling. For the reasons stated above, we neglect that fact in the following discussions, although, in some applications, it is important.

The noise samples are simply as follows:

$$I_{nk} = x(t_k)/\sqrt{2} = x_k/\sqrt{2} \qquad (26)$$

$$Q_{nk} = y(t_k)/\sqrt{2} = y_k/\sqrt{2} \qquad (27)$$

The MAGR samples at a frequency at one-half of the 43.8 MHz clock from the PLL chip shown in Fig. 7, or 21.9 MHz. Given that this sampling is done at baseband, it is slightly more than the Nyquist sampling frequency, with at least two samples per P-code chip. Note that this sample rate is also not an integer multiple of the chipping rate, thus the sampling will never be synchronized with the chip transition times. The A/D threshold control is for CW interference suppression. This is a topic of discussion in Chapter 10, this volume.

2. Precorrelation Filtering

Precorrelation filtering (low-pass filters in Fig. 10 and bandpass filter in Fig. 11) is necessary to prevent aliasing while sampling in a digital receiver. However, this filtering also causes correlation losses, because the sidelobes of the PRN code spectrum are eliminated. Figure 12 illustrates this loss for a sharp cutoff,

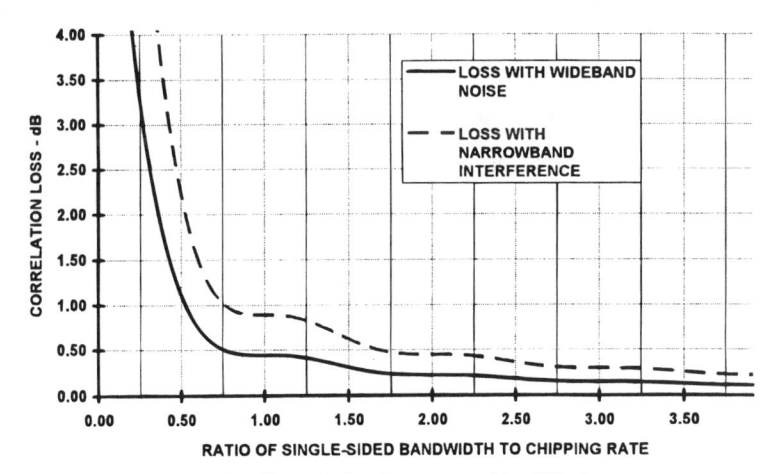

Fig. 12 Correlation loss caused by filtering.

linear phase filter of unity gain as a function of the ratio of its single-sided bandwidth to the PRN code chipping rate. Two cases are plotted. One is in the presence of wide-band noise. In this case, the filter also band-limits the noise, thus reducing the effective correlation loss. The other case is if the noise (interference) is completely within the filters bandpass, in which case, the filter doesn't alter it at all. This latter loss is simply twice the loss (in dB) of the first case.

3. Precorrelation Automatic Gain Control

Anytime multibit quantization is used, a precorrelation AGC (or A/D threshold control, which has a similar effect) is required. Most commercial receivers use 1-bit quantization (or hard-limiting). This implementation does not require an AGC, provided that the front end has the required dynamic range to accommodate gain variations and expected interference. Some hard-limiting receivers don't have the dynamic range for the latter. The AGC is used for three reasons: 1) increased dynamic range; 2) quantization level control; and 3) pulse interference suppression. The latter case requires a wide band fast AGC, which is the case in the MAGR. However, as described later, the effects of a slow AGC coupled with digital sampling on pulse interference is similar. For a receiver used in a military environment, however, the fast AGC is more effective. The MAGR effectively has two AGCs, including the A/D threshold control.

The increased dynamic range prevents strong continuous interference from capturing the sampling process, even in the case of 1-bit sampling (hard-limiting). Without it, 1-bit sampling can be a disaster, because the interference will saturate the front end and keep the signal away from the zero level being sampled.

The GPSCard™ uses the AGC to perform quanitization level control. The MAGR, in a gross sense, does the same, but fine tunes the quantization level control with the A/D level control. Both perform this by processing sample statistics collected in their respective signal-processing chips. The only difference is that one drives the noise (or interference) to the desired A/D threshold levels, while the other drives the A/D threshold levels to the noise (or interference). The effect is the same.

4. Sampling and Quantization Effects

Sampling rate, quantization, and precorrelation bandwidth are all related and have a combined effect on implementation losses in a GPS receiver. There are many papers on the subject, mostly dealing with communications, so the background theory is not detailed here.[13-17] Only some of the effects are presented. These effects differ, depending upon whether the signal is corrupted with Gaussian noise, other types of interference, or both. The effects of interference are discussed in Chapter 10, this volume. Only the Gaussian noise effects are discussed here.

The effects of sample rates are essentially described by the Nyquist theory, which states that the sample rate should be at least twice the signal-plus-noise bandwidth—twice the single-sided bandwidth for baseband sampling, and twice the two-sided IF bandwidth in the case of IF sampling (because both I and Q are sampled). The Nyquist sample rate (f_s) relationship to signal-plus-noise

GPS RECEIVERS 353

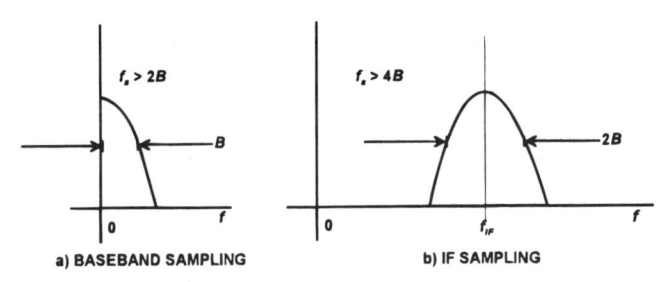

Fig. 13 Nyquist sampling rates.

a) BASEBAND SAMPLING
b) IF SAMPLING

bandwidth for those two cases is illustrated in Fig. 13. This rule applies to both the bandwidth of the signal and the bandwidth of the noise to prevent aliasing. Thus, the bandwidth of the precorrelation filters have a significant effect, even if the bandwidth of the signal is limited. To limit the digital signal-processing rates, sometimes there tends to be a desire to limit the bandwidth below the signal bandwidth. Of course, this results in a filtering implementation (correlation) loss, as described above. (This occurs in analog receivers as well, but the reasons for limiting the bandwidth are not the same.) However, given that the sample rate is fixed to be at least twice the bandwidth of the signal-plus-noise, the results of quantization do not vary much with sample rate, assuming the filter roll-off is sufficient so as to not cause significant aliasing.[13] Even so, the results do vary with respect to the bandwidth. For example, a rule of thumb states that the loss caused by 1-bit (hard-limiting) quantization in Gaussian noise is 1.96 dB. However, that number is only true for the infinite (large with respect to signal bandwidth) bandwidth/infinite sample rate case.

Quantization of signal-plus-noise and the associated digital decoding are shown in Fig. 14. Three quantization cases are shown—2-bit, 2.5-bit, and 3-bit, using three, five, and seven threshold levels, respectively. In the three cases, L is the

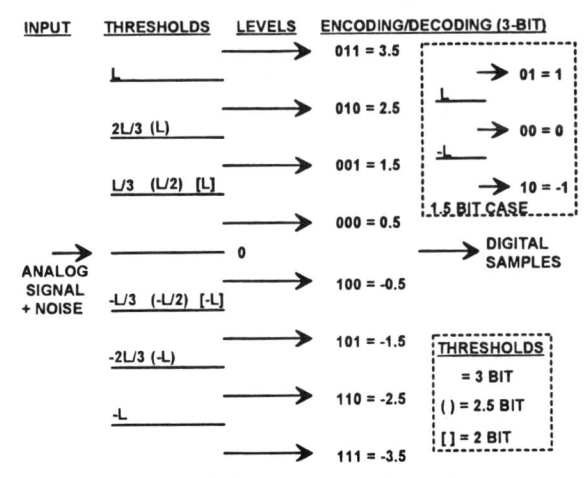

Fig. 14 Quantization mechanization.

highest threshold level. The MAGR quantization is somewhat different, however. Its 1.5-bit quantization has no level at zero, and only two threshold levels, as shown in Fig. 10.[1] The encoding/decoding for that case is shown in the inset of Fig. 14. The GEC Plessey GP1010 chip has the same implementation.[2] This is not always a good idea if the incoming signal-plus-noise is completely (or almost) between the two thresholds, in which case, all is lost. However, the MAGR has the wideband AGC that prevents this.

The degradation caused by quantization for two different bandwidths is shown in Figs. 15 and 16 for the 1-bit, 2-bit, 3-bit, 4-bit, and 5-bit cases as a function of the ratio of the largest threshold L to the rms noise level.[13] The vertical dotted lines show the optimum ratios. The bandwidth relationships in these two figures is close to that of the two receivers that are used as examples. Figure 15 presents the case where the single-sided bandwidth equals the code chipping rate, similar to that of the MAGR for P-code processing. Because the MAGR has no zero threshold, its degradation curve would approach infinity as the level/noise ratio becomes large, instead of being asymptotic to the 1-bit case. However, as stated above, that cannot happen. As it is, however, its degradation is approximately

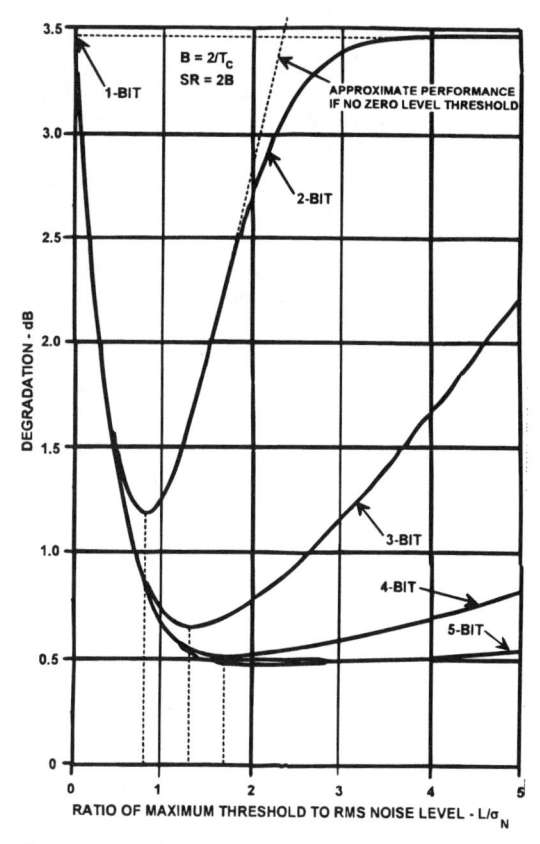

Fig. 15 Quantization degradation in a narrow bandwidth.[13]

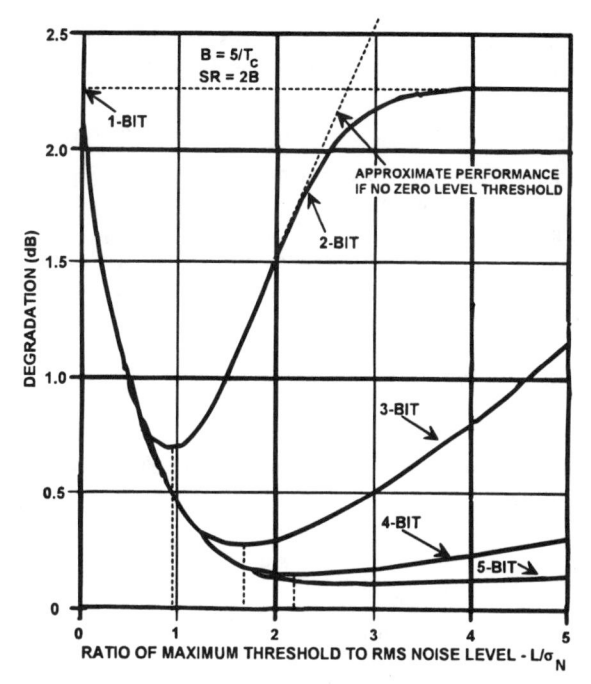

Fig. 16 Quantization degradation in a wide bandwidth.[13]

that of the 2-bit case, somewhere around 1.2 dB. This does include the loss of 0.45 dB due to filter presented in Fig. 12. Note that the minimum degradation caused by quantization is approximately the filter loss shown in Fig. 12.

Figure 16 presents the case where the single-sided bandwidth is five times the chipping rate, close to that of the GPSCard™ for C/A-code processing. Its total loss is less than 0.5 dB. Also shown in Fig. 15 is the case of most low-cost commercial receivers—narrow bandwidth 1-bit processing—nearly 3.5 dB of loss.

The reason why opening the precorrelation bandwidth reduces the quantization loss and its threshold sensitivity is because there is more noise to *dither* the signal about the thresholds. Of course, the filtering loss is also less. In all cases, the GPS signal is well below the noise at this point in the receiver.

5. Multichannel Signal Processing

It is at this point in a modern GPS receiver, after the sampling and quantization functional block in Fig. 2, where the signal processing is split into multiple receiver channels for simultaneous tracking of multiple satellites. The processing that follows is identical in each channel. There is, however, one common thread to all these channels, which is the generation of local clocks and microprocessor interrupts. The local clocks are used to clock the NCOs for the derivation of the reference residual carriers and code generation clocks, and to provide a reference

6. Doppler Removal (Phase Rotation) and Number-Controlled Oscillators

The next signal-processing block shown in Fig. 2 (output at point ①) is Doppler removal or phase rotation, which are one and the same. This process is used by the carrier-tracking loops to track either the phase or the frequency of the incoming signal. This is accomplished with a digital implementation of a single sideband modulator, which is nothing more than the execution of the following trigonometric identities:

$$\begin{aligned}I_{1,sk} &= \frac{A}{\sqrt{2}} C_k D_k \cos(\phi_k - \phi_{rk}) \\ &= \frac{A}{\sqrt{2}} C_k D_k \cos\phi_k \cos\phi_{rk} + \frac{A}{\sqrt{2}} C_k D_k \sin\phi_k \sin\phi_{rk} \\ &= I_{sk} \cos\phi_{rk} + Q_{sk} \sin\phi_{rk}\end{aligned} \quad (28)$$

and

$$\begin{aligned}Q_{1,sk} &= \frac{A}{\sqrt{2}} C_k D_k \sin(\phi_k - \phi_{rk}) \\ &= \frac{A}{\sqrt{2}} C_k D_k \sin\phi_k \cos\phi_{rk} - \frac{A}{\sqrt{2}} C_k D_k \cos\phi_k \sin\phi_{rk} \\ &= Q_{sk} \cos\phi_{rk} - I_{sk} \sin\phi_{rk}\end{aligned} \quad (29)$$

where ϕ_{rk} is the reference phase at time t_k that is generated in an NCO, and ϕ_k is the phase of the I and Q samples in Eqs. (24) and (25). Similarly, the noise components are also phase rotated, but still random, generating the following:

$$I_{1,nk} = x_{1,k}/\sqrt{2} \quad (30)$$

and

$$Q_{1,nk} = y_{1,k}/\sqrt{2} \quad (31)$$

Figure 17 is a block diagram of an NCO. The NCO accumulates phase at its clocking rate based upon a frequency number input. Every time its accumulator rolls over, a new cycle is generated. The time that it takes to do this is a cycle period. The illustrated N_b-bit NCO has the following properties:

1) Frequency output resolution $\Delta f_{min} = 2^{-N_b} f_c$ Hz, where f_c is the clocking rate, which is usually equal to the sample rate entering the Doppler function.

2) The frequency range $\Delta f_{max} = f_c/2$ Hz is governed by the Nyquist criteria. That is, the range of the NCO is one-half the clocking frequency.

3) Phase resolution $\Delta \phi_{min} = \Delta f/f_c$ cycles is the step size of the phase accumulator. At the maximum frequency, this is $1/2$ cycle. Thus, the clocking frequency should be somewhat higher than the Nyquist criteria for good phase resolution, but no better than required for hardware phase tracking of the signal. Normally, the frequency generated in the carrier NCO is that of Doppler plus an intentional

GPS RECEIVERS

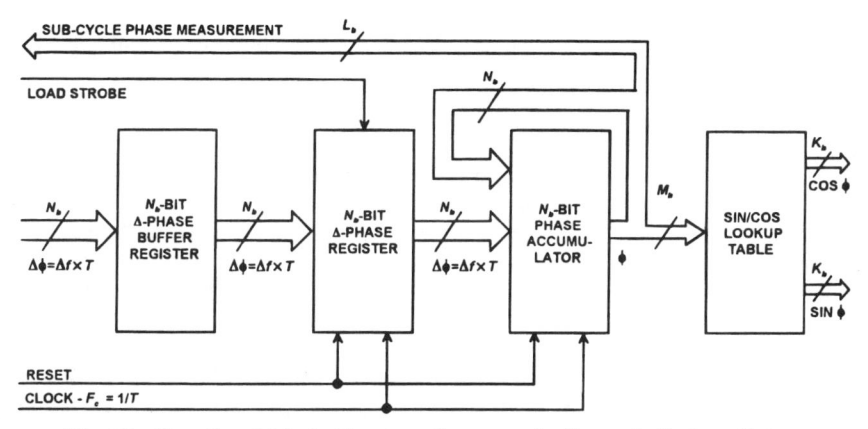

Fig. 17 Functional block diagram of a numerically controlled oscillator.

frequency offset, which is small compared to the clocking frequency of the NCO. Thus, this phase resolution error is negligible. Furthermore, it is random if the clocking frequency is asynchronous with the intentional frequency offset, which is the case in both the MAGR and the GPSCard™.

4) The number of output bits M_b need not be the entire N_b bits and commensurate with the phase resolution for digital receiver applications. All that is required is to generate the desired resolution of the sine and cosine of phase.

5) The number of bits representing the sines and cosines of phase K_b does not have to be any higher than three bits to provide good performance. A single bit (square wave) represents only 0.91 dB loss in the Doppler removal process. Some low-cost commercial receivers use only a single bit. Three bits, which is used in both the MAGR and the GPSCard™ represents only 0.02 dB loss.[1] This requirement relaxes the number of bits M_b required as an input to the sine/cosine lookup table described in 4) above.

6) L_b bits of phase can be latched and used by the microprocessor as a real-time phase measurement, with a resolution of 2^{-L_b} cycles.

7. Code Clock Generation

The code clock is generated in the same way in the code NCO as the reference carrier offset is generated in the carrier NCO, with the exception that sine and cosine components are not generated. The code clock is generated directly from the phase of the code NCO. The same NCO clock/frequency generation properties described above for the carrier NCO also applies to the code NCO, with the exception that the *entire* code clock is generated, not just the Doppler and frequency offsets. Thus, the NCO clock frequency for the code NCO is usually required to be higher than that for the carrier NCO. That is why the MAGR uses a 43.8 MHz clock from its PLL chip (see Fig. 7). This clock provides a 0.2336 P-chip phase resolution, which may seem to be coarse. However, because it is asynchronous with the 10.23 MHz P-code clock, this phase resolution error is random, and it is smoothed in the code-tracking loop. This phase resolution error is even less for the GPSCard™ (in fraction of C/A-chip), because the C/A-code is clocked at only 1.023 MHz, which is also asynchronous with the 20.473 MHz NCO clock.

8. Coder Implementation

The requirements for GPS receiver PN coder implementation are specified in ICD-GPS-200 and in the *Global Positioning System Standard Positioning Service Signal Specification*.[18,19] However, the code generator functional block diagrams provided in that ICD and *Signal Specification* are only for the purpose of describing the requirements. There are other implementation methods. Examples of other methods are given below for C/A coder implementations.

An example functional block diagram of a P-code generator is given in Fig. 18. Note that there are four shift registers, two each for the X1 and X2 code generators. Each of these 12-stage shift registers that have $2^{12} - 1 = 4095$ possible states are short-cycled, either at 4092 or 4093 states, and reset. Both of the X1 shift registers are reset on X1 epochs (every 1.5 s), while the X2 shift registers are reset every 1.5 s plus 37 chip clock cycles. All shift registers are

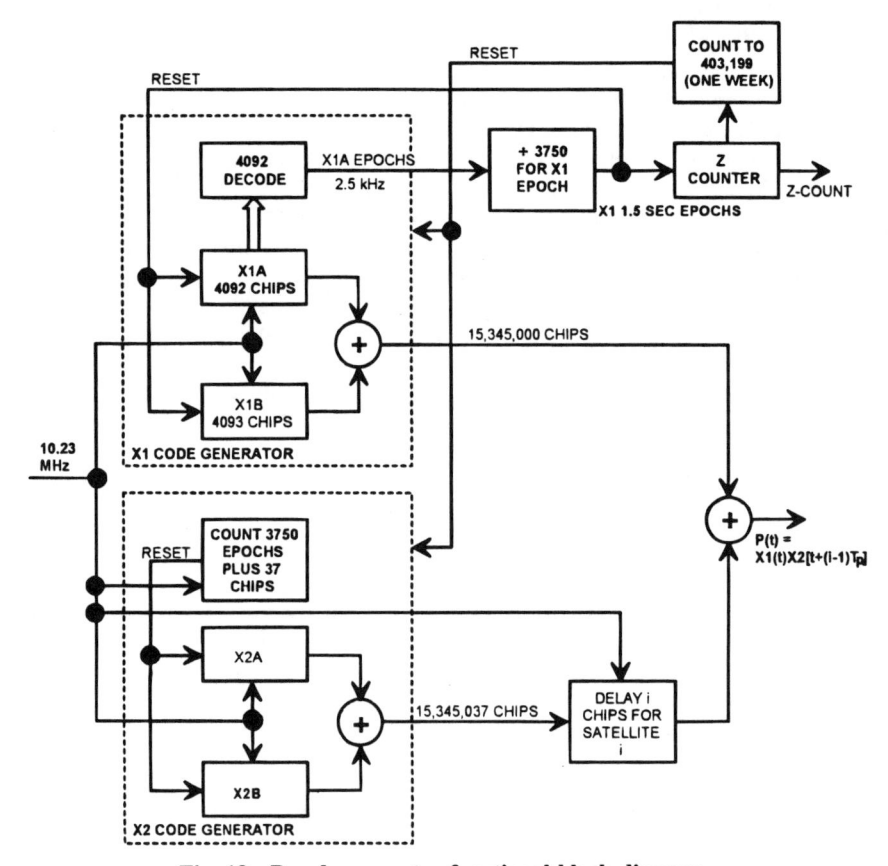

Fig. 18 P-code generator functional block diagram.

GPS RECEIVERS

reset at the end of the week. Although the X1 and X2 coders repeat every 1.5 and 1.5 (+ 37 chip periods) s, the fact that they are running asynchronously at 10.23 MHz, their modulo-2 addition generates an extremely long code. Delaying the X2 code with respect to the X1 code an additional $i - 1$ chips for the ith satellite provides codes for each satellite. The count of the X1 epochs provides a Z-count that is used as basic timing for the system to which the data message and the C/A-coder are synchronized.

Alternatives for the implementation of the C/A-code generator are important, because some implementations cannot be used to generate all possible C/A-codes, although all the GPS C/A-codes can be generated. This is important, because future GPS system augmentations that include other satellites such as the Wide Area Augmentation System (WAAS) geostationary satellites (see Chapter 4, the companion volume), and pseudollites (see Chapter 2, the companion volume) will make use of other C/A-codes in the C/A-code family of 1023 codes.[20] The C/A-coder implementation illustrated in ICD-GPS-200 and in Fig. 19, which uses a selection of two taps of the G2 register to generate a delay of the G2 code, will only generate 45 possible codes, of which 9 are unbalanced codes. The remaining 36 are reserved for GPS.

Two other alternatives for the implementation of the C/A-code generator are illustrated in Figs. 20 and 21. The first initializes the G2 code state at a delayed value, while the second delays the G2 code state, which are equivalent. These two implementations will generate all 1023 codes. Another implementation stores the 1023 code bits of each used code in read-only memory (ROM) and shifts the bits out ROM with the code clock. Although this implementation was used on earlier commercial receiver, it is not flexible, because a new code must be stored each time one is added to the system. Furthermore, with the evolution of application specific integrated circuit (ASIC) technology, ROM storage is inefficient.

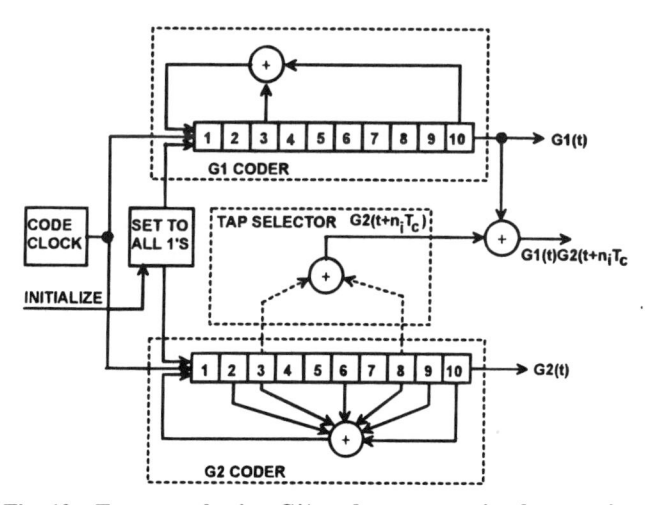

Fig. 19   Two-tap selection C/A-code generator implementation.

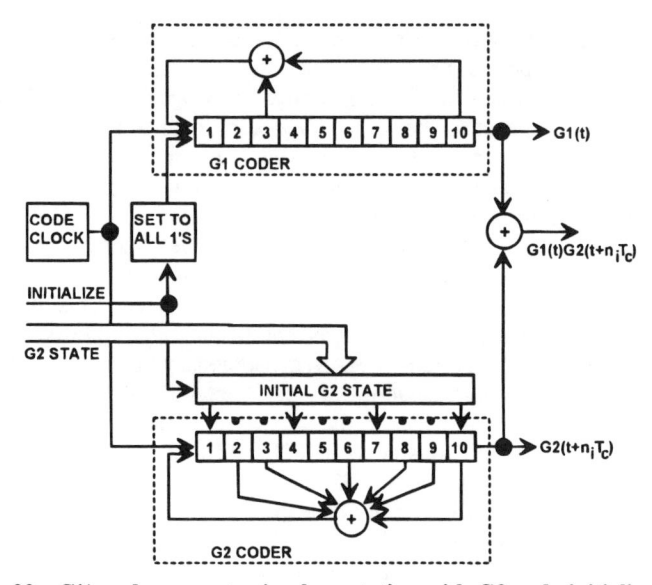

Fig. 20   C/A-code generator implementation with G2 code initialization.

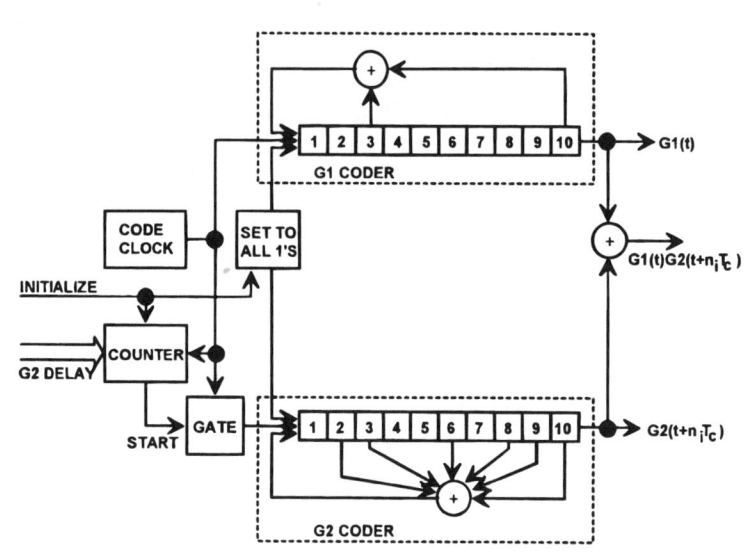

Fig. 21   C/A-code generator implementation with delayed G2 coder initialization.

9. Correlation

The correlation process of the GPSCard™ is illustrated in Fig. 22.[21,22] In this process, early, punctual, and late codes are derived in a shift register that shifts the (early) C/A code from the code generator at a clocking rate defined by the desired early/late correlator spacing (in a fraction of a C/A-code chip). The dual correlation process is realized by performing a multibit *exclusive–or* between the single-bit codes and the multibit I and Q samples. A discriminator selection process allows the selection of either early and late or early-minus-late and punctual correlation. Early and late correlation is used during the signal acquisition process using the maximum (approximately) one chip spacing ($N_{space} = 10$) for rapid acquisition. Early-minus-late and punctual correlation is used during tracking using a 0.1 chip spacing ($N_{space} = 2$) for optimum parallel code (early-minus-late times punctual) and carrier (punctual) tracking. This dynamic spacing concept provides fast acquisition and C/A-code tracking performance that approaches that of conventional P-code tracking performance.

The MAGR correlation process is similar, with two exceptions. First, the MAGR does not allow multiple correlator spacings and is fixed at 1/2 chip, whether processing the P code or the C/A code. That is, the clock feeding the correlator spacing shift register is either twice the C/A-code chipping rate or twice the P-code chipping rate. The second exception is that the MAGR has 10

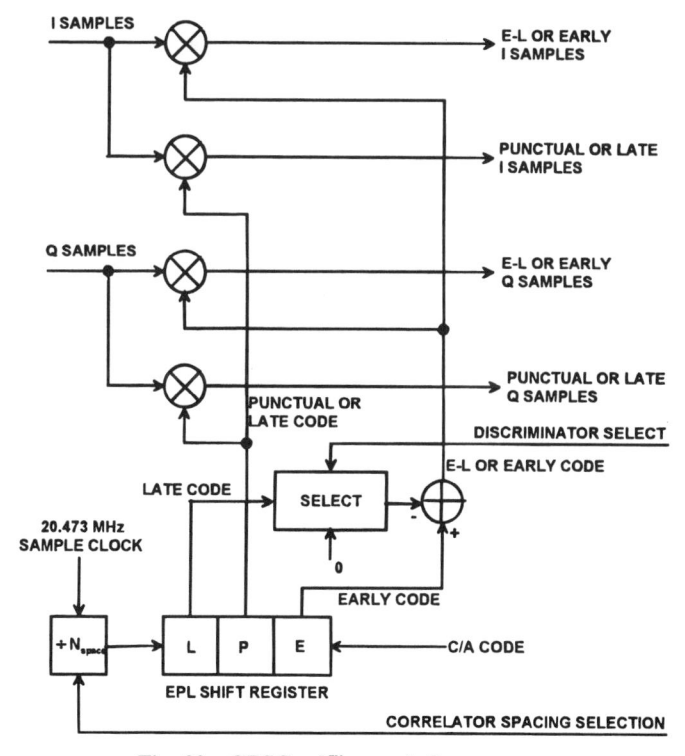

Fig. 22 GPSCard™ correlation process.

correlators instead of 2 for the purposes of rapid acquisition and reacquisition in the presence of jamming.

The reference codes $C_{kr,j}$ for the correlation process at time t_k are outputs of the early-punctual-late (EPL) shift register stages j, $j = e, p$, and l denoting early, punctual, and late. The outputs of correlator j corresponding to each stage j are then obtained by multiplying the I and Q samples of Eqs. (28–31) by the reference codes, resulting in the I and Q samples at point ② of Fig. 2, which are, for the kth sample and the jth shift register stage

$$I_{2,sk,j} = I_{1,sk}C_{kr,j} = \frac{A}{\sqrt{2}} C_k C_{kr,j} D_k \cos(\phi_k - \phi_{rk}) \tag{32}$$

$$Q_{2,sk,j} = Q_{1,sk}C_{kr,j} = \frac{A}{\sqrt{2}} C_k C_{kr,j} D_k \sin(\phi_k - \phi_{rk}) \tag{33}$$

$$I_{1,nk,j} = I_{1,nk}C_{kr,j} = C_{kr,j}x_{1,k}/\sqrt{2} = x_{2,k,j}/\sqrt{2} \tag{34}$$

$$Q_{2,nk,j} = Q_{1,nk}C_{kr,j} = C_{kr,j}y_{1,k}/\sqrt{2} = y_{2,k,j}/\sqrt{2} \tag{35}$$

where the $x_{2,kj}$ and $y_{2,kj}$ are spread versions of $x_{1,k}$ and $y_{1,k}$, and where

$$E(C_k C_{kr,j}) = R(\tau_{kj})$$
$$\approx 1 - |\tau_{kj}|, |\tau_{kj}| \leq 1$$
$$\approx 0, |\tau_{kj}| > 1 \tag{36}$$

where $R(\tau_{kj})$ is the cross-correlation function between the incoming signal code and the reference code j for a delay τ_{kj} at time t_k. This function and the statistics of the noise samples are influenced by the precorrelation bandwidth and possible interference. Discussions of these influences are reserved for the performance discussions of Sec. IV. For now, the assumption is that the precorrelation bandwidth and sample rate are large with respect the code chipping rate, that there is no interference is present and no losses. Then, the correlation process does serve as a noise filter so that the noise samples have the following statistics[22]:

$$E(x_{2,k,j}^2) = E(y_{2,k,j}^2) = N_0/T_s \tag{37}$$

where $T_s = 1/(2B)$ is the inverse of the precorrelation Nyquist sample rate, and

$$E(x_{2,k,j}y_{2,k,j}) = 0 \tag{38}$$

and the time correlation between noise samples in correlators j and m is

$$\rho_{nj,m} = R(\tau_{j,m}) \tag{39}$$

where $\tau_{j,m}$ is the spacing between correlators j and m.

The correlator shown in Fig. 22, as does the MAGR correlator, has the capability of switching one of the correlators to an early-minus-late mode. In this case, the reference code is the difference between the early and late codes j (e) and m (l), while the other reference code is the *punctual* or *prompt* code.[22,23]

$$C_{kr,e-l} = C_{kr,e} - C_{kr,l} \tag{40}$$

and

$$E(C_k C_{kr,e-l}) = E(C_k C_{kr,e}) - E(C_k C_{kr,l}) = R(\tau_{ke}) - R(\tau_{kl}) \qquad (41)$$

and the corresponding noise statistics are as follows:

$$E(x^2_{2,k,e-l}) = E(y^2_{2,k,e-l}) = \frac{2N_0}{T_s}(1 - \rho_{n,e-l}) \qquad (42)$$

$$\rho_{n,e-l,p} = 0 \qquad (43)$$

which is the correlation between the punctual noise (designated by p) and the early-minus-late noise (designated by e-l). This is because both the early and late codes are correlated equally to the punctual code. The variance of the punctual noise samples is the same as indicated above in Eqs. (37) and (38). Note that when the correlation spacing becomes small, the noise variances of Eq. (42) also become small because $\rho_{n,e-l}$ becomes closer to 1.[22]

10. Postcorrelation Processing/Predetection Bandwidth

The function block that follows the correlators in Fig. 2 is that of accumulators, which make up the postcorrelation processing and filter the correlated signal components prior to processing in the microprocessor. The accumulation process is simply the accumulation of the correlated samples over T seconds, where T is usually a C/A-code epoch period of one ms. The accumulations are stored and collected by the microprocessor, and the accumulators are dumped, resulting in an accumulate-and-dump filtering of the signal components. The number of samples accumulated is $M_E = T/T_s$, which may vary some because of code Doppler. Remember, T_s is in local receiver time, while T is derived from the channel's code clock. However, this variation is minimal over 1 ms, and its effect is negligible.

Unless full correlation is achieved, the correlated signal components are still pseudorandom. Thus, this accumulation process serves as a time average of the components in Eqs. (32) and (33). Thus, we can only evaluate the result as an expected value, where, at point ③ of Fig. 2.

$$E[I_{3,si}] = E\left(\sum_{k=1}^{M_E} I_{2,sk}\right) = \frac{A}{\sqrt{2}} R(\tau_i) D_i \sum_{k=1}^{M_E} \cos(\phi_k - \phi_{rk}) \qquad (44)$$

$$E[Q_{3,si}] = E\left(\sum_{k=1}^{M_E} Q_{2,sk}\right) = \frac{A}{\sqrt{2}} R(\tau_i) D_i \sum_{k=1}^{M_E} \sin(\phi_k - \phi_{rk}) \qquad (45)$$

where all M_E samples are within a databit D_i period. Note that the subscripts have changed to i, indicating a sample with an end point at time t_i. Also, no distinction is made as to which correlator j is being processed, because the process applies to all correlators, and the delay τ_i is assumed to be constant over the time interval T.

The sums of the sine and cosine samples can be approximated as numerical integrations, where

$$\frac{1}{T_s} \sum_{k=1}^{M_E} [\cos(\phi_k - \phi_{rk})] T_s \approx \frac{1}{T_s} \int_0^T \cos(2\pi \Delta f_i t + \Delta \phi_i) dt \tag{46}$$

$$\frac{1}{T_s} \sum_{k=1}^{M_E} [\sin(\phi_k - \phi_{rk})] T_s \approx \frac{1}{T_s} \int_0^T \sin(2\pi \Delta f_i t + \Delta \phi_i) dt \tag{47}$$

where

$$(\phi_i - \phi_{ri}) - (\phi_{i-1} - \phi_{r,i-1}) = 2\pi \Delta f_i T + \Delta \phi_i \tag{48}$$

is the change in the difference between the signal phase and the reference phase over the interval T for an assumed constant frequency and phase error Δf_i and $\Delta \phi_i$. The integration of Eqs. (46) and (47) results in Eqs. (44) and (45) becoming the following:

$$I_{3,si} = \frac{A}{\sqrt{2}} M_E \frac{\sin(\pi \Delta f_i T)}{(\pi \Delta f_i T)} R(\tau_i) D_i \cos(\Delta \phi_i) \tag{49}$$

$$Q_{3,si} = \frac{A}{\sqrt{2}} M_E \frac{\sin(\pi \Delta f_i T)}{(\pi \Delta f_i T)} R(\tau_i) D_i \sin(\Delta \phi_i) \tag{50}$$

$$I_{3,ni} = x_{3,i}/\sqrt{2} = \sum_{k=1}^{M_E} x_{3,k}/\sqrt{2} \tag{51}$$

$$Q_{3,ni} = y_{3,i}/\sqrt{2} = \sum_{k=1}^{M_E} y_{3,k}/\sqrt{2} \tag{52}$$

The punctual noise samples now have the following variances:

$$E(x_{3,i}^2) = E(y_{3,i}^2) = \frac{N_0}{T_s} M_E \tag{53}$$

and for the early-minus-late samples corresponding to Eq. (42),

$$E(x_{3,i,e-l}^2) = E(x_{3,i,e-l}^2) = \frac{2N_0}{T_s} M_E (1 - \rho_{n,e-l})$$

Let us now define a new set of I and Q samples in a more convenient form to use for performance analysis. First, let the signal power $S = P_s = A^2/2$. Now, multiply the samples by a factor of

$$\sqrt{\frac{2T}{M_E^s N_0}} = \sqrt{\frac{2T_s}{M_E N_0}} \tag{54}$$

Then,

$$I_i = \sqrt{\frac{2T_s}{M_E N_0}} (I_{3,si} + I_{3,ni}) = \frac{\sin(\pi \Delta f_i T)}{(\pi \Delta f_i T)} \sqrt{2 \frac{S}{N_0} T} \, R(\tau_i) D_i \cos(\Delta \phi_i) + \eta_{Ii} \quad (55)$$

$$Q_i = \sqrt{\frac{2T_s}{M_E N_0}} (Q_{3,si} + Q_{3,ni}) = \frac{\sin(\pi \Delta f_i T)}{(\pi \Delta f_i T)} \sqrt{2 \frac{S}{N_0} T} \, R(\tau_i) D_i \sin(\Delta \phi_i) + \eta_{Qi} \quad (56)$$

This multiplication normalizes the noise power so that

$$E(\eta_{Ii}^2) = E(\eta_{Qi}^2) = 1 \quad (57)$$

and, for the early-minus-late samples,

$$E(\eta_{Ii,e-l}^2) = E(\eta_{Qi,e-l}^2) = 2(1 - \rho_{n,e-l}) \quad (58)$$

and where, for convenience, S/N_0 = signal-to-noise density in ratio-Hz; $C/N_0 = 10 \log_{10} S/N_0$ = carrier-to-noise density in dB-Hz; and, $(S/N_0)T$ = signal-to-noise ratio in a $1/T$ Hz two-sided noise bandwidth. The method for determining signal-to-noise density in the presence of interference is described in Appendix A.

IV. Receiver Software Signal Processing

Receiver software signal processing consists of predetection bandwidth control, signal acquisition, bit synchronization, AGC, carrier tracking, code tracking, lock detection, carrier-to-noise density determination, data demodulation, parity decoding, and measurement processing.

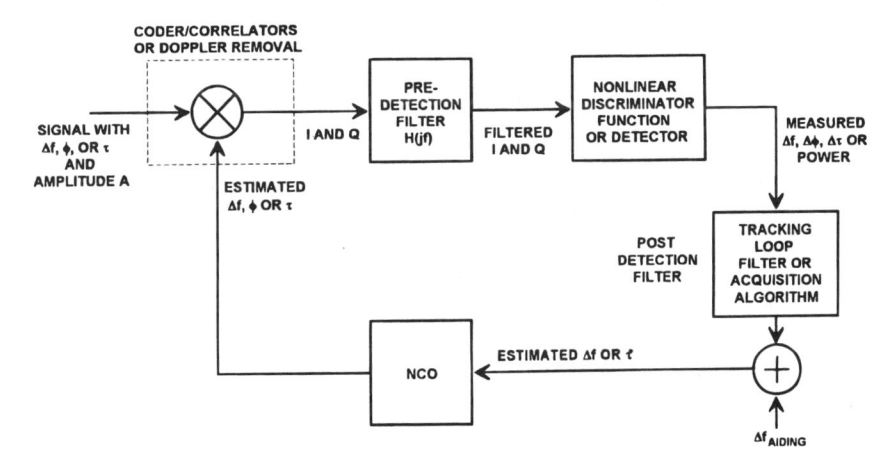

A. A Signal-Processing Model and Noise Bandwidth Concepts

Before launching into the software signal-processing discussions, it is appropriate first to lay the foundation with a generic signal-processing model and the concept of noise bandwidth. Figure 23 illustrates the signal-processing model

Fig. 23 Signal-processing model.

for all types of software signal processing, where the goal is to measure carrier frequency, carrier phase, code phase, or signal power and to perform the appropriate processing of those measurements.

The coder/correlators, Dopper removal, and NCO functions are in hardware and are as described in Sec. III. The predetection filtering consists of the hardware accumulate-and-dump described therein plus any additional accumulation of I and Q samples in software. This predetection filtering establishes the predetection noise bandwidth prior to performing some sort of nonlinear processing of the I and Q samples. The software accumulation process is simply an extension of the accumulation-and-dump process described above. For our purposes here, let us assume that the accumulation process starts with the T second samples, accumulating over M of those samples to an interval of T_I s. Normally, in a GPS receiver, T is one C/A-code epoch period of 1 ms, and M varies between 1 and 20, resulting in a T_I of 1–20 ms, depending upon the mode of operation. Then, the unity dc gain transfer function for this new accumulation-and-dump filter is as follows:

$$H(\omega) = \frac{\sin(\omega T_I/2)}{\omega T_I/2} \tag{59}$$

which is exactly the gain factor of Eqs. (55) and (56) with ω set at the frequency-tracking error $2\pi\Delta f_i$, and T_I set to T.

The equivalent noise bandwidth of a filter is based upon the passing of white noise through the filter, producing an output noise power of[3]:

$$P_n = \frac{N_0}{2} \int_{-\infty}^{\infty} |H(\omega)|^2 \mathrm{d}f \tag{60}$$

where $N_0/2$ is the white noise spectral density. The two-side noise bandwidth is then the noise power if $N_0/2$ is unity, and the dc gain of the filter is also unity. That is, the two-side noise bandwidth, in Hz, is as follows:

$$B_n = \frac{1}{|H(0)|^2} \int_{-\infty}^{\infty} |H(\omega)|^2 \mathrm{d}f \tag{61}$$

Single-sided noise bandwidth is simply half of the two-sided noise bandwidth, where the integration is from 0 to ∞. For the accumulate-and-dump filter transfer function of Eq. (59), Eq. (61) evaluates to $1/T_I$ Hz.

The distinction of two-sided vs single-sided bandwidth, illustrated in Fig. 13, can be confusing. Typically, predetection bandwidth is defined in terms of two-sided noise bandwidth; whereas, postdetection bandwidth (such as tracking-loop bandwidth) is defined in terms of single-sided bandwidth. This is because, in earlier days of communications, predetection filtering was usually accomplished at IF; whereas, postdetection filtering was accomplished at baseband. Furthermore, tracking loops are related more closely to control theory, which is a field that defines processing in terms of single-sided bandwidth. These conventions are maintained throughout this chapter.

Postdetection filtering is any filtering performed after nonlinear processing, such as a tracking-loop filter or the averaging of power in an acquisition algorithm. A tracking-loop filter provides estimates of tracking errors, which are used as

GPS RECEIVERS

inputs to an NCO. In the case of acquisition algorithms, estimates of those inputs are used, but the processing itself is open loop, and the outputs are compared to thresholds. The same is the case for lock detectors. Automatic gain control is a special case in modern receivers and is discussed later.

B. Signal Acquisition

The signal acquisition process consists of a two-dimensional search in time (code phase) and frequency. Figure 24 is a typical C/A-code signal power profile about the signal's main correlation peak when observed through an accumulate-and-dump filter with a response defined in Eq. (59). Note that there are numerous power peaks, only one of which is a valid peak (the largest one in the center). The extra peaks in the frequency direction are attributable to side lobes of the filter. The extra peaks in the code offset direction are attributable to minor correlation peaks. Signal acquisition procedures must be implemented to avoid these invalid peaks. The P-code signal only has the extra peaks in the frequency direction.

The signal acquisition process consists of the following receiver processing tasks:

1) A noise floor is established to set the thresholds for the signal acquisition detector described in Appendix B.

2) Determination of initial estimates of carrier and code Doppler and associated carrier Doppler uncertainty. The initial estimates are set in the carrier and code NCOs. The associated uncertainty is used to set the initial predetection bandwidth and, if required, to determine the number of carrier frequency *bins* that must be observed. A frequency *bin* is at a carrier frequency estimate offset with a width that is a fraction of the predetection bandwidth.

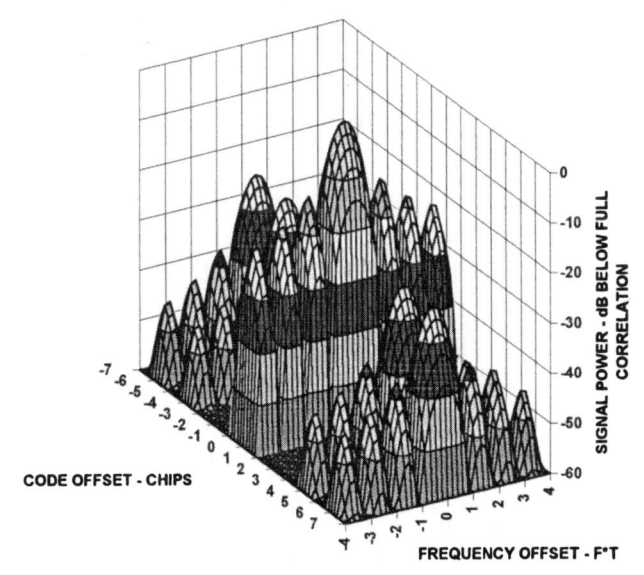

Fig. 24 Two-dimensional search in code phase and frequency.

3) Determination of a received code-phase estimate and associated uncertainty based upon the estimated pseudorange to the satellite being acquired. This estimate is used to set the reference coder phase, while the uncertainty is used to set the code search aperture. Normally, for initial acquisition when time is unknown, the uncertainty is bounded by the entire C/A code, and the coder can be initialized at any phase. If the uncertainty is less than one-half the entire C/A code in subsequent acquisitions, or for P-code acquisitions, the coder phase is usually initialized at the code phase estimate minus some factor times the phase uncertainty (in received time) for the start of the code phase search. A code search is always performed from the *early* side of the estimated location of the correlation peak to prevent the acquisition of a multipath signal.

4) The search for the correlation peak is accomplished by stepping the coder from *early* to *late* through the search aperture and measuring signal-plus-noise power at each code phase position and comparing the power to a predetermined threshold. Normally, these code-phase positions are spaced one-half code chip apart to ensure that near-peak power is encountered during the search. Multiple correlators can be used to speed-up the search by sampling power at multiple positions simultaneously. Descriptions of the power measurements and predetermination of the threshold are given in Appendix B. Special considerations outlined in Appendix B are required for C/A-code signal acquisitions to avoid minor correlation and cross-correlation peak detection.

5) If signal detection fails, a second try may be required, a search through a different frequency *bin* might be required, or the code phase uncertainty might have to be increased. Numerous acquisition strategies are possible.

C. Automatic Gain Control

Precorrelation AGC is required to maintain an optimum A/D conversion, as described previously. Some sort of postcorrelation AGC is also required for optimum signal acquisition and tracking. For example, for the signal acquisition process described above, a noise floor defined by Eqs. (B3) and (B4) of Appendix B must be maintained to obtain the desired false alarm rate. If the precorrelation noise is simply bandlimited white noise, in which case the post correlation noise power can be predicted, no postcorrelation AGC would be required. This is because the precorrelation AGC would maintain the desired noise floor. This is also the case for one-bit sampling without precorrelation AGC. However, if interference or nonwhite noise is present, the postcorrelation noise power is not necessarily predictable.

Postcorrelation AGC can be accomplished a number of ways. For example, for acquisition, we could simply filter the measurement as follows:

$$WBP|_{H_0} = \sum_{i=1}^{KM} (I_i^2 + Q_i^2) \qquad (62)$$

to estimate the wideband noise power in a 1 kHz bandwidth, at a code offset where the signal is not present, prior to and during the acquisition process. This wideband noise power has an expected value equal to the test statistic of Eq. (B4) when the I and Q noise samples have unity variance, but has an expected value of some other value $P_{n,1kHz}$ when they do not. If so, it suffices to alter the threshold of Eq. (B5) to the following new value:

$$TH'' = TH' \cdot \frac{\left[\sum_{i=1}^{KM}(I_i^2 + Q_i^2)\right]_f}{2MK} \tag{63}$$

where $[\ ]_f$ indicates the filtered value. Note that the measurement occurs over the same KM samples as the test statistic, and the filtered value should not be updated until after an H_0 decision has been made, so as not to be influenced by the presence of a signal.

Another approach is to divide the test statistic by the noise power defined in Eq. (62) measured over the *same KM* samples as the test statistic. This defines a *normalized* detector that is useful in a pulsed interference environment, but is only valid when M is greater than 1. Any interference pulses that are spread by the PN code suppress the detector and will not cause false alarms; whereas, the signal causes the narrow-band power of the numerator to exceed the wide-band power of the denominator. Unfortunately, the determination of the threshold to implement this procedure is not straightforward and must be determined by simulation and Monte Carlo analysis.

D. Generic Tracking Loops

The signal-processing model shown in Fig. 23 also applies to code- and carrier-tracking loops, which, in general, use nonlinear error removal and error discriminators or detectors. The simplest of loops is a first-order Costas phase–lock-loop, as shown in Fig. 25 with a linearized phase error removal. Because the signs of the received databits are unknown, a typical discriminator is, for $M = 1$,

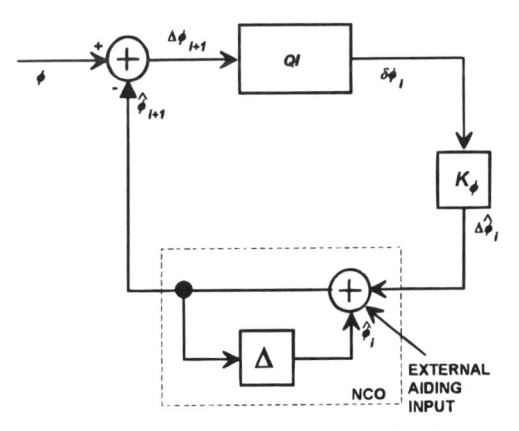

Fig. 25 A generic linearized tracking loop.

$$\delta\phi_i = Q_i I_i \tag{64}$$

The product eliminates dependence on the signs of the databits, because they affect I and Q alike, until their noise components dominate their signal components. This leads to a loss-of-lock.

The expected value of Eq. (64), in terms of Eqs. (55) and (56), yields the following

$$\begin{aligned}E(\delta\phi_i) &= E(I_i)E(Q_i) + E(Q_i)E(x_i) + E(I_i)E(y_i) + E(x_i y_i) \\ &= E(I_i)E(Q_i) = \frac{\sin^2(\pi\Delta f_i T)}{(\pi\Delta f_i T)^2} \frac{S}{N_0} TR^2(\tau_i)\sin(2\Delta\phi_i) \\ &\approx 2\frac{S}{N_0} T\Delta\phi_i \end{aligned} \tag{65}$$

in linearized form for small frequency errors and small code- and phase-tracking errors. If the PLL is truly tracking, the frequency errors have to be small. If the code tracking errors are not small, the amplitude of the above discriminator is simply reduced. This does point out the importance of code and carrier tracking simultaneously.

The variance of that discriminator is the following:

$$\begin{aligned}\sigma^2_{\delta\phi} &= E(I_i^2 Q_i^2) - E^2(I_i Q_i) \\ &= E^2(I_i)E(y_i^2) + E^2(Q_i)E(x_i^2) + E(x_i^2 y_i^2) \\ &= 2\frac{\sin^2(\pi\Delta f_i T)}{(\pi\Delta f_i T)^2} \frac{S}{N_0} T + 1 \\ &\approx 2\frac{S}{N_0}T + 1 = 2\frac{S}{N_0} T\left(1 + \frac{1}{2S/N_0 T}\right) \end{aligned} \tag{66}$$

The second term in the bracket of the variance equation is attributable to squaring loss, which is proportional to the predetection bandwidth $1/T$.

Because the interest is to measure $\delta\phi_i$, note that the gain of the discriminator is $2\,S/N_0 T$. Thus, it varies with signal-to-noise ratio. Because of this, the loop gain must somehow be compensated to maintain a loop bandwidth and, in the case of higher-order loops, to maintain loop dynamic response characteristics. Normally, the optimum closed-loop response (input-to-output estimate of input) takes on the following form[24]:

First order:

$$H(\omega) = \frac{\omega_N}{j\omega + \omega_N}$$

Second order:

$$= \frac{j2\zeta\omega_N\omega + \omega_N^2}{-\omega^2 + j2\zeta\omega_N\omega + \omega_N^2}$$

Third order:

$$= \frac{-2\omega_N\omega^2 + j2\omega_N^2\omega + \omega_N^3}{j\omega^3 - 2\omega_N\omega^2 + j2\omega_N^2\omega + \omega_N^3} \quad (67)$$

resulting single-sided noise bandwidths of

First order:

$$B_L = \frac{\omega_N}{4}$$

Second order:

$$= \frac{\omega_N}{8\zeta}(4\zeta^2 + 1)$$

Third order:

$$= \frac{\omega_N}{1.2} \quad (68)$$

For the linear first-order loop, the closed-loop noise time-update error $\Delta\phi_i$ equation is of the following form:

$$\Delta\phi_{i+1} = (1 - 4B_LT)\Delta\phi_i + \frac{4B_LT}{E(\delta\phi_i)}\delta\phi_i \quad (69)$$

which is the loop error response for the loop shown in Fig. 25 for a loop update rate of $1/T$. The desired gain for this first-order loop is $\omega_N T = 4B_L T$. Thus, to achieve that at a design signal-to-noise density S/N_{od}, using the discriminator gain derived in Eq. (65):

$$K_\phi = \frac{4B_LT\Delta\phi_i}{E(\delta\phi_i)}\bigg|_{\Delta\phi_i=0} = \frac{2B_L}{S/N_{od}} \quad (70)$$

which is already reflected in Eq. (69). This indicates that a gain schedule, AGC, or a normalized discriminator is required to maintain loop performance at all S/N_0. The variance time-update equation is the expected value of the square of both sides of Eq. (69), and accounting for the fact that the noise in $\delta\phi_i$ is independent of that in all previous errors $\Delta\phi_i$, is as follows:

$$\sigma_{\phi,i+1}^2 = (1 - 4B_LT)^2\sigma_{\phi,i}^2 + K_\phi^2\sigma_{\delta\phi}^2 \quad (71)$$

However, in steady state, $\sigma_\phi^2 = \sigma_{\phi,i+1}^2 = \sigma_{\phi,i}^2$ so that (ignoring the aiding input)

$$\sigma_\phi^2 = \frac{K_\phi^2\sigma_{\delta\phi}^2}{8B_LT(1 - 2B_LT)} \approx \frac{B_L}{S/N_0}\left(1 + \frac{1}{2S/N_0T}\right) \quad (72)$$

at the design point, or accounting for the discriminator gain variation with signal-to-noise density. Note that the term $1 - 2B_LT$ is neglected, which assumes that the loop gain is small enough not to cause stability problems; that is, $2B_LT \ll 1$.

It can be shown that the noise performance of higher-order loops with the responses and noise bandwidths shown above in Eqs. (67) and (68) is the same, provided that the loop gains meet the constraints described above.

In general, this analysis can be performed for all of the GPS receiver tracking loops with varying complexity, unless more sophisticated normalized discriminator functions are used. Then, Monte Carlo techniques are usually required to simulate the discriminator statistics to determine the values of those derived in Eqs. (65) and (66). Then, without simulating the closed-loop, the loop noise performance can be determined using Eq. (71), except that the numerator is replaced with the determined statistics. That is, Eq. (70) becomes as follows:

$$K_\phi = \frac{4B_LT}{G_d} \tag{73}$$

where a Monte Carlo estimate of the discriminator's gain is as follows:

$$G_d = \frac{\overline{(\delta\phi)}_{out}|_{\Delta\phi=\Delta\phi_{in}}}{\Delta\phi_{in}} \tag{74}$$

for a small $\Delta\phi_{in}$ in the linear range of the discriminator. The square root of Eq. (72) then becomes the following:

$$\sigma_\phi \approx \frac{\sqrt{2B_LT}}{G_d} \hat{\sigma}_{\delta\phi} \tag{75}$$

where $\hat{\sigma}_{\delta\phi}$ is a Monte Carlo estimate of the square root of Eq. (66). Closed-loop Monte Carlo simulations can be used to achieve the same results. However, if the loop time constant is relatively long, the time to achieve a good estimate of the performance can be quite long. A few closed-loop simulations can be used, however, to substantiate the results.

Aiding of the tracking loop can be achieved by providing an estimate of one of the states of the tracking loop in real time from an external source, as is shown in Fig. 25. In case of a code-tracking (delay-lock) loop (DLL), this external source could be the Doppler estimate from the carrier tracking loop, scaled appropriately. In a jamming scenario, this aiding input might be the output of a corrected inertial sensor measurement, projected along the line-of-sight to the satellite and corrected for satellite Doppler and satellite and receiver clock drift.

E. Delay Lock Loops

Note that Eqs. (55) and (56) contain the cross-correlation function $R(\tau_i)$. These equations represent the I and Q samples from any one of the correlators in Fig. 22. However, the DLL uses outputs of early and late or early-minus-late correlators in its discriminator, depending upon the type of discriminator. Three of the more popular DLL discriminators are the noncoherent discriminators $\delta\tau_i = I_{E,i}^2 + Q_{E,i}^2 - I_{L,i}^2 - Q_{L,i}^2$ (early-minus-late power), and $\delta\tau_i = I_{E-L,i}I_{P,i} + Q_{E-L,i}Q_{P,i}$ (dot-product) where E, L, $E-L$, and P indicate early, late, early-minus-late, and punctual phases of the reference code used in the cross correlation, as illustrated in Fig. 22. A coherent DLL has the form $\delta\tau_i = I_{E-L,i}\,\text{sign}(I_{P,i})$ where sign $(I_{P,i})$ is the sign of the GPS navigation message databit demodulated in the PLL or frequency–lock-loop (FLL). This coherent DLL requires parallel carrier-phase tracking (and thus, the *coherent* qualifier). The noncoherent discriminators require no carrier tracking as long as the frequency offset is small with respect to the predetection bandwidth. This is an important feature in the presence of jamming using an external aiding source.

Each of these three discriminators has its advantages, and each has its disadvantages. The first, the noncoherent early-minus-late power discriminator, is the most robust discriminator. Being noncoherent makes it robust, and it has the largest linear range of the three. It has an expected value of[22]

$$E[\delta\tau_i] = \overline{I}_{E,i}^2 + \overline{Q}_{E,i}^2 - \overline{I}_{L,i}^2 - \overline{Q}_{L,i}^2$$
$$= 2S/N_0 T_I [R^2(\tau_i - d/2) - R^2(\tau_i + d/2)]$$
$$\approx 4S/N_0 T_I (2 - d)\tau_i \tag{76}$$

in the range of $|\tau_i| \leq d/2$, where $d < 2$ is the spacing between the late and early correlators (in units of PN chips), and where

$$T_I = MT \tag{77}$$

is the predetection integration interval after summing M I and Q samples in software, providing a predetection bandwidth of $1/T_I$. The closed-loop noise variance of a DLL using the early-minus-late power discriminator is[22] as follows:

$$\sigma_\tau^2 = \frac{B_L d}{2S/N_0}\left(1 + \frac{2}{(2-d)S/N_0 T_I}\right) \tag{78}$$

for simultaneously measured early and late samples and a precorrelation bandwidth that is wide relative to the chipping rate of the code. The squaring loss term in the brackets is the largest of the three discriminators. To measure early and late power simultaneously, which is important from a noise performance point of view, two correlators are required for this discriminator. Thus, in order to track the carrier in a punctual correlator (at the correlation peak), a third correlator would be required. However, the performance is best when d is small, in which case either the early or late correlator are near the correlation peak.

The noncoherent dot-product DLL discriminator has an expected value of[22]

$$E[\delta\tau_i] = \overline{I}_{E-L,i}\overline{I}_{P,i} + \overline{Q}_{E-L,i}\overline{Q}_{P,i}$$
$$= 2S/N_0 T_I [R(\tau_i - d/2) - R(\tau_i + d/2)]R(\tau_i)$$
$$\approx 4S/N_0 T_I \tau_i (1 - |\tau_i|) \tag{79}$$

in the range of $|\tau_i| \leq d/2$, and a closed-loop noise variance of[22]

$$\sigma_\tau^2 = \frac{B_L d}{2S/N_0}\left(1 + \frac{1}{S/N_0 T_I}\right) \tag{80}$$

for simultaneous measured early-minus-late and punctual samples.

The coherent DLL discriminator has an expected value of[22]

$$E[\delta\tau_i] \approx \bar{I}_{E-L,i}$$
$$= \sqrt{2(S/N_0T_I)}[R(\tau_i - d/2) - R(\tau_i + d/2)]$$
$$\approx 2\sqrt{2S/N_0T_I}\tau_i \qquad (81)$$

in the range of $|\tau_i| \leq d/2$ for low bit error rates as long as phase lock is maintained and a closed-loop noise variance of[22]

$$\sigma_\tau^2 \approx \frac{B_L d}{2S/N_0} \qquad (82)$$

for simultaneous measured early-minus-late and punctual samples under the same conditions. Equations (81) and (82) are not a result of rigorous derivation that would have taken into account bit errors or cycle slips. This is because the coherent DLL only works when phase tracking is successful. The coherent DLL only applies to those applications that also require successful phase tracking. The advantage of its use is that a hardware "Q channel" is never required for the DLL, and thus reducing "gate count" in the receiver's digital signal-processing chip. However, not having this Q channel also reduces the receiver's capabilities for rapid signal acquisition, because the Q-less channel cannot be used when applying Eqs. (B1) and (B2) of Appendix B during acquisition. Also, receivers designed to operate in an interference or scintillating environment would never implement a coherent DLL, because they would be expected to operate when phase tracking is not possible. For example, the MAGR does not even use a PLL. It uses the noncoherent dot-product DLL while tracking only the carrier frequency using an FLL or only using external frequency aiding in times of severe jamming.

Note that all three implementations of the DLL presented here have the same noise performance at higher signal-to-noise conditions at times when squaring loss is minimal in the noncoherent designs. Of course, when the squaring loss in a noncoherent DLL would be significant, the performance of the coherent DLL would also be marginal, and Eq. (82) would not apply.

In general, the DLL is always aided—carrier aided when carrier tracking (phase or frequency) is also being accomplished, or externally aided when carrier tracking in not being accomplished. This has been the case since the beginning of GPS, starting with the X-Set. The aiding input, in the case of carrier aiding, is simply the carrier tracking loops estimate of Doppler, in Hz at L_1 or L_2, divided by the ratio of the carrier frequency to the code chipping rate (1540 in the case of the C/A code and 154 or 120 in the case of the P code).

The general closed-loop implementation is shown in Fig. 26. Note that the loop gains are different, depending upon whether of not a first- or second-order loop is implemented. Because carrier aiding is the norm, a first-order loop is usually sufficient. The loop only has to correct for initial tracking errors, twice the rate of change of the ionosphere or differences in code and carrier multipath. A second-order loop may be required using external aiding, depending upon how accurate the aiding is. Note also that a second-order loop can be used as a first-order loop by simply redefining the loop gains (the integrator gain would be set

GPS RECEIVERS

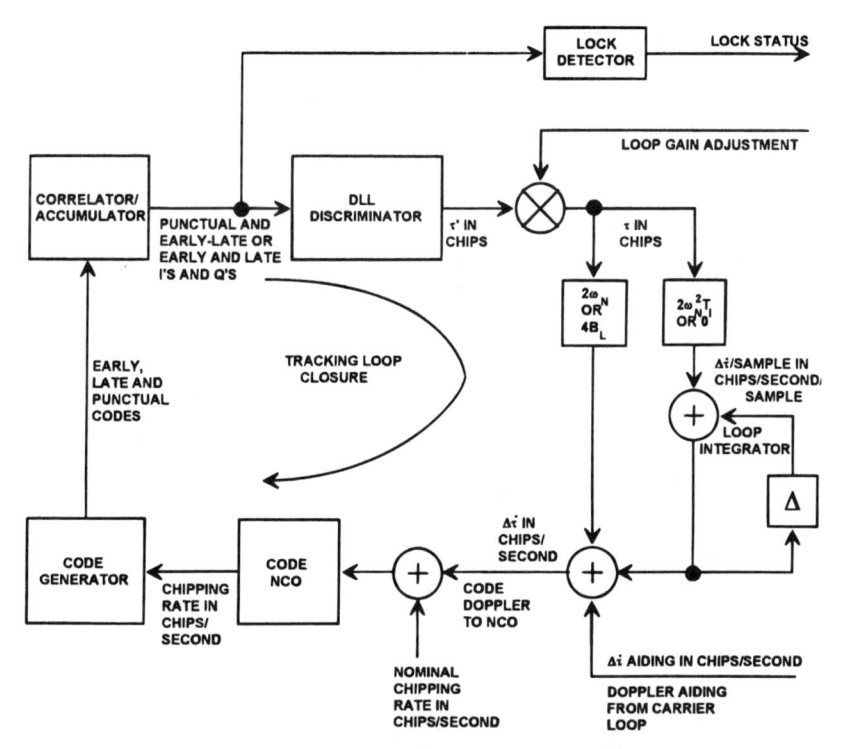

Fig. 26 Delay lock loop Implementation.

to 0). Because of the aiding inputs, typically, DLL loop bandwidths are quite small—on the order of 0.05–1 Hz, depending upon the application. The predetection bandwidth of the discriminator usually takes on the same predetection bandwidth of the simultaneously tracking carrier loop. If externally aided, the predetection bandwidth will be based upon the Doppler aiding error, which, for an interfering environment, should be quite small, or the minimum 50 Hz data bandwidth ($1/T_i$).

1. Example Delay Lock Loop Noise Performance[22]

Equation (75) provides a method of estimating tracking performance that is very useful, because each discriminator output is statistically independent in time. In that way, statistics can be computed on many independent samples in a short period of time, as opposed to using statistically independent closed-loop samples from narrow-band tracking loops.

The performance parameter of Eq. (75), altered for errors in code phase instead of carrier phase, is plotted in Fig. 27 along with the wide-band theoretical case for the simulation and test results for a loop bandwidth of 1 Hz for a dot-product DLL for different correlator spacings. The simulation and test results agree very well, but deviate at the narrow spacings because of precorrelation band limiting. In fact, the results show that a spacing of 0.2 chips performs as good as or better

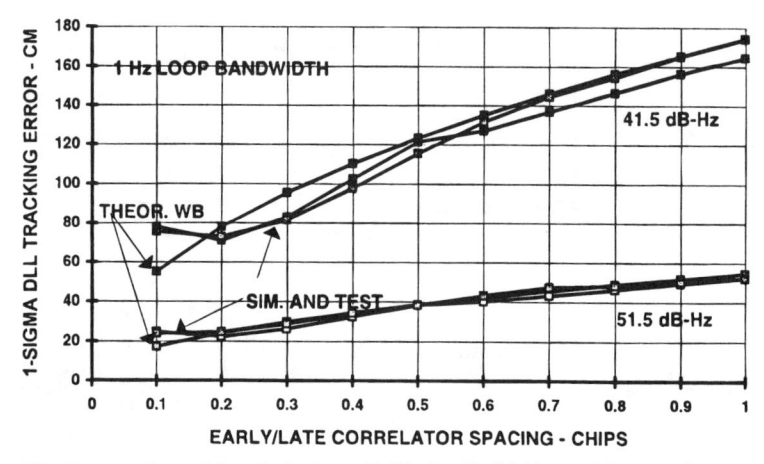

Fig. 27 Dot-product delay lock loop (1 Hz bandwidth) tracking performance vs spacing (noise only).

than a spacing of 0.1 chips for the 8 MHz bandwidth. Both spacings are much better than a 1 chip spacing, however.

To determine the performance for other loop noise bandwidths, we simply multiply the ordinate axis of Fig. 27 by the square root of the loop bandwidth. Thus, for a bandwidth of 0.05 Hz (multiply by 0.2236), better than 10 cm accuracy is achieved at nominal C/N_0 values, which are usually above 45 dB Hz. This bandwidth represents a loop time constant of 5 s, which is more than wide enough to track the code–carrier divergence caused by the ionosphere in the carrier-aided DLL. In fact, further smoothing would result in even better results. However, the overall tracking performance is limited by the effects of multipath.

2. Multipath Rejection

Although the narrow correlator spacing produces superior noise performance in a DLL, multipath effects in a DLL tend to dominate the error budget using the C/A-code. However, the narrow spacing also reduces the effects of multipath. This is because distortion of the cross-correlation function near its peak because of multipath is less severe than at regions away from the peak.

In Hagerman,[25] it was indicated that narrow spacing did not reduce the maximum ranging error caused by multipath. However, that statement was made with respect to *coherent* DLLs, which are more susceptible to carrier-phase tracking error. This is true, because the PLL is tracking the composite true/multipath signal. Thus, *coherent* is not really *coherent* to the true signal, and extreme cross-correlation function distortion can occur, even with narrow spacing. The results presented in Hagerman[25] are true for strong multipath returns.

This does not apply to *noncoherent* DLLs, because those types of discriminators cancel most reliance on carrier phase. Dot-product DLL performance in a multipath environment is essentially the same as the early-minus-late power DLL for what is called Region I tracking.[26] For C/A-code applications, this is the only region of interest, because carrier aiding should never allow a transfer to Region II. Generally speaking, in Region II, the multipath signal itself is being tracked; whereas, in Region I, the multipath is simply causing a signal-tracking error by distorting the discriminator.[25,26]

A two-path signal is given as follows:

$$s_m(t) = AC_f(t)\cos(\omega_0 t + \phi_0) + \alpha AC_f(t - \delta)\cos[\omega_0(t - \delta) + \phi_0] \quad (83)$$

where $C_f(t)$ is the filtered PN code, α is the relative multipath signal amplitude, and δ is the relative time delay of the multipath signal with respect to the receipt of true signal. Theoretically, α can take on values greater than one, but practically, it will be somewhat less than one. Note that the phase of the multipath signal differs from the true signal phase by $\omega_0\delta$.

Equation (83) essentially reflects the signal model used in Eq. (1), with the exception that, here, the datastream is neglected, and the code is filtered. This is important if or when the correlator spacing is narrowed, because the bandwidth affects the multipath distortion.

If we apply the coherent discriminator of Eq. (81), the composite discriminator becomes as follows:

$$E[d\tau_i] \approx \sqrt{2S/N_0T_I}[R_f(\tau_i - d/2) - R_f(\tau_i + d/2)]\cos\phi_i$$
$$+ \alpha\sqrt{2S/N_0T_I}[R_f(\tau_i - \delta - d/2) - R_f(\tau_i - \delta + d/2)]\cos(\phi_m + \phi_i) \quad (84)$$

where ϕ_m is the relative phase between the multipath component and the signal component, and $R_f(\tau_i)$ is the cross correlation between a filtered and an unfiltered PN code. For a ϕ_m of 180 deg and an α of unity, Eq. (84) is identically zero for a large range of τ_i, depending upon the multipath delay and correlator spacing when the spacing is less than one chip. Thus, for the period of time that these conditions occur, the DLL would not be tracking at all. This was the phenomenon that prompted the dishonorable mention for narrow spacing by Hagerman.[25] However, that phenomenon would be rare. In fact, it has been shown that the tracking error attributable to multipath for the coherent DLL is more symmetric and is, thus, easier to smooth away.[27] This potential tracking problem does not occur in the case of noncoherent DLLs. The analysis to follow is only for the noncoherent dot-product case.

By then applying Eq. (83) and normalizing with $2S/N_0T_I$, the resulting dot-product discriminator output becomes[22] the following:

$$E[d\tau_i] = [R_f(\tau_i - d/2) - R_f(\tau_i + d/2)]R_f(\tau_i)$$
$$+ \alpha^2[R_f(\tau_i - \delta - d/2) - R_f(\tau_i - \delta + d/2)]R_f(\tau_i - \delta)$$
$$+ \alpha[R_f(\tau_i - d/2) - R_f(\tau_i + d/2)]R_f(\tau_i - \delta)\cos\phi_m$$
$$+ \alpha[R_f(\tau_i - \delta - d/2) - R_f(\tau_i - \delta + d/2)]R_f(\tau_i)\cos\phi_m \quad (85)$$

When the discriminator output is set to zero as a function of τ_i, the sign of τ_i changes as a function of the multipath phase and varies much more for the larger correlator spacing. The maximum (positive) and minimum (negative) errors occur when the relative multipath phase is 0 and 180 deg, respectively. Thus, it suffices to evaluate the discriminator at those values only, producing an envelope of the multipath error vs multipath delay. Varying relative phase causes the tracking error to take on all values inside of the envelope for a given multipath delay.

Evaluation of the error envelope vs multipath delay and an α of 0.5 is in Fig. 28 for the C/A code with conventional 1 chip spacing in a 2 MHz precorrelation bandwidth vs the C/A code with 0.1 chip spacing in an 8-MHz bandwidth. Also shown is the error envelope for a 20-MHz bandwidth P code using one P-code chip correlator spacing. The evaluation for the P code shows up as a very small region at less that 0.15 C/A-chip multipath delay. The 0.1 chip error envelope is, indeed, much smaller than that for 1 chip spacing, but not as small as the P-code error envelope for two reasons. First of all, the C/A code correlates with the multipath signal with up to 10 times the delay than does the P code. Second, the 8-MHz bandwidth limits the reduction of the multipath effect.

To evaluate what would happen if the bandwidth is opened up to 20 MHz with a spacing of 0.05 chips, the same process is repeated for that case. The results are shown in Fig. 29. For the region of 0.15 chip multipath delay or less, the small C/A-code correlator spacing slightly outperforms the conventional P-code performance. This wider bandwidth makes the receiver more susceptible to what would be out-of-band interference in a conventional spacing receiver. However, the correlator, which is a matched filter, does itself reject interference in this wider bandwidth, but not to the degree that could be achieved with sharp cutoff filters at IF.

Extensive testing has verified this multipath rejection performance.[28–30] Additional discussions of multipath effects appear in Chapter 14, this volume.

F. Carrier Tracking

There are two methods of carrier tracking—carrier-frequency tracking and carrier-phase tracking. The latter method also tracks the carrier frequency. Most

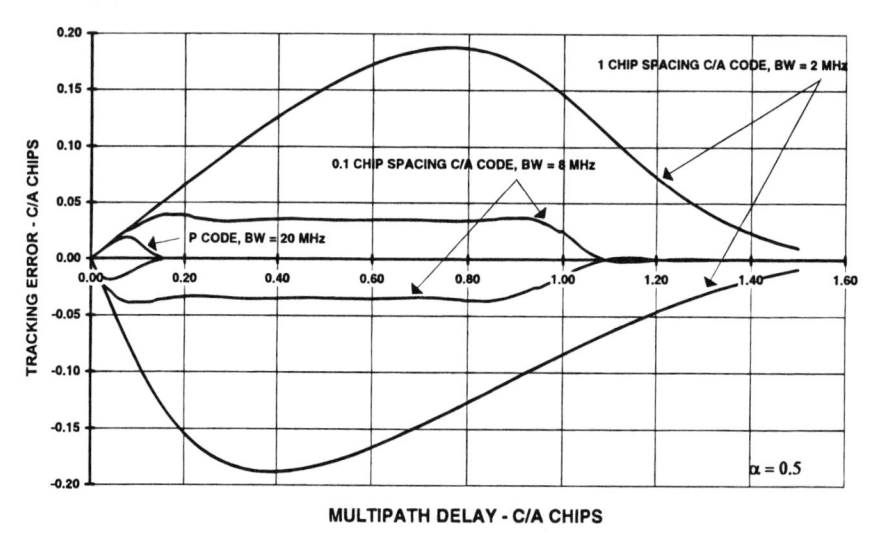

Fig. 28 Multipath error envelopes for 0.1 and 1 chip spacing and P code.

Fig. 29 Multipath error envelopes with a 20-MHz precorrelation bandwidth.

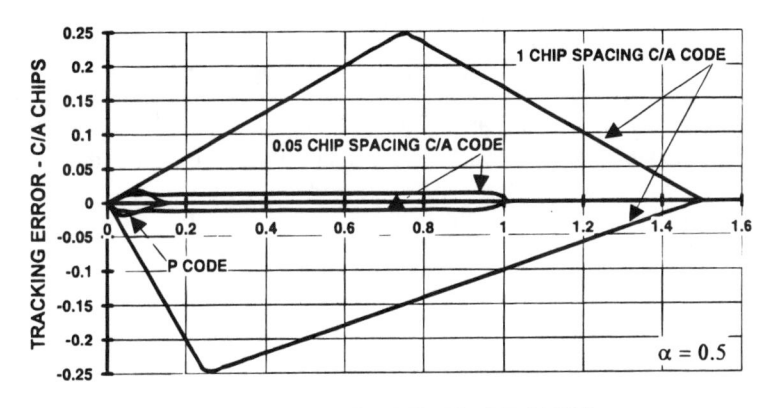

GPS receivers track carrier phase to obtain the very accurate integrated carrier Doppler measurements. However, even some of those use frequency tracking as a transition between signal (code) acquisition and phase tracking. This is because it is very difficult to acquire the phase of the carrier directly because of large frequency uncertainties, and the transition using frequency tracking provides a very efficient means of doing so. Some GPS receivers never track carrier phase. For example, the MAGR tracks only the carrier frequency because its military applications never require integrated Doppler measurements.[23] Short-term *delta-range* measurements are required, but they are obtained by the MAGR as open-loop corrections to the integration of the carrier Doppler.

Frequency tracking is accomplished using an automatic frequency control (AFC) loop, which is sometimes called a frequency–lock-loop.

1. Frequency-Tracking Loops

Frequency tracking is essentially the same as differential carrier-phase tracking. In most cases, the frequency-tracking discriminators measure the change in carrier phase over a finite interval of time. Example discriminators are the following:
Cross-product discriminator:

$$\delta f_i = I_{i-1}Q_i - I_iQ_{i-1} \tag{86}$$

Decision-directed cross-product discriminator:

$$\delta f_i = (I_{i-1}Q_i - I_iQ_{i-1})\text{sign}(I_{i-1}I_i + Q_{i-1}Q_i) \tag{87}$$

Differential arctangent discriminator:

$$\delta f_i = \tan^{-1}\left(\frac{Q_i}{I_i}\right) - \tan^{-1}\left(\frac{Q_{i-1}}{I_{i-1}}\right) \tag{88}$$

All of these discriminators are affected by the databits. For example, the cross-product discriminator must always be computed within a databit period. That is, the previous sample i-1 must be in the same bit period as the current sample i. Otherwise, if the bit sign changes, the discriminator results in the wrong sign. The decision-directed cross product solves that problem by modulating the cross product with the sign of the dot product, which changes sign along with the (noise-free) databits. With the differential arctangent discriminator, the quadrant of the sample must be resolved for the same reason.

Let us assume here that a databit interval is T_B, and its relationship to the predetection interval T_I is as follows:

$$T_B = KT_I \tag{89}$$

Then, the expected value of the cross-product discriminator summed over a databit period, in terms of Eqs. (55–57) and Eq. (77), is as follows:

$$E(\delta f'_k) = E\left(\sum_{i=1}^{K-1} \delta f_i\right)$$

$$= 2\frac{\sin^2(\pi \Delta f_k T_I)}{(\pi \Delta f_k T_I)^2} \frac{S}{N_0} T_I D_k \sum_{i=1}^{K-1} [\cos(\Delta\phi_{i-1})\sin(\Delta\phi_i) - \sin(\Delta\phi_{i-1})\cos(\Delta\phi_i)]$$

$$= 2\frac{\sin^2(\pi \Delta f_k T_I)}{(\pi \Delta f_k T_I)^2} \frac{S}{N_0} T_I D_k \sum_{i=1}^{K-1} \sin(\Delta\phi_i - \Delta\phi_{i-1})$$

$$\approx 2\frac{\sin^2(\pi \Delta f_k T_I)}{(\pi \Delta f_k T_I)^2} \frac{S}{N_0} T_I D_k (K-1)\sin(2\pi \Delta f_k T_I)$$

$$\approx 4\pi S/N_0 T_I^2 D_k (K-1)\Delta f_k \tag{90}$$

for small Δf_i, and assuming that $\Delta\phi_i - \Delta\phi_{i-1} = 2\pi \Delta f_i T_I = 2\pi \Delta f_k T_I$ over $K-1$ samples, with the first sample of the bit period being discarded. Note that the discriminator is proportional to $\sin(2\pi \Delta f_k T_I)$ attenuated by the sinc^2 function and has more than one zero point as a function of Δf_k, some with positive slope and some with negative slope. Thus, care must be taken when initializing the tracking loop with a large initial frequency error. Thus, to start with, T_I should be as small as possible and gradually reduced, for the cross-product discriminator, to a minimum value of 10 ms. (Actually, a minimum value of 5 ms outperforms the discriminator using 10 ms, because K takes on a value of 3 instead of 1.) This gradual narrowing of both the predetection bandwidth and the range of the discriminator is even more important using the other two discriminators, because the decision-directed property causes a discontinuity in the discriminator, narrowing their effective frequency range by two.

The tracking error variance of the cross-product discriminator, in Hz, is[31]

$$\sigma^2_{\Delta f} = \frac{1}{2\pi^2}\frac{K}{K-1}\frac{B_L B_I^2}{S/N_0}\left(\frac{1}{K-1} + \frac{B_I}{2S/N_0}\right) \tag{91}$$

where $B_I = 1/T_I$ is the predetection noise bandwidth with a minimum of 100 Hz, with which K would be 2. Prior to bit synchronization, when the times of bit edges are unknown, a predetection bandwidth of 1 kHz is used. In this case, the first sample per bit cannot be discarded, and the variance becomes as follows:

$$\sigma_{\Delta f}^2 = \frac{1}{2\pi^2} \frac{B_L B_I^2}{S/N_0} \left(\frac{1}{20} + \frac{B_I}{2S/N_0} \right) \quad (92)$$

It is shown later that this variance is dominated by errors caused by *not* discarding the first sample of a databit at high signal-to-noise density.

The decision-directed cross-product discriminator has the same expected value at higher signal to noise ratios. However, its noise variance is[31,32] as follows:

$$\sigma_{\Delta f}^2 = \frac{1}{2\pi^2} \frac{B_L B_I^2}{(1 - 2P_e)^2 S/N_0} \left(\frac{1}{K} + \frac{B_I}{2S/N_0} \right) \left(1 + \frac{16 B_L P_e (1 - 1.5 P_e)}{B_I (1 - 2P_e)^2} \right) \quad (93)$$

where

$$P_e = \frac{1}{2} e^{-S/N_0 T_I} \quad (94)$$

is the differential bit error rate using the dot-product to sense sign changes in the databit stream. These sign changes can be used to demodulate the data differentially (differential phase shift keying (DPSK) demodulation). Note that the variance in Eq. (93) is similar to that of Eqs. (91) and (92) at higher signal-to-noise ratios.

The statistics of the differential arctangent discriminator are derived using Monte Carlo simulations. These statistics compare very well with the equations for the other two discriminators. The computation of the arctangent function is not as intensive as we would think.[23] It can be approximated by breaking up a cycle into octants, and using a linear piece wise function of either Q_i/I_i or I_i/Q_i, depending upon the quadrant of operation, which must be sensed in order to correct for databit sign changes. Samples of Eq. (88) that cross bit boundaries can be discarded or used, using the sensed sign changes. If used, the discriminator's frequency range narrows similar to that of the decision-directed cross-product discriminator. If discarded, the discriminator characteristics are similar to the cross-product discriminator.

Typical discriminators are shown in Figs. 30 and 31 for a medium bandwidth discriminator ($K = 4$, or $B_I = 200$ Hz) with boundary samples discarded and for a narrow bandwidth discriminator ($K = 1$, or $B_I = 50$ Hz) with sign changes sensed and used, respectively. The plots are for positive frequency errors only. The discriminators for negative frequency errors are simply the inverted mirror image. Note that for the medium bandwidth case with the boundary samples discarded, a second apparent tracking point occurs at 200 Hz. This would result in the loss of code lock because of lack of energy in the 200 Hz filter null. However, in the narrow bandwidth case with sign changes sensed and used, a second tracking point occurs at a 25 Hz offset. This results in a false lock. Unfortunately, if transitioning to a PLL, the PLL would could also false lock at this frequency offset. Thus, care must be taken to ensure that this false lock is sensed using appropriate lock detectors.

The general AFC loop implementation is shown in Fig. 32. The loop shown is a second-order loop, which will track a constant range acceleration. The loop can accept external aiding in the form of estimated Doppler. Note that there are two loop integrators (accumulators), because the carrier NCO does not act as an

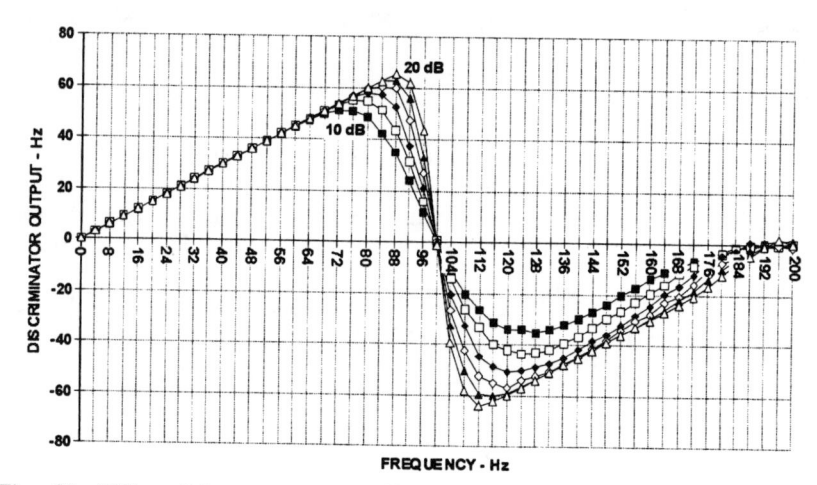

Fig. 30 Differential arctangent medium bandwidth discriminator; bit edges discarded.

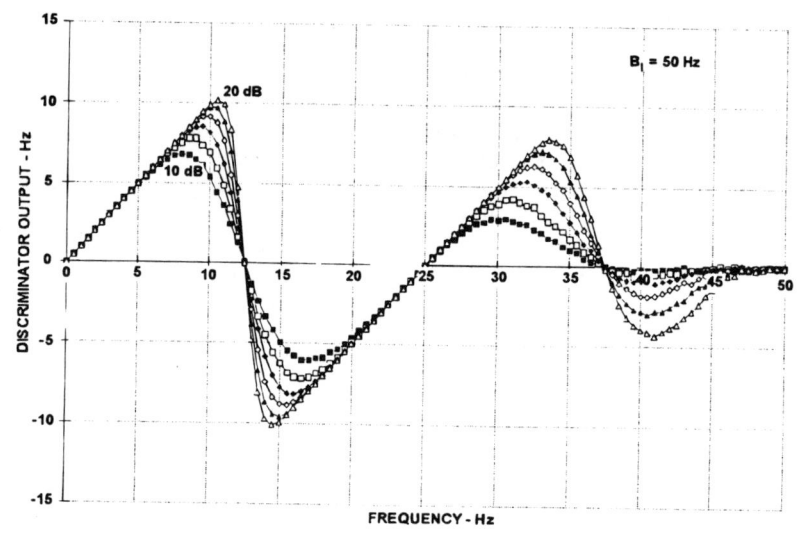

Fig. 31 Differential arctangent narrow bandwidth discriminator; bit edges detected.

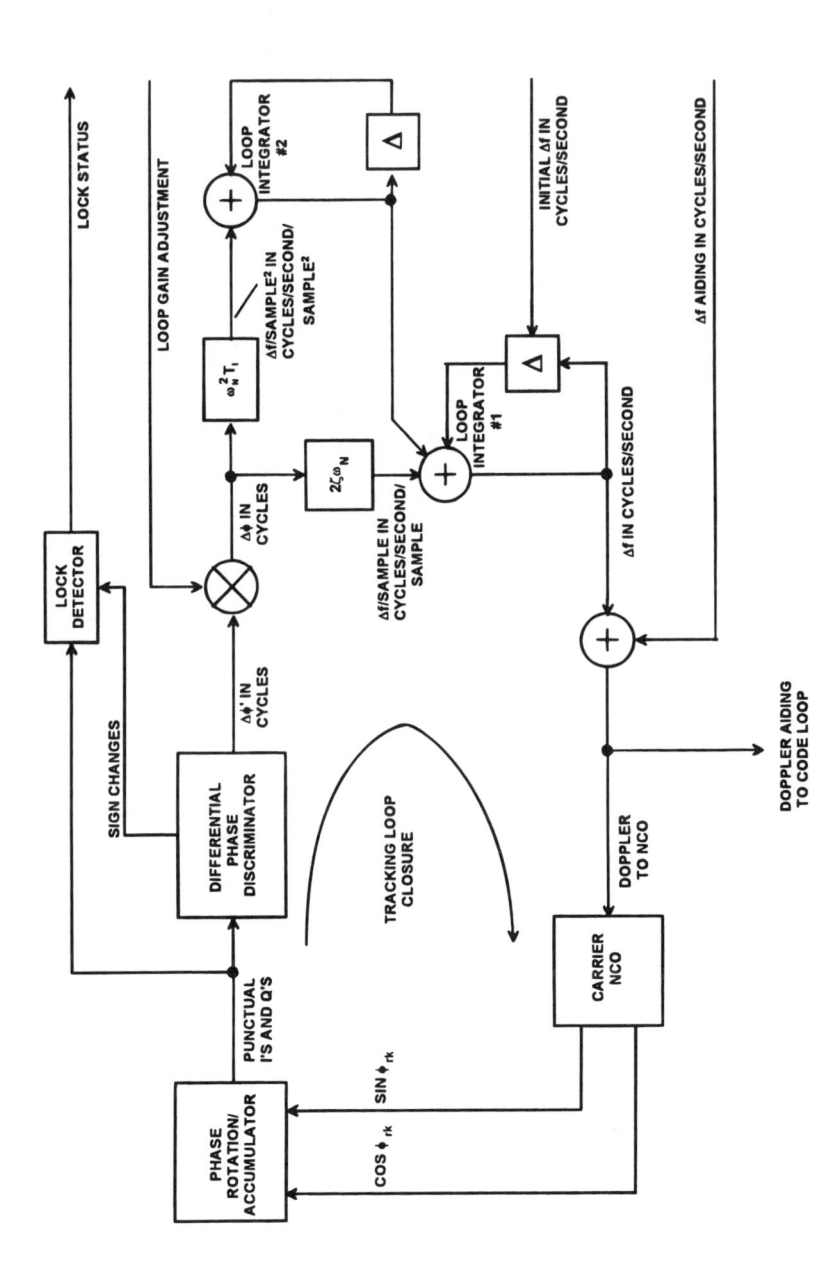

Fig. 32 Automatic frequency control loop implementation.

integrator for an AFC loop. It is simply a means of converting a frequency number to sine and cosine of frequency.

Example automatic frequency control loop noise performance. Typical frequency errors for a medium bandwidth AFC loop using the differential arctangent discriminator in a Monte Carlo simulation (four runs) are shown in Fig. 33. The case presented is that using a discriminator in which the bit edge samples are discarded. The performance is compared to that of the theoretical performance of the cross-product discriminator. Note that the agreement is very good, except at low signal-to-noise ratio where the loop broke lock. Because the theoretical performance is for a linearized loop, it would not indicate loss of lock. However, as a rule-of-thumb, the threshold for a nondecision-directed loop is approximately the predetection bandwidth divided by 12, which is approximately 1/3 the pull-in range (50 Hz at low signal-to-noise ratio). In this case, that is $200/12 \approx 17$ Hz. There is also fairly good agreement with that rule-of-thumb.

Figure 34 shows the performance for a predetection bandwidth of 1 kHz without the knowledge of where the bit edges are (prior to bit synchronization). Note that the error is dominated by the bit edge errors at C/N_0 values above 32 dB-Hz, and the error is always larger than that predicted by the theoretical cross-product performance below that because of the bit edge errors. Note that the loop never broke lock for C/N_0 values above 20 dB-Hz. That is because the discriminator has such a wide range.

The MAGR[23] and the GPSCard™ receivers both use differential arctangent AFC loops. However, the GPSCard™ only uses it for carrier acquisition and

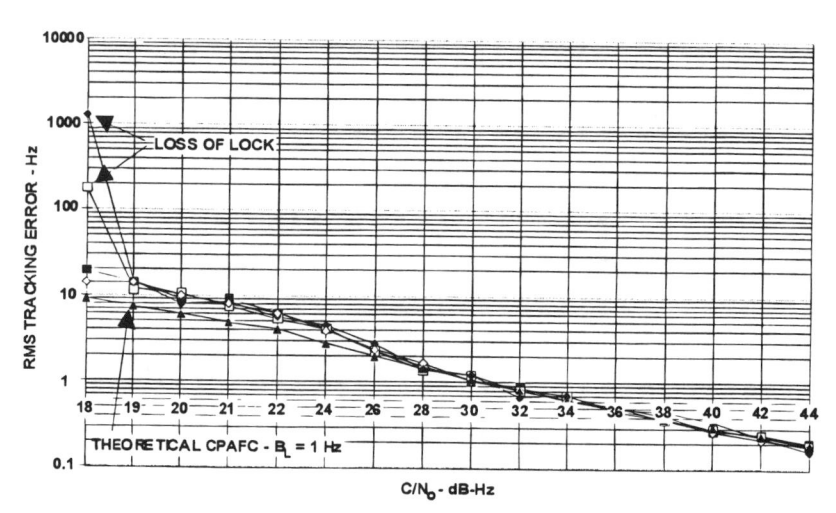

Fig. 33 Medium bandwidth automatic frequency loop performance.

reacquisition and during bit synchronization operations and then switches to a PLL mode. The MAGR uses it always in its unaided mode of operation, although it does make short-term carrier-phase (*delta-range*) measurements. It does so by performing an open-loop accumulation of the differential phase error measurements used in the AFC loop. It uses that accumulation to correct an accumulation of the Doppler applied to the NCO over the same period (less than 1 s).[23]

2. *Phase Lock Loops*

As opposed to the AFC loop, the PLL uses phase tracking discriminators that measure carrier phase. Example discriminators are the following:
Generic Costas discriminator:

$$\delta\phi_i = Q_i I_i \tag{95}$$

which is the one described in the generic tracking loop discussions presented earlier.

Decision-directed Costas discriminator:

$$\delta\phi_i = Q_i \text{sign}(I_i) \tag{96}$$

where sign (I_i) is also the sign of the databit D_i with an ambiguity of 180 deg.

Arctangent discriminator:

$$\delta\phi_i = \tan^{-1}\left(\frac{Q_i}{I_i}\right) \tag{97}$$

All of these discriminators must be computed within a databit period. Except for

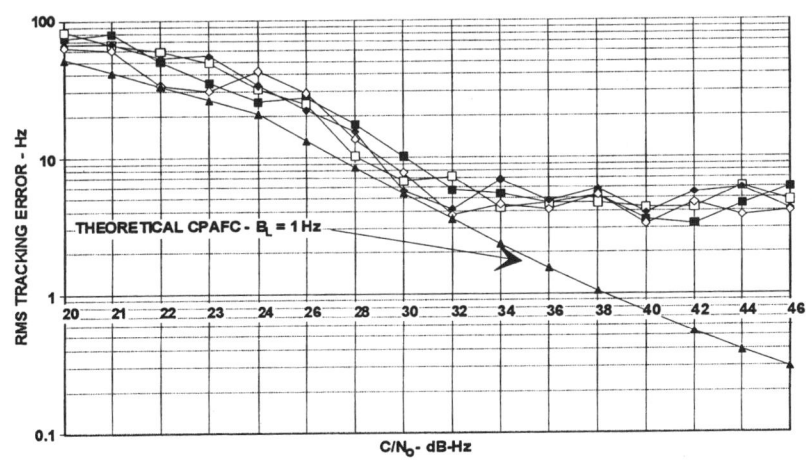

Fig. 34 1-kHz bandwidth automatic frequency control loop performance with bit edges ignored.

possible operation prior to achieving bit synchronization, they are usually computed over the entire bit period.

The statistics for the generic Costas discriminator and closed-loop performance are described above in Eqs. (65–72). Because the other two example discriminators are bit decision-dependent, their performance is best determined using Monte Carlo simulations. However, as shown below, the predicted performance using the equations for the generic Costas discriminator is a good estimation of their performance.

The statistics of the arctangent phase discriminator are derived using Monte Carlo simulations. A typical discriminator is shown in Fig. 35 for a narrow bandwidth ($B_l = 50$ Hz) with sign changes sensed and used. The discriminator for negative phase errors is simply the inverted mirror image. Note that, besides the point at 0 cycles, another tracking point occurs at 0.5 cycles. This is also true for any multiple of 0.5 cycles. Thus, the carrier phase can be tracked at any multiple of 0.5 cycles, creating an abiguity to that level. However, we can change that ambiguity to an integer cycle by observing the demodulated preamble in the system data.

The phase discriminator of Fig. 35 was computed with no frequency error. If there were a frequency error of 25 Hz, a discriminator would exist with phase tracking points at that frequency error, causing a false lock condition. Thus, care must be taken that the AFC loop has converged to the correct frequency before transitioning to PLL operation.

The general PLL loop implementation is shown in Fig. 36. The loop shown is a third-order loop, which will track a constant range acceleration. The loop can accept external aiding in the form of estimated Doppler. Note that there are two loop integrators (accumulators), because the carrier NCO does act as an

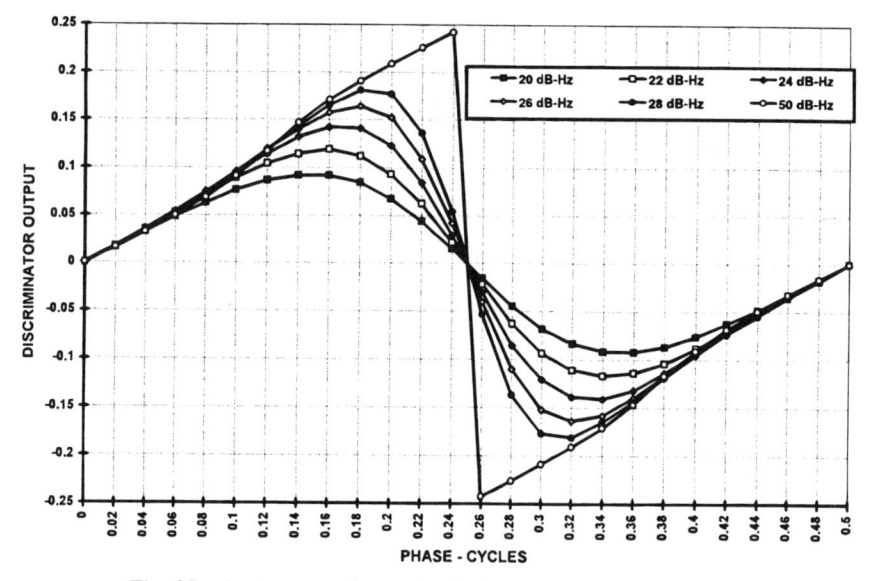

Fig. 35 Arctangent phase discriminator; bit edges detected.

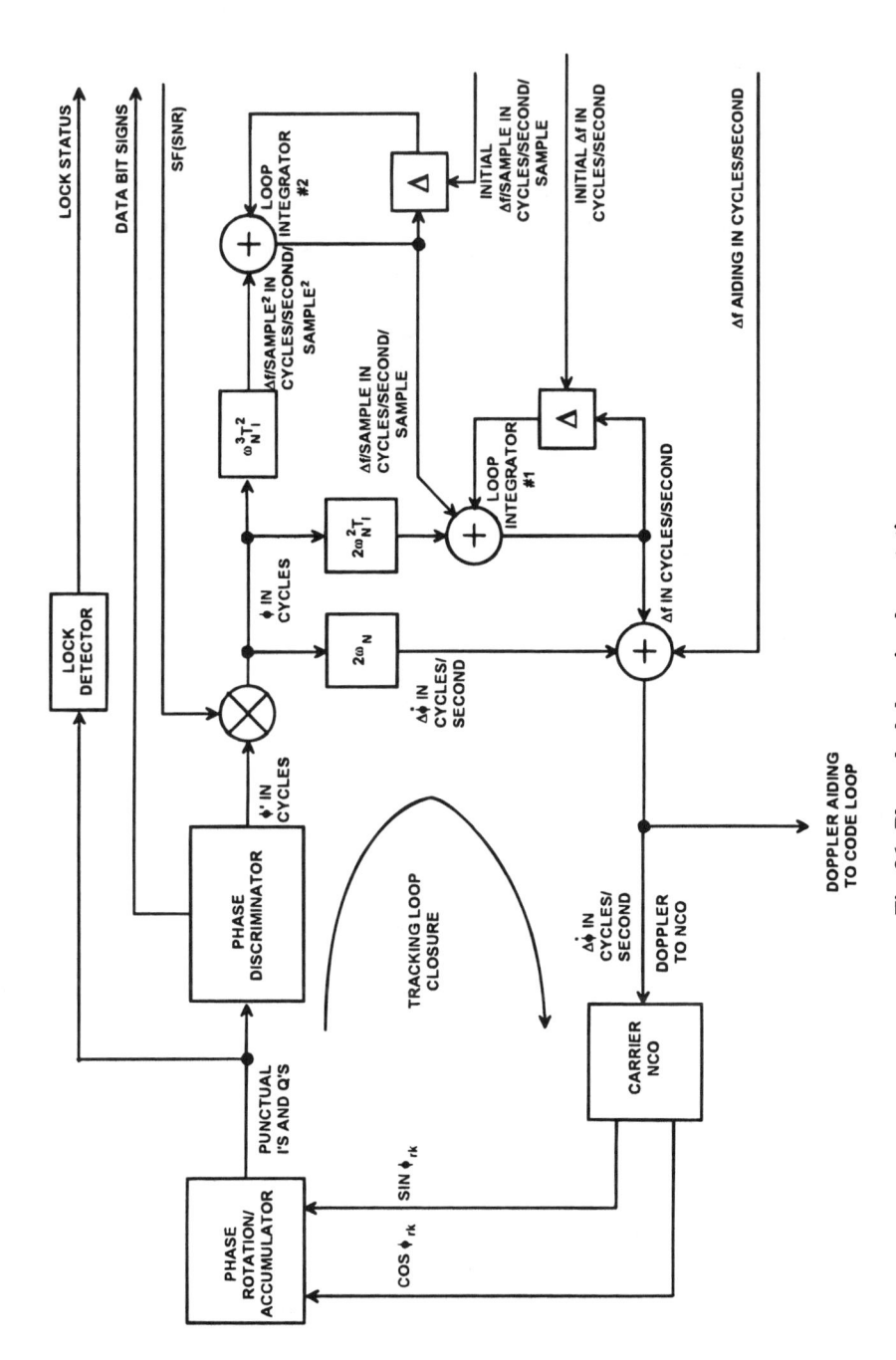

Fig. 36 Phase–lock-loop implementation.

integrator for a PLL loop. Thus, the structure of the loop is very similar to that of the AFC loop in Fig. 32, with the addition of the straight through phase error feedback. This makes for a smooth transition between AFC and PLL loop operations, because only the discriminator and loop gains need changing. The loop integrators and the NCO are naturally initialized.

a. Oscillator phase noise effects. The PLL tracks the phase of the received signal with respect to the receiver's reference oscillator. If either the satellite's or the receiver's oscillator has excessive phase noise, the PLL's tracking performance will be degraded. This phase noise could either be the natural phase noise of the oscillator or that caused by external vibration, because crystal oscillators are basically piezoelectric accelerometers. Thus, they pass along the effects of vibration. Low cost–low performance oscillators can exhibit excessive natural phase noise.

The tracking error of a PLL caused by oscillator phase noise is given as:[33]

$$\sigma_\phi^2 = \frac{1}{2\pi} \int_0^\infty G_\phi(\omega) |1 - H(\omega)|^2 d\omega \quad (98)$$

where $H(\omega)$ is defined in Eq. (67), and $G_\phi(\omega)$ is the single-sideband oscillator phase noise spectral density in radians^2/radian/s. For the transfer functions defined in Eq. (67)

$$|1 - H(\omega)|^2 = \frac{\omega^{2n}}{\omega_N^{2n} + \omega^{2n}} \quad (99)$$

where n is the order of the loop. In the case of phase noise caused by vibration

$$G_\phi(\omega) = (2\pi f_0)^2 k_g^2 \frac{G_g(\omega)}{\omega^2} \quad (100)$$

where $G_g(\omega)$ is the single-sided vibration spectral density in g^2/r/s, k_g is the oscillator's g-sensitivity in parts-per-g, and f_0 is the carrier frequency (L_1 or L_2).

Note that the smaller ω_N is, the larger the phase noise contribution. This is because the PLL cannot follow the oscillator variations. This is the opposite effect caused by thermal noise. Thus, there is a loop bandwidth trade-off between the effects of oscillator phase noise and thermal noise, as well as signal dynamics (discussed below). In general, this trade-off involves quasilinear analysis techniques.[33]

b. Example phase–lock-loop noise performance. Typical phase errors for a 50-Hz predetection bandwidth PLL using the arctangent discriminator in a Monte-Carlo simulation (four runs) are shown in Fig. 37 (with and without oscillator effects). The performance is compared to that of the theoretical performance of the generic Costas discriminator (B_L = 5 Hz). Note that the agreement for the "no clock" case is very good, except at low signal-to-noise ratio where the loop broke lock (below 23 dB-Hz C/N_0). Because the theoretical performance is for a linearized loop, it would not indicate loss of lock. When a "good" low-cost commercial oscillator Monte-Carlo model is added to the simulation, performance

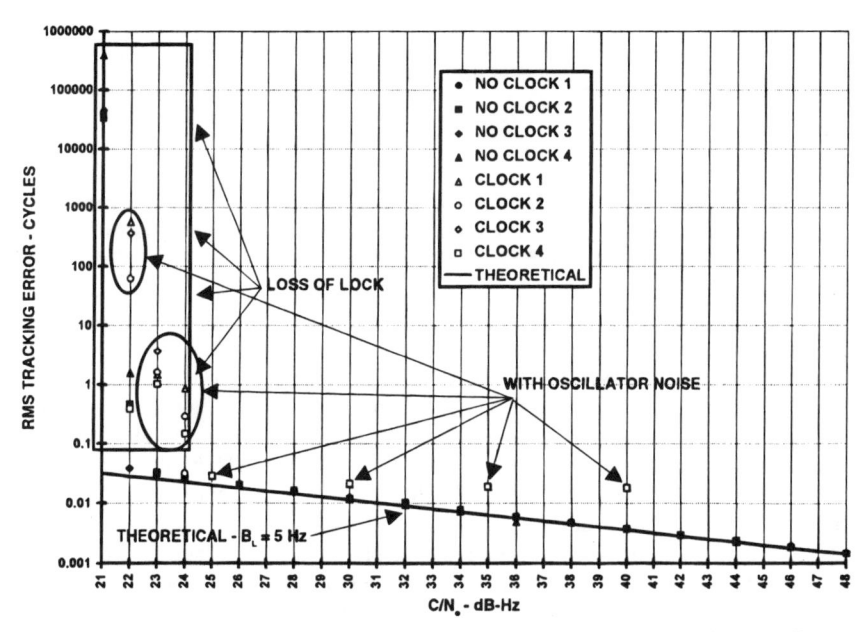

Fig. 37 Typical phase lock loop performance with and without oscillator effects.

degrades as follows: 1) the tracking threshold increases 2 dB; and 2) the high C/N_0 tracking errors are dominated by the effects of the oscillator phase noise.

c. Effects of signal dynamics. We choose ω_N as a function of the desired dynamic response of the tracking loop, but at a minimum value required in order to minimize the noise bandwidth of the loop, maximizing the noise performance of the loop. The value of ω_N affects both the pull-in capability of the loop and the response to dynamic transients. In the case of a third-order loop implementation, the loop will track a constant phase acceleration, but deviates in the presence of a phase jerk. Furthermore, the loop also has a peak dynamic error caused by steps in either frequency or phase acceleration (frequency rate). The peak errors caused by either a frequency, phase acceleration (frequency rate), and phase jerk (frequency acceleration) step in the third-order loop can be approximated by the steady-state response of a first-, second-, or third-order loop caused by a frequency, phase acceleration, or phase jerk step function, respectively, where, these steady-state errors (in cycles) are

First order:
$$\delta\phi_{ss} = \frac{\Delta f}{\omega_N} = \frac{\Delta f}{(4B_L)} \tag{101}$$

Second order:
$$\delta\phi_{ss} = \frac{\Delta \dot{f}}{\omega_N^2} = \frac{\Delta \dot{f}}{(1.885B_L)^2} \tag{102}$$

Third order:

$$\delta\phi_{ss} = \frac{\Delta \ddot{f}}{\omega_N^3} = \frac{\Delta \ddot{f}}{(1.2B_L)^3} \qquad (103)$$

where Δf is in Hz.

The peak error for the third-order loop caused by a step phase acceleration typically will overshoot the steady-state value of Eq. (103) by a small numerical factor, but not of much significance. The second half of Eqs. (101–103)—the half in terms of B_L—can be used to estimate the peak errors of a third-order loop (not steady-state) caused by the applicable step input. For example, the peak response caused by a step-phase acceleration of 4 g ($\gg$ 210 Hz/s) would be $59.12/B_L^2$ or 2.36 cycles for a B_L of 5 Hz. That would obviously cause loss of lock. However, we would never expect a step-phase acceleration, but a jerk that would ramp to the acceleration to that value. For example, assume that the acceleration ramps up to 4 g over 1 s for a jerk of 4 g/s. Then, the peak phase error would be $21.56/B_L^3$ or 0.973 cycles for a B_L of 5 Hz. That is still excessive, but illustrates the improvement. Either the jerk must be reduced, or the bandwidth must be increased. Note that an increase of bandwidth by a factor of 2 reduces the peak error by a factor of 8, producing an acceptable result. One word of caution in increasing the bandwidth—the responses presented here assume continuous analog-type loop responses, or digital loops with update rates on the order of 5–10 times the loop bandwidth.

G. Lock Detectors

If the GPS signal is being tracked, we must know it is being, or *not* being, tracked. Lock detectors are required to perform this function. Lock detectors can be implemented to determine code lock, frequency lock, or phase lock, although, frequency lock and phase lock detection also infer code lock. Representative implementation and performance of these detectors are presented here.

1. Code Lock Detectors and C/N_0 Estimation

Code lock detection is very similar to estimating C/N_0 because code lock is required to achieve good C/N_0, inferring that the receiver is operating on or near the correlation peak. In fact, here a connection is shown between a "good" code lock detector and a "good" C/N_0 estimator.

Knowledge of code lock is obviously the same as the knowledge of received signal power, similar to the power measurements defined in Appendix B. Because a precorrelation AGC could be used in the receiver, knowledge of ambient noise power and interference power are not necessarily known, unless special provisions are made in the receiver hardware. However, the receiver's processor can determine total signal-plus-noise power, and it can do so in different noise bandwidths. In fact, the comparison of total signal-plus-noise power in two different bandwidths can be used to determine signal-to-noise ratio, which is more important that the knowledge of received signal power when inferring lock status.

Consider measurements of total power in $1/T$ (wide-band power) and $1/MT$ (narrow-band power) noise bandwidths of the following form:

$$\text{WBP}_k = \left(\sum_{i=1}^{M} (I_i^2 + Q_i^2)\right)_k \tag{104}$$

and

$$\text{NBP}_k = \left(\sum_{i=1}^{M} I_i\right)_k^2 + \left(\sum_{i=1}^{M} Q_i\right)_k^2 \tag{105}$$

computed over the same M samples. For an unknown postcorrelation noise power, the exact relationship between these measurements and signal-plus-noise power is not known. However, a normalized power defined as follows:

$$\text{NP}_k = \frac{\text{NBP}_k}{\text{WBP}_k} \tag{106}$$

gives us statistics that provide a monatomic function of C/N_0, as shown in Fig. 38. These statistics are based upon the statistics of the ratio of random variables X and Y given, as in Mood et al.[34] and Socci et al.:[35]

$$\mu_{X/Y} \approx \frac{\mu_X}{\mu_Y} - \frac{1}{\mu_Y^2}\text{cov}(X, Y) + \frac{\mu_X}{\mu_Y^3}\sigma_Y^2 \tag{107}$$

and

$$\sigma_{X/Y}^2 \approx \left(\frac{\mu_X}{\mu_Y}\right)^2 \left[\frac{\sigma_X^2}{\mu_X^2} + \frac{\sigma_Y^2}{\mu_Y^2} - \frac{2\,\text{cov}(X, Y)}{\mu_X \mu_Y}\right] \tag{108}$$

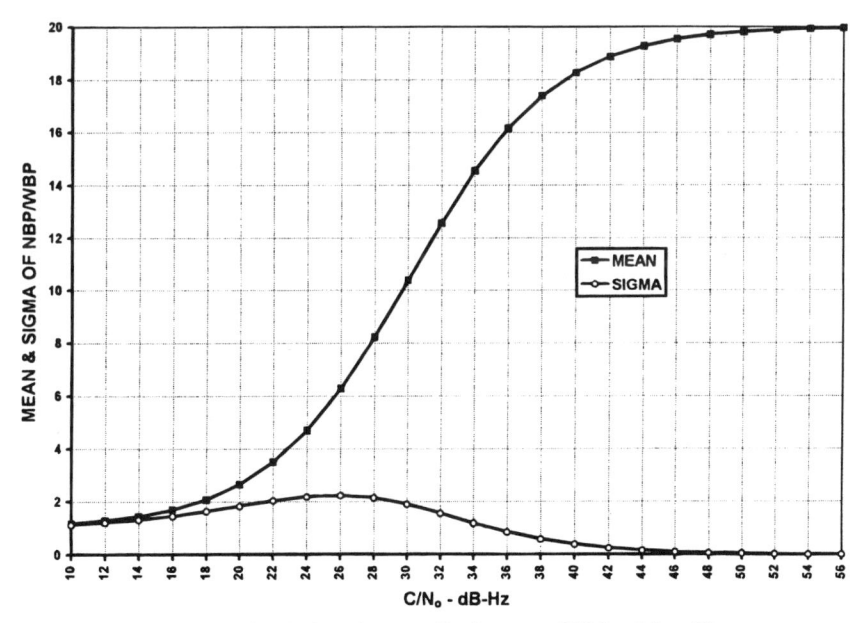

Fig. 38 Statistics of normalized power NP for $M = 20$.

Using the samples defined in Eqs. (55) and (56), and for small frequency and code tracking errors, the statistics of the quantities defined in Eqs. (104) and (105) are

$$\mu_{NBP_k} = 2M[(S/N_0)MT + 1] \tag{109}$$

$$\mu_{WBP_k} = 2M[(S/N_0)T + 1] \tag{110}$$

$$\sigma^2_{NBP_k} = 4M^2[2(S/N_0)MT + 1] \tag{111}$$

$$\sigma^2_{WBP_k} = 4M[2(S/N_0)T + 1] \tag{112}$$

$$\mathrm{cov}[NBP_k, WBP_k] = 4M[2(S/N_0)MT + 1] \tag{113}$$

resulting in the statistics for NP_k shown in Fig. 38 for $M = 20$.
When there is no signal present

$$\mu_{NP_k} | S/N_0 = 0 = 1 \tag{114}$$

and

$$\sigma^2_{NP_k} | S/N_0 = 0 = 1 - \frac{1}{M} \tag{115}$$

The lock detector measurement is as follows:

$$\hat{\mu}_{NP} = \frac{1}{K} \sum_{k=1}^{K} NP_k \tag{116}$$

reduces the standard deviations by a factor of $\sqrt{K}$, and a threshold can be set at a desired minimum C/N_0.

The same measurement can be used in a C/N_0 estimator, given as follows:

$$\frac{\hat{C}}{N_0} = 10 \log_{10} \left(\frac{1}{T} \frac{\hat{\mu}_{NP} - 1}{M - \hat{\mu}_{NP}} \right) \tag{117}$$

The error plots of this estimate for $M = 20$ and $K = 50$ (1 s average) are shown in Fig. 39. These error plots show that longer averaging is required for C/N_0 values less than 25–30 dB-Hz.

2. Frequency Lock Detectors

For a receiver that primarily tracks carrier phase, a frequency lock detector is not required because AFC or FLL loops are only used in transition for fixed periods of time, and code lock is used to determine successful transitioning. Ultimately, carrier lock is determined as phase lock after transitioning to the PLL. Normally, frequency lock detectors are not very sensitive because of predetection bandwidth constraints imposed by bit transitions.

In the case of the MAGR, however, the PLL is never used, and the code lock detector is always used as the lock detector. False lock (25-Hz offset) is detected with a continuously failing parity, because data demodulation results in an apparent bit transition every 20 ms.[23] False lock can also be detected by a discrepancy between carrier and code Doppler.

GPS RECEIVERS

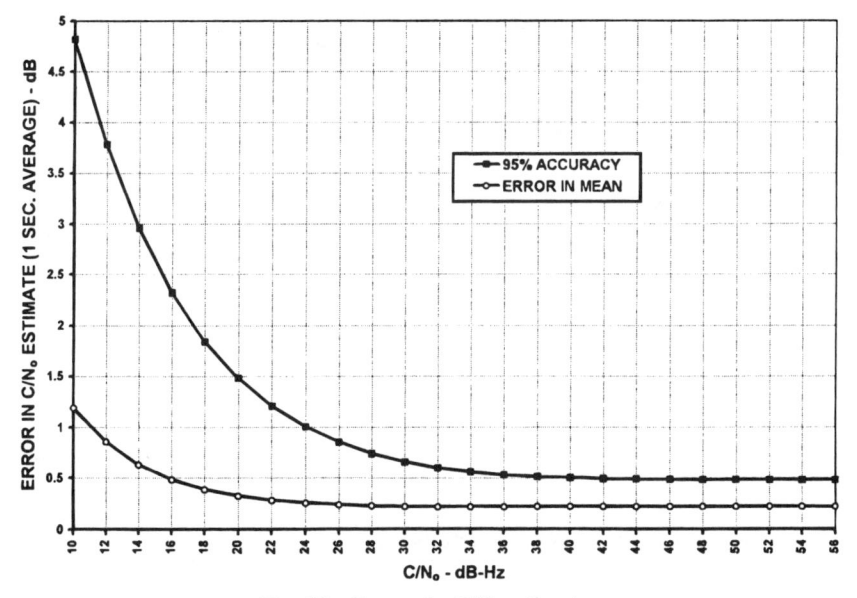

Fig. 39 Errors in C/N_0 estimator.

3. Phase Lock Detectors

Phase lock can be detected using the normalized estimate of the cosine of twice the carrier phase given as follows:

$$C2\phi_k = \frac{\text{NBD}_k}{\text{NBP}_k} \tag{118}$$

where

$$\text{NBD}_k = \left(\sum_{i=1}^{M} I_i\right)_k^2 - \left(\sum_{i=1}^{M} Q_i\right)_k^2 \tag{119}$$

This quantity has the following statistics that are similar to those of Eqs. (109–113).

$$\mu_{\text{NBD}_k} = 2(S/N_0)M^2T \cos 2\phi_k \tag{120}$$

$$\sigma_{\text{NBD}_k}^2 = \sigma_{\text{NBP}_k}^2 = 4M^2[2(S/N_0)MT + 1] \tag{121}$$

$$\text{cov}(\text{NBD}_k, \text{NBP}_k) = 8(S/N_0)M^2T \cos 2\phi \tag{122}$$

Similar analysis used for the code lock detector can be used here to derive the statistics of phase lock detector of Eq. (118). From this analysis, it can be shown that the mean of Eq. (119) is a function of twice the carrier phase. Figure 40 shows Monte Carlo values of such a detector for the simulation runs presented in Fig. 37 (four runs without oscillator effects).

Unfortunately, this phase lock detector (with M of 20) does not detect the 25-Hz offset false lock, because the sum of Q_i integrates to 0, and the sum of I_i

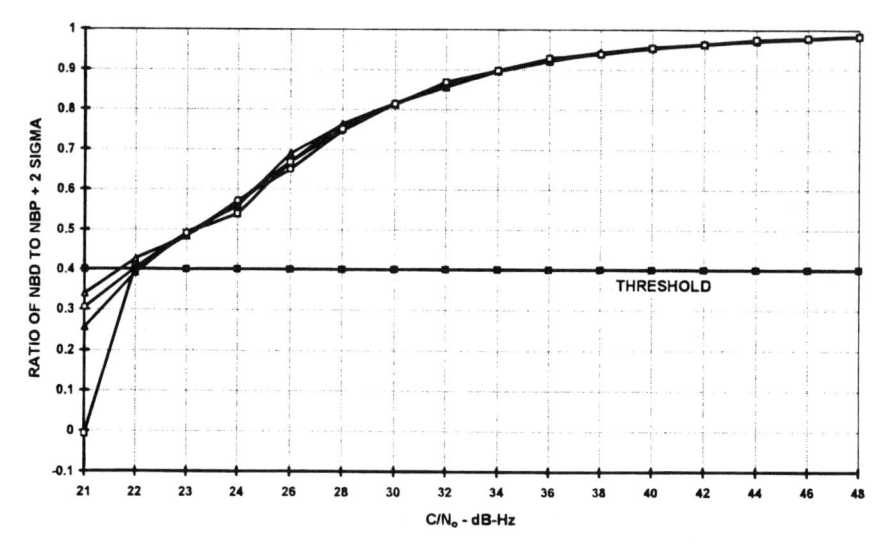

Fig. 40 Carrier-phase lock detection for Monte Carlo simulations.

integrates to a positive or negative value. However, as described above for frequency false lock, this false lock can also be detected here with a continuously failing parity. False lock can also be detected here by a discrepancy between carrier and code Doppler.

Parity can also be used as a general phase lock detector, because large phase tracking errors or cycle slips also cause bit sign detection errors. Figure 41 shows the effect of phase error on bit error rate (BER) as a function of energy-per-bit in dB.

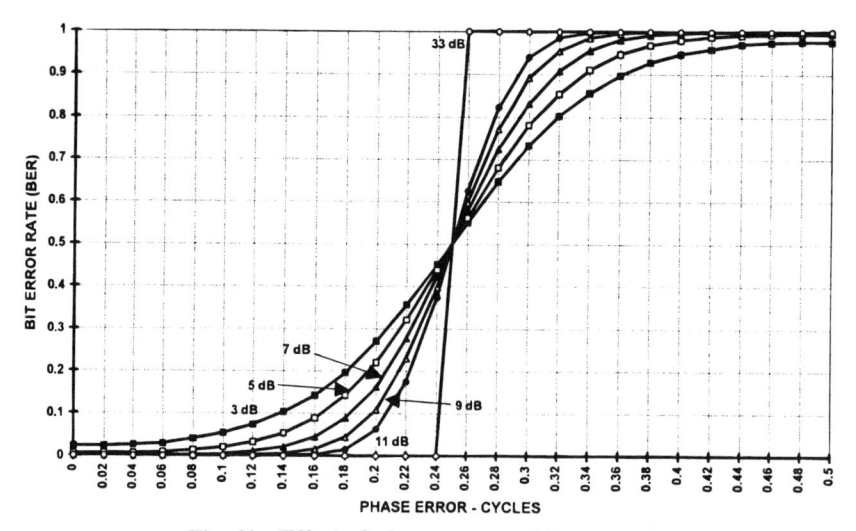

Fig. 41 Effect of phase error on bit error rate.

H. Bit Synchronization

Upon initial wide-band C/A-code acquisition, when position and time uncertainties are large, the C/A-code epoch ambiguity results in the lack of knowledge of databit timing. Thus, bit synchronization is required. There are a number of techniques available to achieve bit synchronization.[33] The MAGR uses a modified in-phase/midphase bit synchronizer.[23,33]

The GPSCard™ uses a histogram approach that was also used by Magnavox in early GPS receivers. This approach breaks an assumed databit period (20 ms) into 20 C/A-code 1-ms epoch periods and senses sign changes between successive epochs. For each sensed sign change, a corresponding histogram cell count is incremented until a count in one specific cell exceeds the other 19 bins by a prespecified amount. An example of one such histogram for success bit synchronization along with count thresholds is shown in Fig. 42. The procedure is as follows:

1) A cell counter K_{cell} is arbitrarily set and runs from 0 to 19.

2) Each sensed sign change is recorded by adding 1 to the histogram cell corresponding to K_{cell}.

3) The process continues until one of the following occurs:
   a) Two cell counts exceed threshold NBS_2.
   b) Loss of lock
   c) One cell count exceeds threshold NBS_1.

4) If a occurs, bit synchronization fails because of low C/N_0 or lack of bit sign transitions, and bit synchronization is reinitialized. If b occurs, lock is reestablished. If c occurs, bit synchronization is successful, and the C/A-code epoch count is reset to the correct value.

The thresholds NBS_1 and NBS_2 are determined as follows:

1) The probability of making an error in determining a sign change at a desired S/N_0 is

$$P_{esc} = 2P_e(1 - P_e) \tag{123}$$

where[33]

$$P_e = \text{erfc}'\left[\sqrt{2(S/N_0)T}\right] \tag{124}$$

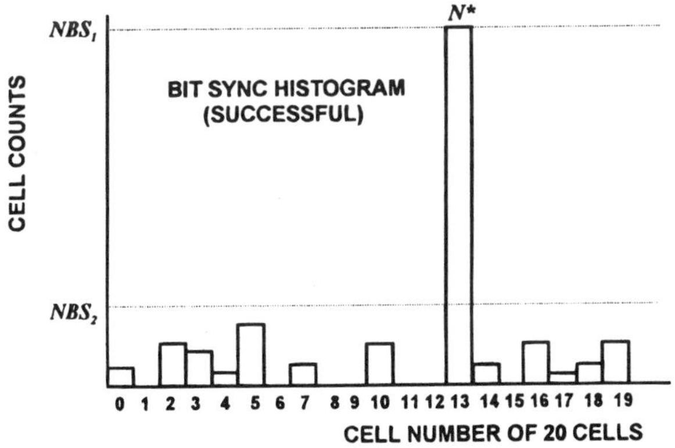

Fig. 42 Example of a successful bit synchronization histogram.

if a PLL is being used, where

$$\text{erfc}'(x) = \frac{1}{\sqrt{2\pi}} \int_x^\infty e^{-y^2/2} dy \tag{125}$$

Equation (121) is replaced with Eq. (92) with $T_I = T$ if an AFC loop, which detects sign changes directly, is being used.

2) The number of entries N_{bs} in a cell has a binomial distribution. Over T_{bs} seconds, the average number of sign changes (bit transitions) is $25T_{bs}$, so that

$$\text{NBS}_1 = \overline{N}^* = 25T_{bs} \tag{126}$$

in the correct cell, and

$$\overline{N}_{bs} = 50T_{bs}P_{esc} \tag{127}$$

in the other cells. The standard deviation of N_{bs} in any cell is

$$\sigma_{N_{bs}} = \sqrt{50T_{bs}P_{esc}(1 - P_{esc})} \tag{128}$$

3) The thresholds, as well as the time interval T_{bs}, are selected to provide a good spread, say three sigma, between them at a desired S/N_0. That is, given Eq. (125) for NBS_1, select NBS_2 and T_{bs} for the desired S/N_0 so that

$$25T_{bs} - 3\sqrt{50T_{bs}P_{esc}(1 - P_{esc})} \geq \text{NBS}_2 \geq 50T_{bs}P_{esc} \tag{129}$$

The use of an AFC loop during bit synchronization proves to be more robust than the use of a PLL, because cycle slipping is avoided, a problem that can exist because of the wide predetection bandwidth. As shown in Fig. 34, the AFC loop with a 1-kHz predetection bandwidth is quite robust.

I. Data Demodulation, Frame Synchronization, and Parity Decoding

1. Data Demodulation

Data demodulation is not only a byproduct of both kinds of carrier tracking, it is usually necessary as part of the frequency and phase discrimination processes, as is illustrated in Figs. 32 and 36. In the case of the PLL, the sign of the bits are determined directly, although there exists a sign ambiguity because the PLL can also track 180 deg out-of-phase. In the case of the AFC discriminator, only sign changes are produced, so that the sign of the very first bit is set arbitrarily. Thus, the result is the same as for the PLL—there exists a sign ambiguity. In both cases, this ambiguity is resolved during frame synchronization and subsequently as part of the parity algorithm.

The performance of data demodulation is measured in terms of bit error rate, where, for the PLL,

$$\text{BER} = \text{erfc}'\left[\sqrt{40(S/N_0)T}\right] \tag{130}$$

and in the case of the AFC loop (differential demodulation)

$$\text{BER} = 0.5e^{-20(S/N_0)T} \tag{131}$$

GPS RECEIVERS

In either case, tracking errors can degrade these theoretical performances. The PLL BER performance is about 1.5 dB better. Furthermore, any differential bit error using the AFC discriminator results in a string of bit errors to the end of a word, which could cause an undetected bit error in the parity decoding. The MAGR, because it doesn't implement a PLL, uses this AFC differential decoding, but, fortunately, there is sufficient margin and data redundancy to overcome these potential problems. The GPSCard™ uses a PLL for data demodulation.

2. Frame Synchronization

Prior to dataframe synchronization, parity decoding cannot be accomplished because of unknown word boundaries when timing uncertainties are large. The GPS navigation message frame structure is shown in Fig. 43.[18,19] Note that at the beginning of each 6-s subframe there is an 8-bit preamble of 10001011, which, because of a sign ambiguity, could be inverted at 01110100. The procedure for positively finding the *correct* preamble is as follows:

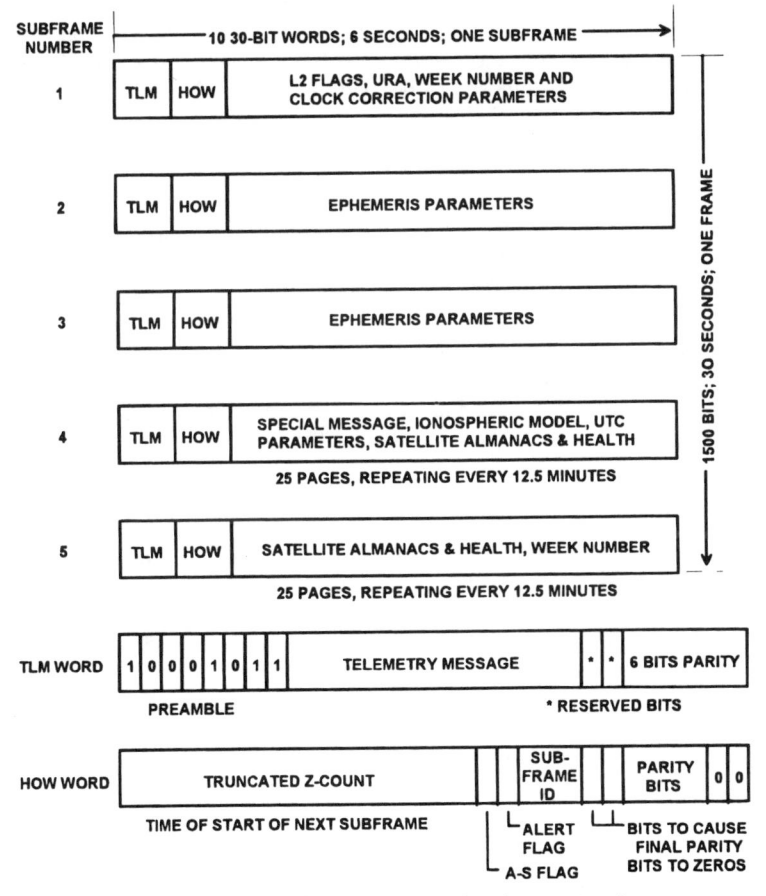

Fig. 43 GPS navigation message dataframe structure.

1) Search for either an upright or inverted preamble.

2) When one is found, which could be a legitimate pattern somewhere else in the data stream, a check is required to see if it is a beginning of a 30-bit word. This is accomplished by collecting the following 22 bits and checking parity. If parity doesn't pass, the candidate preamble is discarded.

3) If parity passes, it verifies that the preamble existed at the beginning of a word. The parity algorithm will also resolve the sign ambiguity. However, there are also legitimate such patterns at beginning of other words, so additional checks are required. If it is the correct [telemetry (TLM)] word, the following word must be a handover word (HOW) that contains a truncated Z-count. The first eight bits of this truncated Z-count can also resemble a preamble.

4) Parity should pass on the HOW word. If not, the frame synchronization procedure should be restarted. Two checks can be made to verify a legitimate HOW word—the Z-count is reasonable, and it agrees with the subframe count.[36] Of course, there is a small probability that these conditions could also occur elsewhere in the message. Thus, further checking is required.

5) If the HOW seems legitimate, provisional demodulation of the other words can commence, and they can be stored in memory. A final check on the next preamble and the next Z-count solidifies the frame synchronization. That is, the preamble is where it is supposed to be, and the Z-count increments by one.

3. Parity Decoding

Although the GPS Signal Specifications[18,19] show otherwise for some of the message words (as is the HOW word in Fig. 43), it is sufficient to assume that the last 6 bits of each 30-bit word are the parity bits for decoding purposes. The parity encoding algorithm is a (32,26) Hamming code, thus the parity overlaps the 30-bit words.[18,19,36] The last 2 raw bits (D29* and D30*) of the previous word are used with the first 24 bits of the 30-bit word to make up a 26-bit vector d, after "exclusive or-ing" the 24 bits with D30* (resolving the sign ambiguity). Parity decoding is accomplished with the matrix "exclusive-or"

$$D = H \oplus d \quad (132)$$

where D is the 6-bit parity vector, and H is a 6×26 mask matrix with 1s in the elements corresponding to the "exclusive ors" of the 6 parity equations of the GPS Signal Specifications.[19] If, after this operation, performed one column at a time for each incoming bit, the parity vector equals the last 6 bits of the 30-bit word, parity passes, and the last 24 bits of d are the decoded bits with the sign ambiguity resolved.

This parity algorithm positively detects up to three simultaneous bit errors and can correct 1-bit errors. However, if error correction is performed, it will no longer detect three simultaneous bit errors. The algorithm is not very strong in the presence of burst errors ($P_s = 1/64$), so care should be taken in believing its results. However, significant redundancy is available in the navigation message, so that cross-checking can be accomplished.

For random bit errors, the undetected BER (UBER) is approximately

$$\text{UBER} \approx \frac{4}{32}\left(\frac{32!}{4!28!}\right)P_e^4(1-P_e)^{28}$$

$$= 4495 P_e^4 (1-P_e)^{28} \tag{133}$$

which provides approximately 5 dB of error detection. The probability of losing a word (word error rate, WER) is given as follows:

$$\text{WER} = 1 - (1-P_e)^{32} \tag{134}$$

Appendix A: Determination of Signal-to-Noise Density

A.1 Introduction

In the various sections of this chapter, only the effects of wideband noise are considered. However, interference, spread by a PRN code, also produces noise that affects the effective signal-to-noise density S/N_0, where, in ratio-Hz,

$$\left(\frac{S}{N_0}\right)^{-1} = \left(\frac{S}{N_{0T}}\right)^{-1} + \left(\frac{S}{N_{0I}}\right)^{-1} \tag{A1}$$

where N_{0T} is the spread thermal noise density, and N_{0I} is the spread interference noise density. The spreading process described in Eqs. (34) and (35) and the statistics defined in Eqs. (37–43) assumed that the precorrelation bandwidth and sampling rate were infinite, or at least large with respect to the code chipping rate. However, in general, the thermal noise is bandlimited at IF, so it is spread to some extent, which is the reason for the reduced correlation loss shown in Fig. 12. In general, the spreading process is defined as a signal (noise or otherwise) being multiplied by the reference code. Thus, the density of noise or interference at the output of the correlator is the convolution of the interference spectral density and the spectral density of the code.[37]

$$N_{0I}(f') = \int_{-\infty}^{\infty} S_C(f) S_I(f' - f) df \tag{A2}$$

where $S_C(f)$ is the spectral density of the reference PRN code and $S_I(f)$ is the density of the interference or noise.

A.2 Pseuorandom Noise Code Spectral Densities

For the long reference P code, $S_C(f)$ is given as follows:

$$S_C(f) = T_p \frac{\sin^2(\pi f T_p)}{(\pi f T_p)^2} \tag{A3}$$

where T_p is $[1/(10.23 \times 10^6)]$ s. However, the reference C/A code has a discrete spectral density that can be described as follows:

$$S_C(f) = \sum_{j=-\infty}^{\infty} c_j \delta(f - 1000j) \tag{A4}$$

where the c_j are spectral line coefficients, $\delta(f)$ is the dirac delta function, and the c_j vary about the envelope

$$c_j = 1000 T_{c/a} \frac{\sin^2(1000\pi j T_{c/a})}{(1000\pi j T_{c/a})^2} \tag{A5}$$

where $T_{c/a}$ is $[1/(1.023 \times 10^6)]$ s. Both of these reference PRN code spectral densities have the property that

$$\int_{-\infty}^{\infty} S_c(f) df = 1 \tag{A6}$$

A.3 Noise Spectral Densities

Bandlimited Noise

First consider bandlimited noise with density N_0 with a two-sided bandwidth of B_I (brick wall filter). Then,

$$N'_{0T} = N_{0T}(0) = N_0 \int_{-B_I/2}^{B_I/2} S_c(f) df \leq N_0 \tag{A7}$$

The reduction of N_0 to N'_{0T} is shown as the bottom curve in Fig. 12. This is the case for both PRN codes. The variations of the C/A-code lines average out over this wide bandwidth.

A similar equation applies for wide- and narrow-band interference with the following spectral density:

$$S_I(f) = \frac{P_I}{f_u - f_l}; f_l \leq f \leq f_u \tag{A8}$$

for upper and lower frequency limits f_u and f_l and total interference power (relative to the signal power S) P_I. Then, with these upper and lower frequencies converted to IF. We have the following.

$$N_{0I}(f') = \frac{P_I}{f_u - f_l} \int_{\max[-B_I/2, f' - f_{UIF}]}^{\min[B_I/2, -f' + f_{UIF}]} S_c(f) df \tag{A9}$$

In the case of the P-code, for narrow bandwidth noise interference centered near the GPS frequencies (i.e., somewhat less than the code chipping rate), this equation becomes

$$N'_{0I} = N_{0I}(0) \approx P_I T_p \tag{A10}$$

If the interference is offset with respect to the center of the GPS frequencies a significant amount, the quantity of Eq. (A10) would be reduced by the ratio of T_p to the magnitude of Eq. (A3) near the center of the narrowband interference.

For the C/A code, Eq. (A9) becomes the following:

$$N_{0I}(f'_j) = \frac{P_I}{f_u - f_l} \sum_{j=\max[-j_{B_{IF}}, j' - j_{UF}]}^{\min[j_{B_{IF}}, -j' + j_{UF}]} C_j \tag{A11}$$

where the upper and lower limits represent the spectral components within the interference band, limited by the IF bandwidth. For narrow-band interference, Eq. (A11) is not generally equal to that of Eq. (A7) because of the variations in the spectral lines of the codes. However, for interference bandwidths on the order of 100 kHz and greater, the summation is over enough lines to average out the variation, and the resulting density approaches the following:

$$N'_{0I} = N_{0I}(0) \approx P_I T_{c/a} \qquad (A12)$$

The noise density will vary (up or down) by approximately 3 dB per decade for interference bandwidths below that down to 10 kHz, and by approximately 6 dB per decade for interference bandwidths between 1 and 10 kHz.

Continuous Wave (CW) Interference

CW interference has a spectral density of the following:

$$S_I(f) = P_I \delta(f - f_I) \qquad (A13)$$

where f_I is the frequency of the interference. Thus, the post correlation noise density is as follows:

$$N_{0I}(f') = P_I S_c(f' - f_I) \qquad (A14)$$

which is the same as Eq. (A10) for the P-code. In the case of the C/A-code, in the worst case, it can take on the value of one of the spectral lines times the interference power, if centered on the line. The resulting spectral density is still a spectral line, because the spreading would repeat every C/A-code period.

$$N_{0I}(f'_j) = P_I c_{j'} \delta(f'_j - f_I) \qquad (A15)$$

which selects the C/A-code spectral line on which the interference lands.

Now, Eq. (A15) is only fully valid if there were no data modulated on the processed signal's C/A code and the spectral density were computed over a relatively long period of time. For the P-code cases, and for relatively wideband interference effects on the C/A code, the above equations are valid, simply because the resulting spread interference, which becomes noise with a bandwidth of the code, is statistically independent over time and can be treated as such. In the case of narrow-band interference; i.e., with somewhat less than a 100-kHz bandwidth, the resulting noise spread with a C/A code is not totally statistically independent over time, and thus, cannot be treated as such. Also, because the signal code is modulated with the 50-Hz data, the resulting spectral density components of the interference can only be computed with a minimum bandwidth of 50 Hz. Thus, the resulting spectral density caused by spreading CW interference is not a spectral line, but a spread spectrum with a minimum bandwidth equal to the postcorrelation receiver-processing bandwidth. CW interference not coinciding with a C/A-code spectral line will also pass through the correlation process, but with a reduced effect. These properties make it very difficult to predict the effects of narrow-band and CW interference on the C/A-code, and results in the specification of *worst-case effects*. These effects are those presented in Eqs. (A12) and (A15) for the largest C/A-code spectral lines, which can be about 9 dB larger than that of the envelope given in Eq. (A5), filtered further with the appropriate postcorrelation (predetection) bandwidth.

Self-Interference

Another type of interference is self-interference caused by signals from other satellites, or, from *pseudolites,* which are GPS-like signal sources located on the surface of the Earth (see Chapter 2, companion volume). In general, the spectral density on the incoming interference from N signals is as follows:

$$S_I(f) = \sum_{k=1}^{N} P_{Ik} S_C(f - f_{Ik}) \tag{A16}$$

where f_{Ik} is the offset frequency of interference k with respect to the reference code, which for the satellites is the Doppler difference between satellite k and the desired satellite, and P_{Ik} is the received power from interference k. Pseudolites can be offset further in frequency.

For the P codes, the frequency offsets are small with respect to the bandwidths of the signals. Thus, the interference density is simply as follows:

$$N'_{0I} = N_{0I}(0) \approx \int_{-\infty}^{\infty} S_C^2(f) df \sum_{k=1}^{N} P_{Ik} = \frac{2}{3} T_p \sum_{k=1}^{N} P_{Ik} \tag{A17}$$

For all but strong pseudolite signals, this would be insignificant interference to a P-code signal because of the factor T_p.

For the satellite C/A codes, for average self-interference, it also suffices to evaluate Eq. (A17), but using $T_{c/a}$ instead of T_p. That is, it suffices to treat the spectrum as being continuous. This average interference, which can be nonnegligible, however small, to a satellite with a low received power when a few satellites with high received powers are visible. However, this is not the most critical problem. The C/A codes also have significant cross-correlation properties, as described in Chapter 3, this volume. This interference density simply describes the average interference.

Appendix B: Acquisition Threshold and Performance Determination

The signal acquisition measurement model for a GPS receiver is illustrated in Fig. B1. At point 1, the samples are those of Eqs. (55) and (56). At point 2, M

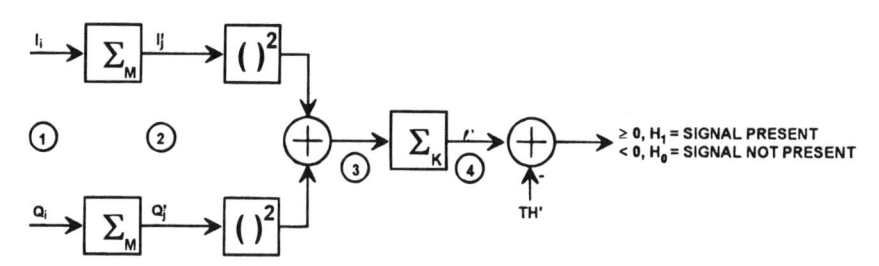

Fig. B1 Signal acquisition model.

of these samples are summed to narrow the predetection bandwidth by a factor of $1/M$. At point 3, we have the sum of the squares of point 2. Then, at point 4, the postdetection bandwidth is narrowed by summing up K of the point 3 samples. This quantity ℓ' is known as the test statistic that is compared against a threshold TH' to determine if the signal is present or not present.

The signal detection problem is set up as a hypothesis test, testing the hypothesis H_1 that the signal is present versus the hypothesis H_0 that the signal is not present. The test statistic for the derivations that follow is defined to be as follows:

$$\ell = \frac{1}{M} \sum_{j=1}^{K} \left[\left(\sum_{i=1}^{M} I_i \right)^2 + \left(\sum_{i=1}^{M} Q_i \right)^2 \right]_j \geq \text{TH} \tag{B1}$$

under hypothesis H_1, and

$$\ell = \frac{1}{M} \sum_{j=1}^{K} \left[\left(\sum_{i=1}^{M} I_i \right)^2 + \left(\sum_{i=1}^{M} Q_i \right)^2 \right]_j < \text{TH} \tag{B2}$$

under hypothesis H_0, where ℓ is the test statistic, TH is the threshold, M is the number of in-phase and quadraphase samples summed prior to squaring, K is the number of samples summed after squaring.

The statistic is $1/M$ of that which is usually used and shown in Fig. B1, but that is for derivation convenience. The resulting threshold can then be multiplied by M to arrive at the desired threshold. The test statistic as defined in Eq. (B1) has an expected value

$$E(\ell \mid H_0) = 2K \tag{B3}$$

for unity variance I and Q noise samples, where, in actual implementations, we would assume that

$$E(\ell' \mid H_0) = 2KM = M \cdot E(\ell \mid H_0) \tag{B4}$$

where ℓ' is the "real" test statistic and

$$\text{TH}' = M \cdot \text{TH} \tag{B5}$$

which can be computed after the threshold TH is determined.

The general probability density function of ℓ under hypothesis H_1, as a function of K, and the following two-sided predetection signal-to-noise ratio:

$$\beta = (S/N_0)MT \tag{B6}$$

is

$$p(y \mid H_1) = \frac{1}{2} \left(\frac{y}{2K\beta} \right)^{\frac{1}{2}(K-1)} e^{-\frac{1}{2}(y - 2K\beta)} I_{K-1}(\sqrt{2K\beta y}); \quad y \geq 0 \tag{B7}$$

where $I_{K-1}(X)$ is a modified Bessel function of the first kind. This probability density function can be derived from that for a single sample K (see Van Trees[38]) through a change of variables (i.e., $y = z^2$) and convoluting K independent densities, and using the properties of the modified Bessel functions that[39]

$$\frac{d^{K-1}[X^{K-1} I_{K-1}(X)]}{dX^{K-1}} = X^{K-1} I_0(X) \tag{B8}$$

and

$$I_{K-1}(X) - I_{K+1}(X) = 2I_K(X)/X \tag{B9}$$

in evaluating the Jacobian.

The corresponding probability of detection for a threshold TH is then as follows:

$$p_d = \int_{TH}^{\infty} p(y|H_1) dy \tag{B10}$$

Likewise, the general probability density of ℓ under hypothesis H_0, as a function of K, is as follows:

$$p(y|H_0) = \frac{1}{2^K(K-1)!} y^{K-1} e^{-y/2}; \, y \geq 0 \tag{B11}$$

which describes a chi-square density function with $2K$ DOF. The Eq. (B11) is a special case of Eq. (B7) with $\beta = 0$, which is, of course, indeterminate with that value. However, using l'Hôpital's Rule $K - 1$ times and the relationship in Eq. (B8), Eq. (B7) does reduce to Eq. (B11) for $\beta = 0$.

The corresponding probability of false alarm (false alarm rate) for a threshold TH is then as follows:

$$p_f = \int_{TH}^{\infty} p(y|H_0) dy \tag{B12}$$

which is only a function of the threshold TH and the postdetection sum K. For a given threshold TH and sum K, this false alarm rate can be determined from tables, or from the following incomplete gamma function:

$$p_f = \left[1 + \frac{TH}{2} + \frac{1}{2!}\left(\frac{TH}{2}\right)^2 + \cdots + \frac{1}{(K-1)!}\left(\frac{TH}{2}\right)^{K-1} \right] e^{TH/2} \tag{B13}$$

Usually, the desired false alarm rate is a given. Thus, the procedure is to use the above equations to solve for the threshold TH for that desired false alarm rate and then evaluate the performance of the detector as a function of C/N_0. If it doesn't perform well enough, either K or M, or both, will have to be increased, which effectively slows down the search rate. Increasing M (decreasing the predetection bandwidth) is more effective, but not always possible because of large Doppler uncertainties, or because of databit edge occurrence. Note that increasing M does not change the threshold TH, only the "real" threshold TH' and the probability of detection performance (because β increases). Unfortunately, the only way that the threshold TH can be determined for a desired p_f is to do it iteratively via trail-and-error, or a method such as the Newton–Rhapson method.[40]

The problem with using Eqs. (B11–B13) for determining the threshold TH for the GPS C/A-code acquisition application is that, because of code cross correlations, the hypothesis that the signal is not present (H_0) using those equations is not correct. This is because there is always "some" signal present in the form of minor correlation peaks or cross-correlation peaks. Thus, a different criteria for finding the threshold TH is required. One that has been found to be successful, is to assume that β (β_N) is not zero, but consists of some signal level that, it is hoped, is much less than the signal of hypothesis H_1. A rule of thumb is to assume that the cross-correlation power levels are approximately 23 dB below the level of a strong GPS signal—say 50 dB Hz—for a power level of approximately 27 dB Hz; that is, set

$$\beta_N = 500 M T_I \qquad (B14)$$

and iteratively solve Eqs. (B7–B10) for the threshold TH, instead of Eqs. (B11–B13).

This does, of course, make the solution for TH much tougher. However, the derivative of Eq. (B10) with respect to the threshold TH is the negative of Eq. (B7) evaluated at the threshold TH. Using this, a Newton–Rhapson procedure can be used to iteratively converge to a solution for TH for a desired p_f.

References

[1] Frank, G. B., and Yakos, M. D., "Collins Next Generation Digital GPS Receiver," *Proceedings, IEEE PLANS 90, Position, Location, and Navigation Symposium*, (Las Vegas, NV) IEEE, New York, March 21–23, 1990, pp. 286–292.

[2] Anon., GEC Plessey Semiconductors Data Sheet DS3076-2.4, "GP1010 Global Positioning Receiver Front End," Oct. 1992.

[3] Schwartz, M., *Information Transmission, Modulation, and Noise*, 4th ed., McGraw-Hill, New York, 1990, chap. 6, pp. 464–470.

[4] Anon., *Critical Item Development Specification for the AS-3822/URN Fixed Reception Pattern Antenna 3 FRPA3 of the User Segment NAVSTAR Global Positioning System*, CI-FRPA-3070A, Nov. 11, 1987.

[5] Anon., Sensor Systems GPS Antennas Data Sheet, Dec. 1992.

[6] Anon., Aeronautical Radio, Inc., *ARINC Characteristic 743A GPS/GLONASS Sensor*, Prepared by the Airlines Electronic Engineering Committee, March 16, 1992.

[7] Anon., *Prime Item Development Specification for the AM-7134/URN Antenna Electronics Amplifier of the User Segment NAVSTAR Global Positioning System*, CI-AE-3061A, 21 March 1988.

[8] Anon., *Specification for NAVSTAR Global Positioning System (GPS) Miniaturized Airborne GPS Receiver (MAGR)*, Final Draft, Specification No. CI-MAGR-300, Code Identification: 07868, 30 March 1990.

[9] Boithias, L., *Radio Wave Propagation*, (trans. to English by D. Beeson) McGraw-Hill, New York, 1987.

[10] Davenport, W. B., Jr., and Root, W. L., *An Introduction to the Theory of Random Signals and Noise*, IEEE Press, New York, 1987, chap. 10, pp. 207–216.

[11] Friis, H. T., "Noise Figures in Radio Receivers," *Proceedings of the IRE*, Vol. 32, July 1944, pp. 419–422.

[12] Spilker, J. J., Jr., *Digital Communications by Satellite*, Prentice-Hall, Englewood Cliffs, NJ, 1977, chap. 6, pp. 174–175.

[13] Chang, H., "Presampling Filtering, Sampling and Quantization Effects on Digital Matched Filter Performance," *Proceedings of the International Telemetering Conference*, (San Diego, CA), 1982, pp. 889–915.

[14] Turin, G. L., "An Introduction to Digital Matched Filters," *Proceedings of the IEEE*, Vol. 64, No. 7, 1976, pp. 1092–1112.

[15]Lim, T. L., "Noncoherent Digital Matched Filters: Multi-Bit Quantization," *IEEE Transactions on Communications,* Vol. 26, No. 4, 1978, pp. 409–419.

[16]Amoroso, F., "Adaptive A/D Converter to Suppress CW Interference in Spread-Spectrum Communications," *IEEE Transactions on Communications,* Vol. 31, No. 10, Oct. 1983, pp. 1117–1123.

[17]Amoroso, F., and Bricker, J. L., "Performance of the Adaptive A/D Converter in Combined CW and Gaussian interference," *IEEE Transactions on Communications,* Vol. 34, No. 3, March 1986, pp. 209–213.

[18]Anon., ARINC Research Corp., *NAVSTAR GPS Space Segment/Navigation User Interfaces* (Public Release Version), ICD-GPS-200, Rev. B-PR, 3 July 1991.

[19]Anon., Department of Defense, *Global Positioning System (GPS) Standard Positioning Service (SPS) Signal Specification,* Dec. 8, 1993.

[20]Nagle, J. R., Van Dierendonck, A. J., and Hua, Q. D., "Inmarsat-3 Navigation Signal C/A-Code Selection and Interference Analysis," *Navigation,* Vol. 39, No. 4, 1992–1993, pp. 445–461.

[21]Fenton, P., Falkenberg, B., Ford, T., Ng K., and Van Dierendonck, A. J., "NovAtel's GPS Receiver—the High Performance OEM Sensor of the Future," *Proceedings of ION GPS-91* (Albuquerque, NM), Institute of Navigation, Washington, DC, Sept. 11–13, 1991, pp. 49–58.

[22]Van Dierendonck, A. J., Fenton, P., and Ford, T., "Theory and Performance of Narrow Correlator Spacing in a GPS Receiver," *Navigation,* Vol. 39, No. 3, 1992, pp. 265–283.

[23]Rambo, J. C., "Receiver Processing Software Design of the Rockwell International DoD Standard GPS Receivers," *Proceedings of ION GPS-89,* (Colorado Springs, CO), Institute of Navigation, Washington, DC, Sept. 27–29, 1989, pp. 217–226.

[24]Chie, C. M., and Lindsey, W. C., "Phase-Locked Loops, Performance Measures, and Summary of Analytical Results," *Phase-Locked Loops,* IEEE Press, New York, 1986, pp. 3–25.

[25]Hagerman, L. L., *Effects of Multipath on Coherent and Noncoherent PRN Ranging Receiver,* Aerospace Corporation Rept. No. TOR-0073(3020-03)-3, May 15, 1973.

[26]Natali, F. D., *Comparison of the Noncoherent Delay-Lock Loop and the Data Estimating Delay-Lock Loop in the Presence of Specular Multipath,* Stanford Telecommunications, Inc., Rept. No. STI-TR-33048, April 5, 1983.

[27]Braasch, M. S., *On the Characterization of Multipath Errors in Satellite-Based Precision Approach and Landing Systems,* Ph.D. Dissertation, Ohio Univ., Athens, OH, June 1992.

[28]Cannon, M. E., and Lachapelle, G., "Analysis of a High-Performance C/A-Code GPS Receiver in Kinematic Mode," *Navigation,* Vol. 39, No. 3, 1992, pp. 285–300.

[29]Hundley, W., Rowson, S., Courtney, G., Wullschleger, V., Velez, R., and O'Donnel, P., "Flight Evaluation of a Basic C/A-Code Differential GPS Landing System for Category I Precision Approach," *Navigation,* Vol. 40, No. 2, 1993, pp. 161–178.

[30]Rowson, S. V., Courtney, G. R., and Hueschen, R. M., "Performance of Category IIIB Automatic Landings Using C/A Code Tracking Differential GPS," *Navigation,* Vol. 41, No. 2, 1994, pp. 127–144.

[31]Natali, F. D., "AFC Tracking Algorithms," *IEEE Transactions on Communications,* Vol. 32, No. 8, Aug. 1984, pp. 935–947.

[32]Natali, F. D., "Noise Performance of a Cross-Product AFC with Decision Feedback for DPSK Signals," *IEEE Transactions on Communications,* Vol. COM-34, No. 3, March 1986, pp. 303–307.

[33]Spilker, J. J., Jr., *Digital Communications by Satellite,* Prentice-Hall, Englewood Cliffs, NJ, 1977, chap. 12, pp. 347–357.

[34]Mood, A. M., Graybill, F. A., and Boes, D. C., *Introduction to the Theory of Statistics,* 3rd ed., McGraw-Hill, New York, 1974, chap. 2, pp. 180–181.

[35]Socci, G., Van Dierendonck, A. J., and Neumann, J., *GPS Satellite Signal Acquisition and Tracking in Extreme Jamming Environments,* Stanford Telecom Rept. STel-TR-89350, April 21, 1989.

[36]Van Dierendonck, A. J., Russell, S. S., Koptizke, E. R., and Birnbaum, M., "The GPS Navigation Message," *Global Positioning System,* Vol. 1, Institute of Navigation, Washington, DC, 1980, pp. 55–73.

[37]Brown, R. G., and Hwang, P. Y. C., *Introduction to Random Signals and Applied Kalman Filtering,* 2nd ed., Wiley, New York, 1992, chap. 2, p. 130.

[38]Van Trees, H. L., *Detection, Estimation and Modulation Theory,* Wiley New York 1968.

[39]Beyer, W.(ed.), *CRC Standard Mathematical Tables,* 25th ed., CRC Press, Boca Raton, FL, 1978, pp. 416–419.

[40]Hildebrand, F. B., *Introduction to Numerical Analysis,* McGraw-Hill, New York 1956, chap. 10, pp. 447–450.

Chapter 9

GPS Navigation Algorithms

P. Axelrad*
University of Colorado, Boulder, Colorado 80309
and
R. G. Brown†
Iowa State University, Ames, Iowa 50010

I. Introduction

THE previous chapters described the hardware and software needed to make GPS observations in the receiver. This chapter focuses on how these observations are processed to form a navigation solution. Fundamentally, a navigation solution is an estimate of the user position plus any other required parameters. The term "state" is used to describe all the parameters to be determined. The typical states in a GPS navigation estimator are three components of position, clock offset, and clock drift. In a moving application, three components of velocity are added. There are many applications described in the companion volume in which GPS is integrated with one or more other sensors, such as an altimeter or an inertial navigation system (INS). In such configurations, the state may be expanded to include specific sensor error states; however, in this chapter, we restrict ourselves to stand-alone GPS navigation estimation.

A navigation algorithm embedded in the GPS receiver combines raw measurements from the signal processor with GPS satellite orbit data to estimate the observer state. This process requires two sets of models—a measurement model and a dynamics or process model. The dynamics model describes the evolution of the system state. The measurement model relates the state to the GPS observations.

Section II describes the GPS measurements and Sec. III shows how they may be combined into a single point navigation estimate. Section IV provides various dynamic models used in GPS. The Kalman filter and some variations are described in Sec. V, and specific numerical examples are given in Sec. VI. For further information on filtering, the reader is advised to refer to Ref. 1, which specifically addresses GPS navigation filters, or more generally, Refs. 2, 3, or 4.

Copyright © 1994 by the authors. Published by the American Institute of Aeronautics and Astronautics, Inc., with permission. Released to AIAA to publish in all forms.
*Assistant Professor, Department of Aerospace Engineering Sciences.
†Distinguished Professor Emeritus, Department of Electrical and Computer Engineering.

II. Measurement Models

We consider three types of GPS measurements—pseudorange, Doppler, and accumulated delta range (ADR). The specifics of how a receiver actually forms these measurements is discussed in Chapters 7 and 8, this volume. Here we concentrate on the mathematical models of how the observations relate to the state of the vehicle.

A. Pseudorange

When the signal processor delay lock loop (DLL) finds the point of maximum correlation with a given GPS satellite signal, it produces an observation of the code phase, or equivalently, signal transmit time t_T for the current local receive time t_R. (In most cases, the observation is not a function of the filter navigation solution; however, the vector delay lock loop described in Chapter 8 of this volume provides a means to integrate the DLL and filter functions for improved performance.) The observed signal propagation delay is $(t_R - t_T)$. The pseudorange observable is merely this time interval scaled by the speed of light in a vacuum: $\rho = c\,(t_R - t_T)$.

The pseudorange observation between a user and satellite i can be related to the user position and clock states as follows:

$$\rho_i = |r_i - r_u| + c \cdot b_u + \epsilon_{\rho_i} \tag{1}$$

Where r_i is the satellite position at transmit time; r_u is the receiver position at receive time; b_u is the bias in the receiver clock (in s), and ϵ_ρ is the composite of errors produced by atmospheric delays, satellite ephemeris mismodeling, selective availability (SA), receiver noise, etc. (in m). Chapter 11, this volume provides an error budget for ϵ_ρ under various conditions.

The state to be estimated, consisting of r_u and $c \cdot b_u$ is embedded in this measurement equation. To extract it we must linearize the measurement equation about some nominal value, for example, about our current best estimate.

Given an a priori estimate of the state $\hat{x} = [\hat{r}_u^T\ c \cdot \hat{b}_u]^T$ and an estimate of the bias contributions caused by ionospheric and tropospheric delay, relativistic effects, satellite clock errors $\hat{\epsilon}_{\rho_i}$, we can predict what the pseudorange measurement should be as follows:

$$\hat{\rho}_i = |r_i - \hat{r}_u| + c \cdot \hat{b}_u + \hat{\epsilon}_{\rho_i} \tag{2}$$

The measurement residual $\Delta\rho$, which is the difference between the predicted and actual measurement, can be modeled as linearly related to the error in the state estimate, $\Delta x \equiv [\Delta r^T\ c \cdot \Delta b]^T$, by performing a Taylor expansion about the current state estimate. The linearized result is given by the following:

$$\Delta\rho_i = \hat{\rho}_i - \rho_i = [-\hat{\mathbf{1}}_i^T\ \ 1]\begin{bmatrix}\Delta r \\ c \cdot \Delta b\end{bmatrix} + \Delta\epsilon_{\rho_i} \tag{3}$$

where

$$\hat{\mathbf{1}}_i \equiv \frac{r_i - \hat{r}_u}{|r_i - \hat{r}_u|}, \quad \Delta r \equiv \hat{r}_u - r_u, \quad \Delta b \equiv \hat{b}_u - b_u, \quad \Delta\epsilon_{\rho_i} \equiv \hat{\epsilon}_{\rho_i} - \epsilon_{\rho_i}$$

$\hat{\mathbf{1}}_i$ is the estimated line of sight unit vector from the user to the satellite; and

$\Delta\epsilon_{\rho_i}$ is the residual error after the known biases have been removed. This linearized model is the fundamental GPS pseudorange measurement equation.

The residual measurement error $\Delta\epsilon_{\rho_i}$ is generally composed of a slowly varying term, usually dominated by SA in civilian receivers, plus random or white noise. The expected variance of the error is required for any weighted navigation solution algorithm. An order of magnitude estimate of the slow terms can be obtained from the user equivalent range error (URE) reported in the Navigation message (see Chapter 4, this volume). The high-frequency error is produced primarily by receiver noise and quantization. For a typical receiver, the standard deviation is about 1/100 of the code chip, or about 3 m for C/A code and 0.3 m for P code. A more precise estimate can be based on the signal-to-noise ratio calculation in the channel, as described in Chapter 8, this volume.

B. Doppler

The numerically controlled oscillator (NCO), which controls the carrier-tracking loop, provides an indication of the observed frequency shift of the received signal. This observed frequency differs from the nominal L_1 or L_2 frequency because of Doppler shifts produced by the satellite and user motion, as well as the frequency error or drift of the satellite and user clocks. The Doppler shift caused by satellite and user motion is the projection of the relative velocities onto the line of sight scaled by the transmitted frequency $L_1 = 1575.42$ MHz divided by the speed of light, as follows:

$$D_i = -\left(\frac{v_i - v_u}{c} \cdot \mathbf{1}_i\right) L_1 \qquad (4)$$

The Doppler can be converted to a pseudorange rate observation given by the following:

$$\dot{\rho}_i = (v_i - v_u) \cdot \frac{r_i - r_u}{|r_i - r_u|} + f + \epsilon_{\dot{\rho}_i} \qquad (5)$$

where f is the receiver clock drift in m/s; and $\epsilon_{\dot{\rho}_i}$ is the error in the observation in m/s. Again, this effect can be predicted, based upon the current estimates of the velocity $\hat{v}_u$; line of sight vector $\hat{\mathbf{1}}_i$, the clock drift estimate in m/s $\hat{f}$; and the known error rates $\hat{\epsilon}_{\dot{\rho}_i}$ as follows:

$$\hat{\dot{\rho}}_i = (v_i - v_u) \cdot \hat{\mathbf{1}}_i + \hat{f} + \hat{\epsilon}_{\dot{\rho}_i} \qquad (6)$$

The linearized Doppler measurement equation is then as follows:

$$\Delta\dot{\rho}_i \equiv \hat{\dot{\rho}}_i - \dot{\rho}_i = [-\hat{\mathbf{1}}_i^T \quad 1] \begin{bmatrix} \Delta v \\ \Delta f \end{bmatrix} + \Delta\epsilon_{\dot{\rho}_i} \qquad (7)$$

Note that the Doppler does depend on the observer position through the line-of-sight unit vector. This dependence can be exploited to perform "Doppler positioning" in which the position is solved for using the Doppler observations and, sometimes, the rate of change of Doppler. This is the positioning method employed with the Navy's Transit satellites. The observation geometry for Doppler positioning is substantially weaker than ranging; thus, it is not used much in GPS, except to set an a priori position estimate.

C. Accumulated Delta Range

The ADR is produced by the signal processor by accumulating the commanded values to the NCO required to maintain lock on the signal. In other words, it keeps track of changes in the observed range to the satellite. Thus, both terms "accumulated delta range" or "integrated Doppler" are appropriate. In the literature (cf. Ref. 5) this measurement has also been called "carrier beat phase," referring to the output of a mixing process between a nominal L_1 carrier signal generated in the receiver, and the received Doppler-shifted version.

The distinction between the ADR and a code-based pseudorange is that the ADR has an ambiguous starting value. Once the phase–lock-loop (PLL) begins to follow the carrier signal, it can keep track of the total change in range; however, there is no way to know the whole number of carrier cycles between the satellite and the user antenna. Thus, for stand-alone navigation it is not possible to use the ADR for absolute estimation of position. (The ADR initial condition problem is similar to that encountered in inertial navigation systems.) However, for differential GPS or attitude determination, it is possible to determine the difference in the ambiguities between two nearby stations and/or two GPS satellites. Thus, for these applications, the precision of the ADR can be fully exploited.

A common use of the ADR in stand-alone navigation is to smooth the noisy pseudorange measurements. A number of techniques are available for doing this. A commonly used technique forms a weighted average of the code and carrier-based measurements.[6] Reference 7 has also suggested an integration scheme where the ADR provides the reference trajectory (in much the same manner as an INS), and then the pseudorange data is used at a slower rate to update the reference trajectory via a Kalman filter. Other ad hoc methods have also been successfully employed for real-time, stand-alone applications. The ADR measurements play only a minor role in many GPS stand-alone navigation applications; however, they are pivotal to kinematic differential operations and surveying, as described in Chapters 15 and 18 in the companion volume.

D. Navigation Data Inputs

To compute the predicted pseudorange and Doppler, the navigation algorithm must have information on the position and velocity of the GPS satellite, as well as error models to correct the satellite clock offset and atmospheric delays. This information is provided via the Navigation message. The satellite positions are computed as described in Chapter 4, this volume; the atmospheric delay models are described in Chapter 8, this volume; and the clock corrections are described in Chapter 4, this volume. This information is also contained in Ref. 8.

III. Single-Point Solution

The physical measurements and equations provided in the previous section are all that is required for a single-point solution or kinematic solution. In this method, the navigation estimate is the least squares solution to the measurement equations made at a single time. For each satellite tracked by the receiver, the predicted pseudorange is formed using Eq. (2), and the linearized observation Eq. (3) is formed. All the measurements are then combined into a set of normal equations:

$$\Delta \rho = G \Delta x + \Delta \epsilon_\rho \tag{8}$$

where

$$\Delta \rho \equiv \begin{bmatrix} \Delta \rho_1 \\ \Delta \rho_2 \\ \vdots \\ \Delta \rho_n \end{bmatrix}, \quad G \equiv \begin{bmatrix} -\hat{\mathbf{1}}_1^T & 1 \\ -\hat{\mathbf{1}}_2^T & 1 \\ \vdots & \vdots \\ -\hat{\mathbf{1}}_n^T & 1 \end{bmatrix}, \quad \Delta x \equiv \begin{bmatrix} \Delta r_u \\ c \cdot \Delta b_u \end{bmatrix}, \quad \Delta \epsilon_\rho \equiv \begin{bmatrix} \Delta \epsilon_{\rho_1} \\ \Delta \epsilon_{\rho_2} \\ \vdots \\ \Delta \epsilon_{\rho_n} \end{bmatrix}$$

which is to be solved for a correction, Δx to the a priori state estimate. To improve the state estimate subtract Δx from the a priori values. In GPS, "G" is frequently referred to as the geometry matrix, and corresponds to the measurement connection matrix, commonly named "H" in the more general literature on filtering.

The $\Delta \epsilon_\rho$ are assumed to be zero mean, so that the least squares solution to the set of normal equations is given by the following:

$$\Delta \hat{x} \equiv (G^T G)^{-1} G^T \Delta \rho \tag{9}$$

or, if a weight R_i^{-1} is assigned to each observation, the weighted least squares estimate is as follows:

$$\Delta \hat{x} \equiv (G^T R^{-1} G)^{-1} G^T R^{-1} \Delta \rho \tag{10}$$

If the a priori estimate used to construct G, is off by a lot (typically more than a few km), the least squares solution may be iterated until the change in the estimate is sufficiently small. Because G only depends upon the line-of-sight unit vector, it is not very sensitive to errors in the observer position. Numerically efficient methods for solving Eqs. (8–10) are well known (c.f. Ref. 9).

A. Solution Accuracy and Dilution of Precision

How accurate is the single-point, least squares solution? The accuracy is decided by two factors, the measurement quality and the user-to-satellite geometry. [Chapter 11, this volume, on errors in GPS details the contributions of both measurement errors and geometry, and Chapter 5, this volume, provides an extensive discussion of geometric dilution of precision (GDOP).] The measurement quality is described by the variance of the measurement error, which for a typical pseudorange is in the range of 0.3 to 30 m, depending on the error conditions. The geometry is described by the "G" matrix, which is composed of line-of-sight vectors and "1s" for the clock states.

The solution error covariance can be expressed as follows:

$$E[\Delta \hat{x} \ \Delta \hat{x}^T] = E[(G^TG)^{-1}G^T\Delta\rho \ \Delta\rho^T G(G^TG)^{-1}]$$
$$= (G^TG)^{-1}G^T R G(G^TG)^{-1} \quad (11)$$

where R is the pseudorange measurement covariance. If we assume (somewhat incorrectly, as described later) that the measurement errors are uncorrelated and have equal variance σ^2 then $R = \sigma^2 I$, and the point solution error covariance reduces to the following:

$$E[\Delta \hat{x} \ \Delta \hat{x}^T] = \sigma^2 (G^TG)^{-1} \quad (12)$$

If the state is parameterized so that $\Delta x = [\Delta E \ \Delta N \ \Delta U \ c \cdot \Delta b]^T$, where ΔE, ΔN, and ΔU, are the east, north, and up position errors, respectively; and $c \cdot \Delta b$ is the clock bias error, then the variance of the state estimates is given by the following:

$$E[\Delta \hat{x} \ \Delta \hat{x}^T] = \begin{bmatrix} E[\Delta E^2] & E[\Delta E \Delta N] & E[\Delta E \Delta U] & E[\Delta E c \cdot \Delta b] \\ E[\Delta N \Delta E] & E[\Delta N^2] & E[\Delta N \Delta U] & E[\Delta N c \cdot \Delta b] \\ E[\Delta U \Delta E] & E[\Delta U \Delta N] & E[\Delta U^2] & E[\Delta U c \cdot \Delta b] \\ E[c \cdot \Delta b \Delta E] & E[c \cdot \Delta b \Delta N] & E[c \cdot \Delta b \Delta U] & E[c \cdot \Delta b^2] \end{bmatrix} \quad (13)$$

Most of the time, we are primarily interested in the diagonal elements. The following DOPs summarize the contribution of the geometry:

$$A \equiv (G^TG)^{-1}$$

$$\begin{aligned}
\text{GDOP} &\equiv \sqrt{\text{trace}(A)} & & \text{geometrical DOP} \\
\text{PDOP} &\equiv \sqrt{A_{11} + A_{22} + A_{33}} & & \text{position DOP} \\
\text{HDOP} &\equiv \sqrt{A_{11} + A_{22}} & & \text{horizontal DOP} \quad (14) \\
\text{VDOP} &\equiv \sqrt{A_{33}} & & \text{vertical DOP} \\
\text{TDOP} &\equiv \sqrt{A_{44}} & & \text{time DOP}
\end{aligned}$$

Thus, the total position error magnitude can be estimated by $\sigma \times \text{PDOP}$, and the vertical position error by $\sigma \times \text{VDOP}$, etc. However, keep in mind that this is only an approximation, because of the assumption that all satellite pseudorange measurements errors are independent and have the same statistics. The equations for a single-point velocity solution are identical with the pseudorange measurements replaced by pseudorange rates.

Typical modern receivers track 5–12 satellites simultaneously. More satellites produce improved geometry, generally leading to a more accurate single-point navigation solution. (An exception can occur if the ranging error to the additional satellite is exceptionally poor.) ADR smoothing can reduce the receiver-induced measurement noise in each observation; however, it cannot eliminate the effects of SA or atmospheric effects. To improve the navigation estimate further, we must tie together measurements over time by including knowledge of the vehicle dynamics in a solution filter.

B. Point Solution Example

As an example of the least squares solution method, assume an observer is actually located on the surface of the Earth at 0° latitude, 0° longitude, and has a clock error from GPS time equivalent to 85,491.5 m. If the observer state is comprised of the WGS-84 (1984 Word Geodetic System) position components x, y, z, and the clock bias in meters, the true state is as follows:

$$x = [6,378,137.0 \text{ m} \quad 0.0 \text{ m} \quad 0.0 \text{ m} \quad 85,000.0 \text{ m}]^T$$

At a certain time there are seven satellites visible above an elevation of 10 deg at the positions shown in Table 1.

If the a priori position and clock estimate is given by

$$\hat{x} = [6,377,000.0 \text{ m} \quad 3,000.0 \text{ m} \quad 4,000.0 \text{ m} \quad 0.0 \text{ m}]^T$$

then the computed range and line-of-sight unit vector to each satellite are as shown in Table 2.

Table 1 Satellite positions for point solution example

Satellite	X position, m	Y position, m	Z position, m
SV 01	22,808,160.9	−12,005,866.6	−6,609,526.5
SV 02	21,141,179.5	−2,355,056.3	−15,985,716.1
SV 08	20,438,959.3	−4,238,967.1	16,502,090.2
SV 14	18,432,296.2	−18,613,382.5	−4,672,400.8
SV 17	21,772,117.8	13,773,269.7	6,656,636.4
SV 23	15,561,523.9	3,469,098.6	−21,303,596.2
SV 24	13,773,316.6	15,929,331.4	−16,266,254.4

Table 2 Computed pseudorange and line-of-sight vectors

Satellite	Computed pseudorange, m	Line-of-sight X	Line-of-sight Y	Line-of-sight Z
SV 01	21,399,408.0	0.767832	−0.561178	−0.309052
SV 02	21,890,921.6	0.674443	−0.107718	−0.730427
SV 08	22,088,910.4	0.636607	−0.192041	0.746895
SV 14	22,666,464.0	0.531856	−0.821318	−0.206314
SV 17	21,699,943.6	0.709454	0.634576	0.306574
SV 23	23,460,242.4	0.391493	0.147744	−0.908243
SV 24	23,938,978.9	0.308965	0.665289	−0.679655

The computed geometry matrix is as follows:

$$G = \begin{bmatrix} -0.767832 & 0.561178 & 0.309052 & 1 \\ -0.674443 & 0.107718 & 0.730427 & 1 \\ -0.636607 & 0.192041 & -0.746895 & 1 \\ -0.531856 & 0.821318 & 0.206314 & 1 \\ -0.709454 & -0.634576 & -0.306574 & 1 \\ -0.391493 & -0.147744 & 0.908243 & 1 \\ -0.308965 & -0.665289 & 0.679655 & 1 \end{bmatrix}$$

Table 3 provides the simulated measured pseudorange (already corrected for known errors such as ionospheric delay, satellite clock, etc.) and the pseudorange residual $\Delta\rho_i \equiv \hat{\rho}_i - \rho_i$. The standard deviation of the pseudorange errors is 6 m. Note that the pseudorange residuals are dominated by the large error in the estimate of the receiver clock.

Solving the normal equations for a correction to the state estimate gives $\Delta x = [-1,131.8 \quad 2,996.8 \quad 3,993.1 \quad -84,996.4]^T$ m. Subtracting this from the a priori estimate gives the improved estimate $\hat{x} = [6,378,131.8 \quad 3.2 \quad 6.9 \quad 84,996.4]^T$ m. This new estimate is closer to the true value of the state; however, it contains errors produced by the pseudorange measurement error as well as the approximation in the line-of-sight vectors caused by the incorrect a priori guess. To see how large the latter effect is, we can redo the least squares solution using the improved estimate to compute the elements of the G matrix. The resulting correction to the state estimate is $\Delta x = [0.3 \quad -0.1 \quad -0.2 \quad 0.6]^T$ m, and the "improved" state estimate (actually worse than the last estimate) is $\hat{x} = [6,378,131.5 \quad 3.3 \quad 7.1 \quad 84,995.8]^T$ m. Thus, it is apparent that the approximation made in computing the G matrix was quite good, and to get a solution at the 1-m level, it is not necessary to iterate if the solution is already known to within a few kilometers.

The final error in the state estimate is $\Delta x = [-5.5 \quad 3.2 \quad 7.1 \quad -4.2]^T$ m. Now let us compare this to the error bound predicted by the GDOP approximation. The A matrix can be computed from the G matrix given in Eq. (14). The DOPS for each of the state components are computed as the square roots of the diagonal elements of A. The measurement standard deviation $\sigma = 6$ m.

Thus, the DOP approximations seem to be valid for this example.

Table 3 Simulated pseudorange and residual

Satellite	Measured pseudorange, m	Pseudorange residual, m
SV 01	21,480,623.2	−81,215.3
SV 02	21,971,919.2	−80,997.6
SV 08	22,175,603.9	−86,693.4
SV 14	22,747,561.5	−81,097.6
SV 17	21,787,252.3	−87,308.8
SV 23	23,541,613.4	−81,371.0
SV 24	24,022,907.4	−83,928.6

Table 4 Actual point solution errors compared to DOP predictions

State component	DOP value	Expected $1-\sigma$ error, $\sigma \times$ DOP	Actual error
X position	3.0	18.0 m	−5.5 m
Y position	0.8	4.8 m	3.2 m
Z position	0.8	4.8 m	7.1 m
Clock bias	1.9	11.4 m	−4.2 m
Total error	3.7	22.2 m	10.4 m

IV. User Process Models

In anticipation of employing Kalman filter methods in the GPS solution (Sec. V), we now look at user "process models." The vehicle dynamics are summarized in the filter process model. The GPS has the capability to provide real-time three-dimensional position, velocity, and time information to any user. However, there are times when all this information is not required or valuable. The degree to which the user dynamics are constrained or predictable dictates the type of process model used.

A. Clock Model

Two states required in any GPS-based navigation estimator are the user clock bias and drift, which represent the phase and frequency errors in the atomic frequency standard or crystal oscillator in the receiver. Within the navigation algorithm the two-state model shown in Fig. 1 is commonly employed.

This model says that we expect both the frequency and phase to random walk over a short period of time. The discrete process equations are given by Ref. 1.

$$x_c(k) = \Phi_c(\Delta t) x_c(k-1) + w_c(k-1) \qquad (15)$$

where

$$x_c \equiv \begin{bmatrix} b \\ f \end{bmatrix}, \qquad \Phi_c(\Delta t) = \begin{bmatrix} 1 & \Delta t \\ 0 & 1 \end{bmatrix}$$

$$Q_c \equiv E[w_c w_c^T] = \begin{bmatrix} s_b \Delta t + S_f \dfrac{\Delta t^3}{3} & S_f \dfrac{\Delta t^2}{2} \\ S_f \dfrac{\Delta t^2}{2} & S_f \Delta t \end{bmatrix}$$

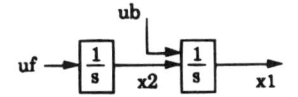

Fig. 1 Dynamic model for GPS clock states.

The white noise spectral amplitudes S_b and S_f can be related to the classical Allan variance parameters. The approximate relation given in Ref. 1 (p. 427) is $S_f = 2\,h_0$, and $S_g = 8\,\pi^2\,h_{-2}$ (see also Ref. 10). Figure 2 shows simulated clock states for a crystal oscillator with $h_0 = 2 \times 10^{-19}$, and $h_{-2} = 2 \times 10^{-20}$ (Ref. 1, Chapter 10).

Two clock states of this type must be included for all types of GPS users. In a time transfer receiver, which is described in greater detail in Chapter 16 of the companion volume, these two clock states are the ones of primary interest. In this case, the receiver position is generally known to at least the level of accuracy of timing information desired (1 m ~ 3 ns). For highest accuracy, the position is held fixed and only the two clock parameters are estimated.

B. Stationary User or Vehicle

If the user antenna is known to be stationary at an unknown location, three position coordinate states may be added to the clock model to form a 5-element state vector. It is assumed that the velocity is zero, thus, the dynamic model for

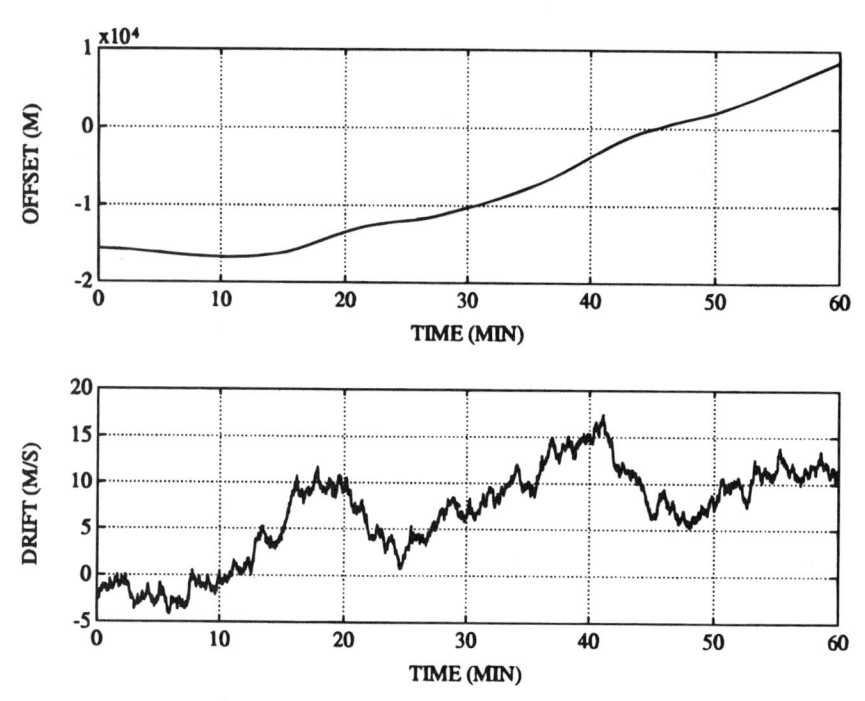

Fig. 2 Simulated GPS clock errors. The bottom graph shows an example of clock frequency drift measured in m/s. The top graph shows the corresponding clock offset in meters. Note, the receiver clock errors can be very large, in this case more than 10 km, and must be estimated along with the position solution.

the stationary user is given by the following:

$$x_s(k) = \Phi_s(\Delta t)x_s(k-1) + w_s(k-1) \qquad (16)$$

where

$$x_s \equiv [x \quad y \quad z \mid b \quad f]^T$$

$$\Phi_s(\Delta t) = \begin{bmatrix} I & 0 \\ \hdashline 0 & \Phi_c(\Delta t) \end{bmatrix}, \quad \text{and } I \text{ is a } 3 \times 3 \text{ identity matrix}$$

$$Q_s \equiv E[w_s w_s^T] = \begin{bmatrix} Q_p & 0 \\ \hdashline 0 & Q_c \end{bmatrix}$$

There are two important things to note. First, it cannot be assumed that the clock state is constant; thus, the frequency error state is required as well as the bias. Second, even for a stationary result, we model the dynamics by a random walk to prevent numerical problems in the navigation algorithm. The process noise covariance Q_p represents the uncertainty in the dynamic model. Thus, for a stationary observer, we would think that it could be set to zero. This is not generally done because it can lead to numerical problems or cause the filter to "go to sleep." In this situation, the estimation error covariance has decreased so far that the estimator gain for new measurements goes to zero—essentially the filter begins to ignore new information. As long as this situation is avoided, Q_p can be set to a small value to maximize the smoothing that will occur.

C. Low Dynamics

The next step up in user dynamics is a low dynamic vehicle, such as a boat or car. In these cases, the position, velocity, and clock terms must be estimated, leading to an 8-element state representation. The discrete model for such a user is shown in Fig. 3. The corresponding dynamic model is given by the following:

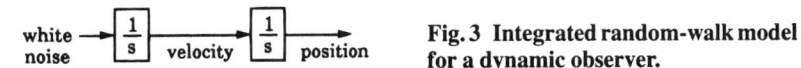

Fig. 3 Integrated random-walk model for a dynamic observer.

$$x_L(k) = \Phi_L(\Delta t)x_L(k-1) + w_L(k-1) \qquad (17)$$

where

$$x_L \equiv [x \ y \ z | \dot{x} \ \dot{y} \ \dot{z} | b \ f]^T$$

$$\Phi_L(\Delta t) = \begin{bmatrix} I & \Delta t I & \vdots & 0 \\ 0 & I & \vdots & 0 \\ \hdashline 0 & 0 & \vdots & \Phi_c(\Delta t) \end{bmatrix}$$

$$Q_L \equiv E[w_L w_L^T] = \begin{bmatrix} Q_p & Q_{pv} & \vdots & 0 \\ Q_{pv} & Q_v & \vdots & 0 \\ \hdashline 0 & 0 & \vdots & Q_c \end{bmatrix}$$

The effect of unknown random accelerations between measurement updates is represented by Q_v. Often, different values are used for horizontal and vertical components; i.e., a car cannot change its vertical velocity substantially; whereas, it can accelerate or decelerate rapidly. If the dynamical uncertainty of the vehicle is large, filtering will not improve the navigation solution.

D. High Dynamics

When the vehicle, such as a fighter aircraft or missile, has the potential for significant accelerations, it is usually necessary to measure and account for the deterministic changes in velocity. This leads to the integrated GPS/INS system, which is discussed in detail in Chapter 2 of the companion volume. A less accurate way of handling the high dynamics problem is to add three acceleration states to the process model and let the stand-alone GPS system estimate the vehicle acceleration in addition to position and velocity. The acceleration states are usually modeled as either random walk or Markov processes. This method of coping with high dynamics is not as good as a full-fledged integrated GPS/INS system, but it is better than treating acceleration as white noise, which is what has to be done if the acceleration states are omitted.

V. Kalman Filter and Alternatives

As mentioned previously, one disadvantage of the point solution approach is that it does not carry any information from one measurement epoch to the next; i.e. it does not include any of the known user dynamics. A second problem is

GPS NAVIGATION ALGORITHMS 421

that the solution accuracy is extremely dependent on the instantaneous satellite geometry. Employing a Kalman filter addresses both of these issues (Ref. 1).

A. Discrete Extended Kalman Filter Formulation

For use with stand-alone GPS, a discrete, extended Kalman filter (EKF) is generally used. This means that measurements are incorporated at discrete intervals, and the measurement models are linearized about the current best estimate of the state. The updated state estimate is formed as a linear blend of the previous estimate (projected forward to the current time) and the current measurement information. The relative weighting in the blend is determined by the a priori error covariance and the measurement error covariance. After updating, the state estimate and its error covariance matrix are projected ahead to the next measurement time via the assumed process dynamics.

The state x includes the two clock components, three position components, and possibly three velocity components, depending on the type of dynamic model used. Associated with the state are four key matrices that must be specified in the discrete EKF:

G—the measurement connection matrix, the elements of which are the partials of the measurement model with respect to each of the states (In most literature on filtering, this matrix is refered to as H; we use G for geometry matrix to be consistent with the GPS literature.)

R—the measurement noise covariance matrix

Φ—the state transition matrix, which is a linearized representation of the process model

Q—the process noise covariance matrix

The selection of R and Q has a significant effect on the convergence and accuracy of the filter solutions. The adjustment of these parameters is refered to as "filter tuning." Tuning is often performed to achieve the best possible performance while avoiding filter divergence in the face of unmodeled errors.

In addition to specifying the four matrices, we must also establish an initial estimate of the state $\hat{x}_0^-$ and the state covariance matrix (P_0^-). The filter proceeds by processing all available measurements at each epoch (the measurement update) and then propagating the state estimate and covariance ahead to the next epoch (the time update). Estimates of the state and covariance after the measurement update are indicated by a superscript "+"; estimates and covariances propagated ahead are indicated by a superscript "−".

The measurement update is summarized as follows:

1) Compute the expected pseudorange $\hat{\rho}_k$ according to Eq. (2) based on the GPS satellite position and the a priori state estimate $\hat{x}_k^-$.

$$\hat{\rho}_k = h(\hat{x}_k^-)$$

2) Construct the measurement connection matrix as follows:

$$G_k(\hat{x}_k^-) \equiv \frac{\partial h(\hat{x}_k^-)}{\partial x}$$

3) Compute the gain matrix K_k according to the following:

$$K_k = P_k^- G_k^T (G_k P_k^- G_k^T + R_k)^{-1} \tag{18}$$

4) Update the state as follows:

$$\hat{x}_k^+ = \hat{x}_k^- + K_k(\rho_{k-\text{measured}} - \hat{\rho}_k) \tag{19}$$

(Note that the quantity in parenthesis is opposite in sign to the measurement residual $\Delta\rho = \hat{\rho}_k - \rho_k$ defined earlier.)

5) and the following covariance matrix*:

$$P_k^+ = (I - K_k G_k)P_k^-(I - K_k G_k)^T + K_k R_k K_k^T \tag{20}$$

The projection steps are as follows:

1) Propagate the covariance matrix to the next measurement epoch as follows:

$$P_{k+1}^- = \Phi_k P_k^+ \Phi_k^T + Q_k \tag{21}$$

2) Propagate the state estimate to the next measurement epoch using the assumed process dynamics. If the dynamic model is linear, the propagation is given by the following:

$$\hat{x}_{k+1}^- = \Phi_k \hat{x}_k^+ \tag{22}$$

In the more general case where the zero-noise dynamic model is given by the following:

$$\frac{d}{dt} x = f(x, t) \tag{23}$$

$\hat{x}_k^+$ is projected forward by numerically integrating the nonlinear dynamic model. These steps are illustrated in Fig. 4.

The standard EKF assumes that the measurement and process noise are not correlated with each other and that each is uncorrelated between time epochs. For GPS, the former assumption is largely valid, but the latter is not. In the next section, methods for dealing with correlated measurement noise are discussed.

B. Steady-State Filter Performance

A key feature of the Kalman filter is that, under many conditions, it quickly converges to a quasi-steady-state condition. In GPS navigation, for example, the geometry matrix G changes rather slowly, and in all of the models considered here, the state transition matrix is only a function of the measurement time interval. Thus, after a short time, the increase in the error covariance caused by state propagation and dynamic uncertainty is matched by the decrease in error covariance caused by the measurement update. The discrete, steady-state Kalman filter can be derived by assuming that both P^+ and P^- are constants in Eqs. (18), (20), and (21). Substitution results in a discrete time Riccati equation that does converge but in general does not afford a closed form solution. This is described in further detail in Refs. 11 and 12. An interesting point is that in GPS, it frequently takes only a few measurement epochs to achieve this steady-state value.

---

* The form of the expression given in Eq. (20) for the covariance measurement update ensures that if P^- is symetric, P^+ will also be symetric. The more commonly used form $P_k^+ = (I - K_k H_k) P_k^-$, does not have this property. In this form, numerical difficulties can result, for example, if the initial covariance is very large and the measurements are very accurate.

```
                    ┌─────────────────────────────┐
                    │      COMPUTE GAIN           │
  Enter x̂₀⁻ and P₀⁻ │                             │
  ─────────────────▶│ Kₖ = Pₖ⁻Gₖᵀ(GₖPₖ⁻Gₖᵀ+Rₖ)⁻¹  │
                    └─────────────────────────────┘
```

(flow chart diagram)

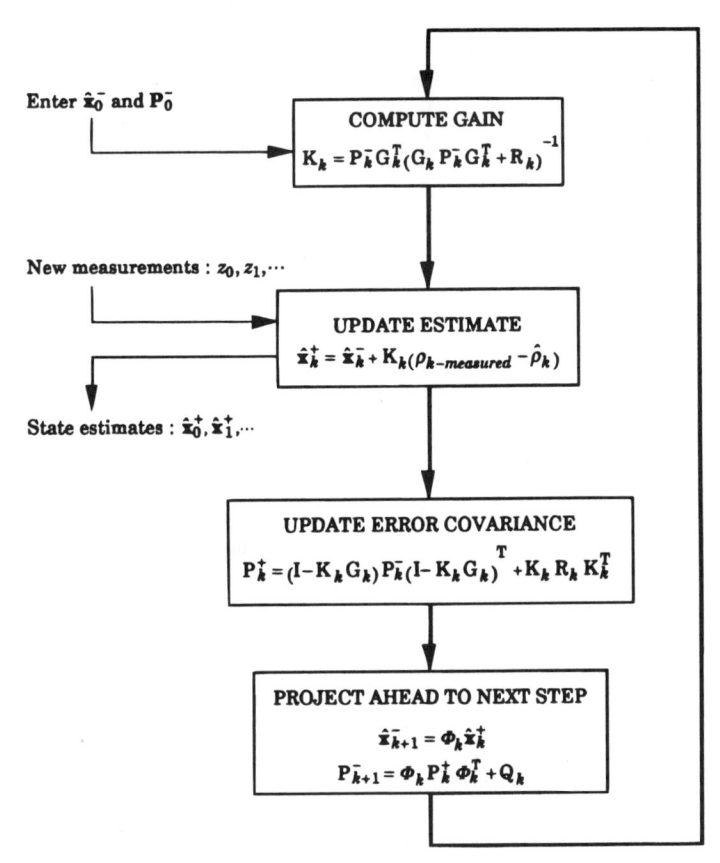

Fig. 4   Flow chart of the discrete Kalman filter solution.

Steps shown in the flow chart:

- **COMPUTE GAIN**: $K_k = P_k^- G_k^T (G_k P_k^- G_k^T + R_k)^{-1}$
- **UPDATE ESTIMATE**: $\hat{x}_k^+ = \hat{x}_k^- + K_k(\rho_{k-measured} - \hat{\rho}_k)$
- New measurements: $z_0, z_1, \ldots$
- State estimates: $\hat{x}_0^+, \hat{x}_1^+, \ldots$
- **UPDATE ERROR COVARIANCE**: $P_k^+ = (I - K_k G_k) P_k^- (I - K_k G_k)^T + K_k R_k K_k^T$
- **PROJECT AHEAD TO NEXT STEP**: $\hat{x}_{k+1}^- = \Phi_k \hat{x}_k^+$; $P_{k+1}^- = \Phi_k P_k^+ \Phi_k^T + Q_k$

C. Alternate Forms of the Kalman Filter

Under special conditions, there may be computational difficulties with the standard Kalman filter formulation shown above. These include numerical instability and divergence produced by round-off errors even when high precision arithmetic is employed. Square-root filters can mitigate these problems to some extent. Perhaps the most widely used square-root filter in GPS is the *U-D* filter. This is described in detail in Ref. 13, and in less detail in Refs. 1 and 4.

In the U-D filter, the covariance matrix P is decomposed into the factored form $P = UDU^T$. The measurement and time updates are formulated directly in terms of the U and D matrices, and conversion to the covariance form is only required for input and output purposes. In brief, the divergence problems encountered in propagating the U and D separately are less severe than they would be if P were propagated in unfactored form.

D. Dual-Rate Filter

A second implementation issue relates to the rate at which the various filter matrices are updated. The computational load of updating the measurement connection matrix, the covariance, and the Kalman gain at each measurement epoch may be a severe burden on the receiver processor. In reality, the measurement geometry changes rather slowly, and once the filter has converged, the covariance matrix tends to a steady value until there is a change in the satellites tracked. Thus, some GPS receivers employ a dual rate or background scheme for updating some of these parameters. This approach may have benefit in reducing the real-time processing load, but also has the potential to increase the software complexity.

E. Correlated Measurement Noise

One of the major difficulties for GPS navigation filters in civilian receivers is selective availability. Recall that the Kalman filter assumes that successive measurement errors are uncorrelated to each other; however, when SA is present, there is an unknown, slowly varying error associated with each satellite, as described in Chapter 16, this volume. Even without SA, measurement errors produced by orbit errors, ionosphere, troposphere, and multipath have time-correlated statistics. This clearly violates one of the basic filter assumptions; thus, even if the measurement noise as described by the R matrix represents the typical error variance, it does not correctly model the effect.

One approach to dealing with this problem would be to add a random walk state to the filter for each satellite tracked. However, for a dynamic user, there is not sufficient information to separate the SA from the vehicle motion effectively, nor is there any assurance that random walk (or any other one-state process) will properly model the SA process in effect at the moment.

Another method for dealing with this unknown signal dynamics is to include a "consider" state for each satellite in a Schmidt filter (Ref. 1, Chap. 9). The Schmidt filter accounts for the covariance of these additional states without trying to estimate the actual parameter values. This improves the overall state covariance estimate and leads to more realistic error bounds and better performance under SA. Of course, the Schmidt filter cannot remove the effect of SA; it does, however, account for it to within the bounds of our uncertain knowledge of SA.

Several methods are described in the literature that can be applied to correlated measurement errors for which the dynamic process governing the errors is known (cf. Refs. 1, 3, and 12). For example, an error that can be modeled by a first order Markov process with known variance and time constant can be readily accomodated in the filter. The key is to construct a new measurement that differences prior measurements to remove the correlated error. This results in modifications to the Kalman filter gain and covariance update equations, which are given in the references mentioned.

VI. GPS Filtering Examples

The following examples illustrate the set up and performance of a Kalman filter under a variety of conditions. All of the results were generated based on a simulation written in MATLAB® that includes the full primary GPS satellite constellation. The first case considered is a stationary buoy with no SA. The second is a vehicle traveling at constant velocity. The last two cases illustrate the effects of unmodeled dynamics and correlated measurement errors on the filter performance.

A. Buoy Example

As a first example of the operation of a Kalman Filter for GPS navigation, we model a fixed buoy located near the surface of the Earth at 0° latitude and 0° longitude. The buoy is nominally not moving, so the stationary dynamic model is used. A relatively large dynamic uncertainty in each position component of $Q_p(i,i) = (10 \text{ cm})^2$ over a 1-s interval is allowed to account for the possibility of random buoy motion on the water. The dynamic model for the clock is the second-order system described in Sec. IV.

Thus, the state is comprised of the three WGS-84 coordinates plus clock bias and drift components $x = [x \ y \ z \ b \ f]^T$. The initial estimate of the state is

$$\hat{x} = [6{,}377{,}000.0 \text{ m} \quad 3{,}000.0 \text{ m} \quad 4{,}000.0 \text{ m} \quad 0.0 \text{ m} \quad 0.0 \text{ m/s}]^T$$

The diagonal elements of the initial covariance matrix are selected to reflect the uncertainty in the a priori state estimates. Generally these can be set very large, because after the first set of measurements is received, the uncertainty will be reduced to approximately the level of the single-point solution. One caveat is that if there are large differences between the initial uncertainties in the states, it can lead to numerical problems if the alternate form of Eq. (20) is used. For this example we use the following:

$$P_0^- = \begin{bmatrix} 25 \text{ km}^2 & & & & 0 \\ & 25 \text{ km}^2 & & & \\ & & 25 \text{ km}^2 & & \\ & & & 10^4 \text{ km}^2 & \\ 0 & & & & \left(c \cdot 10^{-6} \frac{\text{km}}{\text{s}}\right)^2 \end{bmatrix}$$

The receiver takes measurements to all GPS satellites at an elevation above 10 deg, at intervals of 1 s. Each pseudorange measurement is assumed to have a variance $R = 36 \text{ m}^2$.

Figure 5 presents the estimation errors for the x position and clock bias states over a 20 s simulation run. The graphs show the actual error in the estimate as well as the $\pm 1 - \sigma$ bounds computed from the diagonals of the filter covariance matrix. (The initial a priori covariance is off the scale.) This clearly illustrates the convergence of the filter. As each set of measurements is incorporated into the filter, the state uncertainty is decreased by an amount related to the a priori covariance, the measurement variance, and the measurement geometry. When the state is projected ahead to the next measurement epoch, the uncertainty increases because of the limitations of our knowledge of the governing dynamics.

After only about five epochs, the Kalman filter reaches the almost steady-state condition in which the amount of information gained in the measurement step is equal to the loss of information in the projection step. If the satellite geometry

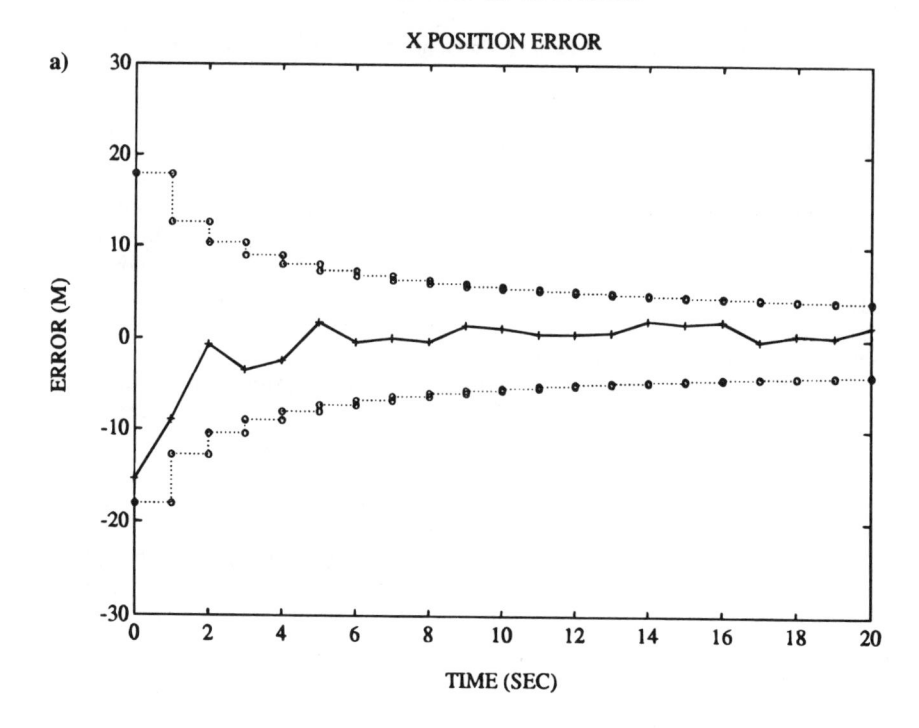

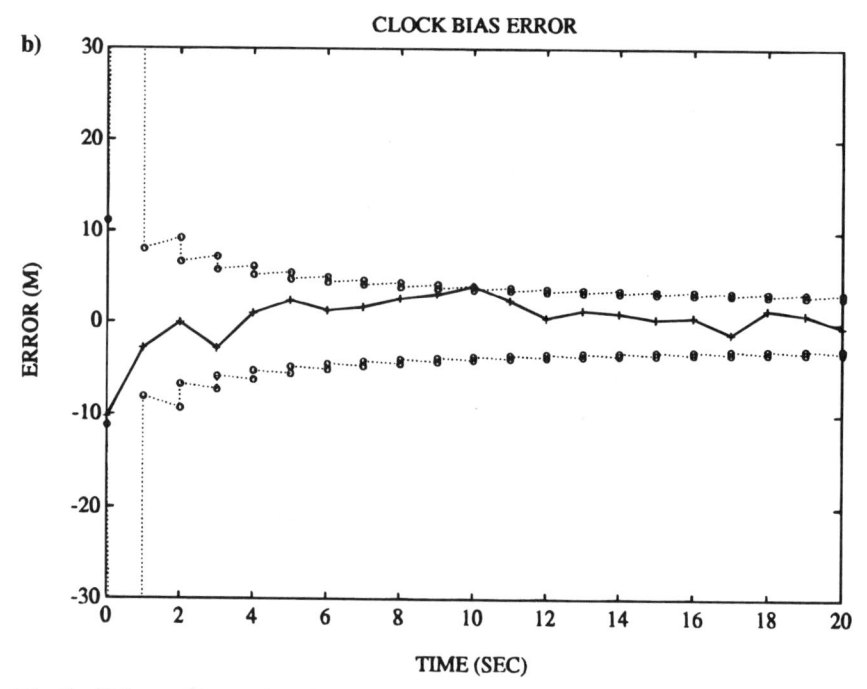

Fig. 5 Kalman filter estimation errors for buoy example. In both graphs the solid line with the "+" symbols represents the actual error. The dotted lines with the "o" symbols represent the filter computed $1 - \sigma$ bounds both before and after the measurement update; a) x position estimation error; b) clock bias estimation error.

remains fairly consistent, this condition will continue. If the geometry degrades or if measurements are missed, the uncertainty will grow at a rate determined by the process noise matrix.

For comparison with the Kalman filter results, Fig. 6 shows the errors in the z position and clock bias point solutions utilizing the same measurement data. Note that although the filter solution uncertainty was reduced to less than 5 m within 20 s, the point solution accuracy does not improve. In this static example, the point solution could be averaged to form an improved estimate that would be at least as accurate as the filter solution.

This example shows good performance of a Kalman filter for GPS navigation solutions because it employs accurate models for both the observations and the observer dynamics. In the next two examples, we look at what happens if these rules are violated.

B. Low Dynamics

The next example is for a user traveling along the Earth surface at 100 m/s in the direction due North from 0° latitude, 0° longitude. The position and velocity components of the state are implemented in WGS-84 coordinates, with the full state given by, $x_L \equiv [x\ y\ z\ |\ \dot{x}\ \dot{y}\ \dot{z}\ |\ b\ f]^T$.

The process model given in Eq. (17) is used in the filter to propagate the state and covariance between measurement epochs. The discrete process noise matrix for a 1-s measurement interval is assumed to be diagonal with uncertainty for the velocity states set to $(1\ \text{cm/s})^2$, and $(10\ \text{cm})^2$ for the position states.

The initial estimate of the state is

$$\hat{x} = [6{,}377{,}000.0\ \text{m}\quad 3{,}000.0\ \text{m}\quad 4{,}000.0\ \text{m}$$
$$\qquad 0.0\ \text{m/s}\quad 0.0\ \text{m/s}\quad 0.0\ \text{m/s}\quad 0.0\ \text{m}\quad 0.0\ \text{m/s}]^T$$

The diagonal elements of the initial covariance matrix for the position states are each set to 25 km^2; for the velocity states to 1 km^2/s^2; for the clock bias state to 10^4 km^2; and for the clock drift state to $(c \cdot 10^{-6})^2$ km^2/s^2. The receiver takes measurements to all GPS satellites above an elevation of 10 deg at intervals of 1 s. Each pseudorange measurement is assumed to have a variance $R = 36$ m^2.

Figure 7 shows the filter estimation error for each of the position and clock bias states. The results are similar to the buoy case in that the filter estimates the trajectory accurately, and the $\pm 1 - \sigma$ uncertainty bounds it computes are reasonable. (In all graphs, the solid line with the "+" symbols represents the actual error. The dotted lines with the "o" symbols represent the filter computed $1 - \sigma$ bounds both before and after each measurement update.)

C. Unmodeled Dynamics

The next step is to introduce unmodeled dynamics. Figure 8 shows the trajectory for a vehicle traveling at a speed of 100 m/s in a 5 km radius circle. Thus, the acceleration is quite mild at 2 m/s^2. The same models were used as in the previous example except that the discrete process noise values were increased to $(1\ \text{m/s})^2$ for each of the velocity states.

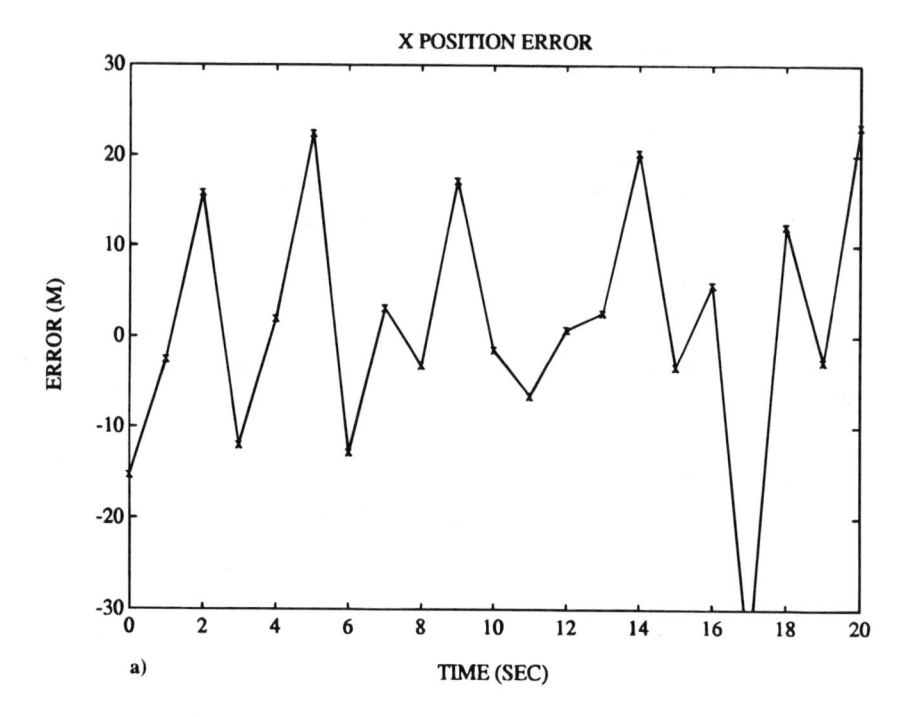

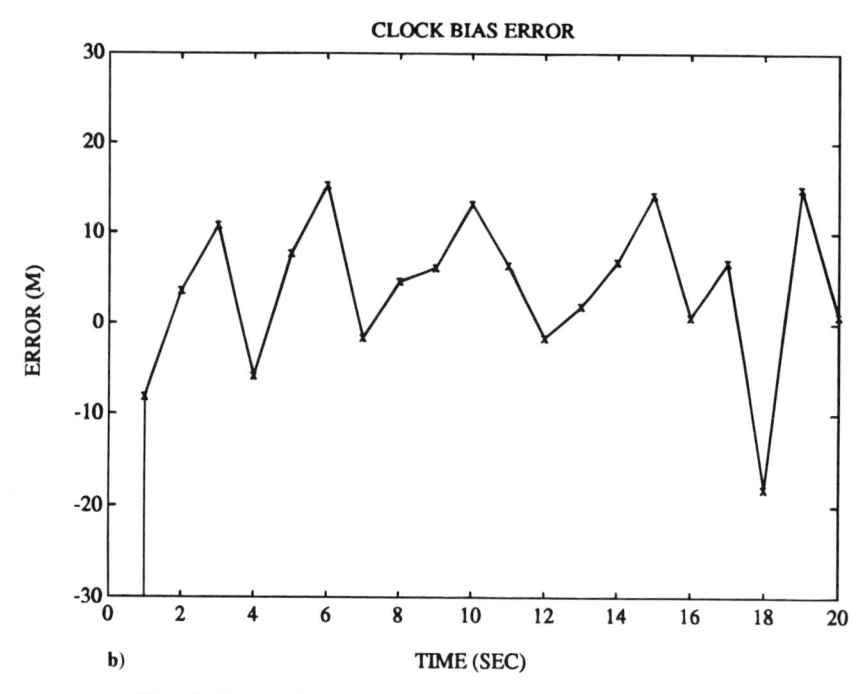

Fig. 6 Point solution estimation errors for buoy example.

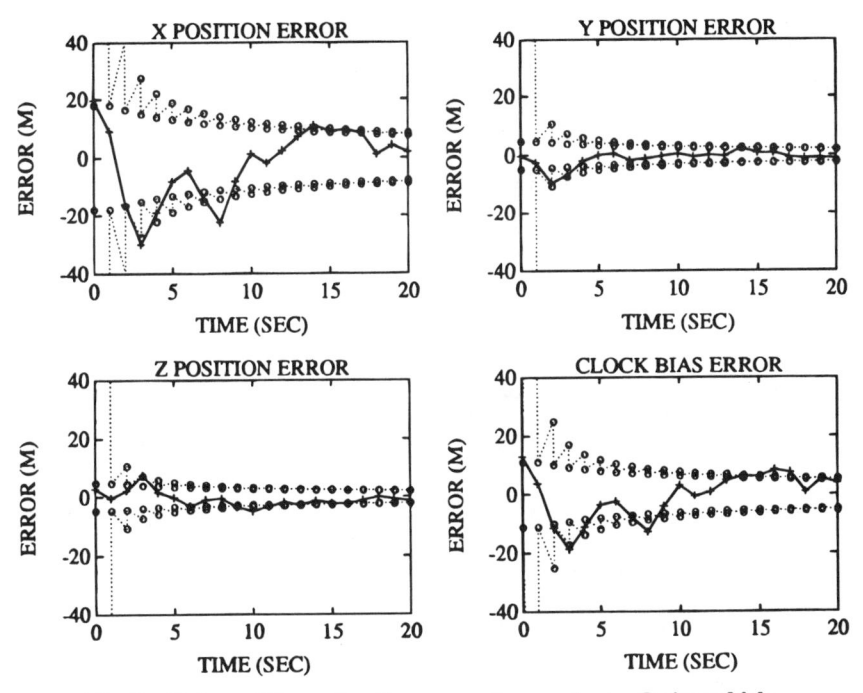

Fig. 7 Kalman filter estimation errors for constant velocity vehicle.

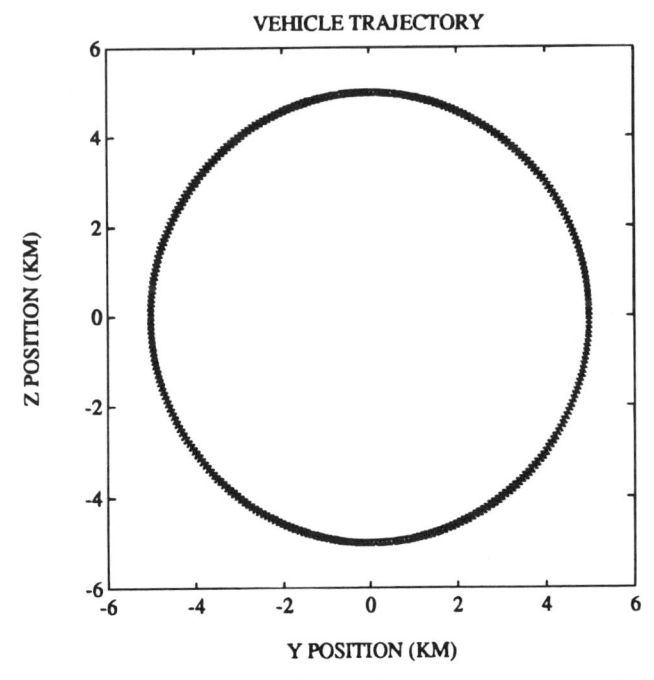

Fig. 8 True Y–Z trajectory for vehicle traveling in a circle at a speed of 100 m/s.

The filter results are shown in Fig. 9. Unlike the previous examples, the position and velocity estimation errors for the Y and Z components have a correlated nature that is not reflected in the filter computed covariance bounds. This illustrates an important limitation of the Kalman filter—it can only work as well as its models. Of course, in this particular case, it would be possible to augment the dynamic model to include acceleration states that would lead to improved performance.

D. Correlated Measurement Errors

As described in Chapter 17 of this volume SA introduces highly correlated errors into the GPS measurements. The figures in Chapter 17 illustrate a variety of observed and simulated SA error profiles. For this example, the model described in Chapter 17, Sec. II was used to generate simulated SA errors for a single satellite. An error profile was created for each GPS satellite, and the buoy filter example was re-run. The only change made to the filter parameters was to increase the measurement variance to $(35 \text{ m})^2$ corresponding to the variance of the actual errors (note that this is not a Schmidt filter). Recall, however, that the filter assumes that the measurement errors are uncorrelated between epochs, which is clearly not the case with SA. Figures 10 and 11 illustrate the filtered and point solution results obtained. Note that in both cases, the solution wanders, following the wandering SA profiles. The filtered solution is smoother, but not more accurate than the point solution.

VII. Summary

This chapter has presented mathematical models for each of the basic GPS measurement types. Pseudoranges were used to form a single point position and clock bias solution. This solution depends only on observations from a single measurement epoch. To improve the navigation accuracy, we must include knowledge of the vehicle dynamics. The extended Kalman filter is an approach commonly used in GPS receivers and data-processing packages. Dynamic models for a stationary, low dynamic, and high dynamic user were given, as well as a useful model for a typical receiver clock. Examples were shown of the performance of the filter as compared to the least squares solution under various conditions, and the consequences of mismodeled dynamics and measurements were described.

In addition to stand-alone navigation, Kalman filtering plays an important role in other GPS applications. Several of the chapters in the companion volume describe models and results obtained for diverse applications such as orbit determination, and aircraft approach and landing. The key to success with this approach is an accurate model of the dynamics and the measurement processes.

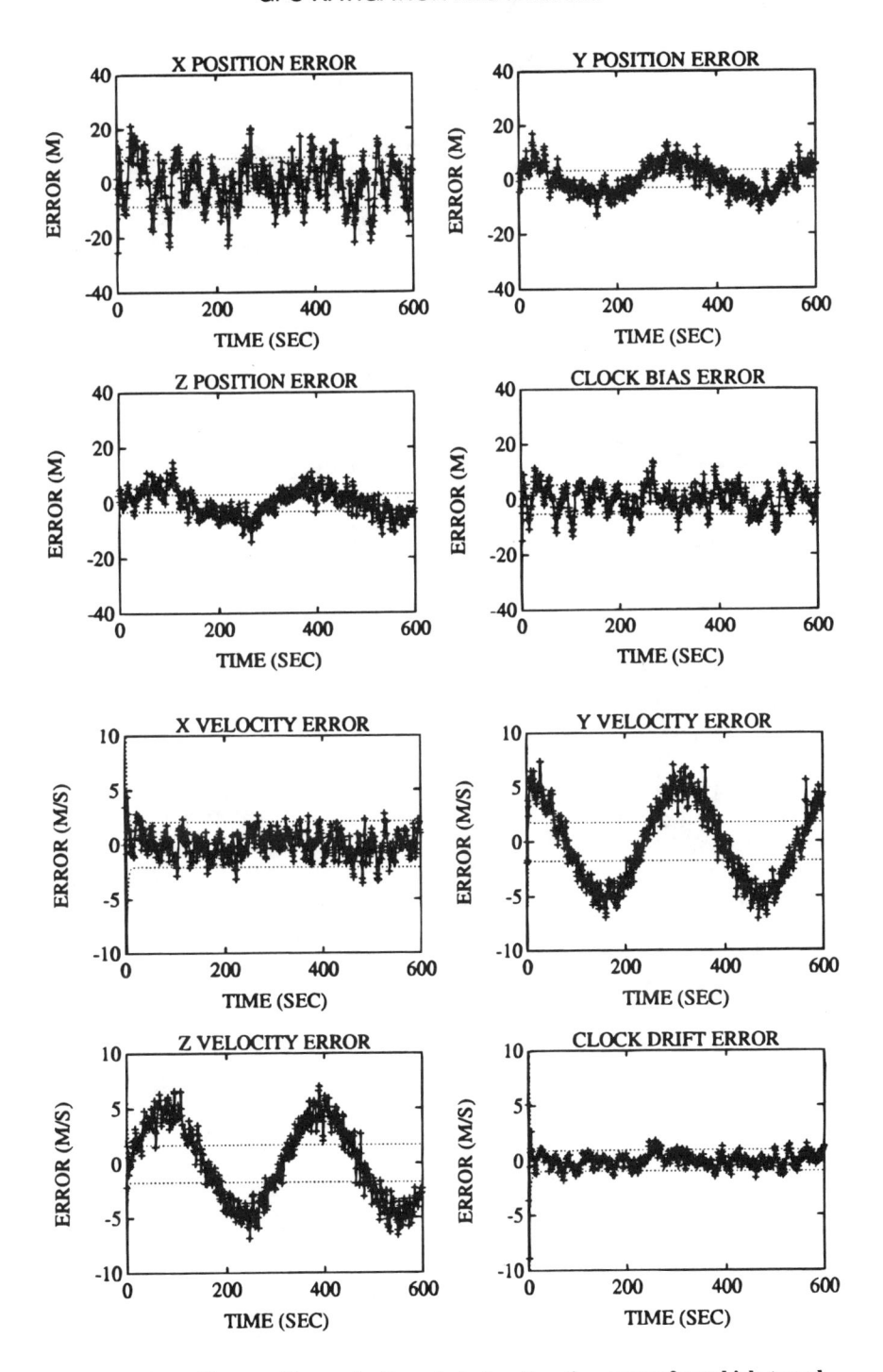

Fig. 9 Kalman filter position, velocity and clock estimation errors for vehicle traveling in a circle. The solid line with the "+" symbols represents the actual error. The dotted lines represent the filter computed $1-\sigma$ bounds after each measurement update.

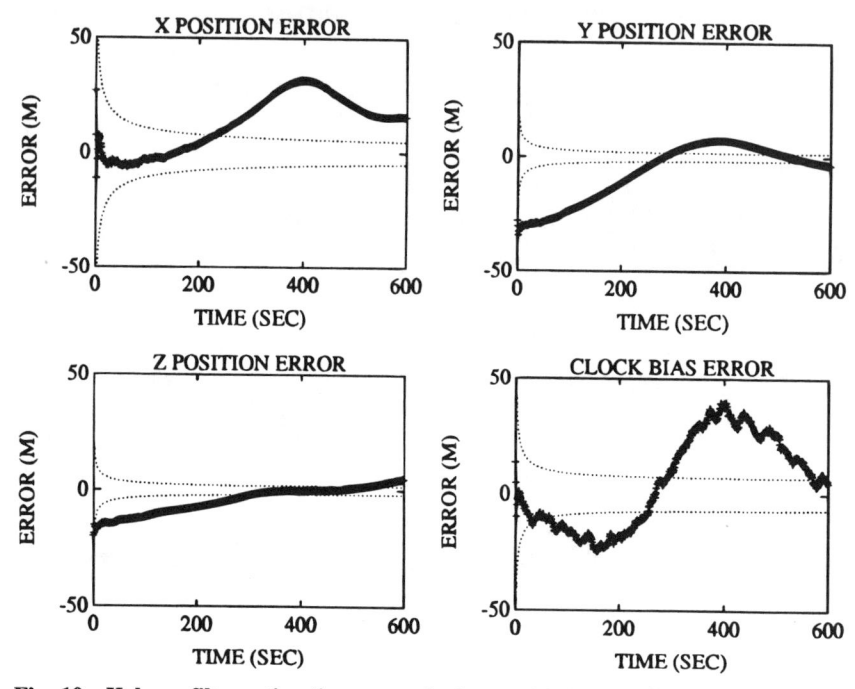

Fig. 10 Kalman filter estimation errors for buoy with SA pseudorange errors. The solid line with the "+" symbols represents the actual error. The dotted lines represent the filter computed $1 - \sigma$ bounds after each measurement update. In this case, the filter bounds are clearly not meaningful.

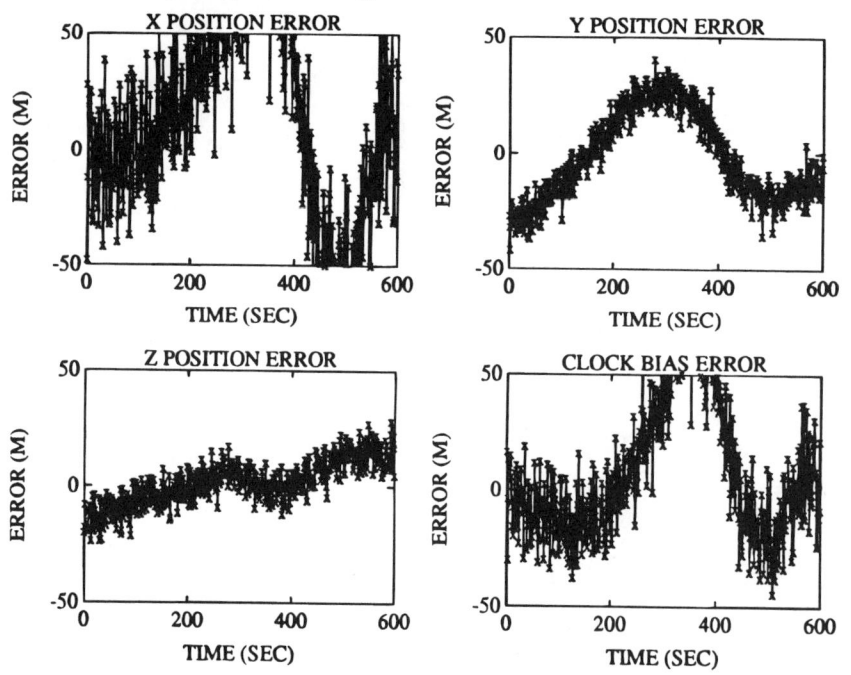

Fig. 11 Point solution estimation errors for buoy with SA pseudorange errors. The solid line with the "+" symbols represents the actual error. The dotted lines represent the filter computed $1 - \sigma$ bounds after each measurement update. In this case, the filter bounds are clearly not meaningful.

References

[1] Brown, R. G., and Hwang, P. Y. C., *Introduction to Random Signals and Applied Kalman Filtering*, 2nd ed., Wiley, New York, 1992.

[2] Gelb, A. (ed.), *Applied Optimal Estimation*, MIT Press, Cambridge, MA, 1984.

[3] Bryson, A. E., and Ho, Y. C., *Applied Optimal Control*, Hemisphere, New York, 1975.

[4] Maybeck, P. S., *Stochastic Models, Estimation and Control*, Vol. 1, Academic Press, New York, 1979.

[5] Wells, D. E., et al., *Guide to GPS Positioning*, Canadian GPS Associates, 1986.

[6] Hatch, R., "The Synergism of GPS Code and Carrier Measurements," *Proceedings of the Third International Symposium on Satellite Doppler Positioning*, (New Mexico State University), Feb. 8–12, 1982, pp. 1213–1231.

[7] Hwang, P. Y. C., and Brown, R. G., "GPS Navigation: Combining Pseudorange with Continuous Carrier Phase Using a Kalman Filter," *Proceedings of ION GPS-89*, (Colorado Springs, CO), Institute of Navigation, Washington, DC, Sept. 27–29, 1989, pp. 185–190.

[8] Anon., ICD-GPS-200, IRN-200B-PR-00J, Rev.B-PR, July 1992.

[9] Golub, G. H., and Van Loan, C. F., *Matrix Computations*, Johns Hopkins University Press, Baltimore, MD, 1989.

[10] Van Dierendonck, A. J., McGraw, J. B., and Brown, R. G., "Relationship Between Allan Variances and Kalman Filter Parameters," *Proceedings of the 16th Annual Precise Time and Time Interval (PTTI) Applications and Planning Meeting*, NASA Goddard Space Flight Center, Nov., 1984, pp. 273–293.

[11] Minkler, G., and Minkler, J., *Theory and Application of Kalman Filtering*, Magellan Book Company, 1993.

[12] Kailath, T. (ed.), *Linear Least-Squares Estimation*, Dowden, Hutchinson & Ross, 1977.

[13] Bierman, G. J., *Factorization Methods for Discrete Sequential Estimation*, Academic Press, Orlando, FL, 1977.

Chapter 10

GPS Operational Control Segment

Sherman G. Francisco*
IBM Federal Systems Company, Bethesda, Maryland 20817

THE GPS user solution is based on Space Vehicle (SV) orbit and time scale information contained in the navigation data sets generated by the operational control segment (OCS).[1] The space assets of GPS are managed and supported by this dedicated ground-based segment, which consists of L-band facilities to continually track the GPS radio frequency navigation signals, a digital computer network to process scientific and system control data, dedicated S-band facilities to conduct duplex information transfer sessions with each individual space vehicle, dedicated secure communications datalinks to couple the globally distributed ground assets into one integrated real-time global system, and a personnel system located at the principal ground facility where dedicated members of the United States Air Force operate the complex system. Functions allocated to this segment of GPS include the operational responsibility to perform the following:

1) Control and maintain the status, health, and configuration of the SV Constellation.

2) Support the user segment with precision predictions of ephemeris and time scale calibration data. Prepare and upload the formatted navigation message data sets to the SV for subsequent metered retransmission to the user community.

3) Monitor the quality of navigation and time transfer services as provided to the end users.

4) Support system interfaces to associated services (i.e., United States Naval Observatory).

5) Manage and schedule the ground assets of the control segment.

This chapter provides a summary description of the present OCS, which includes both the manned master control station (MCS) facility located at Falcon Air Force Base and the automated globally distributed radio frequency station assets deployed to control and continually monitor each satellite of the space vehicle constellation. System architecture, satellite contact coverage, measurement data correction, estimation, and prediction are addressed in this chapter.

Copyright © 1994 by the American Institute of Aeronautics and Astronautics, Inc. International Business Machines Corporation (IBM) has a royalty-free license to exercise all rights under the copyright claimed herein for IBM's purposes. All other rights are reserved by the copyright owner.
*Senior Technical Staff Member.

Figure 1 illustrates the components of this segment and the major system interfaces, including those to other GPS segments and to associated support services.

This OCS architecture provides the GPS operating agency the means for self-sufficient support of the navigation mission once operational space vehicles are stabilized on orbit, and the OCS facilities have ample resources to handle the fully populated constellation supplemented with additional on-orbit spare satellites. External operational interfaces are implemented to the Air Force Satellite Control Facility (AFSCF), to the United States Naval Observatory (USNO), and to the Defense Mapping Agency (DMA) for the respective data exchange functions of SV hand-over in their assigned orbit plane, of coordinating the universal coordinated time (UTC) absolute time scale (which is not visible to GPS instrumentation), and of providing Earth orientation data relative to the existing standard international conventions for inertially fixed coordinate systems. These external support dependencies are strategic in nature because the primary GPS services can autonomously continue service within the established system performance specification for many weeks. Less formal system interfaces also exist to import from the Jet Propulsion Laboratory (JPL) the predicted Sun–Moon position data, to exchange pertinent information with associated user data services, and to support the initial prelaunch activities at Cape Canaveral.

Both the operational procedures and the equipment configuration have been designed to ensure continuous GPS system services. Although the GPS user community is buffered by the distributed store-and-forward data concept implemented in GPS satellite memory, the OCS is designed so that single failures on the ground can be accommodated without disrupting the time-critical functions allocated to this segment. Redundancy of all critical equipment is provided, and

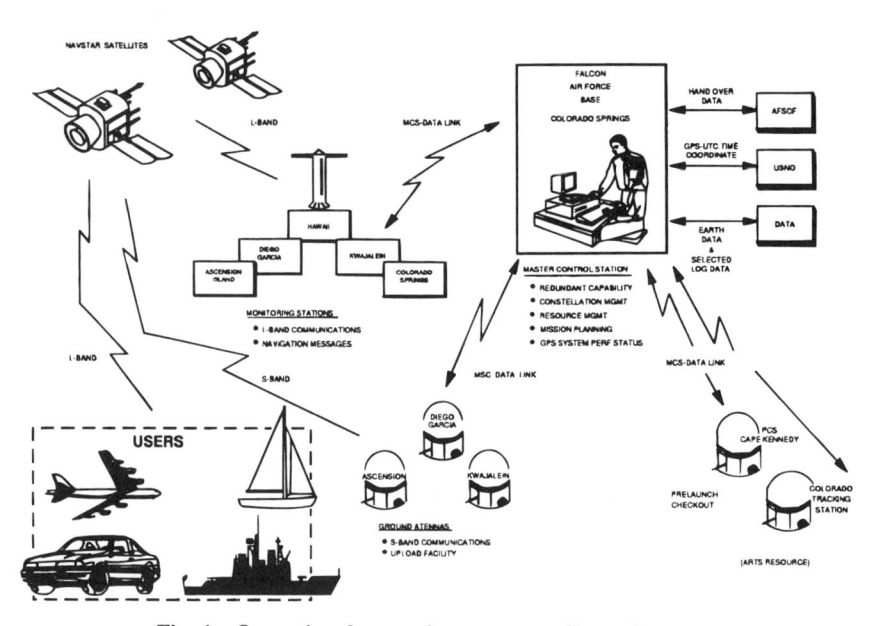

Fig. 1　Operational control segment configuration.

nonreal-time support tasks such as the numerically intensive integration of future reference orbits are identified and judiciously scheduled at OCS convenience on the standby system assets so as to make efficient use of the overall facility investment while meeting the critical data preparation needs. This conservative design philosophy extends to providing database redundancy and to the protection of vital information from contamination. High availability architectural features are cooperatively incorporated in both hardware and software to tolerate single equipment failures, to facilitate equipment reconfiguration and maintenance actions, and to perform (if necessary) process initialization/restart functions without residual corruption of either the database values or of the critical estimation process. Stringent consistency checks are applied to the measurement data, to the estimation process, and to the OCS products generated for internal and for export use. The OCS navigation service product integrity is thus maintained in the presence of normally encountered implementation imperfections.

The robust, underlying GPS service integrity concept of allocating tactical autonomy to each space vehicle and of implementing end-to-end error detection and fail-safe interlocks in the OCS protocols associated with data movement make possible fault-tolerant operational procedures that ensure dataset consistency at the satellite. The occasional OCS process adjustments and equipment reconfigurations encountered during continuous operations are formulated so as to be transparent to users who properly adhere to the total content of the navigation message and the application constraints as documented in the ICD-GPS-200[2] documentation.

Thus, the OCS provides high-integrity navigation data sets at the SV memory. Data errors occurring on the SV–user link are correctable using codes that originated at the ground OCS MCS. Although improbable, an unpredictable flaw in the timing of the GPS signal structure as generated by the SV has occasionally occurred on Block I satellites. The OCS detects and records such events, but has no practical means to provide a service alarm in a timely (10 s) manner. The defensive concept of receiver autonomous integrity monitor (RAIM) has evolved within the community for the user to identify any such SV signal inconsistencies based on the inherent information redundancy of the fully populated constellation. For critical applications such as precision aviation approach navigation, real-time communications can augment the basic GPS service by disseminating both the current status of service and differential corrections to the predictions for each satellite. Such feed-forward error correction overcomes the practical limits of precision time scale and orbit prediction in the baseline GPS concept, but at a cost of service vulnerability.

As space technology continues to advance, some OCS functions of the base system level concept may be reallocated to future space vehicles so as to reduce tactical dependencies on ground assets. Estimation of navigation data (such as SV clock correction term estimation) could be improved by the availability of satellite-to-satellite cross-link measurement capabilities, which makes the on-board refinement of navigation data possible.[3] The precision ranging service from these unaided SVs would also be improved[4] with the potential of a continuously available user range error (URE) of 2 m enabled by the reduced delay in providing precise relative clock corrections to the user sets.

The Consolidated Space Operation Center (CSOC) is located at Falcon Air Force Base (Fig. 2) near Colorado Springs, Colorado. Here the MCS for GPS

Fig. 2 Consolidated space operations center.

resides and operates 24 hours a day with a dedicated United States Air Force staff. They are responsible for the SV constellation and operate the ground support processing that generates the fresh navigation data uploads.

This facility is the overall operations center for GPS with operational responsibility for all OCS functions including navigation information processing, satellite data upload, vehicle command and control, and overall system management. Full responsibility of each on-orbit SV commences once the normal Earth-oriented attitude stabilization is established which enables L-band tracking, an operational capability unique to the GPS control segment (CS). Monitoring the performance of the numerous GPS services and system components by OCS is an essential element of system availability and service quality assurance.

Visibility of the GPS signal structure is provided to the OCS by the five monitor stations (MS). These are remote, unmanned* GPS assets used to passively track the continuous navigation ranging signals of the entire satellite constellation and to acquire the digital data transmitted on L-band by each SV. Both the upload message generation and the GPS service quality monitoring are based on analysis of these tracking data.

Another type of unmanned OCS radio frequency asset is the ground antenna (GA), a full-duplex S-band communications facility that has dedicated command and control sessions with a single SV at a time. Interactive burst-communication sessions are periodically established for navigation data upload, satellite command, and telemetry reception that are essential to constellation operations and navigation service support. S-band ranging services are not available. Both types

---

*On-call maintenance provided by the host site.

of these distributed GPS radio frequency facilities include the functionality to tolerate automatically short disruption of communications with no induced flaw in service and to support remote reconfiguration with diagnostics conducted from the MCS. Location of these radio frequency facilities has been a programmatic compromise between radio frequency contact utility to the GPS mission and the availability of adequate secure physical facilities with on-call technical support.

I. Monitor Stations

Navigation signal visibility is provided to the MCS by the globally distributed unmanned monitor stations. The GPS monitor stations are radio frequency-passive facilities that receive the same signal structure as the user community. Some low-rate telemetry unique to the instantaneous SV state is of operational interest to the OCS. These stations track the apparent pseudoranges and carrier phase between SV–MS pairs and also collect the transmitted navigation messages for two operationally routine purposes:

1) The pseudorange and carrier phase histories are required to derive precisely and provide the GPS user with ephemerous and clock calibration data for each SV.

2) Both the pseudorange histories and the broadcast navigation data are required to monitor the services as provided to the user community.

Note that the technical performance required of OCS monitor stations must exceed that of the conventional high-precision user set for two fundamental reasons:

1) The OCS task of prediction is more demanding of data quality than is the deterministic end user's current position solution. Both the state separation process encountered in estimating current nonorthogonal system states and the time projection process encountered in predicting future SV states are extremely sensitive to noise, bias, and systematic errors in the measurements. Monitor station measurement errors are magnified in their propagation to become user input error components.

2) The MCS must maintain clear, detailed signal visibility and cope with abnormal signal structures both to permit initial SV process alignments and to support overall system diagnostics. The user has no need to deal with out-of-specification signals or malfunctioning system elements.

The GPS is committed to the operational requirement of global continuous navigation service, and monitoring the system service implies that each satellite's signal should be continuously monitored for pseudorange accuracy compliance, message content, and signal health. The ground projected latitude of block II and later satellites will never exceed the orbit inclination angle of 55 deg; therefore, near-equatorial sites that can track the complete 110-deg groundtrack latitude band are the most desirable MS locations for contact efficiency. Although observations from a spread of latitudes is beneficial to geopotential and force anomaly modeling, covariance error analysis shows that they are not critical to observe the GPS states. Scientific modeling efforts external to the operational program are adequate support to achieve the system performance requirements.

Colorado Springs, Ascension Island, Diego Garcia, Kwajalein, and Hawaii have been chosen as the OCS monitor station sites that best meet the overall OCS requirements, and the precise coordinates of the antenna phase center were accurately determined in the WGS84 coordinate system through special surveys and custom off-line tracking data reductions. Figure 3 illustrates the resulting track coverages provided by this selection of monitor station sites for SV lines of sight that are above the horizontal by a practical 5-deg elevation angle at the receiving antenna. This selection of sites achieves a 95.87 % average coverage for the baseline constellation with only five monitor stations. Contact statistics with each of the operational ground tracks is different, varying from 93.63 to 100% of the ground track. Because of the latitude of Colorado Springs, a slight gap in signal monitoring (not navigation service) does occur for satellites when ground tracks are over the open ocean west of southern South America, which is apparent in Fig. 3. A limited L-band capability is also implemented in the eastern launch site at Cape Canaveral for the OCS to verify segment compatibility prior to satellite launch, but use of this facility does not close the slight operational coverage gap.

Regions of tracking coverage overlap (simultaneous L-band contact with the same SV by two monitor stations) are very important in establishing a robust GPS estimation process. Observed residuals in pseudorange measurements must be allocated to probable errors in time and in SV position by the action of the Kalman filter estimator. Solution for the states of an isolated satellite and of the monitor stations is quite fragile because of the extensive linear relationships that prevail in pseudorange-based measurement systems and to the effects of accumulated model uncertainty (process noise) when marginal measurement geometry exists to distinguish the error source. Common view strengthens any solution by enabling direct time transfer between MS sites. Fortunately, the extreme situation of sporadic track contact only occurs during system development (and possibly during recovery from an improbable loss of the OCS database). Although it was deemed impractical at the time of initial OCS development to

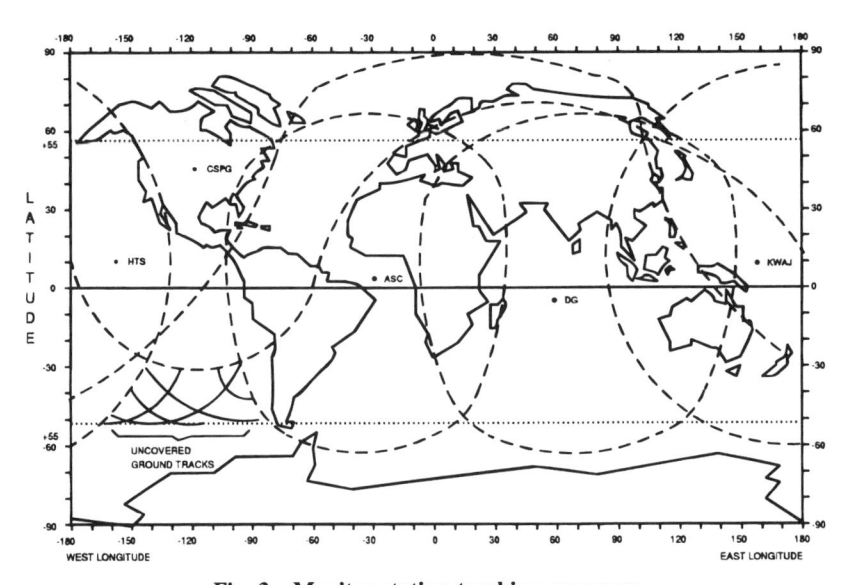

Fig. 3 **Monitor station tracking coverage.**

implement the total GPS state solution in one large filter formulation so as to fully exploit the available measurement geometry (space and time), near optimum performance has been practically achieved at a much lower computer burden by partitioning the constellation and adhering to the following important system design choices that have been found very beneficial in stabilizing the estimation process and achieve a robust solution within estimation filter partitions, or subsets of states.

1) Select the SVs for inclusion in each partition so as to maintain (near) continuous time transfer between monitor stations within each partition through common-view SV observations. This strategy is made possible by the existence of significant regions of tracking coverage overlap provided by the MS site selection, thus greatly strengthening the GPS system solution by effectively decoupling the MS time and the SV state components of pseudorange errors.

2) Adopt the same SV-signal instance for sampling the common view measurements of a common-view sample set to establish cleanly one satellite state for each common view sample set. This strategy, which effectively decouples satellite velocity states, is achieved in the OCS monitor stations by basing the pseudorange measurement instant on the waveform epoch in the received signal structure. Utilization of this sample time convention reduces the terms in the measurement model, and this choice is essential if corrections for selective availability are to be made with precision.

3) Eliminate the interchannel measurement biases that disrupt the clean separation of system ground states. This reduces the true number of time states and is achieved by the OCS monitor station architecture practice of using, whenever possible, the same physical hardware for common MS functions (i.e., downconverter, station time reference).

Thus, the objective of achieving the full GPS performance potential strongly drove both the selection of MS sites and the detailed receiver equipment architecture. The above design guidance, which greatly influences MS implementation, was formulated through extensive error analysis and simulation efforts at the GPS system level, and these advanced principals have been verified through a decade of stable estimator performance. Even in a partitioned estimator solution, exploiting the common-view information potential is possible through advantageous selection of the MS sites, prudent selection of the SV members within a partition, and the adoption of common-view measurement time tagging (adopting the specific source SV code epoch for sampling). This practice of simultaneous measurement instances for a SV significantly stabilized the partitioned estimator formulation and reduced estimation process sensitivity to systematic error stresses. Errors in position and time are more decoupled, which significantly reduces the degree of error source aliasing. Estimation of the individual satellite trajectories is then numerically solid because the large off-diagonal covariances characteristic of weak geometry are avoided. The relatively large clock process noise levels are less coupled into the position solution, and once the MS time-transfer network closes through common-view measurements, robust estimator performance is established. This strategy has had the effect of immediately stabilizing the relative monitor station time scales for each partition.

Obviously, the precision of knowledge estimated by the CS at the beginning of the prediction period must be an order of magnitude better in quality than the end-ranging service specification after a possibly long prediction interval. Prediction magnifies any errors existing at the beginning of the interval, and so errors in MS measurements affect the user solution more than do the errors in user set measurements. Similarly, the measurements to substantiate system performance metrics must be substantially better than the specified ranging service requirement to be a reliable performance indicator.

Two distinct categories of tracking performance requirements exist. The precision range data, which are used to estimate clock and SV trajectory states, must be free of significant error. Systematic errors are especially damaging because they are not countered by averaging the abundant measurement data, and their presence can cause significant spatial biases in the resulting navigation service product. The OCS allocated processes of separating states and of predicting ahead in time greatly amplifies the size of error to the user. Therefore, systematic errors, such as the effects of multipath interference, must be carefully controlled. The ranging measurement data used by the estimation filters is restricted by the OCS operational procedures to track elevation angles of 15 deg or greater above the horizon so as to avoid significant multipath effects and to eliminate the relatively large, uncorrectable measurement errors that could result from excessively long propagation paths in the troposphere. Exceptional multipath rejection ratio characteristics and phase center stability are still required above this mask angle, but this can be achieved with the simplicity of a single-beam antenna that precludes the problem of antenna channel bias. In contrast, measurement data to support navigation service monitoring and subsequent user process emulation for service verification need only be consistent in quality with that of baseline user practice, which is a relaxation from the accuracy required for the OCS estimation data acquisition. Tracking performance monitoring must, however, extend down to 5-deg elevation angle to achieve coverage with a practical number of MS sites while providing range data better than the baseline user requirement and with acceptable bit error rates.

These OCS L-band tracking requirements imposed a formidable antenna design problem. Master station design practice is based on the philosophy that the raw data acquisition performance of an operational system should be consistent, when practical, with the achievable state of the art so as to assure clear visibility of data anomalies for fault analysis. Design studies and then chamber testing of the actual antennas indicate that a phase stability of plus or minus two centimeters is achievable on both L_1 and L_2 without spatial calibration. This uncertainty is consistent with the assessed level of error in the measurement model components.

Thus, a single, unsteered beam fulfilling both the navigation service monitoring and the precision range tracking requirements on both L_1 and L_2 frequencies is implemented at each MS site, and multiple radio frequency channels with resulting equipment duplication and the introduction of interchannel bias problems were avoided. Vega Precision Laboratories provided the high-performance design and developed the production MS antennas (Fig. 4). A shaded, bent turnstile is used to receive dual-frequency, right-hand circularly polarized signal transmission. The conical ground plane with annular chokes at the base produces the specified performance with 14-dB multipath rejection ratio for signal paths above 15-deg elevation angle. The design is optimized for L_2 to compensate for the difference in SV signal transmission strengths. The antenna design contains the low noise

Fig. 4 Monitor station antenna.

amplifier in the pedestal base, as is shown at the Ascension monitor station installation. (An external heated radome with heating is installed at the CSOC location to handle the Colorado winter climate.)

In operational practice, the CS experiences a measurement update innovation level of 2–4 cm for (carrier-aided) 15-min smoothed measurements during normal steady-state operation for each MS. Because this is achieved with estimation filter tuning optimized for the multiday prediction interval performance that demands the use of low but realistic process noise values, the data acquisition and measurement strategy and implementations are verified in practice.

The L-band receivers for the CS must have other additional functionality not found in conventional user sets. They must work with signals not in compliance with the operational specification. Initial acquisition of a new satellite cannot be dependent upon the complex signal alignment promised the user, and no correct prompting is assured. The MS receiver must accommodate and detect abnormal signal structures that naturally occur prior to completion of the navigation payload alignment or when a related equipment fault is experienced. Good visibility of the signal structure is retained by the specialized OCS MS receiver design, and the capability for a technical specialist at the MCS to assist in the initial alignment or SV diagnostic process is provided.

For normal tracking, the OCS simultaneously tracks multiple SVs on both L-band frequencies to permit correction of the ionospheric delay. Contact scheduling for each MS is based on the requirements for each MS to track all visible satellites above the estimator data criterion of 15-deg elevation angle, and to track continually each SV of the constellation by some MS, if anywhere visible to the OCS. This strategy maximizes the common-view contacts and provides the long track histories for carrier-aided smoothing to support the estimation

Fig. 5 Monitor station equipment.

process. It also maintains maximum signal monitor contact with each SV. Adequate channel assets are provided in each monitor station to meet this schedule criteria, to provide an active spare channel, and to include a calibration channel programmed to cycle through all the channels. This innovation permits the MCS to sense and correct for any residual interchannel biases in the receiver correlator instrumentation. All modes of track acquisition can be accomplished so that the operator can verify L-band support of the numerous user strategies and that reliable SV contact in unusual adverse conditions is possible.

The specialized monitor station receiver equipment was developed by Stanford Telecommunications, Inc. and provide accurate multisatellite range data with minimum interchannel biases. The circa 1978 technology, as shown in Fig. 5, provides data of unprecedented quality to support the precision estimation of clock and ephemeris parameters, and measurement performance is comparable to the best of today's receivers. The internal clock circuits and the two down converters are common to all channels, so that any common bias can be absorbed within the estimated monitor station clock states. Critical cables are trimmed to consistent lengths in order to match delay and drift affects. Pseudorange and accumulated delta range measurements are obtained each 1.5 s on the exact received X-1 epoch for both the L_1 and the L_2 signals.[5] This selection meeting the objective of SV-referenced sampling is convenient to the designer, and the high sample rate captures the information content of the noisy pseudorange measurements. The resulting signal visibility not dependent upon SV alignment is essential to support SV diagnostics and the MS–MCS communications is adequate to return all the data to the MCS. Thermal noise for P-code tracking under the worse specification signal conditions is approximately 2.5 m, one sigma, for pseudorange and less than 0.7 cm, one sigma, for accumulated delta range (ADR). These measurements, the received navigation message, and signal-to-noise power data observations are formatted by the receiver and then transferred to the local digital processor for message handling at this autonomous facility.

A test and calibration signal generator, which provides remote station diagnostic capability, is also controlled by this processor. These test signals can be configured and routed to the LNA, down-converter, or correlator by an MCS ground controller. This capability permits precise station-ready verification and is used to support local maintenance whenever a service call is made.

Frequency Electronic, Inc. (FEI) developed the monitor station frequency standard rack, which establishes the precision reference for pseudorange measurements. Although ultrastable atomic time scales at each MS are not critical to GPS navigation mission once a major portion of the constellation is up, this equipment was important to the development program. Two cesium atomic frequency standards are installed at each unmanned monitor station to permit functional substitution of the hot spare from the MCS. These Hewlett Packard 50-61A Options 4 Standards were modified to provide remote indication of clock parameters considered characteristic of atomic standard health. Phase comparison is instrumented in the FEI design so that the MCS can maintain a current frequency calibration of the backup unit and transfer the timescale to the redundant unit with no further time scale calibration. Remote fault sensing with the option of automatic switchover is provided. The battery-powered backup units have been designed for high availability in unmanned applications. They are fully integrated with the overall MS asset monitoring control structure.

Advanced time subsystems are being introduced to give better support to the time transfer community. This extension is intended to improve the coupling of GPS atomic time and the official USNO reference. Support of local time users can be integrated into the local monitor station reference, which is more current than that available from the SVs.

Remote indicating sensors of barometric pressure, temperature and dew point are included in each MS to permit approximate correction for tropospheric delay. The dew-point sensor at unmanned sites has been undependable, and often a default value is substituted to accommodate sensor malfunction.

Tracking orders and equipment configuration commands are received from the MCS over dedicated, secure communication channels. Measurements and status data are forwarded to the MCS over the same duplex channel, which utilizes commercially developed SDLC protocol to provide error detection and block data retransmission features transparent to the application software. Buffer time of 5 min is provided to accommodate communication equipment reconfiguring or other short service outages without loss of data.

II. Master Control Station

The master control station consists of the processing complex and controller facilities necessary to manage completely the operational GPS space assets and to produce the navigation message. Figure 6 provides an overview of the major functions and the flow of information types throughout OCS. Communications and remote processing can be tested remotely and initialized from the MCS to recover from a fault or extended outage of any remote unmanned facilities. The navigation mission requires upload availability of 98% (not to be confused with

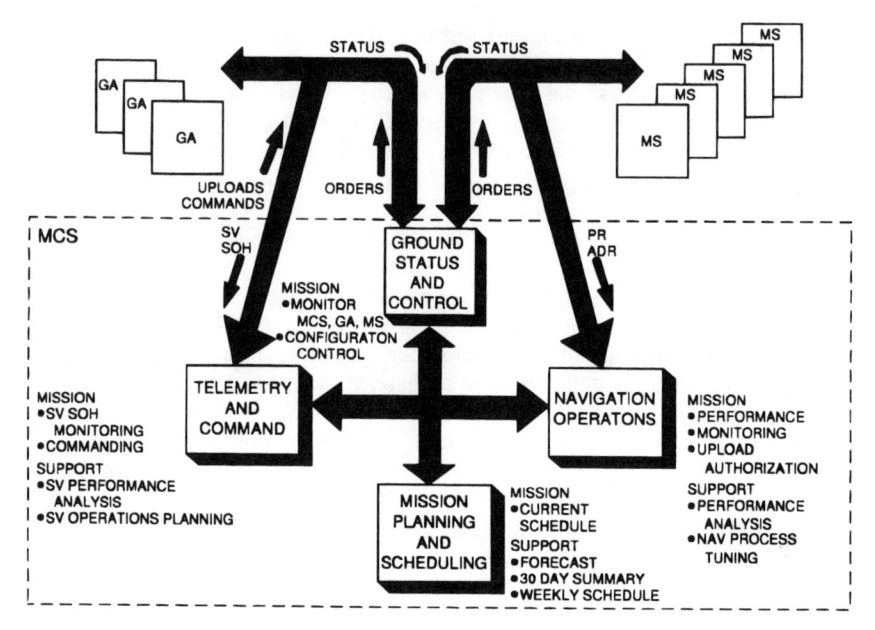

Fig. 6 Operations overview.

navigation service availability), so redundancy is provided for all mission critical equipment. Dual processors, communications controllers, and peripherals are configured to permit processing of the on-line navigation processing and satellite control functions with either processor unit. Many other GPS functions required to maintain the operation are deferrable for short periods of time when necessary to accommodate peak computation loads, fault recovery, and maintenance. Real-time communications are critical, and the OCS has implemented remote initialization and synchronization control from the MCS as well as the diagnostic capabilities to control remotely both digital and analog line looping features. Data Products New England implemented line and service management features of the MCS communications subsystem, which facilitate maintaining communications services. Loop back of the equipment and the control of the communications net is facilitated by the flexibility of this communications interface. All communications with the MCS are encrypted for integrity to prevent spoofing of the system.

Personnel at the MCS control all navigation, processor, constellation, and OCS assets and are responsible for the overall GPS integrity. This requires the establishment of the procedures and efficient access to critical mission data. The MCS provides many identical work consoles initially developed by the Sanders Corporation; each having a keyboard, two-color displays, and a hard copy unit. Air Force controllers are on duty during a full shift. Each shift is under the direction of a senior duty officer, who has overall shift responsibility. This officer is aided by a ground controller, SV engineer, navigation specialist, and multiple pass controllers. The ground controller is responsible for the ground equipment and communications, while the SV engineer is responsible for the overall satellite status and the resolution of problems concerning the SV. The navigation specialist is responsible for the overall navigation mission performance and manages the generation of upload messages. Prime responsibility for the satellite upload transaction, TT&C, and other satellite contacts rests with the pass controllers. Each operator can sign on at any of the consoles; giving his identity, function, and required passwords. The software then configures the display and interface options for that console to meet the specific requirements of the mission function authorized during the sign-on.

The MCS navigation process must generate predicted clock time scale and ephemeris on which to base the data in the navigation message, which is then stored and forwarded to the user by the satellite. Figure 7 illustrates the complex system information flow that is the basis of the GPS concept and integrates the segments to provide the GPS navigation service. As an illustration of the intertwined dependencies within GPS, consider that the pseudorandom radio frequency signal originating at the SV is tracked by the C Segment from which the navigation message is generated and returned to the SV for dissemination to the user, and also back to the CS for verification of the total system performance. The scientific navigation processing that generates the data for the formatted navigation message from measurements acquired by the MS is detailed in a subsequent section of this chapter.

III. Ground Antenna

Navigation message uploads, SV commands, and telemetry data are communicated via S-band sessions between a specific SV and one ground antenna (GA) as scheduled by a pass controller. Such information must be communicated to the SV constellation by the network of globally distributed, unmanned ground antenna facilities. Figure 8 illustrates the contact opportunities provided by the chosen GA sites of Ascension Island, Diego Garcia, and Kwajalein. The regions

	INPUT	FUNCTION	PRODUCT
SPACE	NAVIGATION MESSAGE COMMANDS	PROVIDE ATOMIC TIME SCALE GENERATE PSEUDORANGE SIGNALS STORE AND FORWARD NAVIGATION MESSAGES	PSEUDORANDOM RF SIGNAL NAVIGATION MESSAGE TELEMETRY
CONTROL	PSEUDORANDOM RF SIGNAL TELEMETRY UTC	CALIBRATE TIME SCALE, PREDICT EPHEMERIS MANAGE SPACE ASSETS	NAVIGATION MESSAGE COMMANDS
USER	PSEUDORANDOM RF SIGNAL NAVIGATION MESSAGE	SOLVE NAVIGATION EQUATIONS	POSITION VELOCITY TIME

Fig. 7 System information flow.

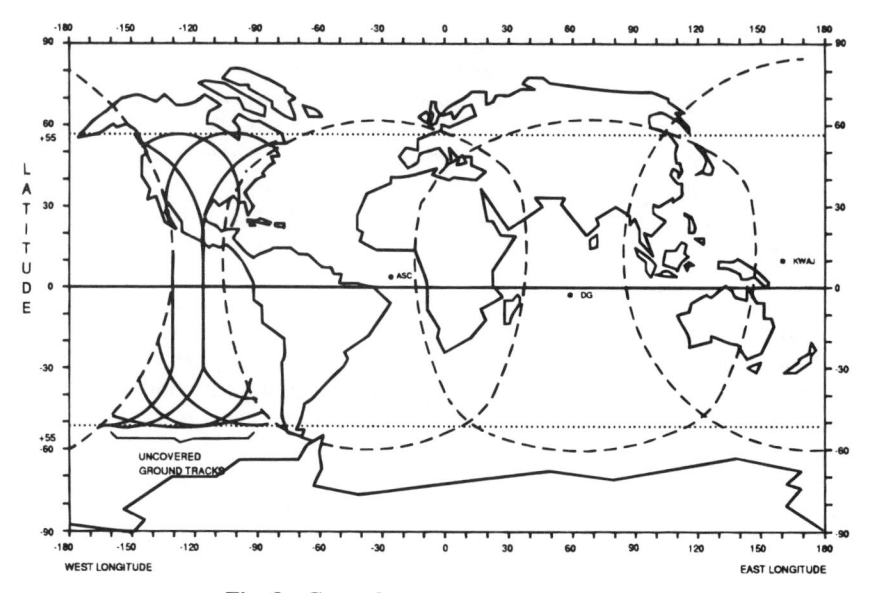

Fig. 8 Ground antenna contact window.

of coverage provide scheduling flexibility to best meet the overall MCS contact requirements. A fourth GA facility owned by GPS is located at the Eastern Launch Site, but is not depicted for coverage because radio frequency transmissions are severely restricted at launch sites, and the prime intent of this equipment is to support segment compatibility verification during prelaunch operations. In the event of a system emergency involving a SV in this contact gap, interoperability with an automated remote tracking station (ARTS) at the Falcon AFB is possible on an emergency basis to provide critical S-band services.

The dedicated GA installations developed by the Harris Corporation consist of a Scientific Atlanta 10-m S-band antenna and extensive Harris electronic equipment assembled in two rack groups according to function. Figure 9 shows the installation of the rack group containing the radio frequency exciter, high-power transmitter, receiver, and the servodrive equipment. Site selection was influenced by the location of existing remote tracking stations and with an objective of establishing the best overall space-tracking capability. Extensive remote readiness testing, fault alarms, and diagnostic features are incorporated to ensure equipment integrity and to support the on-call site maintenance. On-line equipment safety provisions respond to local sensor inputs to protect the GPS assets from thermal or power anomalies, even if communications with the MCS are disrupted. Command, telemetry, and navigation upload traffic is handled by these GA installations. Tracking orders, equipment configuration commands, and data are received over secured, dedicated duplex communications channels. The local GA processors store this transaction information on disks prior to SV contact, which minimizes operational sensitivity to communications anomalies. The command and upload protocol can be automatically maintained by the GA equipment in accordance with a programmable contact protocol decision tree

Fig. 9   Ground antenna electronics.

once the S-band session is initialized, and the conclusion is that a safe system state is ensured. Log files are maintained that unambiguously document the actual transaction so that the responses and final state can be reconstructed. Telemetry data received from the SV are buffered on local disks to accommodate any limitation of MCS communications services and safeguard the historic data. Dual strings of electronics equipment are installed at each of the fully operational sites to achieve the required service availability. Servodrive redundancy has been provided by the Scientific Atlanta pedestal design to improve the fault tolerance with single drive (at reduced slew rates) of the mechanical antenna pedestal drive.

The antenna shown in Fig. 10 is at Cape Canaveral before the radome was installed. All GPS S-band ground antennas have an installed radome to protect the equipment from the elements. All uploads and commands are sent to the satellite through these S-band facilities, utilizing a full duplex S-band protocol with error detection to correct radio frequency link errors and to ensure the delivery of consistent datasets. The timing of uploads to all SVs through these limited GA facilities is determined by the mission scheduler based on individual satellite clock performance, GA equipment status, contact visibility, and the cyclic quality of the navigation message predictions. Special transactions to manage occasional events, such as the eclipse season, attitude control support, station keeping, and vehicle reconfiguration, are merged with the routine navigation payload transaction traffic.

IV.   Navigation Data Processing

The principal navigation product of the CS is formatted information that enables the user to evaluate the SV positions precisely and the time corrections that apply

Fig. 10 GPS ground antenna.

to each SV's atomic time scale. Thus, the OCS must estimate time corrections and positions for each SV in the GPS constellation and project these states forward in time for subsequent application by the user. This is accomplished by scientific analysis of the pseudorange measurements observed by the MS. The CS is required by the system level error budget to generate and maintain the quality of this predicted digital data product so that the user has the following corrected signal service at his or her receiver antenna: 1) pseudorange service has an rms error component no worse than 6 m; and 2) UTC transfer service has an rms error component no worse than 97 ns.

The navigation information from the CS is passed to the user as parameters of GPS standardized mathematical models[6] specified in ICD-GPS-200.[2] This information is packaged by the OCS as datasets valid for a specific time interval and uploaded to the SV memory to be metered out to the user as they become valid. Numerical evaluation of the model equations by the user yields the required information as a function of time to utilize pseudorange measurements. Although this final OCS product is indexed by SV and dataset time window, the solution is derived for a system of many SVs, and the predicted functions are continuous for weeks after upload.

Figure 11 depicts the scientific processing that produces the precise navigation message content derived from the pseudorange tracking information acquired by the MS. The process architecture contains four distinct categories of function encountered in transforming the raw signal tracking data into the formatted datasets containing predicted information suitable for upload to the SV. These categories are measurement data preprocessing, support function preparation, system state estimation, and predicted navigation message parameter generation.

OPERATIONAL CONTROL SEGMENT

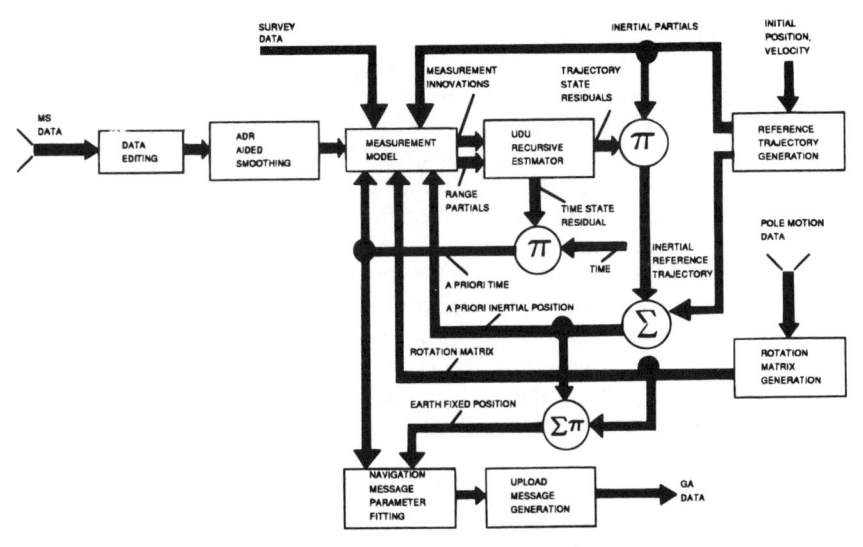

Fig. 11 Navigation data processing.

The measurement data preprocessing consists of correction of receiver measurement anomalies, data editing, correction of the ionospheric delay, smoothing of the high sample rate data, sample realignment from the SV time scale to the GPS time scale, and correction of the tropospheric delay. The raw monitor station track data of pseudorange and carrier phase from each specific SV frequency channel have both been sampled on the received 1.5-s $X1$ epoch to preserve coherence of data across code and carrier. These time history sequences with measurement time tagging referenced to SV—time are examined and correlated with receiver tracking fault indicators to verify sample measurement validity and data continuity. The pseudorange data are corrected for interchannel biases as estimated by analysis of the roving calibration, channel-scanning data, and the accumulated delta range data are adjusted for phase accumulator rollovers encountered during the extended track intervals. If present, the SA effects on the clock are corrected to improve the statistical margin in data editing. Both first and second difference histories are compared with tabulated threshold values to detect any data inconsistencies or gaps that would preclude the application of ionospherically corrected pseudorange or of carrier-aided smoothing. The tabulated criteria for data acceptance are based on analysis of possible range dynamics, expected tracker noise levels, and observed fault signatures. The CS is designed to be data-rich to combat random measurement noise, but must reliably support an estimation process with centimeter-level innovations, so it is wise to discard the occasional marginal measurement points and not risk contamination of the state estimator product.

At the performance level required of the CS, the ionospheric effects on propagation cause a major pseudorange deviation. The actual code delay and carrier phase advance introduced by the ionosphere is caused by the number of electrons that are encountered in the path. This varies by the propagation path, the local time of day, the latitude, and the level of ionosphere excitation caused by such things as radiation and solar activity. Electromagnetic interaction with this column of electrons produces a range deviation up to 15 m for the vertical path and is inversely proportional to the square of the carrier frequency squared of the radio frequency signal. Three times this vertical delay is encountered at the GPS horizon. Although this varying delay is not precisely predictable, its frequency dependency enables nearly complete correction for such demanding applications as the GPS CS. Dual-frequency observations on L_1 and L_2 allow the solution for pseudorange and carrier measurement histories nearly identical to those that would have been observed in an ionosphere-free medium. With f-1 being approximately 1.6 GHz and f-2 being 1.2 GHz, the correction equations are evaluated for each sample to eliminate the effects of the ionosphere, although the random noise component has been amplified by the algorithm. Only single-frequency measurements will be used by the CS to salvage some utility from a failing SV that can no longer provide dual-frequency signals, but the quality service of this SV will be realistically indicated as much less than normal.

The ionospherically corrected pseudorange history has a measurement noise level of over 1 m. This error source can be attenuated by smoothing the many independent data points available during a Kalman period, but detrimental biases will develop if the encountered data curvatures mismatch the polynomial model used for smoothing. Polynomials of high order are less effective in smoothing because of the increased degrees of freedom. In contrast, the ionospherically corrected accumulated delta range has an insignificant subcentimeter measurement noise level, but contains a large unknown bias of integration (or accumulator initiation). Carrier-aided smoothing takes advantage of the code–carrier measurement sample coherence feature of the GPS MS design, which provides data with identical (excluding multipath effects) curvature in ionospherically corrected pseudorange and integrated carrier measurement histories. The difference function of these two measurement types consists of random noise and the unknown ADR bias, but no curvature caused by range geometry. Averaging this difference function produces the required unbiased estimate of the ADR bias to permit utilization of the corrected carrier as an unbiased, low-noise, ionospherically corrected pseudorange measurement function. Other sources of curvature would bias the smoothed measurements.

Multipath is the only signal anomaly that is not coherent across code and carrier, and, fortunately, the MS antenna design has been very successful in rejecting multipath at line-of-sight elevations above the 15 deg criterion of acceptable tropospheric deviations. Thus, the effective residual difference in expected value between the two measurement types is a simple bias with no dynamics during the tracking interval of interest, and solution for unbiased low-noise measurements requires only averaging the difference between the corrected pseudorange history and the ADR history over the Kalman interval. Substitution of low-noise ADR data for the relatively noisy pseudorange data now supports a full solution for the GPS system states, including clocks. This ultrarefinement of measurement data for the estimator update is key in achieving the quality of predictions required of the system, and is practical for the CS because of the reliable availability of long signal-tracking intervals.

The Kalman filter operates periodically on the GPS system time scale, and so interpolation of the SV time aligned measurement samples is required to produce a measurement at the correct sample time. Choice of the SV code epoch as the measurement time for all sampling ensures small interpolation intervals and simplifies the measurement model implementation. Fortunately, the previous smoothing process efficiently yields pseudorange measurement data refined to the carrier noise level at any 1.5-s sample point. A local (six sample) fit is applied in the vicinity of the Kalman cycle time epoch to generate the applied measurement sample properly aligned on the GPS time scale.

The significant pseudorange measurement deviations are listed in Table 1, which tabulates the maximum magnitude encountered to show the potential error risk. The propagation effects due to the troposphere produce variations in the apparent range of up to 15 m for vertical paths, and this is amplified by as much as a factor of 12 when the satellite is tracked at the GPS horizon. Measurement use by the precision estimator has been restricted to those acquired at tracking elevations of 15 deg or higher to protect the estimation from large uncorrected errors. A correction of this tropospheric-induced delay is now made for the Kalman measurement sample using the Hopfield–Black model.[7,8] The effective speed of light encountered along the path results from the change of dialectic constant of the media. This slowing down of the signal is common, and essentially equal at both the L_1 and L_2 frequencies. Eighty percent of the delay component is from the effects of the dry air and is a function of the column of troposphere through which the signal is propagated. The correction model considers the station altitude, atmospheric pressure, and the elevation of the line of sight to account for increased length of tropospheric propagation for nonvertical paths. This dry component of troposphere can be determined and calibrated to within a few centimeters of error.

The remaining wet component accounts for up to 20% of the tropospheric delay and is from the effects of water vapor on the dialectic constant. This component cannot be precisely corrected, but, fortunately, it is smaller. When an inversion layer does not exist, good corrections are possible by using current surface measurements of relative humidity and temperature with the troposphere model, which considers path length and lapse rate. Protective code is provided to substitute nominal values characteristic of the site if any sensor malfunction is detected. When there is an atmospheric inversion condition, this model-based correction is inaccurate. Because of this weakness in measurement correction

Table 1 Pseudorange deviations

Physical phenomenon	Maximum magnitude
Special relativity	2.5×10^{-10}
General relativity	7.0×10^{-10}
Ionosphere	15 meters[a]
Troposphere	2.4 meters[b]
Earth rotation	35 meters

[a]Three times this vertical delay is encountered at the GPS horizon.
[b]Twelve times this vertical delay is encountered at the GPS horizon.

technology, only data observed at or above 15-deg elevation are used by the data-rich operational GPS Kalman estimator. Marginal measurement data may also be excluded by action of a statistical acceptance criterion placed on the magnitude of the innovation to protect the state estimates. The only alternative instrumentation technique that could yield precise correction in the presence of inversion conditions uses radiometers, and this laboratory-grade instrumentation has been deemed impractical and unnecessary for the GPS operational system.

The observed pseudorange measurement data contain components that deviate from the conceptual GPS formulation of ideal atomic time scales and static geometry in inertial space. The SV is moving rapidly through a nonuniform gravity field, which introduces relativistic effects of significance at the accuracy levels required to achieve the GPS system potential. The cesium clock technology adopted for the baseline SV atomic time scale is specified at a fractional frequency stability of two parts in ten to the thirteenth power. Special relativity effects alone are three orders of magnitude greater, and most of this is a clock frequency bias corrected through the combined effects of prelaunch adjustments, orbit control of the clock physics operating point, management of the phase offset instrumentation available on most SVs, and the clock correction terms in the navigation message generated by the CS. General relativity effects include components induced by dynamic changes in both velocity and the encountered gravity field, and the measurement correction model used by the CS and specified for user sets is identical to maintain consistency throughout GPS. The formulation makes use of relations in orbital mechanics for efficient evaluation. Note that the dynamic portion of this pseudorange deviation correction is not included in the correction polynomial distributed as part of the navigation message because of the curvature level that could be encountered for orbits of maximum allowable eccentricity.

Although the user is free to formulate algorithms best suited to his or her instrumentation and application, the GPS fundamental concept remains the same. Based on the code phase of the received signal at pseudorange measurement time, the user knows by code definition the SV time when this signal point was emitted. Applying the (inverse) time calibration information for this SV, the corresponding GPS system time and then the SV position can be evaluated to support solution of the GPS navigation equation. Corrections for relativity, for delays caused by the propagation medium, and for motion of the receiving antenna are allocated to the measurement model computations. The concept of pseudorange is based on the equivalence of time and distance for electromagnetic propagation nominally at the speed of light. Substantial deviations are induced by the signal propagation through the ionosphere and the troposphere, and the maximum magnitude value depicted in Table 1 is for a vertical path. Twelve times this effect is encountered at the GPS horizon of 5 deg because of longer path length in the region of medium deviation.

An update rate of four per hour has been adopted for the Kalman estimator process and provides adequate flexibility in upload scheduling. Note that the choice of a long Kalman period does not introduce bias from the measurement smoothing process because the technique of carrier aiding pseudorange smoothing removed the measurement dynamics up front, and the local data model simplifies to just a constant. Only measurement history segments that are intact for most of a 15-min Kalman interval are retained as significant for adding information to the operational system states. Shorter segments can be processed, but the data would be automatically deweighted because of their higher noise level when less smoothing is provided.

The principal terms of the measurement model relate to the atomic time scale calibrations and the physical geometry of the antennas, both system information components estimated by the CS. The Earth orientation relative to inertial space is not estimated, but is imported from external services. Special transformations are implemented to isolate the nonreal geometric discontinuity at the time of polar motion import from the GPS navigation service. Attention to the measurement model consistency is essential to provide stable performance for all conditions, especially during partial resource outages. The measurement model used by the CS is provided by the following measurement model equations:

$R_{SA} = c\Delta t_{SA}$ Selective availability

$\left. \begin{array}{l} PR_1 = PR_{M1} + R_{SA} - \delta I_{M1} \\ PR_2 = PR_{M2} + R_{SA} - \delta I_{M2} \end{array} \right\}$ Instrumentation error

$\left. \begin{array}{l} \phi_1 = \phi_{M1} + f_1 \Delta t_{SA} + N_1 \phi_{RO} \\ \phi_2 = \phi_{M2} + f_2 \Delta t_{SA} + N_2 \phi_{RO} \end{array} \right\}$ Rollover

$\left. \begin{array}{l} \Delta PR_1(i) = PR_1(i) - PR_1(i-1) \\ \Delta\Delta PR_1(i) = \Delta PR_1(i) - \Delta PR_1(i-1) \end{array} \right\}$ Differences

$$\vdots$$

$\left. \begin{array}{l} -T_{R1} \leq \Delta PR_1 \leq T_{R1} \\ -T_{R2} \leq \Delta\Delta PR_1 \leq T_{R2} \end{array} \right\}$ Edit flags

$$\vdots$$

$\left. \begin{array}{l} PR = \dfrac{PR_2 - \gamma PR_1}{1 - \gamma} \\ \\ ADR = \dfrac{\lambda_2 \phi_2 - \gamma \lambda_1 \phi_1}{1 - \gamma} \end{array} \right\}$ Ionosphere

$B = \overline{ADR - PR}$ ADR Track bias

$SPR = ADR - B$ Smooth data

$\Delta t_{SV} = \left(a_{S0} + a_{S1}(t_R - t_{oc}) + \Delta t_{SA} - \dfrac{2R^R_{SV} \cdot V^R_{SV}}{c^2} \right) \Big/ (1 + a_{S1})$

$\Delta t_{SV} = \Delta t_{SV} + \dfrac{(\Delta t_{SV} - (t_R - t_{oc}))^2 a_{S2}}{1 + a_{S1}}$

$\tilde{\Delta} t_R = 0.07$

$$DPR(d) = \sum_{j=-2}^{2} w_j^d \, SPR \, (t_R + 1.5_j) \quad d = 0, 1, 2 \qquad \text{Model}$$

$$SPRI = DPR(0) - DPR(1)\Delta t_{SV} + DPR(2)\Delta t_{SV}^2 \qquad \text{Interpolation}$$

$$t_R = \left(\frac{SPRI}{c} + \Delta t_{SV} - \Delta t_{SA} - a_{M0}\right)\bigg/(1 + a_{M1}) \qquad \text{Receive Time}$$

$$\left.\begin{array}{l} D = R_{SV} - [R_{MS}(\tilde{\Delta}t_R) + (t_R - \tilde{\Delta}t_R)\mathbf{V}_{MS}] \\ D = |D| \end{array}\right\} \qquad \text{Geometry}$$

$$Z = t_R C = \delta R_{\text{TROPO}} - D \qquad \text{Residual}$$

Table 2 lists the system states for which solutions are supported. Position and velocity states for the space vehicle are epoch values, corresponding to conditions at the reference time adopted when precomputing the nominal trajectory for each satellite. All others are current state values for the integer Kalman points with 15-min spacing on the GPS time scale. Clock phases relative to the GPS time scale are modeled as a linear time function for cesium standards and as a quadratic time function for rubidium standards. These are scaled by the speed of light to form pseudorange terms in the measurement model.

Absolute time is not observable from system pseudorange measurements, which are based on precision time differences. The initial formulation declared one monitor station to be the master time reference, and its clock state values were not altered by the measurement update process. Slaving to the UTC reference is accomplished by inserting a small aging term for the master clock that is limited to values that can propagate throughout the system without introducing significant errors in the prediction process. A change to the clock ensemble formulation is presently implemented where each MS clock is considered a separate statistical process, and thus all GPS clocks are estimated in a consistent fashion. The free GPS time scale is governed by an ensemble of all clocks in the system. This model improvement did necessitate the implementation of constraint equations to maintain time alignment between partitions and with the external absolute UTC time reference.

Fortunately, the satellite will remain close to a precomputed reference trajectory that starts at a nominal epoch, and this model can be used to define a linear

Table 2 Estimated states

SV position at epoch
SV velocity at epoch
SV clock phase
SV clock frequency
SV clock aging (rubidium only)
Solar flux
SV Y-axis acceleration bias
MS clock phase
MS clock frequency
MS wet-troposphere height

region of several kilometers about the nominal reference trajectory that starts at the epoch. Periodic regeneration is performed whenever a force event is introduced that invalidates the model and whenever the residuals from the Kalman estimator indicate that the current states are near the edge of acceptable linearity. Partial derivatives with respect to each epoch state component of inertial position and velocity are integrated along with the position prediction at time of reference trajectory generation to support reducing the estimation and prediction problem to linear mathematical relationships.

Nominal position and velocity values for the SV antenna are precomputed for each integer 15-min point in GPS time, and the component set corresponding to the measurement update time must be updated with the product of the residual states from the Kalman estimator and the trajectory's first partial derivatives (also precomputed) to obtain the a priori values for use by the measurement model (Fig. 11). This mathematical convenience and the control timing conventions adopted for the phase adjustor at the space vehicle were the motivation for adopting the SV referenced measurement sample convention. Correction of the SV position to that of code epoch emission time is simply the product of the clock calibration and the SV velocity.

Position of the MS antenna's phase center at the time of epoch reception is the last component of the measurement model. Up to 35 m of deviation in range is caused by the effects of Earth rotation during the propagation time. This value is pseudorange dependent because the Earth continues to turn during the signal propagation interval, whatever the delay mechanism. The MS antenna site survey data are adjusted for crustal tides and also for the rotation of the Earth. Predictions for both the hour angle and the axis of rotation necessary for inertial coordinate conversion is provided to GPS by the Naval Observatory. The net correction of MS position vector is linearized in time about the center of the possible epoch reception window. Evaluation time for MS antenna position is derived from the observed pseudorange and the a priori MS clock states applicable for the update point. The measurement strategy adopted by the CS again permits a simple precise measurement model.

V. System State Estimation

The OCS estimation errors must be held at the submeter level to support the GPS ranging error budget as provided to the user community. Any mismatch between the measurement data and the a priori measurement model values (termed an innovation) indicates that an adjustment of system states will statistically improve the knowledge of the system. A measurement acceptance criterion based on the a priori measurement variance is implemented to protect the estimator by blocking apparent blunders that would harm the estimator's knowledge base. The Kalman filter provides the estimation technology to distribute the small probable adjustment across the system state residuals based on the partials characteristic of the measurement model, the known measurement noise, and the current state uncertainty, as expressed by the covariance matrix.

Estimation at the GPS CS is based on a $U\text{-}D\text{-}U$ factored formulation of the Kalman Filter. This square root variation was chosen for its superior numerical stability characteristics in the presence of roundoff errors, because the GPS estimator must reliably operate continuously. The overall OCS process was formulated to provide natural state decoupling wherever possible, but significant off-diagonal terms in the covariance matrix are unavoidable for systems based on pseudorange. The GPS estimation problem has many state variables, and mathematical operations on a single large matrix would impose a formable load on computer resources at the time of initial MCS implementation. However, the only coupling between the states of different SVs is through the MS time scale system, and to gain computational efficiency the CS has partitioned the GPS problem into subsets of SVs, each with the common-view measurement connectivity to precisely maintain its MS time scale system. As computer technology advances, all operational SVs will be accommodated in a single partition.

Valid measurements for each applicable MS–SV pair are sequentially processed to gain the information in the current measurement set. The equations for this measurement update[9] are as follows

$v_j^T = A_j S_j$

$\bar{\sigma}_j = 1/(v_j^T v_j + 1)$ — Predicted residual variance inverse

$\bar{K}_j = S_j v_j$ — Unweighted Kalman gain

$\delta_j = z_j - A_j x_j$ — Predicted residuals

$x_{j+1} = x_j + \bar{K}_j(\delta_j \bar{\sigma}_j)$ — Updated state estimate

$\gamma_j = \bar{\sigma}_j(1 + \sqrt{\bar{\sigma}_j})$

$S_{j+1} = S_j - (\gamma_j \bar{K}_j) v_j^T$ — Square root covariance update

Although this conventional filter derivation is for uncorrelated measurement noise, a more complex variation[9] is used in the operational system that accounts for the pair-wise error correlations of ADR data.

Between the Kalman measurement updates that account for the gain of information, a Kalman time update is performed that accounts for the statistical loss of information caused by process model uncertainty and the passage of time. Tuning, or the selection of values for the statistical parameters must be consistent with reality if the GPS performance potential is to be achieved. The process noise for the clock states was derived from the observed Allen variance for each type of clock. Process noise for the trajectory states is applied as part of the time update in current RAC coordinates, with noise values traceable to the actual degree of model decay, as revealed by batch fit studies conducted offline. Following this scientific rationale with noise value traceability to real operational experience, brings statistical realism to the filter tuning, which was found essential to obtain the favorable prediction performance of several weeks duration even when anomalies in the monitor station tracking schedule were encountered.

$$\overline{D}_{j+1} = (v_{j+1}^{(j+1)})^T D v_{j+1}^{(j+1)}$$

$$\overline{U}(k, j+1) = (v_k^{(j+1)})^T D v_{j+1}^{(j+1)} / \overline{D}_{j+1} \qquad k = 1, \ldots, j$$

$$v_k^{(j)} = v_k^{(j+1)} - \overline{U}(k, j+1) v_{j+1}^{(j+1)} \qquad k = 1, \ldots, j$$

and

$$\overline{D} = (v_1^{(1)})^T D v_1^{(1)}$$

The implemented recursive estimator has proved to be computationally efficient and provides a numerically robust system in the operational environment, in part because of the quality of the measurements and the strategic utilization of common-view measurement opportunities. Equipment and communication service outages are occasionally experienced, and although the capability to monitor the navigation service has been occasionally disrupted, the estimation process has remained sound.

Prediction is the real essence of navigation processing performed by the CS, and it requires exceptional state estimate. Prediction also has extensive dependency on precise geophysical models, and those used by the MCS are based on the scientific experience of the entire GPS community. Days of measurement data are required to clearly distinguish all of the system states, and use of this information by the real-time user will be displaced in time from the end of the measurement data acquisition interval by more than the nominal 1-day service life of the upload, which is much longer than the Kalman interval prediction that supports the measurement update process. The baseline GPS system operations concept envisioned that each SV would be uploaded three times a day to achieve the specified ranging performance of 6 m. However, the cesium clocks have been performing on orbit with such stability that the specified performance is routinely achieved with one upload per day. Thus, the user set routinely operates with navigation message sets received up to 26 h after the last measurement, and may continue to navigate with this predicted data for several additional hours. In view of the observation time required to separate GPS states, an information delay is inevitable, and this service time is a reasonable performance trade for ground based estimation.

Generation of the navigation message is based on the existence of models adequate to support predictions meeting the system error budget. This requirement is more stringent than that required of the models just to separate the SV position and clock states. The parameters of the models map into the required states of the estimator, as provided by Table 2. All states associated with the preprepared reference trajectories are values at the trajectory epoch, and the remaining states are current values.

Figure 12 illustrates the clock model, and illustrates its function during both the observation interval and the prediction interval. The calibration, or difference

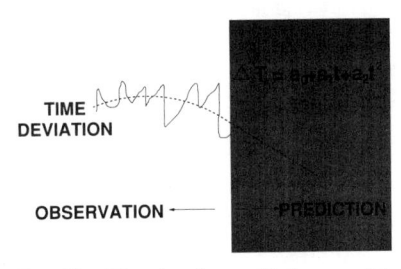

Fig. 12 Time scale prediction model.

between the actual clock phase and an ideal clock phase is a statistical process that contains both deterministic trends and random components. Atomic clock calibrations are routinely modeled as a polynomial that includes a bias term, a linear (or frequency) term, and for rubidium devices a quadratic (or aging) term. During the observation interval, the model definition provides the basis for estimating a priori time values to support the measurement update and provide structure for state separation from the encountered measurement data. Prediction beyond the last estimated time is then accomplished by evaluating the model for times beyond the availability of observations. The random model component, which for clocks is characterized by the Allen variance, interferes with this prediction process in two ways. First, estimation of the model parameters from a finite period of observation will contain error whose effects are amplified as the model projects ahead in time. Second, the random component will continue as a difference between the prediction and that which will be experienced by the system. What can legitimately be considered deterministic trends depends both upon the real process stability and the intervals of time considered, and study of the Allan variance statistic suggests the utilization of effective observation intervals that are no shorter than the desired prediction interval. The Kalman frequency state supports one-day predictions well with tuning derived from the Allan variance. Because of limitations in the process noise modeling, very long time predictions are best supported with a subsequent time projection based on Kalman estimates separated by the prediction interval rather than the instantaneous frequency estimate.

The other critical model to GPS performance provides the trajectory of the SV antenna's phase center. The laws of mechanics provide the solution for the MS center of mass from time of initial conditions based on a knowledge of the forces encountered. Given an initial position and velocity, the satellite's center of gravity nearly follows a free-fall trajectory in the complex gravity field. Forces vary as a function of time and geometry, and extremely precise models are possible for the high altitude orbits of the GPS satellites because the drag and very high spatial frequency geopotential forces are small. Given a force, the acceleration is proportional to the force divided by the estimated mass of the vehicle, which when integrated and added to the initial velocity condition, will provide the velocity of the satellite. A second integration, utilizing the initial position, provides a precision prediction of the position.

This process is illustrated by Fig. 13. The models that create this component of forces acting on the SV are dependent upon both the position and time. The major force component acting on the satellite is caused by the Earth's gravitational field. The Earth is more complex than a point mass, and the geopotential model adopted for the operational GPS is WGS-84 expressed as a high-order spherical harmonic expansion. Dynamic corrections are applied to the first-order terms of the expansion to account for the dynamic plastic deformation of the spheroid resulting from the gravity gradient (a tidal effect) induced by the sun and moon.

The next most significant force components acting on the space vehicle are the gravity potential caused by the Sun and Moon. The Jet Propulsion Laboratory provides GPS with long-term predictions of major bodies within the solar system, which has more than adequate accuracy to support the system requirements.

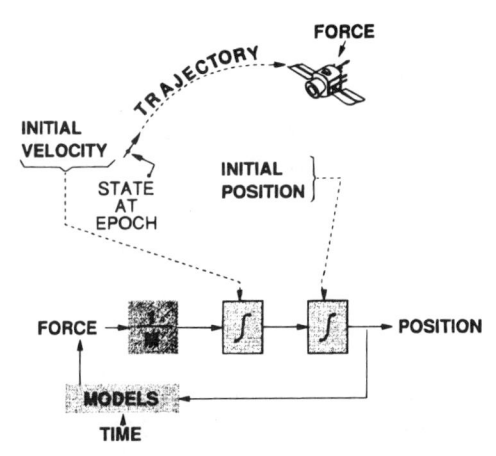

Fig. 13 Ephemeris prediction.

Solar-flux reaction on the satellite are based upon detailed models provided by the space vehicle builder that characterizes each major surface in terms of geometry and finish. The force component is evaluated by algorithms that account for flux modulation while entering and exiting the solar eclipse conditions. Surface properties, attitude of the space vehicle body relative to the sun, and partial shadowing by the large solar panels is accounted for in the force evaluation. Because much of the observed variation in solar force relative to the model is believed due to darkening of solar cells, an effective solar flux level for each SV is estimated as an ad hoc state of the Kalman filter for each SV.

Analysis of residual systematic trends has resulted in adopting an ad hoc acceleration model along the Y-axis of the satellite. The mechanics of this apparent force are not well understood with no consensus as to the cause, but inclusion of this body-aligned force component whose magnitude is estimated by the Kalman filter has clearly improved the performance of trajectory prediction. Another force residue seems to be thermal propulsion during battery conditioning, but no serious attempt has been made to include this in the force model.

The net force acting on the satellite is scaled by the mass to evaluate acceleration in inertial coordinates. Factory records provide the initial vehicle mass, and this value is modified as consumables are used. Double integration from initial position and velocity at epoch provides a reference trajectory for the mass center, and because the Earth–sun stabilization results in a deterministic angular attitude, the position of the antenna phase center is known through a translation. (Angular drift during eclipse is an unmodeled motion that has been evident.) Planned station-keeping maneuvers are modeled, and the predictions are included in the reference trajectory. However, the accuracy of force during thruster activation doesn't support long-time predictions, and as soon as a solid estimation of position and velocity are available, generation of a new reference trajectory is expedited.

The initial position and velocity of the satellite is not precisely known, the force models are not perfect, and thus the trajectory obtained from this integration must be updated as more recent information is collected by the L-band monitor stations. Fortunately, the satellite will remain close to this precomputed reference trajectory, and the model can be used to define a linearized region of several kilometers about the nominal reference trajectory, which starts at the epoch, as is illustrated by Fig. 14.

A mathematical expansion that is linear in dimensions of the Kalman states is adequate for efficiently evaluating predictions of the future SV antenna positions slightly off this precomputed reference trajectory without rerunning the computationally costly integrator process. Partial derivatives of position and velocity with respect to each Kalman state are evaluated by integration right along with the generation of predicted trajectory. Thus, the estimation and prediction update evaluation is reduced to a linear mathematical function of the residual state estimates for the region of immediate GPS interest. The vector product of these partial derivative functions and the estimated residual state values from the Kalman filter provides the update to the reference trajectory functions for preparation of the navigation message and for supporting evaluation of the filter gains necessary to process newer measurement data. Preparation of these reference trajectory position and first partial projections are scheduled at a time when computational resources are plentiful. Periodic regeneration is performed whenever a force event is introduced that invalidates the model and whenever the residuals from the Kalman estimator indicate that the current states are near the edge of acceptable linearity. The bounds of acceptable linearity were evaluated through extensive offline analysis of higher-order terms in the expansion.

Given current values for estimated state residuals, evaluation of the first-order expansion provides the position trajectory used to process monitor station measurements or to generate the predicted navigation message for navigation message upload. The application of the model has been formulated to include compensation for the offset between the satellite center of gravity and antenna phase center. This permits the continued efficient operation of the system with linear modification to the states for several weeks after the computationally burdensome integration has been performed in advance for each satellite. These estimated states allow the Kalman estimator to incorporate new monitor station measurements to improve the predicted trajectory as new information is acquired.

An independent test of the navigation process has been conducted with synthetic data carefully prepared by the Aerospace Corp., which includes random processes

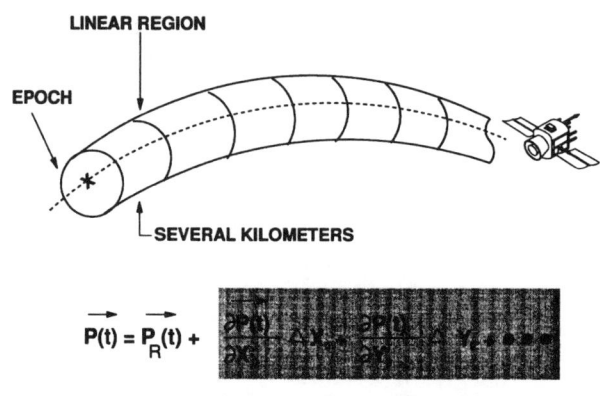

Fig. 14 Reference trajectory expansion.

for clocks and terminal measurement noise. The test data example (Fig. 15) illustrates the results corresponding to an SV equipped with a cesium clock. The URE statistic was evaluated in projections to 32 Earth-fixed user locations after allowing two days for estimator convergence. This independent truth data indicate that the OCS meets performance requirements for orbit and clock prediction with nominal SV clock stabilities of 2E-13. Clock performance better than specification will yield even better system performance and the on-line performance history of the GPS is much better than that predicted with the simulated data. On-orbit clock performance is observed to be better than specification and supports the 6-m URE requirements at a 24-h age of data.

VI. Navigation Message Generation

Preparation of the navigation message is based on the estimated position of the satellite (antenna) as a function of time obtained as an evaluation of the first order Taylor expansion about the reference trajectory. The current Kalman states for satellite position and the prepared partials provide an efficient generation of the most current trajectory knowledge in inertial space. This utilizes inertial to Earth-fixed coordinate transformation matrixes that are consistent with externally supplied Earth rotation data. Continuity constraints are imposed on this system input to maintain navigation service integrity across editors of the coordinate transformation parameters.

These predictions are formatted for the navigation upload message by creating navigation datasets. The predicted trajectory interval consists of a 4-h segment of prediction during the first day and 6 h for days 2–14 of the upload as illustrated in Fig. 16. Weighted least squares fitting is used to evaluate the numerical values of orbit model terms appearing in navigation message. To avoid problems with the mathematical singularities that exist when eccentricity is zero, the actual fitting process is done for the intermediate model value of "Argument of Latitude"

Fig. 15 Test data example 2E-13 clock.

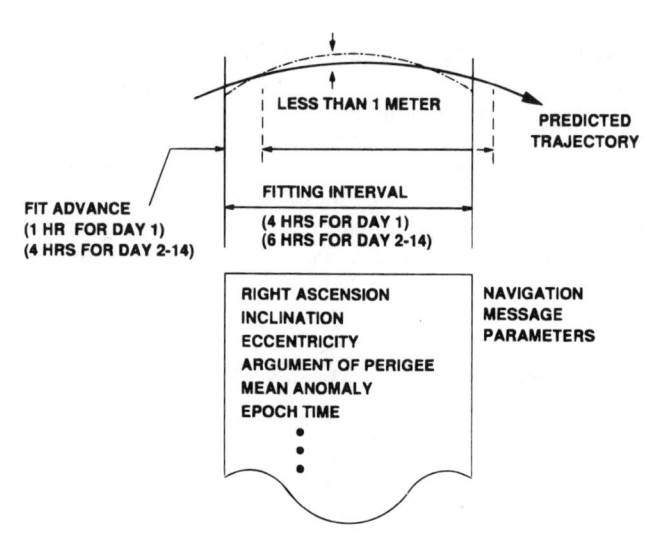

```
                    ↓
              ╱‾‾‾‾‾‾‾‾‾‾╲
             ↑    ↑        ╲→ PREDICTED
        LESS THAN 1 METER      TRAJECTORY

         FITTING INTERVAL
FIT ADVANCE
(1 HR FOR DAY 1)    (4 HRS FOR DAY 1)
(4 HRS FOR DAY 2-14) (6 HRS FOR DAY 2-14)

         RIGHT ASCENSION    NAVIGATION
         INCLINATION        MESSAGE
         ECCENTRICITY       PARAMETERS
         ARGUMENT OF PERIGEE
         MEAN ANOMALY
         EPOCH TIME
```

Fig. 16 Navigation dataset fitting.

and two intermediate product terms containing "eccentricity." This equivalent formulation for the SV position has no singularity, and produces stable values for all terms. The specified navigation message parameters are reliably obtained by algebraic substitution. The fit error weighting is tapered to be heavy at the ends of the dataset interval so as to obtain a uniform error distribution and avoid adverse end effects. This provides concise formatted information to the user that is within 1 m of the predicted trajectory. These are in accordance with the interface ICD GPS 200, which is the governing specification for the interface to the GPS user community. The fit interval advances one hour during day one and four hours for day two, so the user can utilize the transmitted data for a period of several hours after navigation messages have been collected. The upload schedule and length of data buffered at each of the distributed space vehicles is managed to ensure the specified accuracy performance with extended service availability as required of a navigation utility used in critical applications. To permit continued service in the event of a national disaster, 2-week predictions are actually generated for each upload to the SV memory.

VII. Time Coordination

Redundant atomic clocks are provided at each MS, one a hot spare. Transitioning to the alternate MS time reference requires a quick substitution of calibration data. The bias term can be resolved with one good measurement update, but the frequency estimate is dependent upon observations over a considerable period of time. Special clock phase comparison instrumentation is implemented at each MS, and this difference history is used to calibrate indirectly the frequency of the hot spare to support a rapid switchover.

The GPS is based on the measurement of time differences, and the system cannot sense absolute time. For each partition, the Kalman measurement update refines the relative knowledge of the time scales, but one extra degree-of-freedom exists. The initial implementation at the MCS applied the artificial concept of a master MS clock as a reference that was not estimated, and this tied the time scales of partitions together.

Time accuracy of 97 ns, one sigma, at the satellite is required to support the time transformation mission. This is a less demanding performance requirement than that of providing the GPS navigation service, but must rely on the United States Naval Observatory for a measurement link to UTC. Based on their observations, the clock calibration terms for the master clock are adjusted by a time-steering mechanism. Any single clock time stress results in a movement of all estimated time scale calibrations, but it takes a complete constellation upload cycle to refresh the stored data on the SV. For this reason, the dynamics of this GPS time scale alignment to UTC is constrained in rate so as to safeguard the navigation service, which is dependent upon the consistency of all calibrated clocks. Time steering of multiple GPS clocks would distort both the time and orbit estimations, which will produce errors in the user navigation solutions. Considerable time is required for the calibration changes to propagate through the estimation process and correct the inconsistency stored in the many SV memories. The change to the broadcast calibration of the GPS time scale need not be constrained because it doesn't affect navigation.

An improved concept of time has been implemented where the calibration of all clocks are estimated, and the GPS timescale is based on the ensemble. This composite clock[10] formulation relies on the concept of a mean time component in all clocks, and bounds the covariance growth earlier experienced with implementing an overall GPS ensemble. GPS time stability is improved and the mechanics of maintaining UTC alignment are more direct.

VIII. Navigation Product Validation

System performance is monitored by performing repeated navigation solutions with the received ranging and navigation data observed at each of the five monitor stations just as a user might. The nominal signal acquisition steps can be exercised if desired. This navigation solution is an absolute indicator of process health, but does not continually utilize each satellites signal. A continuous monitor implemented for each Kalman partition set of SVs is the evaluation of the observed range residual (ORD) for measurements at each MS using the broadcast navigation data. This is a sensitive continuous indicator of consistency within a partition. Data service integrity is monitored by comparing each bit of the received message with the corresponding upload database to verify proper dissemination.

This performance monitoring does introduce the questions of possible distortions at other geographic points. Such a possibility suggests further off-line analysis by the GPS community with potential for continuing performance enhancements through improved modeling.

References

[1] Parkinson, B. W., "Navstar Global Positioning System (GPS)," *Proceedings of the National Telecommunications Conference,* Nov. 1976.

[2]Among, "Navstar GPS Space Segment/Navigation User Interfaces," ICD-GPS-200, Jan. 25, 1983.

[3]Ananda, M. P., et al., "Autonomous Navigation of the Global Positioning System Satellites," *AIAA Guidance and Control Conference* (Seattle, WA), AIAA, New York, Aug. 1984.

[4]Divine D., III, and Francisco, S. G., "Synchromesh, A Practical Enhancement to GPS Service," *IEEE PLANS '84, IEEE Position, Location, and Navigation Symposium* (San Diego, CA), Institute of Electrical and Electronics Engineers, New York, Nov. 1984

[5]Spilker, J. J., Jr., "GPS Signal Structure and Performance Characteristics," *Navigation,* Vol. 25, No. 2, 1978.

[6]Van Dierendonck, A. J., Russell. E. R., Kopitzke, E. R., and Birnbaum, M., "The GPS Navigation Message," *Navigation,* Vol. 25, No. 2, 1978.

[7]Hopfield, H. S., "Tropospheric Effects on Electromagnetically Measured Range, Prediction from Surface Water Data," *Radio Science,* Vol. 6, No. 3, March 1971, pp. 356–367.

[8]Black, H. D., "An Easily Implemented Algorithm for Tropospheric Range Correction," *Journal of Geophysical Research,* Vol. 83, April 1978, pp. 1825–1828.

[9]Bierman, G. J., "Factorization Methods for Discrete Sequential Estimation," *Mathematics in Science and Engineering,* Vol. 128, Academic Press, New York, 1977.

[10]Brown, K. R., Jr., "The Theory of the GPS Composite Clock," *Proceedings of ION GPS-91,* Institute of Navigation, Washington, DC, Sept. 1991.

Part II. GPS Performance and Error Effects

Chapter 11

GPS Error Analysis

Bradford W. Parkinson*
Stanford University, Stanford, California 94305

I. Introduction

ALTHOUGH the Global Positioning System (GPS) is clearly the most accurate worldwide navigation system yet developed, it still can exhibit significant errors. By understanding these errors, the user can both hope to reduce them and to understand the limitations of the GPS system. This section of the book should help develop that understanding. This chapter provides both an overview of the sources of error and a detailed analysis of the general error equations. It also presents a standard table of errors that should help clarify the impacts of variations in the specific error magnitudes. Later chapters delve into the expected ranges of these errors.

This development assumes that we are dealing with state-of-the-art receiver technology. In general, this requires a six-channel, continuous tracking implementation. Receivers with fewer channels will probably give a significantly degraded performance. In fact, a number of implementation compromises can produce receiver errors that are greater than those presented here.

This chapter first develops the general error equations and then illustrates how the dilution of precision (DOP) caused by satellite geometry can seriously degrade results. Finally, the error budget is summarized.[1]

II. Fundamental Error Equation

A. Overview of Development

As explained in introductory chapters, a GPS receiver fundamentally measures a quantity called pseudorange ρ, which is a raw, one-way range measurement corrupted by a user clock bias.[2] Using either models or measurements, ρ can be corrected for atmospheric and other effects to produce corrected pseudorange ρ_c. With an approximate user location, the receiver can then process these corrected pseudoranges (to four or more satellites) to determine location in a convenient coordinate system. (Various manufacturers have implemented the "anywhere"

Copyright ©1994 by the author. Published by the American Institute of Aeronautics and Astronautics, Inc., with permission. Released to AIAA to publish in all forms.
*Professor, Department of Aeronautics and Astronautics; Director of Stanford GPS Program.

fix, which can start from any location. Such techniques are beyond this text and are not a part of error analysis.) This calculation is developed in this chapter.

For GPS, the underlying coordinate system is currently the 1984 World Geodetic System (WGS-84), which is an accepted worldwide geodetic coordinate system. It is expected that this will be replaced with a more accurate version as satellite geodesy improves. It is usual and convenient for the receiver to perform initial calculations in an Earth-centered, Earth-fixed (ECEF) Cartesian coordinate system. These coordinates can then be converted to any other required reference. For *error analysis,* it is usual to consider a local coordinate frame centered at the user and oriented East, North, and Up. This is convenient because many users have differing sensitivities to vertical errors and horizontal errors.

B. Derivation of the Fundamental Error Equation

This section develops the GPS error equation, beginning with the fundamental measurements and proceeding through analysis of the effects of various error sources. It lays the groundwork for the essential understanding of the dilution of precision (DOP) concept.

1. Ideal Measurement

The "true" or ideal measurement is the GPS signal arrival time. This is equal to the signal transmission time delayed by the vacuum transit time and corrected for the true additional delays caused by the ionosphere and troposphere:

$$t_A = t_T + D/c + T + I \tag{1}$$

where t_A = true arrival time (s); t_T = true transmit time (s); D = true range (m); c = vacuum speed of light (m/s); T = true tropospheric delay (s); and I = true ionospheric delay (s).

2. Measured Arrival Time

The measured arrival time reflects the user's clock bias and other measurement errors:

$$t_{Au} = t_A + b_u + v \tag{2}$$

where t_{Au} = arrival time measured by the user (s); b_u = user clock bias estimate (s); and v = receiver noise, multipath, interchannel error (different for each satellite (s)).

3. Satellite Transmission Time

The satellite clock correction transmitted by the satellite can also be in error (the dominant error may be due to selective availability SA):

$$t_{Ts} = t_T + B \tag{3}$$

where t_{Ts} = value of transmission time in the current satellite message (s); and B = true error in satellite's transmission time (includes SA).

GPS ERROR ANALYSIS

4. True Range

The true range is the absolute value of the vector difference between the true satellite position and the true user position:

$$D = |\bar{r}_s - \bar{r}_u| = \bar{\mathbf{l}}_s \cdot [\bar{r}_s - \bar{r}_u] \tag{4}$$

where $\bar{r}_s$ = true satellite position; $\bar{r}_u$ = true user position; and $\bar{\mathbf{l}}_s$ = true unit vector from user to satellite.

The right-hand expression in Eq. (4), which uses the vector dot product, is a convenient way to calculate the range in the later user equation. In this calculation, the *estimated* user position can be used to find the unit vector from user to satellite. This unit vector need not be exact. Even *errors* of *several hundred meters* in user or satellite location have a very small effect on the dot product that uses this unit vector. Such a position error would produce angular errors of a few arc-seconds (about 10^{-5} rad) in the unit vector. Because the dot product *error* would be proportional to the cosine of this angle, it is of the order (angle)^2/2 or about $10^{-10}/2$ times the range. Because the range is 2×10^7 m, the effect would be less than 1 mm.

5. Pseudorange

The user receiver actually "measures" the "pseudorange" ρ, given by the following:

$$\rho = c \cdot (t_{Au} - t_{Ts}) \tag{5}$$

This is called the pseudorange because it is linearly a function of the range to the satellite, but it is also corrupted by the user's clock bias, which must be estimated and removed. In addition, it must be corrected for the estimated satellite time bias and for variations in the speed of transmission.

Substituting Eqs. (3), (2), and (1) into Eq. (5) gives the following result:

$$\rho = D + c \cdot (b_u - B) + c \cdot (T + I + v) \tag{6}$$

Using Eq. (4) in this expression gives the following:

$$\rho = \bar{\mathbf{l}}_s \cdot [\bar{r}_s - \bar{r}_u] + c \cdot (b_u - B) + c \cdot (T + I + v) \tag{7}$$

To account for the estimated value ($\hat{}$) and the estimate error (Δ), each of the above terms is to be broken into two parts as follows:

$\bar{r}_s = \hat{\bar{r}}_s - \Delta \bar{r}_s$, where $\hat{\bar{r}}_s$ = satellite position reported in the transmitted message (m)

$\bar{r}_u = \hat{\bar{r}}_u - \Delta \bar{r}_u$, where $\hat{\bar{r}}_u$ = user estimated position (m)

$\bar{\mathbf{l}}_s = \hat{\bar{\mathbf{l}}}_s - \Delta \bar{\mathbf{l}}_s$, where $\hat{\bar{\mathbf{l}}}_s$ = unit vector from user to the satellite estimated from $\hat{\bar{r}}_s$ and $\hat{\bar{r}}_u$

$b_u = \hat{b}_u - \Delta b_u$, where $\hat{b}_u$ = user clock bias estimate common to a set of simultaneous measurements (s)

$B = \hat{B} - \Delta B - S$, where $\hat{B}$ = satellite transmitted clock bias (s)

ΔB = the "natural" satellite clock error; that is, the error in control system prediction (s)

$$T = \hat{T} - \Delta T$$
$$I = \hat{I} - \Delta I$$

S = error in transmit time due to SA (s)
$\hat{T}$ = estimated (or modeled) tropospheric delay (s)
$\hat{I}$ = estimated (or modeled) ionospheric delay (s)

Equation (7) can then be modified to account for the estimated values:

$$\rho_j = (\hat{\bar{1}}_{sj} - \Delta \bar{1}_{sj}) \cdot (\hat{\bar{r}}_{sj} - \Delta \bar{r}_{sj} - \hat{\bar{r}}_u + \Delta \bar{r}_u) + c \cdot (\hat{b}_u - \Delta b_u - \hat{B}_j + \Delta B_j + S_j) + c \cdot (\hat{I}_j - \Delta I_j + \hat{T}_j - \Delta T_j + v_j) \qquad (8)$$

where the j subscript is the satellite number and has been added to point out quantities unique to each satellite.

In preparation for conversion to matrix form, Eq. (8) can be rewritten as follows:

$$\underbrace{\hat{\bar{1}}_{sj} \cdot \hat{\bar{r}}_u}_{(b)} \underbrace{- c \cdot \hat{b}_u}_{(a)} \underbrace{- \hat{\bar{1}}_{sj} \cdot \Delta \bar{r}_u + c \cdot \Delta b_u}_{(d)} = \underbrace{\hat{\bar{1}}_{sj} \cdot \hat{\bar{r}}_{sj} - \rho_j + c \cdot [\hat{I}_j + \hat{T}_j - \hat{B}_j]}_{(b)}$$

$$+ \underbrace{-\hat{\bar{1}}_{sj} \cdot \Delta \bar{r}_{sj} - \Delta \bar{1}_{sj} \cdot (\hat{\bar{r}}_{sj} - \hat{\bar{r}}_u) + c \cdot (\Delta B_j + S_j) - c \cdot (\Delta I_j + \Delta T_j) + c \cdot v_j}_{(c)}$$

$$+ \text{ higher order terms} \qquad (9)$$

The large dots represent the dot product. Note that the terms underlined as (a) are the user's position and clock errors to be solved, the terms (b) are estimated or measured by the user, and the terms (c) are unknown errors that produce the solution errors given by (d). The right-hand portion of (b) includes ρ_{cj}, or corrected pseudorange, as: $\rho_{cj} = \rho_j - c \cdot [\hat{I}_j + \hat{T}_j - \hat{B}_j]$. This variable is used below.

Next, define the following matrices for K satellites in view (note that $K = 4$ is normally the minimum number of measurements):

$$G_{K \times 4} \triangleq \begin{bmatrix} \hat{\bar{1}}_{s1}^T & 1 \\ \hat{\bar{1}}_{s2}^T & 1 \\ \vdots & \vdots \\ \hat{\bar{1}}_{sK}^T & 1 \end{bmatrix} \qquad A_{K \times 3K} \triangleq \begin{bmatrix} \hat{\bar{1}}_{s1}^T & & 0 \\ & \hat{\bar{1}}_{s2}^T & \\ & & \ddots \\ 0 & & \hat{\bar{1}}_{sK}^T \end{bmatrix}$$

$$\hat{\bar{x}}_{4 \times 1} \triangleq \begin{bmatrix} \hat{\bar{r}}_u \\ -c \cdot \hat{b}_u \end{bmatrix} \qquad \Delta \bar{x}_{4 \times 1} \triangleq \begin{bmatrix} \Delta \bar{r}_u \\ -c \cdot \Delta b_u \end{bmatrix} \qquad \bar{R}_{3K \times 1} \triangleq \begin{bmatrix} \hat{\bar{r}}_{s1} \\ \hat{\bar{r}}_{s2} \\ \vdots \\ \hat{\bar{r}}_{sK} \end{bmatrix}$$

$$\Delta\overline{R}_{3K\times 1} \triangleq \begin{bmatrix} \Delta\overline{r}_{s1} \\ \Delta\overline{r}_{s2} \\ \vdots \\ \Delta\overline{r}_{sK} \end{bmatrix}, \quad -\hat{\overline{\rho}}_{cK\times 1} \triangleq \begin{bmatrix} -\rho_1 + c\cdot(\hat{I}_1 + \hat{T}_1 - \hat{B}_1) \\ -\rho_2 + c\cdot(\hat{I}_2 + \hat{T}_2 - \hat{B}_2) \\ \vdots \\ -\rho_K + c\cdot(\hat{I}_K + \hat{T}_K - \hat{B}_K) \end{bmatrix} = \begin{bmatrix} -\rho_{c1} \\ -\rho_{c2} \\ \vdots \\ -\rho_{cK} \end{bmatrix}$$

$$\epsilon_{K\times 3K} \triangleq \begin{bmatrix} \Delta\overline{1}_1^T & & 0 \\ & \Delta\overline{1}_2^T & \\ & & \ddots \\ 0 & & \Delta\overline{1}_K^T \end{bmatrix}, \quad \overline{P}_{3K\times 1} \triangleq \begin{bmatrix} \overline{r}_u \\ \overline{r}_u \\ \vdots \\ \overline{r}_u \end{bmatrix}$$

and $\Delta\overline{B}$, $\overline{S}$, $\Delta\overline{I}$, $\Delta\overline{T}$, $\overline{v}$ are all obvious. The error in the unit vector to the satellite is $\Delta\overline{1}_j$. The matrix transpose is a convenient matrix notation for the dot product. Equation (8) then becomes the following (neglecting higher-order terms):

$$\mathbf{G}\cdot\hat{\overline{x}} - \mathbf{G}\cdot\Delta\overline{x} = \mathbf{A}\cdot\overline{R} - \hat{\overline{\rho}}_c - \mathbf{A}\cdot\Delta\overline{R} + c$$
$$\cdot(\Delta\overline{B} + \overline{S} - \Delta\overline{I} - \Delta\overline{T} + \overline{v}) + \epsilon\cdot(\overline{R} - \overline{P}) \qquad (10)$$

The user does not know the last terms of Eq. (10), which are the errors, and calculates position based on the following:

$$\mathbf{G}\cdot\hat{\overline{x}} = \mathbf{A}\cdot\overline{R} - \hat{\overline{\rho}}_c$$

to find the following for $K = 4$:

$$\hat{\overline{x}} = \mathbf{G}^{-1}(\mathbf{A}\cdot\overline{R} - \hat{\overline{\rho}}_c) \qquad (11a)$$

and the following for $K > 4$:

$$\hat{\overline{x}} = (\mathbf{G}^T\mathbf{G})^{-1}\mathbf{G}^T(\mathbf{A}\cdot\overline{R} - \hat{\overline{\rho}}_c) \qquad (11b)$$

using the generalized matrix inverse (or pseudoinverse) of $\mathbf{G}$.

These are the fundamental position calculations. Note that $\mathbf{G}$, the geometry matrix, is constructed from the set of approximate directions to the satellites, as is the matrix $\mathbf{A}$. The vector $\overline{R}$ is constructed from the location of the satellites that has been transmitted and $\hat{\overline{\rho}}_c$ is the corrected pseudorange to each satellite.

Inserting Eq. (11a) back into Eq. (10) cancels appropriate terms and leaves the fundamental error equation*:

$$\mathbf{G}\cdot\Delta\overline{x} = c\cdot(-\Delta\overline{B} - \overline{S} + \Delta\overline{I} + \Delta\overline{T} - \overline{v}) - \epsilon\cdot(\overline{R} - \overline{P}) + \mathbf{A}\cdot\Delta\overline{R} \equiv \Delta\overline{\rho} \qquad (12)$$

Thus, the right-hand side consists of all the ranging and calculation errors expressed in meters as we have defined them. Distance and time can be equated by recalling that light travels one meter in about three ns.

---

* The situation for the generalized inverse, Eq. (11b), is somewhat more complicated and is not presented here. The same expression for $\Delta\overline{x}$ is obtained. The same generalized development can then be used.

$$\text{for } K = 4, \quad \Delta \bar{x} = \mathbf{G}^{-1} \Delta \bar{\rho}_c \qquad (13a)$$

$$\text{for } K > 4, \quad \Delta \bar{x} = (\mathbf{G}^T \mathbf{G})^{-1} \mathbf{G}^T \Delta \bar{\rho}_c \qquad (13b)$$

where $\Delta \bar{x}$ is the positioning error in meters.

III. Geometric Dilution of Precision

It is intuitively obvious that satellite geometry can affect the accuracy of Eq. (13). This section develops the quantitative tools to understand the "dilution" of precision caused by various satellite geometrical configurations.

A. Derivation of the Geometric Dilution of Precision Equation

Now the well-known geometric dilution of precision (GDOP) equation can be easily derived. The covariance of position (m) is calculated as follows:

$$\text{cov(position)} = E(\Delta \bar{x} \cdot \Delta \bar{x}^T) = (\mathbf{G}^T \mathbf{G})^{-1} \mathbf{G}^T \cdot E[(\Delta \bar{\rho} \cdot \Delta \bar{\rho}^T)] \cdot \mathbf{G}(\mathbf{G}^T \mathbf{G})^{-1}, \; K > 4 \qquad (14a)$$

$$\text{cov(position)} = \mathbf{G}^{-1} \cdot E[\Delta \bar{\rho} \cdot \Delta \bar{\rho}^T] \cdot \mathbf{G}^{-T}, \; K = 4 \qquad (14b)$$

where E is the expectation operator. Because the $\mathbf{G}$ matrix does not have a random component, it has been brought outside the expectation operation.

If all ranging errors have the same variance $[\sigma_R^2]$ (m²) and are uncorrelated zero mean $(E[\Delta \rho_i \Delta \rho_j] = 0, i \neq j)$, then the expectation in Eqs. (14a and 14b) becomes $\sigma_R^2 \cdot U$, where U is the 4 × 4 identity matrix. Then both Eqs. (14) collapse to the following:

$$\text{cov(position)} = \sigma_R^2 \cdot [\mathbf{G}^T \mathbf{G}]^{-1}$$

Therefore, $[\mathbf{G}^T \mathbf{G}]^{-1}$ is the matrix of multipliers of *ranging* variance to give *position* variance. It is known as the GDOP or geometric dilution of precision matrix. If the position coordinates are the ordered right-hand set, east, north and up, then the *square root* of the ordered diagonal terms from the upper left are: east DOP, north DOP, vertical DOP, and time DOP. Note that σ_R^2 is expressed in m² rather than s².

cov(Position) =

$$\sigma_R^2(\text{m}^2) \begin{bmatrix} (\text{East DOP})^2 & & & \text{covariance terms} \\ & (\text{North DOP})^2 & & \\ & & (\text{Vertical DOP})^2 & \\ \text{covariance terms} & & & (\text{Time DOP})^2 \end{bmatrix}$$

The scalar GDOP is defined to be the square root of the trace of the GDOP matrix. Also HDOP (horizontal) = $\sqrt{(\text{NorthDOP})^2 + (\text{EastDOP})^2}$, and PDOP-(position) = $\sqrt{(\text{HDOP})^2 + (\text{VDOP})^2}$.

For satellites constrained to be above a minimum elevation angle* (greater than 0 deg), the best GDOP for $K = 4$ is obtained when one satellite is overhead and the other three are equally spaced at the minimum elevation angles around the horizon. In later chapters, we explore the power of pseudolites that provide ranging signals *transmitted from the ground*. These have negative elevation angles for an aircraft and can significantly improve geometry.

B. Power of the GDOP Concept

The concept of GDOP is a powerful tool for GPS. All receivers use some algorithm based on GDOP to select the best set of satellites to track among the group of up to 11 satellites in view. *Positioning* accuracy can then be estimated as the *ranging* accuracy multiplied by a dilution factor. This dilution factor (DOP) depends solely on geometry.

Typically, variations in geometry are far greater than variations in ranging accuracy for the nominal satellite constellation. The GDOP concept also quantizes the effect when the nominal satellites are not in view. Examples include local terrain shading, satellite outages, and user shading caused by vehicle extensions such as aircraft wings, etc. During these circumstances, the GDOP calculation for those satellites still being tracked will give the multiplier on ranging accuracies to yield positioning accuracies.

C. Example Calculations

To gain insight into the GDOP concept, some sample calculations are useful. They reveal the tradeoff of accuracy with satellite location. We define satellite direction as azimuth (Az-measured 360 deg clockwise from true north) and elevation (E-measured up from local horizontal—0–90 deg). This can be translated into the east, north, up coordinate frame, and the G matrix becomes:

$$\begin{bmatrix} (\cos(E_1)*\sin(Az_1)) & (\cos(E_1)*\cos(Az_1)) & \sin(E_1) & 1 \\ (\cos(E_2)*\sin(Az_2)) & (\cos(E_2)*\cos(Az_2)) & \sin(E_2) & 1 \\ (\cos(E_3)*\sin(Az_3)) & (\cos(E_3)*\cos(Az_3)) & \sin(E_3) & 1 \\ (\cos(E_4)*\sin(Az_4)) & (\cos(E_4)*\cos(Az_4)) & \sin(E_4) & 1 \end{bmatrix}$$

Using only satellites as ranging sources, the best accuracy† is found with three satellites equally spaced on the horizon, at minimum elevation angle, and one satellite directly overhead (See Table 1).

---

* Elevation angle is the angle of the satellite above the local horizontal.

† Note that the lower satellite elevation angles tend to have the greater *ranging errors*. The statement is that this configuration offers the best *geometry*. Usually geometry has a larger effect on accuracy than ranging errors.

Table 1 Satellite location

	Sat 1	Sat 2	Sat 3	Sat 4
Elevation, deg	5	5	5	90
Azimuth, deg	0	120	240	0

The following example, for 5-deg minimum satellite elevation, illustrates the technique. Using the best satellite geometry previously described, and the formulas for GDOP we get the following:

$$\mathbf{G} = \begin{bmatrix} 0.000 & 0.996 & 0.087 & 1.000 \\ 0.863 & -0.498 & 0.087 & 1.000 \\ -0.863 & -0.498 & 0.087 & 1.000 \\ 0.000 & 0.000 & 1.000 & 1.000 \end{bmatrix}$$

Recall that the position and time solution (**x**) is simply given by $\mathbf{G}^{-1}$ (for the four-measurement case*) times the column of corrected pseudoranges ρ_c. **G** inverse for the above **G** is listed below. Note that each of the first three rows sums to exactly zero. This implies that any common bias in ranging measurements will not affect the position solution. Note also that the last row sums to one. This is the row that gives the time solution (or bias in the local clock). This implies that any common error shows up solely as an error in the local clock. To see these two results, simply multiply on the right by a column whose elements are all the same value. These two results are always true. They arise from inverting the **G** matrix, which always has a constant last column of ones.

$$\mathbf{G}^{-1} = \begin{bmatrix} 0.000 & 0.580 & -0.580 & 0.000 \\ 0.670 & -0.335 & -0.335 & 0.000 \\ -0.365 & -0.365 & -0.365 & 1.095 \\ 0.365 & 0.365 & 0.365 & -0.095 \end{bmatrix}$$

The GDOP matrix for this example $(\mathbf{G}^T\mathbf{G})^{-1}$ is shown below. Note that, in this case, all off-diagonal terms are zero except for those correlating vertical errors and time. This correlation is negative, which implies that the errors tend to have opposite signs. In general, large correlations will be found between vertical errors and timing errors, and in general, off-diagonal terms will not be zero.

---

*For a greater number of satellites, $\mathbf{G}^{-1}$ is replaced with $(\mathbf{G}^T\mathbf{G})^{-1}\mathbf{G}^T$.

$$(\mathbf{G}^T\mathbf{G})^{-1} = \begin{bmatrix} 0.672 & 0.000 & 0.000 & 0.000 \\ 0.000 & 0.672 & 0.000 & 0.000 \\ 0.000 & 0.000 & 1.600 & -0.505 \\ 0.000 & 0.000 & -0.505 & 0.409 \end{bmatrix}$$

Listed below are the results of taking the square roots of appropriate diagonal terms of the GDOP matrix above:

HDOP (horizontal DOP) = 1.16   TDOP (time DOP) = 0.64

VDOP (vertical DOP) = 1.26   GDOP = 1.83

PDOP (position DOP) = 1.72

A PDOP of 1.72 is very good. In fact, it is the optimum for four satellites with a minimum elevation angle of 5 deg. A more representative median (50th percentile) worldwide result for PDOP is about 2.5. With all satellites of a 24-satellite constellation available, PDOP numbers as high as six or seven will be found. Thus, geometry can affect the results by a factor of as much as five or more.

D. Impact of Elevation Angle on GDOP

Figure 1 illustrates the impact of satellite position on DOP.[3,4] Maintaining the same symmetric spacing in azimuth, the minimum elevation angle is varied, including negative angles that could only be achieved with Earth-based transmitters (pseudolites).

The HDOP is quite flat over the whole range, which reflects the optimum azimuthal configuration. Satellites more concentrated in one-half of the sky than the other would cause greater variation in these results. Another interesting conclusion is that the lower elevation angles are of significant help for the vertical position.

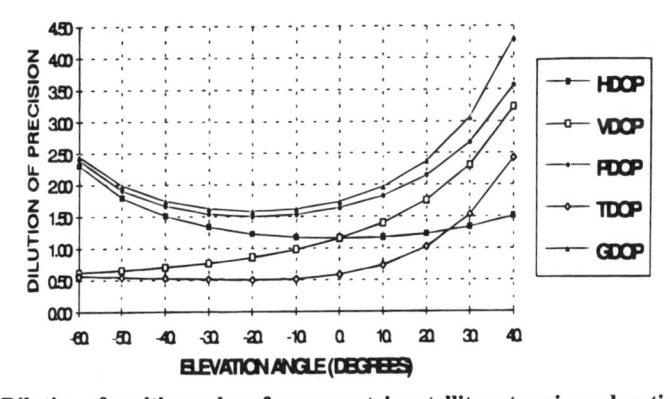

Fig. 1   Dilution of position values for symmetric satellites at various elevation angles above the horizon. The minimum GDOP is at −19.5 deg (i.e., below the local horizontal). This corresponds to locating four transmitters at the apexes of a regular tetrahedron.

IV. Ranging Errors

A. Six Classes of Errors

Ranging errors are grouped into the six following classes:
1) *Ephemeris data*—Errors in the transmitted location of the satellite
2) *Satellite clock*—Errors in the transmitted clock, including SA
3) *Ionosphere*—Errors in the corrections of pseudorange caused by ionospheric effects
4) *Troposphere*—Errors in the corrections of pseudorange caused by tropospheric effects
5) *Multipath*—Errors caused by reflected signals entering the receiver antenna
6) *Receiver*—Errors in the receiver's measurement of range caused by thermal noise, software accuracy, and interchannel biases

Each class is briefly discussed in the following sections. Representative values for these errors are used to construct an error table in a later section of this chapter. A more complete discussion of individual error sources can be found in succeeding chapters.

B. Ephemeris Errors

Ephemeris errors result when the GPS message does not transmit the correct satellite location. It is typical that the radial component of this error is the smallest: the tangential and cross-track errors may be larger by an order of magnitude. Fortunately, the larger components do not affect ranging accuracy to the same degree. This can be seen in the fundamental error Eq. (12). The $\Delta \overline{R}$ represents each satellite position error, but when dot-multiplied by the unit satellite direction vector (in the A matrix), only the projection of satellite positioning error along the line of sight creates a ranging error.

Because satellite errors reflect a position prediction, they tend to grow with time from the last control station upload. It is possible that a portion of the deliberate SA error is added to the ephemeris as well. However, the predictions are long smooth arcs, so all errors in the ephemeris tend to be slowly changing with time. Therefore, their utility in SA is quite limited.

As reported during phase one, (Bowen, 1986) in 1984,[5] *for predictions of up to 24 hours, the rms ranging error attributable to ephemeris was 2.1 m.* These errors were closely correlated with the satellite clock, as we would expect. Note that these errors are the same for both the P- and C/A-codes (see Chapter 16 of this volume for a more detailed discussion of ephemeris and clock errors).

C. Satellite Clock Errors

Fundamental to GPS is the one-way ranging that ultimately depends on satellite clock predictability. These satellite clock errors affect both the C/A- and P-code users in the same way. The error effect can be seen in the fundamental error Eq. (11) as ΔB. This effect is also independent of satellite direction, which is important when the technique of differential corrections is used. All differential stations and users measure an identical satellite clock error.

A major source of apparent clock error is SA, which is varied so as to be unpredictable over periods longer than about 10 minutes. The rms value of SA is typically about 20 m in ranging, but this can change after providing appropriate notice, depending on need. The U.S. Air Force has guaranteed that the two-dimensional rms (2 DRMS) positioning error (approximately 90th percentile) will be kept to less than 100 m. This is now a matter of U.S. federal policy and can only be changed by order of the President of the United States.

More interesting is the underlying accuracy of the system with SA off. The ability to predict clock behavior is a measure of clock quality. GPS uses atomic clocks (cesium and rubidium oscillators),[1] which have stabilities of about 1 part in 10^{13} over a day. If a clock can be predicted to this accuracy, its error in a day ($\sim 10^5$ s) will be about 10^{-8} s or about 3.5 m. The experience reported in 1984 was 4.1 m for 24-hour predictions. *Because the standard deviations of these errors were reported to grow quadratically with time, an average error of 1–2 m for 12-hour updates is the normal expectation.*

D. Ionosphere Errors

Because of free electrons in the ionosphere, GPS signals do not travel at the vacuum speed of light as they transit this region. The modulation on the signal is *delayed* in proportion to the number of free electrons encountered and is also (to first order) proportional to the inverse of the carrier frequency squared ($1/f^2$). The phase of the radio frequency carrier is *advanced* by the same amount because of these effects. Carrier-smoothed receivers should take this into account in the design of their filters. The ionosphere is usually reasonably well-behaved and stable in the temperate zones; near the equator or magnetic poles it can fluctuate considerably. An in-depth discussion of this can be found in Chapter 12, this volume.

All users will correct the raw pseudoranges for the ionospheric delay. The simplest correction will use an internal diurnal model of these delays. The parameters can be updated using information in the GPS communications message (although the accuracy of these updates is not yet clearly established). *The effective accuracy of this modeling is about 2–5 m in ranging for users in the temperate zones.*

A second technique for *dual-frequency P-code* receivers is to measure the signal at both frequencies and directly solve for the delay. The difference between L_1 and L_2 arrival times allows a direct algebraic solution. *This dual-frequency technique should provide 1–2 m of ranging accuracy, due to the ionosphere, for a well-calibrated receiver.*

A third technique is to rely on a near real-time update. An example would be the proposed Wide Area Differential GPS system (WADGPS). *This should also produce corrections with accuracies of 1–2 m or better in the temperate zones of the world.*

E. Troposphere Errors

Another deviation from the vacuum speed of light is caused by the troposphere. Variations in temperature, pressure, and humidity all contribute to variations in the speed of light of radio waves. Both the code and carrier will have the same delays. This is described further in the chapter devoted to these effects, Chapter

13 of this volume. *For most users and circumstances, a simple model should be effectively accurate to about 1 m or better.*

F. Multipath Errors

Multipath is the error caused by reflected signals entering the front end of the receiver and masking the real correlation peak. These effects tend to be more pronounced in a static receiver near large reflecting surfaces, where 15 m or more in ranging error can be found in extreme cases. Monitor or reference stations require special care in siting to avoid unacceptable errors. The first line of defense is to use the combination of antenna cut-off angle and antenna location that minimizes this problem. A second approach is to use so-called "narrow correlator" receivers which tend to minimize the impact of multipath on range tracking accuracies. *With proper siting and antenna selection, the net impact to a moving user should be less than 1 m under most circumstances.* See Chapter 14 of this volume for further discussion of multipath errors.

G. Receiver Errors

Initially most GPS commercial receivers were sequential in that one or two tracking channels shared the burden of locking on to four or more satellites. With modern chip technology, it is common to place three or more tracking channels on a single inexpensive chip. As the size and cost have shrunk, techniques have improved and five- or six-channel receivers are common. Most modern receivers use reconstructed carrier to aid the code tracking loops. This produces a precision of better than 0.3 m. Interchannel bias is minimized with digital sampling and all-digital designs.

The limited precision of the receiver software also contributed to errors in earlier designs, which relied on 8-bit microprocessors. With ranges to the satellites of over 20 million meters, a precision of $1:10^{10}$ or better was required. Modern microprocessors now provide such precision along with the co-requisite calculation speeds. *The net result is that the receiver should contribute less than 0.5 ms error in bias and less than 0.2 m in noise.* Further information on receiver errors is available in Chapters 3, 7, 8, and 9 of this volume.

V. Standard Error Tables

These overview discussions on error sources and magnitudes, as well as the effects of satellite geometry, can be summarized with the following error tables. Each error is described as a bias (persistence of minutes or more) and a random effect that is, in effect "white" noise and exhibits little correlation between samples of range. The total error in each category is found as the root sum square (rss) of these two components.

Each *component* of error is assumed to be statistically uncorrelated with all others, so they are combined as an rss as well. The receiver is assumed to filter the measurements so that about 16 samples are effectively averaged reducing the random content by the square root of 16. Of course, averaging cannot improve the bias-type errors.

GPS ERROR ANALYSIS

Finally, each satellite error is assumed to be uncorrelated and of zero mean, so the application of HDOP and VDOP are justified as the last step. Despite these limiting assumptions, the resulting error model has proved to be surprisingly valid. Of course, the assumptions on uncorrelated errors is almost always violated to some degree. For example, if the estimate of zenith ionosphere delay is in error, a proportional error is induced in all measurements through the obliquity calculation. Clearly, such an error would be correlated. These and other correlations have not caused serious problems in the use of this model.

A. Error Table without SA: Normal Operation for C/A Code

Table 2 assumes that SA is not operating. Consequently, the residual satellite clock error, at 2.1 m, is not the dominant error; in fact, the largest error is expected to be the mismodeling of the ionosphere, at 4.0 m. Thus, the worldwide civilian positioning error for GPS is potentially about 10 m (horizontal), as shown in Table 2.

B. Error Table with SA

A second example shows the impact of SA on these errors. Because the deliberately mismodeled clock so dominates the ranging error, all other effects could be safely ignored in the error budget. The results of Table 3 have been repeatedly corroborated by actual measurements. Note that SA is listed as a bias because it cannot be averaged to zero with a 1 s (or less) filter. Selective availability is expected to be zero mean, but only when averaged over many hours or perhaps days. Of course, such averaging is not practical for a dynamic user who only sees the satellite for a portion of the orbit. If differential corrections are used,

Table 2 Standard error model—no SA

Error source	One-sigma error, m		
	Bias	Random	Total
Ephemeris data	2.1	0.0	2.1
Satellite clock	2.0	0.7	2.1
Ionosphere	4.0	0.5	4.0
Troposphere	0.5	0.5	0.7
Multipath	1.0	1.0	1.4
Receiver measurement	0.5	0.2	0.5
User equivalent range error (UERE), rms[a]	5.1	1.4	5.3
Filtered UERE, rms	5.1	0.4	5.1
Vertical one-sigma errors—VDOP= 2.5			12.8
Horizontal one-sigma errors—HDOP= 2.0			10.2

[a]This is the statistical ranging error (one-sigma) that represents the total of all contributing sources. The dominant error is usually the ionosphere. A horizontal error of 10 m (one-sigma) is the expected performance for the temperate latitudes using civilian (C/A-code) receivers.

Table 3 SA error model

Error source	One-sigma error, m		
	Bias	Random	Total
Ephemeris data	2.1	0.0	2.1
Satellite clock	20.0	0.7	20.0
Ionosphere	4.0	0.5	4.0
Troposphere	0.5	0.5	0.7
Multipath	1.0	1.0	1.4
Receiver measurement	0.5	0.2	0.5
UERE, rms	20.5	1.4	20.6
Filtered UERE, rms	20.5	0.4	20.5
Vertical one-sigma errors—VDOP= 2.5			51.4
Horizontal one-sigma errors—HDOP= 2.0			41.1

they will eliminate the SA error entirely (if corrections are passed at a sufficiently high data rate) as discussed in Chapter 21, this volume.

The 41-m horizontal error is a one-sigma (σ) result; under the existing agreement between the U.S. Department of Transportation (DOT) and the U.S. Department of Defense (DOD), the 2 DRMS horizontal error is to be less than 100 m. The impact on the vertical error is probably greater, because the VDOP value usually exceeds the HDOP value.

C. Error Table for Precise Positioning Service (PPS Dual-Frequency P/Y Code)

The errors for dual-frequency P/Y code are similar to those above except that SA errors are eliminated because the authorized user can decode the magnitude as part of a classified message. An expected horizontal error is less than 10 m. The ionosphere error is reduced to 1-m bias and about 0.7 m of noise by the dual-frequency measurement. The dominant sources are the satellite ephemeris and clocks. This is illustrated in Table 4.

VI. Summary

Excluding the deliberate degradation of SA, the dominant error source for satellite ranging with single frequency receivers is usually the ionosphere. It is on the order of four meters, depending on the quality of the single-frequency model. For dual-frequency (P/Y-code) receivers (which eliminate SA) the Standard Error Model of Table 1 has one principal change (in addition to the elimination of the SA error). The ionospheric error is reduced from four meters to about one meter.

Greater variations in the errors are due to geometry, which are quantified as dilutions of precision or DOPs. While geometric dilutions of 2.5 are about the worldwide average, this factor can range up to 10 or more with poor satellite geometry. Reduced satellite availability (and consequent increases in DOP) could

Table 4 Precise error model, dual-frequency, P/Y code

Error source	One-sigma error, m		
	Bias	Random	Total
Ephemeris data	2.1	0.0	2.1
Satellite clock	2.0	0.7	2.1
Ionosphere	1.0	0.7	1.2
Troposphere	0.5	0.5	0.7
Multipath	1.0	1.0	1.4
Receiver measurement	0.5	0.2	0.5
UERE, rms	3.3	1.5	3.6
Filtered UERE, rms	3.3	0.4	3.3
Vertical one-sigma errors—VDOP= 2.5			**8.3**
Horizontal one-sigma errors—HDOP= 2.0			**6.6**

be caused by satellite outages, local terrain masking, or user antenna tilting (for example due to aircraft banking). Typical normal accuracy (one-sigma) for well-designed civil equipment under nominal operating conditions *with SA off* should be about 10 m horizontal and 13 m vertical.

References

[1]Martin, E. H., "GPS User Equipment Error Models," *Global Positioning System Papers,* Vol. I, Institute of Navigation, Washington, DC, 1980, pp. 109–118.

[2]Milliken, R. J., and Zollar, C. J., "Principle of Operation of NAVSTAR and System Characteristics," *Global Positioning System Papers,* Vol. I, Institute of Navigation, Washington, DC, 1980, pp. 3–14.

[3]Copps, E. M., "An Aspect of the Role of the Clock in a GPS Receiver," *Global Positioning System Papers,* Vol. III, Institute of Navigation, Washington, DC, 1986.

[4]Massat, P., and Rudnick, K., "Geometric Formulas for Dilution of Precision Calculations," *Navigation,* Vol. 37, No. 4, 1990–1991.

[5]Bowen, R., et al., "GPS Control System Accuracies," *Global Positioning System Papers,* Vol. III, Institute of Navigation, Washington, DC, 1986, pp. 241–247.

Chapter 12

Ionospheric Effects on GPS

J. A. Klobuchar*
Hanscom Air Force Base, Massachusetts 01731

I. Introduction

THE ionosphere is an important source of range and range-rate errors for users of the global positioning system (GPS) satellites who require high-accuracy measurements. At times, the range errors of the troposphere and the ionosphere can be comparable, but the variability of the Earth's ionosphere is much larger than that of the troposphere, and it is more difficult to model. The ionospheric range error can vary from only a few meters, to many tens of meters at the zenith, whereas the tropospheric range error at the zenith is generally between two to three meters. Fortunately, the ionosphere is a dispersive medium; that is, the refractive index is a function of the operating frequency, and two-frequency GPS users can take advantage of this property of the ionosphere to measure and correct for the first-order ionospheric range and range-rate effects directly. Unlike the troposphere, the ionosphere can change rapidly in absolute value. Although the range error of the troposphere generally does not change by more than ±10%, even over long periods of time, the ionosphere frequently changes by at least one order of magnitude during the course of each day. The major effects the ionosphere can have on GPS are the following: 1) group delay of the signal modulation, or absolute range error; 2) carrier phase advance, or relative range error; 3) Doppler shift, or range-rate errors; 4) Faraday rotation of linearly polarized signals; 5) refraction or bending of the radio wave; 6) distortion of pulse waveforms; 7) signal amplitude fading or amplitude scintillation; and 8) phase scintillations.

In order to understand the reasons for these potential effects on GPS performance, a brief description of the major characteristics of the ionosphere is necessary.

II. Characteristics of the Ionosphere

To first order, the ionosphere is formed by the ultraviolet (uv) ionizing radiation from the Sun. Different regions of the ionosphere are produced by different

---

This paper is declared a work of the U.S. Government and is not subject to copyright protection in the United States.
*Geophysics Directorate, Ionospheric Effects Division, Phillips Laboratory.

chemical species. The ionosphere is a weakly ionized plasma, or gas, which can affect radiowave propagation in various ways. The electron and ion densities are assumed to be equal in the ionosphere, and the density of the ions is much less than 1% of the neutral density at all heights. When the ionosphere was first discovered, the original regions were named with the alphabetic letters E and F, for electric and field,[1] with the thought that regions of less density, and earlier letters of the alphabet, would eventually be found at lower heights. Today we know that the ionosphere is composed of the D, E, F1, and F2 regions, named in order of increasing height. Figure 1 illustrates the different regions and their electron densities in the ionosphere. The D, E, and F1 regions are closely tied to the uv ionizing daytime radiation from the sun, and are not present at night. The F2 region is present at night, but it is lower in density and generally has its maximum density at a greater height during the night, as compared with daytime.

The various regions of the ionosphere are produced by different wavelengths of radiation from the sun, with the harder solar radiation, namely x rays, penetrating farther into the neutral atmosphere, and the less intense uv radiation being stopped at greater heights where they produce ionization. Much early work was done by Chapman[2] in developing the mathematics of the production of ionization of the atmosphere.

Because the neutral atmosphere is approximately in diffusive equilibrium, the scale height of each neutral atomic and modecular species falls off exponentially with increasing height above the Earth's surface. Thus, the total neutral density is mostly composed of the heavier, molecular species at lower heights, and of the lighter, atomic species at greater heights. Above approximately 180 km, electron diffusion becomes important, and electrons generated by solar uv emissions are free to move to greater heights following the Earth's magnetic lines of force. Additional changes in electron density above approximately 180 km are produced by electric fields that cause electrons to move in a direction perpendicular to the magnetic lines of force, while neutral winds can cause electrons to

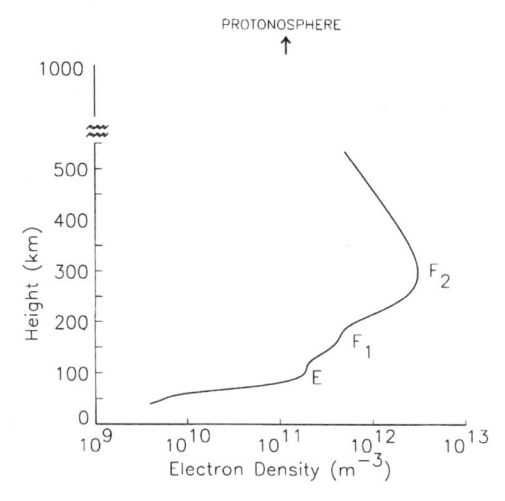

Fig. 1 Electron density of the different regions of the ionosphere vs height for daytime conditions.

flow either up or down the Earth's magnetic field lines, further complicating the specification of electron density at any given height above where diffusion becomes important.

The chemistry of the F2 region is predominately attributable to the ionization of atomic oxygen, whereas the electrons at great heights are attributable to ionized hydrogen gas. The scale height of each species is: $H = kt/mg$, where k is Boltzmann's constant, T is the absolute temperature, m is the mass of the species, and g is the acceleration of gravity.

The scale height of each species is inversely proportional to its atomic weight; thus, the scale height of the electrons associated with H^+ is 16 times greater than that of the electrons due to O^+ ions. Charge neutrality generally is assumed to be the case in the ionosphere, thus the number of electrons always is equal to the number of ions. The scale height of the F2 region is typically 60 km, while that of the protonosphere is over 1,000 km. Thus, although the electron density of the protonosphere is small, the number of electrons does not fall off very fast with increasing height. Therefore, the number of electrons in the protonosphere can be an important fraction of the total, especially during the nighttime periods when the electron density of the F2 region is small, as it normally is during the nighttime.

The parameter of the ionosphere that produces most of the effects on GPS signals is the total number of electrons in the ionosphere. This integrated number of electrons, commonly called the total electron content (TEC), is expressed as the number of electrons in a vertical column having a 1-m^2 cross section, and extending all the way from the GPS satellite to the observer. Details of the behavior of the TEC of the Earth's ionosphere are given in Sec. V of this chapter.

The electron density of the F2 region is not only the highest of the various regions, producing the greatest potential effects on many radiowave systems, but is the most highly variable, safely keeping many ionospheric researchers employed to understand better and be able to predict the physics of its detailed behavior. The major characteristics and importance of each region of the ionosphere for potential effects on GPS signals are summarized as follows (note that heights given are only approximate):

1) D region, 50–90 km: This region, produced by ionization of several molecular species from hard x rays and solar Lyman α radiation, causes absorption of radio signals at frequencies up to the low vhf band, and has no measurable effect on GPS frequencies.

2) E Region, 90–140 km: The normal E region, produced by solar soft x rays, has a minimal effect on GPS. An intense E region, with irregular structure, produced by solar particle precipitation in the auroral region, might cause minor scintillation effects. Sporadic E, still of unknown origin, is very thin and also has a negligible effect at GPS frequencies.

3) F1, 140–210 km: The normal F1 region, combined with the E region, can account for up to 10% of the ionospheric time delay encountered by GPS. Diffusion is not important at F1 region heights, and, as with the normal E region, it has a highly predictable density from known solar emissions. The F1 region is produced through ionization of molecular species, and its electron density nicely merges into the bottomside of the F2 region.

4) F2, 210–1,000 km: The F2 region is the most dense and also has the highest variability, causing most of the potential effects on GPS receiving systems. The height of the peak of the electron density of the F2 region generally varies from 250 to 400 km, but it can be even much higher or somewhat lower under extreme conditions. The F2 region is produced mainly from ionization of atomic oxygen, which is the principal constituent of the neutral atmosphere at those heights. The F2, and to some extent the F1, regions, cause most of the problems for radiowave propagation at GPS frequencies.

5) $H^+ > 1,000$ km: The protonosphere, is a region of ionized hydrogen, with a lesser contribution from helium gas. It is of low density, but extends out to approximately the orbital height of GPS satellites. It can be a significant source of unknown electron density and consequent variability of time delay for GPS users. Estimates of the contribution of the protonosphere vary from 10% of the total ionospheric time delay during daytime hours, when the electron density of the F2 region is highest, to approximately 50% during the nighttime, when the F2 region density is low. The electron content of the protonosphere does not change by a large amount during the day, but is depleted during major magnetic storms and can take several days to recover to prestorm values.

III. Refractive Index of the Ionosphere

In order to quantify the propagation effects on a radio wave traveling through the ionosphere, the refractive index of the medium must be specified. The refractive index of the ionosphere, n, has been derived by Appleton and Hartree,[1] and it can be expressed as

$$n^2 = 1 - \frac{X}{1 - iZ - \frac{Y_T^2}{2(1 - X - iZ)} \pm \left[\frac{Y_T^4}{4(1 - X - iZ)^2} + Y_L^2\right]^{1/2}} \quad (1)$$

where $X = Ne^2/\epsilon_0 m\omega^2 = f_n^2/f^2$, $Y_L = eB_L/m\omega, = f_H \cos\theta/f$, $Y_T = eB_T/m\omega = f_H \sin\theta/f$, $Z = \nu/\omega$, $w = 2\pi f$, where f is the system operating frequency, in Hz and

- e = electron charge, -1.602×10^{-19} coulomb
- ϵ = permittivity of free space, $= 8.854 \times 10^{-12}$ farad/m
- m = rest mass of a electron, $= 9.107 \times 10^{-31}$ kg
- θ = the angle of the ray with respect to the Earth's magnetic field
- ν = the electron-neutral collision frequency
- f_H = the electron gyro frequency

The electron gyro frequency f_H is typically 1.5 MHz; the plasma frequency f_N rarely exceeds 20 MHz; and the collision frequency, f_ν, is approximately 10^4 Hz. Thus, to an accuracy of better than 1%, the refractive index of the ionospheric is given by the following:

$$n = 1 - (X/2) \quad (2)$$

The ionospheric refractive index is the basis for the effects on GPS signals described later, and the first-order form is sufficient for most purposes. Higher-order corrections are described in Sec. IV.D.

IV. Major Effects on Global Positioning Systems Caused by the Ionosphere

Knowing the refractive index of the ionosphere it is possible to derive the group delay or absolute range error; the carrier phase advance or relative range error; and the Doppler shift or range-rate error. It is also possible to calculate the potential effects of Faraday rotation and refraction, or bending, of the radio wave. The distortion of pulse waveforms is described briefly. Finally, the effects of signal fading, or amplitude scintillation, and phase scintillations are described. All of these effects are produced because the refractive index of the ionosphere differs from unity.

A. Ionospheric Group Delay—Absolute Range Error

1. Single-Frequency Group Delay

The group delay of the ionosphere produces range errors, which can be expressed either in units of distance, or in units of time delay, to GPS users. This group delay can be determined by

$$\Delta t = \frac{1}{c} \int (1 - n) dl \tag{3}$$

or

$$\Delta r = \int (1 - n) dl \tag{4}$$

At L-band the first-order refractive index is, $n = 1 - X/2$, where

$$X = \frac{40.3}{f^2} \int N dl$$

and the ionospheric group delay is

$$\Delta t = \frac{40.3}{cf^2} \int N dl, \ldots, \text{seconds} \tag{5}$$

The quantity $\int N dl$ is the TEC, in el/m^2, integrated along the path from observer to each GPS satellite. The temporal and spatial variations of TEC, which are responsible for the variability of ionospheric time delay to GPS users, are described in Sec. V.

2. Dual-Frequency Group Delay

By measuring the group path delay independently at the two, widely spaced GPS frequencies, $L_2 = f_2$ and $L_1 = f_1$, the TEC along the path from satellite to receiver can be measured directly. A dual-frequency GPS receiver measures the difference in ionospheric time delay at $L_2 - L_1$, referred to as $\delta(\Delta t)$. From Eq. (5), we obtain

$$\delta(\Delta t) = (40.3/c) \times \text{TEC} \times [(1/f_2^2) - (1/f_1^2)] = \Delta t_1 [(f_1^2 - f_2^2)/f_2^2] \tag{6}$$

or

$$\Delta t_1 = [f_2^2/(f_1^2 - f_2^2) \times \delta(\Delta t)] \tag{7}$$

where Δt_1 is the ionospheric time delay at L_1.

The value $\delta(\Delta t)$ is obtained from the difference of the simultaneous measurements of the total range, including ionospheric time delay, at the two frequencies f_1 and f_2, because the geometric distance is, of course, the same at all frequencies. The quantity, is $[f_2^2/(f_2^2 - f_2^2)]$ is called the ionospheric scaling factor. For the GPS pair of frequencies, this factor is 1.546.

If the two GPS frequencies had been chosen to be too closely spaced, the differential ionospheric time delay between them would have been very small and would have been masked by the receiving system noise. A wider frequency separation between L_1 and L_2 would have made the measurement capability of absolute ionospheric range error more precise, but probably would have required two separate transmitting and receiving antennas, and more elaborate transmitter and receiver designs. The frequency separation between L_1 and L_2, giving an ionospheric scaling factor of 1.546, is a reasonable compromise between system hardware design and absolute ionospheric range error/time delay requirements. Although Δt_1 is 1.546 times the difference between the two, relatively noisy, pseudorange measurements this differential time delay can be averaged over many samples, and, hence, can be measured to subnanosecond accuracy. The limitations in measuring absolute differential pseudoranges are mostly caused by multipath and lack of a precise knowledge of the differential pseudorange as transmitted from each GPS satellite.

B. Ionospheric Carrier Phase Advance

1. Single-Frequency Ionospheric Carrier Phase Advance

The carrier phase advance, as compared with the received carrier phase in the absence of an ionosphere, can be expressed as

$$\Delta \phi = \frac{1}{\lambda} \int (1 + n) \, dl, \ldots, \text{cycles, or wavelengths} \tag{8}$$

or

$$\Delta \phi = \frac{f}{2c} \int X dl = \frac{40.3}{cf} \int N dl = \frac{1.34 \times 10^{-7}}{f} \int N dl, \ldots, \text{cycles} \tag{9}$$

Remember that $v_g v_\phi = c^2$, where v_g, v_ϕ are the group and phase velocities, respectively. Although the carrier phase travels faster than the velocity of light, it carries no information, and thus, communication does not occur faster than the velocity of light.

2. Differential Carrier Phase Advance

As a radio signal traverses the ionosphere, the phase of the carrier of the radio frequency transmission is advanced from its velocity in free space. In practice, the amount of this phase advance cannot be measured readily on a single frequency unless both the transmitter and the receiver have exceptional oscillator stability and the satellite orbital characteristics are extremely well known. Usually two, coherently derived, frequencies are required for this measurement. In the case of the GPS satellites, the L_1 and L_2 transmitted carriers are phase coherent, both being derived from a common 10.23 MHz oscillator. The differential carrier phase shift (δ_ϕ) between the two frequencies then can be measured. That differential measurement is related to TEC by

$$\Delta\delta_\phi = [(1.34 \times 10^{-7})/f_L \times [(m^2 - 1)/m^2]/\text{TEC}, \ldots, \text{cycles} \qquad (10)$$

where $m = f_1/f_2$. The GPS system uses differential carrier phase to correct automatically for range-rate errors in its system. Differential carrier phase also provides a very precise measure of changes in relative TEC during a satellite pass, but, because of the unknown number of differential cycles of phase, absolute TEC values must be obtained from the differential group delay measurement.

3. Relationship Between Carrier Phase Advance and Group Delay

The relationship between group delay and carrier phase is simply

$$\Delta\phi = -f\Delta t \qquad (11)$$

or, for every cycle of carrier phase advance, there are

$$1/f, \ldots, \text{seconds} \qquad (12)$$

of time delay. In the case of GPS at L_1, one cycle of carrier phase advance is equivalent to 0.635 ns of group delay. The minus sign in Eq. (11) is meant to indicate that the differential code group delay and the differential carrier phase advance move in opposite directions. We must be careful to note this when using carrier-aided code tracking, if the ionosphere changes significantly during the observation period. It is also possible to measure *relative* ionospheric changes using the L_1 C/A code minus L_1 carrier phase during a pass.

4. Useful Numbers for GPS Are Shown in Table 1.

C. Higher-Order Ionospheric Effects

Because we are concerned here only with radiowave propagation at GPS frequencies, the terms X, Y_T, Y_L, and Z in Eq. (1) are all much less than one. Thus, the refractive index of the ionosphere at GPS frequencies, as given by Brunner and Gu,[3] neglecting those terms whose magnitude is less than 10^{-9}, can be expressed as

$$n = 1 - (X/2) \pm (XY/2)\cos\theta - (X^2/8) \qquad (13)$$

As Eq. (13) illustrates, the terms contributing to the refractive index of the ionosphere are A: 1, the free-space velocity; B: $(X/2)$, the first-order, or $(1/f^2)$ term; C: $(XY/2)\cos\theta$, the second-order, or $(1/f^3)$, term; and D: $(X^2/8)$, the third-order, or $(1/f^4)$ term where f is the GPS operating frequency. X and Y are defined in Sec. III. Using the Brunner and Gu[3] derivation, the magnitudes

Table 1 Relationships between the various global positioning system first-order measured parameters and total electron content of Earth's ionosphere

L_2–L_1, differential group delay

360 deg, or 151.098 ns of delay, measured at L_1, or 97.75 ns of differential delay; i.e., 1 code chip
    360 deg = 278.83 × 10^{16} (el/m²);   1 deg = 0.7745 × 10^{16} (el/m²)
1 ns of differential code delay
    = 2.852 × 10^{16} (el/m²)
    = 1.546 ns of delay at L_1
    = 0.464 m of range error at L_1
1 ns of delay, measured at L_1
    = 1.8476 × 10^{16} (el/m²)
    = 0.300 m of range error at L_1
1 cycle, or 1 wavelength, 19.04 cm, of carrier phase advance at L_1 = 1.173 × 10^{16} (el/m²)
1 m of range error:
    measured at L_1 = 6.15 × 10^{16} (el/m²)
    measured at L_2 = 3.73 × 10^{16} (el/m²)
1 TEC unit [1 × 10^{16} (el/m²)]
    = 0.351 ns of differential delay
    = 0.542 ns of delay at L_1
    = 0.163 m of range error at L_1
    = 0.853 cycles of phase advance at L_1

L_2–L_1 differential carrier phase advance, measured at L_2

1 deg = 6.456 × 10^{13} (el/m²)
0.1 rad = 3.699 × 10^{14} (el/m²)
360 deg = 2.324 × 10^{16} (el/m²)
To convert differential carrier phase advance (measured at L_2) to an equivalent single frequency phase change at a specified frequency.

Frequency	Multiply GPS differential carrier phase by
244 MHz	12.81
1 GHz	3.125

of the higher-order terms at GPS frequencies, for maximum worldwide ionospheric conditions, are $B \approx 2 \times 10^{-4}$; $C \approx 2 \times 10^{-7}$; and $D \approx 2 \times 10^{-8}$.

The ratios of the higher-order terms, C and D, to the first order term, B, again under these worst case ionospheric conditions, is $C/B \approx 10^{-3}$; and $D/B \approx 10^{-4}$.

Thus, these higher-order terms are much less than 1% of the first-order term at GPS frequencies, even for the extremely high value of $f_n = 25$ MHz used for the maximum ionospheric plasma frequency, and for the GPS L_2 frequency used in the computation. For a more typical ionospheric maximum plasma frequency of, say, 12 MHz, and for GPS L_1 users, the higher-order terms in ionospheric refraction are even much less than those given here. Thus, within better than 0.1% accuracy, even during worst case ionospheric conditions, the ionospheric refractive index at GPS frequencies can be expressed simply as

$$n = 1 - (X/2) \qquad (14)$$

During times of high TEC, the first-order range error can be a few hundred meters. At these times, higher-order ionospheric effects can be several tens of centimeters of range error, which represent large errors in geodetic measurements. Brunner and Gu[3] have used the full form of the refractive index given in Eq. (1) to calculate the residual range error from the first-order form for refractive index. Their model also includes the geomagnetic field, and the effects of ray bending at both the GPS frequencies, L_1 and L_2. They claim that their improved form of dual-frequency ionospheric correction eliminates the ionospheric higher-order effects to better than 1 mm residual range error. However, in order to achieve this order of ionospheric error correction, they require knowledge of the actual maximum electron density, N_m, an ionospheric electron density profile shape factor they call η, and the average value of the longitudinal component of the Earth's magnetic field along the ray path, $\overline{B(\cos\theta)}$. In a practical case, these parameters are not easy to estimate.

Bassiri and Hajj[4] have done similar work on higher-order ionospheric range errors for GPS. They find the magnitudes of the second- and third-order terms at L_1 to be ~ 1.6 cm and ~ 0.9 mm, respectively, for a TEC of 10^{18} (el/m^2). Rather that requiring a knowledge of the electron density profile shape, as is the case with Brunner and Gu,[3] they use a constant ionospheric shape factor of 0.66, and they assume a constant height for the maximum of the electron density profile. Their form of higher-order ionospheric corrections are not as good as those claimed by Brunner and Gu,[3] but they are much easier to implement, and, in the practical case of actual ionospheric data required, may result in corrections of the same magnitude. GPS users who require ionospheric corrections to centimeter or millimeter accuracy should refer to the works of Brunner and Gu[3] and Bassiri and Hajj[4] for additional details.

D. Obtaining Absolute Total Electron Content from Dual-Frequency GPS Measurements

1. Removing Multipath Effects

If both the differential carrier phase and the differential group delay are measured with a dual-frequency GPS receiver, the user easily can obtain both the absolute TEC and its rate of change. Jorgensen,[5] first showed that the differential group delay could be used to fix the differential carrier phase to an absolute scale after a satellite pass, thus obtaining the best of both measurements; namely, the absolute scale obtained from the differential group delay and the precision of the differential carrier phase. Hatch[6] suggested that combining GPS carrier with code measurements results in better absolute and relative positioning. Because the differential carrier phase is much less sensitive to multipath, the final result, providing there are no unresolved carrier cycle slips during the pass, is a precise, smooth rendition of absolute TEC over an entire pass. The ratio of multipath effects on the rf carrier, as compared with multipath observed on the P-code modulation, is proportional to their respective wavelengths. That is, multipath on the L_2 carrier is 120 times smaller than on the 10.23 MHz modulation.

Figure 2 shows a typical pass of a GPS satellite. The top panel is the L_1 carrier signal strength divided by the receiver noise, on a linear amplitude scale. The middle panel shows the relative differential carrier phase on a relative scale.

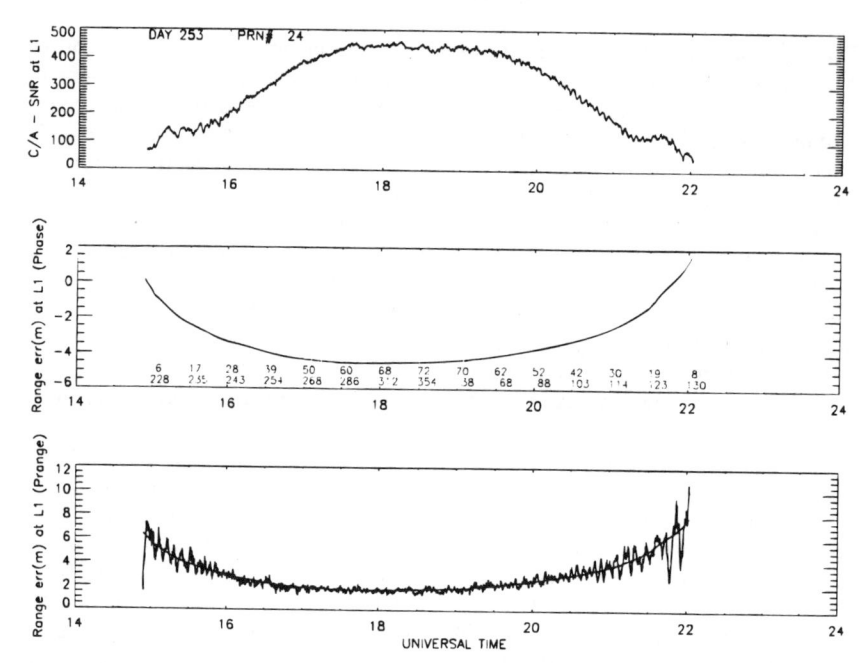

Fig. 2 Recording of a typical global positioning system pass, showing the carrier-assisted relative signal-to-noise ratio at L_1 (top panel), relative range error computed from the differential carrier phase (middle panel), and absolute rate error computed from the differential P-code group delay (bottom panel).

Normally the differential carrier phase is set at zero relative range error at the beginning of each pass. Elevation and azimuth values are printed in the middle panel, just under the relative differential phase record. The bottom panel shows the absolute ionospheric range error, in meters, obtained from differential group delay. Note the large multipath effects in the differential group delay, especially at both ends of the pass. The fact that the differential group variations are attributable to multipath effects becomes obvious if two, or more, successive days of passes are plotted to the same scale. The individual multipath variations agree exactly with a time shift corresponding to the sidereal difference in the time of passage of the GPS satellite along the same track over successive days.

The bottom portion of Fig. 2 also shows the differential carrier phase, now translated to an absolute scale. The process of converting the differential carrier phase to an absolute scale, by fitting to the absolute differential group delay, is done over the higher-elevation portions of each pass, when multipath is generally smallest. Multipath may not be a zero mean process and may not necessarily average out if the fitting procedure were done over the entire pass, or even worse, if the fitting were done only over the low-elevation portions of a pass.

After the differential carrier phase is converted to an absolute scale by fitting it to the differential group delay curve over the desirable, low multipath, portion of each pass, the differential group delay data are simply discarded, because they have done their task. The final TEC values are precise, accurate, and without multipath, unless the multipath environment is really terrible, in which case a small, residual amount of multipath can even be seen in the differential carrier phase.

If a user requires absolute TEC measurements soon after acquisition of each GPS satellite at a low-elevation angle, the fitting of the differential carrier phase to the differential group delay must be done for only a few minutes of data, usually where multipath on the differential group is large. There are several, relatively straightforward, schemes for removing multipath at low-elevation angles. One is to use a high-gain, directional antenna that has smaller multipath effects. An antenna with a large ground plane also can attenuate multipath reflections. Another method is to take advantage of the fact that the orbit of each GPS satellite repeats in its ground track over successive days, thereby giving the stationary user nearly the same multipath conditions. Suitable matching and filtering techniques can be employed to use the multipath for preceeding days to remove its effects in near real time. The user must take care to see that the day-to-day differences in multipath for each GPS satellite really are the same. GPS antennas deployed in locations where the multipath effects may change from day to day, such as near automobile parking lots or aircraft terminals, may not be suitable for this procedure.

2. *Automated Ionospheric Range and Range-Rate Error Corrections*

In using an operational dual-frequency GPS receiver, all the processes described above are done automatically and are transparent to the user. If the ionospheric first-order range and range-rate error corrections are done for each satellite soon after each satellite is acquired, it is likely that only the differential group delay is used to obtain an absolute range error correction. However, the process of determining an absolute ionospheric correction by using both the differential carrier phase and the differential group delay, with an improving fit of the phase data to the group delay data as more data are acquired, could be used routinely.

E. Ionospheric Doppler Shift/Range-Rate Errors

Because frequency is simply the time derivative of phase, an additional contribution to geometric Doppler shift results because of changing TEC. This additional frequency shift is generally small compared with the normal geometric Doppler shift, but can be computed by

$$\Delta f = (d n/dt) = [(1.34 \times 10^{-7})/f] \, (d/dt) \, \text{TEC}, \ldots, \text{Hz} \qquad (15)$$

For high-orbit satellites, such as the GPS satellites, where the apparent satellite motion across the sky is slow, the diurnal changes in TEC are generally greater than the geometric ones. An upper limit to the rate of change of TEC, for a stationary user, is approximately 0.1×10^{16} (el/m^2) per second. This value yields an additional frequency shift of 0.085 Hz at L_1, which would not be significant compared with a typical required receiver carrier tracking loop bandwidth of at least a few Hz. The value of 0.085 Hz at L_1 corresponds to 1.6 cm/s of range-rate error.

Ionospheric range-rate or Doppler shift errors are attributable to the time rate of change of the electron content of the ionosphere, as seen by the observing system. The range-rate error depends upon the diurnal rate of change of the electron content of the ionosphere, the structure of any large-scale irregularities that may exist in the region, and the motion of any vehicle. For instance, a GPS satellite moving up from the horizon will usually encounter fewer electrons as it rises in elevation, simply because of the decrease of the signal path length in the ionosphere. An observer in a high-velocity aircraft, or even more so, in a low-orbit spacecraft, generally will encounter geometric changes far greater than the temporal rate of change of electron content in the ionosphere.

For the slowly moving GPS satellites, the satellite motion, diurnal changes in the ionosphere, and observer platform motion, all contribute to ionospheric range-rate errors. It is difficult to model the relatively high day-to-day variability of range-rate changes attributed to the ionosphere even for a fixed observer because of the large variability in the day-to-day ionospheric rates of change. Thus, for ionospheric range-rate errors, corrections through the use of an ionospheric model, particularly one with a simple representation of diurnal ionospheric changes, such as that in the single-frequency GPS user ionospheric algorithm, are not recommended. The dual-frequency GPS user can, of course, automatically correct for both the first-order range and range-rate ionospheric errors.

F. Faraday Rotation

1. Amount of Faraday Rotation

When a linearly polarized radio wave traverses the ionosphere, the wave undergoes rotation of the plane of this linear polarization. At frequencies of approximately 100 MHz, and higher, the amount of this polarization rotation can be described by

$$\Omega = \frac{k}{f^2} \int B \cos \theta \, Ndl, \ldots, \text{radians} \qquad (16)$$

where the quantity inside the integral is the product of electron density times the longitudinal component of the Earth's magnetic field, integrated along the radio wavepath. Many ionospheric workers have used this effect to make measurements of the TEC of the ionosphere.

Because the longitudinal magnetic field intensity, $B \cos \theta$, changes much slower with height than the electron density of the ionosphere, Eq. (16) can be rewritten as

$$\text{TEC} = (\Omega f^2 / k B_L) \qquad (17)$$

where $B_L = \overline{B \cos(\theta)}$ is taken at a mean ionospheric height, usually near 400 km, $k = 2.36 \times 10^{-5}$, and TEC is simply $\int Ndl$.

Generally, the equivalent vertical TEC is determined by dividing the slant TEC by the secant of the zenith angle at a mean ionospheric height. The equivalent vertical TEC is the one most often used for comparison purposes among sets of TEC data, because these different slant TEC values cannot easily be directly compared. Much of the TEC data available today from stations throughout the world, used in model construction and testing, are from Faraday rotation measurements from vhf telemetry signals of opportunity from various geostationary satellites.

For satellite navigation and communication designers, the Faraday polarization rotation effect is a nuisance. If a linearly polarized wave is transmitted from a satellite to an observer on, or near, the surface of the Earth, the amount of polarization rotation may be nearly an odd integral multiple of 90 deg, thereby giving no signal on the receiver's linearly polarized antenna, unless the user is careful to realign the antenna polarization for maximum receiver signal. The Faraday rotation problem is overcome by the use of circular polarization of the correct sense at both the satellite and at the user's receiver.

2. Faraday Rotation Effects on Global Positioning Systems

GPS signals are transmitted with right-hand circular polarization; thus, Faraday rotation is not a problem with GPS users. The optimum receiving antenna for GPS users also would be one of right-hand circular polarization, to ensure that the receiving antenna polarization matches the characteristics of the received signal. However, it is impossible to design a nearly omnidirectional GPS receiving antenna having circular polarization over most of the entire sky. If a GPS user uses a linearly polarized antenna, the loss will be 3dB, because only one-half of the potential signal energy is being received. Thus, the optimum receiving antenna will exhibit right-hand circular polarization over as much of the sky as possible. Generally the mobile user finds it difficult to utilize circular polarization, because of the continual vehicle directional changes; thus, the user settles for nearly linear polarization. The up to 3-dB loss between transmitted circular polarization and receiver nearly linear polarization is a necessary price GPS users pay for antenna maneuverability and simplicity. The transmitted signal levels from the GPS satellites were designed to provide adequate signal strength for users with linearly polarized antennas.

Had the GPS satellites been transmitting linearly polarized radio waves, the polarization rotation, viewing satellites in various directions from a northern midlatitude station, could be up to 90 deg. For values of polarization rotation near 90 deg, signal loss for two, cross-aligned, linearly polarized antennas can be in excess of 30 dB. Thus, circular polarization was wisely chosen for use on GPS.

G. Angular Refraction

The refractive index of the Earth's ionosphere is responsible for the bending of radio waves from a straight line geometric path between satellite and ground. Normally, for GPS users, the small bending of radio waves is not a problem. The angular refraction, or bending, produces an apparent higher elevation angle than the geometric elevation. Millman and Reinsmith[7] have derived expressions relating the refraction to the resultant angular bending. Perhaps the easiest expressions to use, as given by Millman and Reinsmith,[7] relate the ionospheric range error to angular refraction. This expression is:

$$\Delta E = \{(R + r_0 \sin E_0)(r_0 \cos E_0)/[h_i(2r_0 + h_i) + (r_0^2 \sin^2 E_0)] \times (\Delta R/R)\} \quad (18)$$

where E_0 is the apparent elevation angle; R is the apparent range; ΔR is computed from, $\Delta R = (40.3/f^2) \times$ TEC; r_0 is the Earth's radius; and h_i is the height of the centroid of the TEC distribution, generally taken to be between 300 and 400 km.

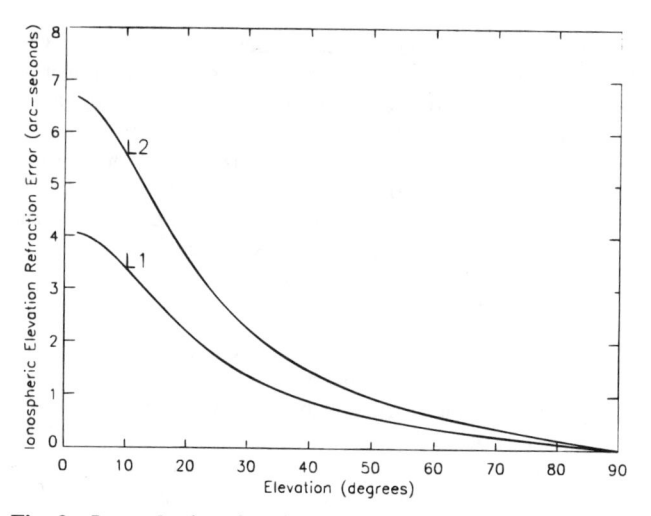

Fig. 3 Ionospheric refraction vs elevation for both L_1 and L_2.

Typical values of elevation refraction error for GPS at L_1 and at L_2, for a TEC of 10^{18} (el/m^2) column are given in Fig. 3.

The refraction, or radio wave bending illustrated in Fig. 3, is generally not a problem for GPS, because the user does not attempt to use GPS satellites at elevation angles lower than approximately 5 deg, due to other effects at low elevation angles, including antenna multipath, and tropospheric delay effects that increase greatly at low elevation angles. Errors in the azimuth of GPS radio signals transmitted through the ionosphere also can occur. They depend upon azimuthal gradients in TEC that generally are smaller than vertical gradients, and they can be neglected in most practical cases.

H. Distortion of Pulse Waveforms

Two characteristics of the ionosphere can produce distortion of pulses of rf energy propagated through it. The GPS signals consist of spread spectrum pseudorandom noise, having bandwidths of approximately 2 MHz and 20 MHz for the C/A and the P codes, respectively. The ionosphere can produce dispersion of the spread spectrum signals from GPS, but this effect is very small. The dispersion, or differential time delay caused by the normal ionosphere, as derived by Millman,[8] produces a difference in pulse arrival time across a bandwidth, Δf, of

$$\Delta t = [(80.6 \times \Delta f)/cf^3] \times \text{TEC}, \ldots, \text{seconds} \qquad (19)$$

where c is the velocity of light in m/s, f and Δf are expressed in Hz, and TEC is in el/m^2 column. The dispersive term for pulse distortion is thus proportional to TEC. When the difference in group delay time across the bandwidth of the pulse is the same magnitude as the width of the pulse, it will be significantly disturbed by the ionosphere. The dispersion across the 20-MHz GPS bandwidth is normally small and can be ignored.

In addition to pulse distortion by the dispersive effects caused by the TEC of the normal background ionosphere radio pulses are also modified by scattering from ionospheric irregularities. Yeh and Yang[9] have computed pulse mean arrival time and mean pulsewidth caused by both dispersion and scattering.

I. Amplitude Scintillation

Irregularities in the Earth's ionosphere produce both diffraction and refraction effects, causing short-term signal fading, which can severely stress the tracking capabilities of a GPS receiver. Signal enhancements also occur, but the GPS user cannot make use of the brief periods of stronger signal in any useful manner. The fading can be so severe that the signal level will drop completely below the receiver lock threshold and must be continually re-acquired. Figure 4 shows an example of several minutes of GPS receiver AGC recordings taken at at Kwajalein Island in the equatorial Pacific Ocean during the maximum of an 11-year solar cycle period in 1980. Note the rapid fading of signal on both the L_1 and the L_2 frequencies. Some of the fades exceeded the average signal-to-noise ratio of the GPS receiver then in use, which was $\approx$ 20 dB. Times when the signal apparently remained at a low, constant level for periods of several seconds, especially evident on the L_2 frequency, are when the receiver either partially or completely lost phase lock, and was attempting to re-acquire. Although Fig. 4 only shows a 6-min sample of strong scintillation fading, in order to show individual fades in detail, the strong fading typically can last for periods of up to several hours in the evening, broken up with varying intervals of time with no fading.

The equatorial scintillation region can encompass nearly 50% of the Earth, but fortunately, the times of strong scintillation effects observed in the near-equatorial regions are generally limited to approximately 1 hour after local sunset to local midnight. Figure 5, taken from Goodman and Aarons,[10] pictorially illustrates the regions and times of strong L-band equatorial scintillations.

There are exceptions to these times, but if precise measurements using GPS can possibly be avoided during the approximate 19–24 local time hours in this region, during years of high solar activity, and during the months of normally high scintillation activity, the chances of encountering strong scintillation effects

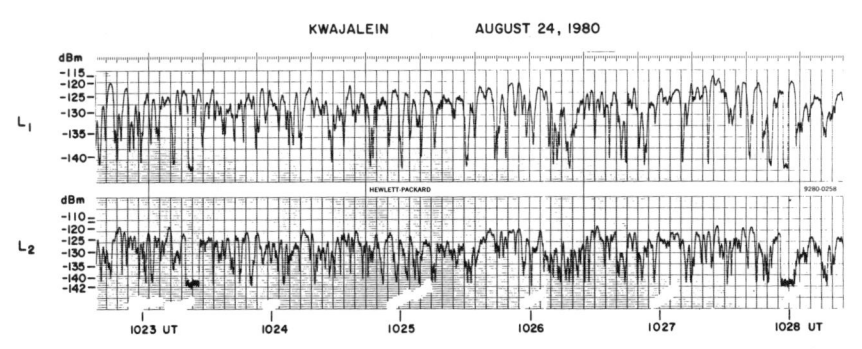

Fig. 4 Strong amplitude scintillation with fading greater than 15 dB measured on both L_1 and L_2 from Kwajalein Island in the Pacific Ocean.

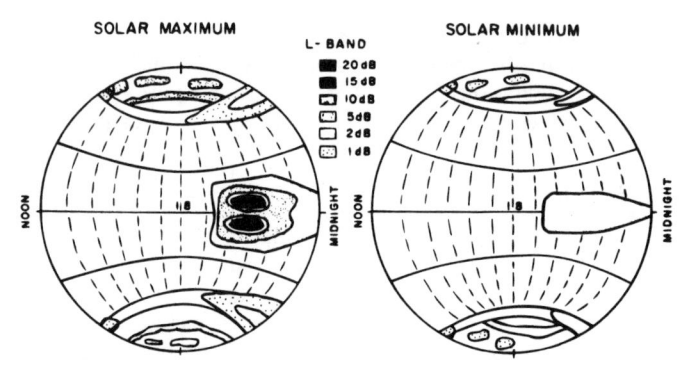

Fig. 5 Pictorial representation of "worst case" *L*-band ionospheric scintillation fading depths at *L*-band.[10]

are small. Furthermore, there are seasonal and solar cycle effects that also reduce the chances of encountering scintillation in the near-equatorial regions. In the months from April through August, chances are small of having significant scintillation in the American, African, and Indian longitude regions. In the Pacific region, scintillation effects maximize during these months. From September through March, the situation is just the reverse. The times and geographic regions of maximum statistical occurrence of equatorial anomaly scintillation should be avoided, if possible.

Fading (and enhancements) in the amplitude of the signal, or scintillation of the amplitude of the received signal, is caused by irregularities of scale size from hundreds of meters to kilometers in the electron density of the ionosphere. The fading effects that these irregularities produce can be characterized by a statistical description of the percentage of time below which the signal fades, and is described by Whitney et al.[11]

The regions of the world having the strongest scintillation effects are illustrated in Fig. 6, which indicates the location of Kwajalein Island in the Pacific Ocean, and Ascension Island in the South Atlantic Ocean.

Basu et al.[12] have published the seasonal and solar cycle dependence of *L*-band amplitude scintillation observed from Ascension Island, which is located at the peak of the equatorial anomaly peak. Their results are summarized in Fig. 7. The mean monthly sunspot number is also plotted in this figure. Note that the occurrence of strong amplitude scintillation is closely related to the sunspot number. During the years of near-minimum solar activity there should be little, if any, strong scintillation effects on GPS in the equatorial and low-latitude region. Strong amplitude scintillations have been found at frequencies at least as high as 10 GHz in Japan[13] and at 4 GHz from India.[14] The only relatively long-term scintillation measurements using GPS signals have been reported by Wanninger,[15] who investigated the occurrence of relatively long period equatorial phase scintillation, by using GPS data available from two stations in the International GPS Geodynamics Service (IGS) network. He used differential carrier phase data, available at 1-min intervals, to make an index of ionospheric phase variations. Although the spectrum of phase scintillations still has significant power down

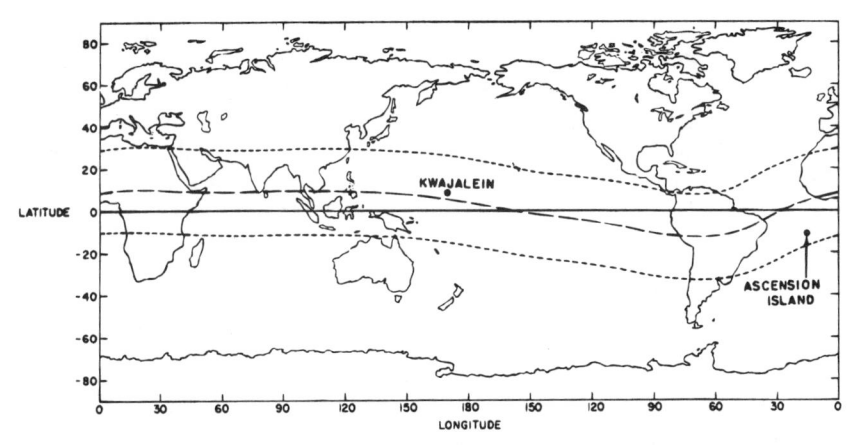

Fig. 6 Regions of maximum scintillation occur near and within the short dashed lines; Earth's magnetic equator is indicated by the long dashed line.

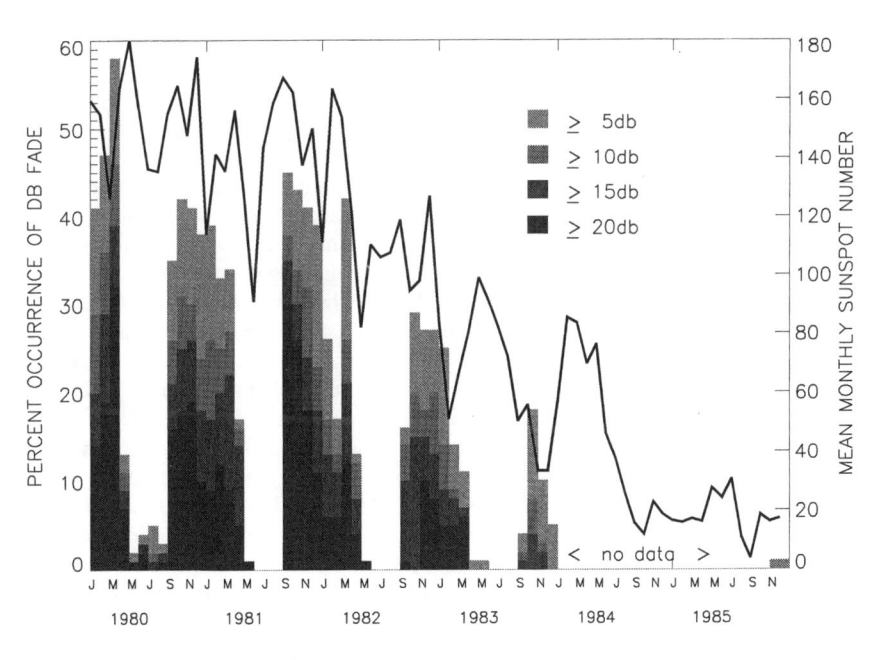

Fig. 7 Occurrence of 1.5 GHz amplitude scintillations of indicated depth observed at Ascension Island during the 2000–2400 local time period.[12]

to frequencies greater than 1 Hz, these faster scintillations are, of course, associated with those that Wanninger[15] studied. He showed the temporal and seasonal variations of the 1-min phase scintillation index for GPS data taken at Kokee, Hawaii for 1992 and basically confirmed previous work on the seasonal dependense of low-latitude scintillation by Basu et al.,[12] Aarons et al.,[16] and others. Wanninger[15] for the first time used a substantial amount of actual GPS differential phase data, and viewed scintillations from all the visible GPS satellites, rather than in only one direction, as has been the case with workers using signals of opportunity available from geostationary satellites.

In the auroral and polar cap latitudes, any significant magnetic storm activity can produce scintillation effects, but, surprisingly, the high-latitude scintillations are not as severe as those that have been measured in the near-equatorial belt. However, they can last for many hours, even days, and are not limited to the local late evening hours, as the near-equatorial scintillation effects are. There is little published L-band amplitude scintillation data available for the auroral and polar cap regions. The maximum fading depth observed on a GPS signal from the north polar cap region (Basu, private communication) was approximately 10 dB, with an uncertain calibration. This fading depth is much less than the >25 dB fades that have been observed in the equatorial anomaly region.

J. Ionospheric Phase Scintillation Effects

During times when the amplitude fading effects (Fig. 4) are strong, the irregularities in the ionosphere that produce range-rate errors are also generally moving across the observer's ray path rapidly, causing rapid receiver carrier phase changes. This is the most severe test of a GPS receiver in the natural environment. Rapid changes in signal phase, called phase scintillations, are attributable largely to rapid, but very small, changes in the electron content of the ionosphere. During those times, the signal amplitude is generally fading also, because of ionospheric scintillation effects, and the receiver lock is being stressed severely.

During times of severe phase scintillation, the phase will not change in a consistent, rapid manner to yield greater ionospheric Doppler shifts, but the phase of the incoming rf signal will have a large random fluctuation superimposed upon the changes associated with the normal rate of change in TEC. This large, random component may spread out the spectrum of the received signal sufficiently to cause a receiver having a bandwidth of only 1 Hz, or narrower, to lose phase lock, as the receiver signal phase may have little energy remaining in the carrier, and instead may be spread over several Hz, with little recognizable carrier remaining. A knowledge of phase scintillation rates is required to determine the spread of the receiver signal phase. GPS receiver phase lock problems can occur if the ionosphere produces a phase change faster than the receiver bandwidth can allow. Typically, a change of only 1 radian of phase is required at L_1, corresponding to $0.19 \times 10^{16} (1/m^2)$ column, or only 0.2% of a typical 10^{18} $(1/m^2)$ column background TEC, within the inverse receiver bandwidth time, to cause problems in maintaining receiver loop lock. If the receiver carrier tracking loop bandwidth is only ≈ 1 Hz, just wide enough to accomodate the geometric Doppler shift, there will be times when the rate of change of the ionosphere can cause apparent Doppler phase changes greater than that, and loss of lock will result. Normally, those regions of the Earth where strong phase effects occur are limited to the near-equatorial latitudes. However, during major magnetic storms, such as that which occurred in March 1989, these effects can occur over the midlatitudes.

V. Total Electron Content

The parameter of the ionosphere that produces most of the effects on GPS signals discussed in the many subsections of Sec. IV above, is the TEC. A value of TEC equal to 1×10^{16} electrons/m^2 is called one TEC unit. Values of TEC from 10^{16} to 10^{19} electrons per m^2, along the radio wavepath, represent the extremes of observed values in the Earth's ionosphere. No value of TEC greater than 10^{19} el/m^2 has ever been measured, even along a slant path in the Earth's ionosphere during a high solar maximum period, so that number is probably a safe maximum design value for systems designers.

Studies of the diurnal, seasonal, and solar cycle behavior of TEC have been conducted by numerous observers, using various measurement techniques, over more than 30 years. See Refs. 17 and 18 for review papers on TEC. Although TEC measurements are generally made from satellite radio signals of opportunity, observed at arbitrary elevation angles, the TEC normally is expressed as an equivalent vertical TEC by dividing the slant TEC by the secant of the elevation angle at a mean ionospheric height, usually taken between 350 and 400 km. Vertical electron content values can more easily be compared than slant values observed at various elevation angles.

Figure 8 shows contours of ionospheric time delay at L_1 in ns for monthly average conditions for 1990, a year of high solar activity. These contours of monthly average values of time delay were obtained from the Bent[19] model of TEC. Note that the regions of highest ionospheric time delay are located, on average, approximately ±15 to ±20 deg either side of the Earth's magnetic

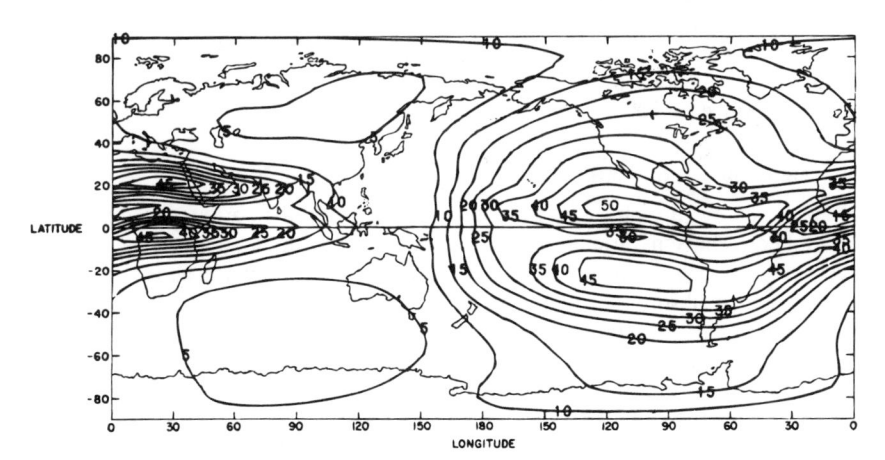

Fig. 8 Contours of zenith ionospheric time delay, in units of ns at L_1, for the March equinox season for a solar maximum year (sunspot number, 153).

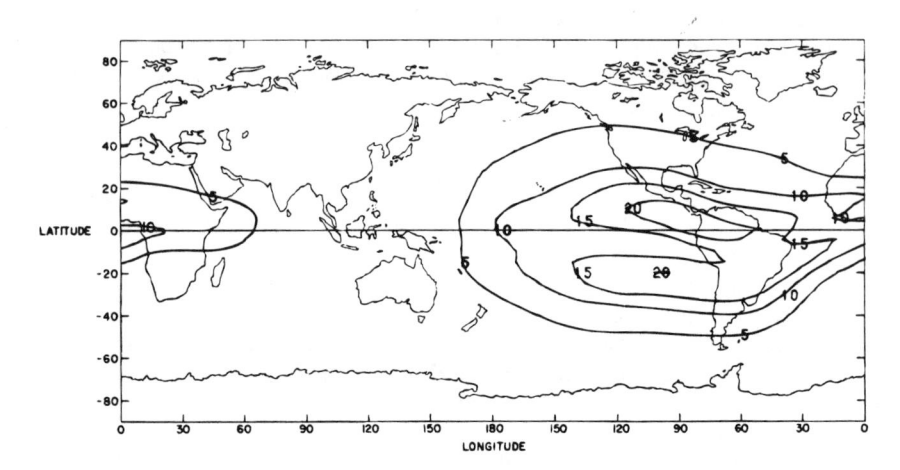

Fig. 9 Contours of ionospheric zenith time delay at L_1 for the March equinox season for a year of low sunspot activity (sunspot number, 10).

equator, and can be > 50 ns as viewed vertically. When we look at a 5-deg elevation angle, the time delay can be three times the vertical value.

Figure 9 shows contours of monthly average ionospheric zenith time delay for 1995, a year of expected solar minimum conditions. These contours also were obtained from the Bent[19] model. In this figure, more than half of the world has a zenith time delay of less that 5 ns, and the maximum value is only 20 ns.

The day-to-day variability of the ionosphere has a standard deviation of ≈ ±20% to 25% about monthly average conditions. The deviations from monthly average conditions are more-or-less normally distributed, but there are occasional days in which the monthly average ionospheric range error is exceeded by a factor of 2 or more. A "worst case" slant range error caused by the Earth's ionosphere, including the factor of 3 conversion from vertical to low-elevation angle viewing of a GPS satellite, during solar maximum conditions, could be as high as 300 ns, or 100 m.

A. Dependence of Total Electron Content on Solar Flux

The sun's uv flux produces the ionization found in the various regions of the ionosphere. The sun varies in its uv energy output with an approximate 11-year cycle. Direct measurements of solar uv flux only can be made above the ionosphere, which absorbs all of the uv flux in producing the ionosphere. Only a limited number of direct measurements of solar uv flux have been made, but a useful surrogate measure of solar ultraviolet activity is the number of sunspots observed on the solar disk. Sunspots have been observed since the early 1700s. Figure 10 illustrates the sunspot number for the last 300 years of solar activity. Note from Fig. 10, that the last 50 years of solar activity have included the highest four cycles that have occurred during the entire 300 years of directly observed sunspot activity.

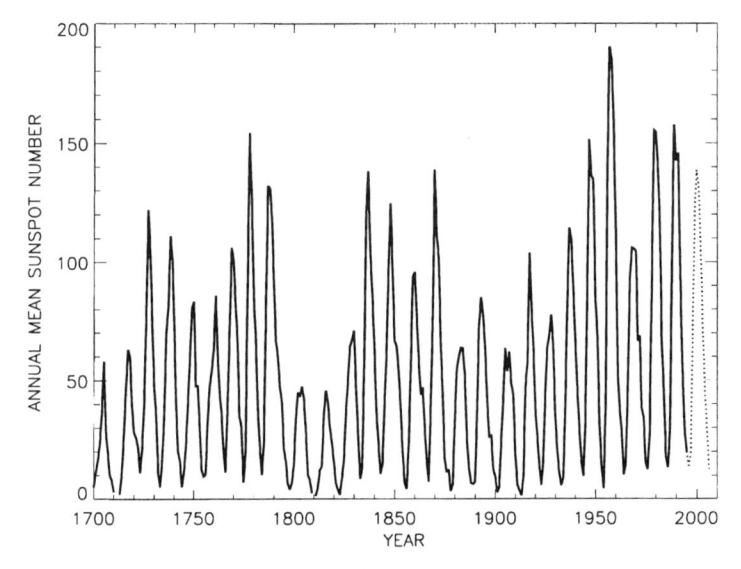

Fig. 10 Mean yearly sunspot numbers since 1700.

Beginning in 1947, changes in solar uv activity also have been inferred by measurements of the solar flux at a radio wavelength of 10.7 cm, or a radio frequency of approximately 2.8 GHz. Solar 10.7-cm radio wavelength flux measurements are a better quantitative measure of solar activity than the sunspot number, which is a more subjective, manual measure of solar variability. Both the observed sunspot activity on the surface of the Sun and the 10.7-cm solar radio flux are approximate measures of solar uv activity, but they are the only continuous measures of solar variability. Measurements of solar flux at 10.7-cm wavelength normally are used as surrogate measurements of uv flux.

The art of predicting long-term solar activity is not very precise; thus, the best guess about the next solar cycle peak, expected to occur near the turn of the century, is that it will be an average one, certainly lower than four of the last five cycles. There is at least one published prediction that the next solar maximum will exceed the maximum of the cycle that peaked in 1989, and perhaps will be the highest cycle ever recorded (Ref. 20). However, Brown[21] has done a study of predictions of solar cycle activity and has found that fewer than 23% of the predictions were within ±10%. Thus, not much faith should be placed on predictions of future long-term solar activity. At the present time, the best guess is that the maximum of the solar cycle, which should have its peak near the year 2000, will be of average activity.

Results of studies of the correlation of solar 10.7-cm wavelength flux against the day-to-day variability of the ionosphere show poor short-term, but good long-term, agreement. Over an 11-year solar cycle, the monthly average daytime TEC correlates well with the monthly average solar radio flux, but TEC variations for individual days during most months, do not correlate well with day-to-day changes in solar 10.7-cm radio flux. Figure 11 illustrates the lack of correlation of daily values of mean daytime TEC when plotted against daily values of the solar

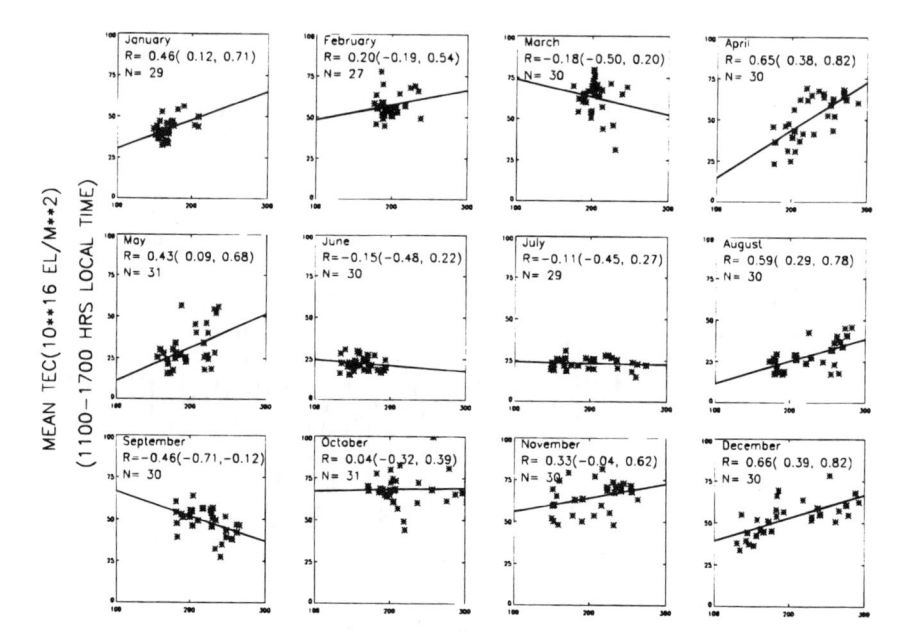

Fig. 11 Mean daytime zenith TEC plotted against daily solar 10.7-cm radio flux (the TEC data are from Hamilton, MA, for 1981, a year of high solar activity).

radio 10.7-cm flux. The correlation coefficients, as well as their 95% confidence intervals are given in each panel of Fig. 11. For most months, the correlation is very small. Thus, the day-to-day values of TEC cannot be predicted by knowing the solar radio flux.

B. Ionospheric Models

1. Empirical and Physical Models of the Ionosphere

Models of the electron density of the Earth's ionosphere fall into two types: 1) empirical models, or those derived from existing data; and 2) physical models, or those derived from physical principles, generally having empirical inputs as required, to describe the behavior of the various forcing functions. The first empirical model of the ionosphere was constructed for the parameter foF2, which is proportional to the electron density at the peak of the F2 region.[22] Various representations of the foF2 model form the basis for all the empirical models of the ionosphere; e.g., the International Reference Ionosphere, or IRI model[23] and the Bent model.[19] Rush[24] and Tascioni et al.[25] discuss some of the many versions of foF2 empirical models.

Those models derived from physical principles[26] attempt to include all the physical processes (i.e., forcing functions) that produce the electron content of the ionosphere. The great strength of physical models lies in their ability to look at the relative influence of various physical input parameters and how they might be expected to vary. The physical models, at present, are much too cumbersome to use directly as a means of determining monthy mean TEC for range error corrections for the GPS user. The GPS user doesn't want to become an expert on the relative merits of ionospheric models, but simply wants to know what models of ionospheric range error are available, how complicated they are to use, and how much of the error they will correct.

The first thing that a GPS user must understand about the ionosphere is its large variability. The day-to-day variability of the TEC from the monthly mean value, at any given time and location, is approximately 20–25%, 1σ. None of the ionospheric models presently available has been very successful in predicting the day-to-day variability of the ionosphere. In fact, the existing models are considered to be excellent if the bias between the prediction of the monthly mean TEC and the actual monthly mean TEC is within $\pm 10\%$. With a model bias of at least 10%, and a day-to-day variability that cannot yet be modeled of 20–25%, the overall error in estimating the ionospheric range error for GPS by using only an ionospheric model is approximately 22–27%, 1σ. Because even current state-of-the-art models have a bias about the true monthly average of at least 10%, and the models cannot reproduce the day-to-day variability, it is reasonable to use a relatively simple model for the monthly average ionospheric time delay that is computationally fast. Of course, the ideal situation is one in which the GPS user can directly make dual-frequency group delay and differential carrier phase measurements to correct automatically for the first-order ionospheric time delay.

2. Single-Frequency User Ionospheric Algorithm

In the mid-1970s, a simple algorithm was developed for the GPS single-frequency user to correct for approximately 50% of the ionospheric range error.[27,28] The 50% correction goal was established because the GPS satellite message had space for only eight coefficients to describe the worldwide behavior of the Earth's ionosphere (see Chapter 4, this volume). Furthermore, these coefficients could not be updated more often than once per day, and generally not even that often. Finally, simple equations had to be used to implement the algorithm to avoid causing undue computational stress on the GPS user. After all, the main computational task of the GPS user was to compute his or her position from the transmitted elements, not to spend too much time and computational assets in computing an ionospheric time delay correction.

It was immediately realized that there was a relation between the percentage ionospheric correction and the complexity of the computations required. Figure 12 illustrates this relation, in qualitative terms. Without near real-time data, it is considered impossible to model the day-to-day behavior of the ionosphere to better than the approximate 20–25% residual error about the monthly median behavior. Accordingly, only the monthly median behavior was modeled.

The median diurnal behavior of the ionosphere at any one location, can be modeled quite well by using an equation having only a few terms. As an example, Fig. 13 illustrates the monthly median behavior of TEC over the island of Jamaica. Note that a constant value of 9.4 ns for that entire month will yield over a 50% correction. A slightly more complicated function, namely a "half cosine" term, as illustrated in Fig. 13, yielded a representation of the median curve with only

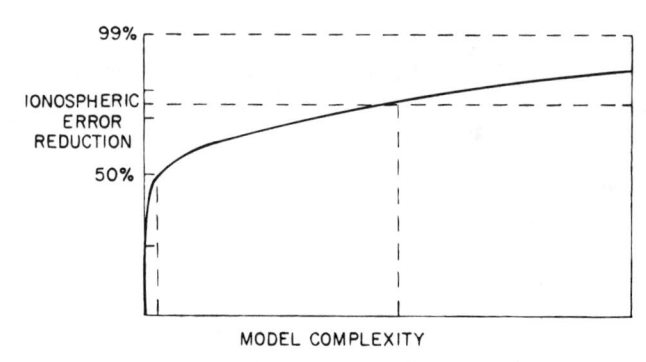

Fig. 12 Relative model complexity vs ionospheric error correction.

a residual 10% error. Thus, the half cosine form was used to represent the diurnal variation of TEC in the single-frequency GPS user algorithm. This half cosine is

$$T_{iono} = F \times \{DC + A \cos [2\pi(t - \phi)/P]\} \quad (20)$$

There are four potential variables in the cosine representation of diurnal TEC. They are: 1) the amplitude; A; 2) the period, P; 3) the DC offset term; and 4) the phase of the maximum with respect to local noon, ϕ. Realizing that only eight coefficients could be used to represent these four variables that describe

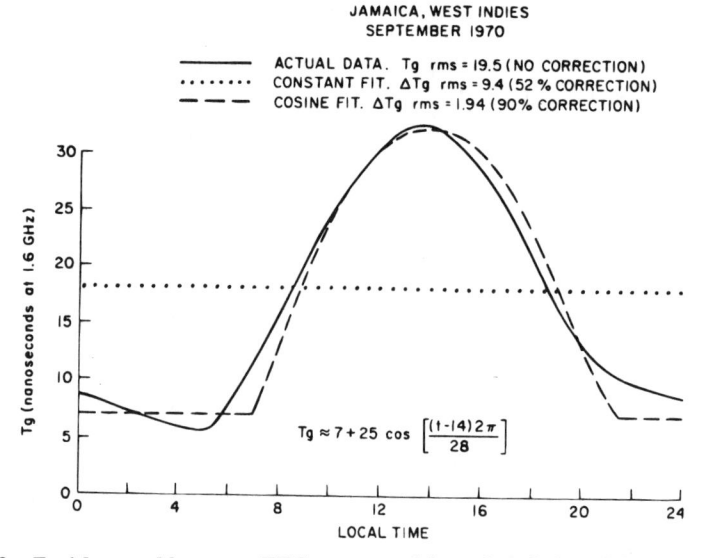

Fig. 13 Zenith monthly mean TEC, expressed in units of time delay of ns at L_1 from Jamaica, West Indies, for an average solar maximum year. Also shown are the rms values of the actual data, the rms errors after using a constant fit for all hours, and after a cosine fit has been applied to the monthly average data.

the diurnal variation of TEC over the entire globe, a study was made to determine which of the four cosine coefficients could be simplified to best use the total of eight terms available in the GPS message. When the original single-frequency GPS user ionospheric correction algorithm was made in 1975, the Bent Model[19] was the best available model of worldwide TEC. Thus, the Bent model was used to produce model diurnal curves for a large range of latitudes using a longitude in the approximate middle of the CONUS region. The variation of the DC and the phase terms in the equivalent cosine representation of diurnal TEC over a wide range of latitudes was much smaller than the variation of the amplitude and the period terms. Accordingly, the DC term was chosen to have a constant value of 5 ns, largely because little data were available on the contribution to TEC from the Earth's protonosphere, that region of electron density above a vertical height of approximately 1,000 km. The phase term was given a constant value of 14 h local time, however, the range of values found for the dirunal peak in TEC varied from 11–17 h for certain seasons, latitudes, and conditions of solar activity. The remaining terms, the amplitude and the period of the equivalent positive "half cosine," were then each allowed to have four coefficients to represent their behavior over the entire globe. The functional form chosen to represent the behavior of the amplitude and the period terms was simply a cubic polynomial in geomagnetic latitude, rather than in geographic latitude, because it had been found many years before that the electron density at the peak of the F region could be fit better by using geomagnetic latitude, rather than geographic latitude. The equations for the ionospheric correction algorithm for the single frequency GPS user are given in Appendix A.

There are several significant limitations of this algorithm, but Feess and Stephens[29] have shown that the residual ionospheric error for the comparisons they did against actual GPS dual-frequency ionospheric data, was approximately 40%. Thus, the correction for ionospheric time delay/range error was approximately 60% rms. There are certainly much better ionospheric models available than the simple, single-frequency GPS user ionospheric correction algorithm. However, going from the simple algorithm of $\approx$ 50–60%, to a state-of-the-art, complex ionospheric model that cannot beat the $\approx$ 75% barrier of the inherent, unknown, (and presently unmodelable) day-to-day variability of the ionosphere is hardly worth the additional trouble. If the GPS user really needs a better correction than that provided by the ionospheric correction algorithm, serious consideration should be given to using either differential GPS, or a codeless, dual-frequency receiver to actually measure the first-order ionospheric differential group delay and differential carrier phase advance.

C. Single-Frequency GPS Ionospheric Corrections

Recent work has been done by Cohen et al.[30] in attempting to measure the ionosphere using a single-frequency GPS receiver by using the L_1 carrier-aided code minus the L_1 carrier phase changes during a GPS pass. Although *relative* changes in the ionosphere over the time of the pass can be obtained in this manner, the *absolute* TEC cannot be determined. Various schemes of fitting the received code minus carrier phase data to the expected zenith angle variation of the TEC and/or other modeled variations in TEC over the pass have been attempted. It must be emphasized that these schemes are only as good as the form of the ionospheric model used. For actual ionospheric variations during a GPS pass that differ from the model, (that is when real ionospheric observations are most needed), these single-frequency GPS ionospheric correction schemes fall short.

D. Magnetic Storms Effects on Global Positioning Systems

Magnetic storms occur because of particles from solar flares arriving at the Earth and causing changes in the Earth's magnetic field. The visible aurora, or "northern lights," are caused by high-energy particles flowing along the Earth's magnetic field lines into the high latitudes, where they interact with the neutral atmosphere to produce excited ions, giving off red and green displays. These particles also produce additional electrons, and are indirectly responsible for strong electric fields, both of which can produce large electron densities and large shear effects which, in turn, produce irregularities. The irregularities cause strong amplitude and phase scintillation fading effects on GPS receivers operating in the auroral and polar cap latitudes.

The strong electric fields generated in the ionosphere during those magnetically disturbed times push electons over the polar cap, and large, rapid changes in ionospheric group delay can move through a GPS signal ray path, greatly changing the ionospheric range and range-rate errors within time periods of the order of 1 min. Klobuchar et al.[31] monitored GPS dual-frequency signals from the polar cap at Thule, Greenland for an approximate 2-week period. During that time, they found many occurrences of rapidly changing TEC when "patches" of additional $F2$ region ionization moved rapidly across the path to the GPS satellite being monitored. These large increases in TEC were accompanied by small amplitude fading and phase scintillation effects. Their observations were taken in 1984, a period of relatively low solar activity.

The effects observed in the auroral and polar cap regions should not be confused with those observed in the equatorial and low-latitude regions that occur during magnetically quiet conditions, and are generally quenched during magnetic distrubances. During the seasons normally having no scintillations in the equational and low latitudes, a magnetic storm can cause strong scintillations.

Whereas these strong ionospheric strom-induced changes in the polar cap and auroral regions can last many hours, or even days, the geographic area involved, compared with the near-equatorial region, is small. The northern polar cap is less than 2% of the globe, and the entire northern auroral and polar cap region comprises less than 7% of the Earth's surface area. On the other hand, the extended equatorial region of potentially strong evening time scintillation effects can extend up to ±50% of the Earth's surface.

During rare, very strong magnetic storms, such as the one that occurred in March 1989, these auroral effects can extend well into the midlatitudes, and can cause unusual effects on GPS receivers. At those times, the single-frequency GPS user ionospheric time delay correction algorithm would be likely to do a poor job of correcting for ionospheric range errors, and the range and range-rate changes would be much larger than during magnetically quiet times. For instance, during the March 1989 great magnetic storm, in which the aurora extended over virtually all of the continental United States, the range-rate errors produced by rapid variations in TEC exceeded a 1 Hz change in 1 s. Thus, receivers with only a 1-Hz bandwidth were continually losing lock during the worst part of this storm, because of their inability to follow changes faster than 1 Hz in 1 s. These rare times of great magnetic storms severely test the capabilities of all GPS receivers, but they do not occur frequently, and, as with hurricanes and tornadoes, they are part of natures's ionospheric "weather'" we must learn to accept. [The National Oceanic and Atmospheric Administration (NOAA) operates the Space Environment Services Center in Boulder, Colorado, which gives forecasts of solar and magnetic activity. If you are planning a major field campaign, it would be good to check with them for their predictions of ionospheric effects on your precise measurements. They can be reached at (303) 497-3171, or via modem, at (303) 497-5000. The service is free.]

V. Differential GPS Positioning

Two navigation services are provided by the GPS satellites. One is the Precise Positioning Service (PPS) offered to authorized military GPS users, which requires the reception of carrier and modulation phase from both the L_1 and the L_2 frequencies. In fact, the only reason the L_2 signal is transmitted from the GPS satellites is to allow qualified military users to correct automatically for the first-order effects of the range and range-rate errors produced by the Earth's ionosphere. After initial system tests, it was found that the L_1 only user was able to attain a better accuracy than that planned by the GPS designers. In order to prevent unauthorized users from having better dynamic positioning capabilities than that intended, the GPS Program Office developed methods for using selective availability, SA, with a resultant 100-m, 2σ positional accuracy for the standard positioning service. These methods include encoding a quasi-random error on each satellite clock, and truncating, or reducing the accuracy of the orbital elements of each GPS satellite.

Several civilian users of the GPS signals found that when the GPS signals were used in a differential carrier mode, that is, two receivers at different locations were used, the differential errors between them could be reduced to submeter accuracy. If the location of one of the receivers is known precisely, then the location of the second receiver could be determined with respect to the known location of the first receiver to decameter accuracy. The deliberate use of SA can be overcome by monitoring GPS signals from a receiver at known location, and transmitting, via a suitable radio link, the instantaneous clock errors for each satellite received at that station to other users within the field of view of the same satellite. The U.S. Coast Guard has plans to transmit corrections for GPS clock errors to civilian users at marine locations along the coastal areas of the United States, as well as in the Great Lakes region.

Normally, the effects of the ionosphere on differential GPS users are small, at least when the relative geographic area of the differential region used is small; for instance, within a single harbor or aircraft approach and landing area. The differential ionospheric electron content, or range error, within a single harbor or aviation precision approach area is generally small. However, differential GPS attempted over a wide area, called, appropriately enough, wide area differential

GPS positioning, or WADGPS, can include the requirement for differential positioning over areas of hundreds, even thousands, of kilometers in radius from a central, known location. In these cases, the differential ionospheric time delay across the region of interest can be a significant limitation to the overall WADGPS positioning accuracy.

A recent study, Klobuchar et al.[32] measured differences in ionospheric range delay for pairs of stations having different separations along generally a north-south direction. They summarized their results for the daytime hours of each season, and plotted the differences from median ionospheric behavior at the 1, 5, 95, and 99% cumulative probability levels vs station separation. Their results are shown in Fig. 14.

The results shown in Fig. 14. indicate that, even at station spacings of 1,000 km, the differential ionospheric range delay between pairs of stations is only a few meters, even at the 1% and the 99% levels. Similar results were found for east–west station spacings and for times encompassing the sunrise–sunset periods. Although their work was not done at solar maximum conditions, a reasonable scaling to a period of high solar activity should still have differential ionospheric range delays from median conditions at each station of less than a few meters between stations spaced several hundred kilometers apart.

The largest gradients in ionospheric range delay between stations in a WADGPS system occur in magnetically disturbed periods. Unfortunately, these times generally cannot be predicted in advance, and the presently available measures of solar/magnetic activity cannot be used to predict when large differences in

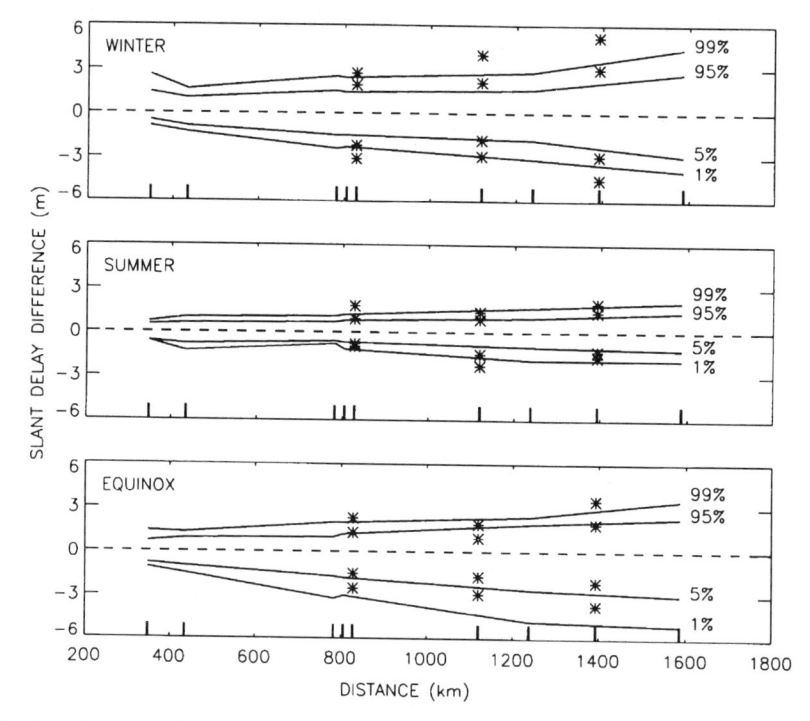

Fig. 14 Ionospheric delay differences vs station spacing for daytime hours; the stations were aligned generally in a north-south direction.

differential ionospheric delay will occur; however, once a magnetic storm is in progress, the actual dual-frequency codeless GPS stations in a WADGPS network will be the best indicators that the differential ionospheric range delays are outside some predetermined confidence interval. When that occurs, the WAGPS user can be informed, generally directly in the same message used to communicate the differential clock and other corrections to the WADGPS user.

Appendix: Ionospheric Correction Algorithm for the Single-Frequency GPS User

The form of the single-frequency GPS user ionospheric correction algorithm requires the user's approximate geodetic latitude Φ_u, longitude λ_u, elevation angle E, and azimuth A to each GPS satellite. The units used for angular measure are semicircles, and time is expressed in seconds. The coefficients α_n and β_n are transmitted as part of the GPS satellite message. The calculation proceeds as follows:

1) Calculate the Earth-centered angle, ψ

$$\psi = 0.0137/(E + 0.11) - 0.022 \quad \text{(semicircles)} \quad \text{(A1)}$$

2) Compute the subionospheric latitude, Φ_I

$$\Phi_I = \Phi_U + \psi \cos A \quad \text{(semicircles)} \quad \text{(A2)}$$

If $\Phi_I > +0.416$, then $\Phi_I = +0.416$. If $\Phi_I < -0.416$, then $\Phi_I = -0.416$.

3) Then, compute the subionospheric longitude, λ_I

$$\lambda_I = \lambda_u + (\psi \sin a / \cos \Phi_I) \quad \text{(semicircles)} \quad \text{(A3)}$$

4) Find the geomagnetic latitude, Φ_m, of the subionospheric location looking toward each GPS satellite. It is found by

$$\Phi_m = \Phi_I + 0.064 \cos (\lambda_I - 1.617) \quad \text{(semicircles)} \quad \text{(A4)}$$

5) Find the local time, t, at the subionospheric point

$$t = 4.32 \times 10^4 \lambda_I + GPS \text{ time (seconds)} \quad \text{(A5)}$$

If $t > 86,400$, use $t = t - 86,400$. If $t < 0$, add 86,400.

6) To convert to slant time delay, compute the slant factor, F

$$F = 1.0 + 16.0 \times (0.53 - E)^3 \quad \text{(A6)}$$

7) Then, compute the ionospheric time delay T_{iono} by first computing x

$$x = \frac{2\pi(t - 50400)}{\sum_{n=0}^{3} \beta_n \Phi^n_m} \quad \text{(A7)}$$

If $|\chi| > 1.57$

$$T_{iono} = F \times (5 \times 10^{-9}) \quad (A8)$$

Otherwise

$$T_{iono} = F \times \left[5 \times 10^{-9} + \sum_{n=0}^{3} \alpha_n \Phi^n{}_m \times \left(1 - \frac{x^2}{2} + \frac{x^4}{24}\right) \right] \quad (A9)$$

References

[1] Davies, K., *Ionospheric Radio*, Peter Peregrinus, Ltd., London, 1989.

[2] Chapman, S., "The Absorption and Dissociative or Ionizing Effect of Monochromatic Radiation in an Atmosphere on a Rotating Earth. Part I, and Part II, Grazing Incidence," Proc. Phys. Soc., Vol. 43, No. 26, 1931, p. 484.

[3] Brunner, F. K., and Gu, M., "An Improved Model for the Dual Frequency Ionospheric Correction of GPS Observations," *Manuscripta Geodaetica*, Vol. 16, 1991, pp. 205–214.

[4] Bassire, S., and Hajj, G. A., "Higher-Order Ionospheric Effects on the Global Positioning System Observables and Means of Modeling Them," *Manuscripta Geodaetica*, Vol. 18, No. 6, 1993, pp. 280–289.

[5] Jorgensen, P. S., "Ionospheric Measurements from NAVSTAR Satellites," SAMSO-TR-79-29, AD A068809, Dec. 1978, available from the Defense Technical Information Center, Cameron Station, Alexandria, VA 22304.

[6] Hatch, R., "The Synergism of GPS Code and Carrier Measurements," Magnavox Technical Paper, MX-TM-3353-82, Jan. 1982.

[7] Millman, G. H., and Reinsmith, G. M., "An Analysis of the Incoherent Scatter-Faraday Rotation Technique for Ionospheric Propagation Error Correction," General Electric Technical Information Series Report, R74EMH2, Feb. 1974.

[8] Millman, G. H., "Atmospheric Effects on Radio Wave Propagation," *Modern Radar, Analysis, Evaluation, and System Design*, edited by R. S. Berkowitz, John Wiley & Sons, New York, 1965.

[9] Yeh, K. C., and Yang, C. C., "Mean Arrival Time and Mean Pulsewidth of Signals Propagating Through a Dispersive and Random Medium," *IEEE Transactions on Antennas and Propagation*, Vol. AP-25, Sept. 1977, pp. 710–713.

[10] Goodman, J. M., and Aarons, J., "Ionospheric Effects on Modern Electronic Systems," *Proceedings of the IEEE*, Vol. 78, 1990, pp. 512–528.

[11] Whitney, H. E., Aarons, J., Allen, R. S., and Seemann, D. R., "Estimation of the Cumulative Amplitude Probability Distribution Function of Ionospheric Scintillations," *Radio Science*, Vol. 7, No. 12, Dec. 1972, pp. 1095–1104.

[12] Basu, S., MacKenzie, E., and Basu, S., "Ionospheric Constraints on VHF/UHF Communications Links during Solar Maximum and Minimum Periods," *Radio Science*, Vol. 23, No. 3, 1988, pp. 363–378.

[13] Minakoshi, H., Nishimuta, I., and Ogawa, T., "Observation of Equatorial Ionospheric Irregularities at Lower Mid-Latitudes," *Proceedings of International Beacon Satellite Symposium*, Massachusetts Institute of Technology, Cambridge, MA, July 1992.

[14] Dabas, R. S., Banerjee, P. K., Bhattacharya, S., Reddy, B. M., and Singh, J., "Gigahertz Scintillation Observations at 22.0° N. Magnetic Latitude in the Indian Zone," *Radio Science*, Vol., 26, No. 3, 1991, pp. 759–771.

[15]Wanninger, L., "Ionospheric Monitoring using IGS Data," Paper presented at the 1993 Berne IGS Workshop, Berne, Switzerland, March 1993.

[16]Aarons, J., "Gigahertz Scintillations Associated with Equatorial Patches," *Radio Science*, Vol. 18, No. 3, 1983, pp. 421–434.

[17]Evans, J. V., "Satellite Beacon Contributions to Studies of the Structure of the Ionosphere," *Reviews of Geophysics*, Vol. 15, 1977, pp. 325–350.

[18]Davies, K., "Recent Progress in Satellite Radio Beacon Studies with Particular Emphasis on the ATS-6 Radio Beacon Experiment," *Space Science Reviews*, Vol. 25, 1980, pp. 357–430.

[19]Llewellyn, S. K., and Bent, R. B., "Documentation and Description of the Bent Ionospheric Model," AFCRL-TR-73-0657, AD 772733, 1973.

[20]Wilson, R. M., "An Early Estimate for the Size of Cycle 23," *Solar Physics*, Vol. 140, 1992, pp. 181–193.

[21]Brown, G. M., "The Peak of Solar Cycle 22: Predictions in Retrospect," *Ann. Geophysicae*, Vol. 10, 1992, pp. 453–461.

[22]Jones, W. B., and Gallet, R. M., "Representation of Diurnal and Geographic Variations of Ionospheric Data by Numerical Methods," *Telecommunication Journal*, Vol. 32, 1965, p. 18.

[23]Bilitza, D., Rawer, K., Bossy, L., and Gulyaeva, T., "International Reference Ionosphere—Past, Present, and Future: I. Electron Density," *Advances in Space Research*, Vol. 13, No. 3, 1993, pp. (3)3–(3)13.

[24]Rush, C. M., "Ionospheric Radio Propagation Models and Prediction Models—A Mini Review," *IEEE Transactions on Antennas and Propagation*, AP-34, 1986, p. 1163.

[25]Tascione, T. F., et al.,"New Ionospheric and Magnetospheric Specification Models," *Radio Science*, Vol. 23, 1988, pp. 211–222.

[26]Crain, D. J., Sojka, J. J., Schunk, R. W., Doherty, P. H., and Klobuchar, J. A., "A First-Principle Derivation of the High-Latitude Total Electron Content Distribution," *Radio Science*, Vol. 28, No. 1, 1993, pp. 49–61.

[27]Klobuchar, J. A., "A First-Order, Worldwide, Ionospheric Time Delay Algorithm," AFCRL-TR-75-0502, AD A018862, available from the Defense Technical Information Center, Cameron Station, Alexandria, VA 22304.

[28]Klobuchar, J. A., "Ionospheric Time-Delay Algorithm for Single-Frequency GPS Users," *IEEE Transactions on Aerospace and Electronic Systems*, Vol. AES-23, No. 3, 1987, pp. 325–331.

[29]Feess, W. A., and Stephens, S. G., "Evaluation of GPS Ionospheric Time-Delay Model," *IEEE Transactions on Aerospace and Electronic Systems*, Vol. AES-23, No. 3, 1987, pp. 332–338.

[30]Cohen, C. E., Pervan, B., and Parkinson, B. W., "Estimation of Absolute Ionospheric Delay Exclusively through Single-Frequency GPS Measurements," presented at the ION-GPS meeting, Albuquerque, NM, Sept. 1992.

[31]Klobuchar, J. A., Bishop, G. J., and Doherty, P. H., "Total Electron Content and L-Band Amplitude and Phase Scintillation Measurements in the Polar Cap Ionosphere," AGARD-CPP-382, May 1985.

[32]Klobuchar, J. A., Doherty, P. H., El-Arini, M. B., "Potential Ionospheric Limitations to Wide-Area Differential GPS," Proceedings of ION GPS-93, Salt Lake City, UT, Sept. 22–24, 1993, pp. 1245–1254.

Chapter 13

Tropospheric Effects on GPS

J. J. Spilker Jr.*
Stanford Telecom, Sunnyvale, California 94089

I. Tropospheric Effects
A. Introduction

THIS chapter discusses the effects of the troposphere on the GPS *L*-band signals and the resulting effect on GPS positioning. The specific effects discussed include tropospheric attenuation, scintillation, and delay. To be precise, the term tropospheric used in this chapter is somewhat of a misnomer because roughly 25% of the delay effect is caused by atmospheric gases above the troposphere, specifically gases in the tropopause and stratosphere as shown in Fig. 1. The troposphere produces attenuation effects that are generally below 0.5 dB and delay effects on the order of 2–25 m. These effects vary with elevation angle because lower elevation angles produce a longer path length through the troposphere and also vary with the detailed atmospheric gas density profile vs altitude.

1. Atmospheric Constituents and Profile

The atmosphere consists of dry gases and water vapor. The wet and dry constituents of the atmosphere affect the propagation delay of the radio frequency signals quite differently, and these constituents have different pressure profiles. Water vapor is confined to the troposphere and generally exists only below altitudes of 12 km above sea level, and most of the water vapor is below 4 km. Water vapor density varies widely with position and time and is much more difficult to predict than the dry atmosphere. For example, significant changes in water vapor can occur over tens of km and hours of time. Fortunately, however, water vapor effects represent only a relatively small fraction (roughly 1/10) of the total. Because the water vapor content is highly nonuniform, a total average fractional volume is not a very meaningful method of description of the water vapor content. Instead, a measure of water vapor often employed is the total amount of water in a vertical column of air one square cm in area:

Copyright © 1994 by the author. Published by the American Institute of Aeronautics and Astronautics, Inc., with permission. Released to AIAA to publish in all forms.
*Ph.D., Chairman of the Board.

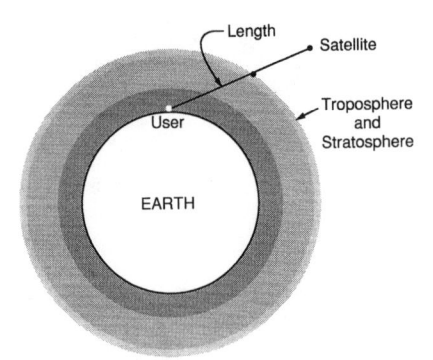

Fig. 1 Propagation of the GPS signal through the troposphere and stratosphere produces attenuation, delay and scintillation effects. The tropopause is the boundary between the troposphere and the higher altitude stratosphere.

$$W = \int_0^\infty \rho_w(h) dh \ g$$

where $\rho_w(h)$ is the water vapor density in g/cm^3. The total integrated water content W can vary enormously with position and ranges from 0.01 to 7.5 g, from the polar region to the tropics.

On the other hand, the dry atmosphere is relatively uniform in its constituents. Typical dry gas constituents have molar weights and fractional densities,* as shown in Table 1. At the GPS frequencies, oxygen is the dominant source of attenuation.

Table 1 Molar weights and approximate fractional volumes[a] of the major constituents of dry air[1]

Constituent	Molar weight, kg/kmol	Fractional volume unitless
N_2	28.0134	0.78084
O_2	31.9988	0.209476
Ar	39.948	0.00934
CO_2	44.00995	0.000314
Ne	20.183	0.00001818
He	4.0026	0.00000524
Kr	83.30	0.00000114
X_e	131.30	0.000000087

[a]For reference, at sea level and at 100% humidity, the water vapor occupies roughly 1.7% by volume, but note that percent humidity varies considerably with both time and position.[2]

---

* The fractional volume of CO_2 is the only major constituent of dry air that shows a significant variation, and it shows an annual variation of 6 ppm out of approximately 300 ppm and is increasing at a rate of 1.7 ppm/year.[3] However, for our purposes, these effects are negligible.

The temperature profile of a highly simplified model of the atmosphere in Fig. 2 shows that the temperature generally decreases with altitude at a constant lapse rate of −5 to −7°C per km of altitude increase from sea level up to the tropopause. At higher latitudes in winter and at nighttime, there is sometimes a temperature inversion layer in the 0.5–2 km region before the constant lapse rate of the troposphere begins. The tropopause is a region of approximately constant temperature and has an altitude of 8–12 km in the winter, or 10–12 km during the summer. The height of the tropopause has a downward slope from the equator toward the poles, and there is a small discontinuity in the tropopause height just above and below the equator. In the tropopause, the temperature rate of change decreases to zero. In the lower stratosphere (just above the tropopause), the rate of change of temperature gradually reverses to a slight +1°C to 2°C/km, and this relatively slow, nonuniform increase continues through the stratosphere up to an altitude of approximately 50 km. At this altitude, the stratopause, the temperature rate begins to reverse, and the temperature is roughly 0°C.[3,4] In the region directly above the stratopause; namely, the mesosphere, the temperature again decreases until it reaches approximately −90°C at the mesopause, which has a height of approximately 90 km. However, the atmospheric pressure in the mesosphere is so small (0.02–1 mb) as to be inconsequential for purposes here.

The atmospheric pressure at sea level is roughly 1013 mb and decreases with altitude to approximately 200–350 mb at the tropopause at the pole, and 70–150

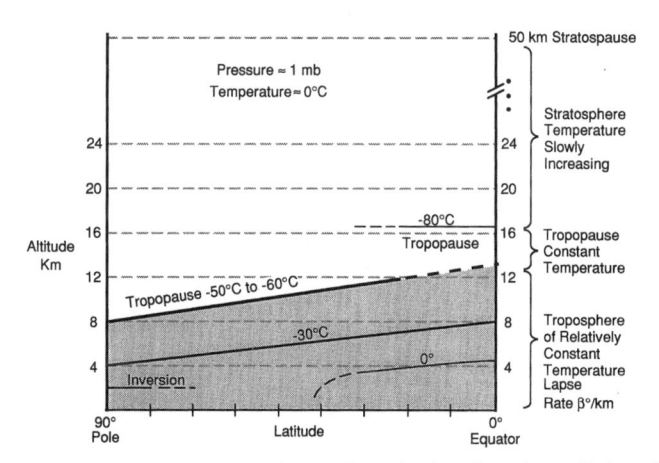

Fig. 2 Simplified model of the isothermal lines during the winter. Below the tropopause, the isotherms are approximately equally spaced at a temperature lapse rate of approximately 5–7°C/km decrease in temperature for every km of altitude increase. At the tropopause, the temperature lapse rate equals zero and then gradually reverses to a small rate of increase of approximately 1°–2°C/km in the stratosphere. Note that the height of the tropopause is roughly 4–8 km lower at the pole than near the equator. Thus, the tropopause has a downward slope toward the poles. During the summer, the tropopause is approximately 2 km higher in altitude at the poles but remains in the range of 17 km at the equator. At the polar regions, there is also an arctic temperature inversion at approximately 2 km, where the temperature lapse rate reverses.[3,4]

mb at the equator. Pressure decreases further to only 30 mb at approximately an altitude of 24 km at the pole and is only 1 mb at the stratopause. One atmosphere of pressure is defined as 1013.25 mb. The water vapor content lies primarily in the region below 4 km.

B. Atmospheric Attenuation

Atmospheric attenuation in the 1–2 GHz frequency band is dominated by oxygen attenuation, but even this effect normally is small. The attenuation is on the order of 0.035 dB for a satellite at zenith.[5,6] However, it can be ten times larger (in dB) at low elevation angles. The effects of water vapor, rain, and nitrogen attenuation at frequencies in the GPS frequency bands are negligible.[3]

Oxygen attenuation $A(E)$ in dB for the 1.5 GHz frequency range is approximately 0.035 dB at zenith ($E = 90$ deg) and varies with elevation angle E in proportion to the tropospheric path length L (obliquity factor or mapping function). If the troposphere is modeled by a simple uniform spherical shell of height h_m above the Earth, as shown in Fig. 3, then the length of the path L varies with elevation angle E, as shown in the fig. Thus, $A(E)$ has the following approximate value:

$$A(E) \cong \frac{2A(90 \text{ deg})(1 + a/2)}{\sin E + \sqrt{\sin^2 E + 2a + a^2}}$$

$$\cong \begin{cases} \dfrac{2A(90 \text{ deg})}{\sin E + 0.043} \text{ dB for small } E \text{ but } > 3 \text{ deg} \\ \dfrac{A(90 \text{ deg})}{\sin E} \text{ dB for } E > 10 \text{ deg} \end{cases} \quad (1)$$

where $a = h_m/R_e \ll 1$ and h_o is the equivalent height for oxygen $h_m = 6$ km, and R_e is the Earth radius $R_e \cong 6378$ km.[7]

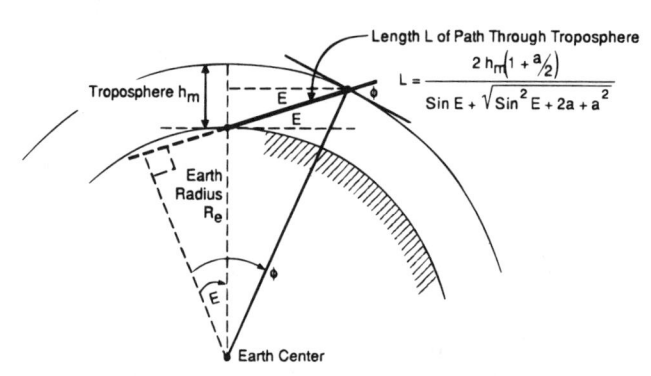

Fig. 3 Path length L through a uniform shell troposphere at elevation angle E where $\cos E = (1 + a)\cos \phi$ where $a \triangleq h_m/R_e$.

The attenuation of Eq. (1) is plotted in Fig. 4 for elevation angles $E > 3$ deg. This expression for $A(E)$ has assumed a spherical troposphere symmetrical in azimuth and uniform in density up to the equivalent height h_o. Note, if the troposphere were simply a planar layer with thickness h over a flat Earth, the length would simply be $h_o \csc E$. Near the horizon; e.g., below 3 deg, the uniform spherical model of Fig. 3 is no longer accurate, and neither Eq. (1) nor Fig. 4 should be used. The complexity of the atmosphere at low elevation angles is discussed later in this chapter. Nonetheless, note from Fig. 4 that the attenuation at 5 deg elevation angles is 0.38 dB, which is approximately ten times the zenith attenuation of 0.035 dB.

The mapping or obliquity function [i.e., the ratio $A(E)/A(90\ \text{deg})$] at low elevation angles for the troposphere is significantly larger than that for the ionosphere (10 compared to 3) because the troposphere extends down to the surface of the Earth. Thus, at low elevation angles, the ray path to the satellite penetrates the lower troposphere in a more nearly horizontal direction than it does at the ionosphere, which exists only above approximately 60 km. In practice, we should avoid using GPS satellites below approximately 5-deg elevation not only because of the lower signal levels associated with tropospheric attenuation, but because of larger uncertainties in tropospheric and ionospheric delay and greater scintillation effects caused by both the troposphere and the ionosphere (discussed later in this chapter and discussed in the previous chapter for the ionosphere). Furthermore, potential signal multipath, refraction, and receiving antenna gain roll-off effects may be magnified at low elevation angles.

C. Rainfall Attenuation

For a frequency of 2 GHz, the attenuation even for dense, 100 mm/h rainfall, is less than 0.01 dB/km; thus, it has a very small effect. Rainfall attenuation below 2 GHz is even less; thus, rain attenuation is of little consequence in the frequency bands of interest for GPS, 1.57542 GHz and 1.2276 GHz.[5]

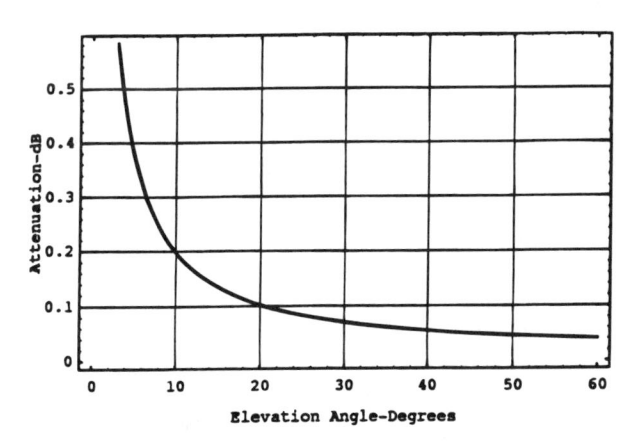

Fig. 4 Atmospheric attenuation vs elevation in degrees. Near the horizon at $E \leq 3$ deg, the model is no longer accurate.

D. Tropospheric Scintillation

Tropospheric scintillation is caused by irregularities and turbulence in the atmospheric refractive index primarily in the first few kilometers above the ground. The satellite–Earth propagation link through the troposphere is affected by a combination of random absorption and scattering from a continuum of signal paths that, in turn, cause random amplitude and phase scintillations in the received waveform. The scintillation effect varies with time and is dependent upon frequency, elevation angle, and weather conditions, especially dense clouds. At elevation angles above 10 deg, the predominant effect is forward scattering caused by atmospheric turbulence. At GPS frequencies, these effects are generally relatively small except for a small fraction of the time and at low elevation angles.

A received carrier from a satellite generally has the form $A(t)\sin(\omega t + \phi)$. Define the scintillation intensity $x(t)$ as the log of the amplitude ratio $A(t)/\overline{A(t)}$:

$$x(t) = 20 \log_{10}[A(t)/\overline{A(t)}] \text{ dB} \qquad (2)$$

where $\overline{A(t)}$ is the mean (short-term) amplitude of the signal. The probability density of $x(t)$ in dB in the short term is Gaussian and has variance σ_x^2 (see Boithias,[5] Ippolito,[8] and Moulsey and Vilar[9]). Thus, $A(t)$ has lognormal statistics. Experimental measurements of received carrier amplitude[8,9] have shown that the statistics of amplitude scintillation $A(t)$, although well represented by a lognormal distribution (Gaussian in dB representation with rms value σ_x) in the short term are not truly stationary and, over a longer observation interval, have significant fluctuations in the σ_x parameter with time caused, for example, by changes in weather. The appearance of large irregularly shaped cumulus clouds in the satellite–user path can change the value of σ_x. In observations taken while a large cumulus cloud passed through the antenna beam, the value of σ_x increased by a factor of six (see Ippolito[8]), although when scaled to L-band these effects are still small.

Moulsey and Vilar[9] represent $A(t)$ with a conditional lognormal distribution. Thus the probability of $x(t)$ conditional on a given σ_x is then normal with a variance σ_x^2 and is then as follows:

$$p(x|\sigma_x) = \frac{1}{\sigma_x\sqrt{2\pi}} \exp\left(-\frac{x^2}{2\sigma_x^2}\right) \qquad (3)$$

The rms value σ_x in dB is itself a random variable with a mean σ_m, and its long-term fluctuations have a probability density that is also lognormal:

$$p(\sigma_x) = \frac{1}{\sigma_\sigma \sigma_x \sqrt{2\pi}} \exp\left[-\frac{(\log \sigma_x - \log \sigma_m)^2}{2\sigma_\sigma^2}\right] \qquad (4)$$

where σ_σ^2 is the variance of this distribution. Thus $p(x|\sigma_x)$ can be viewed as a short-term statistic wherein σ_x itself has a distribution for the long term. The rms value of x in the long term is defined as σ_{xm}, which can be evaluated using $p(x) = \int p(x|\sigma_x)p(\sigma_x)d\sigma_x$ and found to be $\sigma_{xm} = \sigma_m \exp(\sigma_\sigma^2/4)$ dB.

For a value $\sigma_\sigma = 1$, which seems to match measurements by Moulsey and Vilar,[9] the value of $\sigma_{xm} = 1.28 \sigma_m$. The interpretation of this result is that although the long-term mean value of σ_x is σ_m, and σ_m may be small; e.g., on the order of 0.259 dB, the variance $\sigma_\sigma^2 = 1$ can lead to momentary periods of much larger scintillation.

The CCIR (1982) has given an expression for the long-term mean value of σ_x, namely, σ_m. For small antennas such as omnidirectional GPS antennas, the CCIR expression for the long-term rms amplitude scintillation varies with frequency and elevation angle as follows:

$$\sigma_m = 0.025 f^{7/12} (\csc E)^{-0.85} \text{ dB} \qquad (5)$$

where f is in GHz, and the elevation angle is E. (Larger antennas, not of importance for most GPS applications, produce an antenna averaging effect that decreases the value of σ_m). For $L_1 = 1.57542$ GHz we have the following:

$$\sigma_m = 0.0326 (\csc E)^{-0.85} \text{ dB} \qquad (5a)$$

and $\sigma_m = 0.259$ dB at $E = 5$ deg. Over the long term, the value of σ_x can vary substantially. If the mean σ_m of σ_x is taken as $\sigma_m = 0.259$ corresponding to the (worst case) 5 deg elevation angle for the CCIR model, and $\sigma_m = 0.081$ dB for 20-deg elevation angle, the cumulative distribution of σ_x is then as shown in Fig. 5. Thus, roughly 10% of the time the rms scintillation σ_x has a value of 0.9 dB and 0.3 dB for these two elevation angles, respectively, for this model. Thus, for low elevation angles and small fractions of time, tropospheric scintillation can be significant, but otherwise it is quite small.

II. Tropospheric Delay

The signal received from a GPS satellite is refracted by the atmosphere as it travels to the user on or near the Earth's surface. The atmospheric refraction causes a delay that depends upon the actual path (slightly curved) of the ray and

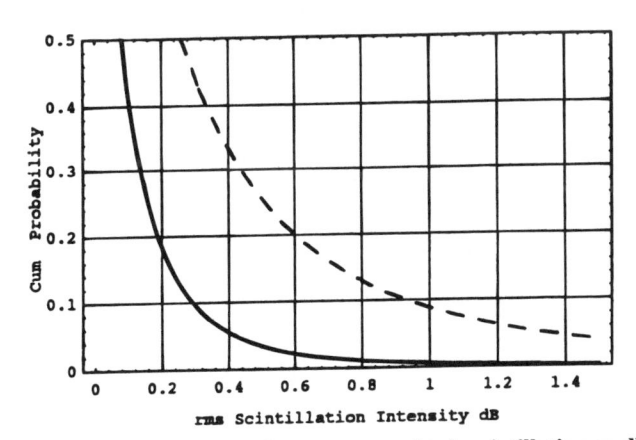

Fig. 5 Cumulative probability for the rms tropospheric scintillation σ_x dB. For this example, the mean value of $\sigma_m = 0.259$ dB for small 5 deg elevation angles (dashed curve) and $\sigma_m = 0.081$ dB for 20 deg elevation angle solid curve. For this lognormal model and a cumulative probability of 10%, the rms scintillation intensity σ_x is equal to 0.9 dB and 0.3 dB, respectively, for 5-deg and 20-deg elevation angles.

the refractive index of the gases along that path. For an atmosphere symmetric in azimuth about the user antenna, the delay depends only upon the vertical profile of the atmosphere and the elevation angle to the satellite.

A. Path Length and Delay

There are two major delay effects of the troposphere. The first and larger effect is a dry atmosphere excess delay caused primarily by N_2 and O_2. The dry atmosphere zenith excess delay corresponds to approximately 2.3 m and varies with local temperature and atmospheric pressure in a reasonably predictable manner.* The dry atmosphere effect varies by less than 1% in a few hours. The second effect—the wet atmosphere or water vapor effect—is generally smaller, 1–80 cm at zenith; i.e., perhaps $1/10$ the size of the dry atmosphere delay effect, but it varies markedly, 10–20%, in a few hours and is less predictable even with surface humidity measurements. Dual frequency 22, 31 GHz radiometer measurements can be made to make more precise predictions, but these measurements are fairly complicated and not feasible for most navigation applications.

The tropospheric delay is caused by the larger refractive index n ($n > 1$) of atmospheric gases than that of free space ($n = 1$), which causes the speed of light (group velocity) in the medium to decrease below its free space value c. Thus, a ray with an infinitesimal path length ds that travels through a medium with a refractive index n has a time delay $d\tau = n \, ds/c$ and has an equivalent distance $c \, d\tau = n \, ds$. In this section, delays are measured in meters rather than time, and the c is omitted after this paragraph. The difference between the actual total path delay $S = c\tau = \int n(s) \times 10^{-6} \, ds$ and the geometrical path distance S_g is the excess tropospheric delay Δ. The difference between the actual refractive index and unity is of the order of magnitude of $n-1 \cong 2.7 \times 10^{-4}$ at sea level and varies with altitude, latitude, and various meteorological conditions.

At the L-band frequencies of GPS signals, the refractive index is essentially constant with frequency, and hence, is nondispersive; i.e., $dn/df = 0$, and thus, the group velocity and phase velocity are the same. As shown in the previous chapter, this equivalence is not true in the ionosphere.

The spatially varying refractive index causes the signal path to have a slight curvature with respect to the geometric straight line path. Thus, the actual satellite-to-user path is a slightly curved path, as shown in Fig. 6 where the path curvature has been greatly exaggerated for clarity. The total length of the actual path from P_1 to P_2 in m is as follows:

$$S = \int_{\text{Actual}} ds \tag{6}$$

and is clearly longer than the straight line geometric path S_g (at elevation angles less than 90 deg) that the ray would take in a vacuum. However, it is well known from Fermat's principle[5,10] that the actual ray path is that path which minimizes the total delay from P_1 to P_2 (omitting c from $c\tau$ from now on):

---

*In the frequency range below 15 GHz, both the wet and dry components have an effect. At optical frequencies, the whole effect is attributable to the dry component.

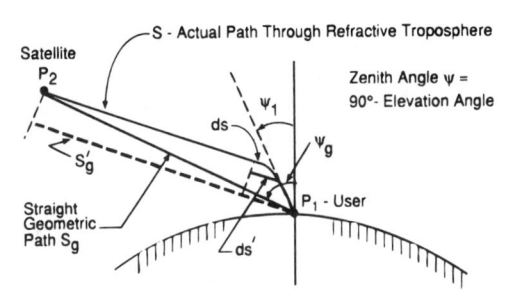

Fig. 6 Actual slightly curved path S and straight geometric paths through the refractive atmosphere. The difference in the actual path distance vs the geometric path distance is $\Delta_g \triangleq S - S_g \cong S - S_g'$ where $\Delta_g \triangleq \int ds - \int ds'$.

$$\tau = \int_{\text{Actual}} n(s)\, ds \tag{7}$$

The actual (curved) path from P_1 proceeds along a path which is initially closer to the zenith so that it passes through the region of high dielectric constant more rapidly. Thus, the actual curved path length S is longer than the straight line geometric path length (a rectilinear chord) S_g in m:

$$S_g \triangleq \int_{\text{geo}} ds \quad \text{and} \quad \tau_{\text{geo}} = \int_{\text{geo}} n(s)\, ds \tag{8}$$

although it is shorter in time delay. The notation τ_{geo} represents the path delay which would be accumulated along the geometric rectilinear chord. The integral path notation geo defines the geometric straight line path. That is, delay τ is shorter than the delay τ_{geo} of the straight line path. The quantity of primary interest is the excess delay caused by the atmosphere $\Delta = \tau - S_g$. The difference between the length of the actual curved path and the straight line path is typically quite small $S - S_g < 0.1$ m[11] except at low elevation angles.

If point s along the ray path is expressed in polar coordinates, then $s = re^{j\phi}$ (two dimensions), where the center of the coordinate system is at the Earth center, then the total delay [Eq. (7)], can be written as follows:

$$\tau = \int_{\text{Actual}} n(s)\, ds = \int_{\text{Actual}} n(r,\phi)\sqrt{1 + (r\phi')^2}\, dr \tag{9}$$

where $ds = dr\sqrt{1 + (r\phi')^2}$ and $\phi' \triangleq d\phi/dr$ for the path. Assume that $n(r,\phi) = n(r)$; i.e., the troposphere is spherically symmetric about the Earth's center. Fermat's principle and the calculus of variations then gives a minimum τ for the path and results in the following relationship:

$$\frac{n(r)r^2\phi'(r)}{\sqrt{1 + (r\phi')^2}} = \text{constant, or } n(r)r \sin \psi(r) = n(0)R_e \sin \psi(0) = \text{constant} \tag{10}$$

which is Snell's Law, and where $\psi(r)$ is the angle between the Earth's center radial vector at (r,ϕ), and the path tangent at height h above the Earth (see Fig.

7). At the user point P_1 on the Earth's surface, the radial distance $r = R_e$, the Earth radius, and the angle $\psi(0)$ is the zenith angle at the user $\psi(0) = 90$ deg $- E$, where E is the elevation angle. Another method of arriving at the same result is to assume a concentric set of spherical shells of height Δr and constant refractive index n_i, as shown in Fig. 7. Thus, we have Snell's law for spherical shells as follows:

$$r_i n_i \sin \psi_i = r_1 n_1 \sin \psi_1 = n(0) R_e \sin \psi(0) = \text{constant} \tag{10a}$$

In summary, the curved path, although physically slightly longer, has a slightly shorter total delay than would a path traveling in a straight line through the troposphere. Thus, the difference between the actual curved delay and straight line geometric distance (rectilinear chord) is as follows:

$$\Delta = \tau - S_g = \int_{\text{Actual}} n(s) \, ds - \int_{\text{geo}} ds \tag{11}$$

The second integral is the delay for the straight line (neglecting relativity) path that the ray would take in a vacuum. This difference can be rewritten as follows:

$$\Delta = \int_{\text{Actual}} (n - 1) \, ds + \left[\int_{\text{Actual}} ds - \int_{\text{geo}} ds \right] = \int_{\text{Actual}} (n - 1) \, ds + \Delta_g \tag{12}$$

where the second term in brackets Δ_g is the difference between the curved and

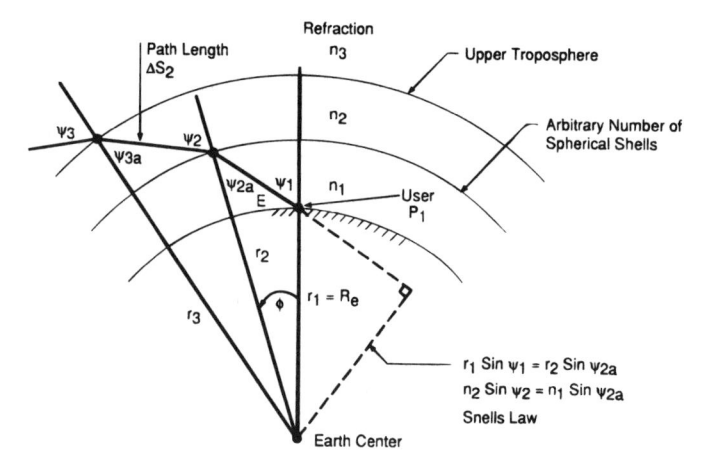

Fig. 7 Diagram of a ray path through sequential spherical shells of uniform refractive index n_i with base shell radi r_i and heights $h_i = r_i - R_e$, respectively. Snells law for spherical shells is $r_1 n_1 \sin \psi_1 = r_2 n_2 \sin \psi_2$. For planar boundaries Snells law has the familar form $n_2 \sin \psi_2 = n_1 \sin \psi_{2a}$. The same relationships hold for an arbitrary number of uniform shells of width Δr. The length of each path within a shell of infinitesimal width dh is related to the shell width by the ratio

$$ds/da = R_e(1 + a \, n(a))/(2a \, n(a) + a^2 n^2(a) + \sin^2 E)^{1/2}$$

where $a \triangleq h/R_e$, $n(a)$ is the refractive index; and a is the normalized height.

free-space paths $S - S_g \triangleq \Delta_g$. Ray-tracing calculations comparing excess delay Δ for a straight line path vs. a curved path from Snell's law have been calculated by Janes et al.[7] for the U.S. Standard atmosphere for various latitudes and elevation angles. The error caused by the neglect of path curvature is less than 3 mm for $E \geq 20$ deg; 2 cm for $E = 10$ deg; and increases to 17 cm at $E = 5$ deg (primarily from dry gases). The initial elevation angle difference between the straight line path and the curved path is on the order of seconds of arc, except for low elevation angles. The actual difference between the actual zenith angle $\psi(0)$ and the geometric straight line zenith angle ψ_g is difficult to compute (we must integrate $\int \tan \psi \, dn/n = \psi_g - \psi(0)$). However, a loose upper bound is easily obtained as follows:

$$\Delta\psi = \psi_g - \psi(0) < N(0) \times 10^{-6} \tan \psi(0) \tag{13}$$

by integrating over a single shell of constant refractive index $N(0)$ (see also Ref. 9). A more precise formula (radio ranging standard formula) from Saastamoinan[11] is as follows:

$$\Delta\psi \text{ in seconds of arc} \cong \frac{16.0 \tan \psi_g}{T}\left(P + \frac{4800e}{T}\right)$$

$$- 0.07(\tan^3 \psi_g + \tan \psi_g)\left(\frac{P}{1000}\right) \tag{14}$$

where P and e are the atmosphere pressure and the partial pressure of water vapor in millibars, respectively, and T is the absolute temperature in °Kelvin. The quantity $\Delta\psi$ is the angle of refraction. The angle $\Delta\psi$ is generally quite small, generally $\Delta\psi < 0.1$ deg.

In general, the estimation models for the actual troposphere excess delay from point P_1 to point P_2 is as follows:

$$\Delta = (10^{-6}) \int_{\text{Actual}} N ds + \Delta_g \tag{15}$$

where the refractivity $N \triangleq (n - 1) \times 10^6$ has both dry and wet terms. The excess delay Δ can be approximated as the sum of wet and dry delay effects in a simplified form similar to the following:

$$\Delta = [\Delta_{zd} m_d(E) + \Delta_{zw} m_w(E)] \tag{16}$$

where Δ represents the excess tropospheric delay; Δ_{zd} represents the dry zenith delay; Δ_{zw} represents the wet zenith delay; and $m_d(E)$ and $m_w(E)$ are the dry and wet mapping functions (obliquity factor) that magnify the tropospheric delay as the elevation angle E decreases. In general, $m_d(E)$ and $m_w(E)$ are really functions of the atmospheric profile as well as E. However, expressing them as a function of E alone is a useful approximation.

To analyze Δ further, it is first necessary to investigate the refractivity profiles of the troposphere with altitude. These results can then be employed to compute the zenith delay terms Δ_{zd} and Δ_{zw}. Finally, models can be developed for the mapping functions $m_d(E)$ and $m_w(E)$. We must be careful to understand exactly what is meant by the wet and dry terms because the actual path of the ray is dependent on all contributions to the index of refraction. As shown below, it is often convenient to break up the excess delay into a hydrostatic component Δ_h, which is dependent only upon the total pressure (and not the mix ratio of dry and wet air) and a wet component Δ_w.

B. Tropospheric Refraction Versus Pressure and Temperature

The general empirical expression for refractivity of nonideal gases including water vapor to be used here is as follows (see Davis et al.[1] and Thayer[12]).

$$N = (n - 1)10^6 = k_1(P_d/T)Z_d^{-1} + k_2(e/T)Z_w^{-1} + k_3(e/T^2)Z_w^{-1}$$

$$= 77.604(P_d/T)Z_d^{-1} + 64.79(e/T)Z_w^{-1} + 377600(e/T^2)Z_w^{-1}$$

$$= 77.604(P_d/T)Z_d^{-1} + (e/TZ_w)(64.79 + 377600/T) \qquad (17)$$

where P_d and e are the partial pressure of dry air and partial pressure of water vapor both in millibars; T is the temperature in °Kelvin; and the compressibility Z factors correct for the small departures of the moist atmosphere from an ideal gas. Ideal gases obey the relationship $PV = RT$; nonideal gases operate as $PV = ZRT$ where R is the universal gas constant. The experimentally derived values used here are: $k_1 = 77.604 \pm 0.0124$; $k_2 = 64.79 \pm 10$; $k_3 = 377600 \pm 3000$. See Table 2 for various values of k_i.

The inverse compressibility expressions are as follows[1]:

$$Z_d^{-1} = 1 + P_d[57.97 \times 10^{-8}(1 + 0.52/T) - 9.4611 \times 10^{-4} T_c/T^2]$$

$$Z_w^{-1} = 1 + 1650 (e/T^3)[1 - 0.01317T_c + 1.75 \times 10^{-4}T_c^2 + 1.44 \times 10^{-6}T_c^3] \qquad (18)$$

where T_c is temperature in °Celsius; T is in °Kelvin; and P_d is the dry pressure in millibars. The molar weight of dry air is $M_d = 28.9644$ kg/kmol, and for water $M_w = 18.0152$ kg/kmol. The ratio $M_w/M_d = 0.621977$.

The task at hand is to define models for the pressure and temperature vs. altitude that will yield models for N vs altitude h using Eqs. (17) and (18). The first term in Eq. (17) is referred to as the dry term, and the second two terms as the wet terms. Typically, the dry term represents perhaps 90% of the total effect and is relatively well predicted by surface pressure measurements. The wet terms, although small, are not easily predicted, and the water vapor effects vary considerably with position, elevation angle, and time. The pressure P and density ρ for the dry and wet air are related by the gas law as follows:

$$P_d = \rho_d \frac{R}{M_d} TZ_d, \qquad P_w = e = \rho_w \frac{R}{M_w} TZ_w \qquad (19)$$

where ρ_d and ρ_w are the dry and wet densities, respectively.

Thus, the refractivity N can be expressed in terms of the densities by substituting Eq. (19) into the first two terms of Eq. (17) as follows:

$$N = k_1 \frac{R\rho_d}{M_d} + k_2 \frac{R\rho_w}{M_w} + k_3 \frac{e}{T^2} Z_w^{-1}$$

$$= k_1 \frac{R\rho}{M_d} + \left(k_2 - k_1 \frac{M_w}{M_d}\right) \frac{e}{T} Z_w^{-1} + k_3(e/T^2)Z_w^{-1}$$

Table 2 Constants used in the refraction Eq. (17) by various sources for microwave frequencies. In this chapter, the Thayer values are employed.

Source	k_1(°K/mb)	k_2(°K/mb)	k_3(°K²/mb)
Thayer[12] (with Davis et al.,[1] uncertainty)	77.604 ± .014	64.79 ± 10	377600 ± 3000
Hill,[13] theoretical	—	98 ± 1	358300 ± 300
Birmbaum and Chatterjee,[14] measurements	—	71.4 ± 5.8	374700 ± 2900
Boudouris,[15] measurements	—	72 ± 11	375000 ± 3000

$$= (77.604 \pm 0.014) \rho \frac{R}{M_d} + k_2' \frac{e}{T} Z_w^{-1} + k_3 \frac{e}{T^2} Z_w^{-1}$$

$$= 22.276 \rho + (16.5 \pm 10) \frac{e}{T} Z_w^{-1} + 377600 \frac{e}{T^2} Z_w^{-1} \quad (20a)$$

where $\rho = \rho_d + \rho_w$ is the total mass density, $k_2' = k_2 - k_1 M_w/M_d$; and, of course, both the ρ_d, ρ_w densities and T vary with height h. Note that there is considerable uncertainty in the coefficient of the $(e/T)Z_w^{-1}$ term.[1] The uncertainty in the k_1 term is ±0.018%. Note that the first term in Eq. (20a) is dependent only on the total density ρ and is independent of the fractional water vapor content. The first term in the last version of Eq. (20) is termed the hydrostatic component. Total pressure is related to total density through the hydrostatic equation $dP/dh = -\rho(h) g(h)$, where $g(h)$ is the acceleration of gravity at the receiver antenna that also has a variation with height (only small variation for the relatively short range of heights for the troposphere). Generally $g(h)$ has very little variation with h, and a good approximation is obtained by replacing $g(h)$ by its weighted mean value over the range of integration $g_m = [\int \rho(h)g(h)dh]/[\int \rho(h)dh]$.

For this reason, we rewrite Eq. (20a) as the sum of hydrostatic and wet refractivity:

$$N = N_h + N_w \quad (20b)$$

where the hydrostatic refractivity is $N_h = (k_1 R/M_d) \rho = 22.276 \rho$, and the wet refractivity is as follows:

$$N_w = \left[k_2 - k_1 \frac{M_w}{M_d} \right] \frac{e}{T} Z_w^{-1} + k_3 \frac{e}{T^2} Z_w^{-1}$$

The hydrostatic refractivity is sometimes referred to as the dry refractivity and differs only by the Z_d^{-1} coefficient.

1. Temperature, Pressure, Refraction Index Models

The variation of the refractivity of the atmosphere vs altitude is exceedingly complex, and the models for refractive index are in the end based on fitting coefficients to match extensive sets of measured data. Nevertheless, it is instructive to review fundamental thermodynamics and ideal gas laws because some of the more complex models bear a close resemblance to these idealized models.

a. Troposphere–ideal gas models—adiabatic process. First consider an atmosphere of a single ideal gas with a pressure vs. altitude $P(h)$. At the lower altitudes within the troposphere, there is continual upward and downward motion of the air masses. Assume as an approximation that the process is assumed to be reversible and adiabatic (no heat entering or leaving the process). Also, assume that the gases are ideal* $PV = RT$. For an ideal gas with constant pressure heat capacity C_p and constant volume heat capacity C_v, where $C_p = C_v + R$, a reversible adiabatic process with no heat transfer yields the following:

$$C_v\, dT + P\, dV = 0 \tag{21a}$$

Because $P = RT/V$ and $R/C_v = \gamma - 1$ where $C_p/C_v = \gamma$ is defined as the ratio of heat capacities. By using $PV = RT$ and $R/C_v = \gamma - 1$, Eq. (21a) becomes the differential equations that follow:

$$dT/T = -R/V\, dV = -(\gamma - 1)\, dV/V, \quad \text{or} \quad dT/T = [(\gamma - 1)/\gamma]\, dP/P \tag{21b}$$

Equation (21b) has the following solutions:

$$\left(\frac{T_2}{T_1}\right) = \left(\frac{V_1}{V_2}\right)^{\gamma-1}, \quad \text{and} \quad \left(\frac{T_2}{T_1}\right) = \left(\frac{P_2}{P_1}\right)^{(\gamma-1)/\gamma} \tag{21c}$$

For a diatomic gas such as O_2, the heat capacity ratio changes slowly with temperature and has an approximate value of $\gamma = 1.4$. For polyatomic gases such as H_2O and CO_2, the value of γ is usually less than 1.3. Define the gas density as $\rho = M/V$, where M is the molar mass, and V is the volume. The hydrostatic relationship for the pressure P is dependent on the height h of the air column. For a differential air column of height dh the hydrostatic pressure increases by $(\rho dh)g\, dP$ at the bottom of the differential element compared to the top by the differential weight of the air per unit area:

$$dP = -\rho(h)g(h)\, dh = -\frac{Mg}{V}\, dh, \quad \text{and} \quad \frac{dP}{P} = \frac{-Mg}{RT}\, dh \tag{22}$$

Assume now that the acceleration of gravity g is constant. Thus, using Eqs. (21) and (22), we have the following:

$$\frac{dT}{T} = \left(\frac{\gamma - 1}{\gamma}\right)\frac{dP}{P} = -\left(\frac{\gamma - 1}{\gamma}\right)\left(\frac{Mg}{RT}\right) dh, \quad \text{or}$$

$$\frac{dT}{dh} = -\left(\frac{\gamma - 1}{\gamma}\right)\frac{Mg}{R} \cong \text{constant} = \beta \tag{23}$$

if γ is also constant with h over the region of interest. For $\gamma = 1.4$, the rate of change of temperature with altitude is then as follows:

---

* Note that real gases do not strictly obey the ideal gas law, and the van der Waals equation $[P + (a/v^2)]\,(v - b) = RT$ is a more accurate representation.[16] In this chapter, the effects of nonideal gases are taken into account by the compressibility factors Z_d and Z_w for the dry gases and water vapor.

TROPOSPHERIC EFFECTS ON GPS

$$\beta = -34.0866(\gamma - 1)/\gamma = -9.74 \text{ deg/km} \quad \text{for} \quad \gamma = 1.4$$

$$M = M_d = 28.9644$$

$$\beta = -7.87 \text{ deg/km} \quad \text{for} \quad \gamma = 1.3$$

Thus, the adiabatic assumption with constant C_v, C_p leads to a constant lapse rate $dT/dh = \beta = -[(\gamma - 1)/\gamma] Mg/R$. See Table 3 for example measured values of β, which are on the order of (-6 ± 0.5) deg/km, and other useful physical constants.

For nitrogen/oxygen atmosphere with $\gamma = 1.4$, the pressure then varies with altitude h as follows:

$$P = P_0\left(1 + \frac{\beta h}{T_0}\right)^{\gamma/(\gamma-1)} = P_0\left(1 + \frac{\beta h}{T_0}\right)^{1.4/(1.4-1)} = P_0\left(1 + \frac{\beta h}{T_0}\right)^{3.5} \quad (24)$$

where the ratio of heat capacities for diatomic gases is $\gamma = 1.4$, and T_0 is the temperature at sea level. The typical observed values of β are negative at approximately -5 to $-7°C$/km in the troposphere. For $\gamma = 1.3$, the exponent would be 4.33. Measurements give a ratio γ of heat capacitances, which corresponds to a somewhat higher exponent of $\gamma/(\gamma - 1) = 5.17$.[17]

We can also simply assume at the outset that the temperature has a constant lapse rate without the adiabatic assumption; i.e., $T = T_0 + \beta h$, where β is expressed in units of deg/km. Then, from the hydrostatic equation, we can integrate $dP/P = -(Mg/RT) dh$ for constant g over the range of heights $h = (0,h)$ to obtain the following:

$$\int_{P_0}^{P} \frac{dP}{P} = \frac{-Mg}{R} \int_0^h \left(\frac{dh}{T_0 + \beta h}\right) \quad \text{yields} \quad P = P_0\left(\frac{T_0 + \beta h}{T_0}\right)^{-Mg/R\beta} \quad (25)$$

where the surface pressure $P = P_0$ at $h = 0$. Thus, because $N_h \sim P_d/T$, the hydrostatic refractivity N_h can be written in terms of the surface value of hydrostatic refractivity N_{do} with a power law dependence on h as follows:

$$N_h = N_{h0}\left(\frac{T_0 + \beta h}{T_0}\right)^\mu \quad (26)$$

where $\mu \triangleq (-Mg/R\beta) - 1$. Thus, the exponent depends on the lapse rate β. If $\beta = -6.81°C$/km, then the exponent $-Mg/R\beta - 1 = 4.0$, a value shown later to be used by Hopfield[18,19] in a two quartic model of the troposphere pressure variation vs. altitude. Other more complex models[11,20] also use a power law model but only for the troposphere, not the stratosphere. (Note that at an altitude $h = -T_0/\beta$, the temperature would have fallen to absolute zero in this model if β remained constant for all h.) The rate of change of N_h at $h = 0$ for this model is then obtained by differentiating Eq. (26) to obtain the following[21]:

$$dN_n/dh = \beta\mu N_{h0}/T_0 = \frac{-N_{h0}Mg}{RT_0}\left(1 + \frac{R\beta}{Mg}\right) \quad \text{at} \quad h = 0$$

See Table 3 for typical values of dN_h/dh (shown as dN/dh).

Table 3 Thermodynamic constants and typical values of refraction variables and height of tropopause[1,11,17]

Typical measured values of tropopause height and lapse rates

Location	Location, north latitude	Tropopause, h_T km	Lapse rate, β °K/km
Onsala, South Sweden	57°	10.5	−5.7
Ettelsburg, Germany	51°	9.6	−5.7
Haystack, East Massachusetts	43°	13.6	−5.6
Owens Valley, Southern California	37°	12.8	−5.6
Fort Davis, Southwest Texas	31°	13.4	−6.3

Typical values of key parameters at surface and tropopause (models 1,2,3)[20]

Zone	Radius, km	P_0 mb	T, °K	$(n-1)$ 10^6	$-(dN/dh)$ 10^6/km	β, °K/km
Tropical	6360 + 0	1010.60	299.85	265.72	24.82	−6.06
Tropical	6360 + 16.8	98.03	198.00	39.06	5.525	
Temperate	6380 + 0	1015.00	285.08	280.87	27.28	−6.45
Temperate	6380 + 10.4	245.33	218.00	88.78	11.28	
Arctic	6400 + 0	1020.00	252.50	318.67	56.96	−6.525
Arctic	6400 + 8.8	303.56	223.60	107.39	13.33	

g_m = weighted mean gravitational constant = $[\int \rho(h)g(h)dh]/[\int \rho(h)dh]$
= $9.8062(1 - 0.0026 \cos 2\lambda - 0.00031 \bar{h}) ms^{-2}$ where $\bar{h}$ is the effective height above sea level of the center of gravity of the atmospheric column. Saastamoinen[20] has shown that $\bar{h} \cong 7.3 + 0.9 h_0$ km where h_0 is the user height and λ is the user latitude.

g_{eff} = effective gravitational constant = $9.784 (1 - 0.0026 \cos 2\lambda - 0.00028 h_0) ms^{-2}$ is the effective value of g for this $\bar{h}$. λ is the user latitude and h_0 is the user height.

1 atmosphere = 1013.25 millibars
M_d = mean molar mass of dry air 28.9644 kg/kmol
M_w = molar mass of water = 18.0152 kg/mol
1 kJ = 10^3 kg m²/s² = 10^{10} erg
R = universal gas constant 8.31434 kJ/(kmol · °K)

$$\frac{M_d g}{R} = \frac{(28.9644)(9.784)}{8.314} \frac{°K}{km} = 34.0866 \frac{°K}{km}.$$ For dry air at sea level $h = 0$ and $\lambda = 45$ deg.

0°C = 273.16°K

Key:
C_v, diatomic gas   Heat capacity, constant volume ≈ $(5/2)R$ at room temperature.
C_p, diatomic gas   Heat capacity, constant pressure ≈ $(7/2)R$; $C_p = C_v + 1$.

b. *Stratosphere*—isothermal model of pressure vs altitude. At higher altitudes from the top of the troposphere (termed the tropopause), at altitude h_T ($h_T \cong$ 12 km), and throughout the stratosphere to the stratopause, a constant temperature (isothermal) condition can be used as an approximation. Except in the tropopause itself, the temperature is not truly constant but varies more slowly than in the troposphere. For a region of ideal gases of constant temperature, the eq. $dP/P = (Mg/RT)\,dh$ can be integrated easily because T is a constant. This isothermal (constant temperature) condition leads to the exponential barometric height equation for pressure at altitudes above h_T:

$$P = P_0 \exp\left[\frac{-Mg(h - h_T)}{RT_T}\right] \qquad (27)$$

where R is the universal gas constant, 8.31434 kJ/(kmol · °K); g is the effective Earth's free fall acceleration in the stratosphere; P_T is the pressure at the tropopause; T_T is the temperature at the tropopause; and M is the molar mass of the air ($M \approx M_d$ in the stratosphere) (see Table 3). Thus, for the isothermal model, pressure decreases exponentially with height.

2. Zenith Delay for Dry Gases—Hydrostatic Component

The exact variation of the refractive index vs altitude is critical if we need to determine the excess tropospheric delay vs elevation angle, especially at the lower elevation angles (large zenith angles). However, for a satellite at the user's zenith, the hydrostatic gas effect of the atmosphere is independent of the shape of the profile with height and the dry air/water vapor mix ratio. The differential dry delay error at zenith Δ_{dz} (for no water vapor) from Eq. (17) is as follows:

$$d\Delta_{dz} = N_d(h) \times 10^{-6}\, dh \quad \text{where} \quad N_d(h) = 77.624\, P_d/T \qquad (28)$$

where $N_d(h)$ is the dry refractivity as a function of altitude h. In this analysis, it is more useful to work with the total hydrostatic delay (which is primarily the dry delay component) where the total hydrostatic refractivity from Eq. (20) is $N_h(h) = 22.276\, \rho(h)$, and the zenith hydrostatic delay is $\Delta_{hz} = 22.276 \int \rho(h)dh$. Note that the surface pressure P_0 is related to the density ρ by the hydrostatic equation $P_0 = \int \rho(h)g(h)dh \cong g_{eff} \int \rho(h)dh$ where $g_{eff} = \int \rho(h)g(h)dh / \int \rho(h)dh$ is the effective gravitational constant and is approximated in Table 3. Thus, the zenith excess hydrostatic delay [see Eq. (20)] is as follows:

$$\Delta_{hz} = 22.276 \times 10^{-3}\, P_0/g_{eff} = 2.2768 \times 10^{-3} P_0/G \text{ m} \qquad (29)$$

where we define $G \stackrel{\Delta}{=} (1 - 0.002626 \cos 2\lambda - 0.00028 h_0) = g_{eff}/9.784$, and P_0 is in millibars.

Water vapor does not conform to the hydrostatic equation because water vapor is subject to myriad effects including condensation and has wide variations in concentration unlike the dry gases that are uniformly mixed. Thus, there is no simple, highly accurate estimate of the wet delay related to the surface vapor pressure. The use of water vapor radiometers is generally impractical for most GPS applications; however, where they are available, Elgered[3] has developed expressions to estimate delay from these measurements to within a few mm for zenith delay.

III. Empirical Models of the Troposphere
A. Saastamoinen Total Delay Model

In a series of papers in 1972 and 1973, Saastamoinen[11,20] presented one of the first models of the refraction of the troposphere that estimates delay vs elevation angle E. In this model, the dry pressure is modeled using the constant lapse rate model for the troposphere and an isothermal model above the tropopause. The vertical gradient of temperature is $T = T_0 + \beta(r - r_0)$, and the resulting pressure profile is $P = P_0(T/T_0)^{-Mg/R\beta}$ where r is the radius from the Earth center ($r = R_e + h$) and r_0 is the user radius (usually $r_0 = R_e$, the Earth radius), and T_0 is the user temperature. The radius r ranges in value from r_0 to r_T which represents the radius to the tropopause. The corresponding dry refractivity, as discussed previously, is then $n - 1 = (n_0 - 1)(T/T_0)^\mu$ where $\mu = -M/R\beta - 1$ is a constant exponent [see Eq. (26)]. Using the isothermal model Eq. (27) above the tropopause, the pressure drops exponentially from its initial value at the tropopause P_T:

$$P = P_T \exp\left[-\frac{gM}{RT_T}(h - h_T)\right]$$

where the subscript T refers to the values at the tropopause.

The wet refraction is dependent on the partial pressure e, which decreases in somewhat the same way as total pressure in the troposphere Eq. (25) although much more rapidly. Saastamoinen[11,20] uses an exponent four times as large as in Eq. (25) to account for this difference, $e = e_0(T/T_0)^{-4gM/R\beta}$.

Saastamoinen[11,20] described both a precision model and a standard model for the tropospheric delay. Only the standard model is given here, and its delay correction for radio frequency ranging for elevation angles $E \geq 10$ deg is as follows[11,20]:

$$\Delta = 0.002277 (1 + D)\sec \psi_0 \left[P_0 + \left(\frac{1255}{T_0} + 0.005\right)e_0 - B \tan^2 \psi_0\right] + \delta_R \, \text{m} \quad (30)$$

where Δ is the delay correction in meters; P_0, e_0 are in millibars; and T_0 is in °K. The correction terms B and δ_R are given in Table 4 for various user heights h. The apparent zenith angle $\psi_0 = 90$ deg $- E$. The value of D in Eq. (30) is $D = 0.0026 \cos 2\phi + 0.00028h$, where ϕ is the local latitude, and h is the station height in km.

B. Hopfield Two Quartic Model

Hopfield[18,19,21] has developed a dual quartic zenith model of the refraction, different quartics for the dry and wet atmospheric profiles. Black[22,23] has extended this zenith model to add the elevation angle mapping function, as shown in a later paragraph.

Figure 8 shows an example of the measured variation in the vertical integrated index of refractivity, the zenith excess delay.[18] As can be seen, the dry component is on the order of 2.3 m and has a relatively small variation, whereas the wet

Table 4 Correction terms for Saastamoinen's[11,20] standard model, Eq. (30)

	Apparent zenith Angle	Station height above sea level							
		0 km	0.5 km	1 km	1.5 km	2 km	3 km	4 km	5 km
	60 deg 00 min,	+0.003	+0.003	+0.002	+0.002	+0.002	+0.002	+0.001	+0.001
	66 deg 00 min,	+0.006	+0.006	+0.005	+0.005	+0.004	+0.003	+0.003	+0.002
	70 deg 00 min,	+0.012	+0.011	+0.010	+0.009	+0.008	+0.006	+0.005	+0.004
	73 deg 00 min,	+0.020	+0.018	+0.017	+0.015	+0.013	+0.011	+0.009	+0.007
	75 deg 00 min,	+0.031	+0.028	+0.025	+0.023	+0.021	+0.017	+0.014	+0.011
δ_R, m	76 deg 00 min,	+0.039	+0.035	+0.032	+0.029	+0.026	+0.021	+0.017	+0.014
	77 deg 00 min,	+0.050	+0.045	+0.041	+0.037	+0.033	+0.027	+0.022	+0.018
	78 deg 00 min,	+0.065	+0.059	+0.054	+0.049	+0.044	+0.036	+0.030	+0.024
	78 deg 30 min,	+0.075	+0.068	+0.062	+0.056	+0.051	+0.042	+0.034	+0.028
	79 deg 00 min,	+0.087	+0.079	+0.072	+0.065	+0.059	+0.049	+0.040	+0.033
	79 deg 30 min,	+0.102	+0.093	+0.085	+0.077	+0.070	+0.058	+0.047	+0.039
	79 deg 45 min,	+0.111	+0.101	+0.092	+0.083	+0.076	+0.063	+0.052	+0.043
	80 deg 00 min,	+0.121	+0.110	+0.100	+0.091	+0.083	+0.068	+0.056	+0.047
B mb		1.156	1.079	1.006	0.938	0.874	0.757	0.654	0.563

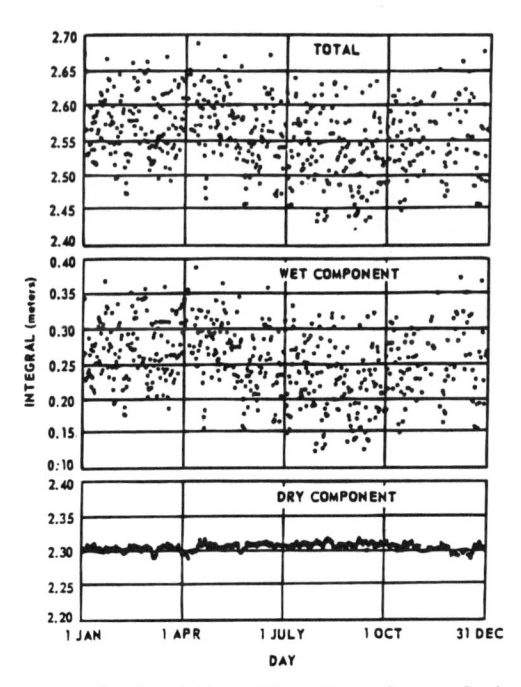

Fig. 8 Vertical integral of refractivity at Pago Pago, Samoa during 1967, balloon data from Hopfield.[18]

component is on the order of 0.25 m but varies ± 40%. Note that the scales used in all three plots are the same 5 cm/division. In data taken overland or in cold climates, the wet component is often much lower, ranging from 0 to 0.1 m.

The two-quartic model for the refractive index vs altitude h is similar in form to that of Eq. (26). Each fourth power term represents the dry and wet refractive indices as follows:

$$N_d(h) = N_{d0}\left(1 - \frac{h}{h_d}\right)^4 \quad \text{for} \quad h \leq h_d = 43 \text{ km} \tag{31a}$$

and

$$N_w(h) = N_{w0}\left(1 - \frac{h}{h_w}\right)^4 \quad \text{for} \quad h \leq h_w = 12 \text{ km} \tag{31b}$$

where N_{d0} and N_{w0} are the respective dry and wet refractive indexes at the surface, and h is the altitude. The zenith delay is then obtained by integrating Eq. (31) as follows:

$$\Delta = 10^{-6} \int_0^{h_d} N_{d0}\left(1 - \frac{h}{h_d}\right)^4 dh + 10^{-6} \int_0^{h_w} N_{w0}\left(1 - \frac{h}{h_w}\right)^4 dh$$

$$= \frac{10^{-6}}{5}[N_{d0}h_d + N_{w0}h_w] = \Delta_d + \Delta_w \tag{32a}$$

Note that h_d and h_w refer to height above the surface level where the surface refractivities N_{d0} and N_{w0} are measured. This quartic relationship is taken from the constant lapse rate model in Eq. (25), which gives an exponent $-(Mg/R\beta) - 1$. As mentioned earlier, this exponent becomes 4 at a lapse rate $\beta = -6.81°K/km$. Hopfield[18,19,21] chose the 4th power for ease of calculation and as a good approximation to the observations, and integrate along the satellite to user ray path for the wet atmosphere (troposphere only) to $h_w = 12$ km and the dry atmosphere to approximately $h_d = 43$ km to cover the troposphere and stratosphere. The values of h_d and h_w are determined by best fit to experimental measurements. As shown earlier in Eq. (29), the zenith integral of dry (really hydrostatic delay) tropospheric delay is equal to a constant times the surface pressure:

$$\Delta_d = kP_0 \tag{32b}$$

where $k = (77.6 \, R/g) \times 10^{-9}$ in cgs units (Ref. 19).

C. Black and Eisner (B&E) Model

Black and Eisner[23] began with a model for excess delay for paths from point P_1 to P_2 at various elevation angles, as follows:

$$\Delta = 10^{-6} \int_{P_1}^{P_2} [N_d(s) + N_w(s)] \, ds \tag{33}$$

where the refraction equation is a simplified version of Eq. (17); namely, $N_d = 77.6\ P/T$ (see Eq. 32), and $N_w = 3.73 \times 10^5\ (e/T^2)$. For a straight geometric line path that neglects ray-bending effects at elevation angle E, the differential ds has been shown [simplified version of Eq. (1) for small a] to be $ds = dh/\sqrt{1 - [\cos E/(1 + a)]^2}$, and the excess delay becomes as follows:

$$\Delta = 10^{-6} \int_0^{h_d} \frac{(N_d(h) + N_w(h))\ dh}{\sqrt{1 - [\cos E/(1 + a)]^2}} \quad (34)$$

where the $h_d = 45$ km; $a \stackrel{\Delta}{=} h/R_e$, and N_w is nonzero only for $h < h_w = 13$ km.

Figure 9 shows the Haystack observatory measurements (in summer) of the profile of water vapor density vs height. Note the very substantial 1σ variation of the water vapor density measurements. The expression for delay error Δ, Eq. (34), is now approximated by the product of the zenith delay and a mapping function $m(E)$. The zenith delay approximation is based on numerical integration of a quartic dry term modeled after Hopfield[18,19,21] and a wet term that is exponential with saturated conditions at the surface and models measured profiles similar to Fig. 9.

The B&E delay error[23] is thus approximated by the following:

$$\Delta = (\Delta_{dz} + \Delta_{wz})\ m(E, T) \quad (35)$$

where B&E use a single mapping function (as opposed to separate mapping functions for wet and dry). The mapping $m(E,T)$ is a function of both the elevation E and has the following small temperature dependence:

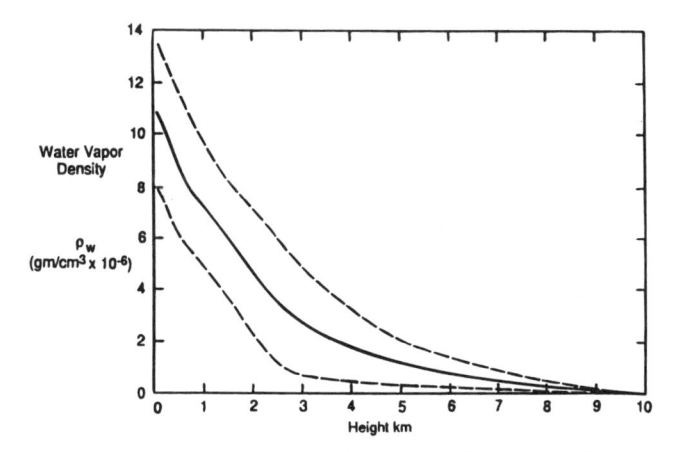

Fig. 9 Mean profile of water vapor density ρ_w taken from 45 radiosonde measurements at Haystack Observatory in August 1973. The dashed lines depict the 1 σ standard deviation. "The profile approximates an exponential dependence with a scale height of 2.2 km although the high part decreases more slowly than an exponential."[23,24]

$$m(E, T) = 1/\sqrt{1 - ((\cos E)/(1 + X_{dw}h_d/R_e))^2} \tag{36}$$

where X_{dw} in general varies with temperature.

For elevation angles in the range 7 deg $< E \le 90$ deg and surface temperatures in the region $-30°C < T_0 < 40°C$, the value of X_{dw} is in the range, $0.00088 \le X_{dw}h_d/R_e \le 0.001$. Thus, B&E approximate $m(E,T) \cong m(E)$

$$m(E) = 1/\sqrt{1 - ((\cos E)/(1 + 0.001))^2} = 1.001/\sqrt{(.001)^2 + 0.002 + \sin^2 E} \tag{37}$$

Compared with the results for more exact models also developed by Black and Eisner[23] this approximate model showed an error of approximately 7 cm at 7-deg elevation angle and 1 cm at 13-deg elevation angle.

D. Water Vapor Zenith Delay Model—Berman

Berman[25] has developed a simple model for water vapor zenith delay based on the correlation between the dry gas and water vapor delay for day and night measurements at Edwards Air Force Base, CA. Thus, if we can estimate the dry zenith delay from surface pressure measurements and estimate the surface wet and dry refractivities, we can then estimate the wet zenith delay as some fraction of that delay. The Berman model[25] gives the ratio of zenith excess delays for wet and dry components, $\tau_w/\tau_d = k N_{\omega 0}/N_{d0}$, as proportional to the ratio of wet and dry surface refractive indexes where the empirically determined scale factor is as follows:

$$k = \begin{cases} 0.2896 & \text{day} \\ 0.3773 & \text{night} \\ 0.3224 & \text{mixed day/night} \end{cases}$$

Note that this empirical result was obtained in a relatively dry region of California.

E. Davis, Chao, and Marini Mapping Functions

As discussed earlier, the mapping function attempts to relate accurately the actual excess path delay $\Delta(E)$ as a function of elevation angle and meteorological conditions to the zenith excess delay Δ_z. The mapping function can be used in conjunction with other models for the zenith delay (e.g., Davis et al.[1] and Hopfield[18]). For elevation angles near 90 deg (the zenith), clearly, a simple $1/\sin E$ approximation is generally sufficient. However, at lower elevation angles, the nonuniform and finite width spherical shell model troposphere make this simple model inadequate.

Marini[26] described a continued fraction version of the mapping function as follows:

$$m(E) = \cfrac{1}{\sin E + \cfrac{a}{\sin E + \cfrac{b}{\sin E + \cfrac{c}{\sin E + \dots}}}} \tag{38}$$

where the a, b, c, ... are constants. Chao[27] employed a model of this form for the excess delay Δ that utilizes separate mapping functions for the wet and dry components. These mapping functions are of the same form as the continued fraction representation except that only the first two terms are used and the second $\sin E$ has been replaced by $\tan E$ so as to make $m(90 \text{ deg}) = 1$. The specific dry and wet mapping functions used by Chao are as follows:

$$m_d(E) = \cfrac{1}{\sin E + \cfrac{0.00143}{\tan E + 0.0445}}$$

$$m_w(E) = \cfrac{1}{\sin E + \cfrac{0.00035}{\tan E + 0.017}} \quad (39)$$

Davis[1] has developed a more sophisticated function for the dry (hydrostatic) mapping function wherein the coefficients a, b, and c are dependent on surface pressures, temperature, lapse rates, and tropospheric height h_T. This representation has been compared with ray-tracing results and found to be accurate to within about 2.5 cm even at elevation angles as low as 5 deg. This model is termed the Davis Cfa 2.2 mapping function and is as follows:

$$m(E) = \cfrac{1}{\sin E + \cfrac{a}{\tan E + \cfrac{b}{\tan E + c}}} \quad (40)$$

where a, b, and c are dependent upon measurements or estimates[1]:

$$a = 0.001185[1 + 0.6071 \times 10^{-4}(P_0 - 1000) - 0.1471 \times 10^{-3} e_0$$
$$+ 0.3072 \times 10^{-2}(T_0 - 20) + 0.1965 \times 10^{-1}(\beta + 6.5)$$
$$- 0.5645 \times 10^{-2}(h_T - 11.231)] \quad (40a)$$

$$b = 0.001144[1 + 0.1164 \times 10^{-4}(P_0 - 1000) - 0.2795 \times 10^{-3} e_0$$
$$+ 0.3109 \times 10^{-2}(T_0 - 20) + 0.3038 \times 10^{-1}(\beta + 6.5)$$
$$- 0.1217 \times 10^{-1}(h_T - 11.231)]$$

$$c = -0.0090$$

and β is the tropospheric temperature lapse rate in °C/km; h_T is the height of the tropopause in km; T_0 is the surface temperature in °C; and P_0 is the surface pressure in mb.

F. Altshuler and Kalaghan Delay Model

This tropospheric excess delay model[28] is easy to use and that can either use monthly average refractivity values or can be updated with real-time corrections. The model, hereafter called the A&K model (note that there is an errata sheet in the published report for this model), consists of empirical expressions for tropospheric range error corrections suitable for GPS users, based upon worldwide statistics of surface refractivity. The model requires that the user input the elevation angle, height above the surface, and the surface refractivity, if known. If the user cannot measure the actual surface refractivity, the A&K model has an expression for the average surface refractivity as a function of height above sea level, latitude, and season of the year. The A&K model also contains a term for off-zenith angle correction, valid down to a viewing angle of 3-deg elevation angle. The A&K model has the following form:

$$\Delta(E, h, N_s) = 2.29286 \, m(E)H(h)F(h, N_s) \quad \text{feet}$$
$$= (0.3048)(2.29286) \, m(E)H(h)F(h, N_s) \quad \text{meter} \quad (41)$$

where $m(E)$ is the A&K mapping function, and H is a function of the user's height (note that one foot is exactly 0.3048m). The function F depends on both the user's height and the surface refractivity N_s. The factors are chosen to be of polynomial form to simplify computation:

$$m(E) = \{(0.1556 + 138.8926/E - 105.0574/E^2 + 31.5070/E^3)$$
$$+ [1.000 + 1.0 \times 10^{-4}(E - 30.00)^2]\}/2.29286 \quad (41a)$$

where E is in deg. [The factor 2.29286 has been broken out from the mapping function in the original A&K formulation, so that $m(90 \text{ deg}) = 1$.] The function $H(h)$ is as follows:

$$H(h) = [(0.00970 - 2.08809(h + 8.6286)^{-1} + 122.73592(h + 8.6286)^{-2}$$
$$- 703.82166(h + 8.6286)^{-3}] \quad (41b)$$

where h is in thousands of feet. The third function $F(h, N_s)$ is as follows:

$$F(h, N_s) = 3.28084 \left[\frac{6.81758}{h + 3.28084} + 0.30480(h + 3.28084) + 0.00423 \, N_s \right.$$
$$\left. - 1.33333 \right] [1 - 1.41723 \times 10^{-6}(N_s - 315.000)] \quad (41c)$$

The surface refractivity N_s, if not known, can be estimated as follows:

$$N_s = \alpha_0 + \alpha_1 h' + \alpha_2 \phi + \alpha_3 h' s^2 + \alpha_4 \phi s^2 + \alpha_5 h' c + \alpha_6 \phi \lambda c \quad (41d)$$

where ϕ is the latitude in degrees, and h' is the height of the surface above sea level in feet. The constants α_i and c are as follows: $\alpha_0 = 369.03$; $\alpha_1 = -0.01553$; $\alpha_2 = -0.92442$; $\alpha_3 = -0.0016$; $\alpha_4 = -0.19361$; $\alpha_5 = 0.00063$; $\alpha_6 = -0.05958$; $s = \sin(\pi M/12)$; and $c = \cos(\pi M/12)$. The value M (month) varies with the season: $M = 1.5$ winter; 4.5 spring; 7.5 summer; and 10.5 fall. The average global surface refractivity[28] is $\overline{N} = 324.8$, and the standard deviation is $\sigma_N = 30.1$.

Altshuler and Kalaghan[28] have estimated errors in their tropospheric range error model to be approximately 3.7% if the surface refractivity is known, and 6% if it is not known. If only a global average surface refractivity of 324.8 N units is used, they estimate the standard error of their model to be approximately 8%.

G. Ray Tracing and Simplified Models

For some purposes, a much simpler model of the tropospheric delay is all that is required. Typical troposphere delays vs elevation angle for a user at sea level have been computed by Janes et al.[7] by ray tracing for U.S. Standard Atmosphere and are shown in Table 5. Notice that the total delay at zenith ranges from a low of 2.316 m at latitude 75°N in January to a high of 2.576 m at 30°N in July. Shown in the lower portion of Table 5, the dry atmosphere delay ranged from 2.297 m in July at 60°N to a high of 2.328 m in January at 30°N for a 3-cm variation. The wet delay at zenith varied from a low of 0.015 m in January at 75°N to a high of 0.263 m in July at 30°N for a 24.8-cm variation. Notice that the ratio of 5-deg elevation angle delay to 90-deg elevation angle delay is fairly constant ranging from 10.183 to 10.231 at 75°N but that at the midlatitude of 30–45°, the ratio ranges only from 10.187 to 10.201 for only a 0.14% variation. Note the ratio $\Delta_w(90°)$ to $\Delta(90°)$ of wet delay to total delay is approximately 7% in summer in the midlatitudes (45–75°) but is only 0.6–2% in the winter.

If a simple model for sea level delay is employed, the excess delay could be modeled as follows:

$$\Delta = 2.47/(\sin E + 0.0121) \qquad (42)$$

which gives a value at zenith of 2.44 that is the average value for users in Table 3 at 30°N, 45°N, 60°N, and gives a mapping function value of approximately 10.2 at a 5-deg elevation angle.

Figure 10 shows this simplified mapping function

$$m(E) = 1.0121/(\sin E + 0.0121)$$

and the mapping functions of Black and Eisner[23]

$$m(E) = 1/\sqrt{1 - (\cos E/1.001)^2},$$

and Altshuler and Kalaghan[28]. They are relatively close to one another on the graph.

If we assume that refractivity varies as $(1 - h/h_d)^4 = (1 - .023h)^4$ for h in km and a scale height $h_d = 43$ km, then the zenith delay for a user at altitude h is computed as follows:

$$\Delta(h) = N_s \int_h^{h_d} (1 - h/h_d)^4 \, dh = \Delta(0)(1 - h/h_d)^5 \qquad (43)$$

As an alternative approach, an exponential profile can be employed to give the following simplified model:

$$\Delta(h) = 2.47 \, e^{-0.133h}/(\sin E + 0.0121) \qquad (44)$$

so that the zenith delay decreases to $e^{-0.133h} = 0.393$ of the sea level value at $h = 7$ km altitude.

H. Lanyi Mapping Function and GPS Control Segment Delay Estimate

Lanyi[29,30] has published one of the more precise but complex estimates of the tropospheric delay vs elevation angle where excess delay is of the form

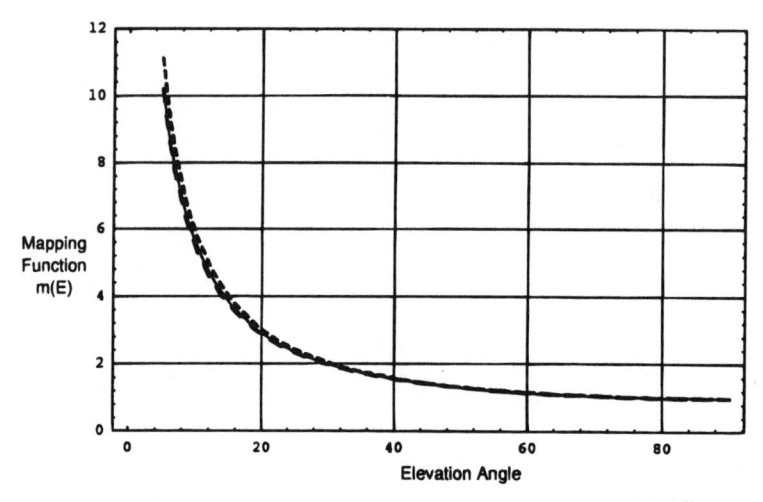

Fig. 10 Mapping functions of the simplified model (dashed), the B&E[23] (fine solid line) model, and the A&K[28] model (fine dashed line) for the range of E $5 < E \leq 90$ deg. The values of $m(5$ deg) are 10.1969, 10.218, and 11.114 for the simplified model, Eq. (44) the B&E model, and the A&K model, respectively.

$$\Delta(E) = F(E)/\sin E \qquad E > 5°$$

where

$$F(E) = \Delta_d F_d(E) + \Delta_w F_w(E) + (\Delta_d^2/D)F_{b1}(E) + (2\Delta_d\Delta_w/D)F_{b2}(E)$$
$$+ \Delta_w^2/D)F_{b3}(E) + (\Delta_d^3/\Delta^2)F_b4(E).$$

The terms Δ_w and Δ_d are the wet and dry zenith delays, respectively, D is a scale height ≈ 8.567 km, and the F_{b1}, F_{b2}, F_{b3}, F_{b4}, are various complicated mapping functions. Suffice it to say that special cases of the Lanyi mapping function yield approximations to the Saastamoinen and Black mapping functions.

The GPS Control Segment also employs a model for tropospheric excess delay to correct the smoothed pseudoranges. The overall equation for tropospheric delay is[31]:

$$\Delta(E) = \frac{(0.02312)P_s[T - 4.11 + 5(r_m - r_a)/148.98]}{T\sqrt{1 - \left[\frac{r_a \cos E}{r_m + (1 - C)(r_d - r_m)}\right]^2}} +$$

$$\frac{(0.0746)eh[1 + 5(r_m - r_a/h]}{T^2\sqrt{1 - \left[\frac{r_a \cos E}{r_m + (1 - C)h}\right]^2}} \qquad E > 5°$$

where P_s = monitor station barometric pressure converted to kilopascals, T = monitor station measured temperature converted to Kelvin, r_m = radial distance

Table 5 Troposphere delay (wet + dry component) in meters vs elevation angle in degrees, latitude, and season developed by ray tracing the U.S. Standard Atmosphere; also shown in the lower portion of the table is the individual wet and dry delay components (using results from Janes et al.[7])

Elevation angle E	15°N annual avg.	30°N July	30°N Jan.	45°N July	45°N Jan.	60°N July	60°N Jan.	75°N July	75°N Jan.
5 deg	26.088	26.279	25.06	25.365	24.239	24.781	23.853	24.409	23.694
10 deg	14.237	14.341	13.689	13.854	13.234	13.352	13.002	13.338	12.911
20 deg	7.416	7.47	7.132	7.219	6.893	7.05	6.769	6.95	6.719
90 deg = zenith	2.558	2.576	2.46	2.49	2.378	2.431	2.334	2.397	2.316
Ratio, Δ(5 deg)/Δ(90 deg)	10.199	10.201	10.187	10.187	10.193	10.194	10.220	10.183	10.231
Ratio, Δ_w(90 deg)/Δ(90 deg)	0.0958	0.1020	0.0537	0.0731	0.0240	0.0551	0.0124	0.0405	0.0065

Dry/wet excess delay, m

Elevation angle	Annual	30°N latitude July	30°N latitude Jan.	45°N latitude July	45°N latitude Jan.	60°N latitude July	60°N latitude Jan.	75°N latitude July	75°N latitude Jan.
90 deg	2.313/.245	2.313/.263	2.328/.132	2.308/.182	2.321/.057	2.297/.134	2.305/.029	2.300/.097	2.301/.015
20 deg	6.702/.714	6.703/.767	6.747/.385	6.688/.531	6.727/.166	6.658/.392	6.683/.086	6.667/.283	6.674/.045
10 deg	12.841/1.396	12.842/1.499	12.936/.753	12.817/1.037	12.910/.324	12.766/.766	12.835/.167	12.786/.552	12.823/.088
5 deg	23.381/2.707	23.379/2.900	23.600/1.460	23.356/2.009	23.612/.627	23.300/1.481	23.530/.323	23.341/1.068	23.523/.171

from the Earth's center to the meterorological sensors, r_a = radial distance from the Earth's center to the monitor station antenna, r_d = tropospheric dry radius, E = satellite elevation, h = wet height of troposphere, e = estimate of the partial water vapor pressure, C = assumed integration constant = 0.85 when $E > 5.0°$.[31]

I. Model Comparisons

Janes et al.[7] have done an extensive comparison of the various models described in the previous paragraphs. A few words of summary are appropriate here. For the zenith delay, both the Saastamoinen and Hopfield models in Refs. 11 and 18–21 gave results for the dry component that were within several mm of the ray trace delay as computed for the U.S. Standard Atmosphere (Table 5). Even at a 5-deg elevation angle, the Hopfield model was within 5 cm, whereas the two-layer model[11,20] gave more accurate results to within 6 mm. At zenith, the water vapor component for the standard model[11,20] gave results mostly within 30 mm, whereas the model errors[18,19,21] were within 20 mm.

Janes also compared the mapping function error $\delta m(E)$ scaled by the total zenith delay; namely, $\delta m(E)\Delta_z$. For the total delay at the 20-deg elevation angle, the Saastamoinen and Black and Eisner models of Refs. 11, 20, and 23 gave results within 8 mm. These errors rose to approximately 50 mm at 10 deg elevation. At 5-deg elevation angle the result[23] was still within 10 cm, whereas the standard model[11,20] produced errors of 1.2 m. For the dry gas component the Davis mapping function was the most accurate, yielding errors of less than 6 cm at 5-deg elevation angle.

J. Tropospheric Delay Errors and GPS Positioning

Errors in the estimates of the tropospheric delay have some amount of cancellation relative to GPS positioning because of the correlation in the errors for different satellites. Clearly, a constant bias for four pseudoranges to four different satellites would not affect position errors at all but would cause a user clock time bias error. As we have seen, the significant tropospheric errors are likely to be those from satellites at low elevation angles. If, as an example, four satellites at low elevation angle all have an identical error bias in the troposphere delay, and a satellite at zenith has no delay bias error, the bias error in the horizontal plane position estimate cancels out, but an altitude error remains.

One of the primary contributors to horizontal position error, in addition to lower elevation angle prediction of the dry tropospheric delay (with satellites of different elevation angles) is the variation in wet low elevation angle errors for satellites at substantially different azimuths. At 5-deg elevation angle, the wet troposphere can contribute widely varying delay errors (relative to the total wet delay contribution ≈ 25 cm). The water vapor zenith correction can also change significantly 2 cm with a 10-km horizontal displacement, or in a few hours of time passage, and these can be magnified by the mapping function at low elevation angles E. Water vapor zenith delay changes of 3 cm have been observed in one hour.[3] There is also no way to predict the wet delay correction change with the azimuth. It is expected, however, that the dry component of delay error would be highly correlated with azimuth; only the wet component is expected to decorrelate in any significant way with azimuth.

References

[1] Davis, J. L., et al., "Geodesy by Radio Interferometry: Effects of Atmospheric Modeling Errors on Estimates of Baseline Length," *Radio Science,* Vol. 20, 1985.

[2] NOAA, "U.S. Standard Atmosphere, 1976," NOAA-S/T76-1562, National Technical Information Service, U.S. Dept. of Commerce, Springfield, VA, 1976.

[3] Elgered, G., "Tropospheric Radio Path Delay," edited by M. A. Jansse, *Atmospheric Remote Sensing by Microwave Radiometry,* Wiley, New York, 1993.

[4] Smith, D. G., (ed.), The Cambridge Encyclopedic of Earth Sciences, Cambridge University Press, New York, 1981.

[5] Boithias, L., *Radio Wave Propagation,* McGraw Hill, New York, 1982.

[6] Spilker, J. J., Jr., "Digital Communications by Satellite," Prentice Hall, Englewood Cliffs, NJ, 1977, 1995.

[7] Janes, H. W., Langley, R. B., and Newby, S. P., "Analysis of Tropospheric Delay Prediction Models," *Bulletin Géodésique,* Vol. 65, 1991.

[8] Ippolito, L. J., Jr., *Radio Wave Propagation in Satellite Communications,* Van Nostrand, New York, 1986.

[9] Moulsley, T. J., and Vilar, E. "Experimental and Theoretical Statistics of Microwave Amplitude Scintillation on Satellite Downlinks," *IEEE Trans. Antennas and Propagation,* Vol. AP-30, pp. 1099–1106, Nov. 1982.

[10] Sears, F. W., *Optics,* Addison-Wesley, Reading, MA, 1949.

[11] Saastamoinen, J., "Contribution to the Theory of Atmospheric Refraction, *Bulletin Géodésique,* Vol. 105, Sept. 1972, Vol. 106, Dec. 1972, Vol. 107, March 1973.

[12] Thayer, G. D., "An Improved Equation for the Radio Refractive Index of Air," *Radio Science,* Vol. 9, 1974.

[13] Hill, R. J., Lawrence, R. S., and Priestly, J. T., "Theoretical and Calculated Aspects of the Radio Refractive Index of Water Vapor," *Radio Science,* Vol. 17, pp. 1251–1257, 1982.

[14] Birmbaum, G., and Chatterjee, S. K., "The Dielectric Constant of Water Vapor in the Microwave Region," *Journal of Applied Physics,* Vol. 23, 1952.

[15] Boudouris, G., "On the Index of Retraction of Air, the Absorption and Dispersion of Centimeter Waves by Gases," *Journal of Research of the National Bureau of Standards,* 1963.

[16] Sears, F. W., *Thermodynamics, The Kinetic Theory of Gases, and Statistical Mechanics,* Addison-Wesley, Reading, MA, 1953.

[17] Abbott, M. M., and Van Ness, H. C., *Thermodynamics,* McGraw Hill, New York, 1989.

[18] Hopfield, H. S., "Tropospheric Effect on Electromagnetically Measured Range: Prediction from Surface Weather Data," Applied Physics Laboratory, Johns Hopkins University, Baltimore, MD, July 1970.

[19] Hopfield, H. S., "Tropospheric Range Error Parameters—Further Studies," Applied Physics Laboratory, Johns Hopkins University, Baltimore, MD, June 1972.

[20] Saastamoinen, J., "Atmospheric Correction for the Troposphere and Stratosphere in Radio Ranging of Satellites," Geophysical Monograph 15, American Geophysical Union, Washington, DC, 1972.

[21] Hopfield, H. S., "Two Quartic Tropospheric Refractivity Profile for Correcting Satellite Data," *Journal of Geophysical Research,* April 1969.

[22] Black, H. D., "An Easily Implemented Algorithm for the Tropospheric Range Correction," *Journal of Geophysical Research,* Vol. 38, No. 4, 1978, pp. 1825–1828.

[23]Black, H. D., and Eisner A., "Correcting Satellite Doppler Data for Tropospheric Effects," *Journal of Geophysical Research,* Vol. 89, 1984.

[24]Moran, J. M., and Rosen, B. R., "The Estimation of the Propagation Delay Through the Troposphere from Microwave Radiometer Data," *Radio Interferometry Techniques for Geodesy,* NASA Conf. Publ. 2125, 1979.

[25]Berman, A. L., "The Prediction of Zenith Refraction from Surface Measurements of Meteorological Parameters," JPL TR-32-1602 California Institute of Technology, Jet Propulsion Laboratory, Pasadena, CA, 1976.

[26]Marini, J. W., "Correction of Satellite Tracking Data for an Arbitrary Tropospheric Profile," *Radio Science,* Vol. 7, 1972.

[27]Chao, C. C., "The Tropospheric Calibration Model for Mariner Mars, 1971," JPL TR 32-1587, Jet Propulsion Laboratory, Pasadena, CA, March 1974.

[28]Altshuler, E. E., and Kalaghan, P. M., "Tropospheric Range Error Corrections for the NAVSTAR System," Microwave Physics Laboratory, Air Force Cambridge Research Laboratories, April 1974.

[29]Lanyi, G., "Tropospheric Delay Effects in Radio Interferometry," Telecommunications and Data Acquisition Progress Rep., Jet Propulsion Lab, Pasadena, CA, April–June 1984.

[30]Sovers, O. J., and Lanyi, G. E., "Evaluation of Current Tropospheric Mapping Functions by Deep Space Network Very Long Baseline Interferometry," The Telecommunications and Data Acquisition Progress Rep. 42–119, Jet Propulsion Laboratory, Pasadena, CA, Nov. 1994.

[31]Master Control Station Kalman Filter–Mission Support Study Guide, 2nd ed. Oct. 1993.

Bibliography

Brunner, F. K., ed., "Atmospheric Effects on Geodetic Space Measurements," Monograph 12, School of Surveying, University of New South Wales, Kinsington, NSW, Australia, 1988.

Hendy, M. R., and Brunner, F. K., "Modeling the Zenith Wet Component of the Tropospheric Path Delay for Microwaves," *Australian Journal of Geodesy, Photogrammetry & Surveying,* Dec. 1990.

Chapter 14

Multipath Effects

Michael S. Braasch*
Ohio University, Athens, Ohio 45701

I. Introduction

THE Global Positioning System (GPS) has been shown to be capable of supporting a wide variety of exciting applications. In addition to the usual functions of position, velocity, and time determination, it is also possible to perform attitude and heading determination of dynamic platforms, measurement of flexing in large space structures, and to provide precision approach and landing guidance. The later applications require high accuracy in real time. However, these technologies face a major stumbling block. It is the effect of multipath. Multipath is the phenomenon whereby a signal arrives at a receiver via multiple paths attributable to reflection and diffraction. Multipath represents the dominant error source in satellite-based precision guidance systems.

Multipath has been cited as a major error source both in differential satellite systems as well as interferometry.[1,2] Multipath distorts the signal modulation and degrades accuracy in conventional and differential systems. Multipath also distorts the phase of the carrier, and hence degrades the accuracy of the interferometric systems. Furthermore, because interferometric systems often employ pseudorange measurements for initialization (ambiguity resolution) purposes, multipath contamination of the pseudorange can increase the time required for initialization.

For standard code-based differential systems, signal degradation attributable to multipath can be severe. This stems from the fact that multipath is a highly localized phenomenon. Multipath sources that affect the ground reference station receiver do not necessarily cause errors in the mobile receiver. Likewise, multipath sources that affect the mobile receiver do not necessarily affect the ground reference station.

Multipath effects in pseudorandom noise (PRN) ranging have been studied since the early 1970s. Hagerman[3] derived relationships involving multipath and PRN code-tracking error. This fundamental work formed the basis for the analysis of GPS code and carrier multipath errors during the field tests at the Yuma Proving Ground (YPG).[4] In the early 1980s, the effect of multipath on short

---

Copyright © 1995 by the American Institute of Aeronautics and Astronautics, Inc. All rights reserved.

*Assistant Professor, Department of Electrical and Computer Engineering.

baseline interferometry was studied at the Massachusetts Institute of Technology[5,6] and at the Charles Stark Draper Laboratory.[7] The studies concluded that the effects of multipath (in this differential carrier-phase tracking system) could be reduced to a few centimeters of error over short baselines if the signals could be averaged over a period of an hour or more. Bletzacker[8] also considered multipath errors in geodetic applications. Performance improvements were obtained by mounting the antenna on rf-absorbing material, thereby improving the characteristics of the antenna pattern. Tranquilla and Carr[9] confirmed this by collecting data in stressful environments using a geodetic antenna with and without an rf-absorbing ground plane.

Falkenberg et al.[10] and Lachapelle et al.[1] describe marine differential GPS experiments in which multipath was mitigated through the use of rf-absorbing ground planes and filtering schemes. Evans[11] demonstrated multipath effects on ionospherically corrected code and carrier measurements from a geodetic GPS receiver. Georgiadou and Kleusberg[12] considered multiple reflections and showed that multipath on short baselines could be detected using dual-frequency measurements. Abidin[13] examined the effects of multipath in dual-frequency-measurement-based ambiguity resolution.

The effect of multipath on ionospheric measurements using GPS was presented by Bishop et al.[14] Their work verified the theoretical multipath relations derived by Hagerman[3] and considered various mitigation schemes for static applications. Sennott and Pietraszewski[15] and Sennot and Spalding[16] have developed state variable models for the estimation and mitigation of multipath in differential GPS ground reference stations.

Van Nee[17] has shown that code-phase multipath error traces tend not to be zero mean and can have periods on the order of an hour. This contradicts the popular notion that code-phase multipath can be eliminated in static applications simply through averaging. Van Nee[18,19] has shown that an exception to this rule occurs when a coherent delay–lock-loop (DLL) receiver is used and the rate-of-change of the multipath relative phase is large compared to the tracking loop bandwidth.

Much of this chapter is based upon Braasch,[20] which examined the characteristics of multipath error in the precision approach and landing environment.

II. Signal and Multipath Error Models

Fundamental to the understanding of the effects of multipath in any given environment is an understanding of PRN ranging receivers and how multipath distortion results in ranging errors. This section derives closed form expressions for code-phase and carrier-phase multipath errors resulting from a single multipath ray entering a stationary receiver. Although most multipath scenarios involve multiple rays, much insight results from the analysis of the single-ray case. It also serves as the starting point from which the multiple-ray case can be considered. The analysis considers the two most prevalent types of receivers: the coherent delay–lock-loop (DLL) and the noncoherent DLL. This section extends the theoretical developments documented by Hagerman[3] to include consideration of code multipath error spectra. It should be noted that although the derivations follow the signal flow of analog receivers, the results are exactly the same for modern digital architectures.

A. Pseudorandom Noise Modulated Signal Description

The signal broadcast from the satellite in a PRN ranging system may be expressed as follows:

$$s_1(t) = A \cos[\omega_o t + p(t)\pi/2] \tag{1}$$

where $0.5*A^2$ is the average signal power into the receiver, ω_o is the frequency of the received signal in radians per second (carrier frequency plus Doppler shift), and $P(t)$ is the PRN code (either $+1$ or -1).

Note that the actual GPS signal is considerably more complicated. The GPS carrier is modulated by two PRN codes (the coarse/acquisition (C/A) code and the precision (P) code) in addition to navigation data. For the present purpose of multipath analysis, however, the model given by Eq. (1) is adequate. By applying trigonometric identities, Eq. (1) may be rewritten as follows:

$$\hat{s}_1(t) = -Ap(t)\sin(\omega_o t) \tag{2}$$

Once inside the receiver, multipath is characterized by four parameters (all of which are relative to the direct signal): 1) amplitude; 2) time delay; 3) phase; 4) phase rate of change. For the present discussion, a stable multipath scenario is assumed; thus, the relative phase rate of change is assumed to be zero. Relative phase of the multipath is a function of the relative time delay and the reflection coefficient of the reflecting object. If the received signal is composed of the direct signal plus a single multipath ray, it may be expressed as follows:

$$s_{1m}(t) = \hat{s}_1(t) + \alpha \hat{s}_1(t + \delta) \tag{3}$$

where α is the multipath relative amplitude and δ is the multipath relative time delay (note: must be negative given the convention used in the equation).

Note that the relative phase of the multipath is not shown. Substitution of Eq. (2) into Eq. (3) and inclusion of relative phase yields the following:

$$s_{1m}(t) = -Ap(t)\sin(\omega_o t) - \alpha Ap(t + \delta)\sin(\omega_o t + \theta_m) \tag{4}$$

where θ_m is the multipath relative phase.

B. Coherent Pseudorandom Noise Receiver

This section derives expressions for the coherent DLL discriminator curve in the absence and presence of multipath. Also derived is the expression for the composite phase of the multipath corrupted signal as it is tracked by the carrier-tracking loop.

1. Coherent Delay Lock Loop Discriminator Curve in the Absence of Multipath

Following the usual signal flow in an analog receiver, the incoming signal $S_1(t)$ is mixed with early and late versions of the PRN code modulated onto a local oscillator frequency:

$$s_E(t) = -p(t + \tau - \tau_d)\sin(\omega_l t + \theta) \quad (5)$$

$$s_L(t) = -p(t + \tau + \tau_d)\sin(\omega_l t + \theta) \quad (6)$$

where τ is the DLL tracking error, τ_d is the time advance of the early code or the time delay of the late code (relative to the on-time code), θ is the tracking error of the phase lock loop (PLL), and ω_l is the local oscillator frequency in rad/s.

The signals output from the mixers are passed through bandpass filters (BPF). It is assumed that the passband of the BPFs is narrow enough to reject the sum-frequency terms. The filters also serve to complete the correlation process. The bars shown over the quantities in Eqs. (7) and (8) denote the operation of the BPFs. The output of the filters is as follows:

$$\overline{s_1(t)s_E(t)} = 1/2\, AR(\tau - \tau_d)\cos(\omega_1 t - \theta) \quad (7)$$

$$\overline{s_1(t)s_L(t)} = 1/2\, AR(\tau + \tau_d)\cos(\omega_1 t - \theta) \quad (8)$$

where ω_1 is the difference-frequency ($\omega_1 = \omega_o - \omega_l$) and $R(\tau)$ is the correlation function of the PRN code.

A sufficient approximation of the PRN code correlation function is given by the following:

$$R(\tau) = 1 - \frac{|\tau|}{T}, \quad |\tau| \leq T$$
$$= 0, \quad |\tau| > T \quad (9)$$

where T is the PRN code bit period (note: a PRN code bit is also known as a "chip." Accordingly, the PRN code bit-rate is also known as the "chipping-rate").

Correlation sidelobes are ignored, and infinite bandwidth is assumed. The finite bandwidth of the BPFs distorts the shape of the PRN code bits and results in a smoothing of the correlation function.[23] Comparison of the results of this analysis (Sec. II. B.) with those obtained when smoothing the correlation function[24] reveals that the smoothing slightly reduces the maximum range error due to a given set of multipath parameters. As a result, Eq. (9) yields slightly conservative results.

The signals out of the BPFs are thus the early and late correlator functions at the intermediate frequency (IF). Low-pass filters can be used to remove the IF term leaving the baseband signal. Assuming a PLL carrier-phase tracking error of zero ($\theta = 0$) then yields the following:

$$\overline{s_1(t)s_E(t)} = 1/2\, AR(\tau - \tau_d) \quad (10)$$

$$\overline{s_1(t)s_L(t)} = 1/2\, AR(\tau + \tau_d) \quad (11)$$

Finally, the discriminator function is formed, for example, by differencing the outputs of the LPFs. The normalized form of the discriminator function is as follows:

$$D_c(\tau) = R(\tau + \tau_d) - R(\tau - \tau_d) \quad (12)$$

where the subscript c denotes coherent DLL operation.

MULTIPATH EFFECTS

Note that the DLL tracks the peak of the correlation function by tracking the zero-crossing of the discriminator function since both occur for $\tau = 0$. As derived in the next section, multipath distorts the discriminator curve so that the zero-crossing occurs for some non-zero τ. Thus, the τ corresponding to the zero-crossing is the DLL tracking error caused by the multipath. Note that although τ is an error in correlation, ranging is a time-domain function. An error in correlation corresponds to an equal but opposite error in timing. Thus, the ranging error is given by the opposite of the DLL tracking error.

2. Coherent Delay Lock Loop Discriminator Curve in the Presence of Multipath

In the presence of multipath, the incoming signal is given by Eq. (4). In the case of the coherent DLL, the early and late codes are modulated onto a local oscillator frequency, which is phase-locked to the incoming signal. However, the incoming signal has been perturbed because of the presence of the multipath. Thus, the PLL tracks the phase of the composite signal and not that of the direct signal. The early and late signals are now given by the following:

$$s_{Em}(t) = -p(t + \tau - \tau_d)\sin(\omega_I t + \theta_c + \theta) \quad (13)$$

$$s_{Lm}(t) = -p(t + \tau + \tau_d)\sin(\omega_I t + \theta_c + \theta) \quad (14)$$

where θ_c is the composite phase of the direct plus multipath signal and θ is the PLL carrier-phase tracking error of the composite signal.

The relations governing the composite phase of the direct plus multipath signal are derived in the next section. The outputs of the mixers are given by multiplying Eq. (4) by Eqs. (13) and (14).

Passing the output of the mixers through the BPFs completes the correlation process and removes the sum frequency terms. Low-pass filters then allow only the baseband signals to pass, yielding the following:

$$\overline{s_{1m}(t)s_{Em}(t)} = 1/2\, AR(\tau - \tau_d)\cos(-\theta_c - \theta)$$
$$+ 1/2\, \alpha AR(\tau - \tau_d - \delta)\cos(\theta_m - \theta_c - \theta) \quad (15)$$

$$\overline{s_{1m}(t)s_{Lm}(t)} = 1/2\, AR(\tau + \tau_d)\cos(-\theta_c - \theta)$$
$$+ 1/2\, \alpha AR(\tau + \tau_d - \delta)\cos(\theta_m - \theta_c - \theta) \quad (16)$$

Assuming a PLL tracking error of zero and taking advantage of the fact that the cosine is an even function yields the following:

$$\overline{s_{1m}(t)s_{Em}(t)} = 1/2\, AR(\tau - \tau_d)\cos(\theta_c)$$
$$+ 1/2\, \alpha AR(\tau - \tau_d - \delta)\cos(\theta_m - \theta_c) \quad (17)$$

$$\overline{s_{1m}(t)s_{Lm}(t)} = 1/2\, AR(\tau + \tau_d)\cos(\theta_c)$$
$$+ 1/2\, \alpha AR(\tau + \tau_d - \delta)\cos(\theta_m - \theta_c) \quad (18)$$

Again, the discriminator function is formed by differencing the outputs of the LPFs. The normalized form of the discriminator function is as follows:

$$D_{cm}(\tau) = R(\tau + \tau_d)\cos(\theta_c) + \alpha R(\tau + \tau_d - \delta)\cos(\theta_m - \theta_c)$$
$$- R(\tau - \tau_d)\cos(\theta_c) - \alpha R(\tau - \tau_d - \delta)\cos(\theta_m - \theta_c)$$
$$= [R(\tau + \tau_d) - R(\tau - \tau_d)]\cos(\theta_c)$$
$$+ \alpha[R(\tau + \tau_d - \delta) - R(\tau - \tau_d - \delta)]\cos(\theta_m - \theta_c) \qquad (19)$$

where the subscript *cm* denotes coherent DLL operation in the presence of multipath.

3. Carrier Phase Lock Loop Operation in the Presence of Multipath

Prior to entering the PLL, the incoming signal is mixed with the on-time code modulated onto the local oscillator frequency. The incoming signal in the presence of a single multipath ray was given by Eq. (4). The on-time signal generated by the receiver is given by the following.

$$s_{Om}(t) = -p(t + \tau)\sin(\omega_l t + \theta_c + \theta) \qquad (20)$$

After mixing the two signals and passing through a BPF, we achieve the following;

$$\overline{s_{1m}(t)s_{Om}(t)} = 1/2\ AR(\tau)\cos((\omega_o - \omega_l)t - \theta_c - \theta)$$
$$+ 1/2\ \alpha AR(\tau - \delta)\cos((\omega_o - \omega_l)t + \theta_m - \theta_c - \theta) \qquad (21)$$

Assuming perfect carrier PLL tracking of the composite signal ($\theta = 0$) and substituting $\omega_1 = \omega_o - \omega_l$, Eq. (21) becomes the following:

$$\overline{s_{1m}(t)s_{Om}(t)} = 1/2\ AR(\tau)\cos(\omega_1 t - \theta_c)$$
$$+ 1/2\ \alpha AR(\tau - \delta)\cos(\omega_1 t + \theta_m - \theta_c) \qquad (22)$$

Using the trigonometric identity for the cosine of a sum gives the following:

$$\overline{s_{1m}(t)s_{Om}(t)} = 1/2\ AR(\tau)[\cos(\omega_1 t)\cos(\theta_c) + \sin(\omega_1 t)\sin(\theta_c)]$$
$$+ 1/2\ \alpha AR(\tau - \delta)[\cos(\omega_1 t)\cos(\theta_c - \theta_m) + \sin(\omega_1 t)\sin(\theta_c - \theta_m)] \qquad (23)$$

Grouping the cosine and sine terms leads to the following:

$$\overline{s_{1m}(t)s_{Om}(t)} = 1/2\ A[R(\tau)\cos(\theta_c)$$
$$+ \alpha R(\tau - \delta)\cos(\theta_c - \theta_m)]\cos(\omega_1 t) + 1/2\ A[R(\tau)\sin(\theta_c) \qquad (24)$$
$$+ \alpha R(\tau - \delta)\sin(\theta_c - \theta_m)]\sin(\omega_1 t)$$

The PLL tracks the composite signal so that the coefficient of the $\sin(\omega_1 t)$ term is nulled:

$$R(\tau)\sin(\theta_c) + \alpha R(\tau - \delta)\sin(\theta_c - \theta_m) = 0 \qquad (25)$$

a. Region I: Absolute value of the DLL tracking error less than one chip. In this region, the correlation function $R(\tau)$ is nonzero. After making trigonometric substitutions and rearranging, Eq. (25) may be rewritten as follows:

$$\tan(\theta_c) = \sin(\theta_c)/\cos(\theta_c)$$
$$= \alpha R(\tau - \delta)\sin(\theta_m)/[R(\tau) + \alpha R(\tau - \delta)\cos(\theta_m)] \quad (26)$$

For region I operation, then, the composite phase of the signal entering the PLL is given by the arctangent of the right-hand side of Eq. (26). The interdependency of the code- and carrier-tracking loops may also be observed in Eq. (26) by noting the presence of the code-correlation function.

Assuming that the multipath strength is always less than or equal to the direct, it can be shown that the carrier-phase error can be no more than 90 deg. At the GPS L_1 band (1575.42 MHz), this corresponds approximately to 4.8 cm. As shown later, code-phase errors can exceed 100 m. Carrier-phase tracking, thus, yields relatively multipath-free measurements.

b. Region II: Absolute value of the DLL tracking error greater than or equal to one chip. In this region, $R(\tau)$ is approximately zero. Although this is not strictly true, it is a reasonable approximation in light of its proximity to zero relative to the peak value of the correlation function. Using this assumption, Eq. (25) simplifies to the following:

$$\alpha R(\tau - \delta)\sin(\theta_c - \theta_m) = 0 \quad (27)$$

therefore

$$\theta_c = \theta_m + 2\pi N \quad (28)$$

For region II operation, then, the composite phase of the signal entering the PLL is simply the multipath phase relative to the direct signal. The direct signal phase is taken to be zero. Conceptually, region II operation involves the DLL tracking the multipath rather than the direct signal. In either operating region, then, the multipath-induced carrier-phase measurement error is given by θ_c.

C. Noncoherent Pseudorandom Noise Receiver

This section details expressions for the noncoherent DLL discriminator curve in the absence and presence of multipath. Because it can be shown that the PLL carrier-phase error attributable to multipath has the same form as for the coherent receiver architecture, the derivation is not given here.

1. Noncoherent Delay Lock Loop Discriminator Curve in the Absence of Multipath

The signal flow for the noncoherent DLL is similar to that of the coherent DLL except the outputs of the early and late correlators (mixers plus bandpass filters) are squared prior to being low-pass filtered and differenced. The signals entering the LPFs are given by squaring the expressions in Eqs. (7) and (8):

$$\overline{\{s_1(t)s_E(t)\}}^2 = 1/4\ A^2R^2(\tau - \tau_d)\cos^2(\omega_1 t - \theta) \qquad (29)$$

$$\overline{\{s_1(t)s_L(t)\}}^2 = 1/4\ A^2R^2(\tau + \tau_d)\cos^2(\omega_1 t - \theta) \qquad (30)$$

Note that because the early and late signals are being generated noncoherently, θ is simply the phase offset of the frequency tracking loop. After Eqs. (29) and (30) have been expanded, the LPFs reject the double-frequency terms leaving the following:

$$\overline{\{s_1(t)s_E(t)\}}^2 = 1/8\ A^2R^2(\tau - \tau_d) \qquad (31)$$

$$\overline{\{s_1(t)s_L(t)\}}^2 = 1/8\ A^2R^2(\tau + \tau_d) \qquad (32)$$

Again, the discriminator function is formed by differencing the outputs of the LPFs. Thus, the normalized form of the noncoherent discriminator function is as follows:

$$D_n(\tau) = R^2(\tau + \tau_d) - R^2(\tau - \tau_d) \qquad (33)$$

where the subscript n denotes noncoherent DLL operation.

2. Noncoherent Delay Lock Loop Discriminator Curve in the Presence of Multipath

Following a procedure similar to that for the analysis of the coherent DLL, the normalized form of the noncoherent discriminator function is given by the following:

$$D_{nm}(\tau) = R^2(\tau + \tau_d) - R^2(\tau - \tau_d) + \alpha^2[R^2(\tau + \tau_d - \delta) - R^2(\tau - \tau_d - \delta)]$$
$$+ 2\alpha\cos(\theta_m)[R(\tau + \tau_d)R(\tau + \tau_d - \delta) - R_,(\tau - \tau_d)R(\tau - \tau_d - \delta)] \qquad (34)$$

where the subscript nm denotes noncoherent DLL operation in the presence of a single multipath ray.

D. Simulation Results

Having derived the multipath error equations, simulations allow for the quantification of the error encountered under various multipath conditions.

1. C/A Code

The pseudorange error envelope as a function of relative multipath amplitude and delay is given in Fig. 1 for the standard coherent DLL. Note that the standard separation of 1 PRN bit period between the early and late correlators is assumed. In the figure, the relative multipath amplitude is constant and the upper curve represents error attributable to a multipath ray that is in phase with the direct signal. The bottom curve represents the out-of-phase case. It is important to remember that the relative phase of the multipath signal is a function of the electromagnetic properties of the reflecting object in addition to the path length difference. A multipath signal with a given delay can, therefore, take an arbitrary relative phase depending upon the reflector. Examination of the plot reveals that even for a weak multipath signal (-20 dB), the peak error is 15 m. It is also

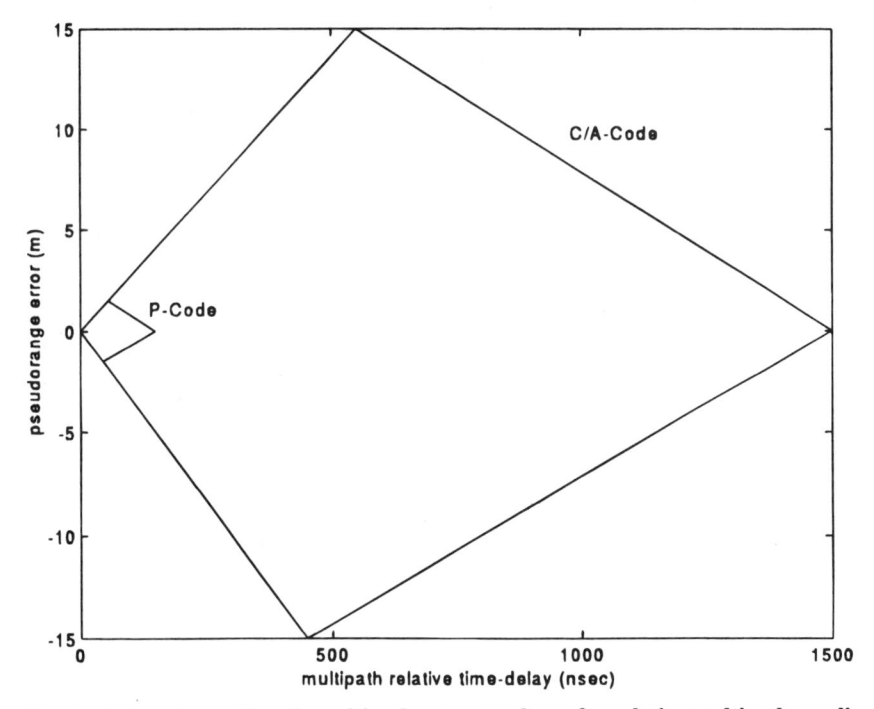

Fig. 1 C/A-code and P-code multipath error envelopes for relative multipath amplitude set at −20 dB.

important to note that the error drops to zero at a delay of 1466 ns (1.5 chips). Furthermore, as shown in Refs. 19 and 20, the error increases when the multipath relative strength increases. Assuming an environment in which the multipath signal strength never exceeds that of the direct, the peak multipath error equals one-half of a chip length.[3] Although not shown, the error envelope for the noncoherent DLL is exactly the same as for the coherent DLL. Their behavior within the envelope is different, however, as is shown later in the plots of average error.

It is important to note that, in reality, multipath signals that are delayed by more than 1.5 chips can induce some error. Recall that in the derivation of the multipath error equations, the sidelobes of the correlation function were ignored. This was not a problem for analyzing multipath with relative time delays of less than 1.5 chips. However, for longer delay multipath, the correlation sidelobes provide the mechanism for interference with the direct signal. The small magnitude of these sidelobes relative to the main lobe means that long delay multipath signals will be attenuated about 24 dB in the correlation process. Although weak, this kind of multipath can induce error on the order of several meters.

For a better understanding of the behavior of the error curve within the envelope, Figs. 2 and 3 show C/A-code pseudorange error over a small range of time delay for different relative multipath amplitudes. For these examples, the multipath relative phase was assumed to be strictly a function of relative path delay. When

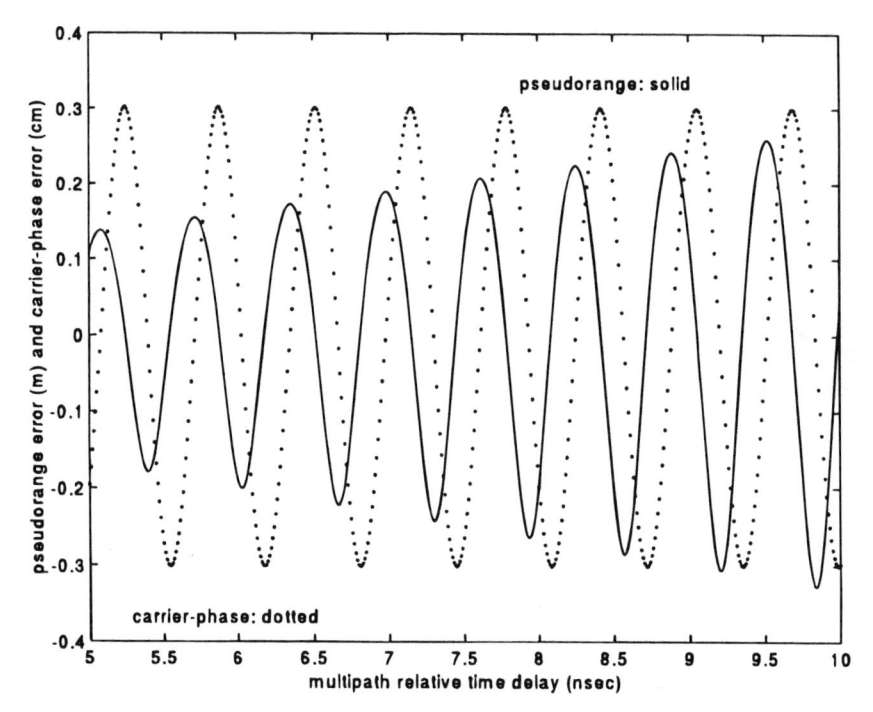

Fig. 2 C/A-code and carrier-phase multipath error for relative multipath amplitude set at −20 dB. Relative phase was calculated strictly from relative path delay.

the relative multipath amplitude is small, the error varies sinusoidally as a function of relative path delay and thus is narrow band. When the relative amplitude is large, however, the error is not sinusoidal and, in fact, it contains sharp discontinuities. In this case, the error signal is wide band.[20,21] This disproves the popular myth that every peak in a multipath error spectrum corresponds to a separate multipath ray. The nonsinusoidal behavior is a result of nonlinearities in the receiver architecture. The effects reach further than simply causing an increased error signal bandwidth. As was first shown by Hagerman[3] and later by Van Nee[17] and Braasch and Van Graas,[21] the error signal is not zero-mean. In situations where the multipath relative phase is fluctuating (nonzero relative phase rate-of-change), the errors will not average out to zero.

Plots of average error vs time delay are given in Fig. 4. Because relative-phase is a complex function of the electromagnetic properties of the reflecting object in addition to the time delay, errors may be computed by holding relative amplitude and phase constant, and varying the time delay. For each of the plots shown, error was averaged over 10 relative-phase values evenly spaced between 0 and 180 deg. As can be seen in the plots, not only are the error traces nonzero mean, the average value can easily be several meters. For example, a relative multipath amplitude of −8 dB yields a peak average error of 8 m for the coherent DLL.

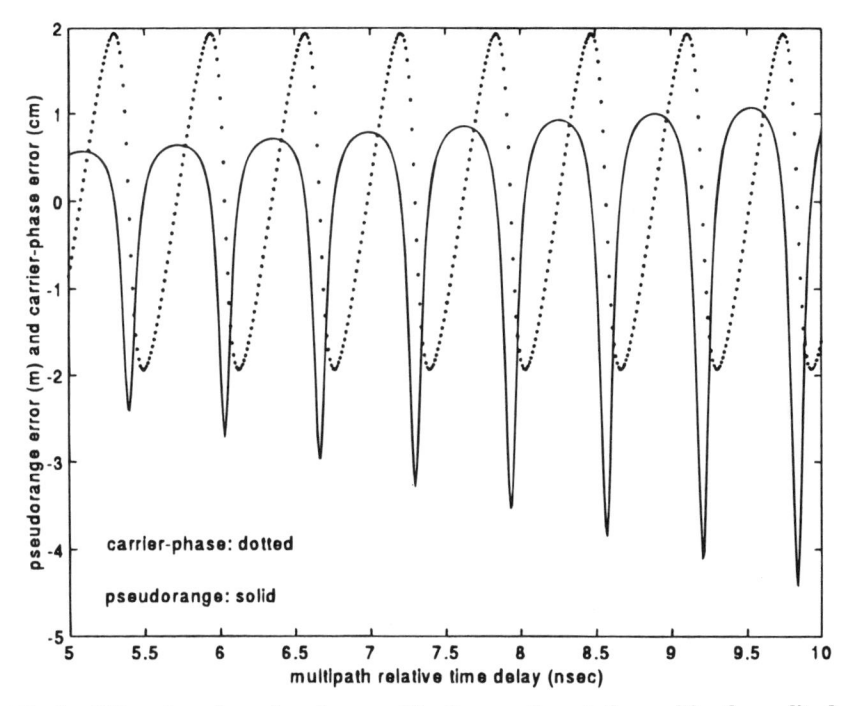

Fig. 3 C/A-code and carrier-phase multipath error for relative multipath amplitude set at −4.4 dB. Relative phase was calculated strictly from relative path delay.

2. P Code

The P code is the second PRN code modulated onto the GPS carrier. Although it was made available to the public during the satellite constellation buildup, DOD policy dictates that this code be encrypted (forming the so-called Y code) so as to be accessible only by authorized users. For those users having access to it, however, considerable multipath reduction or rejection is gained. Because the P code (10.23 MHz) is modulated at a rate ten times higher than the C/A code (1.023 MHz), its chips are thus one-tenth the length. The P code, therefore, is much less sensitive because it is affected only by multipath with relative time delays less than 1.5 times its own chip. Instantaneous and average multipath error curves for the P code have exactly the same shape as for the C/A code. In fact, the only difference is the scale factor of 10. The results for the P code may be obtained from Figs. 2–4 simply by dividing the C/A-code pseudorange numbers on both axes by 10. The P-code error envelope is given in Fig. 1.

3. Carrier Phase

As mentioned earlier, carrier-phase multipath errors typically are on the order of centimeters. This is illustrated in Figs. 2 and 3. Carrier-phase errors are shown with the corresponding pseudorange errors. It is interesting to note that, as the

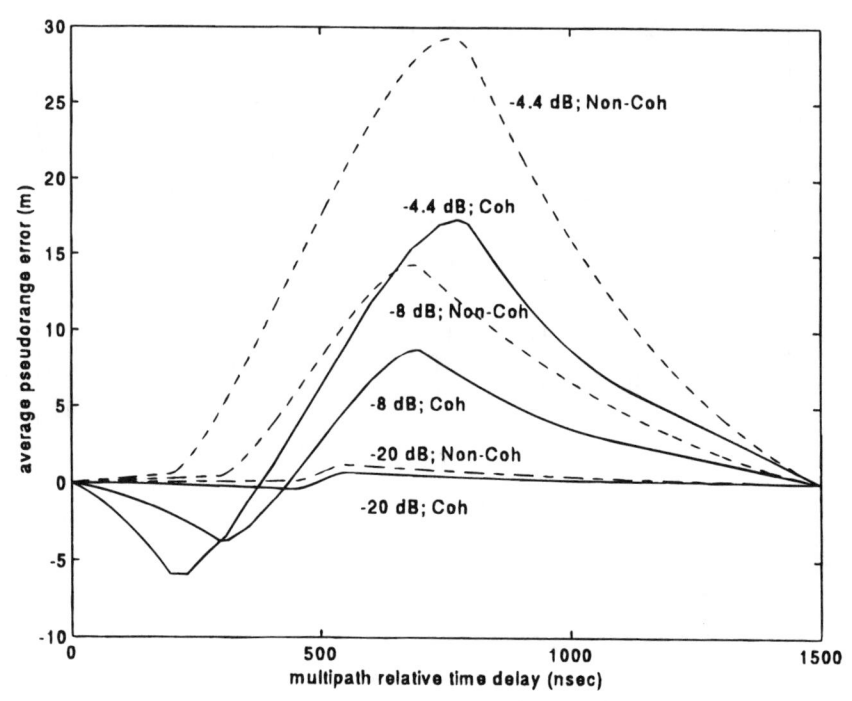

Fig. 4 C/A-code average multipath error. Curves are given for relative multipath amplitudes of −4.4 dB, −8 dB, and −20 dB for both coherent and noncoherent DLLs. For a given time delay for each of the curves, error was averaged over ten relative-phase values evenly spaced between 0 and 180 deg.

theory would predict, the pseudorange errors peak when the carrier-phase errors are zero and vice versa.

III. Aggravation and Mitigation

In the previous section, the basic relationships governing the response of the receiver to multipath were derived. Multipath was parameterized in terms of amplitude, time delay, phase, and phase-rate relative to the direct signal. As was shown, multipath error is proportional to the relative strength of the multipath signal and nonlinearly dependent upon time-delay and phase. The time-delay of a given multipath signal is entirely dependent upon the geometry of the environment in which the receiver is located. The amplitude and phase, however, are dependent upon both the environment and the characteristics of the user's antenna.

A. Antenna Considerations

In Sec. II, the relationships governing the response of the receiver to multipath were derived. Multipath was parameterized in terms of amplitude, time delay, phase and phase-rate relative to the direct signal. Prior to arriving at the receiver tracking loops, multipath must pass through the antenna. Realizable antennas (as opposed to theoretical isotropic radiators) do not receive signals equally from all directions. In fact, partial multipath rejection is built in to some antenna designs by shaping the gain pattern. Because most multipath arrives from angles near the horizon, multipath may be reduced by shaping the pattern to have low gain in these directions. However, extensive shaping of the pattern requires either a large aperture or multiple elements and signal processing. This might be feasible for the reference station in a differential system. However, this is not possible for highly dynamic platforms (i.e., aircraft), which require compact antennas and virtually omnidirectional gain.

Additional multipath attenuation by the antenna results from polarization discrimination. The direct GPS signal incident on the antenna is right-hand circularly polarized. In general, a single reflection from a planar surface will be left-hand elliptically polarized if the angle of incidence is less than the Brewster angle.[22] An ideal GPS antenna would completely reject all signals that are left-hand circularly polarized. In reality, total rejection is not obtained, but attenuation on the order of 10 dB is typical. The situation deteriorates further when we consider reflectors with nonsmooth surfaces. Reflections from very rough surfaces have random polarization characteristics. As a result, typically only 3 dB of attenuation can be achieved.

Having determined the multipath rejection properties of the antenna itself, the issue of antenna siting can be addressed. The multipath error equations (discussed in Sec. II) along with the antenna characteristics (just described) give general guidance regarding the siting of ground reference stations. To minimize the effects of multipath, the antenna should be located as far away from other objects as possible and/or should be located in such a way that multipath arrives from directions in which the antenna has low gain.

Locating the antenna away from objects minimizes signal blockage and reduces the strength of multipath signals. Mere distance, however, does not guarantee multipath error immunity. One of the great GPS multipath myths is the following (C/A-code processing assumed): GPS receivers reject multipath signals that arrive more than 1.5 chips (440 m or 1466 ns) after the direct signal; therefore, objects greater than 440 m from the receiver are not of concern. The error in this statement stems from a misinterpretation of the concept of multipath delay. As shown in sec. II, the 1.5 chips delay is a delay of the multipath *relative to the direct signal.* As was shown in Braasch,[20] the relative delay of the multipath is a function of receiver-to-object distance *and* proximity of the object to the satellite-to-receiver line-of-sight. The closer an object is to the line-of-sight, the greater the receiver-to-object distance can be while still yielding a relative multipath delay of less than 1.5 chips. The conclusion, then, is that increased receiver-to-object distance only guarantees a weakening of the multipath signal.

Maximization of the receiver-to-object distance was the first general siting guideline mentioned above. The second considered the gain pattern of the antenna. Multipath mitigation can be maximized if the antenna is located so that multipath arrives from directions in which the antenna has low gain. Multipath usually arrives from angles near the horizon and below. Multipath mitigation, then, is achieved by using an antenna with low gain in these directions.

B. Receiver Design

As demonstrated in Sec. II, multipath error is highly dependent upon receiver architecture. The following sections expand upon this.

1. Coherent vs Noncoherent Delay Lock Loop Architectures

As the plots in Sec. II demonstrate, for a given set of multipath parameters, average multipath error is smaller for a coherent DLL than for a noncoherent DLL. Furthermore, van Nee[19] has shown that when the relative phase rate-of-change of the multipath is large compared to the code tracking loop bandwidth, the coherent DLL error does average to zero. The primary disadvantage of the coherent DLL is a matter of robustness and not multipath. A coherent DLL will be disrupted by cycle slips, whereas a noncoherent DLL will continue to function independently of the carrier-phase tracking loop.

2. C/A Code with Narrow Correlator Spacing

Fenton et al.[23] and Van Dierendonck et al.[24] describe a patented development in GPS receiver design that lessens the effect of multipath by narrowing the spacing of the early and late correlators in a noncoherent DLL. By using a small portion of the correlation function (around the peak) to form the discriminator, maximum multipath error is reduced by a factor of 10 and multipath with relative delays of approximately 1 chip or greater is rejected entirely (0.1 chip correlator spacing). Braasch[26] and Van Nee[18] showed that the narrow correlator concept applies to the coherent DLL also. Meehan and Young[25] describe variations of the narrow correlator concept which benefit carrier-phase measurements as well as code measurements.

3. P Code

As discussed in Sec. II the P-code multipath performance is far superior to C/A-code tracking. For a given set of multipath parameters, peak error on the C/A code is 10 times larger than on the P code. When access is available, P code is clearly the choice when multipath is considered. It is important to note, however, that the 10:1 multipath error reduction applies to the maximum error and not necessarily the instantaneous error. This is illustrated in Fig. 1 where a relative multipath amplitude of -20 dB has been simulated. Note that for short delay multipath, the P-code and C/A-code multipath error envelopes coincide. Thus, for multipath in this region, the P code will not necessarily perform an order of magnitude better than the C/A code.

IV. Multipath Data Collection

In many instances it is desirable to evaluate the multipath in a given environment. This is particularly true in the siting of a ground reference station. Thus, the situation arises that GPS data are to be collected, and the multipath component of the error is to be isolated. Note that a truth reference system is not useful in this case. A truth reference system allows for the determination of the total system error. However, in this case only the multipath component of the error is desired.

The GPS signal itself may be exploited to isolate the combination of code multipath plus receiver error[4,11]. The GPS code and carrier-phase (integrated Doppler) measurements may be expressed as follows:[26]

$$\rho_{code} = D + c(b_u - B) + c(T + I + M_{code} + HW + \nu_{code}) + URE$$
$$+ SA + MEAS_{code} \tag{35}$$

$$\rho_{phase} = D + c(b_u - B) + c(T - I + M_{phase} + HW + \nu_{phase}) + URE$$
$$+ SA + MEAS_{phase} + \Delta \tag{36}$$

where

B	=	satellite clock offset from system time
b_u	=	receiver clock offset from system time
c	=	speed of light, m/s
D	=	true line-of-sight range from the satellite to the user in meters
HW	=	receiver hardware delay, s
I	=	apparent change in signal path delay due to propagation through the ionosphere, s
M	=	apparent change in signal path delay due to specular (i.e., nondiffuse) multipath, s
MEAS	=	receiver measurement (tracking) errors
SA	=	(selective availability) range error due to the intentional degradation of the satellite clock and orbit information by the DOD (used for security purposes), m
T	=	apparent change in signal path delay due to propagation through the troposphere, s
URE	=	(user range error) range error due to satellite clock and orbit uncertainty, m
ν	=	apparent change in signal path delay due to a combination of receiver noise and diffuse multipath, s
Δ	=	range difference between the code and integrated Doppler measurements due to an integer wavelength ambiguity, m
ρ_{code}	=	code measurement (pseudorange), m
ρ_{phase}	=	carrier-phase (integrated Doppler) measurement, m

Diffuse multipath arises from reflection and diffraction from a group of electrically small objects. Each of these objects individually produces a negligible multipath field, but the sum effect can be on the order of receiver noise values. This effect is lumped in with receiver noise because it is generally uncorrelated over time, and therefore noise-like in behavior. Note that the ionosphere term is equal in magnitude but opposite in sign for the two measurements. More detailed information on the effects of the ionosphere can be found in Chapter 12, this volume. The integer wavelength ambiguity arises in the carrier measurement because the basic measurement is that of integrated Doppler (change-of-range relative to the start of signal tracking). For the moment, consider the observable obtained by differencing the code and carrier-phase measurements:

$$\rho_{code} - \rho_{phase} = 2I + MEAS_{code} - MEAS_{phase} + \nu_{code} - \nu_{phase} + M_{code}$$
$$- M_{phase} - \Delta \tag{37}$$

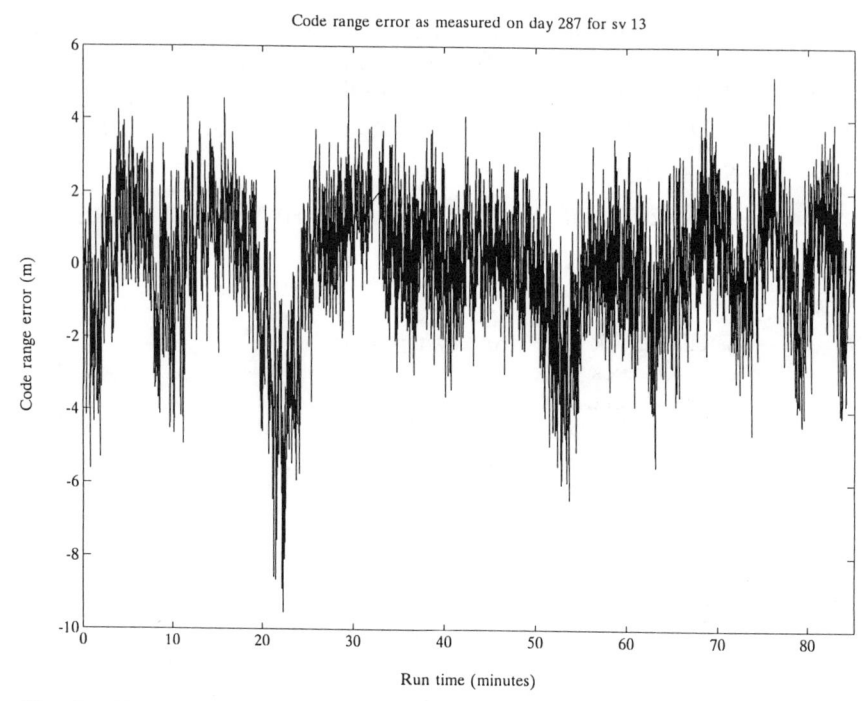

Fig. 5 Ohio University hangar test. Data were collected using an Ashtech 3DF receiver approximately 4 m from the hangar (which is approximately 10 m tall). Note that the curve shows the combined effects of multipath and receiver noise.

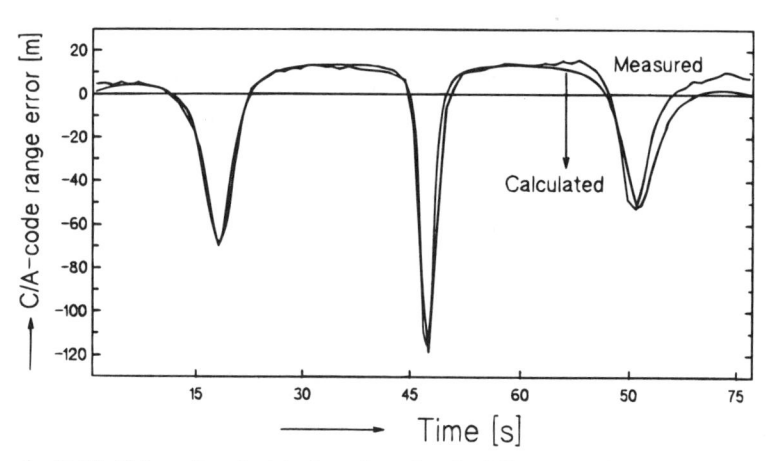

Fig. 6 Delft University electrical engineering building test. Data were collected using a Trimble 4000SST receiver approximately 30 m from the building (which is approximately 100 m tall). The curve labeled "calculated" is the result of fitting a model based on theory described in Sec. II. Reprinted from Van Nee (1992) with permission from R. van Nee and the Delft University of Technology.

MULTIPATH EFFECTS

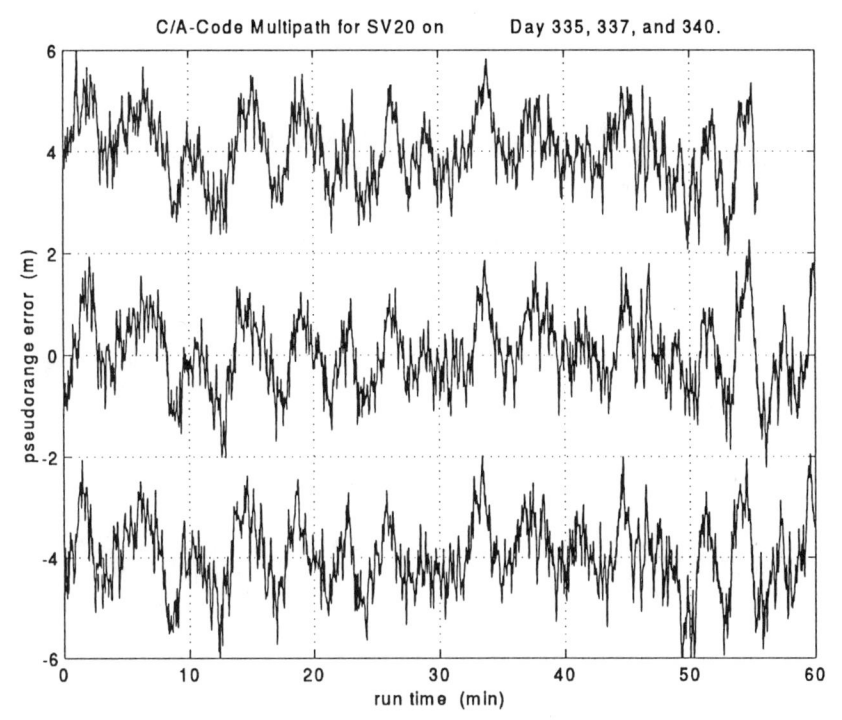

Fig. 7 Day-to-day repeatability of C/A-code multipath. Curves show pseudorange multipath error for three days. The data from day 235 have been offset in amplitude by +4 m and in time by 0 min; data from day 237 have been offset in amplitude by 0 m and in time by 8 min; and data from Day 240 have been offset in amplitude by −4 m and in time by 20 min. Data were collected using a Turborogue receiver on the roof of the University of Colorado, Boulder, Engineering Center.

Phase measurement errors, phase noise, and multipath typically are negligible.[27,28] The usual application of this process involves differencing code and carrier data collected over a given period of time. The integer ambiguity may, thus, be removed by subtracting out the bias. What remains is a combination of multipath, receiver code measurement error, noise, and an ionospheric term:

$$(\rho_{code} - \rho_{phase})_{adj} = 2I + \text{MEAS}_{code} + \nu_{code} + M_{code} \qquad (38)$$

Receiver code measurement errors typically result from dynamics-induced tracking loop lags.[29] Although these are usually correlated over time, receiver noise is not, and therefore it may be reduced through filtering. The filtering scheme smooths the code against the carrier measurements. Typical implementations are the Hatch filter[30] and the complementary Kalman filter.[31] The remaining term besides multipath is then the effect due to the ionosphere. However, measurements from two different carrier frequencies can be used to eliminate the ionospheric term. For those users with single-frequency receivers, the effect of the ionosphere is typically a long-term drift, which can easily be distinguished from the higher-frequency multipath errors.

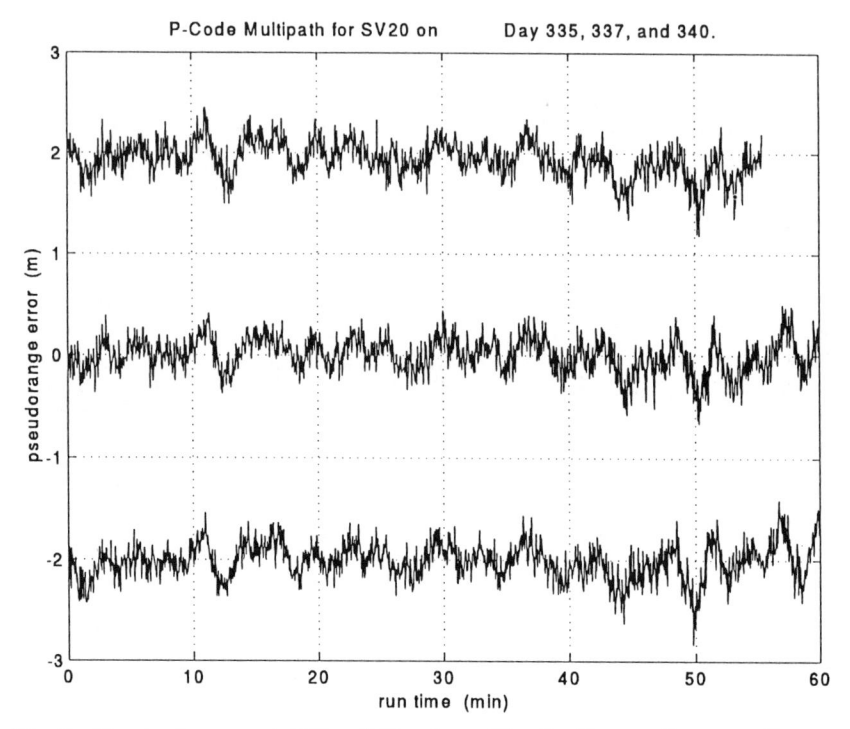

Fig. 8 Day-to-day repeatability of P-code multipath. Curves show pseudorange multipath error for three days. The data from Day 235 have been offset in amplitude by +2 m and in time by 0 min; data from Day 237 have been offset in amplitude by 0 m and in time by 8 min; and data from Day 240 have been offset in amplitude by −2 m and in time by 20 min. Data were collected using a Turborogue receiver on the roof of the University of Colorado, Boulder, Engineering Center.

As a result, the differencing of code and carrier measurements allows for the isolation of pseudorange multipath and receiver code measurement errors. The magnitude and behavior of receiver code measurement errors is, quite obviously, receiver dependent. However, it must be taken into consideration when analyzing the data. Test results indicate that receiver code measurement errors can be on the order of one meter.

For static data collection efforts, the repeating pattern of the satellite orbits can be exploited to gain confidence in the multipath analysis. The question often arises as to whether a given set of error data is truly multipath or not. The answer lies in collecting data on successive days. If, for instance, data are collected from satellite 20 on Wednesday between 2:00 p.m. and 3:00 p.m., data from the same satellite should be collected during the same time on Thursday. Because the satellite orbits the Earth twice every sidereal day (23 h 56 min), it returns to the same location four minutes earlier each day. Thus, errors thought to be multipath can be checked for repeatability. Any residual noise errors and the like will not repeat from day to day. However, because the same antenna, object, and satellite positions are being used, the multipath data should repeat. In this way, confidence

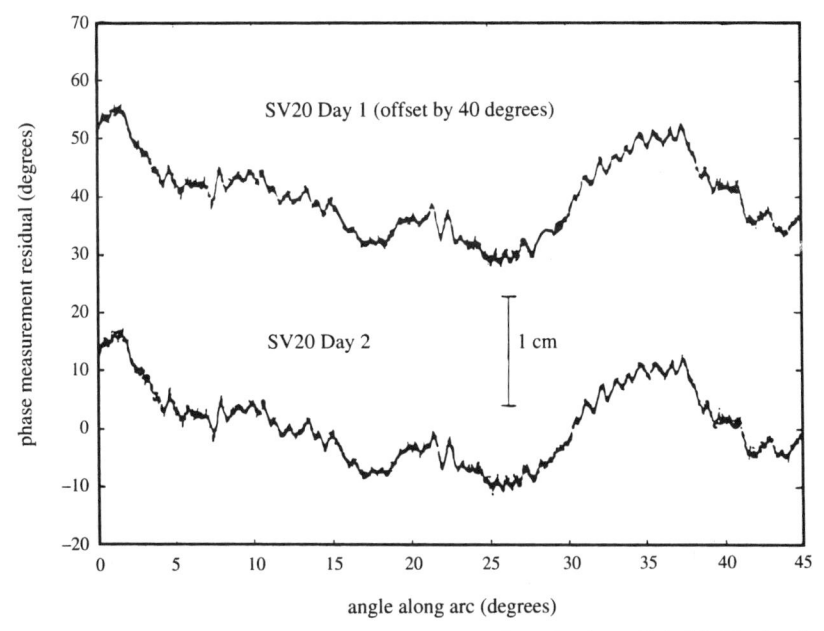

Fig. 9 Carrier-phase multipath error. Data was collected using a Trimble Navigation TANS-based GPS receiver. Multipath was induced by mounting a rectangular sheet metal reflector near one of the interferometer antennas. Reprinted from Cohen (1992) with permission.

can be gained that a set of error data is truly due to multipath. Obviously, this method assumes that the multipath comes from stationary objects.

Examples of GPS code multipath errors are given in Figs. 5–8. For the data shown in Fig. 5, a GPS antenna was placed approximately 4 m in front of a hangar door at the Ohio University airport.[32] In this case, the measurements were not filtered, and thus the graph shows a combination of multipath and receiver noise on the code measurements. The low-frequency trend shows the oscillating error attributable to the phase variation of the multipath. The trend is much clearer in fig. 6. These data were collected by the Electrical Engineering Department at the Delft University of Technology in Delft, The Netherlands.[18] A GPS antenna was placed approximately 30 m away from the Delft University Electrical Engineering building. The building is approximately 100 m high and 70 m wide. This figure illustrates several points about multipath error. First, it is highly nonsinusoidal. This verifies the earlier claim and is a direct result of the nonlinearities in the receiver architecture. Second, the maximum error is −120 m, thus verifying the assertion that code multipath errors can be extremely large. Finally, using the equations described in Sec. II, a model was fit to the data with the assumption that there were two major reflectors. Time delays were calculated from the geometry of the test set-up. The other parameters were adjusted to achieve a good fit. The results show clearly that the theoretical equations are good descriptions of actual multipath error.

Figures 7 and 8 illustrate C/A-code and P-code multipath day-to-day repeatability. The data sets were collected on DOY 335, 337, and 340 of 1993 (Dec. 1, 3, 6) on the roof of the Engineering Center at the University of Colorado, Boulder. Each of the three days of data for each code has been offset from each other for visual clarity (4 m for the C/A code, 2 m for the P code). The lower frequency oscillations due to multipath are apparent as well as the day-to-day repeatability. Sample correlation coefficients between data sets are on the order of 0.50–0.60.

Figure 9 illustrates carrier-phase multipath and its repeatability in a controlled environment. These data were collected at Stanford University using a short baseline GPS interferometer.[33] A metal sheet was intentionally placed near one antenna to induce multipath. The carrier-phase residuals clearly demonstrate the presence of multipath.

Acknowledgments

The author gratefully acknowledges the assistance of the Delft University of Technology, Faculty of Electrical Engineering, Telecommunications and Traffic-Control Systems Group (Delft, The Netherlands). In particular, the help of Edward Breeuwer, Richard van Nee, and Durk van Willigen was invaluable. Penina Axelrad and Christopher Comp of the Colorado Center for Astrodynamics Research, University of Colorado, Boulder, are thanked for their comments on the manuscript and for the collection of the C/A-code and P-code multipath data. Clark Cohen, Department of Aeronautics and Astronautics, Stanford University is thanked for the provision of the carrier-phase multipath error plot.

References

[1]Lachapelle, G., Falkenberg, W., Neufeldt, D., and Keilland, P., "Marine DGPS Using Code and Carrier in a Multipath Environment," *Proceedings of ION GPS-89,* Colorado Springs, CO, Sept. 1989, Institute of Navigation, Washington, DC, pp. 343–347.

[2]Cohen, C., and Parkinson, B., "Mitigating Multipath Error in GPS-Based Attitude Determination," *Guidance and Control 1991,* Vol. 74, *Advances in the Astronautical Sciences,* edited by R. Culp and J. McQuerry, American Astronautical Society, 1991, pp. 53–68.

[3]Hagerman, L., "Effects of Multipath on Coherent and Noncoherent PRN Ranging Receiver," Aerospace Rep. TOR-0073(3020-03)-3, Development Planning Division, The Aerospace Corporation, May 15, 1973.

[4]General Dynamics, Electronics Division, "Final User Field Test Report for the NAVSTAR Global Positioning System Phase I, Major Field Test Objective No. 17: Environmental Effects, Multipath Rejection," Rept. GPS-GD-025-C-US-7008, San Diego, CA, March 28.

[5]Counselman, C., and Gourevitch, S., "Miniature Interferometer Terminals for Earth Surveying: Ambiguity and Multipath with Global Positioning System," *IEEE Transactions on Geoscience and Remote Sensing,* Vol. GE-19, No. 4, 1981, pp. 244–252.

[6]Counselman, C., "Miniature Interferometer Terminals for Earth Surveying (MITES): Geodetic Results and Multipath Effects," Digest of the International Geoscience and Remote Sensing Symposium, Washington, DC, June 1981, Institute of Electrical and Electronics Engineers, New York, pp. 219–224.

[7]Greenspan, R., Ng, A., Przyjemski, J., and Veale, J., "Accuracy of Relative Positioning by Interferometry with Reconstructed Carrier GPS: Experimental Results," *Proceedings of the Third International Geodetic Symposium on Satellite Doppler Positioning,* New Mexico State University, Las Cruces, NM, Feb. 1982, pp. 1177–1195.

[8]Bletzacker, F., "Reduction of Multipath Contamination in a Geodetic GPS Receiver," Proceedings of the First International Symposium on Precise Positioning with the Global Positioning System, U.S. Department of Commerce, National Oceanic and Atmospheric Administration, Rockville, MD, April 1985, pp. 413–422.

[9]Tranquilla, J., and Carr, J., "GPS Multipath Field Observations at Land and Water Sites," *Navigation,* Vol. 37, No. 4, 1990–1991, pp. 393–414.

[10]Falkenberg, W., Kielland, P., and Lachapelle, G., "GPS Differential Positioning Technologies for Hydrographic Surveying," Record of the Position, Location, and Navigation Symposium, PLANS, Orlando, FL, Institute of Electrical and Electronics Engineers, New York, Dec. 1988, pp. 310–317.

[11]Evans, A., "Comparison of GPS Pseudorange and Biased Doppler Range Measurements to Demonstrate Signal Multipath Effects," Proceedings of the International Telemetering Conference, Las Vegas, NV, Instrument Society of America, Research Triangle Park, NC, Oct. 1986, pp. 795–801.

[12]Georgiadou, Y., and Kleusberg, A., "On Carrier Signal Multipath Effects in Relative GPS Positioning," Manuscripta Geodaetica, Vol. 13, 1988, pp. 172–179.

[13]Abidin, H., "Extrawidelaning for 'On the Fly' Ambiguity Resolution: Simulation of Multipath Effects," *Proceedings of ION GPS-90,* Colorado Springs, CO, Institute of Navigation, Washington, DC, Sept. 19–20, 1990, pp. 525–533.

[14]Bishop, G., Klobuchar, J., and Doherty, P., "Multipath Effects on the Determination of Absolute Ionospheric Time Delay From GPS Signals," *Radio Science,* Vol. 20, No. 3, 1985, pp. 388–396.

[15]Sennott, J., and Pietraszewski, D., "Experimental Measurement and Characterization of Ionospheric and Multipath Errors in Differential GPS," *Navigation,* Vol. 34, No. 2, 1987, pp. 160–173.

[16]Sennott, J., and Spalding J., "Multipath Sensitivity and Carrier Slip Tolerance of an Integrated Doppler DGPS Navigation Algorithm," Record of the Position Location and Navigation Symposium (PLANS), Las Vegas, NV, Institute of Electrical and Electronics Engineers, New York, March 1990, pp. 638–644.

[17]van Nee, R., "Multipath Effects on GPS Code Phase Measurements," *Proceedings of ION GPS-91,* Albuquerque, NM, Institute of Navigation, Washington, DC, Sept. 1991, pp. 915–924.

[18]van Nee, R., "GPS Multipath and Satellite Interference," *Proceedings of the Forty-eighth Annual Meeting of the Institute of Navigation,* Institute of Navigation, June 1992, Washington DC, pp. 167–177.

[19]van Nee, R.,"Spread Spectrum Code and Carrier Synchronization Errors Caused by Multipath and Interference," *IEEE Transactions on Aerospace and Electronic Systems,* Vol. 29, No. 4, 1993, pp. 1359–1365.

[20]Braasch, M., "On the Characterization of Multipath in Satellite-Based Precision Approach and Landing Systems," Ph.D. Dissertation, Department of Electrical and Computer Engineering, Ohio University, Athens, OH, June 1992.

[21]Braasch, M., and van Graas, F., "Mitigation of Multipath in DGPS Ground Reference Stations," *Proceedings of the ION National Technical Meeting,* San Diego, CA, Institute of Navigation, Washington, DC, Jan. 1992, pp. 105–114.

[22]Balanis, C., *Advanced Engineering Electromagnetics*, Wiley, New York, 1989, pp. 185–196.

[23]Fenton, P. et al., "Novatel's GPS Receiver: The High Performance OEM Sensor of the Future," *Proceedings of ION GPS-91*, Albuquerque, NM, Institute of Navigation, Washington, DC, Sept. 1991, pp. 49–58.

[24]Van Dierendonck, A., Fenton, P., and Ford, T., "Theory and Performance of Narrow Correlator Spacing in a GPS Receiver," *Navigation*, Vol. 39, No. 3, 1992, pp. 265–283.

[25]Meehan, T., and Young, L., "On-Receiver Signal Processing for GPS Multipath Reduction," Proceedings of the 6th International Geodetic Symposium on Satellite Positioning, Columbus, OH, Defense Mapping Agency and the Ohio State University, Columbus, OH, March 1992, pp. 200–208.

[26]Braasch, M., "A Signal Model for GPS," *Navigation*, Vol. 37, No. 4, 1990–1991, pp. 363–377.

[27]Ferguson, K., et al., "Three-Dimensional Attitude Determination with the Ashtech 3DF 24-Channel GPS Measurement System," *Proceedings of the ION National Technical Meeting*, Phoenix, AR, Institute of Navigation, Washington, DC, Jan. 1991.

[28]Braasch, M., and van Graas, F., "Guidance Accuracy Considerations for Real-Time GPS Interferometry," *Proceedings of ION GPS-91*, Albuquerque, NM, Institute of Navigation, Washington, DC, Sept. 1991, pp. 373–386.

[29]Braasch, M., "Isolation of GPS Multipath and Receiver Tracking Errors," *Proceedings of the ION National Technical Meeting*, San Diego, CA, Institute of Navigation, Washington, DC, Jan. 1994, pp. 511–521.

[30]Hatch, R., "The Synergism of GPS Code and Carrier Measurements," *Proceedings of the Third International Geodetic Symposium on Satellite Doppler Positioning*, New Mexico State University, Las Cruces, NM, Feb. 1982, pp. 1213–1231.

[31]van Graas, F., and Braasch, M., "GPS Interferometric Attitude and Heading Determination: Initial Flight Test Results," *Navigation*, Vol. 38, No. 4, 1991–1992, pp. 297–316.

[32]Breeuwer, E., "Modeling and Measuring GPS Multipath Effects," Master's Thesis, Faculty of Electrical Engineering, Delft University of Technology, Delft, The Netherlands, Jan. 1992.

[33]Cohen, C., "Attitude Determination Using GPS," Ph.D. Dissertation, Department of Aeronautics and Astronautics, Stanford University, Stanford, CA, Dec. 1992.

Chapter 15

Foliage Attenuation for Land Mobile Users

J. J. Spilker Jr.*
Stanford Telecom, Sunnyvale, California 94089

I. Introduction

LAND mobile users are expected to be one of the largest categories of GPS users, and it is important to examine the specific GPS propagation issues for this environment. It has already been shown that the optimum geometric dilution of precision (GDOP) is provided when several of the GPS satellites are at low elevation angle near the horizon. However, the land mobile user environment differs from that of aircraft in flight or ships at sea in that the user driving along a road or freeway is often subject to shadowing, diffraction, and scattering of the satellite signal by trees, utility poles, buildings, or hills. These effects are accentuated by the need for operation at low elevation angles for at least some of the GPS satellites. In addition, the requirement for receiver simplicity, and the need to track several satellites widely spaced in angle simultaneously, generally dictates the use of an omnidirectional or hemispherical antenna. Thus, while receiving the direct line-of-sight ray from the satellite, the user has little means to discriminate against multipath signals scattered from ground reflections, tree limbs and foliage, or other scattering elements. In addition, the direct ray may itself be attenuated by tree foliage.

This chapter describes and models the statistics of the signal environment for the rural land mobile user where tree foliage often shadows the user. The statistical models are then compared with measured data for L-band signals at various elevation angles. Both stationary and mobile users are considered.

The satellite-to-land mobile user links can be categorized in one of three forms: 1) line of sight, unshadowed; 2) shadowed by trees or foliage; and 3) completely obstructed where the link is totally blocked by hills or other major obstructions (see Fig. 1).

The unshadowed link is characterized by a link where there is a clear line of sight to the satellite, but there can also be a scattered and/or specular multipath component in addition to the line-of-sight link. The multipath component can be caused by ground reflections. The unshadowed link can be modeled as the sum

---

Copyright © 1994 by the author. Published by the American Institute of Aeronautics and Astronautics, Inc., with permission. Released to AIAA to publish in all forms.
*Ph.D., Chairman of the Board.

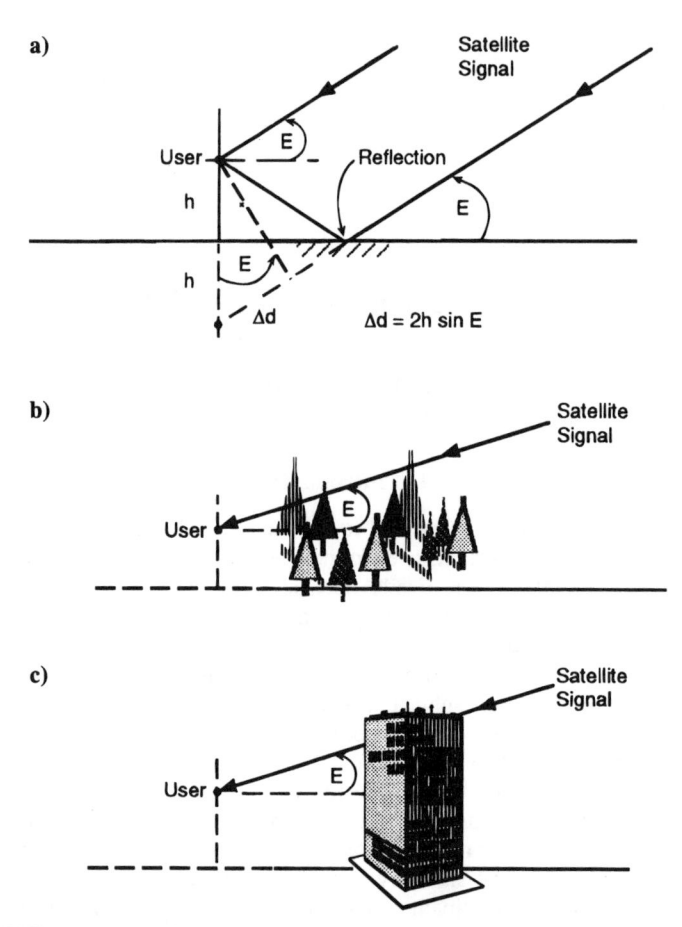

Fig. 1 Different types of links experienced by a mobile user, or fixed user on the ground: a) unshadowed link with multipath reflection a distance h below the user and satellites at elevation angle E; b) link partially shadowed by trees, telephone poles; c) link completely shadowed by buildings, hills, or other obstructions.

of a fixed line-of-sight signal plus a Rayleigh scattered term. The sum of these two components has a distribution alternatively known as a Rician or a Nakagami–Rice distribution.[1]

Partially shadowed links have a randomly attenuated direct link in addition to the scattered component. The attenuation generally is caused by individual trees, utility poles, clusters of trees, or forests along a roadway. The shadowed signal is sometimes modeled by assuming a lognormal distribution for the direct ray. The sum of direct and scattered components has a Rician distribution for any given amplitude direct ray; i.e., the conditional distribution for a given line-of-sight signal amplitude is Rician. The resulting distribution, obtained by averaging over the lognormal direct ray, is known as the Loo distribution.[2]

FOLIAGE ATTENUATION FOR LAND MOBILE USERS

The completely obstructed link may be blocked by hills or large man-made structures and has no useful signal component at all over a substantial period of time.

Discussion in this chapter is limited to the shadowed/unshadowed environment common along road/highways in the rural or suburban environment. Typically, a vehicle moves from the shadowed environment where forests are adjacent to the roadway to regions of relatively open space with line-of-sight propagation and then back to the shadowed region. In a given region, there is a certain fraction of time S that a mobile user is in the shadowed region, and during the remaining fraction of time $1-S$ the user is in the unshadowed region.

The channel model employed in this chapter assumes that the received signal $S_r(t)$ has the form

$$S_r(t) = \alpha(t)s(t - \tau) + \Sigma \beta_i(t)s(t - \tau - \Delta\tau_i) \tag{1}$$

where $\alpha(t)$ represents the random path attenuation caused by the foliage, τ is the delay of the direct line-of-sight satellite user path, and $\beta_i(t)$ represents the random attenuation of the scattered components with additional delay $\Delta\tau_i$. For purposes of this chapter, the delays $\Delta\tau_i$ of the scattered components are considered negligible compared to the inverse bandwidth of the signal. Thus, the channel considered here is strictly limited to the flat fading channel; i.e., the signal spectrum is unchanged by the fading except for a random amplitude. Larger multipath delays; for example, as experienced by an aircraft in flight with sea surface reflections, could produce frequency selective fading. Multipath channels where the delays are not negligible are considered in another chapter.

The discussion that follows first treats the stationary user and views the various GPS satellites through individual trees or groves of trees at different elevation angles. The remainder of the chapter emphasizes mobile users where the link is varying fairly rapidly with time as the vehicle moves past trees and other short duration obstructions.

II. Attenuation of an Individual Tree or Forest of Trees—Stationary User

Consider first the attenuation caused by an individual tree where a GPS satellite is viewed by a stationary user at an elevation angle E deg. Goldhirsh and Vogel[1,3] have published measurements of attenuation of an individual tree vs. elevation angle. The approximate geometry of the tree and the relative position of the receiver are shown in Fig. 2. These measurements were taken at 870 MHz, but the results can be translated to L-band after being scaled up to 1.575 GHz using the following relationship:

$$\text{Attenuation } (f) \sim \sqrt{f}$$
$$\text{Attenuation (1.575 GHz) dB} \approx \sqrt{\frac{1575 \text{ MHz}}{870 \text{ MHz}}} \text{ Attenuation (870 MHz) dB} \tag{2}$$

where the attenuation in dB is scaled by $\sqrt{1575/870} = 1.345$. That is, foliage attenuation varies approximately as the square root of frequency.*

---

*The square root relationship is only a rough approximation. In the frequency range 800 MHz to 2 GHz the rms error has been estimated[1] at 6%.

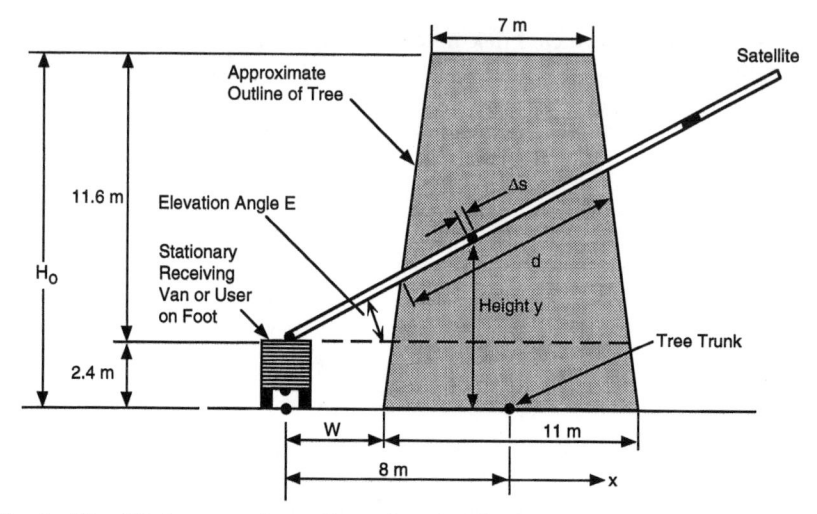

Fig. 2 Simplified geometric configuration of a single callery pear tree vs elevation angle in measurements made by Goldhirsh and Vogel[3] as viewed by a stationary user. The origin of the x, y coordinates used in the text is the base of the tree trunk.

It has also been shown that the attenuation of a tree in full foliage is roughly 35% greater in dB than that of a deciduous tree without foliage. Thus, the bulk of a tree's attenuation is clearly caused by the wood tree limbs, branches, and trunk rather than by the leaves.

Foliage attenuation is often characterized as attenuation in dB/m of foliage penetration. As shown later, these numbers of attenuation per meter tend to vary widely depending on the type of tree and the height of the ray relative to the top of the tree because the foliage density varies with height. The attenuation also varies with the distance of the tree from the user. Fresnel diffraction analyses have been used to explain the average decrease of attenuation with increasing distance of the receiver to the tree.[4]

As an attempt to obtain a simple purely empirical geometric interpretation of the foliage attenuation refer to Fig. 2 again. Consider a ray path through a nonuniform density foliage slab where for each incremental distance Δs along the ray path has some random value of attenuation in dB/m. The mean attenuation coefficient at that position depends on various parameters, such as type of tree, density of trees, position relative to center, and base of tree. The total attenuation then depends on the distance W from the receiver to the edge of the trees, the elevation angle to the satellite, and the geometric extent of the foliage. In effect, we consider the trees to be a slab of foliage with nonuniform attenuation where the attenuation density in dB/m varies with position. The quantity H_o is the height of the foliage slab representing the tree or trees. It is assumed that $A_d(x, y)$ represents the mean attenuation density vs position, and the actual attenuation is either 0 or k_o depending on whether that position in space has foliage or not. The probability of the incremental ray being attenuated at that position cell is p, and all cells are independent. Thus, the mean attenuation for that incremental cell of length Δs is pk_o where k_o varies with position relative to the base of each tree.

It should be emphasized that this representation is only a simple empirical model which we can attempt to fit to the measured data and compare with measured data. If we use the model for attenuation just discussed and we use the law of large numbers for a large number of independent attenuation cells of length Δs, the total path attenuation in dB approximately follows a normal distribution. The distribution of received signal level in absolute terms then follows a lognormal distribution. We return to this distribution later.

We can easily show that the total distance d through the foliage of the model of Fig. 2 does not change substantially as we vary the elevation angle from 0 to 45 deg; d remains at approximately 10 m. Consequently, a uniform attenuation density foliage slab model would show little variation in total attenuation with elevation angle. However, the attenuation measured by Ref. 1 varies considerably from approximately 9.5 dB at 40 deg to approximately 19 dB at 20 deg elevation angle (measured numbers are scaled up to 1.545 GHz). Thus, the attenuation density in dB/m obviously must vary substantially with position relative to this base of the tree if the attenuation vs elevation angle of the model is to match the measurement. That variation is consistent with what we might expect from a simple observation of the wood portion of a typical tree, which has a greater wood density near the base of the trunk.

We have selected an empirical attenuation density profile (in dB/m) that varies with position relative to the base of the tree's center (the trunk), which seems to match fairly well-measured data for the tree of Fig. 2. This attenuation profile is shown in Fig. 3 for this given tree and tree–user distance. The mean attenuation density in dB/m of this empirical model varies as follows:

$$A_d(x, y) = (1 - y/14)^2(1 - |x|/5.5)K \text{ dB/m} \tag{3}$$

This attenuation density profile can be integrated along the path s through the tree at a given elevation angle E to obtain the total attenuation $A_T(E)$ where the following obtains:

$$A_T(E) = \int_{\text{path}} A_d(x, y) \, ds \tag{4}$$

Assume a rectangular tree of 14-m height, 11-m width, a trunk 8 m away from the receiver, and the receiver 2.4 m above ground (see Fig. 2). The mean total attenuation A_T is then

$$A_T(E) \cong \frac{K \sec(E)}{5.5} \left[\int_{2.5}^{8} (1 - y/14)^2(x - 25)dx + \int_{8}^{13.5} (1 - y/14)^2(13.5 - x)dx \right] \tag{5}$$

where $y = 2.4 + x \tan E$. Thus,

$$A_T(E) \approx K[3.78 \sec E - 5.21 \sec E \tan E + 1.937 \sec E \tan^2 E]$$
$$\cong K[3.726 - 0.0670 E] \text{ for a least mean square fit}$$
$$\cong 24.6 - 0.442 E.$$

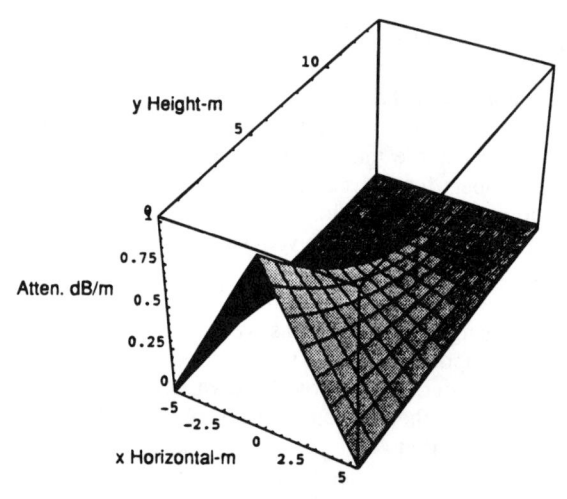

Fig. 3 Simplified mean normalized attenuation density profile of a tree in dB/m that has been adjusted to match measured variation of total attenuation with elevation angle for a callery pear tree of 14-m height and 11-m width at the base. The height is the height above the ground level. The zero point on the x width scale represents the trunk of the tree.

The calculated value of $A_T(E)$ vs elevation angle is shown in Fig. 4 as the dashed curve where K has been selected for a good match to measured data. Also shown in the figure are the measured points from Goldhirsh and Vogel[1] (which have been scaled up to 1.525 from 870 MHz by a factor of 1.345). The solid curve represents the Goldhirsh and Vogel empirical model:

$$A_T(E) = 25.8 - 0.47\ E \tag{6}$$

which is rather close, as can be seen, to the measured points and the linear best fit result using the geometric model of Fig. 2. We can conclude that the attenuation density is clearly not uniform with position, and the empirical model of Fig. 2, at least for this example, gives a good match for the limited range of elevation data available (between 12 and 30 deg).

Other data have been gathered for various types of trees, as shown in Table 1. The table shows measured tree attenuation for various types of trees at 870 MHz. As discussed earlier, in this frequency region, tree foliage attenuation varies approximately as the square root of frequency; i.e., $\sqrt{f_{Li}/870\ \text{MHz}}$ and results are scaled to 1.575 GHz in terms of increased dB/m attenuation. Trees with full foliage have approximately 1.35 times the attenuation in dB/m compared to attenuation for deciduous trees bare of foliage. Note that some trees have as much as 3–4 dB/m of attenuation at 1.5 GHz. Elevation angle information was not available for these measurements. Note also the substantial difference in two attenuation measurements made for pin oak. The measurements were made at two different locations. Thus, we should expect a wide variation in the data from tree to tree, and the data are primarily useful in a statistical sense.

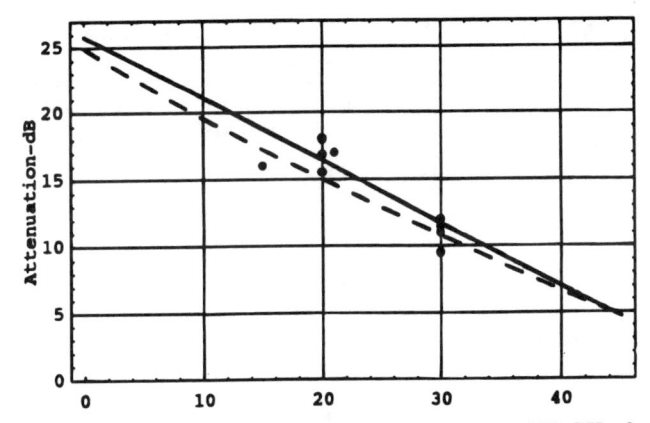

Fig. 4 Measured and approximate attenuation scaled to 1.575 GHz for a callery pear tree of 14-m in height with a base width of 11 m corresponding to Fig. 2. The measured points are shown, the solid curve is the empirical model from Ref. 3, and the dashed curve is the resulting curve from the simple physical model of Fig. 2.

Next, consider the attenuation through a grove of trees, as shown in Fig. 5 where the user–satellite path travels a distance d through the grove. The CCIR[5] has developed an approximate model for mobile user path loss vs the ray foliage penetration distance d of Fig. 5. The CCIR modified exponential decay model gives a path loss $L = \alpha d$ where $\alpha = 1.33 f^{0.284} d^{-0.412}$ dB/m for $14 < d < 400$ m and $\alpha = 0.45 f^{0.284}$ for $0 < d < 14m$ and frequency f in GHz. This model* gives less than a linear dependence over d, which compensates for the Fresnel diffraction effect that gives a decreased attenuation effect with larger user–tree grove distances.[5,6]

A. Foliage Attenuation—Mobile User

Next consider a user who is mobile rather than stationary; i.e., the user is moving rapidly past trees and open spaces alongside the roadway. If these trees and open spaces pass by relatively rapidly compared to the GPS receiver closed loop bandwidth and databit interval of 20 ms = 1/50 bps, then the mean attenuation statistic is more useful than the attenuation of each individual tree.

An empirical model[7] for the fade depth attenuation in dB at 1.5 GHz vs elevation angle E for driving along forested roadways in Maryland is given by the following:

$$F(P, E) = -[a_0 + a_1 E + a_2 E^2] \ln P + bE + c \text{ dB} \qquad (7)$$

where P is the percentage of time this fade depth F is exceeded ($P = 1$–20%); $a_0 = 3.44$; $a_1 = 0.0975$; $a_2 = -0.002$; $b = -0.443$, $c = 34.76$, and E is the elevation angle in degrees to the satellite. A graph of this empirical relationship over the region 20 deg $< E <$ 60 deg for $P = 1\%$, and $P = 10\%$ is shown in Fig. 6.

---
*Note that this CCIR model does not deal with specifics such as the type of foliage, or the distance from user to the grove of trees.

Table 1 Single tree attenuations at $f = 870$ MHz and attenuation coefficient scaled dB/m for 1.5 GHz (scale factor of 1.35)[7]

Tree type	Attenuation dB—870 MHz		Attenuation coefficient dB/m—870 MHz		Attenuation coefficient dB/m—1.575 GHz	
	Largest	Average	Largest	Average	Largest	Average
Burr oak	13.9	11.1	1.0	0.8	1.3	1.1
Callery pear	18.4	10.6	1.7	1.0	2.2	1.3
Holly	19.9	12.1	2.3	1.2	3.0	1.6
Norway maple	10.8	10.0	3.5	3.2	4.6	4.2
Pin oak	8.4	6.3	0.85	0.6	1.1	0.8
Pin oak	18.4	13.1	1.85	1.3	2.4	1.7
Pine cone	17.2	15.4	1.3	1.1	1.7	1.5
Sassafras	16.1	9.8	3.2	1.9	4.2	2.5
Scotch pine	7.7	6.6	0.9	0.7	1.2	0.9
White pine	12.1	10.6	1.5	1.2	2.0	1.6
Overall average	14.3	10.6	1.8	1.3	2.4	1.7

The empirical Eq. (7) for fade depth statistics can be rewritten in terms of the probability of a fade level being exceeded as a function of fade level for various elevation angles. This probability is as follows:

$$P(F, E) = \exp[(-17380 + 500 F + 221.5E)/(-1720 - 48.75E + E^2)] \qquad (8)$$

for $1 < P < 20\%$ and 20 deg $< E >$ 60 deg. The probability of fade exceedance is shown in Fig. 7 for elevation angles of 20, 30, 40, 50, and 60 deg along these forested roadways in Maryland.

B. Probability Distribution Models for Foliage Attenuation—Mobile User

The probability distribution for fading along roadways has one of two different types of models, shadowed or unshadowed. For the unshadowed model, we

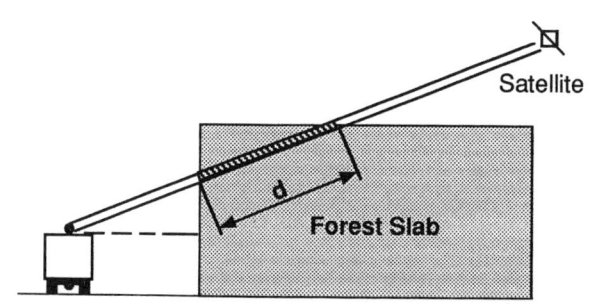

Fig. 5 Model of forest next to roadway. The vegetative path length is d meters.

FOLIAGE ATTENUATION FOR LAND MOBILE USERS

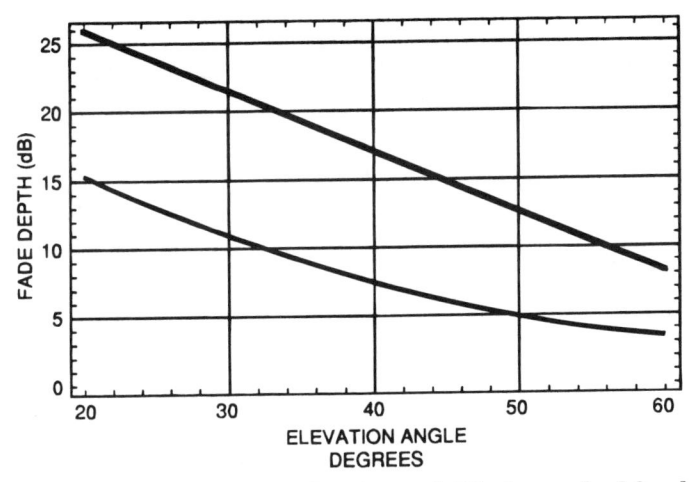

Fig. 6 Fade depth statistics in dB vs elevation angle E in degrees for L-band satellite to mobile users along a forested roadway. The two curves are for 1% (upper heavy curve) and 10% (light curve) probability that the fade is no more than the value shown.

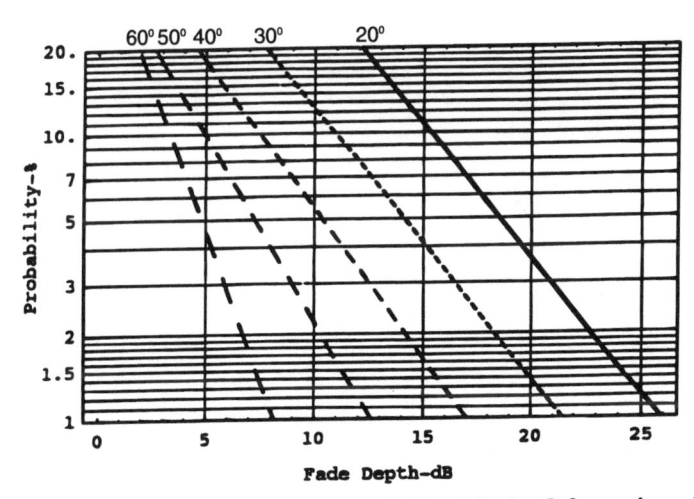

Fig. 7 Probability of fade depth being exceeded vs fade depth for various elevation angles: 20 deg, 30 deg, 40 deg, 50 deg, 60 deg. The solid curve is for an elevation angle of 20 deg, and the curves to the left are for progressively higher elevation angles in 10-deg increments. These empirical data correspond to measurements at L-band taken along forested roadways in Maryland. (data from Ref. 7)

assume that the received signal energy consists of the sum of 1) direct ray (constant amplitude); and 2) Rayleigh scattered signals. The received signal then has the following form:

$$r = a \cos(\omega_o t + \phi) + \omega \cos(\omega_o t + \theta)$$
$$= x_c \cos\omega_o t + x_s \sin \omega_o t \qquad (9)$$

where ω has a Rayleigh distribution with mean square value σ^2 and a is the direct ray signal amplitude. The sum of these two has a Nakagami–Rice distribution for the envelope $z = \sqrt{x_c^2 + x_s^2}$ of the in-phase and quadrature signal components. The probability density of the unshadowed envelope is then as follows:

$$p_u(z) = \frac{z}{\sigma^2} \exp\left[\frac{-z^2 + a^2}{2\sigma^2}\right] I_0\left(\frac{za}{\sigma}\right) \qquad (10)$$

where the normalized line-of-sight signal power is $P_s = a^2/2$, and the power in the Rayleigh scattered signal is $P_R = \sigma^2$, and $I_0(\ )$ is the modified Bessel function of zero order. If the direct line-of-sight a^2 term decreases to zero, we are left with the Rayleigh signal.

In the shadowed environment, we use a distribution proposed by Loo[2] in which the conditional probability density of the total received signal for a given direct signal level a; namely $p(z|a)$, is given by the Nakagami–Rice distribution, and the probability density of the a direct ray $p(a)$ component is lognormal. The lognormal density is given by the following:

$$p(a) = \frac{1}{\sigma_o a \sqrt{2\pi}} \exp\left[\frac{-(\ln a - \mu)^2}{2\sigma_o^2}\right] \qquad (11)$$

as shown in Fig. 8. In essence, the lognormal distribution is the same as a Gaussian

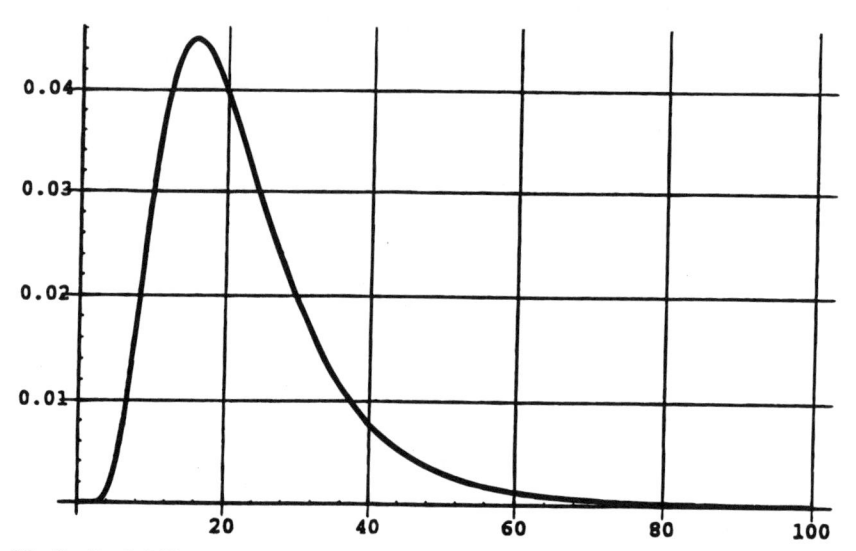

Fig. 8 Probability density functions of the lognormal distribution function for mean $\mu = 3$, and $\sigma_o = 0.5$.

density function where the variable is attenuation or signal level expressed in dB (except for a scale factor to account for the difference between $\log_{10}$ and ℓn).

The shadowed density function formulated by Loo[2] is then the convolution of the Rice and lognormal densities:

$$p_s(z) = \int p(z|a)p(a)\,da \quad (12)$$

Barts and Stutzman[6] have taken this problem a step further by assigning a probability S that the user is shadowed in the course of travel down a road. During the shadowed condition, the probability density is $p_s(z)$. The signal level has an unshadowed probability density $p_u(z)$ for the remaining fraction $1 - S$ of the time. The composite signal probability density is then as follows:

$$p_c(z) = Sp_s(z) + (1 - S)p_u(z) \quad (13)$$

The coefficients of these probability densities can be fit to measured data to give reasonably good models for the measurements.

The Loo probability density from Eq. (12) is as follows:

$$p_s(z) = \frac{2z}{\alpha\sigma\sqrt{2\pi}} \int_0^\infty \frac{1}{a} I_o\left(\frac{2za}{\alpha}\right) \exp\left[\frac{-(\ln a - \mu)^2}{2\sigma^2} - \frac{a^2 + z^2}{\sigma}\right] da \quad (14)$$

where α is the Rayleigh parameter; and μ, σ are the lognormal mean and sigma. This density is rather cumbersome to use. Barts and Stutzman[6] have suggested an approximation to the Loo density for the envelope of the shadowed signal. Their power–law approximation to the shadowed propagation fade distribution is then as follows:

$$P_s(F) \cong \left(\frac{50 - F}{V_1}\right)^{V_2} \quad (15)$$

where

$$V_1 = -0.275\,\overline{K} + 0.723\mu + 0.336\sigma + 56.979$$
$$V_2 = [-0.006\,\overline{K} - 0.008\mu + 0.013\sigma + 0.121]^{-1}$$

where in the expression all parameters $\overline{K}$, μ, σ are expressed in dB, and $P_s(F)$ is the probability that the fade F in dB is exceeded. We have defined $\overline{K} = 10\log(1/\alpha)$ dB.

A similar approximation was given for the unshadowed Ricean distribution:

$$p_u(z) = p_{\text{Rice}}(z) = \frac{2z}{\beta} \exp\left[\frac{-(z^2 + a^2)}{\beta}\right] I_o\left(\frac{2za}{\beta}\right) \quad (16)$$

where B is the mean square value of the Ricean distribution of r, and C is the amplitude of the direct component. Define $K = 10\log(a^2/\beta)$ dB. The approximation of Barts and Stutzman[6] is as follows:

$$P_u(F) \cong \exp\left[\frac{-(F + u_1)}{u_2}\right] \quad (17)$$

where

$$u_1 = 0.01K^2 - 0.378K + 3.98$$
$$u_2 = 331.35\ K^{-2.29}$$

and $P_u(F)$ is the probability or fraction of time that a given fade depth F in dB is exceeded.

III. Measured Models—Satellite Attenuation Data

Fade statistics for mobile users in suburban, rural/forested, rural/farmland, mountainous terrain, and tree-lined roads have been measured by a number of investigators. A set of data taken at the moderately low elevation angle of 19 deg is shown for a variety of environments in Fig. 9. Note again that for GPS applications, we generally want access to satellites at low elevation angles for low GDOP, as well as access to a satellite at zenith. As can be seen, fades of 8 dB or more are possible for a reasonable fraction ($\approx 10\%$) of the time in suburban areas. Fades of 20 dB or more are encountered 1% of the time for suburban and rural forested regions at this low 19-deg elevation angle. It should also be pointed out that there can be a rapid spatial fluctuation in the fading. The deeper fades generally have considerable variation (≈ 5 dB) for movements of the receiver as small as 2 m for forested roadways.

At higher elevation angles where line-of-sight propagation (unshadowed) is maintained, the attenuation (caused by multipath scattering) is significantly less.

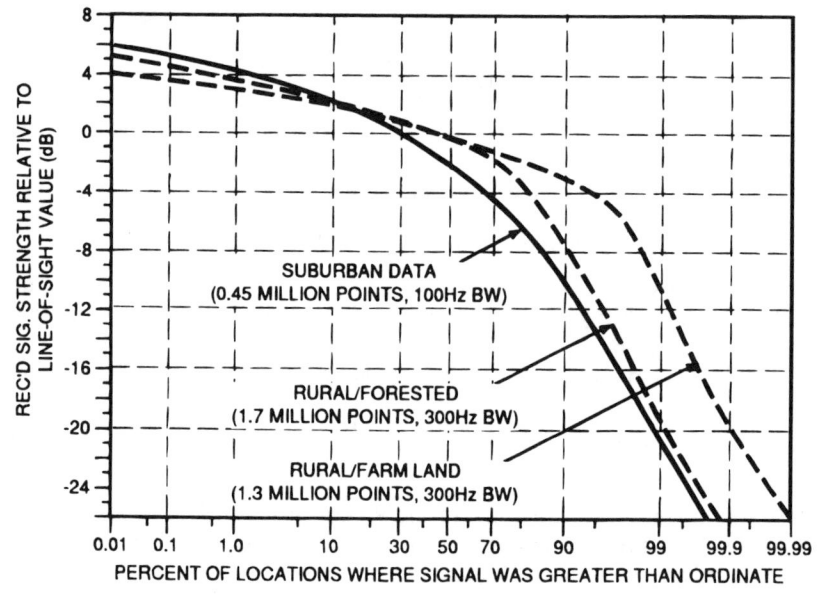

Fig. 9 Measurements of L-band propagation statistics for mobile users on land for various environments at an elevation angle of 19°. These plots are based on data from Kent.[8]

Fig. 10 Unshadowed line-of-sight measurements of *L*-band fade distributions for mobile users on land for various environments and elevation angles. Multipath reflections cause the fading: a) line-of-sight model for mountainous terrain; b) line-of-sight for tree-lined roads. These plots are based on data from Goldhirsh and Vogel.[14]

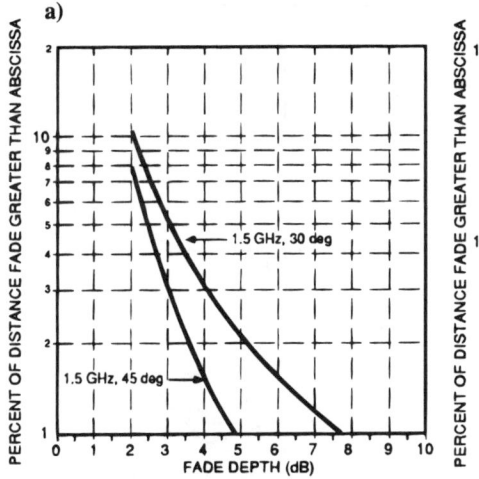

a) Best power curve fit cumulative fade distributions for line-of-sight configurations in mountainous terrain. The cumulative fade distribution probability P for this power law approximation is of the form

$$P = aF^{-b}$$

where F is the fade in dB and where a, b are

	E	
	30°	45°
a	33.19	39.95
b	1.710	2.321

b) Best exponential fit cumulative fade distributions for line-of-sight configurations on tree-lined roads. The best exponential fit is of the form

$$P = u \exp(-vF)$$

where
$$u = 125.6$$
$$v = 1.116$$

are the best u, v values for moderately high elevation angles when averaged over measurements at E = 30°, 45°, 60°

Figure 10 shows best fit empirical cures at the GPS L_1 frequency 1.5 GHz. The 1% fade depth was about 5 dB for either tree-lined roads or mountainous terrain (with an elevation angle of 45 deg).

A. Measured Fading for Tree-lined Roads—Mobile Users

Numerous measurements have been made worldwide on fading depth as a function of elevation angle as a vehicle is driven around various tree-lined roads. Figure 11 shows a summary of a selected set of these measurements. These measurements include the following:

1) Bundrock and Harvey,[9] Melbourne, Australia—1.55 GHz tree-lined road with 85% tree incidence
2) Butterworth and Mott,[10] Ottawa, Canada—1.5 GHz, 19 deg elevation angle, rural, forested, hilly terrain with immature timber with occasional cleared areas, two-lane road with good shoulders

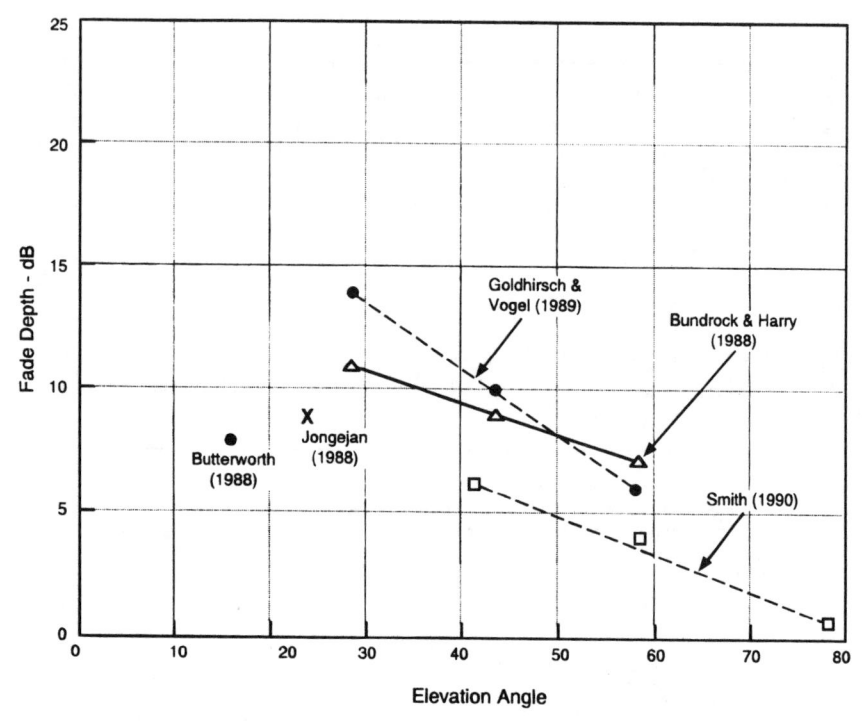

Fig. 11 Summary of various measurements of fading depth for tree-lined roads for 10% probability of the fade level being exceeded.

3) Jongejans et al.,[11] Belgium, 1.5 GHz—hilly Ardennes with roadside lined with bare trees in winter

4) Smith et al.,[12] England—rural, tree-shadowed environment with trees with full leaf cover, L-band

5) Goldhirsh and Vogel[3]—forested roads in Maryland, two- and four-lane highways

In general, the 1% fade points are substantially higher by as much as a factor of two in the dB fade level; i.e., a 10 dB fade for 10% might increase to 20 dB for the 1% mark.

These measured data were generally taken with circularly polarized signals of narrow bandwidth. Reference 13 showed that measurements made in a pine forest for vertical and horizontal polarization showed little difference between the two polarizations and concluded that that type of forest at least is approximately an isotropic medium.

References

[1]Goldhirsh, J., and Vogel, W. J., "Propagation Effects for Land Mobile Satellite System: Overview of Experimental and Modeling Results," NASA Ref. Pub. 1274, Feb. 1992.

[2]Loo, C., "A Statistical Model for a Land Mobile-Satellite Link," *IEEE Transactions on Vehicular Technology,* Aug. 1985.

[3] Goldhirsh, J., and Vogel, W. J., "Mobile Satellite System Fade Statistics for Shadowing and Multipath from Roadside Trace at UHF and L-Band," *IEEE Transactions on Antenna and Propagation,* April 1989.

[4] Yoshikawa, M., and Kagohara, M., "Propagation Characteristics in Land Mobile Satellite Systems," Thirty-Ninth Annual IEEE Vehicular Technical Conference, May 1989.

[5] CCIR Study Group 1990–1994, "Impact of Propagation Impairments on the Design of LED Mobile Satellite System Providing Personal Communication," CCIR, U.S. WP-8D-14 (Rev. 2), October 1992.

[6] Barts, R. M., and Stutzman, W. L., "Modeling and Simulation of Mobile Satellite Propagation," *IEEE Antenna and Propagation,* April 1992.

[7] Goldhirsh, J., and Vogel, W. J., "Mobile Satellite System Propagation Measurements at L-Band Using MARECS B2," IEEE Trans-Antenna and Propagation, February 1990.

[8] Kent, J. D. B., "A Land Mobile Satellite Data System," International Mobile Satellite Conference, Ottawa, 1990.

[9] Bundrock, A., and Harvey, R., "Propagation Measurements for an Australian Land Mobile-Satellite System," *Proceedings of Mobile Satellite Conference,* 1988.

[10] Butterworth, J. S., and Mott, E. E., "Characterization of Propagation Effects for Land Mobile Satellite Services," International Conference Satellite System for Mobile Communication and Navigation, June 1983.

[11] Jongejans, A. A., et al., "PROSAT—Phase I Report," European Space Agency TR, ISA, STR-216, May 1986.

[12] Smith, H., et al., "Assessment of the Channel Offered by a High Elevation Satellite Orbit to Mobiles in Europe," IEEE Conference on Radio Receiver and Associated Systems, July 1990.

[13] Ulaby, F. T., et al., "Measuring the Propagation Propagation Properties of a Forest Canopy Using Polarimetric Scattermeter," *IEEE Transactions on Antenna and Propagation,* Feb., 1990.

[14] Goldhirsh, J., and Vogel, W. J., "An Overview of Results Derived from Mobile-Satellite Propagation Experiments," Inter. Mobile Satellite Conference, Ottawa, 1990.

Chapter 16

Ephemeris and Clock Navigation Message Accuracy

J. F. Zumberge* and W. I. Bertiger†
*Jet Propulsion Laboratory, California Institute of Technology,
Pasadena, California 91109*

IN this chapter, we discuss the accuracy of the ephemeris and clock corrections contained in the Global Positioning System (GPS) navigation message. We first provide a brief description of how the Control Segment generates these quantities. Next, we compare them with results from precise (non-real-time) solutions of satellite parameters derived from the simultaneous analysis of data from a globally distributed network of GPS receivers. Finally, we cast these accuracies into the form of a user equivalent range error.

I. Control Segment Generation of Predicted Ephemerides and Clock Corrections

One of the primary purposes of the Control Segment (Chapter 10, this volume) is to generate predicted satellite ephemerides and clock corrections, which are regularly uploaded to the satellites. The predictions are then included as part of the 50-b/s 1,500-bit navigation message (Chapter 4, this volume) that modulates the transmitted GPS signal. Ground receivers then use the predictions for real-time estimates of satellite coordinates and clock corrections.

Data used for the predictions are acquired from receivers situated at precisely known locations in Hawaii, Colorado, Ascension Island in the Atlantic Ocean, Diego Garcia in the northern Indian Ocean, and Kwajalein in the western Pacific. The distribution in longitude of these sites (Table 1) is reasonably uniform, allowing continuous tracking of all GPS spacecraft. The sites at Ascension, Diego Garcia, and Kwajalein are capable of transmitting computed navigation message updates to the satellites. Receivers at all stations use cesium oscillators for time stability, and measure dual-frequency phase and pseudorange. Meteorological data are acquired at each station and used to aid in estimation of troposphere

---

Copyright © 1994 by the American Institute of Aeronautics and Astronautics, Inc. The U.S. Government has a royalty-free license to exercise all rights under the copyright claimed herein for Governmental purposes. All other rights are reserved by the copyright owner.
*Member Technical Staff, Satellite Geodesy and Geodynamics Systems Group, Tracking Systems and Applications Section.
†Member Technical Staff, Earth Orbiting Systems Group, Tracking Systems and Applications Section.

Table 1 Tracking stations used by the Control Segment and approximate locations

Site	Latitude	Longitude
Hawaii	21°N	158°W
Colorado Springs[a]	39°N	105°W
Ascension Island[b]	8°S	14°W
Diego Garcia[b]	7°S	72°E
Kwajalein[b]	9°N	168°E

[a]Master Control Station.
[b]Can transmit to GPS satellites.

delay. All data are regularly transmitted to the Master Control Station in Colorado Springs.

Only the P-code pseudorange measurements are used as data in the parameter estimation scheme, which is based on a Kalman filter. Estimated satellite parameters include epoch-state position and velocity, solar radiation pressure coefficients, clock bias, drift, and drift rate. Station parameters include similar clock quantities and troposheric delay. The terrestrial coordinate system and gravity field are 1984 World Geodetic System (WGS-84). The reference time is an average of monitor station clocks and a subset of GPS clocks.

Data going back 4 weeks are used to estimate reference satellite trajectories, which are then used to propagate satellite positions and clock corrections into the future. The first 28 h of prediction are divided into overlapping 4-h fit intervals separated by 1 h. The fit results for each such interval are cast in the format of the navigation message (through a fitting procedure), and are uploaded into the satellites once a day, more frequently if required to meet a 10-m user-equivalent range error specification. The daily upload is based on a data window that closed 45 min prior to the upload.[1-3]

Given the daily upload, the satellite broadcasts satellite positions and clock corrections contained in the appropriate 4-h interval. Although predictions beyond 28 h are also uploaded, they are normally not used, because the next day's upload overwrites them with results derived from more current data.

II. Accuracy of the Navigation Message

This section assesses the accuracy of the information broadcast in the navigation message. The "truth cases" to which the navigation messages are compared are daily GPS solutions, from the Jet Propulsion Laboratory (JPL), of satellite positions and clock corrections. First we describe the daily JPL solutions, including estimates of their accuracies. Next, based on the period 1993 July 4–Oct. 22, we compare these daily solutions with their counterparts from the navigation message. Of course, it must be remembered that one of the key differences between the GPS Control Segment solutions and the truth model is that the GPS solutions are predictions of the future based on past data, whereas the "truth" solutions are based on after-the-fact, postfit smoothed estimates.

A. Global Network GPS Analysis at the Jet Propulsion Laboratory

Since June 1992, analysts at the JPL have regularly reduced GPS data from a globally distributed network of 20–40 precision P-code GPS receivers using the GIPSY/OASIS-II software.[4,5] Shown in Fig. 1 are locations of the sites as of fall 1993. In addition to dense coverage in North America and Europe, there is also reasonable coverage elsewhere, including eight sites in the Southern hemisphere.

Receivers at these sites make measurements of the carrier phase and pseudorange observables on both the L_1 and L_2 bands from GPS satellites, at 30-s data intervals. Data are analyzed daily in 30-h batches, centered on GPS noon. The 6-h overlap centered at each GPS midnight allows for consistency checks between solutions from adjacent days. Prior to parameter estimation, data are edited using the TurboEdit alogrithm[6] and decimated to a 10-min interval.

The model used in the analysis corrects for ionospheric delay (through the formation of the ionosphere-free linear combination of phase and pseudorange observables), tropospheric delay (by stochastic estimation of the wet component at each receiver site), transmitter and receiver phase center offsets, Earth orientation (through explicit estimation of pole position and length of day), solid Earth tides, and relativistic effects. Transmitter and receiver clock corrections are estimated as independent parameters at each sample time. [In the case of selective availability (SA)-affected transmitters, this accounts for the dithering of GPS clocks.] The reference clock is a hydrogen maser driving one of the receivers. Satellite parameters include epoch-state position and velocity, and solar radiation pressure; the latter is estimated stochastically. The Earth-fixed reference frame is defined by adopting fixed locations for eight of the receivers, as specified in the international terrestrial reference frame, ITRF-91.[7,8] The carrier phase biases are estimated as piecewise-constant, real-valued parameters.

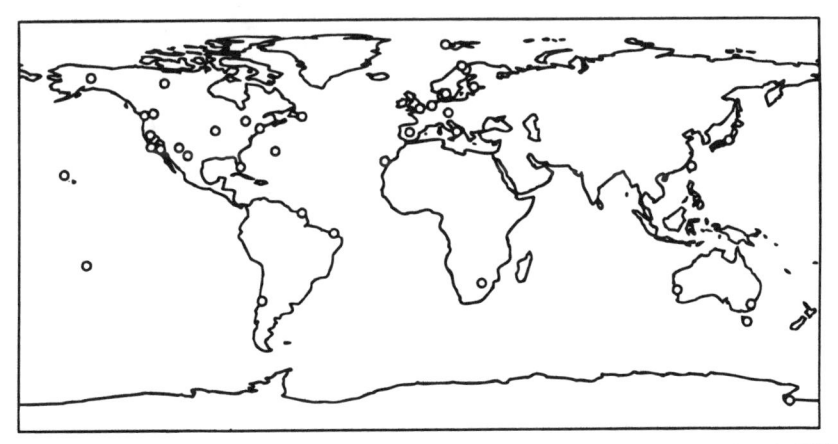

Fig. 1 Global distribution of GPS tracking receivers in the International GPS Service for Geodynamics, fall 1993. In addition to dense coverage in North America and Europe, there is also reasonable coverage elsewhere, including eight sites in the southern hemisphere.

B. Accuracy of the Precise Solution

One assessment of orbit quality from the precise solutions can be made by looking at the continuity of results from adjacent days. For example, the solution for Oct. 7 uses data from 2100 h on Oct. 6 to 0300 h on Oct. 8. Similarly, the solution for Oct. 6 uses data from 2100 h on Oct. 5 to 0300 h on Oct. 7. The difference between these solutions in the position of satellite PRN 12 within ± 3 h of midnight between Oct. 6 and 7 is shown in Fig. 2. The rms variation over the 6-h period is 12 cm, 22 cm, and 25 cm for the radial, cross-track, and along-track components, respectively.

A number of other groups estimate satellite parameters from the same data, using independent software, thus allowing a separate assessment of orbit quality. Shown in Fig. 3 is the comparison of the JPL's orbit solution for PRN12 on Oct. 7, 1993 with that determined by the Center for Orbit Determination in Europe (CODE) at the University of Berne, Switzerland.[9] The vertical scale is the same as that in Fig. 2. The rms difference over the day between the solutions is 7 cm for the radial component, 5 cm for the cross-track, and 9 cm for the along-track. For other days and satellites, this agreement is generally within 20 cm for all components. Comparisons with results from other analysis centers show comparable agreement. To summarize, the precise orbit solutions are typically accurate to 5–30 cm rms, depending on the component and other factors.

A comparison similar to that shown in Fig. 2 is shown for a transmitter clock in Fig. 4. Plotted there is the difference between the Oct. 6 and Oct. 7 solution for the PRN13 clock. The rms difference over the 6-h period is 0.22 ns, of which a portion is attributable to a 0.18-ns bias. This few-tenths-ns rms difference is typical for other satellites and days.

The reference for the precise clock solutions is the maser-based receiver at Algonquin Park, Canada, maintained by the Geodetic Survey Division of Canada's Department of Energy, Mines and Resources. This clock is adjusted periodically,

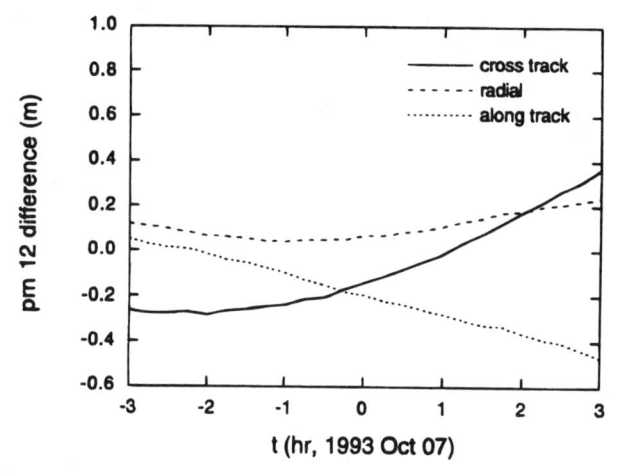

Fig. 2 Difference near the midnight boundary between the JPL's precise solutions of Oct. 6, 1993 and Oct. 7. The rms differences are 0.12 m, 0.22 m, and 0.25 m in the radial, cross-track, and along-track components.

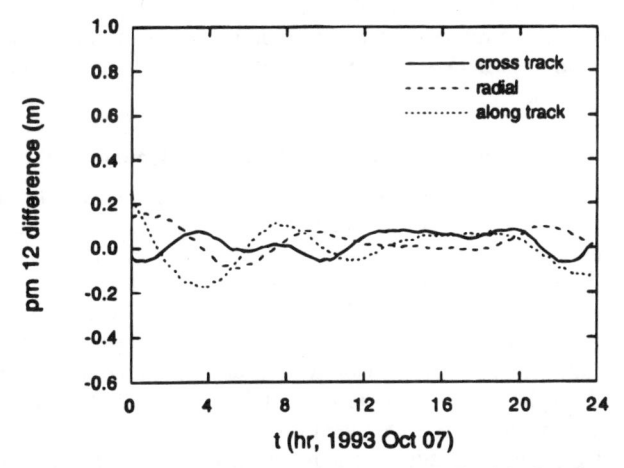

Fig. 3 Comparison of the precise solution from the JPL with that from the CODE (University of Berne), for PRN 12 on Oct. 7, 1993. The rms values over the day (which include both the bias and the variation) are 0.07 m for the radial component, 0.05 m for the cross-track, and 0.09 m for the along-track. The agreement between the JPL and CODE for this satellite and day is somewhat better than typical. However, for other satellites and days the agreement is, with few exceptions, at least as good as a few tens of centimeters.

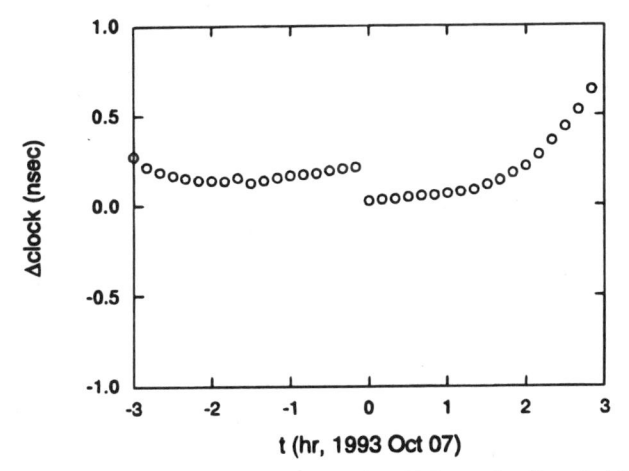

Fig. 4 Difference in the clock solution for PRN 13 from the Oct. 6, 1993 solution and the Oct. 7, 1993 solution. The rms value over the 6-h overlap is 0.22 ns (of which a portion is attributable to a 0.18-ns bias).

and is believed accurate with respect to GPS time to within a few hundred ns, with drift magnitudes of no more than a few tens of ns per day.[10] Of course, biases and drifts in the reference clock will be masked in a comparison such as Fig. 4, but would appear in a comparison between the navigation message and the precise solutions.

C. Comparison of Precise Orbits with Broadcast Ephemerides

Fig. 5 shows the position difference between the JPL precise solution and the broadcast orbit for satellite PRN 12 on Oct. 7, 1993. The interval between updated navigation messages typically is one hour. The satellite and day are the same as in Figs. 2 and 3, although the vertical scale is 10 times larger. The rms values over the day are 0.56 m for the radial component, 1.67 m for the cross-track, and 2.67 m for the along-track.

Similar calculations have been made for all satellites and days over the period July 1, 1993 through Oct. 22, 1993 (a total of 2490 satellite days). For the given satellite and day, the rms difference over the day between the broadcast ephemeris and the precise solution is computed, for each of the three components. The results are summarized in Fig. 6.

Figure 6 contains three histograms, one for each of the position difference components. The median values in the above distributions are 1.3 m for the radial component, 3.6 for the cross-track, and 4.7 m for the along-track. Thus, half of the satellites and days over the ~4-month period had a daily rms agreement between the navigation message and the precise solution of less than 1.3 m in the radial component.

The reference frame ITRF-91 used for the precise solutions differs from the 1984 World Geodetic System (WGS-84) used by the broadcast ephemeris. To test how much this reference-frame difference contributes to the observed

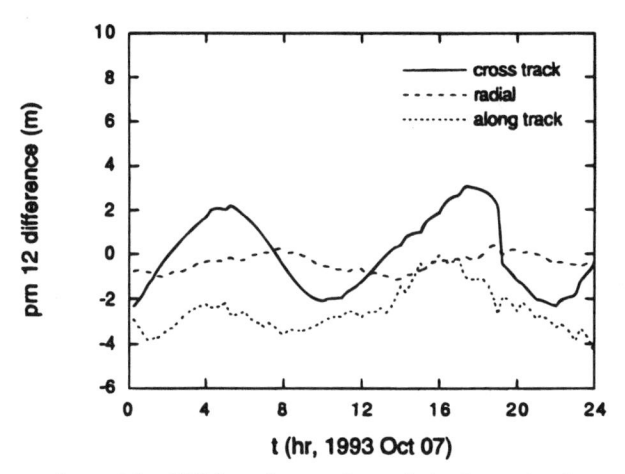

Fig. 5 Comparison of the GPS broadcast ephemeris in the navigation message with the JPL's precise solution, for PRN 12 on Oct. 7, 1993. The rms values over the day are 0.56 m for the radial component, 1.67 m for the cross-track, and 2.67 m for the along-track. The vertical scale is 10× that of Figs. 2 and 3.

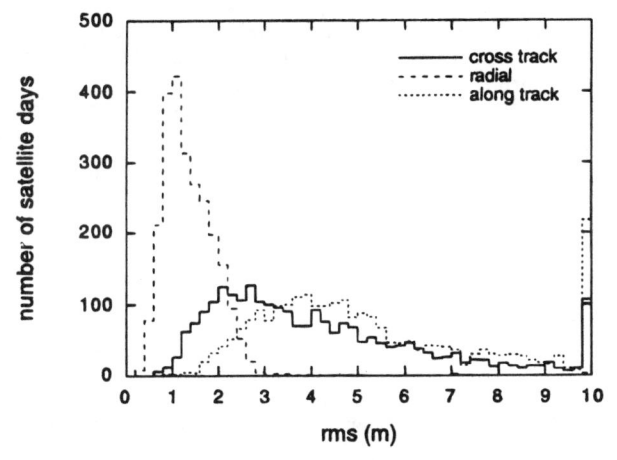

Fig. 6 Comparison of the GPS broadcast ephemeris with precise orbital solutions for the period July 4, 1993 through Oct. 22, 1993. An "event" in one of the three histograms corresponds to a single satellite on a single day (as in Fig. 5). For the given satellite and day, the rms difference over the day between the broadcast ephemeris and the precise solution is computed, for each of the three components. The median values in the above distributions are 1.3 m for the radial component, 3.6 m for the cross-track, and 4.7 m for the along-track. If a daily seven-parameter transformation is applied to align the navigation message reference frame with that of the precise solution, the medians are marginally reduced to 1.2 m, 3.2 m, and 4.5 m.

difference in ephemerides, consider the 7-parameter transformation defined by $x' = (1 + \epsilon)x + T + \Theta x$, where T is a translation vector, ϵ a scale factor, and

$$\Theta \equiv \begin{pmatrix} 0 & \theta_z & \theta_y \\ \theta_z & 0 & -\theta_x \\ -\theta_y & \theta_x & 0 \end{pmatrix}$$

is a rotation matrix. The difference between x and x' can be thought of as arising from 1) a shift or translation T; 2) a change in orientation, characterized by Θ; and 3) an overall expansion or contraction, characterized by ϵ.

The calculations that resulted in Fig. 6 were repeated, but each day a transformation was applied to all coordinates in the broadcast message. The parameters were chosen to minimize $\Sigma_{pct}[\Delta_{pct}^2]$ where $\Delta_{pct} \equiv X_{pct} - x'_{pct}$ is the difference at time t between the precise orbit (X_{pct}) and the transformed broadcast orbit (x'_{pct}) for PRN p and Cartesian component c.

The median values of the daily rms differences in the radial, cross-track, and along-track components are reduced to 1.2 m, 3.2 m, and 4.5 m, respectively, from the values corresponding to Fig. 6. The values of the transformation parameters are given in Table 2, and Fig. 7 shows the daily values of the scale factor.

There are two known effects which would contribute to the scale factor parameter ϵ. The first is that the precise orbits refer to the spacecraft center

Table 2 Parameters in the transformation from the broadcast ephemeris reference frame to that of the precise orbits, based on 120 daily transformations[a]

Parameter	Average	Standard deviation
T_x (cm)	−4.0 ± 2.3	24.7
T_y (cm)	6.3 ± 1.8	19.6
T_z (cm)	−4.6 ± 3.3	36.3
ϵ (ppb)	−12.7 ± 0.2	2.2
θ_x (nrad)	−7.9 ± 3.2	34.9
θ_y (nrad)	−6.8 ± 3.3	35.9
θ_z (nrad)	−125.5 ± 4.3	47.2

[a]The uncertainties in the average values are based on observed daily fluctuations, shown in the third column, divided by $\sqrt{120}$. The most significant parameters are the -12.7×10^{-9} scale factor and the −125.5-nrad rotation around the z-axis.

of mass, whereas the broadcast orbits refer to the spacecraft antenna phase center. Because these points are separated by 0.9519 m, mostly in the radial direction, we could expect a contribution of about $0.9519/(26.55 \times 10^6) \approx$ +35.9 ppb to ϵ in the broadcast-to-precise transformation.

Second, the broadcast orbits use the WGS-84 gravity field, with GM = 3.986005×10^5 km^3 s^{-2}, compared with the JGM2 value (Nerem et al.) of 3.986004415×10^5 km^3 s^{-2} used in the precise solutions. Because the radius varies with (GM)$^{1/3}$, we would expect a contribution to ϵ of $-1/3 \ (3.986005 - 3.986004415)/3.986004415 \approx -48.9$ ppb. (The larger value of WGS-84 would put the satellite out further, requiring a negative value of ϵ to bring it into agreement with the precise orbit.) The sum of these expectations, −13.1 ppb, is remarkably close to the observed value of −12.7 ppb in Table 2.

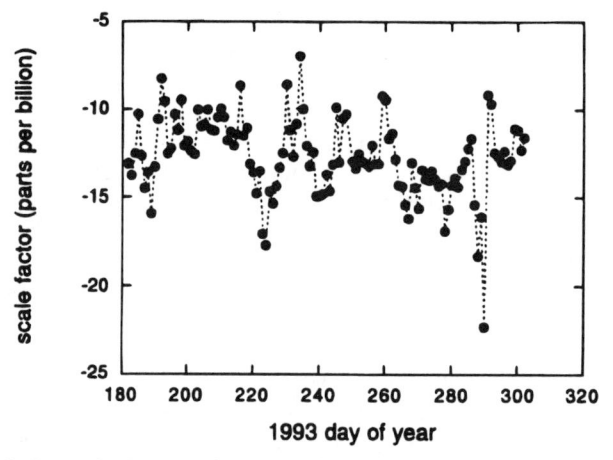

Fig. 7 Scale factor in the transformation from the WGS-84 reference frame to the International Terrestrial Reference Frame. The average value is −12.7 ± 0.2 parts per billion, which corresponds to about −30 cm in the radial direction.

Note that we could use the values in Table 2 to transform WGS-84 coordinates of terrestrial sites to the ITRF-91 reference frame (at epoch midway between July 1 and Oct. 22, 1993), except that ϵ should be taken as $(-12.7 - 35.9)$ ppb $= -48.6$ ppb.

The differences between the precise and broadcast orbits are significantly larger than the accuracies of the precise solutions, as estimated in the previous section. This is not surprising, as the broadcast solutions are, by necessity, the result of an extrapolation in time from hours-old data. The precise solutions, on the other hand, do not have such a real-time constraint. It should also be mentioned that the intentional degradation of broadcast ephemeris quality, one speculated aspect of selectve availability (SA), has not been observed.

D. Comparison of Precise Clocks with Broadcast Clocks

Shown in Fig. 8 are the the clock corrections for PRN 13 on Oct. 7, 1993, as determined by the JPL's precise solution (solid squares) and that from the navigation message (open circles). The interval between points is 30 min. (Points whose estimated uncertainty—"formal error"—exceeds 10 ns in the precise solution are not considered.) The difference between the navigation message and the precise solution is shown in Fig. 9, where the discontinuities in the former are clearly evident.

The GPS satellites on this day can be grouped into two classes depending on the variation over the day in the difference between the precise and broadcast clocks. Figures 9 and 10 contain satellites in the group for which this scatter is of the order of 10 ns. Of the five satellites in this group, three are Block I (PRNs 3, 12, and 13) and two are Block II (PRNs 15 and 20). Note that the time series in Figs. 9 and 10 are reasonably smooth with time.

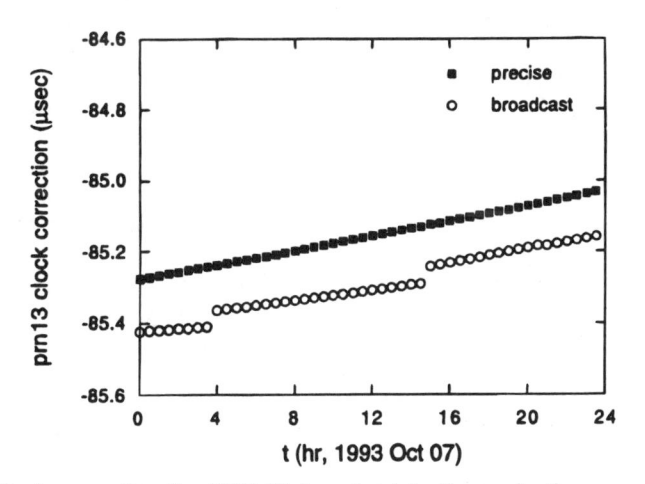

Fig. 8 Clock correction for PRN 13 broadcast in the navigation message (open circles) and determined in JPL's precise solution (solid squares). The precise solution uses a maser-based GPS receiver as its reference. The solid squares describe a slope of about 250 ns/day.

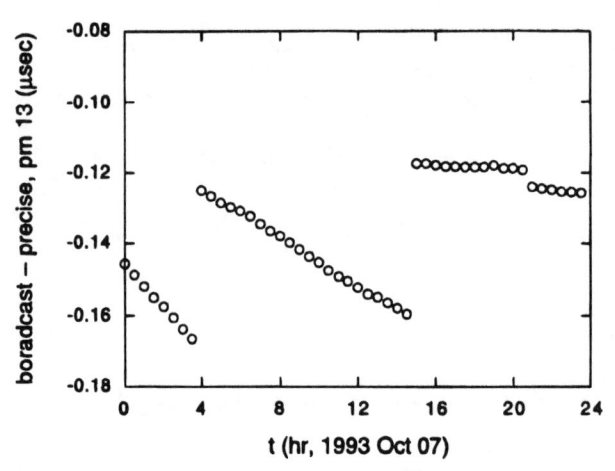

Fig. 9 Difference in the broadcast and precise clock corrections for PRN 13. The discontinuities are attributable to new broadcast messages, as indicated in Fig. 8.

The second class consists of satellites for which the difference is much noisier. All satellites in this group are Block II, and are shown in Fig. 11 as a function of time, together with the histogram that indicates the distribution of the differences. The standard deviation of the distribution is about 80 ns.

The clock dithering component of SA is clearly evident in this second class and absent from the first class. Note that the biases of the distributions in Figs. 9–11 are all about −120 ns, and thus, represent a constant difference between the GPS reference time and that of the precise solution. This could be entirely attributable to the maser-based reference clock used for the latter.

Similar analyses of GPS clocks were made for all satellites and days from July 4, 1993 through Oct. 22, 1993. For each day and satellite, any linear trend in the difference between the precise solution and the broadcast clock over the day was removed. The standard deviation over the day of the detrended difference has been calculated. The distribution of daily standard deviations is shown in Fig. 12. The median of the lower distribution is 4.5 ns, which represents the nonsystematic component of the broadcast clock error for satellites not affected by clock dithering. The median of the upper distribution is 79.9 ns; this quantifies the effect of clock dithering.

A second, systematic component of clock error arises because of differences among satellites each day in the linear trends. (The *average* trend parameters arise from the difference between GPS system time and the time of the Algonquin maser reference; such a difference is not included here because it will have essentially no effect on user position.) Based on the variations in trend parameters among the non-SA satellites over each day, the median value of this systematic component, over all days in the period being studied, has been calculated to be 10.3 ns. The total clock error for non-SA satellites is thus $(4.5^2 + 10.3^2)^{1/2}$ ns ≈ 11.2 ns. For satellites affected by clock dithering it is $(79.9^2 + 10.3^2)^{1/2}$ ns ≈ 80.6 ns.

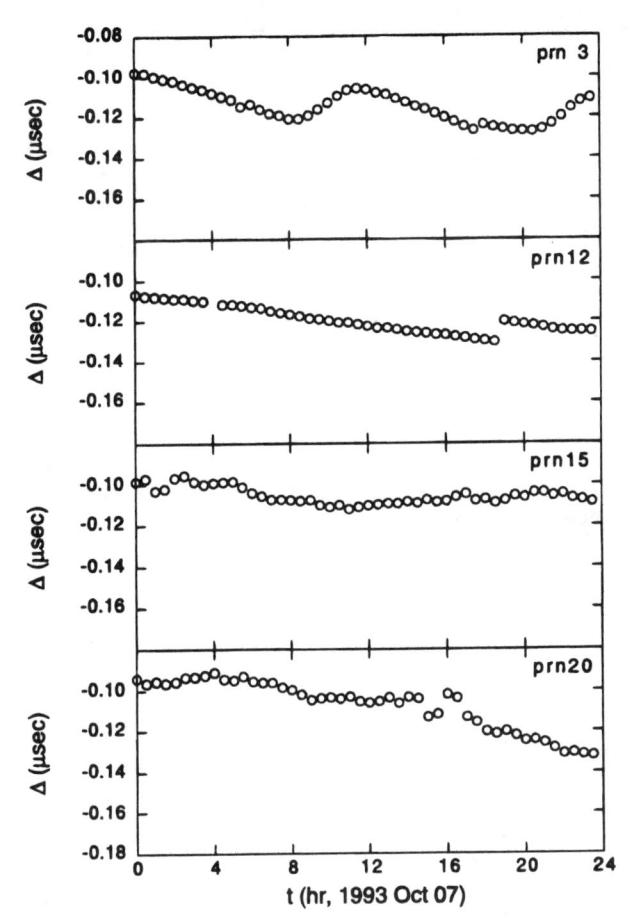

Fig. 10 Difference in the GPS broadcast and precise clock corrections, Δ, for two Block-I spacecraft (PRNs 3 and 12) and two Block-II spacecraft (PRNs 15 and 20). Full scale on each plot is 100 ns. The bias of about −120 ns in all of these (as well as that for PRN 13 in Fig. 9) represents a constant difference between GPS reference time and that of the precise solution.

E. Summary and Discussion

Table 3 compares these results with a prediction[11] of how well the GPS Control Segment would be able to predict clock corrections and GPS ephemerides. The prediction is reasonably consistent with the comparisons between the navigation message and the JPL precise solution, as discussed in this section.

A distribution of differences can be divided into two components: an average (bias) and deviation about that average. Our rms values for orbits include both. Our rms values for orbits include both of these components. Because the transformation from WGS-84 to ITRF will absorb most of the biases, and because the remaining rms values are not reduced much (see Sec. II.C) follow-

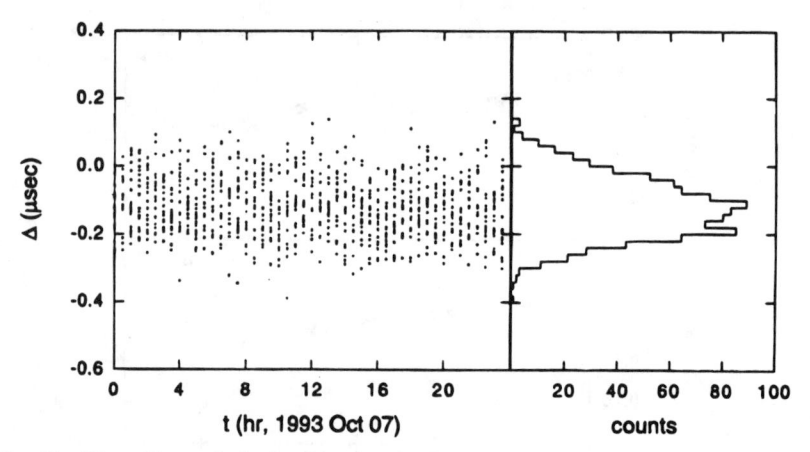

Fig. 11 The effects of clock dithering in SA are shown here, which includes all satellites not shown in Figs. 9 or 10. The distribution of Δ (right) has a standard deviation of about 80 ns and a mean of about -120 ns. (The mean is about the same as those in in Figs. 9 and 10.)

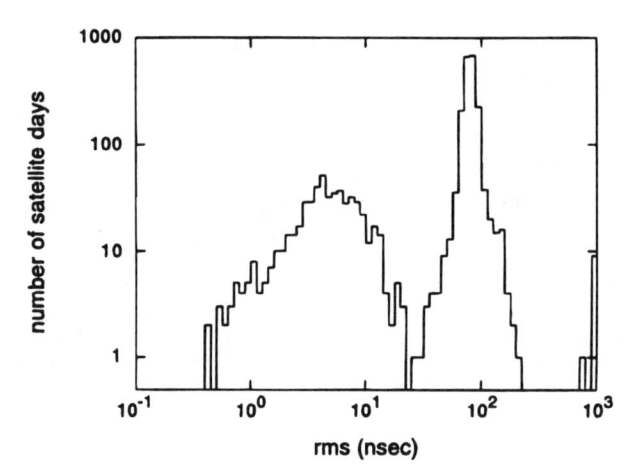

Fig. 12 Comparison of broadcast clock solution with precise solution, for each day and satellite during the period July 4, 1993 through Oct. 22, 1993. The bimodal distribution arises from the effects of clock dithering as part of SA. (There are also a handful of satellites and days when there was rather poor agreement between the precise solution and the broadcast clock.) The median value of the lower distribution (clock dithering presumably not in effect, rms de-trended difference over the day between precise solution and broadcast clock less than 25 ns) is 4.5 ns. The median value of the upper distribution is 79.9 ns.

Table 3 GPS Control Segment performance, predicted and observed

Parameter	Prediction	Observed
Radial (σ_r)	0.8 m	1.2 m
Cross-track ($\sigma_\dagger$)	3.0 m	3.2 m
Along-track (σ_a)	6.3 m	4.5 m
Clock (no SA) (σ_t)	7.7 ns	11.2 ns
Clock (with SA)	—	80.6 ns

ing the transformation, the dominant component is the deviation and not the bias.

The radial uncertainty, just over a meter, is about a factor of 3–4 less than the cross- and along-track uncertainties. This occurs because the range measurement is more sensitive to changes in the radial dimension than to the other dimensions (see Appendix).

We can cast the performances from Table 3 into a user equivalent range error $\sigma_u : \sigma_u^2 = k_r \sigma_r^2 + c^2 \sigma_t^2 + k_\rho \rho_{rt} \sigma_r c \sigma_t + k_{\dagger a}(\sigma_\dagger^2 + \sigma_a^2)$. Here, ρ_{rt} is the correlation coefficient between the radial and clock errors; c is the speed of light; and $\sigma_\dagger$, σ_a are the cross- and along-track errors. For a satellite at zenith, an error in the radial component maps directly into σ_u (this is always true for the clock error), while cross- and along-track errors do not affect the user range. We expect nominally, then, that $k_r = 1$, $k_\rho = 2$, and $k_{\dagger a} = 0$. We show in the Appendix that $k_r \approx 0.959$, $k_\rho \approx 1.959$, and $k_{\dagger a} \approx 0.0204$ account for the average geometry in the relationship between satellite and user positions.

It is expected that the correlation between radial and clock errors is small, in which case $|\rho_{rt}| \ll 1$. For completeness, however, we indicate in Table 4 the contributions to σ_u over the range of possible correlations, that is $\rho_{rt} = 0, \pm 1$.

When SA clock dithering is not in effect, the contributions to σ_u from the ephemeris and clock errors are of the same order, and result in $\sigma_u \approx 3$–4 m. The more usual circumstance, however, has clock dithering in effect, in which case the dominant contribution to σ_u is from the 80-ns noise in the broadcast clock, resulting in $\sigma_u \approx 24$ m.

Appendix: User Equivalent Range Error

Choose a spherical coordinate system with origin at the center of the Earth. Suppose a satellite has the following Cartesian coordinates:

Table 4 User equivalent range error (meters)

					σ_u		
	σ_r	$c\sigma_t$	$\sqrt{(2\sigma_r c\sigma_t)}$	$\sqrt{[k_{\dagger a}(\sigma_\dagger^2 + \sigma_a^2)]}$	$\rho_{rt} = -1$	$\rho_{rt} = 0$	$\rho_{rt} = +1$
No SA	1.2	1.3	1.8	0.8	0.7	1.9	2.6
With SA	1.2	24.0	7.6	0.8	22.8	24.0	25.1

$$\begin{pmatrix} X \\ Y \\ Z \end{pmatrix} = \begin{pmatrix} r \sin\Theta \cos\Phi \\ r \sin\Theta \sin\Phi \\ r \cos\Theta \end{pmatrix}$$

and an observer on the Earth has the following Cartesian coordinates:

$$\begin{pmatrix} x \\ y \\ z \end{pmatrix} = \begin{pmatrix} r_e \sin\theta \cos\phi \\ r_e \sin\theta \sin\phi \\ r_e \cos\theta \end{pmatrix}$$

where $r_e \approx 6370$ km, and $r \approx 26{,}550$ km. We use θ and Θ to indicate polar angle, and ϕ and Φ to indicate azimuth.

The user range is $u = \sqrt{(X-x)^2 + (Y-y)^2 + (Z-z)^2} = u(r, \Theta, \Phi, r_e, \theta, \phi)$. Choose the orientation of the axes so that 1) the z-axis intersects the satellite ($\Theta = 0$, nominally); and 2) an increase in Θ at $\Phi = 0$ corresponds to the direction of the resultant of the cross- and along-track uncertainties in the satellite position. (We assume that errors in different orbit components are not correlated.) Then, the user equivalent range error is as follows:

$$\sigma_u^2 = c^2\sigma_t^2 + \left(\frac{\partial u}{\partial r}\right)^2 \sigma_r^2 + 2\frac{\partial u}{\partial r} \rho_{rt} \sigma_r c \sigma_t + \left(\frac{\partial u}{r\partial\Theta}\right)^2 (\sigma_f^2 + \sigma_a^2)$$

(Note that, at $\Theta = \Phi = 0$, we have $\partial u/\partial X = \partial u/r\partial\Theta$.) Differentiations of $u(r,\Theta,\Phi,r_e,\theta,\phi)$ evaluated at $\Theta = 0$ give the following:

$$\frac{\partial u}{\partial r} = \frac{r - r_e \cos\theta}{\sqrt{r^2 + r_e^2 - 2rr_e\cos\theta}}, \quad \left(\frac{\partial u}{\partial r}\right)^2 = \frac{(r - r_e \cos\theta)^2}{r^2 + r_e^2 - 2rr_e\cos\theta}$$

and

$$\left(\frac{\partial u}{r\partial\Theta}\right)^2 = \frac{r_e^2 \sin^2\theta \cos^2\phi}{r^2 + r_e^2 - 2rr_e\cos\theta}$$

If we assume uniform distribution of positions on the Earth, then we have $p(\theta,\phi)\, d\theta\, d\phi \propto \sin\theta\, d\theta\, d\phi$ as the joint probability distribution of θ and ϕ, for θ less than its maximum value of $\theta_{max} = \cos^{-1}(r_e/r) \approx 76$ deg. If we average $\partial u/\partial r$, $(\partial u/\partial r)^2$, and $(\partial u/r\partial\Theta)^2$ over $0 \le \theta \le \theta_{max}$ and $0 \le \phi \le 2\pi$ with this distribution as the weighing function, we obtain $\sigma_u^2 \approx 0.959\, \sigma_r^2 + c^2 \sigma_t^2 + 1.959\, \rho_{rt}\, \sigma_r\, c\, \sigma_t + 0.0204\, (\sigma_f^2 + \sigma_a^2)$.

References

[1] Shank, C., and Smetek, R., personal communication.

[2] Brown, K. R., Jr., "Characterizations of OCS Kalman Filter Errors," *Proceedings of ION GPS-91*, ION, Washington, DC, Sept. 1991.

[3] Bowen, R., Swanson, P. L., Winn, F. B., Rhodus, N. W., and Feess, W. A., "Global Positioning System Operational Control System Accuracies," *Navigation*, Vol. 32, No. 2, 1985.

[4] Webb, F. H., and Zumberge, J. F., (eds.). "An Introduction to GIPSY/OASIS II," JPL Course Notes, Boulder, CO, JPL D-11088, July 1993.

[5]Wu, S. C., et al., Topex/Poseidon Project: Global Positioning System (GPS) Precision Orbit Determination (POD) Software Design, JPL D-7275, March 1990.

[6]Blewitt, G., "An Automatic Editing Algorithm for GPS Data," *Geophysical Research Letters*, Vol. 17, No. 3, 1990, pp. 199–202.

[7]Boucher, C., Altamimi, Z., Duhem, L., *"ITRF-91 and Its Associated Velocity Field,"* IERS TN 12, Observatoire de Paris, Oct. 1992.

[8]Blewitt, G., Heflin, M. B., Webb, F. H., Lindqwister, U. J., and Malla, R. P., "Global Coordinates with Centimeter Accuracy in the International Terrestrial Reference Frame using GPS," *Geophysical Research Letters,* Vol. 19, No. 9, 1992, pp. 853–856.

[9]Rothacher, M., Beutler, G., Gurtner, W., Brockmann, E., and Mervart, L., "Bernese GPS Software Version 3.4," Documentation May 1993, Druckerei der Universitaet Bern, Bern, Germany, May 1993.

[10]Kouba, J., and Tetreault, P., "International GPS Service for Geodynamics (IGS)," Natural Resource, Canada, Analysis Reports available every week from the IGS Central Bureau at the JPL (access, for example, with lynx http://igscb.jpl.nasa.gov/).

[11]Rusell, S. S., and Schaibly, J. H., "Control Segment and User Performance," *Global Positioning System, Papers published in Navigation, 1980,* The Institute of Navigation, Washington, DC, 1980.

[12]Wells, D., et al., *Guide to GPS Positioning,* Canadian GPS Associates, Fredericton, New Brunswick, Canada, 1986.

Chapter 17

Selective Availability

Frank van Graas* and Michael S. Braasch†
Ohio University, Athens, Ohio 45701

I. Goals and History

SELECTIVE availability (SA) is the intentional degradation of the Global Positioning System (GPS) signal with the objective to deny full position and velocity accuracy to unauthorized users. Selective availability is part of the Standard Positioning Service (SPS), which was formally implemented on 25 March 1990. Although there have been limited periods of time during which the levels of SA were fairly benign, the GPS policy[1] continues to state the following:

> SPS is planned to provide, on a daily basis, the capability to obtain horizontal positioning accuracy within 100 meters (2 drms, 95 percent probability) and 300 meters (99.99 percent probability), vertical positioning accuracy within 140 meters (95 percent probability), and timing accuracy within 340 ns (95 percent probability).

Clearly, this policy indicates that SA will be active, because the performance of GPS without SA would be in the 20-m range for horizontal positioning accuracy (95%).

SA was not part of the experimental design of GPS. However, initial testing of the Coarse/Acquisition (C/A) code during the 1970s revealed accuracies that were much better than anticipated. The C/A code provided 20–30 m position accuracies rather than the predicted accuracy of no better than 100 m.[2] This prompted the DOD to degrade intentionally the accuracy available to unauthorized users. Initially, the level of SA was set at 500 m (95%), but this was changed to 100 m (95%) in 1983. This level of accuracy was chosen because it is comparable with that provided by an on-airport VHF omnidirectional range (VOR) during the nonprecision approach phase of flight.

II. Implementation

The GPS position solution is obtained by solving a set of four or more pseudorange equations ($i = 1$ through the number of measurements)

Copyright © 1994 by the American Institute of Aeronautics and Astronautics, Inc. All rights reserved.
*Associate Professor, Department of Electrical and Computer Engineering.
†Assistant Professor, Department of Electrical and Computer Engineering.

$$\rho_i = \sqrt{(X - X_i)^2 + (Y - Y_i)^2 + (Z - Z_i)^2} + (\Delta t_{\text{RCVR}} - \Delta t_{\text{sv}_i})c \qquad (1)$$

where the unknowns are the receiver position X, Y, Z and the receiver clock with respect to GPS time, Δt_{RCVR}. Propagation delays and other error sources are omitted from Eq. (1). The following parameters are required to solve for the receiver position and clock offset:

ρ_i measured pseudorange
(X_i, Y_i, Z_i) satellite position for satellite number i
Δt_{sv_i} satellite clock offset for satellite i with respect to GPS time

The satellite positions and clock offsets are calculated from the navigation data transmitted by the satellites. Two different methods can be used to deny the full GPS accuracy: 1) manipulation of the navigation message orbit data, also referred to as the ϵ-process; and 2) manipulation of the satellite clock frequency, also referred to as the δ-process or clock dither. Manipulation of the navigation orbit data degrades the accuracy of the calculated satellite positions and results in slowly varying user position errors (periods on the order of hours). Note that the actual satellite orbits are not affected, only the parameters describing the satellite orbits are corrupted. Clock dither, on the other hand, involves the manipulation of the satellite clock itself. This results in fairly rapid errors on the pseudorange measurements with periods on the order of minutes. Because the actual satellite clock is manipulated, clock dither affects both the C/A code and the P code, as well as the integrated Doppler shift measurements.

Although position errors are specified in the GPS policy statement (see previous section), no information is provided with respect to the power spectral density of SA. Therefore, GPS receiver and system designers should anticipate a wide range of possibilities. The next section characterizes SA based on collected data.

III. Characterization of Selective Availability

The SPS provides information on the effects of SA in terms of position and time accuracies, which are summarized in Table 1. Also listed in Table 1 are the corresponding accuracies without SA, which were obtained from the U.S. Naval Observatory Bulletin Board.[3] The bulletin board also provided that frequency stability with SA would be on the order of 1 part in 10^{10}, whereas in the absence of SA, the frequency stability is on the order of 1 part in 10^{12}.

It should be noted that Table 1 provides position and timing accuracies only. No information is provided on the dynamics of the errors. The only way to

Table 1   Standard positioning service position and timing accuracies with and without selective availability

Parameter	With SA	Without SA
Horizontal position	100 m (95%) 300 m (99.99%)	20 m (95%)
Vertical position	140 m (95%)	30 m (95%)
Time	340 ns (95%)	40 ns

determine the dynamics of SA is through actual data collection. Figure 1 represents typical horizontal positioning accuracies of the GPS SPS with and without SA. Without SA, the horizontal position errors are fairly constant over periods on the order of tens of minutes. Position changes without SA are mostly caused by slowly changing propagation delays and satellite clock and ephemeris errors. With SA, the horizontal position "wanders around" within a circle with a radius of approximately 100 m. Successive position errors become uncorrelated after a period of approximately 2–5 min. Similar performance characteristics can be found for vertical positioning and for time transfer.

Although the characterization of SA in the position domain provides helpful information, several shortcomings of just a position domain characterization are quickly recognized.

1) SA is generated in each satellite and seems to be uncorrelated between satellites,[4] which means that the effect on the position accuracy depends on the satellite geometry.

2) The SPS policy assumes that at least 21 satellites are operational; therefore, additional satellite failures could produce larger than normal position errors.

3) No information is provided on velocity and acceleration errors.

4) Differential positioning accuracies are difficult to analyze with only position domain information. These depend on the acceleration of SA, processing time, and the latency of the data link.

5) No information is provided on the power spectral density of SA, which makes it difficult to simulate SA for receiver design and performance analysis purposes.

Because of these shortcomings, a complete analysis of SA-induced errors must start in the range measurement domain. Next, the range domain errors can be converted into positioning and timing errors depending on the application. Figure 2 shows representative, measured SA errors in the range domain. The SA error consists of the sum of a bias component (epsilon error) and a rapidly varying

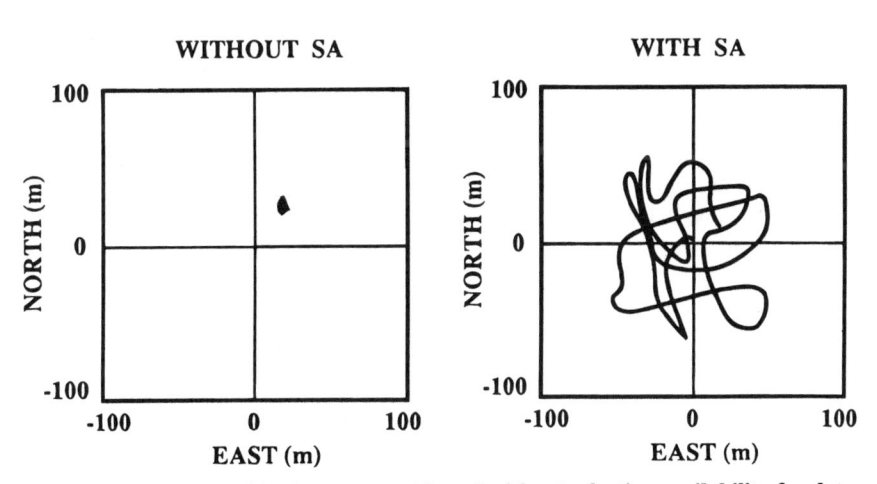

Fig. 1 Horizontal positioning errors with and without selective availability for data collected during a 1-h period.

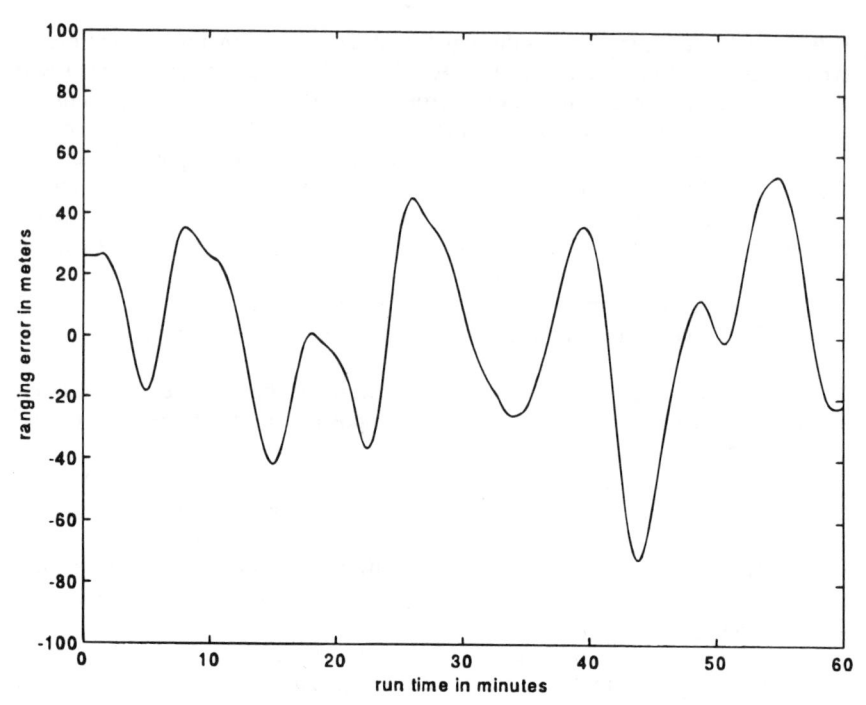

Fig. 2 Representative, measured range selective availability errors.

component (clock dither). The period of the oscillations is on the order of 2–5 min, while the standard deviation is approximately 23 m.

A first-order analysis of SA-induced ranging errors would start with the determination of the one-sigma range, velocity, and acceleration errors. Table 2 provides typical numbers resulting from such an analysis for two cases.[5] Case 1 refers to data collected from PRN 14 on day 124 of 1989, while case 2 represents data collected from both PRN 2 and 14 on day 259 of 1989. The user range accuracies (URA) were set to 32 and 64 m, respectively. The data without SA were collected from PRN 6 (a Block I satellite), while its URA was set to 2–4 m. Kremer et al.[5] gives detailed description of the data-processing techniques used. For both cases, the correlation time of the SA errors was on the order of 180 s. The range biases could either be caused by slow clock dither or by epsilon error.

Table 2 Example of measured line-of-sight range, velocity, and acceleration with and without selective availability

Parameter	One-sigma range	One-sigma velocity	One-sigma acceleration	Range bias
Without SA	1–1.5 m	0.0055 m/s	0.4 mm/s^2	0 m
With SA (case 1)	29 m	0.12 m/s	2 mm/s^2	74 m
With SA (case 2)	38–57 m	0.21 m/s	3.5 mm/s^2	−61 to −38 m

Further characterization of SA usually focuses on the clock dither. Epsilon errors are effectively simulated by adding random biases to the satellite orbit data. It should be noted, however, that SA could consist of either clock dither or epsilon errors, or both. RTCA, Inc. (formerly known as the Radio Technical Commission for Aeronautics), for instance, simulates epsilon error by adding random biases chosen from a Gaussian distribution with zero mean and a standard deviation of 23 m to the satellite range measurements. These biases are held constant during each GPS receiver test. Methods for identifying the epsilon and dither components of SA can be found in Ref. 6. Epsilon error may be found by comparing the orbits obtained from the broadcast ephemeris and from the postfit precise ephemeris. Dither is obtained by processing the corrections generated in a differential ground reference station.

The next step in the characterization of SA would be to determine the actual power spectral density (PSD) of the signal. This involves the postulation of a model for SA, which then can be used to generate statistically equivalent SA. The first SA model was derived by Matchett in 1985.[7] This model was not derived from actual data, but was deduced from a sample probability distribution curve. The GPS Joint Program Office (JPO) generated SA samples and then computed the distribution curve from these samples. A second-order Gauss–Markov process was postulated, and the coefficients were adjusted until its distribution curve matched the one provided by the JPO. A second-order Gauss–Markov process has also been adopted by Special Committee 159 of RTCA for the purpose of simulating SA errors.[8] The first models obtained from actual SA data were time series models derived by Braasch in 1989 using system identification theory.[9] The resulting models were autoregressive moving average (ARMA) models. Later, Chou implemented a second-order Gauss–Markov process based on measured SA data.[10] At the same time, Chou also provided the formulation for a recursive autoregressive model.[11] Other models of interest were published by Lear et al., who presented several time series and analytical models, also based on measured SA data.[12] Four of the above models are discussed in the following sections. These are the second-order Gauss–Markov model, an autoregressive (AR) model, an analytic model, and a recursive AR model (lattice filter). It should be noted that all of these models simulate clock dither only.

A. Second-Order Gauss–Markov Model

The second-order Gauss–Markov model implementation follows the description provided in Ref. 8. The continuous time model of a second-order Gauss–Markov process is given by the following:

$$\ddot{x}_p + 2\beta\omega_0 \dot{x}_p + \omega_0^2 x_p = c \times w \qquad (2)$$

where ω_0 = natural frequency; β = damping factor < 1; w = white Gaussian noise; and power spectral density = 1.

The continuous time model given by Eq. (2) can be expressed in the form of a state-space equation:

$$\begin{bmatrix} \dot{x}_p \\ \dot{x}_v \end{bmatrix} = \begin{bmatrix} 0 & 1 \\ -\omega_0^2 & -2\beta\omega_0 \end{bmatrix} \begin{bmatrix} x_p \\ x_v \end{bmatrix} + \begin{bmatrix} 0 \\ c \end{bmatrix} w \quad (3)$$

where $\dot{x}_v = \dot{x}_p$. The corresponding discrete-time state-space equation is then given by the following:

$$\begin{bmatrix} x_p \\ x_v \end{bmatrix}_{i+1} = \begin{bmatrix} \phi_{11} & \phi_{12} \\ \phi_{21} & \phi_{22} \end{bmatrix} \begin{bmatrix} x_p \\ x_v \end{bmatrix}_i + \begin{bmatrix} u_{11} & u_{12} \\ 0 & u_{22} \end{bmatrix} \begin{bmatrix} w_1 \\ w_2 \end{bmatrix}_i \quad (4)$$

where w_1 and w_2 are white Gaussian noise processes with zero mean and unit variance. The elements of the discrete-time state transition matrix are as follows:

$$\begin{aligned}
\phi_{11} &= e^{-\beta\omega_0 \Delta T}[\cos(\omega_1 \Delta T) + \beta(\omega_0/\omega_1)\sin(\omega_1 \Delta T)] \\
\phi_{12} &= (1/\omega_1)e^{-\beta\omega_0 \Delta T}[\sin(\omega_1 \Delta T)] \\
\phi_{21} &= -\omega_0^2 \phi_{12} \\
\phi_{22} &= e^{-\beta\omega_0 \Delta T}[\cos(\omega_1 \Delta T) - \beta(\omega_0/\omega_1)\sin(\omega_1 \Delta T)]
\end{aligned} \quad (5)$$

where ΔT is in units of seconds.

The error covariance matrix of the white noise is as follows:

$$\begin{bmatrix} Q_{11} & Q_{12} \\ Q_{21} & Q_{22} \end{bmatrix} = \begin{bmatrix} u_{11} & u_{12} \\ 0 & u_{22} \end{bmatrix} \begin{bmatrix} u_{11} & 0 \\ u_{12} & u_{22} \end{bmatrix} \quad (6)$$

in such a way that

$$u_{11} = \sqrt{Q_{11} - Q_{12}^2/Q_{22}} \qquad u_{12} = Q_{12}/\sqrt{Q_{22}} \qquad u_{22} = \sqrt{Q_{22}} \quad (7)$$

The elements of the covariance matrix Q are as follows:

$$Q_{11} = \frac{c^2}{4\beta\omega_0^3}\left[1 - \frac{\omega_0^2}{\omega_1^2} e^{-2\beta\omega_0 \Delta T}\left(1 - \beta^2 \cos(2\omega_1 \Delta T) + \beta \frac{\omega_1}{\omega_0}\sin(2\omega_1 \Delta T)\right)\right]$$

$$Q_{12} = Q_{21} = \frac{c^2}{4\omega_1^2}[e^{-2\beta\omega_0 \Delta T}(1 - \cos(2\omega_1 \Delta T))] \quad (8)$$

$$Q_{22} = \frac{c^2}{4\beta\omega_0}\left[1 - \frac{\omega_0^2}{\omega_1^2} e^{-2\beta\omega_0 \Delta T}\left(1 - \beta^2 \cos(2\omega_1 \Delta T) - \beta \frac{\omega_1}{\omega_0}\sin(2\omega_1 \Delta T)\right)\right]$$

where

$$\omega_0 = \frac{\sigma_v}{\sigma_x} \qquad c^2 = 4\beta \frac{\sigma_v^3}{\sigma_x} \qquad \omega_1 = \frac{\sigma_v}{\sigma_x}\sqrt{1 - \beta^2} \quad (9)$$

time constant: $\tau_d = \dfrac{1}{\beta} \dfrac{\sigma_x}{\sigma_v} = \dfrac{1}{\beta} \dfrac{1}{\omega_0}$

range sigma: $\sigma_x = \sqrt{\dfrac{c^2}{4\beta\omega_0^3}}$ (10)

velocity sigma: $\sigma_v = \sqrt{\dfrac{c^2}{4\beta\omega_0}}$

The power spectral density function specified by RTCA for the above model is given by the following:

$$S(\omega) = c^2/(\omega^4 + \omega_0^4) \text{ m}^2/(\text{rad/s}) \quad (11)$$

where ω is the frequency in rad/s.

The power spectral density is factored into right-half plane and left-half plane components:

$$S(\omega) = c/[\omega^2 - (2/\sqrt{2})(\omega_0\omega) + \omega_0^2] \, c/[\omega^2 + (2/\sqrt{2})(\omega_0\omega) + \omega_0^2] \quad (12)$$

Next, a second-order Gauss–Markov process is created by passing zero mean, unit variance, white Gaussian noise through the filter given by the left-half plane component of the power spectral density. The corresponding realization of the second-order Gauss–Markov process in the time domain is given by Eq. (2), see Ref. 13. From Eq. (2), it is found that the damping factor is given by the following:

$$\beta = 1/\sqrt{2} \quad (13)$$

For simulation purposes, RTCA has proposed the following parameters:

$$\omega_0 = 0.012 \text{ rad/s}$$
$$c^2 = 0.002585 \text{ m}^2 \quad (14)$$

The model then results in the following:

$$\sigma_p = \sqrt{\dfrac{0.002585}{4\dfrac{1}{\sqrt{2}} 0.012^3}} = 23 \text{ m}$$

$$\sigma_v = \sqrt{\dfrac{0.002585}{4\dfrac{1}{\sqrt{2}} 0.012}} = 0.28 \text{ m/s} \quad (15)$$

$$\tau = \sqrt{2}\,\dfrac{23}{0.28} = 118 \text{ s}$$

The process outlined in the preceding equations is to be generated for each satellite. The error x_p is the second-order Gauss–Markov process to be added to the "perfect" pseudorange. The error x_v is the first-order Gauss–Markov process to be added to the instantaneous "perfect" pseudorange rate. Initialization of x_p is Gaussian with σ_p, initialization of x_v is Gaussian with σ_v.

For an update period of 1 s, the matrices in Eq. (4) become the following:

$$\Phi = \begin{bmatrix} \phi_{11} & \phi_{12} \\ \phi_{21} & \phi_{22} \end{bmatrix} = \begin{bmatrix} 0.9999284 & 0.9915387 \\ -0.0001428 & 0.9831014 \end{bmatrix}$$

$$U = \begin{bmatrix} u_{11} & u_{12} \\ 0 & u_{22} \end{bmatrix} = \begin{bmatrix} 0.0146771 & 0.0252060 \\ 0 & 0.0504133 \end{bmatrix}$$

(16)

Figure 3 shows typical results for the second-order Gauss–Markov model. The range SA is slightly noisier than the actual range SA (see Fig. 2). The range rate SA is adequate for simulation purposes, but it is much noisier than actual range rate SA.

B. Autoregressive Model

This section follows the material presented in Ref. 6. In general, time series models are based on the assumption that the data of interest can be modeled as the output of a linear system (pole-zero filter) driven by Gaussian white noise. Conceptually, the derivation of a time series SA model can be thought of as a two-step process. The first step is to send the SA data through a filter and adjust the poles and zeros, or equivalently, the filter coefficients, in such a way that the output is Gaussian white noise with minimum variance (the output is referred

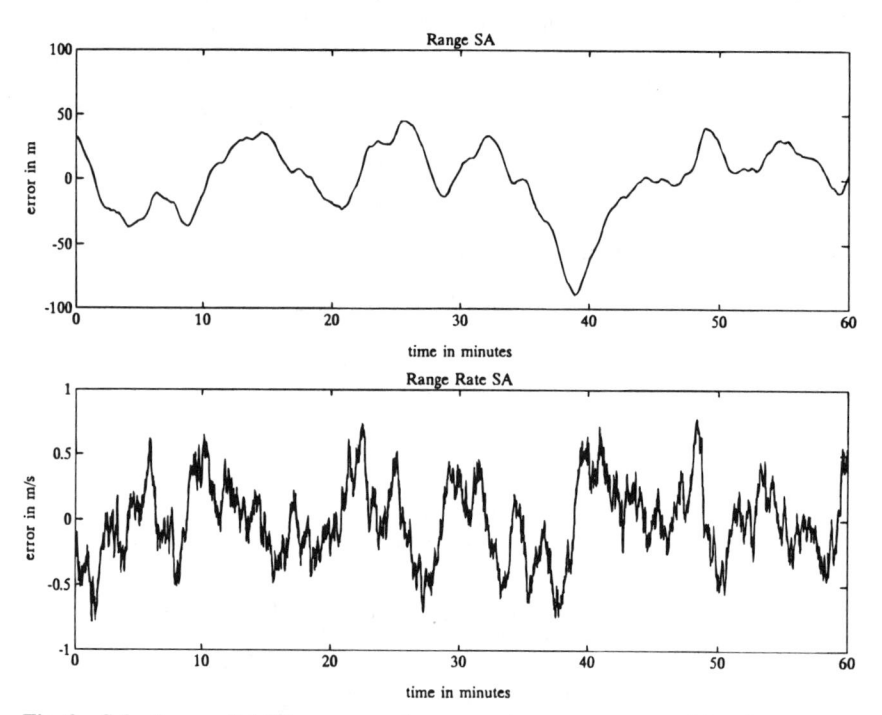

Fig. 3 Selective availability range and range rate errors generated by the second-order Gauss–Markov model.

to as residuals). The second step is to compute the inverse of the filter determined in the first step. Model identification is now complete.[6] Statistically equivalent SA data can then be generated by driving the inverse filter with Gaussian noise, whose variance is equivalent to that of the residuals in the first step. This process is illustrated in Fig. 4. Kelly provides and excellent overview of time series model identification and its application to the problem of microwave landing system (MLS) signal modeling.[14]

Three decisions are inherent in the procedure described in the preceding paragraph. The first is the choice of model (filter) type. Three are possible: 1) a pole-zero filter giving rise to an ARMA model; 2) an all-pole filter yielding an AR model; and 3) an all-zero filter yielding a moving average (MA) model. The second decision is the choice of model order. For example, if an AR model is chosen, how many poles will be used? The third decision is related to the first two and involves determining if a given residual sequence is white.

Because the primary goal is to derive an accurate SA-only model, an AR model type is chosen. This is because ARMA and MA models tend to be noisy. In fact, Braasch[9] concluded that an ARMA model was the best type for the combination of SA and receiver noise. An AR model of order p [referred to as an AR(p)] is defined as follows[15]:

$$x(n) = -\sum_{k=1}^{p} a(k) x(n - k) + e(n) \qquad (17)$$

where x is the model output; n is the time index; $a(k)$ is the kth filter coefficient;

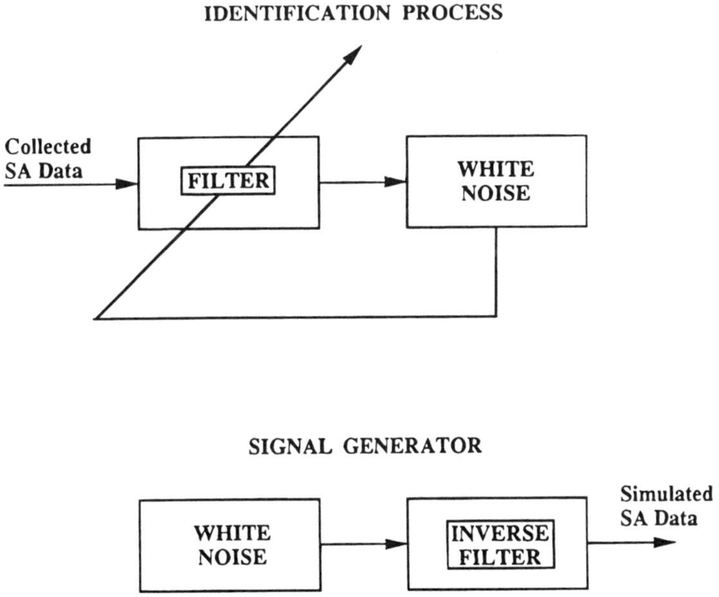

IDENTIFICATION PROCESS

SIGNAL GENERATOR

Fig. 4 Characterization of selective availability using system identification theory.

and e is the input Gaussian white noise. Note that SA models derived from data will operate at the same rate as the data collection rate.

Once having made the decision to use an AR model type, the rest of the process involves finding the optimum model order and coefficients (pole locations). For a given model order, many methods exist for optimizing the coefficients.[15-17] The one chosen is the modified covariance or forward–backward method. The second name stems from the fact that the optimization criterion is the minimization of forward and backward prediction errors. As shown later, this method performs quite well with SA data.

Several methods exist for model order selection. The majority of these have been developed for extremely short data records. The main issue is that we want to derive a model for the underlying statistical process that gave rise to the data. When model orders are selected that are too high (i.e., approaching the number of data points in the sample), the result is a "fit" of the sample data record rather than the underlying statistical process. The model order selection method used in this study is known as the Principle of Parsimony—the simplest acceptable model is the one chosen. An acceptable model is the inverse of the filter that outputs white noise when driven with SA. Note that if the model order is too low, the residuals will not be white, although the coefficients have been optimized.

The model identification, therefore, proceeds as follows. For a given sample of SA data, the coefficient is optimized for a first-order filter, and the residuals

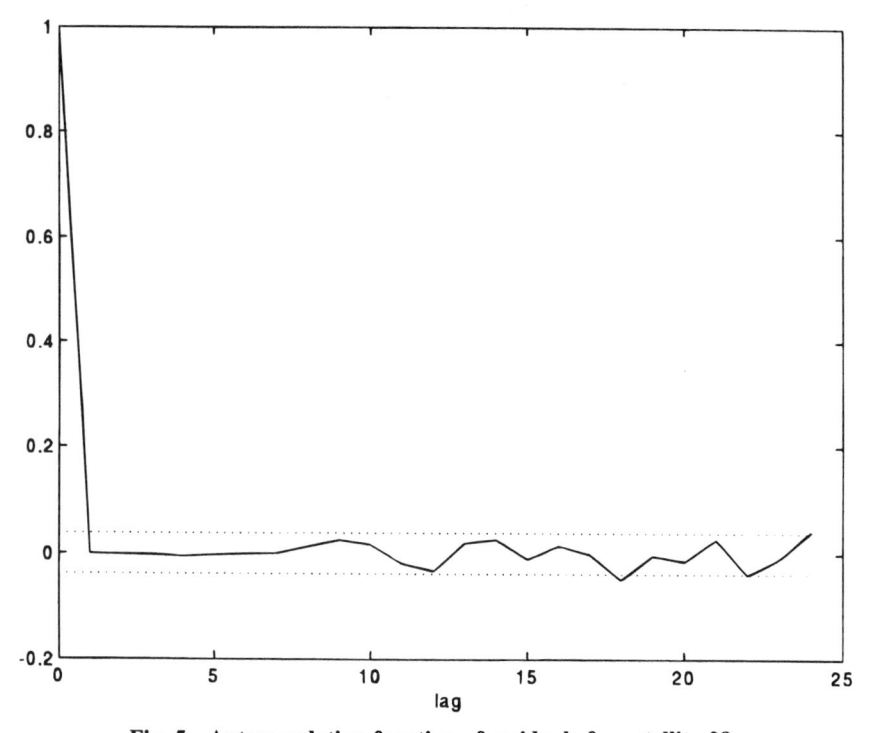

Fig. 5 Autocorrelation function of residuals for satellite 28.

are examined. If they are not white, then the coefficients for a second-order filter are optimized, and the residuals are examined again. The process is repeated until the model order and optimum coefficients are found for which the residuals are white. This process was performed for a variety of SA data sets. Depending upon the data set, models of either 9 or 11 coefficients were derived.

The method for determining whiteness involves examination of the autocorrelation function. An example is given in Fig. 5 where the autocorrelation function is plotted for the residuals from the SA data of satellite 28. Ideally, the autocorrelation function of white noise has a spike at lag zero and is zero everywhere else. However, this can be obtained only for infinite length sequences. As a result some minor "sidelobes" will occur at lags other than zero for white noise sequences that are finite. The dotted lines in the figure represent the 99% confidence intervals for the sidelobes. As can be seen in the plot, the sidelobes lie inside the confidence intervals for the most part, and thus the model is acceptable.

Further validation of the model can be performed by generating some waveforms and comparing the PSDs of the generated and collected data. An example is shown in Figs. 6 and 7. Figure 6 shows the waveform generated by the SA model derived from the SV 28 data. Note that if we compare the waveform to that of the collected data (Fig. 2), they are not the same; however, they are statistically equivalent. That is, the periods and amplitudes of the generated data are the same as for the collected data. This is better illustrated in Fig. 7 where the PSDs of the two waveforms are plotted. The lower line represents the collected

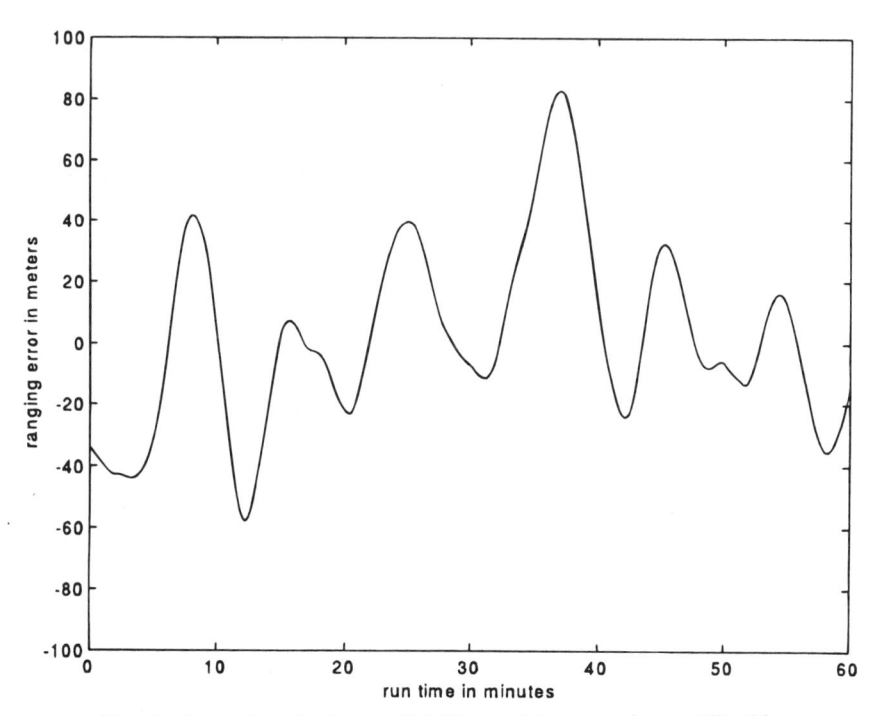

Fig. 6 Example selective availability model output for satellite 28.

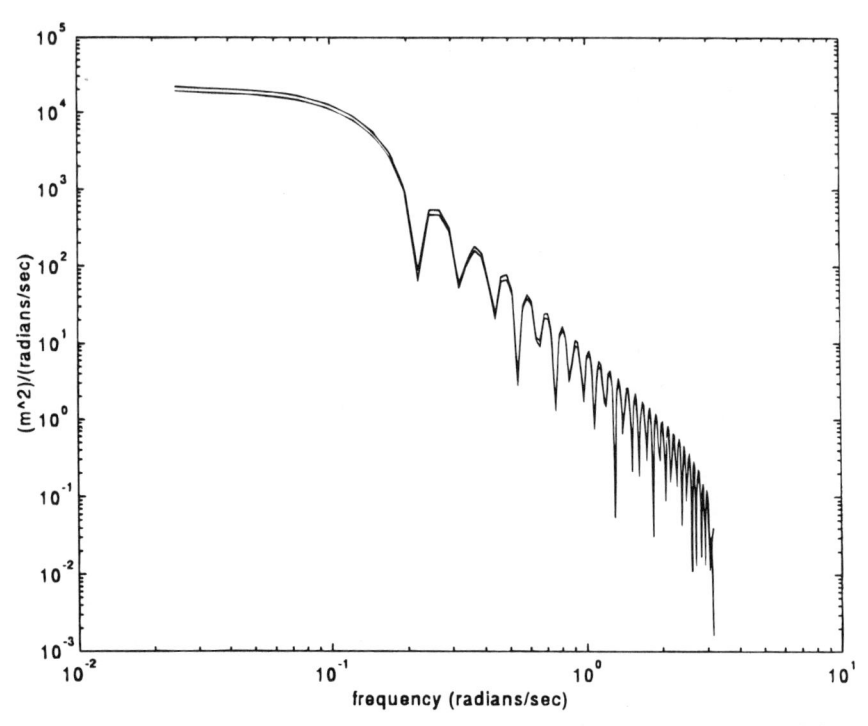

Fig. 7 Power spectral density functions of modeled and measured selective availability for satellite 28.

data, and the upper line represents the generated waveform. Power spectral density comparisons were performed on all of the models derived from the data. In each case, the result was similar to that shown here.

A final step in model validation concerns the power in the residuals. Recall that in step one of the model derivation process, the goal was to find a filter that output white noise (residuals) with minimum variance when driven with SA. The need for minimum variance is important from both a theoretical and practical viewpoint. Theoretically, having residuals with minimum variance means that the filter has been optimized and embodies the structure (i.e., correlation or information) of the SA. Kelly[14] refers to this as the filter "explaining" the data. However, from a practical viewpoint, minimum variance is also required. This is particularly true when trying to model random, yet smooth, waveforms such as SA.

Figures 8 and 9 illustrate the success of the AR model type in this respect. The residuals plotted in Fig. 8 have a standard deviation of 4.12 mm (4.12 × 10^{-3} m). Because this represents the amplitude of the noise driving the model [see Eq. (17)], it follows that any noise-like behavior in the generated SA waveforms will be negligible. This is verified in Fig. 9, which shows the smooth waveform of the generated SA over a short time interval.

A typical set of AR coefficients and the variance of the white noise input for

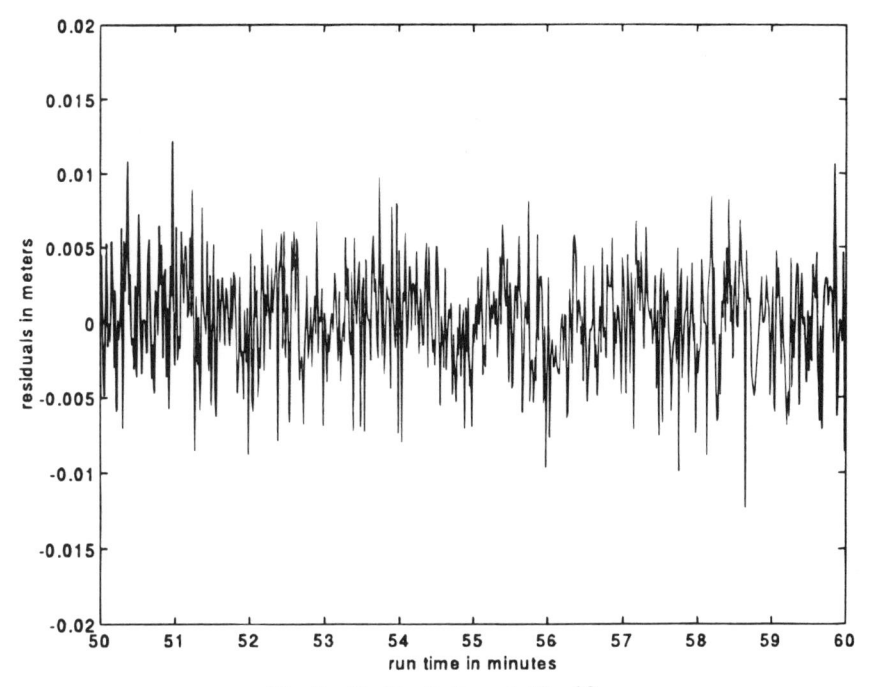

Fig. 8 Residuals for satellite 28.

data collected at a one-second update rate are given below:

$$a(1) = -1.36192741558063$$
$$a(2) = -0.15866710938728$$
$$a(3) = +0.13545921610672$$
$$a(4) = +0.21501267664869$$
$$a(5) = +0.30061078095966$$
$$a(6) = -0.12390183286070$$
$$a(7) = +0.10063573000351$$
$$a(8) = +0.02694677520401$$
$$a(9) = -0.12898590228866$$
$$a(10) = +0.05083106570666$$
$$a(11) = -0.05600186282898$$

$$\sigma_e^2 = 1.6993 \times 10^{-5} \text{ (m}^2\text{)}$$

where σ_e^2 is the variance of the Gaussian white noise input. These particular coefficients were derived from data collected from SV 28 during the first week of December 1992. Data from other satellites were found to produce similar results. The seemingly excessive number of significant figures are required to ensure filter stability.

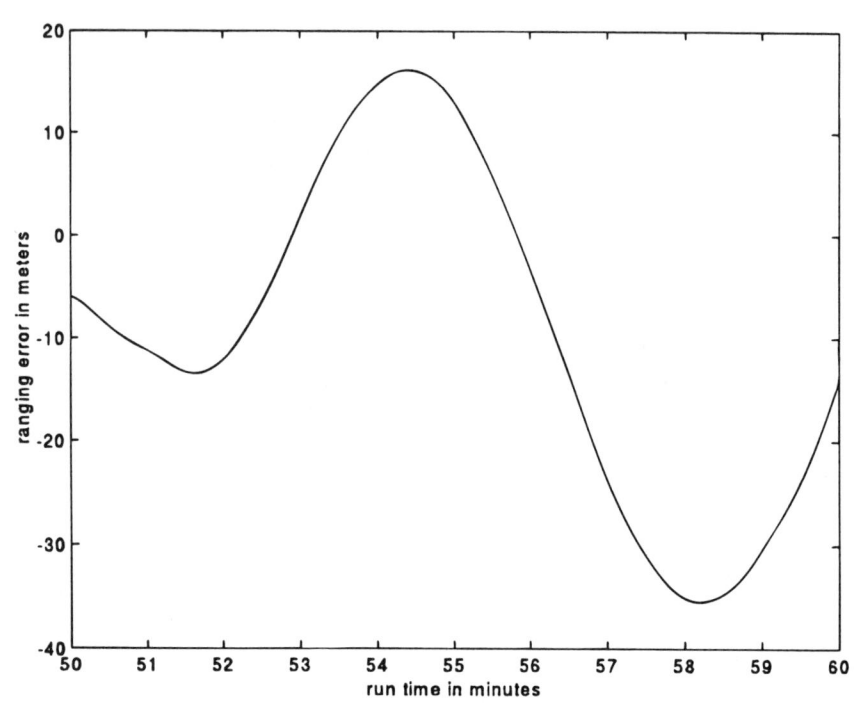

Fig. 9 Modeled selective availability for satellite 28—expanded scale.

C. Analytic Model

Lear et al. present several time series and analytical models in their 1992 paper.[12] Only one of the analytical models is discussed here because of its uniqueness. The term *analytic model* refers to "a piecewise smooth differentiable stochastic process with randomness appearing through the coefficients or model parameters".[12] These parameters are kept constant for randomly selected time intervals. The Rater analytic SA model was developed by observing data from satellite 19 on day 71 of 1992. Using low-noise integrated Doppler shift measurements and ionospheric corrections, the observed raw range was differenced twice. Data differenced once provided smooth pseudorange rate data, while the data differenced twice provided noisy pseudorange acceleration data. Rater observed that the second differences resemble connected ramps of varying slopes. This would imply that the third derivative (jerk) would consist of random steps. The third difference was too noisy, however, to show this behavior.

The Rater analytic SA model uses two randomly selected time intervals from uniform distributions. A constant jerk value is calculated for each time interval based on the length of the interval. The jerk is then integrated three times to obtain the range domain errors. Next, two of these waveforms (independently generated) are added together to obtain the range SA. In detail, the Rater model proceeds as follows:

1) Generate $x_1(t)$: Select the time duration T_1 of x_1 from a uniform distribution (100,700 s). Next, select the time constant τ_1 from a uniform distribution (0,

$T_1/2$ s) where τ is the period of time during which the jerk level is zero. This period is preceded by a jerk level of $+A_1$ and followed by a jerk level of $-A_1$. The jerk level A_1 is calculated from the following:

$$A_1 = -\ddot{x}_1(0)(2T_1 + \tau_1)/T_1(T_1 + \tau_1) \qquad (18)$$

where $x_1(0) = \dot{x}_1(0) = 0$, and $\ddot{x}(0) = 4$ mm/s^2

The following jerk profile is generated:

Time interval	Jerk level
$0 < t \leq (T_1 - \tau_1)/2$	$+A_1$
$(T_1 - \tau_1)/2 < t \leq (T_1 + \tau_1)/2$	0
$(T_1 + \tau_1)/2 < t \leq T_1$	$-A_1$

The jerk profile is integrated three times to obtain $x_1(t)$.

2) Generate $x_2(t)$: Same as the generation of $x_1(t)$, but use new random numbers for T_2 and τ_2 and choose $\ddot{x}_2(0) = -\ddot{x}_1(0)$.

3) Add the two processes $x(t) = x_1(t) + x_2(t)$.

The heart of this model lies in the basic waveform obtained after integrating the jerk profile three times. The result is a Gaussian-like pulse that is entirely non-negative or nonpositive depending upon the sign of $\ddot{x}_1(0)$. Then $x_1(t)$ consists of a series of positive pulses with random lengths and amplitudes, and $s_2(t)$ is a series of negative pulses. Finally, the sum of $x_1(t)$ and $x_2(t)$ results in a SA-like waveform.

Figure 10 shows a typical output of the Rater analytic SA model. As noted by the author of the model, a direct analytic relationship between the choice of parameters and the sigma of the process does not exist.[12] Nevertheless, the model seems to generate representative SA data. In addition, the model generates smooth range rate data, which is not the case for the previous two models.

D. Recursive Autoregressive Model (Lattice Filter)

A recursive AR model allows for the estimation of the AR model coefficients in real time, rather than in (off-line) batch processing. In principle, this would allow the AR model to predict ahead, once the model has "learned" the coefficients. The degradation of the accuracy of the predictions depends on the correlation time of the data. Highly correlated data would allow the AR model to predict ahead for longer periods of time than in the case of data that exhibit a low serial correlation. A recursive least squares (LS) lattice filter was first used by Chou in 1990.[11] The results presented by Chou were very promising, but they have not been repeated with the same level of success for other sets of data (see also Ref. 18). However, the use of a recursive AR model shows significant accuracy improvements in the prediction of the waveform when significant SA accelerations are presents. In the absence of significant SA accelerations, the recursive AR model does not offer any improvement over a simple linear extrapolation.

The LS lattice algorithm lends itself best to the real-time estimation of the AR coefficients because of the inherent stability offered by the lattice structure.[19] Following Refs. 16 and 19, the steps needed to implement the LS lattice filter are given below. The reader is referred to Ref. 19 for a detailed description of LS lattice filters.

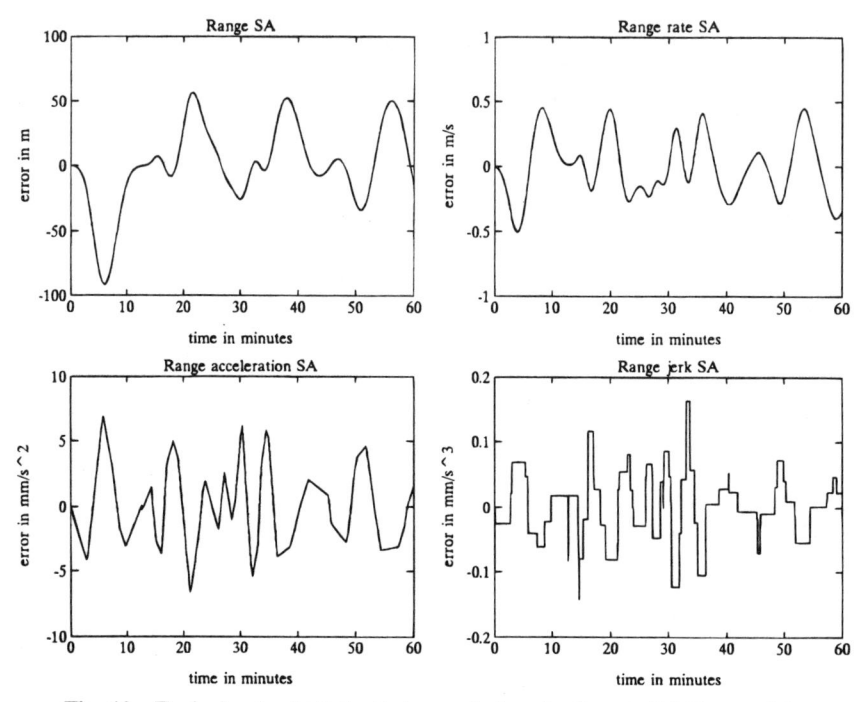

Fig. 10 Typical output of the Rater analytic selective availability model.

Initialization (δ is the steady-state squared prediction error; e.g., $\delta = 10^{-5}$):

$$e_m^b(0) = \Delta_m(0) = 0$$

$$\gamma_m(0) = 1 \qquad (0 \leq m \leq N)$$

$$\epsilon_m^f(0) = \epsilon_m^b(0) = \delta$$

For $n = 1$ to $n = $ final do (n is the current time index; x is the collected SA data)

$$e_0^b(n) = e_0^f(n) = x(n)$$

$$\epsilon_0^b(n) = \epsilon_0^f(n) = \epsilon_0^f(n-1) + x^2(n)$$

$$\gamma_0(n) = 1$$

For $0 \leq m \leq N-1$ do (N is the model order)

The AR filter coefficients are calculated from

$$\Delta_{m+1}(n) = \Delta_{m+1}(n-1) + [e_m^b(n-1)e_m^f(n)/\gamma_m(n-1)]$$

$$e_{m+1}^f(n) = e_m^f(n) - [\Delta_{m+1}(n)e_m^b(n-1)/\epsilon_m^b(n-1)]$$

$$e_{m+1}^b(n) = e_m^b(n-1) - [\Delta_{m+1}(n)e_m^f(n)/\epsilon_m^f(n)]$$

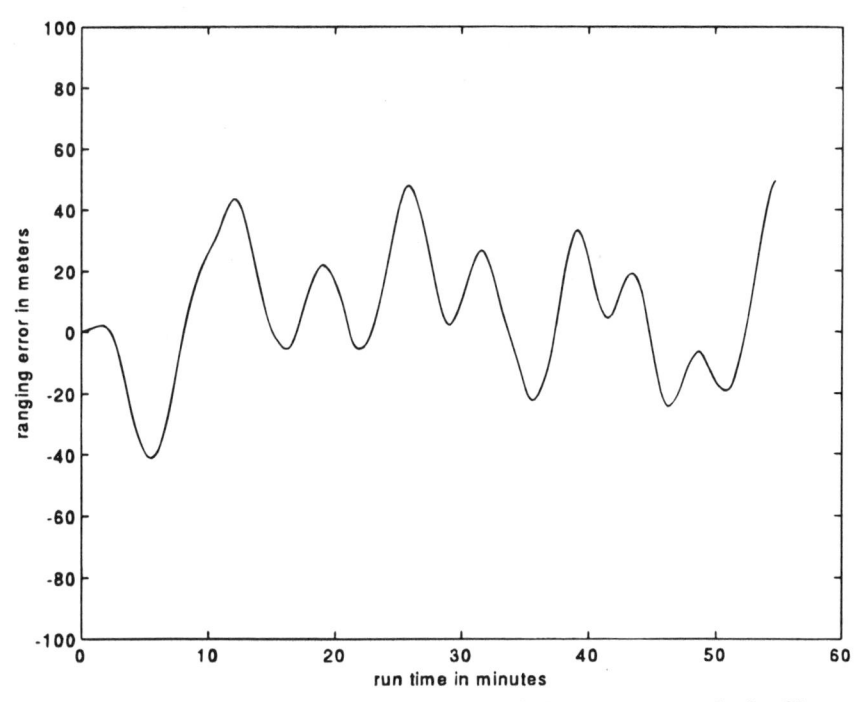

Fig. 11 Selective availability data used to "train" the least squares lattice filter.

$$\epsilon_{m+1}^f(n) = \epsilon_m^f(n) - [\Delta_{m+1}^2(n)/\epsilon_m^b(n-1)]$$

$$\epsilon_{m+1}^b(n) = \epsilon_m^b(n-1) - [\Delta_{m+1}^2(n)/\epsilon_m^f(n)]$$

$$\gamma_{m+1}(n-1) = \gamma_m(n-1) - \{[e_m^b(n-1)]^2/\epsilon_m^b(n-1)\}$$

The reflection coefficient is calculated from

$$k_{m+1}(n) = \Delta_{m+1}(n)/[\epsilon_m^b(n-1)\,\epsilon_m^f(n)]^{1/2}$$

For $1 \leq i \leq m-1$ do

$$\alpha_m(i) = \alpha_{m-1}(i) - k_{m+1}^b \alpha_{m-1}(m-i)$$

End (i do loop)

$$\alpha_m(m) = k_{m+1}^b(n)$$

End (m do loop)

For $1 \leq i \leq N$ do

$$a_i(n) = \alpha_N(i)$$

End (i do loop)

End (n do loop)

The coefficients and parameters used in the LS lattice filter are summarized below:

$e_m^b(n)$ = the backward prediction error for order m at time equals n
$e_m^f(n)$ = the forward prediction error for order m at time equals n
$\Delta_m(n)$ = the partial correlation coefficient between the forward and backward prediction errors for order m at time equals n
$\gamma_m(n)$ = the angle parameter for order m at time equals n
$\epsilon_m^f(n)$ = the squared forward prediction error for order m at time equals n
$\epsilon_m^b(n)$ = the squared backward prediction error for order m at time equals n
δ = the steady-state squared prediction error
$x(n)$ = the collected SA data at time equals n
n = the current time index
N = AR model order
m = model order index (0 through $N-1$)
$k_m(n)$ = the reflection coefficient for order m at time equals n
$\alpha_m(i)$ = the ith prediction coefficient for order m
$a_i(n)$ = the ith AR filter coefficient at time equals n

As an example of the usefulness of the LS lattice filter, an 11th-order filter was "trained" with almost 1 h of simulated SA data generated with the model

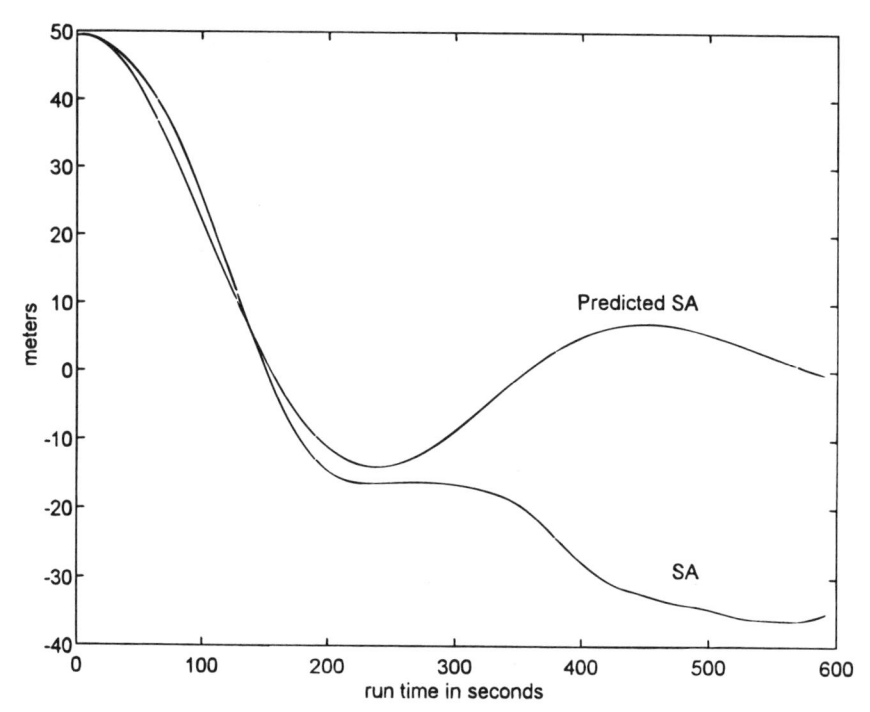

Fig. 12 Comparison of actual selective availability error and selective availability error predicted by the least squares lattice filter.

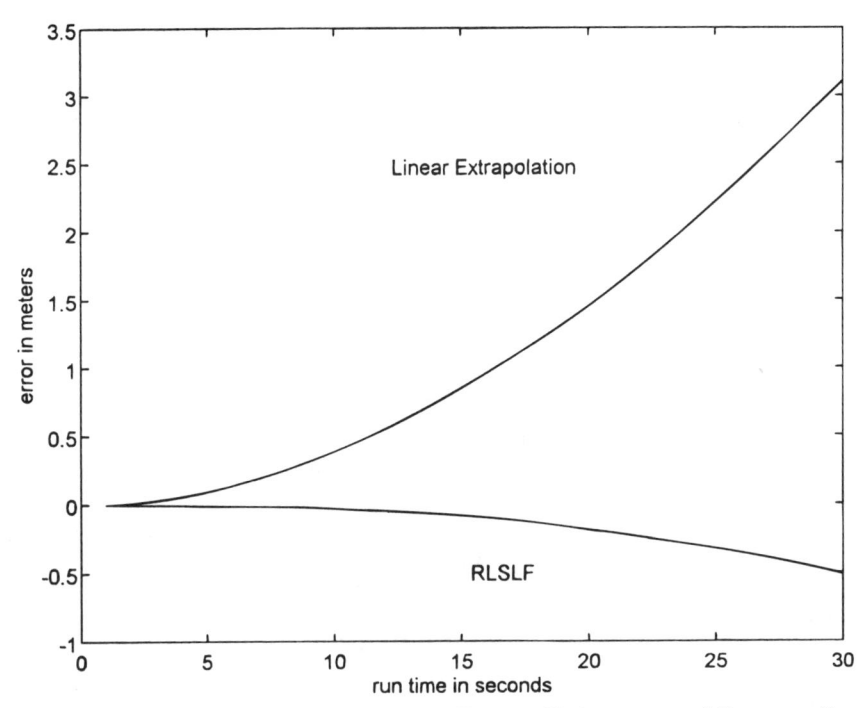

Fig. 13 Comparison of least square lattice filter prediction error and linear prediction error.

described in Sec. III. B., as shown in Fig. 11. At the end of the training period, the SA waveform was just starting to turn around from a positive slope to a negative slope. Figure 12 shows the predicted SA generated by the LS lattice filter, as well as the actual SA. During the first 3 min of the prediction interval, the predicted SA matches the actual SA to within 5 m. For periods of time longer than 3 min, the predicted values go to zero. This is consistent with the anticipated performance: the data can be predicted for a period of time on the order of the correlation time of the data. If a linear extrapolation is used instead of the LS lattice filter, the results are much worse, as illustrated in Fig. 13. After a period of 30 s, the linear extrapolation error is approximately 3 m, while the LS lattice filter prediction error is only -0.5 m. The opposite sign of the prediction errors is caused by the change in the slope of the SA data at the end of the training period. For this particular example, the LS lattice filter order was chosen to best match the SA data. Performance could be degraded if the filter is not adequately matched to the data. It should be noted that in the absence of significant SA acceleration, the performance of the linear extrapolator is statistically similar to the performance of the lattice filter.

E. Selective Availability Model Summary

Three different SA models are presented in this section. The highlights of each of the models are summarized as follows:

1. Second-Order Gauss–Markov Model

This model generates both range and range rate errors. It has been adopted by RTCA for GPS receiver-testing purposes. The main disadvantage of the model is that both the range and the range rate SA are noisier than that observed from actual satellite measurements.

2. Autoregressive Model

This model only generates range errors. Statistically, this is a very powerful model that is closely matched to measured SA data.

3. Analytic Model

This model generates both range and range rate errors. It also closely matches measured SA data. The range rate data represent the measured data better than those generated by the second-order Gauss–Markov model. The main disadvantage of this model is that there is not a direct relation between the choice of the model parameters and the output data. Noise levels must be set empirically.

Because the actual SA algorithms are not known, it is recommended, when required during GPS systems design and performance evaluations, to simulate the effect of SA using a variety of different models.

References

[1] Anon., "1992 Federal Radionavigation Plan," U.S. Department of Transportation and U.S. Department of Defense, Washington, DC, Rep. DOT-VNTSC-RSPA-92-2/DOD-4650.5, Jan. 1993.

[2] Georgiadou, Y., and Doucet, K. D., "The Issue of Selective Availability," *GPS World*, Sept.–Oct. 1990, pp. 53–56.

[3] United States Naval Observatory Electronic Bulletin Board, U.S. Coast Guard, Telephone (703) 866-3890, Aug. 29, 1990.

[4] Misra, P., Bayliss, E., LaFrey, R., and Pratt, M., "Integrated Use of GPS and GLONASS in Civil Aviation Navigation I: Coverage & Data Models," *Proceedings of the ION GPS-90*, Colorado Springs, CO, Institute of Navigation, Washington, DC, Sept. 1990, pp. 425–435.

[5] Kremer, G. T., Kalafus, R. M., Loomis, P. V. W., and Reynolds, J. C., "The Effect of Selective Availability on Differential GPS Corrections," *Navigation*, Vol. 37, No. 1, Spring 1990, pp. 39–52.

[6] Braasch, M. S., Fink, A., and Duffus, K., "Improved Modeling of GPS Selective Availability," *Proceedings of the ION National Technical Meeting*, San Francisco, CA, Institute of Navigation, Washington, DC, Jan. 20–22, 1993, pp. 121–130.

[7] Matchett, G., "Stochastic Simulation of GPS Selective Availability Errors," TM, FAA Contr. DTRS-57-83-C-00077, June 1985.

[8] Studenny, J., "Simulation of a Second-Order Gauss–Markov Process," RTCA Paper 148-93/SC159-424, March 17, 1993.

[9] Braasch, M. S., "A Signal Model for the NAVSTAR Global Positioning System," *Navigation*, Vol. 37, No. 4, Winter 1990–1991, pp. 363–377.

[10]Chou, H., "An Anti-SA Filter for Non-Differential GPS Users," *Proceedings of ION GPS-90*, Colorado Springs, CO, Institute of Navigation, Washington, DC, Sept. 19–21, 1990, pp. 535–542.

[11]Chou, H., "A Robust Filter for DGPS Users," *Proceedings of ION GPS-90*, Colorado Springs, CO, Institute of Navigation, Washington, DC, Sept. 19–21, 1990, pp. 607–612.

[12]Lear, W., Montez, M., Rater, L., and Zyla, L., "The Effect of Selective Availability on Orbit Space Vehicles Equipped with SPS GPS Receivers," *Proceedings of ION GPS-92*, Albuquerque, NM, Institute of Navigation, Washington, DC, Sept. 16–18, 1992, pp. 825–840.

[13]Åström, K., and Wittenmark, B., *Computer Controlled Systems—Theory and Design*, Prentice-Hall, Englewood Cliffs, NJ, 1984, pp. 146–151.

[14]Kelly, R. J., "MLS System Error Model Identification and Synthesis," *IEEE Transactions on Aerospace and Electronic Systems,* Vol. 28, No. 1, Jan. 1992, pp. 164–173.

[15]Marple, S. L., Jr., *Digital Spectral Analysis with Applications,* Prentice-Hall, Englewood Cliffs, NJ, 1988, pp. 172–260.

[16]Kay, S., *Modern Spectral Estimation: Theory and Application,* Prentice-Hall, Englewood Cliffs, NJ, 1987, pp. 106–270.

[17]Ljung, L., *System Identification—Theory for the User,* Prentice-Hall, Englewood Cliffs, NJ, 1987, pp. 169–207.

[18]Chou, H., "An Adaptive Correction Technique For Differential Global Positioning System," Ph.D. Dissertation, W. W. Hansen Experimental Physics Laboratory, Gravity Probe B Program, Stanford Univ., (SUDAAR 613), Stanford, CA, June 1991, pp. 59–82.

[19]Alexander, S. T., *Adaptive Signal Processing—Theory and Applications,* Springer-Verlag, New York, 1986, pp. 34–152.

Chapter 18

Introduction to Relativistic Effects on the Global Positioning System

N. Ashby*
University of Colorado, Boulder, Colorado 80309
and
J. J. Spilker Jr.†
Stanford Telecom, Sunnyvale, California 94089

I. Introduction

A. Objectives

THE GPS is one of the first operational systems, outside of particle accelerators, that has important effects from relativity. The reasons for this are threefold. The GPS satellites have a large velocity, there is a non-negligible gravitational potential difference between that of the satellites and that of the users (usually at or near the Earth's surface), and there are significant Earth rotation effects. These effects of themselves are not necessarily important, clearly, there are other operational satellite systems. However, when coupled with the fact that GPS satellites carry precision atomic frequency standards, that pseudorange measurements can be made to accuracies in the nanosecond range, and that carrier-phase measurements are made to the centimeter level, relativistic effects can, indeed, be significant and must be taken into account (see also Chapter 3, this volume, and Ref. 1).

For a fixed user at sea level on the Earth's surface, there are three primary consequences of relativity effects (see Table 1):

1) There is a fixed frequency offset in the satellite's clock rate when observed from Earth. Most of the effect is purposely removed by slightly offsetting the satellite clocks in frequency prior to launch, the so-called "factory offset" of the clock.

2) The slight eccentricity of each satellite orbit causes an additional periodic clock error effect that varies with the satellite's position in its orbit plane.

Copyright © 1995 by the authors. Published by the American Institute of Aeronautics and Astronautics, Inc., with permission. Released to AIAA to publish in all forms.
*Professor of Physics.
†Ph.D., Chairman of the Board.

Table 1   List of key relativistic effects on GPS

- **Fixed user on the geoid**
  Constant frequency offset (factory offset)
  The net effect of relativity for a zero eccentricity GPS satellite is a combination of effects caused by the satellite's velocity and the Earth's gravitational field, including its quadrupole field. This effect produces a small fixed frequency offset of received signal frequency relative to the satellite transmitted frequency in addition to the classical Doppler shift. This effect is compensated by a prelaunch factory offset in the satellite clock.
  Sinusoidal delay perturbation caused by the eccentricity of the satellite orbit.
  Sagnac delay caused by the Earth's rotation during the time of transit of the satellite signal to the ground user.
- **Moving user or user above the geoid**
  In addition to the above effects, there are effects caused by the user velocity and height of the user above the geoid. Some of these effects can cancel or partially cancel in position estimation. These effects can be significant if the user is another satellite in orbit.
- **Secondary effects**
  There are a host of secondary effects discussed briefly in this chapter that are smaller than the accuracy level required by most users. These include:
  Tidal potential effects on clocks
  The user is on or near the rotating Earth, which in turn, is revolving about the sun with its gravitational field. However, both the satellite and user are in orbit about the sun at approximately the same position so much of this effect cancels.
  Nonspherical gravity potential
  The Earth's gravity potential is slightly nonspherical primarily because of the ellipsoidal shape of the Earth, which causes a currently nonmodeled quadrupole field effect on GPS satellite clocks.
  Shapiro delay
  The Shapiro delay is caused by variations in the apparent velocity of light because of the Earth's gravitational field. Approximate values of the Shapiro delay are calculated and shown to be small.
  Lense-Thirring effect–frame dragging
  This effect is caused by the rotation of the Earth's mass on its axis. This effect, a so-called frame-dragging effect, slightly modifies the solutions to the field equations and generates a slightly different metric with a term related to the Earth's angular momentum, but its effects are negligible for our purposes here.

3) There is also an effect (Sagnac delay) caused by the Earth's rotation during the time of transit of the satellite signal from satellite to ground.

Moving users on or near the Earth's surface or fixed users at an altitude above or below the geoid* may have to make additional corrections caused by their velocity and height above the ground. Satellite users in low-Earth orbit (LEO), geostationary-Earth-orbit (GEO), or other orbit altitudes have additional corrections that may be necessary depending upon the accuracy required. Clearly, the

---

*The geoid is an equipotential surface that differs slightly from the ellipsoidal model of the Earth by geoid undulations. The gravity potential W is the sum of both a gravitational potential Φ and a centrifugal potential $V = -\frac{1}{2}\omega^2(x^2 + y^2)$ and is also affected by nonuniformity in the Earth's density, and topography (see Sec. 3.1).

INTRODUCTION TO RELATIVISTIC EFFECTS 625

Table 2   Objectives of the chapter

- Introduce special and general relativistic effects. For the GPS, most of the significant general relativity effects can be shown from the principle of equivalence, and flat or nearly flat space models are adequate.
- Define the appropriate invariant equations, the metric tensors, and show their application to translational linear motion at constant velocity, rotating coordinates, spherically symmetric and cylindrically symmetric gravitational fields with and without coordinate system rotation.
- Compute the relativistic effects on GPS pseudorange and Doppler frequency measurements on both fixed and moving users, including users in orbit. Discuss effects on the calculation of user position, velocity, and time transfer estimates. Compute both primary and secondary effects.

impact of these relativistic effects can differ depending upon the objective and precision desired; e.g., for GPS and differential GPS (DGPS) position estimates, some common error relativistic effects cancel; whereas, for time transfer, they may not.

This chapter's objective is to describe for the reader the detailed effects of relativity for GPS users, and is summarized in Table 2. For an extensive treatment of relativity, the reader is referred to the texts in the References at the end of the chapter. In this chapter, the effects of the Earth's atmosphere are ignored. These effects have been discussed in Chapters 12 and 13, this volume, and are assumed to have been taken into account already.

B. Statement of the GPS Problem

As already discussed in detail in previous chapters, the GPS receiver makes two types of measurements on the received signal, pseudorange and carrier-phase/Doppler frequency measurements. In this chapter, we focus only on relativistic effects and assume that perfect atomic frequency standards are employed both for the satellite clock and the user clock. We ignore all other error sources covered in other chapters. Ideal atomic clocks moving with the user and satellite are sometimes referred to as *standard clocks*. For our purposes, it is assumed that the satellite transmits coded electromagnetic pulses at time instants as determined by this perfect satellite clock* and transmits these pulses at a precisely determined frequency f_o (see Fig. 1.). The satellite-generated coded pulse—really a sequence of pulses—carry embedded in the code the precise satellite clock time at the precise time of transmission of this coded pulse. The GPS receiver is also assumed to have a perfect atomic clock and measures the time of reception of the satellite-coded pulse waveform and decodes the satellite's clock time word. The receiver then simply subtracts the satellite clock time number (at the time of transmission) from the user clock time number at the time of reception. Corrections must be applied to account for the fact that the user undoubtedly

---

*The GPS clocks prior to launch are set to run at GPS time—an atomic standard time—with a factory offset for relativity. While in orbit, the satellite clocks are periodically corrected by the GPS Control Segment (CS) to match GPS time with both "paper" and, less frequently, physical clock corrections (see Chapters 4 and 10, this volume).

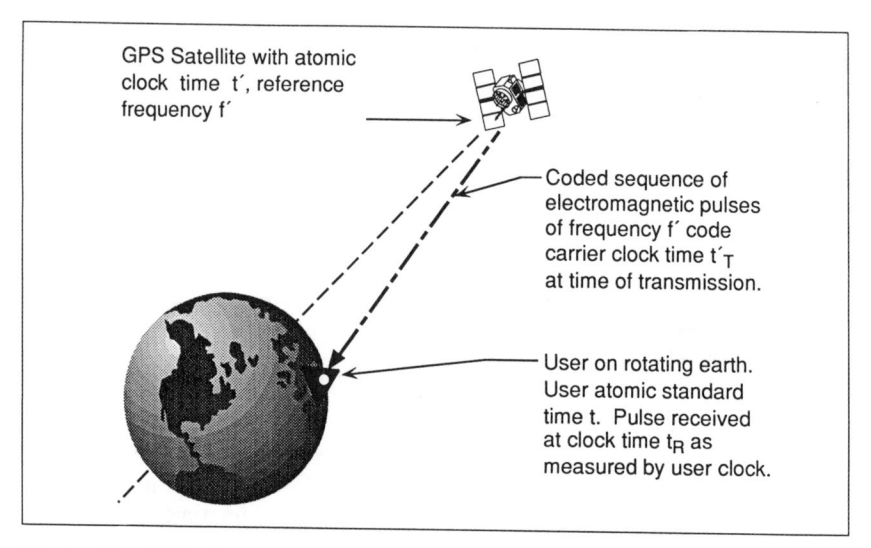

Fig. 1 The GPS relativistic problem. The GPS satellite is assumed to carry a noise-free atomic frequency standard and produces a clock time t' carried on the satellite. The satellite transmits a coded sequence of electromagnetic pulses of frequency f'. The coded pulse sequence represents a number t'_T, which is the satellite standard clock time at the beginning of the coded pulse *transmission*. The GPS user receiver for this example also carries a perfect atomic frequency standard and produces time t. The receiver measures the time of *reception* t_R and then subtracts $t_R - t'_T$ to give pseudorange. The receiver moves during propagation of the signal. The receiver also measures the frequency of the received electromagnetic pulses and compares it with the frequency of its own coordinate clock to measure frequency shift.

has moved during the time of transit, because either the user is on the rotating Earth, in an aircraft slightly above the Earth's surface, or perhaps in another satellite in low-Earth orbit. Finally, the user also compares the satellite frequency and phase changes with its own clock to compute range rate and accumulated delta range (ADR), a carrier-phase measurement. In both instances, pseudorange and carrier frequency are affected by various relativistic effects. The problem is to account for motional Doppler shifts, gravitational frequency shifts, and propagation delays so the user can accurately determine his or her position and time.

The importance of relativistic effects for GPS, and especially for GPS time transfer, is heightened by the fact that today's atomic time standards carried in the GPS satellites and used on the ground are so precise. Today's clocks exhibit long-term Allan variance stability levels (see Chapter 4, this volume) on the order of 10^{-14}, and even better stability is possible for clocks on the ground.

Simultaneity is a crucial concept in the GPS. For users to determine position and time, atomic clocks in the satellites should be synchronized. For users to rendezvous successfully after traversing different paths with different speeds through different gravitational fields, their clocks must be synchronized, and their positions must be accurately determined. When coordinate reference frames are inertial, as in special relativity, the usual Einstein synchronization procedures may be employed.* However, typical users are moving over the rotating Earth and are in the Earth's gravitational field, giving rise to noninertial effects. Proper time elapsed on standard clocks depends upon the clocks' history, and the Einstein procedure is not sufficiently accurate. Instead, we use the concept of a *global coordinate time* wherein there is a hypothetical network of clocks synchronized in an underlying Earth–centered-inertial frame (ECI frame) and running at a rate consistent with atomic time standards on the geoid of the rotating Earth.[2] The definition of simultaneity employed is that of *coordinate simultaneity*,[3] namely, two events with space–time coordinates $\{t_1, x_1, y_1, z_1\}$ and $\{t_2, x_2, y_2, z_2\}$ are simultaneous in the given reference frame if $t_1 = t_2$.

In this chapter, standard time refers to *International Atomic Time* (TAI).†[4] Strictly speaking, GPS time is a coordinate time with rate determined by an independent composite set of clocks (see Chapter 10, this volume) which is indirectly coupled to TAI and Universal Coordinated Time (UTC) (with its leap seconds) through UTC–GPS corrections (see Chapter 4, this volume). Synchronization is achieved by methods discussed in this chapter.

II. Introduction to the Elementary Principles of Relativity

The mathematics of general relativity are extremely complex. However, for this treatment of the relativistic effects on GPS, only a small fraction of the theory is required, and even in this fraction, many simplifications can be made, because the gravitational fields (of the Earth) are relatively weak, and the velocities of the satellites and users are small compared to the velocity of light c. As a result, the curved spaces of general relativity are nearly flat.

Although much of the discussion of the relativistic effects of GPS involves either rotation of the Earth or satellites in orbit, and gravitational fields with some nonuniformity, we stress the admonition of Misner et al.[5] that "physics is simple when analyzed locally. Don't try to describe motion relative to far away objects." Physics is always and everywhere locally Lorentzian (has a local inertial frame). Thus, we begin the discussion by reviewing Euclidean geometry and inertial frames and then introduce the concepts of special relativity, the principle of equivalence and metric tensors. Metric tensors and their associated scalar intervals are the key tools that allow us to define the geometry of the space and interpret the consequences of the theory.

A. Euclidean Geometry and Newtonian Physics

In this section, we begin by considering families of inertial coordinate systems each moving at constant velocity (no acceleration) with respect to the others. To be called *inertial*, a coordinate system must satisfy the properties[6] of Table 3.

---

*Simultaneity can be defined self-consistently for a homogeneous, isotropic inertial frame; if light pulses emitted from two points, A and B, in the frame arrive at the midpoint of A and B at the same time, then the transmit events are simultaneous.

†TAI is a coordinate time scale, defined in a geocentric reference frame with the SI second as scale unit as realized on the rotating geoid. It can be extended to a fixed or moving point in the vicinity of the Earth with sufficient accuracy at the present state of the art by the application of first-order corrections of the General Theory of Relativity; i.e., corrections arising from differences in the gravitational potential and differences of speed, in addition to the rotation of the Earth.

Table 3 Required properties of an inertial coordinate system and related definitions

Event
An event marks an action that takes place at a point in space–time. Events are denoted by ϵ_1, ϵ_2, etc. for this chapter.

Observation by an inertial observer
An observer assigns to an event ϵ, the coordinates x, y, z of the location of the occurrence and the time t read by a clock in that coordinate frame located *exactly* at the position of that event for that frame.

World-line
A world-line is the path taken by a particle or light ray in four-dimensional space–time. The world-line is a line for which x is given as a function of t.

Inertial coordinate frame
The coordinate system must satisfy the three properties:
1) The distance between point P_1 at (x_1, y_1, z_1) and point P_2 at (x_2, y_2, z_2) is a constant independent of time.
2) The clocks that sit at every grid point in the coordinate frame are all synchronized, all run at the same rate, and provide a *coordinate time* for that coordinate frame, the same as proper time measured by synchronized standard clocks. *Proper time* is time measured on a single arbitrarily moving standard clock.
3) The geometry of space at any fixed instant of time is Euclidean, a flat space. Acceleration and gravitational fields violate property 2, except for a very localized region of space.

Much of our analysis focuses on these very localized regions of inertial space that apply to the GPS. Consider two coordinate systems S and S', where S' is moving at constant velocity v in the x direction relative to S, as shown in Fig. 2.

In standard Euclidean geometry, the *Galilean transformation* between coordinates for these Cartesian coordinate systems is as follows:

$$x' = x - vt$$
$$y' = y$$
$$z' = z$$
$$t' = t \tag{1}$$

The very thought that t and t' could be different is not even considered in

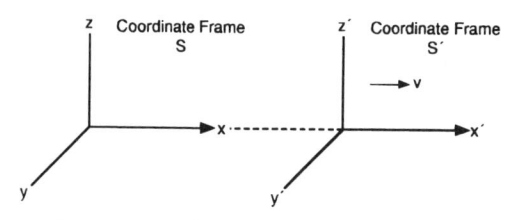

Fig. 2 Two (nonaccelerating) inertial coordinate frames S and S'. Coordinate frame S' is moving in such a way that $x' = x - vt$ where the origins of both frames are coincident at $t = t' = 0$.

Newtonian physics, and there is no constraint on velocity relative to the speed of light. In these Euclidean coordinate systems, the invariant differential interval between two events is the differential distance, given by the following

$$\begin{aligned} d\sigma^2 &= dx^2 + dy^2 + dz^2 & \text{Cartesian coordinates} \\ &= dr^2 + r^2 d\phi^2 + dz^2 & \text{cylindrical coordinates} \\ &= dr^2 + r^2(d\theta^2 + \sin^2\phi\, d\phi^2) & \text{spherical coordinates} \end{aligned} \quad (2)$$

The *spatial metric* in Cartesian coordinates can be written as

$$d\sigma^2 = dx^2 + dy^2 + dz^2 = \rho_{ik} dx^i dx^k = \delta_{ik} dx^i dx^k$$

where ρ_{ik} written in matrix form is the following diagonal matrix:

$$\|\rho_{ik}\| = \begin{bmatrix} 1 & 0 & 0 \\ 0 & 1 & 0 \\ 0 & 0 & 1 \end{bmatrix}$$

See Table 4 for an explanation of the Einstein summation convention. Thus, a differential distance between any two points in Euclidean space is the same regardless of the chosen coordinate system; it is a scalar. This space is a perfectly flat space. The deficiency in this Galilean system is that it makes no allowance for the fundamental fact that the speed of light is a constant c in all inertial frames, nor that time measures could be different in different coordinate frames; i.e., t and t' can be different.

In special and general relativity where there are coordinate systems with different velocities, rotational effects, and gravitational effects, the three-dimensional spatial metric does not suffice. Instead, we deal with a four-dimensional space–time coordinate frame, and the *scalar interval* is of the form, $ds^2 = g_{\alpha\beta} dx^\alpha dx^\beta$ where the individual $g_{\alpha\beta}$ terms can be functions of position and time; i.e., $g_{\alpha\beta}(x^\alpha)$. The set of elements $g_{\alpha\beta}$ together form a *covariant tensor* $g_{\alpha\beta}$ of rank 2. The tensor is symmetric; i.e., $g_{\alpha\beta} = g_{\beta\alpha}$. This *metric tensor* defines the geometry (differential geometry) of the space.* In general, the space–time coordinates are not flat. However, with the GPS and the users in relatively close proximity to the Earth, $r < 40{,}000$ km, the spatial coordinates are very nearly flat.

The fundamental basis for special relativity and its *Lorentz transformation* are contained in Einstein's postulates relating to the constancy of the speed of light in all inertial frames shown in Table 5.

B. Space–Time Coordinates and the Lorentz Transformation

Let us begin by considering four-dimensional space–time coordinates. This coordinate system may be envisioned as a standard Cartesian coordinate system filled with precision standard clocks at every point in the coordinate grid (see

---

*Any symmetric covariant tensor field of rank 2 defines a metric. A manifold with an applicable metric is termed a Riemanian manifold.[7] The concept of a differentiable manifold is abstract; however, it can be described roughly as an n-dimensional space that can be covered by one or more open neighborhoods, each of which has a coordinate system. Pairs of these coordinate systems are related to each other by differentiable coordinate transformations. A smooth surface in three-dimensional Euclidean space is a simple example of a differentiable manifold.[8]

Table 4. General expressions for distance in three-dimensional space and four-dimensional space–time

We introduce a set of three-dimensional spatial coordinates $\{x, y, z\} = \{x^1, x^2, x^3\}$ in frame S and a new set of coordinates $\{x'^1, x'^2, x'^3\}$ in frame S' where $x = x^1(x'^1, x'^2, x'^3)$, $y = x^2(x'^1, x'^2, x'^3)$, $z = x^3(x'^1, x'^2, x'^3)$. A coordinate differential $dx = dx^1$ is related to coordinate differentials dx'^i by the following:

$$dx = dx^1 = \frac{\partial x}{\partial x'^1} dx'^1 + \frac{\partial x}{\partial x'^2} dx'^2 + \frac{\partial x}{\partial x'^3} dx'^3 = \sum_i \frac{\partial x}{\partial x'^i} dx'^i = \frac{\partial x^1}{\partial x'^i} dx'^i$$

Einstein's summation convention is used, where two repeated indices—one subscript and one superscript—automatically imply a summation over the range of the indices. Latin indices range from 1 to 3, while Greek indices range from 0 to 3. Thus, $a_i b^i$ is shorthand for $a_1 b^1 + a_2 b^2 + a_3 b^3$.

Similarly, the differentials for dy and dz may be written in the following compact form:

$$dx^j = \sum_i \frac{\partial x^j}{\partial x'^i} dx'^i = \frac{\partial x^j}{\partial x'^i} dx'^i$$

Quantities that transform under a change of coordinates according to such a rule: $V^j = \partial x^j / \partial x'^i \, V'^i$ are termed *contravariant vectors*. (In three-dimensional Euclidean space, the distinction between contravariant and covariant vectors is not as significant as it is in general space-time.)

The invariant distance measure is as follows:

$$ds^2 = dx^2 + dy^2 + dz^2 = \delta_{ij} dx^i dx^j = \delta_{ij} \frac{\partial x^i}{\partial x'^m} \frac{\partial x^j}{\partial x'^n} dx'^m dx'^n$$

$$= g'_{mn}(x'^1, x'^2, x'^3) dx'^m dx'^n$$

where the *metric tensor* in the new coordinate system $\{x'^1, x'^2, x'^3\}$ is the following:

$$g'_{mn} = \delta_{ij} \frac{\partial x^i}{\partial x'^m} \frac{\partial x^j}{\partial x'^n} = \left[\frac{\partial x}{\partial x'^m} \frac{\partial x}{\partial x'^n} + \frac{\partial y}{\partial x'^m} \frac{\partial y}{\partial x'^n} + \frac{\partial z}{\partial x'^m} \frac{\partial z}{\partial x'^n} \right]$$

The three-dimensional distance measure, thus, simplifies to $ds^2 = g'_{ij} dx'^i dx'^j$. The three-dimensional metric tensor g'_{ij} in an arbitrary coordinate system has nine components but only six are independent because $g'_{ij} = g'_{ji}$.

The metric tensor generalizes to four-dimensional spacetime using the coordinate x^0 for ct where t is the time for that coordinate system, and the spatial coordinates are $\{x^1, x^2, x^3\}$. Then a general *line element* in four dimensions is written

$$ds^2 = g_{\alpha\beta} dx^\alpha dx^\beta$$

The particular dependence of $g_{\alpha\beta}$ on space–time is determined by the physical situation (see Table 10).

In special and general relativity, the increment of *proper time* $d\tau$ elapsed on a standard clock that moves along the space–time path dx^α is just $d\tau = |ds|/c$. Thus, if a clock does not move, $dx^\alpha = 0$ for $\alpha = 1, 2, 3$, and then $d\tau^2 = ds^2/c^2 = g_{00}(dx^0)^2/c^2$, and the proper time elapsed on the clock is $\sqrt{|g_{00}|} dt$.

INTRODUCTION TO RELATIVISTIC EFFECTS

Table 5 Einstein's two postulates[9]

Postulate 1: The Principle of Relativity
"No inertial system is preferred. The equations expressing the laws of physics have the same form in all inertial systems."

Postulate 2: The Principle of the Constancy of the Speed of Light
"The speed of light is a universal constant independent of the state of motion of the source. Any light ray moves in the inertial system of coordinates with constant velocity, c, whether the ray is emitted by a stationary or by a moving source."[10]

Of course, the speed of light c refers to the speed of light in a vacuum. Obviously, the speed of light decreases in a dielectic medium such as air or glass fiber.

Ref. 11). Second, define *events*; for example, a light pulse transmission as, say, ϵ_1, not only by the place of occurrence x, y, z, but also by the time t in that coordinate system at which the event took place. Thus, an event ϵ_1 is labeled in an inertial (nonaccelerating, gravity-free) coordinate frame S as $\epsilon_1(t_1, x_1, y_1, z_1)$ and in coordinate frame S' by $\epsilon_1(t'_1, x'_1, y'_1, z'_1)$ where the clock time is labeled by the time on the clock that is *exactly* at the same location in each coordinate system as the event. The two coordinate frames S and S' are generally in relative motion to one another. Each frame must have its own set of measuring rods and synchronized clocks at rest in that frame. An "observer" is a shorthand way of speaking of the whole ensemble of clocks and measuring rods associated with one coordinate frame, and an observation is the operation of labeling each event with the position coordinates and clock time at that event. It is critical that the clock time label be that associated with the exact position of that event, not a clock time at some distant point. An observation is *not* the act of looking at some clock distant from the event, and attempts to do so will often lead to error and sometimes leads to so-called paradoxes.

1. Time Dilation

Consider next two inertial coordinate frames, as shown in Fig. 3, where the origin of the S' frame moves with respect to reference frame S along the x-axis at velocity v, i.e., $dx/dt = v$. At time $t = 0 = t'$, the origins coincide. Consider an event ϵ_A wherein a light flash is emitted at the origin of both coordinate frames; i.e., ϵ_A ($t = 0$, $x = 0$, $y = 0$, $z = 0$), ϵ_A ($t' = 0$, $x' = 0$, $y' = 0$, $z' = 0$). Event B is the reception of the flash after reflection from a mirror or other reflector at position $y = \Delta L = y'$. (Symmetry considerations imply that the two observers agree on lengths oriented perpendicular to the direction of relative motion.)

Note that the speed of light c is exactly the same in both frames, from Postulate 2 in Table 5. Assume that the reflecting mirror moves with the coordinate frame S' (or equivalently, that the light beam is angled in the S frame so that it returns to the origin of S'). The light pulse in frame S' travels exactly $2\Delta L$ meters, and hence event ϵ_B in S' has coordinates ($t' = 2\Delta L/c$, $x' = 0$, $y' = 0$, $z' = 0$). Thus, the coordinate changes between the two events in S' are $\Delta t' = 2\ \Delta L/c$, $\Delta x' = 0$, $\Delta y' = 0$, $\Delta z' = 0$.

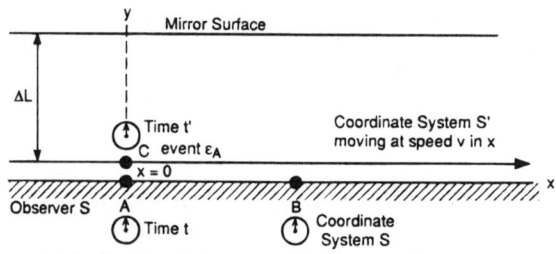

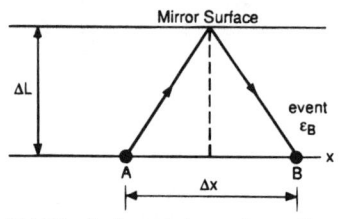

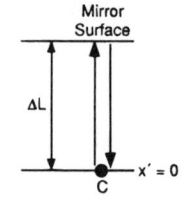

Fig. 3 Example of light pulse transmission in two coordinate frames S and S'. The light pulse is transmitted, event ϵ_A, at position A in frame S and at position C in frame S', which is exactly coincident with A at the time of event ϵ_A. The pulse is received, event ϵ_B, at the same location C in frame S' and at a displaced point B in frame S where $\Delta x = v\Delta t$.

In coordinate frame S, on the other hand, the light pulse arrives, event ϵ_B, at the following time:

$$t = \Delta t = 2\sqrt{(\Delta L)^2 + (\Delta x/2)^2}/c \tag{3}$$

where $\Delta x = v\Delta t$, and $x = \Delta x$, $y = 0 = z$. Thus, in S, event ϵ_B has coordinates $t = \Delta t, x = v\Delta t, y = 0, z = 0$. The coordinate changes between the two events are, therefore, $(\Delta t, \Delta x = v\Delta t, \Delta y = 0, \Delta z = 0)$. Because $\Delta x = v\Delta t$, we can solve Eq. (3) for Δt to obtain the following:

$$\Delta t = \frac{2\Delta L/c}{\sqrt{1 - (v/c)^2}} \tag{4}$$

The ratio of the two time intervals is as follows:

$$\frac{\Delta t}{\Delta t'} = \frac{1}{\sqrt{1 - (v/c)^2}} = \frac{1}{\sqrt{1 - \beta^2}} \stackrel{\Delta}{=} \gamma \geq 1 \tag{5}$$

where γ is the *time dilation* factor, and $\beta \stackrel{\Delta}{=} v/c$ is the normalized speed, with $v = |v|$. Thus, the laboratory reference frame clock accumulates a larger clock count between the same two events than the clock in the moving coordinate system S', and hence, the moving clock seems to be running at a lower frequency clock rate than the laboratory clock. *Moving clocks run slow when compared with a sequence of synchronized clocks at rest.* Of course, if the experiment were reversed so that the light pulse returned to the origin of S, then $\Delta x = 0$, the clock at A in S is moving at velocity $-v$ with respect to S', and it would seem that the S frame clock is running slow.

For the two events ϵ_A, ϵ_B discussed here, the clock in S' moves uniformly from ϵ_A to ϵ_B, and in S', the events occur at the same position in space. Whenever it is possible to introduce a uniformly moving clock whose world line intersects two events, an inertial frame can be constructed with this clock as reference, in which the spatial separation between the events is zero. The space–time separation between the events is then *timelike*, and the proper time elapsed on the clock is a measure of this separation. We can then show that, in general, the space–time separation between events $\Delta s^2 = -(c\Delta t)^2 + \Delta x^2 + \Delta y^2 + \Delta z^2$ is a scalar quantity having the same value in all inertial frames. Indeed, in the present case in S' $-(c\Delta t')^2 + (\Delta x')^2 + (\Delta y')^2 + (\Delta z')^2 = -(c\Delta t')^2 = -4(\Delta L)^2$, while from Eq. (4) in S, $-(c\Delta t)^2 + (\Delta x)^2 + (\Delta y)^2 + (\Delta z)^2 = (c\Delta t)^2(1 - v^2c^2) = [-4(\Delta L)^2/(1 - \beta^2)](1 - \beta^2) = -4(\Delta L)^2$. Thus, because v no longer appears, Δs^2 is an invariant quantity independent of the coordinate frame.

In differential form* we write the following:

$$ds^2 = -c^2 dt^2 + dx^2 + dy^2 + dz^2 \tag{6}$$

This invariant is termed the space–time interval for the *Lorentz space* (inertial space). Notice that if the clock is fixed in the reference frame; i.e., $dx = dy = dz = 0$, then $dt = d\tau$, and dt is equal to proper time. Thus, a fixed (standard) clock in the coordinate system carries proper time (see Table 1). *Proper time* is the time carried by a clock that is transported along a world line in space–time through the events. Figure 4 shows a world-line description of the light pulse transmission. A *world line* is the path taken by a particle or light ray in space–time.

The expression for the invariant ds^2 contains one form of the metric tensor and can be written for this inertial space as the following quadratic form†:

$$ds^2 = \sum_\alpha^3 \sum_\beta^3 g_{\alpha\beta} \, dx^\alpha \, dx^\beta = g_{\alpha\beta} \, dx^\alpha \, dx^\beta \tag{7}$$

For inertial space (Minkowski space), the metric tensor in Cartesian coordinates can be written in matrix form as follows:

---

*Note that many texts use simplifying normalizing notations so that c does not appear, and some use $ds^2 = -dt^2 + (dx^2 + dy^2 + dz^2)/c^2$ and express the line element in time dimensions. In this chapter, ds has the dimensions of meters.

†The metric tensor $g_{\alpha\beta}$ is a tensor of rank 2 that carries out an operation $g(dx, dx) = g_{\alpha\beta} dx^\alpha dx^\beta$ on any two vectors inserted and computes a scalar that has certain invariant properties under coordinate transformation. For more detail refer to Refs. 5, 8, 12, 13. We are using Einstein notation and automatically sum over any index that appears as both a subscript and superscript.

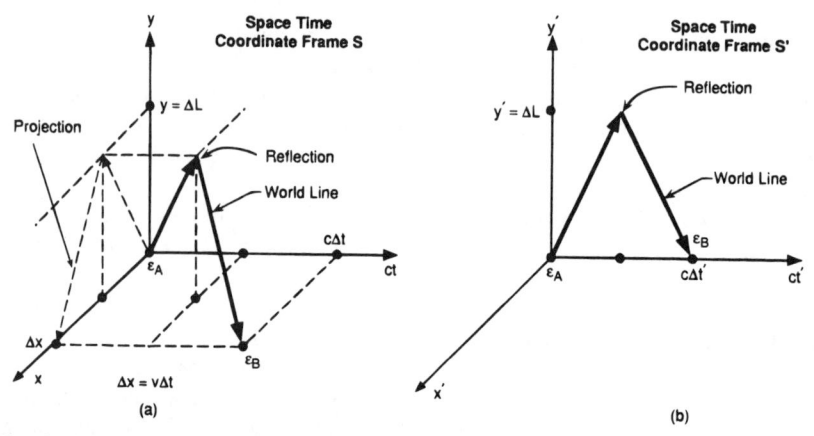

Fig. 4 Space–time coordinate frame world lines in coordinate frames S and S' showing events ϵ_A and ϵ_B. Coordinate frame S' is moving in the x direction with respect to coordinate frame S at speed v. A light pulse is transmitted, event ϵ_A, when the origins at time $t = 0 = t'$ at the origin of S and S'. The pulse is reflected at $y = \Delta L$ and returned to $y = 0$; $y' = 0$, event ϵ_B.

$$\|g_{\alpha\beta}\| = \|\eta_{\alpha\beta}\| \stackrel{\Delta}{=} \begin{bmatrix} -1 & 0 & 0 & 0 \\ 0 & 1 & 0 & 0 \\ 0 & 0 & 1 & 0 \\ 0 & 0 & 0 & 1 \end{bmatrix}$$

where $\eta_{\alpha\beta}$ is defined as the metric tensor for *Minkowski space* and is one form of the generalized Kronecker delta. This scalar invariant, or line element, can also be written in cylindrical and spherical coordinates as follows:

$$ds^2 = -c^2 dt^2 + dr^2 + r^2 d\phi^2 + dz^2$$
$$ds^2 = -c^2 dt^2 + dr^2 + r^2(d\theta^2 + \sin^2 d\phi^2) \tag{8}$$

The line element for any path in space–time is invariant under coordinate transformations between inertial frames. Along the path of a photon, the line element or invariant distance measure, $ds^2 = 0$. Notice that because for photon travel $ds^2 = 0$, the line element yields the result $c^2 dt^2 = dx^2 + dy^2 + dz^2$, which is an expression of the constancy of the speed of light in all inertial frames.*

2. Geodesics

A geodesic is the path in space–time that a body takes when it is free from nongravitational forces. In the flat space of special relativity, inertial space, the body moves with uniform speed along a straight line. In more general spaces,

---

*Note that in three dimensions x, y, t, this equation $c^2 dt^2 = dx^2 + dy^2$ can be visualized as a cone in three dimensions, sometimes referred to as a light cone. Physical objects must travel inside the light cone, and light rays travel on the surface of the cone.

the geodesics* are extremal paths. For physical particles, a geodesic ordinarily corresponds to the shortest path. For light rays, the geodesics have zero space–time length and are termed null geodesics. In general, in a gravity field, the geodesics are curved, at least slightly, in space.

3. Transformations Between Inertial Frames

Consider an inertial system S with $ds^2 = -c^2 dt^2 + dx^2 + dy^2 + dz^2$. This space–time interval is not form-invariant under Galilean coordinate transformations, because if we apply Eqs. (1) to transform ds^2, we obtain

$$ds^2 = -(cdt')^2 (1 - \beta^2) + 2\beta dx' cdt' + dx'^2 + dy'^2 + dz'^2,$$

which is not of the same form as Eq. (6). However, because it is a quadratic form in the primed coordinate increments, by performing additional linear transformations of x' and t' it is possible to reduce this expression to the appropriate form $-(cdt')^2 + dx'^2 + dy'^2 + dz'^2$.

Let us, instead, begin by assuming that x and ct are linear combinations of x' and ct'. For the event ϵ_B of Fig. 3, $x' = 0$, while $ct = \gamma ct'$, and $x = vt = \gamma\beta ct'$. This set of relationships is enough to determine two of the four needed coefficients. Then, for an arbitrary event we must have the following:

$$ct = \gamma ct' + Ax' \qquad (9)$$

$$x = Bx' + \gamma\beta ct' \qquad (10)$$

where A and B are constants that remain to be computed. Consider events at the origin of S, at $x = 0$. These must occur at $x' = -vt'$, so from Eq. (10) $0 = B(-vt') + \gamma\beta ct'$, which gives $B = \gamma$. Invariance of the speed of light implies that if $x' = ct'$, then $x = ct$. That is, the position of a light pulse transmitted from the origins at the moment they coincide must move with speed c in both coordinate systems. Then, substituting $x' = ct'$ in Eqs. (9) and (10) we must have $x = \gamma ct' + \gamma\beta ct' = ct = \gamma ct' + Act'$.

This equation can only be satisfied if $A = \gamma\beta$. The resulting transformations, the so-called Lorentz transformations, can be written as follows:

$$ct = \gamma(ct' + \beta x')$$

$$x = \gamma(x' + \beta ct') \qquad (11)$$

There are many other ways of deriving the Lorentz transformations. Another method is to assume the transformations between coordinates of S' and S are linear, then impose the requirement that ds^2 must have the same form in the two frames. This yields Eqs. (11) after some algebra.

Compare Eq. (11) with the Galilean transformation where we would have $x = x' + vt'$ and $t = t'$. First-order corrections enter in the time transformation; second-order corrections enter through the factor γ.

---

*A geodesic is a curve for which the integral $\int ds$ is an extremum; the curve has zero geodesic curvature.[11]

4. Velocity Addition

Consider a double transformation of velocities where there are three inertial coordinate systems: S_a, S_b, S_c, as shown in Fig. 5. The coordinate frame S_b moves at velocity v_b with respect to S_a, and S_c moves at velocity v_c with respect to S_b. The velocities would, of course, simply add in a Galilean transform with S_c moving with respect to S_a with velocity $v_b + v_c$. Let the three sets of coordinates be $\{ct_a, x_a, y_a, z_a\}$, $\{ct_b, x_b, y_b, z_b\}$, and $\{ct_c, x_c, y_c, z_c\}$.

Using the Lorentz transformation (11), we can write the following:

$$x_a = \gamma_b(x_b + v_b t_b) = \gamma_b[\gamma_c x_c + \gamma_c v_c t_c] + \gamma_b v_b[\gamma_c t_c + v_c \gamma_c x_c]$$

$$= \gamma_b \gamma_c [x_c + v_c t_c + v_b t_c + v_c v_b x_c] = \gamma_b \gamma_c [x_c(1 + v_b v_c) + (v_b + v_c) t_c]$$

$$= \gamma_b \gamma_c (1 + v_b v_c)\left[x_c + \left(\frac{v_b + v_c}{1 + v_b v_c}\right) t_c\right] \tag{12}$$

where $\gamma_b = 1/\sqrt{1 - \beta_b^2}$, and $\beta_b \overset{\Delta}{=} v_b/c$, etc. Thus, the effective velocity of S_c with respect to S_a in the x direction is as follows:

$$v_{\text{eff}} = \frac{v_b + v_c}{1 + v_b v_c} \quad \text{and} \quad \gamma_{\text{eff}} \overset{\Delta}{=} 1/\sqrt{1 - (v_{\text{eff}}/c)^2} = \gamma_b \gamma_c (1 + v_b v_c) \tag{13}$$

Thus, $v_{\text{eff}} \leq c$. This result is generalized in Table 6.

It is important to point out that this velocity addition applies equally when a coordinate system S_b, moving with respect to S_a, has a moving transmitter on it. That is, the transmitter is moving with respect to S_b at velocity v_c and is observed by S_a. We cannot simply use the Galilean velocity addition formulas except as an approximation for small v/c. Tables 7 and 8 summarize the four-velocity and four-momentum characteristics in inertial space. Table 9 summarizes the definition and symbols used herein.

5. Light Pulse Time Delay

Now we consider how different observers, in relative motion, view the propagation of a light pulse from one point to another. This can be approached quite generally from the Lorentz transformations, Eq. (11). Referring to Fig. 6a, the event ϵ_1 is the transmission of a light pulse, viewed in the S' frame, from the point $(x' = -L' \cos\theta', y' = L' \sin\theta', z' = 0)$ at the time $ct' = 0$. The light pulse propagates with speed c to the origin of S', at $x' = y' = z' = 0$, where it

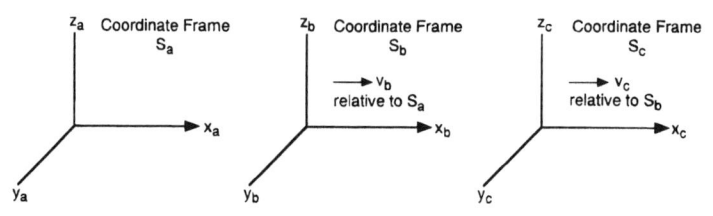

Fig. 5 Three coordinate frames S_a, S_b, S_c where S_b moves at velocity v_b with respect to S_a, and S_c moves at velocity v_c relative to S_b, both in the x direction.

INTRODUCTION TO RELATIVISTIC EFFECTS

Table 6 General Lorentz velocity addition transformation

Consider a particle in an inertial frame S' that moves at velocity $u'_x = \dfrac{dx'}{dt'}$, $u'_y = \dfrac{dy'}{dt'}$, $u'_z = \dfrac{dz'}{dt'}$, i.e., $\mathbf{u}' = (u'_x, u'_y, u'_z)$ with respect to S. Frame S' again moves at velocity v in the x direction with respect to S. The velocity of that particle in the S frame is then obtained from the following differentials:

$$dx = \gamma(dx' + v\,dt') = \gamma(u'_x dt' + v\,dt') = \gamma dt'(u'_x + v)$$

$$dy = dy' = u'_y dt'$$

$$dz = dz' = u'_z dt'$$

$$dt = \gamma(dt' + v\,dx'/c^2) = \gamma(dt' + vu'_x dt'/c^2) = \gamma dt'(1 + u'_x v/c^2)$$

Thus, the velocity as measured in the S frame is $\mathbf{u} = (u_x, u_y, u_z)$ is as follows:

$$u_x = \frac{dx}{dt} = \frac{u'_x + v}{(1 + u'_x v)}$$

$$u_y = \frac{dy}{dt} = u'_y/\gamma(1 + u'_x v/c^2)$$

$$u_z = \frac{dz}{dt} = u'_z/\gamma(1 + u'_x v/c^2)$$

This expression generalizes the velocity addition formula.

---

arrives at the time $t' = L'/c$. The coordinate intervals along the path in frame S' are $c\Delta t' = L'$, $\Delta x' = L'\cos\theta'$, $\Delta y' = -L'\sin\theta'$, and $\Delta z' = 0$.

Figure 6b shows how the path of the light looks in frame S. Because the Lorentz transformations are linear, coordinate intervals along the light path can be calculated by taking differences in Eq. (11):

$$c\Delta t = \gamma(c\Delta t' + (v/c)\Delta x') \tag{14}$$

$$\Delta x = \gamma(\Delta x' + (v/c)c\Delta t') \tag{15}$$

and $\Delta y = \Delta y'$. The last term $\gamma v \Delta x'/c$ in Eq. (14) is an extra delay caused by the distance traveled by S' during the time of transit.

Consider, for example, the special case $\theta' = 0$. The rod of length L' in S' is laid out parallel to the direction of relative motion and, therefore, to S appears to be shorter, by the Lorentz contraction, so $L = L'\sqrt{1 - \beta^2}$. Light propagates toward the front of the rod, which moves away with speed v. The relative speed with which the light catches up to the front of the rod is $c-v$, so the propagation time in S is as follows:

$$\Delta t = \frac{L'\sqrt{1-\beta^2}}{c-v} = \frac{\gamma L'(1-\beta^2)}{c(1-\beta)} = \gamma\Delta t'\left(1 + \frac{v}{c}\right) \tag{16}$$

On the other hand, this result follows directly from the transformation (14), for substituting $\Delta x' = c\Delta t'$ into Eq. (14):

Table 7 Four-velocity and its space–time invariant

A *four-velocity vector* in a local inertial frame is defined by its components:

$$u = \{u^0, u^1, u^2, u^3\} = \left\{\frac{dx^0}{d\tau}, \frac{dx^1}{d\tau}, \frac{dx^2}{d\tau}, \frac{dx^3}{d\tau}\right\} = \left\{\frac{cdt}{d\tau}, \frac{dx}{d\tau}, \frac{dy}{d\tau}, \frac{dz}{d\tau}\right\}$$

where $d\tau$ is the proper time on a hypothetical clock carried along with the particle. $d\tau$ is a scalar invariant, $d\tau = |ds|/c$ along the particle's path. If the particle velocity has magnitude v_p, then

$$d\tau = \sqrt{1 - v_p^2/c^2}\, dt = dt/\gamma(v_P)$$

where

$$v_P^2 = (v_P^1)^2 + (v_P^2)^2 + (v_P^3)^2 \quad \text{and} \quad \gamma(v_P) = (1 - v_P^2/c^2)^{-1/2}$$

In another frame S', the four-velocity vector will have components $u'^\alpha = dx'^\alpha/d\tau$, and under a Lorentz transformation such as Eq. (11) the four-velocity components transform as a contravariant vector (see Table 4):

$$u^\beta = \frac{\partial x^\beta}{\partial x'^\alpha} u'^\alpha \quad \text{or} \quad u^0 = \gamma(v)(u'^0 + vu'^1/c),\ u^1 = \gamma[v](u'^1 + vu'^0/c),\ u^2 = u'^2,\ u^3 = u'^3$$

where $\gamma(v)$ is the time dilation factor of the Lorentz transformation.

Operating on two four-velocity vectors with the metric tensor yields a scalar; e.g.,

$$g(u, v) = g_{\alpha\beta} u^\alpha u^\beta = -u^0 v^0 + u^1 v^1 + u^2 v^2 + u^3 v^3$$

where for any inertial frame with Cartesian coordinates the metric tensor $g_{\alpha\beta} = \eta_{\alpha\beta}$, where $\eta_{\alpha\beta}$ has been defined earlier. The first term with a minus sign arising from η_{00} corresponds to the time variable term in the invariant interval, (6), while the other three terms correspond to the spatial variables. The quantity $g(u, u)$ is invariant under any Lorentz transformation:

$$g(u, u) = -(u^0)^2 + (u^1)^2 + (u^2)^2 + (u^3)^2 = -\gamma[v_P]^2 c^2 + \gamma[v_P]^2 v_P^2 = -c^2$$

because $\gamma(v_P) = (1 - v_P^2/c^2)^{-1/2}$.

$$c\Delta t = \gamma(c\Delta t' + (v/c)c\Delta t') = \gamma c\Delta t'(1 + v/c) \tag{17}$$

In the more general case where the length L' is oriented at a nonzero angle to the x,x'-axes, then $\Delta x' = L' \cos\theta'$, and the Lorentz transformation gives the following:

$$c\Delta t = \gamma c\Delta t' + \gamma(v/c)L' \cos\theta' \tag{18}$$

and the extra time delay is $\gamma v L' \cos\theta'/c$. That is, the correction is proportional to the projection of the vector x' on the x-axis. This correction term can, therefore, be written $\gamma v \cdot x'/c$. The correction survives even if second-order effects such as Lorentz contraction are neglected. As shown later, an analogous effect occurs with movement of the Earth's surface where the delay corresponds to a projection of the light pulse position on the equatorial plane.

6. Drag Effect–Light Velocity in a Moving Dielectric

Einstein's velocity addition formula also applies to the computation of the velocity of light traveling in a moving dielectric. Consider a dielectric with a

Table 8 Four-momentum for particles and photons

The *four-momentum* of a particle with rest mass m and velocity $v = \{v^1, v^2, v^3\}$ and $v = |v|$ is equal to

$$p = \{p^0, p^1, p^2, p^3\} = \{E/c, p^1, p^2, p^3\} = mu$$

where u is the four-velocity and where $p^0 = E/c = m\gamma c = mc/\sqrt{1 - v^2/c^2} \approx mc + 1/2\, mc(v/c)^2$ for small v, and where $p^1 = m\gamma v^1$, $p^2 = m\gamma v^2$, $p^3 = m\gamma v^3$.

For the four-momentum vector p, the quantity $g(p, p)$ is invariant under Lorentz transformation:

$$g(p, p) = -E^2/c^2 + (p^1)^2 + (p^2)^2 + (p^3)^2 = g(mu, mu) = m^2 g(u, u) = -m^2 c^2$$

and is proportional to the squared mass of the particle.

A photon has zero rest mass. Along the path of a photon $ds^2 = 0$; there is no frame S for which a photon is at rest. A photon carries energy and momentum described by its four-momentum vector $p = \{E/c, p^1, p^2, p^3\}$. For a photon $g(p, p) = 0$ and $E^2/c^2 = (p^1)^2 + (p^2)^2 + (p^3)^2$. Energy E and frequency f are related by $E = hf$, where $h = 6.6256 \times 10^{-34}$ Js is Planck's constant. More generally, if the wave number four-vector for photons is $k = \{k^0, k^1, k^2, k^3\} = \{2\pi f/c, k^1, k^2, k^3\}$ then the four-momentum and wave number four-vector are related by $p = hk/2\pi$. Then the Lorentz transformation describes the Doppler shift formula for photons. For example, if in frame S', for which the Lorentz transformations to frame S are given by Eq. (11), a photon has frequency f' and propagates parallel to the x'-axis, then the x'-component of the wave vector is $k'^1 = 2\pi f'/c$, and the frequency f of the photon in S is given by the Lorentz transformation:

$$2\pi f/c = \gamma(2\pi f'/c + \beta 2\pi f'/c) \quad \text{or} \quad f = \gamma(1 + \beta)f' = \sqrt{\frac{1 + \beta}{1 - \beta}} f'$$

Table 9 Key definitions for GPS relativistic effects

x^α	four-vector space–time coordinates $\{x^0, x^1, x^2, x^3\}$, where x^0 is time expressed in meters, $x^0 = ct$
e_0, e_1, e_2, e_3	basis vectors
x	$x^0 e_0 + x^1 e_1 + x^2 e_2 + x^3 e_3$ or $\{x^0, x^1, x^2, x^3\}$
u	four-vector velocity (space–time) $\{u^0, u^1, u^2, u^3\}$, $u = dx/d\tau$ m/s
$a = du/d\tau$	four-acceleration in m/s^2
$c\Delta t$	observed time interval, in meters, measured by clocks at rest in the observer's coordinate frame
β	normalized speed v/c
γ	Lorentz time dilation $\gamma = (1 - \beta^2)^{-1/2} = dt/d\tau$ for inertial frames
τ	proper time is the time measured by a perfect clock moving arbitrarily (measured in seconds)
$g = g_{\alpha\beta}$	metric tensor
ds^2	infinitesimal metric line element $ds^2 = \Sigma\Sigma g_{\alpha\beta} dx^\alpha dx^\beta = g_{\alpha\beta} dx^\alpha dx^\beta$
E	energy
p	four-momentum of a photon of energy $E = -p \cdot u$
S	inertial coordinate frame
$\mathfrak{S}$	general accelerating, gravitational coordinate frame

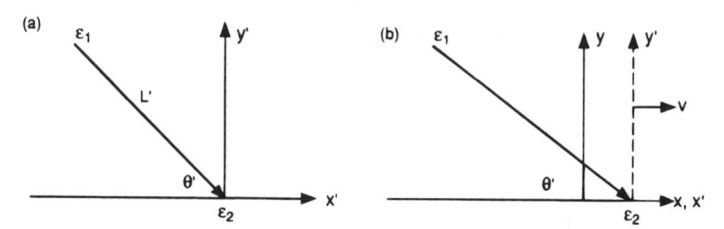

Fig. 6 Transmission of a light pulse from a point originally a distance L' away from the common origin of coordinate frames S and S'. The light pulse is transmitted to the origin of S'.

refractive index n so that the speed of light in this dielectric liquid, fiber, or atmosphere is $u = c/n < c$. Suppose, further, that the dielectric is flowing uniformly past an observer in frame S at velocity v. A light pulse transmitted in the direction of the velocity flow has observed velocity that can be calculated from the velocity addition formula (13). The velocity of the light pulse observed in S is as follows:

$$v_{\text{eff}} = \frac{u+v}{1+uv/c^2} = \frac{(c/n)+v}{1+v/cn} \cong (c/n+v)(1-v/cn) = c/n + v - v/n^2 - v^2/cn$$
$$= c/n + v(1 - 1/n^2) \quad \text{for} \quad v/cn \ll 1 \qquad (19)$$

Thus, in effect, the velocity of the dielectric drags the light with it (or retards it if the dielectric velocity is in the opposite direction from the light). The quantity $1-(1/n^2)$ is Fresnel's "drag coefficient."

7. Flat Space of Special Relativity

The metric tensor of special relativity describes a flat space. That is, two free-particle world lines that begin parallel to each other remain parallel, no matter how far they are extended, exactly the property of Euclidean geometry. The space–time of special relativity (Minkowski's space) is not Euclidean space, because its metric is different; e.g., photons travel on world lines of zero "length." The metric tensor is characterized by having its spatial distance $d\sigma$ given by $d\sigma^2 = dx^2 + dy^2 + dz^2$, or its equivalent in cylindrical or spherical coordinates. Thus, special relativity has a flat, non-Euclidean geometry. As shown later, when gravitational fields are introduced, the spatial coordinates are warped, and the space is no longer flat. For the GPS problem, however, the gravitational field effects are weak, and the space is nearly flat. Corrections to the invariant interval of special relativity can be expanded in powers of the small parameter l/c. Only terms to order $1/c^2$ need be retained.

C. Relativistic Effects of Rotation in the Absence of a Gravitational Field

In this subsection, we examine the relativistic effects on clocks and light pulse transmission on a rotating disk in the absence of gravitation. A clock on the rim of a rotating disk, of course, both has a velocity and undergoes an acceleration.

INTRODUCTION TO RELATIVISTIC EFFECTS

We introduce the concept of a *momentarily co-moving reference frame*[6] (MCRF)—also known as instantaneous rest frame—which allows us to treat this problem with special relativity. These results have been verified experimentally many times. In the next subsection, where the principle of equivalence and gravitation effects are introduced, it is shown that the centrifugal force/acceleration induces an equivalent cylindrically symmetric gravitational field, and identical results are obtained when treating the effects as caused by gravitation.

1. Momentarily Co-Moving Reference Frame

The most useful applications of relativity deal with accelerated (or gravitational) reference frames $\mathfrak{S}$,* and an accelerated particle has no inertial frame in which it is always at rest. For example, rotating frames are accelerated noninertial frames. However, there are inertial frames S, for which some point or region of the accelerated frame is momentarily at rest. In the example of a rotating frame, this momentarily co-moving inertial reference frame is very localized in space. Thus, a *sequence of MCRFs* can allow us to use the Lorentz transformation, or a sequence thereof. This set of localized MCRFs permits the use of special relativity and the associated metric tensor of special relativity for an accelerating frame that can be momentarily treated as a constant velocity frame. These concepts have proved accurate even for highly accelerated frames, such as centrifuges with acceleration of $6 \times 10^4 g$.[14]

2. Relativistic Effects on a Rotating Disc or Centrifuge

There are relativistic effects on clocks on a rotating disc because of the centripetal acceleration and velocity. This discussion provides background for understanding clocks on satellites and the rotating Earth. A clock on the rim of a rotating disc has both a centripetal acceleration $\omega^2 R$ toward the axis, and a velocity ωR, where ω is the angular rate of the disc, and R is the radial distance of the clock from the axis.

3. Clock Synchronization on a Rotating Disk and Angular Rate as Measured by an Inertial Observer

Consider a rotating disk of radius R, which represents a noninertial space $\mathfrak{S}$. Let standard atomic clocks on the rim of the rotating disk at radius R all be synchronized to $t' = 0$ when a momentarily adjacent fixed clock (in the ring of inertial clocks) also reads zero $t = 0$. Alternatively, we can synchronize the clocks at radius R by transmitting a light pulse from the axis of the disk. Clocks on the rotating disk at radius R are then set equal upon reception of the pulse from the axis of the disk. Adaptation of this procedure for synchronization of clocks on the rotating disc means that their synchronization is really determined by Einstein synchronization in the underlying inertial frame. Upon comparing two such adjacent clocks on the rim of the rotating disc by Einstein synchronization performed entirely on the disc, the clocks would be found to be unsynchronized.

---

*In this chapter general noninertial space is referred to by the symbol $\mathfrak{S}$. The inertial frames, including the MCRF, are referred to by symbols S or S'.

Assume that the disk is marked in angular increments of equal size in degrees before rotation began. Obviously, the number of angular increments marked on the disk surface, e.g., degrees, does not change with rotation. There are still the same increments and a total of 360 deg or 2π radians after rotation. (However, as shown later, the relationship between radius and circumference is no longer the same.) Let ϵ_1 denote the event that corresponds to a fixed point and given clock on the rotating disk passing a fixed adjacent point S in inertial nonrotating laboratory space S. On the next revolution, let ϵ_2 denote the event where the same clock fixed to the surface of the rotating disk coincides with the inertial clock at point S. Suppose the time interval between events in nonrotating inertial space is Δt, and the angular rate viewed by the observer in inertial space is $\omega = 2\pi/\Delta t$. Clocks on the rotating disc beat more slowly than clocks in inertial space: $\Delta t' = \sqrt{1 - \beta^2}\,\Delta t$, so the angular rate measured by a clock in the rotating frame is different from ω, and $\omega' = 2\pi/\Delta t' = 2\pi\gamma/\Delta t = \gamma\omega$.

Suppose that a standard clock fixed to the rim of the rotating disc transmits a signal to a receiver at A on the axis of the disc (see Fig. 7a). Let the natural frequency of the signal in the MCRF of the transmitting clock be f_o. The signal travels perpendicular to the direction of relative motion, so there is no longitudinal Doppler shift; only the *redshift* in frequency (time dilation) is significant. In other words, a Lorentz transformation from the MCRF of the transmitting clock to local inertial space shows the received frequency is as follows:

$$f = f_o\sqrt{1 - \beta^2} = f_o/\gamma(\omega R) \qquad (20)$$

The fractional frequency shift is, therefore, as follows:

$$\frac{f - f_o}{f_o} = 1/\gamma(\omega R) - 1 \cong -\tfrac{1}{2}\frac{(\omega R)^2}{c^2} \qquad (21)$$

Turner and Hill[15] employed a radiation source (Co^{57}) near the rim of a

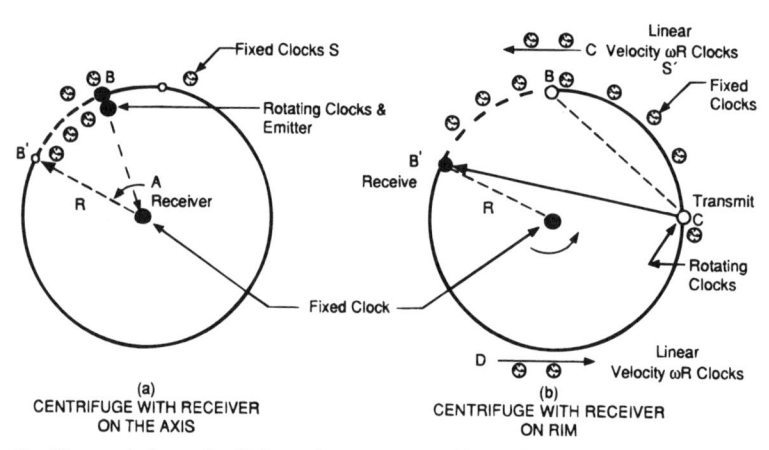

Fig. 7 Transmission of a light pulse on a centrifuge. The centrifuge has radius R and rotates at angular rate ω. In part a, the receiver is on the axis of rotation, and the transmitter is on the rim. In part b, both the transmitter and receiver are on the rim at different locations. For reference, a set of clocks moving at a linear velocity ωR is also shown in part b.

centrifuge and a receiver at the center axis. The frequency shift observed by the receiver on the axis was the same relativistic shift expected for a source with a transverse velocity $v = \omega R$; namely, $\Delta f/f_o = -(1/2)(\omega R/c)^2$, in agreement with Eq. (21).

If both the radiating source and the receiver are on the rim of the centrifuge, as shown in Fig. 7b, the problem is different. Clearly, for the previous example, a fixed inertial observer at the axis (non-rotating) observes that all clocks on the rotating rim have the same velocity $v = \omega R$, and all are observed to have the identical time dilation factor $\gamma(\omega R)$. However, clocks on different parts of the rim have a different instantaneous velocity relative to one another. The question is: "Is there a Doppler shift for a signal transmitted from point C on the rim to point B?"

Place inertial clocks (fixed with respect to the axis of rotation) at point C and point B' (B' is the position of clock B at the time of reception). Both observe the adjacent rotating clocks C and B to have constant time dilation factor $1/\gamma$ as they move. Clearly, the distance between C and B' is constant; i.e., a second light pulse transmitted from C to B' at some time ΔT later arrives with the same delay, at a time $\Delta T + d/c$, where d is the distance the light travels from C to B' observed by S, and d/c is a constant for a constant rotational rate ω.

The rotating clocks (transmitter and receiver) are moving at the same speeds ωR relative to S; also, the direction of propagation of the transmitted photon relative to the transmitter velocity is at the same angle as the propagation direction of the received photon relative to the receiver velocity. Therefore, we might expect that there would be no net frequency shift of the photon when observed in the rotating frame. To show this more rigorously, let the four-momentum of the photon, of frequency f, be as follows:

$$p = \{p^0, p^1, p^2, p^3\} = \{E/c, En^1/c, En^2/c, En^3/c\} = (hf/c)(1, \boldsymbol{n}) \quad (22)$$

where the photon has energy $E = hf$ and momentum $p = hf/c$, and $\boldsymbol{n}$ is the direction of the photon. Let $\boldsymbol{u}$ be the four-velocity of any detector. Then $\boldsymbol{p} \cdot \boldsymbol{u} \equiv g(\boldsymbol{p}, \boldsymbol{u})$ is a scalar invariant that can be evaluated in any reference frame. In particular, in the rest frame of the detector, $\boldsymbol{u} = (c,0,0,0)$, and $\boldsymbol{p} \cdot \boldsymbol{u} = -hf = -E$. Thus, the energy of a photon, hence the frequency, in the rest frame of a transmitter or receiver can be calculated from the invariant $E = hf = -\boldsymbol{p} \cdot \boldsymbol{u}$. Thus, if $\boldsymbol{u}_T$ is the transmitter four-velocity, the energy of the transmitted photon as observed in the rotating frame is the following:

$$E_T = hf_T = -\boldsymbol{p} \cdot \boldsymbol{u}_T = \frac{hf}{c}\gamma(\omega R)(-c + \omega R \cos\theta) \quad (23)$$

where θ is the angle between photon direction and transmitter velocity. Similarly if $\boldsymbol{u}_R$ is the receiver four-velocity, the observed energy of the received photon is as follows:

$$E_R = hf_R = -\boldsymbol{p} \cdot \boldsymbol{u}_R = \frac{hf}{c}\gamma(\omega R)(-c + \omega R \cos\theta) \quad (24)$$

because both velocity vectors intersect the direction of the light ray at the same angle. The energies, and hence, the frequencies, are equal.

Thus, there is no frequency shift at all* in the transmission from C to B at the same radius, as has been proved by experiment.[5] As shown later, the rotating clocks C and B can instead be treated as if each undergoes a radial acceleration and, thus, has an equivalent gravitational potential Φ, which is equal for both clocks: $\Phi(C) = \Phi(B)$.

Distances observed in rotating frame

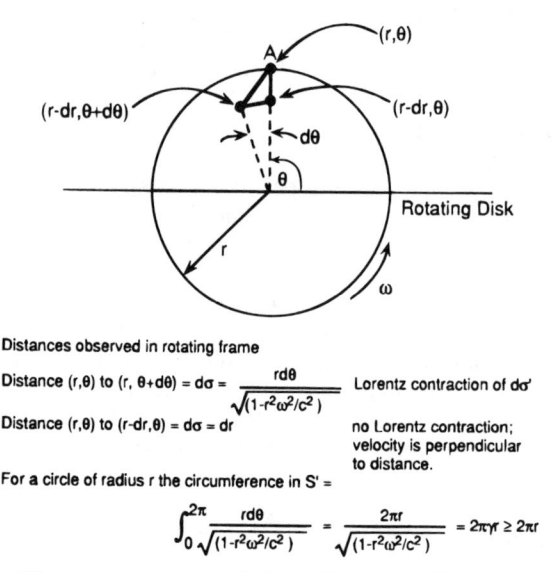

Distance (r,θ) to $(r, \theta+d\theta) = d\sigma = \dfrac{rd\theta}{\sqrt{(1-r^2\omega^2/c^2)}}$ Lorentz contraction of $d\sigma'$

Distance (r,θ) to $(r-dr,\theta) = d\sigma = dr$ no Lorentz contraction; velocity is perpendicular to distance.

For a circle of radius r the circumference in S' =

$$\int_0^{2\pi} \frac{rd\theta}{\sqrt{(1-r^2\omega^2/c^2)}} = \frac{2\pi r}{\sqrt{(1-r^2\omega^2/c^2)}} = 2\pi\gamma r \geq 2\pi r$$

Fig. 8 Distance measures and circumference on the rotating disk.

---

*We might have expected this velocity difference to give a large second-order Doppler effect instead of zero second-order Doppler. For example, examine a clock in an MCRF inertial frame that moves linearly at velocity v in the same direction as clock C relative to a reference inertial frame on the axis and momentarily coincides with the rotating clock C. Consider a second MCRF inertial clock on the opposite side of the rim, which moves at an instantaneous linear velocity $-v$ and is momentarily coincident with clock B'. The relative velocity of the two MCRF inertial frames is $v' = 2v/(1 + v^2/c^2) = 2\omega R/(1 + \omega^2 R^2/c^2)$, and then there is a relativistic Doppler shift of $\gamma(v')$ between these two frames. However, the two rotating clocks on opposite sides of the rotating disk itself, which are *individually* momentarily at rest with the two MCRF inertial frames have no Doppler shift with respect to one another at all. They have a fixed distance between one another; i.e.,

$$\frac{d}{dt}|r_T - r_R| = 0$$

This seeming paradox is related to the Ehrenfest Paradox[16-19] but is not really a paradox at all, it is simply a lapse in the application of the concept of locally inertial coordinate frames. The two clocks on opposite side of the rotating disk are not in any sense local to *one another*. An "observer" in special relativity cannot be "looking" on the opposite side of the disk. The condition where $\gamma(2v)$ does apply corresponds to a rotation where the axis of the disc rotation is shifted from the center of the disc to the receiver location on the rim, and the transmitter on the opposite side of the disc rotates in a circular arc of $2R$ in radius—a completely different motion from the rotation about the center of the disc.

4. Distance Measures on the Rotating Disk

Consider a point A in inertial space near the edge of the disc and a point (r, θ) fixed on the disk, which is momentarily adjacent to point A. Suppose the rotating observer lays down standard rods that are found to have the lengths dr and $rd\theta$ when observed by S, as shown in Fig. 8. Standard rods observed by S have a Lorentz contraction for lengths in line with motion, but no contraction for lengths along the radial line, perpendicular to the motion. Therefore, we have the differential distance measures and circumference, as shown in Fig. 8. The ratio of circumference to radius is no longer $2\pi R$, but $2\pi\gamma R$.

5. Metric Tensor for the Rotating Disk

The metric tensor for the flat space of special relativity can be expressed in nonrotating cylindrical coordinates $\{r, \phi, z\}$:

$$ds^2 = -c^2 dt^2 + dr^2 + r^2 d\phi^2 + dz^2 \qquad (25)$$

The metric tensor for the rotating coordinate system can be expressed using the coordinate transformation[19] $r' = r$, $\phi' = \phi - \omega t$, $t' = t$. Clocks on the rotating disk can no longer be synchronized self-consistently by Einstein's synchronization procedure but instead may be synchronized with a nonrotating inertial clock at the center of the disk where the axis is at rest, as described earlier. The transformation $t' = t$ implies clocks in the underlying inertial frame are used to measure time. The metric line element for the rotating disk then becomes* the following:[7,20]

$$ds^2 = -[1 - (\omega r/c)^2]c^2 dt^2 + dr^2 + r^2 d\phi^2 + 2\omega r^2 d\phi dt + dz^2 \qquad (26)$$

(primes have been dropped). Notice that there is an off-diagonal term in the metric tensor $g_{02} = g_{20} = \omega r^2$.

Consider two rings of standard clocks on the same rotating disk at radii r_1 and r_2. They will beat more slowly than clocks at rest in an adjacent inertial frame by their respective time dilation factors [see Eq. (20)]. Therefore, the ratio of clock frequencies, as observed in S, is as follows:

$$\frac{f_1}{f_2} = \frac{f_0/\gamma(\omega r_1)}{f_0/\gamma(\omega r_2)} = \frac{\gamma(\omega r_2)}{\gamma(\omega r_1)} \qquad (27)$$

and the clock at the larger radius will have a lower frequency.

If we postulate further that there is a vehicle on the rim of a disk at radius r_2, and that vehicle is moving along the circumference of the disk at a velocity v (i.e., ϕ' is not constant), notice that the effective velocity is not precisely equal to $v + \omega r_2$, but rather, we must use the following velocity addition formula:

$$v_{\text{eff}} = \frac{v + \omega r_2}{1 + v\omega r_2/c^2} \qquad (28)$$

Of course, for small velocities v, $\omega r \ll c$, the velocity addition is approximately linear.

---

*This equation is easily obtained from Eq. (6) by substituting $x = r\cos(\omega t + \phi')$, $y = r\sin(\omega t + \phi')$ and differentiating and squaring to obtain

$$dx^2 + dy^2 = dr^2 + r^2(d\phi'^2 + 2\omega \, dt d\phi' + \omega^2 dt^2)$$

6. Transmission Delay for a Receiver on a Rotating Disk

Consider a rotating disk of radius R and angular rate ω. A transmit event ϵ_1 occurs on the rim of the disk when a light pulse is transmitted at time $t' = 0$ on the rotating disk frame $\tilde{S}$ (see Fig. 9) and at time $t = 0$ in the nonrotating frame S. The transmission occurs at point A in frame S and at angle $\phi' = 0$ on the rim of the rotating disk, which is calibrated in uniform angular increments, $(0, 2\pi)$. The origin of the S frame is coincident with the axis of rotation of the disk, and it likewise is calibrated in angle in cylindrical coordinates. The light pulse is received, event ϵ_2, at point B in S at angle ϕ. The total distance traveled in S is $d(\phi) = R\sqrt{(1 - \cos\phi)^2 + \sin^2\phi}$, and the path delay measured in S is then $\Delta t = d(\phi)/c$.

The observer in S would measure the clock in the MCRF local inertial frame at point B' on the disk as running slow by a time dilation by $dt'/dt = 1/\gamma(\omega R)$. During the time of transit, the observer measures an angular increment $\Delta\phi = \omega \Delta t = \omega d/c$. Thus, if the user on the disk wishes to compute angular position, $\phi' = \phi - \Delta\phi$ on the rim of the disk at the instant of transmission, the computation must subtract $\Delta\phi$ from the estimate of ϕ because of the rotation during the time of transit.* An alternative is to subtract the time increment corresponding to the change in delay for the angular increment $\Delta\phi$. This delay increment measured in S is $\delta t = [d(\phi) - d(\phi - \Delta\phi)]/c$, where $d(\phi)$ is the separation distance at angle ϕ.

The difference appearing in the above equation can be computed approximately by differentiation:

$$\delta t = \frac{1}{c} \Delta\phi \frac{d}{d\phi} d(\phi) = \frac{\omega d(\phi)}{c^2} \frac{d}{d\phi} d(\phi) = \frac{\omega R^2}{c^2} \sin\phi \tag{29}$$

where $d(\phi) = R\sqrt{2 - 2\cos\phi}$ has been used. This time difference can be expressed simply in terms of the area of the triangle ABO where O is the axis

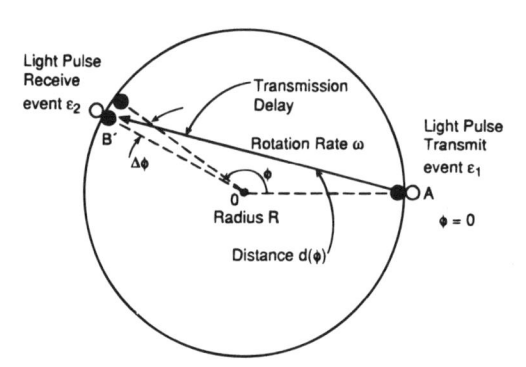

Fig. 9 Transmission delay to a receiver on a rotating disk. The transmit event is ϵ_1, and the receiver event is ϵ_2. The disk radius is R, and the rotation rate is ω. During the time of transit, the disk rotates $\Delta\phi$.

---

*It should be noted that a set of observers on the rotating disk would compute the relationship between radius and circumference as slightly different from $2\pi R$[19] (see Fig. 8).

of rotation. The area is $R^2 \sin\phi/2$ (half the base times the height), so we get the following:

$$\delta t = \frac{2\omega}{c^2} \times \frac{1}{2} R^2 \sin\phi \qquad (30)$$

This delay increment can be related to the delay increment measured by clocks in S' by the usual Lorentz transformation. This delay increment is often referred to as the *Sagnac* Correction.

7. Transmission of Light Around the Perimeter of a Rotating Disk—The Sagnac Correction

Consider a rotating disk of radius R and rotation rate ω as measured by a nonrotating inertial frame S, which is fixed with respect to the axis of rotation. A transmitter at point B in S (see Fig. 10) that is coincident with point A in the MCRF inertial frame S' at $t = 0$ transmits a light pulse (event ϵ_1) completely around the rim of the disk using an infinitely large set of equally spaced reflectors. The total distance around the rim of the disk as measured in the S frame is $2\pi R$, and the light pulse travels from B around the disk and returns to fixed point B in time $t = 2\pi R/c$. However, because of the finite velocity of light during the time of transmission around the disk, the disk has rotated as measured in S by an angle $\omega \Delta t$, where Δt is the total delay. Event ϵ_2 occurs when the light pulse reaches the light transmitter/detector at point A' in the rotating frame. Figure 11 shows the circumference of a cylindrical coordinate system $\{r, \phi, z\}$ unrolled* into a linear coordinate system with only one dimension x and x'.

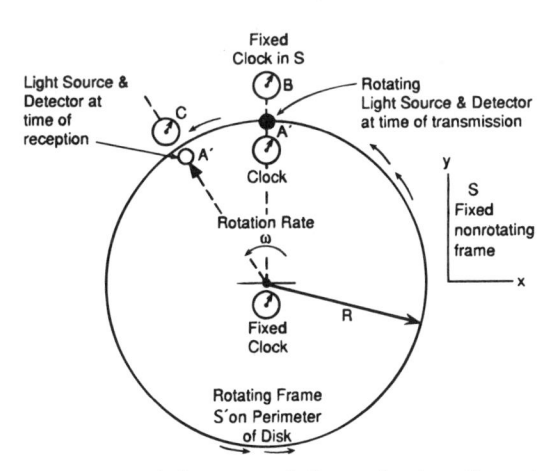

Fig. 10 Light pulse transmission around the perimeter of a rotating disk. The transmit/receive point A' rotates during the time of propagation. Point B is coincident with A' at the time of transmission, and C is coincident with A' at the time of reception.

---

*The cylindrical coordinate at radius R can be unrolled, because we have a flat space in the vicinity of $r = R$, where a local inertial frame exists. Each inertial frame has the same speed relative to S.

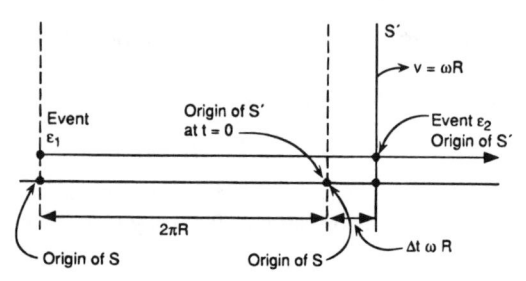

Fig. 11 Light pulse transmission around the rim of the rotating disk laid out on a linear grid by unrolling a cylindrical coordinate system. The origin of S' is moving along the x-axis at velocity $v = \omega R$ relative to frame S.

Event ϵ_1 occurs at $t = t' = 0$ and at $x' = -2\pi\gamma R$, because the disc's circumference in S' is $2\pi\gamma R$. Event ϵ_2 occurs at $t' = 2\pi\gamma R/c$ and $x' = 0$, so $\Delta t' = 2\pi\gamma R/c$, $\Delta x' = 2\pi\gamma R$. Then, from the Lorentz transformation, we get the following:

$$\Delta t = \gamma\left(\frac{2\pi\gamma R}{c} + \frac{\omega R}{c}\frac{2\pi\gamma R}{c}\right) = \frac{2\pi\gamma^2 R}{c}\left(1 + \frac{\omega R}{c}\right)$$

Thus, the extra time caused by the disk rotation is the following:

$$\frac{2\pi\gamma R}{c}\left(1 + \frac{\omega R}{c}\right) - \frac{2\pi R}{c} = \frac{2\omega}{c^2}\pi R^2 \frac{1}{1 - \omega R/c} \tag{31}$$

As an example, if the disk is rotating at the rotation rate of the Earth, and if the radius of the disk is the Earth's equatorial radius, then $\omega = \Omega_e = 7.292115467 \times 10^{-5}$ rad/s, $R = R_e = 6.378137 \times 10^6$ m, and $v = \omega R = 4.651011 \times 10^2$ m/s, $\pi R_e^2 = 1.278019 \times 10^{14}$ m², and $2\Omega_e/c^2 = 1.622714 \times 10^{-21}$ s/m². The value of the extra delay is then as follows:

$$\Delta t - \frac{2\pi R_e}{c} = \frac{207.386 \text{ ns}}{1 - \omega R/c} \cong 207.386 \text{ ns} \tag{32}$$

that is, $\omega R/c = 1.551 \times 10^{-6}$, and the effect of higher-order corrections is negligible.

Thus, suppose we attempted to synchronize a series of clocks by transmitting light pulses from point-to-point westward around the rotating Earth at the equator, each time subtracting the distance $\Delta L/c$ from the clock reading (the Einstein clock synchronization procedure). When the synchronization is completed at the original clock, the time on the last clock synchronized by light pulses leads the time on the original clock. If the light pulses were transmitted in the eastward direction, the last clock would lag the original clock. If transmitted in a westward direction opposite the direction of rotation of the Earth, they would lead the original master clock reference. The same effect (the Sagnac effect*) occurs

---

*This effect is known as the Sagnac effect[22] and must be corrected in clock synchronization around the Earth, as is discussed in more detail in Sec. III. This effect is also the basis for the ring laser and fiber-optic gyroscopes.

if we simply carry a clock around the Earth's equator, as has been verified by experiment.[21]

8. Doppler Effect and Relativity Correction

Consider a light source traveling uniformly through an inertial frame S with constant speed $u = |u|$ with an instantaneous radial velocity u_r relative to the receiver that is at the origin of S, as shown in Fig. 12. Assume successive light pulses are transmitted by a clock on C at proper intervals Δt_0, as measured in the source's rest frame. During this interval between transmit events, the transmitter has increased its distance from the receiver by $\Delta x = \Delta t_0 \gamma(u) u_r$ after accounting for time dilation. Thus, the time difference between pulses received at the origin 0 in S is $\Delta t = \gamma \Delta t_0 + \Delta x/c = \gamma \Delta t_0 + \gamma \Delta t_0 u_r/c = \gamma \Delta t_0 (1 + u_r/c)$, and the relativistic Doppler shift is as follows:

$$f/f_o = \frac{\Delta t_0}{\Delta t} = \frac{1}{\gamma(u)(1 + u_r/c)} = \frac{\sqrt{1 - (u/c)^2}}{1 + u\cos\theta} \quad (33)$$

An example of this effect has been demonstrated in a so-called "thermal" Doppler effect[13] wherein radioactive nuclei bound to a hot crystal are moving in a rapid and random manner. The random direction of their movement causes their first-order Doppler effects $u\cos\theta$ when averaged over all nuclei (and equiprobable angles θ) to average to zero. However, the second-order Doppler shift caused by the time dilation $\gamma(u)$ is related only to the mean square *speed*, not the direction. Hence, the average clock rate of the observed particles is slowed by $\gamma(u)$ is in agreement with measured results.

D. Principle of Equivalence

General Relativity (GR) incorporates gravitation by virtue of the Principle of Equivalence, which was first enunciated by Einstein. Einstein's equivalence principle actually consists of three separate principles, which we now discuss.

The *Weak Principle of Equivalence* states that the trajectory of a test body in a gravitational field is independent of the body's structure and composition. This is just a statement of the proportionality of passive gravitational mass and inertial mass. A "test" body is one small enough not to affect the sources of the gravita-

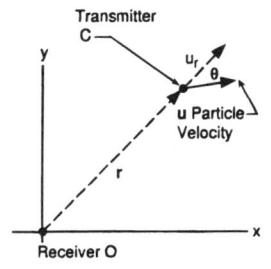

Fig. 12 Particle radiating an electromagnetic signal. The particle velocity in this S frame is u.

tional field. This principle was accurately verified by Eötvös in 1922, and by many others in recent years, to an accuracy of one (1) in 10^{12}. *Local Position Invariance* states that the results of nongravitational experiments must be independent of the space–time position of the reference frame in which they are performed. *Local Lorentz Invariance* states that the results of nongravitational experiments must be independent of the velocity and orientation of the reference frame in which they are performed.

Einstein's equivalence principle includes all of the above principles. There is, in addition, a *Strong Equivalence Principle,* which incorporates Local Position Invariance and Local Lorentz Invariance, and extends the Weak Equivalence Principle by stating that in a gravitational field, the trajectory of a body is even independent of the amount of self-gravitational energy of the body. At present (1995), there is no experimental evidence in conflict with these principles, and for purposes of application in the GPS, they may be adopted together. General Relativity is based on the Strong Equivalence Principle, which states: "*all local freely falling, nonrotating laboratories are fully equivalent for the performance of all physical experiments.*"

Each freely falling laboratory constitutes a *local inertial frame*; "local" means that over a small region of space and time in this frame, the physics can be described by that of special relativity. The gravitational field that causes the laboratory to accelerate (fall), is effectively transformed away by the acceleration. Conversely, because of the equivalence of inertial and passive gravitational mass (weak equivalence) a fictitious gravitational field can be induced in a laboratory by accelerating it and the apparatus within it. The restriction to "local" laboratories means that we must be careful, however, to divide each problem into a network of local problems, and then put these pieces back together.

It is also a tenet of relativity theory, that the lengths of ideal rods (and the rates of ideal clocks, as well) depend only upon their positions and velocities and not on their accelerations. In more mathematical terms, relativistic effects on rods and clocks depend upon first derivatives of position with respect to time and not on any higher derivatives.* In gravitational fields, there will be some dependence upon position in the potential; this effect is discussed later.

Thus, acceleration itself does not affect the rate at which a clock beats. The rate of an accelerated clock relative to an inertial observer is the same as the rate of a clock that is momentarily coincident with and at rest relative to, the accelerated clock (i.e., in the MCRF). This is called the *clock hypothesis.*[7] The clock hypothesis has been verified experimentally by measuring lifetimes of charged elementary particles accelerated in circular paths to speeds approaching c in high-energy accelerators. It has also been verified to high accuracy,[23] by subjecting atomic clocks to large accelerations on a rotating table. Similarly, the

---

*Møller[20] states: "centrifugal forces will then undoubtedly cause a lengthening of the measuring rod. This deformation will, however, depend on the elastic properties of the material from which the measuring rods are made, and all such deformations of the measuring-rod can therefore be corrected for easily.... In general we shall assume that the corrected standard measuring-rods in an accelerated system relative to the measuring rods in a (co-moving) system are subjected to Lorentz contractions only, which means that the lengths of the rods are independent of the accelerations.... In accordance with (this) assumption, it is implied that only the velocity, not the accelerations relative to S, will influence the rate of a standard clock."

rod hypothesis asserts that the length of a standard rod is independent of its state of acceleration, but depends only upon the rod's velocity relative to a local, freely falling (locally inertial) observer.

In the absence of matter, we can create a uniform gravitational field in a laboratory by placing the laboratory in a rocket and accelerating the rocket. The acceleration induces a gravitational field that, over small regions of space and time, is physically indistinguishable from, say, the nearly uniform gravitational field near the Earth's surface. The direction of the induced field is opposite to the acceleration. Conversely, in the presence of the Earth's gravitational field, acceleration in free fall induces an additional gravitational field that precisely cancels the "real" gravitational field.

From Local Lorentz Invariance, a number of extremely important facts about the invariant interval ds^2 can be taken over directly from Special Relativity:

1) Events occur in a four-dimensional space–time manifold, described by four coordinates x^μ. The coordinate x^0 is the *coordinate time variable*. The coordinate time can be defined at all points in space.

2) The quantity ds^2 is a scalar quantity, invariant with respect to arbitrary transformations of coordinates.

3) Along the rays traversed by electromagnetic signals, $ds^2 = 0$. This expresses the principle of constancy of the speed of light.

4) The proper time elapsed on a moving clock may be directly calculated from the following invariant:

$$\Delta\tau = \int_{\text{path}} ds/c = \int_{\text{path}} \frac{1}{c}\sqrt{-g_{\mu\nu}dx^\mu dx^\nu} \qquad (34)$$

5) A test particle in free fall, such as an Earth-orbiting satellite, follows a geodesic path, along which $\int ds$ is an extremum.

Property 1 is important in GPS, because the rate of an atomic clock depends upon its position in a gravitational field and on its velocity, thus, on the history of the clock. Therefore atomic clocks cannot be used directly to transfer time from one station to another; corrections must be applied. From Eq. (34). Obviously the proper time elapsed on an atomic clock will differ from the elapsed coordinate time, $\int dx^0/c$. Because the coordinate time x^0 is a quantity that exists globally, it can be used to establish a network of synchronized clocks. We must keep in mind, however, that coordinate time is only a convenient marker to label events and is *not* the time measured on an arbitrarily moving atomic clock.

Properties 2, 3, and 4 allow us to use Special Relativity, and the scalar ds^2, as tools in interpreting relativity theory. For example, 3 says that along the track of an electromagnetic wave, $ds^2 = 0$. This means that in a local, freely falling reference frame (a local inertial frame), the speed of light as measured with standard rods and standard clocks—as in special relativity—is c. Measurement of the speed of light with coordinate clocks may not yield c. The strong equivalence principle leads to several interesting predictions as freely falling laboratories in a gravitational field accelerate (see Fig. 13). If a light flash is transmitted horizontally across the room in the freely falling laboratory, the light path travels in a straight line when viewed inside the cabin in inertial frame S. However, when viewed from the outside in a nonaccelerating, but gravitational, frame $\mathfrak{S}$, the

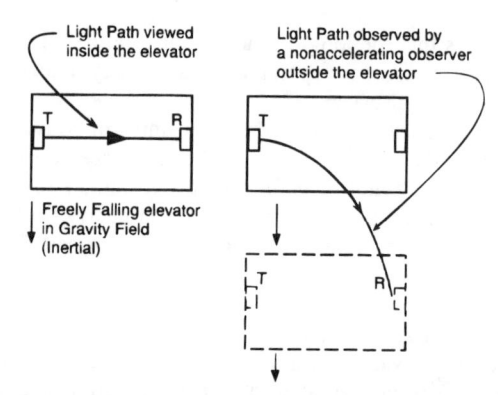

Fig. 13 Transmission of light from transmitter T to receiver R horizontally when viewed inside a freely falling (inertial elevator) in a gravitational field is shown on the left. Because the elevator is accelerating downward toward the gravity source, the light will seem to fall down when observed by an observer fixed in the gravity field as shown on the right.

light path is curved parabolically because of the acceleration. The light ray seems to have weight and be attracted by gravity.

Second, the *strong equivalence principle* implies that light traveling downward in a gravitational field is shifted to a higher frequency; i.e., it is blue shifted and gains energy. As a consequence, atomic clocks at a high elevation in a gravitational field run faster. Some practical examples of this phenomenon are discussed in the following sections.

1. General Relativity and the Gravitational Frequency Shift

Figure 14 shows two infinitesimal elevators, one in free space undergoing a uniform acceleration a, and the other in an equivalent uniform gravitational field of strength $g = -a$.

The principle of equivalence states that the physics inside these two infinitesimal elevators is in every sense the same. Consider the elevator of infinitesimal height dh, in which a photon leaves a transmitter at height dh and falls downward through the induced gravitational field. The light pulse in the freely falling elevator can be viewed as though it were in inertial space, and thus, takes a time $dt = dh/c$ to reach the floor. During this time, the elevator has increased its speed upward by $du = a\,dt = g\,dt = g\,dh/c$. The light pulse then meets the approaching receiver, which measures a longitudinal Doppler shift toward higher frequency; the received frequency will be $f_0 + df$ where the following obtains:

$$f_0 + df = f_0(1 + du/c) = f_0(1 + g\,dh/c^2) \qquad (35)$$

In vector notation, the fractional frequency shift is $df/f_0 = \mathbf{g} \cdot d\mathbf{h}/c^2$, where $d\mathbf{h}$ is the signal path. Define the gravitational potential $\Phi(h)$ with $\mathbf{g} = -\nabla\Phi$. Then we can write the fractional frequency shift as follows:

$$\frac{df}{f_0} = -\frac{d\Phi}{c^2} \qquad (36)$$

For weak fields, this can be integrated to give the following[14]:

INTRODUCTION TO RELATIVISTIC EFFECTS

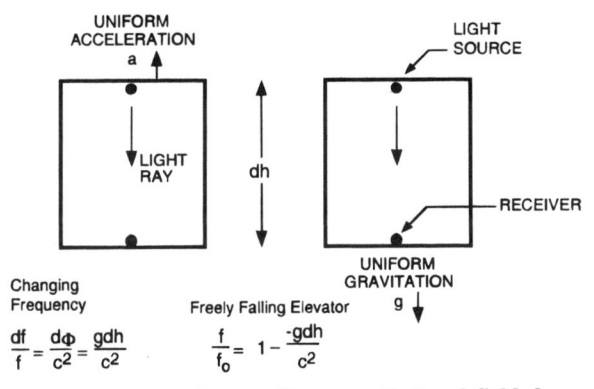

Fig. 14 Infinitesimal elevator in a uniform gravitational field demonstrating the Principle of Equivalence—red shift by static acceleration or gravitation in an elevator of infinitesimal height dh. The elevator is dropped from rest in a gravitational field of strength g.

$$\frac{f}{f_0} = \frac{f_0 + \Delta f}{f_0} \approx 1 - \frac{\Delta \Phi}{c^2} \quad \text{for weak fields where} \quad \left|\frac{\Delta \Phi}{c^2}\right| \ll 1 \quad (37)$$

This argument neglects the distance moved by the receiver during the small time interval dt and is only valid for weak fields. The argument relies on the clock hypothesis, that the Doppler shift is due to relative velocity of the receiver's clock and is independent of the clock's acceleration.

Thus, the gravitational field causes a gravitational red shift or blue shift of the light frequency, depending upon the sign of $\Delta \Phi$. Similarly, an atomic clock at a gravitational potential that differs by $\Delta \Phi$ from the potential of a reference clock, will run faster or slower than the reference clock by the fractional amount $-\Delta \Phi/c^2$. If the observers are near the Earth's surface, at an altitude h above the geoid, where the gravitational field strength is g, then relative to a clock on the geoid $\Delta \Phi/c^2 = gh/c^2$, and the clock frequency will be faster than that of a reference clock on the geoid by the fractional amount gh/c^2. Clocks at the National Institute of Standards and Technology in Boulder, Colorado, at an elevation of about 1650 meters, run faster by approximately 5 μs per year than similar clocks at the Greenwich Observatory, England at an altitude of approximately 24 m.

Pound and Rebka[24] and Pound and Snider[25] accurately demonstrated the gravitational red shift of photons using a 22.5-m helium-filled shaft at Harvard University, in agreement with Eq. (35).

Suppose a uniformly accelerated reference system is created by applying a constant force to each point in the system, in such a way that the induced gravitational field in the accelerated system is static. Suppose, further, that the spatial part of the line element ds^2 is Euclidean. Then it can be shown that the line element must take following the form[18]:

$$ds^2 = -(1 + gz'/c^2)^2(cdt')^2 + (dx')^2 + (dy')^2 + (dz')^2 \quad (38)$$

where t', x', y', z' represent coordinates in the accelerated reference frame.

For weak fields in the neighborhood of the origin, this equation becomes $ds^2 \simeq -(1 + 2gz'/c^2)(cdt')^2 + (dx')^2 + (dy')^2 + (dz')^2$, or, because the gravitational potential in the neighborhood of the origin is $\Phi = gz'$:

$$ds^2 \simeq -(1 + 2\Phi/c^2)(cdt')^2 + (dx')^2 + (dy')^2 + (dz')^2 \qquad (39)$$

Thus, in this case as well, the gravitational potential appears in the component g_{00} of the metric tensor.

2. Spatial Metric

In a general coordinate frame, in which a gravitational field may exist, the metric tensor in spacetime is as follows:

$$ds^2 = g_{\alpha\beta} dx^\alpha dx^\beta \qquad (40)$$

We can then define a three-dimensional spatial metric, in which distances are measured with standard rods, as follows[20]:

$$d\sigma^2 = \rho_{ik} dx^i dx^k, \quad \text{where} \quad i, j = 1, 2, 3 \qquad (41)$$

The spatial metric tensor is $\rho_{ik} = g_{ik} + \rho_i \rho_j$, where the following obtains:

$$\rho_i = \frac{g_{i0}}{\sqrt{-g_{00}}} \qquad (42)$$

The *proper distance* given by the spatial metric (41) is not, in general, equal to the spatial part of the metric tensor, unless the metric tensor is time-orthogonal: $g_{i0} = 0$ for $i \neq 0$; that is, there are no time–space terms in the metric tensor.

For the simple case of isotropic spherical coordinates in General Relativity, for weak fields we have $g_{i0} = 0$, $g_{ij} = \delta_{ij}(1 - 2\Phi/c^2)$. Then the spatial metric (41) becomes $d\sigma^2 = (1 - 2\Phi/c^2)(dr^2 + r^2 d\theta^2 + r^2 \sin^2\theta d\phi^2)$. An observer using standard rods would find the proper spatial interval between two nearby points to be $d\sigma$; this would have to be multiplied by the correction factor $(1 - 2\Phi/c^2)^{-1/2} \approx (1 + \Phi/c^2)$ to obtain the corresponding coordinate distance between the points.

3. Equivalent Scalar Gravitational Potential in a Rotating Reference Frame

Consider the line element on a rotating disk, Eq. (26). Note that the time coordinate t' on the disk is not measured directly by standard clocks at rest on the disk. The transformation $t' = t$ means that coordinate time t' is effectively measured by using clocks in the underlying inertial frame to measure time. A coordinate clock at rest in the rotating frame would have to read, at each instant, the same as a clock at rest in the underlying inertial frame with which it momentarily coincides. In this way, a self-consistent network of synchronized coordinate clocks can be established. Coordinate time in the GPS is established in a similar way.

A standard clock at rest in the rotating frame will beat more slowly than a coordinate clock at the same location by a factor obtained from the line element, Eq. (26), by setting $d\phi = dr = dz = 0$. Thus $d\tau = ds/c = \sqrt{1 - (\omega r/c)^2}\, dt'$. For two clocks at coordinate radii r_1 and r_2, the ratio of standard clock rates is then as follows:

$$\frac{f_2}{f_1} = \frac{\sqrt{1 - (\omega r_1/c)^2}}{\sqrt{1 - (\omega r_2/c)^2}} \tag{43}$$

This ratio could be measured by sending a photon from one clock to the other. Although in the underlying inertial frame, there is no change in frequency of the photon, there is an apparent shift in the rotating frame which the inertial observer interprets as a second-order Doppler shift.

Observers at rest in the rotating frame, however, interpret the shift as caused by the presence of a gravitational field, which is induced locally by the rotational acceleration of the frame. At radius r, the acceleration is $\omega^2 r$ toward the center, so the effective gravitational field strength is $\omega^2 r$ outward. A particle placed at rest in the rotating frame will experience this force as though it were gravitational. The gradient of the gravitational potential is the negative of the force, so we have the following:

$$-\frac{\partial \Phi}{\partial r} = \omega^2 r \tag{44}$$

and, therefore, the effective gravitational potential is as follows:

$$\Phi = -\frac{1}{2}\omega^2 r^2 \tag{45}$$

Therefore, in the line element ds^2, the coefficient of $(cdt)^2$ can be expressed as $g_{00} = -(1 + 2\Phi/c^2)$. In any static gravitational field represented by such a metric tensor coefficient, the ratio of rates of two standard clocks at rest is as follows:

$$\frac{f_2}{f_1} = \frac{\sqrt{g_{00}(1)}}{\sqrt{g_{00}(2)}} = \frac{\sqrt{1 + 2\Phi(1)/c^2}}{\sqrt{1 + 2\Phi(2)/c^2}} \tag{46}$$

Thus, the use of the equivalent potential gives an alternate method for determining the ratio of standard clock rates in the rotating system. Thus, the frequency shift on a rotating disk can be interpreted as either a second-order Doppler shift, using special relativity in the underlying inertial frame, or as an effective gravitational field. In either interpretation, the results are identical.

Note, however, that the gravitational force measured using standard clocks at rest, rather than coordinate clocks, differs from the force discussed above. Because the standard clocks beat more slowly by a factor $\gamma^{-1} = \sqrt{1 - (\omega r/c)^2}$, and the units of acceleration are meters per second squared, the actual gravitational force (the radial four-force) on a unit mass at rest at radius r is $\gamma^2 \omega^2 r$.[14]

4. Spherical Shell Coordinates in a Spherically Symmetric Gravitational Field

Figure 15 shows a series of concentric spheres of radii r_i. On each sphere, suppose there is a network lattice of standard clocks interconnected rigidly along lines of constant longitude ϕ and constant colatitude θ. The coordinate radius r_i is defined by requiring the distance $d\sigma$ measured using standard rods along any arc on the surface to be $d\sigma^2 = r_i^2 (d\theta^2 + \sin^2\theta d\phi^2)$. A sphere of radius r will then have proper surface area $4\pi r^2$. For the moment, assume that there is no mass at the origin and no gravitational field, so $\Phi = 0$. For this

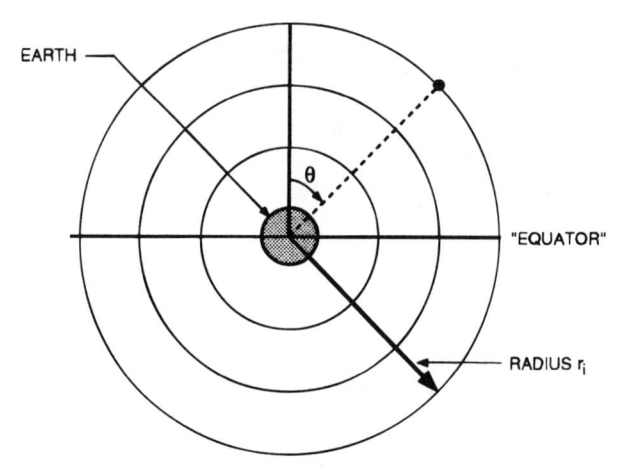

Fig. 15 Set of concentric spherical shells of radius r_i with the Earth at the center. The polar coordinates are r, θ, ϕ corresponding to an Earth-centered inertial frame with nonrotating co-latitude angle θ from the z-axis, and longitude-like angle ϕ.

model, the line element ds^2 for this inertial space in spherical coordinates $\{r, \theta, \phi\}$ is $ds^2 = -c^2 dt^2 + [dr^2 + r^2 d\theta^2 + r^2 \sin^2\theta d\phi^2]$. In this example, of course, the space is uniform, and all clocks run at the same rate.

If now a static mass M is added at the origin, the differential line element changes to the following

$$ds^2 = -\left[1 + \frac{2\Phi}{c^2}\right](cdt)^2 + \frac{dr^2}{[1 + 2\Phi/c^2]} + r^2 d\theta^2 + r^2 \sin^2\theta d\phi^2 \quad (47)$$

where Φ is the Newtonian gravitational potential $\Phi = -GM/r$. The coefficients in Eq. (47) are known as the *Schwarzschild metric tensor*, or more simply as the Schwarzschild metric. This metric is an exact solution of Einstein's field equations for the exterior space of a spherically symmetric, static isolated mass distribution. Thus, the Schwarzschild metric tensor is the following:

$$g_{\alpha\beta} = \begin{bmatrix} -\left[1 + \frac{2\Phi}{c^2}\right] & 0 & 0 & 0 \\ 0 & \left[1 + \frac{2\Phi}{c^2}\right]^{-1} & 0 & 0 \\ 0 & 0 & r^2 & 0 \\ 0 & 0 & 0 & r^2 \sin^2\theta \end{bmatrix} \quad (48)$$

We can easily see from this metric tensor that for a light ray $ds^2 = 0$, the radial and transverse coordinate velocities of light are not equal. For example, solve

INTRODUCTION TO RELATIVISTIC EFFECTS

for dr/dt with $d\Phi$, $d\theta = 0$, and then solve for $rd\theta/dt$ with $dr = 0$. However, the velocity in any direction measured using standard rods and clocks is c. For example, for light traveling through a radial coordinate distance dr in coordinate time dt the standard radial distance is $d\sigma = dr/\sqrt{1 + 2\Phi/c^2}$; whereas, the proper time measured on standard clocks at rest is $d\tau = \sqrt{1 + 2\Phi/c^2}dt$. Thus, the *proper velocity* is as follows:

$$\frac{d\sigma}{d\tau} = \frac{1}{1 + 2\Phi/c^2} \frac{dr}{dt} \tag{49}$$

However, $ds^2 = 0$ gives the following relation between dr and dt:

$$\frac{dr}{dt} = c(1 + 2\Phi/c^2) \tag{50}$$

Therefore, $d\sigma/d\tau = c$. Similar arguments, applied to any path, always give c for the proper velocity of light.

The Schwarzschild metric tensor can also be transformed to a rotating frame, as shown in Table 10. Table 10 also lists the metric tensors for nonuniform gravitational fields that are not spherically symmetric.

In general, the space of General Relativity is described by the *Riemann–Christoffel curvature tensor* $R^\alpha{}_{\mu\rho\sigma}$, which is a tensor of $4^4 = 256$ components (only 21 components are independent).* For flat space, the curvature tensor is zero. For the space of the GPS, as seen in the next section, the gravitational field is weak for users near the Earth's surface, out to at least 100,000 km.

This section has given a brief summary of the metric tensors applicable to the GPS problems. The reader is invited to study Refs. 5–7, 14, 20, 26, and 27.

III. Relativistic Effects in GPS

The major objective of this chapter is to compute the detailed effects of relativity on the various types of GPS measurements. This section employs the metric tensors and other relativistic transformations of the previous section and analyzes three primary applications of the GPS system:

1) Time keeping, time transfer, and clock transport on the Earth's surface. Effects of clock motion and the Earth's geopotential on clock time. Clock synchronization.

2) Relativistic effects on earth-based users of the GPS including: the satellite clock frequency offset, periodic orbit eccentricity effects, time delay effects; so-called Sagnac effects caused by the Earth's rotation during time of transit, and various other effects of user velocity and altitude. In addition, secondary effects such as the Shapiro delay, effects of the Earth's quadrupole moment, and tidal effects from the sun and moon, and Lense–Thirring drag effects are also estimated.

3) Relativistic effects on space-based users of GPS. Effects are computed for low-Earth orbit (LEO) satellites, other GPS satellites, and geostationary satellites.

---

*The elements of the Riemann curvature tensor are functions of the Christoffel symbols $\Gamma^\mu{}_{\mu\sigma}$ (thirteen of these are nonzero for the Schwarzschild metric), which in turn, are, related to the metric tensor components $g_{\alpha\beta}$ and their partial derivatives.

Table 10 List of line elements, Euclidean, Special Relativistic, and General Relativistic metric tensors

Euclidean space

$ds^2 = d\sigma^2 = dx^2 + dy^2 + dz^2$ Cartesian

$\qquad = dr^2 + r^2 d\phi^2 + dz^2$ Cylindrical

$\qquad = dr^2 + r^2(d\theta^2 + \sin^2\theta\, d\phi^2)$ Spherical (θ is angle from the vertical)

Special Relativity–flat space metric tensor

$ds^2 = -c^2 dt^2 + dx^2 + dy^2 + dz^2$ Cartesian ⎫

$\qquad = -c^2 dt^2 + dr^2 + r^2 d\phi^2 + dz^2$ Cylindrical ⎬ Inertial space

$\qquad = -c^2 dt^2 + dr^2 + r^2(d\theta^2 + \sin^2\theta\, d\phi^2)$ Spherical ⎭

$ds^2 = -\left(1 + \dfrac{gz'}{c^2}\right)^2 c^2 dt'^2 + dx'^2 + dy'^2 + dz'^2$ Accelerated frame

Rotating coordinates, $\phi = \omega t + \phi_r$, $d\phi = \omega dt + d\phi_r$, $d\phi^2 = \omega^2 dt^2 + 2\omega\, dt\, d\phi_r + d\phi_r^2$

$ds^2 = (-c^2 + \omega^2 r^2) dt^2 + dr^2 + 2\omega r^2 dt\, d\phi_r + r^2 d\phi_r^2 + dz^2$ ⎫

$\qquad = (-c^2 + \omega^2 r^2) dt^2 + dr^2 + r^2(2\omega\, dt\, \phi_r + d\phi_r^2) + dz^2$ ⎬ Rotating cylindrical

$ds^2 = (-c^2 + \omega^2 r^2 \sin^2\theta) dt^2 + dr^2 + r^2 d\theta^2 + 2\omega\, r^2 \sin^2\theta\, dt\, d\phi_r + r^2 \sin^2\theta\, d\phi_r^2$ ⎫

$\qquad = (-c^2 + \omega^2 r^2 \sin^2\theta) dt^2 + dr^2 + r^2(d\theta^2 + 2\omega \sin^2\theta\, dt\, d\phi_r + \sin^2\theta\, d\phi_r^2)$ ⎭

Rotating spherical

General Relativity–spherically symmetric nonrotating central mass–Schwarzchild metric tensor

$ds^2 = -(1 + 2\Phi/c^2) c^2 dt^2 + (1 + 2\Phi/c^2)^{-1} dr^2 + r^2(d\theta^2 + \sin^2\theta\, d\phi^2)$ Fixed spherical
where $\Phi = -GM/r = -\mu/r$

Spherical rotating coordinates, $\phi = \omega t + \phi_r$

$ds^2 = [-c^2(1 + 2\Phi/c^2) + \omega^2 r^2 \sin^2\theta] dt^2 + (1 + 2\Phi/c^2)^{-1} dr^2 + r^2(d\theta^2 + 2\omega \sin^2\theta\, dt\, d\phi_r$
$\qquad + \sin^2\theta\, d\phi_r^2)$

General Relativity–general weak field approximation not necessarily spherically symmetric (nonrotating mass)–nearly flat space[5,14]

$ds^2 = -(1 + 2\Phi/c^2) c^2 dt^2 + (1 - 2\Phi/c^2)[dr^2 + r^2(d\theta^2 + \sin^2\theta\, d\phi^2)]$ Fixed spherical isotropic

$ds^2 = -(1 + 2\Phi/c^2) c^2 dt^2 + dr^2 + r^2(d\theta^2 + \sin^2\theta\, d\phi^2)$ Flat space spherical

Rotating coordinates (but not a rotating mass), $\phi = \omega t + \phi_r$

$ds^2 = [-c^2(1 + 2\Phi/c^2) + (1 - 2\Phi/c^2)\omega^2 r^2 \sin^2\theta] dt^2 + (1 - 2\Phi/c^2)[dr^2 + r^2(d\theta^2$
$\qquad + 2\omega \sin^2\theta\, dt\, d\phi_r + \sin^2\theta\, d\phi_r^2)]$ Rotating spherical isotropic

It should be pointed out that there can be some cancellation of these effects for GPS position calculations where common range errors to each satellite cancel.

Table 11 lists the key physical constants used in the calculations of relativistic effects. Note that in some cases two values are shown. The first is the WGS-84 (1984 World Geodetic System Standard) value and the second is the more recent JGM-2 value. This section for the most part uses the JGM-2 value. However, WGS-84 values are also given for some of the larger effects in a few cases.

A. Relativistic Effects on Earth-Based Clocks

Clocks on or near the Earth's surface are influenced by several relativistic effects that can cause the clocks to gain or lose time or run at different clock rates depending upon the following:
1) motion/velocity of the clocks
2) change in altitude of the clocks relative to the Earth's reference surface; and
3) direction of motion of the clocks.

Because clocks near Earth's surface experience a gravitational field and may be in motion (in addition to the motion associated with the Earth's rotation), they are subject to gravitational frequency shifts and time dilation. In the GPS, the magnitudes of such effects cannot be neglected. Net time elapsed on a standard clock depends upon the history of the clock; therefore, standard clock readings by themselves cannot be used to establish a network of synchronized clocks. A synchronized clock network satisfies some very important equivalence relations:

- *Reflexivity:* If clock A is synchronized with clock B, then clock B is synchronized with clock A; and
- *Transitivity:* If clock A is synchronized with clock B, and clock B is synchronized with clock C, then clock A is synchronized with clock C.

With such a network, two users could reliably set up a rendezvous; without such a network, little but confusion would result.

In any inertial frame, in the absence of gravitational fields, clocks may be synchronized by the *Einstein synchronization procedure,* based on the constancy of the speed of light. This consists of accounting for the time delay l/c between transmission and reception of a light signal sent between two clocks having spatial separation l. An equivalent procedure uses slow transport of standard clocks, moving so slowly that their time dilation is negligible.

In the GPS, such synchronization is accomplished by adopting a coordinate time t, appropriately scaled, as the measure of time. Suppose for a moment there were no gravitational fields. Then picture an underlying, nonrotating reference frame, a local inertial frame unattached to the spin of the Earth, but with its origin at the center of the Earth. This frame is labeled Earth–centered-inertial (ECI) in Fig. 16. In this nonrotating frame, introduce a fictitious set of standard clocks, available anywhere, all synchronized by the Einstein synchronization procedure, and running at agreed-upon rates so that synchronization is maintained. These clocks read the time t. Next we introduce the rotating Earth with a set of standard clocks distributed around upon it, possibly roving around. We apply to each of the standard clocks a set of corrections based on the known positions and motions of the clocks, to make the corrected clocks read t. This generates

Table 11 Key constants for GPS relativistic effects[a]

a_g	GPS orbit semimajor axis, 26561.75 km = 2.656175 × 10⁷ m
c	speed of light (exact definition), 2.99792458 × 10⁸ m/s
F	Earth flattening = $(R_e - R_{pole})/R_e$, $\begin{cases} 1/298.257223563 \text{ (WGS} - 84) \\ 1/298.1235 \text{ (JGM} - 2) \end{cases}$
g	gravitational acceleration on the Earth's surface (latitude dependent), $g(\lambda)$
G	universal gravitational constant, 6.6732 × 10⁻⁸ cm³/gs² = 6.6732 × 10⁻¹¹ m³/kgs²
h	Planck's constant, 6.626197 × 10⁻²⁷ g cm²/s = 6.626197 × 10⁻³⁴ kg m²/s
J_2	Earth's quadrupole moment, $\begin{cases} 1082.6800 \times 10^{-6} \text{ (WGS} - 84) \\ 1082.6269 \times 10^{-6} \text{ (JGM} - 2) \end{cases}$
$m_e = M_e G/c^2$	ratio for Earth = $\mu/c^2 = m_e$, m_e = 0.44350 cm = 4.435028687 × 10⁻³ m
M_e	mass of Earth, 5.9742 × 10²⁴ kg = 5.9742 × 10²⁷ g
$M_e G/R_e c^2$	dimensionless Earth surface potential, 6.95 × 10⁻¹⁰
M_m/M_e	ratio of mass of moon to that of Earth, 0.01230002
M_s	mass of sun, 1.9891 × 10³⁰ kg = 1.9891 × 10³³ g
$M_s G/c^2$	ratio for sun, 1.47498 × 10³ m
r_{se}	mean Earth–sun distance (1 AU),[b] 1.4959965 × 10¹¹ m
$M_s G/r_{se} c^2$	ratio for sun at Earth orbit radius, 9.85953 × 10⁻¹³
R_e	mean Earth radius at equator, $\begin{cases} 6.378137 \times 10^6 \text{ m (WGS} - 84) \\ 6.3781363 \times 10^6 \text{ m (JGM} - 2) \end{cases}$
R_p	Polar radius of Earth's reference surface, $\begin{cases} 6.356752 \times 10^6 \text{ m (WGS} - 84) \\ 6.3567420 \times 10^6 \text{ m (JGM} - 2) \end{cases}$
r_{me}	mean Earth–moon distance, 3.844 × 10⁸ m
R_s	solar radius, 6.9598 × 10⁸ m
v_e	mean speed of Earth about the sun, 29.784852 km/s = 2.9784852 × 10⁴ ms
v_g	speed of GPS satellite in circular orbit $\sqrt{\mu/a_g}$, 3873.83 m/s
v_g/c	relative speed of GPS satellite in circular orbit to c, 1.29217 × 10⁻⁵
ϵ	satellite orbit eccentricity (for GPS ϵ < 0.005 typical)
ϵ_{es}	eccentricity of Earth's orbit around the sun, 0.017
μ	Earth gravitational constant, $\mu = GM_e$, $\begin{cases} 3.986005 \times 10^{14} \text{ m}^3/\text{s}^2 \text{ (WGS} - 84) \\ 3.986004415 \times 10^{14} \text{ m}^3/\text{s}^2 \text{ (JGM} - 2) \end{cases}$
Ω_e	Earth angular rotation rate (WGS-84), 7.2921151467 × 10⁻⁵ rad/s

[a]References 31, 32, 1984 World Geodetic System Standard, and JGM-2.[41] See also Chapter 4, this volume.
[b]AU, Astronomical Unit.[40]

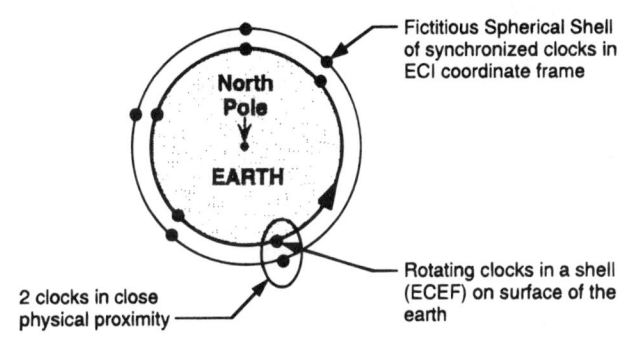

Fig. 16 Global coordinate time. Imagine a grid of fixed clocks on a spherical shell surrounding the Earth and a second grid network of clocks on the surface of the rotating Earth. At each time instant, the network of rotating Earth-based clocks are compared and matched to a network of standard atomic clocks (global coordinate time), which are momentarily at the same physical location.

in the Earth-fixed, rotating system, a coordinate clock time. This time is such that at each instant, a coordinate clock agrees with the fictitious standard clock at rest in the local inertial frame, whose position coincides with the Earth-fixed coordinate clock at that instant. Thus, coordinate time is equivalent to time that would be measured by standard clocks at rest in the local inertial frame.[9,30]

When the Earth's gravitational field is considered, the picture is only a little more complicated. There still exists a coordinate time that can be found by computing a correction for gravitational red shift; this is discussed in the next sections (see Fig. 16).

For a static distribution of mass, and in a local frame of reference that is nonrotating, the metric correct to first order in the small quantity Φ/c^2 is given in Table 10:

$$ds^2 = -\left(1 + \frac{2\Phi}{c^2}\right)(cdt)^2 + \left(1 - \frac{2\Phi}{c^2}\right)(dr^2 + r^2d\theta^2 + r^2\sin^2\theta d\phi^2) \quad (51)$$

The spatial coordinates for this metric are called *isotropic* coordinates. The potential Φ of the Earth, to sufficient accuracy for the calculation of most relativistic effects is as follows:

$$\Phi = -\frac{GM_e}{r}\left[1 - J_2\left(\frac{R_e}{r}\right)^2 P_2(\cos\theta)\right] \quad (52)$$

where G is the Newtonian gravitational constant, M_e is the Earth's mass, r is the distance of the observation point from the Earth's center, J_2 is the Earth's quadrupole moment coefficient, R_e is the Earth's equatorial radius, and $P_2(x) = (3x^2 - 1)/2$ is the Legendre polynomial of second degree. The angle θ is the colatitude, the polar angle of the point of observation, measured down from the axis of rotation. Table 11 gives values for the constants.

One additional expression for the interval is of use, the transformation of Eq. (51) to a rotating Earth-fixed frame of reference. This metric is the last entry in Table 10:

$$ds^2 = -\left(1 + \frac{2\Phi}{c^2} - \left(1 - \frac{2\Phi}{c^2}\right)\frac{\Omega_e^2 r^2 \sin^2\theta}{c^2}\right)(cdt)^2$$

$$+ 2\left(1 - \frac{2\Phi}{c^2}\right)\Omega_e r^2 \sin^2\theta dt\, d\phi_r \quad (53)$$

$$+ \left(1 - \frac{2\Phi}{c^2}\right)(dr^2 + r^2 d\theta^2 + r^2 \sin^2\theta d\phi_r^2)$$

where Ω_e is the Earth's angular rotational speed.

1. Earth's Geoid

The earth's shape is approximately that of an ellipsoid of revolution rather than a perfect sphere (see Fig. 17). For that reason, the Newtonian gravitational potential at Earth's surface is not a constant with geocentric latitude λ (the co-latitude θ is related to geocentric latitude λ by $\cos\theta = \sin\lambda$). On the surface of the ellipsoid, the Newtonian potential is lower at the poles than at the equator where the Earth's radius is larger. On the other hand, the Earth's surface rotational velocity $v(\lambda) = \Omega_e \times r(\lambda)$ causes clocks at the equator to beat more slowly because of time dilation. These effects nearly cancel. When the third effect of gravitational red shift caused by the Earth's quadrupole potential is accounted for, then we can find a surface of constant effective gravitational potential in the Earth-fixed rotating frame, very close to mean sea level. Clocks at rest anywhere on this surface beat at the same rates, to a very high degree of approximation. This calculated surface becomes a surface of reference for determination of the Earth's geoid*—the true surface at mean sea level.[9,28,30]

Consider the line element Eq. (53) in the Earth-fixed rotating frame. Let there be a clock at rest on Earth. Then, keeping only terms of order $1/c^2$, Eq. (53) reduces to the following:

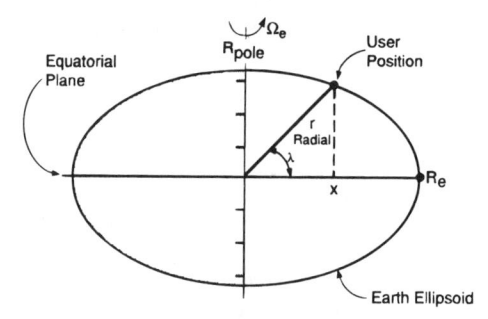

Fig. 17 Earth's ellipsoid shown in geocentric coordinates. The geocentric latitude is λ. The Earth rotates on its axis at angular rate Ω_e.

---

* Sometimes called the u-geoid.[29]

$$-ds^2 = -\left(1 + \frac{2\Phi}{c^2} - \frac{\Omega_e^2 r^2 \sin^2\theta}{c^2}\right)(c\,dt)^2 = -\left(1 + \frac{2\Phi}{c^2} - \frac{\Omega_e^2 r^2 \cos^2\lambda}{c^2}\right)(c\,dt)^2 \tag{54}$$

(t is measured in the ECI frame). Thus, for relativistic purposes, the effective gravitational potential in the rotating frame is as follows:

$$\Phi_{\text{eff}}(r, \lambda) = \Phi(r, \lambda) - \frac{1}{2}\Omega_e^2 r^2 \cos^2\lambda \tag{55}$$

where $\Phi(r, \lambda)$ is given by Eq. (52).

Let us define the reference surface by requiring the effective gravitational potential to have a constant value on this surface, the same value it has on the equator. Substituting values for R_e, GM_e, Ω_e, c, and J_2 from Table 11 using the JGM-2 values gives the following value for the reference potential:

$$\Phi_0 = \Phi_{\text{eff}}(R_e, 0) = -\frac{GM_e}{R_e}(1 + J_2/2) - \frac{1}{2}\Omega_e^2 R_e^2 = -6.2636803 \times 10^7 \text{ m}^2/\text{s}^2 \tag{56}$$

The radius $r(\lambda)$ of the reference surface will vary in such a way that the following obtains:

$$\Phi_0 = -\frac{GM_e}{r(\lambda)}\left(1 - J_2\left(\frac{R_e}{r(\lambda)}\right)^2 \frac{3\sin^2\lambda - 1}{2}\right) - \frac{1}{2}\Omega_e^2 r(\lambda)^2 \cos^2\lambda = \text{constant} \tag{57}$$

Equation (57) determines $r(\lambda)$ implicitly. It can be solved by iteration for $r(\lambda)$, by substituting in increasingly more accurate solutions. The result, accurate to better than a millimeter with the given constants, is the following:

$$r(\lambda) = 6,356,742.0252 + 21353.6419\, x^2 + 39.8320\, x^4 + 0.79780\, x^6 + 0.0030\, x^8 \text{ m} \tag{58}$$

where $x = \cos\lambda$. This radius can differ from that of a true ellipsoid of revolution by as much as 15 meters. At the pole where $\cos\lambda = 0$, the radius is $R_{\text{pole}} = 6{,}356{,}742.0252$ m. This gives a flattening parameter F having the value $F = R_e - R_{\text{pole}}/R_e = 1/298.123510$.

The actual surface of constant effective potential in the rotating frame may differ in height from the above reference surface by as much as 100 meters; calculation of this surface (the Earth's geoid) depends upon detailed knowledge of the many higher multipole moments of the Earth. Figure 18 shows a contour map of Earth's geoid, calculated using the JGM-2 gravity model developed by Goddard Space Flight Center and the University of Texas at Austin.[32]

An estimate of the Earth's gravitational acceleration can be obtained from the acceleration of gravity caused by the reference potential, on this reference surface.

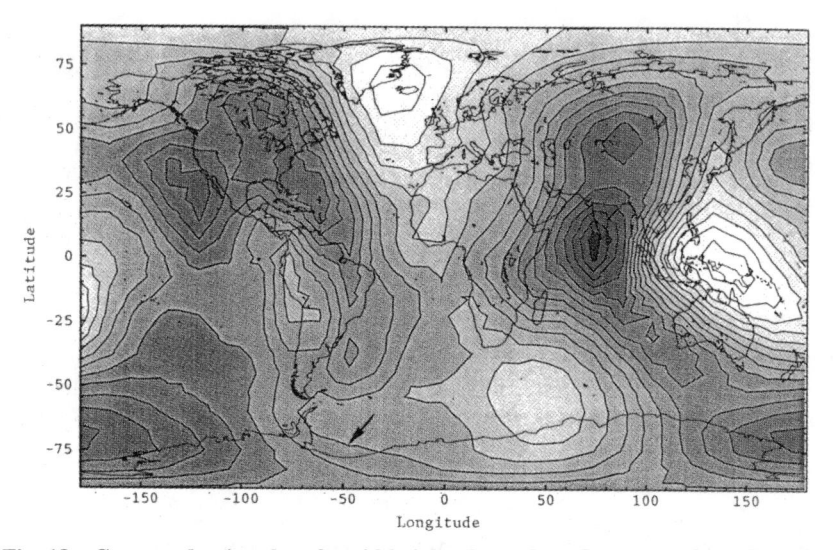

Fig. 18 Contour density plot of geoid height above the reference surface. Contours are given every 10 m. The horizontal scale is longitude in degrees, and the vertical scale is latitude. The contour of zero height is marked with an arrow. The very dark region, which dips to about 100 m below the reference surface, lies near the southern tip of India.

The gradient of the reference potential has two components given by the following:

$$g_r = \frac{\partial \Phi_{\text{eff}}}{\partial r} = \frac{GM_e}{r^2} - \Omega_e^2 r \cos^2\lambda - \frac{3GM_e R_e^2 J_2}{r^4}\left(\frac{3\sin^2\lambda - 1}{2}\right)$$

$$g_\lambda = \frac{1}{r}\frac{\partial \Phi_{\text{eff}}}{\partial \lambda} = \left(\Omega_e^2 r - \frac{3a_1^2 GM_e J_2}{r^4}\right)\sin\lambda \cos\lambda \qquad (59)$$

The acceleration of gravity g_{eff} evaluated on the reference surface, is the magnitude of this gradient:

$$g_{\text{eff}} = \sqrt{g_r^2 + g_\lambda^2}\Big|_{r=r(\lambda)} \qquad (60)$$

with r evaluated on the surface, given by Eq. (58). This expression is very cumbersome; however, the first term GM_e/r^2 in g_r is caused by the main monopole term in the Earth's gravity field and is the largest contribution to the gradient. Furthermore, $\overline{r}$ is close to R_e on the surface. All other terms are very small. This, again, suggests a series expansion in powers of $x = \cos\lambda$. The expressions are too cumbersome to write out symbolically. However, it is a straightforward, but lengthy, calculation to obtain the expansion of Eq. (60) to six decimal places:

$$g_{\text{eff}}(x) = 9.832099 - 0.051038\, x^2 - .000779\, x^4 + .000002\, x^6 \text{ m/s}^2 \qquad (61)$$

Geodetic (geographic) latitude λ' is the angle between the local vertical and the equator.* This differs from geocentric latitude λ because of the Earth's flattened figure. The angles are related by the following:

$$\cos \lambda' = \frac{g_r \cos \lambda - g_\lambda \sin \lambda}{g_{\text{eff}}} \quad (62)$$

Letting $x' = \cos \lambda'$ and evaluating Eq. (62) gives an approximate series expansion for x':

$$x' = 0.993218158\, x + 0.00669351\, x^3 + 0.0000240\, x^5 + 0.00000086\, x^7 \quad (63)$$

which can be inverted approximately to give the following:

$$x = 1.0067639\, x' - 0.0068765\, x'^3 + 0.00011587\, x'^5 - 0.00000091\, x'^7$$
$$(64)$$

These expressions can be used to calculate the radius or acceleration of gravity at the surface in terms of geographic latitude, if needed.

It is useful to estimate the effect of the terms of order x^4 in Eq. (61) on gravitational frequency shifts. At an altitude of $h = 5000$ m above the reference surface, this term would contribute a fractional frequency shift of order

$$.000779 h/c^2 \approx 4.3 \times 10^{-17},$$

compared to a clock on the surface and so could be neglected. In fact, an expansion of $g_{\text{eff}}(x)$ to order x^2 would be sufficiently accurate for many purposes.

Next, it is of interest to see what corrections are needed to obtain a sufficiently accurate expression for g_e on the actual geoid. This is a complex calculation that depends upon numerical analysis of the many Stokes coefficients of the Earth's gravity field.[28] The result is that, accurate to within 0.02 m/s^2, the acceleration of gravity on the actual geoid is essentially equal to the acceleration of gravity attributable to the reference potential on the reference surface at the same geocentric longitude and latitude. Thus, for purposes of computing relativistic frequency shifts, higher terms in the multiple expansion of the Earth's gravity field do not affect users within 5000 m of Earth's geoid. To a fractional accuracy of better than 1×10^{-16}, the acceleration of gravity given above can be used to compute gravitational frequency shifts, $-g_{\text{eff}} h/c^2$, where h is the height of a clock above the geoid. The next most important contributions to the acceleration of gravity on the geoid arise from the radial components of the gradient of the multipole contributions to the potential and are of order 0.0005 m/s^2. These arise primarily from the $l = 2$, $m = 1, 2$ contributions to the multipole potential.†

2. *Change of Coordinate Time Scale*

For a clock at rest on the geoid, Eq. (54) becomes $ds^2 = -(1 + 2\Phi_0/c^2)(c dt)^2$. Taking a square root, expanding as usual keeping only the terms of order $1/c^2$,

---

* There are three commonly used definitions of latitude. Geocentric latitude is measured from the reference ellipsoid's center (λ); geodetic or geographic latitude is measured with respect to the normal to the reference ellipsoid (λ'), and astronomical latitude is measured with respect to the local geoid.

† The multipole coefficients are discussed in Chapter 4, this volume.

and noting that $|ds/c| = d\tau$ is the proper time elapsed on the clock, this becomes the following:

$$d\tau = (1 + \Phi_0/c^2)dt \qquad (65)$$

In this equation, Φ_0 is a constant. The quantity $d\tau$ on the left side can be interpreted as the increment of proper time elapsed on a standard clock at rest, in terms of the elapsed coordinate time dt. Thus, the rather remarkable result has emerged, that ideal clocks at rest on the geoid on the rotating Earth all beat at the same rate. This is reasonable, because the Earth's surface is a gravitational equipotential surface in the rotating frame. Considering clocks at two different latitudes, the one farther north will be closer to the Earth's center because of the flattening—it will, therefore, be more red shifted. However it is also closer to the axis of rotation, and going more slowly, so suffers less second-order Doppler shift. When the quadrupole potential is included, this combination of effects cancels exactly on the geoid. The model explicitly demonstrates this cancellation on the reference surface.

Because all clocks at rest on the geoid beat at the same rate, it is advantageous to exploit this fact to redefine the rate at which coordinate time beats. In fact, the SI second is defined in terms of standard clocks at rest on the Earth's geoid; whereas, the rate of coordinate time in Eq. (51) is defined by standard clocks at rest at infinity. To make the rate of coordinate time the same as that of standard clocks at rest on the Earth's surface, define a new coordinate time t' by means of a constant rate change:

$$t' = (1 + \Phi_0/c^2)t \qquad (66)$$

The correction is a few parts in 10^{10} (see Eq. (56)). Then, because Φ_0 is a constant

$$dt' = (1 + \Phi_0/c^2)dt \qquad (67)$$

When this time scale change is made, keeping only terms of order $1/c^2$ or larger, the metric of Eq. (53) in the Earth-fixed rotating frame becomes the following:

$$-ds^2 = -\left(1 + \frac{2(\Phi_{\text{eff}} - \Phi_0)}{c^2}\right)(cdt')^2 + 2\Omega_e r^2 \sin^2\theta d\phi_r(cdt')/c$$

$$+ \left(1 - \frac{2\Phi}{c^2}\right)[dr^2 + r^2 d\theta^2 + r^2 \sin^2\theta d\phi_r^2] \qquad (68)$$

The same time scale change in the nonrotating ECI metric, Eq. (51), gives the following:

$$ds^2 = -\left(1 + \frac{2(\Phi - \Phi_0)}{c^2}\right)(cdt')^2 + \left(1 - \frac{2\Phi}{c^2}\right)(dr^2 + r^2 d\theta^2$$

$$+ r^2 \sin^2\theta d\phi^2) \qquad (69)$$

To check the consistency of these equations, for a clock at rest on the geoid, $\Phi_{\text{eff}} = \Phi_0$ and the metric of Eq. (68) in the rotating frame reduces simply to $|ds^2| = (cdt')^2$. Thus, the proper time elapsed on clocks at rest on the geoid is identical with the rescaled elapsed coordinate time. This is the correct way to express the fact that ideal clocks at rest on the geoid provide all of our standard reference clocks.[30]

3. Clocks at Rest at Altitude h

Near the geoid, the Earth's potential can be expanded in a Taylor series in distance h normal to the geoid. The effective geopotential to leading order is then as follows:

$$\Phi \cong \Phi_0 + g_{\text{eff}}(\lambda)h \qquad (70)$$

On substituting this approximation into the metric of Eq. (68) and specializing to the case of a clock at rest on the Earth's surface, $|ds^2| = (1 + 2g_{\text{eff}}h/c^2)(cdt')^2$; then taking a square root, expanding in small quantities, and keeping only the leading correction, we have the following:

$$|ds| = (1 + g_{\text{eff}}h/c^2)(cdt') \qquad (71)$$

In this form, the result predicts that a clock at altitude h will beat faster than a clock on the geoid. Solving Eq. (71) approximately for the increment of coordinate time gives the following:

$$dt' = (1 - g_{\text{eff}}h/c^2)|ds/c| \qquad (72)$$

At a latitude of 40°N, the value of the acceleration of gravity from Eq. (61) is 9.801 m/s². The Boulder laboratories of NIST lie at an altitude of about 1650 meters, and the fractional gravitational potential correction term of Eq. (72) is therefore $(-g_{\text{eff}}(40°) \times 1654 \text{ m})/c^2 = -1.80 \times 10^{-13}$. Thus, standard clocks at NIST in Boulder must have their readings reduced by about 15.5 ns per day in order for their rate to agree with clocks in standards laboratories near the geoid. In the GPS, a similar rate correction should, in principle, be made if the master reference clock is located in Colorado Springs; whereas, if the master reference clock is at sea level, no such correction is necessary.

4. Synchronization by Means of Portable Clocks

The invariant interval, Eq. (68), contains all the relativistic effects on a portable clock that were discussed piecemeal in Secs. I and II. We now show how all these effects follow from Eq. (68), in which the variables represent measurements made in the Earth-fixed rotating frame. We consider clocks near the geoid, and insert Eq. (70) into Eq. (68), then factor out $(cdt')^2$.

$$ds^2 = \left[-(1 + 2g_{\text{eff}}h/c^2) + 2\Omega_e r^2 \sin^2\theta \frac{d\phi_r}{(c^2 dt')} \right.$$
$$\left. + \left(1 - \frac{2\Phi}{c^2}\right) \frac{dr^2 + r^2 d\theta^2 + r^2 \sin^2\theta d\phi_r^2}{(cdt')^2} \right] (cdt')^2 \qquad (73)$$

In the last term in Eq. (73), $(dr^2 + r^2 d\theta^2 + r^2 \sin^2\theta d\phi^2)/(cdt')^2$ is just $(v/c)^2$, where v is the ground speed of the moving clock. The product of this factor with $2\Phi/c^2$ is negligible, so the expression reduces to the following:

$$|ds^2| = \left[1 + 2g_{\text{eff}}h/c^2 - 2\Omega_e r^2 \sin^2\theta \frac{d\phi_r}{(c^2 dt')} - \frac{v^2}{c^2} \right] (cdt')^2 \qquad (74)$$

We regard the elapsed proper time $d\tau = |ds/c|$ on the portable clock, and the velocity v, as known or measurable. Solving for the elapsed coordinate time dt' in terms of known quantities, we get the following:

$$(dt')^2 = (d\tau)^2 \left[1 + 2g_{\text{eff}} h/c^2 - 2\Omega_e r^2 \sin^2\theta \frac{d\phi_r}{(c^2 dt')} - \frac{v^2}{c^2} \right]^{-1} \quad (75)$$

Taking square roots of both sides of Eq. (75), expanding the quantity in the square root using the binomial theorem, and keeping only the leading terms in $1/c^2$, gives the following:

$$dt' = (d\tau)\left[1 - g_{\text{eff}} h/c^2 + \Omega_e r^2 \sin^2\theta \frac{d\phi_r}{(c^2 dt')} + \frac{v^2}{2c^2} \right] \quad (76)$$

The three correction terms in Eq. (76) are, respectively, the corrections for gravitational frequency shift, Sagnac effect, and second-order Doppler effect. If a moving clock traverses some path, the corrections can be written in terms of integrals over the path, with respect to proper time on the moving clock:

$$\Delta t' = \int_{\text{path}} d\tau \left[1 - \frac{g_{\text{eff}} h}{c^2} + \Omega_e r^2 \sin^2\theta \frac{d\phi_r}{(c^2 dt')} + \frac{v^2}{2c^2} \right]$$

The Sagnac correction can be written in a simpler form as follows. Let a portable clock move through some arc on the surface of the Earth; the distance of the clock from the rotation axis is $r \sin\theta$, and the projection of the arc on the equatorial plane is of length $r \sin\theta d\phi_r$. Thus, as the clock moves, it sweeps out a series of small triangles projected on the equatorial plane (see Fig. 19). The area of the triangle corresponding to this arc is as follows:

$$dA = \frac{1}{2} r^2 \sin^2\theta d\phi_r \quad (77)$$

Therefore, the contribution from this term in Eq. (76) can be written as follows:

$$\frac{2\Omega_e}{c^2} \int_{\text{path}} dA \frac{d\tau}{dt'} = \frac{2\Omega_e \cdot A}{c^2}$$

because $d\tau/dt' = 1$ to very high order. Hence, the elapsed coordinate time is as follows:

$$\Delta t' = \int_{\text{path}} d\tau (1 - g_{\text{eff}}(\lambda) h/c^2 + v^2/2c^2) + \frac{2\Omega_e \cdot A}{c^2} \quad (78)$$

All quantities on the right side of Eq. (78) are known or measurable; the computation indicated corrects the proper time elapsing on the moving portable clock as it traverses the path, to produce coordinate time. Although the signs of the terms in Eq. (78) may seem incorrect at first glance, they are required in order to produce necessary corrections. Consider, for example, the term $v^2/2c^2$. A moving clock is slowed by time dilation so $\int d\tau$ is smaller than elapsed coordinate time. The correction factor $(1 + v^2/2c^2) > 1$ is applied to correct for

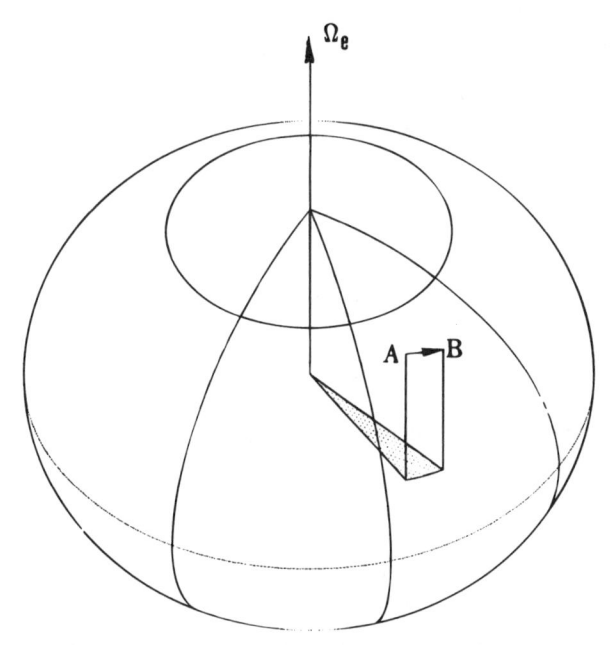

Fig. 19 Path increment of an atomic clock carried slowly over the Earth's surface. The Sagnac effect is proportional to the area swept out by the equatorial projection of a vector from Earth's center to the clock, the shaded area in the figure.

this effect and to yield coordinate time. It must be emphasized, that the application of these corrections to the elapsed proper time

$$\Delta \tau = \int d\tau = \int_{path} ds/c$$

yields elapsed coordinate time in the underlying inertial frame—the ECI frame.

5. Synchronization by Light Signals

In a similar manner, coordinate time may be transferred from one location to another by means of light signals, which satisfy $ds^2 = 0$ along the path. Then Eq. (68) becomes as follows:

$$0 = -\left(1 + \frac{2(\Phi_{eff} - \Phi_0)}{c^2}\right)(cdt')^2 + 2\Omega_e r^2 \sin^2\theta d\phi_r(cdt')/c$$

$$+ \left(1 - \frac{2\Phi}{c^2}\right)[dr^2 + r^2 d\theta^2 + r^2 \sin^2\theta d\phi_r^2] \quad (79)$$

To simplify the computation, let the spatial coordinate distance along the path be represented by dl, where $dl^2 = dr^2 + r^2 d\theta^2 + r^2 \sin^2\theta d\phi_r^2$. Then neglecting all relativistic corrections, the solution of Eq. (79) would yield $dt' = dl/c$. This may be substituted into the relativistic corrections themselves without appreciable error. Thus, as a first step, take the term in $(dt')^2$ to the left side, as shown:

$$\left(1 + \frac{2(\Phi_{\text{eff}} - \Phi_0)}{c^2}\right)(cdt')^2 = 2\Omega_e r^2 \sin^2\theta d\phi_r(dl)/c + \left(1 - \frac{2\Phi}{c^2}\right)dl^2 \quad (80)$$

then to the same order of accuracy, we get the following:

$$(cdt')^2 = 2\Omega_e r^2 \sin^2\theta d\phi_r(dl)/c + \left(1 - \frac{2\Phi + 2\Phi_{\text{eff}} - 2\Phi_0}{c^2}\right)dl^2 \quad (81)$$

Taking square roots of both sides of Eq. (81) and expanding the right side, keeping only the leading relativistic corrections, gives the following:

$$cdt' = \left(1 - \frac{\Phi + \Phi_{\text{eff}} - \Phi_0}{c^2}\right)dl + \Omega_e r^2 \sin^2\theta d\phi_r/c \quad (82)$$

Finally, integrate along the path of the light ray to get the elapsed coordinate time:

$$\Delta t' = \int_{\text{path}} \frac{dl}{c}\left(1 - \frac{(\Phi + \Phi_{\text{eff}} - \Phi_0)}{c^2}\right) + \int_{\text{path}} \Omega_e r^2 \sin^2\theta d\phi_r/c^2 \quad (83)$$

The last term can be expressed in terms of the area swept out by the light ray, projected on the equatorial plane (see Eq. (77)). The result is as follows:

$$\Delta t' = \int_{\text{path}} \frac{dl}{c}\left(1 - \frac{(\Phi + \Phi_{\text{eff}} - \Phi_0)}{c^2}\right) + \frac{2\mathbf{\Omega} \cdot \mathbf{A}_e}{c^2} \quad (84)$$

The area A_e is the total area swept out by the radius vector from the center of the Earth to the light ray while it propagates from transmitter to receiver. Only the area as projected on the equatorial plane will contribute to the Sagnac correction. The potential correction terms in Eq. (84) are very small, but may be significant in the future. The main correction terms is the Sagnac term, the last term in Eq. (84).

6. Sagnac Correction

a. Example 1. Let us neglect the potential term in Eq. (84). Consider light signals sent along a meridian (see Fig. 20). In this case, the radius vector to the light signal sweeps out no area at all, and the Sagnac correction vanishes. Similarly, portable clocks carried from the North pole down along meridians could distribute coordinate time without applying any Sagnac corrections. Thus, time could be transferred from A to B in Fig. 19 along meridians, without applying Sagnac corrections.

b. Example 2. Consider signals or clocks sent eastward in a circle, along a parallel of latitude at geocentric latitude λ, a total distance l. Then $\sin\theta = \cos\lambda$, $\int d\phi_r = l/(r(\lambda)\cos\lambda)$, and the Sagnac correction is as follows:

$$\Delta t' = \frac{2\Omega_e}{c^2} \times \frac{1}{2}(r(\lambda)\cos\lambda)^2 \times \frac{l}{r(\lambda)\cos\lambda} = \frac{\Omega_e r(\lambda) l \cos\lambda}{c^2} \quad (85)$$

If the signals or clocks were sent westward instead of eastward, the Sagnac

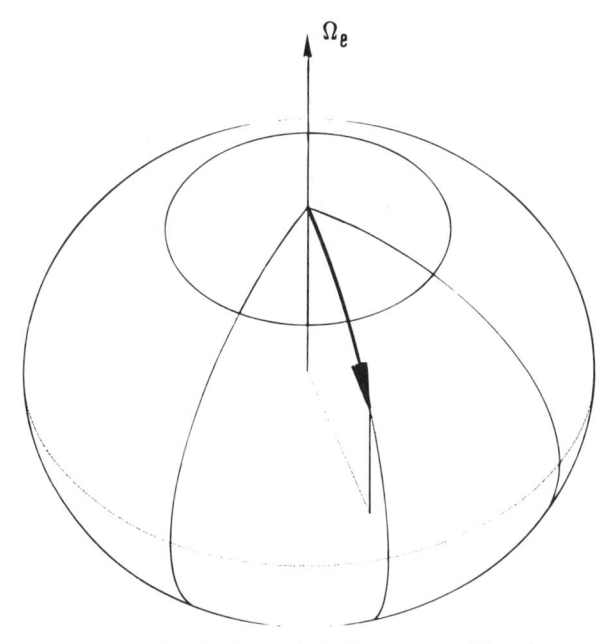

Fig. 20 Path of a portable clock carried along a meridian has zero associated projected area on the equatorial plane.

correction term would change sign. For one eastward circuit around the Earth's equator, the correction is $2\Omega_e \pi R_e^2/c^2 \approx 207$ ns.

   c. *Example 3.* Consider time transfer by portable clock transport from one point to another on the Earth's surface along a great circle. Let the plane of the great circle be inclined at angle I from the Earth's equatorial plane, and let the projection of the path traverse a total geographical longitude $\Delta\phi$, as projected on the equatorial plane. The Sagnac correction for this case is the following:

$$\Delta t' = \frac{2\Omega_e}{c^2} \frac{R_e^2 \cos I}{2} \Delta\phi$$

Figure 21 gives a plot of this correction term as a function of I and $\Delta\phi$. The effect is the same whether light signals or portable clocks are used for the time transfer.

   d. *Example 4.* Consider a receiver at a fixed location, at geographic latitude λ. Calculate the Sagnac correction for a single transmission from a GPS satellite in a circular orbit of radius $a = 26{,}562$ km, with orbit inclination $I = 55$ deg, and line of ascending node Ω on the equator, relative to the longitude of the receiver. The radius vector from the Earth's center to the observer's position is $\boldsymbol{n} = \{\cos \lambda, 0, \sin \lambda\}$. The radius vector $\boldsymbol{r} = \{x, y, z\}$ to the satellite can be expressed as $x = a(\cos \Omega \cos F - \cos I \sin \Omega \sin F)$, $y = a(\sin \Omega \cos F + \cos$

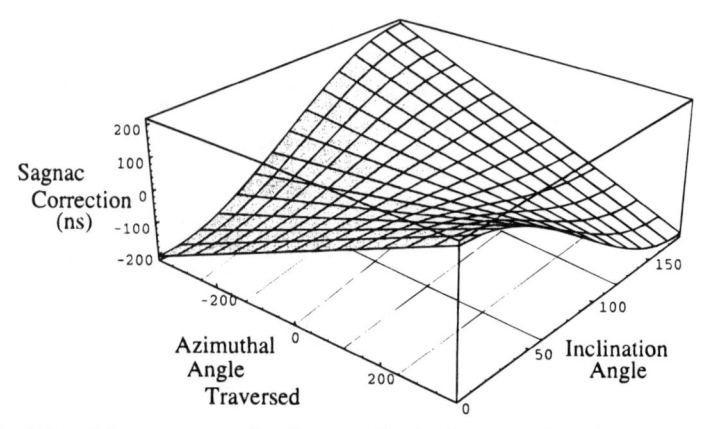

Fig. 21 Plot of Sagnac correction for coordinate time transfer along a great circle on the Earth's surface. The plane of the great circle is inclined at angle I from the Earth's equatorial plane, and the path projected on Earth's equatorial plane traverses longitude difference $\Delta\phi$.

$I \cos \Omega \sin F$), and, $z = a \sin I \sin F$, where $F = \sqrt{\mu/a^3}\, t'$ is the mean anomaly. Then the elevation angle E of the satellite above the observer's horizon is given by the following:

$$\sin E = \frac{\mathbf{r} \cdot \hat{n} - R_e}{\sqrt{r^2 - 2R_e \mathbf{r} \cdot \hat{n} + R_e^2}} \qquad (86)$$

The equatorial projection of area swept out by the signal pulse traveling from transmitter to receiver is the z-component of the following area:

$$A_z = \frac{1}{2} (\mathbf{r} \times (\hat{n} R_e))_z = -\frac{1}{2} R_e y \cos \lambda$$

Therefore, the Sagnac correction is the following:

$$\frac{2\mathbf{\Omega}_e \cdot \mathbf{A}}{c^2} = -\frac{\Omega_e R_e y \cos \lambda}{c^2} \qquad (87)$$

Equations (86) and (87) are used to plot the Sagnac correction in Fig. 22 for a user at geographic latitude 40° as a function of elevation angle E. Corrections are shown for several different nodal angles of the satellite orbit. For comparison of positions determined by GPS, it is the difference between such corrections that is important because some corrections are positive, and some are negative, and the effect can be magnified.

 e. Example 5. Figure 23 shows signal paths from three GPS satellites to three ground stations: National Bureau of Standards in Boulder, CO, USA; Tokyo Astronomical Observatory (TAO); and Physikalisch-Technische-Bundesanstalt (PTB) in Brauschweig, Germany. A given satellite is observed in common view by two ground stations resulting in cancellation of many common range errors. Time transfer around the world is accomplished by this means; the Sagnac correction is proportional to the area of the signal paths, projected onto the equatorial plane.[34]

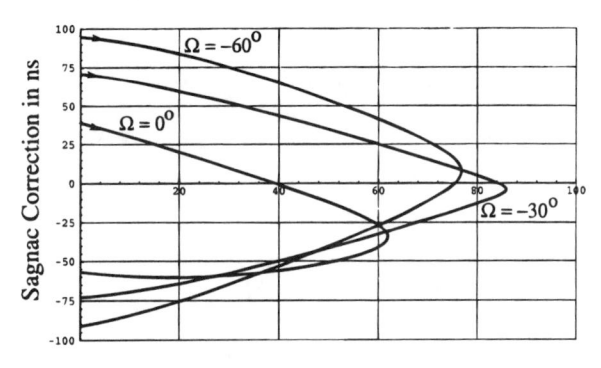

Fig. 22 Sagnac correction in ns for transmission from a GPS satellite to an Earth-fixed receiver at 40° N latitude. The angle Ω is the angle of the line of ascending nodes for the satellite orbit. The independent variable is the satellite elevation angle, viewed from the receiver position.

7. Sagnac Correction Viewed from Earth-Centered Inertial Frame

From the point of view of observers in the ECI frame, the Sagnac correction arises because of the motion of the receiver during propagation of the signal from transmitter to receiver. Similar discussions of the phenomenon on a rotating disc were given in Sec. II. Neglecting gravitational effects, signals propagate with speed c in straight lines in the ECI frame. Referring to Fig. 24, let there be a transmitter at position r_A at the instant a signal is transmitted. The figure shows a signal transmitted from a satellite to a ground-based receiver. At the instant of transmission, let the receiver position be r_B and suppose the receiver has velocity v. (See Table 6 for the net velocity when the receiver moves relative to the Earth's surface.) The total time of propagation of the signal to the receiver is t, and at the instant of reception, the receiver has moved to the position $r_B + vt$. From the point of view of the nonrotating frame, the signal goes in a straight line with speed c from the initial transmitter position r_A to the final receiver position. Thus constancy of c requires the following:

$$ct = |r_B + vt - r_A| \tag{88}$$

Iterative solution of Eq. (88) for t is equivalent to calculating the Sagnac correction. To see this in detail, set $r_B - r_A = \Delta r$, assume that the term involving v on the right-hand side of Eq. (88) is small, and neglect terms that involve v^2. Then, Eq. (88) may be rewritten approximately as $ct = \sqrt{(\Delta r)^2 + 2(\Delta r) \cdot vt}$. The second term under the square root is small compared to the first term, so a Taylor expansion in the small parameter vt gives $ct = |\Delta r| + (\Delta r) \cdot vt/|\Delta r|$. Substitute $|\Delta r|/t = c$ into the right-hand side of the above equation to yield a correction:

$$t = \frac{|\Delta r|}{c} + \frac{(\Delta r) \cdot v}{c^2} \tag{89}$$

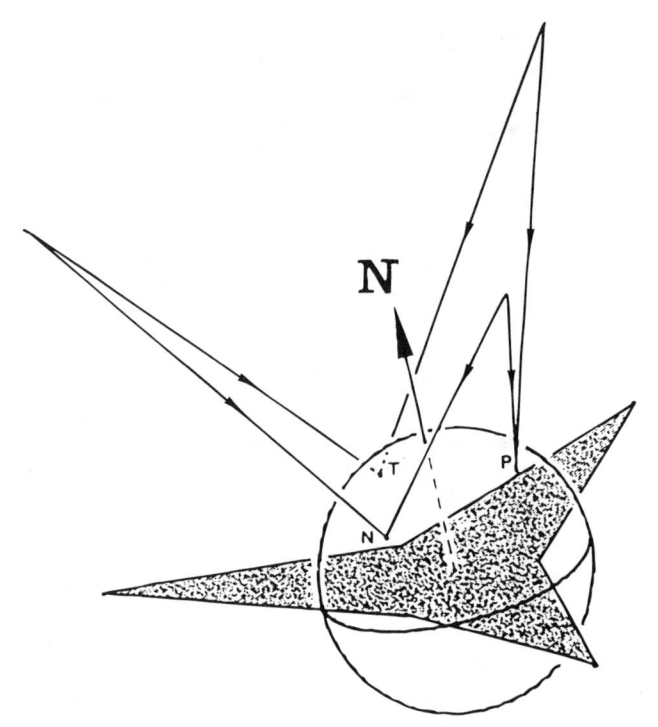

Fig. 23 Around-the-world time transfer using three GPS satellites in common view. The Sagnac correction is proportional to the shaded area outlined by projections of the signal paths on the Earth's equatorial plane.

The last term is the Sagnac correction. This shows analytically that a correction term of first order in the velocity of the moving receiver is obtained by iteratively solving the propagation delay equation for the propagation time and *stopping after a single iteration*. The correction is of the same form as that obtained in the previous section, as can be seen by inserting the expression $v = \Omega_e \times r_B$ for the Earth-fixed receiver. The correction term in Eq. (89) is as follows:

$$\Delta t_{\text{Sagnac}} = \frac{(r_B - r_A) \cdot \Omega_e \times r_B}{c^2} = \frac{r_A \times r_B \cdot \Omega_e}{c^2} = \frac{2\Omega_e}{c^2} \cdot \frac{r_A \times r_B}{2} \qquad (90)$$

In Eq. (90), $r_A \times r_B/2$ is the area of the triangle swept out by an arrow with its tail at Earth's center and its head following the electromagnetic signal pulse. The expression given in Eq. (90) is identical with the last term of Eq. (84).

f. Example 6—A low-Earth orbit satellite. Consider the application of Eq. (89) to a situation in which the receiver is not fixed on the surface of the Earth. Some LEO satellites (e.g., TOPEX) already have GPS receivers; the Space Shuttle is also expected to add such receivers. A good example concerns the proposal to place a GPS receiver in a low-Earth orbiting satellite, using radio occultation

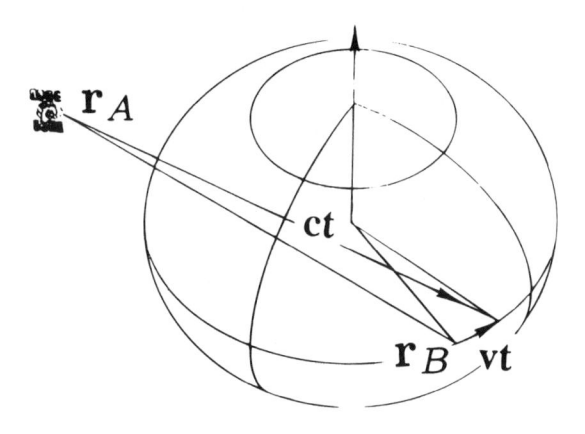

$$\left|\mathbf{r}_B + \mathbf{v}t - \mathbf{r}_A\right| = ct$$

Fig. 24 Time transfer from a GPS satellite to a receiver which moves during the propagation of the signal.

techniques to measure properties of the Earth's atmosphere.[35] The payload will perform ionospheric sounding with positions measured using a GPS receiver in a satellite at about 800-km altitude. The velocity of the satellite is approximately $v \approx 7.5 \times 10^3$ m/s. To estimate the magnitude of the correction given in Eq. (89), assume the receiver is 800-km above Earth's surface, so $r_B \approx 7.178 \times 10^6$m. Also, the distance from Earth's center to the GPS satellite is $r_A = a \approx 2.66 \times 10^7$m. Then, the difference between r_A and r_B could easily be as much as 30,000 km: $|r_A - r_B| \approx 3 \times 10^7$m. Assuming that the cosine that occurs in the dot product of Eq. (89) is unity, the maximum magnitude of the correction is $\Delta t_{\text{Sagnac}} \approx (v/c^2)|r_A - r_B| \approx 2500$ ns. This is a large correction. If, however, the receiver solves the propagation delay equation iteratively, the Sagnac correction will automatically be taken into account.

8. Effect of User Velocity and Altitude

To illustrate these effects consider the application of Eq. (78) to time transfer by a portable clock. For simplicity, suppose a jet aircraft carries a portable atomic clock at an altitude of 10 km, with a constant ground speed of 900 km/h, from the vicinity of Denver, Colorado to Krasnovodsk, Turkmenia, along a parallel of geographic latitude $\lambda' = 40°$. The longitude difference is 157.5°, at a distance from the Earth's rotation axis of 4893 km. The total distance traveled would be 13,500 km during a time 53,798 s. Then $gh/c^2 = 1.09 \times 10^{-12}$, and

$1/2\, v^2/c^2 = 3.48 \times 10^{-13}$. At such low speeds, the gravitational blue shift is larger than slowing because of time dilation. For this trip also, $(2\Omega_e \cdot A)/c^2 = 52.3$ ns. The process of time transfer would require that the following corrections be applied to the portable clock:

First:
$$\int dt\left(-\frac{gh}{c^2}\right) = -58.7 \text{ ns}$$

Second:
$$\int dt\left(+\frac{v^2}{2c^2}\right) = +18.7 \text{ ns}$$

Third:
$$\Delta t_{\text{Sagnac}} = \frac{2\Omega_e \cdot A}{c^2} = +53.4 \text{ ns}$$

The net correction to be applied to the portable atomic clock is 12.7 ns.

9. Summary of Coordinate Time Concept

To generate coordinate time, which is necessary in order to obtain a self-consistent network of synchronized clocks, measurements of position and velocity of a moving atomic clock must be recorded and used to compute corrections to time transferred by moving portable clocks or by electromagnetic signals. Equation (78) for portable clocks and (84) for light signals express the necessary corrections in detail. Thus, referring to Fig. 25, the process of generating a coordinate clock might be described as follows. Theory (General Relativity) is embodied in a computer program, that processes measurements of proper time actually elapsed on a moving clock, or coordinate distance traversed by light signals, to apply corrections that yield coordinate time. This coordinate time has its rate defined by standard clocks at rest on the geoid. Synchronization of the clocks involves application of corrections according to Eqs. (78) or (84) as appropriate; clocks will be synchronized in the underlying ECI frame. Rescaling of the rate of coordinate time to conform to the SI second, which is defined by clocks on the geoid, necessarily involves the correction term Φ_0.

B. Relativistic Effects for Users of the GPS

1. Effects on Clocks in Circular Orbits

The main relativistic effects can be modeled with a spherically symmetric gravitational field with potential $\Phi = -\mu/r$. It is also most convenient to describe satellite motion from the viewpoint of the ECI frame. An appropriate form of

INTRODUCTION TO RELATIVISTIC EFFECTS

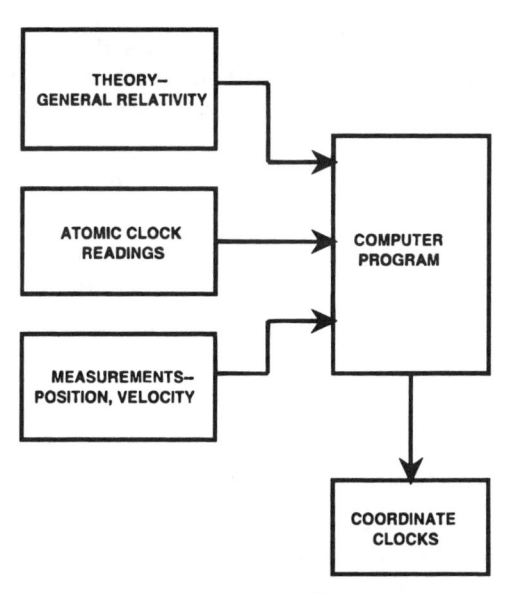

Fig. 25 Schematic diagram of the ingredients in the practical generation of coordinate time. A theory (General Relativity) is embodied in a computation that applies corrections to measured proper times, depending upon position and velocity, to generate coordinate time.

the line element, using Schwarzschild coordinates, is as follows:

$$ds^2 = -\left(1 + \frac{2\Phi}{c^2}\right)(cdt)^2 + \left(1 + \frac{2\Phi}{c^2}\right)^{-1} dr^2 + r^2 d\theta^2 + r^2 \sin^2\theta d\phi^2 \quad (91)$$

where r, θ, ϕ represent the coordinates of the satellite, and t is coordinate time measured by a standard clock at rest at infinity. Note that the time scale change represented by Eq. (66) has not yet been incorporated in Eq. (91).

The JGM-2 value of μ is $\mu = 398600.4415 \times 10^{14}$ m³/s². Define the ratio $m_e \equiv \mu/c^2 = 4.4350280 \times 10^{-3}$ m. Because the potential is spherically symmetric, and the orbit is circular, coordinates can be aligned with the satellite orbit plane so that $\theta = \pi/2$, $dr = d\theta = 0$. In Schwarzschild coordinates, Kepler's third law is of the same form as in Newtonian mechanics. Then setting $r = r_c$ for a circular orbit, the velocity is $v_c^2 = (r_c d\phi/dt)^2 = \mu/r_c = m_e c^2/r_c$. Then, the invariant interval along the satellite path is as follows:

$$\begin{aligned}
ds^2 &= -\left(1 - \frac{2m_e}{r_c}\right)(cdt)^2 + (r_c d\phi)^2 \\
&= -\left(1 - \frac{2m_e}{r_c} - \frac{m_e}{r_c}\right)(cdt)^2 \\
&= -\left(1 - \frac{3m_e}{r_c}\right)(cdt)^2
\end{aligned} \quad (92)$$

Note that higher-order terms in the ratio m_e/r_c can clearly be neglected compared to unity, because for $r_c > R_e = 6.38 \times 10^6$ m, $(m_e/r_c)^2 < 5 \times 10^{-19}$. The proper time elapsed on a satellite clock is, then, the following:

$$d\tau_{sat} = |ds/c| \cong \left(1 - \frac{3m_e}{2r_c}\right)(dt) \tag{93}$$

On the other hand, to the same level of accuracy the proper time elapsed on a MCS clock fixed on the Earth's geoid is given by Eq. (65):

$$d\tau_{ref} \cong (1 + \Phi_0/c^2)(dt) \tag{94}$$

The ratio of elapsed times is the following:

$$\frac{d\tau_{sat}}{d\tau_{ref}} = \frac{1 - 3m_e/2r_c}{1 + \Phi_0/c^2} \cong 1 - \frac{3m_e}{2r_c} - \frac{\Phi_0}{c^2} \tag{95}$$

The value of Φ_0 is given in Eq. (56); the last correction term in Eq. (95), thus, has the value: $-\Phi_0/c^2 = +6.9692842 \times 10^{-10}$. Define the fractional clock rate offset of the satellite clock, relative to reference clocks, as $\delta(r_c) = d\tau_{sat}/d\tau_{ref} - 1$. Figure 26 shows a plot of the fractional clock rate offset $\delta(r_c)$ as a function of the orbit radius r_c. Note that the relativistic clock rate effect takes on both positive and negative values, depending upon the orbit radius. For a satellite clock in very high orbit, the Φ_0 correction, which is dominated by the gravitational potential caused by the Earth, gives the largest contribution; the satellite clock is blue shifted relative to Earth-fixed clocks. A satellite clock in low orbit is slowed because of the second-order Doppler shift (time dilation). There is an intermediate radius at which these effects, as slightly modified by Earth rotation and quadrupole potential, exactly cancel. The effect is zero at

$$r_c = -3c^2 m_e/2\Phi_0 = 3\mu/2[\mu(1 + J_2/2)/R_e + \Omega_e^2 R_e^2/2] \approx 9547 \text{ km}.$$

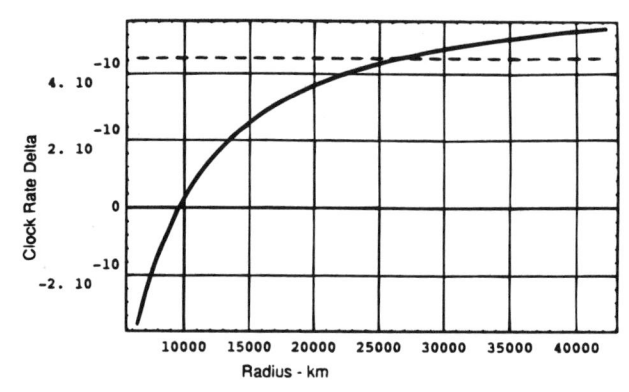

Fig. 26 Relativistic clock rate offset for a satellite with circular orbit in a symmetrical Earth gravitational field plotted vs satellite orbit radius. The relativistic offset of 4.46475×10^{-10} is shown on a dashed line for the GPS satellite radius of 26,562 km.

INTRODUCTION TO RELATIVISTIC EFFECTS

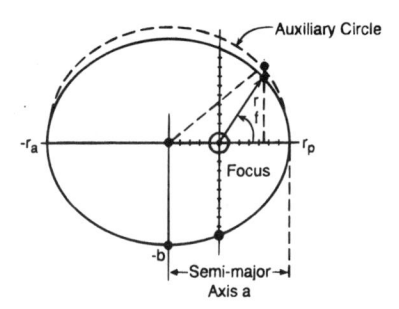

Fig. 27 Key parameters for an elliptical satellite orbit. For the GPS, the orbit eccentricity is generally very small, $e < 0.01$.

At the GPS orbit radius $r_c = 26562$ km, the fractional clock rate offset is as follows:

$$-\frac{3GM}{2ac^2} - \frac{\Phi_0}{c^2} = -2.5045336 \times 10^{-10} + 6.9692842 \times 10^{-10}$$
$$= +4.46475 \times 10^{-10} \quad (96)$$

a fixed constant for a circular orbit system. The relativistic effect would cause the 10.23 MHz clock to be shifted high when viewed from the Earth's surface, by 10.23 MHz $\times$ 4.46475 $\times$ 10^{10} or .004567 Hz. The GPS compensates for this effect by purposely setting the 10.23 MHz clocks lower in frequency at the factory by the same amount, to 10.229 999 999 543 MHz. Note that GPS satellites at the same orbit altitude, in circular orbits, all run at exactly the same rate.

2. Effect of Orbit Eccentricity

Consider next, the effect of orbit eccentricity, as shown in Fig. 27. Key orbital parameters for a Keplerian elliptical orbit are shown in Table 12.

We treat this effect by deriving an expression for the elapsed coordinate time dt' in terms of the elapsed proper time and the velocity, for noncircular orbits.

Table 12 Key parameters of an elliptical satellite orbit

Elliptical orbit of a satellite
    a = semimajor axis
    e = eccentricity = $(r_a - r_b)/2a = \sqrt{1 - b^2/a^2}$
    $p = a(1 - e^2)$ = semilatus rectum
    r = orbit radius, $r = p/(1 + e \cos f) = a(1 - e^2)/(1 + e \cos f)$
    r_a = apogee radius = $a(1 + e)$
    r_p = perigee radius = $a(1 - e)$
    ϵ = specific mechanical energy = $(v^2/2) - \mu/r$
    f = true anomaly, $\cos f = (\cos E - e)/(1 - e \cos E)$
    M = mean anomaly = $n(t' - t'_p)$, $M = E - e \sin E$
    v = speed of satellite = $\sqrt{\mu[(2/r) - (1/a)]}$
    P = period = $2\pi\sqrt{a^3/\mu}$
    $n = 2\pi/P = \sqrt{\mu/a^3}$

An appropriate starting point is the line element in isotropic coordinates, Eq. (51). Factoring $(cdt)^2$ out of the right side of Eq. (51), we have the following:

$$ds^2 = -\left(1 + \frac{2\Phi}{c^2} - \left(1 - \frac{2\Phi}{c^2}\right)\frac{(dr^2 + r^2 d\theta^2 + r^2 \sin^2\theta d\phi^2)}{(cdt)^2}\right)(cdt)^2 \quad (97)$$

Because satellite velocities are small compared to c, the potential term $2\Phi/c^2$ in the last term of Eq. (97) can be neglected. For the same reason, it does not matter whether isotropic coordinates or Schwarzschild coordinates are adopted. To use clocks at rest on the Earth's surface as reference clocks, the scale change in Eq. (66) must be made. Then, because $v^2 = (dr^2 + r^2 d\theta^2 + r^2 \sin^2\theta d\phi^2)/dt^2$, we have the following:

$$ds = \sqrt{\left(1 + \frac{2(\Phi - \Phi_0)}{c^2} - \frac{v^2}{c^2}\right)}(cdt') \quad (98)$$

Eq. (98) may now be solved for dt', expanding in terms of small quantities of order $1/c^2$ as in so many previous examples, and putting $d\tau = |ds/c|$

$$dt' = d\tau\left(1 + \frac{v^2}{2c^2} - \frac{(\Phi - \Phi_0)}{c^2}\right) \quad (99)$$

Here Φ is the Earth's gravitational potential at the satellite's position. The Earth's quadrupole potential falls off so rapidly with distance r from the Earth's center, that its effect on clocks in satellites is small. The quadrupole part of the potential may affect satellite motion slightly; however its effect on the frequency shift of a clock in a GPS satellite is only about one part in 10^{14}; this is discussed later, but is neglected here. The quantity Φ_0, which appears in Eq. (57), represents the frequency shift, relative to a clock at rest at infinity, of a clock on the geoid. The quadrupole contributes to Φ_0 in the term $-GM_e J_2/2R_e$ in Eq. (56); there it contributes a fractional rate correction of -3.77×10^{-13}. This effect is important and is accounted for in the GPS.

Now we discuss the Keplerian orbits, along which the satellites move. This is done here to obtain relations with which to simplify Eq. (99). Because the quadrupole (and higher multipole) parts of the Earth's potential are negligible, in Eq. (99) we can take $\Phi = -\mu/r$. Then, the expressions can be evaluated using what is known about the Newtonian orbital mechanics of the satellites. Let the satellite orbit be a Keplerian orbit of semimajor axis a and eccentricity e (see Fig. 27 and Table 12). Then, the solution of the orbital equations is as follows.[36]

The distance r from the center of the Earth to the satellite is as follows:

$$r = a(1 - e^2)/(1 + e\cos f) \quad (100)$$

The angle f, called the true anomaly, is measured from perigee to the satellite's instantaneous position. The true anomaly can be calculated in terms of another quantity E, called the eccentric anomaly, according to the following relationships:

$$\cos f = \frac{\cos E - e}{1 - e \cos E} \qquad (101)$$

$$\sin f = \sqrt{1 - e^2} \, \frac{\sin E}{1 - e \cos E}$$

Then another way to write the distance r is as follows:

$$r = a(1 - e \cos E) \qquad (102)$$

To find the eccentric anomaly E, we must solve the following transcendental equation:

$$E - e \sin E = \sqrt{\frac{\mu}{a^3}} (t' - t'_p) \qquad (103)$$

where t'_p is the time of perigee passage.

In Newtonian mechanics, the gravitational field is a conservative field, and total energy is conserved. Using the above equations for the Keplerian orbit, we can show that the total energy per unit mass of the satellite is as follows:

$$\frac{1}{2} v^2 - \frac{\mu}{r} = -\frac{\mu}{2a} \qquad (104)$$

Solving Eq. (104) for v^2 and substituting into Eq. (99), then the elapsed coordinate time along the satellite clock path is as follows:

$$\Delta t' = \int_{\text{path}} (d\tau) \left[1 + \frac{3GM}{2ac^2} + \frac{\Phi_0}{c^2} - \frac{2GM}{c^2} \left(\frac{1}{a} - \frac{1}{r} \right) \right] \qquad (105)$$

The first two constant rate correction terms in Eq. (105) have already been discussed.

3. The $e \sin E$ Orbit Eccentricity Effect

The last term in Eq. (105) may be integrated exactly by using the following expression for the rate of change of eccentric anomaly with time, which follows by differentiating Eq. (103):

$$\frac{dE}{dt'} = \frac{\sqrt{\mu/a^3}}{(1 - e \cos E)} \qquad (106)$$

Then

$$\int \left[\frac{2\mu}{c^2} \left(\frac{1}{r} - \frac{1}{a} \right) \right] dt' = \frac{2\mu}{c^2} \int \left(\frac{1}{r} - \frac{1}{a} \right) dt'$$

$$= \frac{2\mu}{ac^2} \int dt' \left(\frac{e \cos E}{1 - e \cos E} \right)$$

$$= \frac{2\sqrt{\mu a}}{c^2} e [\sin E - \sin E_0]$$

$$= +4.4428 \times 10^{-10} \, e\sqrt{a} \, \sin E + \text{constant} \qquad (107)$$

with the semimajor axis a in meters. The constant of integration in Eq. (107) can be dropped, because this term is lumped with other clock offset effects in the Kalman filter computation of the clock correction model. The net clock drift attributable to relativistic effects that vary with distance is as follows:

$$\Delta t' = +4.4428 \times 10^{-10} \frac{1}{\sqrt{m}} e\sqrt{a} \sin E \qquad (108)$$

where a is in meters. This correction must be made by the receiver; it is a correction to the coordinate time as transmitted by the satellite. For a satellite of eccentricity $e = 0.01$, the maximum size of this term is about 23 ns. The correction is needed because of a combination of effects on the satellite clock attributable to gravitational frequency shift, and second-order Doppler shift, which vary because of the orbit eccentricity.

Equation (108) can be expressed without approximation in the following alternative form:

$$\Delta t' = +\frac{2\mathbf{r} \cdot \mathbf{v}}{c^2} \qquad (109)$$

where $\mathbf{r}$ and $\mathbf{v}$ are the position and velocity of the satellite at the instant of transmission. This can be proved using the above expressions for the Keplerian orbits of the satellites. This latter form is usually used in implementations of the receiver software.

Thus, there are two corrections the receiver must apply in order to compute coordinate time correctly. First, the correction of Eq. (108) must be calculated to correct the coordinate time values transmitted from the satellite. Second, the Sagnac correction must be applied to allow for extra signal propagation time, because the receiver moves as the Earth spins; equivalently, four propagation delay equations can be solved iteratively to give the receiver's position and time.

4. Time Transfer in Crosslink Ranging

Next consider the process of transferring time from one satellite to another by crosslink ranging. The standard atomic clock in the satellite suffers a rate adjustment and an "$e \sin E$" correction to get the coordinate time. Then, a signal is sent to the second satellite, which requires calculating a coordinate time of propagation incorporating a relativistic time delay. There is, then, a further transformation of rate and another "$e \sin E$" correction to get the atomic time on the receiving satellite's clock. So that the rate adjustment does not introduce confusion into this process, we assume the rate adjustments are already accounted for and use the subscript S to denote time measurements using rate-adjusted satellite clocks.

Then, let a signal be transmitted from satellite #1, at position X_1 and having velocity V_1 in ECI coordinates, at satellite time $T_S^{(1)}$. The coordinate time at which this occurs, apart from a constant offset, is given by Eq. (107):

$$T^{(1)} = T_S^{(1)} + \frac{2\sqrt{\mu a_1}}{c^2} e_1 \sin E_1 \qquad (110)$$

The coordinate time elapsed during propagation of the signal to the receiver in first approximation l/c, where l is the distance between transmitter at the instant of transmission and receiver at the instant of reception: $\Delta T = T^{(2)} - T^{(1)} = l/c$. The Shapiro time delay corrections to this are discussed in the next section. Finally, the coordinate time of arrival of the signal is related to the time on the receiving satellite's adjusted clock by the inverse of Eq. (110):

$$T_S^{(2)} = T^{(2)} - \frac{2\sqrt{\mu a_2}}{c^2} e_2 \sin E_2 \qquad (111)$$

Collecting all these results, we have the following:

$$T_S^{(2)} = T_S^{(1)} + \frac{l}{c} - \frac{2\sqrt{\mu a_2}}{c_2} e_2 \sin E_2 + \frac{2\sqrt{\mu a_1}}{c^2} e_1 \sin E_1 \qquad (112)$$

In Eq. (112) the distance l is the actual propagation distance, in ECI coordinates, of the signal. If instead, this is expressed, in terms of the distance $|\Delta r|$ between the two satellites at the instant of transmission, then, as in Eq. (89),

$$l = |\Delta r| + \Delta r \cdot v/c$$

and Eq. (112) becomes the following:

$$T_S^{(2)} = T_S^{(1)} + \frac{|\Delta r|}{c} - \frac{2\sqrt{\mu a_2}}{c^2} e_2 \sin E_2 + \frac{2\sqrt{\mu a_1}}{c^2} e_1 \sin E_1 + \frac{\Delta r \cdot v}{c^2} \qquad (113)$$

This result contains all the relativistic corrections that must be considered for direct time transfer by transmission of a time-tagged pulse from one satellite to another. The times $T_S^{(i)}$ are measured on the satellite clocks that have already had their rate adjusted, so the relativistic corrections involved in this rate adjustment have already been accounted for. The term $|\Delta r|/c$, would be the ordinary propagation delay of a signal traveling at the speed of light. There are two "$e \sin E$" correction terms arising, because the in-and-out motion of a satellite whose orbit is eccentric causes both motional and gravitational frequency shifts in the satellite clocks. Finally, there is a term that accounts for the motion of the receiver during the propagation time of the signal.

Because the eccentricity corrections have already been well studied and are normally already incorporated into the GPS receiver software, it should only be necessary to comment that ordinarily the receiver is on the surface of the Earth, and we only have to apply one such correction in the process of calculating the coordinate time at the receiver. In crosslink ranging, however, the receiver may not be simply slaved to a transmitter, but it may have its own very stable atomic clock. Therefore, in comparing the coordinate time—GPS time—to the time on the atomic clock in the receiver, another correction is necessary, the second correction term in Eq. (113), $-2\sqrt{\mu a_2} e_2 \sin E_2/c^2$. We do not discuss these corrections further here, because they are well known. The important relativistic effects are summarized in Table 13.

C. Secondary Relativistic Effects

1. Secondary Periodic Effect of the Earth's Quadrupole Moment on Orbiting Clocks

There is a small secondary effect on orbiting GPS clocks caused by the ellipsoidal shape of the Earth. The resulting quadrupole potential has a small

Table 13 Summary of important relativistic effects in GPS

A. **Effects on ground-based clocks**
  A.1.1. If clocks are at rest on the Earth's surface:
    (a) Fractional frequency shift attributable to Earth's mass: $-GM_e/(c^2 R_e) = -6.9534858 \times 10^{-10}$
    (b) Frequency shift attributable to Earth's quadrupole moment: $-GM_e J_2/(2c^2 R_e) = -3.7640154 \times 10^{-13}$
    (c) Fractional second-order Doppler shift attributable to Earth's spin; at equator $-(1/2)\Omega_e^2 R_e^2/c^2 = -1.2034 \times 10^{-12}$
    (d) Frequency shift attributable to altitude h above the geoid $g_{\text{eff}}(\lambda)h/c^2$, $g_{\text{eff}}/c^2 \approx 1.09 \times 10^{-13}$ km^{-1}
  A.2.2. If clocks are in motion relative to Earth's surface, then in addition to (a, b, c, d) above:
    (e) Fractional second-order Doppler shift attributable to ground speed v, $-(1/2)v^2/c^2$
    (f) Sagnac correction, $2\Omega_e \cdot A/c^2$
B. **Effects on clocks in satellites**
    (g) Fractional frequency shift attributable to Earth's mass, for a circular orbit: $-GM_e/(ac^2) = -1.669689 \times 10^{-10}$
    (h) Fractional second-order Doppler shift attributable to orbital motion: $-(1/2)v^2/c^2 = -8.3485 \times 10^{-11}$
    (a, b, d, g, h.1) constant part–included in frequency offset of transmitter: -0.004567 Hz;
    (a, b, d, g, h.2) $e \sin E$–implemented in receiver; correction is $+4.4428 \times 10^{-10} \, e\sqrt{a} \sin E$
C. **Effects on electromagnetic signals**
    (i) Sagnac correction, $2\Omega_e \cdot A/c^2$

effect on the satellite clock phase. Although this effect is more pronounced at orbit altitudes close to the Earth, there is a small effect, even at GPS altitudes.

The line element for a general gravitational potential, to leading order in $1/c^2$, is given in isotropic coordinates by Eq. (69), in which the time scale change required for clocks at rest on the Earth's surface to serve as references, has been included. Introducing the satellite velocity

$$v^2 = (dr^2 + r^2 \, d\theta^2 + r^2 \sin^2 \theta \, d\phi^2)/dt'^2,$$

the proper time $d\tau = |ds/c|$ elapsed can be expressed for low velocities as follows:

$$|c \, d\tau|^2 = |ds^2| = (c \, dt')^2 \left(1 + \frac{2\Phi - 2\Phi_0}{c^2} - \frac{v^2}{c^2}\right) \quad (114)$$

For satellite motion in a conservative potential, we have the following for the specific energy:

$$\frac{1}{2}v^2 + \Phi = \epsilon = \text{constant} \quad (115)$$

Substituting for v^2 into Eq. (114) gives the following:

$$|ds^2| = (cdt')^2\left(1 + \frac{2\Phi - 2\Phi_0}{c^2} - \frac{2\epsilon - 2\Phi}{c^2}\right) \tag{116}$$

Therefore, solving approximately for dt' in terms of the elapsed proper time $d\tau$ on the orbiting clock, we get the following:

$$dt' = d\tau\left(1 - \frac{2\Phi - \Phi_0 - \epsilon}{c^2}\right) \tag{117}$$

Now, using Eq. (52) for the potential, including the quadrupole contribution, gives the following:

$$dt' = d\tau\left\{1 + \frac{\Phi_0 + \epsilon}{c^2} + \frac{2GM_e}{c^2 r}\left[1 - J_2\left(\frac{R_e}{r}\right)^2\left(\frac{3\sin^2\lambda - 1}{2}\right)\right]\right\} \tag{118}$$

The specific energy will differ slightly from the value given in Eq. (104) because of the additional terms proportional to J_2 in the gravitational potential, which will depend on the particular orbit. This will contribute a small constant rate shift of order $J_2 (R_e/r)^2 \approx 6 \times 10^{-5}$ smaller than the contributions calculated in Eq. (96). These are contributions of the order of 2 parts in 10^{14} and are not currently accounted for.

It is of interest here to examine the contribution to the required time correction, proportional to J_2 in Eq. (118), which have not previously been considered. Call the extra coordinate time correction $\Delta t'_{J_2}$. Then, we have the following:

$$\Delta t'_{J_2} = -\frac{2GM_e J_2 R_e^2}{c^2}\int_{\text{path}}\left(\frac{3\sin^2\lambda - 1}{2r^3}\right)d\tau \tag{119}$$

This correction is very small, so it can be approximated by taking a circular Keplerian orbit in the integrand. Then $r = a$, $d\tau \simeq dt'$ and for an orbit of inclination I, $\sin\lambda$ is as follows:

$$\sin\lambda = \sin I \sin(\omega t') \tag{120}$$

where $\omega = \sqrt{\mu/a^3}$. Then, because $\sin^2 x = (1 - \cos 2x)/2$, the following results:

$$\Delta t'_{J_2} = -\frac{m_e J_2 R_e^2}{a^3}\int_{\text{path}} dt'\left\{\frac{3\sin^2 I}{2}[1 - \cos(2\omega t')] - 1\right\}$$

$$= -\frac{m_e J_2 R_e^2}{a^3}\left(\frac{3\sin^2 I}{2} - 1\right)\Delta t' + \frac{3m_e J_2 R_e^2 \sin^2 I}{4\omega a^3}\sin(2\omega t') + \text{constant}$$

$$\tag{121}$$

The first term in Eq. (121) corresponds to a very small constant rate correction. At the nominal orbit inclination of GPS satellites, $I = 55$ deg, we have the following:

$$\frac{3 \sin^2 I}{2} - 1 = .006515 \qquad (122)$$

The first term, then, corresponds a constant rate correction of a few parts in 10^{17} and is negligible for practical purposes. The amplitude of the time-varying part of Eq. (121) is as follows:

$$\frac{3 m_e J_2 R_e^2 \sin^2 I}{4 \omega a^3} = 7.19 \times 10^{-11} \text{ s} \qquad (123)$$

In terms of an effective path length, this corresponds to a peak-to-peak path delay of 4.31 cm. This is plotted in Fig. 28.

2. Effect of Higher Multipole Moments

Higher-order multipole moment potentials attributable to the Earth fall off more rapidly with distance, and hence, give even smaller contributions than those calculated above. The largest multipole moment coefficients correspond to $l = 2$, $m = 2$ Stokes coefficients having values about 10^{-3} times J_2. Their largest contribution would come on the Earth's surface, where they would contribute a fractional rate effect of order 10^{-16}. This can be neglected.

3. Effect of the Sun and Moon on GPS

The principle of equivalence implies that an observer in free fall in the gravitational field of the solar system cannot sense the presence of external gravitational fields. Although at the instantaneous location of the freely falling observer there is a gravitational field of strength $-\nabla\Phi$ (force per unit mass), this field produces an acceleration $A = -\nabla\Phi$ of the falling observer. Because of this acceleration, an additional fictitious gravitational field $-A$ is induced in the observer's reference frame. The two fields—the real one and the induced one—cancel each

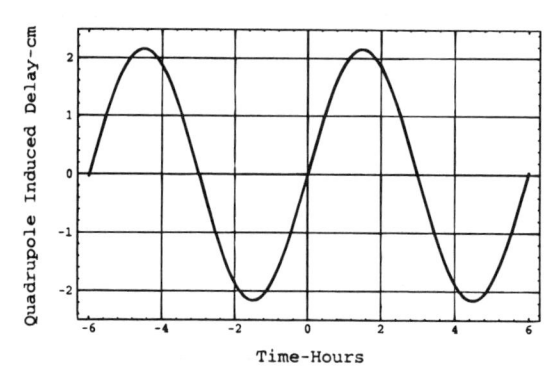

Fig. 28 Quadrupole moment of the Earth's gravitational field causes a small periodic secondary effect on the GPS satellite clock phase.

other; the net field strength at the observer's location is zero. This implies that the gravitational potential in the neighborhood of the freely falling observer cannot have any terms linear in the spatial coordinates. Only quadratic terms can survive—these are tidal terms. The tidal potentials associated with these residual effects are negligible in the GPS. To show this in more detail, we construct a local inertial frame extending over some small region of space in the neighborhood of the observer in free fall. The position of the freely falling observer becomes the origin of local inertial coordinates, and a standard clock at this position defines the rate of coordinate time in the local inertial frame. We apply the line element, Eq. (51), to the motion of this reference clock. Ignoring for the moment the potential attributable to the Earth itself, and retaining only leading relativistic correction terms, the proper time elapsed on the local reference clock is as follows:

$$d\tau = dT\left(1 - \frac{V^2}{2c^2} + \frac{\Phi}{c^2}\right) \qquad (124)$$

where V is the local frame's velocity through the solar system, Φ is the total gravitational potential due to all other solar system bodies, and T is coordinate time in the solar system reference frame, measured by an atomic clock at rest an infinite distance away from the sun, where $\Phi = 0$. The increment of elapsed coordinate time is then, the following to leading order:

$$dT = d\tau\left(1 + \frac{V^2}{2c^2} - \frac{\Phi}{c^2}\right) \qquad (125)$$

This is analagous to Eq. (99) and means that corrections must be applied to the proper time elapsed on the falling clock, which beats more slowly than solar system coordinate clocks because of the second-order Doppler shift (time dilation) and the gravitational frequency shift (always a red shift compared to clocks at infinity.) To help make the distinctions between local inertial coordinates and solar system coordinates, let us use lower case letters $\{t = \tau, r\}$ for local coordinates and upper case letters $\{T, R\}$ for solar system coordinates. Then, when time is distributed throughout the local frame by the Einstein synchronization process, solar system observers will not agree the clocks are synchronized—the breakdown of the concept of synchronization requires an additional term of the form $V \cdot r/c^2$ in the time transformation. Thus, the net coordinate transformation between solar system time T ($X^0 = cT$) and local time t ($x^0 = ct$; t is proper time on the freely falling clock) is as follows:

$$X^0 = cT = \int_{path} dx^0\left(1 + \frac{V^2}{2c^2} - \frac{\Phi(0)}{c^2}\right) + \frac{V \cdot r}{c} \qquad (126)$$

where $\Phi(0)$ is the external potential at the origin of the local frame.

Local spatial coordinates differ from solar system spatial coordinates by two relativistic effects. First, there is Lorentz contraction of the falling measuring rods. Second, the gravitational potential Φ causes the rods to appear to be lengthened. As can be seen from Eq. (41), the standard distance in the presence of a gravitational field Φ is as follows:

$$d\sigma^2 = (1 - 2\Phi/c^2)(dX^2 + dY^2 + dZ^2) \tag{127}$$

which means that a local standard coordinate distance x^i, corresponding to some increment $\Delta\sigma$, needs to be multiplied by a factor $(1 + \Phi/c^2)$ to relate it to solar system coordinate distance. The transformation from local spatial coordinates to solar system spatial coordinates is then of the following form:

$$X^i = X^i(T) + x^i(1 + \Phi(0)/c^2) + \frac{V^i \mathbf{V} \cdot \mathbf{r}}{2c^2}; \quad i = 1, 2, 3 \tag{128}$$

where $X^i(T)$ is the position of the origin of the local freely falling frame, and the last term represents the leading correction for Lorentz contraction in the direction of the velocity.

With these transformations, we can calculate the components of the metric tensor $g_{\alpha\beta}$ in the local frame. The standard transformation equations[27] for a second-rank tensor are as follows.

$$g_{\alpha\beta} = \sum_{\mu,\nu} \frac{\partial X^\mu}{\partial x^\alpha} \frac{\partial X^\nu}{\partial x^\beta} G_{\mu\nu} \tag{129}$$

To leading order, we take the nonzero components of the metric tensor in the solar system to be only those arising from the total potential:

$$G_{00} = -(1 + 2\Phi/c^2), \quad G_{ij} = \delta_{ij}(1 - 2\Phi/c^2) \tag{130}$$

The partial derivatives needed to perform the summation in Eq. (129) are to leading order as follows:

$$\frac{\partial X^0}{\partial x^0} = 1 + \frac{V^2}{2c^2} - \frac{\Phi(0)}{c^2} + \frac{\mathbf{A} \cdot \mathbf{r}}{c^2}$$

$$\frac{\partial X^i}{\partial x^0} = \frac{\partial X^0}{\partial x^i} = \frac{V^i}{c}$$

$$\frac{\partial X^i}{\partial x^j} = \delta_{ij}(1 + \Phi(0)) + \frac{V^i V^j}{2c^2} \tag{131}$$

In writing the above equations, we have used the fact that the time derivative of the position of the origin of the local frame is the frame's velocity, and the time derivative of the frame's velocity is the frame's acceleration $\mathbf{A}$. Then, to lowest order, we derive the following:

$$\begin{aligned}
g_{00} &= -\left(1 + \frac{V^2}{2c^2} - \frac{\Phi(0)}{c^2} + \frac{\mathbf{A} \cdot \mathbf{r}}{c^2}\right)^2 \left(1 + \frac{2\Phi}{c^2}\right) + \frac{V^2}{c^2} \\
&= -\left(1 + \frac{V^2}{c^2} - \frac{2\Phi(0)}{c^2} + \frac{2\mathbf{A} \cdot \mathbf{r}}{c^2} + \frac{2\Phi}{c^2}\right) + \frac{V^2}{c^2} \\
&= -\left(1 + 2\frac{\Phi - \Phi(0) + \mathbf{A} \cdot \mathbf{r}}{c^2}\right)
\end{aligned} \tag{132}$$

Now consider expanding the potential Φ in a Taylor series about the origin of local coordinates:

$$\Phi = \Phi(0) + \nabla\Phi \cdot \mathbf{r} + \frac{1}{2}\sum_{i,j} \frac{\partial^2 \Phi}{\partial X^i \partial X^j} x^i x^j \tag{133}$$

where we have set $X^i - X^i(0) = x^i$, because only small errors are made in a correction that is already small. Thus, we have the following:

$$g_{00} = -\left\{1 + \frac{2}{c^2}\left[(\nabla\Phi + \mathbf{A})\cdot \mathbf{r} + \frac{1}{2}\sum_{i,j} \frac{\partial^2\Phi}{\partial X^i \partial X^j} x^i x^j\right]\right\} \tag{134}$$

The derivatives in this result are evaluated at the origin of the freely falling frame. At that point, however, the acceleration $\mathbf{A}$ precisely cancels the gradient of the potential:

$$\mathbf{A} = -\nabla\Phi \tag{135}$$

The remaining terms in the local metric are tidal terms, depending upon the second derivative of the potential. To estimate the sizes of such terms, consider the solar potential $\Phi = -GM_\odot/R$, where $M_\odot$ is the sun's mass, and $GM_\odot/c^2 = 1477$ m. The tidal contribution to G_{00} is of order $GM_\odot r^2/R^3$, where r is the distance of a clock in the freely falling frame from the origin, and R is the distance of the Earth from the sun. If we take $r \simeq 26{,}572$ km corresponding to a GPS satellite, and $R \simeq 1.5 \times 10^{11}$ m, the fractional frequency shift arising from such tidal terms is approximately $GM_\odot r^2/c^2 R^3 \simeq 3 \times 10^{-16}$; such small contributions are completely negligible.

For the moon, $R = 384{,}000$ km and $GM/c^2 = 5.4 \times 10^{-7}$ m, so $GMr^2/c^2 R^3 \simeq 7 \times 10^{-15}$, which is still negligible.

A similar calculation can easily be carried out for the spatial components of the local metric tensor; however this requires including higher terms in the transformation (128). This problem has been studied in the literature and the reader should consult Ref. 37 for details.

We conclude that although tidal effects may give rise to significant forces on objects near Earth, the associated gravitational frequency shifts are negligible. Note that the acceleration term in Eq. (134) can be traced directly to the resynchronization term—a Lorentz transformation effect—in Eq. (126). This illustrates the importance of properly accounting for the breakdown of simultaneity when discussing relativistic effects.

4. Signal Propagation Delay Caused by the Earth's Gravitational Field

The gravitational field of the Earth causes the coordinate speed of light to depart slightly from its nominal value c, generally resulting in an increase in propagation time. In Schwarzschild coordinates, the coordinate speed along a radial line to the Earth's center is slightly different from the coordinate speed perpendicular to that radial line.[14] In isotropic coordinates, the coordinate speed is independent of direction; thus, great care is needed to interpret such effects properly. The coordinate time delay for Earth–Venus links grazing the sun has been verified experimentally and is often termed the Shapiro time delay effect.[39]

The time delay for GPS is small, as can be seen from the following calculation. Model the Earth's field as spherically symmetric, and use the invariant line element in isotropic coordinates, which is approximately as follows:

$$ds^2 = -\left(1 - \frac{2m_e}{r}\right)(cdt)^2 + \left(1 + \frac{2m_e}{r}\right)(dr^2 + r^2 d\theta^2 + r^2 \sin^2\theta d\phi^2) \quad (136)$$

where $m_e = GM_e/c^2$. Let $dl^2 = dr^2 + r^2 d\theta^2 + r^2 \sin^2\theta d\phi^2$. Along the path of the signal, $ds^2 = 0$, so solving for dt to leading order and integrating along the signal path, the total signal propagation time is as follows:

$$\Delta t = \int_{\text{path}} dt = \frac{1}{c}\int_{\text{path}} dl\left(1 + \frac{2m_e}{r}\right) = \frac{l}{c} + \frac{2m_e}{c}\int_{\text{path}} \frac{dl}{r} \quad (137)$$

The first term l/c would be the ordinary propagation time, except that l is a coordinate path length rather than a proper path length. The second term is the relativistic Shapiro delay. The equivalent path length of this latter delay is of order $2m_e \simeq 0.89$ cm. When the path from transmitter to receiver is radial, the extra relativistic delay is as follows:

$$\Delta t_{\text{delay}} = \frac{2m_e}{c}\int_a^{R_e} \frac{-dr}{r} = \frac{2m_e}{c} \ln\left(\frac{a}{R_e}\right) \quad (138)$$

where $a = 26{,}562$ km is the GPS satellite orbit radius. Then,

$$c\Delta t_{\text{delay}} = .89 \text{ cm} \times 1.427 = 1.26 \text{ cm}.$$

Next consider that the satellite is viewed at an elevation angle E rather than directly overhead. The excess path delay caused by the Earth's gravitational field can then be computed as shown in Fig. 29. The signal path is shown aligned with the x-axis in this figure. Thus, the radius from the Earth's center to a position x on the path has a value as follows:

$$r(x) = [(R_e \sin E + x)^2 + (R_e \cos E)^2]^{1/2} = \sqrt{x^2 + 2xR_e \sin E + R_e^2} \quad (139)$$

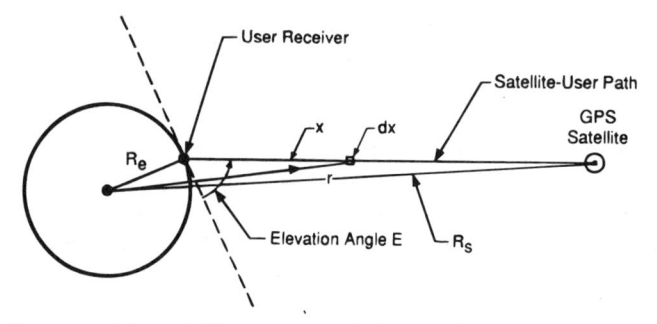

Fig. 29 Diagram of the satellite-to-user path for an elevation angle E. At position x, the radius from the Earth's center is $r(x)$. For simplicity of calculation, the x-axis has been aligned with the path of the satellite–user link.

and x ranges over the path from the transmitter position at the position:

$$x_s = \sqrt{a^2 - R_e^2 \cos^2 E} - R_e \sin E \tag{140}$$

to $x = 0$ at the receiver. Then, the excess delay is as follows:

$$\Delta t_{\text{delay}} = \frac{2m_e}{c} \int_{x_s}^{a} \frac{dx}{\sqrt{x^2 + 2xR_e \sin E + R_e^2}}$$

$$= \frac{2m_e}{c} \ln\left[\frac{\sqrt{a^2 - R_e^2 \cos^2 E} + a}{R_e(1 + \sin E)}\right] \tag{141}$$

The quantity $c\Delta t_{\text{delay}}$ is plotted in Fig. 30 as a function of elevation angle and ranges from 1.88 cm at $E = 0$ to 1.26 cm at $E = 90$ deg. Note that 2 cm of delay corresponds to only 0.0667 ns of delay.

This calculation has not yet accounted for the fact that the rate of coordinate time is actually defined by clocks on the Earth's surface rather than clocks at infinity. Applying the time scale change of Eq. (66), the propagation time of the signal becomes the following:

$$\Delta t' = \left(1 + \frac{\Phi_0}{c^2}\right)\Delta t$$

$$= \left(1 + \frac{\Phi_0}{c^2}\right)\left(\frac{x_s}{c} + \frac{2m_e}{c} \ln\left[\frac{\sqrt{a^2 - R_e^2 \cos^2 E} + a}{R_e(1 + \sin E)}\right]\right)$$

$$\simeq \frac{x_s}{c} + \frac{x_s \Phi_0}{c^3} + \frac{2m_e}{c} \ln\left[\frac{\sqrt{a^2 - R_e^2 \cos^2 E} + a}{R_e(1 + \sin E)}\right] \tag{142}$$

Thus, in terms of coordinate time, the equivalent path length of the time delay is as follows:

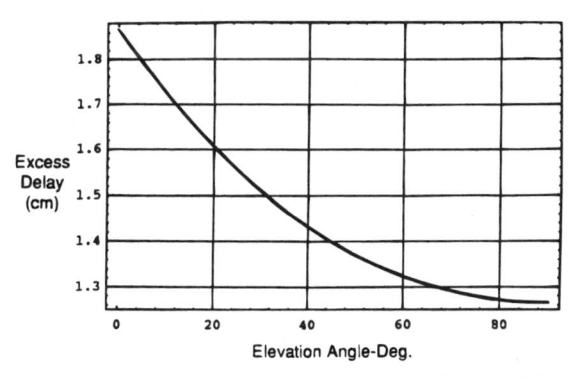

Fig. 30 Plot of relativistic excess propagation delay (Shapiro delay) from a GPS satellite-to-user path vs elevation angle E.

$$c\Delta t'_{\text{delay}} = \frac{\Phi_0}{c^2} \left(\sqrt{a^2 - R_e^2 \cos^2 E} - R_e \sin E \right)$$

$$+ 2m_e \ln \left[\frac{\sqrt{a^2 - R_e^2 \cos^2 E} + a}{R_e(1 + \sin E)} \right] \quad (143)$$

The quantity $c\Delta t'_{\text{delay}}$ is plotted in Fig. 31. The net effect is a fraction of a centimeter and for most purposes can be neglected. We must keep in mind, however, that the main term x_s/c is a coordinate distance, and further small relativistic corrections are required to convert it to a proper distance.

The expression above for coordinate time delay can be converted to an expression for coordinate time delay during crosslink ranging from one GPS satellite to another by specializing as follows. Consider the special case of zero elevation angle. Then, the receiver position a distance R_e from Earth's center is actually the distance of closest approach of the light ray to the Earth's center, and $x_s^2 = a^2 - R_e^2$. If the path of the ray were extended an equal distance in a straight line to another satellite, the net time delay would be double that given in Eq. (142). Let l be the distance between satellites: $x_s = l/2$. Then, for crosslink ranging, the net coordinate time delay $\Delta T'_{\text{delay}}$ is as follows:

$$c\Delta T'_{\text{delay}} = \frac{2x_s \Phi_0}{c^2} + 4m_e \ln \left[\frac{x_s + a}{\sqrt{a^2 - x_s^2}} \right]$$

$$= \frac{l\Phi_0}{c^2} + 2m_e \ln \left[\frac{2a + l}{2a - l} \right] \quad (144)$$

The time delay corrections, while small, may be significant under some circum-

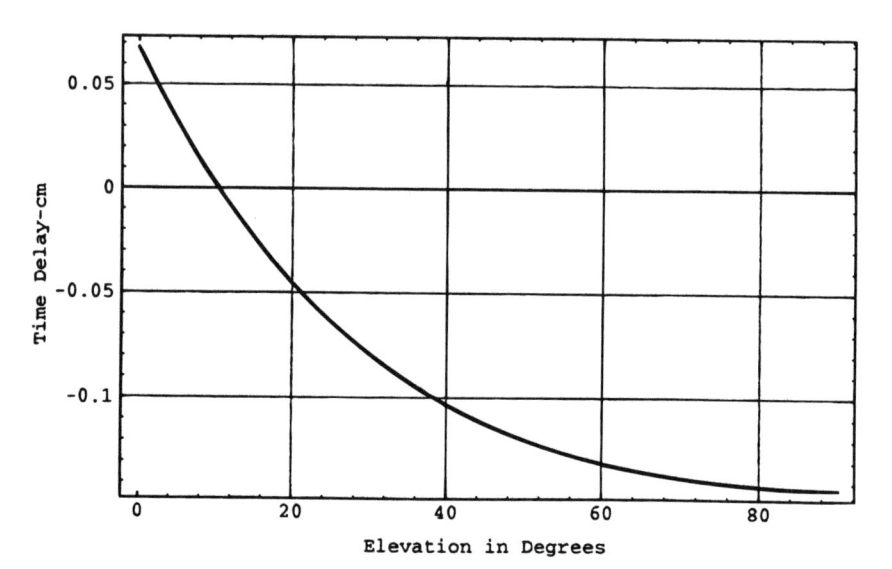

Fig. 31 Time delay vs elevation angle E for a GPS satellite-to-user link, including the time scale change for reference clocks on Earth's surface.

stances. Figure 32 plots the total time delay correction given by the above expression. After the coordinate time scale change is made, the net effect is small. There may be some conditions of autonomous operation, however, in which it will have to be accounted for.

5. Time Transfer in Crosslink Ranging

The Shapiro time delay calculated in Eq. (144) has only a slight effect on time transfer during crosslink ranging. Combining Eqs. (113) and (144) gives the net coordinate time at satellite #2:

$$T_s^{(2)} = T_s^{(1)} + \frac{|\Delta \mathbf{r}|}{c}[1 + \Phi_0/c^2] + \frac{2m_e}{c^3} \ln\left[\frac{2a + l}{2a - l}\right]$$
$$- \frac{2\sqrt{\mu a_2}}{c^2} e_2 \sin E_2 + \frac{2\sqrt{\mu a_1}}{c^2} e_1 \sin E_1 + \frac{\Delta \mathbf{r} \cdot \mathbf{v}}{c^2} \quad (145)$$

There is a tendency for the two terms in Eq. (144) to cancel each other; this is basically because the signals are propagating near the surface of the Earth, where the time unit is defined. The net time delay for a single time transfer process is, then, only a few hundredths of a nanosecond. For many purposes, this effect can be neglected.

The main term determining the propagation time is l/c. The crosslink ranging distance l can vary from a small distance, for satellites in different orbital planes, up to about 51,000 km, requiring a propagation time of about 170 (ms). This crosslink ranging distance is the *coordinate* distance, $|X_2 - X_1|$, between transmitter and receiver using Schwarzschild coordinates described by the metric tensor of Eq. (91). It is not the proper distance between transmitter and receiver. Coordi-

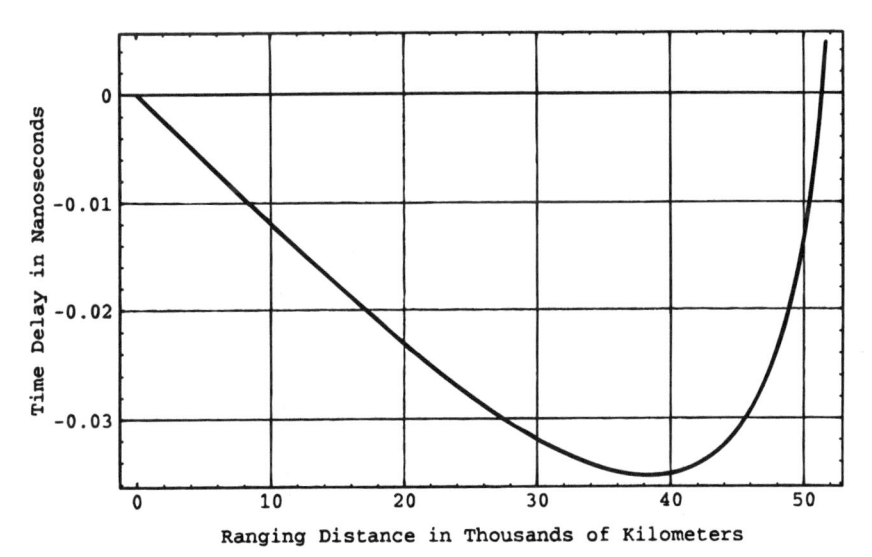

Fig. 32 Plot of crosslink ranging time delay, given by Eq. (144).

nate radii of the satellite orbits can be determined using Kepler's third law, just as in Newtonian orbital dynamics. The difference between proper distance and coordinate distance can be a few cm; this has been discussed elsewhere[1] and is not pursued here; however, it should be kept in mind that this difference can also result in corrections of the order of tenths of nanoseconds, to the propagation delay time.

Another case of interest, with a different geometry, is plotted in Fig. 33. For this figure, Eq. (144) is applied to time transfer between two GPS satellites in orbital planes whose nodal angles differ by 120 deg. The horizontal axis is the altitude in radians of satellite #1 above the Earth's equatorial plane. The different curves correspond to values, differing by 60 deg, of the altitude above the Earth's equatorial plane, of satellite #2. The break in one of the curves corresponds to the Earth's passing between the two satellites.

Thus, for a single time transfer in one direction, the correction is no more than a few hundredths of a nanosecond. This is not currently significant. Under some conceivable conditions of crosslink ranging, however, transmissions between satellites may occur repeatedly and rapidly. The time delay correction might then build up to a significant amount.

To illustrate the buildup of this systematic effect, suppose that signals were transmitted once around a closed planar circuit, in the shape of an equilateral triangle with three GPS satellites involved. Look at only the time delay coming from the logarithm term. It will be $\Delta t_{\text{delay}} = 3 \times [2GM/c^3] \log (2 + \sqrt{3})/(2 - \sqrt{3}) = 0.23$ ns. On the other hand, the time required for signals to propagate around the circuit is $t = 6\sqrt{3}a/2c = 0.46$ s, where a is the radius of the satellite orbit (see Table 11) Thus, the relativistic time delay effect is about 5 parts in 10^{10}. As seen in Figs. 32 and 33, however, this is an overestimate of the size of

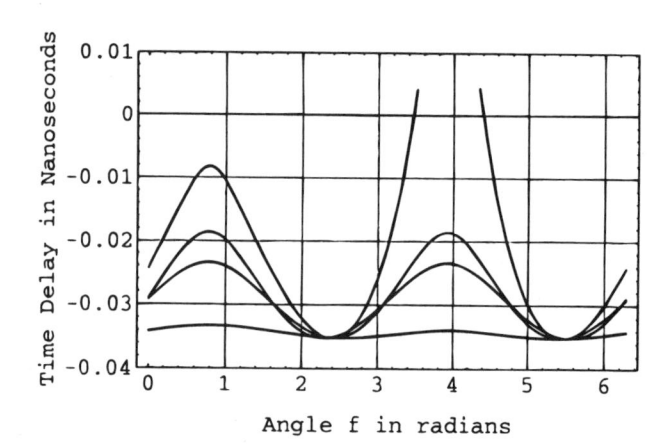

Fig. 33 Plot of total time delay correction, Eq. (144), for a single time transfer between two GPS satellites whose orbital planes are inclined 55 deg from the Earth's equatorial plane and whose orbits lie in planes of nodal angles differing by 120°. The horizontal axis is the altitude (in radians) of satellite #1 above the Earth's equatorial plane. The different curves correspond to values, differing by 60°, of the altitude above the Earth's equatorial plane, of satellite #2.

the total effect because of the tendency for the term coming from $l\Phi_0/c^3$ to cancel the effect coming from the logarithm term.

6. Lense–Thirring Drag

In the neighborhood outside of an uncharged spinning body, space–time is described by the Kerr metric,[5] which in "Boyer–Lindquist" coordinates is as follows:

$$ds^2 = -\frac{\Delta}{\rho^2}(dx^0 - (a/c)\sin^2\theta d\phi)^2 + \frac{\sin^2\theta}{\rho^2}[(r^2 + (a/c)^2)d\phi - (a/c)dx^0]^2 + \frac{\rho^2}{\Delta}dr^2 + \rho^2 d\theta^2 \quad (146)$$

where

$$\Delta \equiv r^2 - 2GMr/c^2 \quad (147)$$

$$\rho^2 \equiv r^2 + (a/c)^2\cos^2\theta \quad (148)$$

and where a is the angular momentum per unit mass of the object. For the Earth, the parameter $a/c \approx 2R_e^2\Omega_e/5c = 3.95$ m. This is sufficiently small that terms in $(a/c)^2$ can be neglected compared to terms of order $R_e^2 \approx 4 \times 10^{13}$ m². Then, the above Kerr metric greatly simplifies to the following:

$$ds^2 = -\left(1 - \frac{2m_e}{r}\right)(dx^0)^2 + \frac{dr^2}{1 - 2m_e/r} + r^2(d\theta^2 + \sin^2\theta d\phi^2) - \frac{4m_e a \sin^2\theta}{cr} dx^0 d\phi \quad (149)$$

This is just the Schwarzschild metric, with an additional term very similar to the term arising from the Sagnac effect. The extra term causes rotation (dragging) of local freely falling inertial frames, and, thence, a precession of orbiting gyroscopes relative to distant stars. This effect becomes more pronounced as r becomes smaller. In the case of the GPS, this term is completely negligible, as we can see by comparing the coefficient of $dx^0 d\phi$ in Eq. (149) with the corresponding coefficient of the Sagnac term in Eq. (68). The ratio for $\theta = \pi/2$, $r = R_e$ is at most

$$\frac{g_{03}(\text{Lense–Thirring})}{g_{03}(\text{Sagnac})} = \left(\frac{4m_e a}{cR_e}\right) \Big/ \left(\frac{2R_e^2\Omega_e}{c}\right) = \frac{4}{5}\frac{m_e}{R_e} \approx 5.6 \times 10^{-10}.$$

Because the Sagnac effect is at most a few hundred nanoseconds, the Lense–Thirring drag effect on time transfer is negligible in the GPS.

References

[1]Ashby, N., "Relativity and the Future of Engineering," *IEEE Transactions on Instrumentation and Measurement,* Vol. 43, Aug. 1994, pp. 505–514.

[2]Ashby, N., "An Earth-Based Coordinate Clock Network," U.S. Dept. of Commerce, NBS TN 659, April 1975, U.S. G.P.O., Washington, DC (S. D. Catalog #C13:46:659).

[3]Klioner, S. A., *Celestial Mechanics and Dynamical Astronomy,* Vol. 53, No. 1, 1992, pp. 81–109.

[4]Winkler, G. M. R., "Synchronization and Relativity," *Proceedings of the IEEE*, Vol. 79(6), June 1991, pp. 1029–1039.

[5]Misner, C. W., Thorne, K. S., and Wheeler, J. A., *Gravitation*, W. H. Freeman, San Francisco, CA, 1973.

[6]Schutz, B. F., *A First Course in General Relativity*, Cambridge University Press, Cambridge, UK, 1990.

[7]d'Inverno, R., *Introducing Einstein's Relativity*, Clarendon Press, Oxford, UK, 1992, p. 33.

[8]Lovelock, D., and Rund, H., *Tensors, Differential Forms and Variational Principles*, Dover, New York, 1989.

[9]Ashby, N., "A Tutorial on Relativistic Effects in the Global Positioning System," NIST Rept. 40RANB9B8112, Boulder, CO, 1990.

[10]Lorentz, H. A., Einstein, A., Minkowski, H., and Weyl, H., *The Principle of Relativity*, Dover, New York, 1923.

[11]Taylor, E. F., and Wheeler, J. A., *Space Time Physics*, W. H. Freeman, New York, 1966.

[12]Danielson, D. A., *Vectors and Tensors in Engineering and Physics*, Addison-Wesley, Menlo Park, CA, 1992.

[13]Aris, R., *Vectors, Tensors, and the Basic Equations of Fluid Mechanics*, Dover, New York, 1989.

[14]Rindler, W., *Essential Relativity*, Springer-Verlag, Berlin, 1977.

[15]Turner, K. C., and Hill, H. A., "New Experimental Limit on Velocity Dependent Interaction with Clocks and Distant Matter," Physical Review B: Solid State, Vol. 134, 1964.

[16]Ehrenfest, P., *Physikalisch Zeitschrift*, Vol. 10, 1909, p. 918.

[17]Sama, N., "On the Ehrenfest Paradox," *American Journal of Physics*, Vol. 40, 1972, pp. 415–418.

[18]Phipps, T. E., "Kinematics of a Rigid Rotor," *Lettere al Nuovo Cimento*, March, 1974.

[19]Grøn, Ø., "Relativistic Description of a Rotating Disc," *American Journal of Physics*, Vol. 43, 1975, pp. 869–876.

[20]Møller, C., *The Theory of Relativity*, 2nd ed., Clarendon Press, Oxford, UK, 1972, pp. 254–256.

[21]Hafele, J. C., and Keating, R. E., "Around the World Atomic Clocks: Predicted Relativistic Time Gains," *Science*, Vol. 177, 1972.

[22]Post, E. J., "Sagnac Effect," *Review of Modern Physics*, April 1967.

[23]Isaak, G. R., *Physics Bulletin*, 1970, pp. 255–257.

[24]Pound, R. V., and Rebka, G. A., "Apparent Weight of Photons," *Physical Review Letters*, Vol. 4, 1960, pp. 337–341.

[25]Pound, R. V., and Snider, J. L., "Effect of Gravity on Gamma Radiation," *Physical Review*, Vol. B140, 1965, pp. 788–803.

[26]S. Weinberg, *Gravitation and Cosmology*, John Wiley, New York, 1972.

[27]Weber, J., *General Relativity and Gravitational Waves*, Interscience, New York, 1961.

[28]Lambeck, K., *Geophysical Geodesy*, Oxford Science Publications, Clarendon Press, Oxford, UK, 1988, pp. 2–20.

[29]Soffel, M. H., *Relativity in Astronomy, Celestial Mechanics, and Geodesy*, Springer-Verlag, Berlin, 1989.

[30]Ashby, N., and Allan, D. W., "Coordinate Time on and Near the Earth," *Physical Review Letters*, Vol. 53, No. 19, 1984, p. 1858.

[31]Chobotov, V. A., (ed.), *Orbital Mechanics,* AIAA Education Series, AIAA, Washington, D.C., 1991.

[32]Larson, W. J., and Wertz, J. R. (eds.), *Space Mission Analysis and Design,* 2nd ed., Kluwer Academic, Dordrecht, The Netherlands, 1992.

[33]Ashby, N., and Allan, D. W., "Practical Implications of Relativity for a Global Coordinate Time Scale," *Radio Science,* Vol. 14(4), July/Aug. 1979, pp. 649–669.

[34]Ashby, N., Allan, D. W., and Weiss, M., "Around-the-World Relativistic Sagnac Experiment," *Science,* Vol. 228, April 5, 1985, pp. 69–70.

[35]Ware, R., "GPS Sounding of the Earth's Atmosphere," *GPS World,* Sept. 1992, pp. 56–57.

[36]Fitzpatrick, P. M., *Principles of Celestial Mechanics,* Academic Press, New York, 1970, p. 33 ff.

[37]Ashby, N., and Bertotti, B., "Relativistic Effects in Local Inertial Frames," *Physical Review,* Vol. D34, 1986, pp. 2246–2258.

[38]Ashby, N., "Relativity and GPS," *GPS World,* Vol. 4, No. 12, Nov. 1993, pp. 42–47; Dec. 1993, p. 44.

[39]Will, C. M., *Theory and Experiment in Gravitational Physics,* Cambridge University Press, Cambridge, UK, 1993.

[40]Kaplan, G. H., (ed.), *The IAU Resolutions on Astronomical Constants, Time Scales, and the Fundamental Reference Frame,* U.S. Naval Observatory Circular No. 163, December, 10, 1981, U.S. Naval Observatory, Washington, DC.

[41]Nerem, R. S., et al., "Gravity Model Development for TOPEX/POSEIDON: Joint Gravity Models 1 and 2," *Journal of Geophysical Research,* Vol. 99, 1994, pp. 24421–24447.

Chapter 19

Joint Program Office Test Results

Leonard Kruczynski*
Ashtech, Sunnyvale, California 94088

I. Introduction

FAVORABLE test results can have a tremendous positive influence on the survivability of a program. This is especially true in the early stages of a program and when a program faces stiff budget battles. A 1979 Comptroller General's report to the Congress was titled "The NAVSTAR Global Positioning System: A Program with Many Uncertainties." The Global Positioning System (GPS) was in a particularly vulnerable position. Two prior reviews by the Comptroller General had made recommendations about the program that, although agreed to by the Department of Defense, had not been implemented. The majority of the questions revolved around funding, schedules, and estimated cost savings. Chapter 4 of the report, however, gave grudging credit to the test results that had been made available. As of October 1978, test data showed that navigation accuracies were better than expected. The favorable test results were one of the bright spots in the Comptroller General's report and undoubtedly contributed to the program's continuation.

In this chapter, we review some of the details of the Joint Program Office (JPO) testing program with an emphasis on the tests conducted at the U.S. Army Yuma Proving Ground (YPG), Arizona. During phase I, Yuma was the site of the majority of tests conducted to support the Concept Validation Phase of GPS. Tests were also conducted by the Navy near San Diego. In phase II, testing expanded worldwide, but YPG remained the heart of JPO testing. U.S. military testers took the GPS receivers to many sites around the world. NATO personnel, an integral part of the JPO, tested GPS equipment in their own countries. Agencies such as the Department of Transportation also conducted tests in support of phase II. The phase III approval carried requirements for equipment modification. Tests continued into phase III at Yuma to validate the modifications that had been made to the equipment.

When not in conflict with GPS user equipment (UE) testing, the JPO supported tests at Yuma that expanded the horizons of GPS. Real-time differential tests were

---

Copyright © 1994 by the American Institute of Aeronautics and Astronautics, Inc. All rights reserved.
* Ph.D.

conducted as early as 1979. The experiences using the ground transmitters (GT) at Yuma provide much information to those who would use pseudolites. Detailed and rigorous testing on the F-4 weapon delivery platform provided proof of the potential weapon delivery accuracy. The results of the F-4 tests became a benchmark for future weapons delivery tests. When selective availability (SA) and antispoofing (AS) were tested early in the program, the YPG GTs were used because Block I satellites cannot implement clock dither or AS. The JPO-developed test capability proved to be a highly valuable resource in the development of GPS.

II. U.S. Army Yuma Proving Ground (YPG)

The U.S. Army Yuma Proving Ground (YPG) is the Army's only general purpose proving ground located in desert terrain (Fig. 1). Located about 25 miles from the city of Yuma, Arizona, YPG covers an area of about 1,400 square miles or 870,000 acres. Originally activated in 1943, the Yuma Test Branch's mission was to test bridges, boats, vehicles, and well-drilling equipment. In 1962, what was then Yuma Test Station was assigned to the U.S. Army Material Development Command and was placed under the immediate control of the U.S. Army Test and Evaluation Command (TECOM). The name was changed to Yuma Proving Ground in 1963 and finally to U.S. Army Yuma Proving Ground in 1973. In 1974, YPG was designated a DoD major test facility partly because of the selection of YPG as the primary test site for GPS user equipment.

Yuma Proving Ground's missions include testing tube artillery systems, aircraft armament systems, air delivery systems, and air movable equipment. The missions also include planning, conduct, and reporting of the results of desert environmental

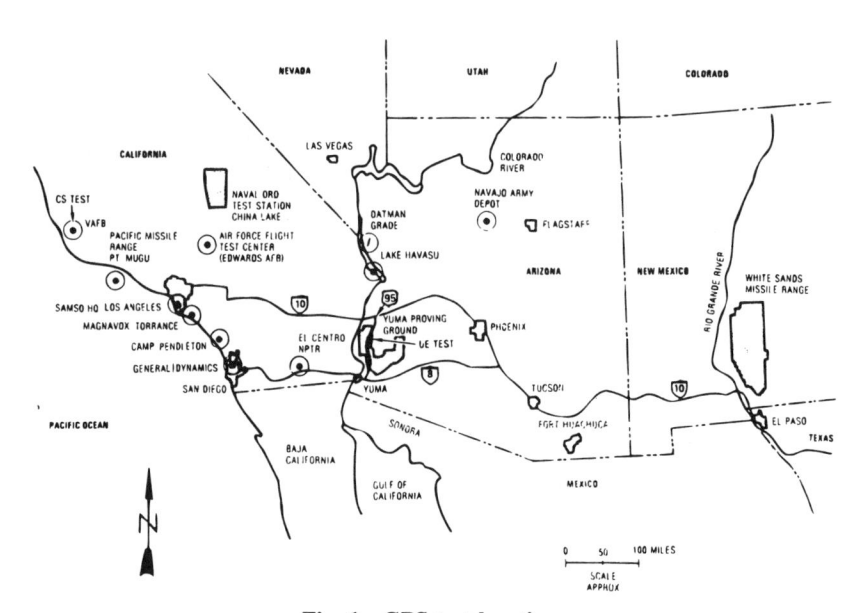

Fig. 1 GPS test locations.

tests of all classes of materiel. The M-1 Abrams tank was a heavy user of the range during the time of GPS testing.

III. Reasons for Selection of Yuma Proving Ground

Yuma Proving Ground has a climate characterized by clear skies, with an average visibility of 25 miles and only three and one-half inches of rain per year. With the extensive real estate and restricted air space, uninterrupted testing can be almost guaranteed. YPG had embarked on a program to use lasers to track artillery shells. The range proposed to use the laser trackers to determine truth trajectories for GPS tests.

IV. Lasers

A total of six lasers were eventually installed at YPG. Figure 2 shows the location of the six lasers and the area of the range covered by at least two lasers. Four lasers were in the southern half of the range and two in the northern half. The tracking lasers are Neodymium YAG lasers built by Sylvania Electronics. The system is designated PATS for precision automated tracking system. The specified radiated power is 50 mJ per pulse, which is automatically attenuated to eye safe levels by measuring the return signal level. Beam divergence, as used in eye safety tests, is less than 2 mrad. Eye protection is required when closer than 2 km to an operating laser.

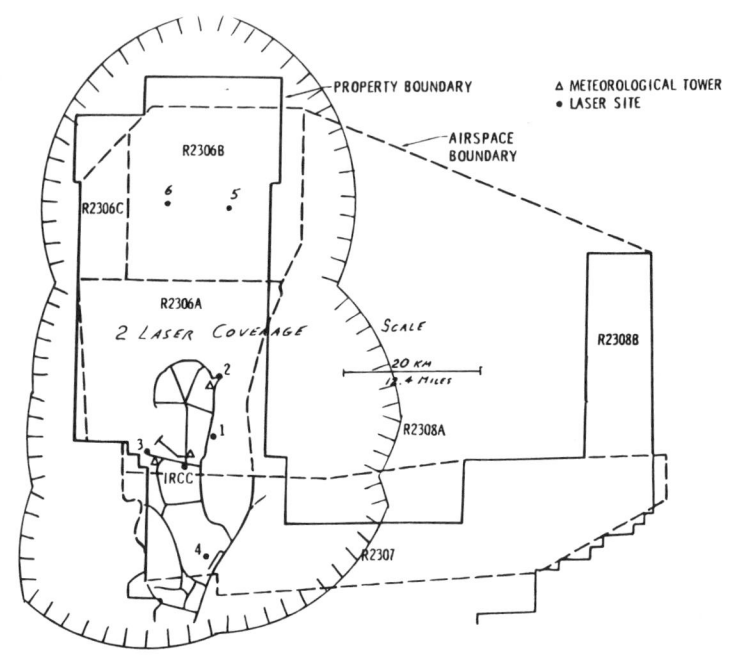

Fig. 2 Yuma Proving Ground area with laser coverage.

Each tracker takes range, azimuth, and elevation measurements at a 100-Hz rate. Specification accuracy for the range measurements is 0.5 m rms from 200 m to 10 km and 1 m rms from 10 km to 30 km. Azimuth and elevation measurement accuracy is 0.1 mrad in each axis. Every fifth measurement is transmitted from the laser site to the Range Operations Center (ROC). No smoothing or editing is performed on the measurements at the laser except to aid in target tracking by the laser. The transmitted measurement is selected only to synchronize the measurements so that a measurement is available on an even second as given by IRIG time. At the laser site, the measurements are recorded at a 100-Hz rate. This field tape is used only for fault isolation. No processing is done on the 100-Hz measurements.

Prior to and immediately following each mission, the laser takes measurements to a series of short targets circling the laser. Typically, there are eight short range targets roughly 1 km from the laser. The short range calibrations are processed by the ROC to provide corrections to tilt (two terms), azimuth, elevation, and range terms.

Each laser also has eight long-range targets at ranges from 1 km to 30 km. The primary use for the long-range calibrations is to determine range bias. Long-range calibrations normally are accomplished monthly or upon request.

"Dump cals" is a third type of calibration for the lasers. To do a dump cal, the laser housing is rotated in elevation through 180 deg, and the short-range targets are tracked. Dump cals are accomplished after maintenance and upon request.

V. Range Space

The Cibola Range covers a ground area approximately 50 km × 100 km and extends to 25 km altitude. Most of the tests were conducted in the southern half of the range. A hilly ridge separates the southern half from the northern half. Two lasers were installed in North Cibola to support aircraft testing that required the entire range. The difficult logistics involved in supporting North Cibola resulted in low utilization of North Cibola. However South Cibola was more than adequate for the majority of testing that had to be accomplished.

Ground vehicle testing was conducted on the dynamometer course. Vehicles on the dynamometer course were tracked by a single laser. Also, the location of the GTs on hills allowed the vehicles to track pseudolite signals when necessary.

VI. Joint Program Office Operating Location

The official presence of the JPO at YPG was the operating location, OL-AA. The operating location normally consisted of Air Force officers and enlisted personnel, civilians assigned by YPG Materiel Test Directorate, military assigned by YPG, Aerospace Corporation personnel, and other agency representation depending on needs. Defense Mapping Agency maintained a slot at OL-AA for several years. General Dynamics Electronics Division and later General Dynamics Services Company provided operations and maintenance support for the operating location.

VII. Satellite Constellation for Test Support

GPS orbital parameters had to satisfy several requirements. High-altitude orbits were desired for survivability and to reduce the number of satellites needed for global coverage. A high inclination was needed to provide coverage in polar areas. A 12-h orbit was selected because it allowed each satellite to be visible over the continental U.S. at least once per day. Twenty-four-hour orbits would not have this feature. In addition, satellites in 24-h orbits were farther from the users, thus requiring greater transmitted power. A critical criterion to support testing was to require that a test window, and preferably an identical test window, exist at a single test site each day. For phase I, the GPS constellation was set up to optimize coverage for testing at YPG.

It is easier to explain GPS test window behavior by assuming that the test site is on the equator and that, on the first day of testing, the satellite passes directly overhead. The requirement for constellation repeatability is that on the next day of testing, the satellite must again be directly overhead. We can further assume that the satellite is crossing the equator in the northerly direction, making this point the ascending node of the orbit.

After two orbits, the satellite will again be at the ascending node over the equator. The key is to require that two orbits of the satellite take the same amount of time as the test site takes to move back under the satellite. Satellites move in space independently of the Earth's rotation. To require that the position of the satellite relative to the ground location repeat requires planning. The two orbits must take 23 hours, 56 minutes and 4 seconds because that is how long it takes the Earth's rotation to bring the test site back to the satellite's ascending node. If this timing is not satisfied, then the location of the satellite with respect to the test site will drift daily, eventually resulting in the lack of a test window at the test site. It is this requirement that gives rise to the 4-min regression of the test window each day.

Had the Earth been perfectly round, the test window would repeat not only in character but also time of day each year. However, the Earth is oblate and the satellite is perturbed. The result is an approximate 50-min shift in the test window each year. This phenomenon helps explain why the GPS test window at Yuma moves from night in the summer of one year to daytime in the summer of a later year.

VIII. Control Segment Responsiveness to Testing Needs

The Control Segment that existed for the testing through phase II consisted of a Master Control Station (MCS) at Vandenberg Air Force Base, California, and monitor stations in Alaska, Hawaii, and Guam. Vandenberg was the launch site for the Block I satellites. The configuration of the monitor stations was especially beneficial to testing at Yuma. The satellites have an eastward motion, so they were generally visible to at least one of the monitor stations prior to rising at Yuma. The MCS was able to upload a satellite either prior to or shortly after the satellite rose at Yuma. By avoiding the use of day-old data, the space and control segment errors seen by receivers under test were minimized.

IX. Trajectory Determination at YPG

A. Real-Time Estimate

The workhorse trajectory at YPG is the real-time estimate (RTE), a position and velocity solution obtained from a single laser. The lasers transmit range, azimuth, and elevation measurements at a 20-Hz rate via microwave links to the ROC. In the ROC, the measurements are preprocessed by a PDP 11/35 and sent to a SEL 32/77 for real-time processing. In the SEL, the measurements are processed first at the incoming 20-Hz rate (every 50 ms) and then at a 5-Hz rate (every 200 ms). The 200-ms task takes every fourth 50 ms measurement on the even 0.2 s, as given by IRIG time.

In the 50-ms task, the measurements are processed using a QD filter with a memory parameter equal to six. Measurement residuals are calculated by subtracting the measurement predicted using a second-order formula from the actual measurements. The output of the 50-ms task is range, azimuth, elevation, and first and second rates of change of range, azimuth, and elevation.

The 200-ms task selects every fourth 50-ms output. Calibration factors obtained from premission tests are now applied. The measurements are then transformed from laser-centered range, azimuth, and elevation to IRCC-centered Cartesian coordinates. The IRCC is the Inverted Range Control Center and is located in the middle of the southern part of the YPG test range. These measurements are processed in the 200-ms task with a QD filter with a memory parameter of 10. QD processing in the 200-ms task is similar to the 50-ms task, except the state consists of IRCC East, North, and Vertical components instead of range, azimuth, and elevation used in the 50-ms task. Following the 200-ms QD, the software attempts to determine which retroreflector is being tracked.

To provide continuous track at all attitudes, most aircraft have two retroreflectors. A retroreflector is an array of corner cubes that reflect light in the direction from which the light comes. A retro mounted on top of the aircraft is needed to maintain track when the aircraft banks. The bottom-mounted retro is the real-time estimate solution point. If the software determines that the upper retro is being tracked by the laser, then lever arm corrections are made to move the solution to the lower retro point.

Retro identification and lever arm adjustment are normally accomplished using one of two methods. If no inertial data are available, a wind–velocity algorithm is used to estimate the attitude of the aircraft. Look angles from the laser to the target are computed and are used to determine which retro is being tracked. In the second method, attitude data from GPS user equipment can be used to provide aircraft attitude information for the look angle computation.

There is a range of look angles for which there is a high degree of uncertainty as to which retro is being tracked. Should the look angles fall into this range, the validity flag for the laser's trajectory estimate is lowered. The laser validity flag can also be lowered if measurements are outside statistical bounds.

If a laser is declared valid, then the solution is a candidate for the RTE. The RTE is the truth trajectory for evaluation of GPS UE. Typically, it is the solution determined by the laser closest to the target if the laser's solution has been declared valid. There is an optional composite laser solution that can be used as the RTE. The composite is a weighted average of the valid laser solutions.

Weighting is a function of range to the target. The closer the laser is to the target, the higher the weighting given the laser solution. There is a maximum gain that can be given any laser.

B. Best Estimate of Trajectory

In critical cases, it is possible to compute a best estimate of trajectory (BET), which considers multiple laser inputs. When requested, BETs are computed postprocessed. The quality of the RTE was sufficient for most tests and BETs were computed sparingly.

C. Validation of Truth Trajectory Accuracy

Performance evaluation of GPS UE must be made using a trajectory estimate that is more accurate than the GPS equipment being tested. Laser tracker specifications indicate that the YPG capabilities were within the requirements to evaluate GPS equipment. To support YPG's ongoing program to improve laser accuracy, validation tests were conducted using the Texas Instruments five-channel phase I High Dynamics User Equipment (HDUE). The HDUE was integrated with a Honeywell inertial measuring unit and installed on a Convair 880. The equipment was operated in a differential navigation mode to remove Space and Control Segment errors. The conclusions reached during the various validation tests was that the RTE was accurate to within 3 m throughout the range, with better accuracy in areas of optimal laser coverage.

Velocity accuracy of the lasers was shown to be about 0.2–0.3 m/s. This is on the order of GPS accuracy, especially when aided by inertial sensors. Evaluation of GPS velocity accuracy using the RTE-computed velocities had to be considered only in the sense of bounding the errors. The source of the velocity error could have been the laser as well as the UE.

D. Ground Truth

Early in the GPS program, it was evident that the accuracy performance of user equipment would vary from day to day, depending on the behavior of the satellites and the ability of the Control Segment to predict satellite orbital and clock behavior. Ionospheric delays can vary considerably over time. Tropospheric delay, although it is a smaller contributor to navigation error, might also be different from one test time to the next. However, it was incumbent on the government to evaluate the performance of the various manufacturers fairly, regardless of the performance of the Space and Control Segments and regardless of atmospheric activity. The solution was to compare the error in the positions calculated by the UE to a "ground truth."

Ground truth is the absolute error in the solutions computed by a "perfect" receiver. The development specification for the IRCC defined ground truth as "a direct real-time estimate of the expected error in an IRCC located user's computation of position and time." It was expected that all receivers under test on the range would see virtually identical errors. Thus, if ground truth was 15 m East of the surveyed location and if the receiver under test was also 15 m East, then the receiver was deemed to have perfect accuracy performance.

To create ground truths, the Magnavox phase I X-set receiver in the IRCC was used. The IRCC was located in the center of the southern part of the YPG test range (Fig. 2). The topography was flat with virtually no blockages except by IRCC equipment itself, such as the lightning rod. To minimize clock errors, a cesium frequency standard was connected to the X-set.

Ground truths were computed by the IRCC computer using measurements from the X-set. The measurements were adjusted by removing estimated range and satellite clock behavior using the satellite navigation data and the IRCC's true position, by using L_1-L_2 data to remove ionospheric delay, and by using locally collected meteorological data to remove tropospheric delays. Receiver clock offsets and drift were removed by using the clock synchronization navigation solution. If the satellite data are accurate, if L_1-L_2 measurements allow complete removal of ionospheric delays, if the tropospheric delays are correctly modeled, if the receiver clock algorithm is accurate, and if other error sources such as multipath and receiver noise are zero, then the corrected pseudorange will be zero. Based on the evaluation of all error sources, corrected pseudorange measurements greater than 1–2 m were most likely caused by satellite navigation data inaccuracies.

The clock synchronization navigation solution was a two-state Kalman filter. Receiver clock offset and drift were the two states. Corrected pseudoranges were input to the Kalman filter and the output was processed until the filter was in steady state. Values for the clock offset could vary wildly on a point-by-point basis, especially if different satellites were used to provide the input data. However, rapidly changing values for the clock offset are neither an accurate depiction of true clock behavior, nor are they desirable for the calculation of position ground truths. The IRCC assumed the clock filter to be in steady state when the best estimate of receiver time was fairly constant, when the average of the pseudorange residuals from all the satellites was zero, and when the differences between the pseudorange residuals of the satellites tended to be constant.

The ground truth navigation solution was calculated using a five-state Kalman filter. The first three states were the three-dimensional position and the other two states were the receiver clock offset and drift as they would be calculated by a navigating receiver. The Kalman filter was tuned to provide an explicit solution with each new set of measurements but with a small amount of filtering to reduce noise.

Ground truths were presented as errors in solutions. To use the ground truth, the receiver under test had to use the same constellation for positioning as was used by the IRCC. Initially, this was no problem because the constellation was limited. Four satellites were launched from February 1978 to December 1978. The fifth and sixth successful launches occurred in 1980. The seventh satellite was launched in 1983. The ground truth computation was modified to allow for the best five GDOP constellations. With five satellites available, only five four-satellite constellation selections are possible. With six satellites, up to 15 different four-satellite constellations are possible. However, many constellations would have geometries beyond acceptable limits.

The IRCC performed the additional function of controlling the ground transmitters (GTs). Using ground transmitters requires that the phase of each GT be accurately synchronized. The GTs themselves did not have atomic clocks.

Table 1 Phase I major field test objectives

Navigation accuracy	Threat performance
Position accuracy	Jamming resistance
Velocity accuracy	Denial of accuracy
Effects of dynamics on accuracy	
	Environmental effects
Demonstrations of military value	Propeller and rotor modulation
Precision weapon delivery	Foliage attenuation
Landing approach	Multipath rejection
Rendezvous	Ionospheric and tropospheric correction
Photomapping	
Nap-of-Earth operations	System characteristics
Static positions	Satellite and clock and ephemeris accuracy
Combined operations	Acquisition and reacquisition time
Crosscountry	Time transfer
Shipboard operations	Signal levels and signal structure

X. Phase I Tests (1972–1979)

After passing DSARC I in December 1973, GPS entered the concept validation phase of development. The major field test objectives for phase I are identified in Table 1. The user equipment tested at YPG included receivers from Magnavox under contract to General Dynamics, Collins, and Texas Instruments. Table 2 lists the basic characteristics of user equipment tested as part of the concept validation phase. User sets were mounted in specially instrumented pallets on several vehicles, such as those listed in Table 3.

XI. Ground Transmitters

Testing GPS at YPG began on March 15, 1977. At that time, there were no satellites in orbit. Instead, four ground transmitters, or pseudolites, were used to transmit a satellite-like signal to provide a simulated constellation. The locations of the ground transmitters are identified in Fig. 3 as IRGT-21 through IRGT-24. The pseudolites transmitted an L_1 signal modulated with C/A and P-codes. The data transmitted was a unique message because it is impossible to model the dynamics of a receiver on the ground in a Keplerian formulation. The pseudolites did not transmit an L_2 signal because ionospheric delay was not an issue. PRN codes from 33 to 36 inclusive have been reserved for GTs.

The four GTs and the IRCC comprised the inverted range (IR). The IR was used in the early days of GPS testing when no satellites were available and later to supplement inadequate coverage, for special tests such as SA/AS, and for real-time differential tests. When four GTs were used, even the best geometry has poor vertical dilution of precision (VDOP) performance. When operating in a hybrid mode, one satellite and three GTs provide an optimal geometry. As more satellites became available, it was advantageous to minimize the use of GTs, even at the expense of larger dilutions of precision (DOPs).

Table 2 Characteristics of concept validation phase user equipment

User equipment	Frequency	Code	Number of channels	Integrated with inertial	Contractor
X	L_1 and L_2	P and C/A	4	Yes or No	General Dynamics/Magnavox
Y	L_1 and L_2	P and C/A	1	Yes or No	General Dynamics/Magnavox
Z	L_1	C/A	1	No	General Dynamics/Magnavox
General development model (GDM)	L_1 and L_2	P and C/A	5	Yes	Collins
High dynamics user equipment (HDUE)	L_1 and L_2	P and C/A	5	No	Texas Instruments
Manpack (MP)	L_1 and L_2	P and C/A	1	No	General Dynamics/Magnavox
Manpack/vehicular equipment (MVUE)	L_1 and L_2	P and C/A	1	No	Texas Instruments

When four pseudolites were used, the DOPs on the range varied significantly and rapidly. Typical horizontal dilutions of precision (HDOPs) at 20,000 ft above the range changed from 2 to 20 in 8 miles. Thus, an aircraft flying at 300 kt would pass through this range of HDOPs in 90 s. VDOPs had several singularities existing over the course. The character of the DOP contours changed when three GTs were used in conjunction with one satellite, but steep gradients were still present.

The IRCC controlled the transmitted power of the GTs during operation. Each pseudolite was equipped with a remotely controlled programmable attenuator. Maintaining proper power levels was necessary to prevent the GT from jamming the receiver. Power levels were adjusted based on the distance between the receiver and the GT. Multiple aircraft tests were not conducted. Power adjustments were especially critical when the receiver lost lock on a satellite because the receiver would switch to C/A code, which was much more susceptible to jamming than the P-code.

When operating in a hybrid mode, the IRCC had to offset the GT clocks and synchronize them to GPS satellite time. The IRCC attempted to keep the GT clocks accurate to within one ns. The IRCC monitored the GT signals and used a uhf radio control system to control the GTs. The pseudolite clocks were driven by temperature stabilized precision oscillators, which were kept in thermal canisters buried six feet under ground at each pseudolite site.

When satellites were used in conjunction with GTs, the aircraft required two antennas; one antenna on the top and a second antenna on the bottom of the aircraft. The upper antenna tracked satellites, while the lower antenna tracked

Table 3 Concept validation phase test vehicles

Vehicle type	GPS set type	Dynamic range
Mobile test van	XU	Static
Man	Manpack, MVUE	Low
Landing craft	XU, YU	Low
Frigate	XU, YU	Low
Armored personnel carrier	Manpack, MVUE	Low
M35 truck	XU, HDUE, Manpack, MVUE	Low to medium
Jeep	Manpack, MVUE	Low to medium
UH-1 helicopter	XU, YU, Manpack, HDUE, MVUE	Medium
C-141 aircraft	XU, XA, YU, YA, GDM, Z, HDUE	High
P3 aircraft	XU, XA, YU, YA, Z	High
F4 aircraft	XA	Very high

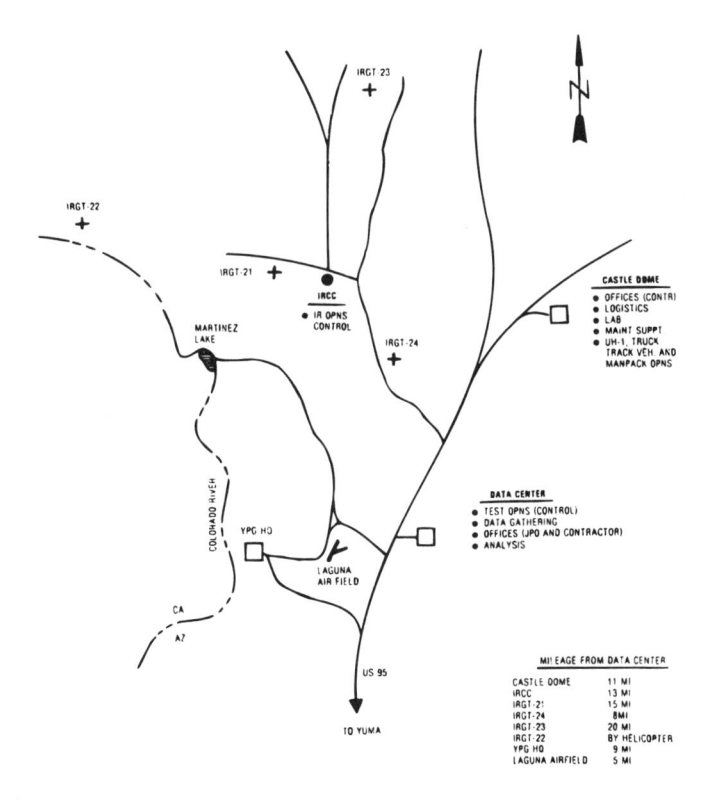

Fig. 3 GPS test area at Yuma Proving Ground showing ground transmitter locations.

the GTs. The phase I UE were designed with two antenna ports. The dual-antenna design also minimized dropouts of GPS satellites during maneuvers. The phase I equipment used inertial attitude data from an external sensor to determine which antenna should be used to track a specific satellite.

The pseudolites on the IR served their assigned purpose of helping to discover and solve user equipment design problems when satellites were not available. Later in the test program, GTs were used to test selective availability and antispoofing performance because the Block I satellites did not have SA/AS capabilities. GTs were also used as the data link in real-time differential tests. The experience with GTs has proved valuable to groups investigating the use of pseudolites such as the Range Applications JPO at Eglin Air Force Base.

Phase I test results were excellent. Figure 4 summarizes the performance of the various equipment tested during the conduct of the phase I program. Three-dimensional position errors at the 90% confidence level ranged from about 35 m for the single-frequency Z-set to 15 m for the multichannel, dual-frequency receivers.

Precise statistical values for weapon delivery performance are classified. However, the JPO often presents graphic depictions of the accuracy of weapon delivery. The results from tests have often been displayed relative to the cupola in the center of the Pentagon or relative to a typical suspension bridge.

XII. Navy Testing for Phase I

The U.S. Navy conducted several tests in support of phase I. Most of the tests were conducted in the waters near San Diego. Figure 5 shows the ground track

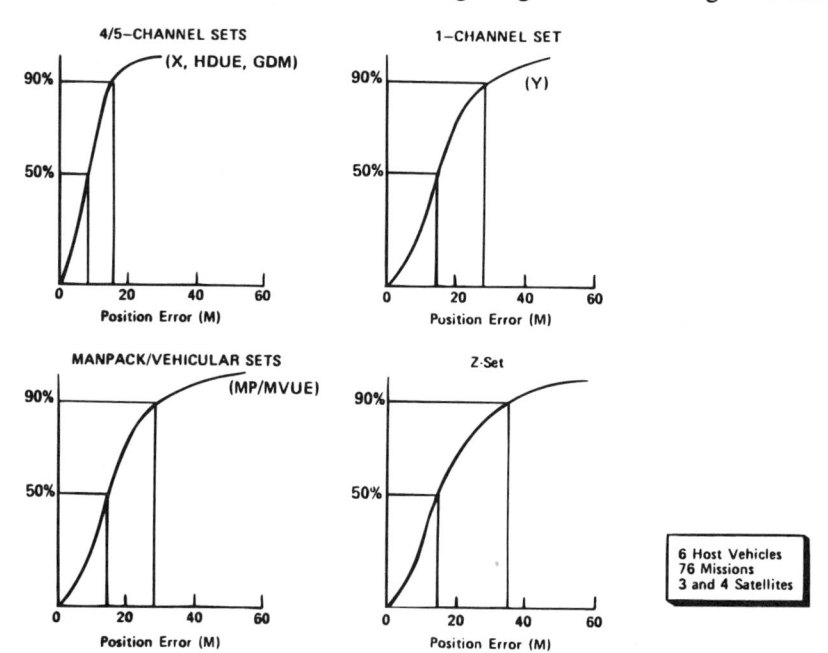

Fig. 4 Phase I results.

HARBOR NAV ABOARD FAST FRIGATE

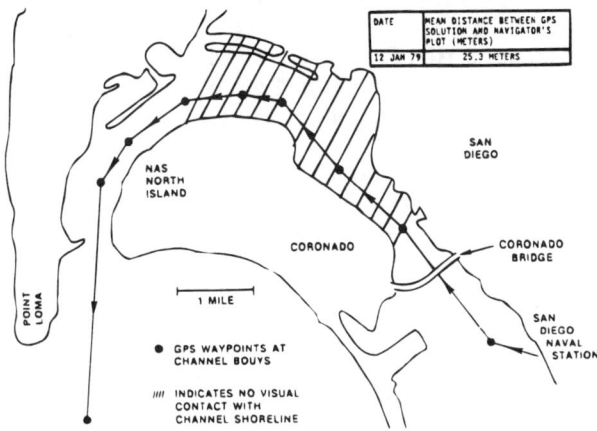

Fig. 5 Navigation aboard a fast frigate in San Diego harbor.

of a fast frigate test in San Diego harbor. The hatched area was fogged in during the test. No laser-accuracy truth solution was available for this test. However, the average difference between the navigator's plot and the GPS solution was about 25 m.

Several Navy tests were aimed at identifying and characterizing multipath effects. Dockside testing was accomplished using a Navy landing craft and a fast frigate. Tests were conducted on two subsequent days for each of the craft using P-code receivers. Signal strength fading and pseudorange errors were analyzed to determine multipath effects. Although frequency standard instabilities made the analysis difficult, the tests identified peak-to-peak pseudorange oscillations of 4–5 m caused by multipath.

XIII. Tests Between Phase I and Phase II (1979–1982)

A. Weapons Delivery

Weapons delivery tests continued while YPG awaited the start of phase II. A Navy F-4 was equipped with a Magnavox phase I X-set and integrated with an IMU. Considerable work was done to tune the pilot's display to inform the pilot better about steering information. The software accounted for aircraft attitude, bomb location, ejection velocities, bomb dynamics, ejection velocities, winds estimates, and so forth. The statistical results are classified, but it was not unusual for the inert bomb to hit the stake used to mark the surveyed target point.

B. Differential Tests

Differential GPS was an integral part of range operations from the beginning. Ground truth was based on the concepts behind differential GPS; that is, the removal of errors common to all users in the vicinity. But differential was not limited to simply laying a ground truth plot over a UE error plot. In 1979, a real-time differential system was established to allow the Air Force to conduct flights to validate the performance of the YPG trajectory estimate.

The Texas Instruments HDUE was a five-channel receiver. In normal operation, four channels tracked satellites on L_1 and the fifth channel sequenced through the satellites on L_2. This allowed the receiver to make real-time ionospheric delay measurements.

To modify the HDUE for real-time differential operation, the fifth receiver channel was used to receive the differential corrections transmitted by a GT. The software was modified to allow proper decoding of the GTs data. At first look this may seem to have the effect of removing the capability to compute ionospheric delay. In reality, the differential message transmitted to the HDUE was computed based on a no-ionosphere position computation, in effect, accounting for ionospheric effects.

Tests of the real-time differential HDUE began on December 13, 1979 at YPG on board a UH-1H helicopter. The initial series of tests clearly demonstrated that nondifferential horizontal errors of approximately 20 m are reduced to 5 m or less upon application of the differential correction.

Table 4 is a summary of horizontal and vertical accuracies reported during a series of differential GPS tests in January 1980. Subsequent analysis of the data showed that the processing was flawed and that actual achieved accuracy was better than indicated in the table. The detailed data analysis determined that the processing had not accounted for the lever arm between the retroreflector and the antenna.

The HDUE in differential mode proved to be very accurate. Range validation tests conducted to test the accuracy of the lasers used the HDUE in differential mode. The tester had become the tested.

The early tests with differential mode were accomplished with the corrections transmitted in terms of adjustments to the position solution. As more satellites were launched and it became impossible to count on the receivers using the same constellation, the HDUE differential mode and the IRCC/GT differential message were modified. The IRCC now transmitted corrections to the individual satellite measurements. The HDUE, after receiving and decoding those corrections, applied them as ionospheric delay estimates to the individual pseudoranges. Tests

Table 4 Differential test summaries, m

Date		Mean	Root sum square	Standard deviation	CEP/ probable error
9 Jan 1980	Horizontal	4.2	4.7	2.1	3.8
	Vertical	−5.2	5.5	1.9	5.2
10 Jan 1980	Horizontal	5.0	5.3	1.8	5.0
	Vertical	3.9	4.5	2.2	4.0
11 Jan 1980	Horizontal	3.8	3.8	0.8	3.7
	Vertical	1.4	2.0	1.5	1.5
16 Jan 1980	Horizontal	3.6	3.8	1.5	3.4
	Vertical	2.1	2.6	1.5	2.1

using the measurement correction method began in 1984, with accuracy results similar to previous differential tests. The system again demonstrated its accuracy by identifying an out-of-calibration laser.

XIV. Phase II: Full-Scale Engineering Development Tests (1982–1985)

Phase II, full-scale engineering development testing of GPS UE began in 1982. Two contractors, Rockwell-Collins and Magnavox, had been selected to compete in the phase II program.

In the concept validation phase, it was appropriate and acceptable to tune equipment to specific situations. Satellite selection algorithms and acquisition techniques were not an issue during phase I. Indeed, with the lack of satellites, it would have been difficult to test such requirements. Environmental requirements for the equipment were also secondary considerations in phase I. In phase II, however, the equipment was to be integrated into several host vehicles and the integrations were to be representative of production requirements. Several other applications of GPS were to be tested, especially those related to the integration into complex avionics systems. Phase II test objectives included the following:

1) Verify reaction time, time to first fix, time to subsequent fix, and reacquisition time.
2) Verify static and dynamic position and velocity accuracies.
3) Evaluate IMU/INS alignment and gyro damping capabilities.
4) Determine susceptibility to electronic warfare and nuclear threats.
5) Evaluate precision weapons delivery using GPS bombing mechanization.

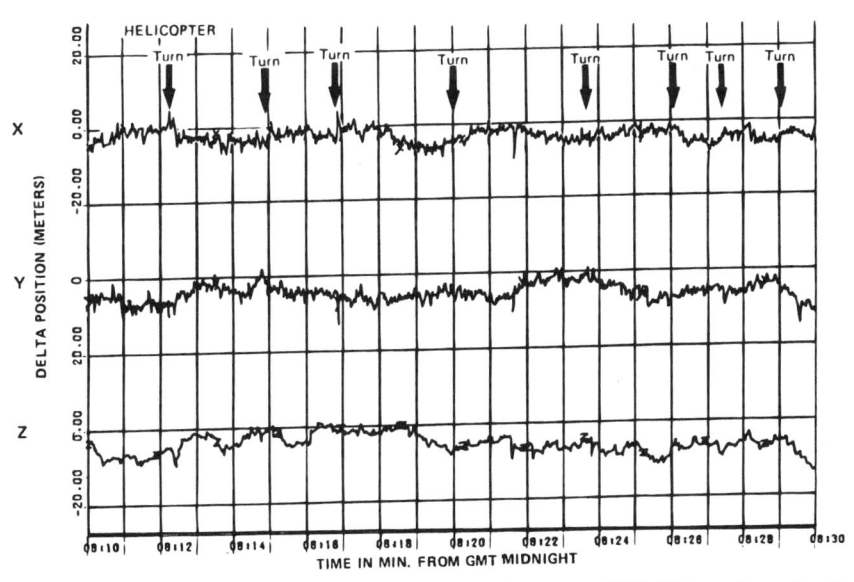

Fig. 6 Example of global positioning system position errors (UH-60 two-channel set).

Chapter 20

Interference Effects and Mitigation Techniques

J. J. Spilker Jr.* and F. D. Natali[†]
Stanford Telecom, Sunnyvale, California 94089

I. Introduction

THE GPS navigation satellite system is designed to serve both military and commercial needs. Because of its military applications, the ability to tolerate significant amounts of interference and jamming was an important consideration in the design of the signal structure. The purpose of this chapter is to describe the effects of interference and to present some design guidelines and mitigation techniques to improve the tolerance to interference of commercial GPS receivers that operate on the C/A signal. The performance of the P(Y)-code signal in the presence of jamming is beyond the scope of this book.

The GPS frequency bands are protected by international and Federal Communication Commission (FCC) frequency assignments. Nonetheless, there is the chance of spurious unintentional interference and even intentional interference. Obviously, any radionavigation system can be disrupted by an interference of sufficiently high power, and the GPS is no exception. Of course, GPS has the obvious limitation that the signal power received has been attenuated by the path loss from the distant satellite. The GPS C/A and P signals are both spread-spectrum signals and, as such, are much less susceptible to interference than a narrow-band signal. For example, a sidetone ranging signal that consists of a pure carrier with tone modulation has a much greater susceptibility to interference. A tone interference could completely disrupt a sidetone- or Doppler-type positioning system if its power simply exceeded that of the desired ranging signal. Interference of even less power could cause significant errors. With the GPS, as described in Chapter 3, this volume, a received tone interference would have to be significantly greater in power than that of the received GPS signal in order to degrade performance appreciably. Thus, the GPS has several advantages over a conventional narrow-band navigation system with respect to purposeful or unintentional interference or disruption for the following reasons:

1) GPS signals, both C/A and P(Y) code, are spread-spectrum signals and permit well-designed receivers to tolerate significantly larger amounts of co-

---

Copyright © 1995 by the authors. Published by the American Institute of Aeronautics and Astronautics, Inc., with permission. Released to AIAA to publish in all forms.
*Ph.D., Chairman of the Board.
† Ph.D., Vice President, Chief Engineer.

channel or adjacent channel interference than can be tolerated by conventional narrowband signals.

2) The GPS position determination is relatively robust, because the system is generally overdetermined; i.e., there are often more satellites visible than the minimum number required for a solution. Furthermore data detection, which has a significantly higher threshold than the tracking functions, is not required continuously.

3) Special receiver design techniques, such as adaptive A/D converters, adaptive frequency notch filters, various types of adaptive delay lock loops (adaptive DLL), and adaptive nulling antennas can further improve receiver performance. An adaptive DLL detects the presence of interference and narrows the tracking loop bandwidth.

4) GPS receivers can be designed to detect interference levels strong enough to disrupt navigation, because the interfering signal must generally be large compared to the relatively stable thermal noise level. Once interference is detected, the receiver can take corrective action.

5) In addition, because the GPS signals are generated on a satellite at a considerable distance from Earth, the satellite signal sources are not easily disrupted by natural disasters. Ground-based transmitters, on the other hand, are sometimes disrupted by violent storms or earthquakes.

The GPS receivers can, by proper design, take maximum advantage of the GPS spread-spectrum signal characteristics. However, beyond a certain interference level, any radionavigation signal can be disrupted. In this situation, the GPS receiver can be designed to detect the presence of the interference and switch to the use of other sensors. This detection process can be similar to that employed in receiver autonomous integrity measurement (RAIM), as discussed in Chapter 5 of the companion volume. However, even in the rare instance when there are an insufficient number of satellites in view to employ RAIM, a high enough level of interference to disrupt GPS navigation can be reliably detected by a properly designed receiver interference detector or by discrete Fourier transform (DFT) processing, and the receiver can alert the operator to take corrective action. Interference mitigation techniques for GPS receivers include the following:

1) Use of adaptive array, multibeam, or adaptive nulling antennas. These techniques can be categorized as spatial signal processing.

2) Radio frequency/intermediate frequency (rf/IF) filtering to reduce adjacent channel and out-of-band interference.

3) Use of sufficient number of bits/sample and AGC to reduce nonlinear signal suppression effects. Adaptive A/D conversion can reduce constant envelope interference effects.

4) Adaptive filtering in the delay lock loop and phase lock tracking filters.

5) Use of the vector delay lock loop described in an earlier chapter on Signal Tracking Theory (Chapter 7, this volume).

6) Detection of interference followed by an appropriate change in the DLL signal processing and aiding from other sensors, such as inertial measurement units (IMUs).

7) Augmentation of the GPS by ground (pseudolite) and space-borne signal sources.

8) Use of Adaptive frequency notch filters to attenuate narrow-band interference.

A. Possible Sources of Interference

Possible sources of unintentional interference to GPS receivers are shown in Fig. 1. None of these is believed to be a real problem to the GPS, but we should be aware of them, nonetheless. Some of these have at times produced interference to GPS receivers. These include the following[1-3]:

1) Out-of-band interference caused by nearby transmitters coupled with inadequate rf filtering in the GPS receiver.

2) Harmonics or intermodulation products of various ground and airborne transmitters. For example, transmitters at 105.028 MHz, 225.06 MHz, 315.084 MHz, 525.14 MHz, and 787.71 MHz have harmonics (the 15th, 7th, 5th, 3rd, and 2nd respectively) that fall within the GPS L_1 frequency band. Normally these transmitters are sufficiently well filtered to avoid radiating interference.

3) Potential active or passive intermodulation products of signals or local oscillators on the same platform as the GPS navigation receiver or from other nearby platforms. We must ensure that these potential intermodulation products are adequately removed by filtering. For example, we must take care to prevent radiation of local oscillator or intermodulation products from an aircraft transmitter that might couple into a GPS receiver on that same aircraft.

4) Pulsed interference from radar signals in nearby frequency bands that are inadequately filtered.

5) Accidental transmission of signals in the wrong frequency band by experimenters.

B. Frequency Allocation in Adjacent and Subharmonic Bands

We first examine the frequency bands near the GPS frequency band and then other bands that could create harmonics or intermodulation products that fall within the GPS bands if the transmitters are improperly filtered. The frequency bands assigned to satellite radio navigation (GPS, GLONASS) are shown in

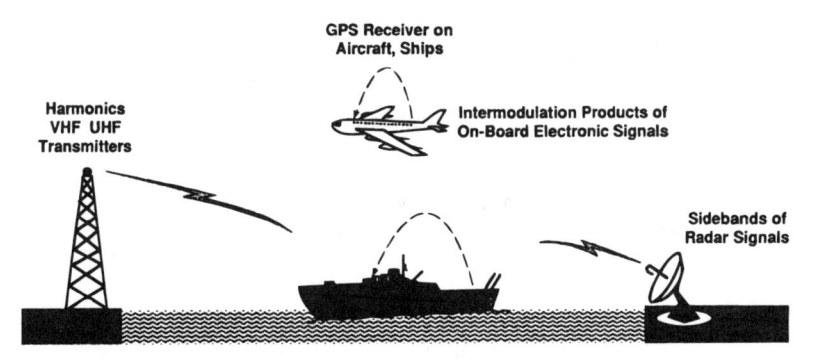

Fig. 1 Possible sources of unintentional interference caused by out-of-band emissions of other signal sources.

Table 1. Other bands in the vicinity of the L_2 signal are presently being considered as a second civil signal. The GPS C/A signal occupies approximately 2.5 MHz of bandwidth centered at 1575.42 MHz (the L_1 frequency).

Nearby bands are assigned for mobile, maritime, aeronautical, satellite communications and satellite-based cellular, as shown in Table 2. The frequency channel just below the GPS L_1 band is a satellite-to-ground link where the power levels received on Earth are expected to be relatively low. On the other hand, the 1610–1626.5 MHz band is a band reserved for hand-held transmitters to transmit to space for satellite-based cellular telephone. Both time division and code division multiple access (TDMA and CDMA) signals are licensed by the FCC for this band. The possible close proximity of these transmitters requires that the GPS receiver have adequate filtering to attenuate these transmissions, which are 34.58 MHz above the GPS L_1 center frequency and 24.58 MHz above the upper (10 MHz) edge of the GPS spectrum.

Another potential source of interference is from satellite communications uplinks at L-band in the 1626.5–1660.5 MHz bands (maritime satrom and aeronautical satrom). If these uplinks are on the same platform as the GPS receiver, there must be enough antenna separation and other isolation to prevent overload of the GPS low-noise amplifier (LNA). In addition one must avoid the effects of active or passive intermodulation products from either satellite communications (SATCOM) multicarriers or from local oscillators in close proximity to the GPS antenna or LNA.

Harmonics of improperly filtered TV channels also can be a potential source of interference in the vicinity of the TV transmission tower. The uhf TV transmission channels are spaced by 6 MHz and extend up to slightly over 800 MHz. Ultrahigh frequency military satellite communication takes place in the 240–400 MHz band with uplinks in the upper portion of the band, primarily in the 290–320 MHz band. Another possible source of interference is a harmonic (12th or 13th) of a VHF civil aviation air-to-ground link in the 118–136 MHz band.

II. Receiver Design for Tolerance to Interference

A. Receiver Systems

A simplified diagram of a GPS receiver that contains both spatial and signal processing for interference suppression is shown in Fig. 2. One or more antennas receive the GPS signal. This signal is rf filtered and downconverted to IF where it is amplified (with AGC) and then sampled and quantized. The quantized signal plus noise is then fed to the parallel code and carrier-tracking loops. All of these

Table 1 Radio navigation frequency bands

Frequency	Bandwidth
1.215–1.240 GHz	25 MHz
1.240–1.260 GHz	20 MHz
1.559–1.610 GHz	51 MHz

INTERFERENCE EFFECTS AND MITIGATION TECHNIQUES 721

Table 2 Frequency bands assigned for mobile satellite communications and satellite-based cellular[a]

Frequency Band	Bandwidth	Usage
1.535–1.559 GHz	24 MHz	Several bands for satellite downlinks to mobile, marine, aeronautical users—space-to-ground
1.610–1.626.5 GHz	16.5 MHz	Uplink based from hand-held units Earth-to-space, satellite-based cellular
782–788 MHz 525 MHz	51 MHz	uhf television—6 MHz channels at 1/2 GPS L_1 frequency uhf television at 1/3 GPS L_1
1626.5–1660.5 MHz	34 MHz	Aeronautical satellite communications uplinks (possible intermodulation products)

[a]Also shown are subharmonic bands that would potentially cause interference if the transmitters are not operating properly.

elements are described in some detail in other chapters. In this chapter, we focus only on the possible effects of an interfering signal, be it broadband or narrowband.

1. Antenna and Adaptive Array Antennas

If only a single GPS antenna is employed, it often is a hemispherical antenna that typically has gain between 0 dB and +6 dB for much of the upper hemisphere. An example antenna pattern is shown in Fig. 3.

Note that at 5 deg elevation angle, the gain for this antenna has dropped to approximately −5 dB. However, as pointed out in the previous chapters, when the elevation angle drops below 5 deg, other effects, such as multipath, physical obstructions, and ionospheric/tropospheric delay uncertainty, may be the dominant error sources. In fact, it may be desirable to attenuate signals received below 5

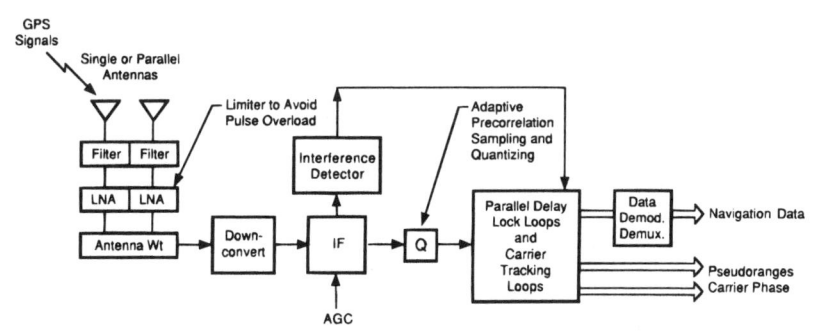

Fig. 2 Simplified diagram of a GPS receiver structure. An antenna array can perform spatial signal processing. An adaptive quantizer can be employed to suppress constant envelope interference. The DLL loop bandwidth can also be adapted. (The LNA is the low noise amplifier.)

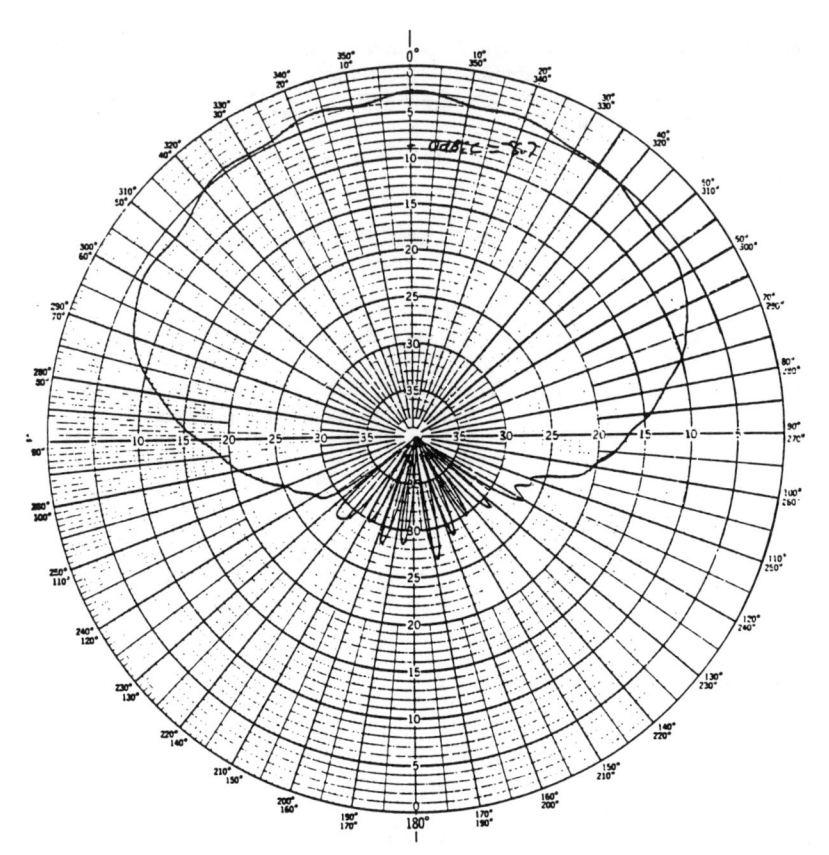

Fig. 3 Typical GPS antenna pattern at L_1 frequency in dB (the gain is 0 dBIC at −8.7 dB on the scale shown) (courtesy of Ball Aerospace).

deg in order to attenuate interfering signals and multipath that might come from the Earth's surface.

Although most GPS receivers employ only a single antenna, some receivers have more than one. For example, two antennas can be employed, one with maximum gain at zenith and another with larger gain at low elevation angles. Satellite receiver channels can be switched between them. Alternatively, a set of antennas can be operated as phased arrays to provide additional antenna gain in the directions to each of the satellites. We can also design an antenna array to null out a single source* of interference by adaptive weighting of multiple antennas, a form of spatial interference suppression. The adaptive nulling of an interfering signal can be performed using an adaptive algorithm very similar to that used in adaptive equalization. Section IV outlines some of the fundamentals of adaptive nulling.

---

*Multiple nulls can also be created for multiple interference sources as long as there is sufficient separation angle between the GPS satellite and the interference source, and there is a sufficient number of antenna elements.

INTERFERENCE EFFECTS AND MITIGATION TECHNIQUES

A multiplicity of antennas can be used in an adaptive array, as shown in Fig. 4.[4,5] The interfering and the satellite signals generally have a significant angular separation if the interference source is on the Earth's surface. Thus, various least mean square and signal-to-noise enhancing adaptive array techniques can be used; e.g., the Applebaum array, to enhance the gain to desired signals and attenuate the interference. Many of these techniques require some knowledge of the signal and the arrival directions, although some adaptive nulling techniques require neither, as discussed in Sec. IV. Constraints can be placed on the adaptive weighting to provide significant gain in the upper hemisphere.

Figure 5 shows an antenna array configured with multiple-beam steering networks. The satellite angle of arrival relative to the user platform must be known to within the beamwidth of the antenna pattern. For fixed ground receiver applications, this information is generally known; for most mobile platforms, however, this operation can be complex.

1. Radio Frequency and Intermediate Frequecy Filtering

Several stages of rf and IF filtering are required. Radio frequency filtering is important to prevent out-of-band interference from causing nonlinear or saturation effects in the LNA and/or intermodulation or image problems in the downconverter mixers. Typically, the user receiver also provides some level of overload protection or limiting prior to the LNA to avoid saturation/burnout of the rf front end caused by high peak power pulses. The rf/IF filter–amplifier combination must be able to reject out-of-band interference and noise, suppress image signal response, prevent intermodulation and spurious effects in the IF amplification and sampling/quantizing operations, while not causing significant group delay variation over the desired signal bandwidth.

The bandwidth of the rf/IF filter for a C/A receiver might be set as wide as 20 MHz in order to get the greatest accuracy in a white thermal noise environment; whereas, this bandwidth could be set as small as 2 MHz (or even 1 MHz) in order to obtain the greater selectivity against out-of-band interference. Figure 6 shows five- and eight-pole Butterworth filter characteristics for bandwidths of

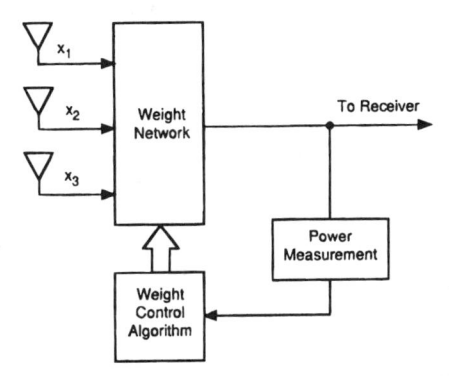

Fig. 4 Adaptive antenna array for nulling interfering signals. The adaptation algorithm can be set to minimize total power because, received power without interference is essentially constant thermal noise.

2 MHz and 20 MHz. Increasing the filter slope selectivity reduces interference effects but increases both the group delay and delay distortion. In summary, although the optimum C/A-code tracking accuracy in thermal noise is obtained by using the full 20 MHz bandwidth, interference rejection and other dynamic range effects may dictate a smaller rf/IF bandwidth.

The AGC sets the signal levels of the receiver so that the performance degradation of the quantizer is minimized. Short duration, high peak power pulses caused by radar sidebands, other pulse interference, or time-gated low duty factor GPS pseudolite transmissions are generally "peak clipped" (limited) or "blanked" (IF disabled for the length of the pulse). The IF or in-phase and quadrature (I,Q) baseband signal is then sampled and quantized. In the next section, the quantizer degradation caused by interference plus thermal noise is analyzed for various numbers of bits per sample and AGC levels. As is shown there, an interfering CW tone with some degree of coherence can cause a 1-bit quantizer to degrade substantially below the performance of that same quantizer with an equal power white thermal noise input. On the other hand, a well-designed multibit A/D converter can actually reduce interference effects for a constant envelope interference.

B. Quantizer Effects in the Presence of Interference

The GPS receiver processing system is assumed to be completely digital; thus, at some point, the received signal plus noise must be sampled and quantized. In this section, the received signal is assumed to contain the desired GPS satellite signal, receiver thermal noise, and an interfering signal. Two forms of the interfering signal are assumed: sinusoidal and Gaussian. A Gaussian interference simply

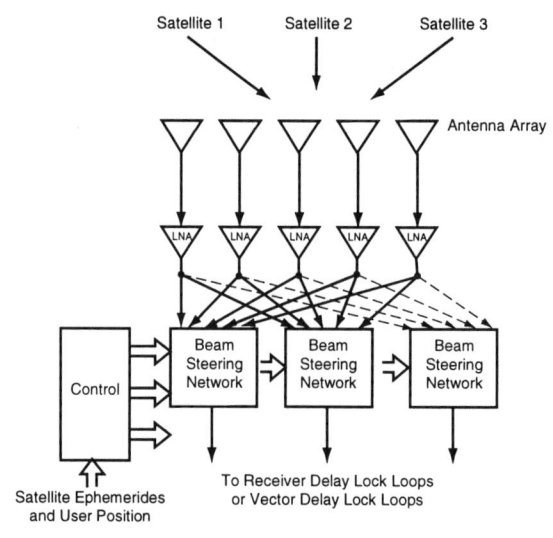

Fig. 5 Multibeam antenna array. This antenna array provides several separate beams, each pointing toward one of the GPS satellites. Approximate knowledge of the satellite angle of arrival is required for this antenna array.

INTERFERENCE EFFECTS AND MITIGATION TECHNIQUES 725

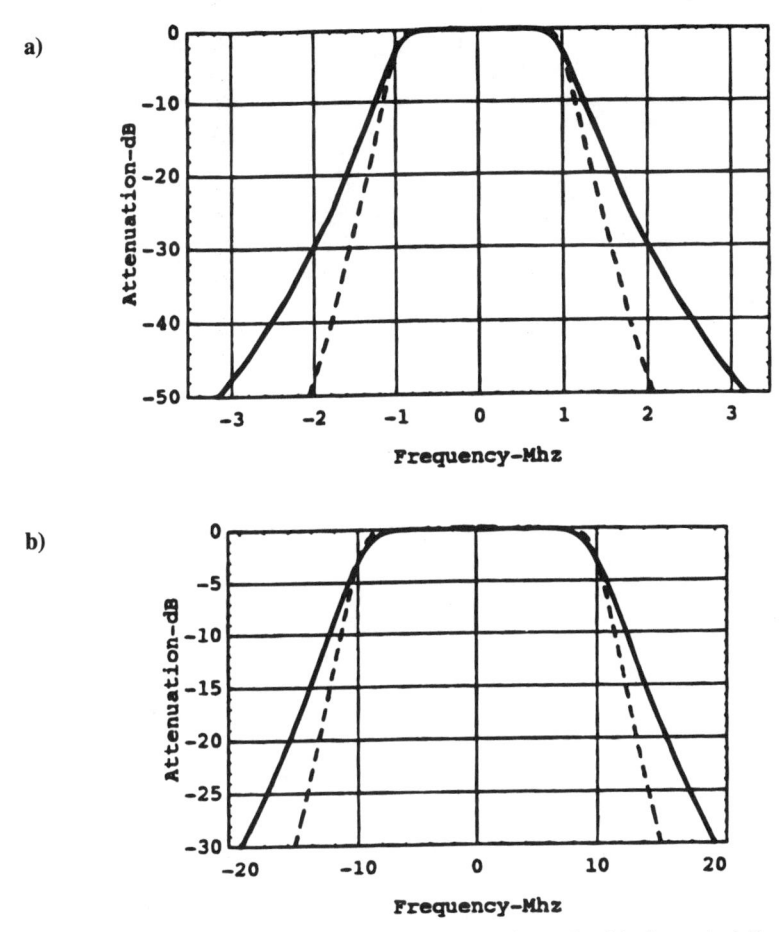

Fig. 6 Example of rf/IF filter frequency response (magnitude) characteristics for five-pole (solid) and eight-pole (dashed) Butterworth filters of 2 MHz and 20 MHz bandwidths.

adds to the Gaussian thermal noise although it may have a different bandwidth. A sinusoidal form of the interference can take on CW, narrowband, and wide bandwidth forms; e.g., an FM signal spread over 100 kHz or 2 MHz. The objective of this section is to analyze the signal-to-noise performance in the output of the quantizer–correlator relative to the input. A sinusoidal interference can have a severe impact on the receiver performance if a 1-bit quantizer is employed or if hard limiting occurs. On the other hand, a well-designed multibit A/D converter can actually suppress a constant envelope interference if the quantizing levels are properly set.

Before we proceed with the detailed discussion of quantizers, it is well to review some results for bandpass limiters. Recall that the GPS signal is a pseudonoise (PN) signal that is biphase modulated on a reference carrier. In a receiver that employs a bandpass limiter,* it is well known that the output signal-to-interference ratio is degraded by 6 dB in the presence of strong sinusoidal interference with significant frequency offset. Frequency offset is defined with respect to the carrier center frequency. That is, the weaker signal is suppressed relative to the strong interference, so it is 6 dB weaker at the output of the bandpass limiter than at the input.[7] If the interference is at exactly the same frequency and in-phase with a desired binary phase-shift keyed (BPSK) PN signal, it can suppress the desired signal by much more than 6 dB and "capture" the receiver. If instead, the interference is a strong Gaussian signal, the suppression of the desired weaker signal by a bandpass hard limiter is $\pi/4$ or -1.05 dB. In this section, we analyze the effects of 1-, 1½-, 2-, and 3-bit quantizers on the correlation performance of a receiver where the signal is received in the presence of various coherently modulated constant envelope interfering signals plus white Gaussian noise.

The GPS received signal power is assumed to range from -157 to -160 dBW for the C/A code for a 0 dB gain antenna. If the thermal noise density is -205.2 dBW/Hz, and a 2 MHz or 63 dB-Hz noise bandwidth is used, the total noise power is $P_n = -205.2 + 63 = -142.2$ dBW. The signal-to-noise ratio (SNR) then ranges from $P_s/P_n = \text{SNR} = -14.8$ dB to -17.8 dB for the C/A code. If a bandwidth of 10 MHz (or 70 dB) is used for higher accuracy, the SNR ranges between -21.8 dB to -24.8 dB. In the analyses that follow, a signal-to-thermal-noise ratio of -23 dB is assumed (the quantizer effects are not very sensitive to the SNR ratio in this range). The signal is assumed to be coherent with the receiver local oscillator so that all of the signal power passes through the in-phase channel; whereas, only half of the noise power is passed in the in-phase channel, the other half being in the quadrature channel. Thus, the ratio of signal amplitude level to rms noise level σ for this example is 1/10.

By way of reference it is well known[7] that the SNR degradation caused by a 1-bit quantizer (with high sampling rate) is a factor of $2/\pi$ or -1.96 dB for a signal plus large Gaussian noise (i.e., low SNR).† This degradation decreases to 0.5495 dB for a 2-bit quantizer for the best quantizer spacing relative to the noise level. For a 3-bit quantizer, the degradation decreases to 0.161 dB for optimum spacing.

Figure 7 shows the quantizer characteristic $Q(r)$ for 2- and 3-bit uniform quantizers where the quantizer interval is Δ, and the output levels are ± 1.5, ± 3.5. The quantizers considered here are all uniform quantizers with equal step sizes. Amoroso[8] shows that there is some advantage in interference suppression if nonuniform steps are employed on the output of a 2-bit quantizer, however, that benefit does not occur with thermal noise, and is not considered here.

1. In-Phase and Quadrature (I,Q) Sampling and Quantizing

In Fig. 8, it is assumed that the C/A signal from one satellite has been coherently downconverted to baseband along with Gaussian noise and sinewave interference.

---

*The bandpass limiter hard limits the envelope of the resultant and passes only the angular information.

†Note that 1-bit quantizing of each of the I and Q channels is not the same as bandpass limiting wherein a narrowband signal plus noise is hard limited and then bandpass filtered to pass frequency components only in the fundamental frequency zone to the output.

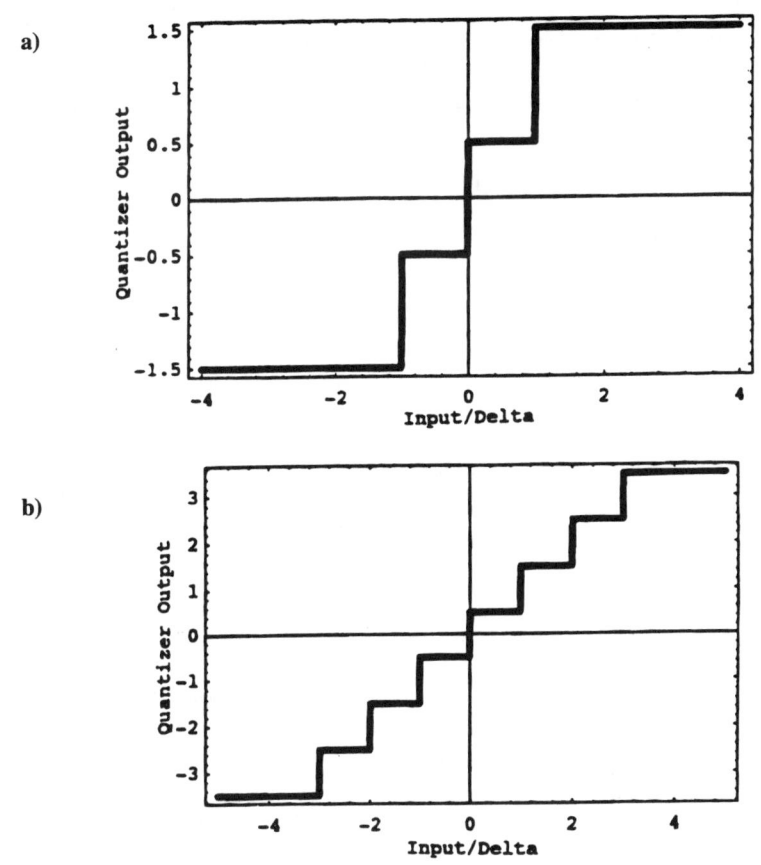

Fig. 7 Quantizer characteristic $Q(r)$ for 2- and 3-bit uniform symmetric quantizers with input quantizer interval Δ.

The interference can be received with or without uncorrelated PN modulation. Assume sufficiently wide bandwidth filters so that filter distortion effects are negligible (or that samples are taken with independent noise samples). The input $r(t)$ to the coherent channel quantizer is then $r(t) = A\, p(t) + K\, q(t) \cos \theta(t) + N_c(t)$, where A is the desired signal amplitude, $p(t)$ is the spreading code, the interference $I(t) = Kq(t) \cos \theta(t)$, and where the PN spreading codes $p(t)$ and $q(t)$ are both assumed to be ± 1. In this model, the interference modulation $q(t)$ may not be at the same rate as the PN code $p(t)$, and, in fact, $q(t)$ may not vary at all. If the interference has no PN modulation and is constant amplitude but offset in frequency by $f = \omega/2\pi$ the interference is $K \cos \omega t$. The bandlimited white noise $N_c(t)$ is assumed to have zero mean, variance σ^2, and Gaussian statistics. The quantizer output is defined as $Q_m[r(t)]$ where Q_m is an m-bit quantizer characteristic. The multiplier (correlator) output is then $c(t) = p(t)\, Q_m[r(t)]$ in the coherent channel.

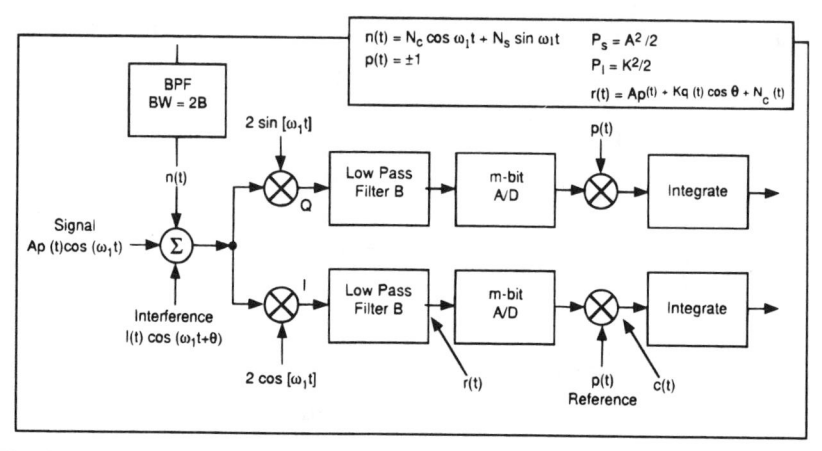

Fig. 8 Digital cross-correlation operation for an input of signal + noise + interference. An *m*-bit A/D converter is used prior to correlation.

The statistics of this multiplier output are examined in the next section. If the signal is sampled at a rate of $2B$ samples per second, the noise samples are independent, because the noise at baseband has bandwidth B. The objective is to compute the ratio of signal power in the correlator output to the variance in the correlator output and compare this ratio with the input signal-to-thermal-noise ratio and the ratio of input signal-to-noise plus interference.

The receiver noise at IF is represented as $n(t) = N_c \cos(\omega_i t + \theta) + N_s \sin(\omega_i t + \theta)$ where $n(t)$ has density N_o (one-sided), and the bandwidth at IF is $2B$. The noise power in that bandwidth is $\sigma^2 = N_o 2B$ (see Fig. 9). The power in $n(t)$ is as follows:

$$E[n^2(t)] = E[N_c^2 \cos^2(\omega_i t + \theta) + N_s^2(t)\sin^2(\omega_i t + \theta)]$$

$$= E\left[\frac{N_c^2}{2} + \frac{N_s^2}{2}\right] = E[N_c^2] = \sigma^2 \qquad (1)$$

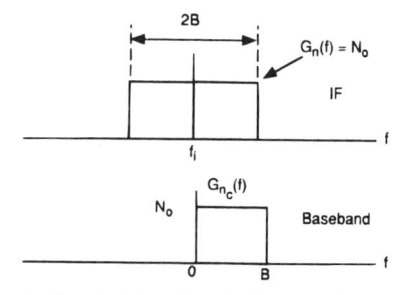

Fig. 9 Noise spectra at rf and at baseband. The baseband spectrum is one-sided.

because $N_c(t)$ and $N_s(t)$ are independent and of equal power, and θ is uniformly distributed; i.e., the variance of $N_c(t)$ is the same as that of $n(t)$. If the signal and noise are coherently mixed with the reference oscillator $2 \cos \omega_1 t$, then the output of the in-phase channel for $\theta = 0$ is the product $2 \cos \omega_1 t \, [Ap(t) \cos \omega_1 t + N_c(t) \cos \omega_1 t + N_s(t) \sin \omega_1 t]$. The low-pass component in this in-phase channel is $Ap(t) + N_c(t)$. If the signal power is P_s and the rf noise power is σ^2, then the SNR at the low-pass filter output of the in-phase channel is $A^2/E(N_c^2) = 2 P_s/\sigma^2$, because $P_s = A^2/2$.

The interfering signal for this example is assumed to have amplitude K and rf power $P_I = K^2/2$. Consider, first, the example where the interference is in-phase with the signal. Other phases and frequencies are considered later for the more general interference model. The total input to the quantizer in the in-phase channel after downconversion is then $r(t) = Ap(t) + I(t) + n(t)$, where $p(t) = \pm 1$ with equal probability, $I(t) = K$, and $n(t)$ has variance σ. The resultant waveform $r(t)$ then has a probability density $\frac{1}{2}$ Normal $(\mu_+, \sigma) + \frac{1}{2}$ Normal (μ_-, σ), where $\mu_+ = K + A$, and $\mu_- = K - A$ are the mean values under these two conditions. The quantizer output $Q[r(t)]$ is then correlated with $p(t)$ to produce the multiplier (correlator) output $c(t)$; i.e.,

$$c(t) = p(t)Q[r(t)] = p(t)Q[Ap(t) + K + n(t)].$$

The correlator-output signal-to-noise ratio is

$$(\text{SNR})_o = E[c(t)]^2/\{E[c^2(t)] - E[c(t)]^2\}.$$

The degradation in quantizer–correlator output SNR vs quantizer input signal-to-thermal-noise ratio* is, then, given by the ratio $R(K,\sigma,A,m,\Delta) = \text{SNR}_o/(A^2/\sigma^2)$, where m is the number of bits in the quantizer, and Δ is the quantizer interval. Define a zero mean Gaussian or normal density function as follows:

$$p(x) = \frac{1}{\sigma\sqrt{2\pi}} e^{-(x^2/2\sigma^2)} \qquad (2)$$

The error function is defined as follows:

$$\text{Erf}(x) \triangleq \frac{2}{\sqrt{\pi}} \int_0^x e^{-t^2} dt, \quad \text{and} \quad \text{Erf}(x_1, x_2) \triangleq \frac{2}{\sqrt{\pi}} \int_{x_1}^{x_2} e^{-t^2} dt \qquad (3)$$

and the normal distribution function is

$$\Phi(y) = \int_{-\infty}^{y} p(x) \, dx$$

Thus, the error function and Gaussian distribution functions are related by the following:

$$\Phi(y) = \frac{1}{2} \{\text{Erf}[y/(\sqrt{2} \, \sigma)] + 1\} \qquad (4)$$

---

*Note that the quantizer input SNR is twice the rf input SNR for a coherent receiver. Note, also, that this degradation is not referenced to the ratio of input signal-to-noise plus interference.

and the probability of a random variable y being in the interval (y_1, y_2); namely, $p(y_1 < y < y_2)$, is as follows:

$$\Phi(y_2) - \Phi(y_1) = \frac{1}{2} \text{Erf}[y_1/(\sqrt{2}\,\sigma),\ y_2/(\sqrt{2}\,\sigma)] \qquad (5)$$

Also, note that for small x, we have the following:

$$\text{Erf}(x) \cong \frac{2}{\sqrt{\pi}}\left[x - \frac{x^3}{3} + O(x^4)\right] \quad \text{for} \quad x \ll 1 \qquad (6)$$

The expected value of the correlator output is the mean value of $Q[r]p$. For a 2-bit quantizer with output levels $\pm 1/2$, $\pm 3/2$, where $p(t) = \pm 1$ with equal probability and all levels normalized to $A = 1$ the expected value of the correlator output is† (see Appendix):

$$E(c) = \frac{1}{2}\left[P\!\left(Q = +\frac{1}{2}\,|\,p = 1\right) - P\!\left(Q = \frac{1}{2}\,|\,p = -1\right)\right]\frac{1}{2}$$

$$+ \frac{3}{2}\left[P\!\left(Q = +\frac{3}{2}\,|\,p = 1\right) - P\!\left(Q = \frac{3}{2}\,p| = -1\right)\right]\frac{1}{2}$$

$$- \frac{1}{2}\left[P\!\left(Q = -\frac{1}{2}\,|\,p = 1\right) - P\!\left(Q = -\frac{1}{2}\,|\,p = -1\right)\right]\frac{1}{2}$$

$$- \frac{3}{2}\left[P\!\left(Q = -\frac{3}{2}\,|\,p = 1\right) - P\!\left(Q = -\frac{3}{2}\,|\,p = -1\right)\right]\frac{1}{2}$$

This expression can be evaluated as follows:

$$E(c) = \frac{1}{8}\left\{\text{Erf}\!\left[\frac{0-(K+1)}{\sigma\sqrt{2}},\ \frac{\Delta-(K+1)}{\sigma\sqrt{2}}\right]\right.$$
$$\left. - \text{Erf}\!\left[\frac{0-(K-1)}{\sigma\sqrt{2}},\ \frac{\Delta-(K-1)}{\sigma\sqrt{2}}\right]\right\}$$
$$+ \frac{3}{8}\left\{\text{Erf}\!\left[\frac{\Delta-(K+1)}{\sigma\sqrt{2}},\ \infty\right] - \text{Erf}\!\left[\frac{\Delta-(K-1)}{\sigma\sqrt{2}},\ \infty\right]\right\}$$
$$- \frac{1}{8}\left\{\text{Erf}\!\left[\frac{-\Delta-(K+1)}{\sigma\sqrt{2}},\ \frac{0-(K+1)}{\sigma\sqrt{2}}\right]\right.$$
$$\left. - \text{Erf}\!\left[\frac{-\Delta-(K-1)}{\sigma\sqrt{2}},\ \frac{0-(K-1)}{\sigma\sqrt{2}}\right]\right\}$$

---

† $P(Q = +1/2\ |\ p = 1)$ is the conditional probability that $Q = +1/2$ given that the PN chip $p = 1$.

$$+ \frac{3}{8} \left\{ \mathrm{Erf}\left[-\infty, \frac{-\Delta - (K+1)}{\sigma\sqrt{2}} \right] - \mathrm{Erf}\left[-\infty, \frac{-\Delta - (K-1)}{\sigma\sqrt{2}} \right] \right\}$$
(7)

The variance of c can be computed in a similar manner by averaging the probabilities that c^2 takes on the values 1/4, 9/4. The variance of c for the 2-bit quantizer is, then, as follows:

$$\mathrm{var}(c) = \left(\frac{1}{2}\right)^2 \left[P\left(Q = +\frac{1}{2} \mid p = 1\right) + P\left(Q = +\frac{1}{2} \mid p = 1\right) \right] \frac{1}{2}$$

$$+ \left(\frac{3}{2}\right)^2 \left[P\left(Q = +\frac{3}{2} \mid p = 1\right) + P\left(Q = +\frac{3}{2} \mid p = -1\right) \right] \frac{1}{2}$$

$$+ \left(\frac{1}{2}\right)^2 \left[P\left(Q = -\frac{1}{2} \mid p = 1\right) + P\left(Q = -\frac{1}{2} \mid p = -1\right) \right] \frac{1}{2}$$

$$+ \left(\frac{3}{2}\right)^2 \left[P\left(Q = -\frac{3}{2} \mid p = 1\right) - P\left(Q = -\frac{3}{2} \mid p = -1\right) \right] \frac{1}{2} \quad (8)$$

2. One-Bit Quantizer—Quantizer–Correlator Degradation

If the quantizer interval is set at $\Delta = 0$, the 2-bit quantizer of Fig. 7a degenerates to a 1-bit quantizer, and the output is then always $\pm 3/2$. Again consider a received input which consists of a unit amplitude biphase modulated signal plus a coherent interference of fixed amplitude K plus Gaussian noise of standard deviation σ. The probability density of the quantizer input given each of the two signs in the biphase modulation then appears as shown in Fig. 10. For this 1-bit quantizer, a coherently related interference of sufficient size can be shown to capture the receiver totally since the quantizer output then has almost no correlation with the input C/A–PN code $p(t)$.

Figure 11 shows the degradation in quantizer output SNR relative to quantizer input SNR plotted in dB for the signal level $Ap = \pm 1$, a noise level $\sigma = 10$, and interference amplitude K. Note that for small coherent interference amplitude $K = 0.1$, the degradation is 1.9548 dB, very close to the small SNR asymptotic value of $2/\pi$ (1.961 dB) for no interference. As the interference amplitude increases to 10, the degradation increases to approximately 6 dB. At larger interference levels for an interference amplitude level of approximately 25, the degradation rapidly increases to approximately 30 dB. This rapid quantizer–correlator degradation with interference levels not much greater than the receiver thermal noise level makes the use of 1-bit quantizing unacceptable if tolerance to coherent constant envelope interference is important. It can easily be seen that if there is no thermal noise, a coherent interference only slightly larger than the input signal amplitude completely suppresses the signal if a 1-bit quantizer is employed.

It is well known that BPSK PN signals are subject to stronger signal capture effects in a bandpass hard limiter, as shown in Fig. 12. A large interference,

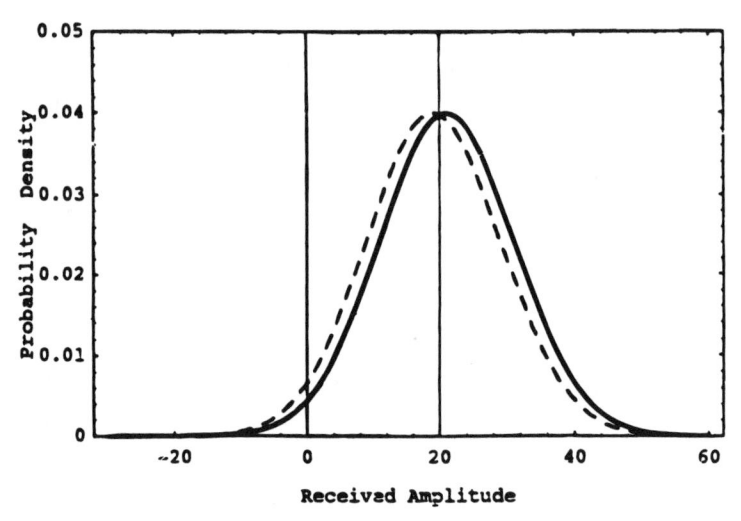

Fig. 10 Probability densities of the resultant of a biphase modulated unit amplitude signal plus coherent interference of level K plus thermal noise with $\sigma = 10$. This resultant signal is the input to a 1-bit quantizer for the two conditions $p = \pm 1$. In this example, $K = 20$, $\sigma = 10$, and $A = 1$. When the interference amplitude $K > \sigma$, the interference level carries the quantizer input to such a level that only very infrequent transitions are made, and the interference begins to capture the receiver totally.

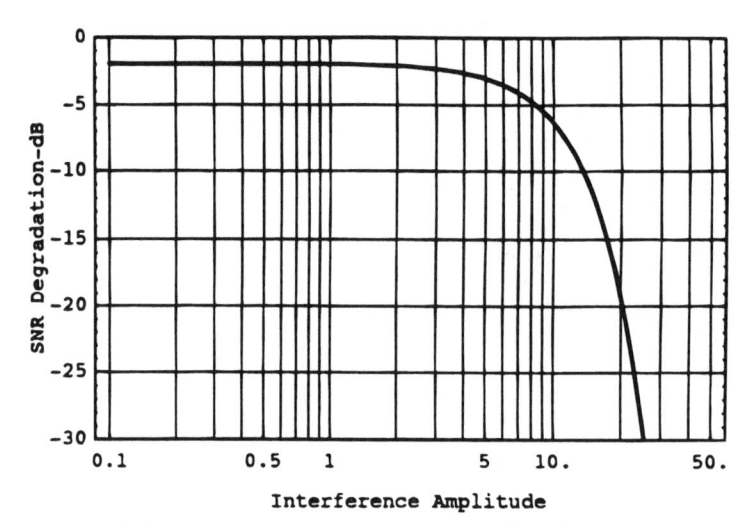

Fig. 11 Degradation in quantizer–correlator output SNR relative to quantizer input SNR caused by a 1-bit quantizer for coherent constant envelope interference of amplitude K relative to the signal amplitude. Gaussian noise is present with $\sigma = 10$. The SNR degradation is defined relative to the input signal to Gaussian noise ratio.

which is momentarily coherent with the signal for that time interval, completely suppresses the signal. Thus, we must avoid bandpass limiting in any GPS receiver with a potential for sinusoidal interference.* These effects with I/Q receiver 1-bit quantization are very similar to those studied by Baer[9] and Aein and Pickholtz.[10] The drastic suppression effects for a 1-bit I,Q quantizer however do not disappear with QPSK PN as they do with bandpass hard-limiter. The interference could be offset by 45 deg from the signal and would affect both I and Q channels equally.

The GPS C/A code, of course, only employs biphase modulation (the quadrature P(Y) code is not generally available to the civil user). Thus, the C/A receiver reference is also biphase modulated. The suppression effect of tone interference on the receiver, however, can be largely avoided through the use of quantizers with two, three, or more bits per sample, and an appropriate AGC, as shown below. Notice also that in the analysis of Fig. 10, the degradation has assumed that the interference is directly in-phase with the signal. If the interference is out of phase by some angle θ, the effective interference amplitude then becomes $K \cos \theta$ instead of K. However, it is clear that, in general, this effect simply means that the interference has to be slightly stronger, and, with θ slowly varying the signal, is suppressed periodically unless a multilevel quantizer is employed.

If the interference is offset in frequency by a large amount compared to the inverse of the averaging time; i.e., the data rate, then the mean and variance of $c(t)$ must be computed by averaging over θ in the $K \cos \theta$ term (with a uniform distribution of θ). This degradation is shown in Fig. 13. Note that for an interference amplitude of 20, the quantizer–correlator degradation is approximately 8.5 dB for the frequency offset interference vs 19 dB for the phase coherent interference, but the degradation is still increasing rapidly with increasing interference, and the 1-bit quantizer performance is unacceptable for this purpose.†

3. Two-Bit Quantizing—Quantizer–Correlator Performance

The performance of 2-bit quantizing is examined for PN signal plus noise alone, signal plus noise plus in-phase interference, and finally signal plus noise

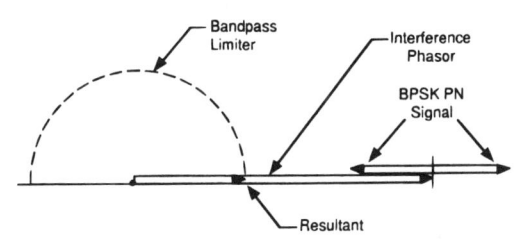

Fig. 12 Sum of BPSK PN signal and a coherent interference of larger amplitude. The resultant phasor is completely independent of the BPSK PN signal for this simple noise-free example, and the signal is completely suppressed. Note that noise is absent in this simple example.

---

*With QPSK PN, high-level sinusoidal interference would completely suppress only the in-phase term; i.e., half the signal power.

†One-bit quantizers are simple and lead to low-cost receivers, but their susceptibility to interference is a serious flaw for some applications.

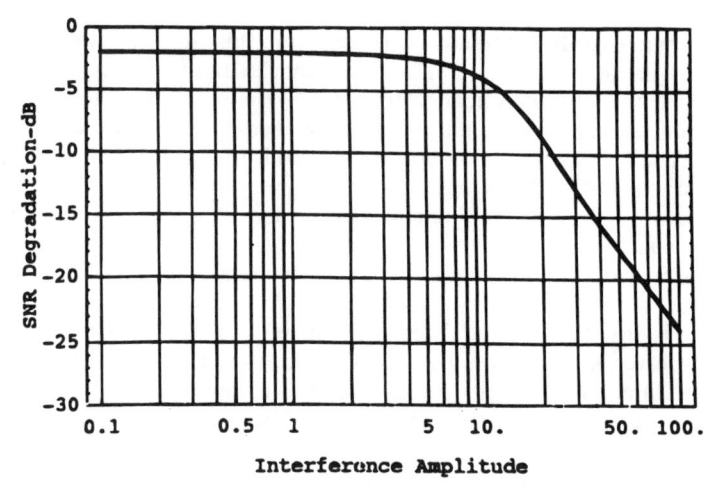

Fig. 13 Quantizer–correlator SNR degradation for a 1-bit quantizer with a frequency offset interfering sine wave $K \sin ax$ of amplitude shown. The signal is received with Gaussian noise ($\sigma = 10$) plus sinusoidal interference. For no interference the SNR degradation is $2/\pi$ or 1.961 dB. For large sinusoidal interference the degradation approaches 6 dB relative to the interference power level.

plus a sinusoidal interference which is offset in frequency from the signal center frequency. Clearly, in the limit for quantizer interval $\Delta = 0$ or ∞, the 2-bit quantizer is either always in saturation, or the received signal plus noise and interference never exceeds the first quantizer magnitude level Δ, and the 2-bit quantizer operates as a 1-bit quantizer. Figure 14 shows the output quantizer–correlator signal-to-noise performance degradation of a 2-bit quantizer relative to the quantizer input signal to thermal noise ratio (or to a receiver with an infinite number of quantizing intervals (linear)). The Gaussian noise has a $\sigma = 10$, and there is no interference. As shown in the figure, the minimum quantizer degradation for this case is 0.5415 dB and occurs at $\Delta = \sigma$. As the value of the input SNR declines to 0; i.e., infinite noise, the degradation increases only slightly to 0.5495 dB. Note that the degradation is still only 1 dB or less for quantizer interval variations of $\Delta = 0.4\,\sigma$ to $1.8\,\sigma$ corresponding to an AGC gain set error of approximately ± 7 dB.

If an interfering signal is present, and it has Gaussian amplitude statistics, the same result of Fig. 13 would also apply to the interfering signal problem. That is, set the AGC levels so that $\Delta = \sqrt{\sigma_n^2 + \sigma_I^2}$ where σ_n^2, σ_I^2 are the noise and interfering power levels, respectively.

Now consider a 2-bit quantizer that is employed with a signal in the presence of coherent interference of amplitude $I_c = K\cos\theta$ (for fixed θ) on the in-phase channel as well as Gaussian noise. The signal is of unit amplitude $Ap = \pm 1$, and the noise has rms value $\sigma = 10$. The quantizer interval is Δ. The quantizer–correlator SNR degradation above that of the quantizer input signal-to-thermal-noise ratio is shown in Fig. 15. Interference levels of $K = 0, 10, 20,$ and 40 correspond to quantizer input interference-to-noise ratios of $-\infty, 0, 6,$ and 12 dB,

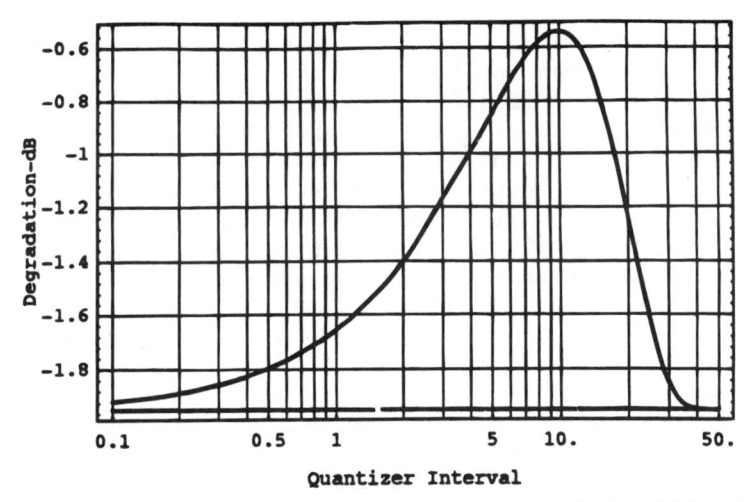

INTERFERENCE EFFECTS AND MITIGATION TECHNIQUES 735

Fig. 14 Quantizer–correlator signal-to-noise degradation in dB for a 2-bit quantizer with normalized quantizer interval Δ for signal plus noise. The degradation is plotted vs quantizer interval Δ. The signal is of amplitude $Ap = \pm 1$, and the Gaussian noise has rms value $\sigma = 10$. There is no interference in this example.

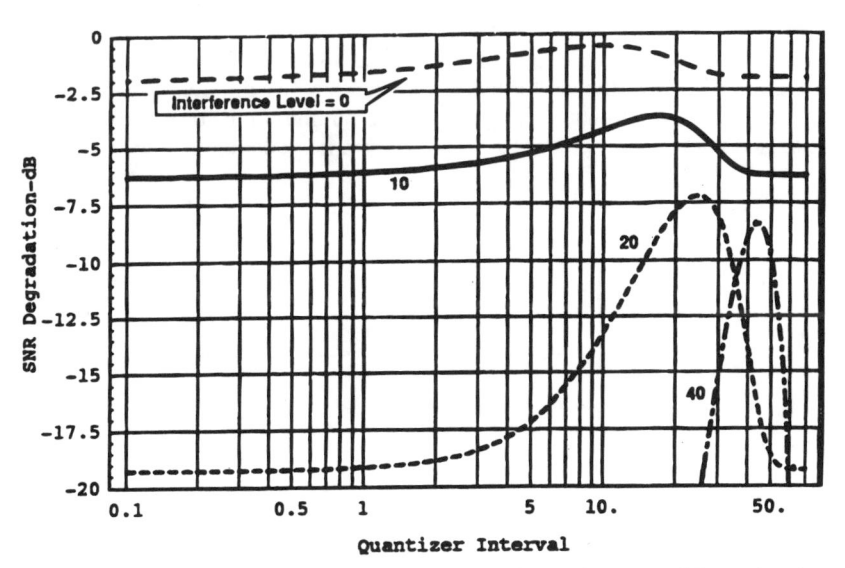

Fig. 15 Quantizer–correlator SNR degradation above that caused by noise alone for the 2-bit quantizer with coherent fixed interference levels of $K = 0, 10, 20$, and 40. If the quantizer was perfectly linear, the added degradation caused by interference would have been 0 dB, 3.0 dB, 7.0 dB, and 12.3 dB, respectively.

respectively. Notice that the range of good quantizer settings grows progressively narrower in percentage as the interference increases. Its width is approximately equal to the standard deviation of the noise. This result implies that the AGC level becomes more critical (when expressed in dB) as the interference level increases.

Note, however, that as the interference increases, the degradation does not increase proportionally with total noise plus interference power, as it would for Gaussian interference. It is clear from these results that the 2-bit quantizer interval Δ (or equivalently the AGC) can be set to attenuate the interference. Amoroso[8] pointed out that it is possible to design the quantizer to be adaptive so as to attenuate a sinusoidal interference. If the quantizer is set at the optimum level for large coherent interference, the input signal plus noise plus interference sits with a peak of the probability density near Δ (recall Fig. 10). Thus, the biphase modulated signal level toggles the magnitude both above and below the quantizer level Δ. In this case, a constant interference in the in-phase channel simply acts as a constant bias, and its effect to a large extent is canceled. Reference 8 showed that this effect can be enhanced by making the quantizer steps nonuniform, in the limit, the quantizer is a $1\frac{1}{2}$-bit null zone quantizer, as shown in Fig. 16. Clearly, in this example, the interference is almost completely canceled for a quantizer setting of $\Delta = K$ for large K.

As is evident from the curves of Figs. 14 and 15, the quantizer setting or AGC can be controlled to make it nearly equally probable to be in the two quantizer amplitude settings. The optimum quantizer setting in Fig. 15 is very close to $\Delta = \Delta_o \triangleq \sqrt{\sigma^2 + K^2}$. For example, at $K = 20$, the value of $\sqrt{10^2 + K^2} = 22.36$.

Note that the fact that the optimum quantizer setting results in equally probable output magnitudes makes it possible to build such an adaptive quantizer relatively easily. We simply have to count the number of occurrences in the two output magnitudes, and adapt the quantizer setting Δ so that they are equal.

Figure 17 shows the performance degradation for different AGC/quantizer settings for the 2-bit quantizer. Note that the optimum quantizer setting for this noise and interference, $\Delta \cong \sqrt{10^2 + K^2}$, gives a degradation that decreases to an

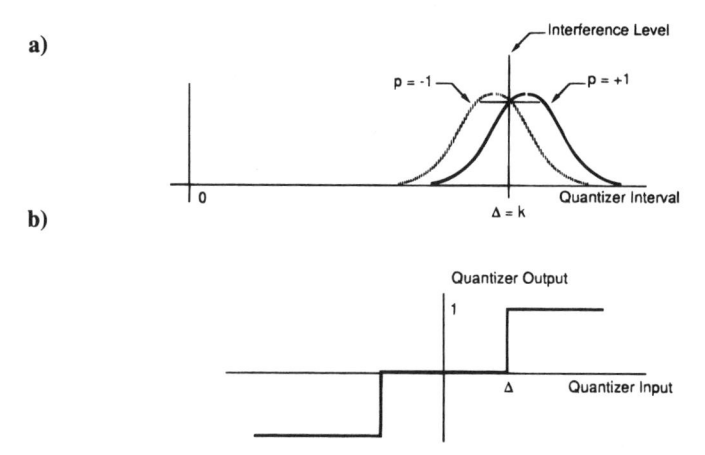

Fig. 16 Null zone quantizer setting with fixed large interference of amplitude K: a) quantizer input probability density; b) $1\frac{1}{2}$-bit, three-level quantizer with a null zone.

INTERFERENCE EFFECTS AND MITIGATION TECHNIQUES 737

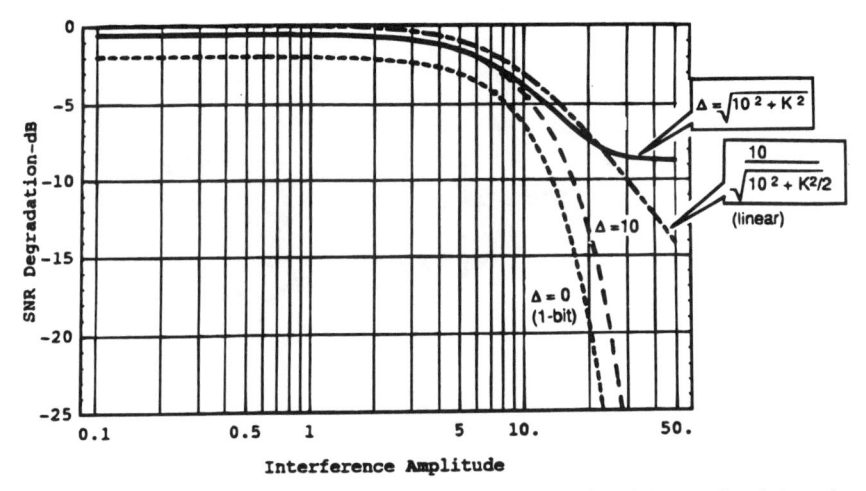

Fig. 17 Signal-to-noise ratio degradation for a 2-bit quantizer for quantizer intervals of $\Delta = \sqrt{10^2 + K^2} = \sigma_{total}$, $\Delta = 10$, and the 1-bit quantizer $\Delta = 0$. The linear reference degradation $10/\sqrt{10^2 + K^2/2}$ is shown as the alternate dash curve.

asymptote of approximately -9 dB. The reason for this behavior is that, as shown earlier in Fig. 5, a fixed interference level will still toggle the quantizer's output with the sum of random noise and signal. Thus, the interference effect itself is suppressed by the quantizer. Obviously, any time variation in the interference amplitude or phase would cause this interference nulling effect to be reduced, as shown below. As shown in Fig. 15, the minimum in quantizer degradation occurs in the region $\Delta = \Delta_0 \pm 10$. Although this result is for perfectly constant coherent interference, it is shown later that even with a sinusoidal offset interference, there is a definite performance improvement for an optimal quantizer compared to a linear quantizer.

A three-dimensional plot of the performance degradation vs both interference level and quantizer interval is shown in Fig. 18. As can be seen from the plot, there is a clearly defined ridge that optimizes the quantizing interval for any given interference level.

If the interference phase θ varies slowly with time relative to the databit rate, but the AGC maintains the quantizer level at a constant setting $\Delta = \sqrt{10^2 + K^2}$, then the quantizer output degradation varies slowly with phase offset (or time if the phase varies linearly with time), as shown in Fig. 19.

4. Two-Bit Quantizer with Frequency Offset Interference

Consider, now, an interference that is frequency offset from the center frequency of the PN signal. Assume that the frequency offset is large compared to the bit rate so that quantizer–correlator performance is determined by the time average of the interference in the in-phase channel ($K \cos \omega t$). Figure 20 shows the output degradation for values of $K = 10, 20, 40, 80$ relative to the signal amplitude of

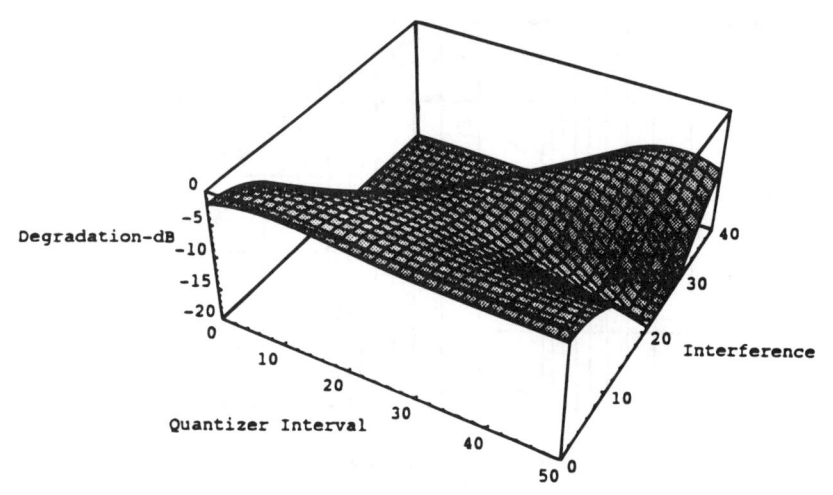

Fig. 18 Quantizer–correlator SNR degradation for the 2-bit quantizer vs both interference level and quantizer interval. The rms noise input is $\sigma = 10$.

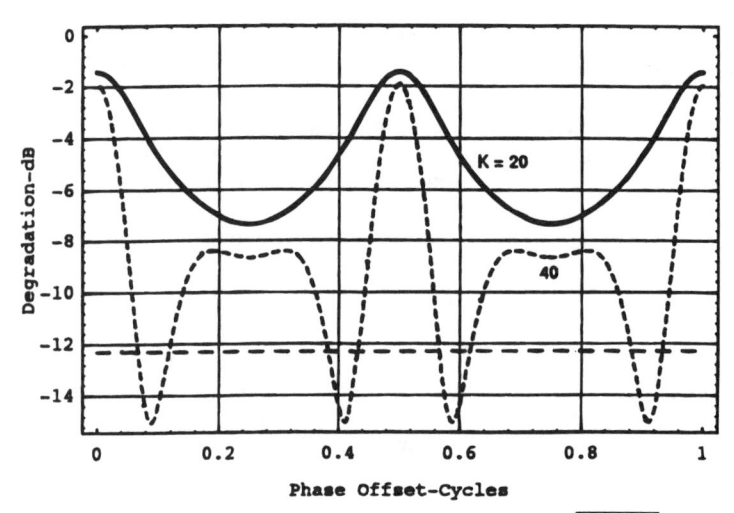

Fig. 19 Quantizer–correlator degradation for a setting of $\Delta = \sqrt{10^2 + K^2}$, with interference levels of $K = 20$ and 40. The horizontal dashed curve corresponds to the ideal quantizer degradation for $K = 40$ and a fixed phase offset $\theta = \pi/4$. The average degradation for this example with the linear quantizer would be 9.6 dB.

INTERFERENCE EFFECTS AND MITIGATION TECHNIQUES 739

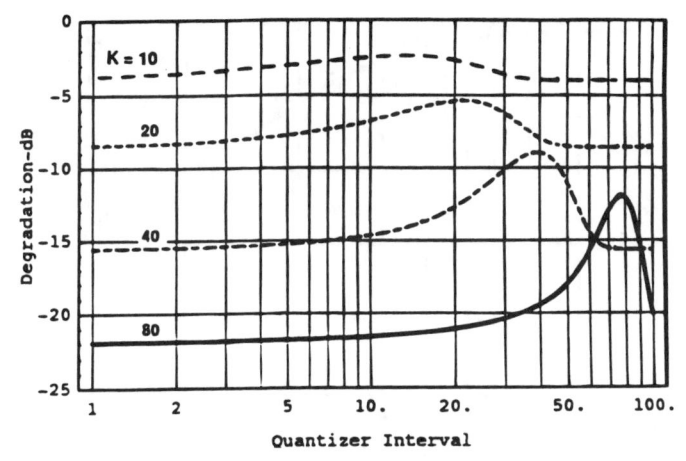

Fig. 20 Two-bit quantizer–correlator degradation in dB for four different values of the interference amplitude with frequency offset interference of the form $K \cos \omega t$. The results are given for $K = 10$, large dashes; $K = 20$, small dashes; $K = 40$, alternate dashes; and $K = 80$, solid curve.

unity and the rms noise $\sigma = 10$. The degradation in the quantizer–correlator output SNR relative to the quantizer input signal to thermal noise ratio varies with quantizer interval, as shown. Note that the optimum quantizer interval, again, has a value of $\Delta \cong \sqrt{\sigma^2 + K^2}$ and corresponds to roughly equal probable magnitude levels in the quantizer output during the interference peaks. Note, further, that the sinusoidal interference causes the quantizer output to have line components in its output. However, these components are spread by the cross-correlation operation with the C/A PN signal.

Figure 21 shows the quantizer–correlator output SNR degradation for the optimum quantizer spacing, $\Delta \cong \sqrt{\sigma^2 + K^2}$ with varying interference levels. Also shown in Fig. 21 is the degradation in quantizer output SNR relative to the quantizer input signal to thermal noise ratio that would be caused by the addition of the interference for a perfectly linear quantizer. Note that the output SNR for the quantizer correlator is improved over that of a perfectly linear quantizer–correlator for constant envelope interference above $K = 30$. At an interference level $K = 100$, the improvement is approximately 7 dB, as shown by the difference between the solid and large dashed curve. Note that the performance with a fixed amplitude interference of Fig. 17 has nearly reached an asymptote at $K = 100$; whereas, with a frequency offset, the performance degradation is still increasing in Fig. 21. Clearly, the optimum quantizer (adaptive AGC) has progressively greater performance compared to the linear quantizer as interference increases.

5. Performance with Three-Bit Quantizing—Quantizer–Correlator Degradation

Figure 22 shows the quantizer–correlator degradation for a 3-bit quantizer with a signal amplitude of $p = \pm 1$ and pure Gaussian noise with an rms value

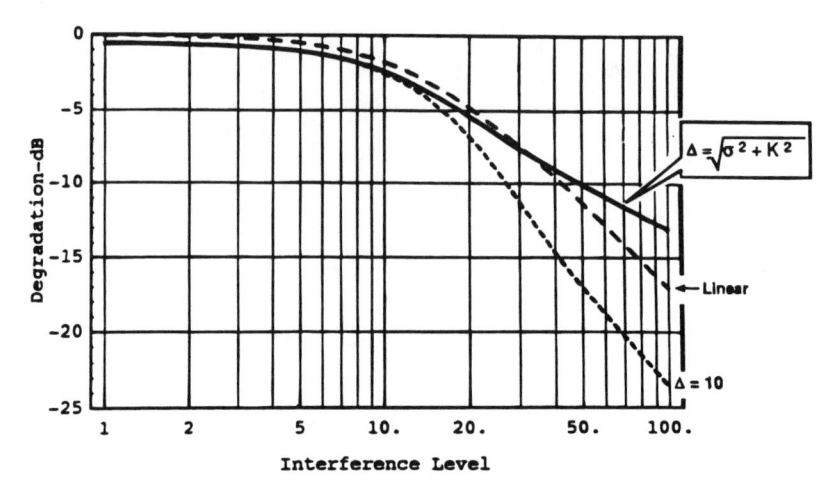

Fig. 21 Two-bit quantizer–correlator output degradation for frequency offset interference $I = K \cos \omega t$ for various quantizer intervals. The solid curve represents near optimum performance with adaptive quantizer spacing $\Delta \cong \sqrt{\sigma^2 + K^2}$; the small dashed curve shows the performance with $\Delta = 10$ fixed at the optimum for no interference. The large dashed curve shows the degradation resulting from a perfectly linear quantizer and is of course equal, to the quantizer input signal to (noise + interference) power ratio.

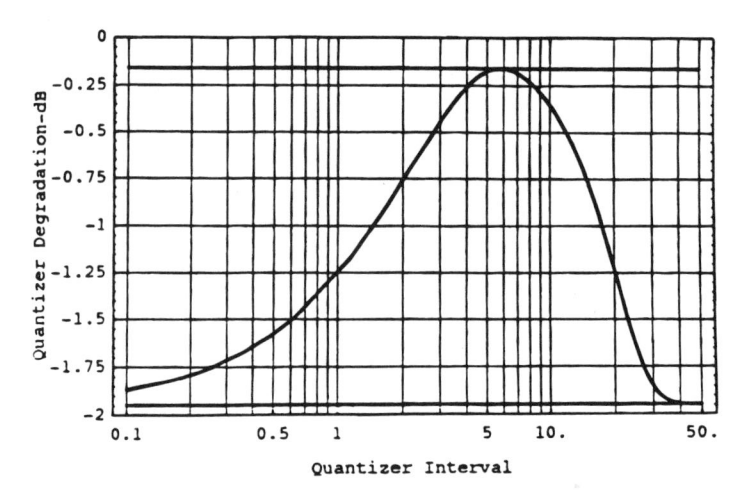

Fig. 22 Three-bit quantizer–correlator degradation in SNR for a signal level $p = \pm 1$, rms noise with $\sigma = 10$ and a quantizer interval Δ. The minimum occurs at $\Delta = 10/\sqrt{3} \approx 5.77$ and is ≤ 0.161 dB degradation. At $\Delta = 0$ or ∞ the degradation increases to -1.96 dB or $2/\pi$, the 1-bit quantizer result.

of $\sigma = 10$. There is no sinusoidal interference in this example. The minimum degradation occurs with a quantizer interval of $\Delta = \sigma/\sqrt{3}$ compared to $\Delta = \sigma$ for the 2-bit quantizer. The minimum degradation is 0.1613 dB vs 0.5415 dB for the 2-bit quantizer. Furthermore, for 3-bit quantization, the quantizing interval for less than 1 dB of degradation can range from $\Delta = 1.5$ to 18 or a 22 dB range. Thus, the AGC setting is considerably less sensitive than for the 2-bit quantizer, which has only a 13 dB range.

Figure 23 shows the cumulative normal probability distribution at the various quantizing intervals in terms of the magnitude probabilities. The probability of being in the magnitude interval 0 to Δ is 0.4363; Δ to 2Δ is 0.3155; 2Δ to $-\infty$ is 0.2482. Thus, the ratio of the probability of being in region 0 to Δ over that of being in Δ to 2Δ is 1.38, a little over unity. The quantizer AGC level, thus, can be set for both the 2-bit and 3-bit quantizers by properly balancing the counts in each magnitude bin (magnitude bits).

The additional quantizer–correlator output SNR degradation, beyond that of noise alone, caused by coherent interference of constant amplitude K is shown in Fig. 24 for $\sigma = 10$. The result for an ideal linear circuit is shown for comparsion. The flattening or double hump behavior appears because of the multiple quantizer magnitude levels in the 3-bit quantizer. Figure 25 gives the excess degradation for the four quantizing intervals $\Delta = \sqrt{\sigma^2 + K^2}/\sqrt{3}$, $\sqrt{\sigma^2 + K^2}$, a constant $\Delta = 10$, and the 1-bit quantizer $\Delta = 0$. For reference, the ratio of the degradation in the quantizer input SNR caused by interference; namely, $10/\sqrt{100 + K^2}$ is also shown. Again, the degradation for $\Delta = \sqrt{\sigma^2 + K^2} \approx K$ for $K \gg 10$ is a special case for which the quantizing interval is almost squarely centered on the peak probability density of the interference plus noise amplitude. The degradation flattens to an asymptote at -9 dB for large K. This quantizer setting is not quite

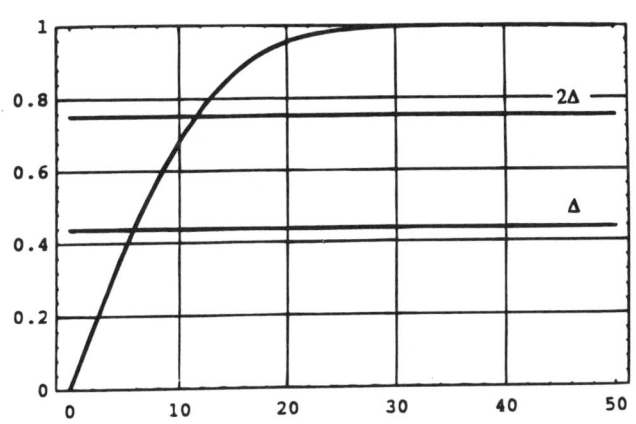

Fig. 23 Cumulative probability at the optimum quantizing interval $\Delta = 10/\sqrt{3}$, and 2Δ. Note that the ratio of the probability of being in the range 0–Δ is 0.44; whereas, the probability of being in the range Δ–2Δ is approximately 0.32. Thus, the ratio of the two probabilities can be used to set AGC level.

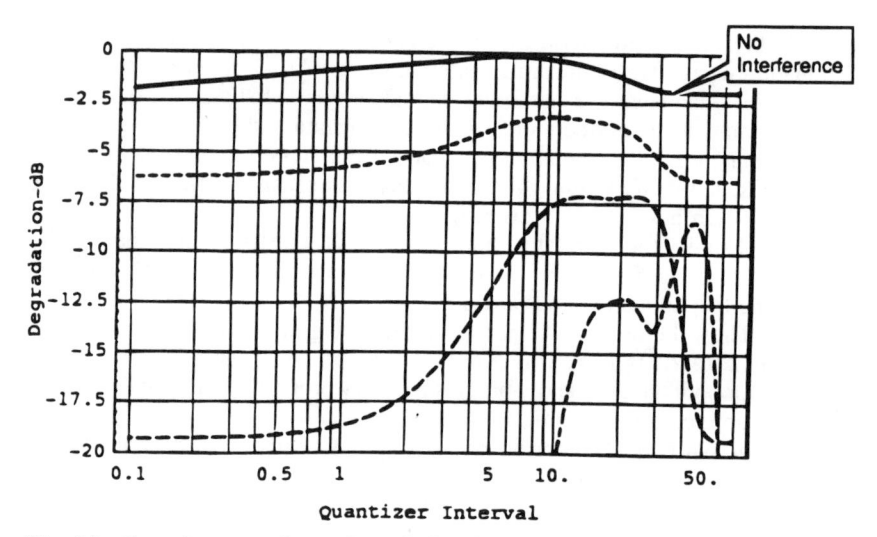

Fig. 24 Quantizer–correlator degradation for a 3-bit quantizer caused by fixed interference of amplitude $K = 0$, solid curve; $K = 10$, small dashes; $K = 20$, large dashes; and $K = 40$, alternate size dashes. The signal is $p = \pm 1$, and the noise has rms value $\sigma = 10$.

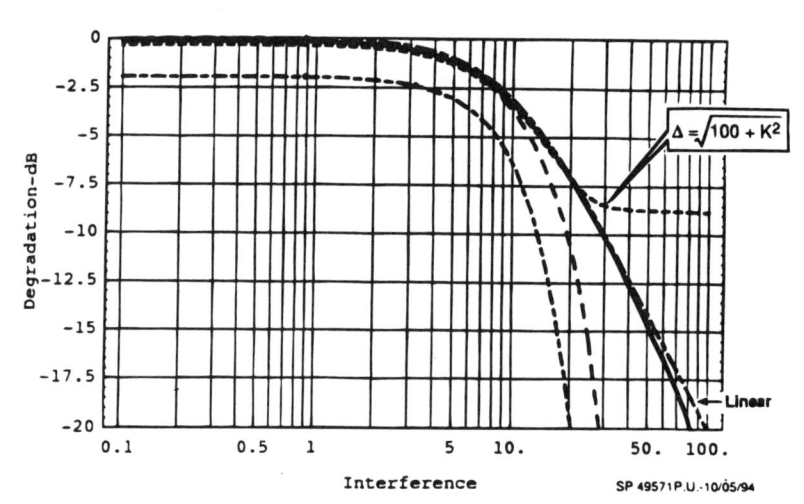

Fig. 25 Quantizer–correlator degradation for the 3-bit quantizer vs interference level for various quantizer intervals, $\Delta = \sqrt{100 + K^2} / \sqrt{3}$, solid curve; $\Delta = \sqrt{100 + K^2}$, small dash curve that levels off; $\Delta = 10$, large dashed curve. The small alternate dashed curve, the curve with the greatest degradation is the 1-bit quantizer $\Delta = 0$ for reference. A reference curve, $10/\sqrt{100 + K^2}$, the linear result, is also shown for reference with large alternate dashes.

as good for noise alone, as shown in Fig. 22. The degradation is about 0.35 dB for $\sigma = 10$, as compared to a minimum of 0.16 dB. The probability of being in the $0-\Delta$ range increases with no interference to 0.6827, but the ratio of being in $0-\Delta$ vs $\Delta-2\Delta$ is approximately unity for high levels of interference. However, the 3-bit quantizer degradation caused by interference with amplitude $K = 10$ is nearly 11 dB less than that for a linear circuit.

Figure 26 shows the three-dimensional plot of degradation of the quantizer–correlator vs both quantizer level and interference amplitude. Again, there is a ridge of optimum quantizer level corresponding to $\Delta \cong \sqrt{K^2 + \sigma^2}$. However, with the 3-bit quantizer, there is a second ridge corresponding to the next quantizing level, and the 3-bit quantizer is not as sensitive to quantizer interval setting as is the two bit quantizer.

Figure 27 shows probability density superimposed on the quantizer intervals for $\Delta = \sqrt{100 + K^2}/\sqrt{3}$, with $K = 20$, i.e., $\Delta = 12.9$, the optimum quantizer setting for noise alone. The recommended quantizer interval for interference plus noise would be $\Delta = 22.36$.

If the coherent interference phase is varied slowly compared to the data bit rate; namely, $I = K \sin \theta$, then the excess degradation for the 3-bit quantizer is as shown in Fig. 28 as a function of the phase. The degradation is given for $K = 20$ and 40. For reference, the peak degradation for the infinite level quantizer is shown as the horizontal dashed curve, and the average degradation is 6 dB above that.

6. Three-Bit Quantizer with Frequency Offset Interference

Consider next a frequency offset interference $K \cos \omega t$, where the offset frequency is larger than the databit rate by several times so that power averaging occurs. Figure 29 shows the quantizer–correlator degradation for values of interference $K = 10$,

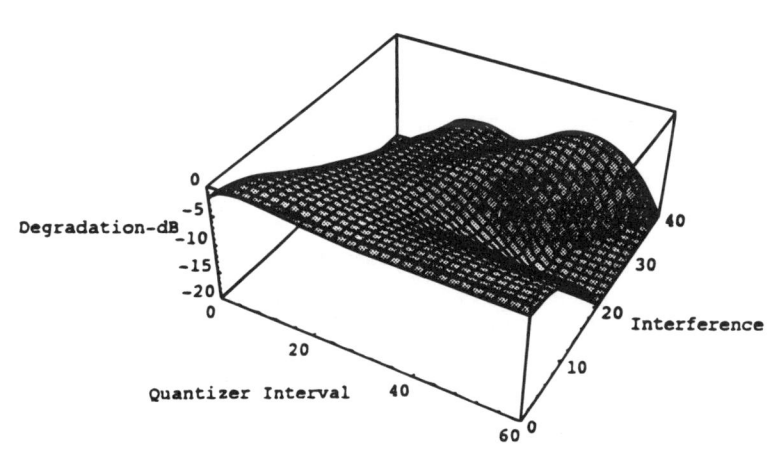

Fig. 26 Excess quantizer–correlator degradation for the 3-bit quantizer vs both quantizer interval and interference level.

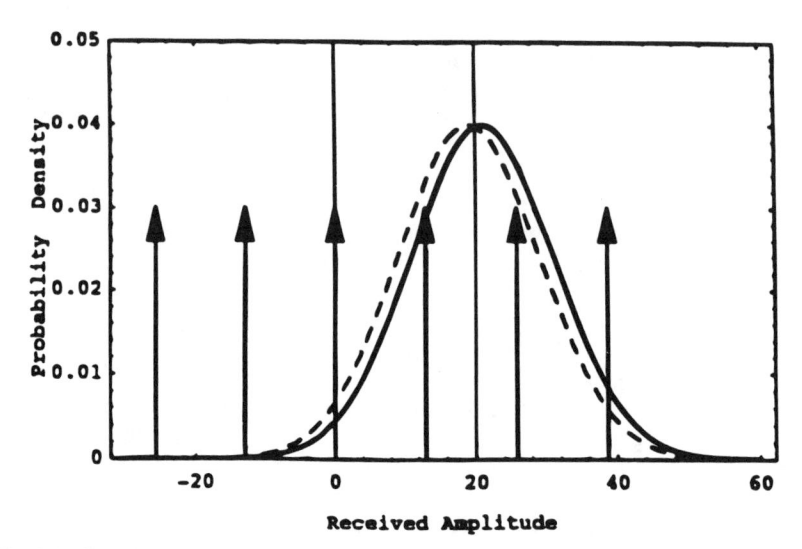

Fig. 27 Signal plus noise probability density and quantizer intervals for the 3-bit quantizer with $K = 20$ and a quantizer interval set by total power /3, i.e., $\Delta = \sqrt{100 + K^2}/\sqrt{3}$. The quantizer intervals are denoted by the arrows.

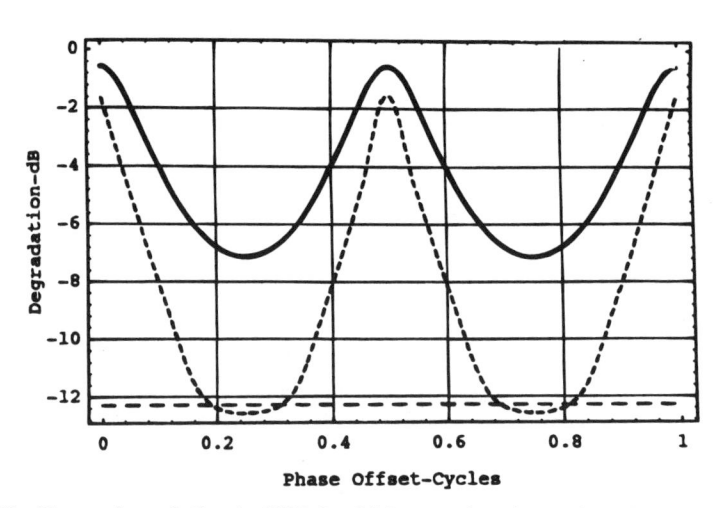

Fig. 28 Excess degradation in SNR for 3-bit quantizer for an interference $K = 20 \sin \theta$, solid curve i and $K = 40 \sin \theta$, small dashed curves. For reference, the peak degradation for an infinite level quantizer is shown as the horizontal dashed curve. The quantizer interval is $\sqrt{\sigma^2 + K^2}$.

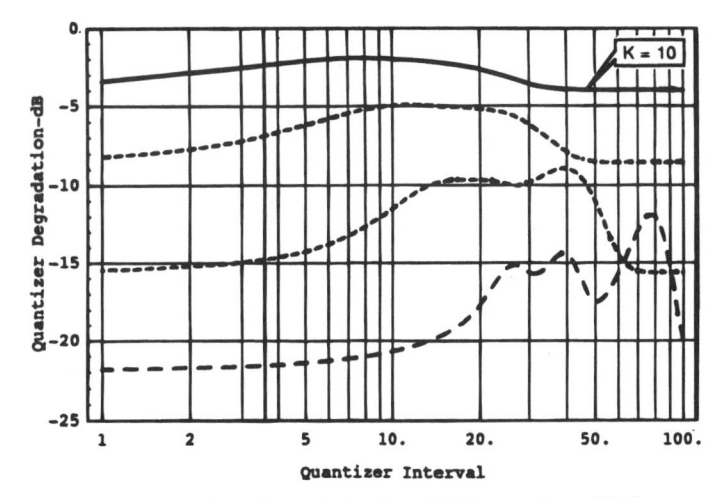

Fig. 29 Quantizer–correlator degradation for a 3-bit quantizer with frequency offset interference $K \cos \omega t$ where $K = 10$, solid curve; $K = 20$, first dashed curve below the solid curve, $K = 40$, second small dashed curve; and $K = 80$, bottom large dashed curve.

20, 40, 80 vs quantizer interval Δ. Again, the optimum quantizer interval is approximately equal to $\Delta \cong \sqrt{K^2 + \sigma^2}$. Note that the peak is generally broad because of the four quantizing magnitude levels. By way of comparison with $K = 40$, the 2-bit quantizer provided a minimum degradation of approximately -9.0 dB relative to that for the thermal noise alone. The 3-bit quantizer for $K = 40$ provides a minimum quantizer degradation of approximately the same value. The difference between the two quantizers is that the performance of the 2-bit quantizer falls off relatively rapidly for decreasing quantizer interval Δ; whereas, the 3-bit quantizer level can decrease by a factor of 3 to 1 and still suffer a penalty of only 1 dB or so. Thus, the 3-bit quantizer is not as sensitive to AGC accuracy.

Figure 30 shows the quantizer–correlator degradation for a 3-bit quantizer vs interference level for a quantizing interval of $\Delta \cong \sqrt{\sigma^2 + K^2}$. The solid curve in the figure shows the degradation for an infinite level quantizer. The optimum 3-bit quantizer exhibits a performance improvement of approximately 4 dB at an interference level of $K = 100$.

The probability density of the amplitude sum of a sine wave interference $K = 40$ with uniform phase and independent Gaussian noise $\sigma = 10$ is shown in Fig. 31. The optimum quantizing interval at $\Delta \cong \sqrt{\sigma^2 + K^2} = 41.23$ is shown as the arrow.

III. Effects of Interference on the GPS C/A Receiver

A. Effects of the C/A-Code Line Components on Narrow-Band Interference Performance

The spectrum of the C/A Gold code consists of line components spaced at 1 kHz because of the 1-ms code period. These line components have approximately,

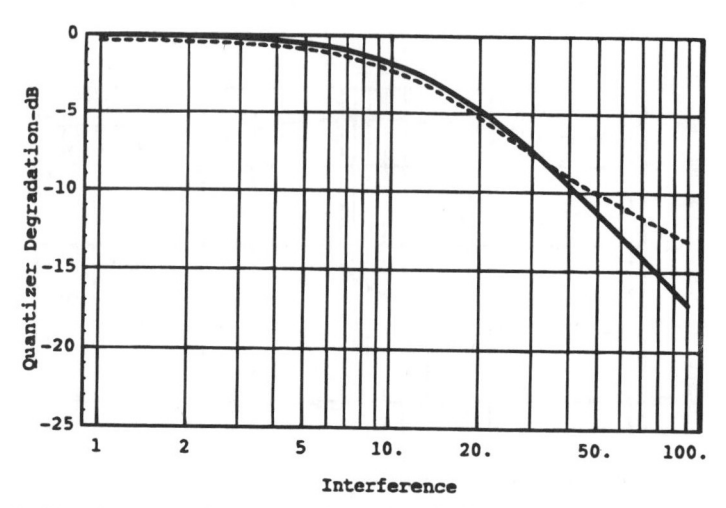

Fig. 30 Quantizer–correlator degradation for the 3-bit quantizer for a quantizing interval $\Delta \cong \sqrt{\sigma^2 + K^2}$ for frequency offset interference $K \cos \omega t$. The dashed curve is the quantizer degradation. The solid curve is the degradation that would accrue for a linear quantizer namely a degradation of $(10/\sqrt{10^2 + K^2/2})$.

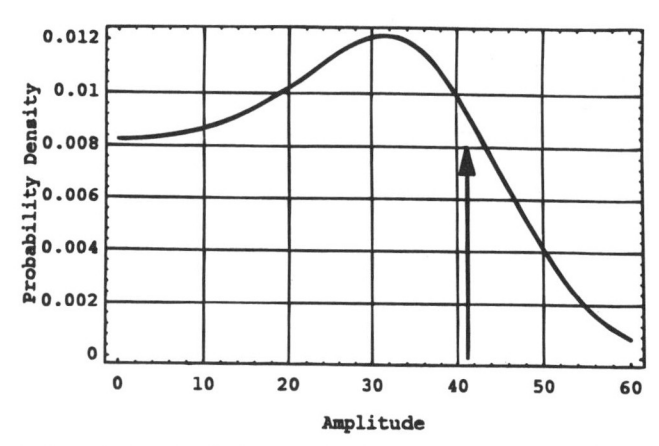

Fig. 31 Probability density of the sum of a sine wave interference $40 \cos \omega t$ + noise of $\sigma = 10$.

INTERFERENCE EFFECTS AND MITIGATION TECHNIQUES   749

$$r(t) = Ap(t)\cos \omega_o t + K \cos(\omega_o + n\Delta \omega t + \theta)$$
$$+ N_c(t)\cos \omega_o t + N_s(t)\sin \omega_o t \qquad (8)$$

where $P_s = A^2/2$ and $P_I = K^2/2$ are the signal and interference powers, respectively. If the noise is bandlimited to an rf bandwidth $2B$ and has power P_n, the noise has one-sided power spectral density $N_o = P_n/2B$. The coherent in-phase channel of the receiver forms the product with $p(t) \cos \omega_o t$ and passes the low-pass frequency components:

$$c(t) = A + Kp(t)\cos(n\Delta \omega t + \theta) + N_c(t) \qquad (9)$$

where $p^2(t) = 1$, $N_c(t)$ has a constant power spectral density N_o, is bandlimited, and has power $P_n/2$. If $n\Delta \omega t + \theta$ is time varying, and $n \neq 0$, then

$$K \cos(n\Delta \omega_o t + \theta)$$

has power $P_I/2$. The purpose of this section is to examine the spectrum of the interference output of the in-phase channel of the correlator.

Figure 34 shows a highly simplified version of the C/A-code spectrum at rf, a narrow-band frequency offset interference spectrum slightly above $f_o + 2000$ Hz and the resulting correlator output spectrum. The baseband spectrum is shown one-sided. Note that both difference terms $f_A - f_B$ and $f_B - f_A$ appear in the output because of the one-sided nature of the output spectrum. Clearly, if the bandwidth of the interference is 500 Hz or more in bandwidth for this offset, the correlator output spectrum would have no gaps.

Now consider a pure CW carrier interference. The GPS C/A reference signal $s(t)$ is periodic, and can be represented as follows:

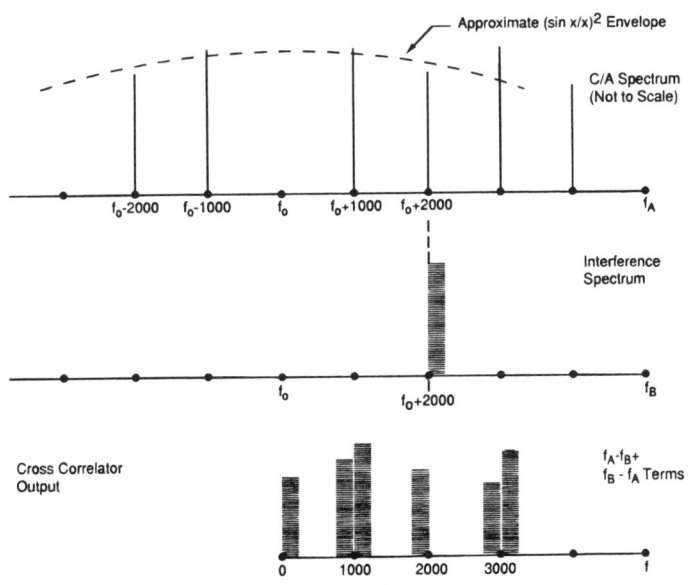

Fig. 34   Interference component at cross-correlator output for a C/A signal.

$$p(t) = \sum_{i=0}^{\infty} a_i \cos(i\Delta\omega t) + b_i \sin(i\Delta\omega t) \qquad (10)$$

where $p(t)$ is taken as $p = \pm 1$, has unit power, and $\Delta\omega = 2\pi \times 1$ kHz (see Figs. 32 and 33). The interference in the in-phase channel is $Kp(t)\cos(n\Delta\omega t + \theta)$, and the product $p(t)I(t)$ has an in-phase component at the origin that acts as co-channel interference on the GPS signal:

$$[2p(t)\cos \omega_o t]I(t)|_{\text{low pass}} = K[a_n \cos\theta + b_n \sin\theta] \qquad (11)$$

where terms of $2\Delta\omega t$ and higher have been removed by filtering. This interference term has power (for θ uniformly distributed and slowly varying with time):

$$K^2\left(\frac{a_n^2 + b_n^2}{2}\right) = K^2\left(\frac{c_n^2}{2}\right) \qquad (12)$$

where $(a_n^2 + b_n^2) = c_n^2$ is the power in the C/A line component at $n\Delta\omega$ and $\Sigma c_n^2 = 1$. Thus, only half the power in the interference falls in the in-phase channel if θ is averaged over time, as we would expect. The other half is in the quadrature channel. The average correlation with the desired signal $Ap(t)$ is then as follows:

$$Ap(t)\cos \omega_o t[p(t)\cos \omega_o t] = \frac{Ap^2}{2} = \frac{A}{2}\Sigma(a_n^2 + b_n^2) = \frac{A}{2}\Sigma c_n^2 \qquad (13)$$

neglecting the sum frequency terms by averaging over time, and where

$$a_n^2 + b_n^2 \triangleq c_n^2.$$

Clearly, if the interference phase is fixed and in-phase with the reference carrier, the full power of this line component appears in the in-phase channel and none in the quadrature channel, just as it does for the desired signal.

As a point of references note that if the C/A code had infinite period and was completely random with rectangular-shaped pulses, the C/A signal would have the following familiar spectrum:

$$G_s(t) = \frac{P_s}{f_s}\left\{\frac{\sin[\pi(f - f_o)/(1.023 \times 10^6)]}{\pi(f - f_o)/(1.023 \times 10^6)}\right\}^2 \qquad (14)$$

where P_s is the signal power, and f_c is the code clock rate. If the spectrum were periodic, but a maximal length sequence, the spectrum would have line components with the same $(\sin x/x)^2$ envelope. The power in a 1-kHz interval is collected in a line component with appropriate power $P_s/f_c \times 1$ kHz near the center frequency; i.e., the spectral line components would be about $-60 + 30$ dB $= -30$ dB down.

Figure 35 shows the spectra of the interference component of the correlator output for various input interference spectra. The interference power in the correlator output is the convolution of the interference spectrum and each of the C/A reference line components. If the interference matches the exact offset and Doppler shift of one of the line components, then the interference translates by convolution to the baseband at $f = 0$, with the same bandwidth as the input

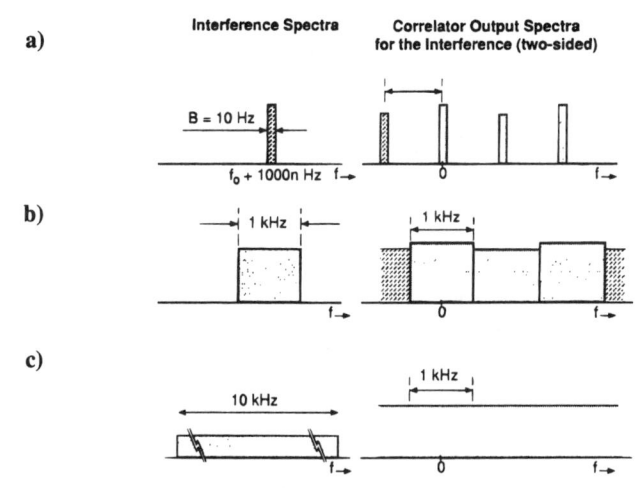

INTERFERENCE EFFECTS AND MITIGATION TECHNIQUES 751

Fig. 35 Spectra of the interference input and the correlator output for various interference bandwidths 10 Hz, 1 kHz, and 10 kHz for the C/A-code correlator. A different C/A signal will have a different set of line components.

interference, as shown in Fig. 35a. Thus, much, or all, of the interference component output can pass through the narrow bandwidth code or carrier-tracking filters. For example, if the line component is 30 dB below the signal power, the interference is only suppressed by 30 dB, not 60 dB. However, note that for most offsets, the interference power will be offset in frequency from $f = 0$, and thus no interference passes through the tracking filters for this very narrow bandwidth interference. For example, if the tracking bandwidth is 10 Hz, 99% of the frequency offsets would produce interference outputs outside the tracking band and be rejected, at least in part.

As a second example, in Fig. 35b, the interference has a bandwidth of 1 kHz, the same bandwidth as the line component spacing. In this example, the interference now is spread continuously over the entire channel. The spectral density levels, however, are not a continuous flat spectrum at -60 dB/Hz but are above and below that level by the same amount that the line components fluctuate up and down below the -30 dB envelope. In this example, however, only a fraction of the signal power passes through the tracking filter as determined by the bandwidth ratio $B_n/10^3$ Hz, where B_n is the closed-loop noise bandwidth of the DLL. The IF tracking filter, however, would generally have 500–2 kHz bandwidth and would be affected by most of this power. Recall from earlier chapters that an IF SNR below 0 dB begins to cause small signal suppression effects.

The third example where the interference bandwidth is 10 kHz (or larger)* is shown in Fig. 35c. In this example, each line component is now spread over the adjacent 10-line component center frequencies, and the line components are, in effect, averaged so that the up and down fluctuation caused by individual line

---

*We assume that the interference bandwidth is less than the PN code chip rate.

Table 3 Effective N_{oI} component caused by the interference[a]

Interference Bandwidth B_I	N_{oI} for unity P_I
0–10 Hz	N/A
1 kHz	$P_I p_n/10^3$
10 kHz	$\approx P_I/10^6$
100 kHz–1 MHz	$P_I/10^6$

[a]The nth line component has relative power p_n. The narrow bandwidth interference (0–10 Hz) produces a narrow-band interference in the correlator output that can fall entirely within the power band of the carrier-tracking loop. However, this is an unlikely effect.

components are smoothed. The resulting spectrum is, then, much closer to the -60 dB/Hz continuous spectrum of the random sequence. In effect, this spectrum can be likened to a C/A spectrum that has been modulated by a 10-kHz random data bit stream, thereby, removing any periodicity in the PN code. As the interference bandwidth increases to 100 kHz, the spectrum is completely smoothed. Table 3 summarizes the effective noise density for the interference of various bandwidths. Table 4 summarizes the effects of various interference bandwidths on the correlator output spectrum.

C. Interference Effects—Effects on Receiver-Tracking Loops

Figure 36 shows a simplified view of the receiver code and carrier-tracking loops as pertains to the interference problem. If the interference is narrow bandwidth in the sense that its bandwidth $B_I \ll 1$ MHz but is many times wider than the 1 kHz spacing between line components; e.g., $B_I > 10$ kHz, then the interference encompasses several line components, and an averaging effect for the line components occurs so that the spectrum of the interference $G_I(t)$ convolved with the C/A spectrum yields a continuous spectrum:

$$G_I(f) \cdot G_{CA}(f) \cong \frac{P_I}{f_{CA}} \left(\frac{\sin \pi f/f_{CA}}{\pi f/f_{CA}} \right)^2 \qquad (15)$$

where $f_{CA} = 1.023$ MHz. Figure 35c shows an approximation to this result for $B_I \geq 10$ kHz.

The effective SNR for the GPS receiver for a receiver noise bandwidth B_n is as follows:

$$\frac{P_s}{P_{n_{\text{eff}}}} = \frac{P_s}{B_n(N_o + P_I/N_o f_{CA})} = \frac{P_s}{B_n N_o (1 + P_I/N_o f_{CA})} \stackrel{\Delta}{=} \frac{P_s}{B_n N_{o_{\text{eff}}}} \qquad (16)$$

If the tracking threshold value for $P_s/P_{n_{\text{eff}}} = R_T$, then the threshold value of P_I is as follows:

Table 4 Generic effects of various bandwidth interference on the correlator output spectrum

Interference Bandwidth	Comment
10 Hz	At the exact frequency of a line component, the interference can pass through all filters. However, as the interference bandwidth widens, only a fraction of the interference will pass through the narrowest code-tracking coherent detector filter. The interference only has a chance of 10 Hz/1 kHz or 1% of matching the frequency of a given satellite. Thus, only one satellite out of those in view is affected, unless two satellites have exactly the same doppler shift and high amplitude line components.
1 kHz	Most of the interference passes through the noncoherent delay lock loop bandpass filter, unless this loop has been narrowed. However, only a fraction of the interference power; e.g., 50 Hz/1 kHz = .05, passes through the data demodulation filter. Carrier- and code-tracking filters receive even a lower fraction. All satellites are affected but to varying degrees depending upon the line component amplitude.
10 kHz	Generally, this bandwidth would provide an averaging of the effects of 10 adjacent line components for each satellite, and then the effective line component strength would generally be closer to -30 dB, rather than the worst case -21.6 dB, and the effective interference noise density is then approximately $P_I/10^6$. In addition, the interference power is spread over the entire C/A-code spectrum.

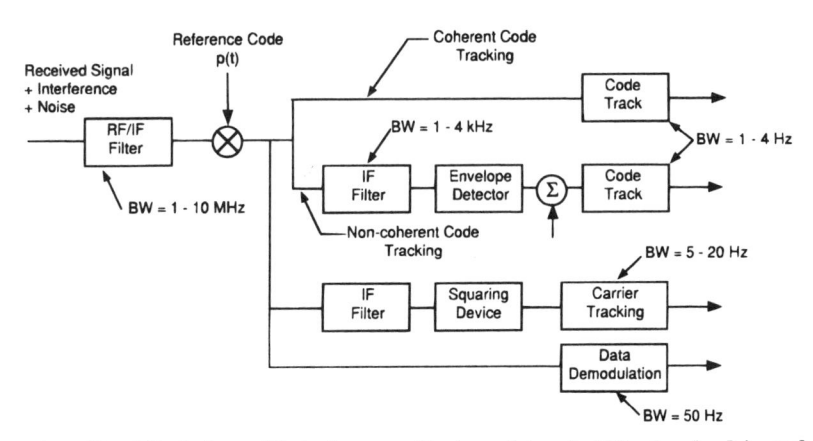

Fig. 36 Simplified view of interference effects and bandwidths involved in code- and carrier-tracking and data demodulation.

$$P_{IT} = \left(\frac{P_s}{B_n R_T N_o} - 1\right) f_{CA} N_o \cong \frac{P_s}{R_T}\left(\frac{f_{CA}}{B_n}\right) \quad \text{for} \quad \frac{P_s}{B_n R_T N_o} \gg 1 \qquad (17)$$

As shown in Chapter 7, this volume, the delay–lock-loop rms delay error σ_ϵ in the quasilinear region is approximated by the following:

$$\left(\frac{\sigma_\epsilon}{T}\right)^2 \cong \frac{1}{\text{SNR}_L} = \left(\frac{N_{o_{\text{eff}}} B_n}{P_s}\right) = \left(\frac{P_{n_{\text{eff}}}}{P_s}\right) \qquad (18)$$

for sufficiently high SNR, and the threshold value of σ_ϵ/T is approximately 0.2, and $R_T = 25$. Thus, the limiting interference level for delay lock tracking is as follows:

$$P_{IT} \cong \frac{P_s}{R_T}\left(\frac{f_{CA}}{B_n}\right) = .04\, P_s\left(\frac{f_{CA}}{B_n}\right) \qquad (19)$$

Table 5 gives an example of the SNR requirement and the noise bandwidth of the various receiver functions.

Similarly, the rms phase error for phase lock tracking of the carrier is as follows:

$$\sigma_\phi \cong \sqrt{\frac{B_n N_{o_{\text{eff}}}}{P_s}} \qquad (20)$$

and $R_T \cong 25$ for above threshold operation.

If the bandwidth of the interference increases above 2 MHz, the convolved signal and interference spectra become wider than the C/A-code bandwidth itself, and these interference effects are decreased.

Table 6 summarizes some of the effects of interference on the GPS receiver for the worst case frequency and sidelobe. For the narrowband interference (10 Hz), one of the interference power components can fall completely within the bandwidth of the carrier-tracking loop (assumed to be 10 Hz.) Neglecting thermal noise effects, if the power in the interference component at the nth 1 kHz frequency offset is P_n, then the required SNR from Table 5 is as follows:

Table 5 Example signal-to-noise ratio requirement for selected bandwidths

	Coherent code track	Carrier track	Data demodulation	IF correlation
Bandwidth Hz/dB-Hz	2 Hz/3 dB-Hz	10 Hz/10 dB-Hz	50 Hz/17 dB-Hz	500 Hz/27 dB-Hz
SNR/dB	25/14 dB	25/14 dB	10/10 dB	1/0 dB
Equipment degradation	2 dB	2 dB	2 dB	2 dB
Total SNR required, dB	16 dB	16 dB	12 dB	2 dB
Total C/N_o required dB-Hz	19 dB-Hz	26 dB-Hz	29 dB-Hz	29 dB-Hz

$$(SNR)_{req} = \frac{P_s}{P_I p_n} = 16 \text{ dB} \quad \text{code and carrier-tracking requirement} \quad (21)$$

where p_n is the nth line component level for the C/A code and $(SNR)_{req}$ is the required SNR.

If the worst case p_n component of -21.6 dBc is assumed (it has a probability of occurrence of only 10^{-3}), the C/A received signal level is $P_s = -157$ dBW, and the interference is offset at exactly the correct frequency (10 Hz/1000 Hz $= 10^{-2}$ for a probability of occurrence of 10^{-5}), then the maximum tolerable interference is -157 dBW $+ 21.6$ dB $- 16$ dB $= P_I = -151.4$ dBW for carrier tracking. The 2-Hz bandwidth of the code loop gives a 7-dB advantage over the carrier loop and so $P_I = -144.4$ dBW for code tracking, or for the average interference component level of $P_n = -30$ dBc, this result becomes $P_I = -143$ dBW for carrier tracking, and $P_I = -136$ dBW for code tracking.

For the 1 kHz interference bandwidth, the result $C/N_o = (P_s/P_I)(p_n/10^3)$ $= 26$ dB $-$ Hz for carrier tracking, and $= 19$ dB $-$ Hz for code tracking. For the worst case, the interference power is $P_I = -157$ dBW $+ 21.6 + 30$ dB $- 26$ dB $= 131.4$ dBW for carrier tracking, and $= -124.4$ dBW for code tracking, and for the average case with an interference bandwidth of 10 kHz or wider, $P_I = -157$ dBW $+ 30$ dB $+ 30$ dB $- 26$ dB $= -123$ dBW for carrier tracking, and $P_I = -116$ dBW for code tracking.

Table 6 shows example interference levels tolerated by a receiver with no adaptive quantizer or interference frequency nulling. It should be emphasized that the tracking bandwidths are only examples. The received C/A-signal level

Table 6 GPS receiver without adaptive bandwidth or adaptive quantizing—maximum interference levels for accurate tracking at the antenna output[a]

Interference bandwidth	Tracking	Threshold interference power dBW			
		Antenna output		Antenna input -5 dB gain to interference	
		Worst case	Average	Worst case	Average
10 Hz	Carrier	-151.4	-143	-146.4	-138
	Code track	-144.4	-136	-139.4	-131
100 Hz	Carrier	-141.4	-133	-136.4	-128
	Code track	-134.4	-126	-129.4	-121
1 kHz	Carrier	-131.4	-123	-126.4	-118
	Code track	-124.4	-116	-119.4	-111
10 kHz	Carrier	-123	-123	-118	-118
	Code track	-116	-116	-111	-111
100 kHz to 1 MHz	Carrier	-123	-123	-118	-118
	Code track	-116	-116	-111	-111

[a]A gain of -5 dB is assumed for the antenna in the direction of the interference source. It is assumed that the satellite signal is acquired and that data demodulation is not required. Results are shown for both carrier and code tracking. Note that with narrow-band interference, the probability that the interference will hit a given satellite is small, and it is not likely that the interference will disrupt more than one satellite at a time.

is assumed to be -157 dBW. In the right-hand columns the interference is assumed to be attenuated by 5 dB relative to the signal by the antenna.

The results of Fig. 6 indicate that the code loop is more robust than the carrier phase locked loop (PLL). Unfortunately, most receivers rely on the PLL to keep the received signal within the receiver pass band filters, and so the PLL determines receiver threshold. One strategy for improving this situation is to incorporate a noncoherent Automatic Frequency Control (AFC) loop to track carrier frequency.[15] Differential data detection can be employed if the PLL is to be eliminated entirely. In this case the receiver threshold is determined by the AFC loop which is generally more robust than the PLL.

By way of reference, if the thermal noise is -205 dBW/Hz and the rf noise bandwidth is 2 MHz = 63 dB-Hz, then the thermal noise power is -142 dBW. Most of the interference levels required to disrupt the carrier and code-tracking loops are well above that level, especially for code tracking. The only exception is for very narrow bandwidth interference, which generally only disrupts one satellite at a time, and the interference frequency must be within 10 Hz of a strong C/A signal spectral line. The higher levels of interference are relatively easy to distinguish from noise, and, as discussed in the next section, the receiver can be implemented so that it will detect and adapt to the interference.

As pointed out in the previous section, an adaptive quantizer can also provide considerable improvement to the performance of any constant envelope interference. Constant envelope interference; e.g., harmonics or intermodulation products of local oscillators, are the most likely of an unlikely event, and adaptive quantizers can attenuate the interference by 10 dB or more for interference above the noise level. As shown in the next section, adaptive null steering antennas and adaptive frequency notch filters can each attenuate the interference by more than 20 dB.

IV. Detection of Interference, Adaptive Delay Lock Loop, Adaptive Frequency Notch Filtering, and Adaptive Null Steering Antennas

The GPS receiver can be designed with the precautions against interference cited earlier; namely, bandpass rf filtering to minimize out-of-band interference, pulse blanking to reduce the effects of high-level pulse interference and prevent LNA overload, adequate number of quantizing levels, and appropriate AGC to ensure full processing gain, and careful design of the code- and carrier-tracking loops.

However, as noted above, sufficiently high levels of interference will overload any type of radionavigation system, and the GPS, even with spread spectrum, is no exception. Note that taking advantage of the GPS signal processing gain by itself is not always the wisest strategy. For example, the results of Table 6 indicate that in interferer with power 14 dB greater than the desired signal will typically disrupt GPS receiver operation. However, if a narrow-band interferer were "notched out" in frequency before correlation, only a modest degradation in tracking accuracy (due to signal distortion) would be noticed. Under most circumstances, narrow-band or broad-band interference of sufficient level to disrupt the GPS receiver is well above the thermal noise level, either in peak power spectral density, total power, or both. Under these circumstances, either a total power

INTERFERENCE EFFECTS AND MITIGATION TECHNIQUES

measurement or a crude spectral density measurement; e.g., using upper band/lower band comparison or more elaborate but still simple, discrete Fourier transform (DFT) processing, can be effective in reliably detecting the presence of interference. When interference is detected, various actions can be taken:

1) Alert the Kalman filter to the interference, give lower weight to the measurements, and use alternate sensors such as an inertial measurement unit during the limited period where interference is present.

2) Reduce the DLL tracking loop filter bandwidths.

3) Alert the integrity-monitoring system of the interference.

4) Utilize adaptive antenna nulling or notch filter techniques (adaptive frequency notching).

A. Adaptation of the Delay Lock Loop and Vector Delay Lock Loop

1. Adaptation of the Tracking Loop Filters

In many circumstances, if interference is detected, the receiver can adapt by reducing the carrier loop bandwidth, reducing the code loop bandwidth, aiding the tracking loops by using information from external sensors (e.g., inertial units), switching antennas, or adapting an antenna array, and eliminating satellites at low elevation angles where there is low antenna gain and lower signal levels. It is sometimes possible to reduce the bandwidth of the tracking filters by a factor of from 2 to 10, with a corresponding ability to reduce the effects of most types of interference.

2. Vector Delay Lock Loop

As discussed in Chapter 7, this volume, if there are more than the required minimum number of satellites, the vector delay lock loop can operate using the total power from all satellites to aid in the tracking operation. It is often possible for a vector delay–lock-loop to operate in the presence of interference and track all satellites under conditions where a set of independent delay lock loops cannot function at all; i.e., none of the code/carrier-tracking loops can operate independently. This advantage accrues when there are many satellites in view, and the total power in all of the satellites is used to track the limited information content of the user position and dynamics. Clearly, when all of the satellite power is employed, the channel capacity is increased; whereas, the information sought (the two- or three-dimensional coordinates of the user) remains fixed. This additional channel capacity can be used to cope with an interfering signal.

B. Rejection of Narrow-Band Interference by Adaptive Frequency Nulling Filters

It is relatively easy to discern a narrow-band interference that is above the power level of the thermal noise. The estimate of the interference level can be used in an adaptive frequency notch filter algorithm that in effect, whitens the spectrum as it enters the delay lock PN-tracking loop. Figure 37 shows one approach to the adaptive interference cancellation where the spectrum is first estimated by a DFT, and then that estimate is compared with the relatively flat spectrum of the thermal noise. The power spectral density in the vicinity of the

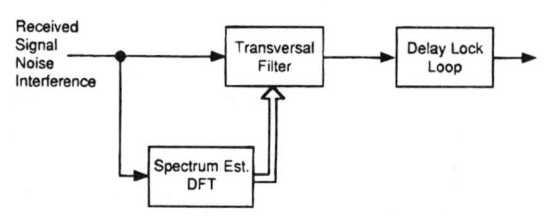

Fig. 37 Adaptive narrow-band interference cancellation using nonparametric spectral estimation using a discrete Fourier transform (DFT) and a transversal filter (linear phase) to whiten the spectrum.

interference is well above that of the thermal noise if the interference is significant. The measurements are then used to adapt the tap coefficients of the K tap transversal filter. If the tap coefficients are denoted by $h(k)$, k = 0,1, ... , K-1 and

$$H(k) = \sum_{n}^{K-1} h(n)e^{-j2\pi nk/K} \quad (22)$$

and power spectral measurements $P(kR_s/K)$ are made (where R_s is the sampling rate), then we get

$$H(k) = \frac{1}{\sqrt{P(kR_s/K)}} e^{-j2\pi(K-1)k/2K} \quad (23)$$

to whiten the output spectrum and $h(k) = h(K-1-k)$ the symmetry condition necessary to provide linear phase (see Ref. 12).

Experimental versions of this system have reduced the interference by 15 dB when it was originally 20 dB above the background spectrum by using a 15-tap suppression filter. Similarly, we can employ the tunable notch filter of Fig. 38a to tune to the frequency of the interference. The notch filter can either be tuned by DFT measurements or simply by scanning the band so as to minimize the total power fed to the correlator if interference is detected. Alternative notch filter frequency control techniques employ automatic frequency control (AFC) or phase–lock-loop (PLL) estimators to tune to the interference frequency. If there is no interference, the notch filter can be simply deleted. Note that if the notch filter is inserted, the effect of the tunable oscillator cancels, because it is used both in the upconversion and downconversion operation.

Another approach employs stronger signal capture as shown in Fig. 38b. This approach employs the stronger signal capture performance of a bandpass hard limiter to capture an interfering sinusoidal signal of any type that is well above the noise level (and signal level). A strong sinusoidal signal suppresses the weaker signal by 6 dB more than the original interference to noise ratio (even if the weaker signal is Gaussian). Thus, if the limiter output is subtracted from the input signal, there is a substantial degree of cancellation of the interference and only a very small cancellation effect in the signal. Experimental tests of this type of canceller have shown good performance with strong narrow-band interference.

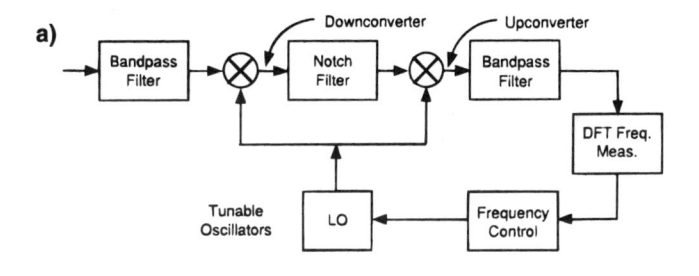

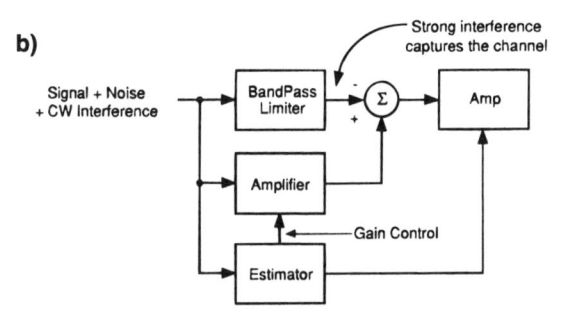

INTERFERENCE EFFECTS AND MITIGATION TECHNIQUES

Fig. 38 Alternative types of interference canceling techniques: a) tunable frequency notch filter designed to minimize total power output by notching narrowband interference; b) excising technique based on stronger signal capture in a bandpass limiter.

C. Adaptive Antennas for Point Source Interference

1. Spatial Signal Processing

Figure 39 shows a simplified sidelobe canceling antenna array that employs a main antenna directed at the upper hemisphere and a number of auxiliary antennas designed to null interference in the sidelobes of the main antenna. By controlling the weights w_1, w_2 on the auxiliary antenna elements, an interference signal in the sidelobes can be effectively nulled. This section briefly outlines some of the concepts for steerable vector array antennas that can be designed to minimize interference by spatially nulling one or more point source interferences.*

Figure 40 shows the weighted output of an N element antenna array where the signal and interference point sources arrive at two different angles from boresight ϕ_s and ϕ_I. Represent the complex signals (in-phase and quadrature) at each of the antenna elements as $x_i(t)$. Then the total received signal can be represented by the complex vector $X(t)$ where

$$\mathbf{X}(t) = [x_1(t), x(t), \ldots, x_N(t)] \qquad (24)$$

and this vector can be split into signal, noise, and interference vectors X_s, X_n, and X_I, respectively:

---

*A detailed treatment of adaptive antennas is beyond the scope of this book. The reader is referred to the excellent work of Compton[5] for a more thorough treatment.

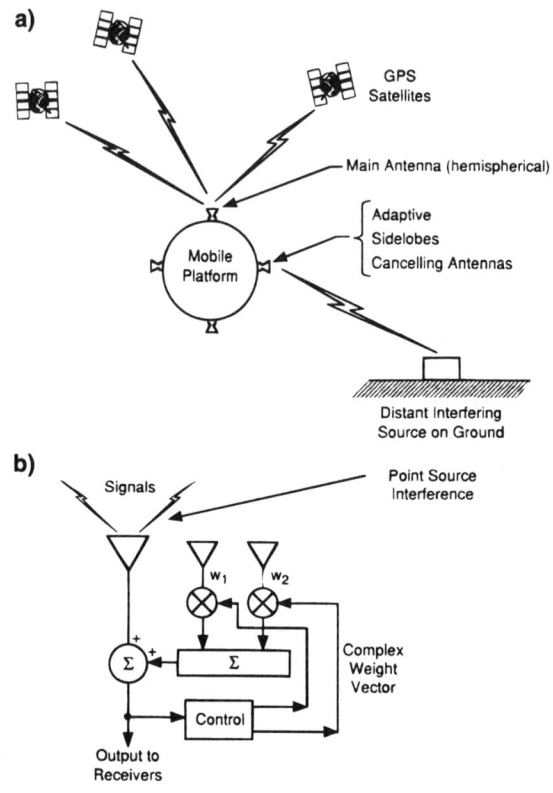

Fig. 39 Adaptive sidelobe cancellation of an interference coming from different directions than the GPS signals. In general, there may be more than one source of interference.

$$\mathbf{X}(t) = \mathbf{X}_s(t) + \mathbf{X}_n(t) + \mathbf{X}_I(t) \tag{25}$$

If each of these received signal vectors are weighted by complex weights w_i and $\mathbf{W} = [w_1, w_2, w_N]$ then the output of the adaptive antenna is as follows:

$$s(t) = \mathbf{W}^T \mathbf{X} = r(t) + n(t) + v(t) \tag{26}$$

which also has been split into desired signal $r(t)$, noise $n(t)$, and interference $v(t)$ components.

For the GPS, there are, in general, M separate satellite signals as opposed to a single signal. Thus, the complete configuration of the adaptive array might appear as shown in Fig. 41, where we show a set of N antenna array elements followed by M adaptive weight matrices and controllers each optimized for a specific satellite. We can also envision a simpler technique in which the antenna array is designed with only a single weight matrix that is designed to place a null in the direction of the interfering signal.

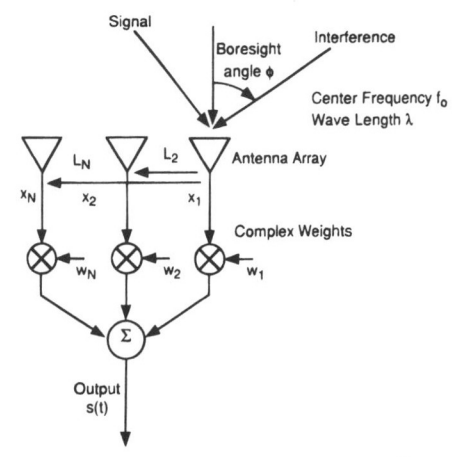

Fig. 40 N-element adaptive antenna array. The antenna elements are separated by length $L_2, L_3, \ldots L_N$ from antenna element 1. If a signal is received at boresight angle ϕ and wavelength λ, the differential phase shift for antenna element 1 is $\theta_i = 2\pi L_i/\lambda \sin \phi$.

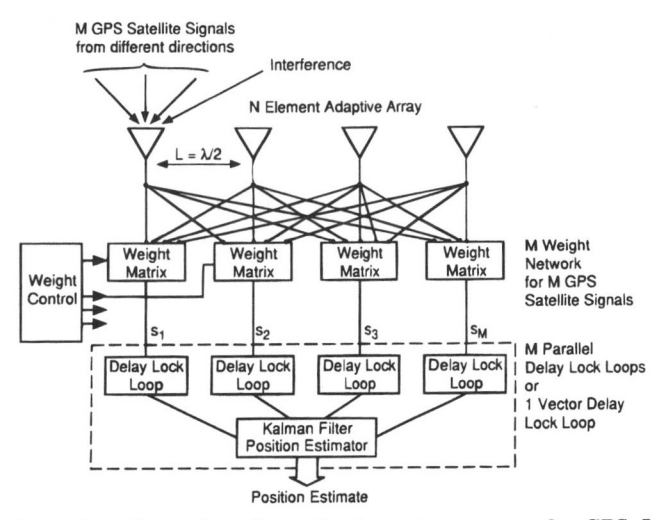

Fig. 41 General configuration of an adaptive antenna array for GPS. In general, there might be a separate weight matrix and control algorithm for each of M GPS satellite signals, each with a different PN code and angle of arrival.

There are a great many different possible types of antenna arrays, and a complete discussion of them is beyond the scope of this book. However, we briefly describe the four types listed in Table 7.

We can form a set of M single-beam antennas, each focused on one of the GPS satellites. If the user platform and orientation are known; the arrival angles to each of the GPS satellites is also known, and a set of programmable M phased arrays can be formed, each pointing its beam to a different GPS satellite. Alternatively, we can form a single multiple-beam antenna with M peaks in a single-antenna pattern. This antenna pattern is time varying, and the antenna element weights are based on the GPS almanac ephemerides.

The first type of adaptive null steering antenna in Table 7 is the *least-mean-square* (LMS) error technique.[13] In this approach, the weight matrix control attempts to minimize the mean-square error between the output of the weight matrix $s(t)$ and a reference waveform $r(t)$, which is a close approximation to the desired signal. For the GPS example, this reference signal can be conveniently generated, because the GPS C/A code is known. We must simply align the signal in time and frequency, which can be done with the aid of the delay lock loop and carrier phase lock loop, and estimate the data. However, this approach is

Table 7 Alternate types of array antennas with application to GPS receivers

Type	Characteristics component
Programmable multibeam antenna array	Array of N antenna elements designed to produce a number of beams (less than N), each pointing in the direction of a GPS satellite. This array is controlled by a deterministic algorithm based on knowledge of satellite orbits.
Least-mean-square (LMS) error adaptive array	The adaptive array weight matrix is designed to minimize the mean-square error in the output signal estimate. This type of adaptive array algorithm requires one to generate a reference signal that can be used to estimate the mean-square error. For the GPS, this reference can be generated by processing the output of a delay–lock-loop. However, this approach in its usual form has definite limitations since because the received signal-to-noise ratio is small.
Applebaum array	This adaption algorithm requires approximate knowledge of the arrival angle of the GPS signal. This arrival angle is then used to generate a steering vector that steers the antenna pattern in the direction of the signal source.
Modified Applebaum array using a power inversion algorithm	This algorithm is designed to minimize the total output power of the weight matrix under the constraint that the weights are not all set to zero. This technique requires neither a reference signal nor knowledge of the signal arrival angle. It is most useful when the interference is large.

vector. One problem with this approach, however, is that the expected value in the above expression is not available in a real-time calculation. Ward and Compton[13] have shown that a good approximation is obtained if we simply substitute the actual value for the expected value, namely, set the derivative of the weight vector as follows:

$$\frac{d\mathbf{W}}{dt} = k\mathbf{X}^*\epsilon \qquad (34)$$

This approximation leads to the block diagram of Fig. 42.

The other key function of Fig. 42 is the generation of the reference signal, $r(t)$ or an approximation thereof. Here the adaptive array output $s(t)$ is correlated with the PN code from the output of the appropriate delay lock loop. The bandwidth compressed IF signal output is then filtered and limited to generate a relatively clean signal with data modulation on the carrier. The properly timed PN code is then remodulated to generate a good approximate reference signal, as shown. As discussed earlier, the received GPS signals are generally received at low SNR, hence the error variance, even with no interference, is quite large unless there is substantial antenna gain. An alternative is to perform the same signal comparison and error estimation, but after PN bandwidth compression.

3. Applebaum Array

In the Applebaum array,[5,6] the weights w_i are adapted so as to maximize the power ratio SINR* = $P_s/(P_I + P_N)$ which is defined as the ratio of the total output signal power to the total of the output interference power P_I plus output noise power P_N. Figure 43 shows one form of the Applebaum array for one of the receiver elements. The key adjustment in the feedback loop is the steering vector $\mathbf{T}^T = [t_1, t_2, \ldots t_N]$. The optimal value of $\mathbf{T} = \mathbf{U}_d$ where $\mathbf{U}_d$ represents a

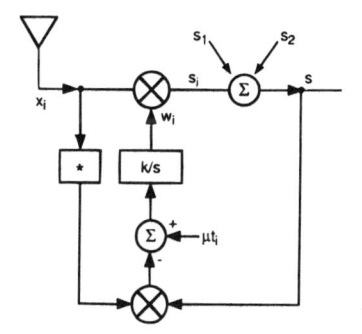

Fig. 43 The Applebaum array with a steering vector T. The optimum SNR is obtained when $T = U_d$ where U_d is a vector representing the interelement phase shifts for a specific signal source direction. Thus, the Applebaum array requires knowledge of the desired signal arrival angle.

---

*Signal-to-interference ratio.

vector of the interelement phase shifts for a narrowband signal arriving from the direction of the actual signal source. Thus, the steering vector requires knowledge of the signal arrival angle.

4. Power Inversion Using the Applebaum Array

An alternative approach is to use the antenna array to invert the power ratio in two signals. That is, the strong signal can be nulled in favor of the weaker signal if a strong interference is detected. The antenna array of Fig. 40 can be formulated to perform power inversion using the processor of Fig. 44. This power inversion array is able to operate when neither the signal nor the arrival angle are known at the receiver; whereas, as discussed above, the LMS and Applebaum adaptive arrays both require knowledge of one or the other. The power inversion array does not always provide as much improvement as adaptive array algorithms, which require additional information. However, when the interference is strong, the power inversion array can still provide a substantial improvement. Furthermore, because the GPS signals are generally received with a low signal-to-thermal-noise ratio, the power inversion array does not null the GPS signals in the absence of interference. In the power inversion array, we select the number of degrees of freedom equal to the number of interferences to be cancelled. An N element antenna array has N-1 degrees of freedom. Thus, a two-element array can provide only one null.

This approach differs from the Applebaum array in that the integrator in the feedback loop is changed to a low-pass filter, and a simple steering vector t_j in Fig. 44 is used to form a weight vector that is different from that normally used in an Applebaum array. The power inversion array weights are designed to minimize total output power under the constraint that the weights are not permitted to go to zero. Thus, for a two-element array, the array automatically directs its only available null to the interference. The differential equation for the tap weights for the jth element is as follows[5]:

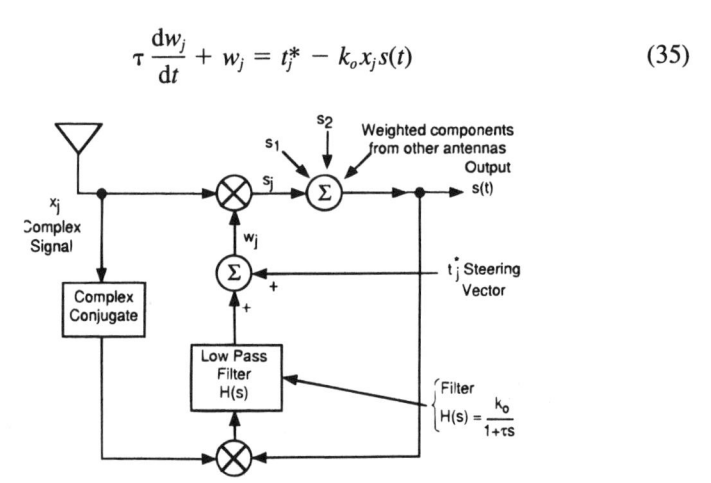

$$\tau \frac{\mathrm{d}w_j}{\mathrm{d}t} + w_j = t_j^* - k_o x_j s(t) \tag{35}$$

Fig. 44 Elements of the antenna array power inversion processor.

where τ is a constant, and t_j^* is the jth component of a steering vector T^* (the asterisk represents complex conjugate). In a power inversion two-element example, assume that the desired GPS signals arrive anywhere in the sector $-\pi \leq \theta_d \leq \pi$, and the antenna elements are spaced a half wavelength apart. The power inversion array can use a very simple steering vector; e.g., $T^* = [1,0]^T$; i.e., one antenna element is given unity weight, and the other is given zero weight, where each antenna pattern is omnidirectional over the half space of interest. The differential equation of the weight vector is as follows:

$$\frac{dW}{dt} + (I + k\Phi)W = T^* \qquad (36)$$

where k is the loop gain. As pointed out previously, the steering vector can weight all elements zero except for one element, which is given unity weight. The loop will still adapt to null out a strong interference.

Compton[5] has shown that we can achieve an output signal to interference plus noise ratio SINR of approximately -22 dB for an input signal-to-interference ratio of -40 dB with an input signal-to-noise ratio of -20 dB and an angular separation between interference and user of somewhat less than 30 deg for a CW interference. Thus, the interference component is almost completely nulled.

Strictly speaking, the above power inversion results apply only to narrow-band signals and interference. However, because the GPS bandwidths are less than 1% of the center frequency for the C/A code, there is negligible degradation from the results for CW signals.

5. Multiple-Beam Antenna Array

Figure 45 shows an adaptive array with M possible sets of adaptive weighting elements for M separate GPS satellite signals, as described earlier in Fig. 41. Each weight matrix, as for example in an Applebaum array, forms a separate beam, which, in turn, can be focused on a separate signal source, such as a separate GPS satellite. A multiple-beam adaptive array of N elements can form N-1 separate beams, and each beam can null N-2 separate interfering sources.[12]

The resolution capability of the array determines the minimum angular separation between the beam maximum and a null. The resolution capability, in turn, is determined by the array aperture size. Resolution is also affected to a smaller extent by the individual element patterns and the number of elements N.

D. Augmentation of the GPS Signals and Constellation

As discussed in Chapters 3 and 5 of this volume, other improvements to the interference performance of the GPS system are possible if the 24-satellite GPS constellation or signals are augmented with additional signal sources. These augmentations include the following possibilities:

1. Increase the L_2 signal power on the GPS satellites to the same level as L_1 and add a C/A-type signal at or near to the L_2 channel. This addition would give the receiver the option of using the L_2 channel if there is significant interference at the L_1 frequency.

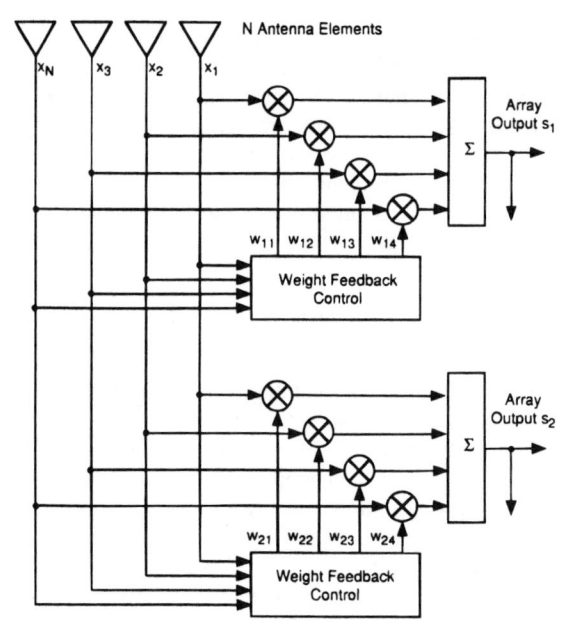

Fig. 45 Multiple-beam adaptive antenna array. An array of N elements has N-1 degrees of freedom and can form N-1 beams (maximum and N-2 nulls). The multibeam array has M simple adaptive arrays in parallel.

2. Add ground pseudolites with GPS-type transmitters in the vicinity of airports to augment the GPS satellites and to provide higher power signals to the GPS receivers. As discussed in Volume II, these signals would generally be time-gated and have a low duty factor in order to avoid interference with the signals from the GPS satellites. Since they are much closer in range, they can be received at much higher power than the satellite signals themselves.

3. Augment the 24-satellite GPS constellation with additional satellites in the same planes, add geostationary satellites which can relay or carry GPS-type payloads themselves or add other medium altitude satellites. See Chapter 5 in this volume and the chapters in Volume II on the Wide Area Augmentation System (WAAS). In addition we can also use the GLONASS signals.

As a final comment, it must be remembered that the large range to the GPS satellites makes even the spread spectrum signal of GPS somewhat susceptible to interference of even moderate power if the interference source is sufficiently close to the receiver. We must therefore consider one or more of the techniques discussed in this chapter as means for coping with the possibility of interference if that interference is sufficiently severe.

Appendix: Mean and Variance of the Correlator Output for an M-Bit Quantizer

In general, an asymmetrical m-bit quantizer has an output $Q(r)$ with value $n + 1/2$ for inputs in the range $-M + 1/2 \leq Q(r) \leq (M - 1) + 1/2$ where

$2^{m-1} = M$. Thus, n ranges from $-M \le n \le M - 1$. The correlation product $c = Q(r) p$ for $Q(r) > 0$ has a positive value when $p = +1$ and a negative value when $p = -1$. Assume a signal level $ap(t) = \pm a$. Assume that $p = \pm 1$ are equally likely with probability 1/2, the expected value of c is easily shown to be the following:

$$E(c) = \sum_{n=-M}^{M-1} \left(n + \frac{1}{2}\right) \left\{ P\left(Q = n + \frac{1}{2} \mid p = +1\right) \right.$$

$$\left. - P\left(Q = n + \frac{1}{2} \mid p = -1\right) \right\} \frac{1}{2} \quad \text{(A1)}$$

where $P(Q = n + 1/2 \mid p = 1)$ is the probability that $Q = n + 1/2$ given that $p = +1$, etc.

These probability difference terms can be written in terms of the error function difference terms, which are defined as follows:

$$P\left(Q = n + \frac{1}{2} \mid p + 1\right) - P\left(Q = n + \frac{1}{2} \mid p - 1\right)$$

$$\stackrel{\Delta}{=} \frac{1}{2} d_e(n, \Delta) \quad \text{for} \quad -M + 1 \le n \le M - 2$$

$$\stackrel{\Delta}{=} \frac{1}{2} d_u(n, \Delta) \quad \text{for} \quad n = M - 1$$

$$\stackrel{\Delta}{=} \frac{1}{2} d_l(n, \Delta) \quad \text{for} \quad n = -M \quad \text{(A2)}$$

where we define Δ as the input quantizing interval, and k is a fixed interference amplitude

$$d_e(n, \Delta) = \text{Erf}\left[\frac{(n-1)\Delta - (k+a)}{\sqrt{2}\,\sigma}, \frac{n\Delta - (k+a)}{\sqrt{2}\,\sigma}\right]$$

$$- \text{Erf}\left[\frac{(n-1)\Delta - (k-a)}{\sqrt{2}\,\sigma}, \frac{n\Delta - (k+a)}{\sqrt{2}\,\sigma}\right]$$

$$d_u(n, \Delta) = \text{Erf}\left[\frac{n\Delta - (k+a)}{\sqrt{2}\,\sigma}, \infty\right] - \text{Erf}\left[\frac{n\Delta - (k-a)}{\sqrt{2}\,\sigma}, \infty\right]$$

$$d_l(n, \Delta) = \text{Erf}\left[-\infty, \frac{n\Delta - (k+a)}{\sqrt{2}\,\sigma}\right] - \text{Erf}\left[-\infty, \frac{n\Delta - (k-a)}{\sqrt{2}\,\sigma}\right] \quad \text{(A3)}$$

Thus, the expected value of c can be written as follows:

$$E(c) = \left(\frac{1}{4}\right) \sum_{n=-M+1}^{M-2} \left(n + \frac{1}{2}\right) d_e(n, \Delta) + \left(M - 1 + \frac{1}{2}\right) d_u[(M-1), \Delta]$$

$$+ \left(-M + \frac{1}{2}\right) d_l[-(M-1), \Delta] \quad (A4)$$

where the quantizer saturates for inputs $r > (M-1)\Delta$ and for $r < -(M-1)\Delta$ with output values of $M - 1 + 1/2 = M - 1/2$ and $-M + 1/2$, respectively.

Likewise, the variance of c is computed as follows:

$$\text{Var}[c] = \sum_{n=-M}^{M-1} \left(n + \frac{1}{2}\right)^2 \left[P\left(Q = n + \frac{1}{2} \mid p = +1\right) \right.$$

$$\left. + P\left(Q = n + \frac{1}{2} \mid p = -1\right) \right] \frac{1}{2} \quad (A5)$$

and the sum of these probabilities can be defined in an analogous manner by computing:

$$P\left(q = n + \frac{1}{2} \mid s + 1\right) + P\left(Q = n + \frac{1}{2} \mid s = -1\right) =$$

$$= \frac{1}{2} P_e(n, \Delta) \quad \text{for} \quad -M + 1 \leq n \leq M - 2$$

$$= \frac{1}{2} P_u(n, \Delta) \quad \text{for} \quad n = M - 1$$

$$= \frac{1}{2} P_l(n, \Delta) \quad \text{for} \quad n = -M \quad (A6)$$

Thus, the variance can be written as follows:

$$\text{Var}[c] = \frac{1}{4} \sum_{n=-M+1}^{m-2} \left(n + \frac{1}{2}\right)^2 P_e[n, \Delta] + \left(M - \frac{1}{2}\right)^2 P_u(M - 1, \Delta)$$

$$+ \left(M - \frac{1}{2}\right)^2 P_l(-(M-1), \Delta) \quad (A7)$$

The ratio of mean to rms variation in the correlator output is then as follows:

$$\frac{E(c)}{\sigma_o} = \frac{E(c)}{\sqrt{\text{Var}(c) - [E(c)]^2}} \quad (A8)$$

For a 3-bit quantizer $M = 2^{3-1} = 4$, and the mean and variance are as follows:

$$E[c] = \frac{1}{4} \sum_{n=-3}^{2} \left(n + \frac{1}{2}\right) d_e(n, \Delta) + \left(3 + \frac{1}{2}\right) d_u[3, \Delta]$$

$$+ \left(-4 + \frac{1}{2}\right) d_l[-3, \Delta] \quad (A9)$$

and

$$\text{Var}[c] = \frac{1}{4} \sum_{n=-3}^{2} \left(n + \frac{1}{2}\right)^2 P_e(n, \Delta)$$

$$+ \left(3\frac{1}{2} \cdot \frac{1}{2}\right)^2 P_u[3, \Delta] + \left(3 + \frac{1}{2}\right)^2 P_l[-3, \Delta] \quad (A10)$$

References

[1] Daher, J. K., Harris, J. M., and Wheeler, "An Evaluation of the Radio Frequency Susceptibility of Commercial GPS Receivers," *IEEE AES Systems Magazine*, Oct. 1994.

[2] Johnson, M. W., and Erlandson, "GNSS Receiver Interference Susceptibility and Civil Aviation Impact," Institute of Navigation, GPS-95 Meeting, Sept. 1995.

[3] Nisner, P., and Owen, J., "Practical Measurement of Radio Frequency Interference to GPS Receivers, "Institute of Navigation, GPS-95 Meeting, Sept. 1995.

[4] Lang, R. H., Bahie El Din, M., and Pickholtz, R. L., "The Effects of Weight Fluctuation in the Performance of Null Steering Antenna Arrays," IEEE Military Communication Conference, Oct. 1984.

[5] Compton, R. T., Jr., *Adaptive Antennas*, Prentice Hall, Englewood Cliffs, NJ, 1988.

[6] Applebaum, A. P., "Adaptive Arrays," *IEEE Transactions on Antennas and Propagation*, Sept. 1976.

[7] Spilker, J. J., Jr., *Digital Communications by Satellite*, Prentice-Hall, New York, 1977, 1995.

[8] Amoroso, F., "Adaptive A/D Converter to Suppress CW Interference in DSPN Spread Spectrum Communication," *IEEE Transaction on Communications,* Oct. 1983.

[9] Baer, H. P., "Interference Effects of Hard Limiting in PN Spread Spectrum Systems," *IEEE Transactions on Communications,* May 1982.

[10] Aein, J. M., and Pickholtz, R. L., "A Simplified Unified Phasor Analysis for PN Multiple Access to Limiting Repeaters," *Spread Spectrum Communications,* edited by C. E. Cook et al., IEEE Press, Piscataway, NJ, 1983.

[11] Holmes, J. K., *Coherent Spread Spectrum Systems*, Krieger Publishing Co., Melbourne, FL, 1990.

[12] Ketchum, J. W., and Proakis, J. G., "Adaptive Algorithms for Estimating and Suppressing Narrow Band Interference in PN Spread Spectrum Systems," *Spread Spectrum Communications,* edited by C. E. Cook et al., IEEE Press, Piscataway, NJ, 1983.

[13] Widrow, B. et al., "Adaptive Antenna System," *Proceedings of the IEEE,* Dec. 1967.

[14] Ward, J., and Compton, R. T., Jr., "High Throughput Slotted Aloha Packet Radio Networks with Adaptive Arrays," *IEEE Transactions on Communications,* March 1993.

[15] Natali, F. D., "AFC Tracking Algorithms," *IEEE Transactions on Communications*, Aug. 1984.

Author Index

Aparicio, M. 209
Ashby, N. 623
Axelrad, P. 409
Bertiger, W. I. 585
Braasch, M. S. 547, 601
Brodie, P. 209
Brown, R. G. 409
Doyle, L. 209
Francisco, S. G. 435
Klobuchar, J. A. 485
Kruczynski, L. 699
Natali, F. D. 717
Parkinson, B. W. 3, 29, 469
Rajan, J. 209
Spilker, J. J., Jr. 29, 57, 121, 177, 245,
　　　　　　　　　　　　517, 569, 623, 717
Torrione, P. 209
Van Dierendonck, A. J. 329
van Graas, F. 601
Zumberge, J. F. 585

Subject Index

A/D (see analog-to-digital converter), 347, 350
Acceleration of gravity (geff), 663–665
Acceleration, centripetal, 641
Accumulate-and-dump, 363–366
Accumulated delta range (ADR), 245, 286, 410, 412
Accuracy of navigation message, 586
Accuracy summary, 18
Acquisition, 254, 267, 273, 286, 297
Adaptive
　adaptive A/D converters, 718
　adaptive antennas, 43
　adaptive array antennas, 721
　adaptive delay lock loops (adaptive DLL), 756
　adaptive frequency notch filters, 756
　adaptive frequency nulling filters, 757
　adaptive Kalman filter, 299
　adaptive null steering antenna, 756
　adaptive nulling antennas, 718
Additive white Gaussian noise (AWGN), 245, 249, 252, 256
Adiabatic process, 529–530
Aerospace Corporation, 6, 27
Aiding, 372, 374
Aircraft navigation, 52
Aliasing, 334, 351, 353
Ambient noise, 338, 340
Analog-to-digital converter (A/D), 347, 350, 725
Analytic mode, 614, 620
Antenna, 82, 84, 88, 332, 340
　antenna gain G, 89
　antennas/low noise amplifiers, 246
　monitoring station, 276–299
　S-band, 276–299
Applebaum array, 722, 764, 765, 766, 767
Applied Physics Laboratory (APL), 4
Arctangent discriminator, 379, 385, 388
ARINC, 341, 343
Atmosphere
　atmospheric attenuation, 520–521
　atmospheric gases, 524
　atmospheric pressure, 519, 527

atmospheric profile, 527
　dry, 52, 518, 524
　ionosphere, 44, 49–51, 54
　troposhere, 34, 44, 49–51, 54
Atmospheric attenuation, 49, 51
Atmospheric sensors, 276–299
Atomic clocks, 5, 14, 15
Atomic frequency standards, 5
Atomic standard time, 152
Attenuation and rejection, 559
Attitude determination, 48
Autocorrelation function, 256, 258, 274, 286, 360
Automatic gain control control (AFC), 273, 379, 382–384, 393, 396
Automatic gain control (AGC), 330–331, 336, 347, 352, 354, 368
Autoregrassive model, 605
Average error, 551

Bessel function, 163
Best Estimate of Trajectory, 705
Block I satellites, 177
Burn-out protection, 332, 343

C/A code generator, 359-360
Carrier aided delay lock loop, 283
Carrier ambiguity, 286
CCIR, 84, 523
Centrer for Orbit Determination, 588
Cesium and rubidium atomic clocks, 152
Channel, 336
Channel capacity, 304
Chip, 73, 337
Christoffel symbols, 657
Clock, 585–586, 594
　atomic, 626
　hypothesis , 650
　reference, 642
　standard, 626
Coarse/acquisition (C/A) code, 57
Code chipping rate, 338

775

SUBJECT INDEX

Code delay lock loop discriminator effects, 285
Code division multiple access (CDMA), 62
Code generator, 355, 356
Code lock, 382
Code phase, 366
Code tracking, 357
Code-carrier divergence, 375
Coherent, 372
Collins radio, 27
Common view time transfer, 46
Compressibility Z factors, 528
Comptroller General's Report on GPS, 699
Control, space vehicle, 276–299
Control segment, 29, 38, 41–42, 47, 276–299
Controlled-reception pattern antenna (CRPA), 332
Coordinate time, 627–628, 651, 659, 661
Coordinates
   Earth-centered inertial (ECI) frame, 31, 627, 659, 669, 676
   Earth-centered Earth-fixed (ECEF) coordinates, 32, 44, 248
   geocentric, 662, 665, 670
   geodetic, 659, 664
   geographic, 664–665, 673, 675
   isotropic, 658, 661, 689–690
   Schwarzschild, 656–658, 676–677
Correleated measurement errors, 203
Correlation, 259–260, 361–363, 367
Correlation loss, 351–353
Correlator, 348, 362–363, 372, 378
Correlator spacing, 269, 351, 361, 363, 375
Costas discriminator, 385, 388
Cramer-Rao lower bound, 110–111
Cross-correlation, 81, 87, 94, 102, 376
   cross-correlation sidelobes, 79, 97
Cross-product discriminator, 379–382
Crosslink ranging, 682–683, 692–694
CW carrier interference, 748
Cycle-and-add property, 78, 113, 116
Cyclotomic cosets, 117

Data bandwidth, 374
Data demodulation, 365, 393, 396
Data modulation, 68, 90, 95–97
Data set fitting, 276–299
Decimations, 117
Decision-directed Costas discriminator, 385
Decision-directed cross-product discriminator, 379–380
Defense Mapping Agency, 158
Defense System Acquistion and Review Council (DSARC), 8, 13, 22
Degraded coverage, 206
Delay lock loop (DLL), 245, 248, 257, 259

   coherent, 256, 261, 275, 278, 372
   noncoherent, 274, 312, 372, 374, 376
   quasicoherent, 267, 279–282
   scalar, 251, 295, 299
   vector, 249–252, 289–290, 297
Delta-range, 384
Demodulate, 381
Desert Storm, 17-18, 24
Dielectric constant, 525
Differential arctangent discriminator, 379, 382
Differential GPS (DGPS), 12, 28
Differential GPS testing, 711, 712
Differential phase shift keying (DPSK), 381
Differential system effects, 547
Differentiated autocorrelation function, 286, 306, 318
Digital signal processing, 345, 348, 353
Dilution of Precision (DOP), 35, 414
Direct Sequence-Spread Sqectrum (DS-SS), 61, 65
Direction cosine matrix, 299
Discriminator characteristic, 274, 279, 282, 381
Doppler removal, 348, 356-357, 375–376
Doppler shift, 53, 97, 103, 411, 642
Dot-product DLL, 374, 376
Downconverter, 334, 337, 345
Dry atmosphere, 49, 518, 524
Dual-frequency receiver, 336
Dual quartic zenith model, 534
Duty factor, 260, 283
Dynamic mode, 419, 431
Dynamic range effects, 352

E sin E effect, 681–683
Early codes, 361
Early-late reference, 256, 263, 265, 269, 262
Earth
   center of mass, 137
   equational radius, 38, 40
   gravitational potential 134, 164–166
   quadrapole moment, 165, 173
Eccentric anomaly, 133, 137, 161, 680
Eccentricity, 179, 623, 660, 679–683
   eccentricity of the orbit, 133, 150, 161, 173
Effective aperture area, 84
Einstein synchronization, 627
Electron content, 49
Elevation angle, 183, 187, 198, 672, 690
Energy, 681
Envelope, 377–378, 401
Envelope correlation, 274–275
Ephemeris errors, 478
Ephemeris time, 135,143, 151–152
Equation of time, 149
Equatorial radius for Earth, 166
Error analysis,11, 470

SUBJECT INDEX

Error covariance matrix, 195-196, 198, 200
Error envelope, 555, 557, 560
Error equation derivation, 470
Errors, C/A code, no S/A, 481
Errors, with S/A, 481
Estimation, 429
Estimation, system states, 276-299
Event, 651
Extended Kalman Filter (EKF), 248, 289, 291

False alarm, 369, 404
False alarm rate, 404
False lock, 382, 386, 393
FCC frequency assignments, 717
Feedback shift registers, 62, 65
Field, Galois, 57, 67, 113-114
Filter, 355, 366
Filtering
  Kalman, 289, 302, 412, 417, 420
Finite field, 113-115
First order loop, 371, 374, 389
Fisher information, 111
Fixed Reception Pattern Antenna (FRPA), 340
Flat space, 640
FLL, 393
Flux density registers, 59
Fokker Planck equation, 321-322
Foliage attenuation, 571-572
Force models, 276-299
Four-momentum, 639, 673
Four-vector, 639
Four-velocity, 639, 643
Fourier transform, 110-110
Frequency division Multiple Access (FDMA), 68-69
Frequency doubler, 337
Frequency Hop Spread Sqectrum (FH-SS), 60-61
Frequency lock, 379, 390, 393
Frequency plan, 333-347
Frequency synthesizer, 331, 333, 337
Fresnel diffraction, 572, 575
Fresnel drag, 640
Friis formula, 344
Front-end, 336
Full-scale engineering development tests, 713

G-sensitivity, 388
Gabor bandwidth, 110
Gain, 374
Galilean transformation, 635
Galosi fields, 57, 113
Gauss-Markov Theorem, 194
Gaussian density, 249, 340, 578
GDOP derivation, 179, 475
General relativity, 637
Generalized inverse matrix, 196, 296
Geocentric latitude, 139

Geodesics, 634
Geoid the earth, 152, 662
Geometric Dilution of Precision (GDOP), 16, 474
  HDOP, 245
  PDOP, 474
  TDOP, 245
  VDOP, 245
Geostationary satellite, 197
GIPSY/OASIS-II, 587
Global Navigation Satellite System (GLONASS)
  clocks, 626
  time, 631
Global Positioning System (GPS), 3, 29
GPS clock correction data formats, 132
GPS ephemeris, 135
GPS Time, 141
GPS-24, 178
GPSCard, 346
Gold code, 78
  balanced Gold Code, 113
Gravitational field, 640
Gravitational frequency shift, 652
Gravity, 17
Ground antenna (GA), 42
Ground traces, 189
Ground transmitter (GT), 706
Ground truth, 705
Group delay, 49

Hamming block code, 130, 398
Hard-limiting, 352
Harmonics, 346
Hemispherical antenna, 569
High Dynamics User Equipment (HDUE), 705
Histogram, 404
History, 601
Horizotal dilution of precision (HDOP), 35
Hybrid GPS receivers, 48
Hybrid GPS/Inertial solution, 53
Hydrostatic refractivity, 533
Hypothesis test, 403

I/Q Sampling, 726
ICD-GPS-200, 121
Ideal gas, 529
IF, 346
IF sampling, 273, 352
Image, 350
In-phase, 365
In-phase/mid-phase bit synchronizer, 365
Inclination angle, 168
Inclined orbits, 159
Incomplete gamma function, 366
Inertial frame, 627
Inertial frame, local, 628

SUBJECT INDEX

Inertial measurement units (IMU's), 248
Information flow, 275–299
Information rate, 254
Initial Operational Capability (IOC), 19
Initial testing, 22
INMARSAT, 82
Intentional (jamming), 54
Interfaces, system, 275–299
Interference rejection, 351, 756
Intermediate frequency, 334
International GPS Service for Geodynamics, 587
International Telecommunication Union (ITU), 59
Inverted Range Control Center (IRCC), 705
IOC, 3
Ionosphere
   ionosphere-free, 16, 374
   ionosphere delay, 50, 305, 396
   ionosphere errors, 478
   ionospheric correction, 144
   ionospheric measurement, 49
   ionospheric obliquity factor, 145
   ionospheric scintillation, 49
Irreducible polynomial, 117
Isolation and evaluation, 547
Isothermal lines, 531
Isothermal model, 531
Issue of Data-Clock (IODC), 136

Jacobian matrix, 291
Jet Propulsion laboratory, 586
JGM-2 gravity model, 663
Joint Program Office (JPO), 24
   motto, 27
   testing, 707

Kalman filter, 248, 290, 296, 409
Kepler's equations, 161
Kerr metric, 687
Kinematic survey, 48

L-band signals, 524
L1 signal, 72,
L2 signal, 84, 338
L3, 71
Land Mobile Navigation, 45
Land mobile users, 569
Laser trackers, 701
Late code, 342
Launch history, 19
Leap seconds, 125, 141
Least square
   least square estimator, 194, 251
   least squares solution, 413
   left-hand-circularly-polarized, 332
Lense-Thirring effect, 624

LHCP, 332
Light, speed of, 631
Lightning protection, 341
Line spectrum 388
Linearized Equivalent Circuit, 267
Linearly-polarized, 332
LMS adaptive antenna, 761
Local inertial frame, 645
Local oscillator (LO), 329, 346
   feed-through, 347
Lock detection, 365, 382
Lognormal distribution, 570
Long-term stability, 333
Loo distribution, 570
Loop bandwidth, 370
Loop gain, 371
Lorentz Invariance, 650
Lorentz Transformation, 629
Loss, 355
Low-noise amplifier (LNA), 329, 332, 341
Low-pass, 347
Lunar gravitation, 169

M-bit quantizer, 726
Magnavox, 27
Magnavox X-set, 706
Mapping function, 521
Marine Vessel Navigation, 45
Maximum a posteriori estimator (MAP), 252
Maximum likelihood estimator, 251
Mean anomaly, 135
Mean solar time, 149
Mean square noise error, 270
Measured pseudorange, 33
Measurement corrections, 276–299
Mesosphere, 519
Metric tensor, 629
Military applications, 24
Miniature airborne GPS receiver (MAGR), 337
Minimal polynominal, 117
Minimum variance, unbiased estimate, 192
Mixer, 347
Mod 2 adders, 65
Model errors, effect of, 416
Model identification, 508
Modified Bessel function, 403
Momentary Comoving Reference Frame (MCRF), 641
Monitor stations, 41
Monte Carlo, 372
Mountainous terrain, 579
Multi-bit quantization, 352
Multipath, 52
   distortion, 376
   errors, 478
   rejection, 332, 376
Multiple access, 36
   noise, 52

SUBJECT INDEX   779

Multiple access channel capacity, 304
Multiple access noise, 52
Multiple beam antenna array, 22

Nakagami-Rice distribution, 570
Narrowband power, 341
Naval Research Lab (NRL), 4, 15
Navigation
  accuracy, 418
  algorithms, 409
  data, 121
Navigation Data Processing, Control Segment, 276–299
Navigation Data Subframe, 123, 132
Navigation Message Generation, 276–299
Navigation Technology Satellite II (NTS-II), 7
Navy Navigation Satellite System (NSSS), 4
Near-far problem, 68
Newton-Rhapson method, 404
Nitrogen/oxygen atmosphere, 524
Noise
  bandwidth, 338, 353
  density, 344, 399
  floor, 367
  temperature, 343
Noncoherent square law detector, 254
Nonideal gases, 528
Nonlinear shift registers, 65
NovAtel, 338
NTS-II, 19
Nudet (Nuclear Detonation) Detection System (NDS), 73
Number-controlled oscillator (NCO), 9, 348, 356
Numerical methods of solving Kepler's equation, 136, 161
Nyquist, 350

Oblateness of the earth, 165, 167
Obliquity factor, 50
Operating Location AA (OL-AA), 702
Operational Control Center, 42
Operations, Control Segment, 275–299
Optimum quantizer, 739
Orbit determination, 36
Oven-controlled crystal oscillator (OCXO), 333
Overdetermined, 194

P code generator, 358
P(Y) code, 44
Parity, 393
Performance Monitoring Service, 276–299
Perigee, 161

Phase distortion, 350
Phase I tests, 699
Phase II tests, 699
Phase lock, 345, 373
Phase locked loop (PLL), 285, 337
Phase noise, 88, 343
Phase rotation, 356
Phased-array, 332
Pioneers, 26
PLL effects, 552
PN chip, 261, 350
Policy, 601
Polynomial, 73, 74, 79, 114, 117
Position, 411, 602, 603
Position dilution of precision (PDOP), 16, 36, 203, 204, 205, 413, 469
  with an accurate clock, 203
Position domain effects, 603
Post-fit, 605
Potential
  gravitational, 654, 663, 678, 680, 684
  quadrupole, 662, 678, 680, 683
Power spectral density, 156, 607, 612, 757
PPS accuracy, 12, 17
Precession, 151, 165, 167
Precise orbits, 590
Precision Automated Tracking System (PATS), 701
Precision P code, 57, 67–68, 73–77
Prediction, 595
  system status, 276–299
Prelaunch facilities, 276–299
Primitive element, 114
Primitive polynomials, 114
Principle of Equivalence, 649
PRN (also PN) coder, 66, 329
Product code, 65, 81
Programmable divider, 346
Prompt code, 362
Propagation, 569
Proper time, 631
Pseudo sampling, 349
Pseudolite, 402
Pseudorandom sequence, 59, 248
Pseudorange, 32, 58, 248, 331, 585
  defined, 33, 469
  deviations, 276–299
  measurements, 410
Punctual code, 363

Quadraphase, 347
Quadrupole moment, of Earth, 686
Quantization, 350
Quantizer
  1-bit quantizer, 731
  2-bit quantizer, 734
  3-bit quantizer, 740
  optimum, 746
  quantizer effects, 746
Quasi-coherent delay lock loop, 245, 280

Radio frequency, 345
Range decorrelation, 87
Range domain effects, 603
Ranging errors, 477
Rayleigh distribution, 53
Rayleigh scattered term, 570
Real-time, 586, 593
Real-Time Estimate of Trajectory (RTE), 704
Receiver, monitoring station, 275–299
Receiver clock mode, 410
Receiver errors, 480
Recursive autoregressive mode, 605
Red shift (or gravitational frequency shift), 642
Reference frame, 587, 590, 593
Reference oscillator, 333
Reflexivity, 659
Refractive index n, 522
Relativity
 general, 627
 principles of, 627
 special, 640
RHCP, 332
Rician, 570
Right-hand-circularly-polarized, 341
Rockwell Collins, 337
Rotating disc, 641

Sagnac effect
 clocks in satellites, 648
 light signals, 648
 portable clocks, 667
Sampling, 246
Satellite
 clock error, 122, 124
 visibility, 181
Scalar invariant, ds 4, 651
Schmidt-Kalman filter, 248
Schwarzschild metric, 656
Scintillation, 49
Second-order Gauss-Markov mode, 605
Second-order loop, 371
Second-order noncoherent delay lock loop, 272
Secular precession rate, 167
Selective availabiltiy (SA), 77, 587, 593
Semi-major axis, 161, 178
Shapiro time delay effect, 624, 689
Shift and add property, 116
Sidereal
 day, 32, 38, 150, 178, 564
 time, 38, 150
Signal processing, 718
Signal-to-noise, 88
 density, 88
 ratio, 88, 254, 729, 766
Single difference, 694
Smoothing, Carrier Aided, 275–299

Snell's Law, 525
Solar day, 149, 150, 178
Solar pressure, 179
Space and Missile Organization (SAMSO), 5, 26
Space segment, 29, 36
Space vehicle health, 139, 141
Spatially nulling, 759
Special realtivity, 627, 640, 641, 651
Spectral density, 317
Specular multipath, 569
Spherical harmonics of the earth's gravitational pull, 166
Spread spectrum signaling, 59
States, system, 276–299
State vectors, 54
Steady state, 390
Stochastic, 587
Stratosphere, 97, 519
Summation convention, Einstein, 629
Sun-Earth gravitation, 150
Surface acoustic wave (SAW), 337
Surface pressure (Po), 97, 531
Synchronization
 by light signals, 669
 by portable clocks, 667
 relativity of, 627
System noise figure, 333

Taylor series, 258
Telemetry, 276–299
 TLM word, 82
Temperature compensated crystal oscillator (TCXO), 333, 388
Tensor
 metric, 640
 rank of, 641
 Riemann-Christoffel curvature, 657
Test statistic, 369
Thermal noise density, 344
Third-order loop, 388
Threshold, 351
Tidal effects, 689
Tides, 587
Time
 dilation, 631
 GPS, 651
 international atomic, 627
 scale, 631
Time frequency search, 254
Toeplitz matrix, 299
Tracking strategy, 276
Transformations, 591
 Galilean, 628
 Lorentz, 635
Transient tracking performance, 268
Transitivity, 659
Trapezoidal pulse sequence, 256
Tropopause, 519

SUBJECT INDEX

Troposphere, 586
   delay, 248
   scintillation, 522
True anomaly, 161
True pseudorange, 33

Universal coordinated time, 152
Universal time, 149
Upload, 276–299
U.S. Naval Observatory, 157
U.S. Standard atmosphere, 526
User clock bias, 124
User range accuracy, 124
User range error, 123
User-equivalent range error, 596-597

Vector delay lock loop, 289-290, 293, 304, 323
Vector, contravariant, 630

Vecror, covariant, 630
Velocity transformation, relativistic, 637
Visibility, space vehicles, 276–299
Voltage control oscillator, 386

Water vapor, 517
Wet refraction, 534
Wide-band power, 390
Wiener filter, 301
World-line, 628

XI epochs, 359

Y code, non-standard codes, 77

Z count, 398, 639

PROGRESS IN ASTRONAUTICS AND AERONAUTICS
SERIES VOLUMES

*1. Solid Propellant Rocket Research (1960)
Martin Summerfield
Princeton University

*2. Liquid Rockets and Propellants (1960)
Loren E. Bollinger
Ohio State University
Martin Goldsmith
The Rand Corp.
Alexis W. Lemmon Jr.
Battelle Memorial Institute

*3. Energy Conversion for Space Power (1961)
Nathan W. Snyder
Institute for Defense Analyses

*4. Space Power Systems (1961)
Nathan W. Snyder
Institute for Defense Analyses

*5. Electrostatic Propulsion (1961)
David B. Langmuir
Space Technology Laboratories, Inc
Ernst Stuhlinger
NASA George C. Marshall Space Flight Center
J.M. Sellen Jr.
Space Technology Laboratories, Inc.

*6. Detonation and Two-Phase Flow (1962)
S.S. Penner
California Institute of Technology
F.A. Williams
Harvard University

*7. Hypersonic Flow Research (1962)
Frederick R. Riddell
AVCO Corp.

*8. Guidance and Control (1962)
Robert E. Roberson
Consultant
James S. Farrior
Lockheed Missiles and Space Co.

*9. Electric Propulsion Development (1963)
Ernst Stuhlinger
NASA George C. Marshall Space Flight Center

*10. Technology of Lunar Exploration (1963)
Clifford I. Cumming
Harold R. Lawrence
Jet Propulsion Laboratory

*11. Power Systems for Space Flight (1963)
Morris A. Zipkin
Russell N. Edwards
General Electric Co.

*12. Ionization in High-Temperature Gases (1963)
Kurt E. Shuler, Editor
National Bureau of Standards
John B. Fenn
Associate Editor
Princeton University

*13. Guidance and Control–II (1964)
Robert C. Langford
General Precision Inc.
Charles J. Mundo
Institute of Naval Studies

*14. Celestial Mechanics and Astrodynamics (1964)
Victor G. Szebehely
Yale University Observatory

*15. Heterogeneous Combustion (1964)
Hans G. Wolfhard
Institute for Defense Analyses
Irvin Glassman
Princeton University
Leon Green Jr.
Air Force Systems Command

*16. Space Power Systems Engineering (1966)
George C. Szego
Institute for Defense Analyses
J. Edward Taylor
TRW Inc.

*17. Methods in Astrodynamics and Celestial Mechanics (1966)
Raynor L. Duncombe
U.S. Naval Observatory
Victor G. Szebehely
Yale University Observatory

*18. Thermophysics and Temperature Control of Spacecraft and Entry Vehicles (1966)
Gerhard B. Heller
NASA George C. Marshall Space Flight Center

*19. Communication Satellite Systems Technology (1966)
Richard B. Marsten
Radio Corporation of America

*20. Thermophysics of Spacecraft and Planetary Bodies: Radiation Properties of Solids and the Electromagnetic Radiation Environment in Space (1967)
Gerhard B. Heller
NASA George C. Marshall Space Flight Center

---

*Out of print.

*21. Thermal Design
Principles of Spacecraft
and Entry Bodies (1969)
Jerry T. Bevans
TRW Systems

*22. Stratospheric
Circulation (1969)
Willis L. Webb
*Atmospheric Sciences
Laboratory, White Sands,
and University of Texas at
El Paso*

*23. Thermophysics:
Applications to Thermal
Design of Spacecraft
(1970)
Jerry T. Bevans
TRW Systems

24. Heat Transfer and
Spacecraft Thermal
Control (1971)
John W. Lucas
Jet Propulsion Laboratory

25. Communication
Satellites for the 70's:
Technology (1971)
Nathaniel E. Feldman
The Rand Corp.
Charles M. Kelly
The Aerospace Corp.

26. Communication
Satellites for the 70's:
Systems (1971)
Nathaniel E. Feldman
The Rand Corp.
Charles M. Kelly
The Aerospace Corp.

27. Thermospheric
Circulation (1972)
Willis L. Webb
*Atmospheric Sciences
Laboratory, White Sands,
and University of Texas at
El Paso*

28. Thermal
Characteristics of the
Moon (1972)
John W. Lucas
Jet Propulsion Laboratory

*29. Fundamentals of
Spacecraft Thermal
Design (1972)
John W. Lucas
Jet Propulsion Laboratory

30. Solar Activity
Observations and
Predictions (1972)
Patrick S. McIntosh
Murray Dryer
*Environmental Research
Laboratories, National
Oceanic and Atmospheric
Administration*

31. Thermal Control and
Radiation (1973)
Chang-Lin Tien
*University of California at
Berkeley*

32. Communications
Satellite Systems (1974)
P.L. Bargellini
COMSAT Laboratories

33. Communications
Satellite Technology (1974)
P.L. Bargellini
COMSAT Laboratories

*34. Instrumentation for
Airbreathing Propulsion
(1974)
Allen E. Fuhs
Naval Postgraduate School
Marshall Kingery
*Arnold Engineering
Development Center*

35. Thermophysics and
Spacecraft Thermal
Control (1974)
Robert G. Hering
University of Iowa

36. Thermal Pollution
Analysis (1975)
Joseph A. Schetz
*Virginia Polytechnic
Institute*
ISBN 0-915928-00-0

*37. Aeroacoustics: Jet
and Combustion Noise;
Duct Acoustics (1975)
Henry T. Nagamatsu, Editor
*General Electric Research
and Development Center*
Jack V. O'Keefe, Associate
Editor
The Boeing Co.
Ira R. Schwartz, Associate
Editor
*NASA Ames Research
Center*
ISBN 0-915928-01-9

*38. Aeroacoustics: Fan,
STOL, and Boundary
Layer Noise; Sonic Boom;
Aeroacoustics
Instrumentation (1975)
Henry T. Nagamatsu, Editor
*General Electric Research
and Development Center*
Jack V. O'Keefe, Associate
Editor
The Boeing Co.
Ira R. Schwartz, Associate
Editor
*NASA Ames Research
Center*
ISBN 0-915928-02-7

39. Heat Transfer with
Thermal Control
Applications (1975)
M. Michael Yovanovich
University of Waterloo
ISBN 0-915928-03-5

*40. Aerodynamics of
Base Combustion (1976)
S.N.B. Murthy, Editor
J.R. Osborn, Associate
Editor
Purdue University
A.W. Barrows
J.R. Ward,
Associate Editors
*Ballistics Research
Laboratories*
ISBN 0-915928-04-3

---
*Out of print.

41. **Communications Satellite Developments: Systems (1976)**
Gilbert E. LaVean
Defense Communications Agency
William G. Schmidt
CML Satellite Corp.
ISBN 0-915928-05-1

42. **Communications Satellite Developments: Technology (1976)**
William G. Schmidt
CML Satellite Corp.
Gilbert E. LaVean
Defense Communications Agency
ISBN 0-915928-06-X

*43. **Aeroacoustics: Jet Noise, Combustion and Core Engine Noise (1976)**
Ira R. Schwartz, Editor
NASA Ames Research Center
Henry T. Nagamatsu, Associate Editor
General Electric Research and Development Center
Warren C. Strahle, Associate Editor
Georgia Institute of Technology
ISBN 0-915928-07-8

*44. **Aeroacoustics: Fan Noise and Control; Duct Acoustics; Rotor Noise (1976)**
Ira R. Schwartz, Editor
NASA Ames Research Center
Henry T. Nagamatsu, Associate Editor
General Electric Research and Development Center
Warren C. Strahle, Associate Editor
Georgia Institute of Technology
ISBN 0-915928-08-6

*45. **Aeroacoustics: STOL Noise; Airframe and Airfoil Noise (1976)**
Ira R. Schwartz, Editor
NASA Ames Research Center
Henry T. Nagamatsu, Associate Editor
General Electric Research and Development Center
Warren C. Strahle, Associate Editor
Georgia Institute of Technology
ISBN 0-915928-09-4

*46. **Aeroacoustics: Acoustic Wave Propagation; Aircraft Noise Prediction; Aeroacoustic Instrumentation (1976)**
Ira R. Schwartz, Editor
NASA Ames Research Center
Henry T. Nagamatsu, Associate Editor
General Electric Research and Development Center
Warren C. Strahle, Associate Editor
Georgia Institute of Technology
ISBN 0-915928-10-8

*47. **Spacecraft Charging by Magnetospheric Plasmas (1976)**
Alan Rosen
TRW Inc.
ISBN 0-915928-11-6

48. **Scientific Investigations on the Skylab Satellite (1976)**
Marion I. Kent
Ernst Stuhlinger
NASA George C. Marshall Space Flight Center
Shi-Tsan Wu
University of Alabama
ISBN 0-915928-12-4

49. **Radiative Transfer and Thermal Control (1976)**
Allie M. Smith
ARO Inc.
ISBN 0-915928-13-2

*50. **Exploration of the Outer Solar System (1976)**
Eugene W. Greenstadt
TRW Inc.
Murray Dryer
National Oceanic and Atmospheric Administration
Devrie S. Intriligator
University of Southern California
ISBN 0-915928-14-0

51. **Rarefied Gas Dynamics, Parts I and II (two volumes) (1977)**
J. Leith Potter
ARO Inc.
ISBN 0-915928-15-9

52. **Materials Sciences in Space with Application to Space Processing (1977)**
Leo Steg
General Electric Co.
ISBN 0-915928-16-7

53. **Experimental Diagnostics in Gas Phase Combustion Systems (1977)**
Ben T. Zinn, Editor
Georgia Institute of Technology
Craig T. Bowman, Associate Editor
Stanford University
Daniel L. Hartley, Associate Editor
Sandia Laboratories
Edward W. Price, Associate Editor
Georgia Institute of Technology
James G. Skifstad, Associate Editor
Purdue University
ISBN 0-915928-18-3

---

*Out of print.

54. Satellite Communication: Future Systems (1977)
David Jarett
TRW Inc.
ISBN 0-915928-18-3

55. Satellite Communications: Advanced Technologies (1977)
David Jarett
TRW Inc.
ISBN 0-915928-19-1

56. Thermophysics of Spacecraft and Outer Planer Entry Probes (1977)
Allied M. Smith
ARO Inc.
ISBN 0-915928-20-5

57. Space-Based Manufacturing from Nonterrestrial Materials (1977)
Gerald K. O'Neill, Editor
Brian O'Leary, Assistant Editor
Princeton University
ISBN 0-915928-21-3

*58. Turbulent Combustion (1978)
Lawrence A. Kennedy
State University of New York at Buffalo
ISBN 0-915928-22-1

*59. Aerodynamic Heating and Thermal Protection Systems (1978)
Leroy S. Fletcher
University of Virginia
ISBN 0-915928-23-X

60. Heat Transfer and Thermal Control Systems (1978)
Leroy S. Fletcher
University of Virginia
ISBN 0-915928-24-8

61. Radiation Energy Conversion in Space (1978)
Kenneth W. Billman
NASA Ames Research Center
ISBN 0-915928-26-4

62. Alternative Hydrocarbon Fuels: Combustion and Chemical Kinetics (1978)
Craig T. Bowman
Stanford University
Jorgen Birkeland
Department of Energy
ISBN 0-915928-25-6

*63. Experimental Diagnostics in Combustion of Solids (1978)
Thomas L. Boggs
Naval Weapons Center
Ben T. Zinn
Georgia Institute of Technology
ISBN 0-915928-28-0

64. Outer Planet Entry Heating and Thermal Protection (1979)
Raymond Viskanta
Purdue University
ISBN 0-915928-29-9

65. Thermophysics and Thermal Control (1979)
Raymond Viskanta
Purdue University
ISBN 0-915928-30-2

66. Interior Ballistics of Guns (1979)
Herman Krier
University of Illinois at Urbana-Champaign
Martin Summerfield
New York University
ISBN 0-915928-32-9

*67. Remote Sensing of Earth from Space: Role of "Smart Sensors" (1979)
Roger A. Breckenridge
NASA Langley Research Center
ISBN 0-915928-33-7

68. Injection and Mixing in Turbulent Flow (1980)
Joseph A. Schetz
Virginia Polytechnic Institute and State University
ISBN 0-915928-35-3

*69. Entry Heating and Thermal Protection (1980)
Walter B. Olstad
NASA Headquarters
ISBN 0-915928-38-8

*70. Heat Transfer, Thermal Control, and Heat Pipes (1980)
Walter B. Olstad
NASA Headquarters
ISBN 0-915928-39-6

*71. Space Systems and Their Interactions with Earth's Space Environment (1980)
Henry B. Garrett
Charles P. Pike
Hanscom Air Force Base
ISBN 0-915928-41-8

*72. Viscous Flow Drag Reduction (1980)
Gary R. Hough
Vought Advanced Technology Center
ISBN 0-915928-44-2

*73. Combustion Experiments in a Zero-Gravity Laboratory (1981)
Thomas H. Cochran
NASA Lewis Research Center
ISBN 0-915928-48-5

74. Rarefied Gas Dynamics, Parts I and II (two volumes) (1981)
Sam S. Fisher
University of Virginia
ISBN 0-915928-51-5

---

*Out of print.

75. **Gasdynamics of Detonations and Explosions (1981)**
J.R. Bowen
University of Wisconsin at Madison
N. Manson
Universite de Poitiers
A.K. Oppenheim
University of California at Berkeley
R. I. Soloukhin
Institute of Heat and Mass Transfer, BSSR Academy of Sciences
ISBN 0-915928-46-9

76. **Combustion in Reactive Systems (1981)**
J.R. Bowen
University of Wisconsin at Madison
N. Manson
Universite de Poitiers
A.K. Oppenheim
University of California at Berkeley
R.I. Soloukhin
Institute of Heat and Mass Transfer, BSSR Academy of Sciences
ISBN 0-915928-47-7

*77. **Aerothermodynamics and Planetary Entry (1981)**
A.L. Crosbie
University of Missouri-Rolla
ISBN 0-915928-52-3

78. **Heat Transfer and Thermal Control (1981)**
A.L. Crosbie
University of Missouri-Rolla
ISBN 0-915928-53-1

*79. **Electric Propulsion and Its Applications to Space Missions (1981)**
Robert C. Finke
NASA Lewis Research Center
ISBN 0-915928-55-8

*80. **Aero-Optical Phenomena (1982)**
Keith G. Gilbert
Leonard J. Otten
Air Force Weapons Laboratory
ISBN 0-915928-60-4

81. **Transonic Aerodynamics (1982)**
David Nixon
Nielsen Engineering & Research, Inc.
ISBN 0-915928-65-5

82. **Thermophysics of Atmospheric Entry (1982)**
T.E. Horton
University of Mississippi
ISBN 0-915928-66-3

83. **Spacecraft Radiative Transfer and Temperature Control (1982)**
T.E. Horton
University of Mississippi
ISBN 0-915928-67-1

84. **Liquid-Metal Flows and Magneto-hydrodynamics (1983)**
H. Branover
Ben-Gurion University of the Negev
P.S. Lykoudis
Purdue University
A. Yakhot
Ben-Gurion University of the Negev
ISBN 0-915928-70-1

85. **Entry Vehicle Heating and Thermal Protection Systems: Space Shuttle, Solar Starprobe, Jupiter Galileo Probe (1983)**
Paul E. Bauer
McDonnell Douglas Astronautics Co.
Howard E. Collicott
The Boeing Co.
ISBN 0-915928-74-4

*86. **Spacecraft Thermal Control, Design, and Operation (1983)**
Howard E. Collicott
The Boeing Co.
Paul E. Bauer
McDonnell Douglas Astronautics Co.
ISBN 0-915928-75-2

87. **Shock Waves, Explosions, and Detonations (1983)**
J.R. Bowen
University of Washington
N. Manson
Universite de Poitiers
A.K. Oppenheim
University of California at Berkeley
R.I. Soloukhin
Institute of Heat and Mass Transfer, BSSR Academy of Sciences
ISBN 0-915928-76-0

88. **Flames, Lasers, and Reactive Systems (1983)**
J.R. Bowen
University of Washington
N. Manson
Universite de Poitiers
A.K. Oppenheim
University of California at Berkeley
R.I. Soloukhin
Institute of Heat and Mass Transfer, BSSR Academy of Sciences
ISBN 0-915928-77-9

*89. **Orbit-Raising and Maneuvering Propulsion: Research Status and Needs (1984)**
Leonard H. Caveny
Air Force Office of Scientific Research
ISBN 0-915928-82-5

---

*Out of print.

90. **Fundamental of Solid-Propellant Combustion (1984)**
Kenneth K. Kuo
Pennsylvania State University
Martin Summerfield
Princeton Combustion Research Laboratories, Inc.
ISBN 0-915928-84-1

91. **Spacecraft Contamination: Sources and Prevention (1984)**
J.A. Roux
University of Mississippi
T.D. McCay
NASA Marshall Space Flight Center
ISBN 0-915928-85-X

92. **Combustion Diagnostics by Nonintrusive Methods (1984)**
T.D. McCay
NASA Marshall Space Flight Center
J.A. Roux
University of Mississippi
ISBN 0-915928-86-8

93. **The INTELSAT Global Satellite System (1984)**
Joel Alper
COMSAT Corp.
Joseph Pelton
INTELSAT
ISBN 0-915928-90-6

94. **Dynamics of Shock Waves, Explosions, and Detonations (1984)**
J.R. Bowen
University of Washington
N. Manson
Universite de Poitiers
A. K. Oppenheim
University of California at Berkeley
R.I. Soloukhin
Institute of Heat and Mass Transfer, BSSR Academy of Sciences
ISBN 0-915928-91-4

95. **Dynamics of Flames and Reactive Systems (1984)**
J.R. Bowen
University of Washington
N. Manson
Universite de Poitiers
A. K. Oppenheim
University of California at Berkeley
R.I. Soloukhin
Institute of Heat and Mass Transfer, BSSR Academy of Sciences
ISBN 0-915928-92-2

96. **Thermal Design of Aeroassisted Orbital Transfer Vehicles (1985)**
H.F. Nelson
University of Missouri-Rolla
ISBN 0-915928-94-9

97. **Monitoring Earth's Ocean, Land, and Atmosphere from Space– Sensors, Systems, and Applications (1985)**
Abraham Schnapf
Aerospace Systems Engineering
ISBN 0-915928-98-1

98. **Thrust and Drag: Its Prediction and Verification (1985)**
Eugene E. Covert
Massachusetts Institute of Technology
C.R. James
Vought Corp.
William F. Kimzey
Sverdrup Technology AEDC Group
George K. Richey
U.S. Air Force
Eugene C. Rooney
U.S. Navy Department of Defense
ISBN 0-930403-00-2

99. **Space Stations and Space Platforms – Concepts, Design, Infrastructure, and Uses (1985)**
Ivan Bekey
Daniel Herman
NASA Headquarters
ISBN 0-930403-01-0

100. **Single- and Multi-Phase Flows in an Electromagnetic Field: Energy, Metallurgical, and Solar Applications (1985)**
Herman Branover
Ben-Gurion University of the Negev
Paul S. Lykoudis
Purdue University
Michael Mond
Ben-Gurion University of the Negev
ISBN 0-930403-04-5

101. **MHD Energy Conversion: Physiotechnical Problems (1986)**
V.A. Kirillin
A.E. Sheyndlin
Soviet Academy of Sciences
ISBN 0-930403-05-3

102. **Numerical Methods for Engine-Airframe Integration (1986)**
S.N.B. Murthy
Purdue University
Gerald C. Paynter
Boeing Airplane Co.
ISBN 0-930403-09-6

103. **Thermophysical Aspects of Re-Entry Flows (1986)**
James N. Moss
NASA Langley Research Center
Carl D. Scott
NASA Johnson Space Center
ISBN 0-930430-10-X

---

*Out of print.

*104. Tactical Missile
Aerodynamics (1986)
M.J. Hemsch
PRC Kentron, Inc.
J.N. Nielson
NASA Ames Research
Center
ISBN 0-930403-13-4

105. Dynamics of
Reactive Systems Part I:
Flames and
Configurations; Part II:
Modeling and
Heterogeneous
Combustion (1986)
J.R. Bowen
University of Washington
J.-C. Leyer
Universite de Poitiers
R.I. Soloukhin
Institute of Heat and Mass
Transfer, BSSR Academy of
Sciences
ISBN 0-930403-14-2

106. Dynamics of
Explosions (1986)
J.R. Bowen
University of Washington
J.-C. Leyer
Universite de Poitiers
R.I. Soloukhin
Institute of Heat and Mass
Transfer, BSSR Academy of
Sciences
ISBN 0-930403-15-0

107. Spacecraft
Dielectric Material
Properties and Spacecraft
Charging (1986)
A.R. Frederickson
U.S. Air Force Rome Air
Development Center
D.B. Cotts
SRI International
J.A. Wall
U.S. Air Force Rome Air
Development Center
F.L. Bouquet
Jet Propulsion Laboratory,
California Institute of
Technology
ISBN 0-930403-17-7

108. Opportunities for
Academic Research in a
Low-Gravity Environment
(1986)
George A. Hazelrigg
National Science
Foundation
Joseph M. Reynolds
Louisiana State University
ISBN 0-930403-18-5

109. Gun Propulsion
Technology (1988)
Ludwig Stiefel
U.S. Army Armament
Research, Development and
Engineering Center
ISBN 0-930403-20-7

110. Commercial
Opportunities in Space
(1988)
F. Shahrokhi
K.E. Harwell
University of Tennessee
Space Institute
C.C. Chao
National Cheng Kung
University
ISBN 0-930403-39-8

111. Liquid-Metal Flows:
Magnetohydrodynamics and
Application (1988)
Herman Branover,
Michael Mond, and
Yeshajahu Unger
Ben-Gurion University of
the Negev
ISBN 0-930403-43-6

112. Current Trends in
Turbulence Research
(1988)
Herman Branover,
Micheal Mond, and
Yeshajahu Unger
Ben-Gurion University of
the Negev
ISBN 0-930403-44-4

113. Dynamics of
Reactive Systems Part I:
Flames; Part II:
Heterogeneous
Combustion and
Applications (1988)
A.L. Kuhl
R & D Associates
J.R. Bowen
University of Washington
J.-C. Leyer
Universite de Poitiers
A. Borisov
USSR Academy of Sciences
ISBN 0-930403-46-0

114. Dynamics of
Explosions (1988)
A.L. Kuhl
R & D Associates
J.R. Bowen
University of Washington
J.-C. Leyer
Universite de Poitiers
A. Borisov
USSR Academy of Sciences
ISBN 0-930403-47-9

115. Machine
Intelligence and Autonomy
for Aerospace (1988)
E. Heer
Heer Associates, Inc.
H. Lum
NASA Ames Research
Center
ISBN 0-930403-48-7

116. Rarefied Gas
Dynamics: Space Related
Studies (1989)
E.P. Muntz
University of Southern
California
D.P. Weaver
U.S. Air Force Astronautics
Laboratory (AFSC)
D.H. Campbell
University of Dayton
Research Institute
ISBN 0-930403-53-3

---

*Out of print.

117. **Rarefied Gas Dynamics: Physical Phenomena (1989)**
E.P. Muntz
University of Southernn California
D.P. Weaver
U.S. Air Force Astronautics Laboratory (AFSC)
D.H. Campbell
University of Dayton Research Institute
ISBN 0-930403-54-1

118. **Rarefied Gas Dynamics: Theoretical and Computational Techniques (1989)**
E.P. Muntz
University of Southernn California
D.P. Weaver
U.S. Air Force Astronautics Laboratory (AFSC)
D.H. Campbell
University of Dayton Research Institute
ISBN 0-930403-55-X

119. **Test and Evaluation of the Tactical Missile (1989)**
Emil J. Eichblatt Jr.
Pacific Missile Test Center
ISBN 0-930403-56-8

120. **Unsteady transonic Aerodynamics (1989)**
David Nixon
Nielsen Engineering & Research, Inc.
ISBN 0-930403-52-5

121. **Orbital Debris from Upper-Stage Breakup (1989)**
Joseph P. Loftus Jr.
NASA Johnson Space Center
ISBN 0-930403-58-4

122. **Thermal-Hydraulics for Space Power, Propulsion and Thermal Management System Design (1989)**
William J. Krotiuk
General Electric Co.
ISBN 0-930403-64-9

123. **Viscous Drag Reduction in Boundary Layers (1990)**
Dennis M. Bushnell
Jerry N. Hefner
NASA Langley Research Center
ISBN 0-930403-66-5

*124. **Tactical and Strategic Missile Guidance (1990)**
Paul Zarchan
Charles Stark Draper Laboratory, Inc.
ISBN 0-930403-68-1

125. **Applied Computational Aerodynamics (1990)**
P.A. Henne
Douglas Aircraft Company
ISBN 0-930403-69-X

126. **Space Commercialization: Launch Vehicles and Programs (1990)**
F. Shahrokhi
University of Tennessee Space Institute
J.S. Greenberg
Princeton Synergetics Inc.
T. Al-Saud
Ministry of Defense and Aviation Kingdom of Saudi Arabia
ISBN 0-930403-75-4

127. **Space Commercialization: Platforms and Processing (1990)**
F. Shahrokhi
University of Tennessee Space Institute
G. Hazelrigg
National Science Foundation
R. Bayuzick
Vanderbilt University
ISBN 0-930403-76-2

128. **Space Commercialization: Satellite Technology (1990)**
F. Shahrokhi
University of Tennessee Space Institute
N. Jasentuliyana
United Nations
N. Tarabzouni
King Abulaziz City for Science and Technology
ISBN 0-930403-77-0

129. **Mechanics and Control of Large Flexible Structures (1990)**
John L. Junkins
Texas A&M University
ISBN 0-930403-73-8

130. **Low-Gravity Fluid Dynamics and Transport Phenomena (1990)**
Jean N. Koster
Robert L. Sani
University of Colorado at Boulder
ISBN 0-930403-74-6

131. **Dynamics of Deflagrations and Reactive Systems: Flames (1991)**
A.L. Kuhl
Lawrence Livermore National Laboratory
J.-C. Leyer
Universite de Poitiers
A. A. Borisov
USSR Academy of Sciences
W.A. Sirignano
University of California
ISBN 0-930403-95-9

---

*Out of print.

132. **Dynamics of Deflagrations and Reactive Systems: Heterogeneous Combustion (1991)**
A.L. Kuhl
Lawrence Livermore National Laboratory
J.-C. Leyer
Universite de Poitiers
A. A. Borisov
USSR Academy of Sciences
W.A. Sirignano
University of California
ISBN 0-930403-96-7

133. **Dynamics of Detonations and Explosions: Detonations (1991)**
A.L. Kuhl
Lawrence Livermore National Laboratory
J.-C. Leyer
Universite de Poitiers
A. A. Borisov
USSR Academy of Sciences
W.A. Sirignano
University of California
ISBN 0-930403-97-5

134. **Dynamics of Detonations and Explosions: Explosion Phenomena (1991)**
A.L. Kuhl
Lawrence Livermore National Laboratory
J.-C. Leyer
Universite de Poitiers
A. A. Borisov
USSR Academy of Sciences
W.A. Sirignano
University of California
ISBN 0-930403-98-3

135. **Numerical Approaches to Combustion Modeling (1991)**
Elaine S. Oran
Jay P. Boris
Naval Research Laboratory
ISBN 1-56347-004-7

136. **Aerospace Software Engineering (1991)**
Christine Anderson
U.S. Air Force Wright Laboratory
Merlin Dorfman
Lockheed Missiles & Space Company, Inc.
ISBN 1-56347-005-0

137. **High-Speed Flight Propulsion Systems (1991)**
S.N.B. Murthy
Purdue University
E.T. Curran
Wright Laboratory
ISBN 1-56347-011-X

138. **Propagation of Intensive Laser Radiation in Clouds (1992)**
O. A. Volkovitsky
Yu. S. Sedenov
L. P. Semenov
Institute of Experimental Meteorology
ISBN 1-56347-020-9

139. **Gun Muzzle Blast and Flash (1992)**
Gunter Klingenburg
Fraunhofer-Institut fur Kurzzeitdynamik, Ernst-Mach-Institut (EMI)
Joseph M. Heimerl
U.S. Army Ballistic Research Laboratory (BRL)
ISBN 1-56347-012-8

140. **Thermal Structures and Materials for High-Speed Flight (1992)**
Earl. A. Thornton
University of Virginia
ISBN 1-56347-017-9

141. **Tactical Missile Aerodynamics: General Topics (1992)**
Michael J. Hemsch
Lockheed Engineering & Sciences Company
ISBN 1-56347-015-2

142. **Tactical Missile Aerodynamics: Prediction Methodology (1992)**
Michael R. Mendenhall
Nielsen Engineering & Research, Inc.
ISBN 1-56347-016-0

143. **Nonsteady Burning and Combustion Stability of Solid Propellants (1992)**
Luigi De Luca
Politecnico di Milano
Edward W. Price
Georgia Institute of Technology
Martin Summerfield
Princeton Combustion Research Laboratories, Inc.
ISBN 1-56347-014-4

144. **Space Economics (1992)**
Joel S. Greenberg
Princeton Synergetics, Inc.
Henry R. Hertzfeld
HRH Associates
ISBN 1-56347-042-X

145. **Mars: Past, Present, and Future (1992)**
E. Brian Pritchard
NASA Langley Research Center
ISBN 1-56347-043-8

146. **Computational Nonlinear Mechanics in Aerospace Engineering (1992)**
Satya N. Atluri
Georgia Institute of Technology
ISBN 1-56347-044-6

147. **Modern Engineering for Design of Liquid-Propellant Rocket Engines (1992)**
Dieter K. Huzel
David H. Huang
ISBN 1-56347-013-6

---

*Out of print.

148. **Metallurgical Technologies, Energy Conversion, and Magnetohydrodynamic Flows (1993)**
Herman Branover
Yeshajahu Unger
Ben-Gurion University of the Negev
ISBN 1-56347-019-5

149. **Advances in Turbulence Studies (1993)**
Herman Branover
Yeshajahu Unger
Ben-Gurion University of the Negev
ISBN 1-56347-018-7

150. **Structural Optimization: Status and Promise (1993)**
Manohar P. Kamat
Georgia Institute of Technology
ISBN 1-56347-56-X

151. **Dynamics of Gaseous Combustion (1993)**
A.L. Kuhl
Lawrence Livermore National Laboratory
J.-C. Leyer
Universite de Poitiers
A. A. Borisov
USSR Academy of Sciences
W. A. Sirignano

152. **Dynamics of Heterogeneous Gaseous Combustion and Reacting Systems (1993)**
A.L. Kuhl
Lawrence Livermore National Laboratory
J.-C. Leyer
Universite de Poitiers
A. A. Borisov
USSR Academy of Sciences
W.A. Sirignano
University of California
ISBN 1-56347-058-6

153. **Dynamic Aspects of Detonations (1993)**
A.L. Kuhl
Lawrence Livermore National Laboratory
J.-C. Leyer
Universite de Poitiers
A. A. Borisov
USSR Academy of Sciences
W.A. Sirignano
University of California
ISBN 1-56347-057-8

154. **Dynamic Aspects of Explosion Phenomena (1993)**
A.L. Kuhl
Lawrence Livermore National Laboratory
J.-C. Leyer
Universite de Poitiers
A. A. Borisov
USSR Academy of Sciences
W.A. Sirignano
University of California
ISBN 1-56347-059-4

155. **Tactical Missile Warheads (1993)**
Joseph Carleone
Aerojet General Corporation
ISBN 1-56347-067-5

156. **Toward a Science of Command, Control, and Communications (1993)**
Carl R. Jones
Naval Postgraduate School
ISBN 1-56347-068-3

157. **Tactical and Strategic Missile Guidance Second Edition (1994)**
Paul Zarchan
Charles Stark Draper Laboratory, Inc.
ISBN 1-56347-077-2

158. **Rarefied Gas Dynamics: Experimental Techniques and Physical Systems (1994)**
Bernie D. Shizgal
University of British Columbia
David P. Weaver
Phillips Laboratory
ISBN 1-56347-079-9

159. **Rarefied Gas Dynamics: Theory and Simulations (1994)**
Bernie D. Shizgal
University of British Columbia
David P. Weaver
Phillips Laboratory
ISBN 1-56347-080-2

160. **Rarefied Gas Dynamics: Space Sciences and Engineering (1994)**
Bernie D. Shizgal
University of British Columbia
David P. Weaver
Phillips Laboratory
ISBN 1-56347-081-0

161. **Teleoperation and Robotics in Space (1994)**
Steven B. Skaar
University of Notre Dame
Carl F. Ruoff
Jet Propulsion Laboratory, California Institute of Technology
ISBN 1-56347-095-0

162. **Progress in Turbulence Research (1994)**
Herman Branover
Yeshajahu Unger
Ben-Gurion University of the Negev
ISBN 1-56347-099-3

---

*Out of print.

163. Global Positioning System: Theory and Applications Volume I (1995)
Bradford W. Parkinson
Stanford University
James J. Spilker Jr.
Stanford Telecom
Penina Axelrad,
Associate Editor
University of Colorado
Per Enge,
Associate Editor
Stanford University
ISBN 1-56347-107-8

164. Global Positioning System: Theory and Applications Volume II (1995)
Bradford W. Parkinson
Stanford University
James J. Spilker Jr.
Stanford Telecom
Penina Axelrad,
Associate Editor
University of Colorado
Per Enge,
Associate Editor
Stanford University
ISBN 1-56347-106-X

---
*Out of print.